EIGHTH EDITION

Mathematical Applications

for the Management, Life, and Social Sciences

EIGHTH EDITION

Mathematical Applications

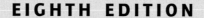

for the Management, Life, and Social Sciences

Ronald J. Harshbarger
University of South Carolina Beaufort

James J. Reynolds
Clarion University of Pennsylvania

HOUGHTON MIFFLIN COMPANY   *Boston   New York*

Vice President, Publisher: Jack Shira
Sponsoring Editor: Leonid Tunik
Development Manager: Maureen Ross
Associate Editor: Jennifer King
Senior Project Editor: Nancy Blodget
Editorial Assistant: Sarah Driver
Composition Buyer: Chuck Dutton
Art and Design Manager: Gary Crespo
Manufacturing Manager: Karen Fawcett
Senior Marketing Manager: Jennifer Jones

Cover photograph: The Starlite Hotel and another Art Deco building in South Beach, Miami, Florida. © James Revillions, Beaford Archive/Corbis. Photographed by Randy Faris.

Photo Credits: **p. 1,** © A.H. Rider/Photo Researchers, Inc.; **p. 58,** © Mark Serota/Reuters/CORBIS; **p. 139,** © Spencer Grant/PhotoEdit, Inc.; **p. 204,** © Royalty-Free/Getty Images; **p. 282,** © Jack Kurtz/The Image Works; **p. 356,** © Ed Kashi/CORBIS; **p. 404,** © Jeff Greenberg/PhotoEdit, Inc.; **p. 466,** © Jeff Greenberg/The Image Works; **p. 534,** © 2004 NBAE/Getty Images; **p. 578,** © Monika Graff/The Image Works; **p. 688,** © Dennis MacDonald/PhotoEdit, Inc.; **p. 762,** © Bernd Obermann/CORBIS; **p. 812,** © Layne Kennedy/CORBIS; **p. 866,** © Sixtus Reimann/VISUM/The Image Works; **p. 950,** © Royalty-Free/CORBIS.

Microsoft Excel screen shots reprinted by permission from Microsoft Corporation.

Printed in the U.S.A.

Library of Congress Control Number: 2005934468

Instructor's exam copy:
ISBN 13: 978-0-618-73162-6
ISBN 10: 0-618-73162-8

For orders, use student text ISBNs:
ISBN 13: 978-0-618-65421-5
ISBN 10: 0-618-65421-6

56789-CRK-09 08 07

Contents

5 Exponential and Logarithmic Functions   356

6 Mathematics of Finance   404

10 Applications of Derivatives   688

11 Derivatives Continued   762

Preface

To paraphrase English mathematician, philosopher, and educator Alfred North Whitehead, the purpose of education is not to fill a vessel but to kindle a fire. In particular, Whitehead encouraged students to be creative and imaginative in their learning and to continually form ideas into new and more exciting combinations. This desirable goal is not always an easy one to realize in mathematics with students whose primary interests are in areas other than mathematics. The purpose of this text, then, is to present mathematical skills and concepts, and to apply them to ideas that are important to students in the management, life, and social sciences. We hope that this look at the relevance of mathematical ideas to a broad range of fields will help inspire the imaginative thinking and excitement for learning that Whitehead spoke of. The applications included allow students to view mathematics in a practical setting relevant to their intended careers. Almost every chapter of this book includes a section or two devoted to the applications of mathematical topics, and every section contains a number of application examples and problems. An index of these applications on the front and back inside covers demonstrates the wide variety used in examples and exercises. Although intended for students who have completed two years of high school algebra or its equivalent, this text begins with a brief review of algebra which, if covered, will aid in preparing students for the work ahead.

Pedagogical Features

In this new edition, we have incorporated many suggestions that reflect the needs and wishes of our users. Important pedagogical features that have characterized previous editions have been retained. They are as follows.

Intuitive Viewpoint.   The book is written from an intuitive viewpoint, with emphasis on concepts and problem solving rather than on mathematical theory. Yet each topic is carefully developed and explained, and examples illustrate the techniques involved.

Flexibility.   At different colleges and universities, the coverage and sequencing of topics may vary according to the purpose of the course and the nature of the student audience. To accommodate alternative approaches, the text has a great deal of flexibility in the order in which topics may be presented and the degree to which they may be emphasized.

Chapter Warm-ups.   With the exception of Chapter 0, a Warm-up appears at the beginning of each chapter and invites students to test themselves on the skills needed for that chapter. The Warm-ups present several prerequisite problem types that are keyed to the appropriate sections in the upcoming chapter where those skills are needed. Students who have difficulty with any particular skill are directed to specific sections of the text for review. Instructors may also find the Warm-ups useful in creating a course syllabus that includes an appropriate scope and sequence of topics.

Application Previews.   Each section begins with an Application Preview that establishes the context and direction for the concepts that will be presented. Each of these Previews contains an example that motivates the mathematics in the section and is then revisited in a completely worked Application Preview example appearing later in the section.

Comprehensive Exercise Sets.   The overall variety and grading of drill and application exercises offer problems for different skill levels, and there are enough challenging problems to stimulate students in thoughtful investigations. Many exercise sets contain critical thinking and thought-provoking multistep problems that extend students' knowledge and skills.

Applications. We have found that integrating applied topics into the discussions and exercises helps provide motivation within the sections and demonstrates the relevance of each topic. Numerous real-life application examples and exercises represent the applicability of the mathematics, and each application problem is identified, so the instructor or student can select applications that are of special interest. In addition, we have found that offering separate lessons on applied topics such as cost, revenue, and profit functions brings the preceding mathematical discussions into clear, concise focus and provides a thread of continuity as mathematical sophistication increases. There are ten such sections in the book and an entire chapter devoted to financial applications. Of the more than 5500 exercises in the book, over 2000 are applied.

Extended Applications and Group Projects. Each of these case studies further illustrates how mathematics can be used in business and personal decision making. In addition, many applications are cumulative in nature in that solutions require students to combine the mathematical concepts and techniques they learned in some of the preceding chapters.

Graphical, Numerical, and Symbolic Methods. A large number of real data and modeling applications are included in the examples and exercises throughout the text and are denoted by the header **Modeling**. Many sections include problems with functions that are modeled from real data and some problems that ask students to model functions from the data given. These problems are solved by using one or more graphical, numerical, or symbolic methods.

Graphing Utilities and Spreadsheets. In the Eighth Edition, we have increased the overall presence of our treatment of graphing utilities and especially spreadsheets. More examples, applications, Technology Notes, Calculator Notes, and Spreadsheet Notes, denoted by the icon , are scattered throughout the text. After the introduction to Excel in Section 0.5 and to graphing calculators in Section 1.4, discussions of the use of technology are placed in subsections and examples in many sections, so that they can be emphasized or de-emphasized at the option of the instructor.

The discussions of graphing utility technology highlight its most common features and uses, such as graphing, window setting, trace, zoom, Solver, tables, finding points of intersection, numerical derivatives, numerical integration, matrices, solving inequalities, and modeling (curve fitting). While technology never replaces the mathematics, it does supplement and extend it by providing opportunities for generalization and alternative ways of understanding, doing, and checking. Some exercises that are better worked with the use of technology, including graphing calculators, computer programs, and computer spreadsheets, are highlighted with the technology icon. Of course, many additional exercises can benefit from the use of technology, at the option of the instructor. Technology can be used to graph functions and to discuss the generalizations, applications, and implications of problems being studied.

In addition, the basics of spreadsheet operations are introduced and opportunities to solve problems with spreadsheets are provided. The Eighth Edition now includes more Excel information specifically relating to solving equations, systems of equations, quadratic equations, matrices, linear programming, output comparisons of $f(x)$, $f'(x)$, and $f''(x)$, and the maxima and minima of functions relative to constraints. Excel is also a particularly useful problem-solving tool when studying the *Mathematics of Finance* in Chapter 6. A separate *Excel Guide* offering examples and user information is available for purchase. Please see the Supplements for the Student section on page xvi of this Preface for more information.

Checkpoints. The Checkpoints ask questions and pose problems within each section's discussion, allowing students to check their understanding of the skills and concepts under discussion before they proceed. Solutions to these Checkpoints appear before the section exercises.

Objective Lists. Every section begins with a brief list of objectives that outline the goals of that section for the student.

Procedure/Example and Property/Example Tables. Appearing throughout the text, these tables aid student understanding by giving step-by-step descriptions of important procedures and properties with illustrative examples worked out beside them.

Boxed Information. All important information is boxed for easy reference, and key terms are highlighted in boldface.

Review Exercises and Chapter Tests. At the end of each chapter, a set of Review Exercises offers students extra practice on topics in that chapter. These Reviews cover each chapter's topics in a section-by-section format, annotated with section numbers, so that any student who has difficulty can turn to the appropriate section for review. A Chapter Test follows each set of Review Exercises. All Chapter Tests provide a mixture of problems that do not directly mirror the order of topics found within the chapter. This organization of the Chapter Test ensures that students have a firm grasp of all material in the chapter.

Changes in the Eighth Edition

In the Eighth Edition, we continue to offer a text characterized by complete and accurate pedagogy, mathematical precision, excellent exercise sets, numerous and varied applications, and student-friendly exposition. There are many changes in the mathematics, prose, art, and design. The more significant ones are as follows.

■ Chapter introductions now include a list of the chapter's sections and featured applications, along with a description of the concepts covered in the chapter.

■ Linear inequalities are now covered in Chapter 1, in the section with linear equations. In addition, they are now used in Chapter 3 to bound the variables in the solution of dependent systems of equations for applied problems.

■ The Simplex Method has been rewritten and is now introduced with an applied example. This gives concrete meaning to slack variables and provides a platform for a new discussion of shadow prices. Also, a new Extended Application, *Slack Variables and Shadow Prices,* allows for an extended study of shadow prices and understanding of the importance of the Simplex Method not only as a means of solving a given linear programming problem, but also as a tool for deciding how to adjust that solution when constraint conditions change.

■ The average rate of change is now introduced with functions and revisited with each new function type. It continues to be used to introduce and motivate the instantaneous rate of change.

■ Data-driven and modeling examples and exercises have been updated or replaced with current applications.

■ The *Mathematics of Finance* (Chapter 6) examples and exercises in most sections have been revised to include multistep (multiple-investment) problems and questions that ask "How long?"

■ An increased number of optional Excel discussions have been inserted where appropriate. Also, larger systems of equations (Chapter 3) and larger linear programming problems (Chapter 4) that benefit from the application of technology, such as Excel, have been added.

Supplements for the Instructor

Eduspace®. Powerful, customizable, and interactive, Eduspace, powered by Blackboard®, is Houghton Mifflin's online learning tool. By pairing the widely recognized tools of Blackboard with high-quality, text-specific content from Houghton Mifflin, Eduspace makes it easy for instructors to create all or part of a course online. Homework exercises, quizzes, tests, tutorials, and supplemental study materials all come ready-to-use or can be modified by the instructor. Visit **www.eduspace.com** for more information.

HM ClassPrep™ with HM Testing (powered by Diploma™). *HM Class Prep* offers text-specific resources for the instructor. *HM Testing* offers instructors a flexible and powerful tool for test generation and test management. Now supported by the Brownstone Research Group's market-leading *Diploma* software, this new version of *HM Testing* significantly improves ease of use.

Blackboard, WebCT®, and Ecollege®. Houghton Mifflin can provide you with valuable content to include in your existing Blackboard, WebCT, and Ecollege systems. This text-specific content enables instructors to teach all or part of their course online.

Supplements for the Student

Eduspace. Powerful, customizable, and interactive, Eduspace, powered by Blackboard, is Houghton Mifflin's online learning tool for instructors and students. Eduspace is a text-specific, web-based learning environment that instructors can use to offer students a combination of practice exercises, multimedia tutorials, video explanations, online algorithmic homework, and more. Specific content is available 24 hours a day to help students succeed in the course.

SMARTHINKING®. Houghton Mifflin's unique partnership with SMARTHINKING brings students real-time, online tutorial support when they need it most. Using state-of-the-art whiteboard technology and feedback tools, students interact and communicate with "e-structors." Tutors guide students through the learning and problem-solving process without providing answers or rewriting students' work. Visit **www.smarthinking.com** for more information.

HM mathSpace® Student Tutorial CD-ROM. For students who prefer the portability of a CD-ROM, this tutorial provides opportunities for self-paced review and practice with algorithmically generated exercises and step-by-step solutions.

Houghton Mifflin Instructional DVDs. Hosted by Dana Mosely and professionally produced, these DVDs cover all sections of the text. Ideal for promoting individual study and review, these comprehensive DVDs also support students in online courses and those who have missed a lecture.

Online Learning Center and Online Teaching Center. The free Houghton Mifflin website at **http://college.hmco.com/PIC/harshbarger8e** contains an abundance of resources. For this edition, we are offering the text-specific *Graphing Calculator Manual* by Ron Harshbarger and Lisa Yocco on the website. This guide gives step-by-step keystrokes and examples for use with TI-82/83/84 calculators.

Student Solutions Guide (ISBN 0-618-67692-9). This guide contains solutions to the odd-numbered exercises from the end-of-section exercise sets, and both even- and odd-numbered solutions to all Chapter Reviews and Chapter Tests.

Excel Guide for Finite Mathematics and Applied Calculus by Revathi Narasimhan (ISBN 0-618-67691-0). This guide provides lists of exercises from the text that can be completed after each step-by-step Excel example. No prior knowledge of Excel is necessary.

Acknowledgments

We would like to thank the many people who helped us at various stages of revising this text. The encouragement, criticism, and suggestions that were offered were invaluable to us. Special thanks go to Jane Upshaw and Jon Beal for their assistance with the Extended Applications. We are especially grateful to Dave Hipfel for his insights regarding slack variables and shadow prices and his help with application exercises and Extended Applications. We also appreciate the ideas for thought-provoking exercises that he and Dip Bhattacharya contributed. In addition, we have been consistently impressed by and are most appreciative of Helen Medley's detailed checking of the chapters and answer section during the page-proof stage.

For their reviews of draft manuscript and the many helpful comments that were offered, we would like to thank

Abdelnour Ahmed-Zaid	University of Houston
Susan P. Bradley	Angelina College
Dwayne C. Brown	University of South Carolina, Lancaster
Mary Jackson	Brookhaven College
Kelly MacArthur	University of Utah
Kathleen Poracky	Monmouth University
Graciela Rodriguez	Laredo Community College
Stan Thompson	Western New Mexico University

Ronald J. Harshbarger

James J. Reynolds

0 Algebraic Concepts

This chapter provides a brief review of the algebraic concepts that will be used throughout the text. You may be familiar with these topics, but it may be helpful to spend some time reviewing them. In addition, each chapter after this one opens with a warm-up page that identifies prerequisite skills needed for that chapter. If algebraic skills are required, the warm-up cites their coverage in this chapter. Thus you will find that this chapter is a useful reference as you study later chapters.

The topics and applications studied in this chapter include the following.

0.1 Sets

A **set** is a well-defined collection of objects. We may talk about a set of books, a set of dishes, a set of students, or a set of individuals with a certain blood type. There are two ways to tell what a given set contains. One way is by listing the **elements** (or **members**) of the set (usually between braces). We may say that a set A contains 1, 2, 3, and 4 by writing $A = \{1, 2, 3, 4\}$. To say that 4 is a member of set A, we write $4 \in A$. Similarly, we write $5 \notin A$ to denote that 5 is not a member of set A.

If all the members of the set can be listed, the set is said to be a **finite set.** $A = \{1, 2, 3, 4\}$ and $B = \{x, y, z\}$ are examples of finite sets. When we do not wish to list all the elements of a finite set, we can use three dots to indicate the unlisted members of the set. For example, the set of even integers from 8 to 8952, inclusive, could be written as

$$\{8, 10, 12, 14, \ldots, 8952\}$$

For an **infinite set,** we cannot list all the elements, so we use the three dots. For example, $N = \{1, 2, 3, 4, \ldots\}$ is an infinite set. This set N is called the set of **natural numbers.**

Another way to specify the elements of a given set is by description. For example, we may write $D = \{x: x \text{ is a Ford automobile}\}$ to describe the set of all Ford automobiles. Furthermore, $F = \{y: y \text{ is an odd natural number}\}$ is read "F is the set of all y such that y is an odd natural number."

● **EXAMPLE 1** *Describing Sets*

Write the following sets in two ways.

(a) The set A of natural numbers less than 6
(b) The set B of natural numbers greater than 10
(c) The set C containing only 3

Solution
(a) $A = \{1, 2, 3, 4, 5\}$ or $A = \{x: x \text{ is a natural number less than 6}\}$
(b) $B = \{11, 12, 13, 14, \ldots\}$ or $B = \{x: x \text{ is a natural number greater than 10}\}$
(c) $C = \{3\}$ or $C = \{x: x = 3\}$

Note that set C of Example 1 contains one member, 3; set A contains five members; and set B contains an infinite number of members. It is possible for a set to contain no members. Such a set is called the **empty set** or the **null set,** and it is denoted by $\varnothing$ or by $\{\ \}$. The set of living veterans of the War of 1812 is empty because there are no living veterans of that war. Thus

$$\{x: x \text{ is a living veteran of the War of 1812}\} = \varnothing$$

Special relations that may exist between two sets are defined as follows.

Relations Between Sets	
Definition	Example
1. Sets X and Y are **equal** if they contain the same elements.	1. If $X = \{1, 2, 3, 4\}$ and $Y = \{4, 3, 2, 1\}$, then $X = Y$.
2. A is called a **subset** of B, which is written $A \subseteq B$, if every element of A is an element of B. The empty set is a subset of every set.	2. If $A = \{1, 2, c, f\}$ and $B = \{1, 2, 3, a, b, c, f\}$, then $A \subseteq B$.
3. If C and D have no elements in common, they are called **disjoint.**	3. If $C = \{1, 2, a, b\}$ and $D = \{3, e, 5, c\}$, then C and D are disjoint.

In the discussion of particular sets, the assumption is always made that the sets under discussion are all subsets of some larger set, called the **universal set** *U*. The choice of the universal set depends on the problem under consideration. For example, in discussing the set of all students and the set of all female students, we may use the set of all humans as the universal set.

We may use **Venn diagrams** to illustrate the relationships among sets. We use a rectangle to represent the universal set, and we use closed figures inside the rectangle to represent the sets under consideration. Figures 0.1–0.3 show such Venn diagrams.

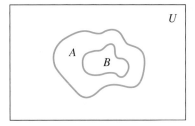

Figure 0.1

B is a subset of *A*; *B* ⊆ *A*.

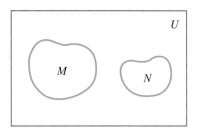

Figure 0.2

M and *N* are disjoint.

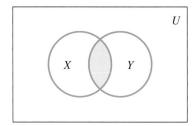

Figure 0.3

X and *Y* are not disjoint.

Set Operations

The shaded portion of Figure 0.3 indicates where the two sets overlap. The set containing the members that are common to two sets is said to be the **intersection** of the two sets.

Set Intersection

The intersection of *A* and *B*, written *A* ∩ *B*, is defined by

$$A \cap B = \{x \colon x \in A \text{ and } x \in B\}$$

● **EXAMPLE 2** *Set Intersection*

If $A = \{2, 3, 4, 5\}$ and $B = \{3, 5, 7, 9, 11\}$, find $A \cap B$.

Solution

$A \cap B = \{3, 5\}$ because 3 and 5 are the common elements of *A* and *B*. Figure 0.4 shows the sets and their intersection.

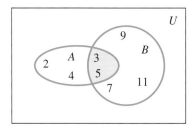

Figure 0.4

● *Checkpoint*

Let $A = \{2, 3, 5, 7, 11\}$, $B = \{2, 4, 6, 8, 10\}$, and $C = \{6, 10, 14, 18, 22\}$. Use these sets to answer the following.
1. (a) Of which sets is 6 an element?   (b) Of which sets is {6} an element?
2. Which of the following are true?
   (a) $2 \in A$ (b) $2 \in B$ (c) $2 \in C$ (d) $5 \notin A$ (e) $5 \notin B$
3. Which pair of *A*, *B*, and *C* is disjoint?
4. Which of *A*, *B*, and *C* are subsets of
   (a) the set *P* of all prime numbers?   (b) the set *M* of all multiples of 2?
5. Which of *A*, *B*, and *C* is equal to
   $D = \{x \mid x = 4n + 2 \text{ for natural numbers } 1 \leq n \leq 5\}$?

The **union** of two sets is the set that contains all members of the two sets.

Set Union

The union of A and B, written $A \cup B$, is defined by

$$A \cup B = \{x: x \in A \text{ or } x \in B \text{ (or both)}\}*$$

We can illustrate the intersection and union of two sets by the use of Venn diagrams. Figures 0.5 and 0.6 show Venn diagrams with universal set U represented by the rectangles and with sets A and B represented by the circles. The shaded region in Figure 0.5 represents $A \cap B$, the intersection of A and B, and the shaded region in Figure 0.6—which consists of all parts of both circles—represents $A \cup B$.

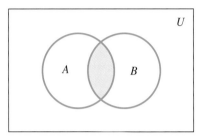

Figure 0.5

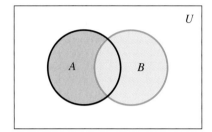

Figure 0.6

● **EXAMPLE 3** *Set Union*

If $X = \{a, b, c, f\}$ and $Y = \{e, f, a, b\}$, find $X \cup Y$.

Solution
$X \cup Y = \{a, b, c, e, f\}$

● **EXAMPLE 4** *Set Intersection and Union*

Let $A = \{x: x \text{ is a natural number less than 6}\}$ and $B = \{1, 3, 5, 7, 9, 11\}$.

(a) Find $A \cap B$. (b) Find $A \cup B$.

Solution
(a) $A \cap B = \{1, 3, 5\}$ (b) $A \cup B = \{1, 2, 3, 4, 5, 7, 9, 11\}$

All elements of the universal set that are not contained in a set A form a set called the **complement** of A.

Set Complement

The complement of A, written A', is defined by

$$A' = \{x: x \in U \text{ and } x \notin A\}$$

We can use a Venn diagram to illustrate the complement of a set. The shaded region of Figure 0.7 represents A', and the *un*shaded region of Figure 0.5 represents $(A \cap B)'$.

*In mathematics, the word *or* means "one or the other or both."

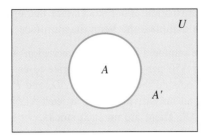

Figure 0.7

● **EXAMPLE 5** *Operations with Sets*

If $U = \{x \in N: x < 10\}$, $A = \{1, 3, 6\}$, and $B = \{1, 6, 8, 9\}$, find the following.

(a) A' (b) B' (c) $(A \cap B)'$ (d) $A' \cup B'$

Solution
(a) $U = \{1, 2, 3, 4, 5, 6, 7, 8, 9\}$ so $A' = \{2, 4, 5, 7, 8, 9\}$
(b) $B' = \{2, 3, 4, 5, 7\}$
(c) $A \cap B = \{1, 6\}$ so $(A \cap B)' = \{2, 3, 4, 5, 7, 8, 9\}$
(d) $A' \cup B' = \{2, 4, 5, 7, 8, 9\} \cup \{2, 3, 4, 5, 7\} = \{2, 3, 4, 5, 7, 8, 9\}$

● **Checkpoint**

Given the sets $U = \{1, 2, 3, 4, 5, 6, 7, 8, 9, 10\}$, $A = \{1, 3, 5, 7, 9\}$, $B = \{2, 3, 5, 7\}$, and $C = \{4, 5, 6, 7, 8, 9, 10\}$, find the following.
6. $A \cup B$ 7. $B \cap C$ 8. A'

● **EXAMPLE 6** *Stocks*

Many local newspapers list "stocks of local interest." Suppose that on a certain day, a prospective investor categorized 23 stocks according to whether

■ their closing price on the previous day was less than \$50/share (set C)

■ their price-to-earnings ratio was less than 20 (set P)

■ their dividend per share was at least \$1.50 (set D)

Of these 23 stocks,

16 belonged to set P 10 belonged to both C and P
12 belonged to set C 7 belonged to both D and P
 8 belonged to set D 2 belonged to all three sets
 3 belonged to both C and D

Draw a Venn diagram that represents this information. Use the diagram to answer the following.

(a) How many stocks had closing prices of less than \$50 per share or price-to-earnings ratios of less than 20?
(b) How many stocks had none of the characteristics of set C, P, or D?
(c) How many stocks had only dividends per share of at least \$1.50?

Solution
The Venn diagram for three sets has eight separate regions (see Figure 0.8(a) on the next page). To assign numbers from our data, we must begin with some information that refers to a single region, namely that two stocks belonged to all three sets (see Figure 0.8(b) on the next page). Because the region common to all three sets is also common to any pair,

we can next use the information about stocks that belonged to two of the sets (see Figure 0.8(c)). Finally, we can complete the Venn diagram (see Figure 0.8(d)).

(a) We need to add the numbers in the separate regions that lie within $C \cup P$. That is, 18 stocks closed under $50 per share or had price-to-earnings ratios of less than 20.
(b) There are 5 stocks outside the three sets C, D, and P.
(c) Those stocks that had only dividends of at least $1.50 per share are inside D but outside both C and P. There are no such stocks.

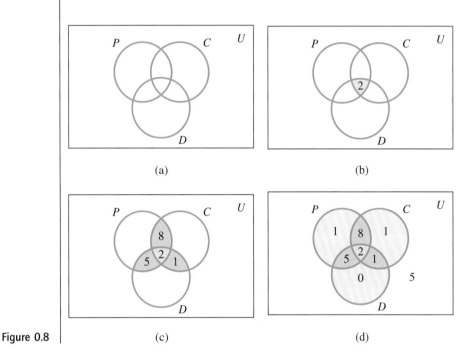

Figure 0.8

(a) (b)

(c) (d)

● **Checkpoint Solutions**

1. (a) Sets B and C have 6 as an element.
   (b) None of A, B, or C has {6} as an element; {6} is itself a set, and the elements of A, B, and C are not sets.
2. (a) True   (b) True   (c) False; $2 \notin C$ (d) False; $5 \in A$ (e) True
3. A and C are disjoint.
4. (a) $A \subseteq P$ (b) $B \subseteq M, C \subseteq M$
5. $C = D$
6. $A \cup B = \{1, 2, 3, 5, 7, 9\}$
7. $B \cap C = \{5, 7\}$
8. $A' = \{2, 4, 6, 8, 10\}$

0.1 Exercises

Use $\in$ or $\notin$ to indicate whether the given object is an element of the given set in the following problems.

1. x $\{x, y, z, a\}$ 2. 3 $\{1, 2, 3, 4, 5, 6\}$
3. 12 $\{1, 2, 3, 4, \ldots\}$
4. 5 $\{x: x$ is a natural number greater than 5$\}$
5. 6 $\{x: x$ is a natural number less than 6$\}$
6. 3 $\varnothing$

In Problems 7–10, write the following sets a second way.
7. $\{x: x$ is a natural number less than 8$\}$
8. $\{x: x$ is a natural number greater than 6, less than 10$\}$
9. $\{3, 4, 5, 6, 7\}$ 10. $\{7, 8, 9, 10, \ldots\}$
11. If $A = \{1, 2, 3, 4\}$ and $B = \{1, 2, 3, 4, 5, 6\}$, is A a subset of B?

12. If $A = \{a, b, c, d\}$ and $B = \{c, d, a, b\}$, is A a subset of B?
13. Is $A \subseteq B$ if $A = \{a, b, c, d\}$ and $B = \{a, b, d\}$?
14. Is $A \subseteq B$ if $A = \{6, 8, 10, 12\}$ and $B = \{6, 8, 10, 14, 18\}$?

Use $\subseteq$ notation to indicate which set is a subset of the other in the following problems.
15. $C = \{a, b, 1, 2, 3\}, D = \{a, b, 1\}$
16. $E = \{x, y, a, b\}, F = \{x, 1, a, y, b, 2\}$
17. $A = \{1, \pi, e, 3\}, D = \{1, \pi, 3\}$
18. $A = \{2, 3, 4, e\}, \varnothing$
19. $A = \{6, 8, 7, 4\}, B = \{8, 7, 6, 4\}$
20. $D = \{a, e, 1, 3, c\}, F = \{e, a, c, 1, 3\}$

In Problems 21–24, indicate whether the following pairs of sets are equal.
21. $A = \{a, b, \pi, \sqrt{3}\}, B = \{a, \pi, \sqrt{3}, b\}$
22. $A = \{x, g, a, b\}, D = \{x, a, b, y\}$
23. $D = \{x : x$ is a natural number less than 4$\}$, $E = \{1, 2, 3, 4\}$
24. $F = \{x : x$ is a natural number greater than 6$\}$, $G = \{7, 8, 9, \ldots\}$
25. From the following list of sets, indicate which pairs of sets are disjoint.
$$A = \{1, 2, 3, 4\}$$
$$B = \{x : x \text{ is a natural number greater than 4}\}$$
$$C = \{4, 5, 6, \ldots\}$$
$$D = \{1, 2, 3\}$$
26. If A and B are disjoint sets, what does $A \cap B$ equal?

In Problems 27–30, find $A \cap B$, the intersection of sets A and B.
27. $A = \{2, 3, 4, 5, 6\}$ and $B = \{4, 6, 8, 10, 12\}$
28. $A = \{a, b, c, d, e\}$ and $B = \{a, d, e, f, g, h\}$
29. $A = \varnothing$ and $B = \{x, y, a, b\}$
30. $A = \{x : x$ is a natural number less than 4$\}$ and $B = \{3, 4, 5, 6\}$

In Problems 31–34, find $A \cup B$, the union of sets A and B.
31. $A = \{1, 2, 4, 5\}$ and $B = \{2, 3, 4, 5\}$
32. $A = \{a, e, i, o, u\}$ and $B = \{a, b, c, d\}$
33. $A = \varnothing$ and $B = \{1, 2, 3, 4\}$
34. $A = \{x : x$ is a natural number greater than 5$\}$ and $B = \{x : x$ is a natural number less than 5$\}$

In Problems 35–46, assume that
$$A = \{1, 3, 5, 8, 7, 2\}$$
$$B = \{4, 3, 8, 10\}$$
$$C = \{2, 4, 6, 8, 10\}$$
and that U is the universal set of natural numbers less than 11. Find the following.
35. A' 36. B' 37. $A \cap B'$
38. $A' \cap B'$ 39. $(A \cup B)'$ 40. $(A \cap B)'$
41. $A' \cup B'$ 42. $(A' \cup B)'$ 43. $(A \cap B') \cup C'$
44. $A \cap (B' \cup C')$ 45. $(A \cap B')' \cap C$
46. $A \cap (B \cup C)$

The difference of two sets, $A - B$, is defined as the set containing all elements of A except those in B. That is, $A - B = A \cap B'$. Find $A - B$ for each pair of sets in Problems 47–50 if $U = \{1, 2, 3, 4, 5, 6, 7, 8, 9\}$.
47. $A = \{1, 3, 7, 9\}$ and $B = \{3, 5, 8, 9\}$
48. $A = \{1, 2, 3, 6, 9\}$ and $B = \{1, 4, 5, 6, 7\}$
49. $A = \{2, 1, 5\}$ and $B = \{1, 2, 3, 4, 5, 6\}$
50. $A = \{1, 2, 3, 4, 5\}$ and $B = \{7, 8, 9\}$

APPLICATIONS

51. *Dow Jones Industrial Average* The following table shows information about yearly lows, highs, and percentage changes for the years 1995 to 2002. Let L be the set of years where the low was greater than 7500. Let H be the set of years where the high was greater than 7500. Let C be the years when the percentage change (from low to high) exceeded 35%.
    (a) List the elements of L, H, and C.
    (b) Is any of L, H, or C a subset of one of the others (besides itself)?
    (c) Write a verbal description of C'.
    (d) Find $H' \cup C'$ and describe it in words.
    (e) Find $L' \cap C$ and describe it in words.

Dow Jones Industrial Average

Year	Low	High	% Change
2002	7286.27	10,635.65	45.9
2001	8235.94	11,332.92	37.6
2000	9796.03	11,722.98	19.7
1999	9120.67	11,497.12	26.1
1998	7359.07	9374.27	27.4
1997	6391.69	8259.31	29.2
1996	5032.94	6569.91	30.5
1995	3832.08	5216.47	36.1

Source: Dow Jones & Company, 2003

52. *Job growth* The number of jobs in 2000, the number projected in 2025, and the projected annual growth rate for jobs in some cities are shown in the table on the next page. Consider the following sets.
    $A =$ set of cities with at least 2,000,000 jobs in 2000 or in 2025
    $B =$ set of cities with at least 1,500,000 jobs in 2000
    $C =$ set of cities with projected annual growth rate of at least 2.5%

(a) List *A*, *B*, and *C* (using the letters to represent the cities).
(b) Is any of *A*, *B*, or *C* a subset of the other?
(c) Find $A \cap C$ and describe the set in words.
(d) Give a verbal description of B'.

Cities	Jobs in 2000 (thousands)	Projected Jobs in 2025 (thousands)	Annual Rates of Increase (%)
O (Orlando)	1098	2207	2.83
M (Myrtle Beach)	133	256	2.64
L (Atlanta)	2715	4893	2.38
P (Phoenix)	1953	3675	2.56
B (Boulder)	233	420	2.38

Source: NPA Data Services, Inc.

National health care   **Suppose that the table below summarizes the opinions of various groups on the issue of national health care. Use this table for Problems 53 and 54.**

	Whites		Nonwhites		
Opinion	Rep.	Dem.	Rep.	Dem.	Total
Favor	100	250	30	200	580
Oppose	250	150	10	10	420
Total	350	400	40	210	1000

53. Identify the number of individuals in each of the following sets.
    (a) Republicans and those who favor national health care
    (b) Republicans or those who favor national health care
    (c) White Republicans or those who oppose national health care
54. Identify the number of individuals in each of the following sets.
    (a) Whites and those who oppose national health care
    (b) Whites or those who oppose national health care
    (c) Nonwhite Democrats and those who favor national health care
55. *Languages*   A survey of 100 aides at the United Nations revealed that 65 could speak English, 60 could speak French, and 40 could speak both English and French.
    (a) Draw a Venn diagram representing the 100 aides. Use *E* to represent English-speaking and *F* to represent French-speaking aides.
    (b) How many aides are in $E \cap F$?
    (c) How many aides are in $E \cup F$?
    (d) How many aides are in $E \cap F'$?

56. *Advertising*   Suppose that a survey of 100 advertisers in *U.S. News, These Times,* and *World* found the following.

    14 advertised in all three
    30 advertised in *These Times* and *U.S. News*
    26 advertised in *World* and *U.S. News*
    27 advertised in *World* and *These Times*
    60 advertised in *These Times*
    52 advertised in *U.S. News*
    50 advertised in *World*

    Draw a Venn diagram representing this information and use it to answer the following.
    (a) How many advertised in none of these publications?
    (b) How many advertised only in *These Times*?
    (c) How many advertised in *U.S. News* or *These Times*?

57. *College enrollments*   Records at a small college show the following about the enrollments of 100 first-year students in mathematics, fine arts, and economics.

    38 take math
    42 take fine arts
    20 take economics
    4 take economics and fine arts
    15 take math and economics
    9 take math and fine arts
    12 take math and economics but not fine arts

    Draw a Venn diagram representing this information and label all the areas. Use this diagram to answer the following.
    (a) How many take none of these three courses?
    (b) How many take math or economics?
    (c) How many take exactly one of these three courses?

58. *Survey analysis*   In a survey of the dining preferences of 110 dormitory students at the end of the spring semester, the following facts were discovered about Adam's Lunch (AL), Pizza Tower (PT), and the Dining Hall (DH).

    30 liked AL but not PT
    21 liked AL only
    63 liked AL
    58 liked PT
    27 liked DH
    25 liked PT and AL but not DH
    18 liked PT and DH

    Draw a Venn diagram representing this survey and label all the areas. Use this diagram to answer the following.
    (a) How many liked PT or DH?
    (b) How many liked all three?
    (c) How many liked only DH?

59. *Blood types* Blood types are determined by the presence or absence of three antigens: A antigen, B antigen, and an antigen called the Rh factor. The resulting blood types are classified as follows:

> *type A* if the A antigen is present
> *type B* if the B antigen is present
> *type AB* if both the A and B antigens are present
> *type O* if neither the A nor the B antigen is present

These types are further classified as *Rh-positive* if the Rh-factor antigen is present and *Rh-negative* otherwise.

(a) Draw a Venn diagram that illustrates this classification scheme.

(b) Identify the blood type determined by each region of the Venn diagram (such as A⁺ to indicate type A, Rh-positive).

(c) Use a library or another source to find what percentage of the U.S. population has each blood type.

0.2 The Real Numbers

This text uses the set of **real numbers** as the universal set. We can represent the real numbers along a line called the **real number line.** This number line is a picture, or graph, of the real numbers. Each point on the real number line corresponds to exactly one real number, and each real number can be located at exactly one point on the real number line. Thus, two real numbers are said to be equal whenever they are represented by the same point on the real number line. The equation $a = b$ (a equals b) means that the symbols a and b represent the same real number. Thus, $3 + 4 = 7$ means that $3 + 4$ and 7 represent the same number. Table 0.1 lists special subsets of the real numbers.

TABLE 0.1 Subsets of the Set of Real Numbers

	Description	Example (some elements shown)
Natural numbers	$\{1, 2, 3, \ldots\}$ The counting numbers.	
Integers	$\{\ldots, -2, -1, 0, 1, 2, \ldots\}$ The natural numbers, 0, and the negatives of the natural numbers.	
Rational numbers	All numbers that can be written as the ratio of two integers, a/b, with $b \neq 0$. These numbers have decimal representations that either terminate or repeat.	
Irrational numbers	Those real numbers that *cannot* be written as the ratio of two integers. Irrational numbers have decimal representations that neither terminate nor repeat.	
Real numbers	The set containing all rational and irrational numbers (the entire number line).	

The properties of the real numbers are fundamental to the study of algebra. These properties follow.

Properties of the Real Numbers

Let a, b, and c denote real numbers.

1. Addition and multiplication are commutative.

$$a + b = b + a \qquad ab = ba$$

2. Addition and multiplication are associative.

$$(a + b) + c = a + (b + c) \qquad (ab)c = a(bc)$$

3. The additive identity is 0.

$$a + 0 = 0 + a = a$$

4. The multiplicative identity is 1.

$$a \cdot 1 = 1 \cdot a = a$$

5. Each element a has an additive inverse, denoted by $-a$.

$$a + (-a) = -a + a = 0$$

Note that there is a difference between a negative number and the negative of a number.

6. Each nonzero element a has a multiplicative inverse, denoted by a^{-1}.

$$a \cdot a^{-1} = a^{-1} \cdot a = 1$$

Note that $a^{-1} = 1/a$.

7. Multiplication is distributive over addition.

$$a(b + c) = ab + ac$$

Note that Property 5 provides the means to subtract by defining $a - b = a + (-b)$ and Property 6 provides a means to divide by defining $a \div b = a \cdot (1/b)$. The number 0 has no multiplicative inverse, so division by 0 is undefined.

Inequalities and Intervals

We say that a is less than b (written $a < b$) if the point representing a is to the left of the point representing b on the real number line. For example, $4 < 7$ because 4 is to the left of 7 on the real number line. We may also say that 7 is greater than 4 (written $7 > 4$). We may indicate that the number x is less than or equal to another number y by writing $x \leq y$. We may also indicate that p is greater than or equal to 4 by writing $p \geq 4$.

● **EXAMPLE 1** *Inequalities*

Use $<$ or $>$ notation to write

(a) 6 is greater than 5.　　　　　　　　(b) 10 is less than 15.
(c) 3 is to the left of 8 on the real number line.　　(d) x is at most 12.

Solution
(a) $6 > 5$　　(b) $10 < 15$　　(c) $3 < 8$
(d) "x is at most 12" means it must be less than or equal to 12. Thus, $x \leq 12$.

The subset of the real numbers consisting of all real numbers x that lie between a and b, excluding a and b, can be denoted by the *double inequality* $a < x < b$ or by the **open interval** (a, b). It is called an open interval because neither of the endpoints is included in the interval. The **closed interval** $[a, b]$ represents the set of all real numbers x satisfying $a \leq x \leq b$. Intervals containing one endpoint, such as $(a, b]$ and $[a, b)$, are called **half-open intervals.**

We can use $[a, +\infty)$ to represent the inequality $x \geq a$ and $(-\infty, a)$ to represent $x < a$. In each of these cases, the symbols $+\infty$ and $-\infty$ are not real numbers but represent the fact that x increases without bound ($+\infty$) or decreases without bound ($-\infty$). Table 0.2 summarizes three types of intervals.

TABLE 0.2 Intervals

Type of Interval	Inequality Notation	Interval Notation	Graph
Open interval	$x > a$	(a, ∞)	
	$x < b$	$(-\infty, b)$	
	$a < x < b$	(a, b)	
Half-open interval	$x \geq a$	$[a, \infty)$	
	$x \leq b$	$(-\infty, b]$	
	$a \leq x < b$	$[a, b)$	
	$a < x \leq b$	$(a, b]$	
Closed interval	$a \leq x \leq b$	$[a, b]$	

● *Checkpoint*

1. Evaluate the following, if possible. For any that are meaningless, so state.
   (a) $\dfrac{4}{0}$ (b) $\dfrac{0}{4}$ (c) $\dfrac{4}{4}$ (d) $\dfrac{4-4}{4-4}$

2. For (a)–(d), write the inequality corresponding to the given interval and sketch its graph on a real number line.
   (a) $(1, 3)$ (b) $(0, 3]$ (c) $[-1, \infty)$ (d) $(-\infty, 2)$

3. Express each of the following inequalities in interval notation and name the type of interval.
   (a) $3 \leq x \leq 6$ (b) $-6 \leq x < 4$

Absolute Value

Sometimes we are interested in the *distance* a number is from the origin (0) of the real number line, without regard to direction. The distance a number a is from 0 on the number line is the **absolute value** of a, denoted $|a|$. The absolute value of any nonzero number is positive, and the absolute value of 0 is 0.

● **EXAMPLE 2 Absolute Value**

Evaluate the following.

(a) $|-4|$ (b) $|+2|$ (c) $|0|$ (d) $|-5 - |-3||$

Solution
(a) $|-4| = +4 = 4$ (b) $|+2| = +2 = 2$
(c) $|0| = 0$ (d) $|-5 - |-3|| = |-5 - 3| = |-8| = 8$

Note that if a is a nonnegative number, then $|a| = a$, *but if a is negative, then $|a|$ is the positive number* $(-a)$. Thus

Absolute Value

$$|a| = \begin{cases} a & \text{if } a \geq 0 \\ -a & \text{if } a < 0 \end{cases}$$

In performing computations with real numbers, it is important to remember the rules for computations.

Operations with Real (Signed) Numbers

Procedure	Example
1. (a) To add two real numbers with the same sign, add their absolute values and affix their common sign.	1. (a) $(+5) + (+6) = +11$
	$\left(-\dfrac{1}{6}\right) + \left(-\dfrac{2}{6}\right) = -\dfrac{3}{6} = -\dfrac{1}{2}$
(b) To add two real numbers with unlike signs, find the difference of their absolute values and affix the sign of the number with the larger absolute value.	(b) $(-4) + (+3) = -1$
	$(+5) + (-3) = +2$
	$\left(-\dfrac{11}{7}\right) + (+1) = -\dfrac{4}{7}$
2. To subtract one real number from another, change the sign of the number being subtracted and proceed as in addition.	2. $(-9) - (-8) = (-9) + (+8) = -1$
	$16 - (8) = 16 + (-8) = +8$
3. (a) The product of two real numbers with like signs is positive.	3. (a) $(-3)(-4) = +12$
	$\left(+\dfrac{3}{4}\right)(+4) = +3$
(b) The product of two real numbers with unlike signs is negative.	(b) $5(-3) = -15$
	$(-3)(+4) = -12$
4. (a) The quotient of two real numbers with like signs is positive.	4. (a) $(-14) \div (-2) = +7$
	$+36/4 = +9$
(b) The quotient of two real numbers with unlike signs is negative.	(b) $(-28)/4 = -7$
	$45 \div (-5) = -9$

When two or more operations with real numbers are indicated in an evaluation, it is important that everyone agree upon the order in which the operations are performed so that a unique result is guaranteed. The following **order of operations** is universally accepted.

Order of Operations

1. Perform operations within parentheses.

2. Find indicated powers ($2^3 = 2 \cdot 2 \cdot 2 = 8$).

3. Perform multiplications and divisions from left to right.

4. Perform additions and subtractions from left to right.

● **EXAMPLE 3 *Order of Operations***

Evaluate the following.

(a) $-4 + 3$ (b) $-4^2 + 3$ (c) $(-4 + 3)^2 + 3$ (d) $6 \div 2(2 + 1)$

Solution

(a) -1

(b) Note that with -4^2 the power 2 is applied only to 4, not to -4 (which would be written $(-4)^2$). Thus $-4^2 + 3 = -(4^2) + 3 = -16 + 3 = -13$

(c) $(-1)^2 + 3 = 1 + 3 = 4$ (d) $6 \div 2(3) = (6 \div 2)(3) = 3 \cdot 3 = 9$

● **Checkpoint**

True or false:
4. $-(-5)^2 = 25$
5. $|4 - 6| = |4| - |6|$
6. $9 - 2(2)(-10) = 7(2)(-10) = -140$

The text assumes that you have a scientific or graphing calculator. Discussions of some of the capabilities of graphing calculators and graphing utilities will be found throughout the text.

Most scientific and graphing calculators use standard algebraic order when evaluating arithmetic expressions. Working outward from inner parentheses, calculations are performed from left to right. Powers and roots are evaluated first, followed by multiplications and divisions and then additions and subtractions.

● **Checkpoint Solutions**

1. (a) Meaningless. A denominator of zero means division by zero, which is undefined.
   (b) $\frac{0}{4} = 0$. A numerator of zero (when the denominator is not zero) means the fraction has value 0.
   (c) $\frac{4}{4} = 1$
   (d) Meaningless. The denominator is zero.
2. (a) $1 < x < 3$
   (b) $0 < x \leq 3$
   (c) $-1 \leq x < \infty$ or $x \geq -1$
   (d) $-\infty < x < 2$ or $x < 2$

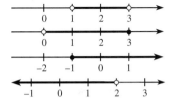

3. (a) $[3, 6]$; closed interval
   (b) $[-6, 4)$; half-open interval
4. False. $-(-5)^2 = (-1)(-5)^2 = (-1)(25) = -25$. Exponentiation has priority and applies only to -5.
5. False. $|4 - 6| = |-2| = 2$ and $|4| - |6| = 4 - 6 = -2$.
6. False. Without parentheses, multiplication has priority over subtraction.
   $9 - 2(2)(-10) = 9 - 4(-10) = 9 + 40 = 49$.

0.2 Exercises

In Problems 1–2, indicate whether the given expression is one or more of the following types of numbers: rational, irrational, integer, natural. If the expression is meaningless, so state.

1. (a) $\dfrac{-\pi}{10}$ (b) -9 (c) $\dfrac{9}{3}$ (d) $\dfrac{4}{0}$

2. (a) $\dfrac{0}{6}$ (b) -1.2916 (c) 1.414 (d) $\dfrac{9}{6}$

Which property of real numbers is illustrated in each part of Problems 3 and 4?

3. (a) $8 + 6 = 6 + 8$
   (b) $5(3 + 7) = 5 \cdot 3 + 5 \cdot 7$
   (c) $-e \cdot 1 = -e$
4. (a) $6(4 \cdot 5) = (6 \cdot 4)(5)$
   (b) $-15 + 0 = -15$
   (c) $\left(\dfrac{3}{2}\right)\left(\dfrac{2}{3}\right) = 1$

Insert the proper sign $<$, $=$, or $>$ to replace $\square$ in Problems 5–10.

5. $-14 \,\square\, -3$ 6. $\pi \,\square\, 3.14$
7. $0.333 \,\square\, \dfrac{1}{3}$ 8. $\dfrac{1}{3} + \dfrac{1}{2} \,\square\, \dfrac{5}{6}$
9. $|-3| + |5| \,\square\, |-3 + 5|$
10. $|-9 - 3| \,\square\, |-9| + |3|$

In Problems 11–22, evaluate each expression.

11. $-3^2 + 10 \cdot 2$ 12. $(-3)^2 + 10 \cdot 2$
13. $\dfrac{4 + 2^2}{2}$ 14. $\dfrac{(4 + 2)^2}{2}$
15. $\dfrac{16 - (-4)}{8 - (-2)}$ 16. $\dfrac{(-5)(-3) - (-2)(3)}{-9 + 2}$
17. $\dfrac{|5 - 2| - |-7|}{|5 - 2|}$ 18. $\dfrac{|3 - |4 - 11||}{-|5^2 - 3^2|}$

19. $\dfrac{(-3)^2 - 2 \cdot 3 + 6}{4 - 2^2 + 3}$

20. $\dfrac{6^2 - 4(-3)(-2)}{6 - 6^2 \div 4}$

21. $\dfrac{-4^2 + 5 - 2 \cdot 3}{5 - 4^2}$

22. $\dfrac{3 - 2(5 - 2)}{(-2)^2 - 2^2 + 3}$

23. What part of the real number line corresponds to the interval $(-\infty, \infty)$?

24. Write the interval corresponding to $x \geq 0$.

In Problems 25–28, express each inequality or graph using interval notation, and name the type of interval.

25. $1 < x \leq 3$

26. $-4 \leq x \leq 3$

27.

28.

In Problems 29–32, write an inequality that describes each interval or graph.

29. $[-3, 5)$

30. $(-2, \infty)$

31.

32.

In Problems 33–40, graph the subset of the real numbers that is represented by each of the following and write your answer in interval notation.

33. $(-\infty, 4) \cap (-3, \infty)$ 34. $[-4, 17) \cap [-20, 10]$

35. $x > 4$ and $x \geq 0$ 36. $x < 10$ and $x < -1$

37. $[0, \infty) \cup [-1, 5]$ 38. $(-\infty, 4) \cup (0, 2)$

39. $x > 7$ or $x < 0$ 40. $x > 4$ and $x < 0$

In Problems 41–46, use your calculator to evaluate each of the following. List all the digits on your display in the answer.

41. $\dfrac{-1}{25916.8}$

42. $\dfrac{51.412}{127.01}$

43. $(3.679)^7$

44. $(1.28)^{10}$

45. $\dfrac{2500}{(1.1)^6 - 1}$

46. $100\left[\dfrac{(1.05)^{12} - 1}{0.05}\right]$

APPLICATIONS

47. *Take-home pay* A sales representative's take-home pay is found by subtracting all taxes and retirement contributions from gross pay (which consists of salary plus commission). Given the following information, complete (a)–(c).

    Salary = $300.00   Commission = $788.91

    Retirement = 5% of gross pay

    Taxes: State = 5% of gross pay,

    Local = 1% of gross pay

    Federal withholding =
       25% of (gross pay less retirement)

    Federal social security and Medicare =
       7.65% of gross pay

    (a) Find the gross pay.
    (b) Find the amount of federal withholding.
    (c) Find the take-home pay.

48. *Tax load* The approximate percent P of average income used to pay federal, state, and local taxes is given by

    $$P = 0.24627t + 25.96473$$

    where t is the number of years past 1950.
    (a) What t-value represents the year 1985?
    (b) The actual tax load for 1985 was 34.8%. What does the formula give as an approximation?
    (c) Approximate the tax load for 2008.
    (d) What P-value signals that this approximation formula is no longer valid? Explain.

49. *Cable rates* By using data from Nielsen Media Research, we can approximate the average monthly cost C for basic cable TV service quite accurately by either
    (1) $C = 1.2393t + 4.2594$ or
    (2) $C = 0.0232t^2 + 0.7984t + 5.8271$
       where t is the number of years past 1980.
    (a) Both (1) and (2) closely fit all the data from 1982 to 1997, but which is more accurate for 1996 and for 1997? Use the fact that average cable rates for those years were as follows:

       1996: $24.41   1997: $26.00

    (b) Use the formula from (a) that was more accurate and predict the average cost for basic cable service for 2008.

50. *Health statistics* From data adapted from the National Center for Health Statistics, the height H in inches and age A in years for boys between 4 and 16 years of age are related according to

    $$H = 2.31A + 31.26$$

    To account for normal variability among boys, normal height for a given age is ±5% of the height obtained from the equation.
    (a) Find the normal height range for a boy who is 10.5 years old, and write it as an inequality.
    (b) Find the normal height range for a boy who is 5.75 years old, and write it as an inequality.

51. *Income taxes*   The 2004 tax brackets for a single person claiming one personal exemption are given in the
following table.

Taxable Income *I*	Tax Due *T*
$0–$7150	10% *I*
$7151–$29,050	$715 + 15%(*I* − 7150)
$29,051–$70,350	$4000 + 25%(*I* − 29,050)
$70,351–$146,750	$14,325 + 28%(*I* − 70,350)
$146,751–$319,100	$35,717 + 33%(*I* − 146,750)
Over $319,100	$92,592.50 + 35%(*I* − 319,100)

Source: 1040 Forms and Instructions, IRS, 2004

(a) Write the last three taxable income ranges as
inequalities.
(b) If an individual has a taxable income of $29,050,
calculate the tax due. Repeat this calculation for a
taxable income of $70,350.
(c) Write an interval that represents the amount of tax
due for a taxable income between $29,050 and
$70,350.

0.3   Integral Exponents

If $1000 is placed in a 5-year savings certificate that pays an interest rate of 10% per year,
compounded annually, then the amount returned after 5 years is given by

$$1000(1.1)^5$$

The 5 in this expression is an **exponent.** Exponents provide an easier way to denote certain multiplications. For example,

$$(1.1)^5 = (1.1)(1.1)(1.1)(1.1)(1.1)$$

An understanding of the properties of exponents is fundamental to the algebra needed to
study functions and solve equations. Furthermore, the definition of exponential and logarithmic functions and many of the techniques in calculus also require an understanding of
the properties of exponents.

For any real number a,

$$a^2 = a \cdot a, \quad a^3 = a \cdot a \cdot a, \quad \text{and} \quad a^n = a \cdot a \cdot a \cdot \cdots \cdot a \quad (n \text{ factors})$$

for any positive integer n. The positive integer n is called the **exponent,** the number a is
called the **base,** and a^n is read "a to the nth power."

Note that $4a^n$ means $4(a^n)$, which is different from $(4a)^n$. The 4 is the coefficient of a^n
in $4a^n$. Note also that $-x^n$ is not equivalent to $(-x)^n$ when n is even. For example,
$-3^4 = -81$, but $(-3)^4 = 81$.

Some of the rules of exponents follow.

Positive Integer Exponents

For any real numbers a and b and positive integers m and n,

1. $a^m \cdot a^n = a^{m+n}$

2. For $a \neq 0$, $\dfrac{a^m}{a^n} = \begin{cases} a^{m-n} & \text{if } m > n \\ 1 & \text{if } m = n \\ 1/a^{n-m} & \text{if } m < n \end{cases}$

3. $(ab)^m = a^m b^m$

4. $\left(\dfrac{a}{b}\right)^m = \dfrac{a^m}{b^m} \quad (b \neq 0)$

5. $(a^m)^n = a^{mn}$

● **EXAMPLE 1** *Positive Integer Exponents*

Use properties of positive integer exponents to rewrite each of the following. Assume all denominators are nonzero.

(a) $\dfrac{5^6}{5^4}$ (b) $\dfrac{x^2}{x^5}$ (c) $\left(\dfrac{x}{y}\right)^4$ (d) $(3x^2y^3)^4$ (e) $3^3 \cdot 3^2$

Solution

(a) $\dfrac{5^6}{5^4} = 5^{6-4} = 5^2$ (b) $\dfrac{x^2}{x^5} = \dfrac{1}{x^{5-2}} = \dfrac{1}{x^3}$ (c) $\left(\dfrac{x}{y}\right)^4 = \dfrac{x^4}{y^4}$

(d) $(3x^2y^3)^4 = 3^4(x^2)^4(y^3)^4 = 81x^8y^{12}$ (e) $3^3 \cdot 3^2 = 3^{3+2} = 3^5$

For certain calculus operations, use of negative exponents is necessary in order to write problems in the proper form. We can extend the rules for positive integer exponents to all integers by defining a^0 and a^{-n}. Clearly $a^m \cdot a^0$ should equal $a^{m+0} = a^m$, and it will if $a^0 = 1$.

Zero Exponent

> For any nonzero real number a, we define $a^0 = 1$. We leave 0^0 undefined.

In Section 0.2, we defined a^{-1} as $1/a$ for $a \neq 0$, so we define a^{-n} as $(a^{-1})^n$.

Negative Exponents

$$a^{-n} = (a^{-1})^n = \left(\frac{1}{a}\right)^n = \frac{1}{a^n} \qquad (a \neq 0)$$

$$\left(\frac{a}{b}\right)^{-n} = \left[\left(\frac{a}{b}\right)^{-1}\right]^n = \left(\frac{b}{a}\right)^n \qquad (a \neq 0, b \neq 0)$$

● **EXAMPLE 2** *Negative and Zero Exponents*

Write each of the following without exponents.

(a) $6 \cdot 3^0$ (b) 6^{-2} (c) $\left(\dfrac{1}{3}\right)^{-1}$ (d) $-\left(\dfrac{2}{3}\right)^{-4}$ (e) $(-4)^{-2}$

Solution

(a) $6 \cdot 3^0 = 6 \cdot 1 = 6$ (b) $6^{-2} = \dfrac{1}{6^2} = \dfrac{1}{36}$

(c) $\left(\dfrac{1}{3}\right)^{-1} = \dfrac{3}{1} = 3$ (d) $-\left(\dfrac{2}{3}\right)^{-4} = -\left(\dfrac{3}{2}\right)^4 = \dfrac{-81}{16}$

(e) $(-4)^{-2} = \dfrac{1}{(-4)^2} = \dfrac{1}{16}$

As we'll see in the chapter on the mathematics of finance (Chapter 6), negative exponents arise in financial calculations when we have a future goal for an investment and want to know how much to invest now. For example, if money can be invested at 9%, compounded annually, then the amount we must invest now (which is called the present value) in order to have $10,000 in the account after 7 years is given by $10,000(1.09)^{-7}$. Calculations such as this are often done directly with a calculator.

Using the definitions of zero and negative exponents enables us to extend the rules of exponents to all integers and to express them more simply.

Rules of Exponents

For real numbers a and b and *integers* m and n,

1. $a^m \cdot a^n = a^{m+n}$
2. $a^m / a^n = a^{m-n}$ $(a \neq 0)$
3. $(ab)^m = a^m b^m$
4. $(a^m)^n = a^{mn}$
5. $(a/b)^m = a^m / b^m$ $(b \neq 0)$
6. $a^0 = 1$ $(a \neq 0)$
7. $a^{-n} = 1/a^n$ $(a \neq 0)$
8. $(a/b)^{-n} = (b/a)^n$ $(a, b \neq 0)$

Throughout the remainder of the text, we will assume all expressions are defined.

● **EXAMPLE 3 *Operations with Exponents***

Use the rules of exponents and the definitions of a^0 and a^{-n} to simplify each of the following with positive exponents.

(a) $2(x^2)^{-2}$ (b) $x^{-2} \cdot x^{-5}$ (c) $\dfrac{x^{-8}}{x^{-4}}$ (d) $\left(\dfrac{2x^3}{3x^{-5}}\right)^{-2}$

Solution

(a) $2(x^2)^{-2} = 2x^{-4} = 2\left(\dfrac{1}{x^4}\right) = \dfrac{2}{x^4}$ (b) $x^{-2} \cdot x^{-5} = x^{-2-5} = x^{-7} = \dfrac{1}{x^7}$

(c) $\dfrac{x^{-8}}{x^{-4}} = x^{-8-(-4)} = x^{-4} = \dfrac{1}{x^4}$ (d) $\left(\dfrac{2x^3}{3x^{-5}}\right)^{-2} = \left(\dfrac{2x^8}{3}\right)^{-2} = \left(\dfrac{3}{2x^8}\right)^2 = \dfrac{9}{4x^{16}}$

● **Checkpoint**

1. Complete the following.

(a) $x^3 \cdot x^8 = x^?$ (b) $x \cdot x^4 \cdot x^{-3} = x^?$ (c) $\dfrac{1}{x^4} = x^?$

(d) $x^{24} \div x^{-3} = x^?$ (e) $(x^4)^2 = x^?$ (f) $(2x^4 y)^3 = ?$

2. True or false:

(a) $3x^{-2} = \dfrac{1}{9x^2}$ (b) $-x^{-4} = \dfrac{-1}{x^4}$ (c) $x^{-3} = -x^3$

3. Evaluate the following, if possible. For any that are meaningless, so state. Assume $x > 0$.

(a) 0^4 (b) 0^0 (c) x^0 (d) 0^x (e) 0^{-4} (f) -5^{-2}

● **EXAMPLE 4 *Rewriting a Quotient***

Write $(x^2 y)/(9wz^3)$ with all factors in the numerator.

Solution

$$\frac{x^2 y}{9wz^3} = x^2 y\left(\frac{1}{9wz^3}\right) = x^2 y\left(\frac{1}{9}\right)\left(\frac{1}{w}\right)\left(\frac{1}{z^3}\right) = x^2 y \cdot 9^{-1} w^{-1} z^{-3}$$
$$= 9^{-1} x^2 y w^{-1} z^{-3}$$

● **EXAMPLE 5 *Rewriting with Positive Exponents***

Simplify the following so all exponents are positive.

(a) $(2^3 x^{-4} y^5)^{-2}$ (b) $\dfrac{2x^4 (x^2 y)^0}{(4x^{-2} y)^2}$

Solution

(a) $(2^3x^{-4}y^5)^{-2} = 2^{-6}x^8y^{-10} = \dfrac{1}{2^6} \cdot x^8 \cdot \dfrac{1}{y^{10}} = \dfrac{x^8}{64y^{10}}$

(b) $\dfrac{2x^4(x^2y)^0}{(4x^{-2}y)^2} = \dfrac{2x^4 \cdot 1}{4^2x^{-4}y^2} = \dfrac{2}{4^2} \cdot \dfrac{x^4}{x^{-4}} \cdot \dfrac{1}{y^2} = \dfrac{2}{16} \cdot \dfrac{x^8}{1} \cdot \dfrac{1}{y^2} = \dfrac{x^8}{8y^2}$

● **Checkpoint Solutions**

1. (a) $x^3 \cdot x^8 = x^{3+8} = x^{11}$ (b) $x \cdot x^4 \cdot x^{-3} = x^{1+4+(-3)} = x^2$

   (c) $\dfrac{1}{x^4} = x^{-4}$ (d) $x^{24} \div x^{-3} = x^{24-(-3)} = x^{27}$

   (e) $(x^4)^2 = x^{(4)(2)} = x^8$ (f) $(2x^4y)^3 = 2^3(x^4)^3y^3 = 8x^{12}y^3$

2. (a) False. $3x^{-2} = 3\left(\dfrac{1}{x^2}\right) = \dfrac{3}{x^2}$

   (b) True. $-x^{-4} = (-1)\left(\dfrac{1}{x^4}\right) = \dfrac{-1}{x^4}$

   (c) False. $x^{-3} = \dfrac{1}{x^3}$

3. (a) $0^4 = 0$ (b) Meaningless. 0^0 is undefined.
   (c) $x^0 = 1$ since $x \neq 0$ (d) $0^x = 0$ because $x > 0$

   (e) Meaningless. 0^{-4} would be $\dfrac{1}{0^4}$, which is not defined.

   (f) $-5^{-2} = (-1)\left(\dfrac{1}{5^2}\right) = \dfrac{-1}{25}$

0.3 Exercises

Evaluate in Problems 1–8. Write all answers without using exponents.

1. $(-4)^4$ 2. -5^3 3. -2^6 4. $(-2)^5$

5. 3^{-2} 6. 6^{-1} 7. $-\left(\dfrac{3}{2}\right)^2$ 8. $\left(\dfrac{2}{3}\right)^3$

In Problems 9–16, use rules of exponents to simplify the expressions. Express answers with positive exponents.

9. $6^5 \cdot 6^3$ 10. $\dfrac{7^8}{7^3}$ 11. $\dfrac{10^8}{10^9}$

12. $\dfrac{5^4}{(5^{-2} \cdot 5^3)}$ 13. $(3^3)^3$ 14. $(2^{-3})^{-2}$

15. $\left(\dfrac{2}{3}\right)^{-2}$ 16. $\left(\dfrac{-2}{5}\right)^{-4}$

In Problems 17–20, simplify by expressing answers with positive exponents $(x, y, z \neq 0)$.

17. $(x^2)^{-3}$ 18. x^{-4} 19. $xy^{-2}z^0$ 20. $(xy^{-2})^0$

In Problems 21–34, use the rules of exponents to simplify so that only positive exponents remain.

21. $x^3 \cdot x^4$ 22. $a^5 \cdot a$ 23. $x^{-5} \cdot x^3$

24. $y^{-5} \cdot y^{-2}$ 25. $\dfrac{x^8}{x^4}$ 26. $\dfrac{a^5}{a^{-1}}$

27. $\dfrac{y^5}{y^{-7}}$ 28. $\dfrac{y^{-3}}{y^{-4}}$ 29. $(x^4)^3$ 30. $(y^3)^{-2}$

31. $(xy)^2$ 32. $(2m)^3$ 33. $\left(\dfrac{2}{x^5}\right)^4$ 34. $\left(\dfrac{8}{a^3}\right)^3$

In Problems 35–46, compute and simplify so that only positive exponents remain.

35. $(2x^{-2}y)^{-4}$ 36. $(-32x^5)^{-3}$

37. $(-8a^{-3}b^2)(2a^5b^{-4})$ 38. $(-3m^2y^{-1})(2m^{-3}y^{-1})$

39. $(2x^{-2}) \div (x^{-1}y^2)$ 40. $(-8a^{-3}b^2c) \div (2a^5b^4)$

41. $\left(\dfrac{x^3}{y^{-2}}\right)^{-3}$ 42. $\left(\dfrac{x^{-2}}{y}\right)^{-3}$

43. $\left(\dfrac{a^{-2}b^{-1}c^{-4}}{a^4b^{-3}c^0}\right)^{-3}$ 44. $\left(\dfrac{4x^{-1}y^{-40}}{2^{-2}x^4y^{-10}}\right)^{-2}$

45. (a) $\dfrac{2x^{-2}}{(2x)^2}$ (b) $\dfrac{(2x)^{-2}}{(2x)^2}$

    (c) $\dfrac{2x^{-2}}{2x^2}$ (d) $\dfrac{2x^{-2}}{(2x)^{-2}}$

46. (a) $\dfrac{2^{-1}x^{-2}}{(2x)^2}$ (b) $\dfrac{2^{-1}x^{-2}}{2x^2}$

    (c) $\dfrac{(2x^{-2})^{-1}}{(2x)^{-2}}$ (d) $\dfrac{(2x^{-2})^{-1}}{2x^2}$

In many applications it is often necessary to write expressions in the form cx^n, where c is a constant and n is an integer. In Problems 47–54, write the expressions in this form.

47. $\dfrac{1}{x}$ 48. $\dfrac{1}{x^2}$ 49. $(2x)^3$ 50. $(3x)^2$

51. $\dfrac{1}{4x^2}$ 52. $\dfrac{3}{2x^4}$ 53. $\left(\dfrac{-x}{2}\right)^3$ 54. $\left(\dfrac{-x}{3}\right)^2$

In Problems 55–58, use a calculator to evaluate the indicated powers.

55. 1.2^4 56. $(-3.7)^3$ 57. $(1.5)^{-5}$ 58. $(-0.8)^{-9}$

APPLICATIONS

Compound interest If $\$P$ is invested for n years at rate i (as a decimal), compounded annually, the future value that accrues is given by $S = P(1 + i)^n$, and the interest earned is $I = S - P$. In Problems 59–62, find S and I for the given P, n, and i.

59. $1200 for 5 years at 12%
60. $1800 for 7 years at 10%
61. $5000 for 6 years at 11.5%
62. $800 for 20 years at 10.5%

Present value If an investment has a goal (future value) of $\$S$ after n years, invested at interest rate i (as a decimal), compounded annually, then the present value P that must be invested is given by $P = S(1 + i)^{-n}$. In Problems 63 and 64, find P for the given S, n, and i.

63. $15,000 after 6 years at 11.5%
64. $80,000 after 20 years at 10.5%

65. *Federal tax per capita* Using Internal Revenue Service data for 1960–2002, the federal tax per capita (in dollars) can be approximated by the formula

$$T = 257.8(1.0693)^t$$

where t is the number of years past 1950.
(a) What t-values correspond to the years 1970, 1995, 2000, and 2002?
(b) The actual federal tax per capita for the years in (a) are as follows:

1970	1995	2000	2002
$955.31	$5216.70	$7595.24	$6967.40

What does the formula predict (to the nearest cent) for these years?
(c) Why do you think the actual tax per capita fell in 2002?
(d) What does the formula predict for the tax per capita in 2010? Do you think this amount will turn out to be too high, too low, or about right?

66. *Stock shares traded* On the New York Stock Exchange (NYSE) for 1970–2002, the average daily shares traded S (in millions of shares) can be approximated by the formula

$$S = 0.4824(1.16345)^t$$

where t is the number of years past 1950. (*Source:* New York Stock Exchange)

(a) What t-values correspond to the years 1980, 1995, and 2002?
(b) For the years in (a), the actual average millions of shares traded on the NYSE were as follows.

1980	1995	2002
44.871	346.101	1441.015

What does the formula predict for these years?
(c) Suppose in 2010 that a stock market average (such as the Dow Jones Industrial Average) dramatically soared or tumbled; do you think this formula's predictions would be accurate, too low, or too high? Explain.

67. *Endangered species* The number of endangered species y can be approximated by the formula

$$y = \dfrac{1209}{1 + 6.46(1.17)^{-t}}$$

where t is the number of years past 1980. (*Source:* U.S. Fish and Wildlife Service)

(a) The actual numbers of endangered species for selected years were as follows.

1980	1990	2003
224	442	987

For each of these years, find the number of endangered species predicted by the formula. Round your answer to the nearest integer.
(b) How many species does the formula estimate will be added to the endangered list between 2000 and 2010?
(c) Why do you think the answer to (b) is smaller than the number of species added from 1980 to 1990?
(d) Why is it reasonable for a formula such as this to have an upper limit that cannot be exceeded? Use large t-values in the formula to discover this formula's upper limit.

68. *Internet users* The percent p of U.S. households with Internet service can be approximated by the equation

$$p = \dfrac{59.57}{1 + 5.22(1.795)^{-t}}$$

where t is the number of years past 1995. (*Source:* U.S. Department of Commerce)

(a) The percents of U.S. households with Internet service for selected years were as follows.

1997	2000	2003
22.2%	44.4%	56.0%

For each of these years, use the equation to find the predicted percent of households with Internet service.

(b) In the three years from 1997 to 2000, the percent of households with Internet service increased by 22.2%. What percent increase does the equation predict from 2006 to 2009? Why do you think the 2006–2009 change is so different from the 1997–2000 change?

(c) Why is it reasonable for a formula such as this to have an upper limit that cannot be exceeded? Use large t-values in the formula to discover this formula's upper limit.

69. *Health care expenditures* Using data from the U.S. Department of Health and Human Services, the national health care expenditure H (in billions of dollars) can be modeled (that is, accurately approximated) by the formula

$$H = 29.57(1.102)^t$$

where t is the number of years past 1960.

(a) What t-value corresponds to 1970?

(b) Approximate the national health care expenditure in 1970.

(c) Approximate the national health care expenditure in 1993.

(d) Estimate the national health care expenditure in 2010.

0.4 Radicals and Rational Exponents

Roots   A process closely linked to that of raising numbers to powers is that of extracting roots. From geometry we know that if an edge of a cube has a length of x units, its volume is x^3 cubic units. Reversing this process, we determine that if the volume of a cube is V cubic units, the length of an edge is the cube root of V, which is denoted

$$\sqrt[3]{V}\ \text{units}$$

When we seek the **cube root** of a number such as 8 (written $\sqrt[3]{8}$), we are looking for a real number whose cube equals 8. Because $2^3 = 8$, we know that $\sqrt[3]{8} = 2$. Similarly, $\sqrt[3]{-27} = -3$ because $(-3)^3 = -27$. The expression $\sqrt[n]{a}$ is called a **radical**, where $\sqrt{\ }$ is the **radical sign,** n the **index,** and a the **radicand.** When no index is indicated, the index is assumed to be 2 and the expression is called a **square root;** thus $\sqrt{4}$ is the square root of 4 and represents the positive number whose square is 4.

Only one real number satisfies $\sqrt[n]{a}$ for a real number a and an odd number n; we call that number the **principal nth root,** or, more simply, the **nth root.**

For an even index n, there are two possible cases:

1. If a is negative, there is no real number equal to $\sqrt[n]{a}$. For example, there are no real numbers that equal $\sqrt{-4}$ or $\sqrt[4]{-16}$ because there is no real number b such that $b^2 = -4$ or $b^4 = -16$. In this case, we say $\sqrt[n]{a}$ is not a real number.

2. If a is positive, there are two real numbers whose nth power equals a. For example, $3^2 = 9$ and $(-3)^2 = 9$. In order to have a unique nth root, we define the (principal) nth root, $\sqrt[n]{a}$, as the *positive* number b that satisfies $b^n = a$.

We summarize this discussion as follows:

nth Root of a   The **(principal) nth root** of a real number is defined as

$$\sqrt[n]{a} = b \quad \text{only if} \quad a = b^n$$

subject to the following conditions:

	$a = 0$	$a > 0$	$a < 0$
n even	$\sqrt[n]{a} = 0$	$\sqrt[n]{a} > 0$	$\sqrt[n]{a}$ not real
n odd	$\sqrt[n]{a} = 0$	$\sqrt[n]{a} > 0$	$\sqrt[n]{a} < 0$

When we are asked for the root of a number, we give the principal root.

● **EXAMPLE 1** *Roots*

Find the roots, if they are real numbers.

(a) $\sqrt[6]{64}$ (b) $-\sqrt{16}$ (c) $\sqrt[3]{-8}$ (d) $\sqrt{-16}$

Solution

(a) $\sqrt[6]{64} = 2$ because $2^6 = 64$ (b) $-\sqrt{16} = -(\sqrt{16}) = -4$ (c) $\sqrt[3]{-8} = -2$
(d) $\sqrt{-16}$ is not a real number because an even root of a negative number is not real.

Fractional Exponents

In order to perform evaluations on a scientific calculator or to perform calculus operations, it is sometimes necessary to rewrite radicals in exponential form with fractional exponents.

We have stated that for $a \geq 0$ and $b \geq 0$,

$$\sqrt{a} = b \quad \text{only if} \quad a = b^2$$

This means that $(\sqrt{a})^2 = b^2 = a$, or $(\sqrt{a})^2 = a$. In order to extend the properties of exponents to rational exponents, it is necessary to define

$$a^{1/2} = \sqrt{a} \quad \text{so that} \quad (a^{1/2})^2 = a$$

Exponent $1/n$

For a positive integer n, we define

$$a^{1/n} = \sqrt[n]{a} \quad \text{if} \quad \sqrt[n]{a} \text{ exists}$$

Thus $(a^{1/n})^n = a^{(1/n)\cdot n} = a$.

Because we wish the properties established for integer exponents to extend to rational exponents, we make the following definitions.

Rational Exponents

For positive integer n and any integer m (with $a \neq 0$ when $m \leq 0$ and with m/n in lowest terms):

1. $a^{m/n} = (a^{1/n})^m = (\sqrt[n]{a})^m$
2. $a^{m/n} = (a^m)^{1/n} = \sqrt[n]{a^m}$ if a is nonnegative when n is even.

Throughout the remaining discussion, we assume all expressions are real.

● **EXAMPLE 2** *Radical Form*

Write the following in radical form and simplify.

(a) $16^{3/4}$ (b) $y^{-3/2}$ (c) $(6m)^{2/3}$

Solution

(a) $16^{3/4} = \sqrt[4]{16^3} = (\sqrt[4]{16})^3 = (2)^3 = 8$

(b) $y^{-3/2} = \dfrac{1}{y^{3/2}} = \dfrac{1}{\sqrt{y^3}}$

(c) $(6m)^{2/3} = \sqrt[3]{(6m)^2} = \sqrt[3]{36m^2}$

● **EXAMPLE 3** *Fractional Exponents*

Write the following without radical signs.

(a) $\sqrt{x^3}$ (b) $\dfrac{1}{\sqrt[3]{b^2}}$ (c) $\sqrt[3]{(ab)^3}$

Solution

(a) $\sqrt{x^3} = x^{3/2}$ (b) $\dfrac{1}{\sqrt[3]{b^2}} = \dfrac{1}{b^{2/3}} = b^{-2/3}$ (c) $\sqrt[3]{(ab)^3} = (ab)^{3/3} = ab$

Note that with many calculators, if we want to evaluate $\sqrt[10]{3.012}$, we must rewrite it as $(3.012)^{1/10}$. Our definition of $a^{m/n}$ guarantees that the rules for exponents will apply to fractional exponents. Thus we can perform operations with fractional exponents as we did with integer exponents.

● **EXAMPLE 4** *Operations with Fractional Exponents*

Simplify the following expressions.

(a) $a^{1/2} \cdot a^{1/6}$ (b) $a^{3/4}/a^{1/3}$ (c) $(a^3 b)^{2/3}$ (d) $(a^{3/2})^{1/2}$ (e) $a^{-1/2} \cdot a^{-3/2}$

Solution
(a) $a^{1/2} \cdot a^{1/6} = a^{1/2+1/6} = a^{3/6+1/6} = a^{4/6} = a^{2/3}$
(b) $a^{3/4}/a^{1/3} = a^{3/4-1/3} = a^{9/12-4/12} = a^{5/12}$
(c) $(a^3 b)^{2/3} = (a^3)^{2/3} b^{2/3} = a^2 b^{2/3}$
(d) $(a^{3/2})^{1/2} = a^{(3/2)(1/2)} = a^{3/4}$
(e) $a^{-1/2} \cdot a^{-3/2} = a^{-1/2-3/2} = a^{-2} = 1/a^2$

● *Checkpoint*

1. Which of the following are *not* real numbers?
    (a) $\sqrt[3]{-64}$ (b) $\sqrt{-64}$ (c) $\sqrt{0}$ (d) $\sqrt[4]{1}$ (e) $\sqrt[5]{-1}$ (f) $\sqrt[8]{-1}$
2. (a) Write as radicals: $x^{1/3}, x^{2/5}, x^{-3/2}$

    (b) Write with fractional exponents: $\sqrt[4]{x^3} = x^?$, $\dfrac{1}{\sqrt{x}} = \dfrac{1}{x^?} = x^?$

3. Evaluate the following.
    (a) $8^{2/3}$ (b) $(-8)^{2/3}$ (c) $8^{-2/3}$ (d) $-8^{-2/3}$ (e) $\sqrt[15]{71}$
4. Complete the following.
    (a) $x \cdot x^{1/3} \cdot x^3 = x^?$ (b) $x^2 \div x^{1/2} = x^?$ (c) $(x^{-2/3})^{-3} = x^?$

    (d) $x^{-3/2} \cdot x^{1/2} = x^?$ (e) $x^{-3/2} \cdot x = x^?$ (f) $\left(\dfrac{x^4}{y^2}\right)^{3/2} = ?$

5. True or false:
    (a) $\dfrac{8x^{2/3}}{x^{-1/3}} = 4x$ (b) $(16x^8 y)^{3/4} = 12x^6 y^{3/4}$

    (c) $\left(\dfrac{x^2}{y^3}\right)^{-1/3} = \left(\dfrac{y^3}{x^2}\right)^{1/3} = \dfrac{y}{x^{2/3}}$

Operations with Radicals

We can perform operations with radicals by first rewriting in exponential form, performing the operations with exponents, and then converting the answer back to radical form. Another option is to apply directly the following rules for operations with radicals.

Rules for Radicals	Examples
Given that $\sqrt[n]{a}$ and $\sqrt[n]{b}$ are real,*	
1. $\sqrt[n]{a^n} = (\sqrt[n]{a})^n = a$	1. $\sqrt[5]{6^5} = (\sqrt[5]{6})^5 = 6$
2. $\sqrt[n]{a} \cdot \sqrt[n]{b} = \sqrt[n]{ab}$	2. $\sqrt[3]{2}\,\sqrt[3]{4} = \sqrt[3]{8} = \sqrt[3]{2^3} = 2$
3. $\dfrac{\sqrt[n]{a}}{\sqrt[n]{b}} = \sqrt[n]{\dfrac{a}{b}}$ $(b \neq 0)$	3. $\dfrac{\sqrt{18}}{\sqrt{2}} = \sqrt{\dfrac{18}{2}} = \sqrt{9} = 3$

*Note that this means $a \geq 0$ and $b \geq 0$ if n is even.

Let us consider Rule 1 for radicals more carefully. Note that if n is even and $a < 0$, then $\sqrt[n]{a}$ is not real, and Rule 1 does not apply. For example, $\sqrt{-2}$ is not a real number, and

$$\sqrt{(-2)^2} \neq -2 \quad \text{because} \quad \sqrt{(-2)^2} = \sqrt{4} = 2 = -(-2)$$

We can generalize this observation as follows: If $a < 0$, then $\sqrt{a^2} = -a > 0$, so

$$\sqrt{a^2} = \begin{cases} a & \text{if } a \geq 0 \\ -a & \text{if } a < 0 \end{cases}$$

This means

$$\sqrt{a^2} = |a|$$

● **EXAMPLE 5** *Simplifying Radicals*

Simplify:

(a) $\sqrt[3]{8^3}$ (b) $\sqrt[5]{x^5}$ (c) $\sqrt{x^2}$ (d) $\left[\sqrt[7]{(3x^2 + 4)^3}\right]^7$

Solution
(a) $\sqrt[3]{8^3} = 8$ by Rule 1 for radicals
(b) $\sqrt[5]{x^5} = x$ (c) $\sqrt{x^2} = |x|$ (d) $\left[\sqrt[7]{(3x^2 + 4)^3}\right]^7 = (3x^2 + 4)^3$

Up to now, to *simplify* a radical has meant to find the indicated root. More generally, a radical expression $\sqrt[n]{x}$ is considered simplified if x has no nth powers as factors. Rule 2 for radicals ($\sqrt[n]{a} \cdot \sqrt[n]{b} = \sqrt[n]{ab}$) provides a procedure for simplifying radicals.

● **EXAMPLE 6** *Simplifying Radicals*

Simplify the following radicals; assume the expressions are real numbers.

(a) $\sqrt{48x^5y^6}$ $(y \geq 0)$ (b) $\sqrt[3]{72a^3b^4}$

Solution
(a) To simplify $\sqrt{48x^5y^6}$, we first factor $48x^5y^6$ into perfect-square factors and other factors. Then we apply Rule 2.

$$\sqrt{48x^5y^6} = \sqrt{16 \cdot 3 \cdot x^4xy^6} = \sqrt{16}\sqrt{x^4}\sqrt{y^6}\sqrt{3x} = 4x^2y^3\sqrt{3x}$$

(b) In this case, we factor $72a^3b^4$ into factors that are perfect cubes and other factors.

$$\sqrt[3]{72a^3b^4} = \sqrt[3]{8 \cdot 9a^3b^3b} = \sqrt[3]{8} \cdot \sqrt[3]{a^3} \cdot \sqrt[3]{b^3} \cdot \sqrt[3]{9b} = 2ab\sqrt[3]{9b}$$

Rule 2 for radicals also provides a procedure for multiplying two roots with the same index.

● **EXAMPLE 7** *Multiplying Radicals*

Multiply the following and simplify the answers, assuming nonnegative variables.

(a) $\sqrt[3]{2xy} \cdot \sqrt[3]{4x^2y}$ (b) $\sqrt{8xy^3z}\sqrt{4x^2y^3z^2}$

Solution

(a) $\sqrt[3]{2xy} \cdot \sqrt[3]{4x^2y} = \sqrt[3]{2xy \cdot 4x^2y} = \sqrt[3]{8x^3y^2} = \sqrt[3]{8} \cdot \sqrt[3]{x^3} \cdot \sqrt[3]{y^2} = 2x\sqrt[3]{y^2}$

(b) $\sqrt{8xy^3z}\sqrt{4x^2y^3z^2} = \sqrt{32x^3y^6z^3} = \sqrt{16x^2y^6z^2}\sqrt{2xz} = 4xy^3z\sqrt{2xz}$

Rule 3 for radicals ($\sqrt[n]{a}/\sqrt[n]{b} = \sqrt[n]{a/b}$) indicates how to find the quotient of two roots with the same index.

● **EXAMPLE 8** *Dividing Radicals*

Find the quotients and simplify the answers, assuming nonnegative variables.

(a) $\dfrac{\sqrt[3]{32}}{\sqrt[3]{4}}$ (b) $\dfrac{\sqrt{16a^3x}}{\sqrt{2ax}}$

Solution

(a) $\dfrac{\sqrt[3]{32}}{\sqrt[3]{4}} = \sqrt[3]{\dfrac{32}{4}} = \sqrt[3]{8} = 2$ (b) $\dfrac{\sqrt{16a^3x}}{\sqrt{2ax}} = \sqrt{\dfrac{16a^3x}{2ax}} = \sqrt{8a^2} = 2a\sqrt{2}$

Rationalizing Occasionally, we wish to express a fraction containing radicals in an equivalent form that contains no radicals in the denominator. This is accomplished by multiplying the numerator *and* the denominator by the expression that will remove the radical. This process is called **rationalizing the denominator.**

● **EXAMPLE 9** *Rationalizing Denominators*

Express each of the following with no radicals in the denominator. (Rationalize each denominator.)

(a) $\dfrac{15}{\sqrt{x}}$ (b) $\dfrac{2x}{\sqrt{18xy}}$ $(x, y > 0)$ (c) $\dfrac{3x}{\sqrt[3]{2x^2}}$ $(x \neq 0)$

Solution

(a) We wish to create a perfect square under the radical in the denominator.

$$\frac{15}{\sqrt{x}} \cdot \frac{\sqrt{x}}{\sqrt{x}} = \frac{15\sqrt{x}}{x}$$

(b) $\dfrac{2x}{\sqrt{18xy}} \cdot \dfrac{\sqrt{2xy}}{\sqrt{2xy}} = \dfrac{2x\sqrt{2xy}}{\sqrt{36x^2y^2}} = \dfrac{2x\sqrt{2xy}}{6xy} = \dfrac{\sqrt{2xy}}{3y}$

(c) We wish to create a perfect cube under the radical in the denominator.

$$\frac{3x}{\sqrt[3]{2x^2}} \cdot \frac{\sqrt[3]{4x}}{\sqrt[3]{4x}} = \frac{3x\sqrt[3]{4x}}{\sqrt[3]{8x^3}} = \frac{3x\sqrt[3]{4x}}{2x} = \frac{3\sqrt[3]{4x}}{2}$$

● *Checkpoint*

6. Simplify:

   (a) $\sqrt[7]{x^7}$ (b) $\left[\sqrt[5]{(x^2 + 1)^2}\right]^5$ (c) $\sqrt{12xy^2} \cdot \sqrt{3x^2y}$

7. Rationalize the denominator of $\dfrac{x}{\sqrt{5x}}$ if $x \neq 0$.

It is also sometimes useful, especially in calculus, to *rationalize the numerator* of a fraction. For example, we can rationalize the numerator of

$$\frac{\sqrt[3]{4x^2}}{3x}$$

by multiplying the numerator and denominator by $\sqrt[3]{2x}$, which creates a perfect cube under the radical:

$$\frac{\sqrt[3]{4x^2}}{3x} \cdot \frac{\sqrt[3]{2x}}{\sqrt[3]{2x}} = \frac{\sqrt[3]{8x^3}}{3x\sqrt[3]{2x}} = \frac{2x}{3x\sqrt[3]{2x}} = \frac{2}{3\sqrt[3]{2x}}$$

● **Checkpoint Solutions**

1. Only *even* roots of negatives are not real numbers. Thus $\sqrt{-64}$ and $\sqrt[8]{-1}$ are not real numbers.

2. (a) $x^{1/3} = \sqrt[3]{x}$, $\quad x^{2/5} = \sqrt[5]{x^2}$, $\quad x^{-3/2} = \dfrac{1}{x^{3/2}} = \dfrac{1}{\sqrt{x^3}}$

   (b) $\sqrt[4]{x^3} = x^{3/4}$, $\quad \dfrac{1}{\sqrt{x}} = \dfrac{1}{x^{1/2}} = x^{-1/2}$

3. (a) $8^{2/3} = (\sqrt[3]{8})^2 = 2^2 = 4$ $\qquad$ (b) $(-8)^{2/3} = (\sqrt[3]{-8})^2 = (-2)^2 = 4$

   (c) $8^{-2/3} = \dfrac{1}{8^{2/3}} = \dfrac{1}{4}$ $\qquad$ (d) $-8^{-2/3} = -\left(\dfrac{1}{8^{2/3}}\right) = -\dfrac{1}{4}$

   (e) $\sqrt[15]{71} = (71)^{1/15} \approx 1.32867$

4. (a) $x \cdot x^{1/3} \cdot x^3 = x^{1+1/3+3} = x^{13/3}$ $\qquad$ (b) $x^2 \div x^{1/2} = x^{2-1/2} = x^{3/2}$

   (c) $(x^{-2/3})^{-3} = x^{(-2/3)(-3)} = x^2$ $\qquad$ (d) $x^{-3/2} \cdot x^{1/2} = x^{-3/2+1/2} = x^{-1}$

   (e) $x^{-3/2} \cdot x = x^{-3/2+1} = x^{-1/2}$ $\qquad$ (f) $\left(\dfrac{x^4}{y^2}\right)^{3/2} = \dfrac{(x^4)^{3/2}}{(y^2)^{3/2}} = \dfrac{x^6}{y^3}$

5. (a) False. $\quad \dfrac{8x^{2/3}}{x^{-1/3}} = 8x^{2/3} \cdot x^{1/3} = 8x^{2/3+1/3} = 8x$

   (b) False. $\quad (16x^8y)^{3/4} = 16^{3/4}(x^8)^{3/4}y^{3/4} = (\sqrt[4]{16})^3x^6y^{3/4} = 8x^6y^{3/4}$

   (c) True.

6. (a) $\sqrt[7]{x^7} = x$

   (b) $\left[\sqrt[5]{(x^2+1)^2}\right]^5 = (x^2+1)^2$

   (c) $\sqrt{12xy^2} \cdot \sqrt{3x^2y} = \sqrt{36x^3y^3} = \sqrt{36x^2y^2 \cdot xy} = \sqrt{36x^2y^2}\sqrt{xy} = 6xy\sqrt{xy}$

7. $\dfrac{x}{\sqrt{5x}} = \dfrac{x}{\sqrt{5x}} \cdot \dfrac{\sqrt{5x}}{\sqrt{5x}} = \dfrac{x\sqrt{5x}}{5x} = \dfrac{\sqrt{5x}}{5}$

0.4 Exercises

Unless stated otherwise, assume all variables are nonnegative and all denominators are nonzero.

In Problems 1–8, find the powers and roots, if they are real numbers.

1. (a) $\sqrt{256/9}$ $\qquad$ (b) $\sqrt{1.44}$
2. (a) $\sqrt[5]{-32^3}$ $\qquad$ (b) $\sqrt[4]{-16^5}$
3. (a) $16^{3/4}$ $\qquad$ (b) $(-16)^{-3/2}$
4. (a) $-27^{-1/3}$ $\qquad$ (b) $32^{3/5}$
5. $\left(\dfrac{8}{27}\right)^{-2/3}$
6. $\left(\dfrac{4}{9}\right)^{3/2}$
7. (a) $8^{2/3}$ $\qquad$ (b) $(-8)^{-2/3}$
8. (a) $8^{-2/3}$ $\qquad$ (b) $-8^{2/3}$

In Problems 9 and 10, rewrite each radical with a fractional exponent, and then approximate the value with a calculator.

9. $\sqrt[9]{(6.12)^4}$ $\qquad$ 10. $\sqrt[12]{4.96}$

In Problems 11–14, replace each radical with a fractional exponent. Do not simplify.

11. $\sqrt{m^3}$ $\quad$ 12. $\sqrt[3]{x^5}$ $\quad$ 13. $\sqrt[4]{m^2n^5}$ $\quad$ 14. $\sqrt[5]{x^3}$

In Problems 15–18, write in radical form. Do not simplify.

15. $x^{7/6}$ $\qquad$ 16. $y^{11/5}$
17. $-(1/4)x^{-5/4}$ $\qquad$ 18. $-x^{-5/3}$

In Problems 19–32, use the properties of exponents to simplify each expression so that only positive exponents remain.

19. $y^{1/4} \cdot y^{1/2}$ 20. $x^{2/3} \cdot x^{1/5}$ 21. $z^{3/4} \cdot z^4$

22. $x^{-2/3} \cdot x^2$ 23. $y^{-3/2} \cdot y^{-1}$ 24. $z^{-2} \cdot z^{5/3}$

25. $\dfrac{x^{1/3}}{x^{-2/3}}$ 26. $\dfrac{x^{-1/2}}{x^{-3/2}}$ 27. $\dfrac{y^{-5/2}}{y^{-2/5}}$

28. $\dfrac{x^{4/9}}{x^{1/12}}$ 29. $(x^{2/3})^{3/4}$ 30. $(x^{4/5})^3$

31. $(x^{-1/2})^2$ 32. $(x^{-2/3})^{-2/5}$

In Problems 33–38, simplify each expression by using the properties of radicals. Assume nonnegative variables.

33. $\sqrt{64x^4}$ 34. $\sqrt[3]{-64x^6y^3}$ 35. $\sqrt{128x^4y^5}$

36. $\sqrt[3]{54x^5x^8}$ 37. $\sqrt[3]{40x^8y^5}$ 38. $\sqrt{32x^5y}$

In Problems 39–46, perform the indicated operations and simplify.

39. $\sqrt{12x^3y} \cdot \sqrt{3x^2y}$ 40. $\sqrt[3]{16x^2y} \cdot \sqrt[3]{3x^2y}$

41. $\sqrt{63x^5y^3} \cdot \sqrt{28x^2y}$ 42. $\sqrt{10xz^{10}} \cdot \sqrt{30x^{17}z}$

43. $\dfrac{\sqrt{12x^3y^{12}}}{\sqrt{27xy^2}}$ 44. $\dfrac{\sqrt{250xy^7z^4}}{\sqrt{18x^{17}y^2}}$

45. $\dfrac{\sqrt[4]{32a^9b^5}}{\sqrt[4]{162a^{17}}}$ 46. $\dfrac{\sqrt[3]{-16x^3y^4}}{\sqrt[3]{128y^2}}$

In Problems 47–50, use properties of exponents and radicals to determine a value for x that makes each statement true.

47. $(A^9)^x = A$ 48. $(B^{20})^x = B$

49. $(\sqrt[7]{R})^x = R$ 50. $(\sqrt{T^3})^x = T$

In Problems 51–56, rationalize each denominator and then simplify.

51. $\sqrt{2/3}$ 52. $\sqrt{5/8}$ 53. $\dfrac{\sqrt{m^2x}}{\sqrt{mx^2}}$

54. $\dfrac{5x^3w}{\sqrt{4xw^2}}$ 55. $\dfrac{\sqrt[3]{m^2x}}{\sqrt[3]{mx^5}}$ 56. $\dfrac{\sqrt[4]{mx^3}}{\sqrt[4]{y^2z^5}}$

In calculus it is frequently important to write an expression in the form cx^n, where c is a constant and n is a rational number. In Problems 57–60, write each expression in this form.

57. $\dfrac{-2}{3\sqrt[3]{x^2}}$ 58. $\dfrac{-2}{3\sqrt[4]{x^3}}$ 59. $3x\sqrt{x}$ 60. $\sqrt{x} \cdot \sqrt[3]{x}$

In calculus problems, the answers are frequently expected to be in a simple form with a radical instead of an exponent. In Problems 61–64, write each expression with radicals.

61. $\dfrac{3}{2}x^{1/2}$ 62. $\dfrac{4}{3}x^{1/3}$ 63. $\dfrac{1}{2}x^{-1/2}$ 64. $\dfrac{-1}{2}x^{-3/2}$

APPLICATIONS

65. *Richter scale* The Richter scale reading for an earthquake measures its intensity (as a multiple of some min-imum intensity used for comparison). The intensity I corresponding to a Richter scale reading R is given by

$$I = 10^R$$

(a) A quake measuring 8.5 on the Richter scale would be severe. Express the intensity of such a quake in exponential form and in radical form.

(b) Find the intensity of a quake measuring 8.5.

(c) The San Francisco quake that occurred during the 1989 World Series measured 7.1, and the 1906 San Francisco quake (which devastated the city) measured 8.25. Calculate the ratio of these intensities (larger to smaller).

66. *Sound intensity* The intensity of sound I (as a multiple of the average minimum threshold of hearing intensity) is related to the decibel level D (or loudness of sound) according to

$$I = 10^{D/10}$$

(a) Express $10^{D/10}$ using radical notation.

(b) The background noise level of a relatively quiet room has a decibel reading of 32. Find the intensity I_1 of this noise level.

(c) A decibel reading of 140 is at the threshold of pain. If I_2 is the intensity of this threshold and I_1 is the intensity found in (b), express the ratio I_2/I_1 as a power of 10. Then approximate this ratio.

67. *Investment* If \$1000 is invested at $r\%$ compounded annually, the future value S of the account after two and a half years is given by

$$S = 1000\left(1 + \frac{r}{100}\right)^{2.5} = 1000\left(1 + \frac{r}{100}\right)^{5/2}$$

(a) Express this equation with radical notation.

(b) Find the value of this account if the interest rate is 6.6% compounded annually.

68. *Global travel spending* Global spending S for travel and tourism (in billions of dollars) for the years 1990–2002 can be approximated by the equation

$$S = 43.3t^{4/5}$$

where t is the number of years past 1980. (*Source:* World Tourism Organization)

(a) Express this equation with radical notation.

(b) Estimate the global spending for travel and tourism for the year 2010.

69. *Population* The population P of India (in billions) for 2000–2050 can be approximated by the equation

$$P = 0.924t^{0.13}$$

where $t > 0$ is the number of years past 2000. (*Source:* United Nations)

(a) Express this equation with radical notation.

(b) Does this equation predict a greater increase from 2005 to 2010 or from 2045 to 2050? What might explain this difference?

70. *Transportation* The percent p of paved roads and streets in the United States can be approximated with the equation

$$p = 6.61t^{0.56}$$

where t is the number of years past 1940. (*Source:* U.S. Department of Transportation)

(a) Express this equation with radical notation.

(b) Does this equation estimate a greater percent change during the decade of the 1970s or during the decade from 2000 to 2010? What might explain this?

(c) When can you be certain this equation is no longer valid?

Half-life **In Problems 71 and 72, use the fact that the quantity of a radioactive substance after t years is given by $q = q_0(2^{-t/k})$, where q_0 is the original amount of radioactive material and k is its half-life (the number of years it takes for half the radioactive substance to decay).**

71. The half-life of strontium-90 is 25 years. Find the amount of strontium-90 remaining after 10 years if $q_0 = 98$ kg.

72. The half-life of carbon-14 is 5600 years. Find the amount of carbon-14 remaining after 10,000 years if $q_0 = 40.0$ g.

73. *Population growth* Suppose the formula for the growth of the population of a city for the next 10 years is given by

$$P = P_0(2.5)^{ht}$$

where P_0 is the population of the city at the present time and P is the population t years from now. If $h = 0.03$ and $P_0 = 30,000$, find P when $t = 10$.

74. *Advertising and sales* Suppose it has been determined that the sales at Ewing Gallery decline after the end of an advertising campaign, with daily sales given by

$$S = 2000(2^{-0.1x})$$

where S is in dollars and x is the number of days after the campaign ends. What are the daily sales 10 days after the end of the campaign?

75. *Company growth* The growth of a company can be described by the equation

$$N = 500(0.02)^{0.7^t}$$

where t is the number of years the company has been in existence and N is the number of employees.

(a) What is the number of employees when $t = 0$? (This is the number of employees the company has when it starts.)

(b) What is the number of employees when $t = 5$?

0.5 Operations with Algebraic Expressions

In algebra we are usually dealing with combinations of real numbers (such as 3, 6/7, and $-\sqrt{2}$) and letters (such as x, a, and m). Unless otherwise specified, the letters are symbols used to represent real numbers and are sometimes called **variables.** An expression obtained by performing additions, subtractions, multiplications, divisions, or extractions of roots with one or more real numbers or variables is called an **algebraic expression.** Unless otherwise specified, the variables represent all real numbers for which the algebraic expression is a real number. Examples of algebraic expressions include

$$3x + 2y, \quad \frac{x^3y + y}{x - 1}, \quad \text{and} \quad \sqrt{x} - 3$$

Note that the variable x cannot be negative in the expression $\sqrt{x} - 3$ and that $(x^3y + y)/(x - 1)$ is not a real number when $x = 1$, because division by 0 is undefined.

Any product of a real number (called the **coefficient**) and one or more variables to powers is called a **term.** The sum of a finite number of terms with nonnegative integer powers on the variables is called a **polynomial.** If a polynomial contains only one variable x, then it is called a polynomial in x.

Polynomial in *x*

The general form of a **polynomial in *x*** is

$$a_n x^n + a_{n-1} x^{n-1} + \cdots + a_1 x + a_0$$

where each coefficient a_i is a real number for $i = 0, 1, 2, \ldots, n$. If $a_n \neq 0$, the **degree** of the polynomial is n, and a_n is called the **leading coefficient.** The term a_0 is called the **constant term.**

Thus $4x^3 - 2x - 3$ is a third-degree polynomial in x with leading coefficient 4 and constant term -3. If two or more variables are in a term, the degree of the term is the sum of the exponents of the variables. The degree of a nonzero constant term is zero. Thus the degree of $4x^2 y$ is $2 + 1 = 3$, the degree of $6xy$ is $1 + 1 = 2$, and the degree of 3 is 0. The **degree of a polynomial** containing one or more variables is the degree of the term in the polynomial having the highest degree. Therefore $2xy - 4x + 6$ is a second-degree polynomial.

A polynomial containing two terms is called a **binomial,** and a polynomial containing three terms is called a **trinomial.** A single-term polynomial is a **monomial.**

Operations with Algebraic Expressions

Because monomials and polynomials represent real numbers, the properties of real numbers can be used to add, subtract, multiply, divide, and simplify polynomials. For example, we can use the Distributive Law to add $3x$ and $2x$.

$$3x + 2x = (3 + 2)x = 5x$$

Similarly, $9xy - 3xy = (9 - 3)xy = 6xy$.

Terms with exactly the same variable factors are called **like terms.** We can add or subtract like terms by adding or subtracting the coefficients of the variables.

● **EXAMPLE 1** *Adding Polynomials*

Compute $(4xy + 3x) + (5xy - 2x)$.

Solution

$$\begin{aligned}(4xy + 3x) + (5xy - 2x) &= 4xy + 3x + 5xy - 2x \\ &= 9xy + x\end{aligned}$$

Subtraction of polynomials uses the Distributive Law to remove the parentheses.

● **EXAMPLE 2** *Subtracting Polynomials*

Compute $(3x^2 + 4xy + 5y^2 + 1) - (6x^2 - 2xy + 4)$.

Solution
Removing the parentheses yields

$$3x^2 + 4xy + 5y^2 + 1 - 6x^2 + 2xy - 4$$

which simplifies to

$$-3x^2 + 6xy + 5y^2 - 3$$

Using the rules of exponents and the Commutative and Associative Laws for multiplication, we can multiply and divide monomials, as the following example shows.

● **EXAMPLE 3** *Products and Quotients*

Perform the indicated operations.

(a) $(8xy^3)(2x^3y)(-3xy^2)$ (b) $-15x^2y^3 \div (3xy^5)$

Solution

(a) $8 \cdot 2 \cdot (-3) \cdot x \cdot x^3 \cdot x \cdot y^3 \cdot y \cdot y^2 = -48x^5y^6$

(b) $\dfrac{-15x^2y^3}{3xy^5} = -\dfrac{15}{3} \cdot \dfrac{x^2}{x} \cdot \dfrac{y^3}{y^5} = -5 \cdot x \cdot \dfrac{1}{y^2} = -\dfrac{5x}{y^2}$

Symbols of grouping are used in algebra in the same way as they are used in the arithmetic of real numbers. We have removed parentheses in the process of adding and subtracting polynomials. Other symbols of grouping, such as brackets, [ ], are treated the same as parentheses.

When there are two or more symbols of grouping involved, we begin with the innermost and work outward.

● **EXAMPLE 4** *Symbols of Grouping*

Simplify $3x^2 - [2x - (3x^2 - 2x)]$.

Solution

$$\begin{aligned}
3x^2 - [2x - (3x^2 - 2x)] &= 3x^2 - [2x - 3x^2 + 2x] \\
&= 3x^2 - [4x - 3x^2] \\
&= 3x^2 - 4x + 3x^2 \\
&= 6x^2 - 4x
\end{aligned}$$

By the use of the Distributive Law, we can multiply a binomial by a monomial. For example,

$$x(2x + 3) = x \cdot 2x + x \cdot 3 = 2x^2 + 3x$$

We can extend the Distributive Law to multiply polynomials with more than two terms. For example,

$$5(x + y + 2) = 5x + 5y + 10$$

● **EXAMPLE 5** *Distributive Law*

Find the following products.

(a) $-4ab(3a^2b + 4ab^2 - 1)$ (b) $(4a + 5b + c)ac$

Solution

(a) $-4ab(3a^2b + 4ab^2 - 1) = -12a^3b^2 - 16a^2b^3 + 4ab$

(b) $(4a + 5b + c)ac = 4a \cdot ac + 5b \cdot ac + c \cdot ac = 4a^2c + 5abc + ac^2$

The Distributive Law can be used to show us how to multiply two polynomials. Consider the indicated multiplication $(a + b)(c + d)$. If we first treat the sum $(a + b)$ as a single quantity, then two successive applications of the Distributive Law gives

$$(a + b)(c + d) = (a + b) \cdot c + (a + b) \cdot d = ac + bc + ad + bd$$

Thus we see that the product can be found by multiplying $(a + b)$ by c, multiplying $(a + b)$ by d, and then adding the products. This is frequently set up as follows.

Product of Two Polynomials	
Procedure	Example
To multiply two polynomials:	Multiply $(3x + 4xy + 3y)$ by $(x - 2y)$.
1. Write one of the polynomials above the other.	1. $3x + 4xy + 3y$ $\underline{\;x - 2y}$
2. Multiply each term of the top polynomial by each term of the bottom one, and write the similar terms of the product under one another.	2. $3x^2 + 4x^2y + 3xy$ $\underline{\; - 6xy - 8xy^2 - 6y^2}$
3. Add like terms to simplify the product.	3. $3x^2 + 4x^2y - 3xy - 8xy^2 - 6y^2$

● **EXAMPLE 6** *The Product of Two Polynomials*

Multiply $(4x^2 + 3xy + 4x)$ by $(2x - 3y)$.

Solution

$$4x^2 + 3xy + 4x$$
$$\underline{2x\;\; - 3y}$$
$$8x^3 + 6x^2y + 8x^2$$
$$\underline{-12x^2y - 9xy^2 - 12xy}$$
$$8x^3 - 6x^2y + 8x^2 - 9xy^2 - 12xy$$

Because the multiplications we must perform often involve binomials, the following special products are worth remembering.

Special Products

A. $(x + a)(x + b) = x^2 + (a + b)x + ab$
B. $(ax + b)(cx + d) = acx^2 + (ad + bc)x + bd$

It is easier to remember these two special products if we note their structure. We can obtain these products by finding the products of the First terms, Outside terms, Inside terms, and Last terms, and then adding the results. This is called the FOIL method of multiplying two binomials.

● **EXAMPLE 7** *Products of Binomials*

Multiply the following.

(a) $(x - 4)(x + 3)$ (b) $(3x + 2)(2x + 5)$

Solution

$$\overset{\text{First}\quad\text{Outside}\quad\text{Inside}\quad\;\text{Last}}{}$$
(a) $(x - 4)(x + 3) = (x^2) + (3x) + (-4x) + (-12) = x^2 - x - 12$
(b) $(3x + 2)(2x + 5) = (6x^2) + (15x) + (4x) + (10) = 6x^2 + 19x + 10$

Additional special products are as follows:

Additional Special Products

C. $(x + a)^2 = x^2 + 2ax + a^2$ binomial squared
D. $(x - a)^2 = x^2 - 2ax + a^2$ binomial squared
E. $(x + a)(x - a) = x^2 - a^2$ difference of two squares
F. $(x + a)^3 = x^3 + 3ax^2 + 3a^2x + a^3$ binomial cubed
G. $(x - a)^3 = x^3 - 3ax^2 + 3a^2x - a^3$ binomial cubed

● **EXAMPLE 8** *Special Products*

Multiply the following.

(a) $(x + 5)^2$
(b) $(3x - 4y)^2$
(c) $(x - 2)(x + 2)$
(d) $(x^2 - y^3)^2$
(e) $(x + 4)^3$

Solution
(a) $(x + 5)^2 = x^2 + 2(5)x + 25 = x^2 + 10x + 25$
(b) $(3x - 4y)^2 = (3x)^2 - 2(3x)(4y) + (4y)^2 = 9x^2 - 24xy + 16y^2$
(c) $(x - 2)(x + 2) = x^2 - 4$
(d) $(x^2 - y^3)^2 = (x^2)^2 - 2(x^2)(y^3) + (y^3)^2 = x^4 - 2x^2y^3 + y^6$
(e) $(x + 4)^3 = x^3 + 3(4)(x^2) + 3(4^2)(x) + 4^3 = x^3 + 12x^2 + 48x + 64$

● *Checkpoint*

1. Remove parentheses and combine like terms: $9x - 5x(x + 2) + 4x^2$
2. Find the following products.
   (a) $(2x + 1)(4x^2 - 2x + 1)$
   (b) $(x + 3)^2$
   (c) $(3x + 2)(x - 5)$
   (d) $(1 - 4x)(1 + 4x)$

All algebraic expressions can represent real numbers, so the techniques used to perform operations on polynomials and to simplify polynomials also apply to other algebraic expressions.

● **EXAMPLE 9** *Operations with Algebraic Expressions*

Perform the indicated operations.

(a) $3\sqrt{3} + 4x\sqrt{y} - 5\sqrt{3} - 11x\sqrt{y} - (\sqrt{3} - x\sqrt{y})$
(b) $x^{3/2}(x^{1/2} - x^{-1/2})$
(c) $(x^{1/2} - x^{1/3})^2$
(d) $(\sqrt{x} + 2)(\sqrt{x} - 2)$

Solution
(a) We remove parentheses and then combine the terms containing $\sqrt{3}$ and the terms containing $x\sqrt{y}$.

$$(3 - 5 - 1)\sqrt{3} + (4 - 11 + 1)x\sqrt{y} = -3\sqrt{3} - 6x\sqrt{y}$$

(b) $x^{3/2}(x^{1/2} - x^{-1/2}) = x^{3/2} \cdot x^{1/2} - x^{3/2} \cdot x^{-1/2} = x^2 - x$
(c) $(x^{1/2} - x^{1/3})^2 = (x^{1/2})^2 - 2x^{1/2}x^{1/3} + (x^{1/3})^2 = x - 2x^{5/6} + x^{2/3}$
(d) $(\sqrt{x} + 2)(\sqrt{x} - 2) = (\sqrt{x})^2 - (2)^2 = x - 4$

In later chapters we will need to write problems in a simplified form so that we can perform certain operations on them. We can often use division of one polynomial by another to obtain the simplification, as shown in the following procedure.

Division of Polynomials

Procedure	Example
To divide one polynomial by another:	Divide $4x^3 + 4x^2 + 5$ by $2x^2 + 1$.
1. Write both polynomials in descending powers of a variable. Include missing terms with coefficient 0 in the dividend.	1. $2x^2 + 1 \overline{)4x^3 + 4x^2 + 0x + 5}$
2. (a) Divide the highest power term of the divisor into the highest power term of the dividend, and write this partial quotient above the dividend. Multiply the partial quotient times the divisor, write the product under the dividend, and subtract, getting a new dividend.	2. (a) $\begin{array}{r} 2x \\ 2x^2 + 1 \overline{)4x^3 + 4x^2 + 0x + 5} \\ \underline{4x^3 \qquad + 2x} \\ 4x^2 - 2x + 5 \end{array}$
(b) Repeat until the degree of the new dividend is less than the degree of the divisor. Any remainder is written over the divisor and added to the quotient.	(b) $\begin{array}{r} 2x + 2 \\ 2x^2 + 1 \overline{)4x^3 + 4x^2 + 0x + 5} \\ \underline{4x^3 \qquad + 2x} \\ 4x^2 - 2x + 5 \\ \underline{4x^2 \qquad + 2} \\ - 2x + 3 \end{array}$ Degree $(-2x + 3) <$ degree $(2x^2 + 1)$ Quotient: $2x + 2 + \dfrac{-2x + 3}{2x^2 + 1}$

● **EXAMPLE 10** *Division of Polynomials*

Divide $(4x^3 - 13x - 22)$ by $(x - 3)$, $\qquad x \neq 3$.

Solution

$$\begin{array}{r} 4x^2 + 12x + 23 \\ x - 3 \overline{)4x^3 + 0x^2 - 13x - 22} \\ \underline{4x^3 - 12x^2} \\ 12x^2 - 13x - 22 \\ \underline{12x^2 - 36x} \\ 23x - 22 \\ \underline{23x - 69} \\ 47 \end{array}$$

$0x^2$ is inserted so that each power of x is present.

The quotient is $4x^2 + 12x + 23$, with remainder 47, or

$$4x^2 + 12x + 23 + \frac{47}{x - 3}$$

● *Checkpoint* 3. Use long division to find $(x^3 + 2x + 7) \div (x - 4)$.

Spreadsheet Note One important use of algebraic expressions is to describe relationships among quantities. For example, the expression "one more than a number" could be written as $n + 1$, where n represents an arbitrary number. This ability to represent quantities or their interrelationships algebraically is one of the keys to using spreadsheets.

Each cell in a spreadsheet has an address based on its row and column (see Table 0.3). These cell addresses can act like variables in an algebraic expression, and the "fill down" or "fill across" capabilities update this cell referencing while maintaining algebraic relationships. For example, we noted previously that if $1000 is invested in an account that earns 10%, compounded annually, then the future value of the account after n years is given by $1000(1.1)^n$. We can track the future value by starting with the spreadsheet shown in Table 0.4.

TABLE 0.3

	A	B	C	D
1	cell A1	cell B1	cell C1	.
2	cell A2	cell B2	cell C2	
3	cell A3	cell B3	cell C3	
4	.			
5	.			

TABLE 0.4

	A	B
1	Year	Future value
2	1	= 1000*(1.1) ^ (A2)
3	=A2+1	

gives →

	A	B
1	Year	Future value
2	1	1100
3	2	

The use of the $=$ sign to begin a cell entry indicates an algebraic expression whose variables are other cells. In cell A3 of Table 0.4, typing the entry $=$A2$+1$ creates a referencing scheme based on the algebraic expression $n + 1$, but with cell A2 acting as the variable. From this beginning, highlighting cell A3 and using the "fill down" command updates the referencing from cell to cell and creates a counter for the number of years (see Table 0.5).

TABLE 0.5

	A	B	C
	Year	Future value	Interest earned
1			
2	1	1100	100
3	2	1210	210
4	3	1331	331
5	4	1464.1	464.1
6	5	1610.51	610.51
7	6	1771.561	771.561
8	7	1948.7171	948.7171
9	8	2143.58881	1143.58881
10	9	2357.947691	1357.947691
11	10	2593.7424601	1593.74246
12	11	2853.1167061	1853.116706
13	12	3138.4283767	2138.428377
14	13	3452.2712144	2452.271214
15	14	3797.4983358	2797.498336
16	15	4177.2481694	3177.248169

The future value is given by a formula, so we use that formula in the cell but replace the variable with the cell name where a value of that variable can be found. The "fill down" command updates the referencing and hence the values used. The first 15 years of the spreadsheet begun in Table 0.4 are shown in Table 0.5, along with a new column for interest earned. Because the interest earned is found by subtracting the original investment of $1000 from the future value, cell C2 contains the entry =B2−1000 and those below it are obtained by using "fill down." The entries for future value and interest earned also could be expressed in dollars and cents by rounding appropriately. ■

● **Checkpoint Solutions**

1. $9x - 5x(x + 2) + 4x^2 = 9x - 5x^2 - 10x + 4x^2$
$$= -x^2 - x$$

Note that without parentheses around $9x - 5x$, multiplication has priority over subtraction.

2. (a) $4x^2 - 2x + 1$

$$\frac{2x + 1}{8x^3 - 4x^2 + 2x}$$
$$\frac{4x^2 - 2x + 1}{8x^3 \qquad\quad + 1}$$

(b) $(x + 3)^2 = x^2 + 2(3x) + 3^2 = x^2 + 6x + 9$

(c) $(3x + 2)(x - 5) = 3x^2 - 15x + 2x - 10 = 3x^2 - 13x - 10$

(d) $(1 - 4x)(1 + 4x) = 1 - 16x^2$ Note that this is different from $16x^2 - 1$.

3.
$$
\begin{array}{r}
x^2 + 4x + 18 \\
x - 4{\overline{\smash{\big)}\,x^3 + 0x^2 + 2x + 7}} \\
\underline{x^3 - 4x^2} \\
4x^2 + 2x + 7 \\
\underline{4x^2 - 16x} \\
18x + 7 \\
\underline{18x - 72} \\
79
\end{array}
$$

The answer is $x^2 + 4x + 18 + \dfrac{79}{x - 4}$.

0.5 Exercises

For each polynomial in Problems 1–4, (a) give the degree of the polynomial, (b) give the coefficient (numerical) of the highest-degree term, (c) give the constant term, and (d) decide whether it is a polynomial of one or several variables.

1. $10 - 3x - x^2$
2. $5x^4 - 2x^9 + 7$
3. $7x^2y - 14xy^3z$
4. $2x^5 + 7x^2y^3 - 5y^6$

The expressions in Problems 5 and 6 are polynomials with the form $a_n x^n + a_{n-1}x^{n-1} + \cdots + a_1 x + a_0$, where n is a positive integer. Complete the following.

5. For $2x^5 - 3x^2 - 5$,
   (a) $2 = a_?$ (b) $a_3 = ?$
   (c) $-3 = a_?$ (d) $a_0 = ?$
6. For $5x^3 - 4x - 17$,
   (a) $a_3 = ?$ (b) $a_1 = ?$
   (c) $a_2 = ?$ (d) $-17 = a_?$

In Problems 7–10, evaluate each algebraic expression at the indicated values of the variables.

7. $4x - x^2$ at $x = -2$
8. $3x^2 - 4y^2 - 2xy$ at $x = 3$ and $y = -4$
9. $\dfrac{2x - y}{x^2 - 2y}$ at $x = -5$ and $y = -3$
10. $\dfrac{16y}{1 - y}$ at $y = -3$
11. Evaluate $1.98T - 1.09(1 - H)(T - 58) - 56.8$ when $T = 74.7$ and $H = 0.80$.
12. Evaluate $R\left[\dfrac{0.083i}{1 - (1 + 0.083i)^{-n}}\right]$ when $R = 100,000$, $i = 0.07$, $n = 360$.

In Problems 13–20, simplify by combining like terms.

13. $(16pq - 7p^2) + (5pq + 5p^2)$
14. $(3x^3 + 4x^2y^2) + (3x^2y^2 - 7x^3)$

15. $(4m^2 - 3n^2 + 5) - (3m^2 + 4n^2 + 8)$
16. $(4rs - 2r^2s - 11rs^2) - (11rs^2 - 2rs + 4r^2s)$
17. $-[8 - 4(q + 5) + q]$
18. $x^3 + [3x - (x^3 - 3x)]$
19. $x^2 - [x - (x^2 - 1) + 1 - (1 - x^2)] + x$
20. $y^3 - [y^2 - (y^3 + y^2)] - [y^3 + (1 - y^2)]$

In Problems 21–58, perform the indicated operations and simplify.

21. $(5x^3)(7x^2)$
22. $(-3x^2y)(2xy^3)(4x^2y^2)$
23. $(39r^3s^2) \div (13r^2s)$
24. $(-15m^3n) \div (5mn^4)$
25. $ax^2(2x^2 + ax + ab)$
26. $-3(3 - x^2)$
27. $(3y + 4)(2y - 3)$
28. $(4x - 1)(x - 3)$
29. $6(1 - 2x^2)(2 - x^2)$
30. $2(x^3 + 3)(2x^3 - 5)$
31. $(4x + 3)^2$
32. $(2y + 5)^2$
33. $\left(x^2 - \dfrac{1}{2}\right)^2$
34. $(x^3y^3 - 0.3)^2$
35. $9(2x + 1)(2x - 1)$
36. $3(5y + 2)(5y - 2)$
37. $(0.1 - 4x)(0.1 + 4x)$
38. $\left(\dfrac{2}{3} + x\right)\left(\dfrac{2}{3} - x\right)$
39. $(0.1x - 2)(x + 0.05)$
40. $(6.2x + 4.1)(6.2x - 4.1)$
41. $(x - 2)(x^2 + 2x + 4)$
42. $(a + b)(a^2 - ab + b^2)$
43. $(x^3 + 5x)(x^5 - 2x^3 + 5)$
44. $(x^3 - 1)(x^7 - 2x^4 - 5x^2 + 5)$
45. (a) $(3x - 2)^2 - 3x - 2(3x - 2) + 5$
    (b) $(3x - 2)^2 - (3x - 2)(3x - 2) + 5$
46. (a) $(2x - 3)(3x + 2) - (5x - 2)(x - 3)$
    (b) $2x - 3(3x + 2) - 5x - 2(x - 3)$
47. $(18m^2n + 6m^3n + 12m^4n^2) \div (6m^2n)$
48. $(16x^2 + 4xy^2 + 8x) \div (4xy)$
49. $(24x^8y^4 + 15x^5y - 6x^7y) \div (9x^5y^2)$
50. $(27x^2y^2 - 18xy + 9xy^2) \div (6xy)$
51. $(x + 1)^3$
52. $(x - 3)^3$
53. $(2x - 3)^3$
54. $(3x + 4)^3$
55. $(x^3 + x - 1) \div (x + 2)$
56. $(x^5 + 5x - 7) \div (x + 1)$
57. $(x^4 + 3x^3 - x + 1) \div (x^2 + 1)$
58. $(x^3 + 5x^2 - 6) \div (x^2 - 2)$

In Problems 59–66, perform the indicated operations with expressions involving fractional exponents and radicals, then simplify.

59. $x^{1/2}(x^{1/2} + 2x^{3/2})$
60. $x^{-2/3}(x^{5/3} - x^{-1/3})$
61. $(x^{1/2} + 1)(x^{1/2} - 2)$
62. $(x^{1/3} - x^{1/2})(4x^{2/3} - 3x^{3/2})$
63. $(\sqrt{x} + 3)(\sqrt{x} - 3)$
64. $(x^{1/5} + x^{1/2})(x^{1/5} - x^{1/2})$
65. $(2x + 1)^{1/2}[(2x + 1)^{3/2} - (2x + 1)^{-1/2}]$
66. $(4x - 3)^{-5/3}[(4x - 3)^{8/3} + 3(4x - 3)^{5/3}]$

APPLICATIONS

67. *Revenue*   A company sells its product for $55 per unit. Write an expression for the amount of money received (revenue) from the sale of x units of the product.

68. *Profit*   Suppose a company's revenue R (in dollars) from the sale of x units of its product is given by

$$R = 215x$$

Suppose further that the total costs C (in dollars) of producing those x units is given by

$$C = 65x + 15{,}000$$

(a) If profit is revenue minus cost, find an expression for the profit from the production and sale of x units.
(b) Find the profit received if 1000 units are sold.

69. *Rental*   A rental truck costs $99.95 for a day plus 21¢ per mile.
(a) If x is the number of miles driven, write an expression for the total cost of renting the truck for a day.
(b) Find the total cost of the rental if it was driven 132 miles.

70. *Cell phones*   Cell Pro makes cell phones and has total weekly costs of $1500 for rent, utilities, and equipment plus labor and material costs of $18.50 for each phone it makes.
(a) If x represents the number of phones produced and sold, write an expression for Cell Pro's weekly costs.
(b) If Cell Pro sells the phones to dealers for $45.50 each, write an expression for the weekly total revenue for the phones.
(c) Cell Pro's weekly profit is the total revenue minus the total cost. Write an expression for Cell Pro's weekly profit.

71. *Investments*   Suppose that you have $4000 to invest, and you invest x dollars at 10% and the remainder at 8%. Write expressions in x that represent
(a) the amount invested at 8%,
(b) the interest earned on the x dollars at 10%,
(c) the interest earned on the money invested at 8%,
(d) the total interest earned.

72. *Medications*   Suppose that a nurse needs 10 cc (cubic centimeters) of a 15.5% solution (that is, a solution that is 15.5% ingredient) of a certain medication, which must be obtained by mixing x cc of a 20% solution and y cc of a 5% solution. Write expressions involving x for
(a) y, the amount of 5% solution,
(b) the amount of ingredient in the x cc of 20% solution,
(c) the amount of ingredient in the 5% solution,
(d) the total amount of ingredient in the mixture.

73. *Package design* The volume of a rectangular box is given by $V = (\text{length})(\text{width})(\text{height})$. If a rectangular piece of cardboard that is 10 in. by 15 in. has a square with sides of length x cut from each corner (see Figure 0.9), and if the sides are folded up along the dotted lines to form a box, what expression of x would represent the volume?

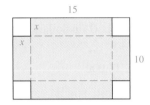

Figure 0.9

![spreadsheet icon] **Problems 74–77 use spreadsheets.**

74. *Coin mix* Suppose an individual has 125 coins, with some nickels and some dimes. We can use a spreadsheet to investigate the values for various numbers of nickels and dimes. Suppose we set up a spreadsheet with column A for the number of nickels, column B for the number of dimes, and column C for the value of the coins. Then the beginning of the spreadsheet might look like this:

	A	B	C
1	nickels	dimes	value
2	0	125	12.5
3	1	124	12.45
4	2	123	12.4

(a) With 0 as the entry in cell A2, what cell A3 entry references A2 and could be used to fill column A with the possible numbers of nickels?

(b) Knowing that the total number of coins is 125, how do we find the number of dimes if we know the number of nickels?

(c) Use the result in (b) to write an entry for cell B2 that is referenced to A2 and that could be used to fill column B with the correct number of dimes for each possible number of nickels in column A.

(d) How do you use the entries in columns A and B to find the corresponding value for column C? Write a rule.

(e) Adapt the rule from (d) to write a C2 entry (referenced appropriately) that could be used to give the correct value for each different number of coins.

(f) Use a spreadsheet to determine how many nickels and how many dimes (with 125 total coins) give the value $7.95.

75. *Area* Suppose we were interested in different areas for rectangles for which the sum of the length and width is 50. If the length and width must be whole numbers, the beginning of a spreadsheet listing the possibilities might look like this:

	A	B	C
1	length	width	area = $(l)(w)$
2	49	1	49
3	48	2	96
4	47	3	141

(a) With cell A2 as 49, what entry in cell A3 (referenced to A2) could be used to fill column A with the possible lengths?

(b) Because the sum of the length and width must be 50, what rule would we use to find the width if we knew the length?

(c) Use your result from (b) to write an entry for cell B2 that is referenced to A2 and that could be used to fill column B with the correct widths for each of the lengths in column A.

(d) What entry in cell C2 (appropriately referenced) could be used to fill column C with the area for any pair of lengths and widths?

(e) Use a spreadsheet to determine the (whole-number) length and width for which we get the smallest area that is still greater than 550.

76. *Cellular telephone subscribers* The number of millions of U.S. cellular telephone subscribers y can be approximated by the equation

$$y = 0.86t^2 + 1.39t + 4.97$$

where t is the number of years past 1990. (*Source:* Cellular Telecommunications and Internet Association) Use a spreadsheet to track the predicted number of U.S. cellular telephone subscribers from 1995 to 2010.

77. *U.S. households with Internet access* By using U.S. Department of Commerce data from 1996 to 2003, the percent P of U.S. households with Internet access can be approximated by

$$P = 5.74t + 14.61$$

where t is the number of years past 1995.
(a) Use a spreadsheet to track the approximate percent from 1995 to 2010.
(b) In what year will the percent first surpass 75%?
(c) For what value of P can you be certain that this formula is no longer valid?

0.6 Factoring

Common Factors

We can factor monomial factors out of a polynomial by using the Distributive Law in reverse; $ab + ac = a(b + c)$ is an example showing that a is a monomial factor of the polynomial $ab + ac$. But it is also a statement of the Distributive Law (with the sides of the equation interchanged). The monomial factor of a polynomial must be a factor of each term of the polynomial, so it is frequently called a **common monomial factor.**

● **EXAMPLE 1** *Monomial Factor*

Factor $-3x^2t - 3x + 9xt^2$.

Solution

1. We can factor out $3x$ and obtain

$$-3x^2t - 3x + 9xt^2 = 3x(-xt - 1 + 3t^2)$$

2. Or we can factor out $-3x$ (factoring out the negative will make the first term of the polynomial positive) and obtain

$$-3x^2t - 3x + 9xt^2 = -3x(xt + 1 - 3t^2)$$

If a factor is common to each term of a polynomial, we can use this procedure to factor it out, even if it is not a monomial. For example, we can factor $(a + b)$ out of the polynomial $2x(a + b) - 3y(a + b)$ and get $(a + b)(2x - 3y)$. The following example demonstrates the **factoring by grouping** technique.

● **EXAMPLE 2** *Factoring by Grouping*

Factor $5x - 5y + bx - by$.

Solution

We can factor this polynomial by the use of grouping. The grouping is done so that common factors (frequently binomial factors) can be removed. We see that we can factor 5 from the first two terms and b from the last two, which gives

$$5(x - y) + b(x - y)$$

This gives us two terms with the common factor $x - y$, so we get

$$(x - y)(5 + b)$$

Factoring Trinomials

We can use the formula for multiplying two binomials to factor certain trinomials. The formula

$$(x + a)(x + b) = x^2 + (a + b)x + ab$$

can be used to factor trinomials such as $x^2 - 7x + 6$.

● **EXAMPLE 3** *Factoring a Trinomial*

Factor $x^2 - 7x + 6$.

Solution

If this trinomial can be factored into an expression of the form

$$(x + a)(x + b)$$

then we need to find a and b such that

$$x^2 - 7x + 6 = x^2 + (a + b)x + ab$$

That is, we need to find a and b such that $a + b = -7$ and $ab = 6$. The two numbers whose sum is -7 and whose product is 6 are -1 and -6. Thus

$$x^2 - 7x + 6 = (x - 1)(x - 6)$$

A similar method can be used to factor trinomials such as $9x^2 - 31x + 12$. Finding the proper factors for this type of trinomial may involve a fair amount of trial and error, because we must find factors a, b, c, and d such that

$$(ax + b)(cx + d) = acx^2 + (ad + bc)x + bd$$

Another technique of factoring is used to factor trinomials such as those we have been discussing. It is useful in factoring more complicated trinomials, such as $9x^2 - 31x + 12$. This procedure for factoring second-degree trinomials follows.

Factoring a Trinomial

Procedure	Example
To factor a trinomial into the product of its binomial factors:	Factor $9x^2 - 31x + 12$.
1. Form the product of the second-degree term and the constant term.	1. $9x^2 \cdot 12 = 108x^2$
2. Determine whether there are any factors of the product of step 1 that will sum to the middle term of the trinomial. (If the answer is no, the trinomial will not factor into two binomials.)	2. The factors $-27x$ and $-4x$ give a sum of $-31x$.
3. Use the sum of these two factors to replace the middle term of the trinomial.	3. $9x^2 - 31x + 12 = 9x^2 - 27x - 4x + 12$
4. Factor this four-term expression by grouping.	4. $9x^2 - 31x + 12 = (9x^2 - 27x) + (-4x + 12)$ $= 9x(x - 3) - 4(x - 3)$ $= (x - 3)(9x - 4)$

In the example just completed, note that writing the middle term $(-31x)$ as $-4x - 27x$ rather than as $-27x - 4x$ (as we did) will also result in the correct factorization. (Try it.)

● **EXAMPLE 4** *Factoring a Trinomial*

Factor $9x^2 - 9x - 10$.

Solution

The product of the second-degree term and the constant is $-90x^2$. Factors of $-90x^2$ that sum to $-9x$ are $-15x$ and $6x$. Thus

$$9x^2 - 9x - 10 = 9x^2 - 15x + 6x - 10$$
$$= (9x^2 - 15x) + (6x - 10)$$
$$= 3x(3x - 5) + 2(3x - 5)$$
$$= (3x - 5)(3x + 2)$$

We can check this factorization by multiplying.

$$(3x - 5)(3x + 2) = 9x^2 + 6x - 15x - 10$$
$$= 9x^2 - 9x - 10$$

Some special products that make factoring easier are as follows.

Special Factorizations

The perfect-square trinomials:

$$x^2 + 2ax + a^2 = (x + a)^2$$
$$x^2 - 2ax + a^2 = (x - a)^2$$

The difference of two squares:

$$x^2 - a^2 = (x + a)(x - a)$$

● **EXAMPLE 5** *Difference of Two Squares*

Factor $25x^2 - 36y^2$.

Solution
The binomial $25x^2 - 36y^2$ is the difference of two squares, so we get

$$25x^2 - 36y^2 = (5x - 6y)(5x + 6y)$$

These two factors are called binomial **conjugates** because they differ in only one sign.

● **EXAMPLE 6** *Perfect Squares*

Factor $4x^2 + 12x + 9$.

Solution
Although we can use the technique we have learned to factor trinomials, the factors come quickly if we recognize that this trinomial is a perfect square. It has two square terms, and the remaining term $(12x)$ is twice the product of the square roots of the squares $(12x = 2 \cdot 2x \cdot 3)$. Thus

$$4x^2 + 12x + 9 = (2x + 3)^2$$

Most of the polynomials we have factored have been second-degree polynomials, or **quadratic polynomials.** Some polynomials that are not quadratic are in a form that can be factored in the same manner as quadratics. For example, the polynomial $x^4 + 4x^2 + 4$ can be written as $a^2 + 4a + 4$, where $a = x^2$.

● **EXAMPLE 7** *Polynomials in Quadratic Form*

Factor (a) $x^4 + 4x^2 + 4$ and (b) $x^4 - 16$.

Solution
(a) The trinomial is in the form of a perfect square, so letting $a = x^2$ will give us

$$x^4 + 4x^2 + 4 = a^2 + 4a + 4 = (a + 2)^2$$

Thus

$$x^4 + 4x^2 + 4 = (x^2 + 2)^2$$

(b) The binomial $x^4 - 16$ can be treated as the difference of two squares, $(x^2)^2 - 4^2$, so

$$x^4 - 16 = (x^2 - 4)(x^2 + 4)$$

But $x^2 - 4$ can be factored into $(x - 2)(x + 2)$, so

$$x^4 - 16 = (x - 2)(x + 2)(x^2 + 4)$$

● *Checkpoint*

1. Factor the following.
   (a) $8x^3 - 12x$ (b) $3x(x^2 + 5) - 5(x^2 + 5)$ (c) $x^2 - 10x - 24$
   (d) $x^2 - 5x + 6$ (e) $4x^2 - 20x + 25$ (f) $100 - 49x^2$
2. Consider $10x^2 - 17x - 20$ and observe that $(10x^2)(-20) = -200x^2$.
   (a) Find two expressions whose product is $-200x^2$ and whose sum is $-17x$.
   (b) Replace $-17x$ in $10x^2 - 17x - 20$ with the two expressions in (a).
   (c) Factor (b) by grouping.
3. True or false:
   (a) $4x^2 + 9 = (2x + 3)^2$ (b) $x^2 + x - 12 = (x - 4)(x + 3)$
   (c) $5x^5 - 20x^3 = 5x^3(x^2 - 4) = 5x^3(x + 2)(x - 2)$

A polynomial is said to be factored completely if all possible factorizations have been completed. For example, $(2x - 4)(x + 3)$ is not factored completely because a 2 can still be factored out of $2x - 4$. The following guidelines are used to factor polynomials completely.

Guidelines for Factoring Completely

Look for: Monomials first.
Then for: Difference of two squares.
Then for: Trinomial squares.
Then for: Other methods of factoring trinomials.

● **EXAMPLE 8** *Factoring Completely*

Factor completely (a) $12x^2 - 36x + 27$ and (b) $16x^2 - 64y^2$.

Solution
(a) $12x^2 - 36x + 27 = 3(4x^2 - 12x + 9)$ Monomial
                  $= 3(2x - 3)^2$ Perfect Square
(b) $16x^2 - 64y^2 = 16(x^2 - 4y^2)$
                $= 16(x + 2y)(x - 2y)$
Factoring the difference of two squares immediately would give $(4x + 8y)(4x - 8y)$, which is not factored completely (because we could still factor 4 from $4x + 8y$ and 4 from $4x - 8y$).

● *Checkpoint Solutions*

1. (a) $8x^3 - 12x = 4x(2x^2 - 3)$
   (b) $3x(x^2 + 5) - 5(x^2 + 5) = (x^2 + 5)(3x - 5)$
   (c) $x^2 - 10x - 24 = (x - 12)(x + 2)$
   (d) $x^2 - 5x + 6 = (x - 3)(x - 2)$
   (e) $4x^2 - 20x + 25 = (2x - 5)^2$
   (f) $100 - 49x^2 = (10 + 7x)(10 - 7x)$
2. (a) $(-25x)(+8x) = -200x^2$ and $-25x + 8x = -17x$
   (b) $10x^2 - 17x - 20 = 10x^2 - 25x + 8x - 20$
   (c) $= (10x^2 - 25x) + (8x - 20)$
                     $= 5x(2x - 5) + 4(2x - 5)$
                     $= (2x - 5)(5x + 4)$
3. (a) False. $4x^2 + 9$ cannot be factored. In fact, sums of squares cannot be factored.
   (b) False. $x^2 + x - 12 = (x + 4)(x - 3)$
   (c) True.

0.6 Exercises

In Problems 1–4, factor by finding the common monomial factor.

1. $9ab - 12a^2b + 18b^2$
2. $8a^2b - 160x + 4bx^2$
3. $4x^2 + 8xy^2 + 2xy^3$
4. $12y^3z + 4yz^2 - 8y^2z^3$

In Problems 5–8, factor by grouping.

5. $7x^3 - 14x^2 + 2x - 4$
6. $5y - 20 - x^2y + 4x^2$
7. $6x - 6m + xy - my$
8. $x^3 - x^2 - 5x + 5$

Factor each expression in Problems 9–20 as a product of binomials.

9. $x^2 + 8x + 12$
10. $x^2 + 6x + 8$
11. $x^2 - x - 6$
12. $x^2 - 2x - 8$
13. $7x^2 - 10x - 8$
14. $12x^2 + 11x + 2$
15. $x^2 - 10x + 25$
16. $4y^2 + 12y + 9$
17. $49a^2 - 144b^2$
18. $16x^2 - 25y^2$
19. (a) $9x^2 + 21x - 8$ (b) $9x^2 + 22x + 8$
20. (a) $10x^2 - 99x - 63$ (b) $10x^2 - 27x - 63$
    (c) $10x^2 + 61x - 63$ (d) $10x^2 + 9x - 63$

In Problems 21–44, factor completely.

21. $4x^2 - x$
22. $2x^5 + 18x^3$
23. $x^3 + 4x^2 - 5x - 20$
24. $x^3 - 2x^2 - 3x + 6$
25. $2x^2 - 8x - 42$
26. $3x^2 - 21x + 36$
27. $2x^3 - 8x^2 + 8x$
28. $x^3 + 16x^2 + 64x$
29. $2x^2 + x - 6$
30. $2x^2 + 13x + 6$
31. $3x^2 + 3x - 36$
32. $4x^2 - 8x - 60$
33. $2x^3 - 8x$
34. $16z^2 - 81w^2$
35. $10x^2 + 19x + 6$
36. $6x^2 + 67x - 35$
37. $9 - 47x + 10x^2$
38. $10x^2 + 21x - 10$
39. $y^4 - 16x^4$
40. $x^8 - 81$
41. $x^4 - 8x^2 + 16$
42. $81 - 18x^2 + x^4$
43. $4x^4 - 5x^2 + 1$
44. $x^4 - 3x^2 - 4$

Use the following factorization formulas involving cubes to factor each expression in Problems 45–52.

Factorizations with Cubes

Perfect cube
$$a^3 + 3a^2b + 3ab^2 + b^3 = (a + b)^3$$

Perfect cube
$$a^3 - 3a^2b + 3ab^2 - b^3 = (a - b)^3$$

Difference of two cubes
$$a^3 - b^3 = (a - b)(a^2 + ab + b^2)$$

Sum of two cubes
$$a^2 + b^3 = (a + b)(a^2 - ab + b^2)$$

---

45. $x^3 + 3x^2 + 3x + 1$
46. $x^3 + 6x^2 + 12x + 8$
47. $x^3 - 12x^2 + 48x - 64$
48. $y^3 - 9y^2 + 27y - 27$

49. $x^3 - 64$
50. $8x^3 - 1$
51. $27 + 8x^3$
52. $a^3 + 216$

In Problems 53–58 determine the missing factor.

53. $x^{3/2} + x^{1/2} = x^{1/2}(?)$
54. $2x^{1/4} + 4x^{3/4} = 2x^{1/4}(?)$
55. $x^{-3} + x^{-2} = x^{-3}(?)$
56. $x^{-1} - x = x^{-1}(?)$
57. $(-x^3 + x)(3 - x^2)^{-1/2} + 2x(3 - x^2)^{1/2}$
    $= (3 - x^2)^{-1/2}(?)$
58. $4x(4x + 1)^{-1/3} - (4x + 1)^{2/3} = (4x + 1)^{-1/3}(?)$

APPLICATIONS

59. *Simple interest*   The future value of a simple-interest investment of P dollars at an annual interest rate r for t years is given by the expression $P + Prt$. Factor this expression.

60. *Reaction to medication*   When medicine is administered, the reaction (measured in change of blood pressure or temperature) can be modeled by (that is, described by)

$$R = \frac{cm^2}{2} - \frac{m^3}{3}$$

where c is a positive constant and m is the amount of medicine absorbed into the blood.* Factor the expression for the reaction.

61. *Sensitivity to medication*   From the formula for reaction to medication given in Problem 60, an expression for sensitivity S can be obtained, where

$$S = cm - m^2$$

Factor this expression for sensitivity.

62. *Volume*   Suppose that squares of side x are cut from four corners of an 8-by-8-inch piece of cardboard and an open-top box is formed (see Figure 0.10). The volume of the box is given by $64x - 32x^2 + 4x^3$. Factor this expression.

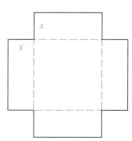

Figure 0.10

*Source: R. M. Thrall et al., *Some Mathematical Models in Biology*, U.S. Department of Commerce, 1967.

63. *Consumer expenditure* The consumer expenditure for a commodity is the product of its market price, p, and the number of units demanded. Suppose that for a certain commodity, the consumer expenditure is given by

$$10,000p - 100p^2$$

(a) Factor this in order to find an expression for the number of units demanded.
(b) Use (a) to find the number of units demanded when the market price is $38.

64. *Power in a circuit* Factor the following expression for the maximum power in a certain electrical circuit.

$$(R + r)^2 - 2r(R + r)$$

65. *Revenue* Revenue R from the sale of x units of a product is found by multiplying the price by the number of items sold.

(a) Factor the right side of $R = 300x - x^2$.
(b) What is the expression for the price of the item?

66. *Poiseuille's law* The expression for the speed of blood through an artery of radius r at a distance x from the artery wall is given by $r^2 - (r - x)^2$. Factor and simplify this expression.

0.7 Algebraic Fractions

Evaluating certain limits and graphing rational functions require an understanding of algebraic fractions. The fraction 6/8 can be reduced to 3/4 by dividing both the numerator and the denominator by 2. In the same manner, the algebraic fraction

$$\frac{(x + 2)(x + 1)}{(x + 1)(x + 3)}$$

can be reduced to

$$\frac{x + 2}{x + 3}$$

by dividing both the numerator and the denominator by $x + 1$, if $x \neq -1$.

Simplifying Fractions

We *simplify* algebraic fractions by factoring the numerator and denominator and then dividing both the numerator and the denominator by any common factors.*

● **EXAMPLE 1** *Simplifying a Fraction*

Simplify $\dfrac{3x^2 - 14x + 8}{x^2 - 16}$ if $x^2 \neq 16$.

Solution

$$\frac{3x^2 - 14x + 8}{x^2 - 16} = \frac{(3x - 2)(x - 4)}{(x - 4)(x + 4)}$$

$$= \frac{(3x - 2)\overset{1}{\cancel{(x - 4)}}}{\underset{1}{\cancel{(x - 4)}}(x + 4)}$$

$$= \frac{3x - 2}{x + 4}$$

---

*We assume that all fractions are defined.

Products of Fractions We can multiply fractions by writing the product as the product of the numerators divided by the product of the denominators. For example,

$$\frac{4}{5} \cdot \frac{10}{12} \cdot \frac{2}{5} = \frac{80}{300}$$

which reduces to $\frac{4}{15}$.

We can also find the product by reducing the fractions before we indicate the multiplication in the numerator and denominator. For example, in

$$\frac{4}{5} \cdot \frac{10}{12} \cdot \frac{2}{5}$$

we can divide the first numerator and the second denominator by 4 and divide the second numerator and the first denominator by 5, which yields

$$\frac{\overset{1}{\cancel{4}}}{\underset{1}{\cancel{5}}} \cdot \frac{\overset{2}{\cancel{10}}}{\underset{3}{\cancel{12}}} \cdot \frac{2}{5} = \frac{1}{1} \cdot \frac{2}{3} \cdot \frac{2}{5} = \frac{4}{15}$$

Product of Fractions

> We *multiply* algebraic fractions by writing the product of the numerators divided by the product of the denominators, and then reduce to lowest terms. We may also reduce prior to finding the product.

● **EXAMPLE 2** *Multiplying Fractions*

Multiply:

(a) $\dfrac{4x^2}{5y} \cdot \dfrac{10x}{y^2} \cdot \dfrac{y}{8x^2}$ (b) $\dfrac{-4x + 8}{3x + 6} \cdot \dfrac{2x + 4}{4x + 12}$

Solution

(a) $\dfrac{4x^2}{5y} \cdot \dfrac{10x}{y^2} \cdot \dfrac{y}{8x^2} = \dfrac{\overset{1}{\cancel{4x^2}}}{\underset{1\cdot1}{\cancel{5y}}} \cdot \dfrac{\overset{2}{\cancel{10x}}}{y^2} \cdot \dfrac{\overset{1}{\cancel{y}}}{\underset{2}{\cancel{8x^2}}} = \dfrac{1}{1} \cdot \dfrac{2x}{y^2} \cdot \dfrac{1}{2} = \dfrac{x}{y^2}$

(b) $\dfrac{-4x + 8}{3x + 6} \cdot \dfrac{2x + 4}{4x + 12} = \dfrac{-4(x - 2)}{3(x + 2)} \cdot \dfrac{2(x + 2)}{4(x + 3)}$

$\qquad\qquad\qquad = \dfrac{\overset{-1}{\cancel{-4}}(x - 2)}{3\underset{1}{(\cancel{x + 2})}} \cdot \dfrac{2\overset{1}{(\cancel{x + 2})}}{\underset{1}{\cancel{4}}(x + 3)}$

$\qquad\qquad\qquad = \dfrac{-2(x - 2)}{3(x + 3)}$

Quotients of Fractions In arithmetic we learned to divide one fraction by another by inverting the divisor and multiplying. The same rule applies to division of algebraic fractions.

● **EXAMPLE 3** *Dividing Fractions*

(a) Divide $\dfrac{a^2b}{c}$ by $\dfrac{ab}{c^2}$. (b) Find $\dfrac{6x^2 - 6}{x^2 + 3x + 2} \div \dfrac{x - 1}{x^2 + 4x + 4}$.

Solution

(a) $\dfrac{a^2b}{c} \div \dfrac{ab}{c^2} = \dfrac{a^2b}{c} \cdot \dfrac{c^2}{ab} = \dfrac{\overset{a\cdot 1}{\cancel{a^2b}}}{\underset{1}{\cancel{c}}} \cdot \dfrac{\overset{c}{\cancel{c^2}}}{\underset{1\cdot 1}{\cancel{ab}}} = \dfrac{ac}{1} = ac$

(b) $\dfrac{6x^2 - 6}{x^2 + 3x + 2} \div \dfrac{x - 1}{x^2 + 4x + 4} = \dfrac{6x^2 - 6}{x^2 + 3x + 2} \cdot \dfrac{x^2 + 4x + 4}{x - 1}$

$\qquad\qquad\qquad\qquad\qquad = \dfrac{6(x - 1)(x + 1)}{(x + 2)(x + 1)} \cdot \dfrac{(x + 2)(x + 2)}{x - 1}$

$\qquad\qquad\qquad\qquad\qquad = 6(x + 2)$

● **Checkpoint**

1. Reduce: $\dfrac{2x^2 - 4x}{2x}$

2. Multiply: $\dfrac{x^2}{x^2 - 9} \cdot \dfrac{x + 3}{3x}$

3. Divide: $\dfrac{5x^2(x - 1)}{2(x + 1)} \div \dfrac{10x^2}{(x + 1)(x - 1)}$

Adding and Subtracting Fractions

If two fractions are to be added, it is convenient that both be expressed with the same denominator. If the denominators are not the same, we can write the equivalents of each of the fractions with a common denominator. We usually use the least common denominator (LCD) when we write the equivalent fractions. The **least common denominator** is the lowest-degree variable expression into which all denominators will divide. If the denominators are polynomials, then the LCD is the lowest-degree polynomial into which all denominators will divide. We can find the least common denominator as follows.

Finding the Least Common Denominator

Procedure	Example
To find the least common denominator of a set of fractions:	Find the LCD of $\dfrac{1}{x^2 - x}, \dfrac{1}{x^2 - 1}, \dfrac{1}{x^2}.$
1. Completely factor each denominator.	1. The factored denominators are $x(x - 1)$, $(x + 1)(x - 1)$, and $x \cdot x$.
2. Identify the different factors that appear.	2. The different factors are x, $x - 1$, and $x + 1$.
3. The LCD is the product of these different factors, with each factor used the maximum number of times it occurs in any one denominator.	3. x occurs a maximum of twice in one denominator, $x - 1$ occurs once, and $x + 1$ occurs once. Thus the LCD is $x \cdot x(x - 1)(x + 1) = x^2(x - 1)(x + 1)$.

The procedure for combining (adding or subtracting) two or more fractions follows.

Adding or Subtracting Fractions	
Procedure	Example
To combine fractions:	Combine $\dfrac{y-3}{y-5}+\dfrac{y-23}{y^2-y-20}$.
1. Find the LCD of the fractions.	1. $y^2-y-20=(y-5)(y+4)$, so the LCD is $(y-5)(y+4)$.
2. Write the equivalent of each fraction with the LCD as its denominator.	2. The sum is $\dfrac{(y-3)(y+4)}{(y-5)(y+4)}+\dfrac{y-23}{(y-5)(y+4)}$.
3. Add or subtract, as indicated, by combining like terms in the numerator over the LCD.	3. $=\dfrac{y^2+y-12+y-23}{(y-5)(y+4)}$ $=\dfrac{y^2+2y-35}{(y-5)(y+4)}$
4. Reduce the fraction, if possible.	4. $=\dfrac{(y+7)(y-5)}{(y-5)(y+4)}=\dfrac{y+7}{y+4}$, if $y\ne 5$.

● **EXAMPLE 4** *Adding Fractions*

Add $\dfrac{3x}{a^2}+\dfrac{4}{ax}$.

Solution
1. The LCD is a^2x.
2. $\dfrac{3x}{a^2}+\dfrac{4}{ax}=\dfrac{3x}{a^2}\cdot\dfrac{x}{x}+\dfrac{4}{ax}\cdot\dfrac{a}{a}=\dfrac{3x^2}{a^2x}+\dfrac{4a}{a^2x}$
3. $\dfrac{3x^2}{a^2x}+\dfrac{4a}{a^2x}=\dfrac{3x^2+4a}{a^2x}$
4. The sum is in lowest terms.

● **EXAMPLE 5** *Combining Fractions*

Combine $\dfrac{y-3}{(y-5)^2}-\dfrac{y-2}{y^2-4y-5}$.

Solution
$y^2-4y-5=(y-5)(y+1)$, so the LCD is $(y-5)^2(y+1)$. Writing the equivalent fractions and then combining them, we get

$$\frac{y-3}{(y-5)^2}-\frac{y-2}{(y-5)(y+1)}=\frac{(y-3)(y+1)}{(y-5)^2(y+1)}-\frac{(y-2)(y-5)}{(y-5)(y+1)(y-5)}$$

$$=\frac{(y^2-2y-3)-(y^2-7y+10)}{(y-5)^2(y+1)}$$

$$=\frac{y^2-2y-3-y^2+7y-10}{(y-5)^2(y+1)}$$

$$=\frac{5y-13}{(y-5)^2(y+1)}$$

Complex Fractions   A fractional expression that contains one or more fractions in its numerator or denominator is called a **complex fraction.** An example of a complex fraction is

$$\frac{\dfrac{1}{3} + \dfrac{4}{x}}{3 - \dfrac{1}{xy}}$$

We can simplify fractions of this type using the property $\dfrac{a}{b} = \dfrac{ac}{bc}$, with c equal to the LCD of *all* the fractions contained in the numerator and denominator of the complex fraction. For example, all fractions contained in

$$\frac{\dfrac{1}{3} + \dfrac{4}{x}}{3 - \dfrac{1}{xy}}$$

have LCD $3xy$. We simplify this complex fraction by multiplying the numerator and denominator as follows:

$$\frac{3xy\left(\dfrac{1}{3} + \dfrac{4}{x}\right)}{3xy\left(3 - \dfrac{1}{xy}\right)} = \frac{3xy\left(\dfrac{1}{3}\right) + 3xy\left(\dfrac{4}{x}\right)}{3xy(3) - 3xy\left(\dfrac{1}{xy}\right)} = \frac{xy + 12y}{9xy - 3}$$

● **EXAMPLE 6** *Complex Fractions*

Simplify $\dfrac{x^{-3} + x^2y^{-3}}{(xy)^{-2}}$ so that only positive exponents remain.

Solution

$$\frac{x^{-3} + x^2y^{-3}}{(xy)^{-2}} = \frac{\dfrac{1}{x^3} + \dfrac{x^2}{y^3}}{\dfrac{1}{(xy)^2}}, \quad \text{LCD} = x^3y^3$$

$$= \frac{x^3y^3\left(\dfrac{1}{x^3} + \dfrac{x^2}{y^3}\right)}{x^3y^3\left(\dfrac{1}{x^2y^2}\right)} = \frac{y^3 + x^5}{xy}$$

● *Checkpoint*

4. Add or subtract:

   (a) $\dfrac{5x - 1}{2x - 5} - \dfrac{x + 9}{2x - 5}$

   (b) $\dfrac{x + 1}{x} + \dfrac{x}{x - 1}$

5. Simplify $\dfrac{\dfrac{y}{x} - 1}{\dfrac{y}{x} - \dfrac{x}{y}}$.

Rationalizing Denominators

We can simplify algebraic fractions whose denominators contain sums and differences that involve square roots by rationalizing the denominators. Using the fact that $(x + y)(x - y) = x^2 - y^2$, we multiply the numerator and denominator of an algebraic fraction of this type by the conjugate of the denominator to simplify the fraction.

● **EXAMPLE 7** *Rationalizing Denominators*

Rationalize the denominators.

(a) $\dfrac{1}{\sqrt{x} - 2}$ (b) $\dfrac{3 + \sqrt{x}}{\sqrt{x} + \sqrt{5}}$

Solution

Multiplying $\sqrt{x} - 2$ by $\sqrt{x} + 2$, its conjugate, gives the difference of two squares and removes the radical from the denominator in (a). We also use the conjugate in (b).

(a) $\dfrac{1}{\sqrt{x} - 2} \cdot \dfrac{\sqrt{x} + 2}{\sqrt{x} + 2} = \dfrac{\sqrt{x} + 2}{(\sqrt{x})^2 - (2)^2} = \dfrac{\sqrt{x} + 2}{x - 4}$

(b) $\dfrac{3 + \sqrt{x}}{\sqrt{x} + \sqrt{5}} \cdot \dfrac{\sqrt{x} - \sqrt{5}}{\sqrt{x} - \sqrt{5}} = \dfrac{3\sqrt{x} - 3\sqrt{5} + x - \sqrt{5x}}{x - 5}$

● *Checkpoint*

6. Rationalize the denominator $\dfrac{\sqrt{x}}{\sqrt{x} - 3}$.

● *Checkpoint Solutions*

1. $\dfrac{2x^2 - 4x}{2x} = \dfrac{2x(x - 2)}{2x} = x - 2$

2. $\dfrac{x^2}{x^2 - 9} \cdot \dfrac{x + 3}{3x} = \dfrac{x^2 \cdot (x + 3)}{(x + 3)(x - 3) \cdot 3x} = \dfrac{x}{3(x - 3)} = \dfrac{x}{3x - 9}$

3. $\dfrac{5x^2(x - 1)}{2(x + 1)} \div \dfrac{10x^2}{(x + 1)(x - 1)} = \dfrac{5x^2(x - 1)}{2(x + 1)} \cdot \dfrac{(x + 1)(x - 1)}{10x^2}$

   $= \dfrac{(x - 1)^2}{4}$

4. (a) $\dfrac{5x - 1}{2x - 5} - \dfrac{x + 9}{2x - 5} = \dfrac{(5x - 1) - (x + 9)}{2x - 5}$

   $= \dfrac{5x - 1 - x - 9}{2x - 5}$

   $= \dfrac{4x - 10}{2x - 5} = \dfrac{2(2x - 5)}{2x - 5} = 2$

   (b) $\dfrac{x + 1}{x} + \dfrac{x}{x - 1}$ has LCD $= x(x - 1)$

   $\dfrac{x + 1}{x} + \dfrac{x}{x - 1} = \dfrac{x + 1}{x} \cdot \dfrac{(x - 1)}{(x - 1)} + \dfrac{x}{x - 1} \cdot \dfrac{x}{x}$

   $= \dfrac{x^2 - 1}{x(x - 1)} + \dfrac{x^2}{x(x - 1)} = \dfrac{x^2 - 1 + x^2}{x(x - 1)}$

   $= \dfrac{2x^2 - 1}{x(x - 1)}$

5. $\dfrac{\dfrac{y}{x} - 1}{\dfrac{y}{x} - \dfrac{x}{y}} = \dfrac{\left(\dfrac{y}{x} - 1\right)}{\left(\dfrac{y}{x} - \dfrac{x}{y}\right)} \cdot \dfrac{xy}{xy} = \dfrac{\dfrac{y}{x} \cdot xy - 1 \cdot xy}{\dfrac{y}{x} \cdot xy - \dfrac{x}{y} \cdot xy}$

$= \dfrac{y^2 - xy}{y^2 - x^2} = \dfrac{y(y - x)}{(y + x)(y - x)} = \dfrac{y}{y + x}$

6. $\dfrac{\sqrt{x}}{\sqrt{x} - 3} = \dfrac{\sqrt{x}}{\sqrt{x} - 3} \cdot \dfrac{\sqrt{x} + 3}{\sqrt{x} + 3} = \dfrac{x + 3\sqrt{x}}{x - 9}$

0.7 Exercises

Simplify the following fractions.

1. $\dfrac{18x^3y^3}{9x^3z}$

2. $\dfrac{15a^4b^5}{30a^3b}$

3. $\dfrac{x - 3y}{3x - 9y}$

4. $\dfrac{x^2 - 6x + 8}{x^2 - 16}$

5. $\dfrac{x^2 - 2x + 1}{x^2 - 4x + 3}$

6. $\dfrac{x^2 - 5x + 6}{9 - x^2}$

7. $\dfrac{6x^3y^3 - 15x^2y}{3x^2y^2 + 9x^2y}$

8. $\dfrac{x^2y^2 - 4x^3y}{x^2y - 2x^2y^2}$

In Problems 9–36, perform the indicated operations and simplify.

9. $\dfrac{6x^3}{8y^3} \cdot \dfrac{16x}{9y^2} \cdot \dfrac{15y^4}{x^3}$

10. $\dfrac{25ac^2}{15a^2c} \cdot \dfrac{4ad^4}{15abc^3}$

11. $\dfrac{8x - 16}{x - 3} \cdot \dfrac{4x - 12}{3x - 6}$

12. $(x^2 - 4) \cdot \dfrac{2x - 3}{x + 2}$

13. $\dfrac{x^2 + 7x + 12}{3x^2 + 13x + 4} \cdot (9x + 3)$

14. $\dfrac{4x + 4}{x - 4} \cdot \dfrac{x^2 - 6x + 8}{8x^2 + 8x}$

15. $\dfrac{x^2 - x - 2}{2x^2 - 8} \cdot \dfrac{18 - 2x^2}{x^2 - 5x + 4} \cdot \dfrac{x^2 - 2x - 8}{x^2 - 6x + 9}$

16. $\dfrac{x^2 - 5x - 6}{x^2 - 5x + 4} \cdot \dfrac{x^2 - x - 12}{x^3 - 6x^2} \cdot \dfrac{x - x^3}{x^2 - 2x + 1}$

17. $\dfrac{15ac^2}{7bd} \div \dfrac{4a}{14b^2d}$

18. $\dfrac{16}{x - 2} \div \dfrac{4}{3x - 6}$

19. $\dfrac{y^2 - 2y + 1}{7y^2 - 7y} \div \dfrac{y^2 - 4y + 3}{35y^2}$

20. $\dfrac{6x^2}{4x^2y - 12xy} \div \dfrac{3x^2 + 12x}{x^2 + x - 12}$

21. $(x^2 - x - 6) \div \dfrac{9 - x^2}{x^2 - 3x}$

22. $\dfrac{2x^2 + 7x + 3}{4x^2 - 1} \div (x + 3)$

23. $\dfrac{2x}{x^2 - x - 2} - \dfrac{x + 2}{x^2 - x - 2}$

24. $\dfrac{4}{9 - x^2} - \dfrac{x + 1}{9 - x^2}$

25. $\dfrac{a}{a - 2} - \dfrac{a - 2}{a}$

26. $x - \dfrac{2}{x - 1}$

27. $\dfrac{x}{x + 1} - x + 1$

28. $\dfrac{x - 1}{x + 1} - \dfrac{2}{x^2 + x}$

29. $\dfrac{4a}{3x + 6} + \dfrac{5a^2}{4x + 8}$

30. $\dfrac{b - 1}{b^2 + 2b} + \dfrac{b}{3b + 6}$

31. $\dfrac{3x - 1}{2x - 4} + \dfrac{4x}{3x - 6} - \dfrac{x - 4}{5x - 10}$

32. $\dfrac{2x + 1}{4x - 2} + \dfrac{5}{2x} - \dfrac{x + 4}{2x^2 - x}$

33. $\dfrac{x}{x^2 - 4} + \dfrac{4}{x^2 - x - 2} - \dfrac{x - 2}{x^2 + 3x + 2}$

34. $\dfrac{3x^2}{x^2 - 4} + \dfrac{2}{x^2 - 4x + 4} - 3$

35. $\dfrac{-x^3 + x}{\sqrt{3 - x^2}} + 2x\sqrt{3 - x^2}$

36. $\dfrac{3x^2(x + 1)}{\sqrt{x^3 + 1}} + \sqrt{x^3 + 1}$

In Problems 37–46, simplify each complex fraction.

37. $\dfrac{3 - \dfrac{2}{3}}{14}$

38. $\dfrac{\dfrac{4}{\frac{1}{4} + \frac{1}{4}}}{}$

39. $\dfrac{x + y}{\dfrac{1}{x} + \dfrac{1}{y}}$

40. $\dfrac{\dfrac{5}{2y} + \dfrac{3}{y}}{\dfrac{1}{4} + \dfrac{1}{3y}}$

41. $\dfrac{2 - \dfrac{1}{x}}{2x - \dfrac{3x}{x + 1}}$

42. $\dfrac{1 - \dfrac{2}{x - 2}}{x - 6 + \dfrac{10}{x + 1}}$

43. $\dfrac{\sqrt{a} - \dfrac{b}{\sqrt{a}}}{a - b}$

44. $\dfrac{\sqrt{x - 1} + \dfrac{1}{\sqrt{x - 1}}}{x}$

45. $\dfrac{\sqrt{x^2+9} - \dfrac{13}{\sqrt{x^2+9}}}{x^2-x-6}$

46. $\dfrac{\sqrt{x^2+3} - \dfrac{x+5}{\sqrt{x^2+3}}}{x^2+5x+4}$

In Problems 47–50, rewrite each of the following so that only positive exponents remain, and simplify.

47. (a) $(2^{-2} - 3^{-1})^{-1}$ (b) $(2^{-1} + 3^{-1})^2$
48. (a) $(3^2 + 4^2)^{-1/2}$ (b) $(2^2 + 3^2)^{-1}$
49. $\dfrac{2a^{-1} - b^{-2}}{(ab^2)^{-1}}$
50. $\dfrac{x^{-2} + xy^{-2}}{(x^2y)^{-2}}$

In Problems 51 and 52, rationalize the denominator of each fraction and simplify.

51. $\dfrac{1 - \sqrt{x}}{1 + \sqrt{x}}$
52. $\dfrac{x - 3}{x - \sqrt{3}}$

In Problems 53 and 54, rationalize the numerator of each fraction and simplify.

53. $\dfrac{\sqrt{x+h} - \sqrt{x}}{h}$
54. $\dfrac{\sqrt{9+2h} - 3}{h}$

APPLICATIONS

55. *Time study*   Workers A, B, and C can complete a job in a, b, and c hours, respectively. Working together, they can complete

$$\frac{1}{a} + \frac{1}{b} + \frac{1}{c}$$

of the job in 1 hour. Add these fractions over a common denominator to obtain an expression for what they can do in 1 hour, working together.

56. *Focal length*   Two thin lenses with focal lengths p and q and separated by a distance d have their combined focal length given by the reciprocal of

$$\frac{1}{p} + \frac{1}{q} - \frac{d}{pq}$$

(a) Combine these fractions.
(b) Use the reciprocal of your answer in (a) to find the combined focal length.

Average cost   **A company's average cost per unit when x units are produced is defined to be**

$$\text{Average cost} = \frac{\text{Total cost}}{x}$$

Use this equation in Problems 57 and 58.

57. Suppose a company's average costs are given by

$$\text{Average cost} = \frac{4000}{x} + 55 + 0.1x$$

(a) Express the average-cost formula as a single fraction.
(b) Write the expression that gives the company's total costs.

58. Suppose a company's average costs are given by

$$\text{Average cost} = \frac{40{,}500}{x} + 190 + 0.2x$$

(a) Express the average-cost formula as a single fraction.
(b) Write the expression that gives the company's total costs.

59. *Advertising and sales*   Suppose that a company's daily sales volume attributed to an advertising campaign is given by

$$\text{Sales volume} = 1 + \frac{3}{t+3} - \frac{18}{(t+3)^2}$$

where t is the number of days since the campaign started. Express the sales volume as a single fraction.

60. *Annuity*   The formula for the future value of an annuity due involves the expression

$$\frac{(1+i)^{n+1} - 1}{i} - 1$$

Write this expression over a common denominator and factor the numerator to simplify.

Key Terms and Formulas

Section	Key Terms	Formulas
0.1	Sets and set membership	
	Natural numbers	$N = \{1, 2, 3, 4 \ldots\}$
	Empty set	$\varnothing$
	Set equality	
	Subset	$A \subseteq B$

(continued)

Section	Key Terms	Formulas
	Universal set	U
	Venn diagrams	
	Set intersection	$A \cap B$
	Disjoint sets	$A \cap B = \varnothing$
	Set union	$A \cup B$
	Set complement	A'
0.2	Real numbers	
	Subsets and properties	
	Real number line	
	Inequalities	
	Intervals and interval notation	
	Closed interval	$a \leq x \leq b$ or $[a, b]$
	Open interval	$a < x < b$ or (a, b)
	Absolute value	
	Order of operations	
0.3	Exponent and base	a^n has base a, exponent n
	Zero exponent	$a^0 = 1, a \neq 0$
	Negative exponent	$a^{-n} = \dfrac{1}{a^n}$
	Rules of exponents	
0.4	Radical	$\sqrt[n]{a}$
	Radicand, index	radicand $= a$; index $= n$
	Principal nth root	$\sqrt[n]{a} = b$ only if $b^n = a$ and $a \geq 0$ and $b \geq 0$ when n is even
	Fractional exponents	$a^{1/n} = \sqrt[n]{a}$ $a^{m/n} = \sqrt[n]{a^m} = (\sqrt[n]{a})^m$
	Properties of radicals	
	Rationalizing the denominator	
0.5	Algebraic expression	
	Variable	
	Constant term	
	Coefficient; leading coefficient	
	Term	
	Polynomial	$a_n x^n + \cdots + a_1 x + a_0$
	Degree	
	Monomial	
	Binomial	
	Trinomial	
	Like terms	
	Distributive Law	$a(b + c) = ab + ac$
	Binomial products	
	Division of polynomials	

Section	Key Terms	Formulas
0.6	Factor	
	Common factor	
	Factoring by grouping	
	Special factorizations	
	Difference of squares	$a^2 - b^2 = (a + b)(a - b)$
	Perfect squares	$a^2 + 2ab + b^2 = (a + b)^2$
		$a^2 - 2ab + b^2 = (a - b)^2$
	Conjugates	$a + b; a - b$
	Quadratic polynomials	$ax^2 + bx + c$
	Factoring completely	
0.7	Algebraic fractions	
	Numerator	
	Denominator	
	Reduce	
	Product of fractions	
	Quotient of fractions	
	Common denominator	
	Least common denominator (LCD)	
	Addition and subtraction of fractions	
	Complex fraction	
	Rationalize the denominator	

Review Exercises

Section 0.1

1. Is $A \subseteq B$, if $A = \{1, 2, 5, 7\}$ and
   $B = \{x: x \text{ is a positive integer}, x \leq 8\}$?
2. Is it true that $3 \in \{x: x > 3\}$?
3. Are $A = \{1, 2, 3, 4\}$ and $B = \{x: x \leq 1\}$ disjoint?

In Problems 4–7, use sets $U = \{1, 2, 3, 4, 5, 6, 7, 8, 9, 10\}$, $A = \{1, 2, 3, 9\}$, and $B = \{1, 3, 5, 6, 7, 8, 10\}$ to find the elements of the sets described.

4. $A \cup B'$ 5. $A' \cap B$
6. $(A' \cap B)'$ 7. Does $(A' \cup B')' = A \cap B$?

Section 0.2

8. State the property of the real numbers that is illustrated in each case.

   (a) $6 + \dfrac{1}{3} = \dfrac{1}{3} + 6$ (b) $2(3 \cdot 4) = (2 \cdot 3)4$

   (c) $\dfrac{1}{3}(6 + 9) = 2 + 3$

9. Indicate whether each given expression is one or more of the following: rational, irrational, integer, natural, or meaningless.

   (a) π (b) $0/6$ (c) $6/0$

10. Insert the proper sign ($<$, $=$, or $>$) to replace each $\square$.
    (a) $\pi \, \square \, 3.14$ (b) $-100 \, \square \, 0.1$ (c) $-3 \, \square \, -12$

For Problems 11–18, evaluate each expression. Use a calculator when necessary.

11. $|5 - 11|$ 12. $44 \div 2 \cdot 11 - 10^2$

13. $(-3)^2 - (-1)^3$ 14. $\dfrac{(3)(2)(15) - (5)(8)}{(4)(10)}$

15. $2 - [3 - (2 - |-3|)] + 11$
16. $-4^2 - (-4)^2 + 3$

17. $\dfrac{4 + 3^2}{4}$ 18. $\dfrac{(-2.91)^5}{\sqrt{3.29^5}}$

19. Write each inequality in interval notation, name the type of interval, and graph it on a real number line.
    (a) $0 \leq x \leq 5$
    (b) $x \geq -3$ and $x < 7$
    (c) $(-4, \infty) \cap (-\infty, 0)$
20. Write an inequality that represents each of the following.
    (a) $(-1, 16)$
    (b) $[-12, 8]$
    (c)

Section 0.3

21. Evaluate each of the following without a calculator.

    (a) $\left(\dfrac{3}{8}\right)^0$ (b) $2^3 \cdot 2^{-5}$

    (c) $\dfrac{4^9}{4^3}$ (d) $\left(\dfrac{1}{7}\right)^3\left(\dfrac{1}{7}\right)^{-4}$

22. Use the rules of exponents to simplify each of the following with positive exponents. Assume all variables are nonzero.

    (a) $x^5 \cdot x^{-7}$ (b) x^8/x^{-2} (c) $(x^3)^3$
    (d) $(y^4)^{-2}$ (e) $(-y^{-3})^{-2}$

For Problems 23–28, rewrite each expression so that only positive exponents remain. Assume all variables are nonzero.

23. $\dfrac{-(2xy^2)^{-2}}{(3x^{-2}y^{-3})^2}$

24. $\left(\dfrac{2}{3}x^2y^{-4}\right)^{-2}$

25. $\left(\dfrac{x^{-2}}{2y^{-1}}\right)^2$

26. $\dfrac{(-x^4y^{-2}z^2)^0}{-(x^4y^{-2}z^2)^{-2}}$

27. $\left(\dfrac{x^{-3}y^4z^{-2}}{3x^{-2}y^{-3}z^{-3}}\right)^{-1}$

28. $\left(\dfrac{x}{2y}\right)\left(\dfrac{y}{x^2}\right)^{-2}$

Section 0.4

29. Find the following roots.
    (a) $-\sqrt[3]{-64}$ (b) $\sqrt{4/49}$ (c) $\sqrt{1.9487171}$

30. Write each of the following with an exponent and with the variable in the numerator.
    (a) $\sqrt{x}$ (b) $\sqrt[3]{x^2}$ (c) $1/\sqrt[4]{x}$

31. Write each of the following in radical form.
    (a) $x^{2/3}$ (b) $x^{-1/2}$ (c) $-x^{3/2}$

32. Rationalize each of the following denominators and simplify.

    (a) $\dfrac{5xy}{\sqrt{2x}}$ (b) $\dfrac{y}{x\sqrt[3]{xy^2}}$

In Problems 33–38, use the properties of exponents to simplify so that only positive exponents remain. Assume all variables are positive.

33. $x^{1/2} \cdot x^{1/3}$ 34. $y^{-3/4}/y^{-7/4}$ 35. $x^4 \cdot x^{1/4}$
36. $1/(x^{-4/3} \cdot x^{-7/3})$ 37. $(x^{4/5})^{1/2}$ 38. $(x^{1/2}y^2)^4$

In Problems 39–44, simplify each expression. Assume all variables are positive.

39. $\sqrt{12x^3y^5}$ 40. $\sqrt{1250x^6y^9}$
41. $\sqrt[3]{24x^4y^4} \cdot \sqrt[3]{45x^4y^{10}}$ 42. $\sqrt{16a^2b^3} \cdot \sqrt{8a^3b^5}$
43. $\dfrac{\sqrt{52x^3y^6}}{\sqrt{13xy^4}}$ 44. $\dfrac{\sqrt{32x^4y^3}}{\sqrt{6xy^{10}}}$

Section 0.5

In Problems 45–62, perform the indicated operations and simplify.

45. $(3x + 5) - (4x + 7)$
46. $x(1 - x) + x[x - (2 + x)]$
47. $(3x^3 - 4xy - 3) + (5xy + x^3 + 4y - 1)$
48. $(4xy^3)(6x^4y^2)$
49. $(3x - 4)(x - 1)$ 50. $(3x - 1)(x + 2)$
51. $(4x + 1)(x - 2)$ 52. $(3x - 7)(2x + 1)$
53. $(2x - 3)^2$ 54. $(4x + 3)(4x - 3)$
55. $(2x^2 + 1)(x^2 + x - 3)$ 56. $(2x - 1)^3$
57. $(x - y)(x^2 + xy + y^2)$ 58. $\dfrac{4x^2y - 3x^3y^3 - 6x^4y^2}{2x^2y^2}$
59. $(3x^4 + 2x^3 - x + 4) \div (x^2 + 1)$
60. $(x^4 - 4x^3 + 5x^2 + x) \div (x - 3)$
61. $x^{4/3}(x^{2/3} - x^{-1/3})$
62. $(\sqrt{x} + \sqrt{a - x})(\sqrt{x} - \sqrt{a - x})$

Section 0.6

In Problems 63–73, factor each expression completely.

63. $2x^4 - x^3$ 64. $4(x^2 + 1)^2 - 2(x^2 + 1)^3$
65. $4x^2 - 4x + 1$ 66. $16 - 9x^2$ 67. $2x^4 - 8x^2$
68. $x^2 - 4x - 21$ 69. $3x^2 - x - 2$
70. $x^2 - 5x + 6$ 71. $x^2 - 10x - 24$
72. $12x^2 - 23x - 24$ 73. $16x^4 - 72x^2 + 81$
74. Factor as indicated: $x^{-2/3} + x^{-4/3} = x^{-4/3}(?)$

Section 0.7

75. Reduce each of the following to lowest terms.

    (a) $\dfrac{2x}{2x + 4}$ (b) $\dfrac{4x^2y^3 - 6x^3y^4}{2x^2y^2 - 3xy^3}$

In Problems 76–82, perform the indicated operations and simplify.

76. $\dfrac{x^2 - 4x}{x^2 + 4} \cdot \dfrac{x^4 - 16}{x^4 - 16x^2}$

77. $\dfrac{x^2 + 6x + 9}{x^2 - 7x + 12} \div \dfrac{x^2 + 4x + 3}{x^2 - 3x - 4}$

78. $\dfrac{x^4 - 2x^3}{3x^2 - x - 2} \div \dfrac{x^3 - 4x}{9x^2 - 4}$ 79. $1 + \dfrac{3}{2x} - \dfrac{1}{6x^2}$

80. $\dfrac{1}{x - 2} - \dfrac{x - 2}{4}$ 81. $\dfrac{x + 2}{x^2 - x} - \dfrac{x^2 + 4}{x^2 - 2x + 1} + 1$

82. $\dfrac{x - 1}{x^2 - x - 2} - \dfrac{x}{x^2 - 2x - 3} + \dfrac{1}{x - 2}$

In Problems 83 and 84, simplify each complex fraction.

83. $\dfrac{x - 1 - \dfrac{x-1}{x}}{\dfrac{1}{x-1} + 1}$ 84. $\dfrac{x^{-2} - x^{-1}}{x^{-2} + x^{-1}}$

85. Rationalize the denominator of $\dfrac{3x - 3}{\sqrt{x} - 1}$ and simplify.

86. Rationalize the numerator of $\dfrac{\sqrt{x} - \sqrt{x-4}}{2}$ and simplify.

APPLICATIONS

Section 0.1

87. *Job effectiveness factors* In an attempt to determine some off-the-job factors that might be indicators of on-the-job effectiveness, a company made a study of 200 of its employees. It was interested in whether the employees had been recognized for superior work by their supervisors within the past year, whether they were involved in community activities, and whether they followed a regular exercise plan. The company found the following.

> 30 answered "yes" to all three
> 50 were recognized and they exercised
> 52 were recognized and were involved in the community
> 77 were recognized
> 37 were involved in the community but did not exercise
> 95 were recognized or were involved in the community
> 95 answered "no" to all three

(a) Draw a Venn diagram that represents this information.
(b) How many exercised only?
(c) How many exercised or were involved in the community?

Section 0.2

88. *Educational spending* Federal spending (in billions of dollars) for public education during the years 1996–2001 is given by the expression $3.32x + 23.16$, where x is the number of years past 1990. Evaluate the expression for $x = 15$ to find the number of billions of dollars spent in 2005.

89. *Poiseuille's law* The expression for the speed of blood through an artery of radius r at a distance x from the artery wall is given by $r^2 - (r - x)^2$. Evaluate this expression when $r = 5$ and $x = 2$.

Section 0.3

90. *Future value* If an individual makes monthly deposits of $100 in an account that earns 9% compounded monthly, then the future value S of the account after n months is given by the formula

$$S = \$100\left[\dfrac{(1.0075)^n - 1}{0.0075}\right]$$

(a) Find the future value after 36 months (3 years).
(b) Find the future value after 20 years.

91. *Head Start program* The amount of federal funds allocated for Head Start between 1975 and 2000 is given by $400(1.107)^x$ million dollars, where x is the number of years past 1975. Evaluate this expression at $x = 0$ and at $x = 25$.

Sections 0.3 and 0.7

92. *Loan payment* Suppose you borrow $10,000 for n months to buy a car at an interest rate of 7.8% compounded monthly. The size of each month's payment R is given by the formula

$$R = \$10,000\left[\dfrac{0.0065}{1 - (1.0065)^{-n}}\right]$$

(a) Rewrite the expression on the right-hand side of this formula as a fraction with only positive exponents.
(b) Find the monthly payment for a 48-month loan. Use both the original formula and your result from (a). (Both formulas should give the same payment.)

Section 0.4

93. *Environment* Suppose that in a study of water birds, the relationship between the number of acres of wetlands A and the number of species of birds S found in the wetlands area was given by

$$S = kA^{1/3}$$

where k is a constant.
(a) Express this formula using radical notation.
(b) If the area is expanded by a factor of 2.25 from 20,000 acres to 45,000 acres, find the expected increase in the number of species (as a multiple of the number of species on the 20,000 acres).

Section 0.5

94. *Profit* Suppose that the total cost of producing and selling x units of a product is $300 + 4x$ and the total revenue from the sale of x units is $30x - 0.001x^2$. Find the difference of these expressions (total revenue minus total cost) to find the profit from the production and sale of x units of the product.

95. *Business loss* Suppose that a commercial building costs \$450,000. After it is placed into service, it loses 0.25% of its original value in each of the x months it is in service. Write an expression for the value of the building after x months.

Section 0.6

96. *Revenue* The revenue for a boat tour is $600 - 13x - 0.5x^2$, where x is the number of passengers above the minimum of 50. This expression will factor into two binomials, with one representing the total number of passengers. Factor this expression.

Section 0.7

97. *Pollution: cost-benefit* Suppose that the cost C (in dollars) of removing $p\%$ of the pollution from the waste water of a manufacturing process is given by

$$C = \frac{540{,}000}{100 - p} - 5400$$

(a) Express the right-hand side of this formula as a single fraction.
(b) Find the cost if $p = 0$. Write a sentence that explains the meaning of what you found.
(c) Find the cost of removing 98% of the pollution.
(d) What happens to this formula when $p = 100$? Explain why you think this happens.

98. *Average cost* The average cost of producing a product is given by the expression

$$1200 + 56x + \frac{8000}{x}$$

Write this expression with all terms over a common denominator.

---

0 Chapter Test

1. Let $U = \{1, 2, 3, 4, 5, 6, 7, 8, 9\}$, $A = \{x : x$ is even and $x \geq 5\}$, and $B = \{1, 2, 5, 7, 8, 9\}$. Complete the following.
   (a) Find $A \cup B'$.
   (b) Find a two-element set that is disjoint from B.
   (c) Find a nonempty subset of A that is not equal to A.

2. Evaluate $(4 - 2^3)^2 - 3^4 \cdot 0^{15} + 12 \div 3 + 1$.

3. Use definitions and properties of exponents to complete the following.
   (a) $x^4 \cdot x^4 = x^?$ (b) $x^0 = ?$, if $x \neq 0$
   (c) $\sqrt{x} = x^?$ (d) $(x^{-5})^2 = x^?$
   (e) $a^{27} \div a^{-3} = a^?$ (f) $x^{1/2} \cdot x^{1/3} = x^?$
   (g) $\dfrac{1}{\sqrt[3]{x^2}} = \dfrac{1}{x^?}$ (h) $\dfrac{1}{x^3} = x^?$

4. Write each of the following as radicals.
   (a) $x^{1/5}$ (b) $x^{-3/4}$

5. Simplify each of the following so that only positive exponents remain.
   (a) x^{-5} (b) $\left(\dfrac{x^{-8}y^2}{x^{-1}}\right)^{-3}$

6. Simplify the following radical expressions, and rationalize any denominators.
   (a) $\dfrac{x}{\sqrt{5x}}$ (b) $\sqrt{24a^2b}\sqrt{a^3b^4}$ (c) $\dfrac{1 - \sqrt{x}}{1 + \sqrt{x}}$

7. Given the expression $2x^3 - 7x^5 - 5x - 8$, complete the following.
   (a) Find the degree.
   (b) Find the constant term.
   (c) Find the coefficient of x.

8. Express $(-2, \infty) \cap (-\infty, 3]$ using interval notation, and graph it.

9. Completely factor each of the following.
   (a) $8x^3 - 2x^2$ (b) $x^2 - 10x + 24$
   (c) $6x^2 - 13x + 6$ (d) $2x^3 - 32x^5$

10. Identify the quadratic polynomial from among (a)–(c), and evaluate it using $x = -3$.
    (a) $2x^2 - 3x^3 + 7$ (b) $x^2 + 3/x + 11$
    (c) $4 - x - x^2$

11. Use long division to find $(2x^3 + x^2 - 7) \div (x^2 - 1)$.

12. Perform the indicated operations and simplify.
    (a) $4y - 5(9 - 3y)$ (b) $-3t^2(2t^4 - 3t^7)$
    (c) $(4x - 1)(x^2 - 5x + 2)$
    (d) $(6x - 1)(2 - 3x)$
    (e) $(2m - 7)^2$ (f) $\dfrac{x^6}{x^2 - 9} \cdot \dfrac{x - 3}{3x^2}$
    (g) $\dfrac{x^4}{3^2} \div \dfrac{9x^3}{x^6}$ (h) $\dfrac{4}{x - 8} - \dfrac{x - 2}{x - 8}$
    (i) $\dfrac{x - 1}{x^2 - 2x - 3} - \dfrac{3}{x^2 - 3x}$

13. Simplify $\dfrac{\dfrac{1}{x} - \dfrac{1}{y}}{\dfrac{1}{x} + y}$.

14. In a nutrition survey of 320 students, the following information was obtained.

> 145 ate breakfast
> 270 ate lunch
> 280 ate dinner
> 125 ate breakfast and lunch

110 ate breakfast and dinner
230 ate lunch and dinner
90 ate all three meals

(a) How many students in the survey ate only breakfast?

(b) How many students skipped breakfast?

15. If \$1000 is invested for x years at 8% compounded quarterly, the future value of the investment is given by $S = 1000\left(1 + \dfrac{0.08}{4}\right)^{4x}$. What will be the future value of this investment in 20 years?

Extended Applications & Group Projects

I. Campaign Management

A politician is trying to win election to the city council, and as his campaign manager, you need to decide how to promote the candidate. There are three ways you can do so: you can send glossy, full-color pamphlets to registered voters of the city, you can run a commercial during the television news on a local cable network, and/or you can buy a full-page ad in the newspaper.

Two hundred fifty thousand voters live in the city, and 36% of them read the newspaper. Fifty thousand voters watch the local cable network news, and 30% of them also read the newspaper.

You also know that the television commercial would cost $40,000, the newspaper ad $27,000, and the pamphlets mailed to voters 90 cents each, including printing and bulk-rate postage.

Suppose the success of the candidate depends on your campaign reaching at least 125,000 voters and that because your budget is limited, you must achieve this goal at a minimum cost. What would be your plan and the cost of that plan?

If you need help devising a method of solution for this problem, try answering the following questions first.

1. How many voters in the city read the newspaper but do not watch the local cable television news?
2. How many voters read the newspaper or watch the local cable television news, or both?
3. Complete the following chart by indicating the number of voters reached by each promotional option, the total cost, and the cost per voter reached.

	Number of Voters Reached	Total Cost	Cost per Voter Reached
Pamphlet			
Television			
Newspaper			

4. Now explain your plan and the cost of that plan.

II. Pricing for Maximum Profit

Overnight Cleaners specializes in cleaning professional offices. The company is expanding into a new city and believes that its staff is capable of handling 300 office units at a weekly cost per unit of $58 for labor and supplies. Preliminary pricing surveys indicate that if Overnight's weekly charge is $100 per unit, then it would have clients for 300 units. However, for each $5 price increase, it could expect to have 10 fewer units. By using a spreadsheet to investigate possible prices per unit and corresponding numbers of units, determine what weekly price Overnight Cleaners should charge in order to have the greatest profit.

The following steps will aid in creating a spreadsheet that can determine the price that gives maximum profit.

(a) Begin with columns for price per unit and number of units, and create a plan by hand to determine what your spreadsheet should look like. Use the information in the problem statement to determine the starting values for price per unit and number of units. Note that a price increase of $5 resulting in a reduction of 10 units to clean can be equivalently expressed as a $0.50 increase resulting in a 1-unit reduction. With this observation, determine the next three entries in the columns for price per unit and number of units. Write the rules used for each.

(b) Profit is found by subtracting expenses (or costs) from income. In the hand-done plan begun in (a), add a column or columns that would allow you to use information about price per unit and number of units to find profit. In your hand-done plan, complete entries for each new column (as you did for price per unit and number of units). Write the rule(s) you used. (At this point, you should have a hand-done plan with at least four entries in each column and a rule for finding more entries in each column.)

(c) Use the starting values from (a) and the rules from (a) and (b) to create a computer spreadsheet (with appropriate cell references) that will duplicate and extend your hand-done plan.

(d) Fill the computer spreadsheet from (c) and find the price per unit that gives Overnight Cleaners its maximum profit. Find its maximum profit. (Print the spreadsheet.)

1 Linear Equations and Functions

A wide variety of problems from business, the social sciences, and the life sciences may be solved using equations. Managers and economists use equations and their graphs to study costs, sales, national consumption, or supply and demand. Social scientists may plot demographic data or try to develop equations that predict population growth, voting behavior, or learning and retention rates. Life scientists use equations to model the flow of blood or the conduction of nerve impulses and to test theories or develop new models by using experimental data.

Numerous applications of mathematics are given throughout the text, but most chapters contain special sections emphasizing business and economics applications. In particular, this chapter introduces two important applications that will be expanded and used throughout the text as increased mathematical skills permit: supply and demand as functions of price (market analysis); and total cost, total revenue, and profit as functions of the quantity produced and sold (theory of the firm).

The topics and applications discussed in this chapter include the following.

Sections	Applications
1.1 Solutions of Linear Equations and Inequalities in One Variable	*Future value of an investment, voting, normal height for a given age, profit*
1.2 Functions **Function notation Operations with functions**	*Income taxes, stock market, mortgage payments*
1.3 Linear Functions **Graphs Slopes Equations**	*Depreciation, gross national product, pricing*
1.4 Graphs and Graphing Utilities **Graphical solutions of linear equations**	*Water purity, women in the work force*
1.5 Solutions of Systems of Linear Equations **Systems of linear equations in three variables**	*Investment mix, medicine concentrations, college enrollment*
1.6 Applications of Functions in Business and Economics	*Total cost, total revenue and profit, break-even, demand and supply, market equilibrium*

Chapter Warm-up

Prerequisite Problem Type	For Section	Answer	Section for Review
Evaluate: (a) $2(-1)^3 - 3(-1)^2 + 1$ (b) $3(-3) - 1$ (c) $14(10) - 0.02(10^2)$ (d) $\dfrac{3-1}{4-(-2)}$ (e) $\dfrac{-1-3}{2-(-2)}$ (f) $\dfrac{3(-8)}{4}$ (g) $2\left(\dfrac{23}{9}\right) - 5\left(\dfrac{2}{9}\right)$	1.1 1.2 1.3 1.5 1.6	(a) -4 (b) -10 (c) 138 (d) $\dfrac{1}{3}$ (e) -1 (f) -6 (g) 4	0.2 Signed numbers
Graph $x \le 3$.	1.1		0.2 Inequalities
(a) $\dfrac{1}{x}$ is *undefined* for which real number(s)? (b) $\sqrt{x-4}$ is a real number for which values of x?	1.1, 1.2	(a) Undefined for $x = 0$ (b) $x \ge 4$	0.2 Real numbers 0.4 Radicals
Identify the coefficient of x and the constant term for: (a) $-9x + 2$ (b) $\dfrac{x}{2}$ (c) $x - 300$	1.1, 1.3 1.5, 1.6	*Coeff.* *Const.* (a) -9 2 (b) $\dfrac{1}{2}$ 0 (c) 1 -300	0.5 Algebraic expressions
Simplify: (a) $4(-c)^2 - 3(-c) + 1$ (b) $[3(x+h) - 1] - [3x - 1]$ (c) $12\left(\dfrac{3x}{4} + 3\right)$ (d) $9x - (300 + 2x)$ (e) $-\dfrac{1}{5}[x - (-1)]$ (f) $2(2y + 3) + 3y$	1.1, 1.2 1.3, 1.5 1.6	(a) $4c^2 + 3c + 1$ (b) $3h$ (c) $9x + 36$ (d) $7x - 300$ (e) $-\dfrac{1}{5}x - \dfrac{1}{5}$ (f) $7y + 6$	0.5 Algebraic expressions
Find the LCD of $\dfrac{3x}{2x + 10}$ and $\dfrac{1}{x + 5}$	1.1	$2x + 10$	0.7 Algebraic fractions

1.1

Solutions of Linear Equations and Inequalities in One Variable

OBJECTIVES

- To solve linear equations in one variable.
- To solve applied problems by using linear equations.
- To solve linear inequalities in one variable.

◗ Application Preview

Using data from 1952–2000, the percent p of the eligible U.S. population voting in presidential elections has been estimated to be

$$p = 63.21 - 0.29x$$

where x is the number of years past 1950. (*Source:* Federal Election Commission) To find the election year in which the percent voting was equal to 53.35%, we solve the equation $53.35 = 63.21 - 0.29x$. (See Example 4.) We will discuss solutions of linear equations and inequalities in this section.

Equations   An **equation** is a statement that two quantities or algebraic expressions are equal. The two quantities on either side of the equal sign are called **members** of the equation. For example, $2 + 2 = 4$ is an equation with members $2 + 2$ and 4; $3x - 2 = 7$ is an equation with $3x - 2$ as its left member and 7 as its right member. Note that the equation $7 = 3x - 2$ is the same statement as $3x - 2 = 7$. An equation such as $3x - 2 = 7$ is known as an equation in one variable. The x in this case is called a **variable** because its value determines whether the equation is true. For example, $3x - 2 = 7$ is true only for $x = 3$. Finding the value(s) of the variable(s) that make the equation true—that is, finding the **solutions**—is called **solving the equation.** The set of solutions of an equation is called a **solution set** of the equation. The variable in an equation is sometimes called the **unknown.**

Some equations involving variables are true only for certain values of the variables, whereas others are true for all values. Equations that are true for all values of the variables are called **identities.** The equation $2(x - 1) = 2x - 2$ is an example of an identity. Equations that are true only for certain values of the variables are called **conditional equations** or simply **equations.**

Two equations are said to be **equivalent** if they have exactly the same solution set. For example,

$$4x - 12 = 16$$
$$4x = 28$$
$$x = 7$$

are equivalent equations because they all have the same solution, namely 7. We can often solve a complicated linear equation by finding an equivalent equation whose solution is easily found. We use the following properties of equality to reduce an equation to a simple equivalent equation.

Properties of Equality	Examples
Substitution Property The equation formed by substituting one expression for an equal expression is equivalent to the original equation.	$3(x - 3) - \frac{1}{2}(4x - 18) = 4$ is equivalent to $3x - 9 - 2x + 9 = 4$ and to $x = 4$. We say the solution set is $\{4\}$, or the solution is 4.
Addition Property The equation formed by adding the same quantity to both sides of an equation is equivalent to the original equation.	$x - 4 = 6$ is equivalent to $x - 4 + 4 = 6 + 4$, or to $x = 10$. $x + 5 = 12$ is equivalent to $x + 5 + (-5) = 12 + (-5)$, or to $x = 7$.

Properties of Equality	Examples
Multiplication Property The equation formed by multiplying both sides of an equation by the same nonzero quantity is equivalent to the original equation.	$\frac{1}{3}x = 6$ is equivalent to $3(\frac{1}{3}x) = 3(6)$, or to $x = 18$. $5x = 20$ is equivalent to $(5x)/5 = 20/5$, or to $x = 4$. (Dividing both sides by 5 is equivalent to multiplying both sides by $\frac{1}{5}$.)

Solving Linear Equations

If an equation contains one variable and if the variable occurs to the first degree, the equation is called a **linear equation in one variable.** The three properties above permit us to reduce any linear equation in one variable (also called an unknown) to an equivalent equation whose solution is obvious. We may solve linear equations in one variable by using the following procedure:

Solving a Linear Equation

Procedure	Example
To solve a linear equation in one variable:	Solve $\dfrac{3x}{4} + 3 = \dfrac{x-1}{3}$.
1. If the equation contains fractions, multiply both sides by the least common denominator (LCD) of the fractions.	1. LCD is 12. $12\left(\dfrac{3x}{4} + 3\right) = 12\left(\dfrac{x-1}{3}\right)$
2. Remove any parentheses in the equation.	2. $9x + 36 = 4x - 4$
3. Perform any additions or subtractions to get all terms containing the variable on one side and all other terms on the other side.	3. $9x + 36 - 4x = 4x - 4 - 4x$ $5x + 36 = -4$ $5x + 36 - 36 = -4 - 36$ $5x = -40$
4. Divide both sides of the equation by the coefficient of the variable.	4. $\dfrac{5x}{5} = \dfrac{-40}{5}$ gives $x = -8$
5. Check the solution by substitution in the original equation.	5. $\dfrac{3(-8)}{4} + 3 \stackrel{?}{=} \dfrac{-8-1}{3}$ gives $-3 = -3$ ✔

● **EXAMPLE 1** *Linear Equations*

(a) Solve for z: $\dfrac{2z}{3} = -6$ (b) Solve for x: $\dfrac{3x+1}{2} = \dfrac{x}{3} - 3$

Solution
(a) Multiply both sides by 3.

$$3\left(\frac{2z}{3}\right) = 3(-6) \quad \text{gives} \quad 2z = -18$$

Divide both sides by 2.

$$\frac{2z}{2} = \frac{-18}{2} \quad \text{gives} \quad z = -9$$

Check: $\dfrac{2(-9)}{3} \stackrel{?}{=} -6$ gives $-6 = -6$ ✔

(b) $\dfrac{3x + 1}{2} = \dfrac{x}{3} - 3$

$6\left(\dfrac{3x + 1}{2}\right) = 6\left(\dfrac{x}{3} - 3\right)$ Multiply both sides by the LCD, 6.

$3(3x + 1) = 6\left(\dfrac{x}{3} - 3\right)$ Simplify the fraction on the left side.

$9x + 3 = 2x - 18$ Distribute to remove parentheses.

$7x = -21$ Add $(-2x) + (-3)$ to both sides.

$x = -3$ Divide both sides by 7.

Check: $\dfrac{3(-3) + 1}{2} \stackrel{?}{=} \dfrac{-3}{3} - 3$ gives $-4 = -4$ ✔

● **EXAMPLE 2** *Future Value of an Investment*

The future value of a simple interest investment is given by $S = P + Prt$, where P is the principal invested, r is the annual interest rate (as a decimal), and t is the time in years. At what simple interest rate r must $P = 1500$ dollars be invested so that the future value is $2940 after 8 years?

Solution
Entering the values $S = 2940$, $P = 1500$, and $t = 8$ into $S = P + Prt$ gives

$$2940 = 1500 + 1500(r)(8)$$

Hence we solve the equation

$2940 = 1500 + 12{,}000r$

$1440 = 12{,}000r$ Subtract 1500 from both sides.

$\dfrac{1440}{12{,}000} = r$ Divide both sides by 12,000.

$0.12 = r$

Thus the interest rate is 0.12, or 12%.
Checking, we see that $2940 = 1500 + 1500(0.12)(8)$. ✔

Fractional Equations

A **fractional equation** is an equation that contains a variable in a denominator. It is solved by first multiplying both sides of the equation by the least common denominator (LCD) of the fractions in the equation. Some fractional equations lead to linear equations. Note that the solution to any fractional equation *must* be checked in the original equation, because multiplying both sides of a fractional equation by a variable expression may result in an equation that is not equivalent to the original equation. If a solution to the fraction-free linear equation makes a denominator of the original equation equal to zero, that value cannot be a solution to the original equation. Some fractional equations have no solutions.

● **EXAMPLE 3** *Solving Fractional Equations*

Solve for x: (a) $\dfrac{3x}{2x + 10} = 1 + \dfrac{1}{x + 5}$ (b) $\dfrac{2x - 1}{x - 3} = 4 + \dfrac{5}{x - 3}$

Solution

(a) First multiply each term on both sides by the LCD, $2x + 10$. Then simplify and solve.

$$(2x + 10)\left(\dfrac{3x}{2x + 10}\right) = (2x + 10)(1) + (2x + 10)\left(\dfrac{1}{x + 5}\right)$$

$3x = (2x + 10) + 2$ gives $x = 12$

Check: $\dfrac{3(12)}{2(12) + 10} \stackrel{?}{=} 1 + \dfrac{1}{12 + 5}$ gives $\dfrac{36}{34} = \dfrac{18}{17}$ ✔

(b) First multiply each term on both sides by the LCD, $x - 3$. Then simplify.

$$(x - 3)\left(\dfrac{2x - 1}{x - 3}\right) = (x - 3)(4) + (x - 3)\left(\dfrac{5}{x - 3}\right)$$

$2x - 1 = (4x - 12) + 5$ or $2x - 1 = 4x - 7$

Add $(-4x) + 1$ to both sides.

$-2x = -6$ gives $x = 3$

Check: The value $x = 3$ gives undefined expressions because the denominators equal 0 when $x = 3$. Hence the equation has no solution. ✔

● *Checkpoint*

1. Solve the following for x, and check.

   (a) $4(x - 3) = 10x - 12$ (b) $\dfrac{5(x - 3)}{6} - x = 1 - \dfrac{x}{9}$ (c) $\dfrac{x}{3x - 6} = 2 - \dfrac{2x}{x - 2}$

When using an equation to describe (model) a set of data, it is sometimes useful to let a variable represent the number of years past a given year. This is called **aligning the data.** Consider the following example.

◗ **EXAMPLE 4** *Voting (Application Preview)*

Using data from 1952–2000, the percent p of the eligible U.S. population voting in presidential elections has been estimated to be

$$p = 63.21 - 0.29x$$

where x is the number of years past 1950. (*Source:* Federal Election Commission) According to this model, in what election year is the percent voting equal to 53.35%?

Solution

To answer this question, we solve

$$53.35 = 63.21 - 0.29x$$
$$-9.86 = -0.29x$$
$$34 = x$$

Checking reveals that $53.35 = 63.21 - 0.29(34)$. ✔

Thus the percent of eligible voters who voted is estimated to be 53.35% in the presidential election year $1950 + 34 = 1984$.

Linear Equations with Two Variables

The steps used to solve linear equations in one variable can also be used to solve linear equations in more than one variable for one of the variables in terms of the other. Solving an equation such as the one in the following example is important when using a graphing utility.

● **EXAMPLE 5** *Solving an Equation for One of Two Variables*

Solve $4x + 3y = 12$ for y.

Solution

No fractions or parentheses are present, so we subtract $4x$ from both sides to get only the term that contains y on one side.

$$3y = -4x + 12$$

Dividing both sides by 3 gives the solution.

$$y = -\frac{4}{3}x + 4$$

Check: $4x + 3\left(\dfrac{-4}{3}x + 4\right) \overset{?}{=} 12$

$$4x + (-4x + 12) = 12 ✔$$

● **Checkpoint**   2. Solve for y: $y - 4 = -4(x + 2)$

● **EXAMPLE 6** *Profit*

Suppose that the relationship between a firm's profit P and the number x of items sold can be described by the equation

$$5x - 4P = 1200$$

(a) How many units must be produced and sold for the firm to make a profit of $150?
(b) Solve this equation for P in terms of x.
(c) Find the profit when 240 units are sold.

Solution

(a) $5x - 4(150) = 1200$
$5x - 600 = 1200$
$5x = 1800$
$x = 360$ units
Check: $5(360) - 4(150) = 1800 - 600 = 1200$ ✔

(b) $5x - 4P = 1200$
$5x - 1200 = 4P$
$P = \dfrac{5x}{4} - \dfrac{1200}{4} = \dfrac{5}{4}x - 300$

(c) $P = \dfrac{5}{4}x - 300$

$P = \dfrac{5}{4}(240) - 300 = 0$

Because $P = 0$ when $x = 240$, we know that profit is $0 when 240 units are sold, and we say that the firm **breaks even** when 240 units are sold.

Stated Problems With an applied problem, it is frequently necessary to convert the problem from its stated form into one or more equations from which the problem's solution can be found. The following guidelines may be useful in solving stated problems.

> **Guidelines for Solving Stated Problems**
>
> 1. Begin by reading the problem carefully to determine what you are to find. Use variables to represent the quantities to be found.
>
> 2. Reread the problem and use your variables to translate given information into algebraic expressions. Often, drawing a figure is helpful.
>
> 3. Use the algebraic expressions and the problem statement to formulate an equation (or equations).
>
> 4. Solve the equation(s).
>
> 5. Check the solution in the problem, not just in your equation or equations. The answer should satisfy the conditions.

● **EXAMPLE 7** *Investment Mix*

Jill Bell has $90,000 to invest. She has chosen one relatively safe investment fund that has an annual yield of 10% and another, riskier one that has a 15% annual yield. How much should she invest in each fund if she would like to earn $10,000 in one year from her investments?

Solution
We want to find the amount of each investment, so we begin as follows:

> Let $x =$ the amount invested at 10%, then
>
> $90,000 - x =$ the amount invested at 15% (because the two investments total $90,000)

If P is the amount of an investment and r is the annual rate of yield (expressed as a decimal), then the annual earnings $I = Pr$. Using this relationship, we can summarize the information about these two investments in a table.

	P	r	I
10% investment	x	0.10	$0.10x$
15% investment	$90,000 - x$	0.15	$0.15(90,000 - x)$
Total investment	$90,000$		$10,000$

The column under I shows that the sum of the earnings is

$$0.10x + 0.15(90,000 - x) = 10,000$$

We solve this as follows.

$$0.10x + 13,500 - 0.15x = 10,000$$
$$-0.05x = -3500 \quad \text{or} \quad x = 70,000$$

Thus the amount invested at 10% is $70,000, and the amount invested at 15% is $90,000 - 70,000 = 20,000$. Check: To check, we return to the problem and note that 10% of $70,000 plus 15% of $20,000 gives a yield of $7000 + $3000 = $10,000. ✔

Linear Inequalities

An **inequality** is a statement that one quantity is greater than (or less than) another quantity. For example, if x represents the number of items a firm produces and sells, x must be a nonnegative quantity. Thus x is greater than or equal to zero, which is written $x \geq 0$. The inequality $3x - 2 > 2x + 1$ is a first-degree (linear) inequality that states that the left member is greater than the right member. Certain values of the variable will satisfy the inequality. These values form the solution set of the inequality. For example, 4 is in the solution set of $3x - 2 > 2x + 1$ because $3 \cdot 4 - 2 > 2 \cdot 4 + 1$. On the other hand, 2 is not in the solution set because $3 \cdot 2 - 2 \not> 2 \cdot 2 + 1$. *Solving* an inequality means finding its solution set, and two inequalities are *equivalent* if they have the same solution set. As with equations, we find the solutions to inequalities by finding equivalent inequalities from which the solutions can be easily seen. We use the following properties to reduce an inequality to a simple equivalent inequality.

Inequalities

Properties	Examples
Substitution Property The inequality formed by substituting one expression for an equal expression is equivalent to the original inequality.	$5x - 4x < 6$ $x < 6$ The solution set is $\{x: x < 6\}$.
Addition Property The inequality formed by adding the same quantity to both sides of an inequality is equivalent to the original inequality.	$2x - 4 > x + 6$ $2x - 4 + 4 > x + 6 + 4$ $2x > x + 10$ $2x + (-x) > x + 10 + (-x)$ $x > 10$
Multiplication Property I The inequality formed by multiplying both sides of an inequality by the same *positive* quantity is equivalent to the original inequality.	$\frac{1}{2}x > 8 \qquad\qquad 3x < 6$ $\frac{1}{2}x(2) > 8(2) \qquad 3x\left(\frac{1}{3}\right) < 6\left(\frac{1}{3}\right)$ $x > 16 \qquad\qquad x < 2$
Multiplication Property II The inequality formed by multiplying both sides of an inequality by the same *negative* number and reversing the direction of the inequality symbol is equivalent to the original inequality.	$-x < 6 \qquad\qquad -3x > -27$ $-x(-1) > 6(-1) \qquad -3x\left(-\frac{1}{3}\right) < -27\left(-\frac{1}{3}\right)$ $x > -6 \qquad\qquad x < 9$

For some inequalities, it requires several operations to find their solution sets. In this case, the order in which the operations are performed is the same as that used in solving linear equations.

● **EXAMPLE 8 *Solution of an Inequality***

Solve the inequality $2(x - 4) < \dfrac{x - 3}{3}$.

Solution

$$2(x - 4) < \frac{x - 3}{3}$$

$6(x - 4) < x - 3$ Clear fractions.

$6x - 24 < x - 3$ Remove parentheses.

$5x < 21$ Perform additions and subtractions.

$x < \dfrac{21}{5}$ Multiply by $\frac{1}{5}$.

Check: Now, if we want to check that this solution is a reasonable one, we can substitute the integer values around 21/5 into the original inequality. Note that $x = 4$ satisfies the inequality because

$$2[(4) - 4] < \frac{(4) - 3}{3}$$

but that $x = 5$ does not because

$$2[(5) - 4] \not< \frac{(5) - 3}{3}$$

Thus $x < 21/5$ is a reasonable solution. ✔

We may also solve inequalities of the form $a \leq b$. This means "*a* is less than *b* or *a* equals *b*." The solution of $2x \leq 4$ is $x \leq 2$, because $x < 2$ is the solution of $2x < 4$ and $x = 2$ is the solution of $2x = 4$.

● **EXAMPLE 9** *Solution of an Inequality*

Solve the inequality $3x - 2 \leq 7$ and graph the solution.

Solution
This inequality states that $3x - 2 = 7$ or that $3x - 2 < 7$. By solving in the usual manner, we get $3x \leq 9$, or $x \leq 3$. Then $x = 3$ is the solution to $3x - 2 = 7$ and $x < 3$ is the solution to $3x - 2 < 7$, so the solution set for $3x - 2 \leq 7$ is $\{x: x \leq 3\}$.

The graph of the solution set includes the point $x = 3$ and all points $x < 3$ (see Figure 1.1).

Figure 1.1

● *Checkpoint*

Solve the following inequalities for y.
3. $3y - 7 \leq 5 - y$ 4. $2y + 6 > 4y + 5$ 5. $4 - 3y \geq 4y + 5$

● **EXAMPLE 10** *Normal Height for a Given Age*

For boys between 4 and 16 years of age, height and age are linearly related. That relation can be expressed as

$$H = 2.31A + 31.26$$

where H is height in inches and A is age in years. To account for natural variation among individuals, normal is considered to be any measure falling within $\pm 5\%$ of the height obtained from the equation.* Write as an inequality the range of normal height for a boy who is 9 years old.

Solution

The boy's height from the formula is $H = 2.31(9) + 31.26 = 52.05$ inches. For a 9-year-old boy to be considered of normal height, H would have to be within $\pm 5\%$ of 52.05 inches. That is, the boy's height H is considered normal if $H \geq 52.05 - (0.05)(52.05)$ and $H \leq 52.05 + (0.05)(52.05)$. We can write this range of normal height by the compound inequality

$$52.05 - (0.05)(52.05) \leq H \leq 52.05 + (0.05)(52.05)$$

or

$$49.45 \leq H \leq 54.65$$

● *Checkpoint Solutions*

1. (a) $4(x - 3) = 10x - 12$
$4x - 12 = 10x - 12$
$-6x = 0$
$x = 0$
Check: $4(0 - 3) \overset{?}{=} 10 \cdot 0 - 12$
$-12 = -12$ ✔

(b) $\dfrac{5(x - 3)}{6} - x = 1 - \dfrac{x}{9}$
LCD of 6 and 9 is 18.
$18\left(\dfrac{5x - 15}{6}\right) - 18(x) = 18(1) - 18\left(\dfrac{x}{9}\right)$
$3(5x - 15) - 18x = 18 - 2x$
$15x - 45 - 18x = 18 - 2x$
$-45 - 3x = 18 - 2x$
$-63 = x$
Check: $\dfrac{5(-63 - 3)}{6} - (-63) \overset{?}{=} 1 - \dfrac{(-63)}{9}$
$5(-11) + 63 \overset{?}{=} 1 + 7$
$8 = 8$ ✔

(c) $\dfrac{x}{3x - 6} = 2 - \dfrac{2x}{x - 2}$
LCD is $3x - 6$ or $3(x - 2)$.
$(3x - 6)\left(\dfrac{x}{3x - 6}\right) = (3x - 6)(2) - (3x - 6)\left(\dfrac{2x}{x - 2}\right)$
$x = 6x - 12 - 3(2x)$ or $x = 6x - 12 - 6x$
$x = -12$
Check: $\dfrac{-12}{3(-12) - 6} \overset{?}{=} 2 - \dfrac{2(-12)}{-12 - 2}$
$\dfrac{-12}{-42} \overset{?}{=} 2 - \dfrac{12}{7}$
$\dfrac{2}{7} = \dfrac{2}{7}$ ✔

2. $y - 4 = -4(x + 2)$
$y - 4 = -4x - 8$
$y = -4x - 4$

3. $3y - 7 \leq 5 - y$
$4y - 7 \leq 5$
$4y \leq 12$
$y \leq 3$

4. $2y + 6 > 4y + 5$
$6 > 2y + 5$
$1 > 2y$
$y < \dfrac{1}{2}$

5. $4 - 3y \geq 4y + 5$
$4 \geq 7y + 5$
$-1 \geq 7y$
$y \leq -\dfrac{1}{7}$

---

*Adapted from data from the National Center for Health Statistics

1.1 Exercises

In Problems 1–12, solve each equation.

1. $4x - 7 = 8x + 2$
2. $3x + 22 = 7x + 2$
3. $x + 8 = 8(x + 1)$
4. $x + x + x = x$
5. $-\dfrac{3}{4}x = 24$
6. $-\dfrac{1}{6}x = 12$
7. $2(x - 7) = 5(x + 3) - x$
8. $3(x - 4) = 4 - 2(x + 2)$
9. $\dfrac{5x}{2} - 4 = \dfrac{2x - 7}{6}$
10. $\dfrac{2x}{3} - 1 = \dfrac{x - 2}{2}$
11. $x + \dfrac{1}{3} = 2\left(x - \dfrac{2}{3}\right) - 6x$
12. $\dfrac{3x}{4} - \dfrac{1}{3} = 1 - \dfrac{2}{3}\left(x - \dfrac{1}{6}\right)$

The equations in Problems 13–18 lead to linear equations. Because not all solutions to the linear equations are solutions to the original equations, be sure to check the solutions in the original equations.

13. $\dfrac{33 - x}{5x} = 2$
14. $\dfrac{3x + 3}{x - 3} = 7$
15. $\dfrac{2x}{2x + 5} = \dfrac{2}{3} - \dfrac{5}{4x + 10}$
16. $\dfrac{3}{x} + \dfrac{1}{4} = \dfrac{2}{3} + \dfrac{1}{x}$
17. $\dfrac{2x}{x - 1} + \dfrac{1}{3} = \dfrac{5}{6} + \dfrac{2}{x - 1}$
18. $\dfrac{2x}{x - 3} = 4 + \dfrac{6}{x - 3}$

In Problems 19–22, use a calculator to solve each equation. Round your answer to three decimal places.

19. $3.259x - 8.638 = -3.8(8.625x + 4.917)$
20. $3.319(14.1x - 5) = 9.95 - 4.6x$
21. $0.000316x + 9.18 = 2.1(3.1 - 0.0029x) - 4.68$
22. $3.814x = 2.916(4.2 - 0.06x) + 5.3$

In Problems 23–26, solve for *y* in terms of *x*.

23. $3x - 4y = 15$
24. $3x - 5y = 25$
25. $9x + \dfrac{3}{2}y = 11$
26. $\dfrac{3x}{2} + 5y = \dfrac{1}{3}$
27. Solve $S = P + Prt$ for t.
28. Solve $\dfrac{y - b}{x - a} = m$ for y.

In Problems 29–34, solve each inequality.

29. $3(x - 1) < 2x - 1$
30. $2(x + 1) > x - 1$
31. $1 - 2x > 9$
32. $17 - x < -4$
33. $\dfrac{3(x - 1)}{2} \leq x - 2$
34. $\dfrac{x - 1}{2} + 1 > x + 1$

In Problems 35–40, solve each inequality and graph the solution.

35. $2(x - 1) - 3 > 4x + 1$
36. $7x + 4 \leq 2(x - 1)$
37. $\dfrac{-3x}{2} > 3 - x$
38. $\dfrac{-2x}{5} \leq -10 - x$
39. $\dfrac{3x}{4} - \dfrac{1}{6} < x - \dfrac{2(x - 1)}{3}$
40. $\dfrac{4x}{3} - 3 > \dfrac{1}{2} + \dfrac{5x}{12}$

APPLICATIONS

41. *Depreciation* A $648,000 property is depreciated for tax purposes by its owner with the straight-line depreciation method. The value of the building, y, after x months of use is given by $y = 648,000 - 1800x$ dollars. After how many months will the value of the building be $387,000?

42. *Depreciation* When an $810,000 building is depreciated for tax purposes (by the straight-line method), its value, y, after x months of use is given by $y = 810,000 - 2250x$. How many months will it be before the building is fully depreciated (that is, its value is $0)? How many years is this?

43. *Credit card debt* High interest rates make it difficult for people to pay off credit card debt in a reasonable period of time. The interest I (in dollars) paid on a $10,000 debt over 3 years when the interest rate is $r\%$ can be approximated by the equation

$$\dfrac{I}{175.393} + 0.663 = r$$

If the credit card interest rate is 19.8%, find the amount of interest paid during the 3 years. (*Source:* Consumer Federation of America)

44. *Seawater pressure* In seawater, the pressure p is related to the depth d according to

$$33p - 18d = 495$$

where d is in feet and p is in pounds per square inch.
(a) Solve this equation for p in terms of d.
(b) The *Titanic* was discovered at a depth of 12,460 ft. Find the pressure at this depth.

45. *Break-even* Burnem, Inc. manufactures blank CDs and sells them to a distributor in packs of 500 CDs. If Burnem's total cost and total revenue (in dollars) for x packs of 500 CDs are given by

Total cost $= 2x + 7920$ and   Total revenue $= 20x$,

how many packs of 500 CDs must Burnem sell to break even?

46. *Break-even* Dish Systems manufactures satellite systems and has its monthly profit P in dollars related to the number of satellite systems, x, by

$$4P = 81x - 29,970$$

Find the number of systems that Dish Systems needs to produce and sell in order to break even.

47. *Profit*   In its second year of operation, a local Internet provider's profits were $170,500. If this amount was 576% of the company's first-year profits, find the first-year profits (to the nearest hundred dollars).

48. *Sales tax*   The total price of a new car (including 6% sales tax) is $21,041. How much of this is tax?

49. *Computers and the Internet*   The percent of various types and locations of U.S. households with Internet service, I, can be approximated by the equation

$$959C - 1000I = 456.8$$

where C is the percent of those households with a computer. (*Source:* National Telecommunications and Information Administration, U.S. Department of Commerce) According to the equation, what percent of households would need to have a computer before the percent with Internet service reached 75%?

50. *Median household income*   The U.S. median household income for blacks can be described by the equation

$$B = 1.172W - 23,760$$

where W is the U.S. median household income for whites. (*Source:* U.S. Bureau of the Census) When black households have a median income of $50,000, what is the predicted median household income for whites?

51. *Course grades*   To earn an A in a course, a student must get at least a 90 average on four tests and a final exam, with the final exam weighted twice that of any one test. If the four test scores are 93, 69, 89, and 97, what is the lowest score the student can earn on the final exam and still get an A in the course?

52. *Course grades*   Suppose a professor counts the final exam as being equal to each of the other tests in her course, and she will also change the lowest test score to match the final exam score if the final exam score is higher. If a student's four test scores are 83, 67, 52, and 90, what is the lowest score the student can earn on the final exam and still obtain at least an 80 average for the course?

53. *Investment mix*   A retired woman has $120,000 to invest. She has chosen one relatively safe investment fund that has an annual yield of 9% and another, riskier one that has a 13% annual yield. How much should she invest in each fund if she would like to earn $12,000 per year from her investments?

54. *Investment yields*   One safe investment pays 10% per year, and a more risky investment pays 18% per year.

A woman who has $145,600 to invest would like to have an income of $20,000 per year from her investments. How much should she invest at each rate?

55. *Salary increases*   A woman making $2000 per month has her salary reduced by 10% because of sluggish sales. One year later, after a dramatic improvement in sales, she is given a 20% raise over her reduced salary. Find her salary after the raise. What percent change is this from the $2000 per month?

56. *Wildlife management*   In wildlife management, the capture-mark-recapture technique is used to estimate the populations of fish or birds in an area or to measure the infestation of insects such as Japanese beetles. Suppose 100 individuals of the species being studied are caught, marked, and released, and one week later 100 more are caught. To estimate the total number of individuals, the following relationship is used:

$$\frac{\text{Total marked found in 2nd capture}}{\text{Total in 2nd capture}} = \frac{\text{Total number marked}}{\text{Total population}}$$

   (a) If in the second capture of 100, it is found that 3 are marked, what is the total population?
   (b) Suppose that 1000 beetles are captured, marked, and released. Suppose further that in the second capture of 1000 it is found that 63 are marked. What is the population estimate?

57. *Profit*   For a certain product, the revenue function is $R(x) = 40x$ and the cost function is $C(x) = 20x + 1600$. To obtain a profit, the revenue must be greater than the cost. For what values of x will there be a profit? Graph the solution.

58. *Car rental*   Thirfty rents a compact car for $33 per day, and Budget rents a similar car for $20 per day plus an initial fee of $78. For how many days would it be cheaper to rent from Budget? Graph the solution.

59. *Purchasing*   Sean can spend at most $900 for a video camera and some video tapes. He plans to buy the camera for $695 and tapes for $5.75 each. Write an inequality that could be used to find the number of tapes (x) that he could buy. How many tapes could he buy?

60. *Taxes*   In Sweetwater, Arizona, water bills are taxed on the basis of the amount of the monthly bill in order to encourage conservation. If the bill is more than $0 but less than $60, the tax is 2% of the bill; if the bill is $60 but less than $80, the tax is 4% of the bill; and if the bill is $80 or more, the tax is 6% of the bill. Write the inequalities that represent the amounts of tax owed in these three cases.

61. *Wireless service spending* The total amount spent in the United States for wireless communication services, S (in billions of dollars), can be modeled (that is, approximated with some accuracy) by

$$S = 6.205 + 11.23t$$

where t is the number of years past 1995. (*Source:* Cellular Telecommunications and Internet Association)
(a) What value of t represents the year 2005?
(b) What values of t give $S > 150$?
(c) In what year does this equation project that spending for wireless communication services will exceed $150 billion?

62. *Cigarette use* The percent p of high school seniors who smoke cigarettes can be modeled by the equation

$$p = 32.88 - 0.03t$$

where t is the number of years past 1975. (*Source:* National Institute of Drug Abuse)
(a) What value of t represents 2007?
(b) What percent does this equation predict for 2007?
(c) This equation ceases to be effective when p becomes negative and perhaps before that. Find the years when $p < 0$.

63. *Heat index* During the summer of 1998, Dallas, Texas endured 29 consecutive days on which the temperature was at least 110°F. On many of those days, the combination of heat and humidity made it feel even hotter than it was. When the temperature is 100°F, the apparent temperature A (or heat index) depends on the humidity h (expressed as a decimal) according to

$$A = 90.2 + 41.3h*$$

(a) For what humidity levels is the apparent temperature at least 110°F? (Note that this answer will be a closed interval. Why?)
(b) For what humidity levels is the apparent temperature less than 100°F?

64. *Wind chill* The combination of cold temperatures and wind speed determine what is called wind chill. The wind chill is a temperature that is the still-air equivalent of the combination of cold and wind. When the wind speed is 25 mph, the wind chill WC depends on the temperature t (in degrees Fahrenheit) according to

$$WC = 1.479t - 43.821$$

For what temperatures does it feel at least 30°F colder than the air temperature? That is, find t such that $WC \leq t - 30$.

*Source: Bosch, W., and C. G. Cobb, "Temperature-Humidity Indices," UMAP Unit 691, *The UMAP Journal,* 10(3), Fall 1989, 237–256.

| 1.2 | # Functions |

OBJECTIVES

- To determine whether a relation is a function
- To state the domains and ranges of certain functions
- To use function notation
- To perform operations with functions
- To find the composite of two functions

◗ Application Preview

The number of individual income tax returns filed electronically has increased in recent years. The relationship between the number of returns filed electronically and the number of years after 1995 can be described by the equation

$$y = 5.064x + 10.067$$

where y is in millions and x is the number of years past 1995. (*Source:* Internal Revenue Service) In this equation, y depends uniquely on x, so we say that y is a **function** of x. Understanding the mathematical meaning of the phrase *function of* and gaining the ability to interpret and apply such relationships are the goals of this section. (The function above will be discussed in Example 4.)

Relations and Functions

An equation or inequality containing two variables expresses a **relation** between those two variables. For example, the inequality $R \geq 35x$ expresses a relation between the two variables x and R, and the equation $y = 4x - 3$ expresses a relation between the two variables x and y.

In addition to defining a relation by an equation, an inequality, or rule of correspondence, we may also define it as any set of **ordered pairs** of real numbers (a, b). For example, the solutions to $y = 4x - 3$ are pairs of numbers (one for x and one for y). We write the pairs (x, y) so that the first number is the x-value and the second is the y-value, and these ordered pairs define the relation between x and y. Some relations may be defined by a table, a graph, or an equation.

Relation

> A **relation** is defined by a set of ordered pairs or by a rule that determines how the ordered pairs are found. It may also be defined by a table, a graph, an equation, or an inequality.

For example, the set of ordered pairs

$$\{(1, 3), (1, 6), (2, 6), (3, 9), (3, 12), (4, 12)\}$$

expresses a relation between the set of first components, $\{1, 2, 3, 4\}$, and the set of second components, $\{3, 6, 9, 12\}$. The set of first components is called the **domain** of the relation, and the set of second components is called the **range** of the relation. Figure 1.2(a) uses arrows to indicate how the inputs from the domain (the first components) are associated with the outputs in the range (the second components). Figure 1.2(b) shows another example of a relation. Because relations can also be defined by tables and graphs, Table 1.1 and Figure 1.3 are examples of relations.

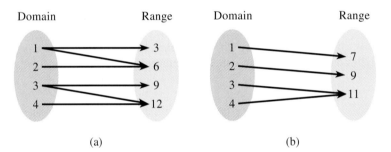

Figure 1.2   (a)   (b)

An equation frequently expresses how the second component (the output) is obtained from the first component (the input). For example, the equation

$$y = 4x - 3$$

expresses how the output y results from the input x. This equation expresses a special relation between x and y, because each value of x that is substituted into the equation results in exactly one value of y. If each value of x put into an equation results in one value of y, we say that the equation expresses y as a **function** of x.

Definition of a Function

> A **function** is a relation between two sets such that to each element of the domain (input) there corresponds exactly one element of the range (output). A function may be defined by a set of ordered pairs, a table, a graph, or an equation.

When a function is defined, the variable that represents the numbers in the domain (input) is called the **independent variable** of the function, and the variable that represents the numbers in the range (output) is called the **dependent variable** (because its values depend on the values of the independent variable). The equation $y = 4x - 3$ defines y as a function of x, because only one value of y will result from each value of x that is substituted

TABLE 1.1
2004 Federal Income Tax

Income	Rate
$0–$7,150	10%
$7,151–$29,050	15%
$29,051–$70,350	25%
$70,351–$146,750	28%
$146,751–$319,100	33%
$319,101+	35%

Source: Internal Revenue Service

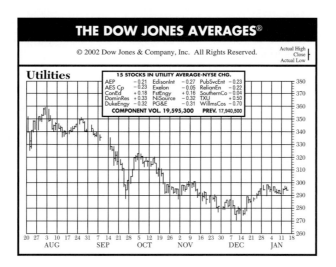

Figure 1.3
Source: The Wall Street Journal, January 17, 2002. Copyright © 2002 Dow Jones & Co. Reprinted by permission of Dow Jones & Co. via Copyright Clearance Center.

into the equation. Thus the equation defines a function in which x is the independent variable and y is the dependent variable.

We can also apply this idea to a relation defined by a table or a graph. In Figure 1.2(b), because each input in the domain corresponds to exactly one output in the range, the relation is a function. Similarly, the data given in Table 1.1 (the tax brackets for U.S. income tax for single wage earners) represents the tax rate as a function of the income. Note in Table 1.1 that even though many different amounts of taxable income have the same tax rate, each amount of taxable income (input) corresponds to exactly one tax rate (output). Thus tax rate (the dependent variable) is a function of a single filer's taxable income (the independent variable). On the other hand, the relation defined in Figure 1.3 is not a function because the graph representing the Dow Jones Average shows that for each day there are at least three different values—the actual high, the actual low, and the close. For example, on September 25, 2001, the average varies from a low of 286 to a high of 298. Graphs such as Figure 1.3 appear in many daily newspapers, so that current versions are readily available. However, this particular figure also has historical interest because it shows a break in the graph when the New York Stock Exchange closed following the terrorist attacks of 9/11/2001.

● **EXAMPLE 1** *Functions*

Does $y^2 = 2x$ express y as a function of x?

Solution
No, because for some values of x there is more than one value of y. In fact, there are two y-values for each $x > 0$. For example, if $x = 8$, then $y = 4$ or $y = -4$, two different y-values for the same x-value. The equation $y^2 = 2x$ expresses a relation between x and y, but y is not a function of x.

Graphs of Functions

It is possible to picture geometrically the relations and functions that we have been discussing by sketching their graphs on a rectangular coordinate system. We construct a rectangular coordinate system by drawing two real number lines (called **coordinate axes**) that are perpendicular to each other and intersect at their origins (called the **origin** of the system).

The ordered pair (a, b) represents the point P that is located a units along the x-axis and b units along the y-axis (see Figure 1.4). Similarly, any point has a unique ordered pair that describes it.

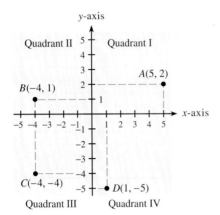

Figure 1.4

The values a and b in the ordered pair associated with the point P are called the **rectangular** (or **Cartesian**) **coordinates** of the point, where a is the **x-coordinate** (or **abscissa**), and b is the **y-coordinate** (or **ordinate**). The ordered pairs (a, b) and (c, d) are equal if and only if $a = c$ and $b = d$.

The **graph** of an equation that defines a function (or relation) is the picture that results when we plot the points whose coordinates (x, y) satisfy the equation. To sketch the graph, we plot enough points to suggest the shape of the graph and draw a smooth curve through the points. This is called the **point-plotting method** of sketching a graph.

● **EXAMPLE 2 *Graphing a Function***

Graph the function $y = 4x^2$.

Solution
We choose some sample values of x and find the corresponding values of y. Placing these in a table, we have sample points to plot. When we have enough to determine the shape of the graph, we connect the points to complete the graph. The table and graph are shown in Figure 1.5(a).

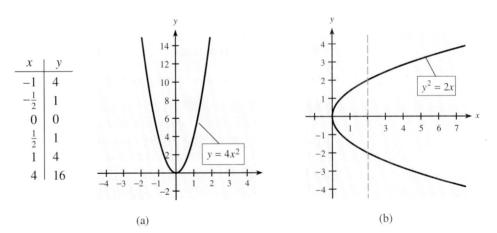

x	y
-1	4
$-\frac{1}{2}$	1
0	0
$\frac{1}{2}$	1
1	4
4	16

Figure 1.5   (a)   (b)

We can determine whether a relation is a function by inspecting its graph. If the relation is a function, then no one input (x-value) has two different outputs (y-values). This means that no two points on the graph will have the same first coordinate (component). Thus no two points of the graph will lie on the same vertical line.

Vertical-Line Test

> If no vertical line exists that intersects the graph at more than one point, then the graph is that of a function.

Performing this test on the graph of $y = 4x^2$ (Figure 1.5(a)), we easily see that this equation describes a function. The graph of $y^2 = 2x$ is shown in Figure 1.5(b), and we can see that the vertical-line test indicates that this is not a function (as we already saw in Example 1). For example, a vertical line at $x = 2$ intersects the curve at (2, 2) and (2, −2).

The graph in Figure 1.6 shows one possible projection of the resources of the Pension Benefit Guaranty Corporation, the agency that insures pensions, (y) versus time (x). It represents a function, because for every x there is exactly one y. On the other hand, as noted previously, the graph in Figure 1.3 (earlier in this section) does not represent a function.

Function Notation

We can use function notation to indicate that y is a function of x. The function is denoted by f, and we write $y = f(x)$. This is read "y is a function of x" or "y equals f of x." For specific values of x, $f(x)$ represents the values of the function (that is, outputs, or y-values) at those x-values. Thus if

$$f(x) = 3x^2 + 2x + 1$$

then

$$f(2) = 3(2)^2 + 2(2) + 1 = 17$$

and

$$f(-3) = 3(-3)^2 + 2(-3) + 1 = 22$$

Figure 1.7 shows the function notation $f(x)$ as (a) an operator on x and (b) a y-coordinate for a given x-value.

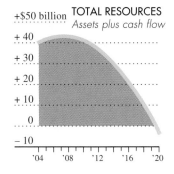

+$50 billion **TOTAL RESOURCES**
Assets plus cash flow

Figure 1.6

Source: The New York Times,
September 14, 2004. Copyright ©
2004 The New York Times Co.
Reprinted by permission.

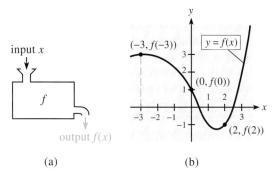

input x

output f(x)

(a)

(−3, f(−3))

y = f(x)

(0, f(0))

(2, f(2))

(b)

Figure 1.7

Letters other than f may also be used to denote functions. For example, $y = g(x)$ or $y = h(x)$ may be used.

● **EXAMPLE 3** *Evaluating Functions*

If $y = f(x) = 2x^3 - 3x^2 + 1$, find the following.
(a) $f(3)$ (b) $f(-1)$ (c) $f(a)$ (d) $f(-a)$

Solution
(a) $f(3) = 2(3)^3 - 3(3)^2 + 1 = 2(27) - 3(9) + 1 = 28$
    Thus $y = 28$ when $x = 3$.

(b) $f(-1) = 2(-1)^3 - 3(-1)^2 + 1 = 2(-1) - 3(1) + 1 = -4$
   Thus $y = -4$ when $x = -1$.
(c) $f(a) = 2a^3 - 3a^2 + 1$
(d) $f(-a) = 2(-a)^3 - 3(-a)^2 + 1 = -2a^3 - 3a^2 + 1$

◑ EXAMPLE 4 *Electronic Income Tax Returns* **(Application Preview)**

The relationship between the number of individual income tax returns filed electronically and the number of years after 1995 can be described by the equation

$$y = f(x) = 5.064x + 10.067$$

where y is in millions and x is the number of years after 1995. (*Source:* Internal Revenue Service)

(a) Find $f(8)$.
(b) Write a sentence that explains the meaning of the result in (a).

Solution
(a) $f(8) = 5.064(8) + 10.067 = 50.579$
(b) The statement $f(8) = 5.064(8) + 10.067 = 50.579$ means that in $1995 + 8 = 2003$, 50.579 million income tax returns were filed electronically.

● EXAMPLE 5 *Mortgage Payment*

TABLE 1.2	
$r(\%)$	$f(r)$
2.6	12
5.2	15
6.3	17
7.4	20
9	30

Table 1.2 shows the number of years that it will take a couple to pay off a $100,000 mortgage at several different interest rates if they pay $800 per month. If r denotes the rate and $f(r)$ denotes the number of years:

(a) What is $f(6.3)$ and what does it mean?
(b) If $f(r) = 30$, what is r?
(c) Does $2 \cdot f(2.6) = f(2 \cdot 2.6)$?

Solution
(a) $f(6.3) = 17$. This means that with a 6.3% interest rate, a couple can pay off the $100,000 mortgage in 17 years by paying $800 per month.
(b) The table indicates that $f(9) = 30$, so $r = 9$.
(c) $2 \cdot f(2.6) = 2 \cdot 12 = 24$ and $f(2 \cdot 2.6) = f(5.2) = 15$, so $2 \cdot f(2.6) \neq f(2 \cdot 2.6)$.

● EXAMPLE 6 *Function Notation*

Given $f(x) = x^2 - 3x + 8$, find $\dfrac{f(x + h) - f(x)}{h}$ and simplify (if $h \neq 0$).

Solution
We find $f(x + h)$ by replacing each x in $f(x)$ with the expression $x + h$.

$$\frac{f(x + h) - f(x)}{h} = \frac{[(x + h)^2 - 3(x + h) + 8] - [x^2 - 3x + 8]}{h}$$

$$= \frac{[(x^2 + 2xh + h^2) - 3x - 3h + 8] - x^2 + 3x - 8}{h}$$

$$= \frac{x^2 + 2xh + h^2 - 3x - 3h + 8 - x^2 + 3x - 8}{h}$$

$$= \frac{2xh + h^2 - 3h}{h} = \frac{h(2x + h - 3)}{h} = 2x + h - 3$$

Domains and Ranges   We will limit our discussion in this text to **real functions,** which are functions whose domains and ranges contain only real numbers. If the domain and range of a function are not specified, it is assumed that the domain consists of all real inputs (*x*-values) that result in real outputs (*y*-values), making the range a subset of the real numbers.

For the types of functions we are now studying, if the domain is unspecified, it will include all real numbers except

1. values that result in a denominator of 0, and
2. values that result in an even root of a negative number.

● **EXAMPLE 7** *Domain and Range*

Find the domain of each of the following functions; find the range for the functions in (a) and (b).

(a) $y = 4x^2$ (b) $y = \sqrt{4 - x}$ (c) $y = 1 + \dfrac{1}{x - 2}$

Solution

(a) There are no restrictions on the numbers substituted for *x*, so the domain consists of all real numbers. Because the square of any real number is nonnegative, $4x^2$ must be nonnegative. Thus the range is $y \geq 0$. If we plot points or use a graphing utility, we will get the graph shown in Figure 1.8(a), which illustrates our conclusions about the domain and range.

(b) We note the restriction that $4 - x$ cannot be negative. Thus the domain consists of only numbers less than or equal to 4. That is, the domain is the set of real numbers satisfying $x \leq 4$. Because $\sqrt{4 - x}$ is always nonnegative, the range is all $y \geq 0$. Figure 1.8(b) shows the graph of $y = \sqrt{4 - x}$. Note that the graph is located only where $x \leq 4$ and on or above the *x*-axis (where $y \geq 0$).

(c) $y = 1 + \dfrac{1}{x - 2}$ is undefined at $x = 2$ because $\dfrac{1}{0}$ is undefined. Hence, the domain consists of all real numbers except 2. Figure 1.8(c) shows the graph of $y = 1 + \dfrac{1}{x - 2}$. The break where $x = 2$ indicates that $x = 2$ is not part of the domain.

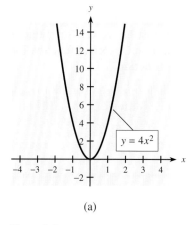

(a)

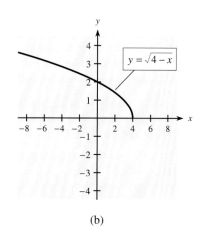

(b)

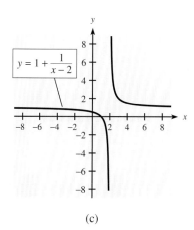

(c)

Figure 1.8

● *Checkpoint*

1. If $y = f(x)$, the independent variable is _____ and the dependent variable is _____.
2. If $(1, 3)$ is on the graph of $y = f(x)$, then $f(1) = ?$
3. If $f(x) = 1 - x^3$, find $f(-2)$.
4. If $f(x) = 2x^2$, find $f(x + h)$.
5. If $f(x) = \dfrac{1}{x + 1}$, what is the domain of $f(x)$?

Operations with Functions

We can form new functions by performing algebraic operations with two or more functions. We define new functions that are the sum, difference, product, and quotient of two functions as follows:

Operations with Functions

Let f and g be functions of x, and define the following.

Sum	$(f + g)(x) = f(x) + g(x)$
Difference	$(f - g)(x) = f(x) - g(x)$
Product	$(f \cdot g)(x) = f(x) \cdot g(x)$
Quotient	$\left(\dfrac{f}{g}\right)(x) = \dfrac{f(x)}{g(x)}$ if $g(x) \neq 0$

● **EXAMPLE 8** *Operations with Functions*

If $f(x) = 3x + 2$ and $g(x) = x^2 - 3$, find the following functions.

(a) $(f + g)(x)$ (b) $(f - g)(x)$ (c) $(f \cdot g)(x)$ (d) $\left(\dfrac{f}{g}\right)(x)$

Solution

(a) $(f + g)(x) = f(x) + g(x) = (3x + 2) + (x^2 - 3) = x^2 + 3x - 1$
(b) $(f - g)(x) = f(x) - g(x) = (3x + 2) - (x^2 - 3) = -x^2 + 3x + 5$
(c) $(f \cdot g)(x) = f(x) \cdot g(x) = (3x + 2)(x^2 - 3) = 3x^3 + 2x^2 - 9x - 6$
(d) $\left(\dfrac{f}{g}\right)(x) = \dfrac{f(x)}{g(x)} = \dfrac{3x + 2}{x^2 - 3}$, if $x^2 - 3 \neq 0$

We now consider a new way to combine two functions. Just as we can substitute a number for the independent variable in a function, we can substitute a second function for the variable. This creates a new function, called a **composite function.**

Composite Functions

Let f and g be functions. Then the **composite functions** g of f (denoted $g \circ f$) and f of g (denoted $f \circ g$) are defined as follows:

$$(g \circ f)(x) = g(f(x))$$
$$(f \circ g)(x) = f(g(x))$$

Note that the domain of $g \circ f$ is the subset of the domain of f for which $g \circ f$ is defined. Similarly, the domain of $f \circ g$ is the subset of the domain of g for which $f \circ g$ is defined.

● **EXAMPLE 9** *Composite Functions*

If $f(x) = 2x^3 + 1$ and $g(x) = x^2$, find the following.
(a) $(g \circ f)(x)$ (b) $(f \circ g)(x)$

Solution

(a) $(g \circ f)(x) = g(f(x))$
$= g(2x^3 + 1)$
$= (2x^3 + 1)^2 = 4x^6 + 4x^3 + 1$

(b) $(f \circ g)(x) = f(g(x))$
$= f(x^2)$
$= 2(x^2)^3 + 1$
$= 2x^6 + 1$

Figure 1.9 illustrates both composite functions found in Example 9.

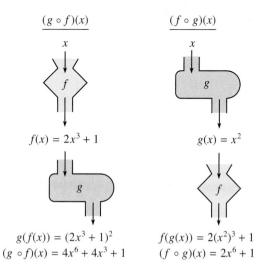

$(g \circ f)(x)$ $(f \circ g)(x)$

x x

f g

$f(x) = 2x^3 + 1$ $g(x) = x^2$

g f

$g(f(x)) = (2x^3 + 1)^2$ $f(g(x)) = 2(x^2)^3 + 1$

Figure 1.9 $(g \circ f)(x) = 4x^6 + 4x^3 + 1$ $(f \circ g)(x) = 2x^6 + 1$

● *Checkpoint*

6. If $f(x) = 1 - 2x$ and $g(x) = 3x^2$, find the following.
   (a) $(g - f)(x)$ (b) $(f \cdot g)(x)$ (c) $(f \circ g)(x)$ (d) $(g \circ f)(x)$ (e) $(f \circ f)(x) = f(f(x))$

● *Checkpoint Solutions*

1. Independent variable is x; dependent variable is y.
2. $f(1) = 3$
3. $f(-2) = 1 - (-2)^3 = 1 - (-8) = 9$
4. $f(x + h) = 2(x + h)^2$
   $= 2(x^2 + 2xh + h^2)$
5. The domain is all real numbers except $x = -1$, because $f(x)$ is undefined when $x = -1$.
6. (a) $(g - f)(x) = g(x) - f(x) = 3x^2 - (1 - 2x) = 3x^2 + 2x - 1$
   (b) $(f \cdot g)(x) = f(x) \cdot g(x) = (1 - 2x)(3x^2) = 3x^2 - 6x^3$
   (c) $(f \circ g)(x) = f(g(x)) = f(3x^2) = 1 - 2(3x^2) = 1 - 6x^2$
   (d) $(g \circ f)(x) = g(f(x)) = g(1 - 2x) = 3(1 - 2x)^2$
   (e) $(f \circ f)(x) = f(f(x)) = f(1 - 2x) = 1 - 2(1 - 2x) = 4x - 1$

1.2 Exercises

In Problems 1–4, use the values in the following table.

x	−7	−1	0	3	4.2	9	11	14	18	22
y	0	0	1	5	9	11	35	22	22	60

1. (a) Explain why the table defines y as a function of x.
   (b) State the domain and range of this function.
2. If the function defined by the table is denoted by f, so that $y = f(x)$, is $f(9)$ an input or an output of f?
3. If the table expresses $y = f(x)$, find $f(0)$ and $f(11)$.
4. Does the table describe x as a function of y? Explain.

In Problems 5 and 6, are the relations defined by the tables functions? Explain why or why not and give the domain and range.

5.
x	1	2	3	8	9
y	−4	−4	5	16	5

6.
x	−1	0	1	3	1
y	0	2	4	6	9

7. Does either of the graphs in Figure 1.10 represent y as a function of x? Explain your choices.

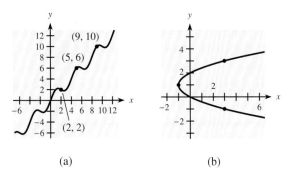

(a)   (b)

Figure 1.10

8. Does either of the graphs in Figure 1.11 represent y as a function of x? Explain your choices.

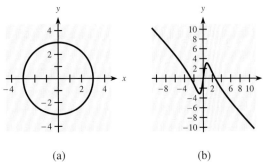

(a)   (b)

Figure 1.11

9. If $y = 3x^3$, is y a function of x?
10. If $y = 6x^2$, is y a function of x?
11. If $y^2 = 3x$, is y a function of x?
12. If $y^2 = 10x^2$, is y a function of x?
13. If $R(x) = 8x - 10$, find the following.
    (a) $R(0)$ (b) $R(2)$ (c) $R(-3)$ (d) $R(1.6)$
14. If $h(x) = 3x^2 - 2x$, find the following.
    (a) $h(3)$ (b) $h(-3)$ (c) $h(2)$ (d) $h(\frac{1}{6})$
15. If $C(x) = 4x^2 - 3$, find the following.
    (a) $C(0)$ (b) $C(-1)$ (c) $C(-2)$ (d) $C(-\frac{3}{2})$
16. If $R(x) = 100x - x^3$, find the following.
    (a) $R(1)$ (b) $R(10)$ (c) $R(2)$ (d) $R(-10)$
17. If $f(x) = x^3 - 4/x$, find the following.
    (a) $f(-\frac{1}{2})$ (b) $f(2)$ (c) $f(-2)$
18. If $C(x) = (x^2 - 1)/x$, find the following.
    (a) $C(1)$ (b) $C(\frac{1}{2})$ (c) $C(-2)$
19. Let $f(x) = 1 + x + x^2$ and $h \neq 0$.
    (a) Is $f(2 + 1) = f(2) + f(1)$?
    (b) Find $f(x + h)$.
    (c) Does $f(x + h) = f(x) + f(h)$?
    (d) Does $f(x + h) = f(x) + h$?
    (e) Find $\dfrac{f(x + h) - f(x)}{h}$ and simplify.
20. Let $f(x) = 3x^2 - 6x$ and $h \neq 0$.
    (a) Is $f(3 + 2) = f(3) + 2$?
    (b) Find $f(x + h)$.
    (c) Does $f(x + h) = f(x) + h$?
    (d) Does $f(x + h) = f(x) + f(h)$?
    (e) Find $\dfrac{f(x + h) - f(x)}{h}$ and simplify.
21. If $f(x) = x - 2x^2$ and $h \neq 0$, find the following and simplify.
    (a) $f(x + h)$ (b) $\dfrac{f(x + h) - f(x)}{h}$
22. If $f(x) = 2x^2 - x + 3$ and $h \neq 0$, find the following and simplify.
    (a) $f(x + h)$ (b) $\dfrac{f(x + h) - f(x)}{h}$
23. If $y = f(x)$ in Figure 1.10(a), find the following.
    (a) $f(9)$ (b) $f(5)$
24. Suppose $y = g(x)$ in Figure 1.11(b).
    (a) Find $g(0)$.
    (b) How many x-values in the domain of this function satisfy $g(x) = 0$?

25. The graph of $y = x^2 - 4x$ is shown in Figure 1.12.

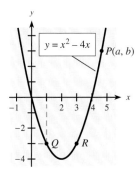

Figure 1.12

(a) If the coordinates of the point P on the graph are (a, b), how are a and b related?
(b) What are the coordinates of the point Q? Do they satisfy the equation?
(c) What are the coordinates of R? Do they satisfy the equation?
(d) What are the x-values of the points on the graph whose y-coordinates are 0? Are these x-values solutions to the equation $x^2 - 4x = 0$?

26. The graph of $y = 2x^2$ is shown in Figure 1.13.

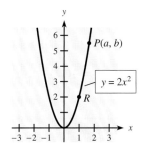

Figure 1.13

(a) If the point P, with coordinates (a, b), is on the graph, how are a and b related?
(b) Does the point $(1, 1)$ lie on the graph? Do the coordinates satisfy the equation?
(c) What are the coordinates of point R? Do they satisfy the equation?
(d) What is the x-value of the point whose y-coordinate is 0? Does this value of x satisfy the equation $0 = 2x^2$?

State the domain and range of each of the functions in Problems 27–30.

27. $y = x^2 + 4$
28. $y = x^2 - 1$
29. $y = \sqrt{x + 4}$
30. $y = \sqrt{x^2 + 1}$

In Problems 31–34, a function and its graph are given. In each problem, find the domain.

31. $f(x) = \dfrac{\sqrt{x - 1}}{x - 2}$

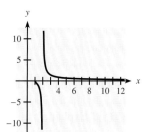

32. $f(x) = \dfrac{x + 1}{\sqrt{x + 3}}$

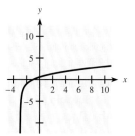

33. $f(x) = 4 + \sqrt{49 - x^2}$

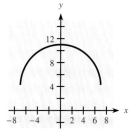

34. $f(x) = -2 - \sqrt{9 - x^2}$

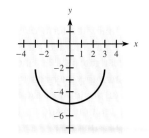

For $f(x)$ and $g(x)$ given in Problems 35–38, find
(a) $(f + g)(x)$
(b) $(f - g)(x)$
(c) $(f \cdot g)(x)$
(d) $(f/g)(x)$

35. $f(x) = 3x \qquad g(x) = x^3$
36. $f(x) = \sqrt{x} \qquad g(x) = 1/x$
37. $f(x) = \sqrt{2x} \qquad g(x) = x^2$
38. $f(x) = (x - 1)^2 \qquad g(x) = 1 - 2x$

For $f(x)$ and $g(x)$ given in Problems 39–42, find
(a) $(f \circ g)(x)$
(b) $(g \circ f)(x)$
(c) $f(f(x))$
(d) $f^2(x) = (f \cdot f)(x)$

39. $f(x) = (x - 1)^3 \qquad g(x) = 1 - 2x$
40. $f(x) = 3x \qquad g(x) = x^3 - 1$
41. $f(x) = 2\sqrt{x} \qquad g(x) = x^4 + 5$
42. $f(x) = \dfrac{1}{x^3} \qquad g(x) = 4x + 1$

APPLICATIONS

43. *Mortgage* A couple seeking to buy a home decides that a monthly payment of $800 fits their budget. Their bank's interest rate is 7.5%. The amount they can borrow, A, is a function of the time t, in years, it will take

to repay the debt. If we denote this function by $A = f(t)$, then the following table defines the function.

t	A	t	A
5	40,000	20	103,000
10	69,000	25	113,000
15	89,000	30	120,000

Source: Comprehensive Mortgage Payment Tables, Publication No. 492, Financial Publishing Co., Boston

(a) Find $f(20)$ and write a sentence that explains its meaning.
(b) Does $f(5 + 5) = f(5) + f(5)$? Explain.
(c) If the couple is looking at a house that requires them to finance \$89,000, how long must they make payments? Write this correspondence in the form $A = f(t)$.

44. *Debt refinancing* When a debt is refinanced, sometimes the term of the loan (that is, the time it takes to repay the debt) is shortened. Suppose the current interest rate is 7%, and the current debt is \$100,000. The monthly payment R of the refinanced debt is a function of the term of the loan, t, in years. If we represent this function by $R = f(t)$, then the following table defines the function.

t	R	t	R
5	1980.12	15	898.83
10	1161.09	20	775.30
12	1028.39	25	706.78

Source: Comprehensive Mortgage Payment Tables, Publication No. 492, Financial Publishing Co., Boston

(a) If they refinance for 20 years, what is the monthly payment? Write this correspondence in the form $R = f(t)$.
(b) Find $f(10)$ and write a sentence that explains its meaning.
(c) Is $f(5 + 5) = f(5) + f(5)$? Explain.

45. *Social Security benefits funding* Social Security benefits paid to eligible beneficiaries are funded by individuals who are currently employed. The following graph, based on known data until 1995 with projections into the future, defines a function that gives the number of workers, n, supporting each retiree as a function of time t (given by calendar year). Let us denote this function by $n = f(t)$.
(a) Find $f(1950)$ and explain its meaning.
(b) Find $f(1990)$.
(c) If, after the year 2050, actual data through 2050 regarding workers per Social Security beneficiary

were graphed, what parts of the new graph *must* be the same as this graph and what parts *might* be the same? Explain.
(d) Find the domain and range of $n = f(t)$ if the function is defined by the graph.

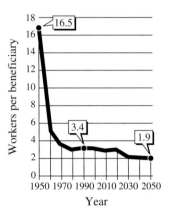

Source: Social Security Administration

46. *Dow Jones Industrial Average* If t represents the number of hours after 9:30 A.M. on Tuesday, October 5, 2004, then the graph defines the Dow Jones Industrial Average D as a function of time t. If we represent this function by $D = f(t)$, use the graph to complete the following.
(a) Find $f(0)$ and $f(6.5)$ and write a sentence that explains what each means.
(b) Find the domain and range for $D = f(t)$ as defined by the graph.
(c) How many t-values satisfy $f(t) = 10,217$? Estimate one such t-value.

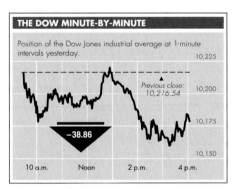

Source: Bloomberg Financial Markets, The New York Times, October 6, 2004. Copyright © 2004 The New York Times Co. Reprinted by permission.

47. *Imprisonment and parole* The figure on the next page shows the number of persons in state prisons at year's end and the number of parolees at year's end, both as functions of the years past 1900. If $y = f(t)$ gives the

number of prisoners and $y = g(t)$ gives the number of parolees, use the figure to complete the following.
(a) Estimate $f(95)$ and $g(95)$.
(b) Find $f(100)$ and explain its meaning.
(c) Find $g(90)$ and explain its meaning.
(d) Find $(f - g)(100)$ and explain its meaning.
(e) Which of $(f - g)(93)$ and $(f - g)(98)$ is greater? Explain.

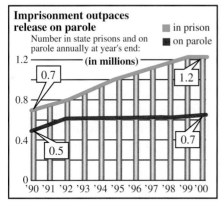

Source: USA Today, January 9, 2000. Copyright © 2000. Reprinted by permission of *USA Today.*

48. ***Women in the work force*** The number (in millions) of women in the work force, given as a function f of the year for selected years from 1920 to 2000, is shown in the figure below.
(a) How many women were in the labor force in 1970?
(b) Estimate $f(1930)$ and write a sentence that explains its meaning.
(c) Estimate $f(1990) - f(1980)$ and explain its meaning.

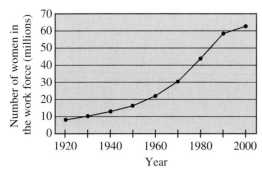

Source: 2004 *World Almanac*

49. ***Wind chill*** Dr. Paul Siple conducted studies testing the effect of wind on the formation of ice at various temperatures and developed the concept of the wind chill factor, which we hear reported during winter weather reports. If the air temperature is $-5°F$, then the wind chill, WC, is a function of the wind speed, s (in mph), and is given by

$$WC = f(s) = 45.694 + 1.75s - 29.26\sqrt{s}$$

(a) Based on the formula for $f(s)$ and the physical context of the problem, what is the domain of $f(s)$?
(b) Find $f(10)$ and write a sentence that explains its meaning.
(c) The working domain of this wind chill function is actually $s \geq 4$. How can you tell that $s = 0$ is not in the working domain, even though it is in the mathematical domain?

50. ***Temperature measurement*** The equation

$$C = \frac{5}{9}F - \frac{160}{9}$$

gives the relation between temperature readings in Celsius and Fahrenheit.
(a) Is C a function of F?
(b) What is the domain?
(c) If we consider this equation as relating temperatures of water in its liquid state, what are the domain and range?
(d) What is C when $F = 40°$?

51. ***Cost*** The total cost of producing a product is given by

$$C(x) = 300x + 0.1x^2 + 1200$$

where x represents the number of units produced. Give
(a) the total cost of producing 10 units
(b) the meaning of $C(100)$
(c) the value of $C(100)$

52. ***Profit*** The profit from the production and sale of a product is $P(x) = 47x - 0.01x^2 - 8000$, where x represents the number of units produced and sold. Give
(a) the profit from the production and sale of 2000 units
(b) the value of $P(5000)$
(c) the meaning of $P(5000)$

53. ***Pollution*** Suppose that the cost C (in dollars) of removing p percent of the particulate pollution from the smokestacks of an industrial plant is given by

$$C(p) = \frac{7300p}{100 - p}$$

(a) Find the domain of this function. Recall that p represents the percent pollution that is removed.
In parts (b)–(e), find the functional values and explain what each means.
(b) $C(45)$ (c) $C(90)$
(d) $C(99)$ (e) $C(99.6)$

54. *Test reliability* If a test that has reliability r is lengthened by a factor n ($n \geq 1$), the reliability R of the new test is given by

$$R(n) = \frac{nr}{1 + (n-1)r} \qquad 0 < r \leq 1$$

If the reliability is $r = 0.6$, the equation becomes

$$R(n) = \frac{0.6n}{0.4 + 0.6n}$$

(a) Find $R(1)$.

(b) Find $R(2)$; that is, find R when the test length is doubled.

(c) What percent improvement is there in the reliability when the test length is doubled?

55. *Area* If 100 feet of fence is to be used to fence in a rectangular yard, then the resulting area of the fenced yard is given by

$$A = x(50 - x)$$

where x is the width of the rectangle.

(a) Is A a function of x?

(b) If $A = A(x)$, find $A(2)$ and $A(30)$.

(c) What restrictions must be placed on x (the domain) so that the problem makes physical sense?

56. *Postal restrictions* If a box with a square cross section is to be sent by a delivery service, there are restrictions on its size such that its volume is given by $V = x^2(108 - 4x)$, where x is the length of each side of the cross section (in inches).

(a) Is V a function of x?

(b) If $V = V(x)$, find $V(10)$ and $V(20)$.

(c) What restrictions must be placed on x (the domain) so that the problem makes physical sense?

57. *Profit* Suppose that the profit from the production and sale of x units of a product is given by

$$P(x) = 180x - \frac{x^2}{100} - 200$$

In addition, suppose that for a certain month the number of units produced on day t of the month is

$$x = q(t) = 1000 + 10t$$

(a) Find $(P \circ q)(t)$ to express the profit as a function of the day of the month.

(b) Find the number of units produced, and the profit, on the fifteenth day of the month.

58. *Fish species growth* For many species of fish, the weight W is a function of the length L that can be expressed by

$$W = W(L) = kL^3 \qquad k = \text{constant}$$

Suppose that for a particular species $k = 0.02$, that for this species the length (in centimeters) is a function of the number of years t the fish has been alive, and that this function is given by

$$L = L(t) = 50 - \frac{(t-20)^2}{10} \qquad 0 \leq t \leq 20$$

Find $(W \circ L)(t)$ in order to express W as a function of the age t of the fish.

59. *Revenue and advertising* Suppose that a company's revenue $R = f(C)$ is a function f of the number of customers C. Suppose also that the amount spent on advertising A affects the number of customers so that $C = g(A)$ is a function g of A.

(a) Is $f \circ g$ defined? Explain.

(b) Is $g \circ f$ defined? Explain.

(c) For the functions in (a) and (b) that are defined, identify the input (independent variable) and the output (dependent variable) and explain what the function means.

60. *Manufacturing* Two of the processes (functions) used by a manufacturer of factory-built homes are sanding (denote this as function s) and painting (denote this as function p). Write a sentence of explanation for each of the following functional expressions involving s and p applied to a door.

(a) $s(\text{door})$ (b) $p(\text{door})$ (c) $(p \circ s)(\text{door})$

(d) $(s \circ p)(\text{door})$ (e) $(p \circ p)(\text{door})$

61. *Fencing a lot* A farmer wishes to fence the perimeter of a rectangular lot with an area of 1600 square feet. If the lot is x feet long, express the amount L of fence needed as a function of x.

62. *Cost* A shipping crate has a square base with sides of length x feet, and it is half as tall as it is wide. If the material for the bottom and sides of the box costs \$2.00 per square foot and the material for the top costs \$1.50 per square foot, express the total cost of material for the box as a function of x.

63. *Revenue* An agency charges \$10 per person for a trip to a concert if 30 people travel in a group. But for each person above the 30, the charge will be reduced by

$0.20. If x represents the number of people above the 30, write the agency's revenue R as a function of x.

64. *Revenue* A company handles an apartment building with 50 units. Experience has shown that if the rent for each of the units is $360 per month, all of the units will be filled, but one unit will become vacant for each $10 increase in the monthly rate. If x represents the number of $10 increases, write the revenue R from the building as a function of x.

1.3

OBJECTIVES

- To find the intercept of graphs
- To graph linear functions
- To find the slope of a line from its graph and from its equation
- To find the rate of change of a linear function
- To graph a line, given its slope and y-intercept or its slope and one point on the line
- To write the equation of a line, given information about its graph

Linear Functions

❍ Application Preview

U.S. Department of Commerce data for selected years from 1980 to 2002 can be used to approximate the gross national product (GNP) of the United States, y (in billions of dollars), as a function of the number of years past 1980, x, by

$$y = 360.74x + 2470.9$$

What does this function tell about how the GNP has changed per year? (See Example 6.) In this section, we will find the slopes and intercepts of graphs of linear functions and apply them.

The function in the Application Preview is an example of a special function, called the **linear function.** A linear function is defined as follows.

Linear Function

> A **linear function** is a function of the form
>
> $$y = f(x) = ax + b$$
>
> where a and b are constants.

Intercepts

Because the graph of a linear function is a line, only two points are necessary to determine its graph. It is frequently possible to use **intercepts** to graph a linear function. The point(s) where a graph intersects the x-axis are called the x-intercept points, and the x-coordinates of these points are the **x-intercepts.** Similarly, the points where a graph intersects the y-axis are the y-intercept points, and the y-coordinates of these points are the **y-intercepts.** Because any point on the x-axis has y-coordinate 0 and any point on the y-axis has x-coordinate 0, we find intercepts as follows:

Intercepts

> (a) To find the **y-intercept(s)** of the graph of an equation, set $x = 0$ in the equation and solve for y. *Note:* A function of x has at most one y-intercept.
> (b) To find the **x-intercept(s),** set $y = 0$ and solve for x.

● EXAMPLE 1 *Intercepts*

Find the intercepts and graph the following.
(a) $3x + y = 9$ (b) $x = 4y$

Solution

(a) To find the *y*-intercept, we set $x = 0$ and solve for *y*. $3(0) + y = 9$ gives $y = 9$, so the *y*-intercept is 9. To find the *x*-intercept, we set $y = 0$ and solve for *x*. $3x + 0 = 9$ gives $x = 3$, so the *x*-intercept is 3. Using the intercepts gives the graph, shown in Figure 1.14.

(b) Letting $x = 0$ gives $y = 0$, and letting $y = 0$ gives $x = 0$, so the only intercept of the graph of $x = 4y$ is at the point (0, 0). A second point is needed to graph the line. Hence, if we let $y = 1$ in $x = 4y$, we get $x = 4$ and have a second point (4, 1) on the graph. It is wise to plot a third point as a check. The graph is shown in Figure 1.15.

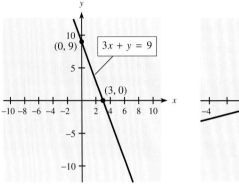

Figure 1.14   Figure 1.15

Note that the equation graphed in Figure 1.14 can be rewritten as

$$y = 9 - 3x \quad \text{or} \quad f(x) = 9 - 3x$$

We see in Figure 1.14 that the *x*-intercept (3, 0) is the point where the function value is zero. The *x*-coordinate of such a point is called a **zero of the function.** Thus we see that *the x-intercepts of a function are the same as its zeros.*

● **EXAMPLE 2** *Depreciation*

A business property is purchased for \$122,880 and depreciated over a period of 10 years. Its value *y* is related to the number of months of service *x* by the equation

$$4096x + 4y = 491{,}520$$

Find the *x*-intercept and the *y*-intercept and use them to sketch the graph of the equation.

Solution

$$x\text{-intercept: } y = 0 \quad \text{gives} \quad 4096x = 491{,}520$$
$$x = 120$$

Thus 120 is the *x*-intercept.

$$y\text{-intercept: } x = 0 \quad \text{gives} \quad 4y = 491{,}520$$
$$y = 122{,}880$$

Thus 122,880 is the *y*-intercept. The graph is shown in Figure 1.16. Note that the units on the *x*- and *y*-axes are different and that the *y*-intercept corresponds to the value of the property 0 months after purchase. That is, the *y*-intercept gives the purchase price. The *x*-intercept corresponds to the number of months that have passed before the value is 0; that is, the property is fully depreciated after 120 months, or 10 years. Note that only positive values for *x* and *y* make sense in this application, so only the Quadrant I portion of the graph is shown.

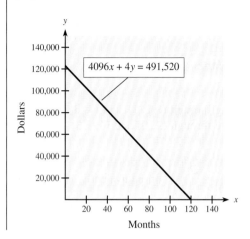

Figure 1.16

Despite the ease of using intercepts to graph linear equations, this method is not always the best. For example, vertical lines, horizontal lines, or lines that pass through the origin may have a single intercept, and if a line has both intercepts very close to the origin, using the intercepts may lead to an inaccurate graph.

Rate of Change; Slope of a Line

Note that in Figure 1.16, as the graph moves from the *y*-intercept point (0, 122,880) to the *x*-intercept point (120, 0), the *y*-value on the line changes $-122,880$ units (from 122,880 to 0), whereas the *x*-value changes 120 units (from 0 to 120). Thus the **rate of change** of the value of the business property is

$$\frac{-122,880}{120} = -1024 \text{ dollars per month}$$

This means that each month the value of the property changes by -1024 dollars, or the value decreases by \$1024/month. This **rate of change** of a linear function is called the **slope** of the line that is its graph (see Figure 1.16). For the graph of a linear function, the ratio of the change in *y* to the corresponding change in *x* measures the slope of the line. For any nonvertical line, the slope can be found by using any two points on the line, as follows:

Slope of a Line

If a nonvertical line passes through the points $P_1(x_1, y_1)$ and $P_2(x_2, y_2)$, its **slope,** denoted by *m*, is found by using either

$$m = \frac{y_2 - y_1}{x_2 - x_1}$$

or, equivalently,

$$m = \frac{y_1 - y_2}{x_1 - x_2}$$

The slope of a vertical line is undefined.

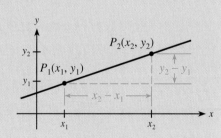

Note that for a given line, the slope is the same regardless of which two points are used in the calculation; this is because corresponding sides of similar triangles are in proportion.

We may also write the slope by using the notation

$$m = \frac{\Delta y}{\Delta x} \quad (\Delta y = y_2 - y_1 \quad \text{and} \quad \Delta x = x_2 - x_1)$$

where Δy is read "delta y" and means "change in y," and Δx means "change in x."

● **EXAMPLE 3** *Slopes*

Find the slope of
(a) line ℓ_1, passing through $(-2, 1)$ and $(4, 3)$
(b) line ℓ_2, passing through $(3, 0)$ and $(4, -3)$

Solution

(a) $m = \dfrac{3 - 1}{4 - (-2)} = \dfrac{2}{6} = \dfrac{1}{3}$ or, equivalently, $m = \dfrac{1 - 3}{-2 - 4} = \dfrac{-2}{-6} = \dfrac{1}{3}$

This means that a point 3 units to the right and 1 unit up from any point on the line is also on the line. Line ℓ_1 is shown in Figure 1.17.

(b) $m = \dfrac{0 - (-3)}{3 - 4} = \dfrac{3}{-1} = -3$

This means that a point 1 unit to the right and 3 units down from any point on the line is also on the line. Line ℓ_2 is also shown in Figure 1.17.

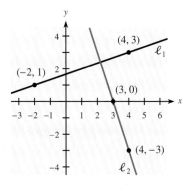

Figure 1.17

From the previous discussion, we see that the slope describes the direction of a line as follows.

Orientation of a Line and Its Slope

1. The slope is *positive* if the line slopes upward toward the right. The function is increasing.

$$m = \frac{\Delta y}{\Delta x} > 0$$

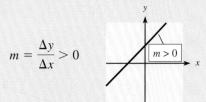

2. The slope is *negative* if the line slopes downward toward the right. The function is decreasing.

$$m = \frac{\Delta y}{\Delta x} < 0$$

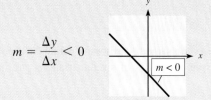

3. The slope of a *horizontal line* is 0, because $\Delta y = 0$. The function is constant.

$$m = \frac{\Delta y}{\Delta x} = 0$$

4. The slope of a *vertical line* is *undefined*, because $\Delta x = 0$.

$$m = \frac{\Delta y}{\Delta x} \text{ is undefined}$$

Two distinct nonvertical lines that have the same slope are parallel, and conversely, two nonvertical parallel lines have the same slope.

Parallel Lines

Two distinct nonvertical lines are *parallel* if and only if their slopes are *equal*.

In general, two lines are perpendicular if they intersect at right angles. However, the appearance of this perpendicularity relationship can be obscured when we graph lines, unless the axes have the same scale. Note that the lines ℓ_1 and ℓ_2 in Figure 1.17 appear to be perpendicular. However, the same lines graphed with different scales on the axes will not look perpendicular. To avoid being misled by graphs with different scales, we use slopes to tell us when lines are perpendicular. Note that the slope of ℓ_1, $\frac{1}{3}$, is the negative reciprocal of the slope of ℓ_2, -3. In fact, as with lines ℓ_1 and ℓ_2, any two nonvertical lines that are perpendicular have slopes that are negative reciprocals of each other.

Slopes of Perpendicular Lines

A line ℓ_1 with slope m, where $m \neq 0$, is *perpendicular* to line ℓ_2 if and only if the slope of ℓ_2 is $-1/m$. (The slopes are *negative reciprocals*.)

Because the slope of a vertical line is undefined, we cannot use slope in discussing parallel and perpendicular relations that involve vertical lines. Two vertical lines are parallel, and any horizontal line is perpendicular to any vertical line.

● *Checkpoint*

1. Find the slope of the line through $(4, 6)$ and $(28, -6)$.
2. If a line has slope $m = 0$, then the line is _____. If a line has an undefined slope, then the line is _____.
3. Suppose that line 1 has slope $m_1 = 5$ and line 2 has slope m_2.
   (a) If line 1 is perpendicular to line 2, find m_2.
   (b) If line 1 is parallel to line 2, find m_2.

Writing Equations of Lines

If the slope of a line is m, then the slope between a fixed point (x_1, y_1) and any other point (x, y) on the line is also m. That is,

$$m = \frac{y - y_1}{x - x_1}$$

Solving for $y - y_1$ gives the point-slope form of the equation of a line.

Point-Slope Form

The equation of the line passing through the point (x_1, y_1) and with slope m can be written in the **point-slope form**

$$y - y_1 = m(x - x_1)$$

● **EXAMPLE 4** *Equations of Lines*

Write the equation for each line that passes through $(1, -2)$ and has

(a) slope $\frac{2}{3}$ (b) undefined slope   (c) point $(2, 3)$ also on the line

Solution

(a) Here $m = \frac{2}{3}$, $x_1 = 1$, and $y_1 = -2$. An equation of the line is

$$y - (-2) = \frac{2}{3}(x - 1)$$

$$y + 2 = \frac{2}{3}x - \frac{2}{3}$$

$$y = \frac{2}{3}x - \frac{8}{3}$$

This equation also may be written in **general form** as $2x - 3y - 8 = 0$. Figure 1.18 shows the graph of this line; the point $(1, -2)$ and the slope are highlighted.

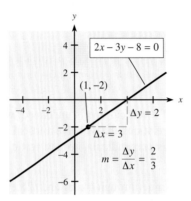

Figure 1.18

(b) Because m is undefined, we cannot use the point-slope form. This line is vertical, so every point on it has x-coordinate 1. Thus the equation is $x = 1$. Note that $x = 1$ is not a function.

(c) First use $(1, -2)$ and $(2, 3)$ to find the slope.

$$m = \frac{3 - (-2)}{2 - 1} = 5$$

Using $m = 5$ and the point $(1, -2)$ (the other point could also be used) gives

$$y - (-2) = 5(x - 1) \quad \text{or} \quad y = 5x - 7$$

The graph of $x = 1$ (from Example 4(b)) is a vertical line, as shown in Figure 1.19(a); the graph of $y = 1$ has slope 0, and its graph is a horizontal line, as shown in Figure 1.19(b).

In general, **vertical lines** have undefined slope and the equation form $x = a$, where a is the x-coordinate of each point on the line. **Horizontal lines** have $m = 0$ and the equation form $y = b$, where b is the y-coordinate of each point on the line.

● **EXAMPLE 5** *Pricing*

U.S. Census Bureau data indicate that the average price p of color television sets can be expressed as a linear function of the number of sets sold N (in thousands). In addition, as N increased by 1000, p dropped by $10.40 and when 6485 (thousand) sets were sold, the average price per set was $504.39. Write the equation of the line determined by this information.

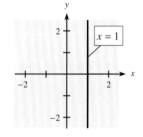

(a)

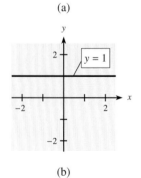

(b)

Figure 1.19

Solution

We see that price p is a function of the number of sets N (in thousands), so the slope is given by

$$m = \frac{\text{change in } p}{\text{change in } N} = \frac{-10.40}{1000} = -0.0104$$

A point on the line is $(N_1, p_1) = (6485, 504.39)$. We use the point-slope form adapted to the variables N and p.

$$p - p_1 = m(N - N_1)$$
$$p - 504.39 = -0.0104(N - 6485)$$
$$p - 504.39 = -0.0104N + 67.444$$
$$p = -0.0104N + 571.834$$

The point-slope form, with the y-intercept point $(0, b)$, can be used to derive a special form for the equation of a line.

$$y - b = m(x - 0)$$
$$y = mx + b$$

Slope-Intercept Form

The **slope-intercept form** of the equation of a line with slope m and y-intercept b is

$$y = mx + b$$

Note that if a linear equation has the form $y = mx + b$, then the coefficient of x is the slope and the constant term is the y-intercept.

◗ EXAMPLE 6 *Gross National Product* **(Application Preview)**

The linear equation $y = 360.74x + 2470.9$ expresses (approximately) the gross national product (GNP) of the United States, y, in billions of dollars, as a function of the number of years past 1980, x.

(a) Find the slope and y-intercept of this linear function and interpret each.
(b) What does this function tell us about how the GNP has changed per year?

Solution

(a) The function $y = 360.74x + 2470.9$ has the form $y = mx + b$. Hence the slope is $m = 360.74$, and the y-intercept is $b = 2470.9$. The slope gives the rate of change of y with respect to x. Thus $m = 360.74$ means that the GNP is changing at the rate of 360.74 billion dollars per year.
  The y-intercept is the value of y when $x = 0$. Thus $b = 2470.9$ means that in 1980 (when $x = 0$), the GNP was (approximately) 2470.9 billion dollars.
(b) As discussed in (a), the rate of change in the GNP per year is given by the slope of this linear function and is 360.74 billion dollars per year.

● EXAMPLE 7 *Writing the Equation of a Line*

Write the equation of the line with slope $\frac{1}{2}$ and y-intercept -3.

Solution

Substituting $m = \frac{1}{2}$ and $b = -3$ in the equation $y = mx + b$ gives $y = \frac{1}{2}x + (-3)$ or $y = \frac{1}{2}x - 3$.

When a linear equation does not appear in slope-intercept form (and does not have the form $x = a$), it can be put into slope-intercept form by solving the equation for y.

● **EXAMPLE 8** *Slope-Intercept Equation Form*

(a) Find the slope and y-intercept of the line whose equation is $x + 2y = 8$.
(b) Use this information to graph the equation.

Solution

(a) To put the equation in slope-intercept form, we must solve it for y.

$$2y = -x + 8 \quad \text{or} \quad y = -\frac{1}{2}x + 4$$

Thus the slope is $-\frac{1}{2}$ and the y-intercept is 4.

(b) First we plot the y-intercept point $(0, 4)$. Because the slope is $-\frac{1}{2} = \frac{-1}{2}$, moving 2 units to the right and down 1 unit from $(0, 4)$ gives the point $(2, 3)$ on the line. A third point (for a check) is plotted at $(4, 2)$. The graph is shown in Figure 1.20.

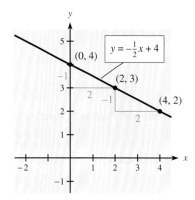

Figure 1.20

It is also possible to graph a straight line if we know its slope and any point on the line; we simply plot the point that is given and then use the slope to plot other points.

The following summarizes the forms of equations of lines.

Forms of Linear Equations

General form: $ax + by + c = 0$
Point-slope form: $y - y_1 = m(x - x_1)$
Slope-intercept form: $y = mx + b$
Vertical line: $x = a$
Horizontal line: $y = b$

● **Checkpoint**

4. Write the equation of the line that has slope $-\frac{3}{4}$ and passes through $(4, -6)$.
5. What are the slope and y-intercept of the graph of $x = -4y + 1$?

● **Checkpoint Solutions**

1. $m = \dfrac{y_2 - y_1}{x_2 - x_1} = \dfrac{-6 - (6)}{28 - (4)} = \dfrac{-12}{24} = -\dfrac{1}{2}$

2. If $m = 0$, then the line is horizontal. If m is undefined, then the line is vertical.

3. (a) $m_2 = \dfrac{-1}{m_1} = -\dfrac{1}{5}$ (b) $m_2 = m_1 = 5$

4. Use $y - y_1 = m(x - x_1)$. Hence $y - (-6) = -\frac{3}{4}(x - 4)$, $y + 6 = -\frac{3}{4}x + 3$, or $y = -\frac{3}{4}x - 3$.

5. If $x = -4y + 1$, then the slope-intercept form is $y = mx + b$ or $y = -\frac{1}{4}x + \frac{1}{4}$. Hence $m = -\frac{1}{4}$ and the y-intercept is $\frac{1}{4}$.

1.3 Exercises

Find the intercepts and graph the following functions.

1. $3x + 4y = 12$
2. $6x - 5y = 90$
3. $5x - 8y = 60$
4. $2x - y + 17 = 0$
5. $3x + 2y = 0$
6. $4x + 5y = 8$

In Problems 7–10, find the slope of the line passing through the given pair of points.

7. $(22, 11)$ and $(15, -17)$
8. $(-6, -12)$ and $(-18, -24)$
9. $(3, 2)$ and $(-1, 2)$
10. $(-4, 2)$ and $(-4, -2)$
11. If a line is horizontal, then its slope is ___?___.
12. If a line is vertical, then its slope is ___?___.
13. What is the rate of change of the function whose graph is a line passing through $(3, 2)$ and $(-1, 2)$?
14. What is the rate of change of the function whose graph is a line passing through $(11, -5)$ and $(-9, -4)$?

In Problems 15 and 16, for each given graph, determine whether each line has a slope that is positive, negative, 0, or undefined.

15.
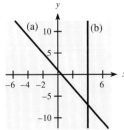

16.
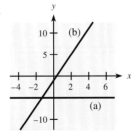

Find the slopes and y-intercepts of the lines whose equations are given in Problems 17–24. Then graph each equation.

17. $y = \dfrac{7}{3}x - \dfrac{1}{4}$
18. $y = \dfrac{4}{3}x + \dfrac{1}{2}$
19. $y = 3$
20. $y = -2$

21. $x = -8$
22. $x = -\dfrac{1}{2}$
23. $2x + 3y = 6$
24. $3x - 2y = 18$

In Problems 25–28, write the equation and sketch the graph of each line with the given slope and y-intercept.

25. Slope $\frac{1}{2}$ and y-intercept -3
26. Slope 4 and y-intercept 2
27. Slope -2 and y-intercept $\frac{1}{2}$
28. Slope $-\frac{2}{3}$ and y-intercept -1

In Problems 29–34, write the equation and graph the line that passes through the given point and has the slope indicated.

29. $(2, 0)$ with slope -5
30. $(1, 1)$ with slope $-\frac{1}{3}$
31. $(-1, 4)$ with slope $-\frac{3}{4}$
32. $(3, -1)$ with slope 1
33. $(-1, 1)$ with undefined slope
34. $(1, 1)$ with 0 slope

In Problems 35–38, write the equation of each line passing through the given pair of points.

35. $(3, 2)$ and $(-1, -6)$
36. $(-4, 2)$ and $(2, 4)$
37. $(7, 3)$ and $(-6, 2)$
38. $(10, 2)$ and $(8, 7)$

In Problems 39–42, determine whether the following pairs of equations represent parallel lines, perpendicular lines, or neither of these.

39. $3x + 2y = 6$; $2x - 3y = 6$
40. $5x - 2y = 8$; $10x - 4y = 8$
41. $6x - 4y = 12$; $3x - 2y = 6$
42. $5x + 4y = 7$; $y = \dfrac{4}{5}x + 7$

43. Write the equation of the line through $(-2, -7)$ that is parallel to $3x + 5y = 11$.
44. Write the equation of the line through $(6, -4)$ that is parallel to $4x - 5y = 6$.

45. Write the equation of the line through (3, 1) that is perpendicular to $5x - 6y = 4$.
46. Write the equation of the line through $(-2, -8)$ that is perpendicular to $x = 4y + 3$.

APPLICATIONS

47. *Depreciation* A \$360,000 building is depreciated by its owner. The value y of the building after x months of use is $y = 360,000 - 1500x$.
    (a) Graph this function for $x \geq 0$.
    (b) How long is it until the building is completely depreciated (its value is zero)?
    (c) The point (60, 270,000) lies on the graph. Explain what this means.

48. *Tax burden* Using data from the Internal Revenue Service, the per capita tax burden T (in hundreds of dollars) can be described by

    $$T(t) = 2.25t + 17.69$$

    where t is the number of years past 1980.
    (a) Find $T(20)$ and write a sentence that explains its meaning.
    (b) Graph the equation for per capita tax burden.

49. *Internet users* The percent of the U.S. population with Internet service can be described by

    $$y = 5.74x + 14.61$$

    where x is the number of years past 1995. (*Source:* Jupiter Media Metrix)
    (a) Find the slope and y-intercept of this equation.
    (b) What interpretation could be given to the y-intercept?
    (c) What interpretation could be given to the slope?

50. *Cigarette use* The percent p of high school seniors who smoke cigarettes can be described by

    $$p = 32.88 - 0.03t$$

    where t is the number of years past 1975. (*Source:* National Institute on Drug Abuse)
    (a) Find the slope and p-intercept of this equation.
    (b) Write a sentence that interprets the meaning of the slope as a rate of change.
    (c) Write a sentence that interprets the meaning of the p-intercept.

51. *Marriage rate* When x is the number of years past 1950, the percent of unmarried women who get married each year is given by $M(x) = -0.762x + 85.284$. (*Source:* Index of Leading Cultural Indicators)

(a) Find the slope and M-intercept for this linear function.
(b) Write a sentence that interprets the M-intercept.
(c) Find the annual rate of change of percent, and write a sentence that explains its meaning.

52. *Temperature-humidity models* Two models for measuring the effects of high temperature and humidity are the Summer Simmer Index and the Apparent Temperature.* For an outside temperature of 100°F, these indices relate the relative humidity, H (expressed as a decimal), to the perceived temperature as follows.

    Summer Simmer: $S = 141.1 - 45.78(1 - H)$

    Apparent Temperature: $A = 91.2 + 41.3H$

    (a) For each index, find the point that corresponds to a relative humidity of 40%.
    (b) For each point in (a), write a sentence that explains its meaning.
    (c) Graph both equations for $0 \leq H \leq 1$.

53. *Earnings and gender* According to the U.S. Bureau of the Census, the relation between the average annual earnings of males and females with various levels of educational attainment can be modeled by the function

    $$F = 0.551M + 6.762$$

    where M and F represent the average annual earnings (in thousands of dollars) of males and females, respectively.
    (a) Viewing F as a function of M, what is the slope of the graph of this function?
    (b) Interpret the slope as a rate of change.
    (c) When the average annual earnings for males reach \$40,000, what does the equation predict for the average annual earnings for females?

54. *Marijuana use* The percent of U.S. high school seniors from 1975 to 2001 who used marijuana can be modeled by

    $$p(x) = 27.085 - 0.196x \quad \text{percent}$$

    where x is the number of years past 1975. (*Source:* National Institute on Drug Abuse)
    (a) What is the slope of the graph of this function?
    (b) Interpret the slope as a rate of change.

55. *Residential electric costs* An electric utility company determines the monthly bill for a residential customer by adding an energy charge of 8.38 cents per kilowatt-hour to its base charge of \$4.95 per month. Write an

*Bosch, W., and C. G. Cobb, "Temperature-Humidity Indices," UMAP Unit 691, *The UMAP Journal,* 10(3), Fall 1989, 237–256.

equation for the monthly charge *y* in terms of *x*, the number of kilowatt-hours used.

56. *Residential heating costs*   Residential customers who heat their homes with natural gas have their monthly bills calculated by adding a base service charge of $5.19 per month and an energy charge of 51.91 cents per hundred cubic feet. Write an equation for the monthly charge *y* in terms of *x*, the number of hundreds of cubic feet used.

57. *Earnings and race*   Data from 2003 for various age groups show that for each $100 increase in the median weekly income for whites, the median weekly income of blacks increases by $61.90. Also, for workers of ages 25 to 54 the median weekly income for whites was $676 and for blacks was $527. (*Source:* U.S. Department of Labor)
    (a) Let *W* represent the median weekly income for whites and *B* the median weekly income for blacks, and write the equation of the line that gives *B* as a linear function of *W*.
    (b) When the median weekly income for whites is $850, what does the equation in (a) predict for the median weekly income for blacks?

58. *Retirement plans*   The retirement plan for Pennsylvania state employees is based on the following formula: "2.5% of average final compensation multiplied by years of credited service." (*Source:* Pennsylvania State Employees Retirement System, 2001) Let *p* represent annual retirement pension, *y* years of service, and *c* average final compensation.
    (a) For someone with average final compensation of $80,000, write the linear equation that gives *p* in terms of *y*.
    (b) For someone intending to retire after 30 years, write the linear equation that gives *p* in terms of *c*.

59. *Consumer price index*   The consumer price index for urban consumers (the CPI-U) for the years 1980–2002 can be accurately approximated by the linear model determined by the line connecting (1980, 82.4) and (2002, 179.9), where the *x*-coordinate is the year and the *y*-coordinate is the price consumers pay in year *x* for goods that cost $100 in 1982. (*Source:* U.S. Bureau of the Census)
    (a) Write the equation of the line connecting these two points to find a linear model for these data.
    (b) Interpret the slope of this line as a rate of change.

60. *Drinking and driving*   The following table gives the number of drinks and the resulting blood alcohol percent for a 180-pound man legally considered driving under the influence (DUI).
    (a) Is the average rate of change of the blood alcohol percent with respect to the number of drinks a constant?
    (b) Use the rate of change and one point determined by a number of drinks and the resulting blood alcohol percent to write the equation of a linear model for these data.
    (c) Verify that the values in the table fit the model.

Number of Drinks	5	6	7	8	9	10
Blood Alcohol Percent	0.11	0.13	0.15	0.17	0.19	0.21

Source: Pennsylvania Liquor Control Board

61. *Pollution effects*   It has been estimated that a certain stream can support 85,000 fish if it is pollution-free. It has further been estimated that for each ton of pollutants in the stream, 1700 fewer fish can be supported. Assuming the relationship is linear, write the equation that gives the population of fish *p* in terms of the tons of pollutants *x*.

62. *Age-sleep relationship*   Each day, a young person should sleep 8 hours plus $\frac{1}{4}$ hour for each year that the person is under 18 years of age. Assuming the relation is linear, write the equation relating hours of sleep *y* and age *x*.

63. *Insulation R-values*   The R-value of insulation is a measure of its ability to resist heat transfer. For fiberglass insulation, $3\frac{1}{2}$ inches is rated at R-11 and 6 inches is rated at R-19. Assuming this relationship is linear, write the equation that gives the R-value of fiberglass insulation as a function of its thickness *t* (in inches).

64. *Depreciation*   Suppose the cost of a business property is $960,000 and a company wants to use a straight-line depreciation schedule for a period of 240 months. If *y* is the value of this property after *x* months, then the company's depreciation schedule will be the equation of a line through (0, 960,000) and (240, 0). Write the equation of this depreciation schedule.

65. *Cholesterol and coronary heart disease risk*   The Seven Countries Study, conducted by Ancel Keys, was a long-term study of the relationship of cholesterol to

coronary heart disease (CHD) mortality in men. The relationship was approximated by the line shown in the accompanying figure. The line passes through the points (200, 25) and (250, 49), which means there were 25 CHD deaths per 1000 among men with 200 mg/dl (milligrams per deciliter) of cholesterol and 49 CHD deaths among men with 250 mg/dl of cholesterol. Using x to represent the cholesterol and y to represent CHD deaths, write the equation that represents this relationship.

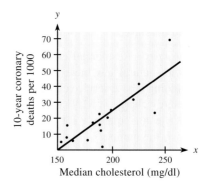

Median cholesterol (mg/dl)

<table>
<tr><td>**1.4**</td><td></td></tr>
</table>

Graphs and Graphing Utilities

OBJECTIVES

- To use a graphing utility to graph equations in the standard viewing window
- To use a graphing utility and a specified range to graph equations
- To use a graphing utility to evaluate functions, to find intercepts, and to find zeros of a function
- To solve linear equations with a graphing utility

◗ Application Preview

Suppose that for a certain city the cost C of obtaining drinking water with p percent impurities (by volume) is given by

$$C = \frac{120{,}000}{p} - 1200$$

where $0 < p \leq 100$. A graph of this equation would illustrate how the purity of water is related to the cost of obtaining it. Because this equation is not linear, graphing it would require many more than two points. Such a graph could be efficiently obtained with a graphing utility. (See Example 3.)

In Section 1.3, "Linear Functions," we saw that a linear equation could easily be graphed by plotting points because only two points are required. When we want the graph of an equation that is not linear, we can still use point plotting, but we must have enough points to sketch an accurate graph.

Some computer software and all graphing calculators have **graphing utilities** (also called *graphics utilities*) that can be used to generate an accurate graph. All graphing utilities use the point-plotting method to plot scores of points quickly and thereby graph an equation.

The computer monitor or calculator screen consists of a fine grid that looks like a piece of graph paper. In this grid, each tiny area is called a pixel. Essentially, a pixel corresponds to a point on a piece of graph paper, and as a graphing utility plots points, the pixels corresponding to the points are lighted and then connected, revealing the graph. The graph is shown in a viewing window or viewing rectangle. The values that define the viewing window or calculator range can be set individually or by using ZOOM keys. The important values are

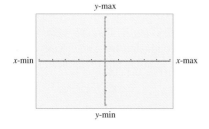

x-min:	the smallest value on the x-axis (the leftmost x-value in the window)
x-max:	the largest value on the x-axis (the rightmost x-value in the window)
y-min:	the smallest value on the y-axis (the lowest y-value in the window)
y-max:	the largest value on the y-axis (the highest y-value in the window)

Most graphing utilities have a *standard viewing window* that gives a window with x-values and y-values between -10 and 10.

$$x\text{-min:} \ -10 \qquad y\text{-min:} \ -10$$
$$x\text{-max:} \ \ \ 10 \qquad y\text{-max:} \ \ \ 10$$

The graph of $y = \frac{1}{3}x^3 - x^2 - 3x + 2$ with the standard viewing window is shown in Figure 1.21.

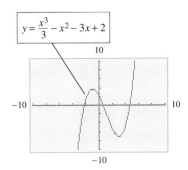

Figure 1.21

With a graphing utility, as with a graph plotted by hand, the appearance of the graph is determined by the part of the graph we are viewing, and the standard window does not always give a **complete graph.** (A complete graph shows the important parts of the graph and suggests the unseen parts.)

For some graphs, the standard viewing window may give a view of the graph that is misleading, incomplete, or perhaps even blank. For example, with the standard viewing window the graph of $y = 9x - 0.1x^2$ looks like a line (see Figure 1.22(a)), but using a different viewing window shows that the standard window is inappropriate for graphing this function (see Figure 1.22(b)). A viewing window can be set manually. For example, the graph shown in Figure 1.22(b) was generated by setting the viewing window (or calculator range) as follows:

$$x\text{-min:} \ -25 \qquad y\text{-min:} \ -50$$
$$x\text{-max:} \ \ 100 \qquad y\text{-max:} \ \ 250$$

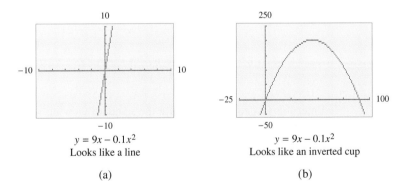

$y = 9x - 0.1x^2$
Looks like a line

$y = 9x - 0.1x^2$
Looks like an inverted cup

Figure 1.22 (a) (b)

Figure 1.23 shows another example in which the standard viewing window does not give the complete graph of a function.

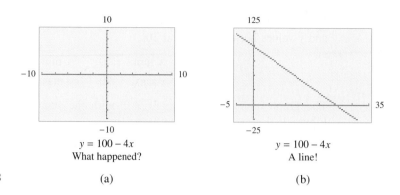

$$y = 100 - 4x$$
What happened?

$$y = 100 - 4x$$
A line!

Figure 1.23   (a)   (b)

The following guidelines may be used for most graphing utilities.

Guidelines for Using a Graphing Utility

To graph an equation in the variables x and y:

1. Solve the equation for y in terms of x.

2. Enter the equation into the graphing utility. Use parentheses as needed to ensure correctness.

3. Determine a viewing window.

4. Activate the graphing utility.

● EXAMPLE 1 *Graphing with Technology*

Graph $2x - 3y = 12$ with a graphing utility.

Solution

First note that this is a linear equation and that the graph is therefore a line. Begin by solving $2x - 3y = 12$ for y.

$$2x - 3y = 12$$
$$-3y = -2x + 12$$
$$y = \frac{-2x}{-3} + \frac{12}{-3}$$
$$y = \frac{2x}{3} - 4$$

Enter this last equation, and use the standard viewing window. The resulting graph is shown in Figure 1.24.

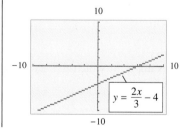

Figure 1.24

● **EXAMPLE 2** *Graphing*

Use a graphing utility with the indicated range to graph $y = x^4 - 8x^2 - 10$.

x-min: -5 y-min: -35

x-max: 5 y-max: 10

Solution

The equation can be entered directly. Set the desired range and activate the graphing utility. The resulting graph is shown in Figure 1.25.

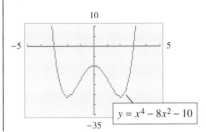

Figure 1.25

For comparison, graph the equation of Example 2 with the standard viewing window. How is it similar to the graph in Figure 1.25? How is it different?

Most graphing utilities have a TRACE capability that allows the user to trace along a graph. Tracing highlights a point on the graph and simultaneously displays its coordinates. These coordinates can be plotted on paper to produce a copy of the graph of the equation. Even if the graph is not visible in the viewing window (because you're at a point beyond the y-range), the coordinates of the trace points are still displayed. These displayed coordinates can be used to help determine a more appropriate viewing window.

For example, if we graph $y = 0.1x^4 - 20x^2 - 12$ with the standard window, none of the graph appears in the window (see Figure 1.26(a)). When we trace along the graph with a standard window, the y-values range from -12 to -1012. Hence we could adjust the y-range to account for these y-values and also expand the x-range. If we use x-min $= -20$, x-max $= 20$, y-min $= -1250$, and y-max $= 500$, we obtain the graph shown in Figure 1.26(b).

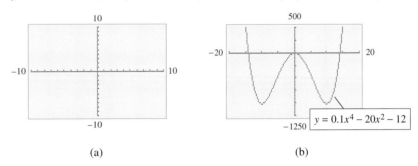

Figure 1.26   (a)   (b)

In addition to using a standard window, we can use windows that are called friendly. A "friendly" window causes TRACE to change each x-value by a "nice" amount, such as 0.1, 0.2, 1, etc., and it will occur if the difference x-max $-$ x-min is nicely divisible by the number of pixels in the x-direction. For example, the TI-83, TI-83 Plus, and TI-84 Plus calculators have 94 pixels in the x-direction, so their windows will be "friendly" if x-max $-$ x-min gives a "nice" number when divided by 94. For example, if we set x-min $= -4.7$ and x-max $= 4.7$ on a T1-83, T1-83 Plus, or TI-84 Plus calculator, each step of TRACE will change x by

$$\frac{4.7 - (-4.7)}{94} = 0.1$$

On these calculators, ZOOM, 4 automatically gives x-min $= -4.7$ and x-max $= 4.7$. Using x-min $= -9.4$ and x-max $= 9.4$ with y-min $= -10$ and y-max $= 10$ gives a window that is "friendly" and close to the standard window.

◖ EXAMPLE 3 *Water Purity (Application Preview)*

For a certain city, the cost C of obtaining drinking water with p percent impurities (by volume) is given by

$$C = \frac{120{,}000}{p} - 1200$$

Because p is the percent impurities, we know that $0 \le p \le 100$. Use the restriction on p and a graphing utility to obtain an accurate graph of the equation.

Solution

To use a graphing utility, we identify C with y and p with x. Thus we enter

$$y = \frac{120{,}000}{x} - 1200$$

We will need an x-range on the graphing utility that corresponds to the p-range in the problem. Because $0 \le p \le 100$, we set the x-range to *include* these values, with the realization that only the portion of the graph above these values applies to our equation and that $p = 0$ (i.e., $x = 0$) is excluded from the domain because it makes the denominator 0. With this x-range we can determine a y-, or C-, range from the equation or by tracing. Figure 1.27(a) shows the graph for x from -25 to 100 and for y from $-25{,}000$ to $125{,}000$. We see that when p is near 0 (the value excluded from the domain), the C-coordinates of the points are very large, indicating that water free of impurities is very costly. However, Figure 1.27(a) does not accurately show what happens for large p-values. Figure 1.27(b) shows another view of the equation with the C-range (that is, the y-range) from -500 to 3500. We see the p-intercept is $p = 100$, which indicates that water containing 100 percent impurities costs nothing ($C = 0$). Note that $p > 100$ has no meaning. Why?

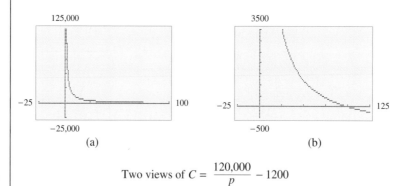

(a)   (b)

Figure 1.27   Two views of $C = \dfrac{120{,}000}{p} - 1200$

● *Checkpoint*   Use a graphing utility and the standard viewing window to graph the following:

1. $x^2 + 4y = 0$ 2. $y = \dfrac{4(x + 1)^2}{x^2 + 1}$

We can evaluate a function $y = f(x)$ with a graphing calculator by using VALUE or TABLE to find the function values corresponding to values of the independent variable. We can also use TRACE to evaluate or approximate function values.

● **EXAMPLE 4** *Evaluating Functions*

For the function $f(x) = x^4 - 8x^2 - 9$, use a graphing utility to

(a) evaluate $f(-2)$ and $f(3)$
(b) determine whether 3 is an x-intercept for the graph of $y = f(x)$
(c) find another x-intercept of the graph of $y = f(x)$

Solution

(a) Graphing $y = x^4 - 8x^2 - 9$ and using VALUE at $x = -2$ or using TABLE gives a y-value of -25, so $f(-2) = -25$. See Figure 1.28(a). Using VALUE at $x = 3$ or using TABLE gives $y = 0$, so $f(3) = 0$.

(b) Because $f(3) = 0$, the point $(3, 0)$ lies on the graph of $y = f(x)$. Thus $x = 3$ is an x-intercept. We also say that 3 is a zero of $f(x)$ and that 3 is a solution of $f(x) = 0$.

(c) It appears that the graph also crosses the x-axis at $x = -3$. Evaluating at $x = -3$ gives $y = 0$, so -3 is another x-intercept of the graph and -3 is a zero of the function.

The table shown in Figure 1.28(b) shows selected values of x and the resulting values of $y = f(x)$. We can also see from the table that $f(-2) = -25$ and that 3 and -3 are zeros of the function and thus x-intercepts of the graph of the function.

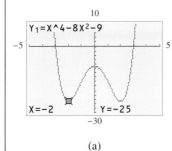

Figure 1.28
(a)   (b)

● **EXAMPLE 5** *Women in the Work Force*

The number y (in millions) of women in the work force is given by the function

$$y = 0.001(6.99x^2 - 101.23x + 6587.43)$$

where x is the number of years past 1900. (*Source:* Bureau of the Census, U.S. Department of Commerce)

(a) Graph this function from $x = 0$ to $x = 105$.
(b) Find the value of y when $x = 44$. Explain what this means.
(c) Use the model to find the number of women in the work force in 2003.

Solution

(a) The graph is found by entering the equation with $x = 0$ to $x = 105$ in the window. Using VALUE for x-values between 0 and 105 helps determine that y-values between 0 and 80 give a complete graph. The graph, using this window, is shown in Figure 1.29(a).

(b) By using VALUE or TABLE, we can find $f(44) \approx 15.666$ (see Figure 1.29(b) and Figure 1.29(c)). This means that there were approximately 15.666 million, or 15,666,000, women in the work force in $1900 + 44 = 1944$.

(c) The year 2003 corresponds to $x = 103$ and $f(103) \approx 70.318$ (see Figure 1.29(d)). Thus in 2003, there were approximately 70,318,000 women in the work force.

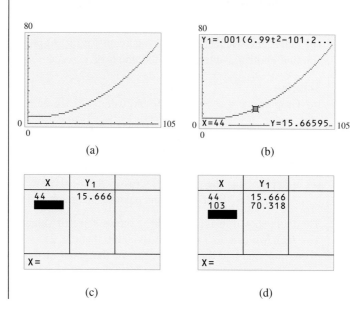

(a)   (b)

Figure 1.29

(c)   (d)

Spreadsheet Note We can use a **spreadsheet,** such as Excel, to find the outputs of a function for given inputs. Table 1.3 shows a spreadsheet for the function $f(x) = 6x - 3$ for the input set $\{-2, -1, 0, 1, 3, 5, 10\}$, with these inputs listed as entries in the first column (column A). When using a formula with a spreadsheet, we use the cell location of the data to represent the variable. Thus if -2 is in cell A2, then $f(-2)$ can be found in cell B2 by typing = 6*A2 − 3 in cell B2. By using the fill-down capability, we can obtain all the function values shown in column B. Spreadsheets such as Excel can also graph this function (see Figure 1.30). ■

TABLE 1.3

	A	B
1	x	f(x) = 6x − 3
2	−2	−15
3	−1	−9
4	0	−3
5	1	3
6	3	15
7	5	27
8	10	57

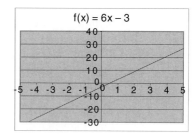

Figure 1.30

Graphical Solutions of Linear Equations

We can use graphing utilities to find the *x*-intercepts of graphs of functions. Because an *x*-intercept is an *x*-value that makes the function equal to 0, the *x*-intercept is also a zero of the function. Thus if an equation is written in the form $0 = f(x)$, the *x*-intercept of $y = f(x)$ is a solution of the equation. This method of solving an equation is called the ***x*-intercept method.**

> ### Solving a Linear Equation Using the *x*-Intercept Method with Graphing Utilities
>
> 1. Rewrite the equation to be solved with 0 (and nothing else) on one side of the equation.
>
> 2. Enter the nonzero side of the equation found in the previous step in the equation editor of your graphing utility, and graph the line in an appropriate viewing window. Be certain that you can see the line cross the horizontal axis on your graph.
>
> 3. Find or approximate the *x*-intercept by inspection with TRACE or by using the graphical solver that is often called ZERO or ROOT. The *x*-intercept is the value of *x* that makes the equation equal to zero, so it is the solution to the equation. The value of *x* found by this method is often displayed as a decimal approximation of the exact solution rather than as the exact solution.
>
> 4. Verify the solution in the original equation.

● **EXAMPLE 6** *Solving an Equation with Technology*

Graphically solve $5x + \dfrac{1}{2} = 7x - 8$ for *x*.

Solution

To solve this equation, we rewrite the equation with 0 on one side.

$$0 = 7x - 8 - \left(5x + \frac{1}{2}\right) \quad \text{or} \quad 0 = 2x - \frac{17}{2}$$

Graphing $y = 2x - \frac{17}{2}$ and finding the *x*-intercept gives *x* approximately equal to $4.25 = \frac{17}{4}$ (see Figure 1.31). Checking $x = \frac{17}{4}$ in the original equation, or solving the equation analytically, shows that $x = \frac{17}{4}$ is the exact solution. ✔

(a)

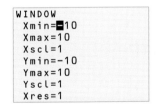

(b)

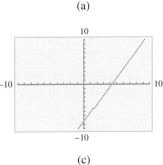

(c)

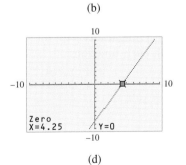

(d)

Figure 1.31

Spreadsheet Note

We can use Excel to solve the equation $0 = 5x - 8$, by finding the value of x that makes $f(x) = 5x - 8$ equal to 0—that is, by finding the zero of $f(x) = 5x - 8$. To find the zero of $f(x)$ with Excel, we enter the formula for $f(x)$ in the worksheet and use Goal Seek under the Tools button.

Under the goal set menu, setting cell B2 to the value 0 and choosing A2 as the changing cell gives $x = 1.6$, which is the solution to the equation. ■

	A	B
1	x	f(x) = 5x − 8
2	1	= 5*A2 − 8

	A	B
1	x	f(x) = 5x − 8
2	1.6	0

● ***Checkpoint Solutions***

1. First solve $x^2 + 4y = 0$ for y.

$$4y = -x^2$$
$$y = \frac{-x^2}{4}$$

The graph is shown in Figure 1.32(a).

2. The equation $y = \dfrac{4(x + 1)^2}{x^2 + 1}$ can be entered directly, but some care with parentheses is needed:

$$y = (4(x + 1)^2)/(x^2 + 1)$$

The graph is shown in Figure 1.32(b).

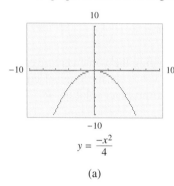

$$y = \frac{-x^2}{4}$$

(a)

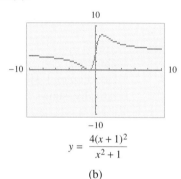

$$y = \frac{4(x + 1)^2}{x^2 + 1}$$

(b)

Figure 1.32

1.4 Exercises

In Problems 1–8, use a graphing utility with the standard viewing window to graph each function.

1. $y = x^2 + 4x + 1$
2. $y = 4 - 3x - x^2$
3. $y = x^3 - 3x$
4. $y = x^3 - 6x^2$
5. $y = 0.1x^3 - 0.3x^2 - 2.4x + 3$
6. $y = \frac{1}{4}x^4 - \frac{1}{3}x^3 - 3x^2 + 8$
7. $y = \frac{12x}{x^2 + 1}$
8. $y = \frac{8}{x^2 + 1}$

In Problems 9 and 10, use a graphing utility with the specified range to graph each equation.

9. $y = x^3 - 12x - 1$
   x-min $= -5$, x-max $= 5$; y-min $= -20$, y-max $= 20$

10. $y = x^4 - 4x^3 + 6$
    x-min $= -4$, x-max $= 6$; y-min $= -40$, y-max $= 10$

In Problems 11 and 12, show the correct way to enter each function in your calculator.

11. $y = \dfrac{3x + 7}{x^2 + 4}$

12. $y = x^2 - \dfrac{4}{x - 1}$

In Problems 13–16, graph each equation with a graphing utility using
(a) the specified range
(b) the standard viewing window (for comparison)

13. $y = 0.01x^3 + 0.3x^2 - 72x + 150$
    x-min $= -100$, x-max $= 80$; y-min $= -2000$, y-max $= 4000$

14. $y = \dfrac{-0.01x^3 + 0.15x^2 + 60x + 700}{100}$

x-min $= -100$, x-max $= 100$; y-min $= -50$,
y-max $= 50$

15. $y = \dfrac{x + 15}{x^2 + 400}$

x-min $= -200$, x-max $= 200$; y-min $= -0.02$,
y-max $= 0.06$

16. $y = \dfrac{x - 80}{x^2 + 1700}$

x-min $= -200$, x-max $= 400$; y-min $= -0.06$,
y-max $= 0.01$

In Problems 17 and 18, do the following.
(a) Algebraically find the x- and y-intercepts of the graph.
(b) Graph each equation with a graphing utility; use a window that includes the intercepts.

17. $y = 0.001x - 0.03$ 18. $y = 50{,}000 - 100x$

In Problems 19–22, a standard viewing window graph of a function is shown. Experiment with the viewing window to obtain a complete graph, adjusting the ranges where necessary.

19. $y = -0.15(x - 10.2)^2 + 10$

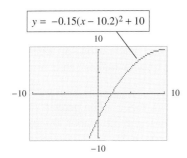

20. $y = x^2 - x - 42$

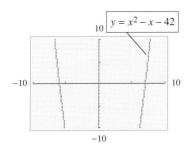

21. $y = \dfrac{x^3 + 19x^2 - 62x - 840}{20}$

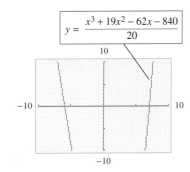

22. $y = \dfrac{-x^3 + 33x^2 + 120x}{20}$

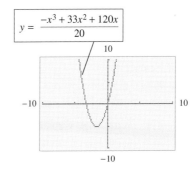

In Problems 23–28, graph the equations with a standard window on a graphing utility. Identify any linear functions.

23. $4x - y = 8$ 24. $5x + y = 8$
25. $4x^2 + 2y = 5$ 26. $6x^2 + 3y = 8$
27. $x^2 + 2y = 6$ 28. $3x^2 - 4y = 8$

Complete Problems 29 and 30 by using VALUE or TABLE.

29. If $f(x) = x^3 - 3x^2 + 2$, find $f(-2)$ and $f(3/4)$.

30. If $f(x) = \dfrac{x^2 - 2x}{x - 1}$, find $f(3)$ and $f(-4)$.

In Problems 31–34, graph each function using a window that gives a complete graph.

31. $y = \dfrac{12x^2 - 12}{x^2 + 1}$ 32. $y = \dfrac{9x^3}{4x^4 + 9}$

33. $y = \dfrac{x^2 - x - 6}{x^2 + 5x + 6}$ 34. $y = \dfrac{x^2 - 4}{x^2 - 9}$

In Problems 35–38, use the *x*-intercept method to find one solution of each equation.

35. $0 = 6x - 21$ 36. $12x + 28 = 0$
37. $10 = x^2 - 3x$ 38. $6x^2 + 4x = 4$

In each of Problems 39 and 40, use a graphing utility to (a) approximate two *x*-intercepts of the graph of the function (to four decimal places) and (b) give the approximate zeros of the function.

39. $y = x^2 - 7x - 9$ 40. $y = 2x^2 - 4x - 11$

APPLICATIONS

41. *Earnings and gender*   With U.S. Census Bureau data, the model that relates the average annual earnings (in thousands of dollars) of females, F, and males, M, with various levels of educational attainment was found to be

    $$F = 0.551M + 6.762$$

    (a) Use a graphing utility to graph this equation for the range M-min $= 0$, M-max $= 35$; F-min $= 0$, F-max $= 30$.
    (b) The point (50, 34.312) lies on the graph of this equation. Explain its meaning.
    (c) Find the average annual female earnings that correspond to male earnings of $62,500.

42. *Earnings and minorities*   With U.S. Census Bureau data, the model that relates the per capita annual income (in thousands of dollars) of blacks, B, and whites, W, was found to be

    $$B = 0.6234W - 0.3785$$

    (a) Use a graphing utility to graph this equation for the range W-min $= 0$, W-max $= 30$; B-min $= 0$, B-max $= 30$.
    (b) The point (40, 24.5575) lies on the graph of this equation. Explain its meaning.
    (c) Find the per capita annual income for blacks that corresponds to a white per capita annual income of $77,500.

43. *Consumer expenditure*   Suppose that the consumer expenditure E (in dollars) depends on the market price p per unit (in dollars) according to

    $$E = 10,000p - 100p^2$$

    (a) Graph this equation with a graphing utility and the range p-min $= -50$, p-max $= 150$; E-min $= -50,000$, E-max $= 300,000$.
    (b) Because E represents consumer expenditure, only values of $E \geq 0$ have meaning. For what p-values is $E \geq 0$?

44. *Rectilinear motion*   The height above ground (in feet) of a ball thrown vertically into the air is given by

    $$S = 112t - 16t^2$$

where t is the time in seconds since the ball was thrown.
    (a) Graph this equation with a graphing utility and the range t-min $= -2$, t-max $= 10$; S-min $= -20$, S-max $= 250$.
    (b) Estimate the time at which the ball is at its highest point, and estimate the height of the ball at that time.

45. *Advertising impact*   An advertising agency has found that when it promotes a new product in a city of 350,000 people, the rate of change R of the number of people who are aware of the product is related to the number of people x who are aware of it and is given by

    $$R = 28,000 - 0.08x$$

where $x \geq 0$ and $R \geq 0$.
    (a) Use the intercepts to determine a window, and then use a graphing utility to graph the equation for $x \geq 0$.
    (b) Is the rate of change increasing or decreasing? Explain why your answer makes sense in the context of the problem.

46. *Learning rate*   In a study using 50 foreign-language vocabulary words, the learning rate L (in words per minute) was found to depend on the number of words already learned, x, according to the equation

    $$L = 20 - 0.4x$$

    (a) Use the intercepts to determine a window, and then use a graphing utility to graph the equation for $x \geq 0$.
    (b) Is the learning rate increasing or decreasing? Explain why your answer makes sense in the context of the problem.

47. *Elderly men in the work force*   The percent P of men 65 years of age or older in the labor force can be modeled by

    $$P = 0.00008x^3 - 0.0128x^2 - 0.0032x + 66.6$$

where x is the number of years past 1890. (*Source:* U.S. Bureau of the Census)
    (a) What x-min and x-max should you use to view this graph for values representing 1890 to 2010?
    (b) Use a table to determine y-min and y-max that show a complete graph of the model.
    (c) Graph this equation.
    (d) If this model is valid until the year 2020, adjust the range to obtain a graph that extends to 2020.
    (e) Write a sentence that compares the percent of men 65 years of age or older in the work force before 2000 with that since 2000.

48. *Cellular subscribers*   The number of millions of U.S. cellular telephone subscribers can be described by

    $$y = 0.86x^2 + 1.39x + 4.97$$

where x is the number of years past 1990. (*Source:* Cellular Telecommunications and Internet Association)

(a) Use a table to determine the viewing window that shows a complete graph of this model for the years from 1990 to 2010. State the window used.

(b) Graph this equation.

(c) For years beyond 2010, does this graph increase or decrease? (Look at this graph in a window that includes these years.) Explain why this means the model eventually will not accurately describe the relationship between the number of cellular subscribers and the year.

49. *Environment* The Millcreek watershed area was heavily strip mined for coal during the late 1960s. Because of the resulting pollution, the streams can't support fish. Suppose the cost C of obtaining stream water that contains p percent of the current pollution levels is given by

$$C = \frac{285{,}000}{p} - 2850$$

Because p is the percent of current pollution levels, $0 \le p \le 100$.

(a) Use the restriction on p and determine a range for C so that a graphing utility can be used to obtain an accurate graph. Then graph the equation.

(b) Describe what happens to the cost as p takes on values near 0.

(c) The point (1, 282,150) lies on the graph of this equation. Explain its meaning.

(d) Explain the meaning of the p-intercept.

50. *Pollution* Suppose the cost C of removing p percent of the particulate pollution from the exhaust gases at an industrial site is given by

$$C = \frac{8100p}{100 - p}$$

Because p is the percent particulate pollution, we know $0 \le p \le 100$.

(a) Use the restriction on p and experiment with a C-range to obtain an accurate graph of the equation with a graphing utility.

(b) Describe what happens to C as p gets close to 100.

(c) The point (98, 396,900) lies on the graph of this equation. Explain the meaning of the coordinates.

(d) Explain the meaning of the p-intercept.

51. *Tax burden* The dollars per capita of federal tax burden T can be described by

$$T = 3.844x^2 - 67.25x + 755.4$$

where x is the number of years past 1950. (*Source:* Internal Revenue Service, *2002 Data Book (2003)*)

(a) Graph this function for values of x that correspond to the years 1950–2010.

(b) Is the graph increasing or decreasing over these years? What does this tell us about the per capita federal tax burden?

52. *Carbon monoxide emissions* The number of millions of short tons of carbon monoxide emissions, y, in the United States can be described by

$$y = 0.0043x^3 - 0.24x^2 + 0.583x + 199.8$$

where x is the number of years past 1970. (*Source:* Environmental Protection Agency)

(a) Graph this function for values of x that represent the years 1970–2010.

(b) According to this function, how many short tons of emissions occurred in 2000?

(c) For years beyond 2000, does the function increase or decrease? If the model is accurate for these future years, what does this tell planners?

1.5

Solutions of Systems of Linear Equations

OBJECTIVES

● To solve systems of linear equations in two variables by graphing
● To solve systems of linear equations by substitution
● To solve systems of linear equations by elimination
● To solve systems of three linear equations in three variables

◗ **Application Preview**

Suppose that a person has $200,000 invested, part at 9% and part at 8%, and that the yearly income from the two investments is $17,200. If x represents the amount invested at 9% and y represents the amount invested at 8%, then to find how much is invested at each rate we must find the values of x and y that satisfy both

$$x + y = 200{,}000 \quad \text{and} \quad 0.09x + 0.08y = 17{,}200$$

(See Example 3.)

Methods of solving systems of linear equations are discussed in this section.

Graphical Solution

In the previous sections, we graphed linear equations in two variables and observed that the graphs are straight lines. Each point on the graph represents an ordered pair of values (x, y) that satisfies the equation, so a point of intersection of two (or more) lines represents a solution to both (or all) the equations of those lines.

The equations are referred to as a **system of equations,** and the ordered pairs (x, y) that satisfy all the equations are the **solutions** (or **simultaneous solutions**) of the system. We can use graphing to find the solution of a system of equations.

● **EXAMPLE 1** *Graphical Solution of a System*

Use graphing to find the solution of the system

$$\begin{cases} 4x + 3y = 11 \\ 2x - 5y = -1 \end{cases}$$

Solution

The graphs of the two equations intersect (meet) at the point $(2, 1)$. (See Figure 1.33.) The solution of the system is $x = 2, y = 1$. Note that these values satisfy both equations.

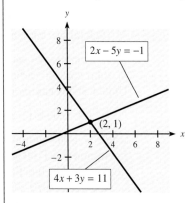

Two distinct nonparallel lines:
one solution

Figure 1.33

If the graphs of two equations are parallel lines, they have no point in common and thus the system has no solution. Such a system of equations is called **inconsistent.** For example,

$$\begin{cases} 4x + 3y = 4 \\ 8x + 6y = 18 \end{cases}$$

is an **inconsistent system** (see Figure 1.34(a)).

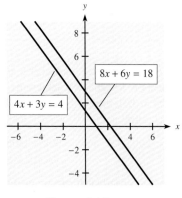

Two parallel lines:
no solution

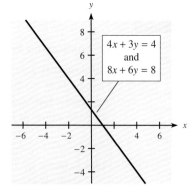

Two coincident lines:
infinitely many solutions

Figure 1.34 (a) (b)

It is also possible that two equations describe the same line. When this happens, the equations are equivalent, and values that satisfy either equation are solutions of the system. For example,

$$\begin{cases} 4x + 3y = 4 \\ 8x + 6y = 8 \end{cases}$$

is called a **dependent system** because all points that satisfy one equation also satisfy the other (see Figure 1.34(b)).

Figures 1.33, 1.34(a), and 1.34(b) represent the three possibilities that can occur when we are solving a system of two linear equations in two variables.

Solution by Substitution
Graphical solution methods may yield only approximate solutions to some systems. Exact solutions can be found using algebraic methods, which are based on the fact that equivalent systems result when any of the following operations are performed.

Equivalent Systems

Equivalent systems result when

1. One expression is replaced by an equivalent expression.
2. Two equations are interchanged.
3. A multiple of one equation is added to another equation.
4. An equation is multiplied by a nonzero constant.

The **substitution method** is based on operation (1).

Substitution Method for Solving Systems

Procedure	Example
To solve a system of two equations in two variables by substitution:	Solve the system $\begin{cases} 2x + 3y = 4 \\ x - 2y = 3 \end{cases}$
1. Solve one of the equations for one of the variables in terms of the other.	1. Solving $x - 2y = 3$ for x gives $x = 2y + 3$.
2. Substitute this expression into the other equation to give one equation in one unknown.	2. Replacing x by $2y + 3$ in $2x + 3y = 4$ gives $2(2y + 3) + 3y = 4$.
3. Solve this linear equation for the unknown.	3. $4y + 6 + 3y = 4$ $$7y = -2 \Rightarrow y = -\frac{2}{7}$$
4. Substitute this solution into the equation in Step 1 or into one of the original equations to solve for the other variable.	4. $x = 2\left(-\frac{2}{7}\right) + 3 \Rightarrow x = \frac{17}{7}$
5. Check the solution by substituting for x and y in both original equations.	5. $2\left(\frac{17}{7}\right) + 3\left(-\frac{2}{7}\right) = 4$ ✔ $$\frac{17}{7} - 2\left(-\frac{2}{7}\right) = 3 ✔$$

● **EXAMPLE 2** *Solution by Substitution*

Solve the system

$$\begin{cases} 4x + 5y = 18 \quad (1) \\ 3x - 9y = -12 \quad (2) \end{cases}$$

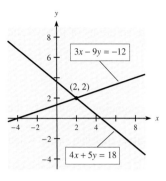

Figure 1.35

Solution

1. $x = \dfrac{9y - 12}{3} = 3y - 4$ Solve for x in equation (2).

2. $4(3y - 4) + 5y = 18$ Substitute for x in equation (1).

3. $12y - 16 + 5y = 18$ Solve for y.

$$17y = 34$$
$$y = 2$$

4. $x = 3(2) - 4$ Use $y = 2$ to find x.

$$x = 2$$

5. $4(2) + 5(2) = 18$ and $3(2) - 9(2) = -12$ ✔   Check.

Thus the solution is $x = 2$, $y = 2$. This means that when the two equations are graphed simultaneously, their point of intersection is $(2, 2)$. See Figure 1.35.

Solution by Elimination   We can also eliminate one of the variables in a system by the **elimination method,** which uses addition or subtraction of equations.

Elimination Method for Solving Systems

Procedure	Example
To solve a system of two equations in two variables by the elimination method:	Solve the system $\begin{cases} 2x - 5y = 4 & (1) \\ x + 2y = 3 & (2) \end{cases}$
1. If necessary, multiply one or both equations by a nonzero number that will make the coefficients of one of the variables identical, except perhaps for signs.	1. Multiply equation (2) by -2. $\quad 2x - 5y = 4$ $\quad -2x - 4y = -6$
2. Add or subtract the equations to eliminate one of the variables.	2. Adding gives $0x - 9y = -2$
3. Solve for the variable in the resulting equation.	3. $y = \dfrac{2}{9}$
4. Substitute the solution into one of the original equations and solve for the other variable.	4. $2x - 5\left(\dfrac{2}{9}\right) = 4$ $\quad 2x = 4 + \dfrac{10}{9} = \dfrac{36}{9} + \dfrac{10}{9}$ $\quad 2x = \dfrac{46}{9} \quad$ so $\quad x = \dfrac{23}{9}$
5. Check the solutions in both original equations.	5. $2\left(\dfrac{23}{9}\right) - 5\left(\dfrac{2}{9}\right) = 4$ ✔ $\quad \dfrac{23}{9} + 2\left(\dfrac{2}{9}\right) = 3$ ✔

◑ EXAMPLE 3 *Investment Mix (Application Preview)*

A person has $200,000 invested, part at 9% and part at 8%. If the total yearly income from the two investments is $17,200, how much is invested at 9% and how much at 8%?

Solution

If x represents the amount invested at 9% and y represents the amount invested at 8%, then $x + y$ is the total investment,

$$x + y = 200{,}000 \qquad (1)$$

and $0.09x + 0.08y$ is the total income earned.

$$0.09x + 0.08y = 17{,}200 \qquad (2)$$

We solve these equations as follows:

$-8x - 8y = -1{,}600{,}000$	(3)	Multiply equation (1) by -8.
$\underline{9x + 8y = 1{,}720{,}000}$	(4)	Multiply equation (2) by 100.
$x = \phantom{-1{,}}120{,}000$		Add (3) and (4).

We find y by using $x = 120{,}000$ in equation (1).

$$120{,}000 + y = 200{,}000 \quad \text{gives} \quad y = 80{,}000$$

Thus \$120,000 is invested at 9%, and \$80,000 is invested at 8%.

As a check, we note that equation (1) is satisfied and

$$0.09(120{,}000) + 0.08(80{,}000) = 10{,}800 + 6400 = 17{,}200 \; ✔$$

● **EXAMPLE 4** *Solution by Elimination*

Solve the systems:

(a) $\begin{cases} 4x + 3y = 4 \\ 8x + 6y = 18 \end{cases}$ 　　 (b) $\begin{cases} 4x + 3y = 4 \\ 8x + 6y = 8 \end{cases}$

Solution

(a) $\begin{cases} 4x + 3y = 4 \\ 8x + 6y = 18 \end{cases}$

Multiply by -2 to get:	$-8x - 6y = -8$	
Leave as is, which gives:	$\underline{8x + 6y = 18}$	
Add the equations to get:	$0x + 0y = 10$	
	$0 = 10$	

The system is solved when $0 = 10$. This is impossible, so there are no solutions of the system. The equations are inconsistent. Their graphs are parallel lines; see Figure 1.34(a) earlier in this section.

(b) $\begin{cases} 4x + 3y = 4 \\ 8x + 6y = 8 \end{cases}$

Multiply by -2 to get:	$-8x - 6y = -8$	
Leave as is, which gives:	$\underline{8x + 6y = 8}$	
Add the equations to get:	$0x + 0y = 0$	
	$0 = 0$	

This is an identity, so the two equations share infinitely many solutions. The equations are dependent. Their graphs coincide, and each point on this graph represents a solution of the system; see Figure 1.34(b) earlier in this section.

● **EXAMPLE 5** *Medicine Concentrations*

A nurse has two solutions that contain different concentrations of a certain medication. One is a 12.5% concentration and the other is a 5% concentration. How many cubic centimeters of each should she mix to obtain 20 cubic centimeters of an 8% concentration?

Solution

Let x equal the number of cubic centimeters of the 12.5% solution, and let y equal the number of cubic centimeters of the 5% solution. The total amount of substance is

$$x + y = 20$$

and the total amount of medication is

$$0.125x + 0.05y = (0.08)(20) = 1.6$$

Solving this pair of equations simultaneously gives

$$
\begin{aligned}
50x + 50y &= 1000 \\
-125x - 50y &= -1600 \\
\hline
-75x \phantom{{}+ 50y} &= -600 \\
x &= 8 \\
8 + y = 20, \text{ so } y &= 12
\end{aligned}
$$

Thus 8 cubic centimeters of a 12.5% concentration and 12 cubic centimeters of a 5% concentration yield 20 cubic centimeters of an 8% concentration.

Checking, we see that $8 + 12 = 20$ and

$$0.125(8) + 0.05(12) = 1 + 0.6 = 1.6 \checkmark$$

● **Checkpoint**   1. Solve by substitution: $\begin{cases} 3x - 4y = -24 \\ x + y = -1 \end{cases}$

2. Solve by elimination: $\begin{cases} 2x + 3y = 5 \\ 3x + 5y = -25 \end{cases}$

3. In Problems 1 and 2, the solution method is given.
   (a) In each case, explain why you think that method is appropriate.
   (b) In each case, would the other method work as well? Explain.

Graphing Utilities   We can check the solution of a system of equations in two variables or solve the system by graphing the two equations on the same screen of a graphing utility and finding a point of intersection of the two graphs. Solutions of systems of equations, if they exist, can be found by using INTERSECT. INTERSECT will give the point of intersection exactly or approximately to a large number of significant digits.   ■

● **EXAMPLE 6** *Solution with Technology*

Solve the system

$$\begin{cases} 3x + 2y = 12 \\ 4x - 3y = -1 \end{cases}$$

Solution

Solving both of these equations for y gives $y = 6 - \dfrac{3x}{2}$ and $y = \dfrac{1}{3} + \dfrac{4x}{3}$.

Entering the equations in a graphing utility, graphing, and using INTERSECT give the intersection of the graphs (see Figure 1.36). The point of intersection is (2, 3), so the solution of the system is $x = 2$, $y = 3$.

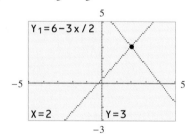

Figure 1.36

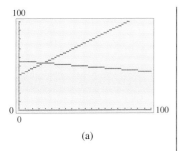

(a)

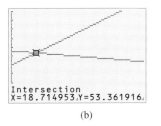

Intersection
X=18.714953 Y=53.361916.

(b)

Figure 1.37

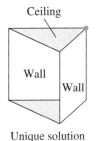

Ceiling

Wall

Wall

Unique solution

Figure 1.38

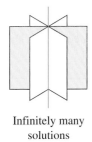

Infinitely many
solutions

Figure 1.39

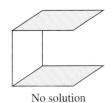

No solution

Figure 1.40

● **EXAMPLE 7** *College Enrollment*

Suppose the percent of males who enrolled in college within 12 months of high school graduation is given by

$$y = -0.126x + 55.72$$

and the percent of females who enrolled in college within 12 months of high school graduation is given by

$$y = 0.73x + 39.7$$

where x is the number of years past 1960. Graphically find the year these models indicate that the percent of females will equal the percent of males.
(*Source: Statistical Abstract of the United States, 2000*)

Solution

Entering the two functions in a graphing utility and setting the window with x-min $= 0$, x-max $= 100$, y-min $= 0$, and y-max $= 100$ gives the graph in Figure 1.37(a). Using INTERSECT (see Figure 1.37(b)) gives a point of intersection of these two graphs at approximately $x = 18.71$, $y = 53.36$. Thus the percent of females will reach the percent of males in 1979. The models indicate that in 1979, the male percent was 53.33% and the female percent was 53.57%.

Solving a system of equations by graphing, whether by hand or with a graphing utility, is limited by two factors. (1) It may be difficult to determine a viewing window that contains the point of intersection and (2) the solution may be only approximate. With *some* systems of equations, the only practical method may be graphical approximation.

However, systems of *linear* equations can be consistently and accurately solved with algebraic methods. Computer algebra systems, some software packages (including spreadsheets), and some graphing calculators have the capability of solving systems of linear equations.

Three Equations in Three Variables

If a, b, c, and d represent constants, then

$$ax + by + cz = d$$

is a first-degree (linear) equation in three variables. When equations of this form are graphed in a three-dimensional coordinate system, their graphs are planes. Two different planes may intersect in a line (like two walls) or may not intersect at all (like a floor and ceiling). Three different planes may intersect in a single point (as when two walls meet the ceiling), may intersect in a line (as in a paddle wheel), or may not have a common intersection. (See Figures 1.38, 1.39, and 1.40.) Thus three linear equations in three variables may have a unique solution, infinitely many solutions, or no solution. For example, the solution of the system

$$\begin{cases} 3x + 2y + z = 6 \\ x - y - z = 0 \\ x + y - z = 4 \end{cases}$$

is $x = 1$, $y = 2$, $z = -1$, because these three values satisfy all three equations, and these are the only values that satisfy them. In this section, we will discuss only systems of three linear equations in three variables that have unique solutions. Additional systems will be discussed in Section 3.3, "Gauss-Jordan Elimination: Solving Systems of Equations."

We can solve three equations in three variables using a systematic procedure called the **left-to-right elimination method.**

Left-to-Right Elimination Method	
Procedure	**Example**
To solve a system of three equations in three variables by the left-to-right elimination method:	Solve: $\begin{cases} 2x + 4y + 5z = 4 \\ x - 2y - 3z = 5 \\ x + 3y + 4z = 1 \end{cases}$
1. If necessary, interchange two equations or use multiplication to make the coefficient of the first variable in equation (1) a factor of the other first variable coefficients.	1. Interchange the first two equations: $$\begin{aligned} x - 2y - 3z &= 5 & (1) \\ 2x + 4y + 5z &= 4 & (2) \\ x + 3y + 4z &= 1 & (3) \end{aligned}$$
2. Add multiples of the first equation to each of the following equations so that the coefficients of the first variable in the second and third equations become zero.	2. Add $(-2) \times$ equation (1) to equation (2) and add $(-1) \times$ equation (1) to equation (3): $$\begin{aligned} x - 2y - 3z &= 5 & (1) \\ 0x + 8y + 11z &= -6 & (2) \\ 0x + 5y + 7z &= -4 & (3) \end{aligned}$$
3. Add a multiple of the second equation to the third equation so that the coefficient of the second variable in the third equation becomes zero.	3. Add $\left(-\frac{5}{8}\right) \times$ equation (2) to equation (3): $$\begin{aligned} x - 2y - 3z &= 5 & (1) \\ 8y + 11z &= -6 & (2) \\ 0y + \frac{1}{8}z &= -\frac{2}{8} & (3) \end{aligned}$$
4. Solve the third equation and *back substitute* from the bottom to find the remaining variables.	4. $z = -2$ from equation (3) $y = \frac{1}{8}(-6 - 11z) = 2$ from equation (2) $x = 5 + 2y + 3z = 3$ from equation (1) so $x = 3, y = 2, z = -2$

● **EXAMPLE 8** *Left-to-Right Elimination*

Solve: $\begin{cases} x + 2y + 3z = 6 & (1) \\ 2x + 3y + 2z = 6 & (2) \\ -x + y + z = 4 & (3) \end{cases}$

Solution

Using equation (1) to eliminate x from the other equations gives the equivalent system:

$$\begin{cases} x + 2y + 3z = 6 & (1) \\ -y - 4z = -6 & (2) \quad (-2) \times \text{equation (1) added to equation (2)} \\ 3y + 4z = 10 & (3) \quad \text{Equation (1) added to equation (3)} \end{cases}$$

Using equation (2) to eliminate y from equation (3) gives

$$\begin{cases} x + 2y + 3z = 6 & (1) \\ -y - 4z = -6 & (2) \\ -8z = -8 & (3) \end{cases}$$

(3) $\times$ equation (2) added to equation (3)

In a system of equations such as this, the first variable appearing in each equation is called the **lead variable.** Solving for each lead variable gives

$$x = 6 - 2y - 3z$$
$$y = 6 - 4z$$
$$z = 1$$

and using **back substitution** from the bottom gives

$$z = 1$$
$$y = 6 - 4 = 2$$
$$x = 6 - 4 - 3 = -1$$

Hence the solution is $x = -1$, $y = 2$, $z = 1$.

Although other methods for solving systems of equations in three variables may be useful, the left-to-right elimination method is important because it is systematic and can easily be extended to larger systems and to systems solved with matrices. (See Section 3.3, "Gauss-Jordan Elimination: Solving Systems of Equations.")

● *Checkpoint*

4. Use left-to-right elimination to solve.

$$\begin{cases} x - y - z = 0 & (1) \\ y - 2z = -18 & (2) \\ x + y + z = 6 & (3) \end{cases}$$

● *Checkpoint Solutions*

1. $x + y = -1$ means $x = -y - 1$, so $3x - 4y = -24$ becomes

$$3(-y - 1) - 4y = -24$$
$$-3y - 3 - 4y = -24$$
$$-7y = -21$$
$$y = 3$$

Hence $x = -3 - 1 = -4$, and the solution is $x = -4$, $y = 3$.

2. Multiply the first equation by (-3) and the second by (2). Add the resulting equations to eliminate the x-variable and solve for y.

$$\begin{array}{r} -6x - 9y = -15 \\ 6x + 10y = -50 \\ \hline y = -65 \end{array}$$

Hence $2x + 3(-65) = 5$ or $2x - 195 = 5$, so $2x = 200$. Thus $x = 100$ and $y = -65$.

3. (a) Substitution works well in Problem 1 because it is easy to solve for x (or for y). Substitution would not work well in Problem 2 because solving for x (or for y) would introduce fractions.

(b) Elimination would work well in Problem 1 if we multiplied the first equation by 4. As stated in (a), substitution wouldn't be as easy in Problem 2.

4.
$$\begin{cases} x - y - z = 0 & (1) \\ y - 2z = -18 & (2) \\ x + y + z = 6 & (3) \end{cases}$$

Add $(-1) \times$ equation (1) to equation (3).

$$\begin{cases} x - y - z = 0 & (1) \\ y - 2z = -18 & (2) \\ 2y + 2z = 6 & (3) \end{cases}$$

Add $(-2) \times$ equation (2) to equation (3).

$$\begin{cases} x - y - z = 0 \\ y - 2z = -18 \\ 6z = 42 \end{cases}$$

Solve the equations for x, y, and z.

$$x = y + z$$
$$y = 2z - 18$$
$$z = 7$$

Back substitution gives $y = 14 - 18 = -4$ and $x = -4 + 7 = 3$. Thus the solution is $x = 3$, $y = -4$, $z = 7$.

1.5 Exercises

In Problems 1–4, the graphs of two equations are shown. Decide whether the system of equations in each problem has one solution, no solution, or an infinite number of solutions. If the system has one solution, estimate it.

1.

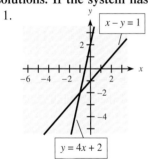

$x - y = 1$

$y = 4x + 2$

2.

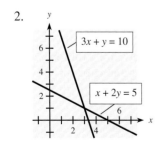

$3x + y = 10$

$x + 2y = 5$

3.

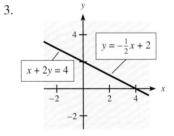

$y = -\frac{1}{2}x + 2$

$x + 2y = 4$

4.

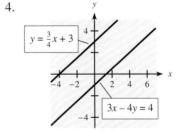

$y = \frac{3}{4}x + 3$

$3x - 4y = 4$

In Problems 5–8, solve the systems of equations by using graphical methods.

5. $\begin{cases} 4x - 2y = 4 \\ x - 2y = -2 \end{cases}$

6. $\begin{cases} x - y = -2 \\ 2x + y = -1 \end{cases}$

7. $\begin{cases} 3x - y = 10 \\ 6x - 2y = 5 \end{cases}$ 8. $\begin{cases} 2x - y = 3 \\ 4x - 2y = 6 \end{cases}$

In Problems 9–12, solve the systems of equations by substitution.

9. $\begin{cases} 3x - 2y = 6 \\ 4y = 8 \end{cases}$ 10. $\begin{cases} 3x = 6 \\ 4x - 3y = 5 \end{cases}$

11. $\begin{cases} 2x - y = 2 \\ 3x + 4y = 6 \end{cases}$ 12. $\begin{cases} 4x - y = 3 \\ 2x + 3y = 19 \end{cases}$

In Problems 13–22, solve each system by elimination or by any convenient method.

13. $\begin{cases} 3x + 4y = 1 \\ 2x - 3y = 12 \end{cases}$ 14. $\begin{cases} 5x - 2y = 4 \\ 2x - 3y = 5 \end{cases}$

15. $\begin{cases} -4x + 3y = -5 \\ 3x - 2y = 4 \end{cases}$ 16. $\begin{cases} x + 2y = 3 \\ 3x + 6y = 6 \end{cases}$

17. $\begin{cases} 0.2x - 0.3y = 4 \\ 2.3x - y = 1.2 \end{cases}$ 18. $\begin{cases} 0.5x + y = 3 \\ 0.3x + 0.2y = 6 \end{cases}$

19. $\begin{cases} \frac{5}{2}x - \frac{7}{2}y = -1 \\ 8x + 3y = 11 \end{cases}$ 20. $\begin{cases} x - \frac{1}{2}y = 1 \\ \frac{2}{3}x - \frac{1}{3}y = 1 \end{cases}$

21. $\begin{cases} 4x + 6y = 4 \\ 2x + 3y = 2 \end{cases}$ 22. $\begin{cases} 6x - 4y = 16 \\ 9x - 6y = 24 \end{cases}$

Using a graphing utility or Excel to find the solution of each system of equations in Problems 23–26.

23. $\begin{cases} y = 8 - \dfrac{3x}{2} \\ y = \dfrac{3x}{4} - 1 \end{cases}$ 24. $\begin{cases} y = 9 - \dfrac{2x}{3} \\ y = 5 + \dfrac{2x}{3} \end{cases}$

25. $\begin{cases} 5x + 3y = -2 \\ 3x + 7y = 4 \end{cases}$ 26. $\begin{cases} 4x - 5y = -3 \\ 2x - 7y = -6 \end{cases}$

Using the left-to-right elimination method to solve the systems in Problems 27–32.

27. $\begin{cases} x + 2y + z = 2 \\ -y + 3z = 8 \\ 2z = 10 \end{cases}$ 28. $\begin{cases} x - 2y + 2z = -10 \\ y + 4z = -10 \\ -3z = 9 \end{cases}$

29. $\begin{cases} x - y - 8z = 0 \\ y + 4z = 8 \\ 3y + 14z = 22 \end{cases}$ 30. $\begin{cases} x + 3y - 8z = 20 \\ y - 3z = 11 \\ 2y + 7z = -4 \end{cases}$

31. $\begin{cases} x + 4y - 2z = 9 \\ x + 5y + 2z = -2 \\ x + 4y - 28z = 22 \end{cases}$ 32. $\begin{cases} x - 3y - z = 0 \\ x - 2y + z = 8 \\ 2x - 6y + z = 6 \end{cases}$

APPLICATIONS

33. *Personal expenditures* For the period 1995–2002, the total personal expenditures (in billions of dollars) in the United States for food, $f(x)$, and for housing, $h(x)$, can be described by

$$f(x) = 40.74x + 742.65 \quad \text{and} \quad h(x) = 47.93x + 725$$

where x is the number of years past 1995. (*Source:* U.S. Department of Commerce) Find the year in which these expenditures were equal and the amount spent on each.

34. *Minority children* In the United States between 1970 and 2002, the numbers of millions of black children $B(x)$ and hispanic children $H(x)$ can be described by

$$B(x) = 0.081x + 8.97 \quad \text{and} \quad H(x) = 0.282x + 3.10$$

where x is the number of years past 1970. (*Source:* U.S. Bureau of the Census)
(a) In what year were the number of children equal? How many of each were there?
(b) Do these models indicate there are more black children or hispanic children now?

35. *Pricing* A concert promoter needs to make $42,000 from the sale of 1800 tickets. The promoter charges $20 for some tickets and $30 for the others.
(a) If there are x of the $20 tickets sold and y of the $30 tickets sold, write an equation that states that the sum of the tickets sold is 1800.
(b) How much money is received from the sale of x tickets for 20 dollars each?
(c) How much money is received from the sale of y tickets for 30 dollars each?
(d) Write an equation that states that the total amount received from the sale is 42,000 dollars.
(e) Solve the equations simultaneously to find how many tickets of each type must be sold to yield the $42,000.

36. *Rental income* A woman has $500,000 invested in two rental properties. One yields an annual return of 10% of her investment, and the other returns 12% per year on her investment. Her total annual return from the two investments is $53,000. Let x represent the amount of the 10% investment and y represent the amount of the 12% investment.
(a) Write an equation that states that the sum of the investments is 500,000 dollars.
(b) What is the annual return on the 10% investment?
(c) What is the annual return on the 12% investment?
(d) Write an equation that states that the sum of the annual returns is 53,000 dollars.
(e) Solve these two equations simultaneously to find how much is invested in each property.

37. *Investment yields* One safe investment pays 10% per year, and a more risky investment pays 18% per year. A woman who has $145,600 to invest would like to have an income of $20,000 per year from her investments. How much should she invest at each rate?

38. *Loans*   A bank lent $118,500 to a company for the development of two products. If the loan for product A was for $34,500 more than that for product B, how much was lent for each product?

39. *Rental income*   A woman has $235,000 invested in two rental properties. One yields 10% on the investment, and the other yields 12%. Her total income from them is $25,500. How much is her income from each property?

40. *Loans*   Mr. Jackson borrowed money from his bank and on his life insurance to start a business. His interest rate on the bank loan was 10%, and his rate on the insurance loan was 12%. If the total amount borrowed was $100,000 and his total yearly interest payment was $10,900, how much did he borrow from the bank?

41. *Nutrition*   Each ounce of substance A supplies 5% of the nutrition a patient needs. Substance B supplies 12% of the required nutrition per ounce. If digestive restrictions require that the ratio of substance A to substance B be 3/5, how many ounces of each should be in the diet to provide 100% of the required nutrition?

42. *Nutrition*   A glass of skim milk supplies 0.1 mg of iron and 8.5 g of protein. A quarter pound of lean red meat provides 3.4 mg of iron and 22 g of protein. If a person on a special diet is to have 7.15 mg of iron and 73.75 g of protein, how many glasses of skim milk and how many quarter-pound servings of meat would provide this?

43. *Bacterial growth*   Bacteria of species A and species B are kept in a single test tube, where they are fed two nutrients. Each day the test tube is supplied with 10,600 units of the first nutrient and 19,650 units of the second nutrient. Each bacterium of species A requires 2 units of the first nutrient and 3 units of the second, and each bacterium of species B requires 1 unit of the first nutrient and 4 units of the second. What populations of each species can coexist in the test tube so that all the nutrients are consumed each day?

44. *Botany*   A biologist has a 40% solution and a 10% solution of the same plant nutrient. How many cubic centimeters of each solution should be mixed to obtain 25 cc of a 28% solution?

45. *Medications*   A nurse has two solutions that contain different concentrations of a certain medication. One is a 20% concentration and the other is a 5% concentration. How many cubic centimeters of each should he mix to obtain 10 cc of a 15.5% solution?

46. *Medications*   Medication A is given every 4 hours and medication B is given twice each day. The total intake of the two medications is restricted to 50.6 mg per day, for a certain patient. If the ratio of the dosage of A to the dosage of B is 5 to 8, find the dosage for each administration of each medication.

47. *Pricing*   A concert promoter needs to take in $380,000 on the sale of 16,000 tickets. If the promoter charges $20 for some tickets and $30 for others, how many of each type must be sold to yield the $380,000?

48. *Pricing*   A nut wholesaler sells a mix of peanuts and cashews. He charges $2.80 per pound for peanuts and $5.30 per pound for cashews. If the mix is to sell for $3.30 per pound, how many pounds each of peanuts and cashews should be used to make 100 pounds of the mix?

49. *Nutrient solutions*   How many cubic centimeters of a 20% solution of a nutrient must be added to 100 cc of a 2% solution of the same nutrient to make a 10% solution of the nutrient?

50. *Mixtures*   How many gallons of washer fluid that is 13.5% antifreeze must a manufacturer add to 200 gallons of washer fluid that is 11% antifreeze to yield washer fluid that is 13% antifreeze?

Application Problems 51–54 require systems of equations in three variables.

51. *Nutrition*   Each ounce of substance A supplies 5% of the nutrition a patient needs. Substance B supplies 15% of the required nutrition per ounce, and substance C supplies 12% of the required nutrition per ounce. If digestive restrictions require that substances A and C be given in equal amounts, and the amount of substance B be one-fifth of either of these other amounts, find the number of ounces of each substance that should be in the meal to provide 100% of the required nutrition.

52. *Dietary requirements*   A glass of skim milk supplies 0.1 mg of iron, 8.5 g of protein, and 1 g of carbohydrates. A quarter pound of lean red meat provides 3.4 mg of iron, 22 g of protein, and 20 g of carbohydrates. Two slices of whole grain bread supply 2.2 mg of iron, 10 g of protein, and 12 g of carbohydrates. If a person on a special diet must have 10.5 mg of iron, 94.5 g of protein, and 61 g of carbohydrates, how many glasses of skim milk, how many quarter-pound servings of meat, and how many two-slice servings of whole grain bread will supply this?

53. *Social services*   A social agency is charged with providing services to three types of clients, A, B, and C. A total of 500 clients are to be served, with $150,000 available for counseling and $100,000 available for emergency food and shelter. Type A clients require an average of $200 for counseling and $300 for emergencies, type B clients require an average of $500 for

counseling and $200 for emergencies, and type C clients require an average of $300 for counseling and $100 for emergencies. How many of each type of client can be served?

54. *Social services* If funding for counseling is cut to $135,000 and funding for emergency food and shelter is cut to $90,000, only 450 clients can be served. How many of each type can be served in this case? (See Problem 53.)

1.6

Applications of Functions in Business and Economics

OBJECTIVES

- To formulate and evaluate total cost, total revenue, and profit functions
- To find marginal cost, revenue, and profit, given linear total cost, total revenue, and profit functions
- To write the equations of linear total cost, total revenue, and profit functions by using information given about the functions
- To find break-even points
- To evaluate and graph supply and demand functions
- To find market equilibrium

◗ Application Preview

Suppose a firm manufactures MP3 players and sells them for $50 each, with costs incurred in the production and sale equal to $200,000 plus $10 for each unit produced and sold. Forming the total cost, total revenue, and profit as functions of the quantity x that is produced and sold (see Example 1) is called the **theory of the firm.** We will also discuss **market analysis,** in which supply and demand are found as functions of price, and market equilibrium is found.

Total Cost, Total Revenue, and Profit

The **profit** a firm makes on its product is the difference between the amount it receives from sales (its revenue) and its cost. If x units are produced and sold, we can write

$$P(x) = R(x) - C(x)$$

where

$$P(x) = \text{profit from sale of } x \text{ units}$$

$$R(x) = \text{total revenue from sale of } x \text{ units}$$

$$C(x) = \text{total cost of production and sale of } x \text{ units*}$$

In general, **revenue** is found by using the equation

$$\text{Revenue} = (\text{price per unit})(\text{number of units})$$

The **cost** is composed of two parts, fixed costs and variable costs. **Fixed costs** *(FC),* such as depreciation, rent, utilities, and so on, remain constant regardless of the number of units produced. **Variable costs** *(VC)* are those directly related to the number of units produced. Thus the cost is found by using the equation

$$\text{Cost} = \text{variable costs} + \text{fixed costs}$$

◗ EXAMPLE 1 *Cost, Revenue, and Profit* **(Application Preview)**

Suppose that a firm manufactures MP3 players and sells them for $50 each. The costs incurred in the production and sale of the MP3 players are $200,000 plus $10 for each player produced and sold. Write the profit function for the production and sale of x players.

Solution

The total revenue for x MP3 players is $50x$, so the total revenue function is $R(x) = 50x$. The fixed costs are $200,000, so the total cost for x players is $10x + 200,000$. Hence, $C(x) = 10x + 200,000$. The profit function is given by $P(x) = R(x) - C(x)$. Thus,

$$P(x) = 50x - (10x + 200,000)$$

$$P(x) = 40x - 200,000$$

Figure 1.41 shows the graphs of $R(x)$, $C(x)$, and $P(x)$.

*The symbols generally used in economics for total cost, total revenue, and profit are *TC, TR,* and π, respectively. In order to avoid confusion, especially with the use of π as a variable, we do not use these symbols.

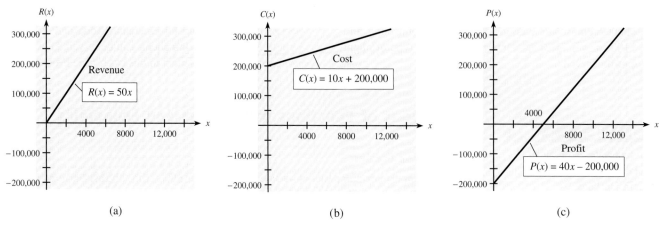

Figure 1.41

By observing the intercepts on the graphs in Figure 1.41, we note the following.

Revenue: 0 units produce 0 revenue.

Cost: 0 units' costs equal fixed costs = $200,000.

Profit: 0 units yield a loss equal to fixed costs = $200,000.
5000 units result in a profit of $0 (no loss or gain).

Marginals

In Example 1, both the total revenue function and the total cost function are linear, so their difference, the profit function, is also linear. The slope of the profit function represents the rate of change in profit with respect to the number of units produced and sold. This is called the **marginal profit** ($\overline{MP}$) for the product. Thus the marginal profit for the MP3 players in Example 1 is $40. Similarly, the **marginal cost** ($\overline{MC}$) for this product is $10 (the slope of the cost function), and the **marginal revenue** ($\overline{MR}$) is $50 (the slope of the revenue function).

● **EXAMPLE 2 Marginal Cost**

Suppose that the cost (in dollars) for a product is $C = 21.75x + 4890$. What is the marginal cost for this product, and what does it mean?

Solution
The equation has the form $C = mx + b$, so the slope is 21.75. Thus the marginal cost is $\overline{MC} = 21.75$ dollars per unit.

Because the marginal cost is the slope of the cost line, production of each additional unit will cost $21.75 more, at any level of production.

Note that when total cost functions are linear, the marginal cost is the same as the variable cost. This is not the case if the functions are not linear, as we shall see later.

● **Checkpoint**

1. Suppose that when a company produces its product, fixed costs are $12,500 and variable costs are $75 per item.
   (a) Write the total cost function if x represents the number of units.
   (b) Are fixed costs equal to $C(0)$?
2. Suppose the company in Problem 1 sells its product for $175 per item.
   (a) Write the total revenue function.
   (b) Find $R(100)$ and give its meaning.
3. (a) Give the formula for profit in terms of revenue and cost.
   (b) Find the profit function for the company in Problems 1 and 2.

● EXAMPLE 3 *Profit*

Suppose the profit function for a product is linear, and the marginal profit is $5. If the profit is $200 when 125 units are sold, write the equation of the profit function.

Solution
The marginal profit gives us the slope of the line representing the profit function. Using this slope ($m = 5$) and the point (125, 200) in the point-slope formula $P - P_1 = m(x - x_1)$ gives

$$P - 200 = 5(x - 125)$$

or

$$P = 5x - 425$$

Break-Even Analysis

We can solve the equations for total revenue and total cost simultaneously to find the point where cost and revenue are equal. This point is called the **break-even point.** On the graph of these functions, we use x to represent the quantity produced and y to represent the dollar value of revenue *and* cost. The point where the total revenue line crosses the total cost line is the break-even point.

● EXAMPLE 4 *Break Even*

A manufacturer sells a product for $10 per unit. The manufacturer's fixed costs are $1200 per month, and the variable costs are $2.50 per unit. How many units must the manufacturer produce each month to break even?

Solution
The total revenue for x units of the product is $10x$, so the equation for total revenue is $R = 10x$. The fixed costs are $1200, so the total cost for x units is $2.50x + 1200$. Thus the equation for total cost is $C = 2.50x + 1200$. We find the break-even point by solving the two equations simultaneously ($R = C$ at the break-even point). By substitution,

$$10x = 2.50x + 1200$$
$$7.5x = 1200$$
$$x = 160$$

Thus the manufacturer will break even if 160 units are produced per month. The manufacturer will make a profit if more than 160 units are produced. Figure 1.42 shows that for $x < 160$, $R(x) < C(x)$ (resulting in a loss) and that for $x > 160$, $R(x) > C(x)$ (resulting in a profit).

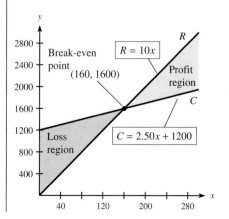

Figure 1.42

Using the fact that the profit function is found by subtracting the total cost function from the total revenue function, we can form the profit function for the previous example. The profit function is given by

$$P(x) = 10x - (2.50x + 1200) \quad \text{or} \quad P(x) = 7.50x - 1200$$

We can find the point where the profit is zero (the break-even point) by setting $P(x) = 0$ and solving for x.

$$0 = 7.50x - 1200$$
$$1200 = 7.50x$$
$$x = 160$$

Note that this is the same break-even point that we found by solving the total revenue and total cost equations simultaneously (see Figure 1.43).

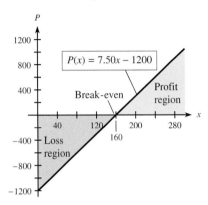

Figure 1.43

● **Checkpoint**   4. Identify two ways in which break-even points can be found.

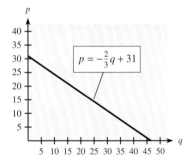

Figure 1.44

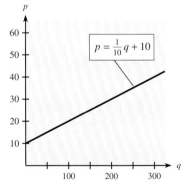

Figure 1.45

Supply, Demand, and Market Equilibrium

Economists and managers also use points of intersection to determine market equilibrium. **Market equilibrium** occurs when the quantity of a commodity demanded is equal to the quantity supplied.

Demand by consumers for a commodity is related to the price of the commodity. The **law of demand** states that the quantity demanded will increase as price decreases, and that the quantity demanded will decrease as price increases. Figure 1.44 shows the graph of a typical linear demand function. Note that although quantity demanded is a function of price, economists have traditionally graphed the demand function with price on the vertical axis. Throughout this text, we will follow this tradition. Linear equations relating price p and quantity demanded q can be solved for either p or q, and we will have occasion to use the equations in both forms.

Just as a consumer's willingness to buy is related to price, a manufacturer's willingness to supply goods is also related to price. The **law of supply** states that the quantity supplied for sale will increase as the price of a product increases. Figure 1.45 shows the graph of a typical linear supply function. As with demand, price is placed on the vertical axis. Note that negative prices and quantities have no meaning, so supply and demand curves are restricted to the first quadrant.

If the supply and demand curves for a commodity are graphed on the same coordinate system, with the same units, market equilibrium occurs at the point where the curves intersect. The price at that point is the **equilibrium price,** and the quantity at that point is the **equilibrium quantity.**

For the supply and demand functions shown in Figure 1.46, we see that the curves intersect at the point (30, 11). This means that when the price is $11, consumers are willing to purchase the same number of units (30) that producers are willing to supply.

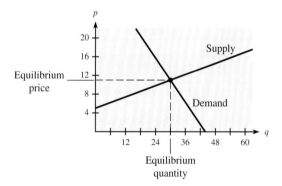

Figure 1.46

In general, the equilibrium price and the equilibrium quantity must both be positive for the market equilibrium to have meaning.

We can find the market equilibrium by graphing the supply and demand functions on the same coordinate system and observing their point of intersection. As we have seen, finding the point(s) common to the graphs of two (or more) functions is called **solving a system of equations** or **solving simultaneously.**

● **EXAMPLE 5** *Market Equilibrium*

Find the market equilibrium point for the following supply and demand functions.

$$\text{Demand:} \quad p = -3q + 36$$
$$\text{Supply:} \quad p = 4q + 1$$

Solution

At market equilibrium, the demand price equals the supply price. Thus,

$$-3q + 36 = 4q + 1$$
$$35 = 7q$$
$$q = 5$$
$$p = 21$$

The equilibrium point is (5, 21).

Checking, we see that

$$21 = -3(5) + 36 \checkmark \qquad \text{and} \qquad 21 = 4(5) + 1 \checkmark$$

Spreadsheet Note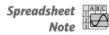

We can use Goal Seek with Excel to find the market equilibrium for given supply and demand functions. To find the equilibrium quantity and price for the supply and demand functions in Example 5, we set up the table with entries for quantity, demand, and supply. We enter the functions as in the following table and add a fourth entry representing demand − supply.

	A	B	C	D
1	q	p: demand	p: supply	demand − supply
2	1	=−3*A2+36	=4*A2+1	=B2−C2

We find the equilibrium quantity and price by using Goal Seek with D2 set to 0, with changing cell A2. The resulting solution gives the equilibrium quantity as 5 and the equilibrium price as 21. ■

	A	B	C	D
1	q	p: demand	p: supply	demand − supply
2	5	21	21	0

● **EXAMPLE 6** *Market Equilibrium*

A group of wholesalers will buy 50 dryers per month if the price is $200 and 30 per month if the price is $300. The manufacturer is willing to supply 20 if the price is $210 and 30 if the price is $230. Assuming that the resulting supply and demand functions are linear, find the equilibrium point for the market.

Solution
Representing price by p and quantity by q, we have
Demand function:

$$m = \frac{300 - 200}{30 - 50} = -5$$

$$p - 200 = -5(q - 50)$$

$$p = -5q + 450$$

Supply function:

$$m = \frac{230 - 210}{30 - 20} = 2$$

$$p - 230 = 2(q - 30)$$

$$p = 2q + 170$$

Because the prices are equal at market equilibrium, we have

$$-5q + 450 = 2q + 170$$

$$280 = 7q$$

$$q = 40$$

$$p = 250$$

The equilibrium point is (40, 250). See Figure 1.47 for the graphs of these functions.

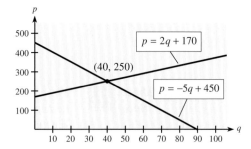

Figure 1.47

Supply and Demand with Taxation

Suppose a supplier is taxed $K per unit sold, and the tax is passed on to the consumer by adding $K to the selling price of the product. If the original supply function is $p = f(q)$, then passing the tax on gives a new supply function, $p = f(q) + K$. Because the value of the product is not changed by the tax, the demand function is unchanged. Figure 1.48 shows the effect that this has on market equilibrium.

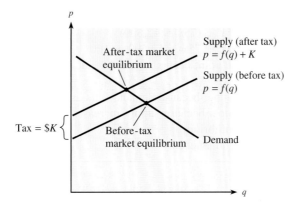

Figure 1.48

Note that the new market equilibrium point is the point of intersection of the original demand function and the new (after-tax) supply function.

● **EXAMPLE 7 *Taxation***

In Example 6 the supply and demand functions for dryers were given as follows.

$$\text{Supply:} \quad p = 2q + 170$$
$$\text{Demand:} \quad p = -5q + 450$$

The equilibrium point was $q = 40$, $p = \$250$. If the wholesaler is taxed $14 per unit sold, what is the new equilibrium point?

Solution
The $14 tax per unit is passed on by the wholesaler, so the new supply function is

$$p = 2q + 170 + 14$$

and the demand function is unchanged. Thus we solve the system

$$\begin{cases} p = 2q + 184 \\ p = -5q + 450 \end{cases}$$
$$2q + 184 = -5q + 450$$
$$7q = 266$$
$$q = 38$$
$$p = 2(38) + 184 = 260$$

The new equilibrium point is $q = 38$, $p = \$260$.
  Checking, we see that

$$260 = 2(38) + 184 \; \checkmark \quad \text{and} \quad 260 = -5(38) + 450 \; \checkmark$$

● **Checkpoint**

5. (a) Does a typical linear demand function have positive slope or negative slope? Why?
   (b) Does a typical linear supply function have positive slope or negative slope? Why?
6. (a) What do we call the point of intersection of a supply function and a demand function?
   (b) What algebraic technique is used to find the point named in (a)?

● **Checkpoint Solutions**

1. (a) $C(x) = 75x + 12{,}500$
   (b) Yes. $C(0) = 12{,}500 = $ Fixed costs. In fact, fixed costs are defined to be $C(0)$.
2. (a) $R(x) = 175x$
   (b) $R(100) = 175(100) = \$17{,}500$, which means that revenue is $17,500 when 100 units are sold.

3. (a) Profit = Revenue − Cost or $P(x) = R(x) − C(x)$
   (b) $P(x) = 175x − (75x + 12{,}500)$
   $$= 175x − 75x − 12{,}500 = 100x − 12{,}500$$
4. The break-even point occurs where revenue equals cost $[R(x) = C(x)]$ or where profit is zero $[P(x) = 0]$.
5. (a) Negative slope, because demand falls as price increases.
   (b) Positive slope, because supply increases as price increases.
6. (a) Market equilibrium
   (b) Solving simultaneously

1.6 Exercises

TOTAL COST, TOTAL REVENUE, AND PROFIT

1. Suppose a calculator manufacturer has the total cost function $C(x) = 17x + 3400$ and the total revenue function $R(x) = 34x$.
   (a) What is the equation of the profit function for the calculator?
   (b) What is the profit on 300 units?
2. Suppose a stereo receiver manufacturer has the total cost function $C(x) = 105x + 1650$ and the total revenue function $R(x) = 215x$.
   (a) What is the equation of the profit function for this commodity?
   (b) What is the profit on 50 items?
3. Suppose a radio manufacturer has the total cost function $C(x) = 43x + 1850$ and the total revenue function $R(x) = 80x$.
   (a) What is the equation of the profit function for this commodity?
   (b) What is the profit on 30 units? Interpret your result.
   (c) How many radios must be sold to avoid losing money?
4. Suppose a computer manufacturer has the total cost function $C(x) = 85x + 3300$ and the total revenue function $R(x) = 385x$.
   (a) What is the equation of the profit function for this commodity?
   (b) What is the profit on 351 items?
   (c) How many items must be sold to avoid losing money?
5. A linear cost function is $C(x) = 5x + 250$.
   (a) What are the slope and the C-intercept?
   (b) What is the marginal cost, and what does it mean?
   (c) How are your answers to (a) and to (b) related?
   (d) What is the cost of producing *one more* item if 50 are currently being produced? What is it if 100 are currently being produced?

6. A linear cost function is $C(x) = 27.55x + 5180$.
   (a) What are the slope and the C-intercept?
   (b) What is the marginal cost, and what does it mean?
   (c) How are your answers to (a) and to (b) related?
   (d) What is the cost of producing *one more* item if 50 are currently being produced? What is it if 100 are currently being produced?
7. A linear revenue function is $R = 27x$.
   (a) What is the slope?
   (b) What is the marginal revenue, and what does it mean?
   (c) What is the revenue received from selling *one more* item if 50 are currently being sold? If 100 are being sold?
8. A linear revenue function is $R = 38.95x$.
   (a) What is the slope?
   (b) What is the marginal revenue, and what does it mean?
   (c) What is the revenue received from selling *one more* item if 50 are currently being sold? If 100 are being sold?
9. Let $C(x) = 5x + 250$ and $R(x) = 27x$.
   (a) Write the profit function $P(x)$.
   (b) What is the slope of the profit function?
   (c) What is the marginal profit?
   (d) Interpret the marginal profit.
10. Given $C(x) = 21.95x + 1400$ and $R(x) = 20x$, find the profit function.
    (a) What is the marginal profit, and what does it mean?
    (b) What should a firm with these cost, revenue, and profit functions do? (*Hint:* Graph the profit function and see where it goes.)
11. A company charting its profits notices that the relationship between the number of units sold, x, and the profit, P, is linear. If 200 units sold results in \$3100 profit and 250 units sold results in \$6000 profit, write the profit function for this company. Find the marginal profit.

12. Suppose that the total cost function for a radio is linear, that the marginal cost is $27, and that the total cost for 50 radios is $4350. Write the equation of this cost function and then graph it.

13. Extreme Protection, Inc. manufactures helmets for skiing and snow boarding. The fixed costs for one model of helmet are $6600 per month. Materials and labor for each helmet of this model are $35, and the company sells this helmet to dealers for $60 each.
    (a) For this helmet, write the function for monthly total costs.
    (b) Write the function for total revenue.
    (c) Write the function for profit.
    (d) Find $C(200)$, $R(200)$, and $P(200)$ and interpret each answer.
    (e) Find $C(300)$, $R(300)$, and $P(300)$ and interpret each answer.
    (f) Find the marginal profit and write a sentence that explains its meaning.

14. A manufacturer of DVD players has monthly fixed costs of $9800 and variable costs of $65 per unit for one particular model. The company sells this model to dealers for $100 each.
    (a) For this model DVD player, write the function for monthly total costs.
    (b) Write the function for total revenue.
    (c) Write the function for profit.
    (d) Find $C(250)$, $R(250)$, and $P(250)$ and interpret each answer.
    (e) Find $C(400)$, $R(400)$, and $P(400)$ and interpret each answer.
    (f) Find the marginal profit and write a sentence that explains its meaning.

BREAK-EVEN ANALYSIS

15. The figure shows graphs of the total cost function and the total revenue function for a commodity.

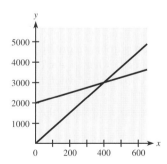

    (a) Label each function correctly.
    (b) Determine the fixed costs.
    (c) Locate the break-even point and determine the number of units sold to break even.
    (d) Estimate the marginal cost and marginal revenue.

16. A manufacturer of shower-surrounds has a revenue function of

$$R(x) = 81.50x$$

and a cost function of

$$C(x) = 63x + 1850$$

Find the number of units that must be sold to break even.

17. A jewelry maker incurs costs for a necklace according to

$$C(x) = 35x + 1650$$

If the revenue function for the necklaces is

$$R(x) = 85x$$

how many necklaces must be sold to break even?

18. A small business recaps and sells tires. If a set of four tires has the revenue function

$$R(x) = 89x$$

and the cost function

$$C(x) = 1400 + 75x$$

find the number of sets of recaps that must be sold to break even.

19. A manufacturer sells belts for $12 per unit. The fixed costs are $1600 per month, and the variable costs are $8 per unit.
    (a) Write the equations of the revenue and cost functions.
    (b) Find the break-even point.

20. A manufacturer sells watches for $50 per unit. The fixed costs related to this product are $10,000 per month, and the variable costs are $30 per unit.
    (a) Write the equations of the revenue and cost functions.
    (b) How many watches must be sold to break even?

21. (a) Write the profit function for Problem 19.
    (b) Set profit equal to zero and solve for x. Compare this x-value with the break-even point from Problem 19(b).

22. (a) Write the profit function for Problem 20.
    (b) Set profit equal to zero and solve for x. Compare this x-value with the break-even point from Problem 20(b).

23. Electronic equipment manufacturer Dynamo Electric, Inc. makes several types of surge protectors. Their base model surge protector has monthly fixed costs of $1045. This particular model wholesales for $10 each and costs $4.50 per unit to manufacture.
    (a) Write the function for Dynamo's monthly total costs.
    (b) Write the function for Dynamo's monthly total revenue.

(c) Write the function for Dynamo's monthly profit.

(d) Find the number of this type of surge protector that Dynamo must produce and sell each month to break even.

24. Financial Paper, Inc. is a printer of checks and forms for financial institutions. For individual accounts, boxes of 200 checks cost $0.80 per box to print and package and sell for $4.95 each. Financial Paper's monthly fixed costs for printing and packaging these checks for individuals are $1245.

(a) Write the function for Financial Paper's monthly total costs.

(b) Write the function for Financial Paper's monthly total revenue.

(c) Write the function for Financial Paper's monthly profit.

(d) Find the number of orders for boxes of checks for individual accounts that Financial Paper must receive and fill each month to break even.

25. A company manufactures and sells bookcases. The selling price is $54.90 per bookcase. The total cost function is linear, and costs amount to $50,000 for 2000 bookcases and $32,120 for 800 bookcases.

(a) Write the equation for revenue.

(b) Write the equation for total costs.

(c) Find the break-even point.

26. A company distributes college logo sweatshirts and sells them for $50 each. The total cost function is linear, and the total cost for 100 sweatshirts is $4360, whereas the total cost for 250 sweatshirts is $7060.

(a) Write the equation for the revenue function.

(b) Write the equation for the total cost function.

(c) Find the break-even point.

In each of Problems 27 and 28, *some* of the graphs of total revenue (*R*), total cost (*C*), variable cost (*VC*), fixed cost (*FC*), and profit (*P*) are shown as functions of the number of units, *x*.

(a) Correctly label the graphs shown.

(b) Carefully sketch and label the graphs of the other functions. Explain your method.

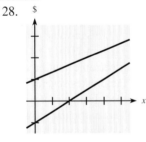

27. 28.

SUPPLY, DEMAND, AND MARKET EQUILIBRIUM

29. As the price of a commodity increases, what happens to demand?

30. As the price of a commodity increases, what happens to supply?

Figure 1.49 is the graph of the demand function for a product, and Figure 1.50 is the graph of the supply function for the same product. Use these graphs to answer the questions in Problems 31 and 32.

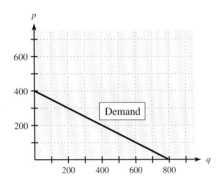

Figure 1.49

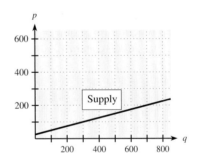

Figure 1.50

31. (a) How many units *q* are *demanded* when the price *p* is $100?

(b) How many units *q* are *supplied* when the price *p* is $100?

(c) Will there be a market surplus (more supplied) or shortage (more demanded) when *p* = $100?

32. (a) How many units *q* are *demanded* when the price *p* is $200?

(b) How many units *q* are *supplied* when the price *p* is $200?

(c) Will there be a market surplus or shortage when the price *p* is $200?

33. If the demand for a pair of shoes is given by $2p + 5q = 200$ and the supply function for it is $p - 2q = 10$, compare the quantity demanded and the quantity supplied when the price is $60. Will there be a surplus or shortfall at this price?

34. If the demand function and supply function for Z-brand phones are $p + 2q = 100$ and $35p - 20q = 350$, respectively, compare the quantity demanded and the quantity supplied when $p = 14$. Are there surplus phones or not enough to meet demand?

35. Suppose a certain outlet chain selling appliances has found that for one brand of stereo system, the monthly demand is 240 when the price is $900. However, when the price is $850, the monthly demand is 315. Assuming that the demand function for this system is linear, write the equation for the demand function. Use p for price and q for quantity.

36. Suppose a certain home improvements outlet knows that the monthly demand for framing studs is 2500 when the price is $1.00 each but that the demand is 3500 when the price is $0.90 each. Assuming that the demand function is linear, write its equation. Use p for price and q for quantity.

37. Suppose the manufacturer of a board game will supply 10,000 games if the wholesale price is $1.50 each but will supply only 5000 if the price is $1.00 each. Assuming that the supply function is linear, write its equation. Use p for price and q for quantity.

38. Suppose a mining company will supply 100,000 tons of ore per month if the price is $30 per ton but will supply only 80,000 tons per month if the price is $25 per ton. Assuming that the supply function is linear, write its equation.

Complete Problems 39–43 by using the accompanying figure, which shows a supply function and a demand function.

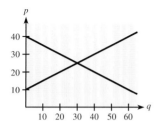

39. (a) Label each function as "demand" or "supply."
    (b) Label the equilibrium point and determine the price and quantity at which market equilibrium occurs.

40. (a) If the price is $30, what quantity is demanded?
    (b) If the price is $30, what quantity is supplied?
    (c) Is there a surplus or shortage when the price is $30? How many units is this surplus or shortage?

41. (a) If the price is $20, what quantity is supplied?
    (b) If the price is $20, what quantity is demanded?
    (c) Is there a surplus or a shortage when the price is $20? How many units is this surplus or shortage?

42. Will a price above the equilibrium price result in a market surplus or shortage?

43. Will a price below the equilibrium price result in a market surplus or shortage?

44. Find the market equilibrium point for the following demand and supply functions.

    Demand: $p = -2q + 320$
    Supply: $p = 8q + 2$

45. Find the market equilibrium point for the following demand and supply functions.

    Demand: $2p = -q + 56$
    Supply: $3p - q = 34$

46. Find the equilibrium point for the following supply and demand functions.

    Demand: $p = 480 - 3q$
    Supply: $p = 17q + 80$

47. Find the equilibrium point for the following supply and demand functions.

    Demand: $p = -4q + 220$
    Supply: $p = 15q + 30$

48. Retailers will buy 45 cordless phones from a wholesaler if the price is $10 each but only 20 if the price is $60. The wholesaler will supply 35 phones at $30 each and 70 at $50 each. Assuming the supply and demand functions are linear, find the market equilibrium point.

49. A group of retailers will buy 80 televisions from a wholesaler if the price is $350 and 120 if the price is $300. The wholesaler is willing to supply 60 if the price is $280 and 140 if the price is $370. Assuming the resulting supply and demand functions are linear, find the equilibrium point for the market.

50. A shoe store owner will buy 10 pairs of a certain shoe if the price is $75 per pair and 30 pairs if the price is $25. The supplier of the shoes is willing to provide 35 pairs if the price is $80 per pair but only 5 pairs if the price is $20. Assuming the supply and demand functions for the shoes are linear, find the market equilibrium point.

Problems 51–58 involve market equilibrium after taxation. Use the figure to answer Problems 51 and 52.

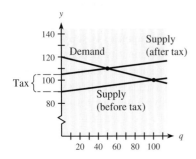

51. (a) What is the amount of the tax?
    (b) What is the original equilibrium price and quantity?
    (c) What is the new equilibrium price and quantity?
    (d) Does the supplier suffer from the tax even though it is passed on?
52. (a) If the tax is doubled, how many units will be sold?
    (b) Can a government lose money by increasing taxes?
53. If a $38 tax is placed on each unit of the product of Problem 47, what are the new equilibrium price and quantity?
54. If a $56 tax is placed on each unit of the product of Problem 46, what is the new equilibrium point?
55. Suppose that a certain product has the following demand and supply functions.

    Demand: $p = -0.05q + 65$
    Supply: $p = 0.05q + 10$

If a $5 tax per item is levied on the supplier and this tax is passed on to the consumer, find the market equilibrium point after the tax.

56. Suppose that a certain product has the following demand and supply functions.

    Demand: $p = -8q + 2800$
    Supply: $p = 3q + 35$

    If a $15 tax per item is levied on the supplier, who passes it on to the consumer as a price increase, find the market equilibrium point after the tax.

57. Suppose that in a certain market the demand function for a product is given by $60p + q = 2100$ and the supply function is given by $120p - q = 540$. Then a tax of $0.50 per item is levied on the supplier, who passes it on to the consumer as a price increase. Find the equilibrium price and quantity after the tax is levied.

58. Suppose that in a certain market the demand function for a product is given by $10p + q = 2300$ and the supply function is given by $45p - q = 360$. If the government levies a tax of $2 per item on the supplier, who passes the tax on to the consumer as a price increase, find the equilibrium price and quantity after the tax is levied.

Key Terms and Formulas

Section	Key Terms	Formulas
1.1	Equation, members; variable; solution	
	Identities; conditional equations	
	Properties of equality	
	Linear equation in one variable	
	Fractional equation	
	Linear equation in two variables	
	Aligning the data	
	Linear inequalities	
	Properties	
	Solutions	
1.2	Relation	
	Function	
	Vertical-line test	
	Domain, range	

Section	Key Terms	Formulas

	Coordinate system
	  Ordered pair, origin, x-axis,
	    y-axis
	Graph
	Function notation
	Operations with functions
	  Composite functions
		$(f \circ g)(x) = f(g(x))$

1.3	Linear function
	Intercepts
	  x-intercept (zero of a function)
	  y-intercept
		$y = ax + b$
		
		where $y = 0$
		where $x = 0$

	Slope of a line	$m = \dfrac{y_2 - y_1}{x_2 - x_1}$
	Rate of change Parallel lines Perpendicular lines Point-slope form Slope-intercept form Vertical line Horizontal line	$m_1 = m_2$ $m_2 = -1/m_1$ $y - y_1 = m(x - x_1)$ $y = mx + b$ $x = a$ $y = b$
1.4	Graphing utilities Standard viewing window Range Evaluating functions x-intercept Zeros of functions Solutions of equations x-intercept method	

1.5	Systems of linear equations
	Solutions
	  Graphical solutions
	Equivalent systems
	Substitution method
	Elimination method
	Left-to-right elimination method
	  Lead variable
	  Back substitution	

1.6	Cost and revenue functions Profit functions Marginal profit Marginal cost Marginal revenue Break-even point Supply and demand functions Market equilibrium	$C(x)$ and $R(x)$ $P(x) = R(x) - C(x)$ Slope of linear profit function Slope of linear cost function Slope of linear revenue function $C(x) = R(x)$ or $P(x) = 0$

Review Exercises

Section 1.1

Solve the equations in Problems 1–6.

1. $3x - 8 = 23$
2. $2x - 8 = 3x + 5$
3. $\dfrac{6x + 3}{6} = \dfrac{5(x - 2)}{9}$
4. $2x + \dfrac{1}{2} = \dfrac{x}{2} + \dfrac{1}{3}$
5. $\dfrac{6}{3x - 5} = \dfrac{6}{2x + 3}$
6. $\dfrac{2x + 5}{x + 7} = \dfrac{1}{3} + \dfrac{x - 11}{2x + 14}$
7. Solve for y: $3(y - 2) = -2(x + 5)$
8. Solve $3x - 9 \le 4(3 - x)$ and graph the solution.
9. Solve $\frac{2}{5}x \le x + 4$ and graph the solution.
10. Solve $5x + 1 \ge \frac{2}{3}(x - 6)$ and graph the solution.

Section 1.2

11. If $p = 3q^3$, is p a function of q?
12. If $y^2 = 9x$, is y a function of x?
13. If $R = \sqrt[3]{x + 4}$, is R a function of x?
14. What are the domain and range of the function $y = \sqrt{9 - x}$?
15. If $f(x) = x^2 + 4x + 5$, find the following.
    (a) $f(-3)$ (b) $f(4)$ (c) $f\left(\dfrac{1}{2}\right)$
16. If $g(x) = x^2 + 1/x$, find the following.
    (a) $g(-1)$ (b) $g\left(\dfrac{1}{2}\right)$ (c) $g(0.1)$
17. If $f(x) = 9x - x^2$, find $\dfrac{f(x + h) - f(x)}{h}$ and simplify.
18. Does the graph in Figure 1.51 represent y as a function of x?

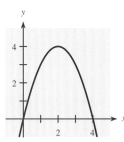

Figure 1.51

19. Does the graph in Figure 1.52 represent y as a function of x?

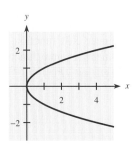

Figure 1.52

20. For the function f graphed in Figure 1.51, what is $f(2)$?
21. For the function f graphed in Figure 1.51, for what values of x does $f(x) = 0$?
22. The following table defines y as a function of x, denoted $y = f(x)$.

x	-2	-1	0	1	3	4
y	8	2	-3	4	2	7

Use the table to complete the following.
    (a) Identify the domain and range of $y = f(x)$.
    (b) Find $f(4)$.
    (c) Find all x-values for which $f(x) = 2$.
    (d) Graph $y = f(x)$.
    (e) Does the table define x as a function of y? Explain.
23. If $f(x) = 3x + 5$ and $g(x) = x^2$, find
    (a) $(f + g)(x)$ (b) $(f/g)(x)$
    (c) $f(g(x))$ (d) $(f \circ f)(x)$

Section 1.3

In Problems 24–26, find the intercepts and graph.

24. $5x + 2y = 10$
25. $6x + 5y = 9$
26. $x = -2$

In Problems 27 and 28, find the slope of the line that passes through each pair of points.

27. $(2, -1)$ and $(-1, -4)$
28. $(-3.8, -7.16)$ and $(-3.8, 1.16)$

In Problems 29 and 30, find the slope and y-intercept of each line.

29. $2x + 5y = 10$
30. $x = -\dfrac{3}{4}y + \dfrac{3}{2}$

In Problems 31–37, write the equation of each line described.

31. Slope 4 and y-intercept 2
32. Slope $-\frac{1}{2}$ and y-intercept 3
33. Through $(-2, 1)$ with slope $\frac{2}{5}$
34. Through $(-2, 7)$ and $(6, -4)$
35. Through $(-1, 8)$ and $(-1, -1)$
36. Through $(1, 6)$ and parallel to $y = 4x - 6$
37. Through $(-1, 2)$ and perpendicular to $3x + 4y = 12$

Section 1.4

In Problems 38 and 39, graph each equation with a graphing utility and the standard viewing window.

38. $x^2 + y - 2x - 3 = 0$
39. $y = \dfrac{x^3 - 27x + 54}{15}$

In Problems 40 and 41, use a graphing utility and
(a) graph each equation in the viewing window given.
(b) graph each equation in the standard viewing window.
(c) explain how the two views differ and why.

40. $y = (x + 6)(x - 3)(x - 15)$ with x-min $= -15$, x-max $= 25$; y-min $= -700$, y-max $= 500$

41. $y = x^2 - x - 42$ with x-min $= -15$, x-max $= 15$; y-min $= -50$, y-max $= 50$

42. What is the domain of $y = \dfrac{\sqrt{x + 3}}{x}$? Check with a graphing utility.

43. Use a graphing utility and the x-intercept method to approximate the solution of $7x - 2 = 0$.

Section 1.5

In Problems 44–50, solve each system of equations.

44. $\begin{cases} 4x - 2y = 6 \\ 3x + 3y = 9 \end{cases}$ 45. $\begin{cases} 2x + y = 19 \\ x - 2y = 12 \end{cases}$

46. $\begin{cases} 3x + 2y = 5 \\ 2x - 3y = 12 \end{cases}$ 47. $\begin{cases} 6x + 3y = 1 \\ y = -2x + 1 \end{cases}$

48. $\begin{cases} 4x - 3y = 253 \\ 13x + 2y = -12 \end{cases}$

49. $\begin{cases} x + 2y + 3z = 5 \\ y + 11z = 21 \\ 5y + 9z = 13 \end{cases}$ 50. $\begin{cases} x + y - z = 12 \\ 2y - 3z = -7 \\ 3x + 3y - 7z = 0 \end{cases}$

APPLICATIONS

Section 1.1

51. *Endangered animals* The number of species of endangered animals, A, can be described by

$$A = 10.06x + 16.73$$

where x is the number of years past 1980. (*Source:* U.S. Fish and Wildlife Service)
(a) To what year does $x = 17$ correspond?
(b) What x-value corresponds to the year 2007?
(c) If this equation remains valid, in what year will the number of species of endangered animals reach 328?

52. *Course grades* In a certain course, grades are based on three tests worth 100 points each, three quizzes worth 50 points each, and a final exam worth 200 points. A student has test grades of 91, 82, and 88, and quiz grades of 50, 42, and 42. What is the lowest percent the student can get on the final and still earn an A (90% or more of the total points) in the course?

53. *Cost analysis* The owner of a small construction business needs a new truck. He can buy a diesel truck for $38,000 and it will cost him $0.24 per mile to operate.

He can buy a gas engine truck for $35,600 and it will cost him $0.30 per mile to operate. Find the number of miles he must drive before the costs are equal. If he normally keeps a truck for 5 years, which is the better buy?

Section 1.2

54. *Heart disease risk* The Multiple Risk Factor Intervention Trial (MRFIT) used data from 356,222 men aged 35 to 57 to investigate the relationship between serum cholesterol and coronary heart disease (CHD) risk. Figure 1.53 shows the graph of the relationship of CHD risk and cholesterol, where a risk of 1 is assigned to 200 mg/dl of serum cholesterol and where the CHD risk is 4 times as high when serum cholesterol is 300 mg/dl.

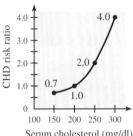

Figure 1.53   Serum cholesterol (mg/dl)

(a) Does this graph indicate that the CHD risk is a function of the serum cholesterol?
(b) Is the relationship a linear function?
(c) If CHD risk is a function f of serum cholesterol, what is $f(300)$?

55. *Mortgage loans* When a couple purchases a home, one of the first questions they face deals with the relationship between the amount borrowed and the monthly payment. In particular, if a bank offers 25-year loans at 7% interest, then the data in the following table would apply.

Amount Borrowed	Monthly Payment
$40,000	$282.72
50,000	353.39
60,000	424.07
70,000	494.75
80,000	565.44
90,000	636.11
100,000	706.78

Source: Comprehensive Mortgage Payment Tables, Publication No. 492, Financial Publishing Co., Boston

Assume that the monthly payment P is a function of the amount borrowed A (in thousands) and is denoted by $P = f(A)$, and answer the following.
(a) Find $f(80)$.
(b) Write a sentence that explains the meaning of $f(70) = 494.75$.

56. *Profit*   Suppose that the profit from the production and sale of x units of a product is given by

$$P(x) = 180x - \frac{x^2}{100} - 200$$

In addition, suppose that for a certain month, the number of units produced on day t of the month is

$$x = q(t) = 1000 + 10t$$

(a) Find $(P \circ q)(t)$ to express the profit as a function of the day of the month.
(b) Find the number of units produced, and the profit, on the fifteenth day of the month.

57. *Fish species growth*   For many species of fish, the weight W is a function of the length L that can be expressed by

$$W = W(L) = kL^3 \qquad k = \text{constant}$$

Suppose that for a particular species $k = 0.02$ and that for this species, the length (in cm) is a function of the number of years t the fish has been alive and that this function is given by

$$L = L(t) = 50 - \frac{(t - 20)^2}{10} \qquad 0 \leq t \leq 20$$

Find $(W \circ L)(t)$ in order to express W as a function of the age t of the fish.

Section 1.3

58. *Distance to a thunderstorm*   The distance d (in miles) to a thunderstorm is given by

$$d = \frac{t}{4.8}$$

where t is the number of seconds that elapse between seeing the lightning and hearing the thunder.
(a) Graph this function for $0 \leq t \leq 20$.
(b) The point $(9.6, 2)$ satisfies the equation. Explain its meaning.

59. *Body-heat loss*   Body-heat loss due to convection depends on a number of factors. If H_c is body-heat loss due to convection, A_c is the exposed surface area of the body, $T_s - T_a$ is skin temperature minus air temperature, and K_c is the convection coefficient (determined by air velocity and so on), then we have

$$H_c = K_c A_c (T_s - T_a)$$

When $K_c = 1$, $A_c = 1$, and $T_s = 90$, the equation is

$$H_c = 90 - T_a$$

Sketch the graph.

60. *Profit*   A company charting its profits notices that the relationship between the number of units sold, x, and the profit, P, is linear.
(a) If 200 units sold results in $3100 profit and 250 units sold results in $6000 profit, write the profit function for this company.
(b) Interpret the slope from part (a) as a rate of change.

61. *Health care costs*   The total cost of health care in the United States as a percent of the gross domestic product (GDP) can be described by

$$p = 0.219x + 2.75$$

where p is the percent of the GDP and x is the number of years past 1950. (*Source:* U.S. Department of Health and Human Services)
(a) Is p a linear function of x?
(b) Find the slope and p-intercept of this function.
(c) Write a sentence that interprets the p-intercept.
(d) Write a sentence that interprets the slope of this function as a rate of change.

62. *Temperature*   Write the equation of the linear relationship between temperature in Celsius (C) and Fahrenheit (F) if water freezes at $0°C$ and $32°F$ and boils at $100°C$ and $212°F$.

Section 1.4

63. *Photosynthesis*   The amount y of photosynthesis that takes place in a certain plant depends on the intensity x of the light present, according to

$$y = 120x^2 - 20x^3 \qquad \text{for } x \geq 0$$

(a) Graph this function with a graphing utility. (Use y-min $= -100$ and y-max $= 700$.)
(b) The model is valid only when $f(x) \geq 0$ (that is, on or above the x-axis). For what x-values is this true?

64. *Flow rates of water*   The speed at which water travels in a pipe can be measured by directing the flow through an elbow and measuring the height to which it spurts out the top. If the elbow height is 10 cm, the equation relating the height h (in centimeters) of the water above the elbow and its velocity v (in centimeters per second) is given by

$$v^2 = 1960(h + 10)$$

(a) Solve this equation for h and graph the result, using the velocity as the independent variable.
(b) If the velocity is 210 cm/sec, what is the height of the water above the elbow?

Section 1.5

65. *Investment mix*   A retired couple has $150,000 to invest and wants to earn $15,000 per year in interest. The safer investment yields 9.5%, but they can supplement their earnings by investing some of their money at 11%. How much should they invest at each rate to earn $15,000 per year?

66. *Botany*   A botanist has a 20% solution and a 70% solution of an insecticide. How much of each must be used to make 4.0 liters of a 35% solution?

Section 1.6

67. *Supply and demand*   A certain product has supply and demand functions $p = 4q + 5$ and $p = -2q + 81$, respectively.
   (a) If the price is $53, how many units are supplied and how many are demanded?
   (b) Does this give a shortfall or a surplus?
   (c) Is the price likely to increase from $53 or decrease from it?

68. *Market analysis*   Of the equations $p + 6q = 420$ and $p = 6q + 60$, one is the supply function for a product and one is the demand function for that product.
   (a) Graph these equations on the same set of axes.
   (b) Label the supply function and the demand function.
   (c) Find the market equilibrium point.

69. *Cost, revenue, and profit*   The total cost and total revenue for a certain product are given by the following:
$$C(x) = 38.80x + 4500$$
$$R(x) = 61.30x$$
   (a) Find the marginal cost.
   (b) Find the marginal revenue.
   (c) Find the marginal profit.
   (d) Find the number of units required to break even.

70. *Cost, revenue, and profit*   A certain commodity has the following costs for a period.

Fixed costs:	$1500
Variable costs:	$22 per unit

The commodity is sold for $52 per unit.
   (a) What is the total cost function?
   (b) What is the total revenue function?
   (c) What is the profit function?
   (d) What is the marginal cost?
   (e) What is the marginal revenue?
   (f) What is the marginal profit?
   (g) What is the break-even point?

71. *Market analysis*   The supply function and the demand function for a product are linear and are determined by the tables that follow. Find the quantity and price that will give market equilibrium.

Supply Function		Demand Function	
Price	Quantity	Price	Quantity
100	200	200	200
200	400	100	400
300	600	0	600

72. *Market analysis*   Suppose that for a certain product the supply and demand functions prior to any taxation are

Supply: $p = \dfrac{q}{10} + 8$

Demand: $10p + q = 1500$

If a tax of $2 per item is levied on the supplier and is passed on to the consumer as a price increase, find the market equilibrium after the tax is levied.

1 Chapter Test

In Problems 1–3, solve the equations.

1. $4x - 3 = \dfrac{x}{2} + 6$

2. $\dfrac{3}{x} + 4 = \dfrac{4x}{x + 1}$ 3. $\dfrac{3x - 1}{4x - 9} = \dfrac{5}{7}$

4. For $f(x) = 7 + 5x - 2x^2$, find and simplify $\dfrac{f(x + h) - f(x)}{h}$.

5. Solve $1 + \dfrac{2}{3}t \le 3t + 22$ and graph the solution.

In Problems 6 and 7, find the intercepts and graph the functions.

6. $5x - 6y = 30$ 7. $7x + 5y = 21$
8. Consider the function $f(x) = \sqrt{4x + 16}$.
   (a) Find the domain and range.
   (b) Find $f(3)$.
   (c) Find the y-coordinate on the graph of $y = f(x)$ when $x = 5$.

9. Write the equation of the line passing through $(-1, 2)$ and $(3, -4)$. Write your answer in slope-intercept form.

10. Find the slope and the *y*-intercept of the graph of $5x + 4y = 15$.

11. Write the equation of the line through $(-3, -1)$ that
    (a) has undefined slope.
    (b) is perpendicular to $x = 4y - 8$.

12. Which of the following relations [(a), (b), (c)] are functions? Explain.

    (a)

    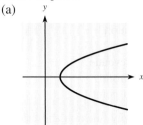

    (b)

    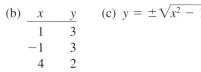

x	y
1	3
−1	3
4	2

    (c) $y = \pm\sqrt{x^2 - 1}$

13. Solve the system

$$\begin{cases} 3x + 2y = -2 \\ 4x + 5y = 2 \end{cases}$$

14. Given $f(x) = 5x^2 - 3x$ and $g(x) = x + 1$, find
    (a) $(fg)(x)$
    (b) $g(g(x))$
    (c) $(f \circ g)(x)$

15. The total cost function for a product is $C(x) = 30x + 1200$, and the total revenue is $R(x) = 38x$, where *x* is the number of units produced and sold.

    (a) Find the marginal cost.
    (b) Find the profit function.
    (c) Find the number of units that gives break-even.
    (d) Find the marginal profit and explain what it means.

16. The selling price for each item of a product is $50, and the total cost is given by $C(x) = 10x + 18,000$, where *x* is the number of items.

    (a) Write the revenue function.
    (b) Find $C(100)$ and write a sentence that explains its meaning.
    (c) Find the number of units that gives break-even.

17. The supply function for a product is $p = 5q + 1500$ and the demand function is $p = -3q + 3100$. Find the quantity and price that give market equilibrium.

18. A building is depreciated by its owner, with the value *y* of the building after *x* months given by $y = 360,000 - 1500x$.

    (a) Find the *y*-intercept of the graph of this function, and explain what it means.
    (b) Find the slope of the graph and tell what it means.

19. An airline has 360 seats on a plane for one of its flights. If 90% of the people making reservations actually buy a ticket, how many reservations should the airline accept to be confident that it will sell 360 tickets?

20. Amanda plans to invest $20,000, part of it at a 9% interest rate and part of it in a safer fund that pays 6%. How much should be invested in each fund to yield an annual return of $1560?

Extended Applications & Group Projects

I. Hospital Administration

Southwest Hospital has an operating room used only for eye surgery. The annual cost of rent, heat, and electricity for the operating room and its equipment is $180,000, and the annual salaries of the people who staff this room total $270,000.

Each surgery performed requires the use of $380 worth of medical supplies and drugs. To promote goodwill, every patient receives a bouquet of flowers the day after surgery. In addition, one-quarter of the patients require dark glasses, which the hospital provides free of charge. It costs the hospital $15 for each bouquet of flowers and $20 for each pair of glasses.

The hospital receives a payment of $1000 for each eye operation performed.

1. Identify the revenue per case and the annual fixed and variable costs for running the operating room.
2. How many eye operations must the hospital perform each year in order to break even?
3. Southwest Hospital currently averages 70 eye operations per month. One of the nurses has just learned about a machine that would reduce by $50 per patient the amount of medical supplies needed. It can be leased for $50,000 annually. Keeping in mind the financial cost and benefits, advise the hospital on whether it should lease this machine.
4. An advertising agency has proposed to the hospital's president that she spend $10,000 per month on television and radio advertising to persuade people that Southwest Hospital is the best place to have any eye surgery performed. Advertising account executives estimate that such publicity would increase business by 40 operations per month. If they are correct and if this increase is not big enough to affect fixed costs, what impact would this advertising have on the hospital's profits?
5. In case the advertising agency is being overly optimistic, how many extra operations per month are needed to cover the cost of the proposed ads?
6. If the ad campaign is approved and subsequently meets its projections, should the hospital review its decision about the machine discussed in Question 3?

II. Fundraising

At most colleges and universities, student organizations conduct fundraising activities, such as selling T-shirts or candy. If a club or organization decided to sell coupons for submarine sandwiches, it would want to find the best deal, based on the amount the club thought it could sell and how much profit it would make. In this project, you'll try to discover the best deal for this type of fundraiser conducted in your area.

1. Contact at least two different sub shops in your area from among local sub shops, a national chain, or a regional or national convenience store chain. From each contact, find the details of selling sub sandwich coupons as a fundraiser. In particular, determine the following for each sub shop, and present your findings in a chart.
   (a) The selling price for each coupon and its value to the coupon holder, including any expiration date
   (b) Your cost for each coupon sold and for each one returned
   (c) The total number of coupons provided
   (d) The duration of the sale
2. For each sub shop, determine a club's total revenue and total costs as linear functions of the number of coupons sold.
3. Form the profit function for each sub shop, and graph the profit functions together. For each function, determine the break-even point. Find the marginal profit for each function, and interpret its meaning.
4. The shop with the best deal will be the one whose coupons will provide the maximum profit for a club.
   (a) For each shop, determine a sales estimate that is based on location, local popularity, and customer value.
   (b) Use each estimate from (a) with that shop's profit function to determine which shop gives the maximum profit.

Fully explain and support your claims; make and justify a recommendation.

2

Quadratic and Other Special Functions

In Chapter 1 we discussed functions in general and linear functions in particular. In this chapter we will discuss quadratic functions and their applications, and we will also discuss other types of functions, including identity, constant, power, absolute value, piecewise defined, and reciprocal functions. Graphs of polynomial and rational functions will also be introduced; they will be studied in detail in Chapter 10. We will pay particular attention to quadratic equations and to quadratic functions, and we will see that cost, revenue, profit, supply, and demand are sometimes modeled by quadratic functions. We also include, in an optional section on modeling, the creation of functions that approximately fit real data points.

The topics and some of the applications discussed in this chapter follow.

Sections	Applications
2.1 Quadratic Equations Factoring methods The quadratic formula	*Social Security benefits, falling objects*
2.2 Quadratic Functions: Parabolas Vertices of parabolas Zeros	*Profit maximization, maximizing revenue*
2.3 Business Applications of Quadratic Functions	*Supply, demand, market equilibrium, break-even points, profit maximization*
2.4 Special Functions and Graphs Basic functions Polynomial functions Rational functions Piecewise defined functions	*Selling price, residential electric costs*
2.5 Modeling: Fitting Curves to Data Linear regression	*Internet access, health care costs, expected life span, consumer price index*

Chapter Warm-up

Prerequisite Problem Type	For Section	Answer	Section for Review
Find $b^2 - 4ac$ if (a) $a = 1, b = 2, c = -2$ (b) $a = -2, b = 3, c = -1$ (c) $a = -3, b = -3, c = -2$	2.1 2.2 2.3	(a) 12 (b) 1 (c) -15	0.2 Signed numbers
(a) Factor $6x^2 - x - 2$. (b) Factor $6x^2 - 9x$.	2.1 2.2 2.3	(a) $(3x - 2)(2x + 1)$ (b) $3x(2x - 3)$	0.6 Factoring
Is $\dfrac{1 + \sqrt{-3}}{2}$ a real number?	2.1	No	0.4 Radicals
(a) Find the y-intercept of $\quad y = x^2 - 6x + 8$. (b) Find the x-intercept of $\quad y = 2x - 3$.	2.2 2.3	(a) 8 (b) $\frac{3}{2}$	1.2 x- and y-intercepts
If $f(x) = -x^2 + 4x$, find $f(2)$. Find the domain of $\quad f(x) = \dfrac{12x + 8}{3x - 9}$.	2.2 2.3 2.4	4 All $x \neq 3$	1.2 Functions
Assume revenue is $R(x) = 500x - 2x^2$ and cost is $C(x) = 3600 + 100x + 2x^2$. (a) Find the profit function $P(x)$. (b) Find $P(50)$.	2.3	 (a) $P(x) = -3600 +$ $\quad 400x - 4x^2$ (b) $P(50) = 6400$	1.6 Cost, revenue, and profit

2.1
OBJECTIVES

- To solve quadratic equations with factoring methods
- To solve quadratic equations with the quadratic formula

Quadratic Equations

◗ Application Preview

With fewer workers, future income from payroll taxes used to fund Social Security benefits won't keep pace with scheduled benefits. The function

$$B = -1.056785714t^2 + 8.259285714t + 74.07142857$$

describes the Social Security trust fund balance B, in billions of dollars, where t is the number of years past the year 2000 (*Source:* Social Security Administration). For planning purposes, it is important to know when the trust fund balance will be 0. That is, for what t-value does

$$0 = -1.056785714t^2 + 8.259285714t + 74.07142857?$$

(See Example 7 for the solution.)

In this section, we learn how to solve equations of this type, using factoring methods and using the quadratic formula.

Factoring Methods

It is important for us to know how to solve equations, such as the equation in the Application Preview. This was true for linear equations and is again true for quadratic equations.

A **quadratic equation** in one variable is an equation that can be written in the *general form*

$$ax^2 + bx + c = 0 \qquad (a \neq 0)$$

where a, b, and c represent constants. For example, the equations

$$3x^2 + 4x + 1 = 0 \qquad \text{and} \qquad 2x^2 + 1 = x^2 - x$$

are quadratic equations; the first of these is in general form, and the second can easily be rewritten in general form.

When we solve quadratic equations, we will be interested only in real number solutions and will consider two methods of solution: factoring and the quadratic formula. We will discuss solving quadratic equations by factoring first. (For a review of factoring, see Section 0.6, "Factoring.")

Solution by factoring is based on the following property of the real numbers.

Zero Product Property

For real numbers a and b, $ab = 0$ if and only if $a = 0$ or $b = 0$ or both.

Hence, to solve by factoring, we must first write the equation with zero on one side.

● **EXAMPLE 1** *Solving Quadratic Equations*

Solve: (a) $6x^2 + 3x = 4x + 2$ \qquad (b) $6x^2 = 9x$

Solution

(a)
$$6x^2 + 3x = 4x + 2$$
$$6x^2 - x - 2 = 0 \qquad \text{Proper form for factoring}$$
$$(3x - 2)(2x + 1) = 0 \qquad \text{Factored}$$
$$3x - 2 = 0 \quad \text{or} \quad 2x + 1 = 0 \qquad \text{Factors equal to zero}$$
$$3x = 2 \qquad\qquad 2x = -1$$
$$x = \frac{2}{3} \qquad\qquad x = -\frac{1}{2} \qquad \text{Solutions}$$

We now check that these values are, in fact, solutions to our original equation.

$$6\left(\frac{2}{3}\right)^2 + 3\left(\frac{2}{3}\right) \stackrel{?}{=} 4\left(\frac{2}{3}\right) + 2 \qquad 6\left(-\frac{1}{2}\right)^2 + 3\left(-\frac{1}{2}\right) \stackrel{?}{=} 4\left(-\frac{1}{2}\right) + 2$$

$$\frac{14}{3} = \frac{14}{3} \; ✔ \qquad\qquad\qquad 0 = 0 \; ✔$$

(b)
$$6x^2 = 9x$$
$$6x^2 - 9x = 0$$
$$3x(2x - 3) = 0$$
$$3x = 0 \quad \text{or} \quad 2x - 3 = 0$$
$$x = 0 \qquad\qquad 2x = 3$$
$$x = \frac{3}{2}$$

Check: $6(0)^2 = 9(0)$ ✔ $\qquad 6\left(\frac{3}{2}\right)^2 = 9\left(\frac{3}{2}\right)$ ✔

Thus the solutions are $x = 0$ and $x = \frac{3}{2}$.

Note that in Example 1(b) it is tempting to divide both sides of the equation by x, but this is incorrect because this results in the loss of the solution $x = 0$. Never divide both sides of an equation by an expression containing the variable.

● EXAMPLE 2 *Solving by Factoring*

Solve: (a) $(y - 3)(y + 2) = -4$ for y (b) $\dfrac{x + 1}{3x + 6} = \dfrac{3}{x} + \dfrac{2x + 6}{x(3x + 6)}$

Solution

(a) Note that the left side of the equation is factored, but the right member is not 0. Therefore, we must multiply the factors before we can rewrite the equation in general form.

$$(y - 3)(y + 2) = -4$$
$$y^2 - y - 6 = -4$$
$$y^2 - y - 2 = 0$$
$$(y - 2)(y + 1) = 0$$
$$y - 2 = 0 \quad \text{or} \quad y + 1 = 0$$
$$y = 2 \qquad\qquad y = -1$$

Check: $(2 - 3)(2 + 2) = -4$ ✔ $\qquad (-1 - 3)(-1 + 2) = -4$ ✔

(b) The LCD of all fractions is $x(3x + 6)$. Multiplying both sides of the equation by this LCD gives a quadratic equation that is equivalent to the original equation for $x \neq 0$ and $x \neq -2$. (The original equation is undefined for these values.)

$$\frac{x + 1}{3x + 6} = \frac{3}{x} + \frac{2x + 6}{x(3x + 6)}$$

$$x(3x + 6)\frac{x + 1}{3x + 6} = x(3x + 6)\left(\frac{3}{x} + \frac{2x + 6}{x(3x + 6)}\right) \qquad \text{Multiply both sides by } x(3x + 6).$$

$$x(x + 1) = 3(3x + 6) + (2x + 6)$$

$$x^2 + x = 9x + 18 + 2x + 6$$

$$x^2 - 10x - 24 = 0$$

$$(x - 12)(x + 2) = 0$$

$$x - 12 = 0 \text{ or } x + 2 = 0$$

$$x = 12 \text{ or } x = -2$$

Checking $x = 12$ and $x = -2$ in the original equation, we see that $x = -2$ does not check because it makes the denominator equal to zero. Hence the only solution is $x = 12$. ✔

● *Checkpoint*

1. The factoring method for solving a quadratic equation is based on the _____ product property. Hence, in order for us to solve a quadratic equation by factoring, one side of the equation must equal _____.
2. Solve the following equations by factoring.
   (a) $x^2 - 19x = 20$ (b) $2x^2 = 6x$

● **EXAMPLE 3** *Falling Object*

A tennis ball is thrown into a swimming pool from the top of a tall hotel. The height of the ball from the pool is modeled by $D(t) = -16t^2 - 4t + 300$ feet, where t is the time, in seconds, after the ball was thrown. How long after the ball was thrown was it 144 feet above the pool?

Solution

To find the number of seconds until the ball is 144 feet above the pool, we solve the equation

$$144 = -16t^2 - 4t + 300$$

for t. The solution follows.

$$144 = -16t^2 - 4t + 300$$

$$0 = -16t^2 - 4t + 156$$

$$0 = 4t^2 + t - 39$$

$$0 = (t - 3)(4t + 13)$$

$$t - 3 = 0 \text{ or } 4t + 13 = 0$$

$$t = 3 \text{ or } t = -13/4$$

This indicates that the ball will be 144 feet above the pool 3 seconds after it is thrown. The negative value for t has no meaning in this application.

The Quadratic Formula

Factoring does not lend itself easily to solving quadratic equations such as

$$x^2 - 5 = 0$$

However, we can solve this equation by writing

$$x^2 = 5$$
$$x = \pm\sqrt{5}$$

In general, we can solve quadratic equations of the form $x^2 = C$ (no x-term) by taking the square root of both sides.

Square Root Property

The solution of $x^2 = C$ is

$$x = \pm\sqrt{C}$$

This property also can be used to solve equations such as those in the following example.

● **EXAMPLE 4** *Square Root Method*

Solve the following equations.

(a) $4x^2 = 5$ (b) $(3x - 4)^2 = 9$

Solution
We can use the square root property for both parts.

(a) $4x^2 = 5$ is equivalent to $x^2 = \frac{5}{4}$. Thus,

$$x = \pm\sqrt{\frac{5}{4}} = \pm\frac{\sqrt{5}}{\sqrt{4}} = \pm\frac{\sqrt{5}}{2}$$

(b) $(3x - 4)^2 = 9$ is equivalent to $3x - 4 = \pm\sqrt{9}$. Thus

$3x - 4 = 3$	$3x - 4 = -3$
$3x = 7$	$3x = 1$
$x = \dfrac{7}{3}$	$x = \dfrac{1}{3}$

The solution of the general quadratic equation $ax^2 + bx + c = 0$, where $a \neq 0$, is called the **quadratic formula.** It can be derived by using the square root property, as follows:

$$ax^2 + bx + c = 0 \qquad \text{Standard form}$$
$$ax^2 + bx = -c \qquad \text{Subtract } c \text{ from both sides.}$$
$$x^2 + \frac{b}{a}x = -\frac{c}{a} \qquad \text{Divide both sides by } a.$$

We would like to make the left side of the last equation a perfect square of the form $(x + k)^2 = x^2 + 2kx + k^2$. If we let $2k = b/a$, then $k = b/(2a)$ and $k^2 = b^2/(4a^2)$. Hence we continue as follows:

$$x^2 + \frac{b}{a}x + \frac{b^2}{4a^2} = \frac{b^2}{4a^2} - \frac{c}{a} \qquad \text{Add } \frac{b^2}{4a^2} \text{ to both sides.}$$
$$\left(x + \frac{b}{2a}\right)^2 = \frac{b^2 - 4ac}{4a^2} \qquad \text{Simplify.}$$
$$x + \frac{b}{2a} = \pm\sqrt{\frac{b^2 - 4ac}{4a^2}} \qquad \text{Square root property}$$
$$x = \frac{-b}{2a} \pm \frac{\sqrt{b^2 - 4ac}}{2|a|} \qquad \text{Solve for } x.$$

Because $\pm 2|a|$ represents the same numbers as $\pm 2a$, we obtain the following:

Quadratic Formula

If $ax^2 + bx + c = 0$, where $a \neq 0$, then

$$x = \frac{-b \pm \sqrt{b^2 - 4ac}}{2a}$$

We may use the quadratic formula to solve all quadratic equations, but especially those in which factorization is difficult or impossible to see. The proper identification of values for a, b, and c to be substituted into the formula requires that the equation be in general form.

● **EXAMPLE 5** *Quadratic Formula*

Use the quadratic formula to solve $2x^2 - 3x - 6 = 0$ for x.

Solution
The equation is already in general form, with $a = 2$, $b = -3$, and $c = -6$. Hence,

$$x = \frac{-b \pm \sqrt{b^2 - 4ac}}{2a}$$

$$= \frac{-(-3) \pm \sqrt{(-3)^2 - 4(2)(-6)}}{2(2)}$$

$$= \frac{3 \pm \sqrt{9 + 48}}{4}$$

$$= \frac{3 \pm \sqrt{57}}{4}$$

Thus the solutions are

$$x = \frac{3 + \sqrt{57}}{4} \quad \text{and} \quad x = \frac{3 - \sqrt{57}}{4}$$

● **EXAMPLE 6** *Nonreal Solution*

Using the quadratic formula, find the (real) solutions to $x^2 = x - 1$.

Solution
We must rewrite the equation in general form before we can determine the values of a, b, and c.

$$x^2 = x - 1$$

$$x^2 - x + 1 = 0 \qquad \text{Note: } a = 1, b = -1, \text{ and } c = 1.$$

$$x = \frac{-(-1) \pm \sqrt{(-1)^2 - 4(1)(1)}}{2(1)}$$

$$x = \frac{1 \pm \sqrt{1 - 4}}{2}$$

$$x = \frac{1 \pm \sqrt{-3}}{2}$$

Because $\sqrt{-3}$ is not a real number, the values of x are not real. Hence there are no real solutions to the given equation.

In Example 6 there were no real solutions to the quadratic equation because the radicand of the quadratic formula was negative. In general, when solving a quadratic equation, we can use the sign of the radicand in the quadratic formula—that is, the sign of $b^2 - 4ac$—to determine how many real solutions there are. Thus we refer to $b^2 - 4ac$ as the **quadratic discriminant.**

Quadratic Discriminant

Given $ax^2 + bx + c = 0$ and $a \neq 0$,

If $b^2 - 4ac > 0$, the equation has 2 distinct real solutions.

If $b^2 - 4ac = 0$, the equation has exactly 1 real solution.

If $b^2 - 4ac < 0$, the equation has no real solutions.

The quadratic formula is especially useful when the coefficients of a quadratic equation are decimal values that make factorization impractical. This occurs in many applied problems.

◗ EXAMPLE 7 *Social Security Trust Fund* (Application Preview)

The function

$$B = -1.056785714t^2 + 8.259285714t + 74.07142857$$

describes the Social Security trust fund balance B, in billions of dollars, where t is the number of years past the year 2000 (*Source:* Social Security Administration). For planning purposes, it is important to know when the trust fund balance will be 0. To find when this occurs, solve

$$0 = -1.056785714t^2 + 8.259285714t + 74.07142857$$

Solution

We use the quadratic formula with

$$a = -1.056785714, \quad b = 8.259285714, \quad c = 74.07142857$$

Thus

$$
\begin{aligned}
t &= \frac{-b \pm \sqrt{b^2 - 4ac}}{2a} \\
&= \frac{-8.259285714 \pm \sqrt{(8.259285714)^2 - 4(-1.056785714)(74.07142857)}}{2(-1.056785714)} \\
&= \frac{-8.259285714 \pm \sqrt{381.3263106}}{-2.113571428} \\
&\approx \frac{-8.259285714 \pm 19.52757821}{-2.113571428}
\end{aligned}
$$

We are interested only in the positive solution, so

$$t = \frac{-8.259285714 - 19.52757821}{-2.113571428} \approx 13.14687393$$

Therefore, the trust fund balance is projected to reach zero slightly more than 13 years after 2000, during the year 2014.

Graphing Utilities We can use graphing utilities to determine (or approximate) the solutions of quadratic equations. The solutions can be found by using commands (such as ZERO, ROOT, or SOLVER) or programs. TRACE can be used to approximate where the graph of the function $y = f(x)$ intersects the x-axis. This **x-intercept** is also the value of x that makes the function zero, so it is called a **zero of the function.** Because this value of x makes $f(x) = 0$, it is also a **solution** (or **root**) of the equation $0 = f(x)$. ■

● **EXAMPLE 8** *Solving with Technology*

Find the solutions of the equation $0 = -7x^2 + 16x - 4$.

Solution

To solve the equation, we find the x-intercepts of the graph of $y = -7x^2 + 16x - 4$. We graph this function and use TRACE to approximate an x-value that gives $y = 0$. Tracing to or evaluating at $x = 2$ gives $y = 0$, so $x = 2$ is an x-intercept. See Figure 2.1(a).

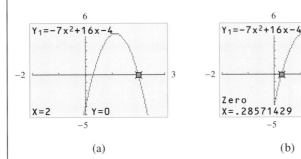

Figure 2.1   (a)   (b)

Tracing near $x = 0.3$ gives a value of y near 0, and an (approximate) x-intercept near 0.3 can be found more accurately by using ZERO (or ROOT). This gives $y = 0$ at $x = 0.28571429$ (approximately). See Figure 2.1(b).

Solving $f(x) = 0$ algebraically gives solutions $x = 2$ and $x = \frac{2}{7}$.

$$0 = (-7x + 2)(x - 2)$$
$$-7x + 2 = 0 \quad \text{or} \quad x - 2 = 0$$
$$x = \frac{2}{7} \qquad\qquad x = 2$$

Spreadsheet Note A spreadsheet such as Excel can also be used to solve a quadratic equation. However, because a quadratic equation $f(x) = 0$ may have two solutions, it is useful to graph the quadratic function $y = f(x)$ before starting Goal Seek to solve the equation with Excel. For example, to solve the equation in Example 8, $0 = -7x^2 + 16x - 4$, we use the graph of $y = f(x)$ to approximate the x-intercepts and thus approximate the solutions to the equation. Graphing the equation shows that the x-intercepts are approximately 2 and 0 (see the graph in Figure 2.1).

Entering x, the function, and values of x near 2 and 0 prepares us to use Goal Seek to find the two solutions.

	A	B
1	x	$-7x^2 + 16x - 4$
2	2	$= -7*A2^2 + 16*A2 - 4$
3	0	$= -7*A3^2 + 16*A3 - 4$

Entering 2 in A2 confirms that $x = 2$ is a solution of the equation because B2 is 0, and using Goal Seek with B3 set to 0 gives the second solution, approximately as 0.285688. This solution is an approximation of the exact solution $x = 2/7$, accurate to four decimal places.

	A	B
1	x	$-7x^2 + 16x - 4$
2	2	0
3	0.285688	-0.00032

● **Checkpoint**

3. The statement of the quadratic formula says that if _____ and $a \neq 0$, then $x =$ _____.

4. Solve $2x^2 - 5x = 9$ using the quadratic formula.

● **Checkpoint Solutions**

1. Zero; zero

2. (a)
$$x^2 - 19x - 20 = 0$$
$$(x - 20)(x + 1) = 0$$

$x - 20 = 0$	$x + 1 = 0$
$x = 20$	$x = -1$

(b)
$$2x^2 - 6x = 0$$
$$2x(x - 3) = 0$$

$2x = 0$	$x - 3 = 0$
$x = 0$	$x = 3$

3. If $ax^2 + bx + c = 0$ with $a \neq 0$, then

$$x = \frac{-b \pm \sqrt{b^2 - 4ac}}{2a}$$

4. $2x^2 - 5x - 9 = 0$, so $a = 2, b = -5, c = -9$

$$x = \frac{-b \pm \sqrt{b^2 - 4ac}}{2a} = \frac{5 \pm \sqrt{25 - 4(2)(-9)}}{4}$$

$$= \frac{5 \pm \sqrt{25 + 72}}{4}$$

$$= \frac{5 \pm \sqrt{97}}{4}$$

$$x = \frac{5 + \sqrt{97}}{4} \approx 3.712 \quad \text{or} \quad x = \frac{5 - \sqrt{97}}{4} \approx -1.212$$

2.1 Exercises

In Problems 1–4, write the equations in general form.

1. $2x^2 + 3 = x^2 - 2x + 4$
2. $x^2 - 2x + 5 = 2 - 2x^2$
3. $(y + 1)(y + 2) = 4$
4. $(z - 1)(z - 3) = 1$

In Problems 5–16, solve each equation by factoring.

5. $9 - 4x^2 = 0$
6. $25x^2 - 16 = 0$
7. $x = x^2$
8. $t^2 - 4t = 3t^2$
9. $x^2 + 5x = 21 + x$
10. $x^2 + 17x = 8x - 14$
11. $4t^2 - 4t + 1 = 0$
12. $49z^2 + 14z + 1 = 0$

13. $\dfrac{w^2}{8} - \dfrac{w}{2} - 4 = 0$ 14. $\dfrac{y^2}{2} - \dfrac{11}{6}y + 1 = 0$

15. $(x - 1)(x + 5) = 7$ 16. $(x - 3)(1 - x) = 1$

In Problems 17–20, multiply both sides of the equation by the LCD, and solve the resulting quadratic equation.

17. $x + \dfrac{8}{x} = 9$ 18. $\dfrac{x}{x - 2} - 1 = \dfrac{3}{x + 1}$

19. $\dfrac{x}{x - 1} = 2x + \dfrac{1}{x - 1}$ 20. $\dfrac{5}{z + 4} - \dfrac{3}{z - 2} = 4$

In Problems 21–28, solve each equation by using the quadratic formula. Give real answers (a) exactly and (b) rounded to two decimal places.

21. $x^2 - 4x - 4 = 0$ 22. $x^2 - 6x + 7 = 0$

23. $2w^2 + w + 1 = 0$ 24. $z^2 + 2z + 4 = 0$

25. $16z^2 + 16z - 21 = 0$ 26. $10y^2 - y - 65 = 0$

27. $5x^2 = 2x + 6$ 28. $3x^2 = -6x - 2$

In Problems 29–34, find the exact real solutions to each equation, if they exist.

29. $y^2 = 7$ 30. $z^2 = 12$

31. $y^2 + 9 = 0$ 32. $z^2 + 121 = 0$

33. $(x + 4)^2 = 25$ 34. $(x + 1)^2 = 2$

In Problems 35 and 36, solve using quadratic methods.

35. $(x + 8)^2 + 3(x + 8) + 2 = 0$

36. $(s - 2)^2 - 5(s - 2) - 24 = 0$

In Problems 37–42, solve each equation by using a graphing utility.

37. $21x + 70 - 7x^2 = 0$ 38. $3x^2 - 11x + 6 = 0$

39. $300 - 2x - 0.01x^2 = 0$

40. $-9.6 + 2x - 0.1x^2 = 0$

41. $25.6x^2 - 16.1x - 1.1 = 0$

42. $6.8z^2 - 4.9z - 2.6 = 0$

APPLICATIONS

43. *Profit*   If the profit from the sale of x units of a product is $P = 90x - 200 - x^2$, what level(s) of production will yield a profit of $1200?

44. *Profit*   If the profit from the sale of x units of a product is $P = 16x - 0.1x^2 - 100$, what level(s) of production will yield a profit of $180?

45. *Profit*   Suppose the profit from the sale of x units of a product is $P = 6400x - 18x^2 - 400$.
    (a) What level(s) of production will yield a profit of $61,800?
    (b) Can a profit of more than $61,800 be made?

46. *Profit*   Suppose the profit from the sale of x units of a product is $P = 50x - 300 - 0.01x^2$.
    (a) What level(s) of production will yield a profit of $250?
    (b) Can a profit of more than $250 be made?

47. *Flight of a ball*   If a ball is thrown upward at 96 feet per second from the top of a building that is 100 feet high, the height of the ball can be modeled by

$$S = 100 + 96t - 16t^2 \text{ feet}$$

where t is the number of seconds after the ball is thrown. How long after it is thrown is the height 100 feet?

48. *Falling object*   A tennis ball is thrown into a swimming pool from the top of a tall hotel. The height of the ball from the pool is modeled by

$$D(t) = -16t^2 - 10t + 300 \text{ feet}$$

where t is the time, in seconds, after the ball is thrown. How long after the ball is thrown is it 4 feet above the pool?

49. *Wind and pollution*   The amount of airborne particulate pollution p from a power plant depends on the wind speed s, among other things, with the relationship between p and s approximated by

$$p = 25 - 0.01s^2$$

    (a) Find the value(s) of s that will make $p = 0$.
    (b) What value of s makes sense in the context of this application? What does $p = 0$ mean in this application?

50. *Drug sensitivity*   The sensitivity S to a drug is related to the dosage size by

$$S = 100x - x^2$$

where x is the dosage size in milliliters.
    (a) What dosage(s) will yield 0 sensitivity?
    (b) Explain what your answer in part (a) might mean.

51. *Corvette acceleration*   The time t, in seconds, that it takes a 2005 Corvette to accelerate to x mph can be described by

$$t = 0.001(0.872x^2 + 8.308x + 711.311)$$

(*Source: Motor Trend*, September 2004). How fast is the Corvette going after 9.03 seconds? Give your answer to the nearest tenth.

52. *Social Security trust fund*   Social Security benefits are paid from a trust fund. As mentioned in the Application Preview, the trust fund balance, B, in billions of dollars, t years past the year 2000 is projected to be described by

$$B = -1.056785714t^2 + 8.259285714t + 74.07142857$$

(*Source:* Social Security Administration projections). Find in what year the trust fund balance is projected to be $1000 billion in the red—that is, when $B = -1000$.

53. *Personal taxes* Suppose that the percent of total personal income that is used to pay personal taxes is given by

$$y = 0.034x^2 - 0.044x + 12.642$$

where x is the number of years past 1990 (*Source:* Bureau of Economic Analysis, U.S. Department of Commerce).
(a) Verify that $x = 7$ is a solution of $14 = 0.034x^2 - 0.044x + 12.642$.
(b) Find the year or years when the percent of total personal income used to pay personal taxes is 14.

54. *Depth of a fissure* A fissure in the earth appeared after an earthquake. To measure its vertical depth, a stone was dropped into it, and the sound of the stone's impact was heard 3.9 seconds later. The distance (in feet) the stone fell is given by $s = 16t_1^2$, and the distance (in feet) the sound traveled is given by $s = 1090t_2$. In these equations, the distances traveled by the sound and the stone are the same, but their times are not. Using the fact that the total time is 3.9 seconds, find the depth of the fissure.

55. *Marijuana use* For the years from 1980 to 2000, the percent p of high school seniors who have tried marijuana can be considered as a function of time t according to

$$p = f(t) = 0.121t^2 - 4.180t + 79.620$$

where t is the number of years past 1980 (*Source:* National Institute on Drug Abuse). In what year after 1980 will the percent predicted by the function reach 44.4%, if this function remains valid?

56. *Projectile motion* Two projectiles are shot into the air over a lake. The paths of the projectiles are given by
(a) $y = -0.0013x^2 + x + 10$ and
(b) $y = -\dfrac{x^2}{81} + \dfrac{4}{3}x + 10$

where y is the height and x is the horizontal distance traveled. Determine which projectile travels farther by substituting $y = 0$ in each equation and finding x.

57. *Percent profit* The Ace Jewelry Store sold a necklace for $144. If the percent profit (based on cost) equals the cost of the necklace to the store, how much did the store pay for it? Use

$$P = \left(\frac{C}{100}\right) \cdot C$$

where P is profit and C is cost.

58. *Tourism spending* The global spending on travel and tourism (in billions of dollars) can be described by the equation

$$y = -1.572x^2 + 53.72x + 20.94$$

where x equals the number of years past 1985 (*Source:* World Tourism Organization). Find the year in which spending is projected to reach $479.9 billion.

59. *Internet use* An equation that models the number of users of the Internet is

$$y = 11.786x^2 - 142.214x + 493 \text{ million users}$$

where x is the number of years since 1990 (*Source: CyberAtlas*, 1999). If the pattern indicated by the model remains valid, when does this model predict there will be 812 million users?

60. *Velocity of blood* Because of friction from the walls of an artery, the velocity of a blood corpuscle in an artery is greatest at the center of the artery and decreases as the distance r from the center increases. The velocity of the blood in the artery can be modeled by the function

$$v = k(R^2 - r^2)$$

where R is the radius of the artery and k is a constant that is determined by the pressure, the viscosity of the blood, and the length of the artery. In the case where $k = 2$ and $R = 0.1$ centimeters, the velocity is $v = 2(0.01 - r^2)$ cm per sec.
(a) What distance r would give a velocity of 0.02 cm/sec?
(b) What distance r would give a velocity of 0.015 cm/sec?
(c) What distance r would give a velocity of 0 cm/sec? Where is the blood corpuscle?

61. *Body-heat loss* The model for body-heat loss depends on the coefficient of convection K, which depends on wind speed v according to the equation

$$K^2 = 16v + 4$$

where v is in miles per hour. Find the positive coefficient of convection when the wind speed is:
(a) 20 mph
(b) 60 mph
(c) What is the change in K for a change in speed from 20 mph to 60 mph?

2.2

OBJECTIVES

- To find the vertex of the graph of a quadratic function
- To determine whether a vertex is a maximum point or a minimum point
- To find the zeros of a quadratic function
- To graph quadratic functions

Quadratic Functions: Parabolas

❍ **Application Preview**

Because additional equipment, raw materials, and labor may cause variable costs of some products to increase dramatically as more units are produced, total cost functions are not always linear functions. For example, suppose that the total cost of producing a product is given by the equation

$$C = C(x) = 300x + 0.1x^2 + 1200$$

where x represents the number of units produced. That is, the cost of this product is represented by a **second-degree function,** or a **quadratic function.** We can find the cost of producing 10 units by evaluating

$$C(10) = 300(10) + 0.1(10)^2 + 1200 = 4210$$

If the revenue function for this product is

$$R = R(x) = 600x$$

then the profit is also a quadratic function:

$$P = R - C = 600x - (300x + 0.1x^2 + 1200)$$
$$P = -0.1x^2 + 300x - 1200$$

We can use this function to find how many units give maximum profit and what that profit is. (See Example 2.) In this section we will describe ways to find the maximum point, or minimum point, for a quadratic function.

Parabolas

In Chapter 1, we studied functions of the form $y = ax + b$, called linear (or first-degree) functions. We now turn our attention to **quadratic** (or second-degree) **functions.** The general equation of a quadratic function has the form

$$y = f(x) = ax^2 + bx + c$$

where a, b, and c are real numbers and $a \neq 0$.

The graph of a quadratic function,

$$y = ax^2 + bx + c \qquad (a \neq 0)$$

has a distinctive shape called a **parabola.**

The basic function $y = x^2$ and a variation of it, $y = -\frac{1}{2}x^2$, are parabolas whose graphs are shown in Figure 2.2.

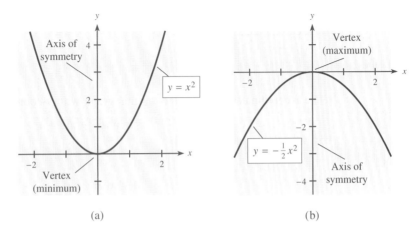

Figure 2.2   (a)   (b)

Vertex of a Parabola

As these examples illustrate, the graph of $y = ax^2$ is a parabola that opens upward if $a > 0$ and downward if $a < 0$. The **vertex,** where the parabola turns, is a **minimum point** if $a > 0$ and a **maximum point** if $a < 0$. The vertical line through the vertex of a parabola is called the **axis of symmetry** because one half of the graph is a reflection of the other half through this line.

The graph of $y = (x - 2)^2 - 1$ is the graph of $y = x^2$ shifted to a new location that is 2 units to the right and 1 unit down; its vertex is shifted from $(0, 0)$ to $(2, -1)$ and its axis of symmetry is shifted 2 units to the right. (See Figure 2.3(a).) The graph of $y = -\frac{1}{2}(x + 1)^2 + 2$ is the graph of $y = -\frac{1}{2}x^2$ shifted 1 unit to the left and 2 units upward, with its vertex at $(-1, 2)$. (See Figure 2.3(b).)

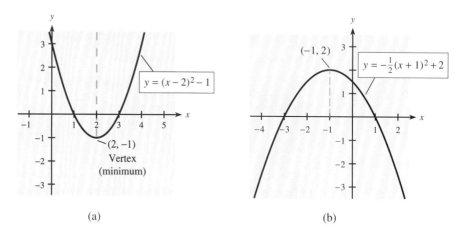

Figure 2.3   (a)   (b)

We can find the x-coordinate of the vertex of the graph of $y = ax^2 + bx + c$ by using the fact that the axis of symmetry of a parabola passes through the vertex. Regardless of the location of the vertex of $y = ax^2 + bx + c$ or the direction it opens, the y-intercept of the graph of $y = ax^2 + bx + c$ is $(0, c)$ and there is another point on the graph with y-coordinate c. See Figure 2.4.

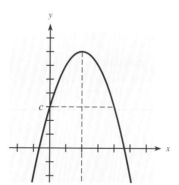

Figure 2.4

The x-coordinates of the points on this graph with y-coordinate c satisfy

$$c = ax^2 + bx + c$$

Solving this equation gives

$$0 = ax^2 + bx$$
$$0 = x(ax + b)$$
$$x = 0 \quad \text{or} \quad x = \frac{-b}{a}$$

The x-coordinate of the vertex is on the axis of symmetry, which is halfway from $x = 0$ to $x = \dfrac{-b}{a}$, at $x = \dfrac{-b}{2a}$. Thus we have the following.

Vertex of a Parabola

The quadratic function $y = f(x) = ax^2 + bx + c$ has its **vertex** at

$$\left(\frac{-b}{2a}, \ f\left(\frac{-b}{2a}\right) \right)$$

The optimum value (either maximum or minimum) of the function occurs at $x = \dfrac{-b}{2a}$ and is $f\left(\dfrac{-b}{2a}\right)$. See Figure 2.5.

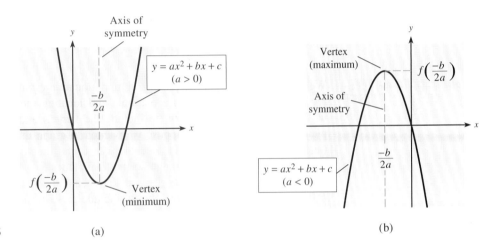

Figure 2.5

(a)

(b)

If we know the location of the vertex and the direction in which the parabola opens, we need very few other points to make a good sketch.

● **EXAMPLE 1 *Vertex and Graph of a Parabola***

Find the vertex and sketch the graph of

$$f(x) = 2x^2 - 4x + 4$$

Solution

Because $a = 2 > 0$, the graph of $f(x)$ opens upward and the vertex is the minimum point. We can calculate its coordinates as follows:

$$x = \frac{-b}{2a} = \frac{-(-4)}{2(2)} = 1$$

$$y = f(1) = 2$$

Thus the vertex is (1, 2). Using x-values on either side of the vertex to plot additional points allows us to sketch the graph accurately. (See Figure 2.6.)

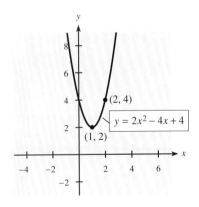

Figure 2.6

We can also use the coordinates of the vertex to find maximum or minimum values without a graph.

◑ EXAMPLE 2 *Profit Maximization* **(Application Preview)**

For the profit function

$$P(x) = -0.1x^2 + 300x - 1200$$

find the number of units that give maximum profit and find the maximum profit.

Solution

$P(x)$ is a quadratic function with $a < 0$. Thus the graph of $y = P(x)$ is a parabola that opens downward, so the vertex is a maximum point. The coordinates of the vertex are

$$x = \frac{-b}{2a} = \frac{-300}{2(-0.1)} = 1500$$

$$P = P(1500) = -0.1(1500)^2 + 300(1500) - 1200 = 223,800$$

Therefore, the maximum profit is $223,800 when 1500 units are sold.

Graphing Utilities   We can also use a graphing utility to graph a quadratic function. Even with a graphing tool, recognizing that the graph is a parabola and locating the vertex are important. They help us to set the viewing window (to include the vertex) and to know when we have a complete graph. For example, suppose we wanted to graph the profit function from Example 2 (and the Application Preview). Because $x \geq 0$ and the vertex and axis of symmetry are at $x = 1500$, we can set the *x*-range so that it includes $x = 0$ and has $x = 1500$ near its center. For the *P*-range (or *y*-range), we know that the maximum is $P = 223,800$ and that $P(0) = -1200$, so the window should include these *y*-values. Figure 2.7 shows the graph. Note that the graph of this function has two *x*-intercepts.

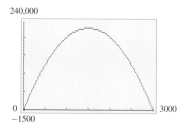

Figure 2.7 ■

Zeros of Quadratic Functions As we noted in Chapter 1, "Linear Equations and Functions," the *x*-intercepts of the graph of a function $y = f(x)$ are the values of x for which $f(x) = 0$, called the **zeros** of the function. As we saw in the previous section, the zeros of the quadratic function $y = f(x) = ax^2 + bx + c$ are the solutions of the quadratic equation

$$ax^2 + bx + c = 0$$

Therefore, the zeros are given by the quadratic formula as

$$x = \frac{-b \pm \sqrt{b^2 - 4ac}}{2a}$$

The information that is useful in graphing quadratic functions is summarized as follows:

Graphs of Quadratic Functions

Form: $y = ax^2 + bx + c$
Graph: parabola (See Figure 2.5 earlier in this section.)

$a > 0$ parabola opens upward; vertex is a minimum point

$a < 0$ parabola opens downward; vertex is a maximum point

Coordinates of vertex: $x = \dfrac{-b}{2a}$, $y = f\left(\dfrac{-b}{2a}\right)$

Axis of symmetry equation: $x = \dfrac{-b}{2a}$

x-intercepts or zeros (if real*):

$$x = \frac{-b + \sqrt{b^2 - 4ac}}{2a}, \quad x = \frac{-b - \sqrt{b^2 - 4ac}}{2a}$$

y-intercept: Let $x = 0$; then $y = c$.

● **EXAMPLE 3** *Graph of a Quadratic Function*

For the function $y = 4x - x^2$, determine whether its vertex is a maximum point or a minimum point and find the coordinates of this point, find the zeros, if any exist, and sketch the graph.

Solution
The proper form is $y = -x^2 + 4x + 0$, so $a = -1$. Thus the parabola opens downward, and the vertex is the highest (maximum) point.

The vertex occurs at $x = \dfrac{-b}{2a} = \dfrac{-4}{2(-1)} = 2$.

The y-coordinate of the vertex is $f(2) = -(2)^2 + 4(2) = 4$.

The zeros of the function are solutions to

$$-x^2 + 4x = 0$$
$$x(-x + 4) = 0$$
$$\text{or } x = 0 \text{ and } x = 4$$

The graph of the function can be found by drawing a parabola with these three points (see Figure 2.8(a)), or by using a graphing utility or a spreadsheet (see Figure 2.8(b)).

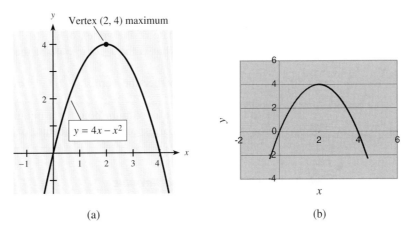

Figure 2.8   (a)   (b)

*If the zeros are not real, the graph does not cross the x-axis.

Example 3 shows how to use the solutions of $f(x) = 0$ to find the x-intercepts of a parabola if $y = f(x)$ is a quadratic function. Conversely, we can use the x-intercepts of $y = f(x)$, if they exist, to find or approximate the solutions of $f(x) = 0$. Exact integer solutions found in this way can also be used to determine factors of $f(x)$.

● **EXAMPLE 4** *Graph Comparison*

Figure 2.9 shows the graphs of two different quadratic functions. Use the figure to answer the following.

(a) Determine the vertex of each function.
(b) Determine the real solutions of $f_1(x) = 0$ and $f_2(x) = 0$.
(c) One of the graphs in Figure 2.9 is the graph of $y = 7 + 6x - x^2$, and one is the graph of $y = x^2 - 6x + 10$. Determine which is which, and why.

Solution
(a) For $y = f_1(x)$, the vertex is the maximum point, at $(3, 16)$.
    For $y = f_2(x)$, the vertex is the minimum point, at $(3, 1)$.
(b) For $y = f_1(x)$, the real solutions of $f_1(x) = 0$ are the zeros, or x-intercepts, at $x = -1$ and $x = 7$. For $y = f_2(x)$, the graph has no x-intercepts, so $f_2(x) = 0$ has no real solutions.
(c) Because the graph of $y = f_1(x)$ opens downward, the coefficient of x^2 must be negative. Hence Figure 2.9(a) shows the graph of $y = f_1(x) = 7 + 6x - x^2$. Similarly, the coefficient of x^2 in $y = f_2(x)$ must be positive, so Figure 2.9(b) shows the graph of $y = f_2(x) = x^2 - 6x + 10$.

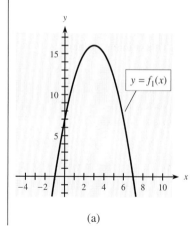

(a)

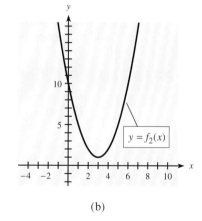

(b)

Figure 2.9

● **Checkpoint**

1. Name the graph of a quadratic function.
2. (a) What is the x-coordinate of the vertex of $y = ax^2 + bx + c$?
   (b) For $y = 12x - \frac{1}{2}x^2$, what is the x-coordinate of the vertex? What is the y-coordinate of the vertex?
3. (a) How can you tell whether the vertex of $f(x) = ax^2 + bx + c$ is a maximum point or a minimum point?
   (b) In 2(b), is the vertex a maximum point or a minimum point?
4. The zeros of a function correspond to what feature of its graph?

● **EXAMPLE 5** *Maximizing Revenue*

Ace Cruise Line offers an inland waterway cruise to a group of 50 people for a price of $30 per person, but it reduces the price per person by $0.50 for each additional person above the 50.

(a) Does reducing the price per person to get more people in the group give the cruise line more revenue?

(b) How many people will provide maximum revenue for the cruise line?

Solution

(a) The revenue to the company if 50 people are in the group and each pays $30 is 50($30) = $1500. The following table shows the revenue for the addition of people to the group. The table shows that as the group size begins to increase past 50 people, the revenue also increases. However, it also shows that increasing the group size too much (to 70 people) reduces revenue.

Increase in Group Size	Number of People	Decrease in Price	New Price ($)	Revenue ($)
0	50	0	30	50(30) = 1500
1	51	0.50(1)	29.50	51(29.50) = 1504.50
2	52	0.50(2) = 1	29	52(29) = 1508
3	53	0.50(3) = 1.50	28.50	53(28.50) = 1510.50
⋮	⋮	⋮	⋮	⋮
20	70	0.50(20) = 10	20	70(20) = 1400
⋮	⋮	⋮	⋮	⋮
x	$50 + x$	$0.50(x)$	$30 - 0.50x$	$(50 + x)(30 - 0.50x)$

(b) The last entry in the table shows the revenue for an increase of x people in the group.

$$R(x) = (50 + x)(30 - 0.50x)$$

Expanding this function gives a form from which we can find the vertex of its graph.

$$R(x) = 1500 + 5x - 0.50x^2$$

The vertex of the graph of this function is $x = \dfrac{-b}{2a} = \dfrac{-5}{2(-0.50)} = 5, R(5) = 1512.50$.

This means that the revenue will be maximized at $1512.50 when 50 + 5 = 55 people are in the group, with each paying 30 - 0.50(5) = 27.50 dollars.

Average Rate of Change

Average Rate of Change Because the graph of a quadratic function is not a line, the rate of change of the function is not constant. We can, however, find the **average rate of change** of a function between two input values if we know how much the function output values change between the two input values.

Average Rate of Change

The average rate of change of $f(x)$ with respect to x over the interval from $x = a$ to $x = b$ (where $a < b$) is calculated as

$$\text{Average rate of change} = \frac{\text{change of } f(x)}{\text{corresponding change in } x\text{-values}} = \frac{f(b) - f(a)}{b - a}$$

The average rate of change is also the slope of the segment (or secant line) joining the points $(a, f(a))$ and $(b, f(b))$.

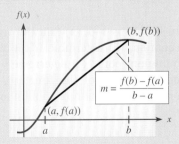

● **EXAMPLE 6** *Average Rate of Change of Revenue*

In Example 5, we found that the revenue for a cruise was defined by the quadratic function

$$R(x) = 1500 + 5x - 0.50x^2$$

where x is the increase in the group size beyond 50 people. What is the average rate of change of revenue if the group increases from 50 to 55 persons?

Solution

The average rate of change of revenue from 50 to 55 persons is the average rate of change of the function from $x = 0$ to $x = 5$.

$$\frac{R(5) - R(0)}{5 - 0} = \frac{1512.50 - 1500}{5} = 2.50 \text{ dollars per person}$$

This average rate of change is also the slope of the **secant line** connecting the points (0, 1500) and (5, 1512.50) on the graph of the function (see Figure 2.10).

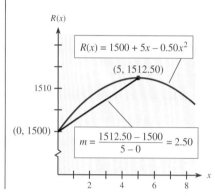

Figure 2.10

● **Checkpoint Solutions**

1. Parabola

2. (a) $x = \dfrac{-b}{2a}$

   (b) $x = \dfrac{-12}{2(-1/2)} = \dfrac{-12}{(-1)} = 12$

   $y = 12(12) - \dfrac{1}{2}(12)^2 = 144 - 72 = 72$

3. (a) Maximum point if $a < 0$; minimum point if $a > 0$.
   (b) (12, 72) is a maximum point because $a = -\frac{1}{2}$.

4. The x-intercepts

2.2 Exercises

In Problems 1–6, (a) find the vertex of the graph of the equation, (b) determine if the vertex is a maximum or minimum point, (c) determine what value of x gives the optimal value of the function, and (d) determine the optimal (maximum or minimum) value of the function.

1. $y = \dfrac{1}{2}x^2 + x$

2. $y = x^2 - 2x$

3. $y = 8 + 2x - x^2$

4. $y = 6 - 4x - 2x^2$

5. $f(x) = 6x - x^2$

6. $f(x) = x^2 + 2x - 3$

In Problems 7–12, determine whether each function's vertex is a maximum point or a minimum point and find the coordinates of this point. Find the zeros, if any exist, and sketch the graph of the function.

7. $y = x - \dfrac{1}{4}x^2$

8. $y = -2x^2 + 18x$

9. $y = x^2 + 4x + 4$

10. $y = x^2 - 6x + 9$

11. $\dfrac{1}{2}x^2 + x - y - 3 = 0$

12. $x^2 + x + 2y = 5$

For each function in Problems 13–16, (a) tell how the graph of $y = x^2$ is shifted, and (b) graph the function.

13. $y = (x - 3)^2 + 1$ 14. $y = (x - 10)^2 + 1$
15. $y = (x + 2)^2 - 2$ 16. $y = (x + 12)^2 - 8$

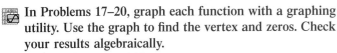 **In Problems 17–20, graph each function with a graphing utility. Use the graph to find the vertex and zeros. Check your results algebraically.**

17. $y = \dfrac{1}{2}x^2 - x - \dfrac{15}{2}$

18. $y = 0.1(x^2 + 4x - 32)$

19. $y = \dfrac{1}{4}x^2 + 3x + 12$

20. $y = x^2 - 2x + 5$

In Problems 21–22, find the average rate of change of the function between the given values of x.

21. $y = -5x - x^2$ between $x = -1$ and $x = 1$.
22. $y = 8 + 3x + 0.5x^2$ between $x = 2$ and $x = 4$.

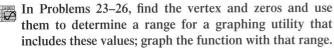 **In Problems 23–26, find the vertex and zeros and use them to determine a range for a graphing utility that includes these values; graph the function with that range.**

23. $y = 63 + 0.2x - 0.01x^2$
24. $y = 0.2x^2 + 16x + 140$
25. $y = 0.0001x^2 - 0.01$
26. $y = 0.01x - 0.001x^2$

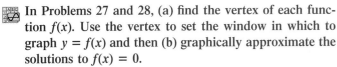 **In Problems 27 and 28, (a) find the vertex of each function $f(x)$. Use the vertex to set the window in which to graph $y = f(x)$ and then (b) graphically approximate the solutions to $f(x) = 0$.**

27. $f(x) = 8x^2 - 16x - 16$
28. $f(x) = 3x^2 - 18x + 16$

 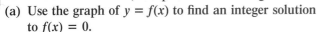 **In Problems 29 and 30, complete the following.**
(a) Use the graph of $y = f(x)$ to find an integer solution to $f(x) = 0$.
(b) Use the solution from (a) to find a factor of $f(x)$.
(c) Factor $f(x)$.
(d) Solve $f(x) = 0$.

29. $f(x) = 3x^2 - 8x + 4$ 30. $f(x) = 5x^2 - 2x - 7$

APPLICATIONS

31. *Profit*   The daily profit from the sale of a product is given by $P = 16x - 0.1x^2 - 100$ dollars.
    (a) What level of production maximizes profit?
    (b) What is the maximum possible profit?

32. *Profit*   The daily profit from the sale of x units of a product is $P = 80x - 0.4x^2 - 200$ dollars.
    (a) What level of production maximizes profit?
    (b) What is the maximum possible profit?

33. *Crop yield*   The yield in bushels from a grove of orange trees is given by $Y = x(800 - x)$, where x is the number of orange trees per acre. How many trees will maximize the yield?

34. *Stimulus-response*   One of the early results in psychology relating the magnitude of a stimulus x to the magnitude of a response y is expressed by the equation

$$y = kx^2$$

where k is an experimental constant. Sketch this graph for $k = 1$, $k = 2$, and $k = 4$.

35. *Drug sensitivity*   The sensitivity S to a drug is related to the dosage x in milligrams by

$$S = 1000x - x^2$$

Sketch the graph of this function and determine what dosage gives maximum sensitivity. Use the graph to determine the maximum sensitivity.

36. *Maximizing an enclosed area*   If 100 feet of fence is used to enclose a rectangular yard, then the resulting area is given by

$$A = x(50 - x)$$

where x feet is the width of the rectangle and $50 - x$ feet is the length. Graph this equation and determine the length and width that give maximum area.

37. *Photosynthesis*   The rate of photosynthesis R for a certain plant depends on the intensity of light x, in lumens, according to

$$R = 270x - 90x^2$$

Sketch the graph of this function, and determine the intensity that gives the maximum rate.

38. *Projectiles*   A ball thrown vertically into the air has its height above ground given by

$$s = 112t - 16t^2$$

where t is in seconds and s is in feet. Find the maximum height of the ball.

39. *Projectiles*   Two projectiles are shot into the air from the same location. The paths of the projectiles are parabolas and are given by
    (a) $y = -0.0013x^2 + x + 10$ and
    (b) $y = \dfrac{-x^2}{81} + \dfrac{4}{3}x + 10$

where x is the horizontal distance and y is the vertical distance, both in feet. Determine which projectile goes higher by locating the vertex of each parabola.

40. *Flow rates of water*   The speed at which water travels in a pipe can be measured by directing the flow through an elbow and measuring the height to which it spurts out the top. If the elbow height is 10 cm, the equation relating the height h (in centimeters) of the water above the elbow and its velocity v (in centimeters per second) is given by

$$v^2 = 1960(h + 10)$$

Solve this equation for h and graph the result, using the velocity as the independent variable.

41. *Cost*   The figure below shows the graph of a total cost function, with x equal to the number of units produced.
   (a) Is the average rate of change of cost greater from $x = a$ to $x = b$ or from $x = b$ to $x = c$? Explain.
   (b) Would the number of units d need to satisfy $d < b$ or $d > b$ for the average rate of change of cost from $x = a$ to $x = d$ to be greater than that from $x = a$ to $x = b$? Explain.

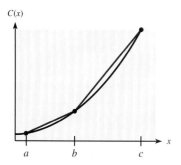

42. *Revenue*   The figure below shows the graph of a total revenue function, with x equal to the number of units sold.
   (a) Is the average rate of change of revenue negative from $x = a$ to $x = b$ or from $x = b$ to $x = c$? Explain.
   (b) Would the number of units d need to satisfy $d < b$ or $d > b$ for the average rate of change of revenue from $x = a$ to $x = d$ to be greater than that from $x = a$ to $x = b$? Explain.

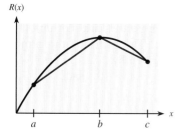

43. *Apartment rental*   The owner of an apartment building can rent all 50 apartments if she charges $600 per

month, but she rents one fewer apartment for each $20 increase in monthly rent.
   (a) Construct a table that gives the revenue generated if she charges $600, $620, and $640.
   (b) Does her revenue from the rental of the apartments increase or decrease as she increases the rent from $600 to $640?
   (c) Write an equation that gives the revenue from rental of the apartments if she makes x increases of $20 in the rent.
   (d) Find the rent she should charge to maximize her revenue.

44. *Revenue*   The owner of a skating rink rents the rink for parties at $600 if 50 or fewer skaters attend, so that the cost per person is $12 if 50 attend. For each 5 skaters above 50, she reduces the price per skater by $0.50.
   (a) Construct a table that gives the revenue generated if 50, 60, and 70 skaters attend.
   (b) Does the owner's revenue from the rental of the rink increase or decrease as the number of skaters increases from 50 to 70?
   (c) Write the equation that describes the revenue for parties with x more than 50 skaters.
   (d) Find the number of skaters that will maximize the revenue.

45. *Pension resources*   The Pension Benefit Guaranty Corporation is the agency that insures pensions. The figure shows one study's projection for the agency's total resources, initially rising (from taking over the assets of failing plans) but then falling (as more workers retire and payouts increase).
   (a) What kind of function might be used to model the agency's total resources?
   (b) If a function of the form $f(x) = ax^2 + bx + c$ were used to model these total resources, would $f(x)$ have $a > 0$ or $a < 0$? Explain.
   (c) If the model from part (b) used x as the number of years past 2004, explain why the model would have $b > 0$ and $c > 0$.

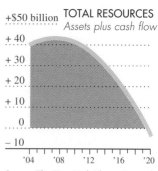

Source: The New York Times, September 14, 2004. Copyright © 2004 The New York Times Co. Reprinted by permission.

46. *Projectile motion*   When a stone is thrown upward, it follows a parabolic path given by a form of the equation

$$y = ax^2 + bx + c$$

If $y = 0$ represents ground level, find the equation of a stone that is thrown from ground level at $x = 0$ and lands on the ground 40 units away if the stone reaches a maximum height of 40 units. (*Hint:* Find the coordinates of the vertex of the parabola and two other points.)

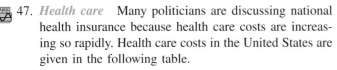 47. *Health care*   Many politicians are discussing national health insurance because health care costs are increasing so rapidly. Health care costs in the United States are given in the following table.

Year	Costs ($ millions)	Year	Costs ($ millions)
1960	26.7	1997	1093.9
1970	73.1	1998	1146.1
1980	245.8	1999	1210.7
1990	695.6	2000	1310.0
1995	987.0	2001	1425.5

Source: U.S. Department of Health and Human Services, Health Care Financing Administration, Office of National Health Statistics

(a) Plot the points from the table. Use x to represent the number of years since 1960 and y to represent costs in millions of dollars.
(b) What type of function appears to be the best fit for these points?
(c) Graph different quadratic functions of the form $y = ax^2 + c$ until you find a curve that is a reasonable fit for the points.

(d) Use your "model" function to predict health care costs in 2010.

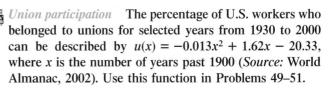 48. *Social Security*   In 1995, America's 45 million Social Security recipients received a 2.6% cost-of-living increase, the second smallest increase in nearly 20 years, a reflection of lower inflation. The percent increase can be described by the function

$$p(t) = -0.3375t^2 + 7.3t - 34.3625$$

where t is the number of years past 1980. (*Source:* Social Security Administration)
(a) Graph the function $y = p(t)$.
(b) From the graph, identify t-values for which $p(t) < 0$ and hence the model is not valid.
(c) From the graph, identify the maximum point on the graph of $p(t)$.
(d) In what year is the cost of living percent increase highest?

Union participation   The percentage of U.S. workers who belonged to unions for selected years from 1930 to 2000 can be described by $u(x) = -0.013x^2 + 1.62x - 20.33$, where x is the number of years past 1900 (*Source:* World Almanac, 2002). Use this function in Problems 49–51.
49. Graph the function $u(x)$.
50. For what year does the function $u(x)$ indicate a maximum percentage of workers belonged to unions?
51. (a) For what years does the function $u(x)$ predict that 0% of U.S. workers will belong to unions?
    (b) When can you guarantee that $u(x)$ can no longer be used to describe the percentage of U.S. workers who belong to unions?

2.3

Business Applications of Quadratic Functions

◗ Application Preview

Suppose that the supply function for a product is given by the quadratic function $p = q^2 + 100$ and the demand function is given by $p = -20q + 2500$. Finding the market equilibrium involves solving a quadratic equation (see Example 1).

In this section, we graph quadratic supply and demand functions and find market equilibrium by solving supply and demand functions simultaneously using quadratic methods. We will also discuss quadratic revenue, cost, and profit functions, including break-even points and profit maximization.

Supply, Demand, and Market Equilibrium

The first-quadrant parts of parabolas or other quadratic equations are frequently used to represent supply and demand functions. For example, the first-quadrant part of $p = q^2 + q + 2$ (Figure 2.11(a) on the next page) may represent a supply curve, whereas the first-quadrant part of $q^2 + 2q + 6p - 23 = 0$ (Figure 2.11(b)) may represent a demand curve.

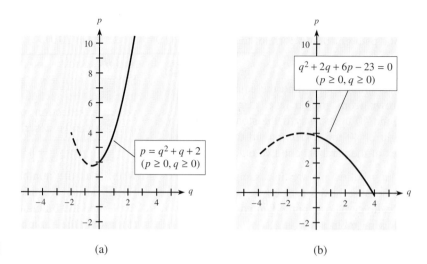

Figure 2.11   (a)   (b)

When quadratic equations are used to represent supply or demand curves, we can solve their equations simultaneously to find the market equilibrium as we did with linear supply and demand functions. As in Section 1.5, we can solve two equations in two variables by eliminating one variable and obtaining an equation in the other variable. When the functions are quadratic, the substitution method of solution is perhaps the best, and the resulting equation in one unknown will usually be quadratic.

◖ EXAMPLE 1 *Supply and Demand (Application Preview)*

If the supply function for a commodity is given by $p = q^2 + 100$ and the demand function is given by $p = -20q + 2500$, find the point of market equilibrium.

Solution
At market equilibrium, both equations will have the same p-value. Thus substituting $q^2 + 100$ for p in $p = -20q + 2500$ yields

$$q^2 + 100 = -20q + 2500$$
$$q^2 + 20q - 2400 = 0$$
$$(q - 40)(q + 60) = 0$$
$$q = 40 \quad \text{or} \quad q = -60$$

Because a negative quantity has no meaning, the equilibrium point occurs when 40 units are sold, at (40, 1700). The graphs of the functions are shown (in the first quadrant only) in Figure 2.12.

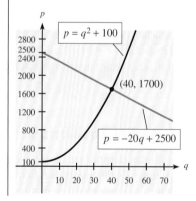

Figure 2.12

● **EXAMPLE 2** *Market Equilibrium*

If the demand function for a commodity is given by $p(q + 4) = 400$ and the supply function is given by $2p - q - 38 = 0$, find the market equilibrium.

Solution

Solving the supply equation $2p - q - 38 = 0$ for p gives $p = \frac{1}{2}q + 19$. Substituting for p in $p(q + 4) = 400$ gives

$$\left(\frac{1}{2}q + 19\right)(q + 4) = 400$$

$$\frac{1}{2}q^2 + 21q + 76 = 400$$

$$\frac{1}{2}q^2 + 21q - 324 = 0$$

Multiplying both sides of the equation by 2 yields $q^2 + 42q - 648 = 0$. Factoring gives

$$(q - 12)(q + 54) = 0$$
$$q = 12 \quad \text{or} \quad q = -54$$

Thus the market equilibrium occurs when 12 items are sold, at a price of $p = \frac{1}{2}(12) + 19 = \$25$ each. The graphs of the demand and supply functions are shown in Figure 2.13(a).

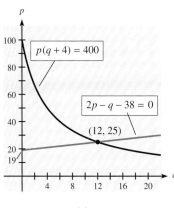

Figure 2.13   (a)

Graphing Utilities

Graphing utilities also can be used to sketch these graphs. A command such as INTERSECT could be used to determine points of intersection that give market equilibrium.

    Figure 2.13(b) shows the graph of the supply and demand functions for the commodity in Example 2. Using the INTERSECT command gives the same market equilibrium point determined in Example 2 (see Figure 2.13(c)).

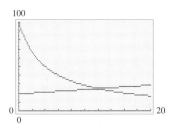 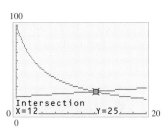

Figure 2.13 (*continued*)   (b)   (c)

● *Checkpoint*

1. The point of intersection of the supply and demand functions is called _____.
2. If the demand and supply functions for a product are

$$p + \frac{1}{10}q^2 = 1000 \quad \text{and} \quad p = \frac{1}{10}q + 10$$

respectively, finding the market equilibrium point requires solution of what equation? Find the market equilibrium.

Break-Even Points and Maximization

In Chapter 1, "Linear Equations and Functions," we discussed linear total cost and total revenue functions. Many total revenue functions may be linear, but costs tend to increase sharply after a certain level of production. Thus functions other than linear functions, including quadratic functions, are used to predict the total costs of products.

For example, the monthly total cost curve for a commodity may be the parabola with equation $C(x) = 360 + 40x + 0.1x^2$. If the total revenue function is $R(x) = 60x$, we can find the break-even point by finding the quantity x that makes $C(x) = R(x)$. (See Figure 2.14.)

Setting $C(x) = R(x)$, we have

$$360 + 40x + 0.1x^2 = 60x$$
$$0.1x^2 - 20x + 360 = 0$$
$$x^2 - 200x + 3600 = 0$$
$$(x - 20)(x - 180) = 0$$
$$x = 20 \quad \text{or} \quad x = 180$$

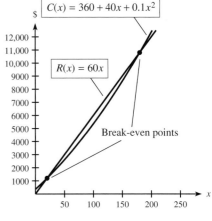

Figure 2.14

Thus $C(x) = R(x)$ at $x = 20$ and at $x = 180$. If 20 items are produced and sold, $C(x)$ and $R(x)$ are both \$1200; if 180 items are sold, $C(x)$ and $R(x)$ are both \$10,800. Thus there are two break-even points.

In a monopoly market, the revenue of a company is restricted by the demand for the product. In this case, the relationship between the price p of the product and the number of units sold x is described by the demand function $p = f(x)$, and the total revenue function for the product is given by

$$R = px = [f(x)]x$$

If, for example, the demand for a product is given by

$$p = 300 - x$$

where x is the number of units sold and p is the price, then the revenue function for this product is the quadratic function

$$R = px = (300 - x)x = 300x - x^2$$

● **EXAMPLE 3 *Break-Even Point***

Suppose that in a monopoly market the total cost per week of producing a high-tech product is given by $C = 3600 + 100x + 2x^2$. Suppose further that the weekly demand function for this product is $p = 500 - 2x$. Find the number of units that will give the break-even point for the product.

Solution

The total cost function is $C(x) = 3600 + 100x + 2x^2$, and the total revenue function is $R(x) = (500 - 2x)x = 500x - 2x^2$.

Setting $C(x) = R(x)$ and solving for x gives

$$3600 + 100x + 2x^2 = 500x - 2x^2$$
$$4x^2 - 400x + 3600 = 0$$
$$x^2 - 100x + 900 = 0$$
$$(x - 90)(x - 10) = 0$$
$$x = 90 \quad \text{or} \quad x = 10$$

Does this mean the firm will break even at 10 units and at 90 units? Yes. Figure 2.15 shows the graphs of $C(x)$ and $R(x)$. From the graph we can observe that the firm makes a profit after $x = 10$ *until* $x = 90$, because $R(x) > C(x)$ in that interval. At $x = 90$, the profit is 0, and the firm loses money if it produces more than 90 units per week.

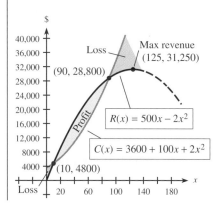

Figure 2.15

● **Checkpoint**

3. The point of intersection of the revenue function and the cost function is called _____ .

4. If $C(x) = 120x + 15,000$ and $R(x) = 370x - x^2$, finding the break-even points requires solution of what equation? Find the break-even points.

Note that for Example 3, the revenue function

$$R(x) = (500 - 2x)x = 500x - 2x^2$$

is a parabola that opens downward. Thus the vertex is the point at which revenue is maximum. We can locate this vertex by using the methods discussed in the previous section.

$$\text{Vertex: } x = \frac{-b}{2a} = \frac{-500}{2(-2)} = \frac{500}{4} = 125 \text{ (units)}$$

It is interesting to note that when $x = 125$, the firm achieves its maximum revenue of

$$R(125) = 500(125) - 2(125)^2 = 31,250 \text{ (dollars)}$$

but the costs when $x = 125$ are

$$C(125) = 3600 + 100(125) + 2(125)^2 = 47,350 \text{ (dollars)}$$

which results in a loss. This illustrates that maximizing revenue is not a good goal. We should seek to maximize profit. Figure 2.15 shows that maximum profit will occur where the distance between the revenue and cost curves (i.e., $R(x) - C(x)$) is largest. This appears to be near $x = 50$, which is midway between the x-values of the break-even points. This is verified in the next example.

● EXAMPLE 4 *Profit Maximization*

For the total cost function $C(x) = 3600 + 100x + 2x^2$ and the total revenue function $R(x) = 500x - 2x^2$ (from Example 3), find the number of units that maximizes profit and find the maximum profit.

Solution

Using Profit = Revenue − Cost, we can determine the profit function:

$$P(x) = (500x - 2x^2) - (3600 + 100x + 2x^2) = -3600 + 400x - 4x^2$$

This profit function is a parabola that opens downward, so the vertex will be the maximum point.

$$\text{Vertex: } x = \frac{-b}{2a} = \frac{-400}{2(-4)} = \frac{-400}{-8} = 50$$

Furthermore, when $x = 50$, we have

$$P(50) = -3600 + 400(50) - 4(50)^2 = 6400 \text{ (dollars)}$$

Thus, when 50 items are produced and sold, a maximum profit of \$6400 is made (see Figure 2.16).

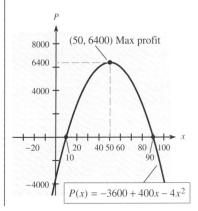

Figure 2.16

Figure 2.16 shows the break-even points at $x = 10$ and $x = 90$, and that the maximum profit occurs at the x-value midway between these x-values. This is reasonable because the graph of the profit function is a parabola, and the x-value of any parabola's vertex occurs midway between its x-intercepts.

It is important to note that the procedures for finding maximum revenue and profit in these examples depend on the fact that these functions are parabolas. Using methods discussed in Sections 2.1 and 2.2, we can use graphing utilities to locate maximum points, minimum points, and break-even points. For more general functions, procedures for finding maximum or minimum values are discussed in Chapter 10, "Applications of Derivatives."

● ***Checkpoint Solutions***

1. The market equilibrium point or market equilibrium

2. Demand: $p = -\frac{1}{10}q^2 + 1000$; supply: $p = \frac{1}{10}q + 10$

   Solution of $-\frac{1}{10}q^2 + 1000 = \frac{1}{10}q + 10$

   $$-q^2 + 10{,}000 = q + 100$$
   $$0 = q^2 + q - 9900$$
   $$0 = (q + 100)(q - 99)$$
   $$q = -100 \quad \text{or} \quad q = 99$$

   Thus market equilibrium occurs when $q = 99$ and $p = 9.9 + 10 = 19.90$.

3. The break-even point

4. Solution of $C(x) = R(x)$. That is, solution of $120x + 15{,}000 = 370x - x^2$, or $x^2 - 250x + 15{,}000 = 0$.

   $$(x - 100)(x - 150) = 0$$
   $$x = 100 \quad \text{or} \quad x = 150$$

   Thus the break-even points are when 100 units and 150 units are produced.

2.3 Exercises

SUPPLY, DEMAND, AND MARKET EQUILIBRIUM

1. Sketch the first-quadrant portions of the following on the same set of axes.
   (a) The supply function whose equation is
      $p = \frac{1}{4}q^2 + 10$
   (b) The demand function whose equation is
      $p = 86 - 6q - 3q^2$
   (c) Label the market equilibrium point.
   (d) Algebraically determine the equilibrium point for the supply and demand functions.

2. Sketch the first-quadrant portions of the following on the same set of axes.
   (a) The supply function whose equation is
      $p = q^2 + 8q + 16$
   (b) The demand function whose equation is
      $p = 216 - 2q$
   (c) Label the market equilibrium point.
   (d) Algebraically determine the equilibrium point for the supply and demand functions.

3. Sketch the first-quadrant portions of the following on the same set of axes.
   (a) The supply function whose equation is
      $p = 0.2q^2 + 0.4q + 1.8$
   (b) The demand function whose equation is
      $p = 9 - 0.2q - 0.1q^2$
   (c) Label the market equilibrium point.
   (d) Algebraically determine the market equilibrium point.

4. Sketch the first-quadrant portions of the following on the same set of axes.
   (a) The supply function whose equation is
      $p = q^2 + 8q + 22$
   (b) The demand function whose equation is
      $p = 198 - 4q - \frac{1}{4}q^2$
   (c) Label the market equilibrium point.
   (d) Algebraically determine the market equilibrium point for the supply and demand functions.

5. If the supply function for a commodity is $p = q^2 + 8q + 16$ and the demand function is $p = -3q^2 + 6q + 436$, find the equilibrium quantity and equilibrium price.

6. If the supply function for a commodity is $p = q^2 + 8q + 20$ and the demand function is $p = 100 - 4q - q^2$, find the equilibrium quantity and equilibrium price.

7. If the demand function for a commodity is given by the equation $p^2 + 4q = 1600$ and the supply function is given by the equation $300 - p^2 + 2q = 0$, find the equilibrium quantity and equilibrium price.

8. If the supply and demand functions for a commodity are given by $4p - q = 42$ and $(p + 2)q = 2100$, respectively, find the price that will result in market equilibrium.

9. If the supply and demand functions for a commodity are given by $p - q = 10$ and $q(2p - 10) = 2100$, what is the equilibrium price and what is the corresponding number of units supplied and demanded?

10. If the supply and demand functions for a certain product are given by the equations $2p - q + 6 = 0$ and $(p + q)(q + 10) = 3696$, respectively, find the price and quantity that give market equilibrium.

11. The supply function for a product is $2p - q - 10 = 0$, while the demand function for the same product is $(p + 10)(q + 30) = 7200$. Find the market equilibrium point.

12. The supply and demand for a product are given by $2p - q = 50$ and $pq = 100 + 20q$, respectively. Find the market equilibrium point.

13. For the product in Problem 11, if a $22 tax is placed on production of the item, then the supplier passes this tax on by adding $22 to his selling price. Find the new equilibrium point for this product when the tax is passed on. (The new supply function is given by $p = \frac{1}{2}q + 27$.)

14. For the product in Problem 12, if a $12.50 tax is placed on production and passed through by the supplier, find the new equilibrium point.

BREAK-EVEN POINTS AND MAXIMIZATION

15. The total costs for a company are given by

$$C(x) = 2000 + 40x + x^2$$

and the total revenues are given by

$$R(x) = 130x$$

Find the break-even points.

16. If a firm has the following cost and revenue functions, find the break-even points.

$$C(x) = 3600 + 25x + \frac{1}{2}x^2,$$

$$R(x) = \left(175 - \frac{1}{2}x\right)x$$

17. If a company has total costs $C(x) = 15,000 + 35x + 0.1x^2$ and total revenues given by $R(x) = 385x - 0.9x^2$, find the break-even points.

18. If total costs are $C(x) = 1600 + 1500x$ and total revenues are $R(x) = 1600x - x^2$, find the break-even points.

19. Given that $P(x) = 11.5x - 0.1x^2 - 150$ and that production is restricted to fewer than 75 units, find the break-even points.

20. If the profit function for a firm is given by $P(x) = -1100 + 120x - x^2$ and limitations on space require that production is less than 100 units, find the break-even points.

21. Find the maximum revenue for the revenue function $R(x) = 385x - 0.9x^2$.

22. Find the maximum revenue for the revenue function $R(x) = 1600x - x^2$.

23. If, in a monopoly market, the demand for a product is $p = 175 - 0.50x$ and the revenue function is $R = px$, where x is the number of units sold, what price will maximize revenue?

24. If, in a monopoly market, the demand for a product is $p = 1600 - x$ and the revenue is $R = px$, where x is the number of units sold, what price will maximize revenue?

25. The profit function for a certain commodity is $P(x) = 110x - x^2 - 1000$. Find the level of production that yields maximum profit, and find the maximum profit.

26. The profit function for a firm making widgets is $P(x) = 88x - x^2 - 1200$. Find the number of units at which maximum profit is achieved, and find the maximum profit.

27. (a) Graph the profit function $P(x) = 80x - 0.1x^2 - 7000$.
    (b) Find the vertex of the graph. Is it a maximum point or a minimum point?
    (c) Is the average rate of change of this function from $x = a < 400$ to $x = 400$ positive or negative?
    (d) Is the average rate of change of this function from $x = 400$ to $x = a > 400$ positive or negative?
    (e) Does the average rate of change of the profit get closer to or farther from 0 when a is closer to 400?

28. (a) Graph the profit function $P(x) = 50x - 0.2x^2 - 2000$.
    (b) Find the vertex of the graph. Is it a maximum point or a minimum point?
    (c) Is the average rate of change of this function from $x = a < 125$ to $x = 125$ positive or negative?
    (d) Is the average rate of change of this function from $x = 125$ to $x = a > 125$ positive or negative?
    (e) Does the average rate of change of the profit get closer to or farther from 0 when a is closer to 125?

29. (a) Form the profit function for the cost and revenue functions in Problem 17, and find the maximum profit.
    (b) Compare the level of production to maximize profit with the level to maximize revenue (see Problem 21). Do they agree?
    (c) How do the break-even points compare with the zeros of $P(x)$?

30. (a) Form the profit function for the cost and revenue functions in Problem 18, and find the maximum profit.

(b) Compare the level of production to maximize profit with the level to maximize revenue (see Problem 22). Do they agree?

(c) How do the break-even points compare with the zeros of $P(x)$?

31. Suppose a company has fixed costs of $28,000 and variable costs of $\frac{2}{5}x + 222$ dollars per unit, where x is the total number of units produced. Suppose further that the selling price of its product is $1250 - \frac{3}{5}x$ dollars per unit.
    (a) Find the break-even points.
    (b) Find the maximum revenue.
    (c) Form the profit function from the cost and revenue functions and find maximum profit.
    (d) What price will maximize the profit?

32. Suppose a company has fixed costs of $300 and variable costs of $\frac{3}{4}x + 1460$ dollars per unit, where x is the total number of units produced. Suppose further that the selling price of its product is $1500 - \frac{1}{4}x$ dollars per unit.
    (a) Find the break-even points.
    (b) Find the maximum revenue.
    (c) Form the profit function from the cost and revenue functions and find maximum profit.
    (d) What price will maximize the profit?

33. The following table gives the total revenues of Cablenet Communications for selected years.

Year	Total Revenues (millions)
1997	$63.13
1998	$69.906
1999	$60.53
2001	$61.1
2002	$62.191
2003	$63.089
2004	$64.904
2005	$67.156

Suppose the data can be described by the equation

$$R(t) = 0.253t^2 - 4.03t + 76.84$$

where t is the number of years past 1992.

(a) Use the function to find the year in which revenue was minimum and find the minimum predicted revenue.

(b) Check the result from (a) against the data in the table.

(c) Graph $R(t)$ and the data points from the table.

(d) Write a sentence to describe how well the function fits the data.

The data in the table give sales revenues and costs and expenses for Continental Divide Mining for various years. Use this table in Problems 34 and 35.

Year	Sales Revenue (millions)	Costs and Expenses (millions)
1995	$2.6155	$2.4105
1996	2.7474	2.4412
1997	2.934	2.6378
1998	3.3131	2.9447
1999	3.9769	3.5344
2000	4.5494	3.8171
2001	4.8949	4.2587
2002	5.1686	4.8769
2003	4.9593	4.9088
2004	5.0913	4.6771
2005	4.7489	4.9025

34. Assume that sales revenues for Continental Divide Mining can be described by

$$R(t) = -0.031t^2 + 0.776t + 0.179$$

where t is the number of years past 1992.

(a) Use the function to determine the year in which maximum revenue occurs and the maximum revenue it predicts.

(b) Check the result from (a) against the data in the table.

(c) Graph $R(t)$ and the data points from the table.

(d) Write a sentence to describe how well the function fits the data.

35. Assume that costs and expenses for Continental Divide Mining can be described by

$$C(t) = -0.012t^2 + 0.492t + 0.725$$

where t is the number of years past 1992.

(a) Use $R(t)$ as given in Problem 34 and form the profit function (as a function of time).

(b) Use the function from (a) to find the year in which maximum profit occurs.

(c) Graph the profit function from (b) and the data points from the table.

(d) Through the decade from 2000 to 2010, does the function project increasing or decreasing profits? Do the data support this trend (as far as it goes)?

(e) How might management respond to this kind of projection?

2.4 Special Functions and Their Graphs

OBJECTIVES

- To graph and apply basic functions, including constant and power functions
- To graph and apply polynomial and rational functions
- To graph and apply absolute value and piecewise defined functions

◑ Application Preview

The average cost per item for a product is calculated by dividing the total cost by the number of items. Hence, if the total cost function for x units of a product is

$$C(x) = 900 + 3x + x^2$$

and if we denote the average cost function by $\overline{C}(x)$, then

$$\overline{C}(x) = \frac{C(x)}{x} = \frac{900 + 3x + x^2}{x}$$

This average cost function is a special kind of function, called a **rational function.** We can use this function to find the minimum average cost. (See Example 5.)

In this section, we discuss polynomial, rational, and other special functions.

Basic Functions

In Section 1.3, "Linear Functions," we saw that linear functions can be written in the form $y = ax + b$. The special linear function

$$y = f(x) = x$$

is called the **identity function** (Figure 2.17(a)), and a linear function defined by

$$y = f(x) = C \qquad C \text{ a constant}$$

is called a **constant function.** Figure 2.17(b) shows the graph of the constant function $y = f(x) = 2$. (Note that the slope of the graph of any constant function is 0.)

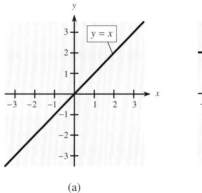
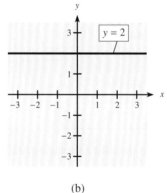

Figure 2.17   (a)   (b)

The functions of the form $y = ax^b$, where $b > 0$, are called **power functions.** Examples of power functions include $y = x^2$, $y = x^3$, $y = \sqrt{x} = x^{1/2}$, and $y = \sqrt[3]{x} = x^{1/3}$. (See Figure 2.18(a)–(d).) The functions $y = \sqrt{x}$ and $y = \sqrt[3]{x}$ are also called **root functions,** and $y = x^2$ and $y = x^3$ are basic **polynomial functions.** The general shape for the power function $y = ax^b$, where $b > 0$, depends on the value of b. Figure 2.19 shows the first-quadrant portions of typical graphs of $y = x^b$ for different values of b. Note how the direction in which the graph bends differs for $b > 1$ and for $0 < b < 1$. Getting accurate graphs of these functions requires plotting a number of points by hand or with a graphing utility. Our goal at this stage is to recognize the basic shapes of certain functions.

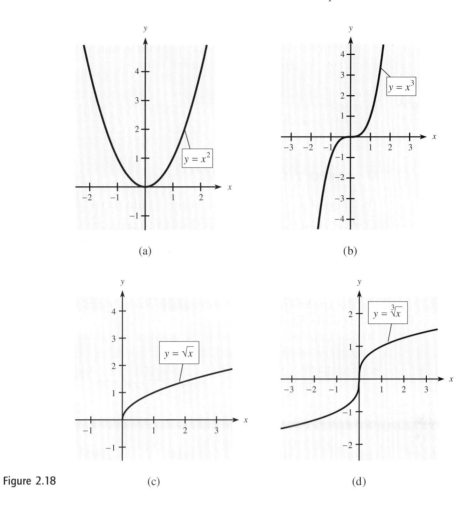

Figure 2.18   (c)   (d)

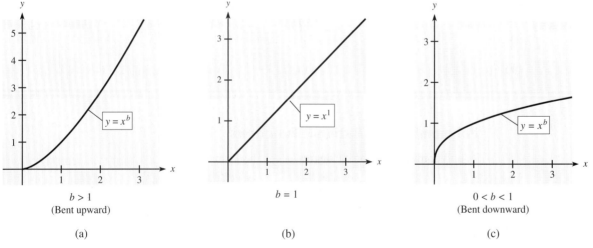

Figure 2.19

Shifts   In Section 2.2, "Quadratic Functions: Parabolas," we noted that the graph of $y = (x - h)^2 + k$ is a parabola that is shifted h units in the x-direction and k units in the y-direction. In general, we have the following:

Shifts of Graphs

> The graph of $y = f(x - h) + k$ is the graph of $y = f(x)$ shifted h units in the x-direction and k units in the y-direction.

● **EXAMPLE 1 Shifted Graph**

The graph of $y = x^3$ is shown in Figure 2.18(b) on the preceding page.

(a) Describe the graph of $y = x^3 - 3$ and graph this function.
(b) Describe the graph of $y = (x - 2)^3$ and graph this function.
(c) Describe the graph of $y = (x - 2)^3 - 3$ and graph this function.

Solution
(a) The graph of $y = x^3 - 3$ is the graph of $y = x^3$ shifted down 3 units. The graph is shown in Figure 2.20(a).
(b) The graph of $y = (x - 2)^3$ is the graph of $y = x^3$ shifted to the right 2 units. The graph is shown in Figure 2.20(b).
(c) The graph of $y = (x - 2)^3 - 3$ is the graph of $y = x^3$ shifted to the right 2 units and down 3 units. The graph is shown in Figure 2.20(c).

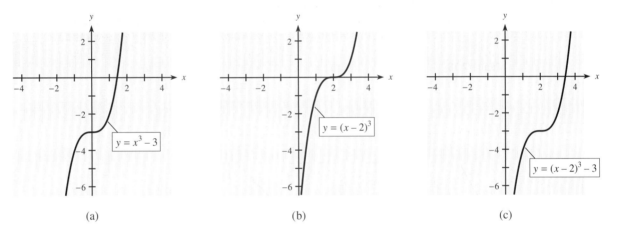

(a)   (b)   (c)

Figure 2.20

Polynomial and Rational Functions

A **polynomial function of degree n** has the form

$$y = a_n x^n + a_{n-1} x^{n-1} + \cdots + a_1 x + a_0$$

where $a_n \neq 0$, and n is an integer, $n \geq 0$.

A **linear function** is a polynomial function of degree 1, and a **quadratic function** is a polynomial function of degree 2.

Accurate graphing of a polynomial function of degree greater than 2 may require the methods of calculus; we will investigate these methods in Chapter 10, "Applications of Derivatives." For now, we will observe some characteristics of the graphs of polynomial functions of degrees 2, 3, and 4; these are summarized in Table 2.1. Using this information and point plotting on a graphing utility yields the graphs of these functions.

TABLE 2.1 Graphs of Some Polynomials

	Degree 2	Degree 3	Degree 4
Turning points	1	0 or 2	1 or 3
x-intercepts	0, 1, or 2	1, 2, or 3	0, 1, 2, 3, or 4
Possible shapes			

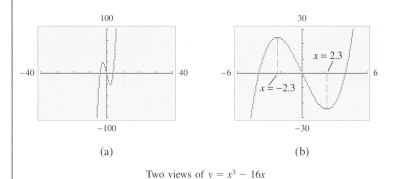

● **EXAMPLE 2** *Cubic Polynomial*

Graph $y = x^3 - 16x$.

Solution

This function is a third-degree (cubic) polynomial function, so it has one of the four shapes shown in the "Degree 3" column in Table 2.1. By graphing the function with a large x-range and y-range, we obtain the graph shown in Figure 2.21(a). We see that the graph has a shape like (b) in Table 2.1, so we know that a smaller viewing window (using a smaller x-range and y-range) will provide a more detailed view of all the graph's turning points (see Figure 2.21(b)). We can use a graphing utility to show that the turning points are at $x = -2.3$ and $x = 2.3$ (to the nearest tenth).

Figure 2.21

Two views of $y = x^3 - 16x$

● **EXAMPLE 3** *Quartic Polynomial*

Graph $y = x^4 - 2x^2$.

Solution

This is a degree 4 (quartic) polynomial function. Thus it has one or three turning points and has one of the six shapes in the "Degree 4" column of Table 2.1. We begin by graphing this function with a large viewing window to obtain the graph in Figure 2.22(a). We see that the graph appears to fit either shape (a) or shape (f) in Table 2.1. Figure 2.22(b) shows the same graph with a smaller viewing window, near $x = 0$. We now see that the graph has shape (f) from Table 2.1, with turning points at $x = -1, x = 0,$ and $x = 1$.

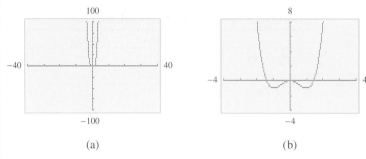

(a)   (b)

Figure 2.22   Two views of $y = x^4 - 2x^2$

● *Checkpoint*   1. All constant functions (such as $f(x) = 8$) have graphs that are _____.
2. Which of the following are polynomial functions?
   (a) $f(x) = x^3 - x + 4$
   (b) $f(x) = \dfrac{x + 1}{4x}$
   (c) $f(x) = 1 + \sqrt{x}$
   (d) $g(x) = \dfrac{1 + \sqrt{x}}{1 + x + \sqrt{x}}$
   (e) $h(x) = 5x$
3. A third-degree polynomial can have at most _____ turning points.

The function $f(x) = \dfrac{1}{x}$ is a special function called a **reciprocal function.** A table of sample values and the graph of $y = \dfrac{1}{x}$ are shown in Figure 2.23. Because division by 0 is not possible, $x = 0$ is not in the domain of the function.

x	y
-3	-0.33
-2	-0.5
-1	-1
-0.1	-10
-0.04	-25
-0.001	-1000
3	0.33
2	0.5
1	1
0.1	10
0.04	25
0.001	1000

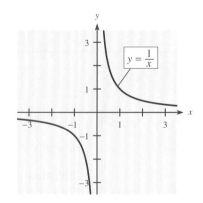

Figure 2.23

The function $y = \dfrac{1}{x}$ is an example of a **rational function,** which is defined below.

Rational Function

A rational function is a function of the form

$$y = \frac{f(x)}{g(x)} \text{ with } g(x) \neq 0$$

where $f(x)$ and $g(x)$ are both polynomials. Its domain is the set of all real numbers for which $g(x) \neq 0$.

Asymptotes

The table of values and the graph in Figure 2.23 show that $|y|$ gets very large as x approaches 0, and that y is undefined at $x = 0$. In this case, we call the line $x = 0$ (that is, the y-axis) a vertical asymptote. On the graphs of polynomial functions, the turning points are usually the features of greatest interest. However, on graphs of rational functions, vertical asymptotes frequently are the most interesting features.

Rational functions sometimes have **horizontal asymptotes** as well as vertical asymptotes. Whenever the values of y approach some finite number b as $|x|$ becomes very large, we say that there is a **horizontal asymptote** at $y = b$.

Note that the graph of $y = \dfrac{1}{x}$ appears to get close to the x-axis as $|x|$ becomes large. Testing values of x for which $|x|$ is large, we see that y is close to 0. Thus, we say that the line $y = 0$ (or the x-axis) is a **horizontal asymptote** for the graph of $y = \dfrac{1}{x}$. The graph in the following example has both a vertical and a horizontal asymptote.

● **EXAMPLE 4** *Rational Functions*

(a) Use values of x from -5 to 5 to develop a table of function values for the graph of

$$y = \frac{12x + 8}{3x - 9}$$

(b) Sketch the graph.

Solution

(a) Because $3x - 9 = 0$ when $x = 3$, it follows that $x = 3$ is not in the domain of this function.

x	-5	-4	-3	-2	-1	0	1
y	2.17	1.90	1.56	1.07	0.33	-0.89	-3.33

2	2.5	2.9	3.1	3.5	4	5
-10.7	-25.33	-142.67	150.67	33.33	18.67	11.33

The values in the table indicate that the graph is approaching the vertical asymptote at $x = 3$.

To see if the function has a horizontal asymptote, we calculate y as $|x|$ becomes larger.

x	$-10,000$	-1000	1000	10,000
y	3.998	3.98	4.01	4.001

This table indicates that the graph is approaching $y = 4$ as $|x|$ increases, so we have a horizontal asymptote at $y = 4$.

(b) Using the information about vertical and horizontal asymptotes and plotting these points give the graph in Figure 2.24.

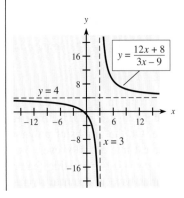

Figure 2.24

A general procedure for finding vertical and horizontal asymptotes of rational functions (see proof in Chapter 9) follows.

Asymptotes

Vertical Asymptote

The graph of the rational function $y = \dfrac{f(x)}{g(x)}$ has a vertical asymptote at $x = c$ if $g(c) = 0$ and $f(c) \neq 0$.

Horizontal Asymptote

Consider the rational function $y = \dfrac{f(x)}{g(x)} = \dfrac{a_n x^n + \cdots + a_1 x + a_0}{b_m x^m + \cdots + b_1 x + b_0}$.

1. If $n < m$ (that is, if the degree of the numerator is less than that of the denominator), a horizontal asymptote occurs at $y = 0$ (the x-axis).
2. If $n = m$ (that is, if the degree of the numerator equals that of the denominator), a horizontal asymptote occurs at $y = \dfrac{a_n}{b_m}$ (the ratio of the leading coefficients).
3. If $n > m$ (that is, if the degree of the numerator is greater than that of the denominator), there is no horizontal asymptote.

Recall that the graph of the function $y = \dfrac{12x + 8}{3x - 9}$ (shown in Figure 2.24) has a vertical asymptote at $x = 3$, and that $x = 3$ makes the denominator 0 but does not make the numerator 0. Observe also that this graph also has a horizontal asymptote at $y = 4$, and that the degree of the numerator is the same as that of the denominator with the ratio of the leading coefficients equal to $12/3 = 4$.

Determining the vertical and/or horizontal asymptotes of a rational function, if any exist, is very useful in determining a window for a graphing utility. A rational function may or may not have turning points. We will see how to find them and use them, when they exist, in Chapter 10.

The graph of the rational function in the following example has a vertical asymptote and a turning point.

◗ EXAMPLE 5 *Total Costs and Average Costs* **(Application Preview)**

Suppose the total cost function for x units of a product is given by

$$C(x) = 900 + 3x + x^2$$

Graph the average cost function for this product, and determine the minimum average cost.

Solution
The average cost per unit is

$$\overline{C}(x) = \frac{C(x)}{x} = \frac{900 + 3x + x^2}{x}$$

or

$$\overline{C}(x) = \frac{900}{x} + 3 + x$$

Because x represents the number of units produced, $x \geq 0$. Because $x = 0$ cannot be in the domain of the function, we choose a window with $x \geq 0$. Figure 2.25 shows the graph of $\overline{C}(x)$. The graph appears to have a minimum near $x = 30$. By plotting points or using MINIMUM on a graphing utility, we can verify that the minimum point occurs at $x = 30$ and $\overline{C} = 63$. Thus the minimum average cost is \$63 per unit when 30 units are produced.

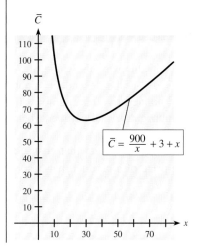

Figure 2.25

● Checkpoint

4. Given $f(x) = \dfrac{3x}{x - 4}$, decide whether the following are true or false.

(a) $f(x)$ has a vertical asymptote at $x = 4$.
(b) $f(x)$ has a horizontal asymptote at $x = 3$.

Piecewise Defined Functions

Another special function comes from the definition of $|x|$. The **absolute value function** can be written as

$$f(x) = |x| \quad \text{or} \quad f(x) = \begin{cases} x & \text{if } x \geq 0 \\ -x & \text{if } x < 0 \end{cases}$$

Note that restrictions on the domain of the absolute value function specify different formulas for different parts of the domain. To graph $f(x) = |x|$, we graph the portion of the line $y = x$ for $x \geq 0$ (see Figure 2.26(a)) and the portion of the line $y = -x$ for $x < 0$ (see Figure 2.26(b)). When we put these pieces on the same graph (Figure 2.26(c)), they give us the graph of $f(x) = |x|$. Because the absolute value function is defined by two equations, we say it is a **piecewise defined function.**

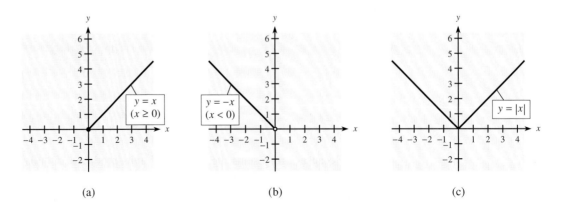

Figure 2.26

(a)   (b)   (c)

It is possible for the selling price S of a product to be defined as a piecewise function of the cost C of the product. For example, the selling price might be defined by two different equations on two different intervals, as follows:

$$S = f(C) = \begin{cases} 3C & \text{if } 0 \leq C \leq 20 \\ 1.5C + 30 & \text{if } C > 20 \end{cases}$$

When we write the equations in this way, the value of S depends on the value of C, so C is the independent variable and S is the dependent variable. The set of possible values of C (the domain) is the set of nonnegative real numbers (because cost cannot be negative), and the range is the set of nonnegative real numbers (because all values of C will result in nonnegative values of S).

Note that the selling price of a product that costs \$15 would be $f(15) = 3(15) = 45$ (dollars) and that the selling price of a product that costs \$25 would be $f(25) = 1.5(25) + 30 = 67.50$ (dollars).

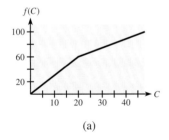

(a)

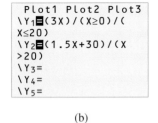

(b)

Figure 2.27

● **EXAMPLE 6** *Selling Price*

Graph the selling price function

$$S = f(C) = \begin{cases} 3C & \text{if } 0 \leq C \leq 20 \\ 1.5C + 30 & \text{if } C > 20 \end{cases}$$

Solution
Each of the two pieces of the graph of this function is a line and is easily graphed. It remains only to graph each in the proper interval. The graph is shown in Figure 2.27(a).

Graphing
Utilities

Graphing utilities can be used to graph piecewise defined functions by entering each piece as a separate equation, along with the interval over which it is defined. Figure 2.27(b) shows how the function from Example 6 could be entered on a T1-84 calculator, using x to represent C.   ■

● **EXAMPLE 7** *Residential Electrical Costs*

The 2006 monthly charge in dollars for x kilowatt hours (kWh) of electricity used by a residential customer of Excelsior Electric Membership Corporation during the months of November through June is given by the function

$$C(x) = \begin{cases} 10 + 0.094x & \text{if } 0 \le x \le 100 \\ 19.40 + 0.075(x - 100) & \text{if } 100 < x \le 500 \\ 49.40 + 0.05(x - 500) & \text{if } x > 500 \end{cases}$$

(a) What is the monthly charge if 1100 kWh of electricity are consumed in a month?
(b) What is the monthly charge if 450 kWh are consumed in a month?
(c) Is there a charge if no electricity is used?

Solution

(a) We need to find $C(1100)$. Because $1100 > 500$, we use it in the third formula line.

$$C(1100) = 49.40 + 0.05(1100 - 500) = \$79.40$$

(b) We evaluate $C(450)$ by using the second formula line for $C(x)$.

$$C(450) = 19.40 + 0.075(450 - 100) = \$45.65$$

(c) If no electricity is used, the charge is $C(0)$; it is found by using the first formula line.

$$C(0) = 10 + 0.094(0) = \$10$$

There is a $10 charge even when no electricity is used. This could be thought of as a service charge or line charge for the privilege and convenience of having electrical service available.

● ***Checkpoint***

5. If $f(x) = \begin{cases} 5 & \text{if } x \le 0 \\ 2x & \text{if } 0 < x < 5 \\ x + 6 & \text{if } x \ge 5 \end{cases}$, find the following.

(a) $f(-5)$ (b) $f(4)$ (c) $f(20)$

● ***Checkpoint Solutions***

1. Horizontal lines
2. Polynomial functions are (a) and (e).
3. Two
4. (a) True.
   (b) False. The horizontal asymptote is $y = 3$.
5. (a) $f(-5) = 5$. In fact, for any negative value of x, $f(x) = 5$.
   (b) $f(4) = 2(4) = 8$
   (c) $f(20) = 20 + 6 = 26$

2.4 Exercises

In Problems 1–12, match each of the functions with one of the graphs labeled (a)–(l) shown following these functions. Recognizing special features of certain types of functions and plotting points for the functions will be helpful.

1. $f(x) = -3$

2. $y = \sqrt{x}$

3. $y = \sqrt[3]{x}$

4. $y = (\sqrt{x})^5$

5. $y = (x + 4)^3 + 1$

6. $y = \dfrac{1}{x}$

7. $y = |x|$

8. $y = |x - 2|$

9. $f(x) = \begin{cases} x^2 & \text{if } x \le 2 \\ 4 & \text{if } x > 2 \end{cases}$

10. $f(x) = \begin{cases} 2 & \text{if } x < 0 \\ x^3 & \text{if } x \ge 0 \end{cases}$

11. $y = \begin{cases} -x & \text{if } x < -1 \\ 1 & \text{if } -1 \le x \le 1 \\ x & \text{if } x > 1 \end{cases}$

12. $y = \sqrt{x - 2}$

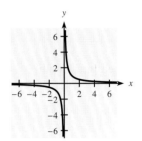

(a)

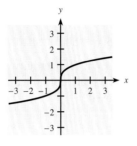

(b)

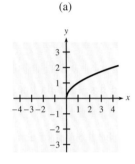

(c)

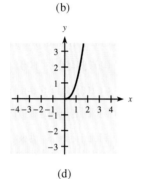

(d)

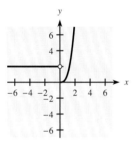

(e)

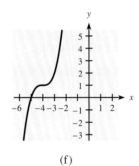

(f)

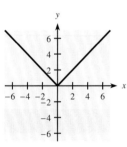

(g)

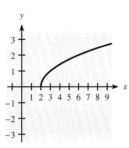

(h)

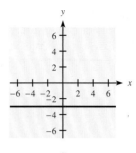

(i)

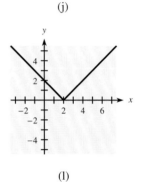

(j)

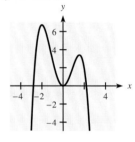

(k)

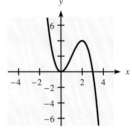

(l)

In Problems 13 and 14, decide whether each function whose graph is shown is the graph of a cubic (third degree) or quartic (fourth degree) function.

13.

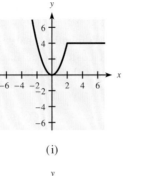

(a)

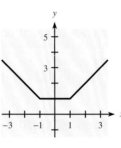

(b)

14.

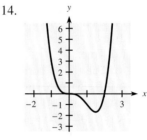

(a)

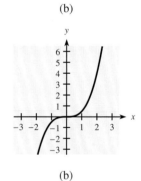

(b)

By recognizing shapes and features of polynomial and rational functions, match each equation in Problems 15–22 with the correct graph among those labeled (a)–(h). Use a graphing utility to confirm your choice.

15. $y = x^3 - x$
16. $y = (x - 3)^2(x + 1)$
17. $y = 16x^2 - x^4$
18. $y = x^4 - 3x^2 - 4$
19. $y = x^2 + 7x$
20. $y = 7x - x^2$
21. $y = \dfrac{x - 3}{x + 1}$
22. $y = \dfrac{1 - 3x}{2x + 5}$

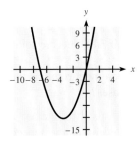

(a)

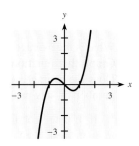

(b)

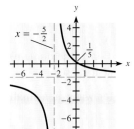

(c)

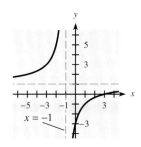

(d)

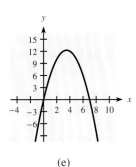

(e)

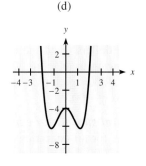

(f)

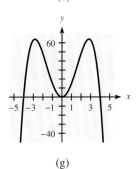

(g)

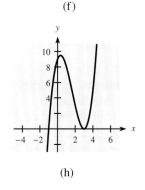

(h)

In Problems 23–28, graph the function.

23. $y = (x + 1)(x - 3)(x - 1)$
24. $y = x^3 - 8x^2 + 19x - 12$
25. $y = \dfrac{1 - 2x}{x}$
26. $y = \dfrac{x}{x - 1}$
27. $y = \begin{cases} x^3 + 2 & \text{if } x < 1 \\ \sqrt{x - 1} & \text{if } x \geq 1 \end{cases}$
28. $y = \begin{cases} x^2 & \text{if } x < 2 \\ 4 - x & \text{if } x \geq 2 \end{cases}$

29. If $F(x) = \dfrac{x^2 - 1}{x}$,

    find the following.
    (a) $F(-\frac{1}{3})$
    (b) $F(10)$
    (c) $F(0.001)$
    (d) Is $F(0)$ defined?

30. If $H(x) = |x - 1|$, find the following.
    (a) $H(-1)$
    (b) $H(1)$
    (c) $H(0)$
    (d) Does $H(-x) = H(x)$?

31. If $f(x) = x^{3/2}$, find the following.
    (a) $f(16)$
    (b) $f(1)$
    (c) $f(100)$
    (d) $f(0.09)$

32. If $k(x) = \begin{cases} 4 - 2x & \text{if } x < 0 \\ |x - 4| & \text{if } 0 < x < 4, \end{cases}$

    find the following.
    (a) $k(-0.1)$
    (b) $k(0.1)$
    (c) $k(3.9)$
    (d) $k(4.1)$

33. If $k(x) = \begin{cases} 2 & \text{if } x < 0 \\ x + 4 & \text{if } 0 \leq x < 1, \\ 1 - x & \text{if } x \geq 1 \end{cases}$

    find the following.
    (a) $k(-5)$
    (b) $k(0)$
    (c) $k(1)$
    (d) $k(-0.001)$

34. If $g(x) = \begin{cases} 0.5x + 4 & \text{if } x < 0 \\ 4 - x & \text{if } 0 \leq x < 4, \\ 0 & \text{if } x > 4 \end{cases}$

    find the following.
    (a) $g(-4)$
    (b) $g(1)$
    (c) $g(7)$
    (d) $g(3.9)$

In Problems 35–40, (a) graph each function (with a graphing utility if one is available); (b) classify each function as a polynomial function, a rational function, or a piecewise defined function; (c) identify any asymptotes; and (d) use the graphs to locate turning points.

35. $f(x) = 1.6x^2 - 0.1x^4$
36. $f(x) = \dfrac{x^4 - 4x^3}{3}$
37. $f(x) = \dfrac{2x + 4}{x + 1}$
38. $f(x) = \dfrac{x - 3}{x + 2}$
39. $f(x) = \begin{cases} -x & \text{if } x < 0 \\ 5x & \text{if } x \geq 0 \end{cases}$
40. $f(x) = \begin{cases} 2x - 1 & \text{if } x < 1 \\ -x & \text{if } x \geq 1 \end{cases}$

APPLICATIONS

41. *Postal restrictions* If a box with a square cross section is to be sent by the postal service, there are restrictions on its size such that its volume is given by $V = x^2(108 - 4x)$, where x is the length of each side of the cross section (in inches).

(a) If $V = V(x)$, find $V(10)$ and $V(20)$.

(b) What restrictions must be placed on x (the domain) so that the problem makes physical sense?

42. *Fixed costs*   Fixed costs FC are business costs that remain constant regardless of the number of units produced. Some items that might contribute to fixed costs are rent and utilities.

$$FC = 2000$$

is an equation indicating that a business has fixed costs of $2000. Graph $FC = 2000$ by putting x (the number of units produced) on the horizontal axis. (Note that FC does not mean the product of F and C.)

 43. *Mutual funds*   The amount of household assets in mutual funds, measured in billions of dollars, for the years 1995 to 1999 was found to be modeled by

$$f(x) = 105.095x^{1.5307} \text{ billion dollars}$$

where x is the number of years past 1990. (*Source:* Federal Reserve, *Time,* April 3, 2000)

(a) Does the graph of this function bend upward or bend downward?

(b) Graph this function.

(c) Use numerical or graphical methods to find when the model predicts that the assets will reach $4,000,000,000,000.

(d) Do you think this model would be accurate through 2002? Why?

 44. *Allometric relationships*   For fiddler crabs, data gathered by Thompson show that the relationship between the weight C of the claw and the weight W of the body is given by

$$C = 0.11W^{1.54}$$

where W and C are in grams. (*Source:* d'Arcy Thompson, *On Growth and Form,* Cambridge University Press, 1961)

(a) Does the graph of this function bend upward or bend downward?

(b) Graph this function.

(c) Use this function to compute the weight of a claw if the weight of the crab's body is 10 grams.

45. *Pollution*   Suppose that the cost C (in dollars) of removing p percent of the particulate pollution from the smokestacks of an industrial plant is given by

$$C(p) = \frac{7300p}{100 - p}$$

(a) Find the domain of this function. Recall that p represents the percent of pollution that is removed. In parts (b)–(e), find the functional values and explain what each means.

(b) $C(45)$ (c) $C(90)$

(d) $C(99)$ (e) $C(99.6)$

(f) What happens to the cost as the percent of pollution removed approaches 100%?

46. *Average cost*   If the weekly total cost of producing 27″ Toshiba television sets is given by $C(x) = 50,000 + 105x$, where x is the number of sets produced per week, then the **average cost** per unit is given by

$$\overline{C}(x) = \frac{50,000 + 105x}{x}$$

(a) What is the average cost per set if 3000 sets are sold?

(b) Graph this function.

(c) Does the average cost per set continue to fall as the number of sets produced increases?

47. *Area*   If 100 feet of fence is to be used to enclose a rectangular yard, then the resulting area of the fenced yard is given by

$$A = x(50 - x)$$

where x is the width of the rectangle.

(a) If $A = A(x)$, find $A(2)$ and $A(30)$.

(b) What restrictions must be placed on x (the domain) so that the problem makes physical sense?

48. *Water usage*   The monthly charge for water in a small town is given by

$$f(x) = \begin{cases} 38 & \text{if } 0 \le x \le 20 \\ 38 + 0.4(x - 20) & \text{if } x > 20 \end{cases}$$

where x is water usage in hundreds of gallons and $f(x)$ is in dollars. Find the monthly charge for each of the following usages: (a) 30 gallons, (b) 3000 gallons, and (c) 4000 gallons. (d) Graph the function for $0 \le x \le 100$.

49. *Federal support for education*   The federal on-budget funds for all educational programs (in billions of constant 2000 dollars) between 1965 and 2000 can be modeled by the function

$$P(t) = \begin{cases} 1.965t - 5.65 & \text{when } 5 \le t \le 20 \\ 0.095t^2 - 2.925t + 54.15 & \text{when } 20 < t \le 40 \end{cases}$$

where t is the number of years past 1960. (*Source:* U.S. Department of Education, National Center for Educational Statistics)

(a) Graph the function P for $5 \le t \le 40$. Describe how the funding for educational programs varied between 1960 and 2000.

(b) What was the amount of funding for educational programs in 1980?

(c) How much federal funding was allotted for educational programs in 1998?

50. *Commercial electrical usage* The monthly charge (in dollars) for x kilowatt hours (kWh) of electricity used by a commercial customer is given by the following function.

$$C(x) = \begin{cases} 7.52 + 0.1079x & \text{if } 0 \le x \le 5 \\ 19.22 + 0.1079x & \text{if } 5 < x \le 750 \\ 20.795 + 0.1058x & \text{if } 750 < x \le 1500 \\ 131.345 + 0.0321x & \text{if } x > 1500 \end{cases}$$

Find the monthly charges for the following usages.
(a) 5 kWh   (b) 6 kWh   (c) 3000 kWh

51. *First-class postage* The postage charged for first-class mail is a function of its weight. The U.S. Postal Service uses the following table to describe the rates.

Weight Increment	Rate
First ounce or fraction of an ounce	0.370
Each additional ounce or fraction	0.230

Source: United States Postal Service, pe.usps.gov/text

(a) Convert this table to a piecewise defined function that represents postage for letters weighing between 0 and 4 ounces, using x as the weight in ounces and P as the postage in cents.
(b) Find $P(1.2)$ and explain what it means.
(c) Give the domain and range of P as it is defined above.
(d) Find the postage for a 2-ounce letter and for a 2.01-ounce letter.

52. *Income tax* The 2004 U.S. federal income tax owed by a married couple filing jointly can be found from the following table. (*Source:* Internal Revenue Service, 2004, Form 1040 Instructions)

Filing Status: Married Filing Jointly				
If Taxable Income is between		Tax Due is		of the amount over
$0 –	$14,300	$0.00 +	10.0%	$0
$14,300 –	$58,100	$1,430.00 +	15.0%	$14,300
$58,100 –	$117,250	$8,000.00 +	25.0%	$58,100
$117,250 –	$178,650	$22,787.50 +	28.0%	$117,250
$178,650 –	$319,100	$39,979.50 +	33.0%	$178,650
$319,100 –	Up	$86,328.00 +	35.0%	$319,100

(a) Write the piecewise defined function T with input x that models the federal tax dollars due as a function of x, the taxable income dollars earned.
(b) Use the function to find $T(70,000)$.
(c) Find the tax due on a taxable income of $150,000.
(d) A friend tells Jack Waddell not to earn any money over $117,250 because it would raise his tax rate to 28% on all of his taxable income. Test this statement by finding the tax due on $117,250 and $117,250 + $1. What do you conclude?

53. *Demand* The demand function for a product is given by

$$p = \frac{200}{2 + 0.1x}$$

where x is the number of units and p is the price in dollars.
(a) Graph this demand function for $0 \le x \le 250$, with x on the horizontal axis.
(b) Does the demand ever reach 0?

54. *Mob behavior* In studying lynchings between 1899 and 1946, psychologist Brian Mullin found that the size of a lynch mob relative to the number of victims predicted the level of brutality. He developed a formula for the other-total ratio (y) that predicts the level of self-attentiveness of people in a crowd of size x with 1 victim.

$$y = \frac{1}{x + 1}$$

The lower the value of y, the more likely an individual is to be influenced by "mob psychology." Graph this function; use positive integers as its domain.

55. *Production costs* A manufacturer estimates that the cost of a production run for a product is

$$C(x) = 30(x - 1) + \frac{3000}{x + 10}$$

where x is the number of machines used.
(a) Graph this total cost function for values of $x \ge 0$.
(b) Interpret any turning points.
(c) Interpret the y-intercept.

OBJECTIVES

- To graph data points in a scatter plot
- To determine the function type that will best model data
- To use a graphing utility to create an equation that models the data
- To graph the data points and model on the same graph

Modeling; Fitting Curves to Data with Graphing Utilities (optional)

◑ Application Preview

When the plot of a set of data points looks like a line, we can use a technique called **linear regression** (or the **least-squares method**) to find the equation of the line that is the best fit for the data points. The resulting equation (**model**) describes the data and gives a formula to plan for the future. We can use technology to fit linear and other functions to sets of data.

The data in Table 2.2 show that the amount of tax paid per capita (per person) has increased every year from 1983 to 2003. In this section we develop a model for these data (see Example 3.)

TABLE 2.2

Year	Federal Tax per Capita	Year	Federal Tax per Capita	Year	Federal Tax per Capita
1983	$2490	1990	$4026	1997	$5716
1984	2738	1991	4064	1998	6142
1985	2982	1992	4153	1999	6443
1986	3090	1993	4381	2000	7044
1987	3414	1994	4710	2001	6865
1988	3598	1995	4985	2002	6326
1989	3884	1996	5315	2003	6282

Source: Tax Foundation

Linear Regression

We saw in Section 1.3, "Linear Functions," that it is possible to write the equation of a straight line if we have two points on the line. Business firms frequently like to treat demand functions as though they are linear, even when they are not exactly linear. They do this because linear functions are much easier to handle than other functions. If a firm has more than two points describing the demand for its product, it is likely that the points will not all lie on the same straight line. However, by using a technique called **linear regression,** the firm can determine the "best line" that fits these points.

Suppose we have the points shown in Figure 2.28. We seek the line that is the "best fit" for the points.

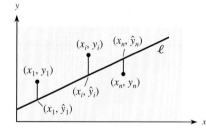

Figure 2.28

Corresponding to each point (x_i, y_i) that we know, we can find a point $(x_i, \hat{y}_i)$ on the line. If the line we have is a good fit for the points, the differences between the y-values of the given points and the corresponding $\hat{y}$-values on the line should be small. The line shown is the best line for the points only if the sum of the squares of these differences is a minimum.*

The formulas that give the best-fitting line for given data are developed in Chapter 14, "Functions of Two or More Variables," but graphing calculators, computer programs, and spreadsheets have built-in formulas or programs that give the equation of the best-fitting line for a set of data.

To find a function that models a set of data, the first step is to plot the data and determine the shape of the curve that would best fit the data points. The plot of the data points is called a **scatter plot.** The scatter plot's shape determines the type of function that will be the best model for the data. The following example shows how a model can be developed for a set of data.

 ● **EXAMPLE 1** *U.S. Households with Internet Access*

The following table gives the percent of U.S. households with internet access in various years.

Year	1996	1997	1998	1999	2000	2001
Percent	8.5	14.3	26.2	28.6	41.5	50.5

Source: U.S. Bureau of the Census

(a) Use a graphing utility to create a scatter plot of the data, with x equal to the number of years after 1995.
(b) Determine what type of function will best model the data.
(c) Use the utility to create an equation that models the data.
(d) Graph the function and the data on the same graph to see how well the function models ("fits") the data.

Solution

(a) When we use x as the number of years past 1995 (so $x = 0$ in 1995) and use y to represent the percent of U.S. households with Internet access, the points representing this function are

$$(1, 8.5) \quad (2, 14.3) \quad (3, 26.2) \quad (4, 28.6) \quad (5, 41.5) \quad (6, 50.5)$$

Plotting these points on a coordinate plane gives the scatter plot shown in Figure 2.29(a). (These points can be plotted by hand or by using a graphing utility.)
(b) It appears that a line could be found that would fit close to the points in Figure 2.29(a). The points do not fit exactly on a line, because the percent of U.S. households with Internet access did not increase by exactly the same amount each year.
(c) The equation of the line that is the best fit for the data can be found by using a graphing utility programmed to perform the linear regression. The linear equation that is the best fit for the points is

$$y = 8.4x - 1.33$$

(d) Figure 2.29(b) shows the graph of the function and the data on the same graph. Note that not all points fit on the graph of the equation, even though this is the line that is the best fit for the data.

Note that in this model the slope of the line is $m = 8.4$ percent per year and represents the (approximate) annual rate of change in the percent of U.S. households with Internet access. Also note that the graph of this equation will rise indefinitely, but the percent of households cannot exceed 100%. Thus, eventually, this model will no longer be valid.

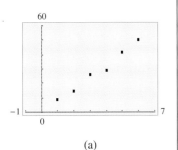

(a)

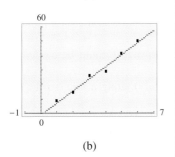

(b)

Figure 2.29

---

*Because linear regression involves minimizing the sum of squared numbers, it is also called the **least-squares method.**

Modeling   Of course, many relations that occur are not linear. The use of graphing utilities permits us to use the same steps as those in Example 1 to model nonlinear relations. Creating an equation, or **mathematical model,** for a set of data is a complex process that involves careful analysis of the data in accordance with a number of guidelines. Our approach to creating a model is much simplified and relies on a knowledge of the appearance of linear, quadratic, power, cubic, and quartic functions and the capabilities of computers and graphing calculators. We should note that the equations the computers and calculators provide are based on sophisticated formulas that are derived using calculus of several variables. These formulas provide the best fit for the data using the function type we choose, but we should keep the following limitations in mind.

1. The computer/calculator will give the best fit for whatever type of function we choose (even if the selected function is a bad fit for the points), so we must choose a function type carefully. To choose a function type, compare the scatter plot with the graphs of the functions discussed earlier in this chapter. The model will not fit the data if we choose a function type whose graph does not match the shape of the plotted data.
2. Some sets of data have no pattern, so they cannot be modeled by functions. Other data sets cannot be modeled by the functions we have studied.
3. Modeling provides a formula that relates the data, but it is not an absolute truth. In fact, even though a model may accurately fit a data set, it may not be a good predictor over a period of time outside the data set.

● **EXAMPLE 2** *Curve Fitting*

Graphs of three sets of data points are shown in Figure 2.30(a–c). Determine what type of function is your choice as the best-fitting curve for each scatter plot. If it is a polynomial function, state the degree.

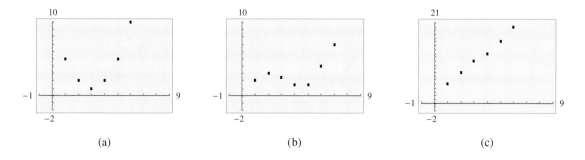

Figure 2.30   (a)   (b)   (c)

Solution
(a) It appears that a parabola will fit the data points in Figure 2.30(a), so the data points can be modeled by a second-degree (quadratic) function.
(b) The two "turns" in the scatter plot in Figure 2.30(b) suggest that it could be modeled by a cubic function.
(c) The data points in Figure 2.30(c) appear to lie (approximately) along a line, so the data can be modeled by a linear function.

After the type of function that gives the best fit for the data is chosen, a graphing utility or spreadsheet can be used to develop the best-fitting equation for the function chosen. The following steps are used to find an equation that models data.

Modeling Data

1. Enter the data and use a graphing utility or spreadsheet to plot the data points. (This gives a scatter plot.)

2. Visually determine what type of function (including degree, if it is a polynomial function) would have a graph that best fits the data.

3. Use a graphing utility or spreadsheet to determine the equation of the chosen type that gives the best fit for the data.

4. To see how well the equation models the data, graph the equation and the data points on the same set of axes. If the graph of the equation does not fit the data points well, another type of function may model the data better.

◗ EXAMPLE 3 Federal Tax per Capita (Application Preview)

The table below gives the amount of federal tax paid per capita (per person) for the years 1983 to 2003.

(a) Create a scatter plot of these data.
(b) Find a cubic model that fits the data points, with $x = 0$ representing 1980.
(c) Use the model to predict the per capita tax for 2006.

Year	Federal Tax per Capita	Year	Federal Tax per Capita	Year	Federal Tax per Capita
1983	$2490	1990	$4026	1997	$5716
1984	2738	1991	4064	1998	6142
1985	2982	1992	4153	1999	6443
1986	3090	1993	4381	2000	7044
1987	3414	1994	4710	2001	6865
1988	3598	1995	4985	2002	6326
1989	3884	1996	5315	2003	6282

Source: Tax Foundation

Solution

(a) The scatter plot is shown in Figure 2.31(a). The "turns" in the graph indicate that a cubic function would be a good fit for the data.

(b) The cubic function that is the best fit for the data is

$$y = -0.848x^3 + 33.166x^2 - 151.360x + 2919.569$$

The graph of the function and the scatter plot are shown in Figure 2.31(b).

(c) The value of y for $x = 26$ (shown in Figure 2.31(c)) predicts $6500 as the tax per capita in 2006.

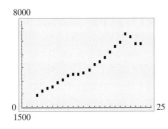

(a)

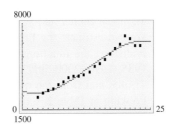

(b)

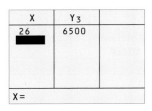

(c)

Figure 2.31

If it is not obvious what model will best fit a given set of data, several models can be developed and compared graphically with the data.

● **EXAMPLE 4** *Expected Life Span*

The expected life span of people in the United States depends on their year of birth.

Year	Life Span (years)	Year	Life Span (years)	Year	Life Span (years)	Year	Life Span (years)
1920	54.1	1975	72.6	1991	75.5	1997	76.5
1930	59.7	1980	73.7	1992	75.5	1998	76.7
1940	62.9	1987	75.0	1993	75.5	1999	76.7
1950	68.2	1988	74.9	1994	75.7	2000	77.0
1960	69.7	1989	75.1	1995	75.8	2001	77.2
1970	70.8	1990	75.4	1996	76.1		

Source: National Center for Health Statistics

(a) Create linear and quadratic models that give life span as a function of birth year with $x = 0$ representing 1900 and, by visual inspection, decide which model gives the better fit.
(b) Use both models to estimate the life span of a person born in the year 2000.
(c) Which model's prediction for the lifespan in 2010 seems better?

Solution

(a) The scatter plot for the data is shown in Figure 2.32(a). It appears that a linear function could be used to model the data. The linear equation that is the best fit for the data is

$$y = 0.250x + 52.642$$

The graph in Figure 2.32(b) shows how well the line fits the data points. The quadratic function that is the best fit for the data is

$$y = -0.00229x^2 + 0.545x + 44.926$$

Its graph is shown in Figure 2.32(c).

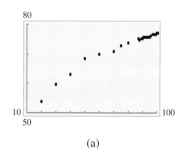

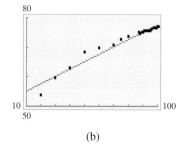

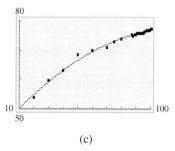

Figure 2.32   (a)   (b)   (c)

X	Y₁	Y₂
85	73.928	74.727
90	75.18	75.45
100	77.685	76.552
110	80.189	77.198

X =

Table 2.3

The quadratic model appears to fit the data points better than the linear model. A table can be used to compare the models (see Table 2.3 with y_1 giving values from the linear model and y_2 giving values from the quadratic model).

(b) We can estimate the life span of people born in the year 2000 with either model by evaluating the functions at $x = 100$. The x-values in Table 2.3 represent the years 1985, 1990, 2000, and 2010. From the linear model the expected life span in 2000 is 77.685, and from the quadratic model the expected life span is 76.552.

(c) Looking at Table 2.3, we see that the linear model may be giving optimistic values in 2010, when $x = 110$, so the quadratic model may be better in the years up to 2010.

● **EXAMPLE 5** *Corvette Acceleration*

The following table shows the times that it takes a 2005 Corvette to reach speeds from 0 mph to 100 mph, in increments of 10 mph after 30 mph.

Time (sec)	Speed (mph)	Time (sec)	Speed (mph)
1.7	30	5.5	70
2.4	40	7.0	80
3.5	50	8.5	90
4.3	60	10.2	100

(a) Use a power function to model the data, and graph the points and the function to see how well the function fits the points.
(b) What does the model indicate the speed is 5 seconds after the car starts to move?
(c) Use the model to determine the number of seconds until the Corvette reaches 110 mph.

Solution
(a) Using x as the time and y as the speed, the scatter plot is shown in Figure 2.33(a). An equation found using the power function $y = ax^b$ (graphed in Figure 2.33(b)) is

$$y = 21.875x^{0.666}$$

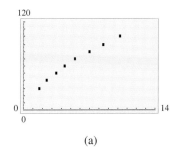
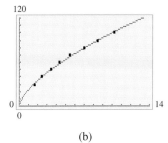

Figure 2.33   (a)   (b)

(b) Using $x = 5$ in this equation gives 63.9 mph as the speed of the Corvette.
(c) Tracing along the graph of $y_1 = 21.875x^{0.666}$ until the y-value is 110 gives 11.3, so the time required to reach 110 mph is 11.3 seconds.

● **EXAMPLE 6** *Consumer Price Index*

The consumer price index urban (CPI-U) measures how prices have changed for urban consumers since 1982. The CPI-U is measured by setting the value of a product or service at 100 in 1982, so that a CPI-U of 300 today would mean that the price is 3 times as high as it was then. The following table gives the CPI-U for the years 1985 through 2003. (*Source:* Bureau of Labor Statistics)

Year	Consumer Price Index	Year	Consumer Price Index
1985	113.2	1995	224.2
1986	121.9	1996	232.4
1987	130.0	1997	239.1
1988	138.3	1998	246.8
1989	148.9	1999	255.1
1990	162.7	2000	266.0
1991	177.1	2001	278.8
1992	190.5	2002	292.9
1993	202.9	2003	306.0
1994	213.4		

(a) Plot the data points, with x equal to the number of years past 1980.

(b) Find a function that models the CPI-U, where x is the number of years past 1980.

(c) Graph this function on the same axes as the scatter plot, for the values of x that represent the years 1980 to 2003. Comment on how well the model fits the data.

(d) Use the model to estimate the CPI-U for 2004.

Solution

(a) The scatter plot of the data points is shown in Figure 2.34(a).

(b) The scatter plot shows that it is reasonable to model the data by a linear function. The linear function that models the consumer price index is

$$y = 10.648x + 58.304$$

where x is the number of years after 1980.

(c) The graph of the data points and the function are shown in Figure 2.34(b).

(d) To estimate the consumer price index for 2004, we evaluate the function at $x = 24$, getting 313.9.

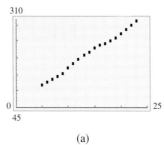
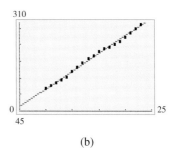

Figure 2.34   (a)   (b)

Spreadsheet Note

The *Excel Chart Wizard* can be used to create scatter plots of data and to find the best-fitting linear or polynomial function for a set of data. The *Excel Guide for Finite Mathematics and Applied Calculus* has detailed instructions for creating scatter plots and for modeling sets of data with functions.

Selecting the Chart Wizard in the Excel program and following the instructions in the Excel Guide will give a scatter plot of the data in Example 6. (See Figure 2.35(a).) After completing the steps to find the scatter plot and selecting ADD TRENDLINE from the Chart menu, choosing linear regression will give the equation of the best-fitting linear function. (See Figure 2.35(b).)

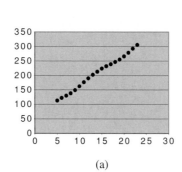
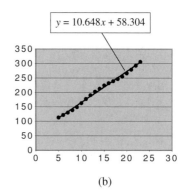

Figure 2.35   (a)   (b)

2.5 Exercises

In Problems 1–8, determine whether the scatter plot should be modeled by a linear, power, quadratic, cubic, or quartic function.

1.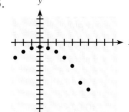

2.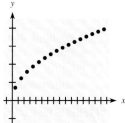

3.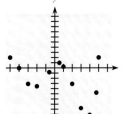

4.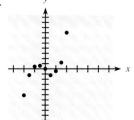

5.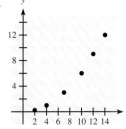

6.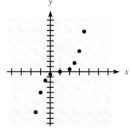

7.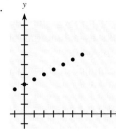

8.

In Problems 9–16, find the equation of the function of the specified type that is the best fit for the given data. Plot the data and the equation.

9. linear

x	y
-2	-7
-1	-5
0	-3
1	-1
2	1
3	3
4	5

10. linear

x	y
-1	-5.5
0	-4
1	-2.5
2	-1
3	.5
4	2
5	3.5

11. quadratic

x	y
-2	7
-1	-0.5
0	-4
1	-3.5
2	1
3	9.5
4	22

12. quadratic

x	y
-4	-8
-3	-4
-2	-2
-1	-2
0	-4
1	-8
2	-14

13. cubic

x	y
-4	-72
-3	-31
-2	-10
-1	-3
0	-4
1	-7
2	-6

14. cubic

x	y
-3	-22
-2	-6
-1	-2
0	-4
1	-6
2	-2
3	14

15. power

x	y
1	2
2	2.8284
3	3.4641
4	4
5	4.4721
6	4.899

16. power

x	y
1	3
2	8.4853
3	15.588
4	24
5	33.541
6	44.091

In Problems 17–24, (a) plot the given points, (b) determine what type of function best models the data, and (c) find the equation that is the best fit for the data.

17.

x	y
-1	-8
0	-3
1	2
2	7
3	12
4	17
5	22

18.

x	y
-1	-2
0	2
1	6
2	10
3	14
4	18
5	22

19.

x	y
−1	2
0	0
1	2
2	3.1748
3	4.1602
4	5.0397
5	5.848

20.

x	y
0.01	0.3
1	3
2	4.2426
3	5.1962
4	6
5	6.7082
6	7.3485

21.

x	y
−2	19
−1	8
0	1
1	−2
2	−1
3	4
4	13

22.

x	y
−4	37
−3	19
−2	7
−1	1
0	1
1	7
2	19

23.

x	y
−3	−11
−2	3
−1	5
0	1
1	−3
2	−1
3	13

24.

x	y
−3	−54
−2	−14
−1	0
0	0
1	−2
2	6
3	36

APPLICATIONS

25. *Earnings and gender*   The table shows the 1999 average earnings of year-round full-time workers by gender and educational attainment.
   (a) Let x represent earnings for males and y represent earnings for females, and create a scatter plot of these data.
   (b) Find a linear model that expresses women's annual earnings as a function of men's.
   (c) Find the slope of the linear model in part (b) and write a sentence that interprets it.

Educational Attainment	Average Annual Earnings (thousands of dollars)	
	Males	Females
Less than 9th grade	18.743	12.392
Some high school	18.908	12.057
High school graduate	30.414	18.092
Some college	33.614	20.241
Associate's degree	40.047	25.079
Bachelor's degree or more	66.810	36.755

Source: U.S. Bureau of the Census

26. *Earnings and race*   The table gives the median household income (in 2003 dollars) for whites and blacks in various years.
   (a) Let x represent the median household income for whites and y represent the corresponding median household income for blacks, and make a scatter plot of these data.
   (b) Find a linear model that expresses the median household income for blacks as a function of the median household income for whites.
   (c) Find the slope of the linear model in (b) and write a sentence that interprets it.

Median Household Income (2003 dollars)		
Year	Whites	Blacks
1981	38,954	21,859
1985	40,614	24,163
1990	42,622	25,488
1995	42,871	26,842
2000	46,910	31,690
2001	46,261	30,625
2002	46,119	29,691
2003	45,631	29,645

Source: U.S. Bureau of the Census

27. *Executions*   The numbers of executions carried out in the United States for the years 1984 through 2003 are given in the table.
   (a) Find the cubic function that best fits the data, using x as the number of years past 1980.
   (b) According to the model developed in part (a), for what year was the number of executions a maximum? Does this result agree with the data?

Year	U.S. Executions	Year	U.S. Executions
1984	21	1994	31
1985	18	1995	56
1986	18	1996	45
1987	25	1997	74
1988	11	1998	68
1989	16	1999	98
1990	23	2000	85
1991	14	2001	66
1992	31	2002	71
1993	38	2003	65

Source: www.cuadp.org

28. *Airport runway near-hits* The chart gives the number of near-collisions on the runways of the nation's airports.
(a) With $x = 0$ representing 1990, find a quadratic function that models the data in the chart.
(b) Use the model to find the year when the number of near-collisions was a minimum. Does this result agree with the data?

Runway near-hits in USA

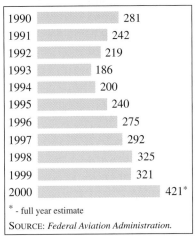

1990	281
1991	242
1992	219
1993	186
1994	200
1995	240
1996	275
1997	292
1998	325
1999	321
2000	421*

* - full year estimate

SOURCE: *Federal Aviation Administration.*

By Julie Snider, *USA Today*

29. *Gross domestic product* The table gives the gross domestic product (GDP, in billions of dollars) of the United States for selected years from 1940 to 2003.
(a) Create a scatter plot of the data, with y representing GDP in billions of dollars and x representing the number of years past 1940.
(b) Find the linear function that best fits the data, with x equal to the number of years past 1940.
(c) Find the cubic function that best fits the data, with x equal to the number of years past 1940.
(d) Graph each of these functions on the same axes as the data points, and visually determine which model is the better fit for the data.

Year	Domestic Gross Product	Year	Domestic Gross Product
1940	837	1980	3,746
1945	1,559	1985	4,207
1950	1,328	1990	4,853
1955	1,700	1995	5,439
1960	1,934	2000	9,817
1965	2,373	2001	10,128
1970	2,847	2002	10,487
1975	3,173	2003	11,004

Source: U.S. Bureau of Economic Analysis

30. *Poverty* The table shows the number of millions of people in the United States who lived below the poverty level for selected years.

(a) Find a model that approximately fits the data, using x as the number of years after 1960.
(b) Use a graph of the model to determine when this model would no longer be valid. Explain why it would no longer be valid.

Year	Persons Living Below the Poverty Level (millions)
1960	39.9
1965	33.2
1970	25.4
1975	25.9
1980	29.3
1986	32.4
1989	31.5
1990	33.6
1991	35.7
1992	38
1993	39.3
1994	38.1
1995	36.4
1996	36.5
1997	35.6
1998	34.5
1999	32.3
2000	31.1
2002	34.6

Source: U.S. Bureau of the Census, U.S. Department of Commerce

31. *Budget deficit* The table gives the yearly budget deficit (as a negative value) or surplus (as a positive value) in billions of dollars for selected years.
(a) Use x as the number of years past 1990 to create a quadratic and a cubic model using the data.
(b) Compare both models with the data to see which is the better fit to the data, and use these models to predict the value for 2005. (The government estimated it to be −$477 billion.)

Year	Yearly Deficit or Surplus	Year	Yearly Deficit or Surplus
1990	−221.2	1998	70.0
1991	−269.4	1999	124.4
1992	−290.4	2000	237.0
1993	−255.0	2001	127.0
1994	−203.1	2002	−159.0
1995	−164.0	2003	−374.0
1996	−107.5	2004	−445.0
1997	−22.0		

Source: Budget of the United States Government

32. *Tax load*   The percent of total personal income used to pay personal taxes is given in the table.
    (a) Using *x* as the number of years past 1990, write the equation of the function that is the best fit for the data.
    (b) When is the model no longer valid?

Year	Percent	Year	Percent
1990	13.0	1996	13.9
1991	12.6	1997	14.6
1992	12.4	1998	14.5
1993	12.6	1999	14.9
1994	12.8	2000	15.5
1995	13.1	2002	17.0

Source: Bureau of Economic Analysis, U.S. Department of Commerce

33. *National health care*   The table below shows the national expenditures for health care in the United States for selected years.
    (a) Use a scatter plot with *x* as the number of years past 1950 and *y* as the total expenditures for health care (in billions) to identify what type (or types) of function(s) would make a good model for these data.
    (b) Find a power model and a quadratic model for the data.
    (c) Which model from (b) more accurately estimates the 2001 expenditures for national health care?
    (d) Use the better model from (c) to estimate the 2010 expenditures for national health care.

Year	National Expenditures for Health Care (in billions)
1960	$26.7
1970	73.1
1980	245.8
1990	695.6
1995	987.0
1999	1210.7
2000	1310.0
2001	1425.5

Source: U.S. Department of Health and Human Services

34. *U.S. population*   The table gives the U.S. population, in millions, for selected years.
    (a) Create a scatter plot for the data in the table, with *x* equal to the number of years past 1960.

    (b) Use the scatter plot to determine the type of function that can be used to model the data, and create a function that best fits the data, with *x* equal to the number of years past 1960.

Year	U.S. Population (millions)
1960	180.671
1965	194.303
1970	205.052
1975	215.973
1980	227.726
1985	238.466
1990	249.948
1995	263.044
1998	270.561
2000	281.422
2003	294.043*
2025	358.030*
2050	408.695*

Source: U.S. Bureau of the Census
*Estimates

35. *Cell phones*   The table gives the number of thousands of U.S. cellular telephone subscriberships.
    (a) Find the power function that is the best fit for these data, with *x* equal to the number of years after 1980 and *y* equal to the number of subscriberships in thousands.
    (b) Use the model to estimate the number of subscriberships in 2005.
    (c) Why does this estimate seem overly optimistic?

Year	Subscriberships (thousands)	Year	Subscriberships (thousands)
1985	340	1994	24,134
1986	682	1995	33,786
1987	1,231	1996	44,043
1988	2,069	1997	55,312
1989	3,509	1998	69,209
1990	5,283	1999	86,047
1991	7,557	2000	109,478
1992	11,033	2001	128,375
1993	16,009	2002	140,767

Source: Semiannual CTIA Wireless Industry Survey

36. *Foreign-born population*   The table gives the percent of the U.S. population that is foreign born.
    (a) Create a scatter plot of the data, with x equal to the number of years after 1900 and y equal to the percent.
    (b) Use the scatter plot to determine the type of function that is the best fit for the data.
    (c) Find the best-fitting function for the data.
    (d) Use the function to estimate the percent in 2005.

Year	Percent Foreign Born	Year	Percent Foreign Born
1900	13.6	1960	5.4
1910	14.7	1970	4.7
1920	13.2	1980	6.2
1930	11.6	1990	8.0
1940	8.8	2000	10.4
1950	6.9	2002	11.5

Source: U.S. Bureau of the Census

Key Terms and Formulas

Section	Key Terms	Formula
2.1	Quadratic equation	$ax^2 + bx + c = 0 \quad (a \neq 0)$
	Square root property	$x^2 = C \Rightarrow x = \pm\sqrt{C}$
	Quadratic formula	$x = \dfrac{-b \pm \sqrt{b^2 - 4ac}}{2a}$
	Quadratic discriminant	$b^2 - 4ac$
2.2	Quadratic function; parabola	$y = f(x) = ax^2 + bx + c$
	Vertex of a parabola	$(-b/2a, f(-b/2a))$ Maximum point if $a < 0$ Minimum point if $a > 0$
	Axis of symmetry	$x = -b/2a$
	Zeros of a quadratic function	$x = \dfrac{-b \pm \sqrt{b^2 - 4ac}}{2a}$
2.3	Supply function Demand function Market equilibrium Cost, revenue, and profit functions	$P(x) = R(x) - C(x)$
	Break-even points	$R(x) = C(x)$ or $P(x) = 0$
2.4	Basic functions	$f(x) = ax + b$ (linear) $f(x) = C, \quad C = \text{constant}$ $f(x) = ax^b$ (power) $f(x) = x$ (identity) $f(x) = x^2, \quad f(x) = x^3, \quad f(x) = 1/x$
	Shifts of graphs	$f(x - h)$ shifts $f(x)$ h units horizontally $f(x) + k$ shifts $f(x)$ k units vertically
	Polynomial function of degree n	$f(x) = a_n x^n + a_{n-1}x^{n-1} + \cdots + a_1 x + a_0$ $a_n \neq 0, n \geq 0, n$ an integer
	Rational function	$f(x) = p(x)/q(x)$, where $p(x)$ and $q(x)$ are polynomials
	Vertical asymptote Horizontal asymptote	
	Absolute value function	$f(x) = |x| = \begin{cases} x & \text{if } x \geq 0 \\ -x & \text{if } x < 0 \end{cases}$
	Piecewise defined functions	

Section	Key Terms	Formula
2.5	Scatter plots Function types Mathematical model Fitting curves to data points Predicting from models	

Review Exercises

Section 2.1

In Problems 1–10, find the real solutions to each quadratic equation.

1. $3x^2 + 10x = 5x$
2. $4x - 3x^2 = 0$
3. $x^2 + 5x + 6 = 0$
4. $11 - 10x - 2x^2 = 0$
5. $(x - 1)(x + 3) = -8$
6. $4x^2 = 3$
7. $20x^2 + 3x = 20 - 15x^2$
8. $8x^2 + 8x = 1 - 8x^2$
9. $7 = 2.07x - 0.02x^2$
10. $46.3x - 117 - 0.5x^2 = 0$

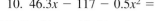 In Problems 11–14, solve each equation by using a graphing utility to find the zeros of the function. Solve the equation algebraically to check your results.

11. $4z^2 + 25 = 0$
12. $z(z + 6) = 27$
13. $3x^2 - 18x - 48 = 0$
14. $3x^2 - 6x - 9 = 0$
15. Solve $x^2 + ax + b = 0$ for x.
16. Solve $xr^2 - 4ar - x^2c = 0$ for r.

In Problems 17 and 18, approximate the real solutions to each quadratic equation to two decimal places.

17. $23.1 - 14.1x - 0.002x^2 = 0$
18. $1.03x^2 + 2.02x - 1.015 = 0$

Section 2.2

For each function in Problems 19–24, find the vertex and determine if it is a maximum or minimum point, find the zeros if they exist, and sketch the graph.

19. $y = \frac{1}{2}x^2 + 2x$
20. $y = 4 - \frac{1}{4}x^2$
21. $y = 6 + x - x^2$
22. $y = x^2 - 4x + 5$
23. $y = x^2 + 6x + 9$
24. $y = 12x - 9 - 4x^2$

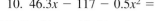 In Problems 25–30, use a graphing utility to graph each function. Use the vertex and zeros to determine an appropriate range. Be sure to label the maximum or minimum point.

25. $y = \frac{1}{3}x^2 - 3$
26. $y = \frac{1}{2}x^2 + 2$
27. $y = x^2 + 2x + 5$
28. $y = -10 + 7x - x^2$
29. $y = 20x - 0.1x^2$
30. $y = 50 - 1.5x + 0.01x^2$
31. Find the average rate of change of $f(x) = 100x - x^2$ from $x = 30$ to $x = 50$.
32. Find the average rate of change of

$$f(x) = x^2 - 30x + 22$$

over the interval $[10, 50]$.

In Problems 33–36, a graph is given. Use the graph to
(a) locate the vertex,
(b) determine the zeros, and
(c) match the graph with one of the equations A, B, C, or D.

A. $y = 7x - \frac{1}{2}x^2$
B. $y = \frac{1}{2}x^2 - x - 4$
C. $y = 8 - 2x - x^2$
D. $y = 49 - x^2$

33.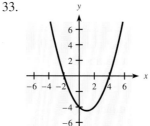

34.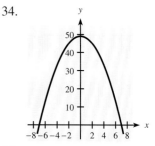

35.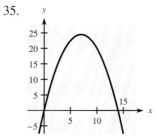

36.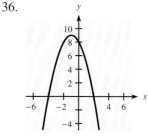

Section 2.4

37. Sketch a graph of each of the following basic functions.
(a) $f(x) = x^2$ (b) $f(x) = 1/x$ (c) $f(x) = x^{1/4}$

38. If $f(x) = \begin{cases} -x^2 & \text{if } x \le 0 \\ 1/x & \text{if } x > 0 \end{cases}$, find the following.
(a) $f(0)$ (b) $f(0.0001)$
(c) $f(-5)$ (d) $f(10)$

39. If $f(x) = \begin{cases} x & \text{if } x \le 1 \\ 3x - 2 & \text{if } x > 1 \end{cases}$, find the following.
(a) $f(-2)$ (b) $f(0)$
(c) $f(1)$ (d) $f(2)$

In Problems 40 and 41, graph each function.

40. $f(x) = \begin{cases} x & \text{if } x \le 1 \\ 3x - 2 & \text{if } x > 1 \end{cases}$
41. (a) $f(x) = (x - 2)^2$ (b) $f(x) = (x + 1)^3$

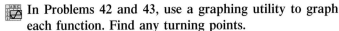 In Problems 42 and 43, use a graphing utility to graph each function. Find any turning points.

42. $y = x^3 + 3x^2 - 9x$
43. $y = x^3 - 9x$

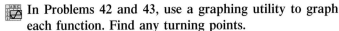 In Problems 44 and 45, graph each function. Find and identify any asymptotes.

44. $y = \dfrac{1}{x - 2}$ 45. $y = \dfrac{2x - 1}{x + 3}$

Section 2.5

46. *Modeling*   Consider the data given in the table.
    (a) Make a scatter plot.
    (b) Fit a linear function to the data and comment on the fit.
    (c) Try other function types and find one that fits better than a linear function.

x	0	4	8	12	16	20	24
y	153	151	147	140	128	115	102

47. *Modeling*   Consider the data given in the table.
    (a) Make a scatter plot.
    (b) Fit a linear function to the data and comment on the fit.
    (c) Try other function types and find one that fits better than a linear function.

x	3	5	10	15	20	25	30
y	35	45	60	70	80	87	95

APPLICATIONS

Section 2.1

48. *Physics*   A ball is thrown into the air from a height of 96 ft above the ground, and its height is given by $S = 96 + 32t - 16t^2$, where t is the time in seconds.
    (a) Find the values of t that make $S = 0$.
    (b) Do both of the values of t have meaning for this application?
    (c) When will the ball strike the ground?

49. *Profit*   The profit for a product is given by $P(x) = 82x - 0.10x^2 - 1600$, where x is the number of units produced and sold. Break-even points will occur at values of x where $P(x) = 0$. How many units will give a break-even point for the product?

Section 2.2

50. *Drug use*   Data indicate that the percent of high school seniors who have tried hallucinogens can be described by the function $f(t) = 0.0241904t^2 - 4.47459t + 216.074$, where t is the number of years since 1900.

(a) For what years does the function predict the percent will be 15%?
(b) For what year does the function predict the percent will be a minimum? Find that minimum predicted percent.

51. *Maximum area*   A rectangular lot is to be fenced in and then divided down the middle to create two identical fenced lots (see the figure). If 1200 ft of fence is to be used, the area of each lot is given by

$$A = x\left(\frac{1200 - 3x}{4}\right)$$

(a) Find the x-value that maximizes this area.
(b) Find the maximum area.

Section 2.3

52. *Supply*   Graph the first-quadrant portion of the supply function

$$p = 2q^2 + 4q + 6$$

53. *Demand*   Graph the first-quadrant portion of the demand function

$$p = 18 - 3q - q^2$$

54. *Market equilibrium*
    (a) Suppose the supply function for a product is $p = 0.1q^2 + 1$ and the demand function is $p = 85 - 0.2q - 0.1q^2$. Sketch the first-quadrant portion of the graph of each function. Use the same set of axes for both and label the market equilibrium point.
    (b) Use algebraic methods to find the equilibrium price and quantity.

55. *Market equilibrium*   The supply function for a product is given by $p = q^2 + 300$, and the demand is given by $p + q = 410$. Find the equilibrium quantity and price.

56. *Market equilibrium*   If the demand function for a commodity is given by the equation $p^2 + 5q = 200$ and the supply function is given by $40 - p^2 + 3q = 0$, find the equilibrium quantity and price.

57. *Break-even points*   If total costs for a product are given by $C(x) = 1760 + 8x + 0.6x^2$ and total revenues are given by $R(x) = 100x - 0.4x^2$, find the break-even points.

58. *Break-even points*   If total costs for a commodity are given by $C(x) = 900 + 25x$ and total revenues are given by $R(x) = 100x - x^2$, find the break-even points.

59. *Maximizing revenue and profit* Find the maximum revenue and maximum profit for the functions described in Problem 58.

60. *Break-even points and profit maximization* Given total profit $P(x) = 1.3x - 0.01x^2 - 30$, find maximum profit and the break-even points and sketch the graph.

61. *Maximum profit* Given $C(x) = 360 + 10x + 0.2x^2$ and $R(x) = 50x - 0.2x^2$, find the level of production that gives maximum profit and find the maximum profit.

62. *Break-even points and profit maximization* A certain company has fixed costs of $15,000 for its product and variable costs given by $140 + 0.04x$ dollars per unit, where x is the total number of units. The selling price of the product is given by $300 - 0.06x$ dollars per unit.
    (a) Formulate the functions for total cost and total revenue.
    (b) Find the break-even points.
    (c) Find the level of sales that maximizes revenue.
    (d) Form the profit function and find the level of production and sales that maximizes profit.
    (e) Find the profit (or loss) at the production levels found in (c) and (d).

Section 2.4

63. *Spread of AIDS* The function $H(t) = 0.099t^{1.969}$, where t is the number of years since 1980 and $H(t)$ is the number of world HIV infections (in hundreds of thousands), has been used as one means of predicting the spread of AIDS.
    (a) What type of function is this?
    (b) What does this function predict as the number of HIV cases in the year 2005?
    (c) Find $H(15)$ and write a sentence that explains its meaning.

 64. *Photosynthesis* The amount y of photosynthesis that takes place in a certain plant depends on the intensity x of the light present, according to
$$y = 120x^2 - 20x^3 \quad \text{for } x \geq 0$$
    (a) Graph this function with a graphing utility. (Use y-min $= -100$ and y-max $= 700$.)
    (b) The model is valid only when $f(x) \geq 0$ (that is, on or above the x-axis). For what x-values is this true?

65. *Cost-benefit* Suppose the cost C, in dollars, of eliminating p percent of the pollution from the emissions of a factory is given by
$$C(p) = \frac{4800p}{100 - p}$$

(a) What type of function is this?
(b) Given that p represents the percent of pollution removed, what is the domain of $C(p)$?
(c) Find $C(0)$ and interpret its meaning.
(d) Find the cost of removing 99% of the pollution.

66. *Municipal water costs* The Borough Municipal Authority of Beaver, Pennsylvania, used the following function to determine charges for water.
$$C(x) = \begin{cases} 1.557x & 0 \leq x \leq 100 \\ 155.70 + 1.04(x - 100) & 100 < x \leq 1000 \\ 1091.70 + 0.689(x - 1000) & x > 1000 \end{cases}$$
where $C(x)$ is the cost in dollars for x thousand gallons of water.
    (a) Find the monthly charge for 12,000 gallons of water.
    (b) Find the monthly charge for 825,000 gallons of water.

Section 2.5

67. *Modeling (Volkswagen Beetle)* The table shows the times that it takes a Volkswagen Beetle to accelerate from 0 mph to speeds of 30 mph, 40 mph, etc., up to 90 mph, in increments of 10 mph.

Time (sec)	Speed (mph)	Time (sec)	Speed (mph)
3.3	30	14.3	70
5.0	40	20.4	80
7.3	50	28.0	90
10.6	60		

Source: Motor Trend

(a) Represent the times by x and the speeds by y, and model the function that is the best fit for the points.
(b) Graph the points and the function to see how well the function fits the points.
(c) What does the model indicate the speed is 5 seconds after the car starts to move?
(d) According to the model, in how many seconds will the car reach 76.5 mph?

68. *Modeling (beach nourishment)* Since the first beach nourishment on Rockaway Beach, New York, in 1923, the amount of sand being pumped onto barrier islands has grown, as shown in the table.
    (a) Using $x = 3$ for the 1920s, $x = 4$ for the 1930s, etc., find a function that fits the data points.
    (b) Predict how much sand will be pumped in the decade 2000–2010.

Decade	Millions of Cubic Yards of Sand
1920s	8
1930s	21
1940s	21
1950s	23
1960s	30
1970s	65
1980s	79
1990s	96

Source: Island Packet, July 13, 1997

69. *Cohabiting households*   The table gives the number of cohabiting (without marriage) households (in thousands) for selected years.
   (a) Find the power function that best fits the data, with your input representing the number of years past 1950.
   (b) Find a quadratic function that is the best fit for the data, using an input equal to the number of years past 1950.

(c) Which model appears to fit the data better, or are they both good fits?

Year	Cohabiting Households (thousands)
1960	439
1970	523
1980	1589
1985	1983
1990	2856
1991	3039
1992	3308
1993	3510
1994	3661
1995	3668
1996	3958
1997	4130
1998	4236
2000	5457

Source: U.S. Bureau of the Census

2   Chapter Test

1. Sketch a graph of each of the following functions.
   (a) $f(x) = x^4$
   (b) $g(x) = |x|$
   (c) $h(x) = -1$
   (d) $k(x) = \sqrt{x}$

2. The figures that follow show graphs of the power function $y = x^b$. Which is the graph for $b > 1$? Which is the graph for $0 < b < 1$?

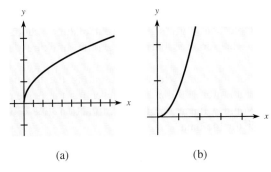

(a)   (b)

3. If $f(x) = ax^2 + bx + c$ and $a < 0$, sketch the shape of the graph of $y = f(x)$.

4. Graph:
   (a) $f(x) = (x + 1)^2 - 1$
   (b) $f(x) = (x - 2)^3 + 1$

5. Which of the following three graphs is the graph of $f(x) = x^3 - 4x^2$? Explain your choice.

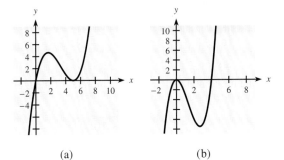

(a)   (b)

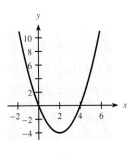

(c)

6. Let $f(x) = \begin{cases} 8x + 1/x & \text{if } x < 0 \\ 4 & \text{if } 0 \le x \le 2 \\ 6 - x & \text{if } x > 2 \end{cases}$

   Find (a) the y-coordinate of the point on the graph of $y = f(x)$ where $x = 16$; (b) $f(-2)$; (c) $f(13)$.

7. Sketch the graph of $g(x) = \begin{cases} x^2 & \text{if } x \le 1 \\ 4 - x & \text{if } x > 1 \end{cases}$

8. Find the vertex and zeros, if they exist, and sketch the graph of $f(x) = 21 - 4x - x^2$.

9. Solve $3x^2 + 2 = 7x$.

10. Solve $2x^2 + 6x = 9$.

11. Solve $\dfrac{1}{x} + 2x = \dfrac{1}{3} + \dfrac{x+1}{x}$.

12. Which of the following three graphs is that of
$g(x) = \dfrac{3x - 12}{x + 2}$? Explain your choice.

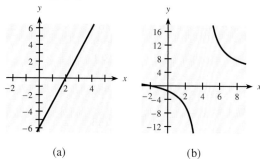

(a)   (b)

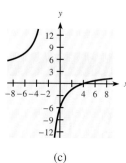

(c)

13. Find the horizontal and vertical asymptotes of the graph of

$$f(x) = \frac{8}{2x - 10}$$

14. Find the average rate of change of $P(x) = 92x - x^2 - 1760$ over the interval from $x = 10$ to $x = 40$.

15. Choose the type of function that models each of the following graphs.

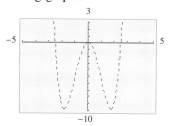

(a)

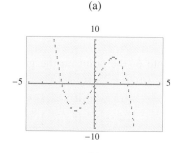

(b)

16. *Modeling*
    (a) Make a scatter plot, then develop a model, for the following data.
    (b) What does the model predict for $x = 40$?
    (c) When does the model predict $f(x) = 0$?

x	0	2	4	6	8	10	12	14	16	18	20
y	20.1	19.2	18.8	17.5	17.0	15.8	16	14.9	13.8	13.7	13.0

17. Suppose the supply and demand functions for a product are given by $6p - q = 180$ and $(p + 20)q = 30,000$, respectively. Find the equilibrium price and quantity.

18. Suppose a company's total cost for a product is given by $C(x) = 15,000 + 35x + 0.1x^2$ and the total revenue for the product is given by $R(x) = 285x - 0.9x^2$, where x is the number of units produced and sold.
    (a) Find the profit function.
    (b) Determine the number of units at which the profit for the product is maximized, and find the maximum possible profit.
    (c) Find the break-even point(s) for the product.

19. The wind chill expresses the combined effects of low temperatures and wind speeds as a single temperature reading. When the outside temperature is 0°F, the wind chill, *WC*, is a function of the wind speed s (in mph) and is given by the following function.*

$$WC = f(s) = \begin{cases} 0 & \text{if } 0 \le s \le 4 \\ 91.4 - 7.46738(5.81 + 3.7\sqrt{s} - 0.25s) \\ & \text{if } 4 < s \le 45 \\ -55 & \text{if } s > 45 \end{cases}$$

    (a) Find $f(15)$ and write a sentence that explains its meaning.
    (b) Find the wind chill when the wind speed is 48 mph.

20. The table below gives the percent of high school students who reported current cigarette smoking during 1991–2001.
    (a) Plot the data, with x representing the number of years after 1990 and y representing the percent.
    (b) Would a linear or quadratic function give a better fit for the data?
    (c) Use the model of your choice in (a) to fit a function to the data.
    (d) Use your model to estimate the percent in 2002.

Year	1991	1993	1995	1997	1999	2001
Percent	27.5	30.5	34.8	36.4	34.8	28.5

Source: www.cdc.gov/tobacco

*Bosch, W., and L. G. Cobb, "Windchill," UMAP Unit 658, *The UMAP Journal*, 5(4), Winter 1984, 477–492.

Extended Applications & Group Projects

I. International Email Usage

The instantaneous nature of telephone communication has always been important for business. However, differences in time zones have caused headaches for those who conduct business internationally. Recently, faxes and email have helped alleviate some of this problem because messages are immediately transferred and made available, even though the recipients may not be there to receive them. Of these, email is faster, more efficient with large documents, and more cost-effective. Thus the growing international economy is stimulating email usage.

The following figure, published in *Newsweek* on January 26, 1998, shows some projections for international email usage. (Data shown are estimates based on the figure.)

No Need to Phone
It's 5:00 in the morning at the Tokyo office
and 8 p.m. in London. The international
economy is creating a boom in e-mail.

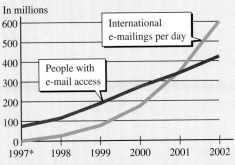

*1997 ESTIMATE. 1998–2002 PROJECTIONS. SOURCE: PIONEER CONSULTING

Source: "No Need to Phone" from *Newsweek,* January 21, 1998,
p. 15. Copyright © 1998 Newsweek, Inc. All rights reserved.
Reprinted by permission.

Year	People with Email Access (millions)	Daily International Emailings (millions)
1997	70	10
1998	130	30
1999	190	80
2000	270	190
2001	350	360
2002	430	600

A. 1. Decide what type of function would be most appropriate to model each of the following.
   (a) International emailings as a function of time in years, $I = I(t)$.
   (b) People with email access as a function of time, $P = P(t)$.
   2. (a) Identify the independent and dependent variables for $I(t)$ and $P(t)$,
   (b) Determine the domain and range for $I(t)$ and $P(t)$.
   3. Develop a model for $I(t)$ and for $P(t)$ by using the given data points for each.

4. (a) Discuss the meaning of intercepts for each model.
   (b) Discuss the meaning of the slope for any linear model.
5. Consider the point of intersection in the figure.
   (a) Discuss its meaning.
   (b) Estimate its coordinates from the figure.
   (c) Calculate its coordinates using the equations from your models.
6. Use your models to find projections for 2005 and 2010, and discuss the implications of your findings.
7. Discuss the limitations of your model.

B. Next investigate international emailings as a function of the number of people with email access.
1. Make a scatter plot of the data points with the number of people as the independent variable. Then decide what type of function best models the data.
2. Develop the function that models these data.
3. What does this new model predict for the number of international emailings when the number of people with email access is
   (a) 190,000,000 (as in 1999);   (b) 650,000,000.
4. Discuss the implications and limitations of this model.

C. Think of one other meaningful way to look at these data. Plot your data points and develop a model. Discuss the implications and limitations of this model.

II. Operating Leverage and Business Risk

Once you have determined the break-even point for your product, you can use it to examine the effects of increasing or decreasing the role of fixed costs in your operating structure. The extent to which a business uses fixed costs (compared to variable costs) in its operations is referred to as **operating leverage.** The greater the use of operating leverage (fixed costs, often associated with fixed assets), the larger the increase in profit as sales rise, and the larger the increase in loss as sales fall (*Source:* Business Owner's Toolkit, June 15, 2002, **www.tookkit.cch.com/text/P06_7540.asp**). The higher the break-even quantity for your product, the greater your **business risk.** To see how operating leverage is related to business risk, complete the following.

1. If x is the number of units of a product that is sold and the price is $\$p$ per unit, write the function that gives the revenue R for the product.
2. (a) If the fixed cost for production of this product is $\$10,000$ and the variable cost is $\$100$ per unit, find the function that gives the total cost C for the product.
   (b) What type of function is this?
3. Write an equation containing only the variables p and x that describes when the break-even point occurs.
4. (a) Solve the equation from 3 for x, so that x is written as a function of p.
   (b) What type of function is this?
   (c) What is the domain of this function?
   (d) What is the domain of this function in the context of this application?
5. (a) Graph the function from 4 for the domain of the problem.
   (b) Does the function increase or decrease as p increases?
6. Suppose that the company has a monopoly for this product so that it can set the price as it chooses. What are the benefit and the danger to the company of pricing its product at $\$1100$ per unit?
7. If the company has a monopoly for this product so that it can set the price, what are the benefit and the danger to the company of pricing its product at $\$101$ per unit?
8. Suppose the company currently is labor-intensive, so that the monthly fixed cost is $\$10,000$ and the variable cost is $\$100$ per unit, but installing modern equipment will reduce the variable cost to $\$50$ per unit while increasing the monthly fixed cost to $\$30,000$. Also suppose the selling price per unit is $p = \$200$.
   (a) Which of these operations will have the highest operating leverage? Explain.
   (b) Which of these operations will have the highest business risk? Explain.
   (c) What is the relation between high operating leverage and risk? Discuss the benefit and danger of the second operating plan to justify your answer.

3

Matrices

A wide variety of application problems can be solved using matrices. A matrix (plural: matrices) is a rectangular array of numbers. In addition to storing data in a matrix and making comparisons of data, we can analyze data and make business decisions by defining the operations of addition, subtraction, scalar multiplication, and matrix multiplication. Matrices can be used to solve systems of linear equations, as we will see in this chapter. Matrices and matrix operations are the basis for computer spreadsheets such as Excel, which are used extensively in education, research, and business. Matrices are also useful in linear programming, which is discussed in Chapter 4.

The topics and applications discussed in this chapter include the following.

Chapter Warm-up

Prerequisite Problem Type	For Section	Answer	Section for Review
Evaluate: (a) $4 - (-2)$ (b) $-4\left(\dfrac{2}{3}\right) + 3$ (c) $(-5)(4) + (8)(1) + (2)(5)$ (d) Multiply each of the numbers $\quad 0, -3, -2, 2$ by $-\dfrac{1}{3}$	3.1 3.2 3.3 3.4 3.5	(a) 6 (b) $\dfrac{1}{3}$ (c) -2 (d) $0, 1, \dfrac{2}{3}, -\dfrac{2}{3}$	0.2 Signed numbers

Write the coefficients of x, y, and z and the constant term in each equation:

(a) $2x + 5y + 4z = 4$
(b) $x - 3y - 2z = 5$
(c) $3x + y = 4$

For Section: 3.3, 3.4, 3.5

Coefficients			Const. term
x	y	z	
2	5	4	4
1	-3	-2	5
3	1	0	4

Section for Review: 0.5 Algebraic expressions

Prerequisite Problem Type	For Section	Answer	Section for Review
Solve: (a) $x_1 - 2x_3 = 2$ for x_1 (b) $x_2 + x_3 = -3$ for x_2 (c) $450{,}000 - 3z \geq 0$ for z	3.3 3.4 3.5	(a) $x_1 = 2 + 2x_3$ (b) $x_2 = -3 - x_3$ (c) $z \leq 150{,}000$	1.1 Linear equations and inequalities
Solve the system: $\begin{cases} x - 0.25y = 11{,}750 \\ -0.20x + y = 10{,}000 \end{cases}$	3.3 3.4 3.5	$x = 15{,}000$ $y = 13{,}000$	1.5 Systems of linear equations

Matrices

- To organize and interpret data stored in matrices
- To add and subtract matrices
- To find the transpose of a matrix
- To multiply a matrix by a scalar (real number)

❶ Application Preview

Data in tabular form can be stored in matrices, which are useful for solving an assortment of problems, including the following. Suppose the purchase prices and delivery charges (in dollars per unit) for wood, siding, and roofing used in construction are given by Table 3.1. If the supplier decides to raise the purchase prices by $0.60 per unit and the delivery charges by $0.05 per unit, the matrix that describes the new unit charges will be the sum of the charges matrix, formed from Table 3.1, and the matrix representing the increases.

TABLE 3.1

	Wood	Siding	Roofing
Purchase	6	4	2
Delivery	1	1	0.5

As we noted in the introduction to this chapter, matrices can be used to store data and to perform operations with the data. The matrix

$$C = \begin{bmatrix} 6 & 4 & 2 \\ 1 & 1 & 0.5 \end{bmatrix}$$

represents the original unit charges; it contains the heart of the information from Table 3.1, without the labels. The matrix representing the increases is given by

$$H = \begin{bmatrix} 0.60 & 0.60 & 0.60 \\ 0.05 & 0.05 & 0.05 \end{bmatrix}$$

We will discuss how to find the sum of these two and other pairs of matrices, how to find the difference of two matrices, and how to multiply a matrix by a real number. (The sum of these two matrices is found in Example 7.)

A business may collect and store or analyze various types of data as a regular part of its record-keeping procedures. The data may be presented in tabular form. For example, a building contractor who builds four different styles of houses may catalog the number of units of certain materials needed to build each style in a table like Table 3.2.

TABLE 3.2

Required Units	House Style			
	Ranch	Colonial	Split Level	Cape Cod
Wood	20	25	22	18
Siding	27	40	31	25
Roofing	16	16	19	16

If we write the numbers from Table 3.2 in the rectangular array below, the *rows* correspond to the types of building materials, and the *columns* correspond to the types of houses. The rows of a matrix are numbered from the top to the bottom and the columns are numbered from left to right.

$$A = \begin{bmatrix} 20 & 25 & 22 & 18 \\ 27 & 40 & 31 & 25 \\ 16 & 16 & 19 & 16 \end{bmatrix} \begin{matrix} \text{Row 1} \\ \text{Row 2} \\ \text{Row 3} \end{matrix}$$

Column 1   Column 2   Column 3   Column 4

Matrices are classified in terms of the numbers of rows and columns they have. Matrix A on the previous page has three rows and four columns, so we say this is a 3×4 (read "three by four") matrix.

The matrix

$$A = \begin{bmatrix} a_{11} & a_{12} & a_{13} & \cdots & a_{1n} \\ a_{21} & a_{22} & a_{23} & \cdots & a_{2n} \\ & \vdots & & & \vdots \\ a_{m1} & a_{m2} & a_{m3} & \cdots & a_{mn} \end{bmatrix}$$

has m rows and n columns, so it is an $m \times n$ matrix. When we designate A as an $m \times n$ matrix, we are indicating the size of the matrix. Two matrices are said to have the same **order** (be the same size) if they have the same number of rows and the same number of columns.

The numbers in a matrix are called its **entries** or **elements.** Note that the subscripts on an entry in matrix A above correspond respectively to the row and column in which the entry is located. Thus a_{23} represents the entry in the second row and the third column, and we refer to it as the "two-three entry." In matrix B below, the entry denoted by b_{23} is 1.

Some matrices take special names because of their size. If the number of rows equals the number of columns, we say the matrix is a **square matrix.** Matrix B is a 3×3 square matrix.

$$B = \begin{bmatrix} 4 & 3 & 0 \\ 0 & 0 & 1 \\ -4 & 0 & 0 \end{bmatrix}$$

● **EXAMPLE 1** *Population and Labor Force*

Figure 3.1 gives the U.S. population and labor force (in millions) for the years 1992 and 2002, and projections for 2012.

Population and labor force, 1992, 2002, and projected 2012

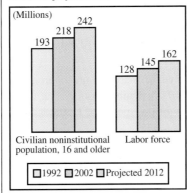

Source: *Occupational Outlook Quarterly,* Winter 2003–2004, p. 43.

Figure 3.1

A partial table of values from Figure 3.1 is shown below. Complete the table that summarizes these data, and use it to create a 2×3 matrix describing the information.

	1992	2002	2012
Population, 16 Plus	193		
Labor Force		145	

Solution

Reading the data from Figure 3.1 and completing the table gives

	1992	2002	2012
Population, 16 Plus	193	218	242
Labor Force	128	145	162

Writing these data in a matrix gives

$$\begin{bmatrix} 193 & 218 & 242 \\ 128 & 145 & 162 \end{bmatrix}$$

A matrix with one row, such as [9 5] or [3 2 1 6], is called a **row matrix,** and a matrix with one column, such as

$$\begin{bmatrix} 1 \\ 3 \\ 5 \end{bmatrix}$$

is called a **column matrix.** Row and column matrices are also called **vectors.**

Any matrix in which *every* entry is zero is called a **zero matrix;** examples include

$$\begin{bmatrix} 0 & 0 \\ 0 & 0 \end{bmatrix}, \quad \begin{bmatrix} 0 & 0 \\ 0 & 0 \\ 0 & 0 \end{bmatrix}, \quad \text{and} \quad \begin{bmatrix} 0 & 0 & 0 & 0 \\ 0 & 0 & 0 & 0 \\ 0 & 0 & 0 & 0 \end{bmatrix}$$

We define two matrices to be **equal** if they are of the same order and if each entry in one equals the corresponding entry in the other.

● **EXAMPLE 2 Matrix Order**

Do the matrices $A = \begin{bmatrix} 1 & 3 & 2 \\ 4 & 1 & 2 \end{bmatrix}$ and $B = \begin{bmatrix} 1 & 4 \\ 2 & 1 \\ 3 & 0 \end{bmatrix}$ have the same order?

Solution

No. A is a 2×3 matrix and B is a 3×2 matrix.

When the columns and rows of matrix A are interchanged to create a matrix B, and vice versa, we say that A and B are **transposes** of each other and write $A^T = B$ and $B^T = A$. We will see valuable uses for transposes of matrices in Chapter 4, "Inequalities and Linear Programming."

● **EXAMPLE 3 Matrices**

(a) Which element of

$$A = \begin{bmatrix} 1 & 0 & 3 \\ 3 & 4 & 2 \\ 7 & 8 & 3 \end{bmatrix} \qquad B = \begin{bmatrix} 1 & 0 & 3 \\ 3 & 4 & 2 \\ 7 & 8 & 4 \end{bmatrix}$$

is represented by a_{32}?
(b) Is A a square matrix?
(c) Find the transpose of matrix A.
(d) Does $A = B$?

Solution

(a) a_{32} represents the element in row 3 and column 2 of matrix A—that is, $a_{32} = 8$.

(b) Yes, it is a 3×3 (square) matrix.

(c)
$$A^T = \begin{bmatrix} 1 & 3 & 7 \\ 0 & 4 & 8 \\ 3 & 2 & 3 \end{bmatrix}$$

(d) No, $a_{33} \neq b_{33}$.

● *Checkpoint*

1. (a) Do matrices A and B have the same order?

$$A = \begin{bmatrix} 1 & 2 & 3 \\ 4 & 5 & 6 \\ 7 & 8 & 9 \end{bmatrix} \quad \text{and} \quad B = \begin{bmatrix} 1 & 4 & 7 \\ 2 & 5 & 8 \\ 3 & 6 & 9 \end{bmatrix}$$

(b) Does matrix A equal matrix B?

(c) Does $B^T = A$?

Addition and Subtraction of Matrices

If two matrices have the same number of rows and columns, we can add the matrices by adding their corresponding entries.

Sum of Two Matrices

If matrix A and matrix B are of the same order and have elements a_{ij} and b_{ij}, respectively, then their **sum** $A + B$ is a matrix C whose elements are $c_{ij} = a_{ij} + b_{ij}$ for all i and j. That is,

$$A + B = \begin{bmatrix} a_{11} & a_{12} & \cdots & a_{1n} \\ a_{21} & a_{22} & \cdots & a_{2n} \\ \vdots & & & \vdots \\ a_{m1} & a_{m2} & \cdots & a_{mn} \end{bmatrix} + \begin{bmatrix} b_{11} & b_{12} & \cdots & b_{1n} \\ b_{21} & b_{22} & \cdots & b_{2n} \\ \vdots & & & \vdots \\ b_{m1} & b_{m2} & \cdots & b_{mn} \end{bmatrix}$$

$$= \begin{bmatrix} a_{11} + b_{11} & a_{12} + b_{12} & \cdots & a_{1n} + b_{1n} \\ a_{21} + b_{21} & a_{22} + b_{22} & \cdots & a_{2n} + b_{2n} \\ \vdots & & & \vdots \\ a_{m1} + b_{m1} & a_{m2} + b_{m2} & \cdots & a_{mn} + b_{mn} \end{bmatrix} = C$$

● **EXAMPLE 4** *Matrix Sums*

Find the sum of A and B if

$$A = \begin{bmatrix} 1 & 2 & 3 \\ 4 & -1 & -2 \end{bmatrix} \quad \text{and} \quad B = \begin{bmatrix} -1 & 2 & -3 \\ -2 & 0 & 1 \end{bmatrix}$$

Solution

$$A + B = \begin{bmatrix} 1 + (-1) & 2 + 2 & 3 + (-3) \\ 4 + (-2) & -1 + 0 & -2 + 1 \end{bmatrix} = \begin{bmatrix} 0 & 4 & 0 \\ 2 & -1 & -1 \end{bmatrix}$$

The matrix $-B$ is called the **negative** of the matrix B, and each element of $-B$ is the negative of the corresponding element of B. For example, if

$$B = \begin{bmatrix} -1 & 2 & -3 \\ -2 & 0 & 1 \end{bmatrix}, \quad \text{then} \quad -B = \begin{bmatrix} 1 & -2 & 3 \\ 2 & 0 & -1 \end{bmatrix}$$

Using the negative, we can define the difference $A - B$ (when A and B have the same order) by $A - B = A + (-B)$, or by subtracting corresponding elements.

● **EXAMPLE 5** *Matrix Differences*

For matrices A and B in Example 4, find $A - B$.

Solution

$A - B$ can be found by subtracting corresponding elements.

$$A - B = \begin{bmatrix} 1 - (-1) & 2 - 2 & 3 - (-3) \\ 4 - (-2) & -1 - 0 & -2 - 1 \end{bmatrix} = \begin{bmatrix} 2 & 0 & 6 \\ 6 & -1 & -3 \end{bmatrix}$$

● *Checkpoint*

2. If $A = \begin{bmatrix} 1 & 2 & 0 \\ 3 & 1 & 2 \end{bmatrix}$, $B = \begin{bmatrix} 3 & 6 & 1 \\ 0 & 2 & 1 \end{bmatrix}$, and $C = \begin{bmatrix} 1 & 0 & 2 \\ 0 & 1 & 1 \end{bmatrix}$, find $A + B - C$.

3. (a) What matrix D must be added to matrix A so that their sum is matrix Z?

$$A = \begin{bmatrix} 1 & -2 & 3 \\ -5 & 1 & 2 \end{bmatrix} \quad \text{and} \quad Z = \begin{bmatrix} 0 & 0 & 0 \\ 0 & 0 & 0 \end{bmatrix}$$

(b) Does $D = -A$?

Graphing Utilities

Graphing calculators and computers have the capability to perform a number of operations on matrices. Computer spreadsheets are also very useful in performing operations with matrices.

  ● **EXAMPLE 6** *Matrix Operations*

Use a graphing calculator or computer to compute $A + B$ and $A - B$ for

$$A = \begin{bmatrix} -1 & 3 & 2 \\ 1 & 0 & 2 \\ 0 & 1 & -1 \\ -4 & 2 & 3 \end{bmatrix} \quad \text{and} \quad B = \begin{bmatrix} -3 & 2 & 1 \\ 0 & -2 & 2 \\ -1 & 6 & 8 \\ 3 & 2 & 4 \end{bmatrix}$$

```
MATRIX[A]   4 ×3
[-1    3     2   ]
[1     0     2   ]
[0     1    -1   ]
[-4    2     3   ]

4,3=3
```

```
MATRIX[B]   4 ×3
[-3    2     1   ]
[0    -2     2   ]
[-1    6     8   ]
[3     2     4   ]

4,3=4
```

Figure 3.2

Solution

On most graphing calculators, you can use MATRIX and EDIT to define a matrix. To define matrix A, we select a name, A, select the dimensions, and enter the elements of matrix A. Note that the address (row and column) of each entry is shown when entering it. Figure 3.2 shows matrix A and matrix B displayed on screens of graphing calculators, with the dimensions of each matrix shown above it. We find the sum and difference of the two matrices by using the regular $+$ and $-$ keys (see Figure 3.3).

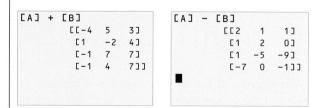

Figure 3.3

Note that if matrix A and matrix B have different orders, they cannot be added or subtracted, and a graphing calculator will give an error statement.

◑ EXAMPLE 7 *Supply Charges* (Application Preview)

Suppose the purchase prices and delivery charges (in dollars per unit) for wood, siding, and roofing used in construction are given by Table 3.3. The supplier decides to raise the purchase prices by $0.60 per unit and the delivery charges by $0.05 per unit. Write the matrix that describes the new unit charges.

TABLE 3.3

	Wood	Siding	Roofing
Purchase	6	4	2
Delivery	1	1	0.5

Solution

The matrix representing the original unit charges is

$$C = \begin{bmatrix} 6 & 4 & 2 \\ 1 & 1 & 0.5 \end{bmatrix}$$

and the matrix representing the increases is given by

$$H = \begin{bmatrix} 0.60 & 0.60 & 0.60 \\ 0.05 & 0.05 & 0.05 \end{bmatrix}$$

The new unit charge for each item is its former charge plus the increase, so the new unit charge matrix is given by

$$C + H = \begin{bmatrix} 6 + 0.60 & 4 + 0.60 & 2 + 0.60 \\ 1 + 0.05 & 1 + 0.05 & 0.5 + 0.05 \end{bmatrix} = \begin{bmatrix} 6.60 & 4.60 & 2.60 \\ 1.05 & 1.05 & 0.55 \end{bmatrix}$$

Note that the sum of C and H in Example 7 could be found by adding the matrices in either order. That is, $C + H = H + C$. This is known as the **commutative law of addition** for matrices. We will see in the next section that multiplication of matrices is *not* commutative.

Scalar Multiplication

Consider the matrix

$$A = \begin{bmatrix} 3 & 2 \\ 1 & 1 \\ 2 & 0 \end{bmatrix}$$

Because $2A$ is $A + A$, we see that

$$2A = \begin{bmatrix} 3 & 2 \\ 1 & 1 \\ 2 & 0 \end{bmatrix} + \begin{bmatrix} 3 & 2 \\ 1 & 1 \\ 2 & 0 \end{bmatrix} = \begin{bmatrix} 6 & 4 \\ 2 & 2 \\ 4 & 0 \end{bmatrix}$$

Note that $2A$ could have been found by multiplying each entry of A by 2. In the same manner,

$$3A = A + A + A = \begin{bmatrix} 3 & 2 \\ 1 & 1 \\ 2 & 0 \end{bmatrix} + \begin{bmatrix} 3 & 2 \\ 1 & 1 \\ 2 & 0 \end{bmatrix} + \begin{bmatrix} 3 & 2 \\ 1 & 1 \\ 2 & 0 \end{bmatrix}$$

$$= \begin{bmatrix} 9 & 6 \\ 3 & 3 \\ 6 & 0 \end{bmatrix} = \begin{bmatrix} 3(3) & 3(2) \\ 3(1) & 3(1) \\ 3(2) & 3(0) \end{bmatrix}$$

We can define **scalar multiplication** as follows.

Scalar Multiplication

Multiplying a matrix by a real number (called a *scalar*) results in a matrix in which each entry of the original matrix is multiplied by the real number. Thus, if

$$A = \begin{bmatrix} a_{11} & a_{12} \\ a_{21} & a_{22} \end{bmatrix}, \quad \text{then} \quad cA = \begin{bmatrix} ca_{11} & ca_{12} \\ ca_{21} & ca_{22} \end{bmatrix}$$

● **EXAMPLE 8** *Scalar Multiplication*

If

$$A = \begin{bmatrix} 4 & 1 & 4 & 0 \\ 2 & -7 & 3 & 6 \\ 0 & 0 & 2 & 5 \end{bmatrix}$$

find $5A$ and $-2A$.

Solution

$$5A = \begin{bmatrix} 5 \cdot 4 & 5 \cdot 1 & 5 \cdot 4 & 5 \cdot 0 \\ 5 \cdot 2 & 5(-7) & 5 \cdot 3 & 5 \cdot 6 \\ 5 \cdot 0 & 5 \cdot 0 & 5 \cdot 2 & 5 \cdot 5 \end{bmatrix} = \begin{bmatrix} 20 & 5 & 20 & 0 \\ 10 & -35 & 15 & 30 \\ 0 & 0 & 10 & 25 \end{bmatrix}$$

$$-2A = \begin{bmatrix} -2 \cdot 4 & -2 \cdot 1 & -2 \cdot 4 & -2 \cdot 0 \\ -2 \cdot 2 & -2(-7) & -2 \cdot 3 & -2 \cdot 6 \\ -2 \cdot 0 & -2 \cdot 0 & -2 \cdot 2 & -2 \cdot 5 \end{bmatrix}$$

$$= \begin{bmatrix} -8 & -2 & -8 & 0 \\ -4 & 14 & -6 & -12 \\ 0 & 0 & -4 & -10 \end{bmatrix}$$

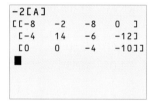

Figure 3.4

Figure 3.4 shows the matrix $-2A$ on a graphing calculator.

● *Checkpoint*

4. (a) Find $3D - D$ if $D = \begin{bmatrix} 1 & 3 & 2 & 10 \\ 20 & 5 & 6 & 3 \end{bmatrix}$

(b) Does $3D - D = 2D$?

● **EXAMPLE 9** *Supply Charges*

Suppose the purchase prices and delivery costs (per unit) for wood, siding, and roofing used in construction are given by Table 3.4. Then the table of unit costs may be represented by the matrix C.

TABLE 3.4

	Wood	Siding	Roofing
Purchase	6	4	2
Delivery	1	1	0.5

$$C = \begin{bmatrix} 6 & 4 & 2 \\ 1 & 1 & 0.5 \end{bmatrix}$$

If the supplier announces a 10% increase on both purchase and delivery of these items, find the new unit cost matrix.

Solution
A 10% increase means that the new unit costs are the former costs plus 0.10 times the former cost. That is, the new costs are 1.10 times the former, so the new unit cost matrix is given by

$$1.10C = 1.10\begin{bmatrix} 6 & 4 & 2 \\ 1 & 1 & 0.5 \end{bmatrix}$$

$$= \begin{bmatrix} 6.60 & 4.40 & 2.20 \\ 1.10 & 1.10 & 0.55 \end{bmatrix}$$

● **EXAMPLE 10** *Power to Influence*

Suppose that in a government agency, paperwork is constantly flowing among offices according to the diagram.

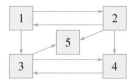

(a) Construct matrix A with elements

$$a_{ij} = \begin{cases} 1 & \text{if paperwork flows directly from } i \text{ to } j. \\ 0 & \text{if paperwork does not flow directly from } i \text{ to } j. \end{cases}$$

(b) Construct matrix B with elements

$$b_{ij} = \begin{cases} 1 & \text{if paperwork can flow from } i \text{ to } j \text{ through at most} \\ & \text{one intermediary, with } i \neq j. \\ 0 & \text{if this is not true.} \end{cases}$$

(c) The person in office i has the most power to influence others if the sum of the elements of row i in the matrix $A + B$ is the largest. What is the office number of this person?

Solution

(a) $A = \begin{bmatrix} 0 & 1 & 1 & 0 & 0 \\ 1 & 0 & 0 & 1 & 1 \\ 0 & 0 & 0 & 1 & 1 \\ 0 & 0 & 1 & 0 & 0 \\ 0 & 0 & 0 & 0 & 0 \end{bmatrix}$ (b) $B = \begin{bmatrix} 0 & 1 & 1 & 1 & 1 \\ 1 & 0 & 1 & 1 & 1 \\ 0 & 0 & 0 & 1 & 1 \\ 0 & 0 & 1 & 0 & 1 \\ 0 & 0 & 0 & 0 & 0 \end{bmatrix}$

(c) $A + B = \begin{bmatrix} 0 & 2 & 2 & 1 & 1 \\ 2 & 0 & 1 & 2 & 2 \\ 0 & 0 & 0 & 2 & 2 \\ 0 & 0 & 2 & 0 & 1 \\ 0 & 0 & 0 & 0 & 0 \end{bmatrix}$

Because the sum of the elements in row 2 is the largest (7), the person in office 2 has the most power to influence others. Because row 5 has the smallest sum (0), the person in office 5 has the least power.

Spreadsheet
Note
Graphing calculators, computer algebra systems, and computer spreadsheets have the capability to perform a number of complex operations on matrices. Many applications in business and industry call for repeated manipulations or operations with very large tables (matrices). Once the data are entered in a graphing calculator, computer program, or spreadsheet, they can be stored and then operated on or combined with other matrices.

The ability to store data in arrays (matrices) and the ability to perform operations with matrices are at the heart of the development of computer spreadsheets such as *Lotus 1-2-3, Quattro Pro,* and *Excel.* Table 3.5(a) shows the financial results (in millions of dollars) from the 2005 Annual Report of the Ace Corporation in a spreadsheet.

TABLE 3.5(a)

	A	B	C	D
1		Financial	Results	($million)
2		2005	2004	2003
3	Earnings from Barrons	41.5	34.1	18.4
4	Dilution gains	2.4	0.9	16.7
5	Earnings from Coors	2.8	−3.0	23.0
6	R&M Holds parent	−11.6	0	0
7	Ace parent	5.0	4.2	3.3
8	Net earnings	40.1	36.2	61.4
9	Operating cash flow	62.3	31.8	91.6
10	Capital expenditures	100.9	81.1	90.7

Among the easiest and most frequently performed operations on a computer spreadsheet is adding rows. For example, the net earnings (row 8 of the spreadsheet) is the sum of the rows 3 through 7 of the spreadsheet. Similarly, the columns of a spreadsheet can be added easily. Table 3.5(b) shows the gains or losses from 2004 to 2005, using column B minus column C, in column D.

TABLE 3.5(b)

	A	B	C	D
1		Financial	Results	($million)
2		2005	2004	Gain (Loss)
3	Earnings from Barrons	41.5	34.1	7.4
4	Dilution gains	2.4	0.9	1.5
5	Earnings from Coors	2.8	−3.0	5.8
6	R&M Holds parent	−11.6	0	−11.6
7	Ace parent	5	4.2	0.8
8	Net earnings	40.1	36.2	3.9
9	Operating cash flow	62.3	31.8	30.5
10	Capital expenditures	100.9	81.1	19.8

● **Checkpoint Solutions**

1. (a) Yes, *A* and *B* have the same order (size).
   (b) No. They have the same elements, but they are not equal because their corresponding elements are not equal.
   (c) Yes.

2. $\begin{bmatrix} 3 & 8 & -1 \\ 3 & 2 & 2 \end{bmatrix}$ 3. (a) $\begin{bmatrix} -1 & 2 & -3 \\ 5 & -1 & -2 \end{bmatrix}$ (b) Yes

4. (a) $3D - D = \begin{bmatrix} 3 & 9 & 6 & 30 \\ 60 & 15 & 18 & 9 \end{bmatrix} - \begin{bmatrix} 1 & 3 & 2 & 10 \\ 20 & 5 & 6 & 3 \end{bmatrix}$

$\qquad\qquad = \begin{bmatrix} 2 & 6 & 4 & 20 \\ 40 & 10 & 12 & 6 \end{bmatrix}$

(b) Yes

3.1 Exercises

Use the following matrices for Problems 1–10.

$A = \begin{bmatrix} 1 & 0 & 2 \\ 3 & 2 & 1 \\ 4 & 0 & 3 \end{bmatrix}$ $B = \begin{bmatrix} 1 & 1 & 3 & 0 \\ 4 & 2 & 1 & 1 \\ 3 & 2 & 0 & 1 \end{bmatrix}$

$C = \begin{bmatrix} 5 & 3 \\ 1 & 2 \end{bmatrix}$ $D = \begin{bmatrix} 4 & 2 \\ 3 & 5 \end{bmatrix}$ $E = \begin{bmatrix} 1 & 0 & 4 \\ 5 & 1 & 0 \end{bmatrix}$

$F = \begin{bmatrix} 1 & 2 & 3 \\ -1 & 0 & 1 \\ 2 & -3 & -4 \end{bmatrix}$ $Z = \begin{bmatrix} 0 & 0 & 0 \\ 0 & 0 & 0 \\ 0 & 0 & 0 \end{bmatrix}$

1. How many rows does matrix B have?
2. What is the order of matrix E?
3. Write the negative of matrix F.
4. Write a zero matrix that is the same order as D.
5. Which of the matrices A, B, C, D, E, F, and Z are square?
6. Write the matrix that is the negative of matrix B.
7. What is the element a_{23}?
8. What is element b_{24}?
9. Write the transpose of matrix A.
10. Write the transpose of matrix F.

Use the matrices M and N in Problems 11 and 12.

$M = \begin{bmatrix} 1 & 0 & 2 \\ 3 & 2 & 1 \\ 4 & 0 & 3 \end{bmatrix}$ and $N = \begin{bmatrix} 1 & 1 & 3 & 0 \\ 4 & 2 & 1 & 1 \\ 3 & 2 & 0 & 1 \end{bmatrix}$

11. What is the sum of matrix M and its negative?
12. If matrix M has elements $m_{3j} = 0$, what is j?

For Problems 13–26, use the following matrices to perform the indicated matrix operations, if possible.

$A = \begin{bmatrix} 1 & 0 & 2 \\ 3 & 2 & 1 \\ 4 & 0 & 3 \end{bmatrix}$ $B = \begin{bmatrix} 1 & 1 & 3 & 0 \\ 4 & 2 & 1 & 1 \\ 3 & 2 & 0 & 1 \end{bmatrix}$

$C = \begin{bmatrix} 5 & 3 \\ 1 & 2 \end{bmatrix}$ $D = \begin{bmatrix} 4 & 2 \\ 3 & 5 \end{bmatrix}$

$E = \begin{bmatrix} 2 & 1 & 1 \\ 1 & 0 & 4 \\ 5 & 1 & 0 \end{bmatrix}$ $F = \begin{bmatrix} 1 & 2 & 3 \\ -1 & 0 & 1 \\ 2 & -3 & -4 \end{bmatrix}$

$Z = \begin{bmatrix} 0 & 0 & 0 \\ 0 & 0 & 0 \\ 0 & 0 & 0 \end{bmatrix}$

13. $C + D$
14. $A + F$
15. $A - F$
16. $Z + E^T$
17. $A + A^T$
18. $A + F^T$
19. $D - E$
20. $B + F$
21. $3B$
22. $4D$
23. $4C + 2D$
24. $8C - 3D$
25. $2A - 3B$
26. $3E + 4F$

In Problems 27–30 find x, y, z, and w.

27. $\begin{bmatrix} x & 1 & 0 \\ 0 & y & z \\ w & 2 & 1 \end{bmatrix} = \begin{bmatrix} 3 & 1 & 0 \\ 0 & 2 & 3 \\ 4 & 2 & 1 \end{bmatrix}$

28. $\begin{bmatrix} 0 & x & 1 \\ 3 & y & y \\ z & 0 & 2 \end{bmatrix} = \begin{bmatrix} 0 & 4 & 1 \\ 3 & 1 & y \\ 1 & 0 & w \end{bmatrix}$

29. $\begin{bmatrix} x & 3 & (2x-1) \\ y & 4 & 4y \end{bmatrix} = \begin{bmatrix} 2x-4 & z & 7 \\ 1 & (w+1) & (3y+1) \end{bmatrix}$

30. $\begin{bmatrix} x & y & (x+3) \\ z & 4 & 4y \end{bmatrix} = \begin{bmatrix} (2x-1) & -1 & w \\ x & (5+y) & -4 \end{bmatrix}$

31. Solve for x, y, and z if

$3\begin{bmatrix} x & y \\ y & z \end{bmatrix} + 2\begin{bmatrix} 2x & -y \\ 3y & -4z \end{bmatrix} = \begin{bmatrix} 14 & 4-y \\ 18 & 15 \end{bmatrix}$

32. Find x, y, z, and w if

$3\begin{bmatrix} x & 4 \\ 4y & w \end{bmatrix} - 2\begin{bmatrix} 4x & 2z \\ -3 & -2w \end{bmatrix} = \begin{bmatrix} 20 & 20 \\ 6 & 14 \end{bmatrix}$

APPLICATIONS

33. *Endangered species* The tables below give the numbers of some species of threatened and endangered wildlife in the United States and in foreign countries in 2003.

United States

	Mammals	Birds	Reptiles	Amphibians	Fishes
Endangered	65	78	14	12	71
Threatened	9	14	22	9	44

Foreign

	Mammals	Birds	Reptiles	Amphibians	Fishes
Endangered	251	175	64	8	11
Threatened	17	6	15	1	0

Source: U.S. Fish and Wildlife Service

(a) Write a matrix A that contains the number of each of these species in the United States in 2003 and a matrix B that contains the number of each of these species outside the United States in 2003.

(b) Find a matrix with the total number of these species. Assume that U.S. and foreign species are different.

(c) Find the matrix $B - A$. What do the negative entries in matrix $B - A$ mean?

34. *Top ten dog breeds* The following tables give the rank and number of registered dogs for the top ten breeds for 1995 and their 2002 data.

(a) Form a matrix for the 1995 data and a matrix for the 2002 data.

(b) Use a matrix operation to find the change from 1995 to 2002 in rank for each breed and in registration numbers for each breed.

(c) Which breed had the greatest improvement in rank?

(d) Which breed had the largest increase in the number of registrations?

1995

Breed	Rank	Number Registered
Labrador retriever	1	132,051
Rottweiler	2	93,656
German shepherd	3	76,088
Golden retriever	4	64,107
Beagle	5	57,063
Poodle	6	54,784
Cocker spaniel	7	48,065
Dachshund	8	44,680
Pomeranian	9	37,894
Yorkshire terrier	10	36,881

2002

Breed	Rank	Number Registered
Labrador retriever	1	154,616
Rottweiler	13	22,196
German shepherd	3	46,963
Golden retriever	2	56,124
Beagle	4	44,610
Poodle	8	33,917
Cocker spaniel	15	20,655
Dachshund	5	42,571
Pomeranian	12	23,061
Yorkshire terrier	6	37,277

Source: American Kennel Club

35. *Pollution abatement* The following tables give the capital expenditures and gross operating costs of manufacturing establishments for pollution abatement, in millions of dollars.

(a) Use a calculator or spreadsheet to find the sum of the matrices containing these data. This will yield the total costs of manufacturing establishments for air, water, and solid contained waste pollution abatement for the years 2002–2005.

(b) Which type of pollution abatement was most expensive in 2005?

Pollution Abatement Capital Expenditures

	Air	Water	Solid Contained Waste
2002	6030.8	2562.0	817.5
2003	3706.3	2814.6	869.1
2004	4403.1	2509.8	953.9
2005	4122.0	2294.9	760.9

Pollution Abatement Gross Operating Costs

	Air	Water	Solid Contained Waste
2002	5010.9	6416.4	5643.5
2003	5033.5	6345.0	6008.2
2004	5395.0	6576.9	5494.5
2005	5574.6	6631.8	5348.6

36. *Death rates* The following tables give the death rates, per 100,000 population, by age for selected years for males and females.

(a) If matrix M gives the male data and matrix F gives the female data, use matrix operations to find the death rate per 100,000 for all people in the age categories given and for the given years.

(b) Find matrix $M - F$ and describe what it means.

Males

	Under 1	1-4	5-14	15-24	25-34	35-44	45-54
1970	2410	93	51	189	215	403	959
1980	1429	73	37	172	196	299	767
1990	1083	52	29	147	204	310	610
1994	899	52	26	151	207	337	585
2001	755	37	20	117	143	259	544

Females

	Under 1	1-4	5-14	15-24	25-34	35-44	45-54
1970	1864	75	32	68	102	231	517
1980	1142	55	24	58	76	159	413
1990	856	41	19	49	74	138	343
1994	719	37	19	47	76	144	326
2001	620	29	15	43	66	148	317

Source: U.S. Department of Health and Human Services, National Center for Health Statistics

37. *Sales* Let matrix A represent the sales (in thousands of dollars) for the Walbash Company in 2004 in various cities, and let matrix B represent the sales (in thousands of dollars) for the same company in 2005 in the same cities.

$$A = \begin{bmatrix} \text{Chicago} & \text{Atlanta} & \text{Memphis} \\ 450 & 280 & 850 \\ 400 & 350 & 150 \end{bmatrix} \begin{matrix} \text{Wholesale} \\ \text{Retail} \end{matrix}$$

$$B = \begin{bmatrix} \text{Chicago} & \text{Atlanta} & \text{Memphis} \\ 375 & 300 & 710 \\ 410 & 300 & 200 \end{bmatrix} \begin{matrix} \text{Wholesale} \\ \text{Retail} \end{matrix}$$

(a) Write the matrix that represents the total sales by type and city for both years.

(b) Write the matrix that represents the change in sales by type and city from 2004 to 2005.

38. *Opinion polls* A poll of 3320 people revealed that of the respondents that were registered Republicans, 843 approved of the president's job performance, 426 did not, and 751 had no opinion. Of the registered Democrats, 257 approved of the president's job performance, 451 did not, and 92 had no opinion. Of those registered as Independents, 135 approved, 127 did not approve, and 38 had no opinion. Of the remaining respondents, who were not registered, 92 approved, 64 did not approve, and 44 had no opinion. Represent these data in a 3 × 4 matrix.

39. *Life expectancy* The following table gives the years of life expected at birth for blacks and whites born in the United States in selected years from 1920 to 2001.

	Whites		Blacks	
	Males	Females	Males	Females
1920	54.4	55.6	45.5	45.2
1940	62.1	66.6	51.5	54.9
1960	67.4	74.1	61.1	67.4
1980	70.7	78.1	63.8	72.5
1990	72.9	79.4	64.5	73.6
2001	75.0	80.2	68.6	75.5

Source: National Center for Health Statistics

Make a matrix A containing the information for whites and a matrix B for blacks. Use these to find how many more years whites in each category are expected to live than blacks.

40. *Imports* (a) From the data in the following table, make a matrix A that gives the value (in billions of dollars) of U.S. imports from the various country groupings in the years 2000 and 2002.

(b) Make a matrix B that gives the value of U.S. exports (in billions of dollars) to the same groupings in the same years.

(c) Find the U.S. trade balance with each country grouping in each year by finding matrix $B - A$.

(d) Find the total U.S. trade with these groupings by finding $A + B$.

Imports to U.S. from Country Groupings ($billions)

	2000	2002
North America	366.8	343.0
Western Europe	240.7	246.0
South/Central America	73.3	69.5
Pacific Rim	418.0	393.5

Exports from U.S. to Country Groupings ($billions)

	2000	2002
North America	290.3	258.4
Western Europe	181.5	157.0
South/Central America	59.3	51.6
Pacific Rim	202.6	178.6

Source: U.S. Commerce Department

41. *Weekly earnings* The following table shows median weekly earnings for men and women of different ages in 2000.

(a) Use the data to make a 6 × 2 matrix with each row representing an age category and each column representing a sex.

(b) If union members' median weekly earnings are 20% more than the earnings given in the table, use a matrix operation to find a matrix giving the results by sex and age for union members.

Age	16–24	25–34	35–44	45–54	55–64	65+
Men	376	603	731	777	738	537
Women	342	493	520	565	505	378

Source: Bureau of Labor Statistics, U.S. Department of Labor

42. *Debt payment* Ace, Baker, and Champ are being purchased by ALCO, Inc., and their outstanding debts must be paid by the purchaser. The matrix below gives the amounts of debt and the terms for the companies being purchased.

$$
\begin{array}{c}
 \\
\text{Ace} \\
\text{Baker} \\
\text{Champ}
\end{array}
\begin{array}{cc}
\text{Due in 30 days} & \text{Due in 60 days} \\
\left[\begin{array}{cc}
\$40{,}000 & \$60{,}000 \\
\$25{,}000 & \$15{,}000 \\
\$35{,}000 & \$58{,}000
\end{array}\right]
\end{array}
$$

(a) If ALCO pays 35% of the amount owed on each debt, write the matrix giving the remaining debts.
(b) Suppose ALCO decides to pay 80% of all debts due in 30 days and to increase the debts due in 60 days by 20%. Write a matrix that gives the debts after these transactions are made.

43. *Expense accounts* A sales associate's expense account for the first week of a certain month has the daily expenses (in dollars) shown in matrix A.

$$
A =
\begin{array}{c}
\\

\end{array}
\begin{array}{cccc}
\text{Meals} & \text{Lodging} & \text{Travel} & \text{Other} \\
\left[\begin{array}{cccc}
44 & 80 & 200 & 10 \\
40 & 80 & 40 & 0 \\
56 & 140 & 90 & 0 \\
30 & 140 & 40 & 20 \\
40 & 0 & 200 & 10
\end{array}\right]
\end{array}
\begin{array}{l}
\text{Monday} \\
\text{Tuesday} \\
\text{Wednesday} \\
\text{Thursday} \\
\text{Friday}
\end{array}
$$

(a) The associate finds that expenses for the second week are 5% more (in each category) than for the first week. Find the expenses matrix for the second week (note that this is $1.05A$).
(b) Find the associate's expenses matrix for the third week if expenses for that week are 10% more (in each category) than they were in the first week.

44. *Debt payment* When a firm buys another company, the company frequently has some outstanding debt that the purchaser must pay. Consider three companies, A, B, and C, that are purchased by Maxx Industries. The following table gives the amount of each company's debt, classified by the number of days remaining until the debt must be paid.

Company	30 Days	60 Days	More than 60 Days
A	$25,000	$26,000	$12,000
B	15,000	52,000	5,000
C	8,000	20,000	120,000

(a) Suppose Maxx Industries pays 20% of the amount owed on each loan. Write the matrix that would give the remaining debt in each category.
(b) What payment plan did Maxx Industries use if the outstanding debt is given by the following matrix?

$$
\begin{bmatrix}
25{,}000 & 26{,}000 & 12{,}000 \\
15{,}000 & 52{,}000 & 5{,}000 \\
8{,}000 & 20{,}000 & 120{,}000
\end{bmatrix}
$$
$$
-\,0.50
\begin{bmatrix}
0 & 26{,}000 & 0 \\
0 & 52{,}000 & 0 \\
0 & 20{,}000 & 0
\end{bmatrix}
-
\begin{bmatrix}
25{,}000 & 0 & 0 \\
15{,}000 & 0 & 0 \\
8{,}000 & 0 & 0
\end{bmatrix}
$$

45. *Management* Management is attempting to identify the most active person in labor's efforts to unionize. The following diagram shows how influence flows from one employee to another among the four most active employees.

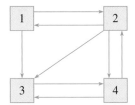

(a) Construct matrix A with elements

$$
a_{ij} =
\begin{cases}
1 & \text{if } i \text{ influences } j \text{ directly.} \\
0 & \text{otherwise.}
\end{cases}
$$

(b) Construct matrix B with elements

$$
b_{ij} =
\begin{cases}
1 & \text{if } i \text{ influences } j \text{ through at most} \\
  & \text{one person, with } i \neq j. \\
0 & \text{otherwise.}
\end{cases}
$$

(c) The person i is most active in influencing others if the sum of the elements in row i of the matrix $A + B$ is the largest. Who is the most active person?

46. *Ranking* In order to rank the five members of its chess team for play against another school, the coach draws the following diagram. An arrow from 1 to 2 means player 1 has defeated player 2.

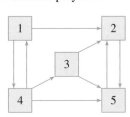

(a) Construct matrix A with elements

$$a_{ij} = \begin{cases} 1 & \text{if } i \text{ defeated } j. \\ 0 & \text{otherwise.} \end{cases}$$

(b) Construct matrix B with elements

$$b_{ij} = \begin{cases} 1 & \text{if } i \text{ defeated someone who} \\ & \text{defeated } j \text{ and } i \ne j. \\ 0 & \text{otherwise.} \end{cases}$$

(c) The player i is the top-ranked player if the sum of row i in the matrix $A + B$ is the largest. What is the number of this player?

47. *Production and inventories* Operating from two plants, the Book Equipment Company (BEC) produces bookcases and filing cabinets. Matrix A summarizes its production for a week, with row 1 representing the number of bookcases and row 2 representing the number of filing cabinets. Matrix B gives the production for the second week, and matrix C that of the third and fourth weeks combined.

$$A = \begin{bmatrix} 50 & 30 \\ 36 & 44 \end{bmatrix} \quad B = \begin{bmatrix} 30 & 45 \\ 22 & 62 \end{bmatrix} \quad C = \begin{bmatrix} 96 & 52 \\ 81 & 37 \end{bmatrix}$$

If column 1 in each matrix represents production from plant 1 and column 2 represents production from plant 2, answer the following.

(a) Write a matrix that describes production for the first 2 weeks.
(b) Find a matrix that describes production for the first 4 weeks.
(c) If matrix D

$$D = \begin{bmatrix} 40 & 26 \\ 29 & 42 \end{bmatrix}$$

describes the shipments made during the first week, write the matrix that describes the units added to the plants' inventories in the first week.

(d) If D also describes the shipments during the second week, describe the change in inventory at the end of two weeks. What happened at plant 1?

48. *Management* The figure that follows depicts the mean (average) flow time for a job when the critical-ratio rule is used to dispatch workers at several machines required to complete the job. In the figure, WHN 1, WHN 2, and WHN 3 represent three different rules for determining when a worker should be transferred to another machine, and Where represents six rules for determining the machine to which the worker is transferred. Construct a 3×6 matrix A to represent these data, with entries rounded to the nearest integer.

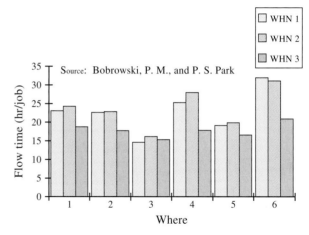

49. *Management* In an evaluation of labor assignment rules when workers are not perfectly interchangeable, Paul M. Bobrowski and Paul Sungchil Park created a dynamic job shop with 9 work centers and 9 workers, both numbered 1–9. The efficiency of each worker is specified in the following labor efficiency matrix, which represents the degree of worker cross-training. (*Source:* Bobrowski, P. M., and P. S. Park, "An evaluation of labor assignment rules," *Journal of Operations Management,* Vol. 11, September 1993)

(a) For what work center(s) is worker 7 least efficient?
(b) For what work center is worker 1 most efficient?

Work Centers

$$\begin{bmatrix} 1.00 & 0.95 & 0.95 & 0.95 & 0.95 & 0.85 & 0.85 & 0.85 & 0.85 \\ 0.85 & 1.00 & 0.95 & 0.95 & 0.95 & 0.95 & 0.85 & 0.85 & 0.85 \\ 0.85 & 0.85 & 1.00 & 0.95 & 0.95 & 0.95 & 0.95 & 0.85 & 0.85 \\ 0.85 & 0.85 & 0.85 & 1.00 & 0.95 & 0.95 & 0.95 & 0.95 & 0.85 \\ 0.85 & 0.85 & 0.85 & 0.85 & 1.00 & 0.95 & 0.95 & 0.95 & 0.95 \\ 0.95 & 0.85 & 0.85 & 0.85 & 0.85 & 1.00 & 0.95 & 0.95 & 0.95 \\ 0.95 & 0.95 & 0.85 & 0.85 & 0.85 & 0.85 & 1.00 & 0.95 & 0.95 \\ 0.95 & 0.95 & 0.95 & 0.85 & 0.85 & 0.85 & 0.85 & 1.00 & 0.95 \\ 0.95 & 0.95 & 0.95 & 0.95 & 0.85 & 0.85 & 0.85 & 0.85 & 1.00 \end{bmatrix} \begin{matrix} \text{W} \\ \text{o} \\ \text{r} \\ \text{k} \\ \text{e} \\ \text{r} \\ \text{s} \end{matrix}$$

50. Suppose that for the workers and the work centers described in the matrix in Problem 49, changes in rules cause efficiency to decrease by .01 at work centers 2 and 5 and to decrease by .02 at work center 7. Write the matrix that describes the new efficiencies.

51. For the data in Problem 49, use a computer spreadsheet to find the average efficiency for each worker over the first four work centers. Which worker is the least efficient? Where does this worker perform best?

52. For the data in Problem 49, use a computer spreadsheet to find the average efficiency for each work center over the first five workers. Which work center is the most efficient? Which work center should be studied for improvement?

3.2

OBJECTIVE

● To multiply two matrices

Multiplication of Matrices

◑ Application Preview

Pentico Industries is a manufacturing company that has two divisions, located at Clarion and Brooks, each of which needs different amounts (units) of production materials as described by the following table.

	Steel	Wood	Plastic
Clarion	20	30	8
Brooks	22	25	15

These raw materials are supplied by Western Supply and Coastal Supply, with prices given in dollars per unit in the following table.

	Western	Coastal
Steel	300	290
Wood	100	90
Plastic	145	180

If one supplier must be chosen to provide all materials for either or both divisions of Pentico, company officials can decide which supplier to choose by constructing matrices from these tables and using **matrix multiplication**. (See Example 3.)

In this section we will discuss finding the **product of two matrices**.

Suppose one store of a firm has 30 washers, 20 dryers, and 10 dishwashers in its inventory. If the value of each washer is $300, that of each dryer is $250, and that of each dishwasher is $350, then the value of this inventory is

$$30 \cdot 300 + 20 \cdot 250 + 10 \cdot 350 = \$17{,}500$$

If we write the value of each of the appliances in the *row matrix*

$$A = \begin{bmatrix} 300 & 250 & 350 \end{bmatrix}$$

and the number of each of the appliances in the *column matrix*

$$B = \begin{bmatrix} 30 \\ 20 \\ 10 \end{bmatrix}$$

then the value of the inventory may be represented by

$$AB = \begin{bmatrix} 300 & 250 & 350 \end{bmatrix} \begin{bmatrix} 30 \\ 20 \\ 10 \end{bmatrix}$$
$$= \begin{bmatrix} 300 \cdot 30 + 250 \cdot 20 + 350 \cdot 10 \end{bmatrix}$$
$$= \begin{bmatrix} \$17{,}500 \end{bmatrix}$$

This useful way of operating with a row matrix and a column matrix is called the product *A* times *B*.

In general, we can multiply the row matrix

$$A = [a_1 \quad a_2 \quad a_3 \quad \ldots \quad a_n]$$

times the column matrix

$$B = \begin{bmatrix} b_1 \\ b_2 \\ b_3 \\ \vdots \\ b_n \end{bmatrix}$$

if A and B have the same number of elements. The product A times B is

$$AB = [a_1 \quad a_2 \quad a_3 \quad \cdots \quad a_n] \begin{bmatrix} b_1 \\ b_2 \\ b_3 \\ \vdots \\ b_n \end{bmatrix} = [a_1b_1 + a_2b_2 + a_3b_3 + \cdots + a_nb_n]$$

Suppose the firm has a second store with 40 washers, 25 dryers, and 5 dishwashers. We can use matrix C to represent the inventories of the two stores.

$$C = \begin{array}{cc} \text{Store I} & \text{Store II} \\ \begin{bmatrix} 30 & 40 \\ 20 & 25 \\ 10 & 5 \end{bmatrix} & \begin{array}{l} \text{Washers} \\ \text{Dryers} \\ \text{Dishwashers} \end{array} \end{array}$$

Suppose the values of the appliances are \$300, \$250, and \$350, respectively. We have already seen that the value of the inventory of store I is \$17,500. The value of the inventory of store II is

$$300 \cdot 40 + 250 \cdot 25 + 350 \cdot 5 = \$20,000$$

Because the value of the inventory of store I can be found by multiplying the row matrix A times the first column of matrix C, and the value of the inventory of store II can be found by multiplying matrix A times the second column of matrix C, we can write this result as

$$AC = [300 \quad 250 \quad 350] \begin{bmatrix} 30 & 40 \\ 20 & 25 \\ 10 & 5 \end{bmatrix} = [17,500 \quad 20,000]$$

The matrix we have represented as AC is called the **product** of the row matrix A and the 3×2 matrix C. We should note that matrix A is a 1×3 matrix, matrix C is a 3×2 matrix, and their product is a 1×2 matrix. In general, we can multiply an $m \times n$ matrix times an $n \times p$ matrix and the product will be an $m \times p$ matrix, as follows:

Product of Two Matrices

Given an $m \times n$ matrix A and an $n \times p$ matrix B, the **matrix product** AB is an $m \times p$ matrix C, with the ij entry of C given by the formula

$$c_{ij} = a_{i1}b_{1j} + a_{i2}b_{2j} + \cdots + a_{in}b_{nj}$$

which is illustrated in Figure 3.5.

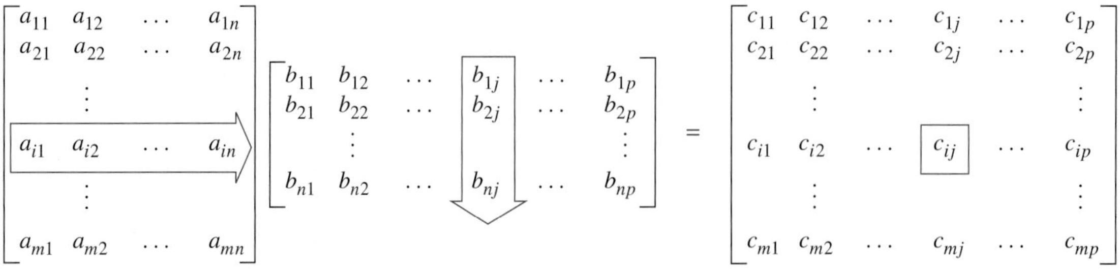

$$
\begin{bmatrix}
a_{11} & a_{12} & \cdots & a_{1n} \\
a_{21} & a_{22} & \cdots & a_{2n} \\
& \vdots & & \\
a_{i1} & a_{i2} & \cdots & a_{in} \\
& \vdots & & \\
a_{m1} & a_{m2} & \cdots & a_{mn}
\end{bmatrix}
\begin{bmatrix}
b_{11} & b_{12} & \cdots & b_{1j} & \cdots & b_{1p} \\
b_{21} & b_{22} & \cdots & b_{2j} & \cdots & b_{2p} \\
& \vdots & & & & \\
b_{n1} & b_{n2} & \cdots & b_{nj} & \cdots & b_{np}
\end{bmatrix}
=
\begin{bmatrix}
c_{11} & c_{12} & \cdots & c_{1j} & \cdots & c_{1p} \\
c_{21} & c_{22} & \cdots & c_{2j} & \cdots & c_{2p} \\
& \vdots & & & & \vdots \\
c_{i1} & c_{i2} & \cdots & c_{ij} & \cdots & c_{ip} \\
& \vdots & & & & \vdots \\
c_{m1} & c_{m2} & \cdots & c_{mj} & \cdots & c_{mp}
\end{bmatrix}
$$

Figure 3.5

● **EXAMPLE 1** *Matrix Product*

Find AB if

$$
A = \begin{bmatrix} 3 & 4 \\ 2 & 5 \\ 6 & 10 \end{bmatrix} \quad \text{and} \quad B = \begin{bmatrix} a & b & c & d \\ e & f & g & h \end{bmatrix}
$$

Solution

A is a 3×2 matrix and B is a 2×4 matrix, so the number of columns of A equals the number of rows of B. Thus we can find the product AB, which is a 3×4 matrix, as follows:

$$
AB = \begin{bmatrix} 3 & 4 \\ 2 & 5 \\ 6 & 10 \end{bmatrix}\begin{bmatrix} a & b & c & d \\ e & f & g & h \end{bmatrix}
$$
$$
= \begin{bmatrix}
3a + 4e & 3b + 4f & 3c + 4g & 3d + 4h \\
2a + 5e & 2b + 5f & 2c + 5g & 2d + 5h \\
6a + 10e & 6b + 10f & 6c + 10g & 6d + 10h
\end{bmatrix}
$$

This example shows that if $AB = C$, element c_{32} is found by multiplying each entry of A's *third* row by the corresponding entry of B's *second* column and then adding these products.

● **EXAMPLE 2** *Matrix Multiplication*

Find AB and BA if

$$
A = \begin{bmatrix} 3 & 2 \\ 1 & 0 \end{bmatrix} \quad \text{and} \quad B = \begin{bmatrix} 1 & 2 \\ 3 & 1 \end{bmatrix}
$$

Solution

Both A and B are 2×2 matrices, so both AB and BA are defined.

$$
AB = \begin{bmatrix} 3 & 2 \\ 1 & 0 \end{bmatrix}\begin{bmatrix} 1 & 2 \\ 3 & 1 \end{bmatrix} = \begin{bmatrix} 3 \cdot 1 + 2 \cdot 3 & 3 \cdot 2 + 2 \cdot 1 \\ 1 \cdot 1 + 0 \cdot 3 & 1 \cdot 2 + 0 \cdot 1 \end{bmatrix} = \begin{bmatrix} 9 & 8 \\ 1 & 2 \end{bmatrix}
$$
$$
BA = \begin{bmatrix} 1 & 2 \\ 3 & 1 \end{bmatrix}\begin{bmatrix} 3 & 2 \\ 1 & 0 \end{bmatrix} = \begin{bmatrix} 1 \cdot 3 + 2 \cdot 1 & 1 \cdot 2 + 2 \cdot 0 \\ 3 \cdot 3 + 1 \cdot 1 & 3 \cdot 2 + 1 \cdot 0 \end{bmatrix} = \begin{bmatrix} 5 & 2 \\ 10 & 6 \end{bmatrix}
$$

Note that for these matrices, the two products AB and BA are matrices of the same size, but they are not equal. That is, $BA \neq AB$. Thus we say that *matrix multiplication is not commutative*.

● *Checkpoint*

1. What is element c_{23} if $C = AB$ with

$$A = \begin{bmatrix} 1 & 2 & 0 \\ 2 & 1 & 3 \end{bmatrix} \quad \text{and} \quad B = \begin{bmatrix} 1 & 0 & 2 \\ 3 & 1 & 1 \\ -2 & 1 & 0 \end{bmatrix}?$$

2. Find the product AB.

◗ EXAMPLE 3 *Material Supply* *(Application Preview)*

Pentico Industries must choose a supplier for the raw materials that it uses in its two manufacturing divisions at Clarion and Brooks. Each division uses different unit amounts of steel, wood, and plastic as shown in the table below.

	Steel	Wood	Plastic
Clarion	20	30	8
Brooks	22	25	15

The two supply companies being considered, Western and Coastal, can each supply all of these materials, but at different prices per unit, as described in the following table.

	Western	Coastal
Steel	300	290
Wood	100	90
Plastic	145	180

Use matrix multiplication to decide which supplier should be chosen to supply (a) the Clarion division, (b) the Brooks division, and (c) both divisions.

Solution

The different amounts of products needed can be placed in matrix A.

$$A = \begin{matrix} & \begin{matrix} \text{Steel} & \text{Wood} & \text{Plastic} \end{matrix} & \\ \begin{bmatrix} 20 & 30 & 8 \\ 22 & 25 & 15 \end{bmatrix} & \begin{matrix} \text{Clarion} \\ \text{Brooks} \end{matrix} \end{matrix}$$

The prices charged by the suppliers are given in matrix B.

$$B = \begin{matrix} & \begin{matrix} \text{Western} & \text{Coastal} \end{matrix} & \\ \begin{bmatrix} 300 & 290 \\ 100 & 90 \\ 145 & 180 \end{bmatrix} & \begin{matrix} \text{Steel} \\ \text{Wood} \\ \text{Plastic} \end{matrix} \end{matrix}$$

The price from each supplier for each division is found from the product AB.

$$AB = \begin{bmatrix} 20 & 30 & 8 \\ 22 & 25 & 15 \end{bmatrix} \begin{bmatrix} 300 & 290 \\ 100 & 90 \\ 145 & 180 \end{bmatrix}$$

$$= \begin{bmatrix} 20 \cdot 300 + 30 \cdot 100 + 8 \cdot 145 & 20 \cdot 290 + 30 \cdot 90 + 8 \cdot 180 \\ 22 \cdot 300 + 25 \cdot 100 + 15 \cdot 145 & 22 \cdot 290 + 25 \cdot 90 + 15 \cdot 180 \end{bmatrix}$$

$$= \begin{matrix} & \begin{matrix} \text{Western} & \text{Coastal} \end{matrix} & \\ \begin{bmatrix} 10{,}160 & 9940 \\ 11{,}275 & 11{,}330 \end{bmatrix} & \begin{matrix} \text{Clarion} \\ \text{Brooks} \end{matrix} \end{matrix}$$

(a) The price for the Clarion division for supplies from Western is in the first row, first column and is $10,160; the price for Clarion for supplies from Coastal is in the first row, second column and is $9940. Thus the best price for Clarion is from Coastal.

(b) The price for the Brooks division for supplies from Western is in the second row, first column and is $11,275; the price for Brooks for supplies from Coastal is in the second row, second column and is $11,330. Thus the best price for Brooks is from Western.

(c) The sum of the first column is $21,435, the price for supplies from Western for both divisions; the sum of the second column is $21,270, the price for supplies from Coastal for both divisions. Thus the best price is from Coastal if one supplier is used for both divisions.

An $n \times n$ (square) matrix (where n is any natural number) that has 1s down its diagonal and 0s everywhere else is called an **identity matrix.** The matrix

$$I = \begin{bmatrix} 1 & 0 & 0 \\ 0 & 1 & 0 \\ 0 & 0 & 1 \end{bmatrix}$$

is a 3×3 identity matrix. The matrix I is called an identity matrix because for any 3×3 matrix A, $AI = IA = A$. That is, if I is multiplied by a 3×3 matrix A, the product matrix is A. Note that when one of two square matrices being multiplied is an identity matrix, the product is *commutative.*

● **EXAMPLE 4 *Identity Matrix***

(a) Write the 2×2 identity matrix.

(b) Given $A = \begin{bmatrix} 4 & -7 \\ 13 & 2 \end{bmatrix}$, show that $AI = IA = A$.

Solution

(a) We denote the 2×2 identity matrix by

$$I = \begin{bmatrix} 1 & 0 \\ 0 & 1 \end{bmatrix}$$

(b) $AI = \begin{bmatrix} 4 & -7 \\ 13 & 2 \end{bmatrix}\begin{bmatrix} 1 & 0 \\ 0 & 1 \end{bmatrix} = \begin{bmatrix} 4+0 & 0-7 \\ 13+0 & 0+2 \end{bmatrix} = \begin{bmatrix} 4 & -7 \\ 13 & 2 \end{bmatrix}$

$IA = \begin{bmatrix} 1 & 0 \\ 0 & 1 \end{bmatrix}\begin{bmatrix} 4 & -7 \\ 13 & 2 \end{bmatrix} = \begin{bmatrix} 4 & -7 \\ 13 & 2 \end{bmatrix}$

Therefore, $AI = IA = A$.

● **EXAMPLE 5 *Encoding Messages***

Messages can be encoded through the use of a code and an encoding matrix. For example, given the code

a	b	c	d	e	f	g	h	i	j	k	l
1	2	3	4	5	6	7	8	9	10	11	12

m	n	o	p	q	r	s	t	u	v	w	x
13	14	15	16	17	18	19	20	21	22	23	24

y	z	blank
25	26	27

and the encoding matrix

$$A = \begin{bmatrix} 3 & 5 \\ 4 & 6 \end{bmatrix}$$

we can encode a message by separating it into number pairs (because A is a 2×2 matrix) and then multiplying each pair by A. Use this code and matrix to encode the message "good job."

Solution

The phrase "good job" is written

g	o	o	d	blank	j	o	b
7	15	15	4	27	10	15	2

and is encoded by writing each pair of numbers as a column matrix and multiplying each matrix by matrix A. Performing this multiplication gives the encoded message.

$$\begin{bmatrix} 3 & 5 \\ 4 & 6 \end{bmatrix}\begin{bmatrix} 7 \\ 15 \end{bmatrix} = \begin{bmatrix} 96 \\ 118 \end{bmatrix}$$

$$\begin{bmatrix} 3 & 5 \\ 4 & 6 \end{bmatrix}\begin{bmatrix} 15 \\ 4 \end{bmatrix} = \begin{bmatrix} 65 \\ 84 \end{bmatrix}$$

$$\begin{bmatrix} 3 & 5 \\ 4 & 6 \end{bmatrix}\begin{bmatrix} 27 \\ 10 \end{bmatrix} = \begin{bmatrix} 131 \\ 168 \end{bmatrix}$$

$$\begin{bmatrix} 3 & 5 \\ 4 & 6 \end{bmatrix}\begin{bmatrix} 15 \\ 2 \end{bmatrix} = \begin{bmatrix} 55 \\ 72 \end{bmatrix}$$

Thus the code sent is 96, 118, 65, 84, 131, 168, 55, 72. Notice that the letter "o" occurred three times in the original message, but was encoded differently each time. We can also get the same code to send by putting the pairs of numbers as columns in one 2×4 matrix and multiplying by A.

$$\begin{bmatrix} 3 & 5 \\ 4 & 6 \end{bmatrix}\begin{bmatrix} 7 & 15 & 27 & 15 \\ 15 & 4 & 10 & 2 \end{bmatrix} = \begin{bmatrix} 96 & 65 & 131 & 55 \\ 118 & 84 & 168 & 72 \end{bmatrix}$$

In this case, the code is sent column by column.

Note in Example 5 that if we used a different encoding matrix, then the same message would be sent with a different sequence of code. We will discuss the methods of decoding messages in Section 3.4, "Inverse of a Square Matrix; Matrix Equations."

● *Checkpoint*

3. (a) Compute AB if $A = \begin{bmatrix} 2 & 4 \\ 3 & 6 \end{bmatrix}$ and $B = \begin{bmatrix} 2 & -6 \\ -1 & 3 \end{bmatrix}$.

   (b) Can the product of two matrices be a zero matrix even if neither matrix is a zero matrix?

   (c) Does BA result in a zero matrix?

Graphing calculators can be used to perform matrix operations.

● EXAMPLE 6 *Matrix Operations*

Use a graphing calculator and the matrices A, B, and C below to perform the operations, if possible.

(a) AB (b) BA (c) BC

$$A = \begin{bmatrix} 2 & 3 & 4 & 5 \\ 3 & 1 & 3 & 5 \\ 3 & -4 & 3 & 1 \end{bmatrix} \quad B = \begin{bmatrix} 2 & 2 & 5 \\ 3 & 5 & 7 \\ 1 & 1 & 2 \\ 5 & 4 & 3 \end{bmatrix} \quad C = \begin{bmatrix} .5 & 3.5 & 2.5 \\ 2 & 7 & .8 \\ 5.5 & 3.5 & 7 \\ 2.5 & 4 & 3 \end{bmatrix}$$

Solution

We begin by entering matrices A, B, and C. The regular operational keys can then be used to perform the required operations.

(a) Figure 3.6(a) shows the product AB. Because A is a 3×4 matrix and B is a 4×3 matrix, the product AB is a 3×3 matrix.

(b) Figure 3.6(b) shows the product BA. Because A is a 3×4 matrix and B is a 4×3 matrix, the product BA is a 4×4 matrix.

(c) Because B and C are both 4×3 matrices, the product BC is not defined. See Figure 3.7.

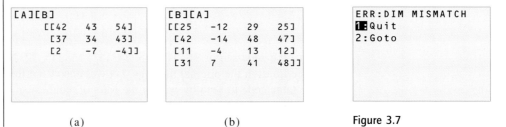

```
[A][B]
    [[42    43    54]
     [37    34    43]
     [2     -7    -4]]
```

```
[B][A]
    [[25    -12    29    25]
     [42    -14    48    47]
     [11    -4     13    12]
     [31    7      41    48]]
```

```
ERR:DIM MISMATCH
1:Quit
2:Goto
```

Figure 3.6   (a)   (b)   Figure 3.7

Like graphing calculators and software programs, computer spreadsheets can be used to multiply two matrices.

Spreadsheet

● EXAMPLE 7 *Venture Capital*

Suppose that a bank has three main sources of income—business loans, auto loans, and home mortgages—and that it draws funds from these sources for venture capital used to provide start-up funds for new businesses. Suppose the income from these sources for each of 3 years is given in Table 3.6, and the bank uses 45% of its income from business loans, 20% of its income from auto loans, and 30% of its income from home mortgages to get its venture capital funds. Write a matrix product that gives the venture capital for these years, and find the available venture capital in each of the 3 years.

TABLE 3.6	Income from Loans		
Year	Business	Auto	Home
2004	63,300	20,024	51,820
2005	48,305	15,817	63,722
2006	55,110	18,621	64,105

Solution

The matrix that describes the sources of income for the 3 years is

$$\begin{bmatrix} 63{,}300 & 20{,}024 & 51{,}820 \\ 48{,}305 & 15{,}817 & 63{,}722 \\ 55{,}110 & 18{,}621 & 64{,}105 \end{bmatrix}$$

and the matrix that describes the percent of each that is used for venture capital is

$$\begin{bmatrix} 0.45 \\ 0.20 \\ 0.30 \end{bmatrix}$$

We can find the product of these matrices with Excel by using the MMULT command as shown in Table 3.7 (see the Excel Guide for details).

TABLE 3.7

= { = MMULT(B3:C5,B11:B13)}				
	A	B	C	D
1		Income from Loans		
2		Business	Auto	Home
3	Matrix A	63,300	20,024	51,820
4		48,305	15,817	63,722
5		55,110	18,621	64,105
6				
7	Matrix B	0.45		
8		0.20		
9		0.30		
10				
11	Product AxB	48,035.80		
12		44,017.25		
13		47,755.20		

Thus the available venture capital for each of the 3 years is as follows.

2004:	$48,035.80
2005:	$44,017.25
2006:	$47,755.20

● **Checkpoint Solutions**

1. $2 \cdot 2 + 1 \cdot 1 + 3 \cdot 0 = 5$

2. $\begin{bmatrix} 7 & 2 & 4 \\ -1 & 4 & 5 \end{bmatrix}$

3. (a) $AB = \begin{bmatrix} 0 & 0 \\ 0 & 0 \end{bmatrix}$ (b) Yes. See (a).   (c) No. $BA = \begin{bmatrix} -14 & -28 \\ 7 & 14 \end{bmatrix}$

3.2 Exercises

In Problems 1–4, multiply the matrices.

1. $\begin{bmatrix} 1 & 2 & 3 \end{bmatrix} \begin{bmatrix} 4 \\ 5 \\ 6 \end{bmatrix}$ 2. $\begin{bmatrix} 2 & 0 & 3 \end{bmatrix} \begin{bmatrix} 0 \\ 1 \\ -3 \end{bmatrix}$ 3. $\begin{bmatrix} 1 & 2 \end{bmatrix} \begin{bmatrix} 3 & 5 \\ 4 & 6 \end{bmatrix}$ 4. $\begin{bmatrix} 3 & 0 \end{bmatrix} \begin{bmatrix} 1 & 2 \\ 4 & 5 \end{bmatrix}$

In Problems 5–24, use matrices A through F.

$$A = \begin{bmatrix} 1 & 0 & 2 \\ 3 & 2 & 1 \\ 4 & 0 & 3 \end{bmatrix} \qquad B = \begin{bmatrix} 1 & 1 & 3 & 0 \\ 4 & 2 & 1 & 1 \\ 3 & 2 & 0 & 1 \end{bmatrix}$$

$$C = \begin{bmatrix} 5 & 3 \\ 1 & 2 \end{bmatrix} \qquad D = \begin{bmatrix} 4 & 2 \\ 3 & 5 \end{bmatrix}$$

$$E = \begin{bmatrix} 1 & 0 & 4 \\ 5 & 1 & 0 \end{bmatrix} \qquad F = \begin{bmatrix} 1 & 0 & -1 & 3 \\ 2 & -1 & 3 & -4 \end{bmatrix}$$

5. *CD*   6. *DC*   7. *DE*
8. *CF*   9. *AB*   10. *EC*
11. *BA*   12. *FB^T*   13. *EB*
14. *BE*   15. *EA^T*   16. *AE^T*
17. *A^2*   18. *A^3*   19. *C^3*
20. *F^2*
21. Does $(AA^T)^T = A^T A$?
22. Are $(CD)E$ and $C(DE)$ equal?
23. Does $CD = DC$? (see Problems 5 and 6)
24. Are $\left(\dfrac{1}{4}A\right)B$ and $\dfrac{1}{4}(AB)$ equal?

In Problems 25–34, use the matrices below. Perform the indicated operations.

$$A = \begin{bmatrix} 2 & 5 & 4 \\ 1 & 4 & 3 \\ 1 & -3 & -2 \end{bmatrix} \qquad B = \begin{bmatrix} -1 & 2 & 1 \\ -5 & 8 & 2 \\ -7 & 11 & -3 \end{bmatrix}$$

$$I = \begin{bmatrix} 1 & 0 & 0 \\ 0 & 1 & 0 \\ 0 & 0 & 1 \end{bmatrix} \qquad C = \begin{bmatrix} 3 & 0 & 4 \\ 1 & 7 & -1 \\ 3 & 0 & 4 \end{bmatrix}$$

$$D = \begin{bmatrix} 4 & 4 & -8 \\ -1 & -1 & 2 \\ -3 & -3 & 6 \end{bmatrix} \qquad F = \begin{bmatrix} 0 & 1 & 2 \\ 0 & 0 & -4 \\ 0 & 0 & 0 \end{bmatrix}$$

$$Z = \begin{bmatrix} 0 & 0 & 0 \\ 0 & 0 & 0 \\ 0 & 0 & 0 \end{bmatrix}$$

25. *AB*   26. *BA*   27. *CD*   28. *DC*
29. *F^3*   30. *AZ*   31. *AI*   32. *IA*
33. *ZCI*   34. *IFZ*

35. Is it true for matrices (as it is for real numbers) that *AB* equals a zero matrix if and only if either *A* or *B* equals a zero matrix? (Refer to Problems 25–34.)
36. Is it true for matrices (as it is for real numbers) that multiplication by a zero matrix gives a result of a zero matrix?
37. (a) Find *AB* and *BA* if

$$A = \begin{bmatrix} a & b \\ c & d \end{bmatrix} \quad \text{and} \quad B = \frac{1}{ad - bc}\begin{bmatrix} d & -b \\ -c & a \end{bmatrix}$$

(b) For *B* to exist, what restriction must $ad - bc$ satisfy?

38. For $F = \begin{bmatrix} 1 & 0 & -1 & 3 \\ 2 & -1 & 3 & -4 \end{bmatrix}$:
   (a) Are FF^T and $F^T F$ defined?
   (b) What size is each product?
   (c) Can $FF^T = F^T F$? Explain.

In each of Problems 39–42, substitute the given values of x, y, and z into the matrix equation and use matrix multiplication to see if the values are the solution of the equation.

39. $\begin{bmatrix} 1 & 2 & 1 \\ 3 & 4 & -2 \\ 2 & 0 & -1 \end{bmatrix}\begin{bmatrix} x \\ y \\ z \end{bmatrix} = \begin{bmatrix} 2 \\ -2 \\ 2 \end{bmatrix}$ $x = 2, y = -1, z = 2$

40. $\begin{bmatrix} 3 & 1 & 0 \\ 2 & -2 & 1 \\ 1 & 1 & 2 \end{bmatrix}\begin{bmatrix} x \\ y \\ z \end{bmatrix} = \begin{bmatrix} 4 \\ 9 \\ 2 \end{bmatrix}$ $x = 2, y = -2, z = 1$

41. $\begin{bmatrix} 1 & 1 & 2 \\ 4 & 0 & 1 \\ 2 & 1 & 1 \end{bmatrix}\begin{bmatrix} x \\ y \\ z \end{bmatrix} = \begin{bmatrix} 5 \\ 5 \\ 5 \end{bmatrix}$ $x = 1, y = 2, z = 1$

42. $\begin{bmatrix} 1 & 0 & 2 \\ 3 & 1 & 0 \\ 1 & 2 & 1 \end{bmatrix}\begin{bmatrix} x \\ y \\ z \end{bmatrix} = \begin{bmatrix} 0 \\ 7 \\ 3 \end{bmatrix}$ $x = 2, y = 1, z = 1$

In Problems 43 and 44, use technology to find the product *AB* of the following matrices.

43. $A = \begin{bmatrix} 0.1 & 0.0 & 0.1 & 0.1 & 0.2 \\ 0.1 & 0.2 & -0.1 & 0.1 & -0.1 \\ 0.1 & -0.1 & 0.1 & -0.2 & 0.1 \\ 0.1 & 0.2 & 0.1 & 0.1 & 0.1 \\ 0.1 & 0.0 & 0.0 & 0.0 & 0.0 \end{bmatrix}$

$B = \begin{bmatrix} 0 & 0 & 0 & 0 & 10 \\ 0 & 5 & \frac{10}{3} & \frac{5}{3} & -10 \\ -10 & -15 & -\frac{20}{3} & \frac{35}{3} & 20 \\ 0 & -5 & -\frac{20}{3} & \frac{5}{3} & 10 \\ 10 & 10 & \frac{20}{3} & -\frac{20}{3} & -20 \end{bmatrix}$

44. $A = \frac{1}{12}\begin{bmatrix} 3 & 15 & -9 & -15 & 6 \\ 3 & -9 & 3 & 9 & -6 \\ -1 & -17 & 3 & 23 & -4 \\ 0 & 12 & 0 & -12 & 0 \\ -1 & -5 & 3 & 5 & 2 \end{bmatrix}$

$B = \begin{bmatrix} 2 & 3 & 0 & 1 & 3 \\ 1 & 0 & 2 & 3 & 1 \\ 0 & 1 & 0 & 2 & 3 \\ 1 & 0 & 2 & 2 & 1 \\ 1 & 0 & 0 & 0 & 3 \end{bmatrix}$

APPLICATIONS

45. *Car pricing*   A car dealer can buy midsize cars for 12% under the list price, and he can buy luxury cars for 15% under the list price. The following table gives the list prices for two midsize and two luxury cars.

Midsize	36,000	42,000
Luxury	50,000	56,000

Write these data in a matrix and multiply it on the left by the matrix

$$\begin{bmatrix} 0.88 & 0 \\ 0 & 0.85 \end{bmatrix}$$

What does each entry in this product matrix represent?

46. *Revenue*   A clothing manufacturer has factories in Atlanta, Chicago, and New York. Sales (in thousands) during the first quarter are summarized in the matrix below.

$$\begin{array}{c} \\ \text{Coats} \\ \text{Shirts} \\ \text{Pants} \\ \text{Ties} \end{array} \begin{array}{ccc} \text{Atl.} & \text{Chi.} & \text{NY} \\ \begin{bmatrix} 40 & 63 & 18 \\ 85 & 56 & 42 \\ 6 & 18 & 8 \\ 7 & 10 & 8 \end{bmatrix} \end{array}$$

During this period the selling price of a coat was $200, of a shirt $40, of a pair of pants $50, and of a tie $30. Use matrix multiplication to find the total revenue received by each factory.

 47. *Egg production*   The following tables give the production of eggs (in millions) for 1995 and 2000 for selected southern states and the average prices in cents per dozen for these states for 1995 and 2000.

(a) Write the production data for 1995 and 2000 in the first table as matrix A. Write the matrix B, which contains the price data in the second table.

(b) What is the order of the product $\frac{1}{12}AB^{T}$?

(c) Where are the entries in the product found in (b) that give the total revenue from the production of eggs for each of these states for the years 1995 and 2000?

Production of Eggs

State	1995	2000
Alabama	2703	2378
Florida	2383	2723
Georgia	4376	5114
Mississippi	1443	1581
North Carolina	3152	2490
South Carolina	1289	1245
Virginia	916	824

Price per Dozen

State	1995	2000
Alabama	96.1	131.0
Florida	48.1	47.8
Georgia	79.4	86.8
Mississippi	99.0	118.0
North Carolina	77.3	107.0
South Carolina	65.8	64.2
Virginia	89.5	96.3

Source: National Agricultural Statistics Service, U.S. Department of Agriculture

48. *Area and population*   Matrix A below gives the fraction of the Earth's area and the projected fraction of its population for five continents in 2050. Matrix B gives the Earth's area (in square miles) and its projected 2050 population. Find the area and population of each given continent by finding AB.

$$A = \begin{array}{c} \\ \begin{array}{cc} \text{Fraction} & \text{Fraction of} \\ \text{of Area} & \text{Population} \end{array} \\ \begin{bmatrix} 0.162 & 0.047 \\ 0.119 & 0.086 \\ 0.066 & 0.065 \\ 0.298 & 0.582 \\ 0.202 & 0.215 \end{bmatrix} \begin{array}{l} \text{North America} \\ \text{South America} \\ \text{Europe} \\ \text{Asia} \\ \text{Africa} \end{array} \end{array}$$

$$B = \begin{array}{c} \begin{array}{cc} \text{Area} & \text{Population} \end{array} \\ \begin{bmatrix} 57,850,000 & 0 \\ 0 & 9,322,000,000 \end{bmatrix} \begin{array}{l} \text{Area} \\ \text{Population} \end{array} \end{array}$$

(*Source:* U.S. Bureau of Census)

49. *Advertising*   A business plans to use three methods of advertising: cable TV, radio, and newspaper. The costs per ad (in thousands of dollars) are given by matrix C.

$$C = \begin{bmatrix} 20 \\ 9 \\ 6 \end{bmatrix} \begin{array}{l} \text{TV} \\ \text{Radio} \\ \text{Newspaper} \end{array}$$

Suppose the business has three target populations for its products: female teenagers, single women 20 to 35, and married women 35 to 60. Matrix T shows the number of ads per month directed at each of these groups.

$$T = \begin{array}{c} \begin{array}{ccc} \text{TV} & \text{Radio} & \text{Papers} \end{array} \\ \begin{bmatrix} 50 & 30 & 5 \\ 40 & 60 & 30 \\ 45 & 40 & 60 \end{bmatrix} \begin{array}{l} \text{Female teens} \\ \text{Single women 20–35} \\ \text{Married women 35–60} \end{array} \end{array}$$

Find the matrix that gives the cost of ads for each target population.

50. *Nutrition* Suppose the weights (in grams) and lengths (in centimeters) of three groups of laboratory animals are given by matrix *A*, where column 1 gives the lengths and each row corresponds to one group.

$$A = \begin{bmatrix} 12.5 & 250 \\ 11.8 & 215 \\ 9.8 & 190 \end{bmatrix}$$

If the increase in both weight and length over the next two weeks is 20% for group I, 7% for group II, and 0% for group III, then the increases in the measures during the 2 weeks can be found by computing *GA*, where

$$G = \begin{bmatrix} 0.20 & 0 & 0 \\ 0 & 0.07 & 0 \\ 0 & 0 & 0 \end{bmatrix}$$

(a) What are the increases in respective weights and measures at the end of these 2 weeks?

(b) Find the matrix that gives the new weights and measures at the end of this period by computing

$$(I + G)A$$

where *I* is the 3 × 3 identity matrix.

Encoding messages **Multiplication by a matrix can be used to encode messages (and multiplication by its inverse can be used to decode messages). Given the code**

a	b	c	d	e	f	g	h	i	j
1	2	3	4	5	6	7	8	9	10

k	l	m	n	o	p	q	r	s	t
11	12	13	14	15	16	17	18	19	20

u	v	w	x	y	z	blank
21	22	23	24	25	26	27

and the code matrix

$$A = \begin{bmatrix} 5 & 9 \\ 6 & 11 \end{bmatrix}$$

complete Problems 51 and 52.

51. Use matrix *A* to encode the message "The die is cast."

52. Use matrix *A* to encode the message "To be or not to be."

53. *Death rates* The following tables give the death rates, per 100,000 population, by age for selected years for males and females. Use matrix operations to find the rate per 100,000 for all people in the age categories given and for the given years.

Males

	Under 1	1–4	5–14	15–24	25–34	35–44	45–54
1970	2410	93	51	189	215	403	959
1980	1429	73	37	172	196	299	767
1990	1083	52	29	147	204	310	610
1994	899	52	26	151	207	337	585
2001	755	37	20	117	143	259	544

Females

	Under 1	1–4	5–14	15–24	25–34	35–44	45–54
1970	1864	75	32	68	102	231	517
1980	1142	55	24	58	76	159	413
1990	856	41	19	49	74	138	343
1994	719	37	19	47	76	144	326
2001	620	29	15	43	66	148	317

Source: U.S. Department of Health and Human Services, National Center for Health Statistics

54. *Manufacturing* Suppose products A and B are made from plastic, steel, and glass, with the number of units of each raw material required for each product given by the table below.

	Plastic	Steel	Glass
Product A	3	1	0.50
Product B	4	0.50	2

Because of transportation costs to the firm's two plants, X and Y, the unit costs for some of the raw materials are different. The table below gives the unit costs for each of the raw materials at the two plants.

	Plant X	Plant Y
Plastic	10	9
Steel	22	26
Glass	14	14

Using the information just given, find the total cost of producing each of the products at each of the factories.

55. *Accounting* The annual budget of the Magnum Company has the following expenses, in thousands of dollars, for selected departments.

	Mfg.	Office	Sales	Shp.	Act.	Mgt.
Supplies	0.7	8.5	10.2	1.1	5.6	3.6
Phone	0.5	0.2	6.1	1.3	0.2	1.0
Transp.	2.2	0.4	8.8	1.2	1.2	4.8
Salaries	251.8	63.4	81.6	35.2	54.3	144.2
Utilities	30.0	1.0	1.0	1.0	1.0	1.0
Materials	788.9	0	0	0	0	0

Mfg. = Manufacturing; Shp. = Shipping;
Act. = Accounting; Mgt. = Management

(a) Write a matrix B for these budget amounts.
(b) Write a matrix A so that BA would contain new budget figures that reflect an 11% increase in manufacturing, sales, and shipping, and a 5% decrease in the other departments.

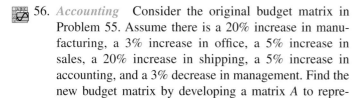

56. *Accounting* Consider the original budget matrix in Problem 55. Assume there is a 20% increase in manufacturing, a 3% increase in office, a 5% increase in sales, a 20% increase in shipping, a 5% increase in accounting, and a 3% decrease in management. Find the new budget matrix by developing a matrix A to represent these departmental increases, then multiplying it from the left by the matrix B in Problem 55.

57. *Management* In an evaluation of labor assignment rules when workers are not perfectly interchangeable, Paul M. Bobrowski and Paul Sungchil Park* created a dynamic job shop with 9 work centers and 9 workers. The efficiency of each worker is specified in the labor efficiency matrix E given here, which represents the degree of worker cross-training. If a worker slowdown causes a 10% decrease in efficiency at all work centers, write the matrix that describes the efficiencies for the 9 workers at the 9 work centers. (Use a graphing calculator, a computer program, or a spreadsheet.)

$$E = \begin{bmatrix} 1.00 & 0.95 & 0.95 & 0.95 & 0.95 & 0.85 & 0.85 & 0.85 & 0.85 \\ 0.85 & 1.00 & 0.95 & 0.95 & 0.95 & 0.95 & 0.85 & 0.85 & 0.85 \\ 0.85 & 0.85 & 1.00 & 0.95 & 0.95 & 0.95 & 0.95 & 0.85 & 0.85 \\ 0.85 & 0.85 & 0.85 & 1.00 & 0.95 & 0.95 & 0.95 & 0.95 & 0.85 \\ 0.85 & 0.85 & 0.85 & 0.85 & 1.00 & 0.95 & 0.95 & 0.95 & 0.95 \\ 0.95 & 0.85 & 0.85 & 0.85 & 0.85 & 1.00 & 0.95 & 0.95 & 0.95 \\ 0.95 & 0.95 & 0.85 & 0.85 & 0.85 & 0.85 & 1.00 & 0.95 & 0.95 \\ 0.95 & 0.95 & 0.95 & 0.85 & 0.85 & 0.85 & 0.85 & 1.00 & 0.95 \\ 0.95 & 0.95 & 0.95 & 0.95 & 0.85 & 0.85 & 0.85 & 0.85 & 1.00 \end{bmatrix}$$

*Bobrowski, P. M., and P. S. Park, "An evaluation of labor assignment rules," *Journal of Operations Management*, Vol. II, September 1993.

3.3

OBJECTIVES

● To use matrices to solve systems of linear equations with unique solutions
● To use matrices to solve systems of linear equations with nonunique solutions

Gauss-Jordan Elimination: Solving Systems of Equations

◗ Application Preview

The Walters Manufacturing Company makes three types of metal storage sheds. The company has three departments: stamping, painting, and packaging. The following table gives the number of hours each division requires for each shed.

Department	Shed		
	Type I	Type II	Type III
Stamping	2	3	4
Painting	1	2	1
Packaging	1	1	2

By using the information in the table and by solving a system of equations, we can determine how many of each type of shed can be produced if the stamping department has 3200 hours available, the painting department has 1700 hours, and the packaging department has 1300 hours. The set of equations used to solve this problem (in Example 3) is as follows:

$$\begin{cases} 2x_1 + 3x_2 + 4x_3 = 3200 \\ x_1 + 2x_2 + x_3 = 1700 \\ x_1 + x_2 + 2x_3 = 1300 \end{cases}$$

In this section, we will solve such systems by using matrices and the Gauss-Jordan elimination method.

Systems with Unique Solutions

In solving systems with the left-to-right elimination method used in Example 8 in Section 1.5, "Solutions of Systems of Linear Equations," we operated on the coefficients of the variables x, y, and z and on the constants. If we keep the coefficients of the variables x, y, and z in distinctive columns, we do not need to write the equations. In solving a system of linear equations with matrices, we first write the coefficients and constants from the system (on the left) in the **augmented matrix** (on the right).

$$\begin{cases} x + 2y + 3z = 6 \\ x \qquad - z = 0 \\ x - y - z = -4 \end{cases} \qquad \left[\begin{array}{ccc|c} 1 & 2 & 3 & 6 \\ 1 & 0 & -1 & 0 \\ 1 & -1 & -1 & -4 \end{array} \right]$$

In the augmented matrix, the numbers on the left side of the solid line form the **coefficient matrix,** with each column containing the coefficients of a variable (0 represents any missing variable). The column on the right side of the line (called the *augment*) contains the constants. Each row of the matrix gives the corresponding coefficients of an equation.

● *Checkpoint*   1. (a) Write the augmented matrix for the following system of linear equations.

$$\begin{cases} 3x + 2y - z = 3 \\ x - y + 2z = 4 \\ 2x + 3y - z = 3 \end{cases}$$

   (b) Write the coefficient matrix for the system.

We can use matrices to solve systems of linear equations by performing the same operations on the rows of a matrix to reduce it as we do on equations in a linear system. The three different operations we can use to reduce the matrix are called **elementary row operations** and are similar to the operations with equations that result in equivalent systems. These operations are

1. Interchange two rows.
2. Add a multiple of one row to another row.
3. Multiply a row by a nonzero constant.

When a new matrix results from one or more of these elementary row operations being performed on a matrix, the new matrix is said to be **equivalent** to the original because both these matrices represent equivalent systems of equations. Thus, if it can be shown that the augmented matrix

$$\left[\begin{array}{ccc|c} 1 & 2 & 3 & 6 \\ 1 & 0 & -1 & 0 \\ 1 & -1 & -1 & -4 \end{array} \right] \text{ is equivalent to } \left[\begin{array}{ccc|c} 1 & 0 & 0 & -0.5 \\ 0 & 1 & 0 & 4 \\ 0 & 0 & 1 & -0.5 \end{array} \right]$$

then the systems corresponding to these matrices have the same solution. That is,

$$\begin{cases} x + 2y + 3z = 6 \\ x \qquad - z = 0 \\ x - y - z = -4 \end{cases} \text{ has the same solution as } \begin{cases} x = -0.5 \\ y = 4 \\ z = -0.5 \end{cases}$$

Thus, if a system of linear equations has a unique solution, we can solve the system by reducing the associated matrix to one whose coefficient matrix is the identity matrix and then "reading" the values of the variables that give the solution.

The process that we use to solve a system of equations with matrices (called the **elimination method** or **Gauss-Jordan elimination method**) is a systematic procedure (similar to the left-to-right elimination method discussed in Section 1.5, "Solutions of Systems of Linear Equations") that uses row operations to attempt to reduce the coefficient matrix to an identity matrix.

● **EXAMPLE 1** *Gauss-Jordan Elimination*

Use matrices to solve the system

$$\begin{cases} 2x + 5y + 4z = 4 \\ x + 4y + 3z = 1 \\ x - 3y - 2z = 5 \end{cases}$$

Solution

The augmented matrix for this system is

$$\begin{bmatrix} 2 & 5 & 4 & | & 4 \\ 1 & 4 & 3 & | & 1 \\ 1 & -3 & -2 & | & 5 \end{bmatrix}$$

Note that this system has 3 equations in 3 variables, but the reduction procedure and row operations apply regardless of the size of the system. The procedure follows.

Gauss-Jordan Elimination Method

Goal (for Each Step)	Row Operation	Equivalent Matrix
1. Get a 1 in row 1, column 1.	Interchange row 1 and row 2.	$\begin{bmatrix} 1 & 4 & 3 & \vert & 1 \\ 2 & 5 & 4 & \vert & 4 \\ 1 & -3 & -2 & \vert & 5 \end{bmatrix}$
2. Add multiples of row 1 *only* to the other rows to get zeros in other entries of column 1.	Add -2 times row 1 to row 2; put the result in row 2. Add -1 times row 1 to row 3; put the result in row 3.	$\begin{bmatrix} 1 & 4 & 3 & \vert & 1 \\ 0 & -3 & -2 & \vert & 2 \\ 0 & -7 & -5 & \vert & 4 \end{bmatrix}$
3. Use rows below row 1 to get a 1 in row 2, column 2.	Multiply row 2 by $-\frac{1}{3}$; put the result in row 2.	$\begin{bmatrix} 1 & 4 & 3 & \vert & 1 \\ 0 & 1 & \frac{2}{3} & \vert & -\frac{2}{3} \\ 0 & -7 & -5 & \vert & 4 \end{bmatrix}$
4. Add multiples of row 2 *only* to the other rows to get zeros as the other entries in column 2.	Add -4 times row 2 to row 1; put the result in row 1. Add 7 times row 2 to row 3; put the result in row 3.	$\begin{bmatrix} 1 & 0 & \frac{1}{3} & \vert & \frac{11}{3} \\ 0 & 1 & \frac{2}{3} & \vert & -\frac{2}{3} \\ 0 & 0 & -\frac{1}{3} & \vert & -\frac{2}{3} \end{bmatrix}$
5. Use rows below row 2 to get a 1 in row 3, column 3.	Multiply row 3 by -3; put the result in row 3.	$\begin{bmatrix} 1 & 0 & \frac{1}{3} & \vert & \frac{11}{3} \\ 0 & 1 & \frac{2}{3} & \vert & -\frac{2}{3} \\ 0 & 0 & 1 & \vert & 2 \end{bmatrix}$
6. Add multiples of row 3 *only* to the other rows to get zeros as the other entries in column 3.	Add $-\frac{1}{3}$ times row 3 to row 1; put the result in row 1. Add $-\frac{2}{3}$ times row 3 to row 2; put the result in row 2.	$\begin{bmatrix} 1 & 0 & 0 & \vert & 3 \\ 0 & 1 & 0 & \vert & -2 \\ 0 & 0 & 1 & \vert & 2 \end{bmatrix}$
7. Repeat the process until it cannot be continued.	All rows have been used. The matrix is in reduced form.	

The previous display shows a series of step-by-step goals for reducing a matrix and shows a series of row operations that reduces this 3×3 matrix.

The system of equations corresponding to this final augmented matrix follows.

$$
\begin{aligned}
x + 0y + 0z &= 3 \\
0x + y + 0z &= -2 \\
0x + 0y + z &= 2
\end{aligned}
$$

Thus the solution is $x = 3$, $y = -2$, and $z = 2$.

When a system of linear equations has a unique solution, the coefficient part of the reduced augmented matrix will be an identity matrix. Note that for this to occur, the coefficient matrix must be square; that is, the number of equations must equal the number of variables. Recall that the Gauss-Jordan elimination method (outlined in Example 1) may be used with systems of any size.

● **EXAMPLE 2** *Using the Gauss-Jordan Method*

Solve the system

$$
\begin{aligned}
x_1 + x_2 + x_3 + 2x_4 &= 6 \\
x_1 + 2x_2 + x_4 &= -2 \\
x_1 + x_2 + 3x_3 - 2x_4 &= 12 \\
x_1 + x_2 - 4x_3 + 5x_4 &= -16
\end{aligned}
$$

Solution

First note that there is a variable missing in the second equation. When we form the augmented matrix, we must insert a zero in this place so that each column represents the coefficients of one variable.

To solve this system, we must reduce the augmented matrix.

$$
\left[\begin{array}{cccc|c}
1 & 1 & 1 & 2 & 6 \\
1 & 2 & 0 & 1 & -2 \\
1 & 1 & 3 & -2 & 12 \\
1 & 1 & -4 & 5 & -16
\end{array}\right]
$$

The reduction process follows. The entry in row 1, column 1 is 1. To get zeros in the first column, add -1 times row 1 to row 2, and put the result in row 2; add -1 times row 1 to row 3 and put the result in row 3; and add -1 times row 1 to row 4 and put the result in row 4. These steps follow.

$$
\left[\begin{array}{cccc|c}
1 & 1 & 1 & 2 & 6 \\
1 & 2 & 0 & 1 & -2 \\
1 & 1 & 3 & -2 & 12 \\
1 & 1 & -4 & 5 & -16
\end{array}\right]
\xrightarrow[\substack{(-1)R_1 + R_3 \to R_3 \\ (-1)R_1 + R_4 \to R_4}]{(-1)R_1 + R_2 \to R_2}
\left[\begin{array}{cccc|c}
1 & 1 & 1 & 2 & 6 \\
0 & 1 & -1 & -1 & -8 \\
0 & 0 & 2 & -4 & 6 \\
0 & 0 & -5 & 3 & -22
\end{array}\right]
$$

The entry in row 2, column 2 is 1. To get zeros in the second column, add -1 times row 2 to row 1, and put the result in row 1.

$$
\xrightarrow{(-1)R_2 + R_1 \to R_1}
\left[\begin{array}{cccc|c}
1 & 0 & 2 & 3 & 14 \\
0 & 1 & -1 & -1 & -8 \\
0 & 0 & 2 & -4 & 6 \\
0 & 0 & -5 & 3 & -22
\end{array}\right]
$$

The entry in row 3, column 3 is 2. To get a 1 in this position, multiply row 3 by $\frac{1}{2}$.

$$\xrightarrow{\frac{1}{2}R_3 \to R_3} \begin{bmatrix} 1 & 0 & 2 & 3 & | & 14 \\ 0 & 1 & -1 & -1 & | & -8 \\ 0 & 0 & 1 & -2 & | & 3 \\ 0 & 0 & -5 & 3 & | & -22 \end{bmatrix}$$

Using row 3, add -2 times row 3 to row 1 and put the result in row 1; add row 3 to row 2 and put the result in row 2; and add 5 times row 3 to row 4 and put the result in row 4.

$$\begin{matrix} (-2)R_3 + R_1 \to R_1 \\ R_3 + R_2 \to R_2 \\ \xrightarrow{5R_3 + R_4 \to R_4} \end{matrix} \begin{bmatrix} 1 & 0 & 0 & 7 & | & 8 \\ 0 & 1 & 0 & -3 & | & -5 \\ 0 & 0 & 1 & -2 & | & 3 \\ 0 & 0 & 0 & -7 & | & -7 \end{bmatrix}$$

Multiplying $-\frac{1}{7}$ times row 4 and putting the result in row 4 gives a 1 in row 4, column 4.

$$\xrightarrow{-\frac{1}{7}R_4 \to R_4} \begin{bmatrix} 1 & 0 & 0 & 7 & | & 8 \\ 0 & 1 & 0 & -3 & | & -5 \\ 0 & 0 & 1 & -2 & | & 3 \\ 0 & 0 & 0 & 1 & | & 1 \end{bmatrix}$$

Adding appropriate multiples of row 4 to the other rows gives

$$\begin{matrix} (-7)R_4 + R_1 \to R_1 \\ 3R_4 + R_2 \to R_2 \\ \xrightarrow{2R_4 + R_3 \to R_3} \end{matrix} \begin{bmatrix} 1 & 0 & 0 & 0 & | & 1 \\ 0 & 1 & 0 & 0 & | & -2 \\ 0 & 0 & 1 & 0 & | & 5 \\ 0 & 0 & 0 & 1 & | & 1 \end{bmatrix}$$

This corresponds to the system

$$\begin{aligned} x_1 &= 1 \\ x_2 &= -2 \\ x_3 &= 5 \\ x_4 &= 1 \end{aligned}$$

Thus the solution is $x_1 = 1, x_2 = -2, x_3 = 5$, and $x_4 = 1$.

Technology Note

Spreadsheets, computer programs, and graphing calculators are especially useful in solving systems of equations, because they can be used to perform the row reductions necessary to obtain a reduced matrix. On some calculators, a direct command will give a reduced form for the augmented matrix. Figure 3.8(a) shows the augmented matrix for the system of equations in Example 2, and Figure 3.8(b) shows the reduced form of the augmented matrix (which is the same as the reduced matrix found in Example 2).

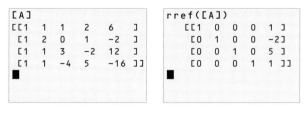

Figure 3.8 (a) (b)

2. Solve the system.

$$\begin{cases} 3x + 2y - \ z = 3 \\ x - \ y + 2z = 4 \\ 2x + 3y - \ z = 3 \end{cases}$$

◗ EXAMPLE 3 *Manufacturing (Application Preview)*

The Walters Manufacturing Company needs to know how best to use the time available within its three manufacturing departments in the construction and packaging of the three types of metal storage sheds. Each one must be stamped, painted, and packaged. Table 3.8 shows the number of hours required for the processing of each type of shed. Using the information in the table, determine how many of each type of shed can be produced if the stamping department has 3200 hours available, the painting department has 1700 hours, and the packaging department has 1300 hours.

TABLE 3.8

Department	Shed		
	Type I	Type II	Type III
Stamping	2	3	4
Painting	1	2	1
Packaging	1	1	2

Solution

If we let x_1 be the number of type I sheds, x_2 be the number of type II sheds, and x_3 be the number of type III sheds, the equation

$$2x_1 + 3x_2 + 4x_3 = 3200$$

represents the hours used by the stamping department. Similarly,

$$1x_1 + 2x_2 + 1x_3 = 1700$$

and

$$1x_1 + 1x_2 + 2x_3 = 1300$$

represent the hours used by the painting department and the packaging department, respectively.

The augmented matrix for this system of equations is

$$\begin{bmatrix} 2 & 3 & 4 & | & 3200 \\ 1 & 2 & 1 & | & 1700 \\ 1 & 1 & 2 & | & 1300 \end{bmatrix}$$

Reducing this augmented matrix proceeds as follows:

$$\begin{bmatrix} 2 & 3 & 4 & | & 3200 \\ 1 & 2 & 1 & | & 1700 \\ 1 & 1 & 2 & | & 1300 \end{bmatrix} \xrightarrow{\text{Switch } R_1 \text{ and } R_3} \begin{bmatrix} 1 & 1 & 2 & | & 1300 \\ 1 & 2 & 1 & | & 1700 \\ 2 & 3 & 4 & | & 3200 \end{bmatrix}$$

$$\xrightarrow[(-2)R_1 + R_3 \to R_3]{(-1)R_1 + R_2 \to R_2} \begin{bmatrix} 1 & 1 & 2 & | & 1300 \\ 0 & 1 & -1 & | & 400 \\ 0 & 1 & 0 & | & 600 \end{bmatrix}$$

$$\begin{array}{c} (-1)R_2 + R_1 \rightarrow R_1 \\ \hline (-1)R_2 + R_3 \rightarrow R_3 \end{array} \left[\begin{array}{ccc|c} 1 & 0 & 3 & 900 \\ 0 & 1 & -1 & 400 \\ 0 & 0 & 1 & 200 \end{array}\right]$$

$$\begin{array}{c} (-3)R_3 + R_1 \rightarrow R_1 \\ \hline R_3 + R_2 \rightarrow R_2 \end{array} \left[\begin{array}{ccc|c} 1 & 0 & 0 & 300 \\ 0 & 1 & 0 & 600 \\ 0 & 0 & 1 & 200 \end{array}\right]$$

Thus the solution to the system is

$$x_1 = 300$$
$$x_2 = 600$$
$$x_3 = 200$$

The company should make 300 type I, 600 type II, and 200 type III sheds.

Systems with Nonunique Solutions

All the systems considered so far had unique solutions, but it is also possible for a system of linear equations to have an infinite number of solutions or no solution at all. Although coefficient matrices for systems with an infinite number of solutions or no solution will not reduce to identity matrices, row operations can be used to obtain a reduced form from which the solutions, if they exist, can be determined.

A matrix is said to be in **reduced form** when it is in the following form:

1. The first nonzero element in each row is 1.
2. Every column containing a first nonzero element for some row has zeros everywhere else.
3. The first nonzero element of each row is to the right of the first nonzero element of every row above it.
4. All rows containing zeros are grouped together below the rows containing nonzero entries.

The following matrices are in reduced form because they satisfy these conditions.

$$\left[\begin{array}{cc|c} 1 & 0 & 4 \\ 0 & 1 & 2 \\ 0 & 0 & 0 \end{array}\right] \qquad \left[\begin{array}{ccc|c} 1 & 0 & 0 & 0 \\ 0 & 1 & 0 & 3 \\ 0 & 0 & 1 & 1 \\ 0 & 0 & 0 & 0 \end{array}\right] \qquad \left[\begin{array}{ccc|c} 1 & 4 & 0 & 1 \\ 0 & 0 & 1 & 2 \\ 0 & 0 & 0 & 0 \end{array}\right]$$

The following matrices are *not* in reduced form.

$$\left[\begin{array}{cc|c} 1 & ① & 0 \\ 0 & 1 & 3 \\ 0 & 0 & 0 \end{array}\right] \qquad \left[\begin{array}{cc|c} 1 & 0 & 1 \\ 0 & ② & 1 \\ 0 & 0 & 0 \end{array}\right] \qquad \left[\begin{array}{cc|c} 0 & 1 & 0 \\ 1 & 0 & 0 \\ 0 & 0 & 1 \end{array}\right]$$

In the first two matrices, the circled element must be changed to obtain a reduced form. Can you see what row operations would transform each of these matrices into reduced form? The third matrix does not satisfy point 3 above.

We can solve a system of linear equations by using row operations on the augmented matrix until the coefficient matrix is transformed to an equivalent matrix in reduced form. We begin by following the goals outlined in Example 1.

● **EXAMPLE 4** *Matrix Solution of a System*

Solve the system

$$\begin{cases} x_1 + x_2 - x_3 + x_4 = 3 \\ x_2 + x_3 + x_4 = 1 \\ x_1 \qquad - 2x_3 + x_4 = 6 \\ 2x_1 - x_2 - 5x_3 - 3x_4 = -5 \end{cases}$$

Solution
To solve this system we must reduce the augmented matrix

$$\begin{bmatrix} 1 & 1 & -1 & 1 & | & 3 \\ 0 & 1 & 1 & 1 & | & 1 \\ 1 & 0 & -2 & 1 & | & 6 \\ 2 & -1 & -5 & -3 & | & -5 \end{bmatrix}$$

Attempting to reduce the coefficient matrix to an identity matrix requires the following.

$$\begin{bmatrix} 1 & 1 & -1 & 1 & | & 3 \\ 0 & 1 & 1 & 1 & | & 1 \\ 1 & 0 & -2 & 1 & | & 6 \\ 2 & -1 & -5 & -3 & | & -5 \end{bmatrix} \xrightarrow[(-2)R_1 + R_4 \to R_4]{(-1)R_1 + R_3 \to R_3} \begin{bmatrix} 1 & 1 & -1 & 1 & | & 3 \\ 0 & 1 & 1 & 1 & | & 1 \\ 0 & -1 & -1 & 0 & | & 3 \\ 0 & -3 & -3 & -5 & | & -11 \end{bmatrix}$$

$$\xrightarrow[\substack{(-1)R_2 + R_1 \to R_1 \\ R_2 + R_3 \to R_3 \\ 3R_2 + R_4 \to R_4}]{} \begin{bmatrix} 1 & 0 & -2 & 0 & | & 2 \\ 0 & 1 & 1 & 1 & | & 1 \\ 0 & 0 & 0 & 1 & | & 4 \\ 0 & 0 & 0 & -2 & | & -8 \end{bmatrix}$$

The entry in row 3, column 3 cannot be made 1 using rows *below* row 2. Moving to column 4, the entry in row 3, column 4 is a 1. Using row 3 gives

$$\xrightarrow[\substack{(-1)R_3 + R_2 \to R_2 \\ 2R_3 + R_4 \to R_4}]{} \begin{bmatrix} 1 & 0 & -2 & 0 & | & 2 \\ 0 & 1 & 1 & 0 & | & -3 \\ 0 & 0 & 0 & 1 & | & 4 \\ 0 & 0 & 0 & 0 & | & 0 \end{bmatrix}$$

The augmented matrix is now reduced; this matrix corresponds to the system

$$\begin{cases} x_1 - 2x_3 = 2 \\ x_2 + x_3 = -3 \\ x_4 = 4 \\ 0 = 0 \end{cases}$$

If we solve each of the equations for the leading variable (the variable corresponding to the first 1 in each row of the reduced form of a matrix) and let any non-leading variables equal any real number (x_3 in this case), we obtain the **general solution** of the system.

$$\begin{aligned} x_1 &= 2 + 2x_3 \\ x_2 &= -3 - x_3 \\ x_4 &= 4 \\ x_3 &= \text{any real number} \end{aligned}$$

The general solution gives the values of x_1 and x_2 dependent on the value of x_3, so we can get many different solutions of the system by specifying different values of x_3. For example, if $x_3 = 1$, then $x_1 = 4$, $x_2 = -4$, $x_3 = 1$, and $x_4 = 4$ is a solution of the system; if we let $x_3 = -2$, then $x_1 = -2$, $x_2 = -1$, $x_3 = -2$, and $x_4 = 4$ is another solution.

We have seen two different possibilities for solutions of systems of linear equations. In Examples 1–3, there was only one solution, whereas in Example 4 we saw that there were infinitely many solutions, one for each possible value of x_3. A third possibility exists: that the system has no solution. Recall that these possibilities also existed for two equations in two unknowns, as discussed in Chapter 1, "Linear Equations and Functions."

● **EXAMPLE 5** *A System with No Solution*

Solve the system

$$\begin{cases} x + 2y - z = 3 \\ 3x + y = 4 \\ 2x - y + z = 2 \end{cases}$$

Solution

The augmented matrix is

$$\begin{bmatrix} 1 & 2 & -1 & | & 3 \\ 3 & 1 & 0 & | & 4 \\ 2 & -1 & 1 & | & 2 \end{bmatrix}$$

This can be reduced as follows:

$$\begin{bmatrix} 1 & 2 & -1 & | & 3 \\ 3 & 1 & 0 & | & 4 \\ 2 & -1 & 1 & | & 2 \end{bmatrix} \xrightarrow[(-2)R_1 + R_3 \to R_3]{(-3)R_1 + R_2 \to R_2} \begin{bmatrix} 1 & 2 & -1 & | & 3 \\ 0 & -5 & 3 & | & -5 \\ 0 & -5 & 3 & | & -4 \end{bmatrix}$$

$$\xrightarrow{-\frac{1}{5}R_2 \to R_2} \begin{bmatrix} 1 & 2 & -1 & | & 3 \\ 0 & 1 & -\frac{3}{5} & | & 1 \\ 0 & -5 & 3 & | & -4 \end{bmatrix}$$

$$\xrightarrow[(5)R_2 + R_3 \to R_3]{(-2)R_2 + R_1 \to R_1} \begin{bmatrix} 1 & 0 & \frac{1}{5} & | & 1 \\ 0 & 1 & -\frac{3}{5} & | & 1 \\ 0 & 0 & 0 & | & 1 \end{bmatrix}$$

The system of equations corresponding to the reduced matrix is

$$\begin{cases} x + \frac{1}{5}z = 1 \\ y - \frac{3}{5}z = 1 \\ 0 = 1 \end{cases}$$

Thus we see that this system has $0 = 1$ as an equation. This is clearly an impossibility, so there is no solution.

Now that we have seen examples of each of the different solution possibilities for n equations in n variables, let us consider what we have learned. In all cases the solution procedure began by setting up and then reducing the augmented matrix. Once the matrix is reduced, we can write the corresponding reduced system of equations. The solutions, if they exist, are easily found from this reduced system. We summarize the possibilities in Table 3.9.

TABLE 3.9

Reduced Form of Augmented Matrix for *n* Equations in *n* Variables	Solution to System
1. Coefficient array is an identity matrix.	Unique solution (see Examples 1, 2, and 3).
2. Coefficient array is *not* an identity matrix with either:	
(a) A row of 0s in the coefficient array with a nonzero entry in the augment.	(a) No solution (see Example 5).
(b) Or otherwise.	(b) Infinitely many solutions. Solve for lead variables in terms of nonlead variables; non-lead variables can equal any real number (see Example 4).

Nonsquare Systems

Finally, we consider systems with fewer equations than variables. Because the coefficient matrix for such a system is not square, it cannot reduce to an identity matrix. Hence these systems of equations cannot have a unique solution but may have no solution or infinitely many solutions. Nevertheless, they are solved in the same manner as those considered so far but with the following two possible results.

1. If the reduced augmented matrix contains a row of 0s in the coefficient matrix with a nonzero number in the augment, the system has no solution.
2. Otherwise, the system has infinitely many solutions.

● **EXAMPLE 6** *Investment*

A trust account manager has $500,000 to be invested in three different accounts. The accounts pay annual interest rates of 8%, 10%, and 14%, and the goal is to earn $49,000 a year. To accomplish this, assume that x dollars is invested at 8%, y dollars at 10%, and z dollars at 14%. Find how much should be invested in each account to satisfy the conditions.

Solution

The sum of the three investments is $500,000, so we have the equation

$$x + y + z = 500,000$$

The interest earned from the 8% investment is $0.08x$, the amount earned from the 10% investment is $0.10y$, and the amount earned from the 14% investment is $0.14z$, so the total amount earned from the investments is given by the equation

$$0.08x + 0.10y + 0.14z = 49,000$$

These two equations represent all the given information, so we attempt to answer the question by solving a system of *two* equations in *three* variables. (Because there are fewer equations than variables, we cannot find a unique solution to the system.) The system is

$$\begin{cases} x + y + z = 500,000 \\ 0.08x + 0.10y + 0.14z = 49,000 \end{cases}$$

We solve this system using matrices, as follows.

$$\begin{bmatrix} 1 & 1 & 1 & | & 500,000 \\ 0.08 & 0.10 & 0.14 & | & 49,000 \end{bmatrix} \xrightarrow{-0.08R_1 + R_2 \to R_2} \begin{bmatrix} 1 & 1 & 1 & | & 500,000 \\ 0 & 0.02 & 0.06 & | & 9,000 \end{bmatrix}$$

$$\xrightarrow{50R_2 \to R_2} \begin{bmatrix} 1 & 1 & 1 & | & 500,000 \\ 0 & 1 & 3 & | & 450,000 \end{bmatrix} \xrightarrow{-R_2 + R_1 \to R_1} \begin{bmatrix} 1 & 0 & -2 & | & 50,000 \\ 0 & 1 & 3 & | & 450,000 \end{bmatrix}$$

This matrix represents the equivalent system

$$\begin{cases} x - 2z = 50{,}000 \\ y + 3z = 450{,}000 \end{cases}$$

Solving each equation for the lead variable gives the following solution.

$$x = 50{,}000 + 2z$$

$$y = 450{,}000 - 3z$$

$$z = \text{any real number}$$

This tells us that there are many possible investments satisfying the conditions given. There are limitations on the solution, however. Because a negative amount cannot be invested, x, y, and z must all be nonnegative. Note that $x \geq 0$ when $z \geq 0$ and that $y \geq 0$ when $450{,}000 - 3z \geq 0$, or when $z \leq 150{,}000$. Thus the amounts invested are:

$$\$z \text{ at } 14\%, \quad \text{where} \quad 0 \leq z \leq 150{,}000$$

$$\$x \text{ at } 8\%, \quad \text{where} \quad x = 50{,}000 + 2z$$

$$\$y \text{ at } 10\%, \quad \text{where} \quad y = 450{,}000 - 3z$$

There are many possible investment plans. One example would be with $z = \$100{,}000$, $x = \$250{,}000$, and $y = \$150{,}000$. Observe that the values for x and y depend on z, and once z is chosen, x will be between \$50,000 and \$350,000, and y will be between \$0 and \$450,000.

● **Checkpoint**

For each system of equations, the reduced form of the augmented matrix is given. Give the solution to the system, if it exists.

3. $\begin{cases} 2x + 3y + 2z = 180 \\ -x + 2y + 4z = 180 \\ 2x + 6y + 5z = 270 \end{cases}$ $\begin{bmatrix} 1 & 0 & 0 & | & 100 \\ 0 & 1 & 0 & | & -80 \\ 0 & 0 & 1 & | & 110 \end{bmatrix}$

4. $\begin{cases} x - 2y + 9z = 12 \\ 2x - y + 3z = 18 \\ x + y - 6z = 6 \end{cases}$ $\begin{bmatrix} 1 & 0 & -1 & | & 8 \\ 0 & 1 & -5 & | & -2 \\ 0 & 0 & 0 & | & 0 \end{bmatrix}$

5. $\begin{cases} x - 4y + 3z = 4 \\ 2x - 2y + z = 6 \\ x + 2y - 2z = 4 \end{cases}$ $\begin{bmatrix} 1 & 0 & -\frac{1}{3} & | & 0 \\ 0 & 1 & -\frac{5}{6} & | & 0 \\ 0 & 0 & 0 & | & 1 \end{bmatrix}$

Technology Note

Whether or not a unique solution exists, we can perform the row operations necessary to reduce an augmented matrix to find any solution that exists. We can perform these row operations with a graphing calculator, and because computer spreadsheets are electronically stored matrices, we can solve systems of linear equations by instructing the spreadsheet to perform these row operations. As expected, the value of using a graphing calculator, a spreadsheet, or a computer algebra system to solve systems of linear equations increases as the size of the system gets larger. Some calculators have commands that will directly give the reduced form of a matrix. ∎

● **EXAMPLE 7** *Solution with Technology*

Solve the system

$$\begin{cases} 2x_1 + 3x_2 - 2x_3 - 2x_4 - 3x_5 = -21 \\ 3x_1 + x_2 - 4x_3 + 3x_4 - x_5 = 0 \\ x_1 + x_2 - 2x_3 - x_4 + x_5 = -2 \\ x_1 + 4x_2 - 4x_3 - 7x_4 + 3x_5 = -16 \\ 2x_1 + 4x_4 - 5x_5 = -7 \end{cases}$$

Solution

The augmented matrix for the system of equations is shown (in two parts, with two columns repeated).

```
[A]
[[2   3   -2   -2   -3...
 [3   1   -4    3   -1...
 [1   1   -2   -1    1...
 [1   4   -4   -7    3...
 [2   0    0    4   -5...
```

```
[A]
...  -2   -2   -3   -21]
...  -4    3   -1    0]
...  -2   -1    1   -2]
...  -4   -7    3  -16]
...   0    4   -5   -7]]
```

The reduced matrix is shown (in two parts, with two columns repeated).

```
...    0            ...
Ans▶Frac
[[1   0   0   0   -5/6 ...
 [0   1   0   0   -2   ...
 [0   0   1   0   -3/2 ...
 [0   0   0   1   -5/6 ...
 [0   0   0   0    0   ...
■
```

```
...    0                ...
Ans▶Frac
...   0   -5/6   -19/6]
...   0   -2     -8   ]
...   0   -3/2   -9/2 ]
...   1   -5/6   -1/6 ]
...   0    0      0   ]]
```

This resulting reduced augmented matrix is

$$\left[\begin{array}{ccccc|c} 1 & 0 & 0 & 0 & -\frac{5}{6} & -\frac{19}{6} \\ 0 & 1 & 0 & 0 & -2 & -8 \\ 0 & 0 & 1 & 0 & -\frac{3}{2} & -\frac{9}{2} \\ 0 & 0 & 0 & 1 & -\frac{5}{6} & -\frac{1}{6} \\ 0 & 0 & 0 & 0 & 0 & 0 \end{array}\right]$$

The resulting equations and the general solution are given below.

$$x_1 - \left(\frac{5}{6}\right)x_5 = -\frac{19}{6} \qquad x_1 = \left(\frac{5}{6}\right)x_5 - \frac{19}{6}$$

$$x_2 - 2x_5 = -8 \qquad x_2 = 2x_5 - 8$$

$$x_3 - \left(\frac{3}{2}\right)x_5 = -\frac{9}{2} \qquad x_3 = \left(\frac{3}{2}\right)x_5 - \frac{9}{2}$$

$$x_4 - \left(\frac{5}{6}\right)x_5 = -\frac{1}{6} \qquad x_4 = \left(\frac{5}{6}\right)x_5 - \frac{1}{6}$$

$$x_5 = \text{any real number}$$

● **Checkpoint Solutions**

1. (a) $\left[\begin{array}{ccc|c} 3 & 2 & -1 & 3 \\ 1 & -1 & 2 & 4 \\ 2 & 3 & -1 & 3 \end{array}\right]$ (b) $\left[\begin{array}{ccc} 3 & 2 & -1 \\ 1 & -1 & 2 \\ 2 & 3 & -1 \end{array}\right]$

$$2. \begin{bmatrix} 3 & 2 & -1 & | & 3 \\ 1 & -1 & 2 & | & 4 \\ 2 & 3 & -1 & | & 3 \end{bmatrix} \xrightarrow[\substack{\text{Switch } R_1 \text{ and } R_2 \\ \text{Then } -3R_1 + R_2 \rightarrow R_2 \\ -2R_1 + R_3 \rightarrow R_3}]{} \begin{bmatrix} 1 & -1 & 2 & | & 4 \\ 0 & 5 & -7 & | & -9 \\ 0 & 5 & -5 & | & -5 \end{bmatrix}$$

$$\xrightarrow[\substack{\text{Switch } R_2 \text{ and } R_3 \\ \text{Then } \frac{1}{5}R_2 \rightarrow R_2 \\ R_2 + R_1 \rightarrow R_1 \\ -5R_2 + R_3 \rightarrow R_3}]{} \begin{bmatrix} 1 & 0 & 1 & | & 3 \\ 0 & 1 & -1 & | & -1 \\ 0 & 0 & -2 & | & -4 \end{bmatrix}$$

$$\xrightarrow[\substack{-\frac{1}{2}R_3 \rightarrow R_3 \\ \text{Then } R_3 + R_2 \rightarrow R_2 \\ -R_3 + R_1 \rightarrow R_1}]{} \begin{bmatrix} 1 & 0 & 0 & | & 1 \\ 0 & 1 & 0 & | & 1 \\ 0 & 0 & 1 & | & 2 \end{bmatrix} \quad \text{Thus } x = 1, y = 1, z = 2.$$

3. $x = 100, y = -80, z = 110$
4. $x - z = 8, y - 5z = -2 \Rightarrow x = z + 8, y = 5z - 2, z = $ any real number
5. No solution

3.3 Exercises

UNIQUE SOLUTIONS

In Problems 1 and 2, use the indicated row operation to change matrix A, where

$$A = \begin{bmatrix} 1 & -2 & -1 & | & -7 \\ 3 & 1 & 2 & | & 0 \\ 4 & 2 & 2 & | & 1 \end{bmatrix}$$

1. Add -3 times row 1 to row 2 of matrix A and place the result in row 2 to get 0 in row 2, column 1.
2. Add -4 times row 1 to row 3 of matrix A and place the result in row 3 to get 0 in row 3, column 1.

In Problems 3 and 4, write the augmented matrix associated with each system of linear equations.

3. $\begin{cases} x - 3y + 4z = 2 \\ 2x \quad\; + 2z = 1 \\ x + 2y + \; z = 1 \end{cases}$ 4. $\begin{cases} x + 2y + 2z = 3 \\ x - 2y \quad\;\; = 4 \\ \quad\;\; y - z = 1 \end{cases}$

In each of Problems 5–8, the given matrix is an augmented matrix used in the solution of a system of linear equations. What is the solution of the system?

5. $\begin{bmatrix} 1 & 0 & 0 & | & 2 \\ 0 & 1 & 0 & | & \frac{1}{2} \\ 0 & 0 & 1 & | & -5 \end{bmatrix}$ 6. $\begin{bmatrix} 1 & 0 & 0 & | & -8 \\ 0 & 1 & 0 & | & 1 \\ 0 & 0 & 1 & | & \frac{1}{3} \end{bmatrix}$

7. $\begin{bmatrix} 1 & 0 & 0 & | & 18 \\ 0 & 1 & 0 & | & 10 \\ 0 & 0 & 1 & | & 0 \end{bmatrix}$ 8. $\begin{bmatrix} 1 & 0 & 0 & | & -1 \\ 0 & 1 & 0 & | & 7 \\ 0 & 0 & 1 & | & 0 \end{bmatrix}$

In Problems 9–12, the given matrix is an augmented matrix representing a system of linear equations. Use the goals of the Gauss-Jordan elimination method (see Example 1) to find the solution of the system.

9. $\begin{bmatrix} 1 & 1 & 2 & | & -1 \\ 0 & 3 & 1 & | & 7 \\ 0 & -2 & 4 & | & 0 \end{bmatrix}$

10. $\begin{bmatrix} 1 & 5 & 2 & | & 6 \\ 0 & -2 & 3 & | & 9 \\ 0 & 1 & 3 & | & 0 \end{bmatrix}$

11. $\begin{bmatrix} 1 & 2 & 5 & | & -4 \\ 2 & -2 & 4 & | & -2 \\ 0 & 1 & -3 & | & 7 \end{bmatrix}$

12. $\begin{bmatrix} 1 & 1 & 1 & | & 3 \\ 3 & -2 & 4 & | & 5 \\ 1 & 2 & 1 & | & 4 \end{bmatrix}$

In Problems 13–22, use row operations on augmented matrices to solve the given systems of linear equations.

13. $\begin{cases} x + y - z = 0 \\ x + 2y + 3z = -5 \\ 2x - y - 13z = 17 \end{cases}$ 14. $\begin{cases} x + 2y - z = 3 \\ 2x + 5y - 2z = 7 \\ -x + y + 5z = -12 \end{cases}$

15. $\begin{cases} 2x - 6y - 12z = 6 \\ 3x - 10y - 20z = 5 \\ 2x \quad\;\; - 17z = -4 \end{cases}$ 16. $\begin{cases} -3x + 6y - 9z = 3 \\ x - y - 2z = 0 \\ 5x + 5y - 7z = 63 \end{cases}$

17. $\begin{cases} x - 3y + 3z = 7 \\ x + 2y - z = -2 \\ 3x + 2y + 4z = 5 \end{cases}$

18. $\begin{cases} 3x + 2y - 4z = -12 \\ 3x - 3y + 2z = -15 \\ 4x + 6y + z = 0 \end{cases}$

19. $\begin{cases} x_1 + 3x_2 + 2x_3 + 2x_4 = 3 \\ x_1 + x_2 + 3x_3 = 4 \\ 2x_1 + 2x_3 - 3x_4 = 4 \\ x_1 - 3x_2 = 1 \end{cases}$

20. $\begin{cases} x_1 + x_2 + x_3 + 2x_4 = 1 \\ x_1 - 3x_3 = 2 \\ x_1 - 3x_2 + x_4 = -2 \\ x_2 - 4x_3 + x_4 = 0 \end{cases}$

21. $\begin{cases} x - 2y + 3z + w = -2 \\ x - 3y + z - w = -7 \\ x - y = -2 \\ x + z + w = 2 \end{cases}$

22. $\begin{cases} x + 4y - 6z - 3w = 3 \\ x - y + 2w = -5 \\ x + z + w = 1 \\ y + z + w = 0 \end{cases}$

NONUNIQUE SOLUTIONS

In each of Problems 23–26, a system of linear equations and a reduced matrix for the system are given. (a) Use the reduced matrix to find the general solution of the system, if one exists. (b) If multiple solutions exist, find two specific solutions.

23. $\begin{cases} x + 2y + 3z = 1 \\ 2x - y = 3 \\ x + 2y + 3z = 2 \end{cases}$ $\begin{bmatrix} 1 & 0 & \frac{3}{5} & | & 0 \\ 0 & 1 & \frac{6}{5} & | & 0 \\ 0 & 0 & 0 & | & 1 \end{bmatrix}$

24. $\begin{cases} 2x + 3y + 4z = 2 \\ x + 2y + 2z = 1 \\ x + y + 2z = 2 \end{cases}$ $\begin{bmatrix} 1 & 0 & 2 & | & 0 \\ 0 & 1 & 0 & | & 0 \\ 0 & 0 & 0 & | & 1 \end{bmatrix}$

25. $\begin{cases} 2x + y - z = 7 \\ x - y - z = 4 \\ 3x + 3y - z = 10 \end{cases}$ $\begin{bmatrix} 1 & 0 & -\frac{2}{3} & | & \frac{11}{3} \\ 0 & 1 & \frac{1}{3} & | & -\frac{1}{3} \\ 0 & 0 & 0 & | & 0 \end{bmatrix}$

26. $\begin{cases} x - y + z = 3 \\ 3x + 2z = 7 \\ x - 4y + 2z = 5 \end{cases}$ $\begin{bmatrix} 1 & 0 & \frac{2}{3} & | & \frac{7}{3} \\ 0 & 1 & -\frac{1}{3} & | & -\frac{2}{3} \\ 0 & 0 & 0 & | & 0 \end{bmatrix}$

27. How can you tell that a system of linear equations has no solution by just looking at its reduced matrix?

28. Describe the procedure for finding the general solution of a system of linear equations if its reduced matrix indicates that it has an infinite number of solutions.

The systems of equations in Problems 29–42 may have unique solutions, an infinite number of solutions, or no solution. Use matrices to find the general solution of each system, if a solution exists.

29. $\begin{cases} x + y + z = 0 \\ 2x - y - z = 0 \\ -x + 2y + 2z = 0 \end{cases}$ 30. $\begin{cases} 2x - y + 3z = 0 \\ x + 2y + 2z = 0 \\ x - 3y + z = 0 \end{cases}$

31. $\begin{cases} 3x + 2y + z = 0 \\ x + y + 2z = 1 \\ 2x + y - z = -1 \end{cases}$ 32. $\begin{cases} x + 3y + 2z = 2 \\ 2x - y - 2z = 1 \\ 3x + 2y = 3 \end{cases}$

33. $\begin{cases} 2x + 2y + z = 2 \\ x - 2y + 2z = 1 \\ -x + 2y - 2z = -1 \end{cases}$ 34. $\begin{cases} 2x + y - z = 2 \\ x - y + 2z = 3 \\ x + y - z = 1 \end{cases}$

35. $\begin{cases} 2x - 5y + z = -9 \\ x + 4y - 6z = 2 \\ 3x - 4y - 2z = -10 \end{cases}$ 36. $\begin{cases} 2x - y + z = 2 \\ 3x + y - 6z = -7 \\ x - y + 2z = 3 \end{cases}$

37. $\begin{cases} x + y + z = 3 \\ x - y + z = 4 \end{cases}$ 38. $\begin{cases} 3x + 2y + z = 3 \\ x - y - z = 2 \end{cases}$

39. $\begin{cases} 3x - 2y + 5z = 14 \\ 2x - 3y + 4z = 8 \end{cases}$ 40. $\begin{cases} x - 4y + z = -4 \\ 2x - 5y - 5z = -9 \end{cases}$

41. $\begin{cases} -0.6x_1 + 0.1x_2 + 0.3x_3 = 0 \\ 0.4x_1 - 0.7x_2 + 0.2x_3 = 0 \\ 0.2x_1 + 0.6x_2 - 0.5x_3 = 0 \end{cases}$

42. $\begin{cases} 0.1x_1 - 0.1x_2 - 0.3x_3 = 0 \\ 0.2x_1 + 0.3x_2 + 0.1x_3 = -0.3 \\ 0.3x_1 + 0.7x_2 + 0.5x_3 = -0.6 \end{cases}$

In Problems 43–48, use technology to solve each system of equations, if a solution exists.

43. $\begin{cases} 3x + 2y + z - w = 3 \\ x - y - 2z + 2w = 2 \\ 2x + 3y - z + w = 1 \\ -x + y + 2z - 2w = -2 \end{cases}$

44. $\begin{cases} 2x_1 + 3x_2 - x_3 + 3x_4 + x_5 = 22 \\ 3x_1 + x_2 - 4x_3 + 3x_4 - x_5 = 0 \\ x_1 + x_2 + 3x_3 - 4x_4 + 2x_5 = 6 \\ x_1 + 2x_2 - 3x_3 + 2x_4 - 2x_5 = -6 \\ 2x_1 + 4x_4 - 5x_5 = -7 \end{cases}$

45. $\begin{cases} x_1 + 2x_2 - x_3 + x_4 = 3 \\ x_1 + 3x_2 + 4x_3 + x_4 = -2 \\ 2x_1 + 5x_2 + 2x_3 + 2x_4 = 1 \\ 2x_1 + 3x_2 - 6x_3 + 2x_4 = 3 \end{cases}$

46. $\begin{cases} x_1 + 2x_2 + 3x_3 + 4x_4 = 2 \\ x_1 + 2x_2 + 4x_3 + 3x_4 = 2 \\ 2x_1 + 4x_2 + 6x_3 + 8x_4 = 4 \\ x_1 + 4x_2 + 3x_3 = 4 \end{cases}$

47. $\begin{cases} x_1 + 3x_2 + 4x_3 - x_4 + 2x_5 = 1 \\ x_1 - x_2 - 2x_3 + x_4 = 3 \\ x_1 + 2x_2 + 3x_3 + x_4 + 4x_5 = 0 \\ 2x_1 + 2x_2 + 2x_3 + 2x_5 = 4 \end{cases}$

48. $\begin{cases} 2x_1 - x_2 + x_3 + 3x_4 + 3x_5 = 7 \\ x_1 + x_2 + x_3 + 2x_4 + 2x_5 = 0 \\ 2x_1 + x_2 - x_3 + x_4 + x_5 = -3 \\ 4x_1 - x_2 - 3x_3 + x_4 + x_5 = 1 \end{cases}$

49. Solve $\begin{cases} a_1x + b_1y = c_1 \\ a_2x + b_2y = c_2 \end{cases}$ for x. This gives a formula for solving two equations in two variables for x.

50. Solve $\begin{cases} a_1x + b_1y = c_1 \\ a_2x + b_2y = c_2 \end{cases}$ for y. This gives a formula for solving two equations in two variables for y.

APPLICATIONS

Problems 51–60 involve systems with unique solutions.

51. *Investment* A man has $235,000 invested in three properties. One earns 12%, one 10%, and one 8%. His annual income from the properties is $22,500 and the amount invested at 8% is twice that invested at 12%.
    (a) How much is invested in each property?
    (b) What is the annual income from each property?

52. *Loans* A bank lent $1.2 million for the development of three new products, with one loan each at 6%, 7%, and 8%. The amount lent at 8% was equal to the sum of the amounts lent at the other two rates, and the bank's annual income from the loans was $88,000. How much was lent at each rate?

53. *Nutrition* A preschool has Campbell's Chunky Beef soup, which contains 4 g of fat and 25 mg of cholesterol per serving (cup), and Campbell's Chunky Sirloin Burger soup, which contains 9 g of fat and 20 mg of cholesterol per serving. By combining the soups, it is possible to get 10 servings of soup that will have 80 g of fat and 210 mg of cholesterol. How many cups of each soup should be used?

54. *Nutrition* A psychologist studying the effects of nutrition on the behavior of laboratory rats is feeding one group a combination of three foods: I, II, and III. Each of these foods contains three additives, A, B, and C, that are being used in the study. Each additive is a certain percentage of each of the foods as follows:

	Foods		
	I	II	III
Additive A	10%	30%	60%
Additive B	0%	4%	5%
Additive C	2%	2%	12%

If the diet requires 53 g per day of A, 4.5 g per day of B, and 8.6 g per day of C, find the number of grams per day of each food that must be used.

55. *Nutrition* The following table gives the calories, fat, and carbohydrates per ounce for three brands of cereal. How many ounces of each brand should be combined to get 443 calories, 5.7 g of fat, and 113.4 g of carbohydrates?

Cereal	Calories	Fat (g)	Carbohydrates (g)
All Fiber	50	0	22.0
Frosted Puffs	108	0.1	25.7
Natural Mixed Grain	127	5.5	18.0

56. *Investment* A brokerage house offers three stock portfolios. Portfolio I consists of 2 blocks of common stock and 1 municipal bond. Portfolio II consists of 4 blocks of common stock, 2 municipal bonds, and 3 blocks of preferred stock. Portfolio III consists of 7 blocks of common stock, 3 municipal bonds, and 3 blocks of preferred stock. A customer wants 21 blocks of common stock, 10 municipal bonds, and 9 blocks of preferred stock. How many units of each portfolio should be offered?

57. *Investment* Suppose that portfolios I and II in Problem 56 are unchanged and portfolio III consists of 2 blocks of common stock, 2 municipal bonds, and 3 blocks of preferred stock. A customer wants 12 blocks of common stock, 6 municipal bonds, and 6 blocks of preferred stock. How many units of each portfolio should be offered?

58. *Nutrition* Each ounce of substance A supplies 5% of the required nutrition a patient needs. Substance B supplies 15% of the required nutrition per ounce and substance C supplies 12% of required nutrition per ounce. If digestive restrictions require that substances A and C be given in equal amounts and that the amount of substance B be one-fifth of these other amounts, find the number of ounces of each substance that should be in the meal to provide 100% nutrition.

59. *Nutrition* A glass of skim milk supplies 0.1 mg of iron, 8.5 g of protein, and 1 g of carbohydrates. A quarter pound of lean red meat provides 3.4 mg of iron, 22 g of protein, and 20 g of carbohydrates. Two slices of whole-grain bread supply 2.2 mg of iron, 10 g of protein, and 12 g of carbohydrates. If a person on a special diet must have 12.1 mg of iron, 97 g of protein, and 70 g of carbohydrates, how many glasses of skim milk, how many quarter-pound servings of meat, and how many two-slice servings of whole-grain bread will supply this?

60. *Transportation* The King Trucking Company has an order for three products for delivery. The following table gives the particulars for the products.

	Type I	Type II	Type III
Unit volume (cubic feet)	10	8	20
Unit weight (pounds)	10	20	40
Unit value (dollars)	100	20	200

If the carrier can carry 6000 cu ft, can carry 11,000 lb, and is insured for $36,900, how many units of each type can be carried?

Application Problems 61–67 involve systems with nonunique solutions.

61. *Nutrition*   A botanist can purchase plant food of four different types, I, II, III, and IV. Each food comes in the same size bag, and the following table summarizes the number of grams of each of three nutrients that each bag contains.

	Foods (grams)			
	I	II	III	IV
Nutrient A	5	5	10	5
Nutrient B	10	5	30	10
Nutrient C	5	15	10	25

The botanist wants to use a food that has these nutrients in a different proportion and determines that he will need a total of 10,000 g of A, 20,000 g of B, and 20,000 g of C. Find the number of bags of each type of food that should be ordered.

62. *Traffic flow*   In the analysis of traffic flow, a certain city estimates the following situation for the "square" of its downtown district. In the following figure, the arrows indicate the flow of traffic. If x_1 represents the number of cars traveling between intersections A and B, x_2 represents the number of cars traveling between B and C, x_3 the number between C and D, and x_4 the number between D and A, we can formulate equations based on the principle that the number of vehicles entering an intersection equals the number leaving it. That is, for intersection A we obtain

$$200 + x_4 = 100 + x_1$$

(a) Formulate equations for the traffic at B, C, and D.

(b) Solve the system of these four equations.

(c) Can the entire traffic flow problem be studied and solved by counting only the x_4 cars that travel between A and D? Explain.

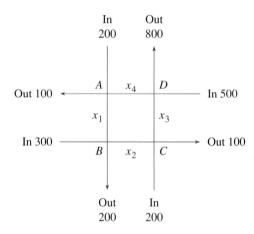

63. *Nutrition*   Three different bacteria are cultured in one dish and feed on three nutrients. Each individual of species I consumes 1 unit of each of the first and second nutrients and 2 units of the third nutrient. Each individual of species II consumes 2 units of the first nutrient and 2 units of the third nutrient. Each individual of species III consumes 2 units of the first nutrient, 3 units of the second nutrient, and 5 units of the third nutrient. If the culture is given 5100 units of the first nutrient, 6900 units of the second nutrient, and 12,000 units of the third nutrient, how many of each species can be supported such that all of the nutrients are consumed?

64. *Irrigation*   An irrigation system allows water to flow in the pattern shown in the figure below. Water flows into the system at A and exits at B, C, D, and E with the amounts shown. Using the fact that at each point the water entering equals the water leaving, formulate an equation for water flow at each of the five points and solve the system.

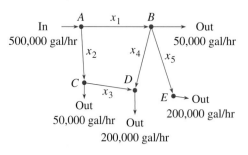

65. *Investment*   An investment club has set a goal of earning 15% on the money they invest in stocks. The members are considering purchasing three possible stocks, with their cost per share (in dollars) and their projected growth per share (in dollars) summarized in the following table.

	Stocks		
	Computer (C)	Utility (U)	Retail (R)
Cost/share	30	44	26
Growth/share	6.00	6.00	2.40

(a) If they have $392,000 to invest, how many shares of each stock should they buy to meet their goal?

(b) If they buy 1000 shares of retail stock, how many shares of the other stocks do they buy? What if they buy 2000 shares of retail stock?

(c) What is the minimum number of shares of computer stock they will buy, and what is the number of shares of the other stocks in this case?

(d) What is the maximum number of shares of computer stock purchased, and what is the number of shares of the other stocks in this care?

66. *Investment*   A trust account manager has $220,000 to be invested. The investment choices have current yields of 8%, 7%, and 10%. Suppose that the investment goal is to earn interest of $16,600, and risk factors make it prudent to invest some money in all three investments.

(a) Find a general description for the amounts invested at the three rates.

(b) If $10,000 is invested at 10%, how much will be invested at each of the other rates? What if $30,000 is invested at 10%?

(c) What is the minimum amount that will be invested at 7%, and in this case how much will be invested at the other rates?

(d) What is the maximum amount that will be invested at 7%, and in this case how much will be invested at the other rates?

67. *Investment*   A brokerage house offers three stock portfolios. Portfolio I consists of 2 blocks of common stock and 1 municipal bond. Portfolio II consists of 4 blocks of common stock, 2 municipal bonds, and 3 blocks of preferred stock. Portfolio III consists of 2 blocks of common stock, 1 municipal bond, and 3 blocks of preferred stock. A customer wants 16 blocks of common stock, 8 municipal bonds, and 6 blocks of preferred stock. If the numbers of the three portfolios offered must be integers, find all possible offerings.

3.4

Inverse of a Square Matrix; Matrix Equations

OBJECTIVES

- To find the inverse of a square matrix
- To use inverse matrices to solve systems of linear equations
- To find determinants of certain matrices

◗ Application Preview

During World War II, the military commanders sent instructions that used simple substitution codes. These codes were easily broken, so they were further coded by using a coding matrix and matrix multiplication. The resulting coded messages, when received, could be unscrambled with a decoding matrix that was the **inverse matrix** of the coding matrix. In Section 3.2, "Multiplication of Matrices," we encoded messages by using the following code:

a	b	c	d	e	f	g	h	i	j	k	l
1	2	3	4	5	6	7	8	9	10	11	12

m	n	o	p	q	r	s	t	u	v	w	x
13	14	15	16	17	18	19	20	21	22	23	24

y	z	blank
25	26	27

and the encoding matrix

$$A = \begin{bmatrix} 3 & 5 \\ 4 & 6 \end{bmatrix}$$

which converts the numbers to the coded message (in pairs of numbers). To decode a message encoded by this matrix A, we find the **inverse** of matrix A, denoted by A^{-1}, and multiply A^{-1} by the coded message. (See Example 4.) In this section, we will discuss how to find and apply the inverse of a given square matrix.

If the product of A and B is the identity matrix, I, we say that B is the inverse of A (and A is the inverse of B). The matrix B is called the **inverse matrix** of A, denoted A^{-1}.

Inverse Matrices

Two square matrices, A and B, are called **inverses** of each other if

$$AB = I \quad \text{and} \quad BA = I$$

In this case, $B = A^{-1}$ and $A = B^{-1}$.

● **EXAMPLE 1** *Inverse Matrices*

Is B the inverse of A if $A = \begin{bmatrix} 1 & 2 \\ 1 & 1 \end{bmatrix}$ and $B = \begin{bmatrix} -1 & 2 \\ 1 & -1 \end{bmatrix}$?

Solution

$$AB = \begin{bmatrix} 1 & 2 \\ 1 & 1 \end{bmatrix} \cdot \begin{bmatrix} -1 & 2 \\ 1 & -1 \end{bmatrix} = \begin{bmatrix} 1 & 0 \\ 0 & 1 \end{bmatrix} = I$$

$$BA = \begin{bmatrix} -1 & 2 \\ 1 & -1 \end{bmatrix} \cdot \begin{bmatrix} 1 & 2 \\ 1 & 1 \end{bmatrix} = \begin{bmatrix} 1 & 0 \\ 0 & 1 \end{bmatrix} = I$$

Thus B is the inverse of A, or $B = A^{-1}$. We also say A is the inverse of B, or $A = B^{-1}$. Thus $(A^{-1})^{-1} = A$.

● *Checkpoint*

1. Are the matrices A and B inverses if

$$A = \begin{bmatrix} 1 & 1 & -1 \\ 1 & -1 & 3 \\ 4 & 2 & 1 \end{bmatrix} \quad \text{and} \quad B = \begin{bmatrix} 3.5 & 1.5 & -1 \\ -5.5 & -2.5 & 2 \\ -3 & -1 & 1 \end{bmatrix}?$$

We have seen how to use elementary row operations on augmented matrices to solve systems of linear equations. We can also find the inverse of a matrix by using elementary row operations.

If the inverse exists for a square matrix A, we find A^{-1} as follows.

Finding the Inverse of a Square Matrix	
Procedure	Example
To find the inverse of the square matrix A:	Find the inverse of matrix $A = \begin{bmatrix} 1 & 2 \\ 1 & 1 \end{bmatrix}$.
1. Form the augmented matrix $[A \mid I]$, where A is the $n \times n$ matrix and I is the $n \times n$ identity matrix.	1. $\begin{bmatrix} 1 & 2 & \vert & 1 & 0 \\ 1 & 1 & \vert & 0 & 1 \end{bmatrix}$
2. Perform elementary row operations on $[A \mid I]$ until we have an augmented matrix of the form $[I \mid B]$— that is, until the matrix A on the left is transformed into the identity matrix.	2. $\begin{bmatrix} 1 & 2 & \vert & 1 & 0 \\ 1 & 1 & \vert & 0 & 1 \end{bmatrix}$ $\xrightarrow{-R_1 + R_2 \to R_2} \begin{bmatrix} 1 & 2 & \vert & 1 & 0 \\ 0 & -1 & \vert & -1 & 1 \end{bmatrix}$ $\xrightarrow[-R_2 \to R_2]{2R_2 + R_1 \to R_1} \begin{bmatrix} 1 & 0 & \vert & -1 & 2 \\ 0 & 1 & \vert & 1 & -1 \end{bmatrix}$
3. The matrix B (on the right) is the inverse of matrix A.	3. The inverse of A is $B = \begin{bmatrix} -1 & 2 \\ 1 & -1 \end{bmatrix}$. We verified this in Example 1.

Note that if A has no inverse, the reduction process on $[A \mid I]$ will yield a row of zeros in the left half.

● **EXAMPLE 2** *Finding an Inverse*

Find the inverse of matrix A, below.

$$A = \begin{bmatrix} 2 & 5 & 4 \\ 1 & 4 & 3 \\ 1 & -3 & -2 \end{bmatrix}$$

Solution

To find the inverse of matrix A, we reduce A in $[A \mid I]$ (the matrix A augmented with the identity matrix I).

$$\left[\begin{array}{ccc|ccc} 2 & 5 & 4 & 1 & 0 & 0 \\ 1 & 4 & 3 & 0 & 1 & 0 \\ 1 & -3 & -2 & 0 & 0 & 1 \end{array} \right]$$

When the left side becomes an identity matrix, the inverse is in the augment.

$$\xrightarrow{\text{Switch } R_1 \text{ and } R_2} \left[\begin{array}{ccc|ccc} 1 & 4 & 3 & 0 & 1 & 0 \\ 2 & 5 & 4 & 1 & 0 & 0 \\ 1 & -3 & -2 & 0 & 0 & 1 \end{array} \right]$$

$$\xrightarrow[{-R_1 + R_3 \rightarrow R_3}]{-2R_1 + R_2 \rightarrow R_2} \left[\begin{array}{ccc|ccc} 1 & 4 & 3 & 0 & 1 & 0 \\ 0 & -3 & -2 & 1 & -2 & 0 \\ 0 & -7 & -5 & 0 & -1 & 1 \end{array} \right]$$

$$\xrightarrow{-\frac{1}{3}R_2 \rightarrow R_2} \left[\begin{array}{ccc|ccc} 1 & 4 & 3 & 0 & 1 & 0 \\ 0 & 1 & \frac{2}{3} & -\frac{1}{3} & \frac{2}{3} & 0 \\ 0 & -7 & -5 & 0 & -1 & 1 \end{array} \right]$$

$$\xrightarrow[{7R_2 + R_3 \rightarrow R_3}]{-4R_2 + R_1 \rightarrow R_1} \left[\begin{array}{ccc|ccc} 1 & 0 & \frac{1}{3} & \frac{4}{3} & -\frac{5}{3} & 0 \\ 0 & 1 & \frac{2}{3} & -\frac{1}{3} & \frac{2}{3} & 0 \\ 0 & 0 & -\frac{1}{3} & -\frac{7}{3} & \frac{11}{3} & 1 \end{array} \right]$$

$$\xrightarrow[\substack{2R_3 + R_2 \rightarrow R_2 \\ -3R_3 \rightarrow R_3}]{R_3 + R_1 \rightarrow R_1} \left[\begin{array}{ccc|ccc} 1 & 0 & 0 & -1 & 2 & 1 \\ 0 & 1 & 0 & -5 & 8 & 2 \\ 0 & 0 & 1 & 7 & -11 & -3 \end{array} \right]$$

The inverse we seek is

$$A^{-1} = \begin{bmatrix} -1 & 2 & 1 \\ -5 & 8 & 2 \\ 7 & -11 & -3 \end{bmatrix}$$

● **Checkpoint**   2. Find the inverse of the matrix

$$A = \begin{bmatrix} 1 & 2 & 1 \\ 0 & 1 & 1 \\ 1 & 2 & 2 \end{bmatrix}$$

The technique of reducing $[A \mid I]$ can be used to find A^{-1} or to find that A^{-1} doesn't exist for any square matrix A. However, a special formula can be used to find the inverse of a 2×2 matrix, if the inverse exists.

Inverse of a 2 × 2 Matrix

If $A = \begin{bmatrix} a & b \\ c & d \end{bmatrix}$, then $A^{-1} = \dfrac{1}{ad - bc} \begin{bmatrix} d & -b \\ -c & a \end{bmatrix}$, provided $ad - bc \neq 0$.

If $ad - bc = 0$, then A^{-1} does not exist.

This result can be verified by direct calculation or by reducing $[A \mid I]$.

● **EXAMPLE 3** *Inverse of a 2 × 2 Matrix*

Find the inverse, if it exists, for each of the following.

(a) $A = \begin{bmatrix} 3 & 7 \\ 2 & 5 \end{bmatrix}$ (b) $B = \begin{bmatrix} 2 & -1 \\ -4 & 2 \end{bmatrix}$

Solution

(a) For A, $ad - bc = (3)(5) - (2)(7) = 1 \neq 0$, so A^{-1} exists.

$$A^{-1} = \frac{1}{1} \begin{bmatrix} 5 & -7 \\ -2 & 3 \end{bmatrix} = \begin{bmatrix} 5 & -7 \\ -2 & 3 \end{bmatrix}$$

(b) For B, $ad - bc = (2)(2) - (-4)(-1) = 4 - 4 = 0$, so B^{-1} does not exist.

◗ **EXAMPLE 4** *Decoding Messages* **(Application Preview)**

In the Application Preview we recalled how to encode messages with

a	b	c	d	e	f	g	h	i	j	k	l
1	2	3	4	5	6	7	8	9	10	11	12

m	n	o	p	q	r	s	t	u	v	w	x
13	14	15	16	17	18	19	20	21	22	23	24

y	z	blank
25	26	27

and with the encoding matrix

$$A = \begin{bmatrix} 3 & 5 \\ 4 & 6 \end{bmatrix}$$

which converts the numbers to the coded message (in pairs of numbers).

To decode any message encoded by A, we must find the inverse of A and multiply A^{-1} by the coded message. Use the inverse of A to decode the coded message 96, 118, 65, 84, 131, 168, 55, 72.

Solution

We first find the inverse of matrix A.

$$A^{-1} = \frac{1}{18 - 20} \begin{bmatrix} 6 & -5 \\ -4 & 3 \end{bmatrix} = -\frac{1}{2} \begin{bmatrix} 6 & -5 \\ -4 & 3 \end{bmatrix} = \begin{bmatrix} -3 & 2.5 \\ 2 & -1.5 \end{bmatrix}$$

We can write the coded message numbers (in pairs) as columns of a matrix B and then find the product $A^{-1}B$.

$$A^{-1}B = \begin{bmatrix} -3 & 2.5 \\ 2 & -1.5 \end{bmatrix} \begin{bmatrix} 96 & 65 & 131 & 55 \\ 118 & 84 & 168 & 72 \end{bmatrix} = \begin{bmatrix} 7 & 15 & 27 & 15 \\ 15 & 4 & 10 & 2 \end{bmatrix}$$

Reading the numbers (down the respective columns) gives the result 7, 15, 15, 4, 27, 10, 15, 2, which is the message "good job."

Calculator Note

Computer software programs and many types of calculators often can be used to find the inverse of a matrix. Because of their ease of usage, it is logical to use graphing utilities to find inverses. For large matrices, computer programs may be necessary. ■

● **EXAMPLE 5** *Finding an Inverse with Technology*

Use technology to find the inverse of the matrix

$$\begin{bmatrix} 1 & 2 & 4 \\ 3 & 1 & 3 \\ 5 & 2 & 4 \end{bmatrix}$$

Solution
Entering this matrix, using the inverse key or command, and converting the entries to fractions gives matrix A^{-1}.

```
[A]-1▶Frac
[[-1/4   0     1/4 ]
 [3/8   -2     9/8 ]
 [1/8    1    -5/8]]
■
```

so $A^{-1} = \begin{bmatrix} -\frac{1}{4} & 0 & \frac{1}{4} \\ \frac{3}{8} & -2 & \frac{9}{8} \\ \frac{1}{8} & 1 & -\frac{5}{8} \end{bmatrix}$

We can verify that the product of matrix A and its inverse A^{-1} is the 3×3 identity matrix.

```
[A]-1[A]
        [[1   0   0]
         [0   1   0]
         [0   0   1]]
```

Spreadsheet Note

We can also find the inverse of a square matrix with Excel. If the inverse exists, it can be found with the command MINVERSE. The inverse of $A = \begin{bmatrix} -1 & 0 & 1 \\ 1 & 4 & -3 \\ 1 & -2 & 1 \end{bmatrix}$ is shown in the following spreadsheet to be $A^{-1} = \begin{bmatrix} 0.5 & 0.5 & 1 \\ 1 & 0.5 & 0.5 \\ 1.5 & 0.5 & 1 \end{bmatrix}$.

==minverse(B1:D3)			
A	**B**	**C**	**D**
1	Matrix A	-1	0	1
2		1	4	-3
3		1	-2	1
4				
5	Inverse A	0.5	0.5	1
6		1	0.5	0.5
7		1.5	0.5	1

It should be noted that roundoff error can occur when technology is used to find the inverse of a matrix and that the product of matrix *A* and its inverse may give a matrix that does not look exactly like the identity matrix, even though it is the identity matrix when the entries are rounded to a reasonable number of decimal places. ■

Matrix Equations

In Example 1 of Section 3.3, "Gauss-Jordan Elimination: Solving Systems of Equations," we solved the system

$$\begin{cases} 2x + 5y + 4z = 4 \\ x + 4y + 3z = 1 \\ x - 3y - 2z = 5 \end{cases}$$

This system can be written as the following matrix equation.

$$\begin{bmatrix} 2x + 5y + 4z \\ x + 4y + 3z \\ x - 3y - 2z \end{bmatrix} = \begin{bmatrix} 4 \\ 1 \\ 5 \end{bmatrix}$$

The matrix equation is often written as follows:

$$\begin{bmatrix} 2 & 5 & 4 \\ 1 & 4 & 3 \\ 1 & -3 & -2 \end{bmatrix} \begin{bmatrix} x \\ y \\ z \end{bmatrix} = \begin{bmatrix} 4 \\ 1 \\ 5 \end{bmatrix}$$

We can see that this second form of a matrix equation is equivalent to the first if we carry out the multiplication. With the system written in this second form, we could also solve the system by multiplying both sides of the equation by the inverse of the coefficient matrix

$$\begin{bmatrix} 2 & 5 & 4 \\ 1 & 4 & 3 \\ 1 & -3 & -2 \end{bmatrix}$$

● **EXAMPLE 6** *Solution with Inverse Matrices*

Solve the system

$$\begin{cases} 2x + 5y + 4z = 4 \\ x + 4y + 3z = 1 \\ x - 3y - 2z = 5 \end{cases}$$

by using inverse matrices.

Solution

To solve the system, we multiply both sides of the associated matrix equation, written above, by the inverse of the coefficient matrix, which we found in Example 2.

$$
\begin{bmatrix} -1 & 2 & 1 \\ -5 & 8 & 2 \\ 7 & -11 & -3 \end{bmatrix} \begin{bmatrix} 2 & 5 & 4 \\ 1 & 4 & 3 \\ 1 & -3 & -2 \end{bmatrix} \begin{bmatrix} x \\ y \\ z \end{bmatrix} = \begin{bmatrix} -1 & 2 & 1 \\ -5 & 8 & 2 \\ 7 & -11 & -3 \end{bmatrix} \begin{bmatrix} 4 \\ 1 \\ 5 \end{bmatrix}
$$

Note that we must be careful to multiply both sides *from the left* because matrix multiplication is not commutative. If we carry out the multiplications, we obtain

$$
\begin{bmatrix} 1 & 0 & 0 \\ 0 & 1 & 0 \\ 0 & 0 & 1 \end{bmatrix} \begin{bmatrix} x \\ y \\ z \end{bmatrix} = \begin{bmatrix} 3 \\ -2 \\ 2 \end{bmatrix}, \quad \text{or} \quad \begin{bmatrix} x \\ y \\ z \end{bmatrix} = \begin{bmatrix} 3 \\ -2 \\ 2 \end{bmatrix}
$$

which yields $x = 3, y = -2, z = 2$, the same solution as that found in Example 1 of the previous section.

Just as we wrote the system of three equations as a matrix equation of the form $AX = B$, this can be done in general. If A is an $n \times n$ matrix and if B and X are $n \times 1$ matrices, then

$$
AX = B
$$

is a **matrix equation.**

If the inverse of a matrix A exists, then we can use that inverse to solve the matrix equation for the matrix X. The general solution method follows.

$$
AX = B
$$

Multiplying both sides of the equation by A^{-1} on the left gives

$$
A^{-1}(AX) = A^{-1}B
$$
$$
(A^{-1}A)X = A^{-1}B
$$
$$
IX = A^{-1}B
$$
$$
X = A^{-1}B
$$

Thus inverse matrices can be used to solve systems of equations. Unfortunately, this method will work only if the solution to the system is unique. In fact, a system $AX = B$ has a unique solution if and only if A^{-1} exists. If the inverse of the coefficient matrix exists, the solution method above can be used to solve the system.

● **EXAMPLE 7** *Solution with Inverse Matrices*

Use an inverse matrix to solve the system of equations.

$$
\begin{cases} -x & + z = 1 \\ x + 4y - 3z = -3 \\ x - 2y + z = 3 \end{cases}
$$

Solution

Writing the system as a matrix equation $AX = B$ gives

$$
\begin{bmatrix} -1 & 0 & 1 \\ 1 & 4 & -3 \\ 1 & -2 & 1 \end{bmatrix} \begin{bmatrix} x \\ y \\ z \end{bmatrix} = \begin{bmatrix} 1 \\ -3 \\ 3 \end{bmatrix}
$$

In the Spreadsheet Note immediately following Example 5, we found that the inverse of the coefficient matrix A is

$$A^{-1} = \begin{bmatrix} 0.5 & 0.5 & 1 \\ 1 & 0.5 & 0.5 \\ 1.5 & 0.5 & 1 \end{bmatrix}$$

Hence the solution is

$$X = A^{-1}B = \begin{bmatrix} 0.5 & 0.5 & 1 \\ 1 & 0.5 & 0.5 \\ 1.5 & 0.5 & 1 \end{bmatrix} \begin{bmatrix} 1 \\ -3 \\ 3 \end{bmatrix} = \begin{bmatrix} 2 \\ 1 \\ 3 \end{bmatrix}$$

Thus $x = 2$, $y = 1$, and $z = 3$.

● **Checkpoint**

3. Use inverse matrices to solve the system.

$$\begin{cases} 7x - 5y = 12 \\ 2x - 3y = 6 \end{cases}$$

Technology
Note

Because technology often can easily find the inverse of a matrix, it is easy to use a graphing utility to solve a system of linear equations by using the inverse of the coefficient matrix. ■

● **EXAMPLE 8** *System Solution with Technology*

Use the inverse of the coefficient matrix to solve the following system of linear equations.

$$\begin{cases} 2x_1 + 3x_2 - x_3 + 3x_4 + x_5 = 22 \\ 3x_1 + x_2 - 4x_3 + 3x_4 - x_5 = 0 \\ x_1 - 2x_2 + 3x_3 - 4x_4 + 2x_5 = 0 \\ x_1 + 2x_2 - 3x_3 + 2x_4 - 2x_5 = -6 \\ 2x_1 + 4x_4 - 5x_5 = -7 \end{cases}$$

Solution
The matrix equation for this system is $AX = B$, as follows:

$$AX = \begin{bmatrix} 2 & 3 & -1 & 3 & 1 \\ 3 & 1 & -4 & 3 & -1 \\ 1 & -2 & 3 & -4 & 2 \\ 1 & 2 & -3 & 2 & -2 \\ 2 & 0 & 0 & 4 & -5 \end{bmatrix} \begin{bmatrix} x_1 \\ x_2 \\ x_3 \\ x_4 \\ x_5 \end{bmatrix} = \begin{bmatrix} 22 \\ 0 \\ 0 \\ -6 \\ -7 \end{bmatrix}$$

We use technology to find the inverse of the coefficient matrix. The inverse of the coefficient matrix may be difficult to read on a small screen. However, we need not record the inverse. If we multiply both sides by A^{-1}, we get the solution, $X = A^{-1}B$ (see Figure 3.9).

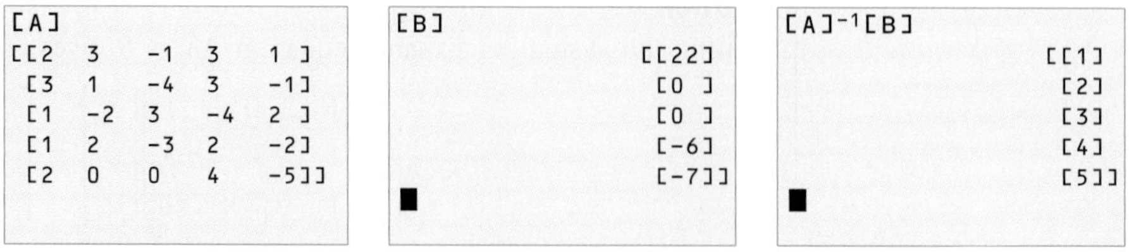

```
[A]
[[2   3   -1   3    1 ]
 [3   1   -4   3   -1]
 [1  -2    3  -4    2 ]
 [1   2   -3   2   -2]
 [2   0    0   4   -5]]
```

```
[B]
          [[22]
           [0 ]
           [0 ]
           [-6]
           [-7]]
■
```

```
[A]⁻¹[B]
          [[1]
           [2]
           [3]
           [4]
           [5]]
■
```

Figure 3.9

Thus the solution is $x_1 = 1$, $x_2 = 2$, $x_3 = 3$, $x_4 = 4$, $x_5 = 5$.

 *Spreadsheet Note* We can also use Excel to solve a system of linear equations, if a unique solution exists. We can solve such a system by finding the inverse of the coefficient matrix and multiplying this inverse times the matrix containing the constants. The spreadsheet that follows shows the solution of Example 7 found with Excel.

	A	B	C	D
1	Matrix A	−1	0	1
2		1	4	−3
3		1	−2	1
4				
5	Inverse A	0.5	0.5	1
6		1	0.5	0.5
7		1.5	0.5	1
8				
9	Matrix B	1		
10		−3		
11		3		
12				
13	Solution X	2		
14		1		
15		3		

Spreadsheet

● **EXAMPLE 9** *Venture Capital*

Suppose a bank draws its venture capital funds from three main sources of income—business loans, auto loans, and home mortgages. The income from these sources for each of 3 years is given in Table 3.10, and the venture capital for these years is given in Table 3.11. If the bank uses a fixed percent of its income from each of the business loans (x), auto loans (y), and home mortgages (z) to get its venture capital funds, find the percent of income used for each of these sources of income.

TABLE 3.10   Income from Loans

	Business	Auto	Home
2004	68,210	23,324	57,234
2005	43,455	19,335	66,345
2006	58,672	15,654	69,334

TABLE 3.11

Venture	Capital
2004	53,448.74
2005	43,447.50
2006	50,582.88

Solution
The matrix equation describing this problem is $AX = V$, or

$$\begin{bmatrix} 68{,}210 & 23{,}324 & 57{,}234 \\ 43{,}455 & 19{,}335 & 66{,}345 \\ 58{,}672 & 15{,}654 & 69{,}334 \end{bmatrix} \begin{bmatrix} x \\ y \\ z \end{bmatrix} = \begin{bmatrix} 53{,}448.74 \\ 43{,}447.50 \\ 50{,}582.88 \end{bmatrix}$$

Finding the inverse of matrix A with Excel and multiplying both sides of the matrix equation (from the left) by A^{-1} gives the solution $X = A^{-1}V$:

$$\begin{bmatrix} x \\ y \\ z \end{bmatrix} = \begin{bmatrix} 0.47 \\ 0.23 \\ 0.28 \end{bmatrix}$$

Thus 47%, 23%, and 28% come from business, auto, and home loans, respectively.

Determinants

Recall that the inverse of $A = \begin{bmatrix} a & b \\ c & d \end{bmatrix}$ can be found with the formula

$$A^{-1} = \frac{1}{ad - bc} \begin{bmatrix} d & -b \\ -c & a \end{bmatrix}$$

if $ad - bc \neq 0$.

The value $ad - bc$ used in finding the inverse of this 2×2 matrix is used so frequently that it is given a special name, the **determinant** of A (denoted *det A* or $|A|$).

**Determinant of a
2 × 2 Matrix**

$$\det \begin{bmatrix} a & b \\ c & d \end{bmatrix} = \begin{vmatrix} a & b \\ c & d \end{vmatrix} = ad - bc$$

● **EXAMPLE 10** *Determinant of a 2 × 2 Matrix*

Find det A if $A = \begin{bmatrix} 2 & 4 \\ 3 & -4 \end{bmatrix}$.

Solution

$$\det A = (2)(-4) - (3)(4) = -8 - 12 = -20$$

There are formulas for finding the determinants of square matrices of orders larger than 2×2, but for the matrices used in this text, calculators and computers make it easy to find determinants. We have already seen that the inverse of a 2×2 matrix A does not exist if and only if det $A = 0$. In general, the inverse of an $n \times n$ matrix B does not exist if det $B = 0$. This also means that if the coefficient matrix of a system of linear equations has a determinant equal to 0, the system does not have a unique solution.

● **EXAMPLE 11** *Solution with Technology*

(a) Use a calculator to find the determinant of

$$A = \begin{bmatrix} 2 & 3 & 0 & 1 & 3 \\ 1 & 0 & 2 & 3 & 1 \\ 0 & 1 & 0 & 2 & 3 \\ 1 & 0 & 2 & 2 & 1 \\ 1 & 0 & 0 & 0 & 3 \end{bmatrix}$$

(b) Does the following system have a unique solution?

$$\begin{cases} 2x_1 + 3x_2 & + x_4 + 3x_5 = 16 \\ x_1 & + 2x_3 + 3x_4 + x_5 = 2 \\ & x_2 & + 2x_4 + 3x_5 = 9 \\ x_1 & + 2x_3 + 2x_4 + x_5 = 3 \\ x_1 & + 3x_5 = 10 \end{cases}$$

Solution

(a) Entering matrix A and using the determinant command gives det $A = 24$.

(b) Because A is the coefficient matrix for the system and det $A = 24$, the system has a unique solution. Using the inverse matrix gives the solution $x_1 = 1$, $x_2 = 2$, $x_3 = 0.5$, $x_4 = -1$, $x_5 = 3$.

● **Checkpoint Solutions**

1. Yes, because both AB and BA give the 3×3 identity matrix.

2.
$$\left[\begin{array}{ccc|ccc} 1 & 2 & 1 & 1 & 0 & 0 \\ 0 & 1 & 1 & 0 & 1 & 0 \\ 1 & 2 & 2 & 0 & 0 & 1 \end{array}\right] \xrightarrow{-R_1 + R_3 \to R_3} \left[\begin{array}{ccc|ccc} 1 & 2 & 1 & 1 & 0 & 0 \\ 0 & 1 & 1 & 0 & 1 & 0 \\ 0 & 0 & 1 & -1 & 0 & 1 \end{array}\right]$$

$$\xrightarrow{-2R_2 + R_1 \to R_1} \left[\begin{array}{ccc|ccc} 1 & 0 & -1 & 1 & -2 & 0 \\ 0 & 1 & 1 & 0 & 1 & 0 \\ 0 & 0 & 1 & -1 & 0 & 1 \end{array}\right]$$

$$\xrightarrow[-R_3 + R_2 \to R_2]{R_3 + R_1 \to R_1} \left[\begin{array}{ccc|ccc} 1 & 0 & 0 & 0 & -2 & 1 \\ 0 & 1 & 0 & 1 & 1 & -1 \\ 0 & 0 & 1 & -1 & 0 & 1 \end{array}\right]$$

The inverse is $A^{-1} = \begin{bmatrix} 0 & -2 & 1 \\ 1 & 1 & -1 \\ -1 & 0 & 1 \end{bmatrix}$.

3. This system can be written

$$\begin{bmatrix} 7 & -5 \\ 2 & -3 \end{bmatrix}\begin{bmatrix} x \\ y \end{bmatrix} = \begin{bmatrix} 12 \\ 6 \end{bmatrix}$$

and for the coefficient matrix A, $ad - bc = -21 - (-10) = -11$. Thus A^{-1} exists,

$$A^{-1} = -\frac{1}{11}\begin{bmatrix} -3 & 5 \\ -2 & 7 \end{bmatrix} = \begin{bmatrix} \frac{3}{11} & -\frac{5}{11} \\ \frac{2}{11} & -\frac{7}{11} \end{bmatrix}$$

so multiplying by A^{-1} *from the left* gives the solution.

$$\begin{bmatrix} \frac{3}{11} & -\frac{5}{11} \\ \frac{2}{11} & -\frac{7}{11} \end{bmatrix}\begin{bmatrix} 7 & -5 \\ 2 & -3 \end{bmatrix}\begin{bmatrix} x \\ y \end{bmatrix} = \begin{bmatrix} \frac{3}{11} & -\frac{5}{11} \\ \frac{2}{11} & -\frac{7}{11} \end{bmatrix}\begin{bmatrix} 12 \\ 6 \end{bmatrix}$$

$$\begin{bmatrix} 1 & 0 \\ 0 & 1 \end{bmatrix}\begin{bmatrix} x \\ y \end{bmatrix} = \begin{bmatrix} \frac{6}{11} \\ -\frac{18}{11} \end{bmatrix}$$

Thus the solution is $x = \frac{6}{11}, y = -\frac{18}{11}$.

3.4 Exercises

1. If A is a 3×3 matrix and B is its inverse, what does the product AB equal?

2. If $C = \begin{bmatrix} 2 & -4 & 12 \\ 0 & 6 & -12 \\ 1 & -2 & 3 \end{bmatrix}$ and

$D = \frac{1}{6}\begin{bmatrix} 1 & 2 & 4 \\ 2 & 1 & -4 \\ 1 & 0 & -2 \end{bmatrix}$, are C and D inverse matrices?

3. If $A = \begin{bmatrix} 1 & 2 & 1 \\ 0 & 0 & 3 \\ 1 & 0 & 1 \end{bmatrix}$ and $B = \begin{bmatrix} 0 & -\frac{1}{3} & 1 \\ \frac{1}{2} & 0 & -\frac{1}{2} \\ 0 & \frac{1}{3} & 0 \end{bmatrix}$,

does $B = A^{-1}$?

4. If $D = \begin{bmatrix} 0 & 2 & -6 \\ -3 & 0 & 3 \\ 0 & -2 & 0 \end{bmatrix}$ and

$C = -\frac{1}{3}\begin{bmatrix} 2 & 4 & 2 \\ 0 & 0 & 6 \\ 2 & 0 & 2 \end{bmatrix}$, does $C = D^{-1}$?

In Problems 5–12, find the inverse matrix for each matrix that has an inverse.

5. $\begin{bmatrix} 4 & 7 \\ 1 & 2 \end{bmatrix}$ 6. $\begin{bmatrix} 4 & 5 \\ 7 & 9 \end{bmatrix}$ 7. $\begin{bmatrix} 2 & -1 \\ -4 & 2 \end{bmatrix}$

8. $\begin{bmatrix} 3 & 12 \\ 1 & 4 \end{bmatrix}$ 9. $\begin{bmatrix} 2 & 2 \\ 4 & 5 \end{bmatrix}$ 10. $\begin{bmatrix} 6 & -4 \\ -1 & 1 \end{bmatrix}$

11. $\begin{bmatrix} 4 & 7 \\ 2 & 1 \end{bmatrix}$ 12. $\begin{bmatrix} 1 & 2 \\ 3 & 0 \end{bmatrix}$

In Problems 13–16, find the inverse of matrix A and check it by calculating AA^{-1}.

13. $A = \begin{bmatrix} 3 & 0 & 0 \\ 0 & 3 & 0 \\ 0 & 0 & 3 \end{bmatrix}$ 14. $A = \begin{bmatrix} 4 & 0 & 0 \\ 0 & 3 & 0 \\ 0 & 0 & 3 \end{bmatrix}$

15. $A = \begin{bmatrix} 0 & 1 & 0 \\ 1 & 1 & 0 \\ 0 & 1 & 1 \end{bmatrix}$ 16. $A = \begin{bmatrix} 0 & 2 & 1 \\ 3 & 0 & 1 \\ 1 & 1 & 1 \end{bmatrix}$

In Problems 17–22, find the inverse matrix for each matrix that has an inverse.

17. $\begin{bmatrix} 3 & 1 & 2 \\ 1 & 2 & 3 \\ 1 & 1 & 1 \end{bmatrix}$ 18. $\begin{bmatrix} 1 & 2 & 3 \\ -1 & 5 & 6 \\ -1 & 3 & 3 \end{bmatrix}$

19. $\begin{bmatrix} 1 & 3 & 5 \\ -1 & -1 & 2 \\ 1 & 5 & 12 \end{bmatrix}$ 20. $\begin{bmatrix} 1 & -1 & 4 \\ -1 & 0 & -2 \\ -1 & -3 & 4 \end{bmatrix}$

21. $\begin{bmatrix} 1 & 2 & 4 \\ 1 & -1 & -3 \\ 2 & 1 & 1 \end{bmatrix}$ 22. $\begin{bmatrix} 3 & 4 & -1 \\ 4 & 2 & 2 \\ 2 & 6 & -4 \end{bmatrix}$

23. Find the inverse of

$$C = \begin{bmatrix} 0 & -1 & 0 & 1 & 0 \\ 1 & 1.5 & -1 & -1.5 & -0.5 \\ 0.5 & 1.75 & -1 & -1.25 & -0.25 \\ -0.5 & -1 & 1 & 1 & 0 \\ 0 & 0.5 & 0 & -0.5 & 0.5 \end{bmatrix}$$

24. Find the inverse of $B = \begin{bmatrix} 2 & 3 & 0 & 1 & 3 \\ 1 & 0 & 2 & 3 & 1 \\ 0 & 1 & 0 & 2 & 3 \\ 1 & 0 & 2 & 2 & 1 \\ 1 & 0 & 0 & 0 & 3 \end{bmatrix}$.

In each of Problems 25–28, the inverse of matrix A is given. Use the inverse to solve for X.

25. $A^{-1} = \begin{bmatrix} 3 & 2 \\ 1 & 1 \end{bmatrix}$ Solve $AX = \begin{bmatrix} 3 \\ 2 \end{bmatrix}$.

26. $A^{-1} = \begin{bmatrix} 1 & -2 \\ 1 & 3 \end{bmatrix}$ Solve $AX = \begin{bmatrix} 5 \\ -1 \end{bmatrix}$.

27. $A^{-1} = \begin{bmatrix} 3 & 2 & 1 \\ 1 & 1 & 2 \\ 1 & 2 & 1 \end{bmatrix}$ Solve $AX = \begin{bmatrix} 3 \\ -1 \\ 2 \end{bmatrix}$.

28. $A^{-1} = \begin{bmatrix} 3 & 0 & -3 \\ 1 & 1 & 2 \\ 1 & 2 & 1 \end{bmatrix}$ Solve $AX = \begin{bmatrix} -1 \\ 2 \\ 3 \end{bmatrix}$.

29. Use the inverse found in Problem 15 to solve

$$\begin{bmatrix} 0 & 1 & 0 \\ 1 & 1 & 0 \\ 0 & 1 & 1 \end{bmatrix} \begin{bmatrix} x \\ y \\ z \end{bmatrix} = \begin{bmatrix} 1 \\ 2 \\ 3 \end{bmatrix}$$

30. Use the inverse found in Problem 16 to solve

$$\begin{bmatrix} 0 & 2 & 1 \\ 3 & 0 & 1 \\ 1 & 1 & 1 \end{bmatrix} \begin{bmatrix} x \\ y \\ z \end{bmatrix} = \begin{bmatrix} 1 \\ 0 \\ 4 \end{bmatrix}$$

In Problems 31–34, use inverse matrices to solve each system of linear equations.

31. $\begin{cases} x + 2y = 4 \\ 3x + 4y = 10 \end{cases}$ 32. $\begin{cases} 3x - 4y = 11 \\ 2x + 3y = -4 \end{cases}$

33. $\begin{cases} 2x + y = 4 \\ 3x + y = 5 \end{cases}$ 34. $\begin{cases} 5x - 2y = 6 \\ 3x + 3y = 12 \end{cases}$

In Problems 35–40, use inverse matrices to find the solution of the systems of equations.

35. $\begin{cases} x + y + z = 3 \\ 2x + y + z = 4 \\ 2x + 2y + z = 5 \end{cases}$ 36. $\begin{cases} 2x - y - 2z = 2 \\ 3x - y + z = -3 \\ x + y - z = 7 \end{cases}$

37. $\begin{cases} x + y + 2z = 8 \\ 2x + y + z = 7 \\ 2x + 2y + z = 10 \end{cases}$ 38. $\begin{cases} x - 2y + z = 0 \\ 2x + y - 2z = 2 \\ 3x + 2y - 3z = 2 \end{cases}$

39. $\begin{cases} -x_2 + x_4 = 0.7 \\ x_1 + x_2 - x_3 - 1.5x_4 - 0.5x_5 = -1.6 \\ 0.5x_1 + 1.75x_2 - x_3 - 1.25x_4 - 0.25x_5 = 1.275 \\ -0.5x_1 - x_2 + x_3 + x_4 = 1.15 \\ 0.5x_2 - 0.5x_4 + 0.5x_5 = -0.15 \end{cases}$

40. $\begin{cases} 2x_1 + 3x_2 + x_4 + 3x_5 = 0 \\ x_1 + 2x_3 + 3x_4 + x_5 = 27 \\ x_2 + 2x_4 + 3x_5 = -15 \\ x_1 + 2x_3 + 2x_4 + x_5 = 24 \\ x_1 + 3x_5 = -24 \end{cases}$

In Problems 41–44, evaluate each determinant.

41. $\begin{vmatrix} 1 & 2 \\ 3 & 4 \end{vmatrix}$ 42. $\begin{vmatrix} 4 & 5 \\ -1 & -2 \end{vmatrix}$

43. $\begin{vmatrix} 3 & -1 \\ 2 & 4 \end{vmatrix}$ 44. $\begin{vmatrix} -1 & 2 \\ 4 & -8 \end{vmatrix}$

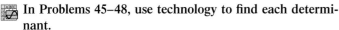
In Problems 45–48, use technology to find each determinant.

45. $\begin{vmatrix} 3 & 2 & 1 \\ -1 & 0 & 2 \\ 0 & 1 & 1 \end{vmatrix}$

46. $\begin{vmatrix} 1 & 2 & 2 \\ 3 & 0 & 1 \\ -1 & 0 & -2 \end{vmatrix}$

47. $\begin{vmatrix} 0 & 1 & 2 \\ 3 & 1 & 1 \\ 4 & -1 & 3 \end{vmatrix}$

48. $\begin{vmatrix} 3 & 1 & 4 & 2 \\ -1 & 2 & 0 & 5 \\ 4 & 3 & 0 & -1 \\ 0 & 4 & 2 & 1 \end{vmatrix}$

In Problems 49–52, use determinants to decide whether each matrix has an inverse.

49. $\begin{bmatrix} 2 & 3 \\ -1 & 4 \end{bmatrix}$

50. $\begin{bmatrix} 3 & 3 \\ 1 & -1 \end{bmatrix}$

51. $\begin{bmatrix} 1 & 3 & -2 \\ 2 & -1 & 5 \\ 3 & 2 & 3 \end{bmatrix}$

52. $\begin{bmatrix} 4 & 1 & 0 \\ 0 & 5 & 0 \\ 0 & 0 & -1 \end{bmatrix}$

APPLICATIONS

Decoding We have encoded messages by assigning the letters of the alphabet the numbers 1–26, with a blank assigned the number 27, and by using an encoding matrix A that converts the numbers to the coded message. To decode any message encoded by A, we must find the inverse of A, denoted by A^{-1}, and multiply A^{-1} times the coded message. In Problems 53–56, use this code and the given encoding matrix A to decode the given message.

53. If the code matrix is

$$A = \begin{bmatrix} 5 & 9 \\ 6 & 11 \end{bmatrix}$$

decode the message

49, 59, 133, 161, 270, 327, 313, 381

54. If the code matrix is

$$A = \begin{bmatrix} 3 & 5 \\ 1 & 2 \end{bmatrix}$$

decode the message

157, 59, 73, 29, 147, 58, 63, 24, 119, 44

55. If the code matrix is

$$A = \begin{bmatrix} 0 & 2 & 1 \\ 3 & 0 & 1 \\ 1 & 1 & 1 \end{bmatrix}$$

decode the message

47, 22, 34, 28, 87, 46, 63, 66, 55, 56, 44, 43, 17, 14, 15

(Because A and A^{-1} are 3×3 matrices, use triples of numbers.)

56. If the code matrix is

$$A = \begin{bmatrix} 1 & 2 & 1 \\ 5 & 8 & 2 \\ 7 & 11 & 3 \end{bmatrix}$$

decode the message

49, 165, 231, 49, 154, 220, 39, 162, 226

57. *Competition* A product is made by only two competing companies. Suppose Company X retains two-thirds of its customers and loses one-third to Company Y each year, and Company Y retains three-quarters of its customers and loses one-quarter to Company X each year. We can represent the number each company had last year by

$$\begin{bmatrix} x_0 \\ y_0 \end{bmatrix}$$

where x_0 is the number Company X had and y_0 is the number Company Y had. Then the number that each will have this year can be represented by

$$\begin{bmatrix} x \\ y \end{bmatrix} = \begin{bmatrix} \frac{2}{3} & \frac{1}{4} \\ \frac{1}{3} & \frac{3}{4} \end{bmatrix} \begin{bmatrix} x_0 \\ y_0 \end{bmatrix}$$

If Company X has 1900 customers and Company Y has 1700 customers this year, how many customers did each have last year?

58. *Demographics* Suppose that a government study showed that 90% of urban families remained in an urban area in the next generation (and 10% moved to a rural area), whereas 70% of rural families remained in a rural area in the next generation (and 30% moved to an urban area). This means that if $\begin{bmatrix} u_0 \\ r_0 \end{bmatrix}$ represents the current percents of urban families, u_0, and of rural families, r_0, then

$$\begin{bmatrix} u \\ r \end{bmatrix} = \begin{bmatrix} 0.90 & 0.30 \\ 0.10 & 0.70 \end{bmatrix} \begin{bmatrix} u_0 \\ r_0 \end{bmatrix}$$

represents the percents of urban families, u, and rural families, r, one generation from now. Currently the population is 60% urban and 40% rural. Find the percents in each group one generation before this one.

In Problems 59–69, set up each system of equations and then solve it by using inverse matrices.

59. *Medication* Medication A is given every 4 hours and medication B is given twice per day, and the ratio of the dosage of A to the dosage of B is always 5 to 8.
   (a) For patient I, the total intake of the two medications is 50.6 mg per day. Find the dosage of each administration of each medication for patient I.

(b) For patient II, the total intake of the two medications is 92 mg per day. Find the dosage of each administration of each medication for patient II.

60. *Transportation* The Ace Freight Company has an order for two products to be delivered to two stores of a company. The table below gives information regarding the two products.

	Product I	Product II
Unit volume (cu ft)	20	30
Unit weight (lb)	100	400

(a) If truck A can carry 2350 cu ft and 23,000 lb, how many of each product can it carry?
(b) If truck B can carry 2500 cu ft and 24,500 lb, how many of each product can it carry?

61. *Investment* One safe investment pays 10% per year, and a more risky investment pays 18% per year. A woman has $145,600 to invest and would like to have an income of $20,000 per year from her investments. How much should she invest at each rate?

62. *Nutrition* A biologist has a 40% solution and a 10% solution of the same plant nutrient. How many cubic centimeters of each solution should be mixed to obtain 25 cc of a 28% solution?

63. *Manufacturing* A manufacturer of table saws has three models, Deluxe, Premium, and Ultimate, which must be painted, assembled, and packaged for shipping. The table gives the number of hours required for each of these operations for each type of table saw. If the manufacturer has 96 hours available per day for painting, 156 hours for assembly, and 37 hours for packaging, how many of each type of saw can be produced each day?

	Deluxe	Premium	Ultimate
Painting	1.6	2	2.4
Assembly	2	3	4
Packaging	0.5	0.5	1

64. *Transportation* Ace Trucking Company has an order for three products, A, B, and C, for delivery. The table gives the volume in cubic feet, the weight in pounds, and the value for insurance in dollars for a unit of each of the products. If the carrier can carry 8000 cu ft and 12,400 lb and is insured for $52,600, how many units of each product can be carried?

	Product A	Product B	Product C
Unit volume (cu ft)	24	20	40
Weight (lb)	40	30	60
Value ($)	150	180	200

65. *Investment* A trust account manager has $1,000,000 to be invested in three different accounts. The accounts pay 6%, 8%, and 10%, and the goal is to earn $86,000 with the amount invested at 10% equal to the sum of the other two investments. To accomplish this, assume that x dollars are invested at 8%, y dollars at 10%, and z dollars at 6%. Find how much should be invested in each account to satisfy the conditions.

66. *Investment* A company offers three mutual fund plans for its employees. Plan I consists of 4 blocks of common stock and 2 municipal bonds. Plan II consists of 8 blocks of common stock, 4 municipal bonds, and 6 blocks of preferred stock. Plan III consists of 14 blocks of common stock, 6 municipal bonds, and 6 blocks of preferred stock. An employee wants to combine these plans so that she has 84 blocks of common stock, 40 municipal bonds, and 36 blocks of preferred stock. How many units of each plan does she need?

67. *Attendance* Suppose that during the first 2 weeks of a controversial art exhibit, the daily number of visitors to the gallery turns out to be the sum of the numbers of visitors on the previous 3 days. We represent the numbers of visitors on 3 successive days by

$$N = \begin{bmatrix} n_t \\ n_{t+1} \\ n_{t+2} \end{bmatrix}$$

(a) Find the matrix M such that

$$MN = \begin{bmatrix} n_{t+1} \\ n_{t+2} \\ n_t + n_{t+1} + n_{t+2} \end{bmatrix}$$

(b) If the numbers of visitors on 3 successive days are given by

$$\begin{bmatrix} 191 \\ 346 \\ 645 \end{bmatrix}$$

use M^{-1} to find the number of visitors 1 day earlier (before the 191-visitor day).

68. *Bee ancestry* Because a female bee comes from a fertilized egg and a male bee comes from an unfertilized egg, the number (n_{t+2}) of ancestors of a male bee $t + 2$ generations before the present generation is the sum of

the number of ancestors t and $t + 1$ generations (n_t and n_{t+1}) before the present. If the numbers of ancestors of a male bee in a given generation t and in the previous generation are given by

$$N = \begin{bmatrix} n_t \\ n_{t+1} \end{bmatrix}$$

then there is a matrix M such that the numbers of ancestors in the two generations preceding generation t are given by

$$MN = \begin{bmatrix} 0 & 1 \\ 1 & 1 \end{bmatrix} \begin{bmatrix} n_t \\ n_{t+1} \end{bmatrix} = \begin{bmatrix} n_{t+1} \\ n_t + n_{t+1} \end{bmatrix}$$

For a given male bee, the numbers of ancestors 5 and 6 generations back are given by

$$\begin{bmatrix} 8 \\ 13 \end{bmatrix}$$

Find the numbers of ancestors 4 and 5 generations back by multiplying both sides of

$$MN = \begin{bmatrix} 8 \\ 13 \end{bmatrix}$$

by the inverse of M.

69. *Bacteria growth* The population of a colony of bacteria grows in such a way that the population size at any hour t is the sum of the populations of the 3 previous hours. Suppose that the matrix

$$N = \begin{bmatrix} n_t \\ n_{t+1} \\ n_{t+2} \end{bmatrix}$$

describes the population sizes for 3 successive hours.

(a) Write matrix M so that MN gives the population sizes for 3 successive hours beginning 1 hour later—that is, such that

$$MN = \begin{bmatrix} n_{t+1} \\ n_{t+2} \\ n_t + n_{t+1} + n_{t+2} \end{bmatrix}$$

(b) If the populations at the ends of 3 successive 1-hour periods were 200 at the end of the first hour, 370 at the end of the second hour, and 600 at the end of the third hour, what was the population 1 hour before it was 200? Use M^{-1}.

3.5

Applications of Matrices: Leontief Input-Output Models

◑ Application Preview

Suppose we consider a simple economy as being based on three commodities: agricultural products, manufactured goods, and fuels. Suppose further that production of 10 units of agricultural products requires 5 units of agricultural products, 2 units of manufactured goods, and 1 unit of fuels; that production of 10 units of manufactured goods requires 1 unit of agricultural products, 5 units of manufactured products, and 3 units of fuels; and that production of 10 units of fuels requires 1 unit of agricultural products, 3 units of manufactured goods, and 4 units of fuels.

Table 3.12 summarizes this information in terms of production of 1 unit. The first column represents the units of agricultural products, manufactured goods, and fuels, respectively, that are needed to produce 1 unit of agricultural products. Column 2 represents the units required to produce 1 unit of manufactured goods, and column 3 represents the units required to produce 1 unit of fuels.

TABLE 3.12

	Outputs		
Inputs	Agricultural Products	Manufactured Goods	Fuels
Agricultural products	0.5	0.1	0.1
Manufactured goods	0.2	0.5	0.3
Fuels	0.1	0.3	0.4

From Table 3.12 we can form matrix *A*, which is called a **technology matrix** or a **Leontief matrix.**

$$A = \begin{bmatrix} 0.5 & 0.1 & 0.1 \\ 0.2 & 0.5 & 0.3 \\ 0.1 & 0.3 & 0.4 \end{bmatrix}$$

The Leontief input-output model, named for Wassily Leontief, is an interesting application of matrices. Just as an airplane model approximates the real object, a *mathematical model* attempts to approximate a real-world situation. Models that are carefully constructed from past information may often provide remarkably accurate forecasts. The model Leontief developed was useful in predicting the effects on the economy of price changes or shifts in government spending.*

Leontief's work divided the economy into 500 sectors, which, in a subsequent article in *Scientific American* (October 1951), were reduced to a more manageable 42 departments of production. We can examine the workings of input-output analysis with a very simplified view of the economy. (We will do this in Example 2 with the economy described above.)

In this section, we will interpret Leontief technology matrices for simple economies and will solve input-output problems for these economies by using Leontief models.

In Table 3.12, we note that the sum of the units of agricultural products, manufactured goods, and fuels required to produce 1 unit of fuels (column 3) does not add up to 1 unit. This is because not all commodities or industries are represented in the model. In particular, it is customary to omit labor from models of this type.

● EXAMPLE 1 *Leontief Matrix*

Use Table 3.12 to answer the following questions.

(a) How many units of agricultural products and of fuels are required to produce 100 units of manufactured goods?
(b) Production of which commodity is least dependent on the other two?
(c) If fuel costs rise, which two industries will be most affected?

Solution
(a) Referring to column 2, manufactured goods, we see that 1 unit requires 0.1 unit of agricultural products and 0.3 unit of fuels. Thus 100 units of manufactured goods require 10 units of agricultural products and 30 units of fuels.
(b) Looking down the columns, we see that 1 unit of agricultural products requires 0.3 unit of the other two commodities; 1 unit of manufactured goods requires 0.4 unit of the other two; and 1 unit of fuels requires 0.4 unit of the other two. Thus production of agricultural products is least dependent on the others.
(c) A rise in the cost of fuels would most affect those industries that use the larger amounts of fuels. One unit of agricultural products requires 0.1 unit of fuels, whereas a unit of manufactured goods requires 0.3 unit, and a unit of fuels requires 0.4 of its own units. Thus manufacturing and the fuel industry would be most affected by a cost increase in fuels.

● *Checkpoint* The following technology matrix for a simple economy describes the relationship of certain industries to each other in the production of one unit of product.

---

*Leontief's work dealt with a massive analysis of the American economy. He was awarded a Nobel Prize in economics in 1973 for his study.

	Ag	Mfg	Fuels	Util
Agriculture	0.32	0.06	0.09	0.06
Manufacturing	0.11	0.41	0.19	0.31
Fuels	0.31	0.14	0.18	0.20
Utilities	0.18	0.26	0.21	0.25

1. Which industry is most dependent on its own goods for its operation?
2. How many units of fuels are required to produce 100 units of utilities?

Open Leontief Model

For a simplified model of the economy, such as that described with Table 3.12, not all information is contained within the technology matrix. In particular, each industry has a gross production. The **gross production matrix** for the economy can be represented by the column matrix

$$X = \begin{bmatrix} x_1 \\ x_2 \\ x_3 \end{bmatrix}$$

where x_1 is the gross production of agricultural products, x_2 is the gross production of manufactured goods, and x_3 is the gross production of fuels.

The amounts of the gross productions used within the economy by the various industries are given by AX. Those units of gross production not used by these industries are called **final demands** or **surpluses** and may be considered as being available for consumers, the government, or export. If we place these surpluses in a column matrix D, then they can be represented by the equation

$$X - AX = D, \quad \text{or} \quad (I - A)X = D$$

where I is an identity matrix. This matrix equation is called the **technological equation** for an **open Leontief model.** This is called an open model because some of the goods from the economy are "open," or available to those outside the economy.

Open Leontief Model

The technological equation for an open Leontief model is

$$(I - A)X = D$$

◑ EXAMPLE 2 *Gross Outputs* *(Application Preview)*

Technology matrix A represents a simple economy with an agricultural industry, a manufacturing industry, and a fuels industry (as introduced in Table 3.12). If we wish to have surpluses of 85 units of agricultural products, 65 units of manufactured goods, and 0 units of fuels, what should the gross outputs be?

$$A = \begin{bmatrix} 0.5 & 0.1 & 0.1 \\ 0.2 & 0.5 & 0.3 \\ 0.1 & 0.3 & 0.4 \end{bmatrix}$$

Solution

Let X be the matrix of gross outputs for the industries and let D be the column matrix of each industry's surpluses. Then the technological equation is $(I - A)X = D$. We begin by finding the matrix $I - A$.

$$I - A = \begin{bmatrix} 1 & 0 & 0 \\ 0 & 1 & 0 \\ 0 & 0 & 1 \end{bmatrix} - \begin{bmatrix} 0.5 & 0.1 & 0.1 \\ 0.2 & 0.5 & 0.3 \\ 0.1 & 0.3 & 0.4 \end{bmatrix} = \begin{bmatrix} 0.5 & -0.1 & -0.1 \\ -0.2 & 0.5 & -0.3 \\ -0.1 & -0.3 & 0.6 \end{bmatrix}$$

Hence we must solve the matrix equation

$$\begin{bmatrix} 0.5 & -0.1 & -0.1 \\ -0.2 & 0.5 & -0.3 \\ -0.1 & -0.3 & 0.6 \end{bmatrix} \begin{bmatrix} x_1 \\ x_2 \\ x_3 \end{bmatrix} = \begin{bmatrix} 85 \\ 65 \\ 0 \end{bmatrix}$$

The augmented matrix is

$$\begin{bmatrix} 0.5 & -0.1 & -0.1 & | & 85 \\ -0.2 & 0.5 & -0.3 & | & 65 \\ -0.1 & -0.3 & 0.6 & | & 0 \end{bmatrix}$$

If we reduce by the methods of Section 3.3, "Gauss-Jordan Elimination: Solving Systems of Equations," we obtain

$$\begin{bmatrix} 1 & 0 & 0 & | & 300 \\ 0 & 1 & 0 & | & 400 \\ 0 & 0 & 1 & | & 250 \end{bmatrix}$$

so the gross outputs for the industries are

Agriculture: $x_1 = 300$

Manufacturing: $x_2 = 400$

Fuels: $x_3 = 250$

In Example 2 we solved the system of equations by using row operations on the augmented matrix. In Example 2 and in general, the technological equation for the open Leontief model can be solved by using the inverse of $I - A$, if the inverse exists. That is,

$$(I - A)X = D \quad \text{has the solution} \quad X = (I - A)^{-1}D$$

Calculator Note
If we have the technology matrix A and the surplus matrix D, we can solve the technological equation for an open Leontief model with a calculator or computer by defining matrices I, A, and D and evaluating $X = (I - A)^{-1}D$. Figure 3.10 shows the solution of Example 2 using a graphing calculator.

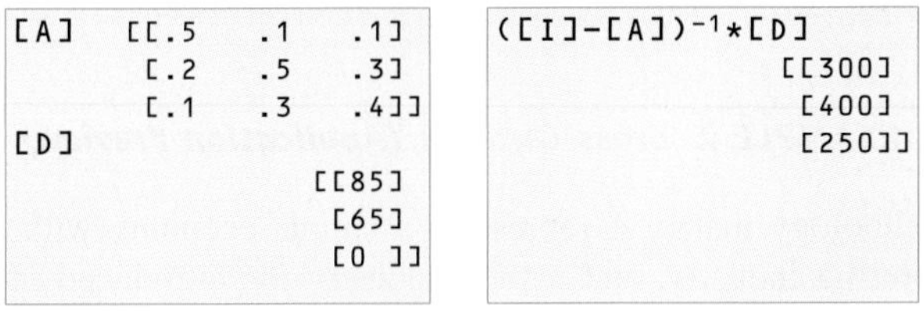

```
[A]   [[.5    .1    .1]        ([I]-[A])⁻¹*[D]
       [.2    .5    .3]                   [[300]
       [.1    .3    .4]]                   [400]
[D]                                        [250]]
              [[85]
               [65]
               [0 ]]
```

Figure 3.10

● **EXAMPLE 3** *Production*

A primitive economy with a lumber industry and a power industry has the following technology matrix.

$$A = \begin{array}{cc} & \begin{array}{cc} \text{Lumber} & \text{Power} \end{array} \\ \begin{bmatrix} 0.1 & 0.2 \\ 0.2 & 0.4 \end{bmatrix} & \begin{array}{c} \text{Lumber} \\ \text{Power} \end{array} \end{array}$$

If surpluses of 30 units of lumber and 70 units of power are desired, find the gross production of each industry.

Solution

Let x be the gross production of lumber, and let y be the gross production of power. Then

$$I - A = \begin{bmatrix} 0.9 & -0.2 \\ -0.2 & 0.6 \end{bmatrix}, \quad X = \begin{bmatrix} x \\ y \end{bmatrix}, \quad \text{and} \quad D = \begin{bmatrix} 30 \\ 70 \end{bmatrix}$$

We must solve the matrix equation $(I - A)X = D$ for X, and we can do this by using

$$X = (I - A)^{-1}D$$

We can find $(I - A)^{-1}$ with the formula from Section 3.4: "Inverse of a Square Matrix; Matrix Equations" or with a graphing utility.

$$(I - A)^{-1} = \frac{1}{0.54 - 0.04} \begin{bmatrix} 0.6 & 0.2 \\ 0.2 & 0.9 \end{bmatrix} = 2 \begin{bmatrix} 0.6 & 0.2 \\ 0.2 & 0.9 \end{bmatrix} = \begin{bmatrix} 1.2 & 0.4 \\ 0.4 & 1.8 \end{bmatrix}$$

Then we can find the gross production of each industry:

$$X = (I - A)^{-1}D = \begin{bmatrix} 1.2 & 0.4 \\ 0.4 & 1.8 \end{bmatrix} \begin{bmatrix} 30 \\ 70 \end{bmatrix} = \begin{bmatrix} 64 \\ 138 \end{bmatrix}$$

Hence the gross productions are

Lumber: $x = 64$

Power: $y = 138$

● **EXAMPLE 4** *Open Economy*

The economy of a developing nation is based on agricultural products, steel, and coal. An output of 1 ton of agricultural products requires inputs of 0.1 ton of agricultural products, 0.02 ton of steel, and 0.05 ton of coal. An output of 1 ton of steel requires inputs of 0.01 ton of agricultural products, 0.13 ton of steel, and 0.18 ton of coal. An output of 1 ton of coal requires inputs of 0.01 ton of agricultural products, 0.20 ton of steel, and 0.05 ton of coal. Write the technology matrix for this economy. Find the necessary gross productions to provide surpluses of 2350 tons of agricultural products, 4552 tons of steel, and 911 tons of coal.

Solution

The technology matrix for the economy is

$$\begin{array}{ccc} \text{Ag} & \text{Steel} & \text{Coal} \end{array}$$
$$\begin{bmatrix} 0.1 & 0.01 & 0.01 \\ 0.02 & 0.13 & 0.20 \\ 0.05 & 0.18 & 0.05 \end{bmatrix} \begin{array}{l} \text{Ag} \\ \text{Steel} \\ \text{Coal} \end{array}$$

The technological equation for this economy is $(I - A)X = D$, or

$$\left(\begin{bmatrix} 1 & 0 & 0 \\ 0 & 1 & 0 \\ 0 & 0 & 1 \end{bmatrix} - \begin{bmatrix} 0.1 & 0.01 & 0.01 \\ 0.02 & 0.13 & 0.20 \\ 0.05 & 0.18 & 0.05 \end{bmatrix} \right) X = \begin{bmatrix} 2350 \\ 4552 \\ 911 \end{bmatrix}$$

Thus the solution is

$$X = \left(\begin{bmatrix} 1 & 0 & 0 \\ 0 & 1 & 0 \\ 0 & 0 & 1 \end{bmatrix} - \begin{bmatrix} 0.1 & 0.01 & 0.01 \\ 0.02 & 0.13 & 0.20 \\ 0.05 & 0.18 & 0.05 \end{bmatrix} \right)^{-1} \begin{bmatrix} 2350 \\ 4552 \\ 911 \end{bmatrix}$$

Note that we can use technology to find the gross productions.

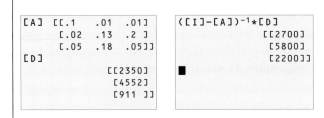

Thus the necessary gross productions of the industries are

$$X = \begin{bmatrix} 2700 \\ 5800 \\ 2200 \end{bmatrix} \begin{matrix} \text{Ag} \\ \text{Steel} \\ \text{Coal} \end{matrix}$$

● **Checkpoint**

3. Suppose a primitive economy has a wood products industry and a minerals industry, with technology matrix A (open Leontief model).

$$A = \begin{matrix} \text{W} \quad\ \text{M} \\ \begin{bmatrix} 0.12 & 0.11 \\ 0.09 & 0.15 \end{bmatrix} \end{matrix} \begin{matrix} \text{Wood products} \\ \text{Minerals} \end{matrix}$$

If surpluses of 1738 units of wood products and 1332 units of minerals are desired, find the gross production of each industry.

A few comments are in order. First, the surplus matrix must contain positive or zero entries. If this is not the case, then the gross production of one or more industries is insufficient to supply the needs of the others. That is, the economy cannot use more of a certain commodity than it produces.

Second, the model we have considered is referred to as an **open Leontief model** because not all inputs and outputs were incorporated within the technology matrix. In particular, because labor was omitted, some (or all) of the surpluses must be used to support this labor. Problems that fit this open Leontief model have unique solutions, so the matrix $(I - A)^{-1}$ can be used to solve the technological equation.

Closed Leontief Model

If a model is developed in which all inputs and outputs are used within the system, then such a model is called a **closed Leontief model.** In such a model, labor (and perhaps other factors) must be included. Labor is included by considering a new industry, households, which produces labor. When such a closed model is developed, all outputs are used within the system, and the sum of the entries in each column equals 1. In this case, there is no surplus, so $D = 0$, and we have the following:

Closed Leontief Model

The technological equation for a closed Leontief model is

$$(I - A)X = 0$$

where 0 is a zero column matrix.

For closed models, the technological equation does *not* have a unique solution, so the matrix $(I - A)^{-1}$ does *not* exist and hence cannot be used to find the solution.

● **EXAMPLE 5** *Closed Economy*

The following closed Leontief model with technology matrix A might describe the economy of the entire country, with x_1 equal to the government's budget, x_2 the value of industrial output (profit-making organizations), x_3 the budget of nonprofit organizations, and x_4 the households budget.

$$A = \begin{bmatrix} & G & I & N & H \\ 0.4 & 0.2 & 0.1 & 0.3 \\ 0.2 & 0.2 & 0.2 & 0.1 \\ 0.2 & 0 & 0.2 & 0.1 \\ 0.2 & 0.6 & 0.5 & 0.5 \end{bmatrix} \begin{matrix} \text{Government} \\ \text{Industry} \\ \text{Nonprofit} \\ \text{Households} \end{matrix}$$

Find the total budgets (or outputs) x_1, x_2, x_3, and x_4.

Solution

This is a closed Leontief model; therefore, we solve the technological equation $(I - A)X = 0$.

$$\left(\begin{bmatrix} 1 & 0 & 0 & 0 \\ 0 & 1 & 0 & 0 \\ 0 & 0 & 1 & 0 \\ 0 & 0 & 0 & 1 \end{bmatrix} - \begin{bmatrix} 0.4 & 0.2 & 0.1 & 0.3 \\ 0.2 & 0.2 & 0.2 & 0.1 \\ 0.2 & 0 & 0.2 & 0.1 \\ 0.2 & 0.6 & 0.5 & 0.5 \end{bmatrix} \right) X = \begin{bmatrix} 0 \\ 0 \\ 0 \\ 0 \end{bmatrix}$$

or

$$\begin{bmatrix} 0.6 & -0.2 & -0.1 & -0.3 \\ -0.2 & 0.8 & -0.2 & -0.1 \\ -0.2 & 0 & 0.8 & -0.1 \\ -0.2 & -0.6 & -0.5 & 0.5 \end{bmatrix} \begin{bmatrix} x_1 \\ x_2 \\ x_3 \\ x_4 \end{bmatrix} = \begin{bmatrix} 0 \\ 0 \\ 0 \\ 0 \end{bmatrix}.$$

The augmented matrix for this system is

$$\begin{bmatrix} 0.6 & -0.2 & -0.1 & -0.3 & | & 0 \\ -0.2 & 0.8 & -0.2 & -0.1 & | & 0 \\ -0.2 & 0 & 0.8 & -0.1 & | & 0 \\ -0.2 & -0.6 & -0.5 & 0.5 & | & 0 \end{bmatrix}$$

If we multiply each entry by -5 and rearrange the rows by moving row 3 to row 1, row 1 to row 2, and row 2 to row 3, we have the equivalent matrix

$$\begin{bmatrix} 1 & 0 & -4 & 0.5 & | & 0 \\ -3 & 1 & 0.5 & 1.5 & | & 0 \\ 1 & -4 & 1 & 0.5 & | & 0 \\ 1 & 3 & 2.5 & -2.5 & | & 0 \end{bmatrix}$$

Then, if we reduce this matrix by the methods of Section 3.3, "Gauss-Jordan Elimination: Solving Systems of Equations," or with a graphing utility, we obtain

$$\begin{bmatrix} 1 & 0 & 0 & -\frac{55}{82} & | & 0 \\ 0 & 1 & 0 & -\frac{15}{41} & | & 0 \\ 0 & 0 & 1 & -\frac{12}{41} & | & 0 \\ 0 & 0 & 0 & 0 & | & 0 \end{bmatrix}$$

The system of equations that corresponds to this augmented matrix is

$$\begin{aligned} x_1 - \tfrac{55}{82}x_4 &= 0 & & & x_1 &= \tfrac{55}{82}x_4 \\ x_2 - \tfrac{15}{41}x_4 &= 0 & \text{or} & & x_2 &= \tfrac{15}{41}x_4 \\ x_3 - \tfrac{12}{41}x_4 &= 0 & & & x_3 &= \tfrac{12}{41}x_4 \\ & & & & x_4 &= \text{households budget} \end{aligned}$$

Note that the economy satisfies the given equation if the government's budget is $\frac{55}{82}$ times the households budget, if the value of industrial output is $\frac{15}{41}$ times the households budget, and if the budget of nonprofit organizations is $\frac{12}{41}$ times the households budget. The dependency here is expected. The fact that the system is closed suggests the dependency, but even more obvious is the fact that industrial output is limited by labor supply.

Calculator Note As noted previously, for a closed Leontief model, the inverse of $I - A$ does not exist. Thus, in Example 5, we must reduce $I - A$ by hand or with a calculator or computer (see Figure 3.11). Note that the existence or nonexistence of $(I - A)^{-1}$ dictates how the technology is used.

```
rref([I]-[A])▸Fr
ac
[[1   0   0    -55/82]
 [0   1   0    -15/41]
 [0   0   1    -12/41]
 [0   0   0   0      ]]
```

Figure 3.11

● Checkpoint   4. For a closed Leontief model:
(a) What is the sum of the column entries of the technology matrix?
(b) State the technological equation.

We have examined two types of input-output models as they pertain to the economy as a whole. Since Leontief's original work with the economy, various other applications of input-output models have been developed. Consider the following parts-listing problem.

● EXAMPLE 6 *Manufacturing*

A storage shed consists of 4 walls and a roof. The walls and roof are made from stamped aluminum panels and are reinforced with 4 braces, each of which is held with 6 bolts. The final assembly joins the walls to each other and to the roof, using a total of 20 more bolts. The parts listing for these sheds can be described by the following matrix.

$$Q = \begin{bmatrix} & S & W & R & P & Br & Bo \\ & 0 & 0 & 0 & 0 & 0 & 0 \\ & 4 & 0 & 0 & 0 & 0 & 0 \\ & 1 & 0 & 0 & 0 & 0 & 0 \\ & 0 & 1 & 1 & 0 & 0 & 0 \\ & 0 & 4 & 4 & 0 & 0 & 0 \\ & 20 & 24 & 24 & 0 & 0 & 0 \end{bmatrix} \begin{matrix} \text{Sheds} \\ \text{Walls} \\ \text{Roofs} \\ \text{Panels} \\ \text{Braces} \\ \text{Bolts} \end{matrix}$$

This matrix is a type of input-output matrix, with the rows representing inputs that are used directly to produce the items that head the columns. We note that the entries here are quantities used rather than proportions used. Also, the columns containing all zeros indicate primary assembly items because they are not made from other parts.

Thus to produce a shed requires 4 walls, 1 roof, and 20 bolts, and to produce a roof requires 1 aluminum panel, 4 braces, and 24 bolts.

Suppose that an order is received for 4 completed sheds and spare parts including 2 walls, 1 roof, 8 braces, and 24 bolts. How many of each of the assembly items are required to fill the order?

Solution

If matrix D represents the order and matrix X represents the gross production, we have

$$D = \begin{bmatrix} 4 \\ 2 \\ 1 \\ 0 \\ 8 \\ 24 \end{bmatrix} \begin{matrix} \text{Sheds} \\ \text{Walls} \\ \text{Roofs} \\ \text{Panels} \\ \text{Braces} \\ \text{Bolts} \end{matrix} \qquad X = \begin{bmatrix} x_1 \\ x_2 \\ x_3 \\ x_4 \\ x_5 \\ x_6 \end{bmatrix} = \begin{bmatrix} \text{total sheds required} \\ \text{total walls required} \\ \text{total roofs required} \\ \text{total panels required} \\ \text{total braces required} \\ \text{total bolts required} \end{bmatrix}$$

Then X must satisfy $X - QX = D$, which is the technological equation for an open Leontief model.

We can find X by solving the system $(I - Q)X = D$, which is represented by the augmented matrix

$$\begin{bmatrix} 1 & 0 & 0 & 0 & 0 & 0 & \bigm| & 4 \\ -4 & 1 & 0 & 0 & 0 & 0 & \bigm| & 2 \\ -1 & 0 & 1 & 0 & 0 & 0 & \bigm| & 1 \\ 0 & -1 & -1 & 1 & 0 & 0 & \bigm| & 0 \\ 0 & -4 & -4 & 0 & 1 & 0 & \bigm| & 8 \\ -20 & -24 & -24 & 0 & 0 & 1 & \bigm| & 24 \end{bmatrix}$$

Although this matrix is quite large, it is easily reduced and yields

$$\begin{bmatrix} 1 & 0 & 0 & 0 & 0 & 0 & \bigm| & 4 \\ 0 & 1 & 0 & 0 & 0 & 0 & \bigm| & 18 \\ 0 & 0 & 1 & 0 & 0 & 0 & \bigm| & 5 \\ 0 & 0 & 0 & 1 & 0 & 0 & \bigm| & 23 \\ 0 & 0 & 0 & 0 & 1 & 0 & \bigm| & 100 \\ 0 & 0 & 0 & 0 & 0 & 1 & \bigm| & 656 \end{bmatrix}$$

This matrix tells us that the order will have 4 complete sheds. The 18 walls from the matrix include the 16 from the complete sheds, and the 5 roofs include the 4 from the complete sheds. The 23 panels, 100 braces, and 656 bolts are the total number of *primary assembly items* (bolts, braces, and panels) required to fill the order.

The inverse of $I - Q$ exists, so it could be used directly or with technology to find the solution to the parts-listing problem of Example 6.

● **Checkpoint Solutions**

1. Manufacturing
2. Under the "Util" column and across from the "Fuels" row is 0.20, indicating that producing 1 unit of utilities requires 0.20 unit of fuels. Thus producing 100 units of utilities requires 20 units of fuels.
3. Let x = the gross production of wood products

   y = the gross production of minerals

   We must solve $(I - A)X = D$.

$$\left(\begin{bmatrix} 1 & 0 \\ 0 & 1 \end{bmatrix} - \begin{bmatrix} 0.12 & 0.11 \\ 0.09 & 0.15 \end{bmatrix} \right) \begin{bmatrix} x \\ y \end{bmatrix} = \begin{bmatrix} 1738 \\ 1332 \end{bmatrix}$$

$$\begin{bmatrix} 0.88 & -0.11 \\ -0.09 & 0.85 \end{bmatrix} \begin{bmatrix} x \\ y \end{bmatrix} = \begin{bmatrix} 1738 \\ 1332 \end{bmatrix}$$

$$\begin{bmatrix} 0.88 & -0.11 \\ -0.09 & 0.85 \end{bmatrix}^{-1} = \frac{1}{(0.88)(0.85) - (0.09)(0.11)} \begin{bmatrix} 0.85 & 0.11 \\ 0.09 & 0.88 \end{bmatrix}$$

Hence

$$\begin{bmatrix} x \\ y \end{bmatrix} = \frac{1}{0.7381} \begin{bmatrix} 0.85 & 0.11 \\ 0.09 & 0.88 \end{bmatrix} \begin{bmatrix} 1738 \\ 1332 \end{bmatrix} = \begin{bmatrix} 2200 \\ 1800 \end{bmatrix}$$

4. (a) The sum of the column entries is 1.
   (b) $(I - A)X = 0$.

3.5 Exercises

1. The following technology matrix for a simple economy describes the relationship of certain industries to each other in the production of 1 unit of product.

$$\begin{array}{cccc} A & M & F & U \end{array}$$
$$\begin{bmatrix} 0.36 & 0.03 & 0.10 & 0.04 \\ 0.06 & 0.42 & 0.25 & 0.33 \\ 0.18 & 0.15 & 0.10 & 0.41 \\ 0.10 & 0.20 & 0.31 & 0.15 \end{bmatrix} \begin{array}{l} \text{Agriculture} \\ \text{Manufacturing} \\ \text{Fuels} \\ \text{Utilities} \end{array}$$

   (a) For each 100 units of manufactured products produced, how many units of fuels are required?
   (b) How many units of utilities are required to produce 40 units of agricultural products?

2. For the economy in Problem 1, what industry is most dependent on utilities?

The following technology matrix describes the relationship of certain industries within the economy to each other. (A&F, agriculture and food; RM, raw materials; M, manufacturing; F, fuels industry; U, utilities; SI, service industries)

$$\begin{array}{cccccc} \text{A\&F} & \text{RM} & \text{M} & \text{F} & \text{U} & \text{SI} \end{array}$$
$$\begin{bmatrix} 0.410 & 0.008 & 0 & 0.002 & 0 & 0.006 \\ 0.025 & 0.493 & 0.190 & 0.024 & 0.030 & 0.150 \\ 0.015 & 0.006 & 0.082 & 0.009 & 0.001 & 0.116 \\ 0.097 & 0.096 & 0.040 & 0.053 & 0.008 & 0.093 \\ 0.028 & 0.129 & 0.039 & 0.058 & 0.138 & 0.409 \\ 0.043 & 0.008 & 0.010 & 0.012 & 0.002 & 0.095 \end{bmatrix} \begin{array}{l} \text{A\&F} \\ \text{RM} \\ \text{M} \\ \text{F} \\ \text{U} \\ \text{SI} \end{array}$$

Use this matrix in Problems 3–10.

3. For each 1000 units of raw materials produced, how many units of agricultural and food products were required?
4. For each 1000 units of raw materials produced, how many units of service were required?
5. How many units of fuels were required to produce 1000 units of manufactured goods?
6. How many units of fuels were required to produce 1000 units of power (utilities' goods)?
7. Which industry is most dependent on its own goods for its operations? Which industry is least dependent on its own goods?

8. Which industry is most dependent on the fuels industry?
9. Which three industries would be most affected by a rise in the cost of raw materials?
10. Which industry would be most affected by a rise in the cost of manufactured goods?
11. Given is the partially solved matrix for the gross production of each industry of an open Leontief model. Find the gross production for each industry.

$$\begin{array}{ccc} \text{U} & \text{A} & \text{M} \end{array}$$
$$\left[\begin{array}{ccc|c} 1 & -2 & -1 & 80 \\ 0 & 8 & -4 & 60 \\ 0 & 0 & 10 & 50 \end{array}\right] \begin{array}{l} \text{Utilities} \\ \text{Agriculture} \\ \text{Manufacturing} \end{array}$$

12. Given is the partially solved matrix for the gross production of each industry of a closed Leontief model. Find the gross production for each industry.

$$\begin{array}{ccc} \text{U} & \text{A} & \text{H} \end{array}$$
$$\left[\begin{array}{ccc|c} 3 & -10 & -7 & 0 \\ 0 & 20 & -10 & 0 \\ 0 & 0 & 0 & 0 \end{array}\right] \begin{array}{l} \text{Utilities} \\ \text{Agriculture} \\ \text{Households} \end{array}$$

13. Suppose a primitive economy consists of two industries, farm products and farm machinery. Suppose also that its technology matrix is

$$A = \begin{array}{cc} & \begin{array}{cc} \text{P} & \text{M} \end{array} \\ & \begin{bmatrix} 0.5 & 0.1 \\ 0.1 & 0.3 \end{bmatrix} \begin{array}{l} \text{Products} \\ \text{Machinery} \end{array} \end{array}$$

If surpluses of 96 units of farm products and 8 units of farm machinery are desired, find the gross production of each industry.

14. Suppose an economy has two industries, agriculture and minerals. Suppose further that the technology matrix for this economy is A.

$$A = \begin{array}{cc} & \begin{array}{cc} \text{A} & \text{M} \end{array} \\ & \begin{bmatrix} 0.3 & 0.1 \\ 0.1 & 0.2 \end{bmatrix} \begin{array}{l} \text{Agriculture} \\ \text{Minerals} \end{array} \end{array}$$

If surpluses of 60 agricultural units and 70 mineral units are desired, find the gross production for each industry.

15. Suppose the economy of an underdeveloped country has an agricultural industry and an oil industry, with technology matrix A.

$$A = \begin{bmatrix} 0.3 & 0.1 \\ 0.2 & 0.1 \end{bmatrix} \begin{matrix} \text{Agricultural products} \\ \text{Oil products} \end{matrix}$$

with column headers A O

If surpluses of 0 units of agricultural products and 610 units of oil products are desired, find the gross production of each industry.

16. Suppose a simple economy with only an agricultural industry and a steel industry has the following technology matrix.

$$A = \begin{bmatrix} 0.5 & 0.2 \\ 0.1 & 0.6 \end{bmatrix} \begin{matrix} \text{Agriculture} \\ \text{Steel} \end{matrix}$$

with column headers A S

If surpluses of 15 units of agricultural products and 33 units of steel are desired, find the gross production of each industry.

17. An economy is based on two industries: utilities and manufacturing. The technology matrix for these industries is A.

$$A = \begin{bmatrix} 0.3 & 0.15 \\ 0.3 & 0.4 \end{bmatrix} \begin{matrix} \text{Utilities} \\ \text{Manufacturing} \end{matrix}$$

with column headers U M

If surpluses of 80 units of utilities output and 180 units of manufacturing are desired, find the gross production of each industry.

18. A primitive economy has a mining industry and a fishing industry, with technology matrix A.

$$A = \begin{bmatrix} 0.25 & 0.05 \\ 0.05 & 0.40 \end{bmatrix} \begin{matrix} \text{Mining} \\ \text{Fishing} \end{matrix}$$

with column headers M F

If surpluses of 147 units of mining output and 26 units of fishing output are desired, find the gross production of each industry.

19. One sector of an economy consists of a mining industry and a manufacturing industry, with technology matrix A.

$$A = \begin{bmatrix} 0.2 & 0.4 \\ 0.3 & 0.3 \end{bmatrix} \begin{matrix} \text{Mining} \\ \text{Manufacturing} \end{matrix}$$

with column headers M Mfg

Surpluses of 36 units of mining output and 278 units of manufacturing output are desired. Find the gross production of each industry.

20. An underdeveloped country has an agricultural industry and a manufacturing industry, with technology matrix A.

$$A = \begin{bmatrix} 0.20 & 0.10 \\ 0.25 & 0.45 \end{bmatrix} \begin{matrix} \text{Agriculture} \\ \text{Manufacturing} \end{matrix}$$

with column headers A M

Surpluses of 8 units of agricultural products and 620 units of manufactured products are desired. Find the gross production of each industry.

21. A simple economy has an electronic components industry and a computers industry. Each unit of electronic components output requires inputs of 0.3 unit of electronic components and 0.2 unit of computers. Each unit of computer industry output requires inputs of 0.6 unit of electronic components and 0.2 unit of computers.
    (a) Write the technology matrix for this simple economy.
    (b) If surpluses of 648 units of electronic components and 16 units of computers are desired, find the gross production of each industry.

22. An economy has an agricultural industry and a textile industry. Each unit of agricultural output requires 0.4 unit of agricultural input and 0.1 unit of textiles input. Each unit of textiles output requires 0.1 unit of agricultural input and 0.2 unit of textiles input.
    (a) Write the technology matrix for this economy.
    (b) If surpluses of 5 units of agricultural products and 195 units of textiles are desired, find the gross production of each industry.

23. A primitive economy consists of a fishing industry and an oil industry. A unit of fishing industry output requires 0.30 unit of fishing industry input and 0.35 unit of oil industry input. A unit of oil industry output requires inputs of 0.04 unit of fishing products and 0.10 unit of oil products.
    (a) Write the technology matrix for this primitive economy.
    (b) If surpluses of 20 units of fishing products and 1090 units of oil products are desired, find the gross production of each industry.

24. An economy has a manufacturing industry and a banking industry. Each unit of manufacturing output requires inputs of 0.5 unit of manufacturing and 0.2 unit of banking. Each unit of banking output requires inputs of 0.3 unit each of manufacturing and banking.
    (a) Write the technology matrix for this economy.
    (b) If surpluses of 141 units of manufacturing and 106 units of banking are desired, find the gross production of each industry.

Interdepartmental costs **Within a company there is a (micro)economy that is monitored by the accounting procedures. In terms of the accounts, the various departments "produce" costs, some of which are internal and some of which are direct costs. Problems 25–28 show how an open Leontief model can be used to determine departmental costs.**

25. Suppose the development department of a firm charges 10% of its total monthly costs to the promotional department, and the promotional department charges 5% of its total monthly costs to the development department. The direct costs of the development department are $20,400, and the direct costs of the promotional department are $9,900. The solution to the matrix equation

$$\begin{bmatrix} 20,400 \\ 9,900 \end{bmatrix} + \begin{bmatrix} 0 & 0.05 \\ 0.1 & 0 \end{bmatrix} \begin{bmatrix} x \\ y \end{bmatrix} = \begin{bmatrix} x \\ y \end{bmatrix}$$

for the column matrix $\begin{bmatrix} x \\ y \end{bmatrix}$ gives the total costs for each department. Note that this equation can be written in the form

$$D + AX = X, \quad \text{or} \quad X - AX = D$$

which is the form for an open economy. Find the total costs for each department in this (micro)economy.

26. Suppose the shipping department of a firm charges 20% of its total monthly costs to the printing department and that the printing department charges 10% of its total monthly costs to the shipping department. If the direct costs of the shipping department are $16,500 and the direct costs of the printing department are $11,400, find the total costs for each department.

27. Suppose the engineering department of a firm charges 20% of its total monthly costs to the computer department, and the computer department charges 25% of its total monthly costs to the engineering department. If during a given month the direct costs are $11,750 for the engineering department and $10,000 for the computer department, what are the total costs of each department?

28. The sales department of an auto dealership charges 10% of its total monthly costs to the service department, and the service department charges 20% of its total monthly costs to the sales department. During a given month, the direct costs are $88,200 for sales and $49,000 for service. Find the total costs of each department.

29. Suppose that an economy has three industries, fishing, agriculture, and mining, and that matrix A is the technology matrix for this economy.

$$A = \begin{bmatrix} F & A & M \\ 0.5 & 0.1 & 0.1 \\ 0.3 & 0.5 & 0.2 \\ 0.1 & 0.3 & 0.4 \end{bmatrix} \begin{matrix} \text{Fishing} \\ \text{Agriculture} \\ \text{Mining} \end{matrix}$$

If surpluses of 110 units of fishing output and 50 units each of agricultural and mining goods are desired, find the gross production of each industry.

30. Suppose that an economy has the same technology matrix as the economy in Problem 29. If surpluses of 180 units of fishing output, 90 units of agricultural goods, and 40 units of mining goods are desired, find the gross production of each industry.

31. Suppose that the economy of a small nation has an electronics industry, a steel industry, and an auto industry, with the following technology matrix.

$$A = \begin{bmatrix} E & S & A \\ 0.6 & 0.2 & 0.2 \\ 0.1 & 0.4 & 0.5 \\ 0.1 & 0.2 & 0.2 \end{bmatrix} \begin{matrix} \text{Electronics} \\ \text{Steel} \\ \text{Autos} \end{matrix}$$

If the nation wishes to have surpluses of 100 units of electronics production, 272 units of steel production, and 200 automobiles, find the gross production of each industry.

32. Suppose an economy has the same technology matrix as that in Problem 31. If surpluses of 540 units of electronics, 30 units of steel, and 140 autos are desired, find the gross production for each industry.

33. Suppose that a simple economy has three industries, service, manufacturing, and agriculture, and that matrix A is the technology matrix for this economy.

$$A = \begin{bmatrix} S & M & A \\ 0.4 & 0.1 & 0.1 \\ 0.2 & 0.5 & 0.2 \\ 0.2 & 0.1 & 0.3 \end{bmatrix} \begin{matrix} \text{Service} \\ \text{Manufacturing} \\ \text{Agriculture} \end{matrix}$$

If surpluses of 24 units of service output, 62 units of manufactured goods, and 32 units of agricultural goods are desired, find the gross production of each industry.

34. Suppose that an economy has the same technology matrix as the economy in Problem 33. If surpluses of 14 units of manufactured goods and 104 units of agricultural goods are desired, find the gross production of each industry.

Problems 35–40 refer to closed Leontief models.

35. Suppose the technology matrix for a closed model of a simple economy is given by

$$A = \begin{bmatrix} P & M & H \\ 0.5 & 0.1 & 0.2 \\ 0.1 & 0.3 & 0 \\ 0.4 & 0.6 & 0.8 \end{bmatrix} \begin{matrix} \text{Products} \\ \text{Machinery} \\ \text{Households} \end{matrix}$$

Find the gross productions for the industries.

36. Suppose the technology matrix for a closed model of a simple economy is given by matrix A. Find the gross productions for the industries.

$$\begin{array}{c} \\ A = \end{array} \begin{array}{c} \begin{matrix} \text{G} & \text{I} & \text{H} \end{matrix} \\ \begin{bmatrix} 0.4 & 0.1 & 0.3 \\ 0.4 & 0.3 & 0.2 \\ 0.2 & 0.6 & 0.5 \end{bmatrix} \end{array} \begin{array}{l} \\ \text{Government} \\ \text{Industry} \\ \text{Households} \end{array}$$

37. Suppose the technology matrix for a closed model of a simple economy is given by matrix A. Find the gross productions for the industries.

$$\begin{array}{c} \\ A = \end{array} \begin{array}{c} \begin{matrix} \text{G} & \text{I} & \text{H} \end{matrix} \\ \begin{bmatrix} 0.4 & 0.2 & 0.2 \\ 0.2 & 0.3 & 0.3 \\ 0.4 & 0.5 & 0.5 \end{bmatrix} \end{array} \begin{array}{l} \\ \text{Government} \\ \text{Industry} \\ \text{Households} \end{array}$$

38. Suppose the technology matrix for a closed model of an economy is given by matrix A. Find the gross productions for the industries.

$$\begin{array}{c} \\ A = \end{array} \begin{array}{c} \begin{matrix} \text{S} & \text{M} & \text{H} \end{matrix} \\ \begin{bmatrix} 0.2 & 0.1 & 0.1 \\ 0.6 & 0.5 & 0.1 \\ 0.2 & 0.4 & 0.8 \end{bmatrix} \end{array} \begin{array}{l} \\ \text{Shipping} \\ \text{Manufacturing} \\ \text{Households} \end{array}$$

39. A closed model for an economy has a manufacturing industry, utilities industry, and households industry. Each unit of manufacturing output uses 0.5 unit of manufacturing input, 0.4 unit of utilities input, and 0.1 unit of households input. Each unit of utilities output requires 0.4 unit of manufacturing input, 0.5 unit of utilities input, and 0.1 unit of households input. Each unit of household output requires 0.3 unit each of manufacturing and utilities input and 0.4 unit of households input.
 (a) Write the technology matrix for this closed model of the economy.
 (b) Find the gross production for each industry.
40. A closed model for an economy identifies government, the profit sector, the nonprofit sector, and households as its industries. Each unit of government output requires 0.3 unit of government input, 0.2 unit of profit sector input, 0.2 unit of nonprofit sector input, and 0.3 unit of households input. Each unit of profit sector output requires 0.2 unit of government input, 0.3 unit of profit sector input, 0.1 unit of nonprofit sector input, and 0.4 unit of households input. Each unit of nonprofit sector output requires 0.1 unit of government input, 0.1 unit of profit sector input, 0.2 unit of nonprofit sector input, and 0.6 unit of households input. Each unit of households output requires 0.05 unit of government input, 0.1 unit of profit sector input, 0.1 unit of nonprofit sector input, and 0.75 unit of households input.
 (a) Write the technology matrix for this closed model of the economy.
 (b) Find the gross production for each industry.
41. For the storage shed in Example 6, find the number of each primary assembly item required to fill an order for 24 sheds, 12 braces, and 96 bolts.
42. Card tables are made by joining 4 legs and a top using 4 bolts. The legs are each made from a steel rod. The top has a frame made from 4 steel rods. A cover and four special clamps that brace the top and hold the legs are joined to the frame using a total of 8 bolts. The parts-listing matrix for the card table assembly is given by

$$\begin{array}{c} \\ A = \end{array} \begin{array}{c} \begin{matrix} \text{CT} & \text{L} & \text{T} & \text{R} & \text{Co} & \text{Cl} & \text{B} \end{matrix} \\ \begin{bmatrix} 0 & 0 & 0 & 0 & 0 & 0 & 0 \\ 4 & 0 & 0 & 0 & 0 & 0 & 0 \\ 1 & 0 & 0 & 0 & 0 & 0 & 0 \\ 0 & 1 & 4 & 0 & 0 & 0 & 0 \\ 0 & 0 & 1 & 0 & 0 & 0 & 0 \\ 0 & 0 & 4 & 0 & 0 & 0 & 0 \\ 4 & 0 & 8 & 0 & 0 & 0 & 0 \end{bmatrix} \end{array} \begin{array}{l} \\ \text{Card table} \\ \text{Legs} \\ \text{Top} \\ \text{Rods} \\ \text{Cover} \\ \text{Clamps} \\ \text{Bolts} \end{array}$$

If an order is received for 10 card tables, 4 legs, 1 top, 1 cover, 6 clamps, and 12 bolts, how many of each primary assembly item are required to fill the order?
43. A sawhorse is made from 2 pairs of legs and a top joined with 4 nails. The top is a 2×4-in. board 3 ft long. Each pair of legs is made from two 2×4-in. boards 3 ft long, a brace, and a special clamp, with the brace and the clamp joined using 8 nails. The parts-listing matrix for the sawhorse is given by

$$\begin{array}{c} \\ A = \end{array} \begin{array}{c} \begin{matrix} \text{SH} & \text{T} & \text{LP} & 2\times 4 & \text{B} & \text{C} & \text{N} \end{matrix} \\ \begin{bmatrix} 0 & 0 & 0 & 0 & 0 & 0 & 0 \\ 1 & 0 & 0 & 0 & 0 & 0 & 0 \\ 2 & 0 & 0 & 0 & 0 & 0 & 0 \\ 0 & 1 & 2 & 0 & 0 & 0 & 0 \\ 0 & 0 & 1 & 0 & 0 & 0 & 0 \\ 0 & 0 & 1 & 0 & 0 & 0 & 0 \\ 4 & 0 & 8 & 0 & 0 & 0 & 0 \end{bmatrix} \end{array} \begin{array}{l} \\ \text{Sawhorse} \\ \text{Top} \\ \text{Leg pair} \\ 2 \times 4\text{s} \\ \text{Brace} \\ \text{Clamp} \\ \text{Nails} \end{array}$$

How many of each primary assembly item are required to fill an order for 10 sawhorses, 6 extra 2×4s, 6 extra clamps, and 100 nails?
44. A log carrier has a body made from a 4-ft length of reinforced cloth having a patch on each side and a dowel slid through each end to act as handles. The parts-listing matrix for the log carrier is given by

$$\begin{array}{c} \\ A = \end{array} \begin{array}{c} \begin{matrix} \text{LC} & \text{B} & \text{H} & \text{C} & \text{P} \end{matrix} \\ \begin{bmatrix} 0 & 0 & 0 & 0 & 0 \\ 1 & 0 & 0 & 0 & 0 \\ 2 & 0 & 0 & 0 & 0 \\ 0 & 1 & 0 & 0 & 0 \\ 0 & 2 & 0 & 0 & 0 \end{bmatrix} \end{array} \begin{array}{l} \\ \text{Log carrier} \\ \text{Body} \\ \text{Handles} \\ \text{Cloth} \\ \text{Patch} \end{array}$$

How many of each primary assembly item are required to fill an order for 500 log carriers and 20 handles?

Key Terms and Formulas

Section	Key Terms	Formulas

3.1	Matrices
	  Entries
	  Order
	  Square, row, column, and zero matrices
	Transpose
	Sum of two matrices
	Negative of a matrix
	Scalar multiplication
		A^T

3.2	Matrix product Identity matrix	2×2 is $\begin{bmatrix} 1 & 0 \\ 0 & 1 \end{bmatrix}$

3.3	Augmented matrix
	Coefficient matrix
	Elementary row operations
	Solving systems of equations
	  Gauss-Jordan elimination method
	Nonunique solutions
	  Reduced form	

3.4	Inverse matrices Matrix equations Determinants	$AX = B$ $\begin{vmatrix} a & b \\ c & d \end{vmatrix} = ad - bc$

3.5	Technology matrix
	Leontief input-output model
	Gross production matrix
	  Final demands (surpluses)
	Leontief model
	Technological equation
	  Open
	  Closed
		
		$(I - A)X = D$
		$(I - A)X = 0$

Review Exercises

Use the matrices below as needed to complete Problems 1–25.

$$A = \begin{bmatrix} 4 & 4 & 2 & -5 \\ 6 & 3 & -1 & 0 \\ 0 & 0 & -3 & 5 \end{bmatrix} \quad B = \begin{bmatrix} 2 & -5 & -11 & 8 \\ 4 & 0 & 0 & 4 \\ -2 & -2 & 1 & 9 \end{bmatrix}$$

$$C = \begin{bmatrix} 4 & -2 \\ 5 & 0 \\ 6 & 0 \\ 1 & 3 \end{bmatrix} \quad D = \begin{bmatrix} 3 & 5 \\ 1 & 2 \end{bmatrix} \quad E = \begin{bmatrix} 1 & 1 \\ 1 & 1 \\ 4 & 6 \\ 0 & 5 \end{bmatrix}$$

$$F = \begin{bmatrix} -1 & 6 \\ 4 & 11 \end{bmatrix} \quad G = \begin{bmatrix} 2 & -5 \\ -1 & 3 \end{bmatrix} \quad I = \begin{bmatrix} 1 & 0 \\ 0 & 1 \end{bmatrix}$$

Section 3.1

1. Find a_{12} in matrix A.
2. Find b_{23} in matrix B.
3. Which of the matrices, if any, are 3×4?

4. Which of the matrices, if any, are 2 × 4?

5. Which matrices are square?

6. Write the negative of matrix B.

7. If a matrix is added to its negative, what kind of matrix results?

8. Two matrices can be added if they have the same _____.

Perform the indicated operations in Problems 9–25.

9. $A + B$ 10. $C - E$ 11. $D^T - I$

12. $3C$ 13. $4I$ 14. $-2F$

15. $4D - 3I$ 16. $F + 2D$ 17. $3A - 5B$

Section 3.2

18. AC 19. CD 20. DF

21. FD 22. FI 23. IF

24. DG^T 25. $(DG)F$

Section 3.3

In Problems 26–28, the reduced matrix for a system of equations is given. (a) Identify the type of solution for the system (unique solution, no solution, or infinitely many solutions). Explain your reasoning. (b) For each system that has a solution, find it. If the system has infinitely many solutions, find two different specific solutions.

26. $\begin{bmatrix} 1 & 0 & -2 & | & 6 \\ 0 & 1 & 3 & | & 7 \\ 0 & 0 & 0 & | & 0 \end{bmatrix}$ 27. $\begin{bmatrix} 1 & 0 & -4 & | & 0 \\ 0 & 1 & 3 & | & 0 \\ 0 & 0 & 0 & | & 1 \end{bmatrix}$

28. $\begin{bmatrix} 1 & 0 & 0 & | & 0 \\ 0 & 1 & 0 & | & -10 \\ 0 & 0 & 1 & | & 14 \end{bmatrix}$

29. Given the following augmented matrix representing a system of linear equations, find the solution.

$$\begin{bmatrix} 1 & 1 & 2 & | & 5 \\ 4 & 0 & 1 & | & 5 \\ 2 & 1 & 1 & | & 5 \end{bmatrix}$$

In Problems 30–36, solve each system using matrices.

30. $\begin{cases} x - 2y = 4 \\ -3x + 10y = 24 \end{cases}$

31. $\begin{cases} x + y + z = 4 \\ 3x + 4y - z = -1 \\ 2x - y + 3z = 3 \end{cases}$

32. $\begin{cases} -x + y + z = 3 \\ 3x - z = 1 \\ 2x - 3y - 4z = -2 \end{cases}$

33. $\begin{cases} x + y - 2z = 5 \\ 3x + 2y + 5z = 10 \\ -2x - 3y + 15z = 2 \end{cases}$

34. $\begin{cases} x - y = 3 \\ x + y + 4z = 1 \\ 2x - 3y - 2z = 7 \end{cases}$

35. $\begin{cases} x - 3y + z = 4 \\ 2x - 5y - z = 6 \end{cases}$

36. $\begin{cases} x_1 + x_2 + x_3 + x_4 = 3 \\ x_1 - 2x_2 + x_3 - 4x_4 = -5 \\ x_1 - x_3 + x_4 = 0 \\ x_2 + x_3 + x_4 = 2 \end{cases}$

Section 3.4

37. Are D and G inverse matrices if

$$D = \begin{bmatrix} 3 & 5 \\ 1 & 2 \end{bmatrix} \quad \text{and} \quad G = \begin{bmatrix} 2 & -5 \\ -1 & 3 \end{bmatrix}?$$

In Problems 38–40, find the inverse of each matrix.

38. $\begin{bmatrix} 7 & -1 \\ -10 & 2 \end{bmatrix}$ 39. $\begin{bmatrix} 1 & 0 & 2 \\ 3 & 4 & -1 \\ 1 & 1 & 0 \end{bmatrix}$

40. $\begin{bmatrix} 3 & 3 & 2 \\ -1 & 4 & 2 \\ 2 & 5 & 3 \end{bmatrix}$

In Problems 41–43, solve each system of equations by using inverse matrices.

41. $\begin{cases} x + 2z = 5 \\ 3x + 4y - z = 2 \\ x + y = -3 \end{cases}$ (See Problem 39.)

42. $\begin{cases} 3x + 3y + 2z = 1 \\ -x + 4y + 2z = -10 \\ 2x + 5y + 3z = -6 \end{cases}$ (See Problem 40.)

43. $\begin{cases} x + 3y + z = 0 \\ x + 4y + 3z = 2 \\ 2x - y - 11z = -12 \end{cases}$

44. Does the following matrix have an inverse?

$$\begin{bmatrix} 1 & 2 & 4 \\ 2 & 0 & -4 \\ 1 & 0 & -2 \end{bmatrix}$$

In Problems 45 and 46,

(a) Find the determinant of the matrix.

(b) Use it to decide whether the matrix has an inverse.

45. $\begin{bmatrix} 4 & 4 \\ -2 & 2 \end{bmatrix}$

46. $\begin{bmatrix} 1 & 2 & 3 \\ 4 & -1 & 8 \\ 6 & 3 & 14 \end{bmatrix}$

APPLICATIONS

The Burr Cabinet Company manufactures bookcases and filing cabinets at two plants, A and B. Matrix M gives the production for the two plants during June, and matrix N gives the production for July. Use them in Problems 47–49.

$$M = \begin{bmatrix} 150 & 80 \\ 280 & 300 \end{bmatrix} \begin{matrix} \text{Bookcases} \\ \text{Files} \end{matrix} \qquad N = \begin{bmatrix} 100 & 60 \\ 200 & 400 \end{bmatrix}$$

$$S = \begin{bmatrix} 120 & 80 \\ 180 & 300 \end{bmatrix} \qquad P = \begin{bmatrix} 1000 & 800 \\ 600 & 1200 \end{bmatrix}$$

(columns labeled A and B)

Section 3.1

47. *Production*   Write the matrix that represents total production at the two plants for the 2 months.

48. *Production*   If matrix P represents the inventories at the plants at the beginning of June and matrix S represents shipments from the plants during June, write the matrix that represents the inventories at the end of June.

Section 3.2

49. *Production*   If the company sells its bookcases to wholesalers for $100 and its filing cabinets for $120, for which month was the value of production higher: (a) at plant A? (b) at plant B?

A small church choir is made up of men and women who wear choir robes in the sizes shown in matrix A.

$$A = \begin{bmatrix} 1 & 14 \\ 12 & 10 \\ 8 & 3 \end{bmatrix} \begin{matrix} \text{Small} \\ \text{Medium} \\ \text{Large} \end{matrix}$$

(columns labeled Men, Women)

$$B = \begin{bmatrix} 25 & 40 & 45 \\ 10 & 10 & 10 \end{bmatrix} \begin{matrix} \text{Robes} \\ \text{Hoods} \end{matrix}$$

(columns labeled S, M, L)

Matrix B contains the prices (in dollars) of new robes and hoods according to size. Use these matrices in Problems 50 and 51.

50. *Cost*   Find the product BA, and label the rows and columns to show what each entry represents.

51. *Cost*   To find a matrix that gives the cost of new robes and the cost of new hoods, find

$$BA \begin{bmatrix} 1 \\ 1 \end{bmatrix}$$

52. *Manufacturing*   Two departments of a firm, A and B, need different amounts of the same products. The fol-

lowing table gives the amounts of the products needed by the two departments.

	Steel	Plastic	Wood
Department A	30	20	10
Department B	20	10	20

These three products are supplied by two suppliers, Ace and Kink, with the unit prices given in the following table.

	Ace	Kink
Steel	300	280
Plastic	150	100
Wood	150	200

(a) Use matrix multiplication to find how much these two orders will cost at the two suppliers. The result should be a 2×2 matrix.

(b) From which supplier should each department make its purchase?

53. *Investment*   Over the period of time 1995–2004, Maura's actual return (per dollar of stock) was 0.013469 per month, Ward's actual return was 0.013543 per month, and Goldsmith's actual return was 0.006504. Under certain assumptions, including that whatever underlying process generated the past set of observations will continue to hold in the present and near future, we can estimate the expected return for a stock portfolio containing these assets in some combination.

Suppose we have a portfolio that has 20% of our total wealth in Maura, 30% in Ward, and 50% in Goldsmith stock. Use the following steps to estimate the portfolio's historical return, which can be used to estimate the future return.

(a) Write a 1×3 matrix that defines the decimal part of our total wealth in Maura, in Ward, and in Goldsmith.

(b) Write a 3×1 matrix that gives the monthly returns from each of these companies.

(c) Find the historical return of the portfolio by computing the product of these two matrices.

(d) What is the expected monthly return of the stock portfolio?

Sections 3.3 and 3.4

54. *Investment*   A woman has $50,000 to invest. She has decided to invest all of it by purchasing some shares of stock in each of three companies: a fast-food chain that sells for $50 per share and has an expected growth of 11.5% per year, a software company that sells for $20 per share and has an expected growth of 15% per year, and a pharmaceutical company that sells for $80 per

share and has an expected growth of 10% per year. She plans to buy twice as many shares of stock in the fast-food chain as in the pharmaceutical company. If her goal is 12% growth per year, how many shares of each stock should she buy?

55. *Nutrition*   A biologist is growing three different types of slugs (types A, B, and C) in the same laboratory environment. Each day, the slugs are given a nutrient mixture that contains three different ingredients (I, II, and III). Each type A slug requires 1 unit of I, 3 units of II, and 1 unit of III per day. Each type B slug requires 1 unit of I, 4 units of II, and 2 units of III per day. Each type C slug requires 2 units of I, 10 units of II, and 6 units of III per day.
   (a) If the daily mixture contains 2000 units of I, 8000 units of II, and 4000 units of III, find the number of slugs of each type that can be supported.
   (b) Is it possible to support 500 type A slugs? If so, how many of the other types are there?
   (c) What is the maximum number of type A slugs possible? How many of the other types are there in this case?

56. *Transportation*   An airline company has three types of aircraft that carry three types of cargo. The payload of each type is summarized in the table below.

	Plane Type		
Units Carried	Passenger	Transport	Jumbo
First-class mail	100	100	100
Passengers	150	20	350
Air freight	20	65	35

Suppose that on a given day the airline must move 1100 units of first-class mail, 460 units of air freight, and 1930 passengers. How many aircraft of each type should be scheduled?

Section 3.5

57. *Economy models*   An economy has a shipping industry and an agricultural industry with technology matrix A.

$$A = \begin{bmatrix} \overset{S}{0.2} & \overset{A}{0.1} \\ 0.1 & 0.2 \end{bmatrix} \begin{matrix} \text{Shipping} \\ \text{Agriculture} \end{matrix}$$

Surpluses of 4720 tons of shipping output and 40 tons of agricultural output are desired. Find the gross production of each industry.

58. *Economy models*   A simple economy has a shoe industry and a cattle industry. Each unit of shoe output requires inputs of 0.1 unit of shoes and 0.2 unit of cattle products. Each unit of cattle products output requires inputs of 0.1 unit of shoes and 0.05 unit of cattle products.
   (a) Write the technology matrix for this simple economy.
   (b) If surpluses of 850 units of shoes and 275 units of cattle products are desired, find the gross production of each industry.

59. *Economy models*   A look at the industrial sector of an economy can be simplified to include three industries: the mining industry, the manufacturing industry, and the fuels industry. The technology matrix for this sector of the economy is given by

$$A = \begin{bmatrix} \overset{M}{0.4} & \overset{Mf}{0.4} & \overset{F}{0.2} \\ 0.2 & 0.4 & 0.2 \\ 0.1 & 0.2 & 0.4 \end{bmatrix} \begin{matrix} \text{Mining} \\ \text{Manufacturing} \\ \text{Fuels} \end{matrix}$$

Find the gross production of each industry if surpluses of 8 units of mined goods, 40 units of manufactured goods, and 140 units of fuels are desired.

60. *Economy models*   Suppose a closed Leontief model for a nation's economy has the following technology matrix.

$$A = \begin{bmatrix} \overset{G}{0.4} & \overset{A}{0.2} & \overset{M}{0.2} & \overset{H}{0.2} \\ 0.2 & 0.4 & 0.1 & 0.2 \\ 0.2 & 0.1 & 0.3 & 0.1 \\ 0.2 & 0.3 & 0.4 & 0.5 \end{bmatrix} \begin{matrix} \text{Government} \\ \text{Agriculture} \\ \text{Manufacturing} \\ \text{Households} \end{matrix}$$

Find the gross production of each industry.

3 Chapter Test

In Problems 1–6, perform the indicated matrix operations with the following matrices.

$$A = \begin{bmatrix} 1 & -2 \\ 3 & 2 \\ 4 & 1 \end{bmatrix} \quad B = \begin{bmatrix} 2 & -2 & 1 \\ 3 & 1 & 5 \end{bmatrix}$$

$$C = \begin{bmatrix} 3 & -4 & -1 \\ 2 & 2 & -1 \end{bmatrix} \quad D = \begin{bmatrix} 1 & 2 & 4 \\ 3 & 5 & 41 \\ 3 & 2 & 3 \end{bmatrix}$$

1. $A^T + B$ 2. $B - C$ 3. CD
4. DA 5. BA 6. ABD

7. Find the inverse of the matrix $\begin{bmatrix} 1 & 3 \\ 2 & 4 \end{bmatrix}$.

8. Find the inverse of the matrix

$$\begin{bmatrix} 1 & 2 & 4 \\ 1 & 2 & 2 \\ 1 & 1 & 4 \end{bmatrix}$$

9. If $AX = B$ and

$$A^{-1} = \begin{bmatrix} 1 & 2 & 0 \\ 3 & 1 & 2 \\ 4 & 1 & 1 \end{bmatrix} \quad \text{and} \quad B = \begin{bmatrix} 3 \\ 1 \\ 2 \end{bmatrix}$$

find the matrix X.

10. Solve

$$\begin{cases} x - y + 2z = 4 \\ x + 4y + z = 4 \\ 2x + 2y + 4z = 10 \end{cases}$$

11. Solve

$$\begin{cases} x - y + 2z = 4 \\ x + 4y + z = 4 \\ 2x + 3y + 3z = 8 \end{cases}$$

12. Solve

$$\begin{cases} x - y + 3z = 4 \\ x + 5y + 2z = 3 \\ 2x + 4y + 5z = 8 \end{cases}$$

13. Use an inverse matrix to solve

$$\begin{cases} x + y + 2z = 4 \\ x + 2y + z + w = 4 \\ 2x + 5y + 4z + 2w = 10 \\ 2y + z + 2w = 0 \end{cases}$$

14. Solve

$$\begin{cases} x - y + 2z - w = 4 \\ x + 4y + z + w = 4 \\ 2x + 2y + 4z + 2w = 10 \\ - y + z + 2w = 2 \end{cases}$$

15. Suppose that the solution of an investment problem involving a system of linear equations is given by

$$B = 75,000 - 3H \quad \text{and} \quad E = 20,000 + 2H$$

where B represents the dollars invested in Barton Bank stocks, H is the dollars invested in Heath Healthcare stocks, and E is the dollars invested in Electronics Depot stocks.

(a) If \$10,000 is invested in the healthcare stocks, how much is invested in the other two stocks?
(b) What is the dollar range that could be invested in the healthcare stocks?
(c) What is the minimum amount that could be invested in the electronics stocks? How much is invested in the other two stocks in this case?

16. In an ecological model, matrix A gives the fractions of several types of plants consumed by the herbivores in the ecosystem. Similarly, matrix B gives the fraction of each type of herbivore that is consumed by the carnivores in the system. Each row of matrix A refers to a type of plant, and each column refers to a kind of herbivore. Similarly, each row of matrix B is a kind of herbivore and each column is a kind of carnivore.

$$A = \begin{array}{c} \text{Herbivores} \\ \begin{bmatrix} 0.2 & 0.1 & 0.4 \\ 0.3 & 0.3 & 0.1 \\ 0.2 & 0.1 & 0.1 \\ 0.1 & 0.2 & 0.5 \\ 0.4 & 0.4 & 0 \end{bmatrix} \end{array} \text{Plants}$$

$$B = \begin{array}{c} \text{Carnivores} \\ \begin{bmatrix} 0.1 & 0.1 & 0.2 \\ 0.2 & 0 & 0.4 \\ 0.1 & 0.5 & 0.1 \end{bmatrix} \end{array} \text{Herbivores}$$

The matrix AB gives the fraction of each plant that is consumed by each carnivore.

(a) Find AB.
(b) What fraction of plant type 1 (row 1) is consumed by each carnivore?
(c) Identify the plant type that each carnivore consumes the most.

17. A furniture manufacturer produces four styles of chairs, and has orders for 1000 of style A, 4000 of style B, 2000 of style C, and 1000 of style D. The following table lists the numbers of units of raw materials the manufacturer needs for each style. Suppose wood costs \$5 per unit, nylon costs \$3 per unit, velvet costs \$4 per unit, and springs cost \$4 per unit.

	Wood	Nylon	Velvet	Springs
Style A	10	5	0	0
Style B	5	0	20	10
Style C	5	20	0	10
Style D	5	10	10	10

(a) Form a 1×4 matrix containing the number of units of each style ordered.

(b) Use matrix multiplication to determine the total number of units of each raw material needed.

(c) Write a 4 × 1 matrix representing the costs of the raw materials.

(d) Use matrix multiplication to find the 1 × 1 matrix that represents the total investment to fill the orders.

(e) Use matrix multiplication to determine the cost of materials for each style of chair.

18. Complete (a) and (b) by using the code discussed in Sections 3.2 and 3.4 (where *a* is 1, ..., *z* is 26, and blank is 27), and the coding matrix is

$$A = \begin{bmatrix} 8 & 5 \\ 3 & 2 \end{bmatrix}$$

(a) Encode the message "Less is more," the motto of architect Mies van der Rohe.

(b) Decode the message 138, 54, 140, 53, 255, 99, 141, 54, 201, 76, 287, 111.

19. A young couple with a $120,000 inheritance wants to invest all this money. They plan to diversify their investments, choosing some of each of the following types of stock.

Type	Cost/Share	Expected Total Growth/Share
Growth	$ 30	$ 4.60
Blue-chip	100	11.00
Utility	50	5.00

Their strategy is to have the total investment in growth stocks equal to the sum of the other investments, and their goal is a 13% growth for their investment. How many shares of each type of stock should they purchase?

20. Suppose an economy has two industries, agriculture and minerals, and the economy has the technology matrix

$$A = \begin{matrix} & \begin{matrix} Ag & M \end{matrix} \\ \begin{matrix} Ag \\ M \end{matrix} & \begin{bmatrix} 0.4 & 0.2 \\ 0.1 & 0.3 \end{bmatrix} \end{matrix}$$

If surpluses of 140 agriculture units and 140 minerals units are desired, find the gross production of each industry.

21. Suppose the technology matrix for a closed model of a simple economy is given by matrix *A*. Find the gross productions of the industries.

$$A = \begin{matrix} & \begin{matrix} P & NP & H \end{matrix} \\ \begin{bmatrix} 0.4 & 0.3 & 0.4 \\ 0.2 & 0.4 & 0.2 \\ 0.4 & 0.3 & 0.4 \end{bmatrix} & \begin{matrix} Profit \\ Nonprofit \\ Households \end{matrix} \end{matrix}$$

Use the following information in Problems 22 and 23. The national economy of Swiziland has four products: agricultural products, machinery, fuel, and steel. Producing 1 unit of agricultural products requires 0.2 unit of agricultural products, 0.3 unit of machinery, 0.2 unit of fuel, and 0.1 unit of steel. Producing 1 unit of machinery requires 0.1 unit of agricultural products, 0.2 unit of machinery, 0.2 unit of fuel, and 0.4 unit of steel. Producing 1 unit of fuel requires 0.1 unit of agricultural products, 0.2 unit of machinery, 0.3 unit of fuel, and 0.2 unit of steel. Producing 1 unit of steel requires 0.1 unit of agricultural products, 0.2 unit of machinery, 0.3 unit of fuel, and 0.2 unit of steel.

22. Create the technology matrix for this economy.

23. Determine how many units of each product will give surpluses of 1700 units of agriculture products, 1900 units of machinery, 900 units of fuel, and 300 units of steel.

24. A simple closed economy has the technology matrix *A*. Find the total outputs of the four sectors of the economy.

$$A = \begin{matrix} & \begin{matrix} Ag & S & F & H \end{matrix} \\ \begin{bmatrix} 0.25 & 0.25 & 0.3 & 0.1 \\ 0.3 & 0.25 & 0.2 & 0.4 \\ 0.15 & 0.2 & 0.1 & 0.3 \\ 0.3 & 0.3 & 0.4 & 0.2 \end{bmatrix} & \begin{matrix} Agriculture \\ Steel \\ Fuel \\ Households \end{matrix} \end{matrix}$$

Extended Applications & Group Projects

I. Taxation

Federal income tax allows a deduction for any state income tax paid during the year. In addition, the state of Alabama allows a deduction from its state income tax for any federal income tax paid during the year. The federal corporate income tax rate is equivalent to a flat rate of 34% if the taxable federal income is between $335,000 and $10,000,000, and the Alabama rate is 5% of the taxable state income.*

Both the Alabama and federal taxable income for a corporation is $1,000,000 *before* either tax is paid. Because each tax is deductible on the other return, the taxable income will differ for the state and federal taxes. One procedure often used by tax accountants to find the tax due in this and similar situations is called *iteration* and is described by the first five steps below.[†]

1. Make an estimate of the federal taxes due by assuming no state tax is due, deduct this estimated federal tax due from the state taxable income, and calculate an estimate of the state taxes due on the basis of this assumption.
2. Deduct the estimated state tax due as computed in (1) from the federal taxable income, and calculate a new estimate of the federal tax due under this assumption.
3. Deduct the new estimated federal income tax due as computed in (2) from the state taxable income, and calculate a new estimate of the state taxes due on the basis of these calculations.
4. Repeat the process in (2) and (3) to get a better estimate of the taxes due to both governments.
5. Continue this process until the federal tax changes from one repetition to another by less than $1. What federal tax is due? What state tax is due?

To find the tax due each government directly, we can solve a matrix equation.

6. Create two linear equations that describe this taxation situation, convert the system to a matrix equation, and solve the system to see exactly what tax is due to each government.
7. What is the effective rate that this corporation pays to each government?

Source: Federal Income Tax Tables, Alabama Income Tax Instructions.
[†]Performing these steps with an electronic spreadsheet is described by Kenneth H. Johnson in "A Simplified Calculation of Mutually Dependent Federal and State Tax Liabilities," in *The Journal of Taxation,* December 1994.

II. Company Profits After Bonuses and Taxes

A company earns $800,000 profit before paying bonuses and taxes. Suppose that a bonus of 2% is paid to each employee after all taxes are paid, that state taxes of 6% are paid on the profit after the bonuses are paid, and that federal taxes are 34% of the profit that remains after bonuses and state taxes are paid.

1. How much profit remains after all taxes and bonuses are paid? Give your answer to the nearest dollar.
2. If the employee bonuses are increased to 3%, employees would receive 50% more bonus money. How much more profit would this cost the company than the 2% bonuses?
3. Use a calculator or spreadsheet to determine the loss of profit associated with bonuses of 4%, 5%, etc., up to 10%, and make a management decision about what bonuses to give in order to balance employee morale with loss of profit.
4. Use technology and the data points found above to create an equation that models the company's profit after expenses as a function of the employee bonus percent.

Inequalities and Linear Programming

Most companies seek to maximize profits subject to the limitations imposed by product demand and available resources (such as raw materials and labor) or to minimize production costs subject to the need to fill customer orders. If the relationships among the various resources, production requirements, costs, and profits are all linear, then these activities may be planned (or programmed) in the best possible (optimal) way by using **linear programming.** Because linear programming provides the best possible solution to problems involving allocation of limited resources among various activities, its impact has been tremendous.

In this chapter, we introduce and discuss this important method. The chapter topics and applications include the following.

Sections	Applications
4.1 Linear Inequalities in Two Variables	*Land management, manufacturing constraints*
4.2 Linear Programming: Graphical Methods	*Profit, revenue, production costs*
4.3 The Simplex Method: Maximization **Nonunique solutions** **Shadow prices**	*Profit, manufacturing*
4.4 The Simplex Method: Duality and Minimization	*Purchasing, nutrition*
4.5 The Simplex Method with Mixed Constraints	*Production scheduling*

Chapter Warm-up

Prerequisite Problem Type	For Section	Answer	Section for Review
(a) Solve $3x - 2y \geq 4$ for y.	4.1	(a) $y \leq \frac{3}{2}x - 2$	1.1 Linear equations and inequalities
(b) Solve $2x + 4y = 60$ for y.		(b) $y = 15 - \frac{1}{2}x$	
Graph the equation $y = \frac{3}{2}x - 2$.	4.1		1.3 Graphing linear equations
Solve the systems: (a) $\begin{cases} x + 2y = 10 \\ 2x + y = 14 \end{cases}$ (b) $\begin{cases} x + 0.5y = 16 \\ x + y = 24 \end{cases}$	4.1 4.2	(a) $x = 6, y = 2$ (b) $x = 8, y = 16$	1.5 Systems of linear equations
Write the system in an augmented matrix: $\begin{cases} x + 2y + s_1 = 10 \\ 2x + y + s_2 = 14 \\ -2x - 3y + f = 0 \end{cases}$	4.3 4.4 4.5	$\begin{bmatrix} 1 & 2 & 1 & 0 & 0 & \vert & 10 \\ 2 & 1 & 0 & 1 & 0 & \vert & 14 \\ -2 & -3 & 0 & 0 & 1 & \vert & 0 \end{bmatrix}$	3.3 Gauss-Jordan elimination
Write a matrix equivalent to matrix A with the entry in row 1, column 2 equal to 1 and with all other entries in column 2 equal to 0. First multiply row 1 by $\frac{1}{2}$. $A = \begin{bmatrix} 1 & 2 & 1 & 0 & 0 & \vert & 10 \\ 2 & 1 & 0 & 1 & 0 & \vert & 14 \\ -2 & -3 & 0 & 0 & 1 & \vert & 0 \end{bmatrix}$	4.3 4.4 4.5	$\begin{bmatrix} \frac{1}{2} & 1 & \frac{1}{2} & 0 & 0 & \vert & 5 \\ \frac{3}{2} & 0 & -\frac{1}{2} & 1 & 0 & \vert & 9 \\ -\frac{1}{2} & 0 & \frac{3}{2} & 0 & 1 & \vert & 15 \end{bmatrix}$	3.3 Gauss-Jordan elimination

4.1

Linear Inequalities in Two Variables

◗ Application Preview

A farm co-op has 6000 acres available to plant with corn and soybeans. Each acre of corn requires 9 gallons of fertilizer/herbicide and 3/4 hour of labor to harvest. Each acre of soybeans requires 3 gallons of fertilizer/herbicide and 1 hour of labor to harvest. The co-op has available at most 40,500 gallons of fertilizer/herbicide and at most 5250 hours of labor for harvesting. The number of acres of each crop is limited (constrained) by the available resources: land, fertilizer/herbicide, and labor for harvesting. Finding the values that satisfy all these constraints at the same time is called solving the system of inequalities. (See Example 4.)

One Linear Inequality in Two Variables

Before we look at systems of inequalities, we will discuss solutions of one inequality in two variables, such as $y < x$. The solutions of this inequality are the ordered pairs (x, y) that satisfy the inequality. Thus $(1, 0)$, $(3, 2)$, $(0, -1)$, and $(-2, -5)$ are solutions of $y < x$, but $(3, 7)$, $(-4, -3)$ and $(2, 2)$ are not.

The graph of $y < x$ consists of all points in which the *y*-coordinate is less than the *x*-coordinate. The graph of the region $y < x$ can be found by graphing the line $y = x$ (as a dashed line, because the given inequality does not include $y = x$). This line separates the *xy*-plane into two **half-planes,** $y < x$ and $y > x$. We can determine which half-plane is the solution region by selecting as a **test point** any point **not on the line;** let's choose $(2, 0)$. Because the coordinates of this test point satisfy the inequality $y < x$, the half-plane containing this point is the solution region for $y < x$. (See Figure 4.1.) If the coordinates of the test point do not satisfy the inequality, then the other half-plane is the solution region. For example, say we had chosen $(0, 4)$ as our test point. Its coordinates do not satisfy $y < x$, so the half-plane that does *not* contain $(0, 4)$ is the solution region. (Note that we get the same region.)

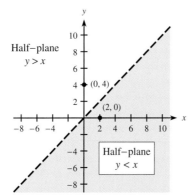

Figure 4.1

● EXAMPLE 1 *Graphing Inequalities*

Graph the inequality $4x - 2y \le 6$.

Solution

First we graph the line $4x - 2y = 6$, or (equivalently) $y = 2x - 3$, as a solid line, because points lying on the line satisfy the given inequality. Next we pick a test point that is not on the line. If we use $(0, 0)$, we see that its coordinates satisfy $4x - 2y \le 6$—that is,

$y \geq 2x - 3$. Hence the solution region is the line $y = 2x - 3$ and the half-plane that contains the test point $(0, 0)$. See Figure 4.2.

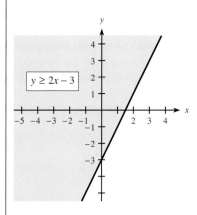

Figure 4.2

Technology Note

Graphing calculators can also be used to shade the solution region of an inequality. Figure 4.3 shows a graphing calculator window for the solution of $4x - 2y \leq 6$, which is equivalent to $4x - 6 \leq 2y$, or $y \geq 2x - 3$.

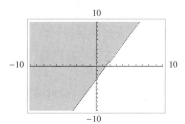

Figure 4.3   ■

Systems of Linear Inequalities

If we have two inequalities in two variables, we can find the solutions that satisfy both inequalities. We call the inequalities a **system of inequalities,** and the solution of the system can be found by finding the intersection of the solution sets of the two inequalities.

The solution set of the system of inequalities can be found by graphing the inequalities on the same set of axes and noting their points of intersection.

● EXAMPLE 2 *Graphical Solution of a System of Inequalities*

Graph the solution of the system

$$\begin{cases} 3x - 2y \geq 4 \\ x + y - 3 > 0 \end{cases}$$

Solution
Begin by graphing the equations $3x - 2y = 4$ and $x + y = 3$ (from $x + y - 3 = 0$) by the intercept method: find y when $x = 0$ and find x when $y = 0$.

$3x - 2y = 4$			$x + y = 3$	
x	y		x	y
0	-2		0	3
4/3	0		3	0

We graph $3x - 2y = 4$ as a solid line and $x + y = 3$ as a dashed line (see Figure 4.4(a)). We use any point not on either line as a test point; let's use $(0, 0)$. Note that the coordinates of $(0, 0)$ do not satisfy either $3x - 2y \geq 4$ or $x + y - 3 > 0$. Thus the solution region for each individual inequality is the half-plane that does not contain the point $(0, 0)$. Figure 4.4(b) indicates the half-plane solution for each inequality with arrows pointing from the line into the desired half-plane (away from the test point). The points that satisfy both of these inequalities lie in the intersection of the two individual solution regions, shown in Figure 4.4(c). This **solution region** is the graph of the solution of this system of inequalities.

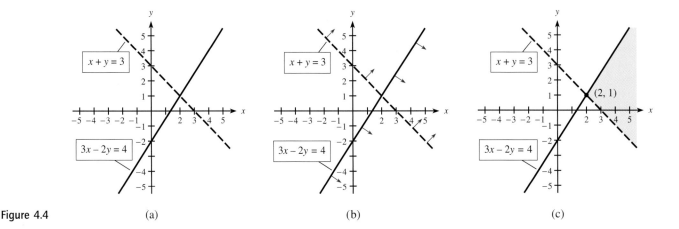

Figure 4.4   (a)   (b)   (c)

The point $(2, 1)$ in Figure 4.4(c), where the two regions form a "corner," is found by solving the *equations* $3x - 2y = 4$ and $x + y = 3$ simultaneously.

● **EXAMPLE 3 Graphical Solution of a System of Inequalities**

Graph the solution of the system

$$\begin{cases} x + 2y \leq 10 \\ 2x + y \leq 14 \\ x \geq 0, y \geq 0 \end{cases}$$

Solution
The two inequalities $x \geq 0$ and $y \geq 0$ restrict the solution to Quadrant I (and the axes bounding Quadrant I).

We seek points in the first quadrant (on or above $y = 0$ and on or to the right of $x = 0$) that satisfy $x + 2y \leq 10$ *and* $2x + y \leq 14$. We can write these inequalities in their equivalent forms $y \leq 5 - \frac{1}{2}x$ and $y \leq 14 - 2x$. The points that satisfy these inequalities (in the first quadrant) are shown by the shaded area in Figure 4.5. We can observe from the graph that the points $(0, 0)$, $(7, 0)$, and $(0, 5)$ are corners of the solution region. The corner $(6, 2)$ is found by solving the *equations* $y = 5 - \frac{1}{2}x$ and $y = 14 - 2x$ simultaneously as follows:

$$5 - \frac{1}{2}x = 14 - 2x$$

$$\frac{3}{2}x = 9$$

$$x = 6$$

$$y = 2$$

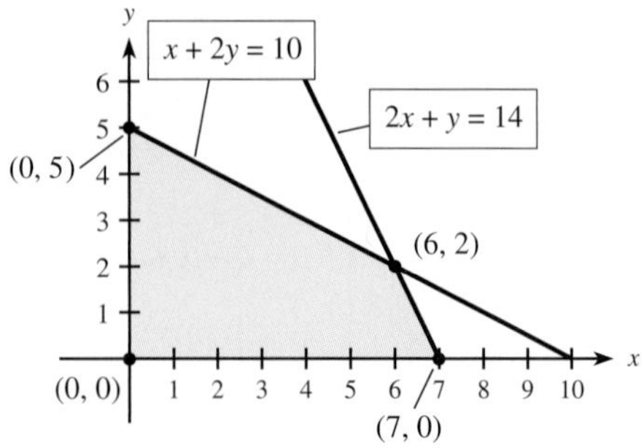

Figure 4.5

We will see that the corners of the solution region are important in solving linear programming problems.

Many applications restrict the variables to be nonnegative (such as $x \geq 0$ and $y \geq 0$ in Example 3). As we noted, the effect of this restriction is to limit the solution to Quadrant I and the axes bounding Quadrant I.

◑ EXAMPLE 4 *Land Management* **(Application Preview)**

A farm co-op has 6000 acres available to plant with corn and soybeans. Each acre of corn requires 9 gallons of fertilizer/herbicide and 3/4 hour of labor to harvest. Each acre of soybeans requires 3 gallons of fertilizer/herbicide and 1 hour of labor to harvest. The co-op has available at most 40,500 gallons of fertilizer/herbicide and at most 5250 hours of labor for harvesting. The number of acres of each crop is limited (constrained) by the available resources: land, fertilizer/herbicide, and labor for harvesting. Write the system of inequalities that describes the constraints and graph the solution region for the system.

Solution
If x represents the number of acres of corn and y represents the number of acres of soybeans, then the total acres planted is limited by $x + y \leq 6000$. The limitation on fertilizer/herbicide is given by $9x + 3y \leq 40,500$, and the labor constraint is given by $\frac{3}{4}x + y \leq 5250$. Because the number of acres planted in each crop must be nonnegative, the system of inequalities that describes the constraints is as follows.

$$x + y \leq 6,000 \qquad (1)$$
$$9x + 3y \leq 40,500 \qquad (2)$$
$$\frac{3}{4}x + y \leq 5,250 \qquad (3)$$
$$x \geq 0, y \geq 0$$

The intercepts of the lines associated with the inequalities are:

(6000, 0) and (0, 6000) for the line $x + y = 6000$ ⠀⠀(from inequality (1))

(4500, 0) and (0, 13,500) for the line $9x + 3y = 40,500$ ⠀⠀(from inequality (2))

(7000, 0) and (0, 5250) for the line $\frac{3}{4}x + y = 5250$ ⠀⠀(from inequality (3))

The solution region is shaded in Figure 4.6, with three of the corners at (0, 0), (4500, 0), and (0, 5250). The corner (3750, 2250) is found by solving simultaneously, as follows.

$$\begin{cases} 9x + 3y = 40,500 & (4) \\ x + y = 6,000 & (5) \end{cases}$$

$$\begin{cases} 9x + 3y = 40,500 & (4) \\ -(3x + 3y = 18,000) & [3 \times \text{Eq. (5)}] \end{cases}$$
$$\overline{\quad 6x \quad\quad = 22,500}$$

$$x = \frac{22,500}{6} = 3750$$

$$y = 6000 - x = 2250$$

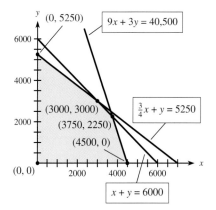

Figure 4.6

The corner (3000, 3000) is found similarly by solving simultaneously with $x + y = 6000$ and $\frac{3}{4}x + y = 5250$.

Any point in the shaded region of Figure 4.6 represents a possible number of acres of corn and of soybeans that the co-op could plant, treat with fertilizer/herbicide, and harvest.

1. Graph the region determined by the inequalities

$$2x + 3y \leq 12$$
$$4x + 2y \leq 16$$
$$x \geq 0, y \geq 0$$

2. Determine the corners of the region.

● EXAMPLE 5 *Manufacturing Constraints*

CDF Appliances has assembly plants in Atlanta and Fort Worth, where they produce a variety of kitchen appliances, including a 12-cup coffee maker and a cappuccino machine. At the Atlanta plant, 160 of the 12-cup models and 200 of the cappuccino machines can be assembled each hour. At the Fort Worth plant, 800 of the 12-cup models and 200 of the cappuccino machines can be assembled each hour. CDF Appliances expects orders for at least 64,000 of the 12-cup models and at least 40,000 of the cappuccino machines. At each plant, the number of assembly hours available for these two appliances is constrained by each plant's capacity and the need to fill the orders. Write the system of inequalities that describes these assembly plant constraints, and graph the solution region for the system.

Solution
Let x be the number of assembly hours at the Atlanta plant, and let y be the number of assembly hours at the Fort Worth plant. The production capabilities of each facility and the anticipated orders are summarized in the following table.

	Atlanta	Fort Worth	Needed
12-cup	160/hr.	800/hr.	At least 64,000
Cappuccino machine	200/hr.	200/hr.	At least 40,000

This table and the fact that both x and y must be nonnegative gives the following constraints.

$$160x + 800y \geq 64,000$$
$$200x + 200y \geq 40,000$$
$$x \geq 0, y \geq 0$$

As before, the solution is restricted to Quadrant I and the axes that bound Quadrant I. The equation $160x + 800y = 64,000$ can be graphed with x- and y-intercepts. The test point $(0, 0)$ does not satisfy $160x + 800y \geq 64,000$, so the region lies above the line (see Figure 4.7(a)). In like manner, the region satisfying $200x + 200y \geq 40,000$ lies above the line $200x + 200y = 40,000$, and the solution region for this system is shown in Figure 4.7(b).

The corner $(150, 50)$ is found by solving simultaneously as follows.

$$\begin{cases} 160x + 800y = 64,000 \\ 200x + 200y = 40,000 \end{cases} \quad \text{is equivalent to} \quad \begin{cases} x + 5y = 400 & (1) \\ x + y = 200 & (2) \end{cases}$$

Finding equation (1) minus equation (2) gives one equation in one variable and allows us to complete the solution.

$$4y = 200, \quad \text{so} \quad y = 200/4 = 50 \quad \text{and} \quad x = 200 - y = 150$$

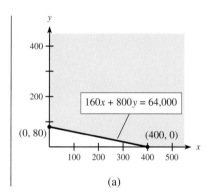
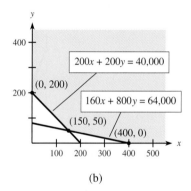

Figure 4.7   (a)   (b)

Many computer programs save considerable time and energy in determining regions that satisfy a system of inequalities. Graphing calculators can also be used in the graphical solution of a system of inequalities.

● **EXAMPLE 6** *Solution Region with Technology*

Use a graphing utility to find the following.

(a) Find the region determined by the inequalities below.

$$5x + 2y \leq 54$$
$$2x + 4y \leq 60$$
$$x \geq 0, y \geq 0$$

(b) Find the corners of this region.

Solution

(a) We write the inequalities above as equations, solved for y.

$$5x + 2y = 54 \Rightarrow 2y = 54 - 5x$$
$$\Rightarrow y = 27 - \frac{5}{2}x$$
$$2x + 4y = 60 \Rightarrow 4y = 60 - 2x$$
$$\Rightarrow y = 15 - \frac{1}{2}x$$

Graphing these equations with a graphing calculator and using shading shows the region satisfying the inequalities (see Figure 4.8).

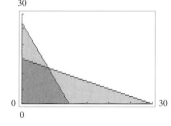

Figure 4.8

(b) Some graphing utilities can determine the points of intersection of two equations with an INTERSECT command. This command can be used to find the intersection of the pair of equations from (a). With other graphing utilities, using the command SOLVER or INTERSECT with pairs of lines that form the borders of this region will give the points where the boundaries intersect. These points, $(0, 0)$, $(0, 15)$, $(6, 12)$, and $(10.8, 0)$, can also be found algebraically. These points are the corners of the solution region. These corners will be important to us in the graphical solutions of linear programming problems in the next section.

● *Checkpoint Solutions*

1.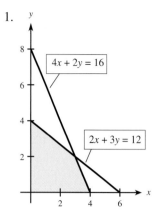

2. The graph shows corners at $(0, 0)$, $(0, 4)$, and $(4, 0)$. A corner also occurs where $2x + 3y = 12$ and $4x + 2y = 16$ intersect. The point of intersection is $x = 3$, $y = 2$, so $(3, 2)$ is a corner.

4.1 Exercises

In Problems 1–6, graph each inequality.

1. $y \le 2x - 1$

2. $y \ge 4x - 5$

3. $\dfrac{x}{2} + \dfrac{y}{4} < 1$

4. $x - \dfrac{y}{3} < \dfrac{-2}{3}$

5. $0.4x \ge 0.8$

6. $\dfrac{-y}{8} > \dfrac{1}{4}$

In Problems 7–12, the graph of the boundary equations for each system of inequalities is shown with that system. Locate the solution region and find the corners.

7. $\begin{cases} x + 4y \le 60 \\ 4x + 2y \le 100 \\ \qquad x \ge 0 \\ \qquad y \ge 0 \end{cases}$

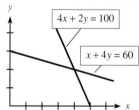

8. $\begin{cases} 4x + 3y \le 240 \\ 5x - \ y \le 110 \\ \qquad x \ge 0 \\ \qquad y \ge 0 \end{cases}$

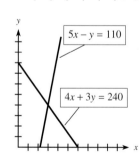

9. $\begin{cases} -3x + 4y \le 18 \\ \ 2x + \ y \ge 10 \\ \ \ x + \ y \le 15 \\ \ x \ge 0, y \ge 0 \end{cases}$

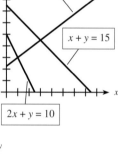

10. $\begin{cases} x + \ y \le 25 \\ 4x + \ y \le 40 \\ 3x + 8y \ge 88 \\ x \ge 0, y \ge 0 \end{cases}$

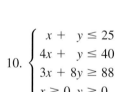

11. $\begin{cases} x + 3y \ge 6 \\ 2x + 4y \ge 10 \\ 3x + y \ge 5 \\ x \ge 0 \\ y \ge 0 \end{cases}$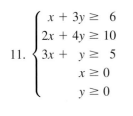

12. $\begin{cases} x + 4y \ge 10 \\ 4x + 2y \ge 10 \\ x + y \ge 4 \\ x \ge 0 \\ y \ge 0 \end{cases}$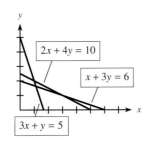

In Problems 13–26, graph the solution of each system of inequalities.

13. $\begin{cases} y < 2x \\ y > x - 1 \end{cases}$

14. $\begin{cases} y > 3x - 4 \\ y < 2x + 3 \end{cases}$

15. $\begin{cases} 2x + y < 3 \\ x - 2y \ge -1 \end{cases}$

16. $\begin{cases} 3x + y > 4 \\ x - 2y < -1 \end{cases}$

17. $\begin{cases} x + 5y \le 200 \\ 2x + 3y \le 134 \\ x \ge 0, y \ge 0 \end{cases}$

18. $\begin{cases} -x + y \le 2 \\ x + 2y \le 10 \\ 3x + y \le 15 \\ x \ge 0, y \ge 0 \end{cases}$

19. $\begin{cases} x + 2y \le 48 \\ x + y \le 30 \\ 2x + y \le 50 \\ x \ge 0, y \ge 0 \end{cases}$

20. $\begin{cases} 3x + y \le 9 \\ 3x + 2y \le 12 \\ x + 2y \le 8 \\ x \ge 0, y \ge 0 \end{cases}$

21. $\begin{cases} x + 2y \ge 19 \\ 3x + 2y \ge 29 \\ x \ge 0, y \ge 0 \end{cases}$

22. $\begin{cases} 4x + y \ge 12 \\ x + y \ge 9 \\ x + 3y \ge 15 \\ x \ge 0, y \ge 0 \end{cases}$

23. $\begin{cases} x + 3y \ge 3 \\ 2x + 3y \ge 5 \\ 2x + y \ge 3 \\ x \ge 0, y \ge 0 \end{cases}$

24. $\begin{cases} x + 2y \ge 10 \\ 2x + y \ge 11 \\ x + y \ge 9 \\ x \ge 0, y \ge 0 \end{cases}$

25. $\begin{cases} x + 2y \ge 20 \\ -3x + 2y \le 4 \\ x \ge 12 \\ x \ge 0, y \ge 0 \end{cases}$

26. $\begin{cases} 3x + 2y \ge 75 \\ -3x + 5y \ge 30 \\ y \le 40 \\ x \ge 0, y \ge 0 \end{cases}$

APPLICATIONS

27. *Management* The Wellbuilt Company produces two types of wood chippers, economy and deluxe. The deluxe model requires 3 hours to assemble and 1/2 hour to paint, and the economy model requires 2 hours to assemble and 1 hour to paint. The maximum number of assembly hours available is 24 per day, and the maximum number of painting hours available is 8 per day.
    (a) Write the system of inequalities that describes the constraints on the number of each type of wood chipper produced. Begin by identifying what x and y represent.
    (b) Graph the solution of the system of inequalities and find the corners of the solution region.

28. *Learning environments* An experiment that involves learning in animals requires placing white mice and rabbits into separate, controlled environments, environment I and environment II. The maximum amount of time available in environment I is 500 minutes, and the maximum amount of time available in environment II is 600 minutes. The white mice must spend 10 minutes in environment I and 25 minutes in environment II, and the rabbits must spend 15 minutes in environment I and 15 minutes in environment II.
    (a) Write a system of inequalities that describes the constraints on the number of each type of animal used in the experiment. Begin by identifying what x and y represent.
    (b) Graph the solution of the system of inequalities and find the corners of the solution region.

29. *Manufacturing* A company manufactures two types of electric hedge trimmers, one of which is cordless. The cord-type trimmer requires 2 hours to make, and the cordless model requires 4 hours. The company has only 800 work hours to use in manufacturing each day, and the packaging department can package only 300 trimmers per day.
    (a) Write the inequalities that describe the constraints on the number of each type of hedge trimmer produced. Begin by identifying what x and y represent.
    (b) Graph the region determined by these constraint inequalities.

30. *Manufacturing* A firm manufactures bumper bolts and fender bolts for antique cars. One machine can produce 130 fender bolts per hour, and another machine can produce 120 bumper bolts per hour. The combined

number of fender bolts and bumper bolts the packaging department can handle is 230 per hour.

   (a) Write the inequalities that describe the constraints on the number of each type of bolt produced. Begin by identifying what x and y represent.

   (b) Graph the region determined by these constraint inequalities.

31. *Advertising* Apex Motors manufactures luxury cars and sport utility vehicles. The most likely customers are high-income men and women, and company managers want to initiate an advertising campaign targeting these groups. They plan to run 1-minute spots on business/investment programs, where they can reach 7 million women and 4 million men from their target groups. They also plan 1-minute spots during sporting events, where they can reach 2 million women and 12 million men from their target groups. Apex feels that the ads must reach at least 30 million women and at least 28 million men who are prospective customers.

   (a) Write the inequalities that describe the constraints on the number of each type of 1-minute spots needed to reach these target groups.

   (b) Graph the region determined by these constraint inequalities.

32. *Manufacturing* The Video Star Company makes two different models of DVD players, which are assembled on two different assembly lines. Line 1 can assemble 30 units of the Star model and 40 units of the Prostar model per hour, and Line 2 can assemble 150 units of the Star model and 40 units of the Prostar model per hour. The company needs to produce at least 270 units of the Star model and 200 units of the Prostar model to fill an order.

   (a) Write the inequalities that describe the production constraints on the number of each type of DVD player needed to fill the order.

   (b) Graph the region determined by these constraint inequalities.

33. *Politics* A candidate wishes to use a combination of radio and television advertisements in her campaign. Research has shown that each 1-minute spot on television reaches 0.09 million people and each 1-minute spot on radio reaches 0.006 million. The candidate feels she must reach at least 2.16 million people, and she must buy a total of at least 80 minutes of advertisements.

   (a) Write the inequalities that relate the number of each type of advertising to her needs.

   (b) Graph the region determined by these constraint inequalities.

34. *Nutrition* In a hospital ward, the patients can be grouped into two general categories depending on their condition and the amount of solid foods they require in their diet. A combination of two diets is used for solid foods because they supply essential nutrients for recovery. The table below summarizes the patient groups and their minimum daily requirements.

	Diet A	Diet B	Daily Requirement
Group 1	4 oz per serving	1 oz per serving	26 oz
Group 2	2 oz per serving	1 oz per serving	18 oz

   (a) Write the inequalities that describe how many servings of each diet are needed to provide the nutritional requirements.

   (b) Graph the region determined by these constraint inequalities.

35. *Manufacturing* A sausage company makes two different kinds of hot dogs, regular and all beef. Each pound of all-beef hot dogs requires 0.75 lb of beef and 0.2 lb of spices, and each pound of regular hot dogs requires 0.18 lb of beef, 0.3 lb of pork, and 0.2 lb of spices. Suppliers can deliver at most 1020 lb of beef, at most 600 lb of pork, and at least 500 lb of spices.

   (a) Write the inequalities that describe how many pounds of each type of hot dog can be produced.

   (b) Graph the region determined by these constraint inequalities.

36. *Manufacturing* A cereal manufacturer makes two different kinds of cereal, Senior Citizen's Feast and Kids Go. Each pound of Senior Citizen's Feast requires 0.6 lb of wheat and 0.2 lb of vitamin-enriched syrup, and each pound of Kids Go requires 0.4 lb of wheat, 0.2 lb of sugar, and 0.2 lb of vitamin-enriched syrup. Suppliers can deliver at most 2800 lb of wheat, at most 800 lb of sugar, and at least 1000 lb of the vitamin-enriched syrup.

   (a) Write the inequalities that describe how many pounds of each type of cereal can be made.

   (b) Graph the region determined by these constraint inequalities.

4.2

Linear Programming: Graphical Methods

OBJECTIVE

● To use graphical methods to find the optimal value of a linear function subject to constraints

◗ Application Preview

Many practical problems in business and economics involve complex relationships among capital, raw materials, labor, and so forth. Consider the following example.

A farm co-op has 6000 acres available to plant with corn and soybeans. Each acre of corn requires 9 gallons of fertilizer/herbicide and 3/4 hour of labor to harvest. Each acre of soybeans requires 3 gallons of fertilizer/herbicide and 1 hour of labor to harvest. The co-op has available at most 40,500 gallons of fertilizer/herbicide and at most 5250 hours of labor for harvesting. If the profits per acre are \$60 for corn and \$40 for soybeans, how many acres of each crop should the co-op plant in order to maximize their profit? What is the maximum profit? (See Example 1.)

The farm co-op's maximum profit can be found by using a mathematical technique called **linear programming.** Linear programming can be used to solve problems such as this if the limits on the variables (called **constraints**) can be expressed as linear inequalities and if the function that is to be maximized or minimized (called the **objective function**) is a linear function.

Feasible Regions and Solutions

Linear programming is widely used by businesses for problems that involve many variables (sometimes more than 100). In this section we begin our study of this important technique by considering problems involving two variables. With two variables we can use graphical methods to help solve the problem. The constraints form a system of linear inequalities in two variables that we can solve by graphing. The solution of the system of constraint inequalities determines a region, any point of which may yield the *optimal* (maximum or minimum) value for the objective function.* Hence any point in the region determined by the constraints is called a **feasible solution,** and the region itself is called the **feasible region.** In a linear programming problem, we seek the feasible solution that maximizes (or minimizes) the objective function.

◗ EXAMPLE 1 *Profit Maximization* (Application Preview)

A farm co-op has 6000 acres available to plant with corn and soybeans. Each acre of corn requires 9 gallons of fertilizer/herbicide and 3/4 hour of labor to harvest. Each acre of soybeans requires 3 gallons of fertilizer/herbicide and 1 hour of labor to harvest. The co-op has available at most 40,500 gallons of fertilizer/herbicide and at most 5250 hours of labor for harvesting. If the profits per acre are \$60 for corn and \$40 for soybeans, how many acres of each crop should the co-op plant in order to maximize their profit? What is the maximum profit?

Solution

Recall from Example 4 in Section 4.1 that if x is the number of acres of corn and y is the number of acres of soybeans, then the constraints for this farm co-op application can be written as the following system of inequalities.

$$x + y \leq 6,000$$
$$9x + 3y \leq 40,500$$
$$\frac{3}{4}x + y \leq 5,250$$
$$x \geq 0, y \geq 0$$

---

*The region determined by the constraints must be *convex* for the optimal to exist. A convex region is one such that for any two points in the region, the segment joining those points lies entirely within the region. We restrict our discussion to convex regions.

Because the profit P is \$60 for each of the x acres of corn and \$40 for each of the y acres of soybeans, the function that describes the co-op's profit is given by

$$\text{Profit} = P = 60x + 40y \quad \text{(in dollars)}$$

Thus the linear programming problem for the co-op can be stated as follows.

$$\text{Maximize Profit} \quad P = 60x + 40y$$
$$\text{Subject to:} \quad x + y \le 6{,}000$$
$$9x + 3y \le 40{,}500$$
$$\frac{3}{4}x + y \le 5{,}250$$
$$x \ge 0, y \ge 0$$

The solution of the system of inequalities, or constraints (found in Example 4 in Section 4.1), forms the feasible region shaded in Figure 4.9(a). *Any* point inside the shaded region or on its boundary is a feasible (possible) solution of the problem. For example, point A (1000, 2000) is in the feasible region, and at this point the profit is $P = 60(1000) + 40(2000) = 140{,}000$ dollars.

To find the maximum value of $P = 60x + 40y$, we cannot possibly evaluate P at every point in the feasible region. However, many points in the feasible region may correspond to the same value of P. For example, at point A in Figure 4.9(a), the value of P is 140,000 and the profit function becomes

$$60x + 40y = 140{,}000$$

or

$$y = \frac{140{,}000 - 60x}{40} = 3500 - \frac{3}{2}x$$

The graph of this function is a line with slope $m = -3/2$ and y-intercept 3500. Many points in the feasible region lie on this line (see Figure 4.9(b)), and their coordinates all result in a profit $P = 140{,}000$. Any point in the feasible region that results in profit P satisfies

$$P = 60x + 40y$$

or

$$y = \frac{P - 60x}{40} = \frac{P}{40} - \frac{3}{2}x$$

In this form we can see that different P-values change the y-intercept for the line but the slope is always $m = -3/2$, so the lines for different P-values are parallel. Figure 4.9(b) shows the feasible region and the lines representing the objective function for $P = 140{,}000$, $P = 315{,}000$, and $P = 440{,}000$. Note that the line corresponding to $P = 315{,}000$ intersects the feasible region at the point (3750, 2250). Values of P less than 315,000 give lines that pass through the feasible region, but represent a profit less than \$315,000 (such as the line for $P = 140{,}000$ through point A). Similarly, values of P greater than 315,000 give lines that miss the feasible region, and hence cannot be solutions of the problem.

Thus, the farm co-op's maximum profit, subject to the constraints (i.e., the solution of the co-op's linear programming problem), is $P = \$315{,}000$ when $x = 3750$ and $y = 2250$. That is, when the co-op plants 3750 acres of corn and 2250 acres of soybeans, it achieves maximum profit of \$315,000.

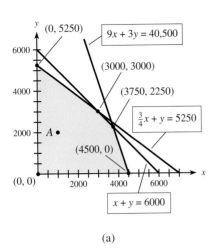
(a)

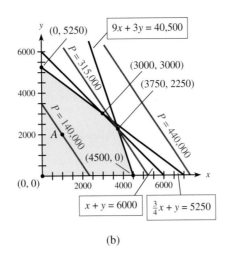
(b)

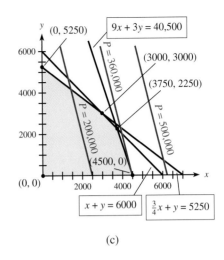
(c)

Figure 4.9

Notice in Example 1 that the objective function was maximized at one of the "corners" (vertices) of the feasible region. This is not a coincidence; it turns out that a corner will always lie on the profit line corresponding to the maximum profit.

Suppose in Example 1 that the objective had been to maximize $P = 80x + 20y$ over the same constraint region. Then the graphs of $P = 80x + 20y$, or $y = \dfrac{P}{20} - 4x$, for $P = 200{,}000$, $P = 360{,}000$, and $P = 500{,}000$ would result in the lines shown in Figure 4.9(c). Notice that the corner $(4500, 0)$ is the only feasible point where $P = 360{,}000$. This figure also shows that any P-value greater than $360{,}000$ will result in a line that misses the feasible region and that any other point inside the feasible region corresponds to a P-value less than $360{,}000$. Thus the maximum value is $P = 360{,}000$ and it occurs at the corner $(4500, 0)$. Note in both cases the objective function was maximized at one of the "corners" of the feasible region.

Solving Graphically

The feasible region in Figure 4.9 is an example of a **closed and bounded region** because it is entirely enclosed by, and includes, the lines associated with the constraints.

Solutions of Linear Programming Problems

1. When the feasible region for a linear programming problem is closed and bounded, the objective function has a maximum value and a minimum value.
2. When the feasible region is not closed and bounded, the objective function may have a maximum only, a minimum only, or no solution.
3. If a linear programming problem has a solution, then the optimal (maximum or minimum) value of an objective function occurs at a corner of the feasible region determined by the constraints.
4. If the objective function has its optimal value at two corners, then it also has that optimal value at any point on the line (boundary) connecting those two corners.

Thus, for a closed and bounded region, we can find the maximum or minimum value of the objective function by evaluating the function at each of the corners of the feasible region formed by the solution of the constraint inequalities. If the feasible region is not closed and bounded, we must check to make sure the objective function has an optimal value.

The steps involved in solving a linear programming problem are as follows.

Linear Programming (Graphical Method)

Procedure	Example
To find the optimal value of a function subject to constraints:	Find the maximum and minimum values of $C = 2x + 3y$ subject to the constraints $$\begin{cases} x + 2y \le 10 \\ 2x + y \le 14 \\ x \ge 0, y \ge 0 \end{cases}$$
1. Write the objective function and constraint inequalities from the problem.	1. Objective function: $C = 2x + 3y$ Constraints: $x + 2y \le 10$ $2x + y \le 14$ $x \ge 0, y \ge 0$
2. Graph the solution of the constraint system. (a) If the feasible region is closed and bounded, proceed to Step 3. (b) If the region is not closed and bounded, check whether an optimal value exists. If not, state this. If so, proceed to Step 3.	2. The constraint region is closed and bounded. See Figure 4.10. 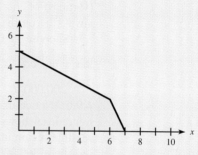 **Figure 4.10**
3. Find the corners of the resulting feasible region. This may require simultaneous solution of two or more pairs of boundary equations.	3. Corners are $(0, 0)$, $(0, 5)$, $(6, 2)$, $(7, 0)$.
4. Evaluate the objective function at each corner of the feasible region determined by the constraints.	4. At $(0, 0)$, $C = 2x + 3y = 2(0) + 3(0) = 0$ At $(0, 5)$, $C = 2x + 3y = 2(0) + 3(5) = 15$ At $(6, 2)$, $C = 2x + 3y = 2(6) + 3(2) = 18$ At $(7, 0)$, $C = 2x + 3y = 2(7) + 3(0) = 14$
5. If two corners give the optimal value of the objective function, then all points on the boundary line joining these two corners also optimize the function.	5. The function is maximized at $x = 6$, $y = 2$. The maximum value is $C = 18$. The function is minimized at $x = 0$, $y = 0$. The minimum value is $C = 0$.

● **EXAMPLE 2** *Maximizing Revenue*

Chairco manufactures two types of chairs, standard and plush. Standard chairs require 2 hours to construct and finish, and plush chairs require 3 hours to construct and finish. Upholstering takes 1 hour for standard chairs and 3 hours for plush chairs. There are 240 hours per day available for construction and finishing, and 150 hours per day are available for upholstering. If the revenue for standard chairs is $89 and for plush chairs is $133.50, how many of each type should be produced each day to maximize revenue?

Solution

Let x be the number of standard chairs produced each day, and let y be the number of plush chairs produced. Then the daily revenue function is given by $R = 89x + 133.5y$. There are constraints for construction and finishing (no more than 240 hours/day) and for upholstering (no more than 150 hours/day). Thus we have the following.

$$\text{Construction/finishing constraint:} \quad 2x + 3y \le 240$$
$$\text{Upholstering constraint:} \quad x + 3y \le 150$$

Because all quantities must be nonnegative, we also have the constraints $x \ge 0$ and $y \ge 0$. Thus we seek to solve the following problem.

Maximize $R = 89x + 133.5y$ subject to

$$\begin{cases} 2x + 3y \le 240 \\ x + 3y \le 150 \\ x \ge 0, y \ge 0 \end{cases}$$

The feasible set is the closed and bounded region shaded in Figure 4.11. The corners of the feasible region are $(0, 0)$, $(120, 0)$, $(0, 50)$, and $(90, 20)$. All of these are obvious except $(90, 20)$, which can be found by solving $2x + 3y = 240$ and $x + 3y = 150$ simultaneously. Testing the objective function at the corners gives the following.

At $(0, 0)$, $R = 89x + 133.5y = 89(0) + 133.5(0) = 0$

At $(120, 0)$, $R = 89(120) + 133.5(0) = 10{,}680$ ⎫

At $(0, 50)$, $R = 89(0) + 133.5(50) = 6675$ ⎬ Maximum at two corners

At $(90, 20)$, $R = 89(90) + 133.5(20) = 10{,}680$ ⎭

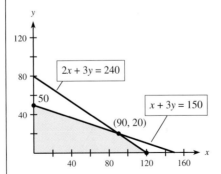

Figure 4.11

Thus the maximum revenue of \$10,680 occurs at either the point $(120, 0)$ or the point $(90, 20)$. This means that the revenue function will be maximized not only at these two corner points but also at any point on the segment joining them. Since the number of each type of chair must be an integer, Chairco has maximum revenue of \$10,680 at any pair of integer values along the segment joining $(120, 0)$ and $(90, 20)$. For example, the point $(105, 10)$ is on this segment, and the revenue at this point is also \$10,680:

$$89x + 133.5y = 89(105) + 133.5(10) = 10{,}680$$

● **Checkpoint**

1. Find the maximum and minimum values of the objective function $f = 4x + 3y$ on the shaded region in the figure, determined by the following constraints.

$$\begin{cases} 2x + 3y \leq 12 \\ 4x - 2y \leq 8 \\ x \geq 0, y \geq 0 \end{cases}$$

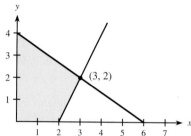

Although the examples so far have all involved closed and bounded regions, similar procedures apply for an unbounded region, although optimal solutions are no longer guaranteed.

● **EXAMPLE 3** *Minimization*

If possible, find the maximum and minimum values of $C = x + y$ subject to the constraints

$$\begin{cases} 3x + 2y \geq 12 \\ x + 3y \geq 11 \\ x \geq 0, y \geq 0 \end{cases}$$

Solution

The graph of the constraint system is shown in Figure 4.12(a). Note that the feasible region is not closed and bounded, so we must check whether optimal values exist. This check is done by graphing $C = x + y$ for selected values of C and noting the trend. Figure 4.12(b) shows the solution region with graphs of $C = x + y$ for $C = 3, C = 5,$ and $C = 8$. Note that the objective function has a minimum but no maximum.

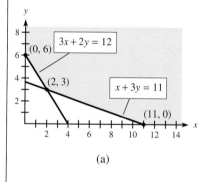

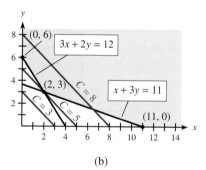

Figure 4.12

(a)

(b)

The corners $(0, 6)$ and $(11, 0)$ can be identified from the graph. The third corner, $(2, 3)$, can be found by solving the equations $3x + 2y = 12$ and $x + 3y = 11$ simultaneously:

$$\begin{array}{r} 3x + 2y = 12 \\ -3x - 9y = -33 \\ \hline -7y = -21 \\ y = 3 \\ x = 2 \end{array}$$

Examining the value of C at each corner point, we have

$$\text{At } (0, 6), \quad C = x + y = 6$$
$$\text{At } (11, 0), \quad C = x + y = 11$$
$$\text{At } (2, 3), \quad C = x + y = 5$$

Thus, we conclude the following:

$$\text{Minimum value of } C = x + y \text{ is 5 at } (2, 3).$$
$$\text{Maximum value of } C = x + y \text{ does not exist.}$$

Note that C can be made arbitrarily large in the feasible region.

● *Checkpoint*

2. Find the maximum and minimum values (if they exist) of the objective function $g = 3x + 4y$ subject to the following constraints.

$$x + 2y \geq 12, \quad x \geq 0$$
$$3x + 4y \geq 30, \quad y \geq 2$$

● **EXAMPLE 4** *Minimizing Production Costs*

Two chemical plants, one at Macon and one at Jonesboro, produce three types of fertilizer, low phosphorus (LP), medium phosphorus (MP), and high phosphorus (HP). At each plant, the fertilizer is produced in a single production run, so the three types are produced in fixed proportions. The Macon plant produces 1 ton of LP, 2 tons of MP, and 3 tons of HP in a single operation, and it charges $600 for what is produced in one operation, whereas one operation of the Jonesboro plant produces 1 ton of LP, 5 tons of MP, and 1 ton of HP, and it charges $1000 for what it produces in one operation. If a customer needs 100 tons of LP, 260 tons of MP, and 180 tons of HP, how many production runs should be ordered from each plant to minimize costs?

Solution

If x represents the number of operations requested from the Macon plant and y represents the number of operations requested from the Jonesboro plant, then we seek to minimize cost

$$C = 600x + 1000y$$

The following table summarizes production capabilities and requirements.

	Macon Plant	Jonesboro Plant	Requirements
Units of LP	1	1	100
Units of MP	2	5	260
Units of HP	3	1	180

Using the number of operations requested and the fact that requirements must be met or exceeded, we can formulate the following constraints.

$$\begin{cases} x + y \geq 100 \\ 2x + 5y \geq 260 \\ 3x + y \geq 180 \\ x \geq 0, y \geq 0 \end{cases}$$

Graphing this system gives the feasible set shown in Figure 4.13. The objective function has a minimum even though the feasible set is not closed and bounded. The corners are $(0, 180)$, $(40, 60)$, $(80, 20)$, and $(130, 0)$, where $(40, 60)$ is obtained by solving $x + y = 100$

and $3x + y = 180$ simultaneously, and where $(80, 20)$ is obtained by solving $x + y = 100$ and $2x + 5y = 260$ simultaneously.

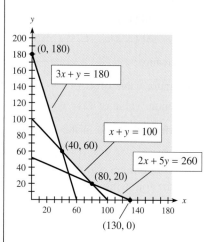

Figure 4.13

Evaluating $C = 600x + 1000y$ at each corner, we obtain

$$
\begin{aligned}
&\text{At } (0, 180), &&C = 180{,}000 \\
&\text{At } (40, 60), &&C = 84{,}000 \\
&\text{At } (80, 20), &&C = 68{,}000 \\
&\text{At } (130, 0), &&C = 78{,}000
\end{aligned}
$$

Thus, by placing orders requiring 80 production runs from the Macon plant and 20 production runs from the Jonesboro plant, the customer's needs will be satisfied at a minimum cost of $68,000.

● EXAMPLE 5 *Maximization Subject to Constraints*

Use a graphing utility to maximize $f = 5x + 11y$ subject to the constraints

$$
\begin{aligned}
5x + 2y &\le 54 \\
2x + 4y &\le 60 \\
x \ge 0,\ y &\ge 0
\end{aligned}
$$

Solution

We write the inequalities above as equations, solved for y. Graphing these equations with a graphing calculator and using shading show the closed and bounded region satisfying the inequalities (see Figure 4.14). By using an INTERSECT command with pairs of lines that form the borders of this region, we see that the boundaries intersect at $(0, 0)$, $(0, 15)$, $(6, 12)$, and $(10.8, 0)$. These points can also be found algebraically. Testing the objective function at each of these corners gives the following values of f:

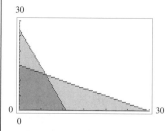

Figure 4.14

$$
\begin{aligned}
&\text{At } (0, 0), &&f = 5x + 11y = 0 \\
&\text{At } (0, 15), &&f = 5x + 11y = 165 \\
&\text{At } (6, 12), &&f = 5x + 11y = 162 \\
&\text{At } (10.8, 0), &&f = 5x + 11y = 54
\end{aligned}
$$

The maximum value is $f = 165$ at $x = 0$, $y = 15$.

● **Checkpoint Solutions**

1. The values of f at the corners are found as follows.

$$\text{At } (0, 0), \quad f = 0$$
$$\text{At } (2, 0), \quad f = 8$$
$$\text{At } (3, 2), \quad f = 12 + 6 = 18$$
$$\text{At } (0, 4), \quad f = 12$$

The maximum value of f is 18 at $x = 3, y = 2$, and the minimum value is $f = 0$ at $x = 0, y = 0$.

2. The graph of the feasible region is shown in Figure 4.15. The values of g at the corners are found as follows:

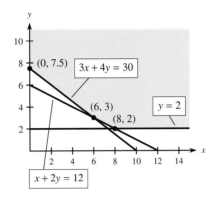

$$\text{At } (0, 7.5), \quad g = 30$$
$$\text{At } (6, 3), \quad g = 18 + 12 = 30$$
$$\text{At } (8, 2), \quad g = 24 + 8 = 32$$

Figure 4.15

The minimum value of g is 30 at both $(0, 7.5)$ and $(6, 3)$. Thus any point on the border joining $(0, 7.5)$ and $(6, 3)$ will give the minimum value 30. For example, $(2, 6)$ is on this border and gives the value $6 + 24 = 30$. The maximum value of g does not exist; g can be made arbitrarily large on this feasible region.

4.2 Exercises

In Problems 1–6, use the given feasible region determined by the constraint inequalities to find the maximum and minimum of the given objective function (if they exist).

1. $C = 2x + 3y$

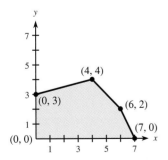

2. $f = 6x + 4y$

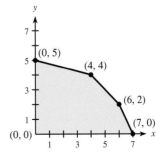

3. $C = 5x + 2y$

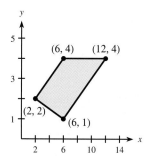

4. $C = 4x + 7y$

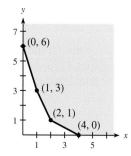

5. $f = 3x + 4y$

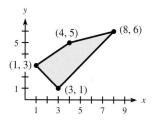

6. $f = 4x + 5y$

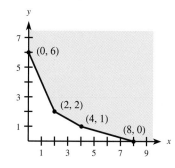

In each of Problems 7–10, the graph of the feasible region is shown. Find the corners of each feasible region, and then find the maximum and minimum of the given objective function (if they exist).

7. $f = 3x + 2y$

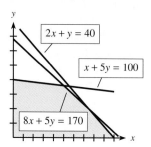

8. $f = 5x + 8y$

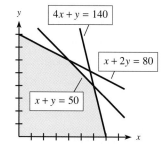

9. $g = 3x + 2y$

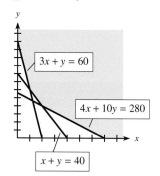

10. $g = x + 3y$

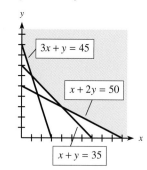

In Problems 11–14, find the indicated maximum or minimum value of the objective function in the linear programming problem. Note that the feasible regions for these problems are found in the answers to Problems 19, 20, 23, and 24 in the 4.1 Exercises.

11. Maximize $f = 3x + 2y$ subject to

$$x + 2y \le 48$$
$$x + y \le 30$$
$$2x + y \le 50$$
$$x \ge 0, y \ge 0$$

12. Maximize $f = 7x + 10y$ subject to

$$3x + y \le 9$$
$$3x + 2y \le 12$$
$$x + 2y \le 8$$
$$x \ge 0, y \ge 0$$

13. Minimize $g = 12x + 48y$ subject to

$$x + 3y \ge 3$$
$$2x + 3y \ge 5$$
$$2x + y \ge 3$$
$$x \ge 0, y \ge 0$$

14. Minimize $g = 12x + 8y$ subject to

$$x + 2y \ge 10$$
$$2x + y \ge 11$$
$$x + y \ge 9$$
$$x \ge 0, y \ge 0$$

In Problems 15–26, solve the following linear programming problems.

15. Maximize $f = 3x + 4y$ subject to

$$x + y \le 6$$
$$2x + y \le 10$$
$$y \le 4$$
$$x \ge 0, y \ge 0$$

16. Maximize $f = x + 3y$ subject to

$$x + 4y \le 12$$
$$y \le 2$$
$$x + y \le 9$$
$$x \ge 0, y \ge 0$$

17. Maximize $f = 2x + 6y$ subject to

$$x + y \le 7$$
$$2x + y \le 12$$
$$x + 3y \le 15$$
$$x \ge 0, y \ge 0$$

18. Maximize $f = 4x + 2y$ subject to

$$x + 2y \le 20$$
$$x + y \le 12$$
$$4x + y \le 36$$
$$x \ge 0, y \ge 0$$

19. Minimize $g = 7x + 6y$ subject to

$$5x + 2y \ge 16$$
$$3x + 7y \ge 27$$
$$x \ge 0, y \ge 0$$

20. Minimize $g = 22x + 17y$ subject to

$$8x + 5y \ge 100$$
$$12x + 25y \ge 360$$
$$x \ge 0, y \ge 0$$

21. Minimize $g = 3x + y$ subject to

$$4x + y \ge 11$$
$$3x + 2y \ge 12$$
$$x \ge 0, y \ge 0$$

22. Minimize $g = 50x + 70y$ subject to

$$11x + 15y \ge 225$$
$$x + 3y \ge 27$$
$$x \ge 0, y \ge 0$$

23. Maximize $f = x + 2y$ subject to

$$x + y \ge 4$$
$$2x + y \le 8$$
$$y \le 4$$

24. Maximize $f = 3x + 5y$ subject to

$$2x + 4y \ge 8$$
$$3x + y \le 7$$
$$x \ge 0, y \le 4$$

25. Minimize $g = 40x + 25y$ subject to

$$x + y \ge 100$$
$$-x + y \le 20$$
$$-2x + 3y \ge 30$$
$$x \ge 0, y \ge 0$$

26. Minimize $g = 3x + 8y$ subject to

$$4x - 5y \ge 50$$
$$-x + 2y \ge 4$$
$$x + y \le 80$$
$$x \ge 0, y \ge 0$$

APPLICATIONS

27. *Manufacturing*   The Wellbuilt Company produces two types of wood chippers, economy and deluxe. The deluxe model requires 3 hours to assemble and $\frac{1}{2}$ hour to paint, and the economy model requires 2 hours to assemble and 1 hour to paint. The maximum number of assembly hours available is 24 per day, and the maximum number of painting hours available is 8 per day. If the profit on the deluxe model is $15 per unit and the profit on the economy model is $12 per unit, how many units of each model will maximize profit? (See Problem 27 in the 4.1 Exercises.)

28. *Learning environments*   An experiment involving learning in animals requires placing white mice and rabbits into separate, controlled environments, environment I and environment II. The maximum amount of time available in environment I is 500 minutes, and the maximum amount of time available in environment II is 600 minutes. The white mice must spend 10 minutes in environment I and 25 minutes in environment II, and the rabbits must spend 15 minutes in environment I and 15 minutes in environment II. Find the maximum possible number of animals that can be used in the experiment, and find the number of white mice and the number of rabbits that can be used. (See Problem 28 in the 4.1 Exercises.)

29. *Manufacturing*   A company manufactures two types of electric hedge trimmers, one of which is cordless. The cord-type trimmer requires 2 hours to make, and the cordless model requires 4 hours. The company has only 800 work hours to use in manufacturing each day, and the packaging department can package only 300 trimmers per day. If the company sells the cord-type model for $22.50 and the cordless model for $45.00, how many of each type should it produce per day to maximize its sales? (See Problem 29 in the 4.1 Exercises.)

30. *Manufacturing*   A firm manufactures bumper bolts and fender bolts for antique cars. One machine can produce 130 fender bolts per hour, and another machine can produce 120 bumper bolts per hour. The combined number of fender bolts and bumper bolts the packaging department can handle is 230 per hour. How many of each type of bolt should the firm produce hourly to maximize its sales if fender bolts sell for $1 and bumper bolts sell for $2? (See Problem 30 in the 4.1 Exercises.)

31. *Politics*   A candidate wishes to use a combination of radio and television advertisements in her campaign. Research has shown that each 1-minute spot on television reaches 0.09 million people and that each 1-minute spot on radio reaches 0.006 million. The candidate feels she must reach at least 2.16 million people, and she must buy a total of at least 80 minutes of advertisements. How many minutes of each medium should be used to minimize costs if television costs $500/minute and radio costs $100/minute? (See Problem 33 in the 4.1 Exercises.)

32. *Nutrition*   In a hospital ward, the patients can be grouped into two general categories depending on their condition and the amount of solid foods they require in their diet. A combination of two diets is used for solid foods because they supply essential nutrients for recovery, but each diet has an amount of a substance deemed detrimental. The table summarizes the patient group, minimum diet requirements, and the amount of the detrimental substance. How many servings from each diet should be given each day in order to minimize the intake of this detrimental substance? (See Problem 34 in the 4.1 Exercises.)

	Diet A	Diet B	Daily Requirement
Group 1	4 oz per serving	1 oz per serving	26 oz
Group 2	2 oz per serving	1 oz per serving	18 oz
Detrimental substance	0.18 oz per serving	0.09 oz per serving	

33. *Production scheduling*   Newjet, Inc. manufactures inkjet printers and laser printers. The company has the capacity to make 70 printers per day and it has 120 hours of labor per day available. It takes 1 hour to make an inkjet printer and 3 hours to make a laser printer. The profits are $40 per inkjet printer and $60 per laser printer. Find the number of each type of printer that should be made to give maximum profit, and find the maximum profit.

34. *Production scheduling*   At one of its factories, a jeans manufacturer makes two styles: #891 and #917. Each pair of style-891 takes 10 minutes to cut out and 20 minutes to assemble and finish. Each pair of style-917 takes 10 minutes to cut out and 30 minutes to assemble and finish. The plant has enough workers to provide at most 7500 minutes per day for cutting and at most 19,500 minutes per day for assembly and finishing. The profit on each pair of style-891 is $6.00 and the profit on each pair of style-917 is $7.50. How many pairs of each style should be produced per day to obtain maximum profit? Find the maximum daily profit.

35. *Nutrition*   A privately owned lake contains two types of game fish, bass and trout. The owner provides two types of food, A and B, for these fish. Bass require 2 units of food A and 4 units of food B, and trout require 5 units of food A and 2 units of food B. If the owner has 800 units of each food, find the maximum number of fish that the lake can support.

36. *Nutrition*   In a zoo, there is a natural habitat containing several feeding areas. One of these areas serves as a feeding area for two species, I and II, and it is supplied each day with 120 pounds of food A, 110 pounds of food B, and 57 pounds of food C. Each individual of species I requires 5 lb of A, 5 lb of B, and 2 lb of C, and each individual of species II requires 6 lb of A, 4 lb of B, and 3 lb of C. Find the maximum number of these species that can be supported.

Shadow prices—land management   **For Problems 37 and 38, refer to the farm co-op application in Example 1 and rework that linear programming problem with the indicated changes in one constraint, then answer the questions.**

37. (a) If one more acre of land became available (for a total of 6001 acres), how would this change the co-op's planting strategy and its maximum profit?
    (b) Repeat part (a) if 8 more acres of land were available.
    (c) Based on parts (a) and (b), what is the profit value of each additional acre of land? This value is called the **shadow price** of an acre of land.

38. (a) If one more gallon of fertilizer/herbicide became available (for a total of 40,501 gallons), how would this change the co-op's planting strategy and its maximum profit?
    (b) Repeat part (a) if 8 more gallons of fertilizer/herbicide were available.
    (c) Based on parts (a) and (b), what is the profit value of each additional gallon of fertilizer/herbicide (that is, the **shadow price** of a gallon of fertilizer/herbicide)?

39. *Manufacturing*   Two factories produce three different types of kitchen appliances. The following table summarizes the production capacity, the number of each type of appliance ordered, and the daily operating costs for the factories. How many days should each factory operate to fill the orders at minimum cost? Find the minimum cost.

	Factory 1	Factory 2	Number Ordered
Appliance 1	80/day	20/day	1600
Appliance 2	10/day	10/day	500
Appliance 3	20/day	70/day	2000
Daily cost	$10,000	$20,000	

40. *Nutrition*   In a laboratory experiment, two separate foods are given to experimental animals. Each food contains essential ingredients, A and B, for which the animals have a minimum requirement, and each food also has an ingredient C, which can be harmful to the animals. The table below summarizes this information.

	Food 1	Food 2	Required
Ingredient A	10 units/g	3 units/g	49 units
Ingredient B	6 units/g	12 units/g	60 units
Ingredient C	3 units/g	1 unit/g	

How many grams of foods 1 and 2 should be given to the animals in order to satisfy the requirements for A and B while minimizing the amount of ingredient C ingested?

41. *Manufacturing*   The Janie Gioffre Drapery Company makes three types of draperies at two different locations. At location I, it can make 10 pairs of deluxe drapes, 20 pairs of better drapes, and 13 pairs of standard drapes per day. At location II, it can make 20 pairs of deluxe, 50 pairs of better, and 6 pairs of standard per day. The company has orders for 2000 pairs of deluxe drapes, 4200 pairs of better drapes, and 1200 pairs of standard drapes. If the daily costs are $500 per day at location I and $800 per day at location II, how many days should Janie schedule at each location in order to fill the orders at minimum cost? Find the minimum cost.

42. *Nutrition*   Two foods contain only proteins, carbohydrates, and fats. Food A costs $1 per pound and contains 30% protein and 50% carbohydrates. Food B costs $1.50 per pound and contains 20% protein and 75% carbohydrates. What combination of these two foods provides at least 1 pound of protein, $2\frac{1}{2}$ pounds of carbohydrates, and $\frac{1}{4}$ pound of fat at the lowest cost?

43. *Manufacturing*   A sausage company makes two different kinds of hot dogs, regular and all beef. Each pound of all-beef hot dogs requires 0.75 lb of beef and 0.2 lb of spices, and each pound of regular hot dogs

requires 0.18 lb of beef, 0.3 lb of pork, and 0.2 lb of spices. Suppliers can deliver at most 1020 lb of beef, at most 600 lb of pork, and at least 500 lb of spices. If the profit is $0.60 on each pound of all-beef hot dogs and $0.40 on each pound of regular hot dogs, how many pounds of each should be produced to obtain maximum profit? What is the maximum profit?

44. *Manufacturing*   A cereal manufacturer makes two different kinds of cereal, Senior Citizen's Feast and Kids Go. Each pound of Senior Citizen's Feast requires 0.6 lb of wheat and 0.2 lb of vitamin-enriched syrup, and each pound of Kids Go requires 0.4 lb of wheat, 0.2 lb of sugar, and 0.2 lb of vitamin-enriched syrup. Suppliers can deliver at most 2800 lb of wheat, at most 800 lb of sugar, and at least 1000 lb of the vitamin-enriched syrup. If the profit is $0.90 on each pound of Senior Citizen's Feast and $1.00 on each pound of Kids Go, find the number of pounds of each cereal that should be produced to obtain maximum profit. Find the maximum profit.

45. *Shipping costs*   TV Circuit has 30 large-screen televisions in a warehouse in Erie and 60 large-screen televisions in a warehouse in Pittsburgh. Thirty-five are needed in a store in Blairsville, and 40 are needed in a store in Youngstown. It costs $18 to ship from Pittsburgh to Blairsville and $22 to ship from Pittsburgh to Youngstown, whereas it costs $20 to ship from Erie to Blairsville and $25 to ship from Erie to Youngstown. How many televisions should be shipped from each warehouse to each store to minimize the shipping cost? *Hint:* If the number shipped from Pittsburgh to Blairsville is represented by x, then the number shipped from Erie to Blairsville is represented by $35 - x$.

46. *Construction*   A contractor builds two types of homes. The Carolina requires one lot, $160,000 capital, and 160 worker-days of labor, whereas the Savannah requires one lot, $240,000 capital, and 160 worker-days of labor. The contractor owns 300 lots and has $48,000,000 available capital and 43,200 worker-days of labor. The profit on the Carolina is $40,000 and the profit on the Savannah is $50,000. List the corner points of the feasible region and find how many of each type of home should be built to maximize profit. Find the maximum possible profit.

47. *Management*   A bank has two types of branches. A satellite branch employs 3 people, requires $100,000 to construct and open, and generates an average daily revenue of $10,000. A full-service branch employs 6 people, requires $140,000 to construct and open, and generates an average daily revenue of $18,000. The bank has up to $2.98 million available to open new branches, and has decided to limit the new branches to a maximum of 25 and to hire at most 120 new employees.

(a) How many branches of each type should the bank open in order to maximize the average daily revenue? Find the maximum average daily revenue.

(b) At the optimal solution from part (a), analyze the bank's constraints (number of new branches, number of new employees, and budget) to determine the "Amount Available," "Amount Used," and "Amount Not Used (Slack)."

(c) Obtaining additional quantities of which constraint items would have the potential to increase the bank's average daily revenue? Explain.

(d) Obtaining more of which constraint item would not increase average daily revenue? Explain.

48. *Manufacturing*   A company manufactures two different sizes of boat lifts. Each size requires some time in the welding and assembly department and some time in the parts and packaging department. The smaller lift requires $\frac{3}{4}$ hour in welding and assembly and $1\frac{2}{3}$ hours in parts and packaging. The larger lift requires $1\frac{1}{2}$ hours in welding and assembly and 1 hour in parts and packaging. The factory has 156 hours/day available in welding and assembly and 174 hours/day available in parts and packaging. Furthermore, daily demand for the lifts is at most 90 large and at most 100 small, and profits are $50 for each large lift and $30 for each small lift.

(a) How many of each type of boat lift should be manufactured each day in order to maximize the profit? Find the maximum profit.

(b) At the optimal solution in part (a), analyze the company's constraints (welding and assembly hours, parts and packaging hours, and demand) to determine the "Amount Available," "Amount Used," and "Amount Not Used (Slack)."

(c) Obtaining additional quantities of which constraint items would have the potential to increase the company's profit? Explain.

(d) Obtaining more of which constraint item would not increase profit? Explain.

The Simplex Method: Maximization

● To use the simplex method to maximize functions subject to constraints

◑ Application Preview

The Solar Technology Company manufactures three different types of hand calculators and classifies them as scientific, business, and graphing according to their calculating capabilities. The three types have production requirements given by the following table.

	Scientific	Business	Graphing
Electronic circuit components	5	7	10
Assembly time (hours)	1	3	4
Cases	1	1	1

The firm has a monthly limit of 90,000 circuit components, 30,000 hours of labor, and 9000 cases. If the profit is $6 for each scientific, $13 for each business, and $20 for each graphing calculator, the number of each that should be produced to yield maximum profit and the amount of that maximum possible profit can be found by using the **simplex method** of linear programming. (See Example 3.)

The Simplex Method   The graphical method for solving linear programming problems is practical only when there are two variables. If there are more than two variables, we could still find the corners of the convex region by simultaneously solving the equations of the boundaries. The objective function could then be evaluated at these corners. However, as the number of variables and the number of constraints increases, it becomes increasingly difficult to discover the corners and extremely time-consuming to evaluate the function. Furthermore, some problems have no solution, and this cannot be determined by using the method of solving the equations simultaneously.

The method discussed in this section is called the **simplex method** for solving linear programming problems. Basically, this method gives a systematic way of moving from one feasible corner of the convex region to another in such a way that the value of the objective function increases until an optimal value is reached or it is discovered that no solution exists.

The simplex method was developed by George Dantzig in 1947. One of the earliest applications of the method was to the scheduling problem that arose in connection with the Berlin airlift, begun in 1948. There the objective was to maximize the amount of goods delivered, subject to such constraints as the number of personnel, the number and size of available aircraft, and the number of runways. Since then, linear programming and the simplex method have been used to solve optimization problems in a wide variety of businesses. In fact, in a recent survey of Fortune 500 firms, 85% of the respondents indicated that they use linear programming.

In 1984, N. Karmarkar, a researcher at Bell Labs, developed a new method (that uses interior points rather than corner points) for solving linear programming problems. When the number of variables is very large, Karmarkar's method is more efficient than the simplex method. For example, the Military Airlift Command and Delta Airlines have realized enormous savings by using Karmarkar's method on problems with thousands of variables. Despite the efficiency of Karmarkar's method with huge linear programming problems, the simplex method is just as effective for many applications and is still widely used.

In discussing the simplex method, we initially restrict ourselves to linear programming problems that satisfy the following conditions, called **standard maximization problems.**

1. The objective function is to be maximized.
2. All variables are nonnegative.

3. The constraints are of the form

$$a_1x_1 + a_2x_2 + \cdots + a_nx_n \le b$$

where $b > 0$.

These conditions may seem to restrict the types of problems unduly, but in applied situations where the objective function is to be maximized, the constraints often satisfy conditions 2 and 3.

Before outlining a procedure for the simplex method, let us develop the procedure and investigate its rationale by seeing how the simplex method compares with the graphical method.

● EXAMPLE 1 *Maximum Profit*

A farm co-op has 6000 acres available to plant with corn and soybeans. Each acre of corn requires 9 gallons of fertilizer/herbicide and 3/4 hour of labor to harvest. Each acre of soybeans requires 3 gallons of fertilizer/herbicide and 1 hour of labor to harvest. The co-op has available at most 40,500 gallons of fertilizer/herbicide and at most 5250 hours of labor for harvesting. If the profits per acre are $60 for corn and $40 for soybeans, how many acres of each crop should the co-op plant in order to maximize their profit? What is the maximum profit?

Solution
This application was solved graphically in Example 1 in Section 4.2. As we saw there, with $x =$ the number of acres of corn and $y =$ the number of acres of soybeans, the farm co-op linear programming problem is

Maximize profit	$P = 60x + 40y$	
subject to	$x + y \le 6{,}000$	(Land constraint)
	$9x + 3y \le 40{,}500$	(Fertilizer/herbicide constraint)
	$\frac{3}{4}x + y \le 5{,}250$	(Labor hours constraint)

Note that $x \ge 0$ and $y \ge 0$. Because variables must be nonnegative in the simplex method, we shall no longer state these conditions.

The simplex method uses matrix methods on systems of equations, so we convert each constraint inequality to an equation by using a new variable, called a **slack variable.** We can think of each slack variable as representing the amount of the constraint left unused. In the land constraint, $x + y$ represents the total number of acres planted, and this number cannot exceed the 6000 acres available. Thus, if we let s_1 represent the unused (unplanted) acres of land, then $s_1 \ge 0$ and

$$x + y \le 6000 \qquad \text{becomes} \qquad x + y + s_1 = 6000$$

Similarly, if s_2 represents the unused gallons of fertilizer/herbicide, then $s_2 \ge 0$ and

$$9x + 3y \le 40{,}500 \qquad \text{becomes} \qquad 9x + 3y + s_2 = 40{,}500$$

And, if s_3 represents the unused hours of labor, then $s_3 \ge 0$ and

$$\frac{3}{4}x + y \le 5250 \qquad \text{becomes} \qquad \frac{3}{4}x + y + s_3 = 5250$$

Notice in each constraint that the expressions in x and y represent the amount of that resource that is used and the slack variable represents the amount not used.

Previously, when we used matrices with systems of equations we arranged the equations with all the variables on the left side of the equation and the constants on the right. If we write the objective function in this form

$$P = 60x + 40y \qquad \text{becomes} \qquad -60x - 40y + P = 0$$

and then the following system of equations describes this problem.

$$\begin{cases} x & + & y & + & s_1 & & & & & & = & 6{,}000 \\ 9x & + & 3y & & & + & s_2 & & & & = & 40{,}500 \\ \frac{3}{4}x & + & y & & & & & + & s_3 & & = & 5{,}250 \\ -60x & - & 40y & & & & & & & + & P & = & 0 \end{cases}$$

We seek to maximize P in the last equation subject to the constraints in the first three equations.

We can place this system of equations in a matrix called a **simplex matrix** or a **simplex tableau.** Note that the objective function is in the last (bottom) row of this matrix with all variables on the left side and that the rows above the function row correspond to the constraints.

$$A = \begin{bmatrix} x & y & s_1 & s_2 & s_3 & P & \\ 1 & 1 & 1 & 0 & 0 & 0 & 6{,}000 \\ 9 & 3 & 0 & 1 & 0 & 0 & 40{,}500 \\ \frac{3}{4} & 1 & 0 & 0 & 1 & 0 & 5{,}250 \\ \hline -60 & -40 & 0 & 0 & 0 & 1 & 0 \end{bmatrix} \begin{array}{l} \text{Land constraint} \\ \text{Fertilizer/herbicide constraint} \\ \text{Labor hours constraint} \\ \text{Objective function (maximum profit)} \end{array}$$

From the graph in Figure 4.16, we know that $(0, 0)$ is a feasible solution giving a value of 0 for P. Because the origin is always a feasible solution of the type of linear programming problems we are considering, the simplex method begins there and systematically moves to new corners while increasing the value of P.

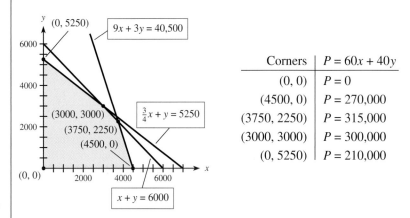

Corners	$P = 60x + 40y$
$(0, 0)$	$P = 0$
$(4500, 0)$	$P = 270{,}000$
$(3750, 2250)$	$P = 315{,}000$
$(3000, 3000)$	$P = 300{,}000$
$(0, 5250)$	$P = 210{,}000$

Figure 4.16

Note that if $x = 0$ and $y = 0$, then all 6000 acres of land are unused ($s_1 = 6000$), all 40,500 gallons of fertilizer/herbicide are unused ($s_2 = 40{,}500$), all 5250 labor hours are unused ($s_3 = 5250$), and profit is $P = 0$. From rows 1, 2, 3, and 4 of matrix A, we can read these values of s_1, s_2, s_3, and P, respectively.

Recall that the last row still corresponds to the objective function $P = 60x + 40y$. If we seek to improve (that is, increase the value of) $P = 60x + 40y$ by changing the value of only one of the variables, then because the coefficient of x is larger, the rate of increase is greatest by increasing x.

In general, we can find the variable that will improve the objective function most rapidly by looking for the *most negative* value in the last row (the objective function row) of the simplex matrix. The column that contains this value is called the **pivot column.**

The simplex matrix A is shown again below with the x-column indicated as the pivot column.

$$A = \begin{bmatrix} \begin{array}{cccccc|c} x & y & s_1 & s_2 & s_3 & P & \\ \boxed{1} & 1 & 1 & 0 & 0 & 0 & 6{,}000 \\ \boxed{9} & 3 & 0 & 1 & 0 & 0 & 40{,}500 \\ \frac{3}{4} & 1 & 0 & 0 & 1 & 0 & 5{,}250 \\ \boxed{-60} & -40 & 0 & 0 & 0 & 1 & 0 \end{array} \end{bmatrix} \quad \begin{array}{l} 6000 \div 1 = 6000 \\ 40{,}500 \div 9 = 4500 \quad \text{smallest} \\ 5250 \div \left(\frac{3}{4}\right) = 5250\left(\frac{4}{3}\right) = 7000 \end{array}$$

↑—— Most negative entry; pivot column

The amount by which x can be increased is limited by the constraining equations (rows 1, 2, and 3 of the simplex matrix). As the graph in Figure 4.16 shows, the constraints limit x to no larger than $x = 4500$. Note that when the positive coefficients in the pivot column (the x-column) of the simplex matrix are divided into the constants in the augment, the *smallest* quotient is $40{,}500 \div 9 = 4500$. This smallest quotient identifies the **pivot row.** The **pivot entry** (or **pivot**) is the entry in both the pivot column and the pivot row. The 9 (circled) in row 2, column 1 of matrix A is the pivot entry. Notice that the smallest quotient, 4500, represents the amount that x can be increased (while $y = 0$), and the point $(4500, 0)$ corresponds to a corner adjoining $(0, 0)$ on the constraint region (see Figure 4.16).

To continue this solution method, we use row operations with the pivot row of the simplex matrix to simulate moving to another corner of the constraint region. We multiply the second row by $\frac{1}{9}$ to convert the pivot entry to 1, and then add multiples of row 2 to each of the remaining rows to make all elements of the pivot column, except the pivot element, equal to 0.

$$A = \begin{bmatrix} \begin{array}{cccccc|c} x & y & s_1 & s_2 & s_3 & P & \\ 1 & 1 & 1 & 0 & 0 & 0 & 6{,}000 \\ \boxed{9} & 3 & 0 & 1 & 0 & 0 & 40{,}500 \\ \frac{3}{4} & 1 & 0 & 0 & 1 & 0 & 5{,}250 \\ -60 & -40 & 0 & 0 & 0 & 1 & 0 \end{array} \end{bmatrix} \xrightarrow{\frac{1}{9}R_2 \to R_2} \begin{bmatrix} \begin{array}{cccccc|c} x & y & s_1 & s_2 & s_3 & P & \\ 1 & 1 & 1 & 0 & 0 & 0 & 6{,}000 \\ \boxed{1} & \frac{1}{3} & 0 & \frac{1}{9} & 0 & 0 & 4{,}500 \\ \frac{3}{4} & 1 & 0 & 0 & 1 & 0 & 5{,}250 \\ -60 & -40 & 0 & 0 & 0 & 1 & 0 \end{array} \end{bmatrix}$$

$$\xrightarrow[\substack{-\frac{3}{4}R_2 + R_3 \to R_3 \\ 60R_2 + R_4 \to R_4}]{-R_2 + R_1 \to R_1} \begin{bmatrix} \begin{array}{cccccc|c} x & y & s_1 & s_2 & s_3 & P & \\ 0 & \frac{2}{3} & 1 & -\frac{1}{9} & 0 & 0 & 1{,}500 \\ 1 & \frac{1}{3} & 0 & \frac{1}{9} & 0 & 0 & 4{,}500 \\ 0 & \frac{3}{4} & 0 & -\frac{1}{12} & 1 & 0 & 1{,}875 \\ 0 & -20 & 0 & \frac{20}{3} & 0 & 1 & 270{,}000 \end{array} \end{bmatrix} = B$$

This "transforms" the simplex matrix A into the simplex matrix B. Recall that this transformation was based on holding $y = 0$ and increasing x as much as possible so the resulting point would be feasible; that is, moving from $(0, 0)$ to $(4500, 0)$ on the graph in Figure 4.16. At the point $(4500, 0)$ the profit is $P = 60x + 40y = 60(4500) + 40(0) = 270{,}000$, and the resources used (and not used, or slack) are as follows.

Land: $\qquad\qquad\qquad x + y + s_1 = 6000 \Rightarrow 4500 + 0 + s_1 = 6000 \Rightarrow s_1 = 1500$

Fertilizer/herbicide: $\quad 9x + 3y + s_2 = 40{,}500 \Rightarrow 40{,}500 + 0 + s_2 = 40{,}500 \Rightarrow s_2 = 0$

Labor hours: $\qquad\quad \dfrac{3}{4}x + y + s_3 = 5250 \Rightarrow 3375 + 0 + s_3 = 5250 \Rightarrow s_3 = 1875$

Let's see how these values ($x = 4500$, $y = 0$, $s_1 = 1500$, $s_2 = 0$, $s_3 = 1875$, and $P = 270{,}000$) can be found in matrix B.

Notice that the columns of B that correspond to the variables x, s_1, and s_3 and to the objective function P are all different, but each contains a single entry of 1 and the other entries are 0. Variables with columns of this type (but *not* the objective function) are called

basic variables. Similarly, the columns for variables y and s_2 do not have this special form, and these columns correspond to **nonbasic variables.** Notice also that the bottom row of B represents

$$-20y + \frac{20}{3}s_2 + P = 270,000, \quad \text{so} \quad P = 270,000 + 20y - \frac{20}{3}s_2$$

depends on y and s_2 (the nonbasic variables). If these nonbasic variables equal 0 (that is, $y = 0$ and $s_2 = 0$), then rows 1, 2, 3, and 4 of matrix B allow us to read the values of s_1, x, s_3, and P, respectively.

$$s_1 = 1500, \quad x = 4500, \quad s_3 = 1875, \quad P = 270,000$$

These solution values correspond to the feasible solution at the corner $(4500, 0)$ in Figure 4.16 and are called a **basic feasible solution.**

We also see that the value of $P = 270,000 + 20y - \frac{20}{3}s_2$ can be increased by increasing the value of y while holding the value of s_2 at 0. In matrix B, we can see this fact by observing that the most negative number in the last row (objective function row) occurs in the y-column. This means the y-column is the new pivot column.

The amount of increase in y is limited by the constraints. In matrix B we can discover the limitations on y by dividing the coefficient of y (if positive) into the constant term in the augment of each row and choosing the smallest nonnegative quotient. This identifies the pivot row, and hence the pivot entry, circled in matrix B.

$$B = \begin{array}{c} \\ \\ \\ \\ \end{array} \begin{array}{cccccc} x & y & s_1 & s_2 & s_3 & P \\ \end{array}$$

$$B = \begin{bmatrix} 0 & \tfrac{2}{3} & 1 & -\tfrac{1}{9} & 0 & 0 & 1,500 \\ 1 & \tfrac{1}{3} & 0 & \tfrac{1}{9} & 0 & 0 & 4,500 \\ 0 & \tfrac{3}{4} & 0 & -\tfrac{1}{12} & 1 & 0 & 1,875 \\ 0 & -20 & 0 & \tfrac{20}{3} & 0 & 1 & 270,000 \end{bmatrix}$$

$1500 \div (\tfrac{2}{3}) = 1500(\tfrac{3}{2}) = 2250$ smallest

$4500 \div (\tfrac{1}{3}) = 4500(\tfrac{3}{1}) = 13,500$

$1875 \div (\tfrac{3}{4}) = 1875(\tfrac{4}{3}) = 2500$

Largest negative; pivot column

The smallest quotient is 2250; note that if we move to a corner adjoining $(4500, 0)$ on the feasible region in Figure 4.16, the largest y-value is 2250 [at the point $(3750, 2250)$].

Let us transform the simplex matrix B by multiplying row 1 by $\frac{3}{2}$ to convert the pivot entry to 1, and then adding multiples of row 1 to the other rows to make all other entries in the y-column (pivot column) equal to 0. The new simplex matrix is C.

$$B = \begin{bmatrix} 0 & \tfrac{2}{3} & 1 & -\tfrac{1}{9} & 0 & 0 & 1,500 \\ 1 & \tfrac{1}{3} & 0 & \tfrac{1}{9} & 0 & 0 & 4,500 \\ 0 & \tfrac{3}{4} & 0 & -\tfrac{1}{12} & 1 & 0 & 1,875 \\ 0 & -20 & 0 & \tfrac{20}{3} & 0 & 1 & 270,000 \end{bmatrix} \xrightarrow{\tfrac{3}{2}R_1 \to R_1} \begin{bmatrix} 0 & 1 & \tfrac{3}{2} & -\tfrac{1}{6} & 0 & 0 & 2,250 \\ 1 & \tfrac{1}{3} & 0 & \tfrac{1}{9} & 0 & 0 & 4,500 \\ 0 & \tfrac{3}{4} & 0 & -\tfrac{1}{12} & 1 & 0 & 1,875 \\ 0 & -20 & 0 & \tfrac{20}{3} & 0 & 1 & 270,000 \end{bmatrix}$$

$$\begin{array}{c} -\tfrac{1}{3}R_1 + R_2 \to R_2 \\ -\tfrac{3}{4}R_1 + R_3 \to R_3 \\ \hline 20R_1 + R_4 \to R_4 \end{array} \xrightarrow{} \begin{bmatrix} 0 & 1 & \tfrac{3}{2} & -\tfrac{1}{6} & 0 & 0 & 2,250 \\ 1 & 0 & -\tfrac{1}{2} & \tfrac{1}{6} & 0 & 0 & 3,750 \\ 0 & 0 & -\tfrac{9}{8} & \tfrac{1}{24} & 1 & 0 & 187.5 \\ 0 & 0 & 30 & \tfrac{10}{3} & 0 & 1 & 315,000 \end{bmatrix} = C$$

In this new simplex matrix C, the basic variables (corresponding to the columns with a single entry of 1) are x, y, and s_3, the nonbasic variables are s_1 and s_2, and the last row (the objective function row) corresponds to

$$30s_1 + \frac{10}{3}s_2 + P = 315,000 \quad \text{or} \quad P = 315,000 - 30s_1 - \frac{10}{3}s_2$$

From the objective function we see that any increase in s_1 or s_2 will cause a *decrease* in P. Observing this or noting that all entries in the last row are nonnegative tells us that we have found the optimal value of the objective function and it occurs when $s_1 = 0$ and $s_2 = 0$. Using this, rows 1, 2, 3, and 4 of matrix C give the values for y, x, s_3, and P, respectively.

$$y = 2250, \quad x = 3750, \quad s_3 = 187.5, \quad P = 315,000, \quad \text{with} \quad s_1 = 0 \quad \text{and} \quad s_2 = 0$$

This corresponds to the corner (3750, 2250) on the graph and to the maximum value of $P = 315,000$ at this corner (see Figure 4.16).

Just as we found graphically, these values mean that the co-op should plant $x = 3750$ acres of corn and $y = 2250$ acres of soybeans for a maximum profit of $315,000. And this planting scheme results in $s_1 = 0$ unused acres of land, $s_2 = 0$ unused gallons of fertilizer/herbicide, and $s_3 = 187.5$ unused hours of labor.

As we review the simplex solution in Example 1, observe that each of the basic feasible solutions corresponded to a corner (x, y) of the feasible region and that two of the five variables x, y, s_1, s_2, and s_3 were equal to 0 at each feasible solution.

At (0, 0):	x and y were equal to 0.
At (4500, 0):	y and s_2 were equal to 0.
At (3750, 2250):	s_1 and s_2 were equal to 0.

See Figure 4.17 for an illustration of how the simplex method moved us from corner to corner until the optimal solution was obtained.

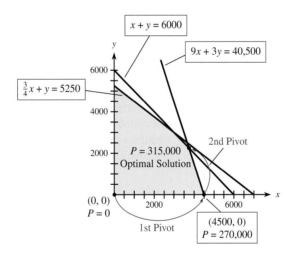

Figure 4.17

Recall from Example 1 that at each step in the simplex method we can identify the **basic variables** directly from the simplex matrix by noting that their columns contain a *single* entry of 1 with all other entries being zero. (Check this.) The remaining columns identify the **nonbasic variables,** and at each feasible solution (that is, at each step in the simplex method) the nonbasic variables were set equal to zero. With the nonbasic variables equal to zero, the values of the basic variables and the objective function can be read from the rows of the simplex matrix.

Simplex Method: Tasks and Procedure

The simplex method involves a series of decisions and operations using matrices. It involves three major tasks.

Simplex Method Tasks

Task A: Setting up the matrix for the simplex method.
Task B: Determining necessary operations and implementing those operations to reach a solution.
Task C: Reading the solution from the simplex matrix.

From Example 1, it should be clear that the pivot selection and row operations of Task B are designed to move the system to the corner that achieves the most rapid increase in the function and at the same time is sensitive to the nonnegative limitations on the variables. The procedure that combines these three tasks to maximize a function is given below.

Linear Programming (Simplex Method)

Procedure	Example

To use the simplex method to solve linear programming problems:

Maximize $f = 2x + 3y$ subject to
$$x + 2y \leq 10$$
$$2x + y \leq 14$$

1. Use a different slack variable to write each constraint as an equation, with positive constants on the right side.

1. $x + 2y \leq 10$ becomes $x + 2y + s_1 = 10$.
$2x + y \leq 14$ becomes $2x + y + s_2 = 14$.
Maximize f in $-2x - 3y + f = 0$.

2. Set up the simplex matrix. Put the objective function in the last row with all variables on the left and coefficient 1 for f.

2.
$$\begin{array}{ccccc} x & y & s_1 & s_2 & f \\ \end{array}$$
$$\left[\begin{array}{ccccc|c} 1 & 2 & 1 & 0 & 0 & 10 \\ 2 & 1 & 0 & 1 & 0 & 14 \\ \hline -2 & -3 & 0 & 0 & 1 & 0 \end{array}\right]$$

3. Find the pivot entry:
   (a) The pivot column has the most negative number in the last row. (If a tie occurs, use either column.)

3. (a)
$$\left[\begin{array}{ccccc|c} 1 & \boxed{2} & 1 & 0 & 0 & 10 \\ 2 & 1 & 0 & 1 & 0 & 14 \\ \hline -2 & \boxed{-3} & 0 & 0 & 1 & 0 \end{array}\right]$$
↑— Most negative entry; pivot column

   (b) The positive coefficient in the pivot column that gives the smallest nonnegative quotient when divided into the constant in the constraint rows of the augment is the pivot entry. If there is a tie, either coefficient may be chosen. If there are no positive coefficients in the pivot column, no solution exists. Stop.

(b) Row 1 quotient: $\dfrac{10}{2} = 5$

Row 2 quotient: $\dfrac{14}{1} = 14$

Therefore, pivot row is row 1.
$$\left[\begin{array}{ccccc|c} 1 & ② & 1 & 0 & 0 & 10 \\ 2 & 1 & 0 & 1 & 0 & 14 \\ \hline -2 & -3 & 0 & 0 & 1 & 0 \end{array}\right]$$
Pivot entry is 2, circled.

4. Use row operations with only the row containing the pivot entry to make the pivot entry a 1 and all other entries in the pivot column zeros. This makes the variable of the pivot column a basic variable. This process is called **pivoting.**

4. Row operations:
Multiply row 1 by $\frac{1}{2}$.
Add -1 times new row 1 to row 2.
Add 3 times new row 1 to row 3.

Result:
$$\left[\begin{array}{ccccc|c} \frac{1}{2} & 1 & \frac{1}{2} & 0 & 0 & 5 \\ \frac{3}{2} & 0 & -\frac{1}{2} & 1 & 0 & 9 \\ \hline -\frac{1}{2} & 0 & \frac{3}{2} & 0 & 1 & 15 \end{array}\right]$$
‾‾‾‾‾‾
Indicators

(continued)

Linear Programming (Simplex Method) *(continued)*

Procedure	Example

5. The numerical entries in the last row are the indicators of what to do next.
   (a) If there is a negative indicator, return to step 3.
   (b) If all indicators are positive or zero, an optimum value has been obtained for the objective function.
   (c) In a complete solution, the variables that have positive entries in the last row are nonbasic variables.

5. $-\frac{1}{2}$ is negative, so we identify a new pivot column and reduce again.

$$\begin{bmatrix} \frac{1}{2} & 1 & \frac{1}{2} & 0 & 0 & | & 5 \\ \textcircled{\frac{3}{2}} & 0 & -\frac{1}{2} & 1 & 0 & | & 9 \\ -\frac{1}{2} & 0 & \frac{3}{2} & 0 & 1 & | & 15 \end{bmatrix}$$

$5 \div \frac{1}{2} = 10$

$9 \div \frac{3}{2} = 6^*$

Pivot column   *6 is smaller quotient.

The new pivot entry is $\frac{3}{2}$ (circled). Now reduce, using the new pivot row, to obtain:

$$\begin{bmatrix} 0 & 1 & \frac{2}{3} & -\frac{1}{3} & 0 & | & 2 \\ 1 & 0 & -\frac{1}{3} & \frac{2}{3} & 0 & | & 6 \\ 0 & 0 & \frac{4}{3} & \frac{1}{3} & 1 & | & 18 \end{bmatrix}$$

Indicators all 0 or positive, so the solution is complete. Nonbasic variables: s_1 and s_2. Basic variables: x and y

6. After setting the nonbasic variables equal to 0, read the values of the basic variables and the objective function from the rows of the matrix.

6. f is maximized at 18 when $y = 2$, $x = 6$, $s_1 = 0$, and $s_2 = 0$. (This same example was solved graphically in the Procedure/Example in Section 4.2.)

As we look back over our work, we can observe the following:

1. Steps 1 and 2 complete Task A.
2. Steps 3–5 complete Task B.
3. Step 6 completes Task C.

● **EXAMPLE 2** *Simplex Method Tasks*

Complete Tasks A, B, and C to find the maximum value of $f = 4x + 3y$ subject to

$$x + 2y \le 8$$
$$2x + y \le 10$$

Solution

Task A: Introducing the slack variables gives the system of equations

$$\begin{cases} x + 2y + s_1 & = 8 \\ 2x + y + s_2 & = 10 \\ -4x - 3y + f & = 0 \end{cases}$$

Writing this in a matrix with the objective function in the last row gives

$$A = \begin{bmatrix} x & y & s_1 & s_2 & f & & \\ 1 & 2 & 1 & 0 & 0 & | & 8 \\ \textcircled{2} & 1 & 0 & 1 & 0 & | & 10 \\ -4 & -3 & 0 & 0 & 1 & | & 0 \end{bmatrix}$$

$8/1 = 8$

$10/2 = 5^*$

↑ Most negative   *Smallest quotient

Note that in this initial simplex matrix A, variables x and y are nonbasic (so set to 0) and this corresponds to the basic feasible solution $x = 0$, $y = 0$ and $s_1 = 8$, $s_2 = 10$, $f = 0$.

Task B: The most negative entry in the last row is -4 (indicated by the arrow), so the pivot column is column 1. The smallest quotient formed by dividing the positive coefficients from column 1 into the constants in the augment is 5 (indicated by the asterisk), so row 2 is the pivot row. Thus the pivot entry is in row 2, column 1 (circled in matrix A above).

Next use row operations with row 2 only to make the pivot entry equal 1 and all other entries in column 1 equal zero. The row operations are

(a) Multiply row 2 by $\frac{1}{2}$ to get a 1 as the new pivot entry.
(b) Add -1 times the new row 2 to row 1 to get a 0 in row 1 above the pivot entry.
(c) Add 4 times the new row 2 to row 3 to get a 0 in row 3 below the pivot entry.

This gives the simplex matrix B, which has nonbasic variables $y = 0$ and $s_2 = 0$ and corresponds to the basic feasible solution $y = 0$, $s_2 = 0$ and $x = 5$, $s_1 = 3$, $f = 20$.

$$
B = \begin{array}{c} \\ \\ \\ \\ \end{array}
\begin{array}{ccccc}
x & y & s_1 & s_2 & f \\
\end{array}
\left[\begin{array}{ccccc|c}
0 & \tfrac{3}{2} & 1 & -\tfrac{1}{2} & 0 & 3 \\
1 & \tfrac{1}{2} & 0 & \tfrac{1}{2} & 0 & 5 \\
\hline
0 & -1 & 0 & 2 & 1 & 20
\end{array}\right]
\begin{array}{l}
3 \div \tfrac{3}{2} = 2* \\
5 \div \tfrac{1}{2} = 10 \\
\\
\end{array}
$$

↑
Most negative   *Smallest quotient

The last row contains a negative entry, so the solution is not optimal and we locate another pivot column to continue. The pivot column is indicated with an arrow, and the smallest quotient is indicated with an asterisk. The new pivot entry is circled.

The row operations using this pivot row are

(a) Multiply row 1 by $\frac{2}{3}$ to get a 1 as the new pivot element.
(b) Add $-\frac{1}{2}$ times the new row 1 to row 2 to get a 0 in row 2 below the pivot element.
(c) Add the new row 1 to row 3 to get a 0 in row 3 below the pivot element.

This gives

$$
C = \begin{array}{c} \\ \\ \\ \\ \end{array}
\begin{array}{ccccc}
x & y & s_1 & s_2 & f \\
\end{array}
\left[\begin{array}{ccccc|c}
0 & 1 & \tfrac{2}{3} & -\tfrac{1}{3} & 0 & 2 \\
1 & 0 & -\tfrac{1}{3} & \tfrac{2}{3} & 0 & 4 \\
\hline
0 & 0 & \tfrac{2}{3} & \tfrac{5}{3} & 1 & 22
\end{array}\right]
$$

Because there are no negative indicators in the last row, the solution is complete.

Task C: s_1 and s_2 are nonbasic variables, so they equal zero. Thus f is maximized at 22 when $y = 2$, $x = 4$, $s_1 = 0$, and $s_2 = 0$.

● EXAMPLE 3 *Manufacturing (Application Preview)*

The Solar Technology Company manufactures three different types of hand calculators and classifies them as scientific, business, and graphing according to their calculating capabilities. The three types have production requirements given by the following table.

	Scientific	Business	Graphing
Electronic circuit components	5	7	10
Assembly time (hours)	1	3	4
Cases	1	1	1

The firm has a monthly limit of 90,000 circuit components, 30,000 hours of labor, and 9000 cases. If the profit is $6 for each scientific, $13 for each business, and $20 for each graphing calculator, how many of each should be produced to yield the maximum profit? What is the maximum profit?

Solution

Let x_1 be the number of scientific calculators produced, x_2 the number of business calculators produced, and x_3 the number of graphing calculators produced. Then the problem is to maximize the profit $f = 6x_1 + 13x_2 + 20x_3$ subject to

Inequalities	Equations with Slack Variables
$5x_1 + 7x_2 + 10x_3 \leq 90{,}000$	$5x_1 + 7x_2 + 10x_3 + s_1 \qquad\quad = 90{,}000$
$x_1 + 3x_2 + 4x_3 \leq 30{,}000$	$x_1 + 3x_2 + 4x_3 + \quad s_2 \qquad = 30{,}000$
$x_1 + x_2 + x_3 \leq 9{,}000$	$x_1 + x_2 + x_3 + \qquad\quad s_3 = 9{,}000$

The slack variables can be interpreted as follows: s_1 is the number of unused circuit components, s_2 is the number of unused hours of labor, and s_3 is the number of unused calculator cases. The simplex matrix, with the first pivot entry circled, is shown below.

Slack variables

$$
\begin{array}{ccccccc}
x_1 & x_2 & x_3 & s_1 & s_2 & s_3 & f \\
\end{array}
$$

$$
\left[
\begin{array}{ccccccc|c}
5 & 7 & 10 & 1 & 0 & 0 & 0 & 90{,}000 \\
1 & 3 & ④ & 0 & 1 & 0 & 0 & 30{,}000 \\
1 & 1 & 1 & 0 & 0 & 1 & 0 & 9{,}000 \\
\hline
-6 & -13 & -20 & 0 & 0 & 0 & 1 & 0
\end{array}
\right]
\quad
\begin{array}{l}
90{,}000/10 = 9000 \\
30{,}000/4 = 7500* \\
9000/1 = 9000 \\
\\
\end{array}
$$

↑ Most negative   *Smallest quotient

We now use row operations to change the pivot entry to 1 and create zeros elsewhere in the pivot column. The row operations are

1. Multiply row 2 by $\frac{1}{4}$ to convert the pivot entry to 1.
2. Using the new row 2,
   (a) Add -10 times new row 2 to row 1.
   (b) Add -1 times new row 2 to row 3.
   (c) Add 20 times new row 2 to row 4.

The result is the second simplex matrix.

Slack variables

$$
\begin{array}{ccccccc}
x_1 & x_2 & x_3 & s_1 & s_2 & s_3 & f \\
\end{array}
$$

$$
\left[
\begin{array}{ccccccc|c}
\frac{5}{2} & -\frac{1}{2} & 0 & 1 & -\frac{5}{2} & 0 & 0 & 15{,}000 \\
\frac{1}{4} & \frac{3}{4} & 1 & 0 & \frac{1}{4} & 0 & 0 & 7{,}500 \\
③\frac{3}{4} & \frac{1}{4} & 0 & 0 & -\frac{1}{4} & 1 & 0 & 1{,}500 \\
\hline
-1 & 2 & 0 & 0 & 5 & 0 & 1 & 150{,}000
\end{array}
\right]
\quad
\begin{array}{l}
15{,}000 \div \frac{5}{2} = 6000 \\
7500 \div \frac{1}{4} = 30{,}000 \\
1500 \div \frac{3}{4} = 2000* \\
\\
\end{array}
$$

↑ Most negative   *Smallest quotient

We check the last row indicators. Because there is an entry that is negative, we repeat the simplex process. Again the pivot entry is circled, and we apply row operations similar to those above. What are the row operations this time? The resulting matrix is

$$
\begin{array}{ccccccc}
x_1 & x_2 & x_3 & s_1 & s_2 & s_3 & f \\
\end{array}
$$

$$
\left[
\begin{array}{ccccccc|c}
0 & -\frac{4}{3} & 0 & 1 & -\frac{5}{3} & -\frac{10}{3} & 0 & 10{,}000 \\
0 & \frac{2}{3} & 1 & 0 & \frac{1}{3} & -\frac{1}{3} & 0 & 7{,}000 \\
1 & \frac{1}{3} & 0 & 0 & -\frac{1}{3} & \frac{4}{3} & 0 & 2{,}000 \\
\hline
0 & \frac{7}{3} & 0 & 0 & \frac{14}{3} & \frac{4}{3} & 1 & 152{,}000
\end{array}
\right]
$$

Checking the last row, we see that all the entries in this row are 0 or positive, so the solution is complete. From the final simplex matrix we see that the nonbasic variables are

x_2, s_2, and s_3, and with these equal to zero we determine all the solution values:

$$x_2 = 0, s_2 = 0, s_3 = 0 \quad \text{and} \quad x_1 = 2000, x_3 = 7000, s_1 = 10,000, f = 152,000$$

Thus the numbers of calculators that should be produced are

$$x_1 = 2000 \text{ scientific calculators}$$
$$x_2 = 0 \text{ business calculators}$$
$$x_3 = 7000 \text{ graphing calculators}$$

in order to obtain a maximum profit of $152,000 for the month. Note also that the slack variables tell us that with this optimal solution there were $s_1 = 10,000$ unused circuit components, $s_2 = 0$ unused labor hours, and $s_3 = 0$ unused calculator cases.

● **Checkpoint**

1. Write the following constraints as equations by using slack variables.

$$3x_1 + x_2 + x_3 \leq 9$$
$$2x_1 + x_2 + 3x_3 \leq 8$$
$$2x_1 + x_2 \qquad \leq 5$$

2. Set up the simplex matrix to maximize $f = 2x_1 + 5x_2 + x_3$ subject to the constraints discussed in Problem 1.

Using the simplex matrix from Problem 2, perform the following.

3. Find the first pivot entry.
4. Use row operations to make the variable of the first pivot entry basic.
5. Find the second pivot entry and make a second variable basic.
6. Find the solution.

Spreadsheet Note

As our discussion indicates, the simplex method is quite complicated, even when the number of variables is relatively small (2 or 3). For this reason, computer software packages are often used to solve linear programming problems.

All of the most popular spreadsheet programs (such as *Excel, Lotus 1-2-3,* and *Quattro Pro*) offer a linear programming solver as a feature or as an add-in. Also, the mathematics software packages MATLAB and MAPLE have linear programming solvers. Excel screens are shown throughout the remainder of this chapter for anyone interested in using Excel to solve linear programming problems.

The Excel screen below shows entries for beginning to solve the linear programming problem that was solved "by hand" with the simplex method in Example 3.

	A	B	C
1	Variables		
2			
3	# scientific calculators (x)	0	
4	# business calculators (y)	0	
5	# graphing calculators (z)	0	
6			
7	Objectives		
8			
9	Maximize profit	=6*B3+13*B4+20*B5	
10			
11	Constraints		
12		Amount used	Maximum
13	Circuit components	=5*B3+7*B4+10*B5	90000
14	Labor hours	=B3+3*B4+4*B5	30000
15	Cases	=B3+B4+B5	9000

Solving this linear programming problem involves using the Excel Solver. (For details of the solution with Excel Solver, see the *Excel Guide for Finite Mathematics and Applied Calculus*.)

The screen below gives the solution.

	A	B	C
1	Variables		
2			
3	# scientific calculators (x)	2000	
4	# business calculators (y)	0	
5	# graphing calculators (z)	7000	
6			
7	Objectives		
8			
9	Maximize profit	152000	
10			
11	Constraints		
12		Amount used	Maximum
13	Circuit components	80000	90000
14	Labor hours	30000	30000
15	Cases	9000	9000

This screen indicates that the profit is maximized at $152,000 when 2000 scientific and 7000 graphing calculators are produced. Note that under "Constraints" the difference between the columns labeled "Maximum" and "Amount used" gives the amount of each resource that was unused—that is, the value of each slack variable: $s_1 = 10{,}000$, $s_2 = 0$, and $s_3 = 0$. Note that this is the same solution found in Example 3. ■

Nonunique Solutions: Multiple Solutions and No Solution

As we've seen with graphical methods, a linear programming problem can have multiple solutions when the optimal value for the objective function occurs at two adjacent corners of the feasible region. Consider the following problem.

Maximize $f = 2x + y$ subject to

$$x + \frac{1}{2}y \le 16$$

$$x + y \le 24$$

$$x \ge 0$$

$$y \ge 0$$

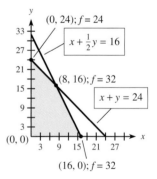

Figure 4.18

From Figure 4.18, we can see graphically that this problem has infinitely many solutions on the segment joining the points (8, 16) and (16, 0). Let's examine how we can discover these multiple solutions with the simplex method.

The simplex matrix for this problem (with the slack variables introduced) is

$$\begin{array}{ccccc} x & y & s_1 & s_2 & f \\ \left[\begin{array}{ccccc|c} ① & \frac{1}{2} & 1 & 0 & 0 & 16 \\ 1 & 1 & 0 & 1 & 0 & 24 \\ \hline -2 & -1 & 0 & 0 & 1 & 0 \end{array}\right] \end{array}$$

The most negative value in the last row is -2, so the x-column is the pivot column. The smallest quotient occurs in row 1; the pivot entry (circled) is 1. Making x a basic variable gives the following transformed simplex matrix.

$$\begin{array}{ccccc} x & y & s_1 & s_2 & f \\ \left[\begin{array}{ccccc|c} 1 & \frac{1}{2} & 1 & 0 & 0 & 16 \\ 0 & \frac{1}{2} & -1 & 1 & 0 & 8 \\ \hline 0 & 0 & 2 & 0 & 1 & 32 \end{array}\right] \end{array}$$

This simplex matrix has no negative indicators, so the optimal value of f has been found ($f = 32$ when $x = 16$, $y = 0$). Note that one of the nonbasic variables (y) has a zero indicator. When this occurs, we can use the column with the 0 indicator as our pivot column and often obtain a new solution that has the same value for the objective function. In this final tableau, if we used the y-column as the pivot column, then we could find the second solution (at $x = 8$, $y = 16$) that gives the same optimal value of f. You will be asked to do this in Problem 39 of the exercises for this section.

Multiple Solutions

> When the simplex matrix for the optimal value of f has a nonbasic variable with a zero indicator in its column, there may be multiple solutions giving the *same* optimal value for f. We can discover whether another solution exists by using the column of that nonbasic variable as the pivot column.

Note that with Excel, multiple solutions may be found by using different initial values for the variables when the problem is entered into Excel.

No Solution   It is also possible for a linear programming problem to have an unbounded solution (and thus no maximum value for f).

● **EXAMPLE 4 Simplex Method When No Maximum Exists**

Maximize $f = 2x_1 + x_2 + 3x_3$ subject to

$$x_1 - 3x_2 + x_3 \le 3$$
$$x_1 - 6x_2 + 2x_3 \le 6$$

if a maximum exists.

Solution
The simplex matrix for this problem (after introduction of slack variables) is

$$\begin{array}{cccccc} x_1 & x_2 & x_3 & s_1 & s_2 & f \\ \left[\begin{array}{ccccc|c} 1 & -3 & ① & 1 & 0 & 0 & 3 \\ 1 & -6 & 2 & 0 & 1 & 0 & 6 \\ \hline -2 & -1 & -3 & 0 & 0 & 1 & 0 \end{array}\right] \end{array}$$

We see that the x_3 column has the most negative number, so we use that column as the pivot column.

We see that the smallest quotient is 3. Both coefficients give this quotient, so we can use either coefficient as the pivot entry. Using the element in row 1, we can make x_3 basic. The new simplex matrix is

$$\begin{array}{ccccccc} & x_1 & x_2 & x_3 & s_1 & s_2 & f \\ & \left[\begin{array}{cccccc|c} 1 & -3 & 1 & 1 & 0 & 0 & 3 \\ -1 & 0 & 0 & -2 & 1 & 0 & 0 \\ \hline 1 & -10 & 0 & 3 & 0 & 1 & 9 \end{array}\right] \end{array}$$

The most negative value in the last row of this matrix is -10, so the pivot column is the x_2-column. But there are no positive coefficients in the x_2-column; what does this mean? Even if we could pivot using the x_2-column, the variables x_1 and s_1 would remain equal to 0. And the relationships among the remaining variables x_2, x_3, and s_2 would be the same.

$$-3x_2 + x_3 = 3 \quad \text{or} \quad x_3 = 3 + 3x_2$$
$$s_2 = 0$$

As x_2 is increased, x_3 also increases because all variables are nonnegative. This means that no matter how much we increase x_2, all other variables remain nonnegative. Furthermore, as we increase x_2 (and hence x_3), the value of f also increases. That is, there is no maximum value for f.

This example indicates that when a linear programming problem has an unbounded solution (and thus no maximum value for f), this is identified by the following condition in the simplex method.

No Solution

> If, after the pivot column has been found, there are no positive coefficients in that column, no maximum solution exists.

Some difficulties can arise in using the simplex method to solve linear programming problems. In the solution process, it is possible for the same simplex matrix to recur before an optimal value for the objective function is obtained. These "degeneracies" are discussed in more advanced courses. They rarely occur in applied linear programming problems, and we shall not encounter them in this text.

Shadow Prices

Finally, we note that one of the most powerful aspects of the simplex method is that it not only solves the given problem but also tells how to adapt the given solution if the situation changes. In particular, from the final simplex matrix (final tableau), the solution to the problem can be read, and there is information about the amount of change in the objective function if 1 additional unit of a resource becomes available (this change is called the **shadow price** of the resource) and how this additional resource changes the optimal solution values.

For example, if one more acre of land became available in the Example 1 application (for a total of 6001 acres), it can be shown that when the co-op plants 3749.5 acres of corn and 2251.5 acres of soybeans, the profit would increase by $30. That is, the shadow price of land is $30/acre (or the marginal profit value of the land is $30/acre). The simplex matrix C below is the final one from Example 1 (where x = acres of corn, y = acres of soybeans, and s_1 = the slack variable for the land constraint). The bottom entry in the s_1-column of C is the shadow price for land. Furthermore, if the s_1-column is added to the last column in C, then matrix D results. The solution obtained from D gives the new planting scheme and the new maximum profit if one more acre of land is planted (x = 3749.5 acres of corn and y = 2251.5 acres of soybeans for a new maximum profit of P = $315,030).

$$C = \begin{bmatrix} 0 & 1 & \frac{3}{2} & -\frac{1}{6} & 0 & 0 & 2{,}250 \\ 1 & 0 & -\frac{1}{2} & \frac{1}{6} & 0 & 0 & 3{,}750 \\ 0 & 0 & -\frac{9}{8} & \frac{1}{24} & 1 & 0 & 187.5 \\ 0 & 0 & 30 & \frac{10}{3} & 0 & 1 & 315{,}000 \end{bmatrix}$$

$$D = \begin{bmatrix} 0 & 1 & \frac{3}{2} & -\frac{1}{6} & 0 & 0 & 2251.5 \\ 1 & 0 & -\frac{1}{2} & \frac{1}{6} & 0 & 0 & 3749.5 \\ 0 & 0 & -\frac{9}{8} & \frac{1}{24} & 1 & 0 & 186\frac{3}{8} \\ 0 & 0 & 30 & \frac{10}{3} & 0 & 1 & 315{,}030 \end{bmatrix}$$

(column headers: $x \quad y \quad s_1 \quad s_2 \quad s_3 \quad P$)

Additional material on shadow prices appears in the second Extended Application/Group Project at the end of this chapter.

● **Checkpoint Solutions**

1. $\begin{aligned} 3x_1 + x_2 + x_3 + s_1 &= 9 \\ 2x_1 + x_2 + 3x_3 + s_2 &= 8 \\ 2x_1 + x_2 + s_3 &= 5 \end{aligned}$

2. $\begin{bmatrix} 3 & 1 & 1 & 1 & 0 & 0 & 0 & 9 \\ 2 & 1 & 3 & 0 & 1 & 0 & 0 & 8 \\ 2 & 1 & 0 & 0 & 0 & 1 & 0 & 5 \\ -2 & -5 & -1 & 0 & 0 & 0 & 1 & 0 \end{bmatrix}$

3. The first pivot entry is the 1 in row 3, column 2.

4. $\begin{bmatrix} 1 & 0 & 1 & 1 & 0 & -1 & 0 & 4 \\ 0 & 0 & 3 & 0 & 1 & -1 & 0 & 3 \\ 2 & 1 & 0 & 0 & 0 & 1 & 0 & 5 \\ 8 & 0 & -1 & 0 & 0 & 5 & 1 & 25 \end{bmatrix}$

5. The second pivot entry is the 3 in row 2, column 3.

$\begin{bmatrix} 1 & 0 & 0 & 1 & -\frac{1}{3} & -\frac{2}{3} & 0 & 3 \\ 0 & 0 & 1 & 0 & \frac{1}{3} & -\frac{1}{3} & 0 & 1 \\ 2 & 1 & 0 & 0 & 0 & 1 & 0 & 5 \\ 8 & 0 & 0 & 0 & \frac{1}{3} & \frac{14}{3} & 1 & 26 \end{bmatrix}$

6. All indicators are 0 or positive, so f is maximized at 26 when $x_1 = 0$, $x_2 = 5$, and $x_3 = 1$. Note also that $s_1 = 3$, $s_2 = 0$, and $s_3 = 0$.

4.3 Exercises

1. Convert the constraint inequalities $3x + 5y \le 15$ and $3x + 6y \le 20$ to equations containing slack variables.
2. Convert the constraint inequalities $2x + y \le 4$ and $3x + 2y \le 8$ to equations containing slack variables.

In Problems 3–6, set up the simplex matrix used to solve each linear programming problem. Assume all variables are nonnegative.

3. Maximize $f = 4x + 9y$ subject to

$$\begin{aligned} x + 5y &\le 200 \\ 2x + 3y &\le 134 \end{aligned}$$

4. Maximize $f = x + 3y$ subject to

$$\begin{aligned} 5x + y &\le 50 \\ x + 2y &\le 50 \end{aligned}$$

5. Maximize $f = 2x + 5y + 2z$ subject to

$$\begin{aligned} 2x + 7y + 9z &\le 100 \\ 6x + 5y + z &\le 145 \\ x + 2y + 7z &\le 90 \end{aligned}$$

6. Maximize $f = 3x + 5y + 11z$ subject to

$$\begin{aligned} 2x + 3y + 4z &\le 60 \\ x + 4y + z &\le 48 \\ 5x + y + z &\le 48 \end{aligned}$$

7. Just by looking at a simplex matrix, how can you tell the number of slack variables?
8. In a simplex matrix, what tells you whether or not the solution is complete?

In Problems 9–18, a simplex matrix for a standard maximization problem is given.

(a) Write the values of *all* the variables (use $x_1, x_2, x_3, \ldots$ and $s_1, s_2, s_3, \ldots$) and of the objective function f.

(b) Indicate whether or not the solution from part (a) is complete (optimal).

(c) If the solution is not complete, find the next pivot or indicate that no solution exists. When a new pivot can be found, state *all* row operations with that pivot (that is, row operations that make that pivot equal to 1, and then make other entries in the pivot column equal to 0). Do not perform the row operations.

9. $\begin{bmatrix} 10 & 27 & 1 & 0 & 0 & 0 & | & 200 \\ 4 & 51 & 0 & 1 & 0 & 0 & | & 400 \\ 15 & 27 & 0 & 0 & 1 & 0 & | & 350 \\ \hline -6 & -7 & 0 & 0 & 0 & 1 & | & 0 \end{bmatrix}$

10. $\begin{bmatrix} 5 & 5 & 7 & 1 & 0 & 0 & 0 & | & 12 \\ 4 & 0 & 6 & 0 & 1 & 0 & 0 & | & 48 \\ 0 & 1 & 1 & 0 & 0 & 1 & 0 & | & 8 \\ \hline -1 & -3 & -1 & 0 & 0 & 0 & 1 & | & 0 \end{bmatrix}$

11. $\begin{bmatrix} 2 & 0 & 1 & -\frac{3}{4} & 0 & | & 12 \\ 3 & 1 & 0 & \frac{1}{3} & 0 & | & 15 \\ \hline -4 & 0 & 0 & 3 & 1 & | & 15 \end{bmatrix}$

12. $\begin{bmatrix} 0 & 4 & 1 & -\frac{1}{5} & 0 & | & 2 \\ 1 & 2 & 0 & 1 & 0 & | & 4 \\ \hline 0 & 2 & 0 & 2 & 1 & | & 12 \end{bmatrix}$

13. $\begin{bmatrix} 4 & 1 & 0 & 0 & \frac{3}{4} & 4 & 0 & | & 14 \\ -2 & 0 & 0 & 1 & -\frac{5}{8} & -2 & 0 & | & 6 \\ 3 & 0 & 1 & 0 & 2 & 6 & 0 & | & 11 \\ \hline 4 & 0 & 0 & 0 & 2 & \frac{1}{2} & 1 & | & 525 \end{bmatrix}$

14. $\begin{bmatrix} 2 & 1 & 1 & 1 & 0 & 0 & 0 & | & 12 \\ -2 & 0 & -1 & -2 & 0 & 1 & 0 & | & 5 \\ 4 & 0 & 2 & -1 & 1 & 0 & 0 & | & 6 \\ \hline -2 & 0 & 5 & 7 & 0 & 0 & 1 & | & 30 \end{bmatrix}$

15. $\begin{bmatrix} 4 & 4 & 1 & 0 & 0 & 2 & 0 & | & 12 \\ 2 & 4 & 0 & 1 & 0 & -1 & 0 & | & 4 \\ -3 & -11 & 0 & 0 & 1 & -1 & 0 & | & 6 \\ \hline -3 & -3 & 0 & 0 & 0 & 4 & 1 & | & 150 \end{bmatrix}$

16. $\begin{bmatrix} 1 & 0 & -3 & 4 & -4 & 0 & 0 & | & 12 \\ 0 & 1 & -4 & 2 & 5 & 0 & 0 & | & 100 \\ 0 & 0 & -1 & -6 & -3 & 1 & 0 & | & 40 \\ \hline 0 & 0 & 6 & -1 & -3 & 0 & 1 & | & 380 \end{bmatrix}$

17. $\begin{bmatrix} 4 & -1 & 0 & 1 & -5 & 0 & 0 & | & 5 \\ 1 & 0 & 1 & 0 & 1 & 0 & 0 & | & 12 \\ 3 & -3 & 0 & 0 & -2 & 1 & 0 & | & 6 \\ \hline -2 & -5 & 0 & 0 & 10 & 0 & 1 & | & 120 \end{bmatrix}$

18. $\begin{bmatrix} 0 & 20 & -6 & 1 & 0 & -3 & 0 & | & 4 \\ 0 & 5 & -6 & 0 & 1 & -2 & 0 & | & 2 \\ 1 & -1 & 0 & 0 & 0 & 4 & 0 & | & 6 \\ \hline 0 & -9 & -12 & 0 & 0 & 10 & 1 & | & 20 \end{bmatrix}$

In Problems 19–24, a simplex matrix is given. In each case the solution is complete, so identify the maximum value of f and a set of values of the variables that gives this maximum value. If multiple solutions may exist, indicate this and locate the next pivot.

19. $\begin{bmatrix} 1 & 0 & \frac{3}{4} & -\frac{3}{4} & 0 & | & 11 \\ 0 & 1 & \frac{5}{12} & \frac{11}{12} & 0 & | & 9 \\ \hline 0 & 0 & 2 & 4 & 1 & | & 20 \end{bmatrix}$

20. $\begin{bmatrix} 0 & 4 & 1 & -\frac{1}{5} & 0 & | & 2 \\ 1 & 2 & 0 & 1 & 0 & | & 4 \\ \hline 0 & 2 & 0 & 2 & 1 & | & 12 \end{bmatrix}$

21. $\begin{bmatrix} 4 & 1 & 0 & 0 & \frac{3}{4} & 4 & 0 & | & 14 \\ -2 & 0 & 0 & 1 & -\frac{8}{5} & -2 & 0 & | & 6 \\ 3 & 0 & 1 & 0 & 2 & 6 & 0 & | & 11 \\ \hline 4 & 0 & 0 & 0 & 2 & \frac{1}{2} & 1 & | & 525 \end{bmatrix}$

22. $\begin{bmatrix} 0 & 1 & 0 & -3 & -1 & -4 & 0 & | & 12 \\ 1 & 0 & 0 & -2 & -1 & -5 & 0 & | & 16 \\ 0 & 0 & 1 & -1 & 2 & 1 & 0 & | & 22 \\ \hline 0 & 0 & 0 & 4 & 1 & 2 & 1 & | & 150 \end{bmatrix}$

23. $\begin{bmatrix} 1 & 0 & 3 & 0 & 6 & 0 & | & 50 \\ 0 & 0 & 4 & 1 & -4 & 0 & | & 6 \\ 0 & 1 & -2 & 0 & 2 & 0 & | & 10 \\ \hline 0 & 0 & 9 & 0 & 0 & 1 & | & 100 \end{bmatrix}$

24. $\begin{bmatrix} 0 & 1 & 2 & 1 & 0 & 0 & | & 20 \\ 1 & 0 & 6 & 1 & 0 & 0 & | & 50 \\ 0 & 0 & -1 & -2 & 1 & 0 & | & 10 \\ \hline 0 & 0 & 4 & 0 & 0 & 1 & | & 88 \end{bmatrix}$

In Problems 25–38, use the simplex method to maximize the given functions. Assume all variables are nonnegative.

25. Maximize $f = 3x + 10y$ subject to

$$14x + 7y \leq 35$$
$$5x + 5y \leq 50$$

26. Maximize $f = 7x + 10y$ subject to

$$14x + 14y \leq 98$$
$$8x + 10y \leq 100$$

27. Maximize $f = 2x + 3y$ subject to

$$x + 2y \le 10$$
$$x + y \le 7$$

28. Maximize $f = 5x + 30y$ subject to

$$2x + 10y \le 96$$
$$x + 10y \le 90$$

29. Maximize $f = 2x + y$ subject to

$$-x + y \le 2$$
$$x + 2y \le 10$$
$$3x + y \le 15$$

30. Maximize $f = 2x + 6y$ subject to

$$2x + 5y \le 30$$
$$x + 5y \le 25$$
$$x + y \le 11$$

31. Maximize $f = 7x + 10y + 4z$ subject to

$$3x + 5y + 4z \le 30$$
$$3x + 2y \le 4$$
$$x + 2y \le 8$$

32. Maximize $f = 64x + 36y + 38z$ subject to

$$4x + 2y + 2z \le 40$$
$$8x + 10y + 5z \le 90$$
$$4x + 4y + 3z \le 30$$

33. Maximize $f = 20x + 12y + 12z$ subject to

$$x + z \le 40$$
$$x + y \le 30$$
$$y + z \le 40$$

34. Maximize $f = 10x + 5y + 4z$ subject to

$$2x + y + z \le 50$$
$$x + 3y \le 20$$
$$y + z \le 15$$

35. Maximize $f = 10x + 8y + 5z$ subject to

$$2x + y + z \le 40$$
$$x + 2y \le 10$$
$$y + 3z \le 80$$

36. Maximize $f = x + 3y + z$ subject to

$$x + 4y \le 12$$
$$3x + 6y + 4z \le 48$$
$$y + z \le 8$$

37. Maximize $f = 24x_1 + 30x_2 + 18x_3 + 18x_4$ subject to

$$12x_1 + 16x_2 + 20x_3 + 8x_4 \le 880$$
$$5x_1 + 10x_2 + 10x_3 + 5x_4 \le 460$$
$$10x_1 - 5x_2 + 8x_3 + 5x_4 \le 280$$

38. Maximize $f = 8x_1 + 17x_2 + 24x_3 + 31x_4 + 36x_5$ subject to

$$x_1 + x_2 + 5x_3 + 3x_4 + 2x_5 \le 5000$$
$$2x_1 + x_2 + 4x_3 + 4x_4 + 2x_5 \le 6000$$
$$3x_1 + x_2 + 5x_3 + 2x_4 + 3x_5 \le 7000$$

Problems 39–44 involve linear programming problems that have nonunique solutions.

The simplex matrices shown in Problems 39 and 40 indicate that an optimal solution has been found but that a second solution is possible. Find the second solution.

39. $\begin{bmatrix} 1 & \frac{1}{2} & 1 & 0 & 0 & | & 16 \\ 0 & \frac{1}{2} & -1 & 1 & 0 & | & 8 \\ 0 & 0 & 2 & 0 & 1 & | & 32 \end{bmatrix}$

40. $\begin{bmatrix} 1 & 2 & 0 & 1 & 0 & | & 40 \\ 0 & 1 & 1 & 2 & 0 & | & 15 \\ 0 & 0 & 0 & 4 & 1 & | & 90 \end{bmatrix}$

In Problems 41–44, use the simplex method to maximize each function (whenever possible) subject to the given constraints. If there is no solution, indicate this; if multiple solutions exist, find two of them.

41. Maximize $f = 3x + 2y$ subject to

$$x - 10y \le 10$$
$$-x + y \le 40$$

42. Maximize $f = 10x + 15y$ subject to

$$-5x + y \le 24$$
$$2x - 3y \le 9$$

43. Maximize $f = 3x + 12y$ subject to

$$2x + y \le 120$$
$$x + 4y \le 200$$

44. Maximize $f = 8x + 6y$ subject to

$$4x + 3y \le 24$$
$$10x + 9y \le 198$$

APPLICATIONS

45. *Manufacturing* Newjet Inc. manufactures two types of printers, an inkjet printer and a laser printer. The company can make a total of 60 printers per day, and it has 120 labor-hours per day available. It takes 1 labor-hour

to make an inkjet printer and 3 labor-hours to make a laser printer. The profits are $40 per inkjet printer and $60 per laser printer.

(a) Write the simplex matrix to maximize the daily profit.

(b) Find the maximum profit and the number of each type of printer that will give the maximum profit.

46. *Construction*   A contractor builds two types of homes. The Carolina requires one lot, $160,000 capital, and 160 worker-days of labor, and the Savannah requires one lot, $240,000 capital, and 160 worker-days of labor. The contractor owns 300 lots and has $48,000,000 available capital and 43,200 worker-days of labor. The profit on the Carolina is $40,000 and the profit on the Savannah is $50,000.

(a) Write the simplex matrix to maximize the profit.

(b) Use the simplex method to find how many of each type of home should be built to maximize profit, and find the maximum possible profit.

47. *Production scheduling*   At one of its factories, a jeans manufacturer makes two styles: #891 and #917. Each pair of style-891 takes 10 minutes to cut out and 20 minutes to assemble and finish. Each pair of style-917 takes 10 minutes to cut out and 30 minutes to assemble and finish. The plant has enough workers to provide at most 7500 minutes per day for cutting and at most 19,500 minutes per day for assembly and finishing. The profit on each pair of style-891 is $6.00 and the profit on each pair of style-917 is $7.50. How many pairs of each style should be produced per day to obtain maximum profit? Find the maximum daily profit.

48. *Budget utilization*   A car rental agency has a budget of $1.8 million to purchase at most 100 new cars. The agency will purchase either compact cars at $15,000 each or luxury cars at $30,000 each. From past rental patterns, the agency decides to purchase at most 50 luxury cars and expects an annual profit of $7500 per compact car and $11,000 per luxury car. How many of each type of car should be purchased in order to obtain the maximum profit while satisfying budgetary and other planning constraints? Find the maximum profit.

49. *Manufacturing*   The Standard Steel Company produces two products, railroad car wheels and axles. Production requires processing in two departments: the smelting department (department A) and the machining department (department B). Department A has 50 hours available per week, and department B has 43 hours available per week. Manufacturing one axle requires 4 hours in department A and 3 hours in department B, and manufacturing one wheel requires 3 hours in department A and 5 hours in department B. The profits

are $300 per axle and $300 per wheel. Find the maximum possible profit and the number of axles and wheels that will maximize the profit.

50. *Experimentation*   An experiment involves placing the males and females of a laboratory animal species in two separate controlled environments. There is a limited time available in these environments, and the experimenter wishes to maximize the number of animals subject to the constraints described.

	Males	Females	Time Available
Environment A	20 min	25 min	800 min
Environment B	20 min	15 min	600 min

How many males and how many females will maximize the total number of animals?

51. *Shipping*   A produce wholesaler has determined that it takes $\frac{1}{2}$ hour of labor to sort and pack a crate of tomatoes and that it takes $1\frac{1}{4}$ hours to sort and pack a crate of peaches. The crate of tomatoes weighs 60 pounds, and the crate of peaches weighs 50 pounds. The wholesaler has 2500 hours of labor available each week and can ship 120,000 pounds per week. If profits are $1 per crate of tomatoes and $2 per crate of peaches, how many crates of each should be sorted, packed, and shipped to maximize overall profit? What is the maximum profit?

52. *Manufacturing*   A manufacturer makes house gutters and downspouts. Each standard gutter piece uses 10 units of aluminum and requires 0.05 hour of labor to make. Each downspout uses 10 units of aluminum and requires 0.1 hour of labor to make. During a typical day, the company has available at most 32 labor hours and at most 5000 units of aluminum. If the profits are $1.50 on each gutter and $2.25 on each downspout, how many of each item should be produced each day to maximize profit, and what is the maximum profit?

53. *Advertising*   Tire Corral has $6000 available per month for advertising. Newspaper ads cost $100 each and can occur a maximum of 21 times per month. Radio ads cost $300 each and can occur a maximum of 28 times per month at this price. Each newspaper ad reaches 6000 men over 20 years of age, and each radio ad reaches 8000 of these men. The company wants to maximize the number of ad exposures to this group. How many of each ad should it purchase? What is the maximum possible number of exposures?

54. *Manufacturing*   A bicycle manufacturer makes a ten-speed and a regular bicycle. The ten-speed requires 2 units of steel and 6 units of aluminum in its frame and 12 special components for the hub, sprocket, and gear assembly. The regular bicycle requires 5 units each of steel and aluminum for its frame and 5 of the special components. Shipments are such that steel is limited to 100 units per day, aluminum is limited to 120 units per day, and the special components are limited to 180 units per day. If the profits are $30 on each ten-speed and $20 on each regular bike, how many of each should be produced to yield the maximum profit? What is the maximum profit?

🖳 **Problems 55–64 involve 3 or more variables. Solve them with the simplex method, Excel, or some other technology.**

55. *Advertising*   The total advertising budget for a firm is $200,000. The following table gives the costs per ad package for each medium and the number of exposures per ad package (with all numbers in thousands).

	Medium 1	Medium 2	Medium 3
Cost/package	10	4	5
Exposures/ package	3100	2000	2400

If the maximum numbers of medium 1, medium 2, and medium 3 packages that can be purchased are 18, 10, and 12, respectively, how many of each ad package should be purchased to maximize the number of ad exposures?

56. *Advertising*   The Laposata Pasta Company has $12,000 available for advertising. The following table gives the costs per ad and the numbers of people exposed to its ads for three different media (with numbers in thousands).

Ad Packages	Newspaper	Radio	TV
Cost	2	2	4
Total audience	30	21	54
Working mothers	6	12	8

If the total available audience is 420,000, and if the company wishes to maximize the number of exposures to working mothers, how many ads of each type should it purchase?

57. *Manufacturing*   A firm has decided to discontinue production of an unprofitable product. This will create excess capacity, and the firm is considering one or more of three possible new products, A, B, and C. The available weekly hours in the plant will be 477 hours in tool

and die, 350 hours on the drill presses, and 150 hours on lathes. The hours of production required in each of these areas are as follows for each of the products.

	Tool and Die	Drill Press	Lathe
A	9	5	3
B	3	4	0
C	0.5	0	2

Furthermore, the sales department foresees no limitations on the sale of products A and C but anticipates sales of only 20 or fewer per week for product B. If the unit profits expected are $30 for A, $9 for B, and $15 for C, how many of each should be produced to maximize overall profit? What is the maximum profit?

58. *Medicine*   A medical clinic performs three types of medical tests that use the same machines. Tests A, B, and C take 15 minutes, 30 minutes, and 1 hour, respectively, with respective profits of $30, $50, and $100. The clinic has 4 machines available. One person is qualified to do test A, two to do test B, and one to do test C. If the clinic has a rush of customers for these tests, how many of each type should it schedule in a 12-hour day to maximize its profit?

59. *Manufacturing*   A manufacturer of blenders produces three different sizes: Regular, Special, and Kitchen Magic. The production of each type requires the materials and amounts given in the table below. Suppose the manufacturer is opening a new plant and anticipates weekly supplies of 90,000 electrical components, 12,000 units of aluminum, and 9000 containers. The unit profits are $3 for the Regular, $4 for the Special, and $8 for the Kitchen Magic. How many of each type of blender should be produced to maximize overall profit? What is the maximum possible profit?

	Regular	Special	Kitchen Magic
Electrical components	5	12	24
Units of aluminum	1	2	4
Blending containers	1	1	1

60. *Construction*   A contractor builds three types of houses, the Aries, the Belfair, and the Wexford. Each house requires one lot, and the following table gives the number of labor-hours and the amount of capital needed for each type of house, as well as the profit on the sale of each house. There are 12 lots, 47,500 labor-hours, and $3,413,000 available for his use.

(a) Building how many of each type of house will maximize his profit?

(b) What is the maximum possible profit?

	Aries	Belfair	Wexford
Labor-hours	3,000	3,700	5,000
Capital	$205,000	$279,600	$350,000
Profit	$20,000	$25,000	$30,000

61. *Revenue* A woman has a building with 26 one-bedroom apartments, 40 two-bedroom apartments, and 60 three-bedroom apartments available to rent to students. She has set the rent at $500 per month for the one-bedroom units, $800 per month for the two-bedroom units, and $1150 per month for the three-bedroom units. She must rent to one student per bedroom, and zoning laws limit her to at most 250 students in this building. There are enough students available to rent all the apartments.

(a) How many of each type of apartment should she rent to maximize her revenue?

(b) What is the maximum possible revenue?

62. *Manufacturing* Patio Iron makes wrought iron outdoor dining tables, chairs, and stools. Each table uses 8 feet of a standard width wrought iron, 4 hours of labor for cutting and assembly, and 2 hours of labor for detail and finishing work. Each chair uses 6 feet of the wrought iron, 2 hours of cutting and assembly labor, and 1.5 hours of detail and finishing labor. Each stool uses 1 foot of the wrought iron, 1.5 hours for cutting and assembly, and 0.5 hour for detail and finishing work, and the daily demand for stools is at most 16. Each day Patio Iron has available at most 108 feet of wrought iron, 50 hours for cutting and assembly, and 40 hours for detail and finishing. If the profits are $60 for each dining table, $48 for each chair, and $36 for each stool, how many of each item should be made each day to maximize profit? Find the maximum profit.

63. *Manufacturing* At one of its factories, a manufacturer of television sets makes one or more of four models of color sets: a 19-inch portable, a 27-inch table model, a 35-inch table model, and a 35-inch floor model. The assembly and testing time requirements for each model are shown in the following table, together with the maximum amounts of time available per week for assembly and testing. In addition to these constraints, the supplier of picture tubes indicated that it would supply no more than 200 picture tubes per week and that of these, no more than 40 could be 19-inch picture tubes.

	19-Inch Portable	27-Inch Table	35-Inch Table	35-Inch Floor	Total Available
Assembly time (hours)	7	10	12	15	2000
Test time (hours)	2	2	4	5	500
Profit (dollars)	46	60	75	100	

Use the profit for each television shown in the table to find the number of sets of each type that should be produced to obtain the maximum profit for the week. Find the maximum profit.

64. *Investment analysis* Conglomo Corporation is considering investing in other smaller companies, and it can purchase any fraction of each company. Each investment by Conglomo requires a partial payment now and a final payment 1 year from now. For each investment, the table below summarizes the amount of each payment (in millions of dollars) and the projected 5-year profit (also in millions of dollars).

	Company 1	Company 2	Company 3	Company 4	Company 5
Paid now	13.2	63.6	6.0	6.0	34.8
Paid in 1 year	3.6	7.2	6.0	1.2	40.8
Projected profit	15.6	19.2	19.2	16.8	46.8

This table may be interpreted as follows: if Conglomo purchases one-fifth of Company 4, then it pays $\frac{1}{5}(6.0) = $1.2 million now and $\frac{1}{5}(1.2) = $0.24 million = $240,000 after 1 year, and the one-fifth share has a projected 5-year profit of $\frac{1}{5}(16.8) = $3.36 million. Conglomo has available $48 million for investment now and $24 million for investment 1 year from now. What fraction of each smaller company should Conglomo purchase in order to maximize the projected 5-year profit? What is the maximum profit? (*Hint:* If x_1 represents the fraction of Company 1 that is purchased, then $x_1 \leq 1$ is a constraint.)

The Simplex Method: Duality and Minimization

4.4

- To formulate the dual for minimization problems
- To solve minimization problems using the simplex method on the dual

◗ Application Preview

A beef producer is considering two different types of feed. Each feed contains some or all of the necessary ingredients for fattening beef. Brand 1 feed costs 20 cents per pound and brand 2 costs 30 cents per pound. Table 4.1 contains all the relevant data about nutrition and cost of each brand and the minimum requirements per unit of beef. The producer would like to determine how much of each brand to buy in order to satisfy the nutritional requirements for Ingredients A and B at minimum cost. (See Example 3.)

TABLE 4.1

	Brand 1	Brand 2	Minimum Requirement
Ingredient A	3 units/lb.	5 units/lb.	40 units
Ingredient B	4 units/lb.	3 units/lb.	46 units
Cost per pound	20¢	30¢	

In this section, we will use the simplex method to solve minimization problems such as this one.

Dual Problems

We have solved both maximization and minimization problems using graphical methods, but the simplex method, as discussed in the previous section, applies only to maximization problems. Now let us turn our attention to extending the use of the simplex method to solve minimization problems. Such problems might arise when a company seeks to minimize its production costs yet fill customers' orders or purchase items necessary for production.

As with maximization problems, we shall consider only **standard minimization problems,** in which the constraints satisfy the following conditions.

1. All variables are nonnegative.
2. The constraints are of the form

$$a_1 y_1 + a_2 y_2 + \cdots + a_n y_n \geq b$$

where b is positive.

In this section, we will show how standard minimization problems can be solved using the simplex method.

● EXAMPLE 1 *A Minimization Problem and Its Dual*

Minimize $g = 11y_1 + 7y_2$ subject to

$$y_1 + 2y_2 \geq 10$$
$$3y_1 + y_2 \geq 15$$

Solution

Because the simplex method specifically seeks to increase the objective function, it does not apply to this minimization problem. Rather than solving this problem, let us solve a different but related problem called the **dual problem** or **dual.**

In forming the dual problem, we write a matrix A that has a form similar to the simplex matrix, but without slack variables and with positive coefficients for the variables in the last row (and with the function to be minimized in the augment).

The matrix for the dual problem is formed by interchanging the rows and columns of matrix A. Matrix B is the result of this procedure, and recall that B is the **transpose** of A.

$$A = \begin{bmatrix} 1 & 2 & | & 10 \\ 3 & 1 & | & 15 \\ 11 & 7 & | & g \end{bmatrix} \qquad B = \begin{bmatrix} 1 & 3 & | & 11 \\ 2 & 1 & | & 7 \\ 10 & 15 & | & g \end{bmatrix}$$

Note that row 1 of A becomes column 1 of B, row 2 of A becomes column 2 of B, and row 3 of A becomes column 3 of B.

In the same way that we formed matrix A, we can now "convert back" from matrix B to the maximization problem that is the dual of the original minimization problem. We state this dual problem, using different letters to emphasize that this is a different problem from the original.

$$\text{Maximize } f = 10x_1 + 15x_2 \text{ subject to}$$
$$x_1 + 3x_2 \leq 11$$
$$2x_1 + x_2 \leq 7$$

The simplex method does apply to this maximization problem, with the following simplex matrix.

$$
\begin{array}{c}
\begin{array}{ccccc} x_1 & x_2 & s_1 & s_2 & f \end{array} \\
\left[
\begin{array}{ccccc|c}
1 & 3 & 1 & 0 & 0 & 11 \\
2 & 1 & 0 & 1 & 0 & 7 \\
\hline
-10 & -15 & 0 & 0 & 1 & 0
\end{array}
\right]
\end{array}
$$

Successive matrices that occur using the simplex method on the maximization problem are shown in the table below with each pivot entry circled. Beside the maximization problem is the solution of the original minimization problem, obtained using the graphical method.

Comparison of a Minimization Problem and Its Dual

Maximization Problem—Simplex Method	Minimization Problem—Graphical Method

Maximization Problem—Simplex Method

1.
$$
\begin{array}{c}
\begin{array}{ccccc} x_1 & x_2 & s_1 & s_2 & f \end{array} \\
\left[
\begin{array}{ccccc|c}
1 & ③ & 1 & 0 & 0 & 11 \\
2 & 1 & 0 & 1 & 0 & 7 \\
\hline
-10 & -15 & 0 & 0 & 1 & 0
\end{array}
\right]
\end{array}
$$

2.
$$
\left[
\begin{array}{ccccc|c}
\frac{1}{3} & 1 & \frac{1}{3} & 0 & 0 & \frac{11}{3} \\
⑤\!\!\frac{5}{3} & 0 & -\frac{1}{3} & 1 & 0 & \frac{10}{3} \\
\hline
-5 & 0 & 5 & 0 & 1 & 55
\end{array}
\right]
$$

3.
$$
\left[
\begin{array}{ccccc|c}
0 & 1 & \frac{2}{5} & -\frac{1}{5} & 0 & 3 \\
1 & 0 & -\frac{1}{5} & \frac{3}{5} & 0 & 2 \\
\hline
0 & 0 & 4 & 3 & 1 & 65
\end{array}
\right]
$$

Maximum $f = 65$ occurs at $x_1 = 2$, $x_2 = 3$.

Minimization Problem—Graphical Method

Corners:

$A = (0, 15)$
$B = (4, 3)$ $\left.\begin{array}{l} \\ \\ \end{array}\right\}$ obtained by solving simultaneously
$C = (10, 0)$

$g = 11y_1 + 7y_2$;
At A, $g = 105$
At B, $g = 65$
At C, $g = 110$

Minimum $g = 65$ occurs at $y_1 = 4$, $y_2 = 3$.

In comparing the solutions of the previous two problems, the first thing to note is that the maximum value of f in the maximization problem and the minimum value of g in the minimization problem are the same. Furthermore, looking at the final simplex matrix, we can also find the values of y_1 and y_2 that give the minimum value of g by looking at the last (bottom) entries in the slack variable columns of the matrix.

$$
\begin{array}{cccccc}
x_1 & x_2 & s_1 & s_2 & f & \\
\begin{bmatrix}
0 & 1 & \frac{2}{5} & -\frac{1}{5} & 0 & 3 \\
1 & 0 & -\frac{1}{5} & \frac{3}{5} & 0 & 2 \\
0 & 0 & 4 & 3 & 1 & 65
\end{bmatrix}
\end{array}
$$

These values give the values of y_1
and y_2 in the minimization problem.

Duality and Solving

From this example we see that the given minimization problem and its dual maximization problem are very closely related. Furthermore, it appears that when problems enjoy this relationship, the simplex method might be used to solve them both. The question is whether the simplex method will always solve them both. The answer is yes. This fact was proved by John von Neumann, and it is summarized by the Principle of Duality.

Principle of Duality

1. When a standard minimization problem and its dual have a solution, the maximum value of the function to be maximized is the same value as the minimum value of the function to be minimized.
2. When the simplex method is used to solve the maximization problem, the values of the variables that solve the corresponding minimization problem are *the last entries in the columns corresponding to the slack variables.*

● **EXAMPLE 2** *Minimization with Simplex Method*

Given the problem

$$\text{Minimize } 18y_1 + 12y_2 = g \text{ subject to}$$
$$2y_1 + y_2 \geq 8$$
$$6y_1 + 6y_2 \geq 36$$

(a) State the dual of this minimization problem.
(b) Use the simplex method to solve both the given problem and its dual.

Solution
(a) The dual of this minimization problem will be a maximization problem. To form the dual, first write the matrix A (without slack variables and with positive coefficients for the variables in the last row) for the given problem.

$$
A = \begin{bmatrix}
2 & 1 & 8 \\
6 & 6 & 36 \\
\hline
18 & 12 & g
\end{bmatrix}
$$

Next, transpose matrix A to yield matrix B.

$$
B = \begin{bmatrix}
2 & 6 & 18 \\
1 & 6 & 12 \\
\hline
8 & 36 & g
\end{bmatrix}
$$

Then write the dual maximization problem, renaming the variables and the function.

Dual problem

Maximize $f = 8x_1 + 36x_2$ subject to

$$2x_1 + 6x_2 \le 18$$

$$x_1 + 6x_2 \le 12$$

(b) To solve the given minimization problem, we use the simplex method on its dual maximization problem. The complete simplex matrix for each step is given, and each pivot entry is circled.

1. $\begin{bmatrix} 2 & 6 & 1 & 0 & 0 & | & 18 \\ 1 & ⑥ & 0 & 1 & 0 & | & 12 \\ -8 & -36 & 0 & 0 & 1 & | & 0 \end{bmatrix}$ 2. $\begin{bmatrix} ① & 0 & 1 & -1 & 0 & | & 6 \\ \frac{1}{6} & 1 & 0 & \frac{1}{6} & 0 & | & 2 \\ -2 & 0 & 0 & 6 & 1 & | & 72 \end{bmatrix}$

3. $\begin{array}{ccccc} x_1 & x_2 & s_1 & s_2 & f \end{array}$
$\begin{bmatrix} 1 & 0 & 1 & -1 & 0 & | & 6 \\ 0 & 1 & -\frac{1}{6} & \frac{2}{6} & 0 & | & 1 \\ 0 & 0 & 2 & 4 & 1 & | & 84 \end{bmatrix}$

For the given minimization problem, the last entries in the slack variable columns give the values of y_1 and y_2. Thus $y_1 = 2$, $y_2 = 4$, and the minimum is $g = 84$.

For the dual maximization problem, the final simplex matrix gives $x_1 = 6$, $x_2 = 1$, and the maximum is $f = 84$.

● *Checkpoint*

Perform the steps below to begin the process of finding the minimum value of $g = y_1 + 4y$, subject to the following constraints.

$$2y_1 + 4y_2 \ge 18$$

$$y_1 + 5y_2 \ge 15$$

1. Form the matrix associated with the minimization problem.
2. Find the transpose of this matrix.
3. State the dual problem by using the matrix from Problem 2.
4. Write the simplex matrix for the dual of this problem.
5. The dual of this minimization problem has the following solution matrix. What is the solution of the minimization problem?

$$\begin{bmatrix} 1 & 0 & \frac{5}{6} & -\frac{1}{6} & 0 & | & \frac{1}{6} \\ 0 & 1 & -\frac{2}{3} & \frac{1}{3} & 0 & | & \frac{2}{3} \\ 0 & 0 & 5 & 2 & 1 & | & 13 \end{bmatrix}$$

◗ EXAMPLE 3 Purchasing (Application Preview)

A beef producer is considering two different types of feed. Each feed contains some or all of the necessary ingredients for fattening beef. Brand 1 feed costs 20 cents per pound and brand 2 costs 30 cents per pound. How much of each brand should the producer buy in order to satisfy the nutritional requirements for ingredients A and B at minimum cost? Table 4.2 contains all the relevant data about nutrition and cost of each brand and the minimum requirements per unit of beef.

TABLE 4.2

	Brand 1	Brand 2	Minimum Requirement
Ingredient A	3 units/lb	5 units/lb	40 units
Ingredient B	4 units/lb	3 units/lb	46 units
Cost per pound	20¢	30¢	

Solution

Let y_1 be the number of pounds of brand 1, and let y_2 be the number of pounds of brand 2. Then we can formulate the problem as follows:

$$\text{Minimizing cost } C = 20y_1 + 30y_2 \text{ subject to}$$
$$3y_1 + 5y_2 \geq 40$$
$$4y_1 + 3y_2 \geq 46$$

The original linear programming problem is usually called the **primal problem.** If we write this in an augmented matrix, without the slack variables, then we can use the transpose to find the dual problem.

$$
\begin{array}{cc}
\text{Primal} & \begin{array}{c}\text{Dual}\\ \text{(with function renamed)}\end{array}
\end{array}
$$

$$
\left[\begin{array}{cc|c}
3 & 5 & 40 \\
4 & 3 & 46 \\
\hline
20 & 30 & C
\end{array}\right]
\qquad
\left[\begin{array}{cc|c}
3 & 4 & 20 \\
5 & 3 & 30 \\
\hline
40 & 46 & f
\end{array}\right]
$$

Thus the dual maximization problem is

$$\text{Maximize } f = 40x_1 + 46x_2 \text{ subject to}$$
$$3x_1 + 4x_2 \leq 20$$
$$5x_1 + 3x_2 \leq 30$$

The simplex matrix for this maximization problem, with slack variables included, is

1.
$$
\left[\begin{array}{ccccc|c}
3 & ④ & 1 & 0 & 0 & 20 \\
5 & 3 & 0 & 1 & 0 & 30 \\
\hline
-40 & -46 & 0 & 0 & 1 & 0
\end{array}\right]
$$

Solving the problem gives

2.
$$
\left[\begin{array}{ccccc|c}
\frac{3}{4} & 1 & \frac{1}{4} & 0 & 0 & 5 \\
⓸\frac{11}{4} & 0 & -\frac{3}{4} & 1 & 0 & 15 \\
\hline
-\frac{11}{2} & 0 & \frac{23}{2} & 0 & 1 & 230
\end{array}\right]
$$
3.
$$
\left[\begin{array}{ccccc|c}
0 & 1 & \frac{5}{11} & -\frac{3}{11} & 0 & \frac{10}{11} \\
1 & 0 & -\frac{3}{11} & \frac{4}{11} & 0 & \frac{60}{11} \\
\hline
0 & 0 & 10 & 2 & 1 & 260
\end{array}\right]
$$

The solution to this problem is

$$y_1 = 10 \text{ lb of brand 1}$$
$$y_2 = 2 \text{ lb of brand 2}$$
$$\text{Minimum cost} = 260 \text{ cents or \$2.60 per unit of beef}$$

In a more extensive study of linear programming, the duality relationship that we have used in this section as a means of solving minimization problems can be shown to have more properties than we have mentioned.

Many linear programming solvers (some of which are available on the Internet) require the submission of both a primal problem and its dual. Thus, if the primal problem is minimization, the dual will be maximization, and vice versa. Furthermore, we've seen that this duality relationship requires the constraints for both the maximization and minimization problems to have a special form. Of course, in applications the constraints may not be uniformly "≤" for a maximization problem or "≥" for a minimization problem; rather, they may be mixed. The next section discusses how to handle mixed constraints.

Spreadsheet Note

When we use Excel, minimization problems are solved in a manner similar to maximization problems. The spreadsheet below is the input for the minimization problem in Example 3.

	A	B	C
1	Variables		
2			
3	Pounds Brand 1	0	
4	Pounds Brand 2	0	
5			
6	Objective		
7			
8	Minimize cost	=20*B3+30*B4	
9			
10	Constraints		
11		Amount in feeds	Required
12	Ingredient A	=3*B3+5*B4	40
13	Ingredient B	=4*B3+3*B4	46

Because of the constraints, the Solver for this problem has different directions on the inequality symbols; Solver is also set to find the minimum (min) rather than the maximum (max) value of the objective function. (For details on using the Excel Solver, see the *Excel Guide for Finite Mathematics and Applied Calculus*.)

The solution found with Solver, shown below, is that cost is minimized at 260 cents when 10 lb of brand 1 and 2 lb of brand 2 are used. This is the same solution as that found "by hand" with the simplex method in Example 3.

	A	B	C
1	Variables		
2			
3	Pounds Brand 1	10	
4	Pounds Brand 2	2	
5			
6	Objective		
7			
8	Minimize cost	260	
9			
10	Constraints		
11		Amount in feeds	Required
12	Ingredient A	40	40
13	Ingredient B	46	46

● **Checkpoint Solutions**

1. $\begin{bmatrix} 2 & 4 & | & 18 \\ 1 & 5 & | & 15 \\ 1 & 4 & | & g \end{bmatrix}$ 2. $\begin{bmatrix} 2 & 1 & | & 1 \\ 4 & 5 & | & 4 \\ 18 & 15 & | & g \end{bmatrix}$

3. Find the maximum value of $f = 18x_1 + 15x_2$ subject to the constraints

$$2x_1 + x_2 \le 1$$
$$4x_1 + 5x_2 \le 4$$

4. The simplex matrix for this dual problem is

$$\begin{bmatrix} 2 & 1 & 1 & 0 & 0 & | & 1 \\ 4 & 5 & 0 & 1 & 0 & | & 4 \\ -18 & -15 & 0 & 0 & 1 & | & 0 \end{bmatrix}$$

5. The minimum value is $g = 13$ when $y_1 = 5$ and $y_2 = 2$.

4.4 Exercises

In Problems 1–4, complete the following.
(a) Form the matrix associated with each given minimization problem and find its transpose.
(b) Write the dual maximization problem. Be sure to rename the variables.

1. Minimize $g = 3y_1 + y_2$ subject to

$$5y_1 + 2y_2 \ge 10$$
$$y_1 + 2y_2 \ge 6$$

2. Minimize $g = 3y_1 + 2y_2$ subject to

$$3y_1 + y_2 \ge 6$$
$$3y_1 + 4y_2 \ge 12$$

3. Minimize $g = 5y_1 + 2y_2$ subject to

$$y_1 + y_2 \ge 9$$
$$y_1 + 3y_2 \ge 15$$

4. Minimize $g = 9y_1 + 10y_2$ subject to

$$y_1 + 2y_2 \ge 21$$
$$3y_1 + 2y_2 \ge 27$$

In Problems 5 and 6, suppose a primal minimization problem and its dual maximization problem were solved by using the simplex method on the dual problem, and the final simplex method is given.
(a) Find the solution of the minimization problem. Use y_1, y_2, y_3 as the variables and g as the function.
(b) Find the solution of the maximization problem. Use x_1, x_2, x_3 as the variables and f as the function.

5. $\begin{bmatrix} 1 & -\frac{2}{5} & 0 & 3 & -\frac{3}{5} & 0 & 0 & | & 5 \\ 0 & -\frac{3}{5} & 0 & -2 & -\frac{11}{5} & 1 & 0 & | & 3 \\ 0 & -\frac{4}{5} & 1 & -1 & -\frac{2}{5} & 0 & 0 & | & 9 \\ \hline 0 & 2 & 0 & 8 & 2 & 0 & 1 & | & 252 \end{bmatrix}$

6. $\begin{bmatrix} 0 & 0 & 1 & \frac{2}{5} & \frac{5}{3} & \frac{4}{15} & 0 & | & 16 \\ 1 & 0 & 0 & -\frac{1}{5} & \frac{11}{3} & \frac{1}{5} & 0 & | & 19 \\ 0 & 1 & 0 & \frac{3}{5} & -\frac{2}{3} & \frac{2}{3} & 0 & | & 22 \\ \hline 0 & 0 & 0 & 12 & 15 & 20 & 1 & | & 554 \end{bmatrix}$

In Problems 7–10, write the dual maximization problem, and then solve both the primal and dual problems with the simplex method.

7. Minimize $g = 2y_1 + 10y_2$ subject to

$$2y_1 + y_2 \ge 11$$
$$y_1 + 3y_2 \ge 11$$
$$y_1 + 4y_2 \ge 16$$

8. Minimize $g = 8y_1 + 4y_2$ subject to

$$3y_1 - 2y_2 \ge 6$$
$$2y_1 + y_2 \ge 11$$

9. Minimize $g = 3y_1 + y_2$ subject to

$$4y_1 + y_2 \ge 11$$
$$3y_1 + 2y_2 \ge 12$$
$$3y_1 + y_2 \ge 6$$

10. Minimize $g = 2y_1 + 5y_2$ subject to

$$4y_1 + y_2 \ge 12$$
$$y_1 + y_2 \ge 9$$
$$y_1 + 3y_2 \ge 15$$

Use the simplex method in Problems 11–14.

11. Minimize $g = 8x + 7y + 12z$ subject to

$$x + y + z \geq 3$$
$$y + 2z \geq 2$$
$$x \geq 2$$

12. Minimize $g = 12y_1 + 8y_2 + 10y_3$ subject to

$$y_1 + 2y_3 \geq 10$$
$$y_1 + y_2 \geq 12$$
$$2y_1 + 2y_2 + y_3 \geq 8$$

13. Minimize $g = 12y_1 + 48y_2 + 8y_3$ subject to

$$y_1 + 3y_2 \geq 1$$
$$4y_1 + 6y_2 + y_3 \geq 3$$
$$4y_2 + y_3 \geq 1$$

14. Minimize $g = 20x + 30y + 36z$ subject to

$$x + 2y + 3z \geq 48$$
$$2x + 2y + 3z \geq 70$$
$$2x + 3y + 4z \geq 96$$

In Problems 15 and 16, a primal maximization problem is given.
(a) Form the dual minimization problem.
(b) Solve both the primal and dual problems with the simplex method.

15. Maximize $f = 40x_1 + 20x_2$ subject to

$$3x_1 + 2x_2 \leq 120$$
$$x_1 + x_2 \leq 50$$

16. Maximize $f = 28x_1 + 12x_2$ subject to

$$7x_1 + 12x_2 \leq 50$$
$$2x_1 + 6x_2 \leq 100$$

In Problems 17–20, you may want to use Excel or another technology.

17. Minimize $g = 40y_1 + 90y_2 + 30y_3$ subject to

$$y_1 + 2y_2 + y_3 \geq 16$$
$$y_1 + 5y_2 + 2y_3 \geq 18$$
$$2y_1 + 5y_2 + 3y_3 \geq 38$$

18. Minimize $w = 48y_1 + 20y_2 + 8y_3$ subject to

$$4y_1 + 2y_2 + y_3 \geq 30$$
$$12y_1 + 4y_2 + 3y_3 \geq 60$$
$$2y_1 + 3y_2 + y_3 \geq 40$$

19. Minimize $g = 50y_1 + 20y_2 + 30y_3 + 80y_4$ subject to

$$40y_1 + 20y_2 + 15y_3 + 50y_4 \geq 50$$
$$3y_1 + 2y_2 + 5y_4 \geq 6$$
$$2y_1 + 2y_2 + 4y_3 + 4y_4 \geq 10$$
$$2y_1 + 4y_2 + y_3 + 5y_4 \geq 8$$

20. Minimize $g = 200y_1 + 100y_2 + 200y_3 + 40y_4$ subject to

$$3y_1 + 2y_2 + y_3 + y_4 \geq 88$$
$$2y_1 + 2y_2 + y_3 \geq 86$$
$$2y_1 + 4y_2 + y_3 \geq 100$$
$$5y_1 + 5y_2 + y_3 \geq 100$$

APPLICATIONS

21. *Production scheduling* CDF Appliances has assembly plants in Atlanta and Fort Worth where they produce a variety of kitchen appliances, including a 12-cup coffee maker and a cappuccino machine. In each hour at the Atlanta plant, 160 of the 12-cup models and 200 of the cappuccino machines can be assembled and the hourly cost is $700. In each hour at the Fort Worth plant, 800 of the 12-cup models and 200 of the cappuccino machines can be assembled and the hourly cost is $2100. CDF Appliances expects orders each week for at least 64,000 of the 12-cup models and at least 40,000 of the cappuccino machines. How many hours per week should each plant be operated in order to provide inventory for the orders at minimum cost? Find the minimum cost.

22. *Manufacturing* Nekita Corporation assembles cell phones and camera cell phones at two different factories within the same city. During each hour at the first factory, 15 cell phones and 30 camera cell phones can be assembled at a cost of $100/hour. During each hour at the second factory, 10 cell phones and 60 camera cell phones can be assembled at a cost of $150/hour. If Nekita expects weekly orders for at least 15,000 cell phones and at least 45,000 camera cell phones, how many hours per week should it schedule at each location to be able to fill the orders at minimum cost? What is the minimum cost?

23. *Manufacturing* The Video Star Company makes two different models of VCRs, which are assembled on two different assembly lines. Line 1 can assemble 30 units of the Star model and 40 units of the Prostar model per hour, and line 2 can assemble 150 units of the Star model and 40 units of the Prostar model per hour. The company needs to produce at least 270 units of the Star model and 200 units of the Prostar model to fill an order. If it costs $200 per hour to run line 1 and $400 per hour to run line 2, how many hours should each line be run to fill the order at the minimum cost? What is the minimum cost?

24. *Nutrition*   A pork producer is considering two types of feed that contain the necessary ingredients for the nutritional requirements for fattening hogs. Red Star Brand contains 9 units of ingredient A and 12 units of ingredient B, and Blue Chip Brand contains 15 units of ingredient A and 9 units of ingredient B. The nutritional requirements for the hogs are at least 120 units of ingredient A and at least 138 units of ingredient B.
    (a) If Red Star Brand costs 50 cents per pound and Blue Chip Brand costs 75 cents, how many pounds of each brand should be bought to satisfy the nutritional requirements at the minimum cost?
    (b) What is the minimum cost?

25. *Production*   A small company produces two products, I and II, at three facilities, A, B, and C. They have orders for 2000 of product I and 1200 of product II. The production capacities and costs per week to operate the three facilities are summarized in the following table.

	A	B	C
I	200	200	400
II	100	200	100
Cost/week	$1000	$3000	$4000

How many weeks should each facility operate to fill the company's orders at a minimum cost, and what is the minimum cost?

26. *Nutrition*   In a hospital ward, the patients can be grouped into two general categories depending on their conditions and the amounts of solid foods they require in their diet. A combination of two diets is used for solid foods because they supply essential nutrients for recovery, but each diet has an amount of a substance deemed detrimental. For each patient group, the table below summarizes the diet requirements and the amounts of the detrimental substance. How many servings from each diet should be given each day in order to minimize the intake of this detrimental substance?

	Diet A	Diet B	Required Daily
Group 1	4 oz per serving	1 oz per serving	26 oz
Group 2	2 oz per serving	1 oz per serving	18 oz
Detrimental substance	0.18 oz per serving	0.07 oz per serving	

27. *Production*   Two factories produce three different types of kitchen appliances. The following table summarizes the production capacities, the numbers of each type of appliance ordered, and the daily operating costs for the factories. How many days should each factory operate to fill the orders at minimum cost?

	Factory 1	Factory 2	Number Ordered
Appliance 1	80/day	20/day	1600
Appliance 2	10/day	10/day	500
Appliance 3	50/day	20/day	1900
Daily cost	$10,000	$20,000	

28. *Nutrition*   In a laboratory experiment, two separate foods are given to experimental animals. Each food contains essential ingredients, A and B, for which the animals have minimum requirements, and each food also has an ingredient C, which can be harmful to the animals. The table below summarizes this information.

	Food 1	Food 2	Required
Ingredient A	10 units/g	3 units/g	49 units
Ingredient B	6 units/g	12 units/g	60 units
Ingredient C	3 units/g	1 unit/g	

How many grams of foods 1 and 2 will satisfy the requirements for A and B and minimize the amount of ingredient C that is ingested?

29. *Politics*   A political candidate wishes to use a combination of radio and TV advertisements in her campaign. Research has shown that each 1-minute spot on TV reaches 0.9 million people and that each 1-minute spot on radio reaches 0.6 million. The candidate feels she must reach 63 million people, and she must buy at least 90 minutes of advertisements. How many minutes of each medium should be used if TV costs $500 per minute, radio costs $100 per minute, and the candidate wishes to minimize costs?

30. *Production*   The James MacGregor Mining Company owns three mines, I, II, and III. Three grades of ore, A, B, and C, are mined at these mines. For each grade of ore, the number of tons per week available from each mine and the number of tons per week required to fill orders are given in the following table.

		I	II	III	Required Tons/Week
Ore Grades	A	10	10	10	90
	B	0	10	10	50
	C	10	0	10	60
Cost/day		$6000	$8000	$12,000	

Find the number of days the company should operate each mine so that orders are filled at minimum cost. Find the minimum cost.

31. *Supply costs* DeTurris Office Supplies presently carries reams of laser printer paper in three different grades: recycled, satin white, and glossy white. Three wholesalers offer special purchase packages consisting of different numbers of reams of these three grades, as shown in the following table. DeTurris Office Supplies needs at least 230 reams of the recycled paper, 240 reams of the satin white paper, and 210 reams of the glossy white paper.
(a) How many of each package should DeTurris buy to minimize its cost?
(b) What is the minimum possible cost?

	Recycled	Satin White	Glossy White	Package Cost
Georgia package	10	5	6	$120
Union package	5	8	6	140
Pacific package	6	6	6	126

32. *Dieting* A dieting company offers three foods, A, B, and C, and groups its customers into two groups according to their nutritional needs. The following table gives the percent of the daily nutritional requirements that a serving of each food provides and the number of ounces of detrimental substances in each food. Determine the combination of food types that will provide at least 100% of the daily requirements *and* will minimize the detrimental substances. What is the minimum amount of the detrimental substance?

	Food A	Food B	Food C	Daily Requirement
Group I	30% per serving	10% per serving	20% per serving	100%
Group II	10% per serving	20% per serving	40% per serving	100%
Detrimental substances	0.1 oz per serving	0.2 oz per serving	0.25 oz per serving	

33. *Nutrition* A hospital wishes to provide at least 24 units of nutrient A and 16 units of nutrient B in a meal, while minimizing the cost of the meal. If three types of food are available, with the nutritional values and costs (per ounce) given in the following table,

(a) What is the minimum possible cost?
(b) How many ounces of each food should be served to minimize cost? Find two solutions that give the minimum cost.

	Units of Nutrient A	Units of Nutrient B	Cost
Food I	2	1	1
Food II	2	5	5
Food III	2	1	2

34. *Pollution* Three factories each dump waste water containing three different types of pollutants into a river. State regulations require the factories to treat their waste in order to reduce pollution levels. The table shows the possible percent reduction of each pollutant at each site and the cost per ton to process the waste.

	Factory 1	Factory 2	Factory 3
Pollutant 1	75%	45%	20%
Pollutant 2	65%	30%	15%
Pollutant 3	10%	15%	5%
Cost/ton	$50	$20	$8

If the state requires a reduction of at least 65 tons per day of pollutant 1, at least 40 tons per day of pollutant 2, and at least 20 tons per day of pollutant 3, find the number of tons of waste that must be treated each day at each site so that the state's requirements are satisfied and the treatment costs are minimized. Find the minimum cost.

35. *Scheduling* Package Express delivers packages and overnight mailers. At each of several decentralized delivery depots in one district, Package Express hires a number of full- and part-time unskilled employees (so one-half of an employee is an individual who works half a day for 5 days). Each employee works the same shift (or part of a shift) for 5 consecutive days. Thus, if x_1 is the number of employees who begin work on Monday, then those same x_1 employees work Monday through Friday. The table shows the minimum number of unskilled workers needed at a depot on each day of the week.

Mon.	Tues.	Wed.	Thurs.	Fri.	Sat.	Sun.
20	16	16	20	22	12	17

How many unskilled employees should begin their shifts on each day of the week in order to staff a depot properly but minimize the total number of these workers that Package Express needs? (*Hint:* The total number of workers on Monday is the sum of those who begin their shifts on any of the days Thursday through Monday. Total numbers of workers on other days are found similarly.)

36. *Scheduling*   Each nurse works 8 consecutive hours at the Beaver Medical Center. The center has the following staffing requirements for each 4-hour work period.

Work Period	Nurses Needed
1 (7–11)	40
2 (11–3)	20
3 (3–7)	30
4 (7–11)	40
5 (11–3)	20
6 (3–7)	10

(a) If y_1 represents the number of nurses starting in period 1, y_2 the number starting in period 2, and so on, write the linear programming problem that will minimize the total number of nurses needed. (Note that the nurses who begin work in period 6 work periods 6 and 1 for their 8-hour shift.)

(b) Solve the problem in (a).

4.5

The Simplex Method with Mixed Constraints

◗ Application Preview

The Laser Company manufactures two models of stereo systems, the Star and the Allstar, at two plants, located in Ashville and in Cleveland. The maximum daily output at the Ashville plant is 900, and the profits there are $200 per unit of the Allstar and $100 per unit of the Star. The maximum daily output at the Cleveland plant is 800, and the profits there are $210 per unit of the Allstar and $80 per unit of the Star. In addition, restrictions at the Ashville plant mean that the number of units of the Star model cannot exceed 100 more than the number of units of the Allstar model produced. If the company gets a rush order for 800 units of the Allstar model and 600 units of the Star model, finding the number of units of each model that should be produced at each location to fill the order and obtain the maximum profit is a **mixed constraint linear programming** problem (see Example 3). The term *mixed constraint* is used because some inequalities describing the constraints contain "≤" and some contain "≥" signs.

Mixed Constraints and Maximization

In the previous two sections, we used the simplex method to solve two different types of standard linear programming problems. These types are summarized as follows:

	Maximization Problems	Minimization Problems
Function	Maximized	Minimized
Variables	Nonnegative	Nonnegative
Constraints	$a_1x_1 + a_2x_2 + \cdots + a_nx_n \le b \ (b \ge 0)$	$c_1y_1 + c_2y_2 + \cdots + c_ny_n \ge d \ (d \ge 0)$
Simplex method	Applied directly	Applied to dual

In this section we will show how to apply the simplex method to mixed constraint linear programming problems that do not exactly fit either of these types.

We begin by considering maximization problems with constraints that have a form different from those noted in the table above. This can happen in one of the following ways:

1. If a constraint has the form $a_1x_1 + a_2x_2 + \cdots + a_nx_n \le b$, where $b < 0$, such as with $x - 2y \le -8$
2. If a constraint has the form $a_1x_1 + a_2x_2 + \cdots + a_nx_n \ge b$, where $b \ge 0$, such as with $2x - 3y \ge 6$

Note in the latter case that multiplying both sides of the inequality by (-1) changes it as follows:

$$a_1x_1 + a_2x_2 + \cdots + a_nx_n \ge b$$

$$\text{becomes} \quad -a_1x_1 - a_2x_2 - \cdots - a_nx_n \le -b$$

That is,

$$2x - 3y \geq 6 \quad \text{becomes} \quad -2x + 3y \leq -6$$

Thus we see that the two possibilities for mixed constraints are actually one: constraints that have the form $a_1x_1 + a_2x_2 + \cdots + a_nx_n \leq b$, where b is any constant. We call these **less than or equal to constraints** and denote them as **≤ constraints.**

Let's consider an example and see how we might proceed with a problem of this type.

● **EXAMPLE 1** *Maximizing with Mixed Constraints*

Maximize $f = x + 2y$ subject to

$$
\begin{aligned}
x + y &\leq 13 \\
2x - 3y &\geq 6 \\
x \geq 0, \; y &\geq 0
\end{aligned}
$$

Solution

We begin by expressing all constraints as "≤" constraints. In this case, we multiply $2x - 3y \geq 6$ by (-1) to obtain $-2x + 3y \leq -6$. We can now use slack variables to write the inequalities as equations and form the simplex matrix.

$$
\begin{aligned}
x + y + s_1 &= 13 \\
-2x + 3y + s_2 &= -6 \\
-x - 2y + f &= 0
\end{aligned}
\qquad \text{gives} \qquad
\begin{bmatrix}
x & y & s_1 & s_2 & f & \\
1 & 1 & 1 & 0 & 0 & 13 \\
-2 & 3 & 0 & 1 & 0 & -6 \\
\hline
-1 & -2 & 0 & 0 & 1 & 0
\end{bmatrix}
$$

The -6 in the upper portion of the last column is a problem. If we compute values for the variables x, y, s_1, and s_2 that are associated with this matrix, we get $s_2 = -6$. This violates the condition of the simplex method that all variables be nonnegative. In order to use the simplex method, we must change the sign of any negative entry that appears in the upper portion of the last column (above the line separating the objective function from the constraints). If the problem has a solution, there will always be another negative entry in the same row as this negative entry, but in a different column. We choose this column as the pivot column, because pivoting with it will give a positive entry in the last column.

$$
\begin{bmatrix}
x & y & s_1 & s_2 & f & \\
1 & 1 & 1 & 0 & 0 & 13 \\
-2 & 3 & 0 & 1 & 0 & -6 \\
\hline
-1 & -2 & 0 & 0 & 1 & 0
\end{bmatrix}
$$
← Negative in last column

↑ Pivot column (negative entry in the same row as the negative entry in the last column)

Once we have identified the pivot column, we choose the pivot entry by forming *all* quotients and choosing the entry that gives the *smallest positive quotient*. Often this will mean pivoting by a negative number, which is the case in this problem.

$$
\begin{bmatrix}
x & y & s_1 & s_2 & f & \\
1 & 1 & 1 & 0 & 0 & 13 \\
\boxed{-2} & 3 & 0 & 1 & 0 & -6 \\
\hline
-1 & -2 & 0 & 0 & 1 & 0
\end{bmatrix}
$$

$13/1 = 13$

$(-6)/(-2) = 3*$

↑ Pivot entry circled

*Smallest positive quotient

Pivoting with this entry gives the following matrix.

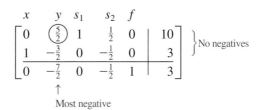

$$\begin{array}{c} \begin{array}{ccccc} x & y & s_1 & s_2 & f \end{array} \\ \left[\begin{array}{ccccc|c} 0 & \frac{5}{2} & 1 & \frac{1}{2} & 0 & 10 \\ 1 & -\frac{3}{2} & 0 & -\frac{1}{2} & 0 & 3 \\ 0 & -\frac{7}{2} & 0 & -\frac{1}{2} & 1 & 3 \end{array}\right] \end{array} \right\}\text{No negatives}$$

Most negative

This new matrix has no negatives in the upper portion of the last column, so we proceed with the simplex method as we have used it previously. The new pivot column is found from the most negative entry in the last row (indicated), and the pivot is found from the smallest positive quotient (circled). Using this pivot completes the solution.

$$\begin{array}{c} \begin{array}{ccccc} x & y & s_1 & s_2 & f \end{array} \\ \left[\begin{array}{ccccc|c} 0 & 1 & \frac{2}{5} & \frac{1}{5} & 0 & 4 \\ 1 & 0 & \frac{3}{5} & -\frac{1}{5} & 0 & 9 \\ 0 & 0 & \frac{7}{5} & \frac{1}{5} & 1 & 17 \end{array}\right] \end{array}$$

We see that when $x = 9$ and $y = 4$, then the maximum value of $f = x + 2y$ is 17.

● **Checkpoint**

1. Write the simplex matrix to maximize $f = 4x + y$ subject to the constraints

$$x + 2y \leq 12$$
$$3x - 4y \geq 6$$
$$x \geq 0, y \geq 0$$

2. Using the matrix from Problem 1, find the maximum value of $f = 4x + y$ subject to the constraints.

Mixed Constraints and Minimization

If mixed constraints occur in a minimization problem, the simplex method does not apply to the dual problem. However, the same techniques used in Example 1 can be slightly modified and applied to minimization problems with mixed constraints. If our objective is to minimize f, then we alter the problem so as to maximize $-f$ subject to "$\leq$ constraints" and then proceed as in Example 1.

● **EXAMPLE 2 *Minimizing with Mixed Constraints***

Minimize the function $f = 3x + 4y$ subject to the constraints

$$x + y \geq 20$$
$$x + 2y \geq 25$$
$$-5x + y \leq 4$$
$$x \geq 0, y \geq 0$$

Solution
Because of the mixed constraints, we seek to maximize $-f = -3x - 4y$ subject to

$$-x - y \leq -20$$
$$-x - 2y \leq -25$$
$$-5x + y \leq 4$$

The simplex matrix for this problem is

$$
\begin{array}{cccccc}
x & y & s_1 & s_2 & s_3 & -f \\
\end{array}
$$
$$
\begin{bmatrix}
-1 & -1 & 1 & 0 & 0 & 0 & \bigm| & -20 \\
-1 & -2 & 0 & 1 & 0 & 0 & \bigm| & -25 \\
-5 & 1 & 0 & 0 & 1 & 0 & \bigm| & 4 \\
3 & 4 & 0 & 0 & 0 & 1 & \bigm| & 0
\end{bmatrix} \leftarrow \text{Negatives}
$$

In this case, there are two negative entries in the upper portion of the last column. When this happens, we can start with either one; we choose -20 (row 1). In row 1 (to the left of -20), we find negative entries in columns 1 and 2. We can choose our pivot column as either of these columns; we choose column 1. Once this choice is made, the pivot is determined from the quotients.

$$
\begin{array}{cccccc}
x & y & s_1 & s_2 & s_3 & -f \\
\end{array}
$$
$$
\begin{bmatrix}
\boxed{-1} & -1 & 1 & 0 & 0 & 0 & \bigm| & -20 \\
-1 & -2 & 0 & 1 & 0 & 0 & \bigm| & -25 \\
-5 & 1 & 0 & 0 & 1 & 0 & \bigm| & 4 \\
3 & 4 & 0 & 0 & 0 & 1 & \bigm| & 0
\end{bmatrix}
\begin{array}{l}
-20/(-1) = 20* \\
-25/(-1) = 25 \\
4/(-5) = -4/5 \\
\end{array}
$$

↑
Pivot column (pivot entry circled)   *Smallest positive quotient

Pivoting with this entry gives the following matrix.

$$
\begin{array}{cccccc}
x & y & s_1 & s_2 & s_3 & -f \\
\end{array}
$$
$$
\begin{bmatrix}
1 & 1 & -1 & 0 & 0 & 0 & \bigm| & 20 \\
0 & \boxed{-1} & -1 & 1 & 0 & 0 & \bigm| & -5 \\
0 & 6 & -5 & 0 & 1 & 0 & \bigm| & 104 \\
0 & 1 & 3 & 0 & 0 & 1 & \bigm| & -60
\end{bmatrix} \leftarrow \text{Negative}
$$

↑
Pivot column

Now there is only one negative in the upper portion of the last column. Looking for negatives in this same row, we see that we can choose either column 2 or column 3 for our pivot column. We choose column 2, and the pivot is circled. Pivoting gives the following matrix.

$$
\begin{array}{cccccc}
x & y & s_1 & s_2 & s_3 & -f \\
\end{array}
$$
$$
\begin{bmatrix}
1 & 0 & -2 & 1 & 0 & 0 & \bigm| & 15 \\
0 & 1 & 1 & -1 & 0 & 0 & \bigm| & 5 \\
0 & 0 & -11 & 6 & 1 & 0 & \bigm| & 74 \\
0 & 0 & 2 & 1 & 0 & 1 & \bigm| & -65
\end{bmatrix}
$$

At this point we have all positives in the upper portion of the last column, so the simplex method can be applied. (The negative value in the lower portion is acceptable because $-f$ is our function.) Looking at the indicators, we see that the solution is complete.

From the matrix we have $x = 15$, $y = 5$, and $-f = -65$. Thus the solution is $x = 15$ and $y = 5$, which gives the minimum value for the function $f = 3x + 4y = 65$.

A careful review of these examples allows us to summarize the key procedural steps for solving linear programming problems with mixed constraints.

Summary: Simplex Method for Mixed Constraints

1. If the problem is to minimize f, then maximize $-f$.

2. a. Make all constraints "$\leq$" contraints by multiplying both sides of any "$\geq$" constraints by (-1).
   b. Use slack variables and form the simplex matrix.

3. In the simplex matrix, scan the *upper portion* of the last column for any negative entries.
   a. If there are no negative entries, apply the simplex method.
   b. If there are negative entries, go to step 4.

4. When there is a negative value in the upper portion of the last column, proceed as follows:
   a. Select any negative entry in the same row and use its column as the pivot column.
   b. In the pivot column, compute all quotients for the entries above the line and determine the pivot from the *smallest positive quotient*.
   c. After completing the pivot operations, return to step 3.

● **EXAMPLE 3** *Production Scheduling for Maximum Profit* **(Application Preview)**

The Laser Company manufactures two models of stereo systems, the Star and the Allstar, at two plants, located in Ashville and in Cleveland. The maximum daily output at the Ashville plant is 900, and the profits there are $200 per unit of the Allstar and $100 per unit of the Star. The maximum daily output at the Cleveland plant is 800, and the profits there are $210 per unit of the Allstar and $80 per unit of the Star. In addition, restrictions at the Ashville plant mean that the number of units of the Star model cannot exceed 100 more than the number of units of the Allstar model produced. If the company gets a rush order for 800 units of the Allstar model and 600 units of the Star model, how many units of each model should be produced at each location to fill the order and obtain the maximum profit?

Solution
The following table identifies our variables x and y and relates other important facts in the problem. In particular, we see that only variables x and y are required.

	Ashville Plant	Total Needed	Cleveland Plant
Allstar	x produced (profit = $200 each)	800	$800 - x$ produced (profit = $210 each)
Star	y produced (profit = $100 each)	600	$600 - y$ produced (profit = $80 each)
Capacity	900	—	800

We wish to maximize profit, and from the table we see that total profit is given by

$$P = 200x + 100y + 210(800 - x) + 80(600 - y)$$

or

$$P = 216{,}000 - 10x + 20y$$

We can read capacity constraints from our table.

$$x + y \leq 900 \quad \text{(Ashville)}$$

$$(800 - x) + (600 - y) \leq 800 \quad \text{or} \quad x + y \geq 600 \quad \text{(Cleveland)}$$

An additional Ashville plant constraint from the statement of the problem is

$$y \leq x + 100 \quad \text{or} \quad -x + y \leq 100$$

Thus our problem is to maximize

$$P = 216{,}000 - 10x + 20y$$

subject to

$$x + y \leq 900$$
$$x + y \geq 600$$
$$-x + y \leq 100$$
$$x \geq 0, y \geq 0$$

We must express $x + y \geq 600$ as a "$\leq$" constraint; multiplying both sides by (-1) gives $-x - y \leq -600$. The simplex matrix is

$$
\begin{array}{ccccccc}
x & y & s_1 & s_2 & s_3 & P & \\
\end{array}
$$

$$
\left[
\begin{array}{cccccc|c}
1 & 1 & 1 & 0 & 0 & 0 & 900 \\
\boxed{-1} & -1 & 0 & 1 & 0 & 0 & -600 \\
-1 & 1 & 0 & 0 & 1 & 0 & 100 \\
\hline
10 & -20 & 0 & 0 & 0 & 1 & 216{,}000
\end{array}
\right] \quad \leftarrow \text{Negative}
$$

↑
Pivot column

The negative in the upper portion of the last column is indicated. In row 2 we find negatives in both column 1 and column 2, so either of these may be our pivot column. Our choice is indicated, and the pivot entry is circled. The matrix that results from the pivot operations follows.

$$
\begin{array}{ccccccc}
x & y & s_1 & s_2 & s_3 & P & \\
\end{array}
$$

$$
\left[
\begin{array}{cccccc|c}
0 & 0 & 1 & 1 & 0 & 0 & 300 \\
1 & 1 & 0 & -1 & 0 & 0 & 600 \\
0 & \boxed{2} & 0 & -1 & 1 & 0 & 700 \\
\hline
0 & -30 & 0 & 10 & 0 & 1 & 210{,}000
\end{array}
\right]
$$

↑
Pivot column

No entry in the upper portion of the last column is negative, so we can proceed with the simplex method. From the indicators we see that column 2 is the pivot column (indicated above), and the pivot entry is circled. The simplex matrix that results from pivoting follows.

$$
\begin{array}{ccccccc}
x & y & s_1 & s_2 & s_3 & P & \\
\end{array}
$$

$$
\left[
\begin{array}{cccccc|c}
0 & 0 & 1 & \boxed{1} & 0 & 0 & 300 \\
1 & 0 & 0 & -\frac{1}{2} & -\frac{1}{2} & 0 & 250 \\
0 & 1 & 0 & -\frac{1}{2} & \frac{1}{2} & 0 & 350 \\
\hline
0 & 0 & 0 & -5 & 15 & 1 & 220{,}500
\end{array}
\right]
$$

↑
Pivot column

The new pivot column is indicated, and the pivot entry is circled. The pivot operation yields the following matrix.

$$
\begin{array}{ccccccc}
x & y & s_1 & s_2 & s_3 & P & \\
\end{array}
$$

$$
\left[
\begin{array}{cccccc|c}
0 & 0 & 1 & 1 & 0 & 0 & 300 \\
1 & 0 & \frac{1}{2} & 0 & -\frac{1}{2} & 0 & 400 \\
0 & 1 & \frac{1}{2} & 0 & \frac{1}{2} & 0 & 500 \\
\hline
0 & 0 & 5 & 0 & 15 & 1 & 222{,}000
\end{array}
\right]
$$

This matrix shows that the solution is complete. We see that $x = 400$ (so $800 - x = 400$), $y = 500$ (so $600 - y = 100$), and $P = \$222{,}000$.

Thus the company should operate the Ashville plant at capacity and produce 400 units of the Allstar model and 500 units of the Star model. The remainder of the order, 400 units of the Allstar model and 100 units of the Star model, should be produced at the Cleveland plant. This production schedule gives maximum profit $P = \$222{,}000$.

Up to this point, we have discussed ways in which any problem can be expressed as a maximization problem with "$\leq$" constraints. If we have problems in this "standard" form, then we can form the dual problem. This means that we are now equipped to transform any linear programming problem so that we can identify both a primal and a dual problem, and take advantage of any solver that requires both problems.

Spreadsheet Note — When using Excel to solve a linear programming problem with mixed constraints, it is not necessary to convert the form of the inequalities. They can simply be entered in their "mixed" form, and the objective function can be maximized or minimized with Solver. ■

● **Checkpoint Solutions**

1. $$\begin{bmatrix} 1 & 2 & 1 & 0 & 0 & | & 12 \\ \boxed{-3} & 4 & 0 & 1 & 0 & | & -6 \\ -4 & -1 & 0 & 0 & 1 & | & 0 \end{bmatrix}$$

2. $$\begin{bmatrix} 0 & \boxed{\frac{10}{3}} & 1 & \frac{1}{3} & 0 & | & 10 \\ 1 & -\frac{4}{3} & 0 & -\frac{1}{3} & 0 & | & 2 \\ 0 & -\frac{19}{3} & 0 & -\frac{4}{3} & 1 & | & 8 \end{bmatrix} \rightarrow \begin{bmatrix} 0 & 1 & \frac{3}{10} & \boxed{\frac{1}{10}} & 0 & | & 3 \\ 1 & 0 & \frac{2}{5} & -\frac{1}{5} & 0 & | & 6 \\ 0 & 0 & \frac{19}{10} & -\frac{7}{10} & 1 & | & 27 \end{bmatrix}$$

$$\rightarrow \begin{bmatrix} 0 & 10 & 3 & 1 & 0 & | & 30 \\ 1 & 2 & 1 & 0 & 0 & | & 12 \\ 0 & 7 & 4 & 0 & 1 & | & 48 \end{bmatrix}$$

The maximum value is $f = 48$ at $x = 12$, $y = 0$, $s_1 = 0$, and $s_2 = 30$.

4.5 Exercises

In Problems 1–4, express each inequality as a "$\leq$" constraint.

1. $3x - y \geq 5$
2. $4x - 3y \geq 6$
3. $y \geq 40 - 6x$
4. $x \geq 60 - 8y$

In Problems 5–8, complete both of the following.
(a) State the given problem in a form from which the simplex matrix can be formed (that is, as a maximization problem with "$\leq$" constraints).
(b) Form the simplex matrix, and circle the first pivot entry.

5. Maximize $f = 2x + 3y$ subject to
$$7x + 4y \leq 28$$
$$-3x + y \geq 2$$
$$x \geq 0, y \geq 0$$

6. Maximize $f = 5x + 11y$ subject to
$$x - 3y \geq 3$$
$$-x + y \leq 1$$
$$x \leq 10$$
$$x \geq 0, y \geq 0$$

7. Minimize $g = 3x + 8y$ subject to
$$4x - 5y \leq 50$$
$$x + y \leq 80$$
$$-x + 2y \geq 4$$
$$x \geq 0, y \geq 0$$

8. Minimize $g = 40x + 25y$ subject to
$$x + y \leq 100$$
$$-x + y \leq 20$$
$$-2x + 3y \geq 6$$
$$x \geq 0, y \geq 0$$

In Problems 9 and 10, a final simplex matrix for a minimization problem is given. In each case, find the solution.

9.
$$
\begin{array}{cccccccc}
x & y & z & s_1 & s_2 & s_3 & -f & \\
\left[\begin{array}{ccccccc|c}
0 & 1 & 0 & \frac{7}{3} & \frac{1}{3} & -\frac{2}{3} & 0 & 8 \\
0 & 0 & 1 & \frac{2}{3} & -\frac{1}{3} & \frac{8}{3} & 0 & 12 \\
1 & 0 & 0 & -\frac{4}{3} & \frac{2}{3} & \frac{4}{3} & 0 & 6 \\
\hline
0 & 0 & 0 & 4 & \frac{4}{3} & 2 & 1 & -120
\end{array}\right]
\end{array}
$$

10.
$$
\begin{array}{cccccccc}
x & y & z & s_1 & s_2 & s_3 & -f & \\
\left[\begin{array}{ccccccc|c}
1 & 0 & 0 & \frac{2}{5} & -\frac{2}{5} & 0 & 0 & 50 \\
0 & 0 & 1 & \frac{1}{5} & \frac{1}{5} & \frac{11}{5} & 0 & 12 \\
0 & 1 & 0 & 1 & -\frac{8}{5} & \frac{14}{5} & 0 & 30 \\
\hline
0 & 0 & 0 & 3 & 1 & 4 & 1 & -1200
\end{array}\right]
\end{array}
$$

In Problems 11–18, use the simplex method to find the optimal solution.

11. Maximize $f = 4x + y$ subject to

$$5x + 2y \le 84$$
$$-3x + 2y \ge 4$$
$$x \ge 0, y \ge 0$$

12. Maximize $f = x + 2y$ subject to

$$-x + 2y \le 60$$
$$-7x + 4y \ge 20$$
$$x \ge 0, y \ge 0$$

13. Minimize $f = 2x + 3y$ subject to

$$x \ge 5$$
$$y \le 13$$
$$-x + y \ge 2$$
$$x \ge 0, y \ge 0$$

14. Minimize $f = 3x + 2y$ subject to

$$x \le 20$$
$$y \le 20$$
$$x + y \ge 21$$
$$x \ge 0, y \ge 0$$

15. Minimize $f = 3x + 2y$ subject to

$$-x + y \le 10$$
$$x + y \ge 20$$
$$x + y \le 35$$
$$x \ge 0, y \ge 0$$

16. Minimize $f = 4x + y$ subject to

$$-x + y \le 4$$
$$3x + y \ge 12$$
$$x + y \le 20$$
$$x \ge 0, y \ge 0$$

17. Maximize $f = 2x + 5y$ subject to

$$-x + y \le 10$$
$$x + y \le 30$$
$$-2x + y \ge -24$$
$$x \ge 0, y \ge 0$$

18. Maximize $f = 3x + 2y$ subject to

$$-x + 2y \le 20$$
$$-3x + 2y \le -36$$
$$x + y \le 22$$
$$x \ge 0, y \ge 0$$

In Problems 19–26, use the simplex method, Excel, or another technology. Assume all variables are nonnegative.

19. Minimize $f = x + 2y + 3z$ subject to

$$x + z \le 20$$
$$x + y \ge 30$$
$$y + z \le 20$$
$$x \ge 0, y \ge 0, z \ge 0$$

20. Maximize $f = -x + 2y + 4z$ subject to

$$x + y + z \le 40$$
$$-x + y + z \ge 20$$
$$x + y - z \ge 10$$
$$x \ge 0, y \ge 0, z \ge 0$$

21. Maximize $f = 2x - y + 4z$ subject to

$$x + y + z \le 8$$
$$x - y + z \ge 4$$
$$x + y - z \ge 2$$
$$x \ge 0, y \ge 0, z \ge 0$$

22. Maximize $f = 5x + 2y + z$ subject to

$$2x + 3y + z \le 30$$
$$x - 2y + z \ge 20$$
$$2x + 5y + 2z \ge 25$$
$$x \ge 0, y \ge 0, z \ge 0$$

23. Minimize $f = 10x + 30y + 35z$ subject to

$$x + y + z \le 250$$
$$x + y + 2z \ge 150$$
$$2x + y + z \le 180$$
$$x \ge 0, y \ge 0, z \ge 0$$

24. Minimize $f = x + 2y + z$ subject to

$$x + 3y + 2z \ge 40$$
$$x + y + z \ge 30$$
$$x + y + z \le 100$$
$$x \ge 0, y \ge 0, z \ge 0$$

25. Maximize $f = 25x_1 + 25x_2 + 25x_3 + 20x_4 + 20x_5 + 20x_6$ subject to

$$x_1 + x_4 \leq 100$$
$$x_2 + x_5 \leq 20$$
$$x_3 + x_6 \leq 30$$
$$0.8x_2 \geq 0.2x_1 + 0.2x_3$$
$$0.9x_5 \geq 0.1x_4 + 0.1x_6$$
$$0.9x_6 \geq 0.1x_4 + 0.1x_5$$

26. Minimize $g = 480x_1 + 460x_2 + 440x_3 + 420x_4 + 530y_1 + 510y_2 + 490y_3 + 470y_4 - 9500$ subject to

$$x_1 + y_1 \geq 30$$
$$x_1 + x_2 + y_1 + y_2 \geq 90$$
$$x_1 + x_2 + x_3 + y_1 + y_2 + y_3 \geq 165$$
$$x_1 + x_2 + x_3 + x_4 + y_1 + y_2 + y_3 + y_4 \geq 190$$
$$x_1 \leq 40, x_2 \leq 40, x_3 \leq 40, x_4 \leq 40$$

APPLICATIONS

27. *Water purification* Nolan Industries manufactures water filters/purifiers that attach to a kitchen faucet. Each purifier consists of a housing unit that attaches to the faucet and a 60-day filter that is inserted into the housing. Past records indicate that on average the number of filters produced plus the number of housing units produced should be at least 500 per week. It takes 20 minutes to make and assemble each filter and 40 minutes for each housing. The manufacturing facility has up to 20,000 minutes per week for making and assembling these units, but due to certain parts supply constraints, the number of housing units per week can be at most 400. If manufacturing costs (for material and labor) are $6.60 for each filter and $8.35 for each housing unit, how many of each should be produced to minimize weekly costs? Find the minimum cost.

28. *Manufacturing* Johnson City Cooperage manufactures 30-gallon and 55-gallon fiber drums. Each 30-gallon drum takes 30 minutes to make, each 55-gallon drum takes 40 minutes to make, and the company has at most 10,000 minutes available each week. Also, workplace limitations and product demand indicate that the number of 55-gallon drums produced plus half the number of 30-gallon drums produced should be at least 160, and the number of 30-gallon drums should be at least twice the number of 55-gallon drums. If Johnson City Cooperage's manufacturing costs are $4 for each 30-gallon drum and $6 for each 55-gallon drum, how many of each drum should be made each week to satisfy the contraints at minimum cost? Find the minimum cost.

29. *Production* A sausage company makes two different kinds of hot dogs, regular and all beef. Each pound of all-beef hot dogs requires 0.75 lb of beef and 0.2 lb of spices, and each pound of regular hot dogs requires 0.18 lb of beef, 0.3 lb of pork, and 0.2 lb of spices. Suppliers can deliver at most 1020 lb of beef, at most 600 lb of pork, and at least 500 lb of spices. If the profit is $0.60 on each pound of all-beef hot dogs and $0.40 on each pound of regular hot dogs, how many pounds of each should be produced to obtain maximum profit? What is the maximum profit?

30. *Manufacturing* A cereal manufacturer makes two different kinds of cereal, Senior Citizen's Feast and Kids Go. Each pound of Senior Citizen's Feast requires 0.6 lb of wheat and 0.2 lb of vitamin-enriched syrup, and each pound of Kids Go requires 0.4 lb of wheat, 0.2 lb of sugar, and 0.2 lb of vitamin-enriched syrup. Suppliers can deliver at most 2800 lb of wheat, at most 800 lb of sugar, and at least 1000 lb of the vitamin-enriched syrup. If the profit is $0.90 on each pound of Senior Citizen's Feast and $1.00 on each pound of Kids Go, find the number of pounds of each cereal that should be produced to obtain maximum profit. Find the maximum profit.

31. *Manufacturing* A company manufactures commercial heating system components and domestic furnaces at its factories in Monaca, PA, and Hamburg, NY. At the Monaca plant, no more than 1000 units per day can be produced, and the number of commercial components cannot exceed 100 more than half the number of domestic furnaces. At the Hamburg plant, no more than 850 units per day can be produced. The profit on each commercial component is $400 at the Monaca plant and $390 at the Hamburg plant. The profit on each domestic furnace is $200 at the Monaca plant and $215 at the Hamburg plant. If there is a rush order for 500 commercial components and 750 domestic furnaces, how many of each should be produced at each plant in order to maximize profits? Find the maximum profit.

32. *Production scheduling* A manufacturer makes Portable Satellite Radios and Auto Satellite Radios at plants in Lakeland and Rockledge. At the Lakeland plant, at most 1800 radios can be produced, and the production of the Auto Satellite Radios can be at most 200 fewer than the production of the Portable Satellite Radios. At the Rockledge plant, at most 1200 radios can be produced. The profits on the Portable Satellite Radios are $100 at Lakeland and $90 at Rockledge, and the profits on the Auto Satellite Radios are $70 at Lakeland and $75 at Rockledge. If the manufacturer gets a rush order for 1500 Portable Satellite Radios and 1300 Auto Satellite Radios, how many of each should be produced at each location so as to maximize profits? Find the maximum profit.

33. *Manufacturing*   Refer to Problem 31. Suppose that the cost of each commercial component is $380 at the Monaca plant and $400 at the Hamburg plant. The cost of each domestic furnace is $200 at the Monaca plant and $185 at the Hamburg plant. If the same rush order for 500 commercial components and 750 domestic furnaces is received, how many of each should be produced at each location to minimize cost? Find the minimum cost.

34. *Manufacturing*   Refer to Problem 32. Suppose the costs associated with Portable Satellite Radios are $50 at Lakeland and $60 at Rockledge, and the costs associated with Auto Satellite Radios are $40 at Lakeland and $25 at Rockledge. If the manufacturer gets the same rush order for 1500 units of Portable Satellite Radios and 1300 units of Auto Satellite Radios, how many units of each should be produced at each location in order to minimize costs? Find the minimum cost.

35. *Water purification*   Three water purification facilities can handle at most 10 million gallons in a certain time period. Plant 1 leaves 20% of certain impurities, and costs $20,000 per million gallons. Plant 2 leaves 15% of these impurities and costs $30,000 per million gallons. Plant 3 leaves 10% impurities and costs $40,000 per million gallons. The desired level of impurities in the water from all three plants is at most 15%. If Plant 1 and Plant 3 combined must handle at least 6 million gallons, find the number of gallons each plant should handle so as to achieve the desired level of purity at minimum cost. Find the minimum cost.

36. *Chemical mixture*   A chemical storage tank has a capacity of 200 tons. Currently the tank contains 50 tons of mixture that has 10% of a certain active chemical and 1.8% of other inert ingredients. The owners of the tank want to replenish the supply in the tank and will purchase some combination of two available mixes. Mix 1 contains 70% of the active chemical and 3% of the inert ingredients; its cost is $100 per ton. Mix 2 contains 30% of the active chemical and 1% of the inert ingredients; its cost is $40 per ton. The desired final mixture should have at least 40% of the active chemical and at most 2% of the inert ingredients. How many tons of each mix should be purchased to obtain the desired final mixture at minimum cost? Find the minimum cost. Note that at least 40% of the active chemical means

$$70\% \ (\text{mix 1}) + 30\% \ (\text{mix 2}) + 10\% \ (\text{mix on hand})$$
$$\geq 40\% \ (\text{mix 1} + \text{mix 2} + \text{mix on hand})$$

37. *Manufacturing*   A ball manufacturer produces soccer balls, footballs, and volleyballs. The manager feels that restricting the types of balls produced could increase revenue. The following table gives the price of each ball, the raw materials cost, and the profit on each ball. The monthly profit must be at least $10,000, and the raw materials costs must be no more than $20,000. How many of each type of ball should be produced to maximize the revenue, and what is the maximum revenue?

	Raw Materials Cost	Price per Ball	Profit per Ball
Footballs	$10	$30	$10
Soccer balls	12	25	8
Volleyballs	8	20	8

38. *Land management*   A farm co-op has two farms, one at Spring Run and one at Willow Bend, where they grow corn and soybeans. Differences in the farms affect the associated costs and yield of each crop, as shown in the table.

	Corn		Soybeans	
	Yield (bu/acre)	Cost/acre	Yield (bu/acre)	Cost/acre
Spring Run	450	$110	360	$99
Willow Bend	585	$132	315	$88

Each farm has 100 acres available to plant, and the total desired yields are at least 46,800 bushels of corn and at least 42,300 bushels of soybeans. How many acres of each crop should be planted at each farm in order to achieve the desired crop yields at minimum cost? What is the minimum cost?

Key Terms

Section	Key Terms
4.1	Linear inequalities in two variables
	Graphs
	Systems
	Solution region

Section	Key Terms
4.2	Linear programming Objective function Constraints Optimum values Feasible region Graphical solution
4.3	Simplex method (maximization) Slack variables Simplex matrix Pivot entry Basic and nonbasic variables Pivoting Basic feasible solutions Nonunique solutions Shadow prices
4.4	Simplex method (minimization) Dual problem Transpose Primal problem
4.5	Simplex method with mixed constraints Creating $\leq$ Constraints

Review Exercises

Section 4.1

In Problems 1–4, graph the solution set of each inequality or system of inequalities.

1. $2x + 3y \leq 12$

2. $4x + 5y \geq 100$

3. $\begin{cases} x + 2y \leq 20 \\ 3x + 10y \leq 80 \\ x \geq 0, \ y \geq 0 \end{cases}$

4. $\begin{cases} 3x + y \geq 4 \\ x + y \geq 2 \\ -x + y \leq 4 \\ x \qquad \leq 5 \end{cases}$

Section 4.2

In Problems 5–8, a function and the graph of a feasible region are given. In each case, find both the maximum and minimum values of the function, if they exist, and the point at which each occurs.

5. $f = -x + 3y$

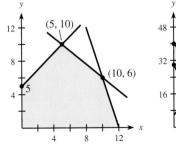

6. $f = 6x + 4y$

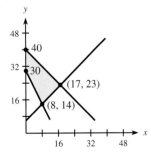

7. $f = 7x - 6y$

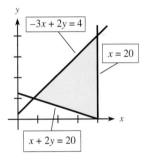

8. $f = 9x + 10y$

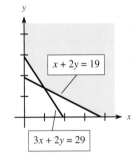

In Problems 9–14, solve the linear programming problems using graphical methods. Restrict $x \geq 0$ and $y \geq 0$.

9. Maximize $f = 5x + 6y$ subject to

$$x + 3y \leq 24$$
$$4x + 3y \leq 42$$
$$2x + y \leq 20$$

10. Maximize $f = x + 4y$ subject to

$$7x + 3y \leq 105$$
$$2x + 5y \leq 59$$
$$x + 7y \leq 70$$

11. Minimize $g = 5x + 3y$ subject to

$$3x + y \geq 12$$
$$x + y \geq 6$$
$$x + 6y \geq 11$$

12. Minimize $g = x + 5y$ subject to

$$8x + y \geq 85$$
$$x + y \geq 50$$
$$x + 4y \geq 80$$
$$x + 10y \geq 104$$

13. Maximize $f = 5x + 2y$ subject to

$$y \leq 20$$
$$2x + y \leq 32$$
$$-x + 2y \geq 4$$
$$x \geq 0, y \geq 0$$

14. Minimize $f = x + 4y$ subject to

$$y \leq 30$$
$$3x + 2y \geq 75$$
$$-3x + 5y \geq 30$$
$$x \geq 0, y \geq 0$$

Section 4.3

In Problems 15–20, use the simplex method to solve the linear programming problems. Assume all variables are nonnegative.

15. Maximize $f = 7x + 12y$ subject to the conditions in Problem 10.

16. Maximize $f = 3x + 4y$ subject to

$$x + 4y \leq 160$$
$$x + 2y \leq 100$$
$$4x + 3y \leq 300$$

17. Maximize $f = 3x + 8y$ subject to the conditions in Problem 16.

18. Maximize $f = 3x + 2y$ subject to

$$x + 2y \leq 48$$
$$x + y \leq 30$$
$$2x + y \leq 50$$
$$x + 10y \leq 200$$

19. Maximize $f = 39x + 5y + 30z$ subject to

$$x + z \leq 7$$
$$3x + 5y \leq 30$$
$$3x + y \leq 18$$

20. Maximize $f = 88x_1 + 86x_2 + 100x_3 + 100x_4$ subject to

$$3x_1 + 2x_2 + 2x_3 + 5x_4 \leq 200$$
$$2x_1 + 2x_2 + 4x_3 + 5x_4 \leq 100$$
$$x_1 + x_2 + x_3 + x_4 \leq 200$$
$$x_1 \qquad\qquad\qquad \leq 40$$

Problems 21 and 22 have nonunique solutions. If there is no solution, indicate this; if there are multiple solutions, find two different solutions. Use the simplex method, with $x \geq 0, y \geq 0$.

21. Maximize $f = 4x + 4y$ subject to

$$x + 5y \leq 500$$
$$x + 2y \leq 230$$
$$x + y \leq 160$$

22. Maximize $f = 2x + 5y$ subject to

$$-4x + y \leq 40$$
$$x - 7y \leq 70$$

Section 4.4

In Problems 23–28, form the dual and use the simplex method to solve the minimization problem. Assume all variables are nonnegative.

23. Minimize $g = 7y_1 + 6y_2$ subject to

$$5y_1 + 2y_2 \geq 16$$
$$3y_1 + 7y_2 \geq 27$$

24. Minimize $g = 3y_1 + 4y_2$ subject to

$$3y_1 + y_2 \geq 8$$
$$y_1 + y_2 \geq 6$$
$$2y_1 + 5y_2 \geq 18$$

25. Minimize $g = 2y_1 + y_2$ subject to the conditions in Problem 24.

26. Minimize $g = 12y_1 + 11y_2$ subject to

$$y_1 + y_2 \geq 100$$
$$2y_1 + y_2 \geq 140$$
$$6y_1 + 5y_2 \geq 580$$

27. Minimize $g = 12y_1 + 5y_2 + 2y_3$ subject to

$$y_1 + 2y_2 + y_3 \geq 60$$
$$12y_1 + 4y_2 + 3y_3 \geq 120$$
$$2y_1 + 3y_2 + y_3 \geq 80$$

28. Minimize $g = 25y_1 + 10y_2 + 4y_3$ subject to

$$7.5y_1 + 4.5y_2 + 2y_3 \geq 650$$
$$6.5y_1 + 3y_2 + 1.5y_3 \geq 400$$
$$y_1 + 1.5y_2 + 0.5y_3 \geq 200$$

Section 4.5

In Problems 29–34, use the simplex method. Assume all variables are nonnegative.

29. Maximize $f = 3x + 5y$ subject to

$$x + y \geq 19$$
$$-x + y \geq 1$$
$$-x + 10y \leq 190$$

30. Maximize $f = 4x + 6y$ subject to

$$2x + 5y \leq 37$$
$$5x - y \leq 34$$
$$-x + 2y \geq 4$$

31. Minimize $f = 10x + 3y$ subject to

$$-x + 10y \geq 5$$
$$4x + y \geq 62$$
$$x + y \leq 50$$

32. Minimize $f = 4x + 3y$ subject to

$$-x + y \geq 1$$
$$x + y \leq 45$$
$$10x + y \geq 45$$

33. Maximize $f = 8x_1 + 10x_2 + 12x_3 + 14x_4$ subject to

$$6x_1 + 3x_2 + 2x_3 + x_4 \geq 350$$
$$3x_1 + 2x_2 + 5x_3 + 6x_4 \leq 300$$
$$8x_1 + 3x_2 + 2x_3 + x_4 \leq 400$$
$$x_1 + x_2 + x_3 + x_4 \leq 100$$

34. Minimize $g = 10x_1 + 9x_2 + 12x_3 + 8x_4$ subject to

$$45x_1 + 58.5x_3 \geq 4680$$
$$36x_2 + 31.5x_4 \geq 4230$$
$$x_1 + x_2 \leq 100$$
$$x_3 + x_4 \leq 100$$

APPLICATIONS

Section 4.2

35. *Manufacturing* A company manufactures backyard swing sets of two different sizes. The larger set requires 5 hours of labor to complete, the smaller set requires 2 hours, and there are 700 hours of labor available each week. The packaging department can package at most 185 swing sets per week. If the profit is $100 on each larger set and $50 on each smaller set, how many of each should be produced to yield the maximum profit? What is the maximum profit? Use graphical methods.

36. *Production* A company produces two different grades of steel, A and B, at two different factories, 1 and 2. The table below summarizes the production capabilities of the factories, the cost per day, and the number of units of each grade of steel that is required to fill orders.

	Factory 1	Factory 2	Required
Grade A steel	1 unit	2 units	80 units
Grade B steel	3 units	2 units	140 units
Cost per day	$5000	$6000	

How many days should each factory operate in order to fill the orders at minimum cost? What is the minimum cost? Use graphical methods.

Use the simplex method to solve Problems 37–45.

Section 4.3

37. *Production* A small industry produces two items, I and II. It operates at capacity and makes a profit of $6 on each item I and $4 on each item II. The following table gives the hours required to produce each item and the hours available per day.

	I	II	Hours Available
Assembly	2 hours	1 hour	100
Packaging & inspection	1 hour	1 hour	60

Find the number of items that should be produced each day to maximize profits, and find the maximum daily profit.

38. *Production* Pinnochio Crafts makes two types of wooden crafts: Jacob's ladders and locomotive engines. The manufacture of these crafts requires both carpentry and finishing. Each Jacob's ladder requires 1 hour of finishing and $\frac{1}{2}$ hour of carpentry. Each locomotive engine requires 1 hour of finishing and 1 hour of carpentry. Pinnochio Crafts can obtain all the necessary raw materials, but only 120 finishing hours and 75 carpentry hours per week are available. Also, demand for Jacob's ladders is limited to at most 100 per week. If Pinnochio Crafts makes a profit of $3 on each Jacob's ladder and $5 on each locomotive engine, how many of each should it produce each week to maximize profits? What is the maximum profit?

39. *Profit* At its Jacksonville factory, Nolmaur Electronics manufactures 4 models of TV sets: HD models in 27-, 32-, and 42-in. sizes and a 42-in. plasma model. The manufacturing and testing hours required for each

model and available at the factory each week as well as each model's profit are shown in the table.

	27-in. HD	32-in. HD	42-in. HD	42-in. Plasma	Available Hours
Manufacturing (hr)	8	10	12	15	1870
Testing (hr)	2	4	4	4	530
Profit	$80	$120	$160	$200	

In addition, the supplier of the amplifier units can provide at most 200 units per week with at most 100 of these for both types of 42-in. models. The weekly demand for the 32-in. sets is at most 120. Nolmaur wants to determine the number of each type of set that should be produced each week to obtain maximum profit.
(a) Carefully identify the variables for Nolmaur's linear programming problem.
(b) Carefully state Nolmaur's linear programming problem.
(c) Solve this linear programming problem to determine Nolmaur's manufacturing plan and maximum profit.

Section 4.4

40. *Nutrition*   A nutritionist wants to find the least expensive combination of two foods that meet minimum daily vitamin requirements, which are 5 units of A and 30 units of B. Each ounce of food I provides 2 units of A and 1 unit of B, and each ounce of food II provides 10 units of A and 10 units of B. If food I costs 30 cents per ounce and food II costs 20 cents per ounce, find the number of ounces of each food that will provide the required vitamins and minimize the cost.

41. *Nutrition*   A laboratory wishes to purchase two different feeds, A and B, for its animals. The following table summarizes the nutritional contents of the feeds, the required amounts of each ingredient, and the cost of each type of feed.

	Feed A	Feed B	Required
Carbohydrates	1 unit/lb	4 units/lb	40 units
Protein	2 units/lb	1 unit/lb	80 units
Cost	14¢/lb	16¢/lb	

How many pounds of each type of feed should the laboratory buy in order to satisfy its needs at minimum cost?

42. *Production*   A company makes three products, I, II, and III, at three different factories. At factory A, it can make 10 units of each product per day. At factory B, it can make 20 units of II and 20 units of III per day.

At factory C, it can make 20 units of I, 20 units of II, and 10 units of III per day. The company has orders for 200 units of I, 500 units of II, and 300 units of III. If the daily costs are $200 at A, $300 at B, and $500 at C, find the number of days that each factory should operate in order to fill the company's orders at minimum cost. Find the minimum cost.

Section 4.5

43. *Profit*   A company makes pancake mix and cake mix. Each pound of pancake mix uses 0.6 lb of flour and 0.1 lb of shortening. Each pound of cake mix uses 0.4 lb of flour, 0.1 lb of shortening, and 0.4 lb of sugar. Suppliers can deliver at most 6000 lb of flour, at least 500 lb of shortening, and at most 1200 lb of sugar. If the profit per pound is $0.35 for pancake mix and $0.25 for cake mix, how many pounds of each mix should be made to earn maximum profit? What is the maximum profit?

44. *Manufacturing*   A company manufactures desks and computer tables at plants in Texas and Louisiana. At the Texas plant, production costs are $12 for each desk and $20 for each computer table, and the plant can produce at most 120 units per day. At the Louisiana plant, costs are $14 for each desk and $19 per computer table, and the plant can produce at most 150 units per day. The company gets a rush order for 130 desks and 130 computer tables at a time when the Texas plant is further limited by the fact that the number of computer tables it produces must be at least 10 more than the number of desks. How should production be scheduled at each location in order to fill the order at minimum cost? What is the minimum cost?

45. *Cost*   Armstrong Industries makes two different grades of steel at its plants in Midland and Donora. Weekly demand is at least 500 tons for grade 1 steel and at least 450 tons for grade 2. Due to differences in the equipment and the labor force, production times and costs per ton for each grade of steel differ slightly at each location, as shown in the table.

	Grade 1		Grade 2		
	Min/ Ton	Cost/ Ton	Min/ Ton	Cost/ Ton	Furnace Time/Week (in Minutes)
Midland	40	$100	42	$120	19,460
Donora	44	$110	45	$90	21,380

Find the number of tons of each grade of steel that should be made each week at each plant so that Armstrong meets demand at minimum cost. Find the minimum cost.

4 Chapter Test

1. Maximize $f = 3x + 5y$ subject to the constraints determined by the shaded region shown in the figure.

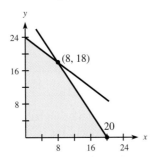

2. Following are three simplex matrices in various stages of the simplex method solution of a maximization problem.
   (a) Identify the matrix for which multiple solutions exist. Circle the pivot and specify all row operations with that pivot that are needed to find a second solution.
   (b) Identify the matrix for which no solution exists. Explain how you can tell there is no solution.
   (c) For the remaining matrix, give the values of the variables (use $x_1, x_2, \ldots$), the slack variables (use $s_1, s_2, \ldots$), and the objective function f.

$$A = \begin{bmatrix} 2 & 0 & -4 & 1 & -2 & 0 & 0 & | & 6 \\ -1 & 1 & 0 & 0 & 1 & 0 & 0 & | & 3 \\ 4 & 0 & -6 & 0 & -1 & 1 & 0 & | & 10 \\ -3 & 0 & -8 & 0 & 4 & 0 & 1 & | & 50 \end{bmatrix}$$

$$B = \begin{bmatrix} 0 & 1 & -2 & \frac{1}{4} & 0 & \frac{1}{5} & 0 & | & 12 \\ 0 & 0 & 4 & -\frac{3}{4} & 1 & -\frac{2}{5} & 0 & | & 20 \\ 1 & 0 & 3 & -\frac{5}{4} & 0 & \frac{4}{5} & 0 & | & 40 \\ 0 & 0 & 4 & 6 & 0 & -\frac{1}{5} & 1 & | & 170 \end{bmatrix}$$

$$C = \begin{bmatrix} 1 & 2 & 0 & 1 & 0 & -\frac{3}{2} & 0 & | & 40 \\ 0 & 1 & 0 & -2 & 1 & \frac{1}{2} & 0 & | & 15 \\ 0 & 3 & 1 & -1 & 0 & \frac{1}{4} & 0 & | & 60 \\ 0 & 0 & 0 & 4 & 0 & 6 & 1 & | & 220 \end{bmatrix}$$

3. Graph the solution region for each of the following.
   (a) $3x - 5y \le 30$ (b) $x > 2$ (c) $\begin{cases} 5y \ge 2x \\ x + y > 7 \\ 2y \le 5x \end{cases}$

4. Formulate the dual maximization problem associated with the following.
   Minimize $g = 2y_1 + 3y_2 + 5y_3$ subject to
   $$3y_1 + 5y_2 + y_3 \ge 100, \quad 4y_1 + 6y_2 + 3y_3 \ge 120,$$
   $$y_1 \ge 0, y_2 \ge 0, y_3 \ge 0$$

5. Find the maximum and minimum values of $f = 5x + 2y$ (if they exist) subject to the constraints determined by the shaded region in the figure.

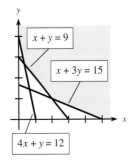

6. Use graphical methods to minimize $g = 3x + y$ subject to $5x - 4y \le 40$, $x + y \ge 80$, and $2x - y \ge 4$.

7. Express the following linear programming problem as a maximization problem with "$\le$" constraints. Assume $x \ge 0, y \ge 0$.
   Minimize $g = 7x + 3y$ subject to
   $$\begin{cases} -x + 4y \ge 4 \\ x - y \le 5 \\ 2x + 3y \le 30 \end{cases}$$

8. The final simplex matrix for a problem is given below. Give the solutions to *both* the maximization problem and the dual minimization problem. Carefully label each solution.

$$\begin{bmatrix} 0 & 1 & -0.5 & 3 & -0.7 & 0 & 0 & | & 15 \\ 1 & 0 & -0.8 & -2 & -0.5 & 0 & 0 & | & 17 \\ 0 & 0 & -0.6 & -1 & -2.4 & 1 & 0 & | & 11 \\ 0 & 0 & 12 & 4 & 18 & 0 & 1 & | & 658 \end{bmatrix}$$

9. Use the simplex method to maximize $f = 70x + 5y$ subject to $x + 1.5y \le 150$, $x + 0.5y \le 90$, $x \ge 0$, and $y \ge 0$.

10. Maximize $f = 60x + 48y + 36z$ subject to
    $$\begin{aligned} 8x + 6y + z &\le 108 \\ 4x + 2y + 1.5z &\le 50 \\ 2x + 1.5y + 0.5z &\le 40 \\ z &\le 16 \end{aligned}$$
    $$x \ge 0, \quad y \ge 0, \quad z \ge 0$$

11. River Brewery is a micro-brewery that produces an amber lager and an ale. Producing a barrel of lager requires 3 lb of corn and 2 lb of hops. Producing a barrel of ale requires 2 lb each of corn and hops. Profits are $35 from each barrel of lager and $30 from each

barrel of ale. Suppliers can provide at most 1200 lb of corn and at most 1000 lb of hops per month. Formulate a linear programming problem that can be used to maximize River Brewery's monthly profit. Then solve it with the simplex method.

12. A marketing research group conducting a telephone survey must contact at least 150 wives and 120 husbands. It costs $3 to make a daytime call and (because of higher labor costs) $4 to make an evening call. On average, daytime calls reach wives 30% of the time, husbands 10% of the time, and neither of these 60% of the time, whereas evening calls reach wives 30% of the time, husbands 30% of the time, and neither of these 40% of the time. Staffing considerations mean that daytime calls must be less than or equal to half of the total calls made. Formulate a linear programming problem that can be used to minimize the cost of completing the survey. Then solve it with any method from this chapter.

13. Lawn Rich, Inc. makes four different lawn care products that have fixed percents of phosphate, nitrogen, potash, and other minerals. The table shows the percent of each of the 4 ingredients present in each of Lawn Rich's products and the profit per ton for each product.

	Product 1	Product 2	Product 3	Product 4
% Phosphate	60	30	20	10
% Nitrogen	20	20	20	20
% Potash	9	6	15	18
% Other minerals	8	3	2	1
Profit/ton	$40	$50	$60	$70

A lawn care company buys in bulk from Lawn Rich and then blends the products for its own purposes. For the blend it desires, the lawn care company wants its Lawn Rich order to contain at least 35 tons of phosphate, at most 20 tons of nitrogen, at most 9 tons of potash, and at most 4 tons of other minerals. How many tons of each product should Lawn Rich produce to fulfill the customer's needs and maximize its own profit? What is the maximum profit?

Extended Applications & Group Projects

I. Transportation

The Kimble Firefighting Equipment Company has two assembly plants, plant A in Atlanta and plant B in Buffalo. It has two distribution centers, center I in Pittsburgh and center II in Savannah. The company can produce at most 800 alarm valves per week at plant A and at most 1000 per week at plant B. Center I must have at least 900 alarm valves per week, and center II must have at least 600 alarm valves per week. The costs per unit for transportation from the plants to the centers follow.

Plant A to center I	$2
Plant A to center II	3
Plant B to center I	4
Plant B to center II	1

What weekly shipping plan will meet the market demands and minimize the total cost of shipping the valves from the assembly plants to the distribution centers? What is the minimum transportation cost? Should the company eliminate the route with the most expensive unit shipment cost?

1. To assist the company in making these decisions, use linear programming with the following variables.

y_1 = the number of units shipped from plant A to center I

y_2 = the number of units shipped from plant A to center II

y_3 = the number of units shipped from plant B to center I

y_4 = the number of units shipped from plant B to center II

2. Form the constraint inequalities that give the limits for the plants and distribution centers, and form the total cost function that is to be minimized.
3. Solve the mixed constraint problem for the minimum transportation cost.
4. Determine whether any shipment route is unnecessary.

II. Slack Variables and Shadow Prices

Part A: With linear programming problems, one set of interesting questions deals with how the optimal solution changes if the constraint quantities change. Surprisingly, answers to these questions can be found in the final simplex tableau. In this investigation, we restrict ourselves to standard maximization problems and begin by reexamining the farm co-op application discussed in Example 1 in Section 4.3.

Farm Co-op Problem Statement
Maximize profit $P = 60x + 40y$ subject to

$$x + y \le 6000 \qquad \text{(Land constraint, in acres)}$$

$$9x + 3y \le 40{,}500 \qquad \text{(Fertilizer/herbicide constraint, in gallons)}$$

$$\frac{3}{4}x + y \le 5250 \qquad \text{(Labor constraint, in hours)}$$

Initial Tableau

$$
A =
\begin{array}{c}
\begin{array}{cccccc}
x & y & s_1 & s_2 & s_3 & P
\end{array} \\
\left[
\begin{array}{cccccc|c}
1 & 1 & 1 & 0 & 0 & 0 & 6{,}000 \\
9 & 3 & 0 & 1 & 0 & 0 & 40{,}500 \\
\frac{3}{4} & 1 & 0 & 0 & 1 & 0 & 5{,}250 \\
\hline
-60 & -40 & 0 & 0 & 0 & 1 & 0
\end{array}
\right]
\end{array}
$$

Final Tableau

$$
C =
\begin{array}{c}
\begin{array}{cccccc}
x & y & s_1 & s_2 & s_3 & P
\end{array} \\
\left[
\begin{array}{cccccc|c}
0 & 1 & \frac{3}{2} & -\frac{1}{6} & 0 & 0 & 2{,}250 \\
1 & 0 & -\frac{1}{2} & \frac{1}{6} & 0 & 0 & 3{,}750 \\
0 & 0 & -\frac{9}{8} & \frac{1}{24} & 1 & 0 & 187.5 \\
\hline
0 & 0 & 30 & \frac{10}{3} & 0 & 1 & 315{,}000
\end{array}
\right]
\end{array}
$$

Notice in the initial tableau that the columns for the slack variables and the objective function form an identity matrix (here a 4×4 matrix), and the row operations of the simplex method transform this identity matrix into another matrix in the same columns of the final tableau. Call this matrix S.

1. (a) Identify S and calculate SA, where A is the initial tableau above. What do you notice when you compare the result of SA with the final tableau, C?

   (b) Look at the Solar Technology manufacturing application in Example 3 in Section 4.3 and identify the corresponding matrix S there. Then find S times the initial tableau and compare your result with the final tableau. What do you notice?

2. (a) In the farm co-op application above, if the number of gallons of fertilizer/herbicide were increased by h to $40{,}500 + h$, then the row 2 entry in the last column of A would become $40{,}500 + h$. Call this new initial tableau A_h. Find the effect this change in fertilizer/herbicide would have on the optimal solution by calculating SA_h. Describe the differences between C (the final tableau of the original co-op problem) and SA_h.

   (b) Repeat part (a) if the number of acres of land is changed by k.

   (c) Use the results of parts (a) and (b) to complete the following general statement, A.
   Statement A: If constraint i has slack variable s_i and the quantity of constraint i is changed by h, then the new optimal solution can be found by using

   $h \cdot$ (the _____-column) plus the _____ column as the new _____ column

(d) Based on your results so far (and similar calculations with other examples, if needed), decide whether the following statement is true or false.

"The process described in statement A for determining the new optimal solution applies for any increment of change in any constraint or combination of constraints as long as the last column of the new final tableau has *all* nonnegative entries."

(e) Apply these results to find the new optimal solution to the farm co-op application if the co-op obtained 16 additional acres of land and 48 more gallons of fertilizer/herbicide.

3. (a) In part 2(a), how much is the objective function changed in the new optimal solution?

(b) If $h = 1$ in part 2(a), how much would the objective function change and where in the final tableau is the value that gives this amount of change?

(c) Recall that the **shadow price** for a constraint measures the change in the objective function if one additional unit of the resource for that constraint becomes available. What is the shadow price for fertilizer/herbicide, and what does it tell us?

(d) Find the shadow prices for acres of land and labor hours and tell what each means.

(e) Complete the following general statement.

Statement B: The shadow price for constraint i is found in the final tableau as the ____ entry in the column for ____.

(f) From this example (and others if you check), notice that if a slack variable is nonzero, then the shadow price of the associated constraint will equal zero. Why do you think this is true?

Part B: Each week the Plush Seating Co. makes stuffed chairs (x) and love seats (y) by using hours of carpentry time (s_1), hours of assembly time (s_2), and hours of upholstery time (s_3) in order to maximize profit P, in dollars. Plush Seating's final tableau is shown below; use it and the results from part A to answer the questions that follow.

$$
\begin{array}{cccccc}
x & y & s_1 & s_2 & s_3 & P \\
\end{array}
$$

$$
\left[
\begin{array}{cccccc|c}
1 & 0 & \frac{1}{3} & -1 & 0 & 0 & 28 \\
0 & 1 & -\frac{1}{3} & 2 & 0 & 0 & 8 \\
0 & 0 & \frac{4}{3} & -10 & 1 & 0 & 10 \\
\hline
0 & 0 & \frac{8}{3} & 64 & 0 & 1 & 2816 \\
\end{array}
\right]
$$

1. Producing how many chairs and love seats will yield a maximum profit? Find the maximum profit.

2. How many hours of assembly time are left unused?

3. How many hours of upholstery time are left unused?

4. How much will profit increase if an extra hour of carpentry time becomes available? What is this amount of increase called?

5. How much will profit increase if an extra hour of upholstery time becomes available?

6. If 1 more hour of labor were available, where should Plush Seating use it? Explain why.

7. If each chair takes 2 hours of upholstery time and each love seat takes 6 hours, how many hours of upholstery time are available each week?

8. If Plush Seating could hire a part-timer to do 6 hours of carpentry and 1 hour of assembly, determine the new optimal solution.

9. Suppose that each worker can do any of the three jobs, and rather than hire a part-timer, management wants to reassign the extra upholstery hours. If they use 9 of those hours in carpentry and the other 1 in assembly, find the new optimal solution.

10. Find a way to reassign any number of the upholstery hours in a way different from that in part 9, so that profit is improved even more than in part 9.

5

Exponential and Logarithmic Functions

In this chapter we study exponential and logarithmic functions, which provide models for many applications that at first seem remote and unrelated. In our study of these functions, we will examine their descriptions, their properties, their graphs, and the special inverse relationship between these two functions. We will see how exponential and logarithmic functions are applied to some of the concerns of social scientists, business managers, and life scientists.

The chapter topics and applications include the following.

Chapter Warm-up

Prerequisite Problem Type	For Section	Answer	Section for Review
Write the following with positive exponents: (a) x^{-3} (b) $\dfrac{1}{x^{-2}}$ (c) $\sqrt{x}$	5.1 5.2 5.3	 (a) $\dfrac{1}{x^3}$ (b) x^2 (c) $x^{1/2}$	0.3, 0.4 Exponents and radicals
Simplify: (a) 2^0 (b) $x^0 \quad (x \neq 0)$ (c) $49^{1/2}$ (d) 10^{-2}	5.1 5.2 5.3	 (a) 1 (b) 1 (c) 7 (d) $\dfrac{1}{100}$	0.3, 0.4 Exponents
Answer true or false: (a) $\left(\dfrac{1}{2}\right)^x = 2^{-x}$ (b) $\sqrt{50} = 50^{1/2}$ (c) If $8 = 2^y$, then $y = 4$. (d) If $x^3 = 8$, then $x = 2$.	5.1 5.2	 (a) True (b) True (c) False; $y = 3$ (d) True	0.3, 0.4 Exponents and radicals
(a) If $f(x) = 2^{-2x}$, what is $f(-2)$? (b) If $f(x) = 2^{-2x}$, what is $f(1)$? (c) If $f(t) = (1 + 0.02)^t$, what is $f(0)$? (d) If $f(t) = 100(0.03)^{0.02t}$, what is $f(0)$? (e) If $f(x) = \dfrac{9.46}{1 + 53.08e^{-1.28x}}$, what is $f(5)$?	5.1 5.2 5.3	(a) 16 (b) $\dfrac{1}{4}$ (c) 1 (d) 3 (e) ≈ 8.693	1.2 Function notation

5.1

OBJECTIVES

- To graph exponential functions
- To evaluate exponential functions
- To model with exponential functions

Exponential Functions

❶ **Application Preview**

If $10,000 is invested at 6%, compounded monthly, then the future value S of the investment after x years is given by the exponential function

$$S = 10,000(1.005)^{12x}$$

(We will find the future value of the investment after 5 years in Example 2.)

In this section we will evaluate and graph exponential functions, and we will model data with exponential functions.

TABLE 5.1

Minutes Passed	Number of Organisms
0	1
1	2
2	4
3	8
4	16

Suppose a culture of bacteria has the characteristic that each minute, every microorganism splits into two new organisms. We can describe the number of bacteria in the culture as a function of time. That is, if we begin the culture with one microorganism, we know that after 1 minute we will have two organisms, after 2 minutes, four, and so on. Table 5.1 gives a few of the values that describe this growth. If x represents the number of minutes that have passed and y represents the number of organisms, the points (x, y) lie on the graph of the function with equation

$$y = 2^x$$

The equation $y = 2^x$ is an example of a special group of functions called **exponential functions,** defined as follows.

Exponential Functions

If a is a real number with $a > 0$ and $a \neq 1$, then the function

$$f(x) = a^x$$

is an **exponential function** with base a.

A table of some values satisfying $y = 2^x$ and the graph of this function are given in Figure 5.1. This function is said to model the growth of the number of organisms in the previous discussion, even though some points on the graph do not correspond to a time and a number of organisms. For example, time x could not be negative, and the number of organisms y could not be fractional.

We defined rational powers of x in terms of radicals in Chapter 0, so 2^x makes sense for any rational power x. It can also be shown that the laws of exponents apply for irrational numbers. We will assume that if we graphed $y = 2^x$ for irrational values of x, those

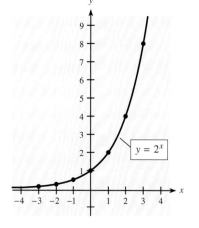

x	$y = 2^x$
-3	$2^{-3} = \frac{1}{8}$
-2	$2^{-2} = \frac{1}{4}$
-1	$2^{-1} = \frac{1}{2}$
0	$2^0 = 1$
1	$2^1 = 2$
2	$2^2 = 4$
3	$2^3 = 8$

Figure 5.1

points would lie on the curve in Figure 5.1. Thus, in general, we can graph an exponential function by plotting easily calculated points, such as those in the table in Figure 5.1, and drawing a smooth curve through the points.

● **EXAMPLE 1** *Graph of an Exponential Function*

Graph $y = 10^x$.

Solution
A table of values and the graph are given in Figure 5.2.

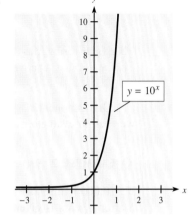

x	y
-3	$10^{-3} = 1/1000$
-2	$10^{-2} = 1/100$
-1	$10^{-1} = 1/10$
0	$10^0 = 1$
1	$10^1 = 10$
2	$10^2 = 100$
3	$10^3 = 1000$

Figure 5.2

Calculator Note   We also could have used a graphing calculator to graph $y = 2^x$ and $y = 10^x$. Figure 5.3 shows graphs of $f(x) = 1.5^x$, $f(x) = 4^x$, and $f(x) = 20^x$ that were obtained using a graphing calculator. Note the viewing window for each graph in the figure.

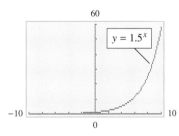

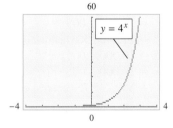

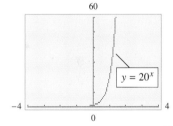

Figure 5.3

Note that the three graphs in the figure and the graphs of $y = 2^x$ and $y = 10^x$ are similar. The graphs of $y = 2^x$ and $y = 10^x$ clearly approach, but do not touch, the negative *x*-axis. However, the graphs in Figure 5.3 appear as if they might eventually merge with the negative *x*-axis. By adjusting the viewing window, however, we would see that these graphs also approach, but do not touch, the negative *x*-axis. That is, the negative *x*-axis is an asymptote for these functions.

In fact, the shapes of the graphs of functions of the form $y = f(x) = a^x$, with $a > 1$, are similar to those in Figures 5.1, 5.2, and 5.3. Exponential functions of this type, and more generally of the form $f(x) = C(a^x)$, where $C > 0$ and $a > 1$, are called **exponential growth functions** because they are used to model growth in diverse applications. Their graphs have the basic shape shown in the following box.

Graphs of Exponential Growth Functions

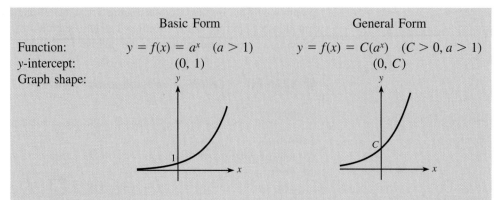

	Basic Form	General Form
Function:	$y = f(x) = a^x \quad (a > 1)$	$y = f(x) = C(a^x) \quad (C > 0, a > 1)$
y-intercept:	$(0, 1)$	$(0, C)$
Graph shape:		

For both forms, the domain is all real numbers, and the range is $y > 0$. The asymptote is the x-axis (negative half). The graphs are similar in shape.

One exponential growth model concerns money invested at compound interest. Often we seek to evaluate rather than graph these functions.

◗ **EXAMPLE 2** *Future Value of an Investment* **(Application Preview)**

If $10,000 is invested at 6%, compounded monthly, then the future value of the investment S after x years is given by

$$S = 10,000(1.005)^{12x}$$

Find the future value of the investment after (a) 5 years and (b) 30 years.

Solution
These future values can be found with a calculator.

(a) $S = 10,000(1.005)^{12(5)}$
$\quad = 10,000(1.005)^{60}$
$\quad = \$13,488.50$ (nearest cent)

(b) $S = 10,000(1.005)^{12(30)}$
$\quad = 10,000(1.005)^{360}$
$\quad = \$60,225.75$ (nearest cent)

Note that the amount after 30 years is significantly more than the amount after 5 years, a result consistent with exponential growth models.

● *Checkpoint*

1. Can any value of x give a negative value for y if $y = a^x$ and $a > 1$?
2. If $a > 1$, what asymptote does the graph of $y = a^x$ approach?

A special function that occurs frequently in economics and biology is $y = e^x$, where e is a fixed irrational number (approximately $2.71828 \ldots$). We will see how e arises when we discuss interest that is compounded continuously, and we will formally define e in Section 9.2, "Continuous Functions; Limits at Infinity."

Because $e > 1$, the graph of $y = e^x$ will have the same basic shape as other growth exponentials. We can calculate the y-coordinate for points on the graph of this function with a calculator. A table of some values (with y-values rounded to two decimal places) and the graph are shown in Figure 5.4.

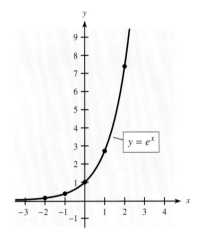

x	$y = e^x$
-2	0.14
-1	0.37
0	1.00
1	2.72
2	7.39

Figure 5.4

Exponentials whose bases are between 0 and 1, such as $y = \left(\frac{1}{2}\right)^x$, have graphs different from those of the exponentials just discussed. Using the properties of exponents, we have

$$y = \left(\frac{1}{2}\right)^x = (2^{-1})^x = 2^{-x}$$

By using this technique, all exponentials of the form $y = b^x$, where $0 < b < 1$, can be rewritten in the form $y = a^{-x}$, where $a = \frac{1}{b} > 1$. Thus, graphs of equations of the form $y = a^{-x}$, where $a > 1$, and $y = b^x$, where $0 < b < 1$, will have the same shape.

● **EXAMPLE 3** *Graphing an Exponential Function*

Graph $y = 2^{-x}$.

Solution

A table of values and the graph are given in Figure 5.5.

x	$y = 2^{-x}$
-3	$2^3 = 8$
-2	$2^2 = 4$
-1	$2^1 = 2$
0	$2^0 = 1$
1	$2^{-1} = \frac{1}{2}$
2	$2^{-2} = \frac{1}{4}$
3	$2^{-3} = \frac{1}{8}$

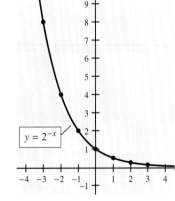

Figure 5.5

● **EXAMPLE 4** *Exponential Function with Base e*

Graph $y = e^{-2x}$.

Solution
Using a calculator to find the values of powers of e (to two decimal places), we get the graph shown in Figure 5.6.

x	$y = e^{-2x}$
-3	$e^6 = 403.43$
-2	$e^4 = 54.60$
-1	$e^2 = 7.39$
0	$e^0 = 1.00$
1	$e^{-2} = 0.14$
2	$e^{-4} = 0.02$

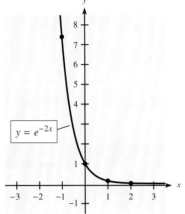

Figure 5.6

We can also use a graphing utility to graph exponential functions such as those in Examples 3 and 4. Additional examples of functions of this type would yield graphs with the same shape. **Exponential decay functions** have the form $y = a^{-x}$, where $a > 1$, or, more generally, the form $y = C(a^{-x})$, where $C > 0$ and $a > 1$. They model decay for various phenomena, and their graphs have the characteristics and shape shown in the following box.

Graphs of Exponential Decay Functions

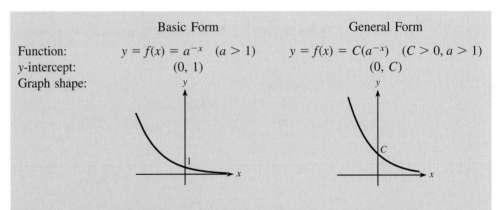

	Basic Form	General Form
Function:	$y = f(x) = a^{-x}$ $(a > 1)$	$y = f(x) = C(a^{-x})$ $(C > 0, a > 1)$
y-intercept:	$(0, 1)$	$(0, C)$
Graph shape:		

For both forms, the domain is all real numbers, the range is $y > 0$, and the asymptote is the x-axis (positive half). The graphs are similar in shape.

Exponential decay functions can be used to describe various physical phenomena, such as the number of atoms of a radioactive element (radioactive decay), the lingering effect of an advertising campaign on new sales after the campaign ends, and the effect of inflation on the purchasing power of a fixed income.

● **EXAMPLE 5** *Purchasing Power and Inflation*

The purchasing power P of a fixed income of $30,000 per year (such as a pension) after t years of 4% inflation can be modeled by

$$P = 30{,}000e^{-0.04t}$$

Find the purchasing power after (a) 5 years and (b) 20 years.

Solution
We can use a calculator to answer both parts.

(a) $P = 30{,}000e^{-0.04(5)} = 30{,}000e^{-0.2} = \$24{,}562$ (nearest dollar)
(b) $P = 30{,}000e^{-0.04(20)} = 30{,}000e^{-0.8} = \$13{,}480$ (nearest dollar)

Note that the impact of inflation over time significantly erodes purchasing power and provides some insight into the plight of elderly people who live on fixed incomes.

Exponential functions with base e often arise in natural ways. As we will see in Section 6.2, the growth of money that is compounded continuously is given by $S = Pe^{rt}$, where P is the original principal, r is the annual interest rate, and t is the time in years. Certain populations (of insects, for example) grow exponentially, and the number of individuals can be closely approximated by the equation $y = P_0 e^{ht}$, where P_0 is the original population size, h is a constant that depends on the type of population, and y is the population size at any instant t. Also, the amount y of a radioactive substance is modeled by the exponential decay equation $y = y_0 e^{-kt}$, where y_0 is the original amount and k is a constant that depends on the radioactive substance.

There are other important exponential functions that use base e but whose graphs are different from those we have discussed. For example, the standard normal probability curve (often referred to as a bell-shaped curve) is the graph of an exponential function with base e (see Figure 5.7). Later in this chapter we will study other exponential functions that model growth but whose graphs are also different from those discussed previously.

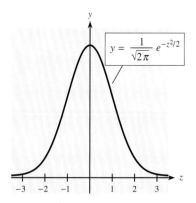

$$y = \frac{1}{\sqrt{2\pi}}\, e^{-z^2/2}$$

Figure 5.7   Standard normal probability curve

● *Checkpoint*

3. True or false: The graph of $y = \left(\dfrac{1}{a}\right)^x$, with $a > 1$, is the same as the graph of $y = a^{-x}$.

4. True or false: The graph of $y = a^{-x}$, with $a > 1$, approaches the positive x-axis as an asymptote.

5. True or false: The graph of $y = 2^{-x}$ is the graph of $y = 2^x$ reflected about the y-axis.

Calculator
Note

Graphing calculators and other graphing utilities allow us to investigate variations of the growth and decay exponentials. For example, we can examine the similarities and differences between the graphs of $y = f(x) = a^x$ and those of $y = mf(x)$, $y = f(kx)$, $y = f(x + h)$, and $y = f(x) + C$ for constants m, k, h, and C. We also can explore these similarities and differences for $y = f(x) = a^{-x}$. ■

● **EXAMPLE 6** *Translations of Exponential Functions*

(a) Use a graphing utility to graph $y = f(x) = e^x$ and $y = mf(x) = me^x$ for $m = -6, -3,$ 2, and 10.

(b) Explain the effects that m has on the shape of the graph, the asymptote, and the y-intercept.

Solution

(a) The graphs are shown in Figure 5.8.

(b) From the graphs, we see that when $m > 0$, the shape of the graph is still that of a growth exponential. When $m < 0$, the shape is that of a growth exponential turned upside down, or reflected through the x-axis (making the range $y < 0$). In all cases, the asymptote is unchanged and is the negative x-axis. In each case, the y-intercept is $(0, m)$.

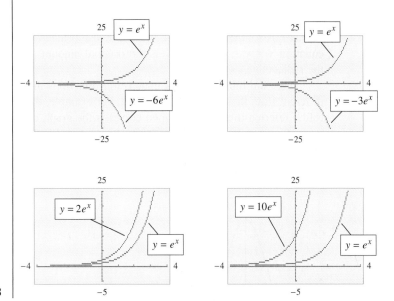

Figure 5.8

Because the outputs of exponential functions grow very rapidly, it is sometimes hard to find an appropriate window in which to show the graph of an exponential function on a graphing utility. One way to find appropriate values of y-min and y-max when x-min and x-max are given is to use TABLE to find values of y for selected values of x. For example, Figure 5.9 shows how we can determine a window suitable for graphing

$$y = 1000e^{0.12x}$$

Figure 5.9(a) shows a table of values of y that correspond to certain values of x, and Figure 5.9(b) shows the graph in a window that uses the table values to set y-max equal to 3500, with y-min $= 0$.

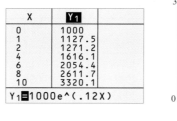

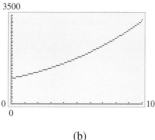

Figure 5.9   (a)   (b)

Modeling with Exponential Functions

Many types of data can be modeled using an exponential growth function. Figure 5.10(a) shows a graph of amounts of carbon in the atmosphere due to emissions from the burning of fossil fuels. With curve-fitting tools available with some computer software or on graphing calculators, we can develop an equation that models, or approximates, these data. The equation model and its graph are shown in Figure 5.10(b).

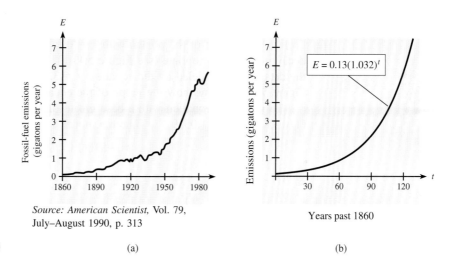

Source: American Scientist, Vol. 79, July–August 1990, p. 313

Figure 5.10   (a)   (b)

The model in Figure 5.10(b) is an exponential growth function. As we have noted, functions of this type have the general form $y = C(a^x)$ with $a > 1$, and exponential decay functions have the general form $y = C(a^x)$ with $0 < a < 1$ [or $y = C(a^{-x})$ with $a > 1$]. Note that when technology is applied to model exponentials (either growth or decay), the form $y = C(a^x)$ is used, but it is written in the equivalent form $y = a*b^x$.

Modeling

● **EXAMPLE 7** *Consumer Price Index*

The purchasing power of $1 is calculated by assuming that the consumer price index (CPI) was $1 during 1983. It is used to show the values of goods that could be purchased for $1 in 1983. Table 5.2 gives the purchasing power of a 1983 dollar for the years 1968–2002.

(a) Use these data, with $x = 0$ in 1960, to find an exponential function that models the decay of the dollar.
(b) What does this model predict as the purchasing power of $1 in 2010?
(c) When will the purchasing power of a dollar fall below $0.20?

TABLE 5.2

Year	Purchasing Power of $1	Year	Purchasing Power of $1	Year	Purchasing Power of $1
1968	2.873	1980	1.215	1992	0.713
1969	2.726	1981	1.098	1993	0.692
1970	2.574	1982	1.035	1994	0.675
1971	2.466	1983	1.003	1995	0.656
1972	2.391	1984	0.961	1996	0.637
1973	2.251	1985	0.928	1997	0.623
1974	2.029	1986	0.913	1998	0.613
1975	1.859	1987	0.880	1999	0.600
1976	1.757	1988	0.846	2000	0.581
1977	1.649	1989	0.807	2001	0.565
1978	1.532	1990	0.766	2002	0.556
1979	1.380	1991	0.734		

Source: U.S. Bureau of Labor Statistics

Solution

(a) From a scatter plot of the points, with $x = 0$ in 1960, it appears that an exponential decay model is appropriate (see Figure 5.11(a)). Using the function of the form $a*b^x$, the data can be modeled by the equation $y = 3.818(0.95066^x)$ with $x = 0$ in 1960. Figure 5.11(b) shows how the graph of the equation fits the data.

  To write this equation with a base greater than 1, we can write 0.95066 as $(1/0.95066)^{-1} = (1.0519)^{-1}$. Thus the model also could be written as

$$y = 3.818(1.0519)^{-x}$$

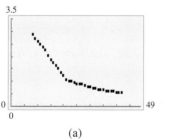

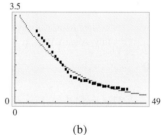

Figure 5.11

(a)   (b)

(b) For the year 2010, we use $x = 50$. Evaluating the function for $x = 50$ gives the purchasing power of $0.3042.

(c) Evaluation from the graph or a table shows that the purchasing power is below $0.20 when x is 59, or in the year 2019, if the model remains valid.

Spreadsheet Note

A spreadsheet also can be used to create a scatter plot of the data in Table 5.2 and to develop an exponential model. Figure 5.12(a) shows an Excel scatter plot of the Table 5.2 data, and Figure 5.12(b) shows the scatter plot with Excel's base e exponential model. (Procedural details for using Excel are in the supplement *Excel Guide for Finite Mathematics and Applied Calculus*.)

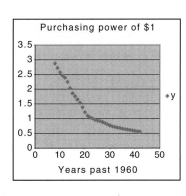

Figure 5.12   (a)   (b)

● **Checkpoint Solutions**
1. No, all values of y are positive.
2. The left side of the x-axis ($y = 0$)
3. True
4. True
5. True

5.1 Exercises

In Problems 1–8, use a calculator to evaluate each expression.

1. $10^{0.5}$
2. $10^{3.6}$
3. $5^{-2.7}$
4. $8^{-2.6}$
5. $3^{1/3}$
6. $2^{11/6}$
7. e^2
8. e^{-3}

In Problems 9–18, graph each function.

9. $y = 4^x$
10. $y = 8^x$
11. $y = 2(3^x)$
12. $y = 3(2^x)$
13. $y = 5^{-x}$
14. $y = 3^{x-1}$
15. $y = 2e^x$
16. $y = 3^{-x}$
17. $y = 3^{-2x}$
18. $y = 5e^{-x}$

19. (a) Find $b > 1$ to express $y = 3(\frac{2}{5})^x$ in the form $y = 3(b^{-x})$.
   (b) Are these functions growth exponentials or decay exponentials? Explain.
   (c) Check your result by graphing both functions.

20. (a) Find $b > 1$ to express $y = 8(\frac{5}{7})^x$ in the form $y = 8(b^{-x})$.
   (b) Are these functions growth exponentials or decay exponentials? Explain.
   (c) Check your result by graphing both functions.

21. (a) Graph $y = 2(1.5)^{-x}$.
   (b) Graph $y = 2(\frac{2}{3})^x$.
   (c) Algebraically show why these graphs are identical.

22. (a) Graph $y = 2.5(3.25)^{-x}$.
   (b) Graph $y = 2.5(\frac{4}{13})^x$.
   (c) Algebraically show why these graphs are identical.

In Problems 23–26, use a graphing utility to graph the functions.

23. Given $f(x) = e^{-x}$. Graph $y = f(x)$ and $y = f(kx) = e^{-kx}$ for each k, where $k = 0.1, 0.5, 2$, and 5. Explain the effect that different values of k have on the graphs.

24. Given $f(x) = 2^{-x}$. Graph $y = f(x)$ and $y = mf(x) = m(2^{-x})$ for each m, where $m = -7, -2, 3$, and 8. Explain the effect that different values of m have on the graphs.

25. Given $f(x) = 4^x$. Graph $y = f(x)$ and $y = f(x) + C = 4^x + C$ for each C, where $C = -5, -2, 3$, and 6. Explain the effect that C has on the graphs.

26. Given $f(x) = 3^x$. Graph $y = f(x)$ and $y = f(x - h) = 3^{x-h}$ for each h, where $h = 4, 1, -2$, and -5. Explain the effect that h has on the graphs.

For Problems 27 and 28, let $f(x) = c(1 + e^{-ax})$ with $a > 0$. Use a graphing utility to graph the functions.

27. (a) Fix $a = 1$ and graph $y = f(x) = c(1 + e^{-x})$ for $c = 10, 50$, and 100.
   (b) What effect does c have on the graphs?

28. (a) Fix $c = 50$ and graph $y = f(x) = 50(1 + e^{-ax})$ for $a = 0.1, 1$, and 10.
   (b) What effect does a have on the graphs?

APPLICATIONS

29. *Compound interest*   If $1000 is invested for x years at 8%, compounded quarterly, the future value that will result is

$$S = 1000(1.02)^{4x}$$

What amount will result in 8 years?

30. *Compound interest*   If $3200 is invested for x years at 8%, compounded quarterly, the interest earned is

$$I = 3200(1.02)^{4x} - 3200$$

What interest is earned in 5 years?

31. *Compound interest*   We will show in the next chapter that if P is invested for n years at 10% compounded continuously, the future value of the investment is given by

$$S = Pe^{0.1n}$$

Use $P = 1000$ and graph this function for $0 \leq n \leq 20$.

32. *Compound interest*   If $1000 is invested for x years at 10%, compounded continuously, the future value that results is

$$S = 1000e^{0.10x}$$

What amount will result in 5 years?

33. *Drug in the bloodstream*   The percent concentration y of a certain drug in the bloodstream at any time t in minutes is given by the equation

$$y = 100(1 - e^{-0.462t})$$

Graph this equation for $0 \leq t \leq 10$. Write a sentence that interprets the graph.

34. *Bacterial growth*   A single bacterium splits into two bacteria every half hour, so the number of bacteria in a culture quadruples every hour. Thus the equation by which a colony of 10 bacteria multiplies in t hours is given by

$$y = 10(4^t)$$

Graph this equation for $0 \leq t \leq 8$.

35. *Product reliability*   A statistical study shows that the fraction of television sets of a certain brand that are still in service after x years is given by $f(x) = e^{-0.15x}$. Graph this equation for $0 \leq x \leq 10$. Write a sentence that interprets the graph.

36. *National health care expenditures*   Using data from the U.S. Department of Health and Human Services, the national health care expenditures H can be modeled by

$$H = 29.57e^{0.097t}$$

where t is the number of years after 1960 and H is in billions of dollars. Graph this equation with a graphing utility to show the graph through the year 2010.

Population growth   **Use the following information to answer Problems 37–40. World population can be considered as growing according to the equation**

$$N = N_0(1 + r)^t$$

where N_0 is the number of individuals at time $t = 0$, r is the yearly rate of growth, and t is the number of years.

37. Sketch the graph for $t = 0$ to $t = 10$ when the growth rate is 2% and N_0 is 4.1 billion.

38. Sketch the graph for $t = 0$ to $t = 10$ when the growth rate is 3% and N_0 is 4.1 billion.

39. Repeat Problem 37 when the growth rate is 5%.

40. Repeat Problem 38 when the growth rate is 7%.

41. *Wireless subscribers*   By using data from The C.T.I.A. Semi-Annual Wireless Survey, the number of wireless subscribers (in millions) can be modeled by

$$N = 12.58e^{0.2096t}$$

where t is the number of years past 1990.
   (a) Is this model one of exponential growth or exponential decay? Explain.
   (b) Graph this equation with a graphing utility to show the graph from 1990 through the year 2010.

42. *Advertising and sales*   Suppose that sales are related to advertising expenditures according to one of the following two models, where S_1 and S_n are sales and x is advertising, all in millions of dollars.

$$S_1 = 30 + 20x - 0.4x^2$$
$$S_n = 24.58 + 325.18(1 - e^{-x/14})$$

   (a) Graph both of these functions on the same set of axes. Use a graphing utility.
   (b) Do these two functions give approximately the same sales per million dollars of advertising for $0 \leq x \leq 20$?
   (c) How do these functions differ for $x > 20$? Which more realistically represents the relationship between sales and advertising expenditures after $20 million is spent on advertising? Why?

43. *Industrial consolidation*   The following figure, from *Investor's Business Daily* (March 5, 1998), shows how quickly the U.S. metal processing industry is consolidating. The linear equation that is the best fit for the number of metal processors as a function of years after 1990 is $y_1 = -329.6970x + 5392.1212$, and the best exponential fit is $y_2 = 9933.7353e^{-0.1672x}$. The linear equation gives a much better fit for the data points than the exponential equation. Why, then, is the exponential

equation a more useful model to predict the number of metal processors in 2010?

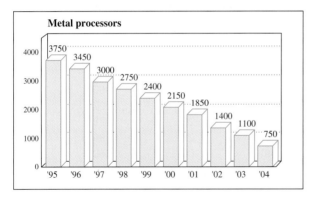

44. **Modeling** *National debt* The table below gives the U.S. national debt for selected years from 1900 to 2003.
    (a) Using a function of the form $y = a*b^x$, with $x = 0$ in 1900 and y equal to the national debt in billions, model the data.
    (b) Use the model to predict the debt in 2010.
    (c) Predict when the debt will be $25 trillion ($25,000 billion).

Year	U.S. Debt ($billions)	Year	U.S. Debt ($billions)
1900	1.2	1990	3233.3
1910	1.1	1992	4064.6
1920	24.2	1994	4692.8
1930	16.1	1996	5224.8
1940	43.0	1998	5526.2
1945	258.7	1999	5656.3
1955	272.8	2000	5674.2
1965	313.8	2001	5807.5
1975	533.2	2002	6228.2
1985	1823.1	2003	6783.2

Source: Bureau of Public Debt, U.S. Treasury

(d) Look at a graph of both the data and the model. What events may affect the accuracy of this model as a predictor of future public debt? Explain.

45. **Modeling** *Personal income* Total personal income in the United States (in billions of dollars) for selected years from 1960 to 2002 is given in the following table.

Year	1960	1970	1980	1990	2000	2002
Personal Income	411.7	836.1	2285.7	4791.6	8406.6	8929.1

Source: Bureau of Economic Analysis, U.S. Department of Commerce

(a) These data can be modeled by an exponential function. Write the equation of this function, with x as the number of years past 1960.
(b) If this model is accurate, what will be the total U.S. personal income in 2010?

46. **Modeling** *Daily shares traded on the NYSE* Selected data for the average daily shares traded (in thousands) on the New York Stock Exchange from 1900 to 2002 are given in the following table.

Year	Daily Average	Year	Daily Average
1900	505	1985	109,169
1920	828	1990	156,777
1940	751	1995	346,101
1950	1,980	1998	673,590
1960	3,042	1999	809,183
1970	11,564	2000	1,041,578
1980	44,871	2002	1,441,015

Source: New York Stock Exchange, *Fact Book, 2002*

(a) Use technology to create an exponential model for the data (with $x = 0$ in 1900).
(b) Use the model to estimate the average daily shares traded in 2000.
(c) Examine a graph of both the data and the model. Comment on the fit of the model to the data.

47. **Modeling** *Consumer Price Index* The consumer price index (CPI) is calculated by averaging the prices of various items after assigning a weight to each item. The following table gives the consumer price indexes for selected years from 1940 through 2002, reflecting buying patterns of all urban consumers, with x representing years past 1900.
    (a) Find an equation that models these data.
    (b) Use the model to predict the consumer price index in 2010.
    (c) According to the model, during what year will the consumer price index pass 300?

Year	Consumer Price Index	Year	Consumer Price Index
1940	14	1980	82.4
1950	24.1	1990	130.7
1960	29.6	2000	172.2
1970	38.8	2002	179.9

Source: U.S. Bureau of the Census

48. **Modeling** *Compound interest* The following table gives the value of an investment, after intervals ranging from 0 to 7 years, of $20,000 invested at 10%, compounded annually.
    (a) Develop an exponential model for these data, accurate to four decimal places, with x in years and y in dollars.
    (b) Use the model to find the amount to which $20,000 will grow in 30 years if it is invested at 10%, compounded annually.

Year	Amount Investment Grows to
0	20,000
1	22,000
2	24,200
3	26,620
4	29,282
5	32,210.20
6	35,431.22
7	38,974.34

49. **Modeling** *Students per computer* The following table gives the average number of students per computer in public schools for the school years that ended in 1985 through 2002.
  (a) Find an exponential model for these data. Let x be the number of years past 1980.
  (b) Is this model an exponential growth function or an exponential decay function? Explain how you know.

(c) How many students per computer in public schools does this model predict for 2010?

Year	Students per Computer	Year	Students per Computer
1985	75	1994	14
1986	50	1995	10.5
1987	37	1996	10
1988	32	1997	7.8
1989	25	1998	6.1
1990	22	1999	5.7
1991	20	2000	5.4
1992	18	2001	5.0
1993	16	2002	4.9

Source: Quality Education Data, Inc., Denver, Colorado

5.2 Logarithmic Functions and Their Properties

OBJECTIVES

- To convert equations for logarithmic functions from logarithmic to exponential form, and vice versa
- To evaluate some special logarithms
- To graph logarithmic functions
- To model logarithmic functions
- To use properties of logarithmic functions to simplify expressions involving logarithms
- To use the change-of-base formula

◑ Application Preview

If P dollars is invested at an annual rate r, compounded continuously, then the future value of the investment after t years is given by

$$S = Pe^{rt}$$

A common question with investments such as this is "How long will it be before the investment doubles?" That is, when does $S = 2P$? (See Example 4.) The answer to this question gives an important formula, called a "doubling-time" formula. This requires the use of **logarithmic functions.**

Logarithmic Functions and Graphs

Before the development and easy availability of calculators and computers, certain arithmetic computations, such as $(1.37)^{13}$ and $\sqrt[16]{3.09}$, were difficult to perform. The computations could be performed relatively easily using **logarithms,** which were developed in the 17th century by John Napier, or by using a slide rule, which is based on logarithms. The use of logarithms as a computing technique has all but disappeared today, but the study of **logarithmic functions** is still very important because of the many applications of these functions.

For example, let us again consider the culture of bacteria described at the beginning of the previous section. If we know that the culture is begun with one microorganism and that each minute every microorganism present splits into two new ones, then we can find the number of minutes it takes until there are 1024 organisms by solving

$$1024 = 2^y$$

The solution of this equation may be written in the form

$$y = \log_2 1024$$

which is read "y equals the logarithm of 1024 to the base 2."

In general, we may express the equation $x = a^y$ $(a > 0, a \neq 1)$ in the form $y = f(x)$ by defining a **logarithmic function.**

Logarithmic Function

> For $a > 0$ and $a \neq 1$, the **logarithmic function**
>
> $$y = \log_a x \quad \text{(logarithmic form)}$$
>
> has domain $x > 0$, base a, and is defined by
>
> $$a^y = x \quad \text{(exponential form)}$$

TABLE 5.3

Logarithmic Form	Exponential Form
$\log_{10} 100 = 2$	$10^2 = 100$
$\log_{10} 0.1 = -1$	$10^{-1} = 0.1$
$\log_2 x = y$	$2^y = x$
$\log_a 1 = 0 \ (a > 0)$	$a^0 = 1$
$\log_a a = 1 \ (a > 0)$	$a^1 = a$

From the definition, we know that $y = \log_a x$ means $x = a^y$. This means that $\log_3 81 = 4$ because $3^4 = 81$. In this case the logarithm, 4, is the exponent to which we have to raise the base 3 to obtain 81. In general, if $y = \log_a x$, then y is the exponent to which the base a must be raised to obtain x.

The a is called the **base** in both $\log_a x = y$ and $a^y = x$, and y is the *logarithm* in $\log_a x = y$ and the *exponent* in $a^y = x$. Thus **a logarithm is an exponent.**

Table 5.3 shows some logarithmic equations and their equivalent exponential forms.

● **EXAMPLE 1** *Logarithms and Exponential Forms*

(a) Write $64 = 4^3$ in logarithmic form.
(b) Write $\log_4 \left(\frac{1}{64}\right) = -3$ in exponential form.
(c) If $4 = \log_2 x$, find x.

Solution

(a) In $64 = 4^3$, the base is 4 and the exponent (or logarithm) is 3. Thus $64 = 4^3$ is equivalent to $3 = \log_4 64$.
(b) In $\log_4 \left(\frac{1}{64}\right) = -3$, the base is 4 and the logarithm (or exponent) is -3. Thus $\log_4 \left(\frac{1}{64}\right) = -3$ is equivalent to $4^{-3} = \frac{1}{64}$.
(c) If $4 = \log_2 x$, then $2^4 = x$ and $x = 16$.

● **EXAMPLE 2** *Evaluating Logarithms*

Evaluate:

(a) $\log_2 8$
(b) $\log_3 9$
(c) $\log_5 \left(\frac{1}{25}\right)$

Solution

(a) If $y = \log_2 8$, then $8 = 2^y$. Because $2^3 = 8$, $\log_2 8 = 3$.
(b) If $y = \log_3 9$, then $9 = 3^y$. Because $3^2 = 9$, $\log_3 9 = 2$.
(c) If $y = \log_5 \left(\frac{1}{25}\right)$, then $\frac{1}{25} = 5^y$. Because $5^{-2} = \frac{1}{25}$, $\log_5 \left(\frac{1}{25}\right) = -2$.

● **EXAMPLE 3** *Graphing a Logarithmic Function*

Graph $y = \log_2 x$.

Solution

We may graph $y = \log_2 x$ by graphing $x = 2^y$. The table of values (found by substituting values of y and calculating x) and the graph are shown in Figure 5.13.

$x = 2^y$	y
$\frac{1}{8}$	-3
$\frac{1}{4}$	-2
$\frac{1}{2}$	-1
1	0
2	1
4	2
8	3

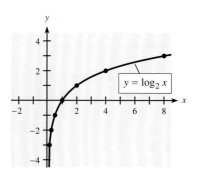

Figure 5.13

From the definition of logarithms, we see that every logarithm has a base. Most applications of logarithms involve logarithms to the base 10 (called **common logarithms**) or logarithms to the base e (called **natural logarithms**). In fact, logarithms to the base 10 and to the base e are the only ones that have function keys on calculators. Thus it is important to be familiar with their names and designations.

Common and Natural Logarithms

Common logarithms:	$\log x$	means	$\log_{10} x$.
Natural logarithms:	$\ln x$	means	$\log_e x$.

Values of common and natural logarithmic functions are usually found with a calculator. For example, a calculator gives $\log 2 \approx 0.301$ and $\ln 2 \approx 0.693$.

◗ EXAMPLE 4 Doubling Time for an Investment (Application Preview)

If \$$P$ is invested for t years at interest rate r, compounded continuously, then the future value of the investment is given by $S = Pe^{rt}$. The doubling time for this investment can be found by solving for t in $S = Pe^{rt}$ when $S = 2P$. That is, we must solve $2P = Pe^{rt}$, or (equivalently) $2 = e^{rt}$.

(a) Express $2 = e^{rt}$ in logarithmic form and then solve for t to find the doubling-time formula.
(b) If an investment earns 10% compounded continuously, in how many years will it double?

Solution
(a) In logarithmic form, $2 = e^{rt}$ is equivalent to $\log_e 2 = rt$. Solving for t gives the doubling-time formula

$$t = \frac{\log_e 2}{r} = \frac{\ln 2}{r}$$

(b) If the interest rate is $r = 10\%$, compounded continuously, the time required for the investment to double is

$$t = \frac{\ln 2}{0.10} \approx 6.93 \text{ years}$$

Note that we could write the doubling time for this problem as

$$t = \frac{\ln 2}{0.10} \approx \frac{0.693}{0.10} = \frac{69.3}{10}$$

In general we can approximate the doubling time for an investment at $r\%$, compounded continuously, with $\frac{70}{r}$. (In economics, this is called the Rule of 70.)

● **EXAMPLE 5** *Market Share*

Suppose that after a company introduces a new product, the number of months m before its market share is s percent can be modeled by

$$m = 20 \ln \left(\frac{40}{40 - s} \right)$$

When will its product have a 35% share of the market?

Solution

A 35% market share means $s = 35$. Hence

$$m = 20 \ln \left(\frac{40}{40 - s} \right)$$

$$= 20 \ln \left(\frac{40}{40 - 35} \right) = 20 \ln \left(\frac{40}{5} \right) = 20 \ln 8 \approx 41.6$$

Thus, the market share will be 35% after about 41.6 months.

● **EXAMPLE 6** *Natural Logarithm*

Graph $y = \ln x$.

Solution

We can graph $y = \ln x$ by evaluating $y = \ln x$ for $x > 0$ (including some values $0 < x < 1$) with a calculator. The graph is shown in Figure 5.14.

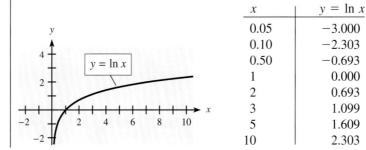

x	$y = \ln x$
0.05	-3.000
0.10	-2.303
0.50	-0.693
1	0.000
2	0.693
3	1.099
5	1.609
10	2.303

Figure 5.14

Note that because $\ln x$ is a standard function on a graphing utility, such a utility could be used to obtain its graph. Note also that the graphs of $y = \log_2 x$ (Figure 5.13) and $y = \ln x$ (Figure 5.14) are very similar. The shapes of graphs of equations of the form $y = \log_a x$ with $a > 1$ are similar to these two graphs.

Graphs of Logarithmic Functions

Equation: $y = \log_a x$ $(a > 1)$
x-intercept: $(1, 0)$
Domain: All positive reals
Range: All reals
Asymptote: y-axis (negtative half)

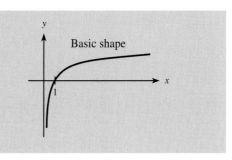

● *Checkpoint*

1. What asymptote does the graph of $y = \log_a x$ approach when $a > 1$?
2. For $a > 1$, does the equation $y = \log_a x$ represent the same function as the equation $x = a^y$?
3. For what values of x is $y = \log_a x, a > 0, a \neq 1$, defined?

Modeling with Logarithmic Functions

The basic shape of a logarithmic function is important for two reasons. First, when we graph a logarithmic function, we know that the graph should have this shape. Second, when data points have this basic shape, they suggest a logarithmic model.

Modeling

● EXAMPLE 7 *Life Expectancy*

The expected life span of people in the United States depends on their year of birth (see the table and scatter plot, with $x = 0$ in 1900, in Figure 5.15). In Section 2.5, "Modeling; Fitting Curves to Data with Graphing Utilities," we modeled life span with a linear function and with a quadratic function. However, the scatter plot suggests that the best model may be logarithmic.

(a) Use technology to find a logarithmic equation that models the data.
(b) The National Center for Health Statistics projects the expected life span for people born in 2000 to be 77.0 and that for those born in 2010 to be 77.9. Use your model to project the life spans for those years.

Year	Life Span (years)	Year	Life Span (years)
1920	54.1	1989	75.2
1930	59.7	1990	75.4
1940	62.9	1992	75.8
1950	68.2	1994	75.7
1960	69.7	1996	76.1
1970	70.8	1998	76.7
1975	72.6	1999	76.7
1980	73.7	2000	77.0
1987	75.0	2001	77.2
1988	74.9		

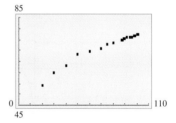

Source: National Center for Health Statistics

Figure 5.15

Solution

(a) A graphing calculator gives the logarithmic model

$$L(x) = 11.307 + 14.229 \ln x$$

where x is the number of years since 1900. Figure 5.16 shows the scatter plot and graph of the model.

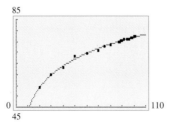

Figure 5.16

(b) For the year 2000, the value of x is 100, so the projected life span is $L(100) = 76.8$. For 2010, the model gives 78.2 as the expected life span. These calculations closely approximate the projections from the National Center for Health Statistics.

Properties of Logarithms

The definition of the logarithmic function and the previous examples suggest a special relationship between the logarithmic function $y = \log_a x$ and the exponential function $y = a^x$ ($a > 0$, $a \neq 1$). Because we can write $y = \log_a x$ in exponential form as $x = a^y$, we see that the connection between

$$y = \log_a x \quad \text{and} \quad y = a^x$$

is that x and y have been interchanged from one function to the other. This is true for the functional description and hence for the ordered pairs or coordinates that satisfy these functions. This is illustrated in Table 5.4 for the functions $y = \log x$ ($y = \log_{10} x$) and $y = 10^x$.

TABLE 5.4

$y = \log_{10} x = \log x$		$y = 10^x$	
Coordinates	Justification	Coordinates	Justification
$(1000, 3)$	$3 = \log 1000$	$(3, 1000)$	$1000 = 10^3$
$(100, 2)$	$2 = \log 100$	$(2, 100)$	$100 = 10^2$
$\left(\frac{1}{10}, -1\right)$	$-1 = \log \frac{1}{10}$	$\left(-1, \frac{1}{10}\right)$	$\frac{1}{10} = 10^{-1}$

In general we say that $y = f(x)$ and $y = g(x)$ are **inverse functions** if, whenever the pair (a, b) satisfies $y = f(x)$, then the pair (b, a) satisfies $y = g(x)$. Furthermore, because the values of the x- and y-coordinates are interchanged for inverse functions, their graphs are reflections of each other about the line $y = x$.

Thus for $a > 0$ and $a \neq 1$, the logarithmic function $y = \log_a x$ (also written $x = a^y$) and the exponential function $y = a^x$ are inverse functions.

The logarithmic function $y = \log x$ is the inverse of the exponential function $y = 10^x$. Thus the graphs of $y = \log x$ and $y = 10^x$ are reflections of each other about the line $y = x$. Some values of x and y for these functions are given in Table 5.4, and their graphs are shown in Figure 5.17.

This inverse relationship is a consequence of the definition of the logarithmic function. We can use this definition to discover several other properties of logarithms.

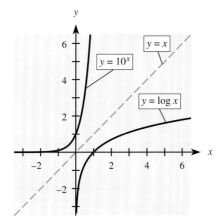

Figure 5.17

Because logarithms are exponents, the properties of logarithms can be derived from the properties of exponents. (The properties of exponents are discussed in Chapter 0, "Algebraic Concepts.") The following properties of logarithms are useful in simplifying expressions that contain logarithms.

Logarithmic Property I

> If $a > 0, a \neq 1$, then $\log_a a^x = x$, for any real number x.

To prove this result, note that the exponential form of $y = \log_a a^x$ is $a^y = a^x$, so $y = x$. That is, $\log_a a^x = x$.

● **EXAMPLE 8** *Logarithmic Property I*

Use Property I to simplify each of the following.

(a) $\log_4 4^3$ (b) $\ln e^x$

Solution
(a) $\log_4 4^3 = 3$
Check: The exponential form is $4^3 = 4^3$. ✔

(b) $\ln e^x = \log_e e^x = x$
Check: The exponential form is $e^x = e^x$. ✔

We note that two special cases of Property I are used frequently; these are when $x = 1$ and $x = 0$.

Special Cases of Logarithmic Property I

> Because $a^1 = a$, we have $\log_a a = 1$.
> Because $a^0 = 1$, we have $\log_a 1 = 0$.

The logarithmic form of $y = a^{\log_a x}$ is $\log_a y = \log_a x$, so $y = x$. This means that $a^{\log_a x} = x$ and proves Property II.

Logarithmic Property II

> If $a > 0, a \neq 1$, then $a^{\log_a x} = x$, for any positive real number x.

● **EXAMPLE 9** *Logarithmic Property II*

Use Property II to simplify each of the following.

(a) $2^{\log_2 4}$ (b) $e^{\ln x}$

Solution
(a) $2^{\log_2 4} = 4$
Check: The logarithmic form is $\log_2 4 = \log_2 4$. ✔

(b) $e^{\ln x} = x$
Check: The logarithmic form is $\log_e x = \ln x$. ✔

If $u = \log_a M$ and $v = \log_a N$, then their exponential forms are $a^u = M$ and $a^v = N$, respectively. Thus

$$\log_a (MN) = \log_a (a^u \cdot a^v) = \log_a (a^{u+v}) = u + v = \log_a M + \log_a N$$

and Property III is established.

Logarithmic Property III

If $a > 0$, $a \neq 1$, and M and N are positive real numbers, then

$$\log_a (MN) = \log_a M + \log_a N$$

● **EXAMPLE 10** *Logarithmic Property III*

(a) Find $\log_2 (4 \cdot 16)$, if $\log_2 4 = 2$ and $\log_2 16 = 4$.
(b) Find $\log_{10} (4 \cdot 5)$ if $\log_{10} 4 = 0.6021$ and $\log_{10} 5 = 0.6990$ (to four decimal places).
(c) Find $\ln 77$, if $\ln 7 = 1.9459$ and $\ln 11 = 2.3979$ (to four decimal places).

Solution
(a) $\log_2 (4 \cdot 16) = \log_2 4 + \log_2 16 = 2 + 4 = 6$
Check: $\log_2 (4 \cdot 16) = \log_2 (64) = 6$, because $2^6 = 64$. ✔

(b) $\log_{10} (4 \cdot 5) = \log_{10} 4 + \log_{10} 5 = 0.6021 + 0.6990 = 1.3011$
Check: Using a calculator, we can see that $10^{1.3011} \approx 20 = 4 \cdot 5$. ✔

(c) $\ln 77 = \ln (7 \cdot 11) = \ln 7 + \ln 11 = 1.9459 + 2.3979 = 4.3438$
Check: Using a calculator, we can see that $e^{4.3438} \approx 77$. ✔

Logarithmic Property IV

If $a > 0$, $a \neq 1$, and M and N are positive real numbers, then

$$\log_a (M/N) = \log_a M - \log_a N$$

The proof of this property is left to the student in Problem 49.

● **EXAMPLE 11** *Logarithmic Property IV*

(a) Evaluate $\log_3 \left(\frac{9}{27} \right)$.
(b) Find $\log_{10} \left(\frac{16}{5} \right)$, if $\log_{10} 16 = 1.2041$ and $\log_{10} 5 = 0.6990$ (to four decimal places).
(c) Find $\ln \left(\frac{3}{8} \right)$, if $\ln 3 = 1.0986$ and $\ln 8 = 2.0794$ (to four decimal places).

Solution
(a) $\log_3 \left(\frac{9}{27} \right) = \log_3 9 - \log_3 27 = 2 - 3 = -1$
Check: $\log_3 \left(\frac{1}{3} \right) = -1$ because $3^{-1} = \frac{1}{3}$. ✔

(b) $\log_{10} \left(\frac{16}{5} \right) = \log_{10} 16 - \log_{10} 5 = 1.2041 - 0.6990 = 0.5051$
Check: Using a calculator, we can see that $10^{0.5051} \approx 3.2 = \frac{16}{5}$. ✔

(c) $\ln \left(\frac{3}{8} \right) = \ln 3 - \ln 8 = 1.0986 - 2.0794 = -0.9808$
Check: Using a calculator, we can see that $e^{-0.9808} \approx 0.375 = \frac{3}{8}$. ✔

Logarithmic Property V

If $a > 0$, $a \neq 1$, M is a positive real number, and N is any real number, then

$$\log_a (M^N) = N \log_a M$$

The proof of this property is left to the student in Problem 50.

● **EXAMPLE 12** *Logarithmic Property V*

(a) Simplify $\log_3 (9^2)$.

(b) Simplify $\ln 8^{-4}$, if $\ln 8 = 2.0794$ (to four decimal places).

Solution

(a) $\log_3 (9^2) = 2 \log_3 9 = 2 \cdot 2 = 4$

Check: $\log_3 81 = 4$ because $3^4 = 81$. ✔

(b) $\ln 8^{-4} = -4 \ln 8 = -4(2.0794) = -8.3176$

Check: Using a calculator, we can see that $8^{-4} \approx 0.000244$ and $\ln (0.000244) \approx -8.3176$. ✔

● **Checkpoint**

4. Simplify:

    (a) $6^{\log_6 x}$ (b) $\log_7 7^3$ (c) $\ln \left(\dfrac{1}{e^2} \right)$ (d) $\log 1$

5. If $\log_7 20 = 1.5395$ and $\log_7 11 = 1.2323$, find the following.

    (a) $\log_7 220 = \log_7 (20 \cdot 11)$ (b) $\log_7 \left(\dfrac{11}{20} \right)$ (c) $\log_7 20^4$

Change of Base

By using a calculator, we can directly evaluate only those logarithms with base 10 or base e. Also, logarithms with base 10 or base e are the only ones that are standard functions on a graphing calculator. Thus, if we had a way to express a logarithmic function with any base in terms of a logarithm with base 10 or base e, we would be able to evaluate the original logarithmic function and graph it with a graphing calculator.

In general, if we use the properties of logarithms, we can write

$$y = \log_b x \quad \text{in the form} \quad b^y = x$$

If we take the base a logarithm of both sides, we have

$$\log_a b^y = \log_a x$$
$$y \log_a b = \log_a x$$
$$y = \frac{\log_a x}{\log_a b}$$

This gives us the **change-of-base formula** from base b to base a.

Change-of-Base Formulas

If $a \neq 1, b \neq 1, a > 0, b > 0$, then

$$\log_b x = \frac{\log_a x}{\log_a b}$$

For the purposes of calculation, we usually convert logarithms to base e.

$$\log_b x = \frac{\ln x}{\ln b}$$

● **EXAMPLE 13** *Change-of-Base Formula*

Evaluate $\log_7 15$ by using a change-of-base formula.

Solution

$$\log_7 15 = \frac{\ln 15}{\ln 7} = \frac{2.70805}{1.94591} = 1.39166 \text{ (approximately)}$$

As a check, a calculator shows that $7^{1.39166} \approx 15$. ✔

● **EXAMPLE 14** *Graphing with Technology*

Use a change-of-base formula to rewrite each of the following logarithms in base e. Then graph each with a graphing utility.

(a) $y = \log_2 x$ (b) $y = \log_7 x$

Solution

(a) $y = \log_2 x = \dfrac{\ln x}{\ln 2}$. See Figure 5.18(a) for the graph. Note that it is exactly the same as the graph in Figure 5.13 in Example 3.

(b) $y = \log_7 x = \dfrac{\ln x}{\ln 7}$. See Figure 5.18(b) for the graph.

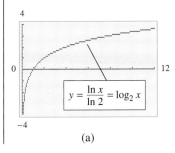
(a)

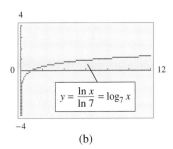
(b)

Figure 5.18

Spreadsheet Note The graph of a logarithm with any base b can be created with Excel by entering the formula "= log (x, b)." ■

Natural logarithms, $y = \ln x$ (and the inverse exponentials with base e), have many practical applications, some of which are considered in the next section. Common logarithms, $y = \log x$, were widely used for computation before computers and calculators became popular. They also have several applications to scaling variables, where the purpose is to reduce the scale of variation when a natural physical variable covers a wide range.

● **EXAMPLE 15** *Richter Scale*

The Richter scale is used to measure the intensity of an earthquake. The magnitude on the Richter scale of an earthquake of intensity I is given by

$$R = \log (I/I_0)$$

where I_0 is a certain minimum intensity used for comparison.

(a) Find R if I is 3,160,000 times as great as I_0.
(b) The 1964 Alaskan earthquake measured 8.5 on the Richter scale. Find the intensity of the 1964 Alaskan earthquake.

Solution

(a) If $I = 3{,}160{,}000 I_0$, then $I/I_0 = 3{,}160{,}000$. Hence

$$R = \log (3{,}160{,}000)$$
$$= 6.5 \text{ (approximated to one decimal place)}$$

(b) For $R = 8.5$, it follows that

$$8.5 = \log (I/I_0)$$

Rewriting this in exponential form gives

$$10^{8.5} = I/I_0$$

and from a calculator we obtain

$$I/I_0 = 316{,}000{,}000 \text{ (approximately)}$$

Thus the intensity is $316{,}000{,}000$ times I_0.

Note that a Richter scale measurement that is 2 units larger means that the intensity is $10^2 = 100$ *times* greater.

● **Checkpoint Solutions**

1. The (negative) y-axis
2. Yes
3. $x > 0$ is the domain of $\log_a x$.
4. (a) x, by Property II
   (b) 3, by Property I
   (c) $\ln \left(\dfrac{1}{e^2} \right) = \ln (e^{-2}) = -2$, by Property I
   (d) 0, because $\log_a 1 = 0$ for any base a.
5. (a) By Property III,
   $\log_7 (20 \cdot 11) = \log_7 20 + \log_7 11 = 1.5395 + 1.2323 = 2.7718$
   (b) By Property IV,
   $\log_7 \left(\frac{11}{20}\right) = \log_7 11 - \log_7 20 = 1.2323 - 1.5395 = -0.3072$
   (c) By Property V,
   $\log_7 20^4 = 4 \log_7 20 = 4(1.5395) = 6.1580$

5.2 Exercises

In Problems 1–4, write each equation in exponential form.

1. $4 = \log_2 16$

2. $4 = \log_3 81$

3. $\dfrac{1}{2} = \log_4 2$

4. $-2 = \log_3 \left(\dfrac{1}{9} \right)$

In Problems 5–10, solve for x by writing the equation in exponential form.

5. $\log_2 x = 3$

6. $\log_4 x = -2$

7. $\log_8 x = -\dfrac{1}{3}$

8. $\log_{25} x = \dfrac{1}{2}$

9. $\log_5 (2x + 1) = 2$

10. $\log_3 (7 - x) = 3$

In Problems 11–14, write each equation in logarithmic form.

11. $2^5 = 32$

12. $5^3 = 125$

13. $4^{-1} = \dfrac{1}{4}$

14. $9^{1/2} = 3$

In Problems 15–20, graph each function.

15. $y = \log_3 x$

16. $y = \log_4 x$

17. $y = \ln x$

18. $y = \log_9 x$

19. $y = \log_2 (-x)$

20. $y = \ln (-x)$

In Problems 21 and 22, use properties of logarithms or a definition to simplify each expression. Check each result with a change-of-base formula.

21. (a) $\log_3 27$
    (b) $\log_5 \left(\dfrac{1}{5} \right)$

22. (a) $\log_4 16$
    (b) $\log_9 3$

23. If $f(x) = \ln (x)$, find $f(e^x)$.

24. If $f(x) = \ln (x)$, find $f(\sqrt{e})$.

25. If $f(x) = e^x$, find $f(\ln 3)$.

26. If $f(x) = 10^x$, find $f(\log 2)$.

In Problems 27 and 28, evaluate each logarithm by using properties of logarithms and the following facts.

$$\log_a x = 3.1 \qquad \log_a y = 1.8 \qquad \log_a z = 2.7$$

27. (a) $\log_a (xy)$ (b) $\log_a \left(\dfrac{x}{z}\right)$

    (c) $\log_a (x^4)$ (d) $\log_a \sqrt{y}$

28. (a) $\log_a (yz)$ (b) $\log_a \left(\dfrac{z}{y}\right)$

    (c) $\log_a (y^6)$ (d) $\log_a \sqrt[3]{z}$

Write each expression in Problems 29–32 as the sum or difference of two logarithmic functions containing no exponents.

29. $\log \left(\dfrac{x}{x+1}\right)$ 30. $\ln \left[(x+1)(4x+5)\right]$

31. $\log_7 (x\sqrt[3]{x+4})$ 32. $\log_5 \left(\dfrac{x^2}{\sqrt{x+4}}\right)$

Use the properties of logarithms to write each expression in Problems 33–36 as a single logarithm.

33. $\ln x - \ln y$ 34. $\log_3 (x+1) + \log_3 (x-1)$

35. $\log_5 (x+1) + \dfrac{1}{2}\log_5 x$

36. $\log (2x+1) - \dfrac{1}{3}\log (x+1)$

In Problems 37–40, use a calculator to determine whether expression (a) is equivalent to expression (b). If they are equivalent, state what properties are being illustrated. If they are not equivalent, rewrite expression (a) so that they are equivalent.

37. (a) $\ln \sqrt{4 \cdot 6}$ (b) $\dfrac{1}{2}(\ln 4 + \ln 6)$

38. (a) $\log \dfrac{\sqrt{56}}{23}$ (b) $\dfrac{1}{2}\log 56 - \log 23$

39. (a) $\log \sqrt[3]{\dfrac{8}{5}}$ (b) $\dfrac{1}{3}\log 8 - \log 5$

40. (a) $\dfrac{\log_2 34}{17}$ (b) $\log_2 34 - \log_2 17$

 41. (a) Given $f(x) = \ln x$, use a graphing utility to graph $f(x) = \ln x$ with $f(x-c) = \ln (x-c)$ for $c = -4$, -2, 1, and 5.
    (b) For each c-value, identify the vertical asymptote and the domain of $y = \ln (x - c)$.
    (c) For each c-value, find the x-intercept of $y = \ln (x - c)$.
    (d) Write a sentence that explains precisely how the graphs of $y = f(x)$ and $y = f(x - c)$ are related.

 42. Given $f(x) = \ln x$, use a graphing utility to graph $f(ax) = \ln (ax)$ for $a = -2, -1, -0.5, 0.2$, and 3 with $f(x) = \ln x$. Explain the differences between the graphs
    (a) when $a > 0$ and when $a < 0$.
    (b) when $|a| > 1$ and when $0 < |a| < 1$.

In Problems 43 and 44, use a change-of-base formula to evaluate each logarithm.

43. (a) $\log_2 17$ (b) $\log_5 (0.78)$
44. (a) $\log_3 12$ (b) $\log_8 (0.15)$

In Problems 45–48, use a change-of-base formula to rewrite each logarithm. Then use a graphing utility to graph the function.

45. $y = \log_5 x$ 46. $y = \log_6 x$
47. $y = \log_{13} x$ 48. $y = \log_{16} x$
49. Prove logarithmic Property IV.
50. Prove logarithmic Property V.
51. The function $y = 3.818(1.0519)^{-x}$ was used to model the purchasing power of \$1 in Example 7 in the previous section. Find k and write this function as an exponential function with base e, so that it has the form $y = 3.818e^{-kx}$. *Hint:* It must be true that $(1.0519)^{-x} = e^{-kx} = (e^k)^{-x}$. Thus $e^k = 1.0519$; write this in logarithmic form.
52. Write the function $y = 4(2.334)^x$ as an exponential function of the form $y = 4e^{kx}$.

APPLICATIONS

Richter scale   Use the formula $R = \log (I/I_0)$ in Problems 53–56.

53. In October 2004, an earthquake measuring 6.8 on the Richter scale occurred in Japan. The largest quake in Japan since 1990 was one in 1993 that registered 7.7. How many times more severe was the 1993 shock than the one in 2004?
54. The strongest earthquake ever to strike Japan occurred in 1933 and measured 8.9 on the Richter scale. How many times more severe was this 1933 quake than the one in October 2004? (See Problem 53.)
55. The San Francisco earthquake of 1906 measured 8.25 on the Richter scale, and the San Francisco earthquake of 1989 measured 7.1. How much more intense was the 1906 quake?
56. The largest earthquakes ever recorded measured 8.9 on the Richter scale; the San Francisco earthquake of 1906 measured 8.25. Calculate the ratio of their intensities. (These readings correspond to devastating quakes.)

Decibel readings   In Problems 57–60, use the fact that the loudness of sound (in decibels) perceived by the human ear depends on intensity levels according to

$$L = 10 \log (I/I_0)$$

where I_0 is the threshold of hearing for the average human ear.

57. Find the loudness when I is 10,000 times I_0. This is the intensity level of the average voice.
58. A sound that causes pain has intensity about 10^{14} times I_0. Find the decibel reading for this threshold.

59. Graph the equation for loudness. Use I/I_0 as the independent variable.

60. The background noise level of a relatively quiet room is about $L_1 = 32$ decibels and of a heated argument is about $L_2 = 66$ decibels. Find the ratio I_2/I_1 of the associated intensities.

pH levels In Problems 61–64, use the following information. Chemists use the pH (hydrogen potential) of a solution to measure its acidity or basicity. The pH is given by the formula

$$\text{pH} = -\log [H^+]$$

where [H^+] is the concentration of hydrogen ions in moles per liter.

61. Most common solutions have pH ranges between 1 and 14. What values of H^+ are associated with these extremes?

62. Find the approximate pH of each of the following.
    (a) blood: $[H^+] = 3.98 \times 10^{-8}$
    $\quad\quad\quad\quad = 0.0000000398$
    (b) beer: $[H^+] = 6.31 \times 10^{-5} = 0.0000631$
    (c) vinegar: $[H^+] = 6.3 \times 10^{-3} = 0.0063$

63. Sometimes pH is defined as the logarithm of the reciprocal of the concentration of hydrogen ions. Write an equation that represents this sentence, and explain how it and the equation given in the information preceding Problem 61 can both represent pH.

64. Find the approximate hydrogen ion concentration $[H^+]$ for each of the following.
    (a) apples: pH = 3.0   (b) eggs: pH = 7.79
    (c) water (neutral): pH = 7.0

Doubling time In Problems 65 and 66, use the formula

$$2 = \left(1 + \frac{r}{100n}\right)^{nt}$$

to find the doubling time t, in years, for an investment at $r\%$ compounded n times per year. Write each exponential statement in logarithmic form. Then use a change-of-base formula to find the doubling time.

65. 8% compounded quarterly

66. 7.2% compounded monthly

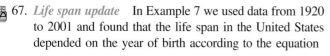 67. *Life span update* In Example 7 we used data from 1920 to 2001 and found that the life span in the United States depended on the year of birth according to the equation

$$L(x) = 11.307 + 14.229 \ln x$$

where x is the number of years after 1900. Using data from 1920 to 1989, the model

$$\ell(x) = 11.6164 + 14.1442 \ln x$$

where x is the number of years after 1900, predicts life span as a function of birth year. Use both models to predict the life spans for people born in 1999 and in 2012. Did adding data from 1990 to 2001 give predictions that were quite different from or very similar to predictions based on the model found earlier?

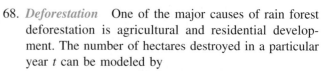 68. *Deforestation* One of the major causes of rain forest deforestation is agricultural and residential development. The number of hectares destroyed in a particular year t can be modeled by

$$y = -3.91435 + 2.62196 \ln t$$

where $t = 0$ in 1950.
    (a) Graph this function.
    (b) Use the model to predict the number of hectares that will be destroyed in the year 2010 because of agricultural and residential development.

69. *Women in the work force* Between the years 1960 and 2002, the percent of women in the work force is given by

$$w(x) = 2.552 + 14.57 \ln x$$

where x is the number of years past 1950. (*Source:* U.S. Bureau of Labor Statistics)
    (a) Graph this function.
    (b) What does this model estimate to be the percent of women in the work force in 2010?
    (c) Use the graph drawn in part (a) to estimate the year in which the percent reached 60.

70. **Modeling** *Poverty threshold* The following table gives the average poverty thresholds for one person for the years 1987 to 2002. Use a logarithmic equation to model these data, with x as the number of years past 1980. What does this model predict for the poverty threshold in 2010?

Year	Poverty Threshold
1987	$5778
1988	6022
1989	6310
1990	6652
1991	6932
1992	7143
1993	7363
1994	7547
1995	7763
1996	7995
1997	8183
1998	8316
1999	8501
2000	8794
2001	9039
2002	9182

Source: U.S. Bureau of the Census

71. **Modeling** *U.S. households with cable TV* The following data represent the percent of U.S. households with cable TV from 1981 to 2002.

(a) Let $x = 0$ in 1980 and find a logarithmic equation (base e) that models the data.

(b) If this model remains valid, use it to predict the percent of U.S. households with cable TV in the year 2010.

(c) What condition for the percent of households must be satisfied for the model to be valid?

Year	Percent	Year	Percent
1981	28.3	1992	61.5
1982	35.0	1993	62.5
1983	40.5	1994	63.4
1984	43.7	1995	65.7
1985	46.2	1996	66.7
1986	48.1	1997	67.3
1987	50.5	1998	67.4
1988	53.8	1999	68.0
1989	57.1	2000	68.0
1990	59.0	2001	69.2
1991	60.6	2002	68.9

Source: Nielson Media Research

5.3   Solutions of Exponential Equations: Applications of Exponential and Logarithmic Functions

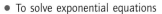

OBJECTIVES

- To solve exponential equations
- To solve exponential growth or decay equations when sufficient data are known
- To solve exponential equations representing demand, supply, total revenue, or total cost when sufficient data are known
- To model and solve problems involving Gompertz curves and logistic functions

◑ Application Preview

In this section, we discuss applications of exponential and logarithmic functions to the solutions of problems involving exponential growth and decay and supply and demand. With exponential decay, we consider various applications, including decay of radioactive materials and decay of sales following an advertising campaign. For example, suppose a company's sales decline following an advertising campaign so that the number of sales is $S = 2000(2^{-0.1x})$, where x is the number of days after the campaign ends. This company would be interested in determining when to begin a new campaign by finding how many days will elapse before sales drop below 350 (see Example 2).

Exponential growth focuses on population growth and growth that can be modeled by Gompertz curves and logistic functions. Finally, we consider applications of demand functions that can be modeled by exponential functions.

Solving Exponential Equations Using Logarithmic Properties

Questions such as the one in the Application Preview can be answered by solving an equation in which the variable appears in the exponent, called an **exponential equation.** Logarithmic properties are useful in converting an equation of this type to one that can be solved easily. In this section we limit our discussion and applications to equations involving a single exponential.

Solving Exponential Equations

Procedure	Example
To solve an exponential equation by using properties of logarithms:	Solve: $4(25^{2x}) = 312{,}500$
1. Isolate the exponential by rewriting the equation with a base raised to a power on one side.	1. $\dfrac{4(25^{2x})}{4} = \dfrac{312{,}500}{4}$ $\qquad 25^{2x} = 78{,}125$
2. Take the logarithm (either base e or base 10) of both sides.	2. We choose base e, although base 10 would work similarly. $\qquad \ln{(25^{2x})} = \ln{(78{,}125)}$
3. Use a property of logarithms to remove the variable from the exponent.	3. Using logarithmic Property V gives $\qquad 2x \ln 25 = \ln{(78{,}125)}$
4. Solve for the variable.	4. Dividing both sides by $2 \ln 25$ gives $\qquad x = \dfrac{\ln{(78{,}125)}}{2 \ln 25} = 1.75$

In Step 2 we noted that using a logarithm base 10 would work similarly, and it gives the same result (try it). Unless the exponential in the equation has base e or base 10, the choice of a logarithm in Step 2 makes no difference. However, if the exponential in the equation has base e, then choosing a base e logarithm is slightly easier (see Examples 1 and 3), and similarly for choosing a base 10 logarithm when the equation has a base 10 exponential.

Growth and Decay

Recall that **exponential decay models** are those that can be described by a function of the following form.

Decay Models $f(x) = C(a^{-x})$ with $a > 1$ and $C > 0$ or $f(x) = C(b^x)$ with $0 < b < 1$ and $C > 0$

We have already mentioned that radioactive decay has this form, as do the demand curves discussed later. Another example of this phenomenon occurs in the life sciences when the valves to the aorta are closed, and blood flows into the heart. During this period of time, the blood pressure in the aorta falls exponentially. This is illustrated in the following example.

● **EXAMPLE 1 Pressure in the Aorta**

Medical research has shown that over short periods of time when the valves to the aorta of a normal adult close, the pressure in the aorta is a function of time and can be modeled by the equation

$$P = 95e^{-0.491t}$$

where t is in seconds. How long will it be before the pressure reaches 80?

Solution

Setting $P = 80$ and solving for t will give us the length of time before the pressure reaches 80.

$$80 = 95e^{-0.491t}$$

To solve this equation for t, we first isolate the exponential containing t by dividing both sides by 95.

$$\frac{80}{95} = e^{-0.491t}$$

Because e is the base of the exponential, we take the logarithm, base e, of both sides.

$$\ln\left(\frac{80}{95}\right) = \ln e^{-0.491t}$$

In this case, logarithmic Property I removes the variable from the exponent.

$$\ln\left(\frac{80}{95}\right) = -0.491t$$

Using a calculator, we have $-0.172 = -0.491t$. Thus,

$$\frac{-0.172}{-0.491} = t$$

$$t = 0.35 \text{ second} \qquad \text{(approximately)}$$

In the exponential equations so far, we have taken base e logarithms of both sides. Next we use base 10 logarithms.

◑ EXAMPLE 2 *Sales Decay* **(Application Preview)**

A company finds that its daily sales begin to fall after the end of an advertising campaign, and the decline is such that the number of sales is $S = 2000(2^{-0.1x})$, where x is the number of days after the campaign ends.

(a) How many sales will be made 10 days after the end of the campaign?
(b) If the company does not want sales to drop below 350 per day, when should it start a new campaign?

Solution

(a) If $x = 10$, sales are given by $S = 2000(2^{-1}) = 1000$.
(b) Setting $S = 350$ and solving for x will give us the number of days after the end of the campaign when sales will reach 350.

$$350 = 2000(2^{-0.1x})$$

$$\frac{350}{2000} = 2^{-0.1x} \qquad \text{Isolate the exponential.}$$

$$0.175 = 2^{-0.1x}$$

With the base 2 exponential isolated, we take logarithms of both sides. This time we choose base 10 logarithms and complete the solution as follows.

$$\log 0.175 = \log (2^{-0.1x})$$
$$\log 0.175 = (-0.1x)(\log 2) \qquad \text{Property V}$$
$$\frac{\log 0.175}{\log 2} = -0.1x$$
$$\frac{-0.7570}{0.3010} = -0.1x$$
$$-2.515 = -0.1x$$
$$x = 25.15$$

Thus sales will be 350 at day 25.15, or during the 26th day after the end of the campaign. If a new campaign isn't begun on or before the 26th day, sales will drop below 350. (See Figure 5.19.)

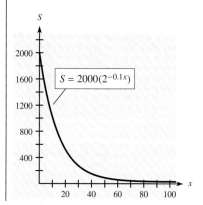

Figure 5.19

$$S = 2000(2^{-0.1x})$$

In business, economics, biology, and the social sciences, the growth of money, bacteria, or population is frequently of interest. Recall that **exponential growth models** are those that can be described by a function of the following form.

Growth Models

$$f(x) = C(a^x) \qquad \text{with } a > 1 \text{ and } C > 0$$

As we mentioned in Section 5.1, "Exponential Functions," the curve that models the growth of some populations is given by $y = P_0e^{ht}$, where P_0 is the population size at a particular time t, h is a constant that depends on the population involved, and y is the total population at time t. This function may be used to model population growth for humans, insects, or bacteria.

● **EXAMPLE 3 *Population***

The population of a certain city was 30,000 in 1980 and 40,500 in 1990. If the formula $y = P_0e^{ht}$ applies to the growth of the city's population, what should the population be in the year 2010?

Solution

We can first use the data from 1980 ($t = 0$) and 1990 ($t = 10$) to find the value of h in the formula. Letting $P_0 = 30,000$ and $y = 40,500$, we get

$$40,500 = 30,000e^{h(10)}$$

$$1.35 = e^{10h} \qquad \text{Isolate the exponential.}$$

Taking the natural logarithms of both sides and using logarithmic Property I give

$$\ln 1.35 = \ln e^{10h} = 10h$$

$$0.3001 = 10h$$

$$h = 0.0300 \qquad \text{(approximately)}$$

Thus the formula for this population is $y = P_0 e^{0.03t}$. To predict the population for the year 2010, we use $P_0 = 40,500$ (for 1990) and $t = 20$. This gives

$$y = 40,500e^{0.03(20)}$$

$$= 40,500e^{0.6}$$

$$= 40,500(1.8221)$$

$$= 73,795 \qquad \text{(approximately)}$$

● **Checkpoint**
1. Suppose the sales of a product, in dollars, are given by $S = 1000e^{-0.07x}$, where x is the number of days after the end of an advertising campaign.
   (a) What are sales 2 days after the end of the campaign?
   (b) How long will it be before sales are $300?

Economic and Management Applications

In Section 2.3, "Business Applications of Quadratic Functions," we discussed quadratic cost, revenue, demand, and supply functions. But cost, revenue, demand, and supply may also be modeled by exponential or logarithmic equations. For example, suppose the demand for a product is given by $p = 30(3^{-q/2})$, where q is the number of thousands of units demanded at a price of p dollars per unit. Then the graph of the demand curve is as given in Figure 5.20.

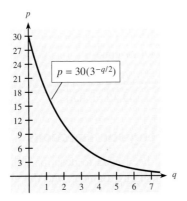

Figure 5.20

● **EXAMPLE 4** *Demand*

Suppose the demand function for q thousand units of a certain commodity is given by $p = 30(3^{-q/2})$.

(a) At what price per unit will the demand equal 4000 units?
(b) How many units, to the nearest thousand units, will be demanded if the price is $17.32?

Solution

(a) If 4000 units are demanded, then $q = 4$ and

$$p = 30(3^{-4/2})$$
$$= 30(0.1111)$$
$$= 3.33 \text{ dollars} \quad \text{(approximately)}$$

(b) If $p = 17.32$, then $17.32 = 30(3^{-q/2})$

$$0.5773 = 3^{-q/2} \qquad \text{Isolate the exponential.}$$
$$\ln 0.5773 = \ln 3^{-q/2}$$
$$\ln 0.5773 = -\frac{q}{2}\ln 3$$
$$\frac{-2\ln 0.5773}{\ln 3} = q$$
$$1.000 \approx q$$

The number of units demanded would be approximately 1000 units.

● *Checkpoint*   2. Suppose the monthly demand for a product is given by $p = 400e^{-0.003x}$, where p is the price in dollars and x is the number of units. How many units will be demanded when the price is $100?

● **EXAMPLE 5** *Total Revenue*

Suppose the demand function for a commodity is given by $p = 100e^{-x/10}$, where p is the price per unit when x units are sold.

(a) What is the total revenue for the commodity?
(b) What would be the total revenue if 30 units were demanded and supplied?

Solution

(a) The total revenue can be computed by multiplying the quantity sold and the price per unit. The demand function gives the price per unit when x units are sold, so the total revenue for x units is $R(x) = x \cdot p = x(100e^{-x/10})$. Thus the total revenue function is $R(x) = 100xe^{-x/10}$.

(b) If 30 units are sold, the total revenue is

$$R(30) = 100(30)e^{-30/10} = 100(30)(0.0498)$$
$$= 149.40 \text{ (dollars)}$$

Gompertz Curves and Logistic Functions

Gompertz Curves   One family of curves that has been used to describe human growth and development, the growth of organisms in a limited environment, and the growth of many types of organizations is the family of **Gompertz curves.** These curves are graphs of equations of the form

$$N = Ca^{R^t}$$

where t represents the time, R ($0 < R < 1$) is a constant that depends on the population, a represents the proportion of initial growth, C is the maximum possible number of individuals, and N is the number of individuals at a given time t.

For example, the equation $N = 100(0.03)^{0.2^t}$ could be used to predict the size of a deer herd introduced on a small island. Here the maximum number of deer C would be 100, the proportion of the initial growth a is 0.03, and R is 0.2. For this example, t represents time, measured in decades. The graph of this equation is given in Figure 5.21.

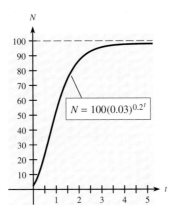

Figure 5.21

● **EXAMPLE 6** *Organizational Growth*

A hospital administrator predicts that the growth in the number of hospital employees will follow the Gompertz equation

$$N = 2000(0.6)^{0.5^t}$$

where t represents the number of years after the opening of a new facility.

(a) What is the number of employees when the facility opens?
(b) How many employees are predicted after 1 year of operation?
(c) Graph the curve.
(d) What is the maximum value of N that the curve will approach?

Solution
(a) The facility opens when $t = 0$, so $N = 2000(0.6)^{0.5^0} = 2000(0.6)^1 = 1200$
(b) In 1 year, $t = 1$, so $N = 2000(0.6)^{0.5} = 2000\sqrt{0.6} = 1549$ (approximately)
(c) The graph is shown in Figure 5.22.
(d) From the graph we can see that as larger values of t are substituted in the function, the values of N approach, but never reach, 2000. We say that the line $N = 2000$ (dashed) is an **asymptote** for this curve and that 2000 is the maximum possible value.

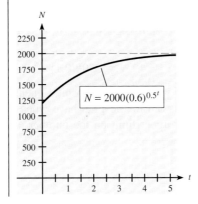

Figure 5.22

● **EXAMPLE 7** *Wildlife Populations*

The Gompertz equation

$$N = 100(0.03)^{0.2^t}$$

predicts the size of a deer herd on a small island t decades from now. During what year will the deer population reach or exceed 70?

Solution
We solve the equation with $N = 70$.

$$70 = 100(0.03)^{0.2^t}$$

$$0.7 = 0.03^{0.2^t} \qquad \text{Isolate the exponential.}$$

$$\ln 0.7 = \ln 0.03^{0.2^t} = 0.2^t \ln 0.03$$

$$\frac{\ln 0.7}{\ln 0.03} = 0.2^t \qquad \text{Again, isolate the exponential.}$$

$$\ln\left(\frac{\ln 0.7}{\ln 0.03}\right) = t \ln 0.2$$

$$\ln(0.10172) = t(-1.6094)$$

$$\frac{-2.2855}{-1.6094} = t$$

$$t = 1.42 \text{ decades} \qquad \text{(approximately)}$$

The population will exceed 70 in just over 14 years, or during the 15th year.

Calculator Note 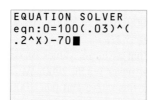 Some graphing utilities have a SOLVER feature or program that can be used to solve an equation. Solution with such a feature or program may not be successful unless a reasonable guess at the solution (which can be seen from a graph) is chosen to start the process used to solve the equation. Figure 5.23 shows the equation from Example 7 and its solution with the SOLVER feature of a graphing calculator.

```
EQUATION SOLVER
eqn:0=100(.03)^(
.2^X)-70■
```

```
100(.03)^(.2^...=0
X=1.4201015868...
bound=(-1E99,1...
```

Figure 5.23

● *Checkpoint*   3. Suppose the number of employees at a new regional hospital is predicted by the Gompertz curve

$$N = 3500(0.1)^{0.5^t}$$

where t is the number of years after the hospital opens.
(a) How many employees did the hospital have when it opened?
(b) What is the expected upper limit on the number of employees?

Logistic Functions   Gompertz curves describe situations in which growth is limited. There are other equations that model this phenomenon under different assumptions. Two examples are

(a) $y = c(1 - e^{-ax})$, $a > 0$ See Figure 5.24(a).

(b) $y = \dfrac{A}{1 + ce^{-ax}}$, $a > 0$ See Figure 5.24(b).

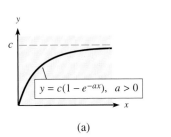

Figure 5.24   (a)   (b)

These equations have many applications. In general, both (a) and (b) can be used to model learning, sales of new products, and population growth, and (b) can be used to describe the spread of epidemics. The equation

$$y = \frac{A}{1 + ce^{-ax}} \qquad a > 0$$

is called a **logistic function**, and the graph in Figure 5.24(b) is called a **logistic curve.**

● **EXAMPLE 8** *DVD Players*

According to the August 20, 2001 issue of *Newsweek*, DVD players were entering U.S. households faster than any other piece of home electronics equipment in history. The sales revenues from 1999 to 2004 can be modeled by the logistic function

$$y = \frac{9.46}{1 + 53.08e^{-1.28x}}$$

where x is the number of years past 1998 and y is in billions of dollars.

(a) Graph this function.
(b) Use the function to estimate the sales revenue for 2004.

Solution
(a) The graph of this function is shown in Figure 5.25 on the following page. Note that the graph is an S-shaped curve with a relatively slow initial rise, then a steep climb, followed by a leveling off. It is reasonable that sales would eventually level off, even if everyone in the United States bought a DVD player. Very seldom will exponential growth continue indefinitely, so logistic functions often better represent many types of sales and organizational growth.
(b) For the 2004 estimate, use $x = 6$ in the function.

$$y = \frac{9.46}{1 + 53.08e^{-1.28(6)}} = \frac{9.46}{1 + 53.08e^{-7.68}}$$

$$\approx \frac{9.46}{1 + 0.0245216276} \approx 9.234$$

Thus the model estimates that in 2004, the sales revenue from DVD players was about $9.234 billion.

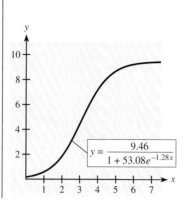

Figure 5.25

● *Checkpoint Solutions*

1. (a) $S = 1000e^{-0.07(2)} = 1000e^{-0.14} \approx \$869.36$ Nearest cent

   (b) $\quad 300 = 1000e^{-0.07x}$

   $\quad\quad 0.3 = e^{-0.07x}$

   $\quad \ln(0.3) = -0.07x$

   $\quad \dfrac{\ln(0.3)}{-0.07} = x \approx 17$ Nearest day

2. $\quad\quad 100 = 400e^{-0.003x}$

   $\quad\quad 0.25 = e^{-0.003x}$

   $\quad \ln(0.25) = -0.003x$

   $\quad \dfrac{\ln(0.25)}{-0.003} = x \approx 462$ Nearest unit

3. (a) $3500(0.1) = 350$

   (b) 3500

5.3 Exercises

In Problems 1–10, solve each exponential equation. Give answers correct to 3 decimal places.

1. $8^{3x} = 32{,}768$

2. $7^{2x} = 823{,}543$

3. $0.13P = P(2^{-x})$

4. $0.05A = A(1.06)^{-x}$

5. $10{,}000 = 1500e^{0.10x}$

6. $2500 = 600e^{0.05x}$

7. $78 = 100 - 100e^{-0.01x}$

8. $500 = 600 - 600e^{-0.4x}$

9. $55 = \dfrac{60}{1 + 5e^{-0.6x}}$

10. $150 = \dfrac{200}{1 + 30e^{-0.3x}}$

For Problems 11 and 12, let $f(x) = \dfrac{A}{1 + ce^{-x}}$. **Use a graphing utility to make the requested graphs.**

11. (a) Fix $A = 100$ and graph $y = f(x) = \dfrac{100}{1 + ce^{-x}}$

    for $c = 0.25, 1, 9,$ and 49.

    (b) What effect does c have on the graphs?

12. (a) Fix $c = 1$ and graph $y = f(x) = \dfrac{A}{1 + e^{-x}}$ for

    $A = 50, 100,$ and 150.

    (b) What effect does A have on the graphs?

GROWTH AND DECAY APPLICATIONS

13. *Sales decay* The sales decay for a product is given by $S = 50{,}000e^{-0.8x}$, where S is the monthly sales and x is the number of months that have passed since the end of a promotional campaign.

    (a) What will be the sales 4 months after the end of the campaign?

    (b) How many months after the end of the campaign will sales drop below 1000, if no new campaign is initiated?

14. *Sales decay*   The sales of a product decline after the end of an advertising campaign, with the sales decay given by $S = 100,000e^{-0.5x}$, where S represents the weekly sales and x represents the number of weeks since the end of the campaign.
    (a) What will be the sales for the tenth week after the end of the campaign?
    (b) During what week after the end of the campaign will sales drop below 400?

15. *Inflation*   The purchasing power P (in dollars) of an annual amount of A dollars after t years of 5% inflation decays according to

$$P = Ae^{-0.05t}$$

    (*Source: Viewpoints,* VALIC)
    (a) How long will it be before a pension of $60,000 per year has a purchasing power of $30,000?
    (b) How much pension A would be needed so that the purchasing power P is $50,000 after 15 years?

16. *Product reliability*   A statistical study shows that the fraction of television sets of a certain brand that are still in service after x years is given by $f(x) = e^{-0.15x}$.
    (a) What fraction of the sets are still in service after 5 years?
    (b) After how many years will the fraction still in service be 1/10?

17. *Radioactive half-life*   An initial amount of 100 g of the radioactive isotope thorium-234 decays according to

$$Q(t) = 100e^{-0.02828t}$$

    where t is in years. How long before half of the initial amount has disintegrated? This time is called the half-life of this isotope.

18. *Radioactive half-life*   A breeder reactor converts stable uranium-238 into the isotope plutonium-239. The decay of this isotope is given by

$$A(t) = A_0e^{-0.00002876t}$$

    where $A(t)$ is the amount of the isotope at time t, in years, and A_0 is the original amount.
    (a) If $A_0 = 500$ lb, how much will be left after a human lifetime (use $t = 70$ years)?
    (b) Find the half-life of this isotope.

19. *Population growth*   If the population of a certain county was 100,000 in 1990 and 110,517 in 2000, and if the formula $y = P_0e^{ht}$ applies to the growth of the county's population, estimate the population of the county in 2015.

20. *Population growth*   The population of a certain city grows according to the formula $y = P_0e^{0.03t}$. If the pop-

ulation was 250,000 in 2000, estimate the year in which the population reaches 350,000.

21. *Health care*   For selected years from 1960 to 2001, the national health care expenditures H, in billions of dollars, can be modeled by

$$H = 29.57e^{0.097t}$$

    where t is the number of years past 1960. (*Source:* U.S. Department of Health and Human Services)
    (a) If this model remains accurate, how long will it be before national health care expenditures reach $4.0 trillion (that is, $4000 billion)?
    (b) How do you think a national health care plan would alter this model?

22. *U.S. debt*   For selected years from 1900 to 2003, the national debt d, in billions of dollars, can be modeled by

$$d = 1.68e^{0.0828t}$$

    where t is the number of years past 1900 (*Source:* Bureau of Public Debt, U.S. Treasury). How long will it be before the debt is predicted to reach $20 trillion (that is, $20,000 billion)? Do you think the War on Terror and the war in Iraq will affect this model? Explain.

ECONOMIC AND MANAGEMENT APPLICATIONS

23. *Demand*   The demand function for a certain commodity is given by $p = 100e^{-q/2}$.
    (a) At what price per unit will the quantity demanded equal 6 units?
    (b) If the price is $1.83 per unit, how many units will be demanded, to the nearest unit?

24. *Demand*   The demand function for a product is given by $p = 3000e^{-q/3}$.
    (a) At what price per unit will the quantity demanded equal 6 units?
    (b) If the price is $149.40 per unit, how many units will be demanded, to the nearest unit?

25. *Supply*   If the supply function for a product is given by $p = 100e^q/(q + 1)$, where q represents the number of hundreds of units, what will be the price when the producers are willing to supply 300 units?

26. *Supply*   If the supply function for a product is given by $p = 200(2^q)$, where q represents the number of hundreds of units, what will be the price when the producers are willing to supply 500 units?

27. *Total cost*   If the total cost function for a product is $C(x) = e^{0.1x} + 400$, where x is the number of items produced, what is the total cost of producing 30 units?

28. *Total cost* If the total cost for x units of a product is given by $C(x) = 400 \ln (x + 10) + 100$, what is the total cost of producing 100 units?

29. *Total revenue* If the demand function for a product is given by $p = 200e^{-0.02x}$, where p is the price per unit when x units are demanded, what is the total revenue when 100 units are demanded and supplied?

30. *Total revenue* If the demand function for a product is given by $p = 4000/\ln (x + 10)$, where p is the price per unit when x units are demanded, what is the total revenue when 40 units are demanded and supplied?

31. *Compound interest* If $8500 is invested at 11.5% compounded continuously, the future value S at any time t (in years) is given by

$$S = 8500e^{0.115t}$$

   (a) What is the amount after 18 months?
   (b) How long before the investment doubles?

32. *Compound interest* If $1000 is invested at 10% compounded continuously, the future value S at any time t (in years) is given by $S = 1000e^{0.1t}$.
   (a) What is the amount after 1 year?
   (b) How long before the investment doubles?

33. *Compound interest* If $5000 is invested at 9% per year compounded monthly, the future value S at any time t (in months) is given by $S = 5000(1.0075)^t$.
   (a) What is the amount after 1 year?
   (b) How long before the investment doubles?

34. *Compound interest* If $10,000 is invested at 1% per month, the future value S at any time t (in months) is given by $S = 10,000(1.01)^t$.
   (a) What is the amount after 1 year?
   (b) How long before the investment doubles?

Securities industry profits The securities industry experienced dramatic growth in the last two decades of the 20th century. The following models for the industry's revenue R and expenses or costs C (both in billions of dollars) were developed as functions of the years past 1980 with data from the U.S. Securities and Exchange Commission's 2000 *Annual Report* (2001).

$$R(t) = 21.4e^{0.131t} \quad \text{and} \quad C(t) = 18.6e^{0.131t}$$

Use these models in Problems 35 and 36.
35. (a) Use the models to predict the profit for the securities industry in 2005.
   (b) How long before the profit found in (a) is predicted to double?
   (c) What happened since 2000 that probably would affect this prediction?

36. Use the models to find how long before the securities industry's profit reaches $100 billion.

37. *Consumer price index (CPI)* By using data from the U.S. Bureau of Labor Statistics for the years 1968–2002, the purchasing power P of a 1983 dollar can be modeled with the function

$$P(t) = 3.818e^{-0.0506t}$$

   where t is the number of years past 1960.
   (a) Find $P(10)$ and $P(70)$, and for each, write a sentence that interprets its meaning.
   (b) How long before the purchasing power of a 1983 dollar is $0.25?

38. *Daily shares traded on the NYSE* For selected years from 1970 to 2002, the average daily shares traded (in millions) on the New York Stock Exchange can be modeled by

$$N(t) = 0.4842e^{0.1514t}$$

   where t is the number of years past 1950. (*Source:* New York Stock Exchange, *Fact Book, 2002*)
   (a) Use the model to estimate the average daily shares traded in 2000.
   (b) Use the model to estimate when the average daily shares traded will reach 5 billion (or 5000 million).

39. *Supply* Suppose the supply of x units of a product at price p dollars per unit is given by

$$p = 10 + 5 \ln (3x + 1)$$

   How many units would be supplied when the price is $50 each?

40. *Demand* Say the demand function for a product is given by $p = 100/\ln (q + 1)$.
   (a) What will be the price if 19 units are demanded?
   (b) How many units, to the neareast unit, will be demanded if the price is $29.40?

GOMPERTZ CURVES AND LOGISTIC FUNCTIONS

41. *Sales growth* The president of a company predicts that sales will increase after she assumes office and that the number of monthly sales will follow the curve given by $N = 3000(0.2)^{0.6^t}$, where t represents the months since she assumed office.
   (a) What will be the sales when she assumes office?
   (b) What will be the sales after 3 months?
   (c) What is the expected upper limit on sales?
   (d) Graph the curve.

42. *Organizational growth* Because of a new market opening, the number of employees of a firm is expected

to increase according to the equation $N = 1400(0.5)^{0.3t}$, where t represents the number of years after the new market opens.
(a) What is the level of employment when the new market opens?
(b) How many employees should be working at the end of 2 years?
(c) What is the expected upper limit on the number of employees?
(d) Graph the curve.

43. *Organizational growth*   Suppose that the equation $N = 500(0.02)^{0.7t}$ represents the number of employees working t years after a company begins operations.
(a) How many employees are there when the company opens (at $t = 0$)?
(b) After how many years will at least 100 employees be working?

44. *Sales growth*   A firm predicts that sales will increase during a promotional campaign and that the number of daily sales will be given by $N = 200(0.01)^{0.8t}$, where t represents the number of days after the campaign begins. How many days after the beginning of the campaign would the firm expect to sell at least 60 units per day?

45. *Drugs in the bloodstream*   The concentration y of a certain drug in the bloodstream t hours after an oral dosage (with $0 \le t \le 15$) is given by the equation

$$y = 100(1 - e^{-0.462t})$$

(a) What is y after 1 hour ($t = 1$)?
(b) How long does it take for y to reach 50?

46. *Population growth*   Suppose that the number y of otters t years after otters were reintroduced into a wild and scenic river is given by

$$y = 2500 - 2490e^{-0.1t}$$

(a) Find the population when the otters were reintroduced (at $t = 0$).
(b) How long will it be before the otter population numbers 1500?

47. *Spread of disease*   On a college campus of 10,000 students, a single student returned to campus infected by a disease. The spread of the disease through the student body is given by

$$y = \frac{10,000}{1 + 9999e^{-0.99t}}$$

where y is the total number infected at time t (in days).
(a) How many are infected after 4 days?
(b) The school will shut down if 50% of the students are ill. During what day will it close?

48. *Spread of a rumor*   The number of people $N(t)$ in a community who are reached by a particular rumor at time t is given by the equation

$$N(t) = \frac{50,500}{1 + 100e^{-0.7t}}$$

(a) Find $N(0)$.
(b) What is the upper limit on the number of people affected?
(c) How long before 75% of the upper limit is reached?

49. *Market share*   Suppose that the market share y (as a percent) that a company expects t months after a new product is introduced is given by $y = 40 - 40e^{-0.05t}$.
(a) What is the market share after the first month (to the nearest percent)?
(b) How long (to the nearest month) before the market share is 25%?

50. *Advertising*   An advertising agency has found that when it promotes a new product in a certain market of 350,000, the number of people x who are aware of the product t days after the ad campaign is initiated is given by

$$x = 350,000(1 - e^{-0.077t})$$

(a) How many people (to the nearest thousand) are aware after 1 week?
(b) How long (to the nearest day) before 300,000 are aware of the new product?

51. *Pollution*   Pollution levels in Lake Erie have been modeled by the equation

$$x = 0.05 + 0.18e^{-0.38t}$$

where x is the volume of pollutants (in cubic kilometers) and t is the time (in years). (Adapted from R. H. Rainey, *Science* 155 (1967), 1242–1243)
(a) Find the initial pollution levels; that is, find x when $t = 0$.
(b) How long before x is 30% of that initial level?

52. *Fish length*   Suppose that the length x (in centimeters) of an individual of a certain species of fish is given by

$$x = 50 - 40e^{-0.05t}$$

where t is its age in months.
(a) Find the length after 1 year.
(b) How long (to the nearest month) will it be until the length is 45 cm?

53. *Mutual funds*   For selected years from 1978 to 2002, the number of mutual funds N, excluding money market funds, can be modeled by

$$N = \frac{8750}{1 + 80.8e^{-0.2286t}}$$

where t is the number of years past 1975. (*Source: Investment Company Institute, Mutual Fund Fact Book (2003)*)

(a) Use the model to estimate the number of mutual funds in 2000.

(b) Use the model to estimate the year when the number of mutual funds will reach 8500.

54. *Chemical reaction*   When two chemicals, A and B, react to form another chemical C (such as in the digestive process), this is a special case of the **law of mass action,** which is fundamental to studying chemical reaction rates. Suppose that chemical C is formed from A and B according to

$$x = \frac{120[1 - (0.6)^{3t}]}{4 - (0.6)^{3t}}$$

where x is the number of pounds of C formed in t minutes.

(a) How much of C is present when the reaction begins?

(b) How much of C is formed in 4 minutes?

(c) How long does it take to form 10 lb of C?

55. **Modeling**   *Internet users*   The following table gives the percent of the U.S. population with Internet connections for the years 1997 to 2003.

(a) Find the logistic function that models these data. Use x as the number of years past 1995.

(b) Use the model to predict the percent of the U.S. population with Internet connections in 2003.

(c) When does the model predict that 59.5% of the U.S. population will have Internet connections?

Year	1997	1998	1999	2000	2001	2002	2003
Percent with Internet	22.2	32.7	39.1	44.4	53.9	55.0	56.0

Source: U.S. Department of Commerce

56. **Modeling**   *Endangered species*   The following table gives the numbers of species of plants that were endangered in various years from 1980 to 2003.

(a) Find the logistic function that models these data. Use x as the number of years past 1975.

(b) Use the model to predict the number of species of endangered plants in 2010.

(c) When does the model predict that 640 plant species will be endangered?

Year	Endangered Plant Species
1980	50
1985	93
1990	179
1995	432
1998	567
1999	581
2000	593
2001	595
2002	596
2003	599

Source: U.S. Fish and Wildlife Service

Key Terms and Formulas

Section	Key Terms	Formulas
5.1	Exponential functions	
	Growth functions	$f(x) = a^x$ $(a > 1)$
		$f(x) = C(a^x)$ $(C > 0, a > 1)$
	Decay functions	$f(x) = C(a^{-x})$ $(C > 0, a > 1)$
		$f(x) = C(b^x)$ $(C > 0, 0 < b < 1)$
	e	$e \approx 2.71828$
5.2	Logarithmic function	$y = \log_a x$, defined by $x = a^y$
	Common logarithm	$\log x = \log_{10} x$
	Natural logarithm	$\ln x = \log_e x$
	Inverse functions	
	Logarithmic Properties I–V	$\log_a a^x = x; \ a^{\log_a x} = x;$
		$\log_a (MN) = \log_a M + \log_a N;$
		$\log_a (M/N) = \log_a M - \log_a N;$
		$\log_a (M^N) = N(\log_a M)$

Section	Key Terms	Formulas
	Change-of-base formulas	$\log_b x = \dfrac{\log_a x}{\log_a b}$ $\log_b x = \dfrac{\ln x}{\ln b}$
5.3	Solving exponential equations	Isolate the exponential; take logarithm of both sides.
	Exponential growth and decay	
	Decay models	$f(x) = Ca^{-x} \quad (a > 1, C > 0)$
	Growth models	$f(x) = Ca^{x} \quad\ (a > 1, C > 0)$
	Gompertz curves	$N = Ca^{R^t}$
	Logistic functions	$y = \dfrac{A}{1 + ce^{-ax}} \quad (a > 0)$

Review Exercises

Sections 5.1 and 5.2

1. Write each statement in logarithmic form.
   (a) $2^x = y$ (b) $3^y = 2x$
2. Write each statement in exponential form.
   (a) $\log_7 \left(\dfrac{1}{49} \right) = -2$ (b) $\log_4 x = -1$

Graph the following functions.

3. $y = e^x$ 4. $y = e^{-x}$
5. $y = \log_2 x$ 6. $y = 2^x$
7. $y = \frac{1}{2}(4^x)$ 8. $y = \ln x$
9. $y = \log_4 x$ 10. $y = 3^{-2x}$
11. $y = \log x$ 12. $y = 10(2^{-x})$

In Problems 13–20 evaluate each logarithm without using a calculator. Check with the change-of-base formula in Problems 13–17.

13. $\log_5 1$ 14. $\log_2 8$
15. $\log_{25} 5$ 16. $\log_3 \left(\dfrac{1}{3} \right)$
17. $\log_3 3^8$ 18. $\ln e$
19. $e^{\ln 5}$ 20. $10^{\log 3.15}$

In Problems 21–24, if $\log_a x = 1.2$ and $\log_a y = 3.9$, find each of the following by using the properties of logarithms.

21. $\log_a \left(\dfrac{x}{y} \right)$ 22. $\log_a \sqrt{x}$
23. $\log_a (xy)$ 24. $\log_a (y^4)$

In Problems 25 and 26, use the properties of logarithms to write each expression as the sum or difference of two logarithmic functions containing no exponents.

25. $\log (yz)$ 26. $\ln \sqrt{\dfrac{x+1}{x}}$

27. Is it true that $\ln x + \ln y = \ln (x + y)$ for all values of x?
28. If $f(x) = \ln x$, find $f(e^{-2})$
29. If $f(x) = 2^x + \log (7x - 4)$, find $f(2)$.
30. If $f(x) = e^x + \ln (x + 1)$, find $f(0)$.
31. If $f(x) = \ln (3e^x - 5)$, find $f(\ln 2)$.

In Problems 32 and 33, use a change-of-base formula to evaluate each logarithm.

32. $\log_9 2158$
33. $\log_{12} (0.0195)$

 In Problems 34 and 35, rewrite each logarithm by using a change-of-base formula, then graph the function with a graphing utility.

34. $y = \log_{\sqrt{3}} x$
35. $f(x) = \log_{11} (2x - 5)$

Section 5.3

In Problems 36–39, solve each equation.
36. $6^{4x} = 46{,}656$
37. $8000 = 250(1.07)^x$
38. $11{,}000 = 45{,}000e^{-0.05x}$
39. $312 = 300 + 300e^{-0.08x}$

APPLICATIONS

Section 5.1

40. *Medicare spending* The graph in the following figure shows the projected federal spending for Medicare as a percent of the gross domestic product. If these expenditures were modeled as a function of time, which of a growth exponential, a decay exponential, or a logarithm would be the best model? Justify your answer.

A Consuming Problem

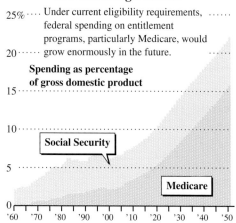

25% ···· Under current eligibility requirements, ······
federal spending on entitlement
programs, particularly Medicare, would
20 ······ grow enormously in the future. ······

**Spending as percentage
of gross domestic product**

Source: Congressional Budget Office, *The New York Times*

41. *Students per computer* The following figure shows the history of the number of students per computer in U.S. public schools. If this number of students were modeled by a function, do you think the best model would be linear, quadratic, exponential, or logarithmic? Explain.

Students per Computer in U.S. Public Schools

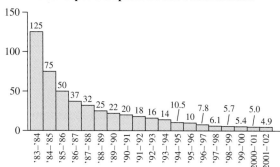

Source: Quality Education Data, Inc., Denver, CO. Reprinted by permission.

42. *Inflation* The purchasing power P of a $60,000 pension after t years of 3% annual inflation is modeled by

$$P(t) = 60,000(0.97)^t$$

(a) What is the purchasing power after 20 years?
(b) Graph this function for $t = 0$ to $t = 25$ with a graphing utility.

43. **Modeling** *Consumer credit* The data in the table show the total consumer credit (in billions of dollars) that was outstanding in the United States for various years from 1980 to 2002.
(a) Find an exponential function that models these data. Use $x = 0$ in 1975.
(b) What does the model estimate for the total consumer credit outstanding in 2002?
(c) What does the model predict for the total consumer credit outstanding in 2010?

(d) Would you think the estimate in (c) is too high or too low? Explain.

Year	Amount (in billions)
1980	$349.4
1985	$593.2
1990	$789.1
1995	$1095.8
2000	$1560.2
2001	$1666.9
2002	$1725.7

Source: Federal Reserve

Section 5.2

44. *Poverty threshold* The average poverty threshold for 1987–2002 for a single individual can be modeled by

$$y = -195.59 + 2975.5 \ln(x)$$

where x is the number of years past 1980 and y is the annual income in dollars. (*Source:* U.S. Bureau of the Census)
(a) What does the model estimate as the poverty threshold in 2015?
(b) Graph this function for $x = 5$ to $x = 40$ with a graphing utility.

45. **Modeling** *Percent of paved roads* The following table shows various years from 1950 to 2000 and, for those years, the percents of U.S. roads and streets that were paved.
(a) Find a logarithmic equation that models these data. Use $x = 0$ in 1940.
(b) What percent of paved roads does the model predict for 2010?
(c) When does the model predict that 75% of U.S. roads and streets will be paved?

Year	Percent	Year	Percent
1950	23.5	1990	58.4
1960	34.7	1995	60.8
1970	44.5	1999	62.4
1980	53.7	2000	63.4

Source: U.S. Department of Transportation, *Highway Statistics Summary to 1995* and *Highway Statistics,* 2000

46. *Stellar magnitude* The stellar magnitude M of a star is related to its brightness B as seen from earth according to

$$M = -\frac{5}{2} \log(B/B_0)$$

where B_0 is a standard level of brightness (the brightness of the star Vega).
(a) Find the magnitude of Venus if its brightness is 36.3 times B_0.
(b) Find the brightness (as a multiple of B_0) of the North Star if its magnitude is 2.1.
(c) If the faintest stars have magnitude 6, find their brightness (as a multiple of B_0).
(d) Is a star with magnitude -1.0 brighter than a star with magnitude $+1.0$?

Section 5.3

47. *Sales decay*   The sales decay for a product is given by $S = 50{,}000e^{-0.1x}$, where S is the weekly sales (in dollars) and x is the number of weeks that have passed since the end of an advertising campaign.
(a) What will sales be 6 weeks after the end of the campaign?
(b) How many weeks will pass before sales drop below 15,000?

48. *Sales decay*   The sales decay for a product is given by $S = 50{,}000e^{-0.6x}$, where S is the monthly sales (in dollars) and x is the number of months that have passed since the end of an advertising campaign. What will sales be 6 months after the end of the campaign?

49. *Compound interest*   If $1000 is invested at 12%, compounded monthly, the future value S at any time t (in years) is given by

$$S = 1000(1.01)^{12t}$$

How long will it take for the amount to double?

50. *Compound interest*   If $5000 is invested at 13.5%, compounded continuously, then the future value S at any time t (in years) is given by $S = 5000e^{0.135t}$.
(a) What is the amount after 9 months?
(b) How long will it be before the investment doubles?

51. *World population*   By using data from the International Data Base of the U.S. Bureau of the Census, the accompanying figure shows a scatter plot of actual or projected world population (in billions) in 5-year increments from 1950 to 2050 ($x = 0$ in 1945). Which of the following functions would be the best model: exponential growth, exponential decay, logarithmic, or logistic? Explain.

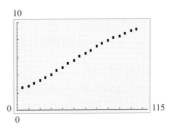

52. *Advertising and sales*   Because of a new advertising campaign, a company predicts that sales will increase and that the yearly sales will be given by the equation

$$N = 10{,}000(0.3)^{0.5t}$$

where t represents the number of years after the start of the campaign.
(a) What are the sales when the campaign begins?
(b) What are the predicted sales for the third year?
(c) What are the maximum predicted sales?

53. *Spread of a disease*   The spread of a highly contagious virus in a high school can be described by the logistic function

$$y = \frac{5000}{1 + 1000e^{-0.8x}}$$

where x is the number of days after the virus is identified in the school and y is the total number of people who are infected by the virus in the first x days.
(a) Graph the function for $0 \le x \le 15$.
(b) How many students had the virus when it was first discovered?
(c) What is the total number infected by the virus during the first 15 days?
(d) In how many days will the total number infected reach 3744?

5 Chapter Test

In Problems 1–4, graph the functions.
1. $y = 5^x$
2. $y = 3^{-x}$
3. $y = \log_5 x$
4. $y = \ln x$

In Problems 5–8, use technology to graph the functions.
5. $y = 3^{0.5x}$
6. $y = \ln(0.5x)$
7. $y = e^{2x}$
8. $y = \log_7 x$

In Problems 9–12, use a calculator to give a decimal approximation of the numbers.
9. e^4
10. $e^{-2.1}$
11. $\ln 4$
12. $\ln 21$
13. Write $\log_{17} x = 3.1$ in exponential form and find x to three decimal places.
14. Write $3^{2x} = 27$ in logarithmic form.

In Problems 15–18, simplify the expressions, using properties or definitions of logarithms.

15. $\log_2 8$ 16. $e^{\ln x^4}$ 17. $\log_7 7^3$ 18. $\ln e^{x^2}$

19. Write $\ln(M \cdot N)$ as a sum involving M and N.

20. Write $\ln\left(\dfrac{x^3 - 1}{x + 2}\right)$ as a difference involving two binomials.

21. Write $\log_4(x^3 + 1)$ as a base e logarithm using a change-of-base formula.

22. Solve for x: $47{,}500 = 1500(1.015)^{6x}$

23. The graph in the following figure shows the number of active workers who will be (or have been) supporting each Social Security beneficiary. What type of function might be an appropriate model for this situation?

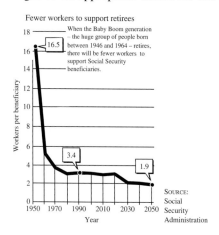

Fewer workers to support retirees

When the Baby Boom generation – the huge group of people born between 1946 and 1964 – retires, there will be fewer workers to support Social Security beneficiaries.

SOURCE: Social Security Administration

24. The graph in the figure shows the history of U.S. cellular telephone subscribership from 1985 to 2002. If these data were modeled by a smooth curve with years on the horizontal axis, what type of function might be appropriate? Explain.

U.S. Cellular Telephone Subscribership, 1985-2002

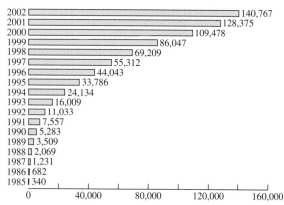

Source: The CTIA Semi-Annual Wireless Industry Indices, December 2004. Used with the permission of CTIA—The Wireless Association.*

*Data may differ slightly from other sources.

25. The total market revenue R (in billions of U.S. dollars) for worldwide telecommunications for the years 1990–2003 can be modeled with the equation

$$R = 330.42e^{0.08333t}$$

where t is the number of years past 1985. (*Source:* International Telecommunications Union)

(a) Find the predicted total market revenue for 2010.

(b) According to the model, how long will it take for the 2010 total market revenue to double?

26. A company plans to phase out one model of its product and replace it with a new model. An advertising campaign for the product being replaced just ended, and typically after such a campaign, monthly sales volume S (in dollars) decays according to

$$S = 22{,}000e^{-0.35t}$$

where t is in months. When the monthly sales volume for this product reaches \$2500, the company plans to discontinue production and launch the new model. How long will it be until this happens?

27. The total U.S. personal income I (in billions of dollars) from 1960 to 2002 can be modeled by

$$I = \frac{14{,}892.86}{1 + 42.1036e^{-0.0993t}}$$

where t is the number of years past 1960. (*Source:* Bureau of Economic Analysis, U.S. Department of Commerce)

(a) What does the model predict for the total U.S. personal income in 2012?

(b) How long will it be until the total U.S. personal income reaches \$12,500 billion?

28. *Modeling*   The table shows the receipts (in billions of dollars) for world tourism for the years 1990–2002.

(a) From a scatter plot of the data (with $x = 0$ in 1985), describe which of exponential, logarithmic, or logistic models is *not* appropriate. Explain.

(b) Find models for the other two types of functions.

(c) Which model from (b) is more optimistic for 2010 world tourism receipts?

Year	Receipts (in billions)	Year	Receipts (in billions)
1990	\$264	1997	\$443
1991	\$278	1998	\$445
1992	\$ 317	1999	\$455
1993	\$323	2000	\$473
1994	\$356	2001	\$459
1995	\$405	2002	\$474
1996	\$439		

Source: World Tourism Organization

Extended Applications & Group Projects

I. Agricultural Business Management

A commercial vegetable and fruit grower carefully observes the relationship between the amount of fertilizer used on a certain variety of pumpkin and the extra revenue made from the sales of the resulting pumpkin crop. These observations are recorded in the table below.

Amount of fertilizer (pounds/acre)	0	250	500	750	1000	1250	1500	1750	2000
Extra revenue earned (dollars/acre)	0	96	145	172	185	192	196	198	199

The fertilizer costs 14¢ per pound. What would you advise the grower is the most profitable amount of fertilizer to use?

Check your advice by answering the following questions.

1. Is each pound of fertilizer equally effective?
2. Estimate the maximum amount of extra revenue that can be earned per acre by observing the data.
3. Graph the data points giving the extra revenue made from improved pumpkin sales as a function of the amount of fertilizer used. What does the graph tell you about the growth of the extra revenue as a function of the amount of fertilizer used?
4. Which of the following equations best models the graph you obtained in Question 3?

   (a) $N = Ca^{Rx}$ (b) $y = \dfrac{A}{1 + ce^{-ax}}$ (c) $y = c(1 - e^{-ax})$

   Determine a specific equation that fits the observed data. That is, find the value of the constants in the model you have chosen.
5. Graph the total cost of the fertilizer as a function of the amount of fertilizer used on the graph you created for Question 3. Determine an equation that models this relationship. For what amount of fertilizer is the extra revenue earned from pumpkin sales offset by the cost of the fertilizer?
6. Use the graphs or the equations you have created to estimate the maximum extra profit that can be made and the amount of fertilizer per acre that would result in that extra profit.

II. National Debt (Modeling)

For years, politicians have talked about a balanced federal budget, and success in this effort was proclaimed in 1998. However, does this mean that there is no longer a national debt? Does it mean that the national debt is getting smaller? The table below gives the U.S. national debt for selected years from 1900 to 1995.

Year	U.S. Debt ($billions)	Year	U.S. Debt ($billions)
1900	1.2	1985	1823.1
1910	1.1	1987	2350.3
1920	24.2	1989	2857.4
1930	16.1	1990	3233.3
1940	43.0	1991	3665.3
1945	258.7	1992	4064.6
1955	272.8	1993	4411.5
1965	313.8	1994	4692.8
1975	533.2	1995	4974.0

Source: Bureau of Public Debt, U.S. Treasury

To investigate this issue, complete the following.

1. Create a scatter plot of these data, and determine the type of function that would best model these data. (Let $x = 0$ in 1900.)
2. Model the data. Rewrite the equation with base e.
3. How accurately does the model predict the debt for 1995? What does it predict for 1998 and 2001?
4. Use this model to predict when the debt will be $50 trillion ($50,000 billion).
5. Compare the results from Questions 3 and 4 with the actual results in the following table, which gives the national debt through 2004. Does the model you created predict the debt for these years fairly well?
6. Create a scatter plot that includes the data in the table below.
7. Graph the model developed in Question 2 on the same set of axes as the scatter plot for the new data. Does the model still represent the data?
8. Create an equation that models the data in the table below. Rewrite the equation with base e.
9. Look at the equations of the two models and graph both models on the same set of axes to determine whether the predictions generated by the new model differ greatly from those of the first model.
10. With the War on Terror and the War in Iraq, these exponential models may no longer be appropriate and may not accurately represent future trends. Use the Internet or a library to find the national debt for the last 3 years and determine whether it is falling or is still growing. Do the debts for these years fit either of the models you created? Try different models, such as cubic or logistic, with the new data. Comment on the appropriateness of these new models.

Year	U.S. Debt ($billions)	Year	U.S. Debt ($billions)
1900	1.2	1992	4064.6
1910	1.1	1993	4411.5
1920	24.2	1994	4692.8
1930	16.1	1995	4974.0
1940	43.0	1996	5224.8
1945	258.7	1997	5413.1
1955	272.8	1998	5526.2
1965	313.8	1999	5656.3
1975	533.2	2000	5674.2
1985	1823.1	2001	5807.5
1987	2350.3	2002	6228.2
1989	2857.4	2003	6783.2
1990	3233.3	2004	7379.1
1991	3665.3		

Source: Bureau of Public Debt, U.S. Treasury

6

Mathematics of Finance

Regardless of whether or not your career is in business, understanding how interest is computed on investments and loans is important to you as a consumer. The proliferation of personal finance and money management software attests to this importance. The goal of this chapter is to provide some understanding of the methods used to determine the interest and future value (principal plus interest) resulting from savings plans and the methods used in repayment of debts.

The topics and applications discussed in this chapter include the following.

Sections	Applications
6.1 Simple Interest; Sequences Arithmetic sequences	*Loans, return on an investment, duration of an investment*
6.2 Compound Interest; Geometric Sequences	*529 plans, future value, present value, comparing investments, doubling time*
6.3 Future Values of Annuities Ordinary annuities Annuities due	*Comparing investment plans, regular investments to reach a goal, time to reach a goal*
6.4 Present Values of Annuities Ordinary annuities Annuities due Deferred annuities	*Retirement planning, lottery prizes, court settlements*
6.5 Loans and Amortization Unpaid balance of a loan	*Home mortgages, buying a car, effect of paying an extra amount*

Chapter Warm-up

Prerequisite Problem Type	For Section	Answer	Section for Review
What is $\dfrac{(-1)^n}{2n}$, if (a) $n = 1$? (b) $n = 2$? (c) $n = 3$?	6.1	(a) $-\frac{1}{2}$ (b) $\frac{1}{4}$ (c) $-\frac{1}{6}$	0.2 Signed numbers
(a) What is $(-2)^6$? (b) What is $\dfrac{\frac{1}{4}\left[1 - \left(\frac{1}{2}\right)^6\right]}{1 - \frac{1}{2}}$?	6.2	(a) 64 (b) $\frac{63}{128}$	0.2 Signed numbers
If $f(x) = \dfrac{1}{2x}$, what is (a) $f(2)$? (b) $f(4)$?	6.1	(a) $\frac{1}{4}$ (b) $\frac{1}{8}$	1.2 Function notation
Evaluate $e^{(0.08)(20)}$.	6.2	4.95303	5.1 Exponential functions
Evaluate: (a) $\dfrac{1 - (1.05)^{-16}}{0.05}$ (b) $5000 + 5000\left[\dfrac{1 - (1.059)^{-9}}{0.059}\right]$	6.3 6.4	(a) 10.83777 (b) 39,156.96575	0.3 Integral exponents
Solve: (a) $1100 = 1000 + 1000(0.058)t$ (b) $12{,}000 = P(1.03)^6$ (c) $850 = A_n\left[\dfrac{0.0065}{1 - (1.0065)^{-360}}\right]$	6.1–6.5	(a) $t \approx 1.72$ (b) $P \approx 10{,}049.81$ (c) $A_n \approx 118{,}076.79$	1.1 Linear equations
Solve: (a) $2 = (1.08)^n$ (b) $20{,}000 = 10{,}000e^{0.08t}$	6.2	(a) $n \approx 9.01$ (b) $t \approx 8.66$	5.3 Solutions of exponential equations

Simple Interest; Sequences

- To find the future value and the amount of interest for a simple interest loan and an investment
- To find the simple interest rate earned on an investment
- To find the time required for a simple interest investment to reach a goal
- To write a specified number of terms of a sequence
- To find specified terms and sums of specified numbers of terms of arithmetic sequences

◊ Application Preview

In this section we begin our study of the mathematics of finance by considering **simple interest**. Simple interest forms the basis for all calculations involving interest that is paid on an investment or that is due on a loan. Also, we can evaluate short-term investments (such as in stocks or real estate) by calculating their simple interest rates. For example, Mary Spaulding purchased some Wind-Gen Electric stock for $6125.00. After 6 months, the stock had risen in value by $138.00 and had paid dividends totaling $144.14. One way to evaluate this investment and compare it with a bank savings plan is to find the simple interest rate that these gains represent. (See Example 3.)

Simple Interest

If a sum of money P (called the **principal**) is invested for a time period t (frequently in years) at an interest rate r per period, the **simple interest** is given by the following formula.

Simple Interest

The simple interest I is given by

$$I = Prt$$

where
I = interest (in dollars)
P = principal (in dollars)
r = annual interest rate (written as a decimal)
t = time (in years)*

Note that the time measurements for r and t must agree.

Simple interest is paid on investments involving time certificates issued by banks and on certain types of bonds, such as U.S. government series H bonds and municipal bonds. The interest for a given period is paid to the investor, and the principal remains the same.

If you borrow money from a friend or a relative, interest on your loan might be calculated with the simple interest formula. We'll consider some simple interest loans in this section, but interest on loans from banks and other lending institutions is calculated using methods discussed later.

● EXAMPLE 1 *Simple Interest*

(a) If $8000 is invested for 2 years at an annual interest rate of 9%, how much interest will be received at the end of the 2-year period?
(b) If $4000 is borrowed for 39 weeks at an annual interest rate of 15%, how much interest is due at the end of the 39 weeks?

Solution
(a) The interest is $I = Prt = \$8000(0.09)(2) = \$1440$.
(b) Use $I = Prt$ with $t = 39/52 = 0.75$ year. Thus

$$I = \$4000(0.15)(0.75) = \$450$$

---

*Periods of time other than years can be used, as can other monetary systems.

Future Value   The **future amount of an investment,** or its **future value,** at the end of an interest period is the sum of the principal and the interest. Thus, in Example 1(a), the future value is

$$S = \$8000 + \$1440 = \$9440$$

Similarly, the **future amount of a loan,** or its **future value,** is the amount of money that must be repaid. In Example 1(b), the future value of the loan is the principal plus the interest, or

$$S = \$4000 + \$450 = \$4450$$

Future Value

If we use the letter S to denote the future value of an investment or a loan, then we have

Future value of investment or loan: $S = P + I$

where P is the principal (in dollars) and I is the interest (in dollars).

The principal P of a loan is also called the **face value** or the **present value** of the loan.

● **EXAMPLE 2** *Future Value and Present Value*

(a) If $2000 is borrowed for one-half year at a simple interest rate of 12% per year, what is the future value of the loan at the end of the half-year?
(b) An investor wants to have $20,000 in 9 months. If the best available interest rate is 6.05% per year, how much must be invested now to yield the desired amount?

Solution
(a) The interest for the half-year period is $I = \$2000(0.12)(0.5) = \$120$. Thus the future value of the investment for the period is

$$S = P + I = \$2000 + \$120 = \$2120$$

(b) We know that $S = P + I = P + Prt$. In this case, we must solve for P, the present value. Also, the time 9 months is $(9/12)$ of a year.

$$\$20,000 = P + P(0.0605)(9/12) = P + 0.045375P$$

$$\$20,000 = 1.045375P$$

$$\frac{\$20,000}{1.045375} = P \quad \text{so} \quad P = \$19,131.89$$

◗ **EXAMPLE 3** *Return on an Investment* **(Application Preview)**

Recall that Mary Spaulding bought Wind-Gen Electric stock for $6125.00 and that after 6 months, the value of her shares had risen by $138.00 and dividends totaling $144.14 had been paid. Find the simple interest rate she earned on this investment if she sold the stock at the end of the 6 months.

Solution
To find the simple interest rate that Mary earned on this investment, we find the rate that would yield an amount of simple interest equal to all of Mary's gains (that is, equal to the rise in the stock's price plus the dividends she received). Thus the principal is $6125.00,

the time is 1/2 year, and the interest earned is the total of all gains (that is, interest $I = \$138.00 + \$144.14 = \$282.14$). Hence

$$I = Prt$$
$$\$282.14 = (\$6125)r(0.5)$$
$$\$282.14 = \$3062.5r$$
$$r = \frac{\$282.14}{\$3062.5} \approx 0.092 = 9.2\%$$

Thus Mary's return was equivalent to an annual simple interest rate of about 9.2%.

● **EXAMPLE 4** *Duration of an Investment*

If $1000 is invested at 5.8% simple interest, how long will it take to grow to $1100?

Solution
With $P = \$1000$, $S = \$1100$, and $r = 0.058$, we solve for t.

$$S = P + Prt$$
$$\$1100 = \$1000 + \$1000(0.058)t$$
$$\$100 = \$58t$$
$$\frac{\$100}{\$58} = t \quad \text{so} \quad t \approx 1.72 \text{ years}$$

● **Checkpoint**

1. What is the simple interest formula?
2. If $8000 is invested at 6% simple interest for 9 months, find the future value of the investment.
3. If a $2500 investment grows to $2875 in 15 months, what simple interest rate was earned?

Sequences

Let us look at the monthly future values of a $2000 investment that earns 1% simple interest for each of 5 months.

Month	Interest ($I = Prt$)	Future Value of the Investment
1	($2000)(0.01)(1) = $20	$2000 + $20 = $2020
2	($2000)(0.01)(1) = $20	$2020 + $20 = $2040
3	($2000)(0.01)(1) = $20	$2040 + $20 = $2060
4	($2000)(0.01)(1) = $20	$2060 + $20 = $2080
5	($2000)(0.01)(1) = $20	$2080 + $20 = $2100

These future values are outputs that result when the inputs are positive integers that correspond to the number of months of the investment. Outputs (such as these future values) that arise uniquely from positive integer inputs define a function whose domain is the positive integers. Such a function is called a **sequence function.** Because the domain of a sequence function is the positive integers, the outputs form an ordered list called a **sequence.**

Sequence

The set of functional values $a_1, a_2, a_3, \ldots$ of a sequence function is called a **sequence.** The values $a_1, a_2, a_3, \ldots$ are called **terms** of the sequence, with a_1 the first term, a_2 the second term, and so on.

Because calculations involving interest often result from using positive integer inputs, sequences are the basis for most of the financial formulas derived in this chapter.

● **EXAMPLE 5** *Terms of a Sequence*

Write the first four terms of the sequence whose nth term is $a_n = (-1)^n/(2n)$.

Solution
The first four terms of the sequence are as follows:

$$a_1 = \frac{(-1)^1}{2(1)} = -\frac{1}{2} \qquad a_2 = \frac{(-1)^2}{2(2)} = \frac{1}{4}$$

$$a_3 = \frac{(-1)^3}{2(3)} = -\frac{1}{6} \qquad a_4 = \frac{(-1)^4}{2(4)} = \frac{1}{8}$$

We usually write these terms in the form $-\frac{1}{2}, \frac{1}{4}, -\frac{1}{6}, \frac{1}{8}$.

Arithmetic Sequences

On the previous page, we described the sequence

$$2020, 2040, 2060, 2080, 2100, \ldots$$

This sequence can also be described in the following way:

$$a_1 = 2020, \quad a_n = a_{n-1} + 20 \quad \text{for } n > 1$$

This sequence is an example of a special kind of sequence called an **arithmetic sequence.** In such a sequence, each term after the first can be found by adding a constant to the preceding term. Thus we have the following definition.

Arithmetic Sequence

A sequence is called an **arithmetic sequence** (progression) if there exists a number d, called the **common difference,** such that

$$a_n = a_{n-1} + d \quad \text{for } n > 1$$

● **EXAMPLE 6** *Arithmetic Sequences*

Write the next three terms of each of the following arithmetic sequences.

(a) $1, 3, 5, \ldots$
(b) $9, 6, 3, \ldots$
(c) $\frac{1}{2}, \frac{5}{6}, \frac{7}{6}, \ldots$

Solution
(a) The common difference is 2, so the next three terms are 7, 9, 11.
(b) The common difference is -3, so the next three terms are $0, -3, -6$.
(c) The common difference is $\frac{1}{3}$, so the next three terms are $\frac{3}{2}, \frac{11}{6}, \frac{13}{6}$.

Because each term after the first term in an arithmetic sequence is obtained by adding d to the preceding term, the second term is $a_1 + d$, the third is $(a_1 + d) + d = a_1 + 2d, \ldots,$ and the nth term is $a_1 + (n - 1)d$. Thus we have the following formula.

nth Term of an Arithmetic Sequence

The **nth term of an arithmetic sequence** (progression) is given by

$$a_n = a_1 + (n - 1)d$$

where a_1 is the first term and d is the common difference between successive terms.

● **EXAMPLE 7 *nth Term of an Arithmetic Sequence***

Find the 11th term of the arithmetic sequence with first term 3 and common difference -2.

Solution
The 11th term is $a_{11} = 3 + (11 - 1)(-2) = -17$.

● **EXAMPLE 8 *Arithmetic Sequence***

If the first term of an arithmetic sequence is 4 and the 9th term is 20, find the 75th term.

Solution
Substituting the values $a_1 = 4$, $a_n = 20$, and $n = 9$ in $a_n = a_1 + (n - 1)d$ gives $20 = 4 + (9 - 1)d$. Solving this equation gives $d = 2$. Therefore, the 75th term is $a_{75} = 4 + (75 - 1)(2) = 152$.

Sum of an Arithmetic Sequence

Consider the arithmetic sequence with first term a_1, common difference d, and nth term a_n. The first n terms of an arithmetic sequence can be written from a_1 to a_n as

$$a_1, a_1 + d, a_1 + 2d, a_1 + 3d, \ldots, a_1 + (n - 1)d$$

or backwards from a_n to a_1 as

$$a_n, a_n - d, a_n - 2d, a_n - 3d, \ldots, a_n - (n - 1)d$$

If we let s_n represent the sum of the first n terms of the sequence described above, then we have the following equivalent forms for the sum.

$$s_n = a_1 + (a_1 + d) + (a_1 + 2d) + \cdots + [a_1 + (n - 1)d] \tag{1}$$
$$s_n = a_n + (a_n - d) + (a_n - 2d) + \cdots + [a_n - (n - 1)d] \tag{2}$$

If we add equations (1) and (2) term by term, we obtain

$$2s_n = (a_1 + a_n) + (a_1 + a_n) + (a_1 + a_n) + \cdots + (a_1 + a_n)$$

in which $(a_1 + a_n)$ appears as a term n times. Thus,

$$2s_n = n(a_1 + a_n)$$

$$s_n = \frac{n}{2}(a_1 + a_n)$$

We may state this result as follows.

Sum of an Arithmetic Sequence

The **sum of the first n terms of an arithmetic sequence** is given by the formula

$$s_n = \frac{n}{2}(a_1 + a_n)$$

where a_1 is the first term of the sequence and a_n is the nth term.

● **EXAMPLE 9** *Sum of an Arithmetic Sequence*

Find the sum of the first 10 terms of the arithmetic sequence with first term 2 and common difference 4.

Solution
We are given the values $n = 10$, $a_1 = 2$, and $d = 4$. Thus the 10th term is $a_{10} = 2 + (10 - 1)4 = 38$, and the sum of the first 10 terms is

$$s_{10} = \frac{10}{2}(2 + 38) = 200$$

● **EXAMPLE 10** *Sum of an Arithmetic Sequence*

Find the sum of the first 91 terms of the arithmetic sequence $\frac{1}{4}, \frac{7}{12}, \frac{11}{12}, \ldots$.

Solution
The first term is $\frac{1}{4}$ and the common difference is $\frac{1}{3}$. Therefore, the 91st term is $a_{91} = \frac{1}{4} + (91 - 1)(\frac{1}{3}) = 30\frac{1}{4} = \frac{121}{4}$. The sum of the first 91 terms is $s_{91} = \frac{91}{2}(\frac{1}{4} + \frac{121}{4}) = \frac{5551}{4} = 1387\frac{3}{4}$.

● **Checkpoint**

4. Of the following sequences, identify the arithmetic sequences.
   (a) 1, 4, 9, 16, . . .
   (b) 1, 4, 7, 10, . . .
   (c) 1, 2, 4, 8, . . .
5. Given the arithmetic sequence $-10, -6, -2, \ldots$, find
   (a) the 51st term
   (b) the sum of the first 51 terms

Finally, we should note that many graphing calculators, some graphics software packages, and spreadsheets have the capability of defining sequence functions and then operating on them by finding additional terms, graphing the terms, or summing a fixed number of terms.

● **Checkpoint Solutions**

1. $I = Prt$
2. $S = P + I = 8000 + 8000(0.06)(\frac{9}{12}) = \$8360$
3. $S = P + I$, so

$$2875 = 2500 + 2500r\left(\frac{15}{12}\right)$$

$$375 = 3125r$$

$$r = \frac{375}{3125} = 0.12, \quad \text{or} \quad 12\%$$

4. Only (b) is an arithmetic sequence; the common difference is 3.
5. (a) The common difference is $d = 4$. The 51st term ($n = 51$) is
   $a_n = a_1 + (n - 1)d = -10 + 50(4) = -10 + 200 = 190$
   (b) For $a_1 = -10$ and $a_{51} = 190$, the sum of the first 51 terms is
   $s_{51} = \frac{51}{2}(-10 + 190) = 4590$.

6.1 Exercises

SIMPLE INTEREST

1. $10,000 is invested for 6 years at an annual simple interest rate of 16%.
   (a) How much interest will be earned?
   (b) What is the future value of the investment at the end of the 6 years?

2. $800 is invested for 5 years at an annual simple interest rate of 14%.
   (a) How much interest will be earned?
   (b) What is the future value of the investment at the end of the 5 years?

3. $1000 is invested for 3 months at an annual simple interest rate of 12%.
   (a) How much interest will be earned?
   (b) What is the future value of the investment after 3 months?

4. $1800 is invested for 9 months at an annual simple interest rate of 15%.
   (a) How much interest will be earned?
   (b) What is the future value of the investment after 9 months?

5. If you borrow $800 for 6 months at 16% annual simple interest, how much must you repay at the end of the 6 months?

6. If you borrow $1600 for 2 years at 14% annual simple interest, how much must you repay at the end of the 2 years?

7. If you lend $3500 to a friend for 15 months at 8% annual simple interest, find the future value of the loan.

8. Mrs. Gonzalez lent $2500 to her son Luis for 7 months at 9% annual simple interest. What is the future value of this loan?

9. A couple bought some stock for $30 per share that pays an annual dividend of $0.90 per share. After 1 year the price of the stock was $33. Find the simple interest rate on the growth of their investment.

10. Jenny Reed bought SSX stock for $16 per share. The annual dividend was $1.50 per share, and after 1 year SSX was selling for $35 per share. Find the simple interest rate of growth of her money.

11. (a) To buy a Treasury bill (T-bill) that matures to $10,000 in 6 months, you must pay $9750. What rate does this earn?
    (b) If the bank charges a fee of $40 to buy a T-bill, what is the actual interest rate you earn?

12. Janie Christopher lent $6000 to a friend for 90 days at 12%. After 30 days, she sold the note to a third party for $6000. What interest rate did the third party receive? Use 360 days in a year.

13. A firm buys 12 file cabinets at $140 each, with the bill due in 90 days. How much must the firm deposit now to have enough to pay the bill if money is worth 12% per year? Use 360 days in a year.

14. A student has a savings account earning 9% simple interest. She must pay $1500 for first-semester tuition by September 1 and $1500 for second-semester tuition by January 1. How much must she earn in the summer (by September 1) in order to pay the first-semester bill on time and still have the remainder of her summer earnings grow to $1500 between September 1 and January 1?

15. If you want to earn 15% annual simple interest on an investment, how much should you pay for a note that will be worth $13,500 in 10 months?

16. What is the present value of an investment at 6% annual simple interest if it is worth $832 in 8 months?

17. If $5000 is invested at 8% annual simple interest, how long does it take to be worth $9000?

18. How long does it take for $8500 invested at 11% annual simple interest to be worth $13,000?

19. A retailer owes a wholesaler $500,000 due in 45 days. If the payment is 15 days late, there is a 1% penalty charge. The retailer can get a 45-day certificate of deposit (CD) paying 6% or a 60-day certificate paying 7%. Is it better to take the 45-day certificate and pay on time or to take the 60-day certificate and pay late with the penalty?

20. An investor owns several apartment buildings. The taxes on these buildings total $30,000 per year and are due before April 1. The late fee is 1/2% per month up to 6 months, at which time the buildings are seized by the authorities and sold for back taxes. If the investor has $30,000 available on March 31, will he save money by paying the taxes at that time or by investing the money at 8% and paying the taxes and the penalty on September 30?

21. Bill Casler bought a $2000, 9-month certificate of deposit (CD) that would earn 8% annual simple interest. Three months before the CD was due to mature, Bill needed his CD money, so a friend agreed to lend him money and receive the value of the CD when it matured.
    (a) What is the value of the CD when it matures?
    (b) If their agreement allowed the friend to earn a 10% annual simple interest return on his loan to Bill, how much did Bill receive from his friend?

22. Suppose you lent $5000 to friend 1 for 18 months at an annual simple interest rate of 9%. After 1 year you need money for an emergency and decide to sell the note to friend 2.

(a) How much does friend 1 owe when the loan is due?

(b) If your agreement with friend 2 means she earns simple interest at an annual rate of 12%, how much did friend 2 pay you for the note?

SEQUENCES

23. Write the first ten terms of the sequence defined by $a_n = 3n$.

24. Write the first seven terms of the sequence defined by $a_n = 2/n$.

25. Write the first six terms of the sequence whose nth term is $(-1)^n/(2n + 1)$.

26. Write the first five terms of the sequence whose nth term is

$$a_n = \frac{(-1)^n}{n^2}$$

27. Write the first four terms and the tenth term of the sequence whose nth term is

$$a_n = \frac{n - 4}{n(n + 2)}$$

28. Write the sixth term of the sequence whose nth term is

$$a_n = \frac{n(n - 1)}{n + 3}$$

ARITHMETIC SEQUENCES

In Problems 29–32, (a) identify d and a_1 and (b) write the next three terms.

29. 2, 5, 8, . . .

30. 3, 9, 15, . . .

31. $3, \frac{9}{2}, 6, \ldots$

32. 2, 2.75, 3.5, . . .

33. Find the 83rd term of the arithmetic sequence with first term 6 and common difference $-\frac{1}{2}$.

34. Find the 66th term of the arithmetic sequence with first term $\frac{1}{2}$ and common difference $-\frac{1}{3}$.

35. Find the 100th term of the arithmetic sequence with first term 5 and eighth term 19.

36. Find the 73rd term of the arithmetic sequence with first term 20 and tenth term 47.

37. Find the sum of the first 38 terms of the arithmetic sequence with first term 2 and 38th term 113.

38. Find the sum of the first 56 terms of the arithmetic sequence with first term 6 and 56th term 226.

39. Find the sum of the first 70 terms of the arithmetic sequence with first term 10 and common difference $\frac{1}{2}$.

40. Find the sum of the first 80 terms of the arithmetic sequence with first term 12 and common difference -3.

41. Find the sum of the first 150 terms of the arithmetic sequence $6, \frac{9}{2}, 3, \ldots$.

42. Find the sum of the first 200 terms of the arithmetic sequence 12, 9, 6, ...

APPLICATIONS

43. *Bee reproduction* A female bee hatches from a fertilized egg, whereas a male bee hatches from an unfertilized egg. Thus a female bee has a male parent and a female parent, whereas a male bee has only a female parent. Therefore, the number of ancestors of a male bee follows the *Fibonacci sequence*

$$1, 2, 3, 5, 8, 13, \ldots$$

Observe the pattern and write three more terms of the sequence.

44. *Salaries* Suppose you are offered a job with a relatively low starting salary but with a $1500 raise for each of the next 7 years. How much more than your starting salary would you be making in the eighth year?

45. *Profit* A new firm loses $2000 in its first month, but its profit increases by $400 in each succeeding month for the next year. What is its profit in the twelfth month?

46. *Pay raises* If you make $27,000 and get $1800 raises each year, in how many years will your salary double?

47. *Salaries* Suppose you are offered two identical jobs: one paying a starting salary of $20,000 with yearly raises of $1000 and one paying a starting salary of $18,000 with yearly raises of $1200. Which job will pay you more for your tenth year on the job?

48. *Profit* A new firm loses $2000 in its first month, but its profit increases by $400 in each succeeding month for the next year. What is its profit for the year?

49. *Pay raises* If you are an employee, would you rather be given a raise of $1000 at the end of each year (plan I) or a raise of $300 at the end of each 6-month period (plan II)? Consider the following table for an employee whose base salary is $20,000 per year (or $10,000 per 6-month period), and answer parts (a)–(g).

Period (months)	Salary Received per 6-Month Period	
	Plan I	Plan II
0–6	$10,000	$10,000
6–12	10,000	10,300
12–18	10,500	10,600
18–24	10,500	10,900
24–30	11,000	11,200
30–36	11,000	11,500

(a) Find the sum of the raises for plan I for the first 3 years.

(b) Find the sum of the raises for plan II for the first 3 years.

(c) Which plan is better, and by how much?

(d) Find the sum of the raises in plan I for 5 years.

(e) Find the sum of the raises in plan II for 5 years.

(f) Which plan is better, and by how much?

(g) Do you want plan I or plan II?

50. *Pay raises*   As an employee, would you prefer being given a \$1200 raise each year for 5 years or a \$200 raise each quarter for 5 years?

6.2   Compound Interest; Geometric Sequences

OBJECTIVES

● To find the future value of a compound interest investment and the amount of interest earned when interest is compounded at regular intervals or continuously

● To find the annual percentage yield (APY) or the effective annual interest rate of money invested at compound interest

● To find the time it takes for an investment to reach a specified amount

● To find specified terms, and sums of specified numbers of terms, of geometric sequences

◗ Application Preview

A common concern for young parents is how they will pay for their children's college educations. To address this, there are an increasing number of college savings plans, such as the so-called "529 plans," which are tax-deferred.

Suppose Jim and Eden established such a plan for their baby daughter Maura. If this account earns 9.8% compounded quarterly and if their goal is to have \$200,000 by Maura's eighteenth birthday, what would be the impact of having \$10,000 in the account by Maura's first birthday? (See Example 3.)

A compound interest investment (such as Jim and Eden's account) is one in which interest is paid into the account at regular intervals. In this section we consider investments of this type and develop formulas that enable us to determine the impact of making a \$10,000 investment in Maura's college tuition account by her first birthday.

Compound Interest

In the previous section we discussed simple interest. A second method of paying interest is the **compound interest** method, where the interest for each period is added to the principal before interest is calculated for the next period. With compound interest, both the interest added and the principal earn interest for the next period. With this method, the principal grows as the interest is added to it. This method is used in investments such as savings accounts and U.S. government series E bonds.

An understanding of compound interest is important not only for people planning careers with financial institutions but also for anyone planning to invest money. To see how compound interest is computed, consider the following table, which tracks the annual growth of \$20,000 invested for 3 years at 10% compounded annually. (Notice that the ending principal for each year becomes the beginning principal for the next year.)

Year	Beginning Principal = P	10% Annual Interest = I	Ending Principal = $P + I$
1	\$20,000	0.10(\$20,000) = \$2000	\$22,000
2	\$22,000	0.10(\$22,000) = \$2200	\$24,200
3	\$24,200	0.10(\$24,200) = \$2420	\$26,620

Note that the future value for each year can be found by multiplying the beginning principal for that year by $1 + 0.10$, or 1.10. That is,

First year: $\qquad \$20,000(1.10) = \$22,000$

Second year: $\quad [\$20,000(1.10)](1.10) = \$20,000(1.10)^2 = \$24,200$

Third year: $\quad [\$20,000(1.10)^2](1.10) = \$20,000(1.10)^3 = \$26,620$

This suggests that if we maintained this investment for n years, the future value at the end of this time would be $\$20,000(1.10)^n$. Thus we have the following general formula.

If $P is invested at an interest rate of r per year, compounded annually, the future value S at the end of the nth year is

$$S = P(1 + r)^n$$

● **EXAMPLE 1** *Annual Compounding*

If $3000 is invested for 4 years at 9% compounded annually, how much interest is earned?

Solution
The future value is

$$S = \$3000(1 + 0.09)^4$$
$$= \$3000(1.4115816)$$
$$= \$4234.7448$$
$$= \$4234.74, \text{ to the nearest cent}$$

Because $3000 of this amount was the original investment, the interest earned is $4234.74 − $3000 = $1234.74.

Some accounts have the interest compounded semiannually, quarterly, monthly, or daily. Unless specifically stated otherwise, a stated interest rate, called the **nominal annual rate,** is the rate per year and is denoted by r. The interest rate *per period,* denoted by i, is the nominal rate divided by the number of interest periods per year. The interest periods are also called *conversion periods,* and the number of periods is denoted by n. Thus, if $100 is invested for 5 years at 6% compounded semiannually (twice a year), it has been invested for $n = 10$ periods (5 years × 2 periods per year) at $i = 3\%$ per period (6% per year ÷ 2 periods per year). The future value of an investment of this type is found using the following formula.

Future Value (Periodic Compounding)

If $P is invested for t years at a nominal interest rate r, compounded m times per year, then the total number of compounding periods is

$$n = mt$$

the interest rate per compounding period is

$$i = \frac{r}{m} \quad \text{(expressed as a decimal)}$$

and the future value is

$$S = P(1 + i)^n = P\left(1 + \frac{r}{m}\right)^{mt}$$

● **EXAMPLE 2** *Periodic Compounding*

For each of the following investments, find the interest rate per period, i, and the number of compounding periods, n.

(a) 12% compounded monthly for 7 years
(b) 7.2% compounded quarterly for 11 quarters

Solution

(a) If the compounding is monthly and $r = 12\% = 0.12$, then $i = 0.12/12 = 0.01$. The number of compounding periods is $n = (7 \text{ yr})(12 \text{ periods/yr}) = 84$.

(b) $i = 0.072/4 = 0.018$, $n = 11$ (the number of quarters given)

Once we know i and n, we can calculate the future value from the formula using a calculator.

◐ EXAMPLE 3 *Future Value (Application Preview)*

Jim and Eden want to have $200,000 in Maura's college fund on her eighteenth birthday, and they want to know the impact on this goal of having $10,000 invested at 9.8%, compounded quarterly, on her first birthday. To advise Jim and Eden regarding this, find

(a) the future value of the $10,000 investment,
(b) the amount of compound interest that the investment earns, and
(c) the impact this would have on their goal.

Solution

(a) For this situation, $i = 0.098/4 = 0.0245$ and $n = 4(17) = 68$. Thus the future value of the $10,000 is given by

$$S = P(1 + i)^n = \$10,000(1 + 0.0245)^{68} = \$10,000(5.18577) = \$51,857.70$$

(b) The amount of interest earned is $51,857.70 - \$10,000 = \$41,857.70$.

(c) Thus $10,000 invested by Maura's first birthday grows to an amount that is slightly more than 25% of their goal. This rather large early investment has a substantial impact on their goal.

We saw previously that compound interest calculations are based on those for simple interest, except that interest payments are added to the principal. In this way, interest is earned on both principal and previous interest payments. Let's examine the effect of this compounding by comparing compound interest and simple interest. If the investment in Example 3 had been at simple interest, the interest earned would have been $Prt = \$10,000(0.098)(17) = \$16,660$. This is almost $25,200 less than the amount of compound interest earned. And this difference would have been magnified over a longer period of time. Try reworking Example 3's investment over 30 years and compare the compound interest earned with the simple interest earned. This comparison begins to shed some light on why Einstein characterized compound interest as "the most powerful force in the Universe."

● EXAMPLE 4 *Present Value*

What amount must be invested now in order to have $12,000 after 3 years if money is worth 6% compounded semiannually?

Solution

We need to find the present value P, knowing that the future value is $S = \$12,000$. Use $i = 0.06/2 = 0.03$ and $n = 3(2) = 6$.

$$S = P(1 + i)^n$$
$$\$12,000 = P(1 + 0.03)^6 = P(1.03)^6$$
$$\$12,000 = P(1.1940523)$$
$$P = \frac{\$12,000}{1.1940523} = \$10,049.81, \text{ to the nearest cent}$$

● **EXAMPLE 5** *Rate Earned*

The events of September 11, 2001, sent stocks into a deep decline. Since then, the war in Iraq, global uncertainty, and increased deficit spending have made recovery slow and unsteady. However, during the 1990s the stock market experienced unprecedented growth, and mutual fund advertisements often had graphs similar to the one in Figure 6.1. For the investment in Figure 6.1, $10,000 would have grown to $30,118 in 10 years. What interest rate, compounded annually, would be earned?

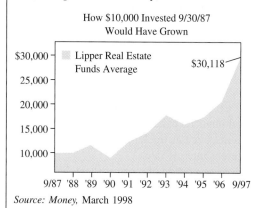

How $10,000 Invested 9/30/87
Would Have Grown

Figure 6.1 *Source: Money,* March 1998

Solution

We use $P = \$10{,}000$, $S = \$30{,}118$, and $n = 10$ in the formula $S = P(1 + i)^n$, and solve for i.

$$\$30{,}118 = \$10{,}000(1 + i)^{10}$$

$$\frac{30{,}118}{10{,}000} = (1 + i)^{10}$$

$$3.0118 = (1 + i)^{10}$$

At this point we take the tenth root of both sides (or, equivalently, raise both sides to the $1/10 = 0.1$ power).

$$3.0118^{0.1} = \left[(1 + i)^{10}\right]^{0.1}$$

$$1.11656 = 1 + i$$

$$0.11656 = i$$

Thus, this investment would earn about 11.66% compounded annually.

● **Checkpoint**

1. If $5000 is invested at 6%, compounded quarterly, for 5 years, find
   (a) the number of compounding periods per year, m,
   (b) the number of compounding periods for the investment, n,
   (c) the interest rate for each compounding period, i,
   (d) the future value of the investment.
2. Find the present value of an investment that is worth $12,000 after 5 years at 9% compounded monthly.

If we invest a sum of money, say $100, then the higher the interest rate, the greater the future value. Figure 6.2 on the following page shows a graphical comparison of the future values when $100 is invested at 5%, 8%, and 10%, all compounded annually over a period of 30 years. Note that higher interest rates yield consistently higher future values and have dramatically higher future values after 15–20 years. Note also that these graphs of the future values are growth exponentials.

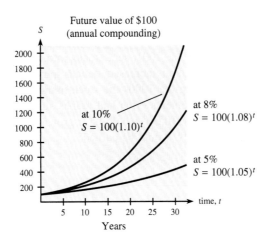

Figure 6.2

Spreadsheet Note Spreadsheets and financial calculators can also be used to investigate and analyze compound interest investments, such as tracking the future value of an investment or comparing investments with different interest rates.

Table 6.1 shows a portion of a spreadsheet that tracks the monthly growth of two investments, both compounded monthly and both with $P = \$1000$, but one at 6% (giving $i = 0.005$) and the other at 6.9% (giving $i = 0.00575$).

TABLE 6.1 Second 12 Months of Two Investments at Different Interest Rates

	A	B	C
		S = 1000(1.005)^n	S = 1000(1.00575)^n
1	Month #		
2	13	$1066.99	$1077.38
3	14	$1072.32	$1083.58
4	15	$1077.68	$1089.81
5	16	$1083.07	$1096.08
6	17	$1088.49	$1102.38
7	18	$1093.93	$1108.72
8	19	$1099.40	$1115.09
9	20	$1104.90	$1121.50
10	21	$1110.42	$1127.95
11	22	$1115.97	$1134.44
12	23	$1121.55	$1140.96
13	24	$1127.16	$1147.52

■

Continuous Compounding Because more frequent compounding means that interest is paid more often (and hence more interest on interest is earned), it would seem that the more frequently the interest is compounded, the larger the future value will become. In order to determine the interest that results from *continuous* compounding (compounding every instant), consider an investment of $1 for 1 year at a 100% interest rate. If the interest is compounded m times per year, the future value is given by

$$S = \left(1 + \frac{1}{m}\right)^m$$

Table 6.2 shows the future values that result as the number of compounding periods increases.

TABLE 6.2

Compounded	Number of Periods per Year	Future Value
Annually	1	$(1 + \frac{1}{1})^1 = 2$
Monthly	12	$(1 + \frac{1}{12})^{12} = 2.6130\ldots$
Daily	360 (business year)	$(1 + \frac{1}{360})^{360} = 2.7145\ldots$
Hourly	8640	$(1 + \frac{1}{8640})^{8640} = 2.71812\ldots$
Each minute	518,400	$(1 + \frac{1}{518,400})^{518,400} = 2.71827\ldots$

Table 6.2 shows that as the number of periods per year increases, the future value increases, although not very rapidly. In fact, no matter how often the interest is compounded, the future value will never exceed $2.72. We say that as the number of periods increases, the future value approaches a limit, which is the number e:

$$e = 2.7182818 \ldots.$$

We discussed the number e and the function $y = e^x$ in Chapter 5, "Exponential and Logarithmic Functions." The discussion here shows one way the number we call e may be derived. We will define e more formally later.

Future Value (Continuous Compounding)

In general, if P is invested for t years at a nominal rate r compounded continuously, then the future value is given by the exponential function

$$S = Pe^{rt}$$

● **EXAMPLE 6** *Continuous Compounding*

(a) Find the future value if $1000 is invested for 20 years at 8%, compounded continuously.
(b) What amount must be invested at 6.5%, compounded continuously, so that it will be worth $25,000 after 8 years?

Solution
(a) The future value is

$$S = \$1000e^{(0.08)(20)} = \$1000e^{1.6}$$
$$= \$1000(4.95303) \quad \text{(because } e^{1.6} \approx 4.95303)$$
$$= \$4953.03$$

(b) Solve for the present value P in $25,000 = Pe^{(0.065)(8)}$.

$$\$25,000 = Pe^{(0.065)(8)} = Pe^{0.52}$$
$$\$25,000 = P(1.68202765)$$
$$\frac{\$25,000}{1.68202765} = P \quad \text{so} \quad P = \$14,863.01, \text{ to the nearest cent}$$

● **EXAMPLE 7** *Comparing Investments*

How much more will you earn if you invest $1000 for 5 years at 8% compounded continuously instead of at 8% compounded quarterly?

Solution

If the interest is compounded continuously, the future value at the end of the 5 years is

$$S = \$1000e^{(0.08)(5)} = \$1000e^{0.4} \approx \$1000(1.49182)$$
$$= \$1491.82$$

If the interest is compounded quarterly, the future value at the end of the 5 years is

$$S = \$1000(1.02)^{20} \approx \$1000(1.485947)$$
$$= \$1485.95, \text{ to the nearest cent}$$

Thus the extra interest earned by compounding continuously is

$$\$1491.82 - \$1485.95 = \$5.87$$

Annual Percentage Yield

As Example 7 shows, when we invest money at a given compound interest rate, the method of compounding affects the amount of interest we earn. As a result, a rate of 8% can earn more than 8% interest if compounding is more frequent than annually.

For example, suppose $1 is invested for 1 year at 8%, compounded semiannually. Then $i = 0.08/2 = 0.04, n = 2$, the future value is $S = \$1(1.04)^2 = \$1.0816$, and the interest earned for the year is $\$1.0816 - \$1 = \$0.0816$. Note that this amount of interest represents an annual percentage yield of 8.16%, so we say that 8% compounded semiannually has an **annual percentage yield (APY), or effective annual rate,** of 8.16%. Similarly, if $1 is invested at 8% compounded continuously, then the interest earned is $\$1(e^{0.08}) - \$1 = \$0.0833$, for an APY of 8.33%.

Banks acknowledge this difference between stated nominal interest rates and annual percentage yields by posting both rates for their investments. Note that the annual percentage yield is equivalent to the stated rate when compounding is annual. In general, the annual percentage yield equals I/P, or just I if $P = \$1$. Hence we can calculate the APY with the following formulas.

Annual Percentage Yield (APY)

Let r represent the annual (nominal) interest rate for an investment. Then the **annual percentage yield (APY)*** is found as follows.

Periodic Compounding. If m is the number of compounding periods per year, then $i = r/m$ is the interest rate per period, and

$$\text{APY} = \left(1 + \frac{r}{m}\right)^m - 1 = (1 + i)^m - 1 \quad \text{(as a decimal)}$$

Continuous Compounding

$$\text{APY} = e^r - 1 \quad \text{(as a decimal)}$$

Thus, although we cannot directly compare two nominal rates with different compounding periods, we can compare their corresponding APYs.

● **EXAMPLE 8** *Comparing Yields*

Suppose a young couple such as Jim and Eden from our Application Preview found three different investment companies that offered college savings plans: (a) one at 10% compounded annually, (b) another at 9.8% compounded quarterly, and (c) a third at 9.65%

*Note that the annual percentage yield is also called the **effective annual rate.**

compounded continuously. Find the annual percentage yield (APY) for each of these three plans in order to discover which plan is best.

Solution

(a) For annual compounding, the stated rate is the APY.

$$APY = (1.10)^1 - 1 = 0.10 = 10\%$$

(b) Because the number of periods per year is $m = 4$ and the nominal rate is $r = 0.098$, the rate per period is $i = r/m = 0.098/4 = 0.0245$. Thus,

$$APY = (1 + 0.0245)^4 - 1 = 1.10166 - 1 = 0.10166 = 10.166\%.$$

(c) For continuous compounding and a nominal rate of 9.65%, we have

$$APY = e^{0.0965} - 1 = 1.10131 - 1 = 0.10131 = 10.131\%.$$

Hence we see that of these three choices, 9.8% compounded quarterly is best. Furthermore, even 9.65% compounded continuously has a higher APY than 10% compounded annually.

● **Checkpoint**

3. Which future value formula below is used for interest that is compounded periodically, and which is used for interest that is compounded continuously?
   (a) $S = P(1 + i)^n$ (b) $S = Pe^{rt}$
4. If $5000 is invested at 6% compounded continuously for 5 years, find the future value of the investment.
5. Find the annual percentage yield of an investment that earns 7% compounded semiannually.

● **EXAMPLE 9 Doubling Time**

How long does it take an investment of $10,000 to double if it is invested at

(a) 8%, compounded annually?
(b) 8%, compounded continuously?

Solution

(a) We solve for n in $20,000 = 10,000(1 + 0.08)^n$.

$$2 = 1.08^n$$

Taking the logarithm, base e, of both sides of the equation gives

$$\ln 2 = \ln 1.08^n$$
$$\ln 2 = n \ln 1.08 \quad \text{Logarithm Property V}$$
$$n = \frac{\ln 2}{\ln 1.08} \approx 9.0 \text{ (years)}$$

(b) Solve for t in $20,000 = 10,000e^{0.08t}$.

$$2 = e^{0.08t} \quad \text{Isolate the exponential.}$$
$$\ln 2 = \ln e^{0.08t} \quad \text{Take "ln" of both sides.}$$
$$\ln 2 = 0.08t \quad \text{Logarithm Property I}$$
$$t = \frac{\ln 2}{0.08} \approx 8.7 \text{ (years)}$$

Our examples and discussion have shown that investments with continuous compounding earn more interest if the times and rates are the same. Also, our development of the formula for future value of an investment with continuous compounding indicates that when the number of compounding periods is very large, there is little difference between periodic and continuous compounding.

Spreadsheet  *Note*
To understand better the effect of compounding periods on an investment, we can use a spreadsheet to compare different compounding schemes. Table 6.3 shows a spreadsheet that tracks the yearly growth of a $100 investment for quarterly compounding and continuous compounding (both at 10%).

TABLE 6.3 First 20 Years of Two Investments with Different Compounding Schemes

	A	B	C
1	Year	S=100(1.025)^(4n)	S=100*exp(.10n)
2	1	$110.38	$110.52
3	2	$121.84	$122.14
4	3	$134.49	$134.99
5	4	$148.45	$149.18
6	5	$163.86	$164.87
7	6	$180.87	$182.21
8	7	$199.65	$201.38
9	8	$220.38	$222.55
10	9	$243.25	$245.96
11	10	$268.51	$271.83
12	11	$296.38	$300.42
13	12	$327.15	$332.01
14	13	$361.11	$366.93
15	14	$398.60	$405.52
16	15	$439.98	$448.17
17	16	$485.65	$495.30
18	17	$536.07	$547.39
19	18	$591.72	$604.96
20	19	$653.15	$668.59
21	20	$720.96	$738.91

Also, most spreadsheets (including Excel) and some calculators have built-in finance packages that can be used to solve the compound interest problems of this section and later sections of this chapter. ■

● *Checkpoint*
6. How long does it take $5000 to double if it is invested at 9% compounded monthly?

Geometric Sequences

If $P is invested at an interest rate of i per period, compounded at the end of each period, the future value at the end of each succeeding period is

$$P(1 + i), P(1 + i)^2, P(1 + i)^3, \ldots, P(1 + i)^n, \ldots$$

The future values for each of the succeeding periods form a sequence in which each term (after the first) is found by multiplying the previous term by the same number. Such a sequence is called a **geometric sequence.**

Geometric Sequence

A sequence is called a **geometric sequence** (progression) if there exists a number r, called the **common ratio,** such that

$$a_n = ra_{n-1} \quad \text{for } n > 1$$

Geometric sequences form the foundation for other applications involving compound interest.

● **EXAMPLE 10** *Geometric Sequence*

Write the next three terms of the following geometric sequences.

(a) 1, 3, 9, . . .   (b) 4, 2, 1, . . .   (c) 3, −6, 12, . . .

Solution
(a) The common ratio is 3, so the next three terms are 27, 81, 243.
(b) The common ratio is $\frac{1}{2}$, so the next three terms are $\frac{1}{2}, \frac{1}{4}, \frac{1}{8}$.
(c) The common ratio is −2, so the next three terms are −24, 48, −96.

Because each term after the first in a geometric sequence is obtained by multiplying the previous term by r, the second term is $a_1 r$, the third is $a_1 r^2, \ldots$ and the nth term is $a_1 r^{n-1}$. Thus we have the following formula.

nth Term of a Geometric Sequence

The **nth term of a geometric sequence** (progression) is given by

$$a_n = a_1 r^{n-1}$$

where a_1 is the first term of the sequence and r is the common ratio.

● **EXAMPLE 11** *nth Term of a Geometric Sequence*

Find the seventh term of the geometric sequence with first term 5 and common ratio −2.

Solution
The seventh term is $a_7 = 5(-2)^{7-1} = 5(64) = 320$.

● **EXAMPLE 12** *Ball Rebounding*

A ball is dropped from a height of 125 feet. If it rebounds $\frac{3}{5}$ of the height from which it falls every time it hits the ground, how high will it bounce after it strikes the ground for the fifth time?

Solution
The first rebound is $\frac{3}{5}(125) = 75$ feet; the second rebound is $\frac{3}{5}(75) = 45$ feet. The heights of the rebounds form a geometric sequence with first term 75 and common ratio $\frac{3}{5}$. Thus the fifth term is

$$a_5 = 75\left(\frac{3}{5}\right)^4 = 75\left(\frac{81}{625}\right) = \frac{243}{25} = 9\frac{18}{25} \text{ feet}$$

Next we develop a formula for the sum of a geometric sequence, a formula that is important in our study of annuities. The sum of the first n terms of a geometric sequence is

$$s_n = a_1 + a_1r + a_1r^2 + \cdots + a_1r^{n-1} \tag{1}$$

If we multiply equation (1) by r, we have

$$rs_n = a_1r + a_1r^2 + a_1r^3 + \cdots + a_1r^n \tag{2}$$

Subtracting equation (2) from equation (1), we obtain

$$s_n - rs_n = a_1 + (a_1r - a_1r) + (a_1r^2 - a_1r^2) + \cdots + (a_1r^{n-1} - a_1r^{n-1}) - a_1r^n$$

Thus

$$s_n(1 - r) = a_1 - a_1r^n$$

$$s_n = \frac{a_1 - a_1r^n}{1 - r} \quad \text{if } r \neq 1$$

This gives the following.

Sum of a Geometric Sequence

The **sum of the first n terms of the geometric sequence** with first term a_1 and common ratio r is

$$s_n = \frac{a_1(1 - r^n)}{1 - r} \quad \text{provided that } r \neq 1$$

● **EXAMPLE 13** *Sums of Geometric Sequences*

(a) Find the sum of the first five terms of the geometric progression with first term 4 and common ratio -3.
(b) Find the sum of the first six terms of the geometric sequence $\frac{1}{4}, \frac{1}{8}, \frac{1}{16}, \ldots$.

Solution
(a) We are given that $n = 5$, $a_1 = 4$, and $r = -3$. Thus

$$s_5 = \frac{4[1 - (-3)^5]}{1 - (-3)} = \frac{4[1 - (-243)]}{4} = 244$$

(b) We know that $n = 6$, $a_1 = \frac{1}{4}$, and $r = \frac{1}{2}$. Thus

$$s_6 = \frac{\frac{1}{4}[1 - (\frac{1}{2})^6]}{1 - \frac{1}{2}} = \frac{\frac{1}{4}(1 - \frac{1}{64})}{\frac{1}{2}} = \frac{1 - \frac{1}{64}}{2} = \frac{64 - 1}{128} = \frac{63}{128}$$

● *Checkpoint*

7. Identify any geometric sequences among the following.
   (a) 1, 4, 9, 16, . . .
   (b) 1, 4, 7, 10, . . .
   (c) 1, 4, 16, 64, . . .
8. (a) Find the 40th term of the geometric sequence $8, 6, \frac{9}{2}, \ldots$.
   (b) Find the sum of the first 20 terms of the geometric sequence $2, 6, 18, \ldots$.

● *Checkpoint Solutions*

1. (a) $m = 4$ (b) $n = m \cdot t = 4 \cdot 5 = 20$ (c) $i = \dfrac{r}{m} = \dfrac{0.06}{4} = 0.015$

   (d) $S = P(1 + i)^n = \$5000(1 + 0.015)^{20} \approx \$6734.28$, to the nearest cent

2. $i = 0.09/12 = 0.0075$ and $n = (5)(12) = 60$

$$\$12,000 = P(1 + 0.0075)^{60} = P(1.5656810270)$$

$$P = \$12,000/1.565681027 = \$7664.40, \text{ to the nearest cent}$$

3. (a) Periodic compounding
   (b) Continuous compounding

4. $S = Pe^{rt} = \$5000e^{(0.06)(5)} = \$5000e^{0.3} \approx \$6749.29$, to the nearest cent

5. $i = 0.07/2 = 0.035$ and $m = 2$

$$\text{APY} = (1 + 0.035)^2 - 1 \approx 0.0712, \text{ or } 7.12\%$$

6. With $i = 0.09/12 = 0.0075$, we solve for n in

$$\$10,000 = \$5000(1 + 0.0075)^n$$

$$2 = (1.0075)^n$$

$$\ln(2) = \ln(1.0075)^n = n\ln(1.0075)$$

$$\frac{\ln(2)}{\ln(1.0075)} = n \quad \text{so} \quad n \approx 92.8 \text{ months}$$

7. Only sequence (c) is geometric with common ratio $r = 4$.

8. (a) $a_{40} = a_1 r^{40-1} = 8\left(\dfrac{3}{4}\right)^{39} \approx 0.0001$

   (b) $s_{20} = \dfrac{a_1(1 - r^{20})}{1 - r} = \dfrac{2(1 - 3^{20})}{1 - 3} = -(1 - 3^{20})$
   $$= 3^{20} - 1 = 3,486,784,400$$

6.2 Exercises

COMPOUND INTEREST

For each investment situation in Problems 1–4, identify (a) the annual interest rate, (b) the length of the investment in years, (c) the periodic interest rate, and (d) the number of periods of the investment.

1. 8% compounded quarterly for 7 years
2. 12% compounded monthly for 3 years
3. 9% compounded monthly for 5 years
4. 10% compounded semiannually for 8 years
5. Find the future value if $8000 is invested for 10 years at 12% compounded annually.
6. What is the future value if $8600 is invested for 8 years at 10% compounded semiannually?
7. What is the future value if $3200 is invested for 5 years at 8% compounded quarterly?
8. What interest will be earned if $6300 is invested for 3 years at 12% compounded monthly?

9. Find the interest that will be earned if $10,000 is invested for 3 years at 9% compounded monthly.
10. Find the future value if $3500 is invested for 6 years at 8% compounded quarterly.
11. What lump sum do parents need to deposit in an account earning 10%, compounded monthly, so that it will grow to $40,000 for their son's college tuition in 18 years?
12. What lump sum should be deposited in an account that will earn 9%, compounded quarterly, to grow to $100,000 for retirement in 25 years?
13. What present value amounts to $10,000 if it is invested for 10 years at 6% compounded annually?
14. What present value amounts to $300,000 if it is invested at 7%, compounded semiannually, for 15 years?
15. Find the future value if $5100 is invested for 4 years at 9% compounded continuously.

16. Find the interest that will result if $8000 is invested at 7%, compounded continuously, for 8 years.

17. What is the compound interest if $410 is invested for 10 years at 8% compounded continuously?

18. If $8000 is invested at 8.5% compounded continuously, find the future value after $4\frac{1}{2}$ years.

19. Grandparents want to make a gift of $100,000 for their grandchild's 20th birthday. How much would have to be invested on the day of their grandchild's birth if their investment could earn:
    (a) 10.5% compounded continuously?
    (b) 11% compounded continuously?
    (c) Describe the effect that this slight change in the interest rate makes over the 20 years of this investment.

20. Suppose an individual wants to have $200,000 available for her child's education. Find the amount that would have to be invested at 12%, compounded continuously, if the number of years until college is:
    (a) 7 years   (b) 14 years
    (c) Does leaving the money invested twice as long mean that only half as much is needed initially? Explain why or why not.

21. Which investment will earn more money, a $1000 investment for 5 years at 8% compounded annually or a $1000 investment for 5 years compounded continuously at 7%?

22. How much more interest will be earned if $5000 is invested for 6 years at 7% compounded continuously, instead of at 7% compounded quarterly?

23. Find the annual percentage yield for an investment at
    (a) 7.3% compounded monthly.
    (b) 6% compounded continuously.

24. What is the annual percentage yield (or effective annual rate) for a nominal rate of (a) 8.4% compounded quarterly and (b) 10% compounded continuously?

In Problems 25 and 26, rank each interest rate and compounding scheme in order from highest yield to lowest yield.

25. 8% compounded quarterly, 8% compounded monthly, 8% compounded annually

26. 6% compounded continuously, 6% compounded semi-annually, 6% compounded monthly

27. Two different investment companies offer college savings plans, one at 8.2% compounded continuously and the other at 8.4% compounded quarterly. Which is the better investment?

28. For life insurance policies, some of the premium pays for the cost of the insurance, and the remainder goes toward the cash value of the policy and earns interest like a savings account. Suppose that, on the cash value of their policies, one insurance company pays 4.8% compounded monthly and another pays 4.82% compounded semiannually. Which company offers a higher yield?

29. The following figure shows a graph of the future value of $100 at 8% compounded annually, along with the graph of $100 at 8% compounded continuously. Which is which? Explain.

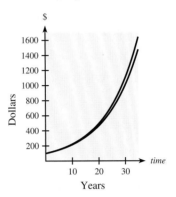

30. The following figure shows a graph of the future value of $350 at 6% compounded monthly, along with the graph of $350 at 6% compounded annually. Which is which? Explain.

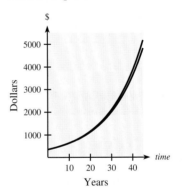

31. If $10,000 had been invested in the Fidelity Select Multimedia Fund when it opened on June 30, 1986, then on June 30, 2004 the investment would have been worth $93,904.80. (*Source:* Fidelity.com, August 2004) What interest rate compounded annually would this investment have earned?

32. If $10,000 had been invested in the Fidelity New Millennium Fund on June 30, 1994, then on June 30, 2004 the investment would have been worth $56,660.36. (*Source:* Fidelity.com, August 2004) What interest rate compounded annually would this investment have earned?

33. How long (in years) would $700 have to be invested at 11.9%, compounded continuously, to earn $300 interest?

34. How long (in years) would $600 have to be invested at 8%, compounded continuously, to amount to $970?

35. At what nominal rate, compounded quarterly, would $20,000 have to be invested to amount to $26,425.82 in 7 years?

36. At what nominal rate, compounded annually, would $10,000 have to be invested to amount to $14,071 in 7 years?

37. For her first birthday, Polly's grandparents invested $1000 in an 18-year certificate for her that pays 8% compounded annually. How much will the certificate be worth on Polly's 19th birthday?

38. To help their son buy a car on his 16th birthday, a boy's parents invest $1500 on his 10th birthday. If the investment pays 9% compounded continuously, how much is available on his 16th birthday?

39. (a) A 40-year-old man no longer qualifies for additional IRAs. If his present IRAs have a balance of $112,860 and if he expects them to earn interest at 7.5% compounded annually, how much can he expect to have when he retires at age 62?
    (b) How much more money would the man have if his investments earned 8.5% compounded annually?

40. (a) The purchase of Alaska cost the United States $7 million in 1869. If this money had been placed in a savings account paying 6% compounded annually, how much money would be available from this investment in 2010?
    (b) If the $7 million earned 7% compounded annually since 1869, how much would be available in 2010?
    (c) Do you think either amount would purchase Alaska in 2010? Explain in light of the value of Alaska's resources or perhaps the price per acre of land.

41. A couple needs $15,000 as a down payment for a home. If they invest the $10,000 they have at 8% compounded quarterly, how long will it take for the money to grow into $15,000?

42. How long does it take for an account containing $8000 to be worth $15,000 if the money is invested at 9% compounded monthly?

43. Mary Stahley invested $2500 in a 36-month certificate of deposit (CD) that earned 8.5% annual simple interest. When the CD matured, she invested the full amount in a mutual fund that had an annual growth equivalent to 18% compounded annually. How much was the mutual fund worth after 9 years?

44. Suppose Patrick Goldsmith deposited $1000 in an account that earned simple interest at an annual rate of 7% and left it there for 4 years. At the end of the 4 years, Patrick deposited the entire amount from that account into a new account that earned 7% compounded quarterly. He left the money in this account for 6 years. How much did he have after the 10 years?

In Problems 45 and 46, use a spreadsheet or financial program on a calculator or computer.

45. Track the future values of two investments of $5000, one at 6.3% compounded quarterly and another at 6.3% compounded monthly for each interest payment period for 10 years.
    (a) How long does it take each investment to be worth more than $7500?
    (b) What are the values of each investment after 3 years, 7 years, and 10 years?

46. Track the future values of two investments of $1000, one at 6.0% compounded semiannually and one at 6.6% compounded semiannually for each interest payment period for 25 years.
    (a) How long before the difference between these investments is $50?
    (b) How much sooner does the 6.6% investment reach $1500?

GEOMETRIC SEQUENCES

For each geometric sequence given in Problems 47 and 48, write the next three terms.

47. (a) 3, 6, 12, . . .
    (b) 81, 54, 36, . . .

48. (a) 4, 12, 36, . . .
    (b) 32, 40, 50, . . .

In Problems 49–58, write an expression that gives the requested term or sum.

49. The 13th term of the geometric sequence with first term 10 and common ratio 2

50. The 11th term of the geometric sequence with first term 6 and common ratio 3

51. The 16th term of the geometric sequence with first term 4 and common ratio $\frac{3}{2}$

52. The 20th term of the geometric sequence with first term 3 and common ratio -2

53. The sum of the first 17 terms of the geometric sequence with first term 6 and common ratio 3

54. The sum of the first 14 terms of the geometric sequence with first term 3 and common ratio 4

55. The sum of the first 35 terms of the geometric sequence 1, 3, 9, . . .

56. The sum of the first 14 terms of the geometric sequence 16, 64, 256, . . .

57. The sum of the first 18 terms of the geometric sequence $6, 4, \frac{8}{3}, \ldots$

58. The sum of the first 31 terms of the geometric sequence $9, -6, 4, \ldots$

APPLICATIONS OF SEQUENCES

59. *Inflation*   A house that 20 years ago was worth $160,000 has increased in value by 4% each year because of inflation. What is its worth today?

60. *Inflation*   If inflation causes the cost of automobiles to increase by 3% each year, what should a car cost today if it cost $25,000 6 years ago?

61. *Population growth*   Suppose a country has a population of 20 million and projects a growth rate of 2% per year for the next 20 years. What will the population of this country be in 10 years?

62. *Spread of AIDS*   Suppose a country is so devastated by the AIDS epidemic that its population decreases by 0.5% each year for a 4-year period. If the population was originally 10 million, what is the population at the end of the 4-year period?

63. *Population growth*   If the rate of growth of a population continues at 2%, in how many years will the population double?

64. *Population*   If a population of 8 million begins to increase at a rate of 0.1% each month, in how many months will it be 10 million?

65. *Ball rebounding*   A ball is dropped from a height of 128 feet. If it rebounds $\frac{3}{4}$ of the height from which it falls every time it hits the ground, how high will it bounce after it strikes the ground for the fourth time?

66. *Water pumping*   A pump removes $\frac{1}{3}$ of the water in a container with every stroke. What amount of water is still in a container after 5 strokes if it originally contained 81 cm^3?

67. *Depreciation*   A machine is valued at $10,000. If the depreciation at the end of each year is 20% of its value at the beginning of the year, find its value at the end of 4 years.

68. *Profit*   Suppose a new business makes a $1000 profit in its first month and has its profit increase by 10% each month for the next 2 years. How much profit will the business earn in its 12th month?

69. *Bacterial growth*   The size of a certain bacteria culture doubles each hour. If the number of bacteria present initially is 5000, how many would be present at the end of 6 hours?

70. *Bacterial growth*   If a bacteria culture increases by 20% every hour and 2000 are present initially, how many will be present at the end of 10 hours?

71. *Profit*   If changing market conditions cause a company earning $8,000,000 in 2005 to project a loss of 2% of its profit in each of the next 5 years, what profit does it project in 2010?

72. *Profit*   Suppose a new business makes a $1000 profit in its first month and has its profit increase by 10% each month for the next 2 years. How much profit will it earn in its first year?

73. *Loans*   An interest-free loan of $12,000 requires monthly payments of 10% of the outstanding balance. What is the outstanding balance after 18 payments?

74. *Loans*   If Sherri must repay a $9000 interest-free loan by making monthly payments of 15% of the unpaid balance, what is the unpaid balance after 1 year?

75. *Chain letters*   Suppose you receive a chain letter with six names on it, and to keep the chain unbroken, you are to mail a dime to the person whose name is at the top, cross out the top name, add your name to the bottom, and mail it to five friends. If your friends mail out five letters each, and no one breaks the chain, you will eventually receive dimes. How many sets of mailings before your name is at the top of the list to receive dimes? How many dimes would you receive? (This is a geometric sequence with first term 5.)

76. *Chain letters*   Mailing chain letters that involve sending money has been declared illegal because most people would receive nothing while a comparative few would profit. Suppose the chain letter in Problem 75 were to go through 12 unbroken progressions.
   (a) How many people would receive money?
   (b) How much money would these people receive as a group?

77. *Chain letters*   How many letters would be mailed if the chain letter in Problem 75 went through 12 unbroken progressions?

6.3

OBJECTIVES

- To compute the future values of ordinary annuities and annuities due
- To compute the payments required in order for ordinary annuities and annuities due to have specified future values
- To compute the payment required to establish a sinking fund
- To find how long it will take to reach a savings goal

Future Values of Annuities

◗ Application Preview

Twins graduate from college together and start their careers. Twin 1 invests $2000 at the end of each of 8 years in an account that earns 10%, compounded annually. After the initial 8 years, no additional contributions are made, but the investment continues to earn 10%, compounded annually. Twin 2 invests no money for 8 years but then contributes $2000 at the end of each year for a period of 36 years (to age 65) to an account that pays 10%, compounded annually. How much money does each twin have at age 65? (See Example 2 and Example 3.)

Each twin's contributions form an **annuity.** An annuity is a financial plan characterized by regular payments. We can view an annuity as a savings plan in which the regular payments are contributions to the account, and then we can ask what the total value of the account will become (as in the Application Preview). Also, we can view an annuity as a payment plan (such as for retirement) in which regular payments are made from an account, often to an individual.

In this section we consider ordinary annuities and annuities due, and we develop a formula for each type of annuity that allows us to find its **future value**—that is, a formula for the value of the account after regular deposits have been made over a period of time.

Ordinary Annuities

In the previous section, we learned how to compute the future value and amount of interest if a fixed sum of money is deposited in an account that pays interest that is compounded periodically or continuously. But not many people are in a position to deposit a large sum of money at one time in an account. Most people save (or invest) money by depositing relatively small amounts at different times. If a depositor makes equal deposits at regular intervals, he or she is contributing to an annuity. The payments (deposits) may be made weekly, monthly, quarterly, yearly, or at any other interval of time. The sum of all payments plus all interest earned is called the **future amount of the annuity** or its **future value.**

Annuities may be classified into two categories—annuities certain and contingent annuities. An **annuity certain** is one in which the payments begin and end on fixed dates. In a **contingent annuity** the payments are related to events that cannot be paced regularly, so the payments are not regular. We will deal with annuities certain in this text, and we will deal first with annuities in which the payments are made at the end of each of the equal payment intervals. This type of annuity certain is called an **ordinary annuity** (and also an **annuity immediate**). The ordinary annuities we will consider have payment intervals that coincide with the compounding period of the interest.

Suppose you invested $100 at the end of each year for 5 years in an account that paid interest at 10%, compounded annually. How much money would you have in the account at the end of the 5 years?

Because you are making payments at the end of each period (year), this annuity is an ordinary annuity.

To find the future value of your annuity at the end of the 5 years, we compute the future value of each payment separately and add the amounts (see Figure 6.3 on the next page). The $100 invested *at the end* of the first year will draw interest for 4 years, so it will amount to $100(1.10)^4$. Figure 6.3 shows similar calculations for years 2–4. The $100 invested at the end of the fifth year will draw no interest, so it will amount to $100.

Thus the future value of the annuity is given by

$$S = 100 + 100(1.10) + 100(1.10)^2 + 100(1.10)^3 + 100(1.10)^4$$

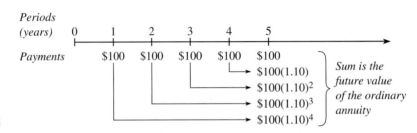

Figure 6.3

The terms of this sum are the first five values of the geometric sequence having $a_1 = 100$ and $r = 1.10$. Thus

$$S = \frac{100[1 - (1.10)^5]}{1 - 1.10} = \frac{100(-0.61051)}{-0.10}$$

$$= 610.51$$

Thus the investment ($100 at the end of each year for 5 years, at 10%, compounded annually) would return $610.51.

Because every such annuity will take the same form, we can state that if a periodic payment R is made for n periods at an interest rate i *per period,* the **future amount of the annuity,** or its **future value,** will be given by

$$S = R \cdot \frac{1 - (1 + i)^n}{1 - (1 + i)}$$

This simplifies to the following.

Future Value of an Ordinary Annuity

> If $R is deposited at the end of each period for n periods in an annuity that earns interest at a rate of i per period, the future value of the annuity will be
>
> $$S = R \cdot s_{\overline{n}|i} = R \cdot \left[\frac{(1 + i)^n - 1}{i}\right]$$
>
> where $s_{\overline{n}|i}$ is read "s, n angle i" and represents the future value of an ordinary annuity of $1 per period for n periods with an interest rate of i per period.

The value of $s_{\overline{n}|i}$ can be computed directly with a calculator or found in Table I in the Appendix.

● **EXAMPLE 1** *Future Value*

Richard Lloyd deposits $200 at the end of each quarter in an account that pays 4%, compounded quarterly. How much money will he have in his account in $2\frac{1}{4}$ years?

Solution
The number of periods is $n = (4)(2.25) = 9$, and the rate *per period* is $i = 0.04/4 = 0.01$. At the end of $2\frac{1}{4}$ years the future value of the annuity will be

$$S = \$200 \cdot s_{\overline{9}|0.01} = \$200\left[\frac{(1 + 0.01)^9 - 1}{0.01}\right]$$

$$= \$200(9.368527) = \$1873.71, \text{ to the nearest cent}$$

If we had used Table I in the Appendix, we would have seen that $s_{\overline{9}|0.01}$, the future value of an ordinary annuity of $1 per period for 9 periods with an interest rate of $0.01 = 1\%$ per period, is 9.368527, and we would have obtained the same result, $200(9.368527) = \$1873.71$.

In the Application Preview, savings plans for twins were described. In the next two examples, we want to answer the questions posed in the Application Preview.

◗ EXAMPLE 2 *Future Value for Twin 2 (Application Preview)*

Twin 2 in the Application Preview invests $2000 at the end of each year for 36 years (until age 65) in an account that pays 10%, compounded annually. How much does twin 2 have at age 65?

Solution
This savings plan is an ordinary annuity with $i = 0.10$, $n = 36$, and $R = \$2000$. The future value (to the nearest dollar) is

$$S = R\left[\frac{(1 + i)^n - 1}{i}\right] = \$2000\left[\frac{(1.10)^{36} - 1}{0.10}\right] = \$598{,}254$$

Let's now complete the solution to the twin problem.

◗ EXAMPLE 3 *Future Value for Twin 1 (Application Preview)*

Twin 1 invests $2000 at the end of each of 8 years in an account that earns 10%, compounded annually. After the initial 8 years, no additional contributions are made, but the investment continues to earn 10%, compounded annually, for 36 more years (until twin 1 is age 65). How much does twin 1 have at age 65?

Solution
We seek the future values of two different investments. The first is an ordinary annuity with $R = \$2000$, $n = 8$ periods, and $i = 0.10$. The second is a compound interest investment with $n = 36$ periods and $i = 0.10$ and whose principal (that is, its present value) is the future value of this twin's ordinary annuity.

We first find the future value of the ordinary annuity.

$$S = R\left[\frac{(1 + i)^n - 1}{i}\right] = \$2000\left[\frac{(1 + 0.10)^8 - 1}{0.10}\right]$$

$$= \$22{,}871.78, \text{ to the nearest cent}$$

This amount is the principal of the compound interest investment. If no deposits or withdrawals were made for the next 36 years, the future value of this investment would be

$$S = P(1 + i)^n = \$22{,}871.78(1 + 0.10)^{36} = \$707{,}028.03, \text{ to the nearest cent}$$

Thus, at age 65, twin 1's investment is worth about $707,028.

Looking back at Examples 2 and 3, we can extract the following summary and see which twin was the wiser.

	Contributions	Account Value at Age 65
Twin 1	$2000/year for 8 years = $16,000	$707,028
Twin 2	$2000/year for 36 years = $72,000	$598,254

Note that twin 1 contributed $56,000 less than twin 2 but had $108,774 more at age 65. This illustrates the powerful effect that time and compounding have on investments (and on loans, as we'll see in Section 6.5, "Loans and Amortization").

Figure 6.4 shows a comparison (each graphed as a smooth curve) of the future values of two annuities that are invested at 6% compounded monthly. One annuity has monthly payments of $100, and one has payments of $125. Note the impact of a slightly larger contribution on the future value of the annuity after 5 years, 10 years, 15 years, and 20 years. Of course, a slightly higher interest rate also can have a substantial impact over time; try a graphical comparison yourself.

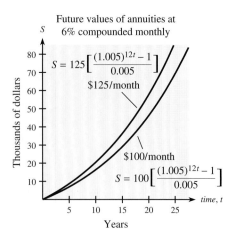

Figure 6.4

Spreadsheet Note A spreadsheet also can be used to compare investments such as those compared graphically in Figure 6.4. Table 6.4 shows the future values at the end of each of the first 10 years of an ordinary annuity of $100 per month at two different rates, 6% and 6.3%, compounded monthly.

TABLE 6.4

	A	B	C
1		Future Value for R = $100	Future Value for R = $100
2	End of Year #	at 6% compounded monthly	at 6.3% compounded monthly
3	1	$1233.56	$1235.26
4	2	$2543.20	$2550.64
5	3	$3933.61	$3951.31
6	4	$5409.78	$5442.82
7	5	$6977.00	$7031.06
8	6	$8640.89	$8722.30
9	7	$10407.39	$10523.21
10	8	$12282.85	$12440.92
11	9	$14273.99	$14483.00
12	10	$16387.93	$16657.50

Sometimes we want to know how long it will take for an annuity to reach a desired future value.

● **EXAMPLE 4 Time to Reach a Goal**

A small business invests $1000 at the end of each month in an account that earns 6% compounded monthly. How long will it take until the business has $100,000 toward the purchase of its own office building?

Solution
This is an ordinary annuity with $S = \$100{,}000$, $R = \$1000$, $i = 0.06/12 = 0.005$, and $n =$ the number of months. To answer the question of how long, solve for n.

Use $S = R\left[\dfrac{(1 + i)^n - 1}{i}\right]$ and solve for n in $100{,}000 = 1000\left[\dfrac{(1 + 0.005)^n - 1}{0.005}\right]$.

This is an exponential equation. Hence, we isolate $(1.005)^n$, take the natural logarithm of both sides, and then solve for n as follows.

$$100,000 = \frac{1000}{0.005}\left[(1.005)^n - 1\right]$$

$$0.5 = (1.005)^n - 1$$

$$1.5 = (1.005)^n$$

$$\ln(1.5) = \ln\left[(1.005)^n\right] = n\left[\ln(1.005)\right]$$

$$n = \frac{\ln(1.5)}{\ln(1.005)} \approx 81.3$$

Because this investment is monthly, after 82 months the company will be able to purchase its own office building.

We can also find the periodic payment needed to obtain a specified future value.

● **EXAMPLE 5** *Payment for an Ordinary Annuity*

A young couple wants to save $15,000 over the next 5 years and then to use this amount as a down payment on a home. To reach this goal, how much money must they deposit at the end of each quarter in an account that earns interest at a rate of 5%, compounded quarterly?

Solution
This plan describes an ordinary annuity with a future value of $15,000 whose payment size, R, is to be determined. Quarterly compounding gives $n = 5(4) = 20$ and $i = 0.05/4 = 0.0125$.

$$S = R\left[\frac{(1 + i)^n - 1}{i}\right]$$

$$\$15,000 = R\left[\frac{(1 + 0.0125)^{20} - 1}{0.0125}\right] = R(22.56297854)$$

$$R = \frac{\$15,000}{22.56297854} = \$664.81, \text{ to the nearest cent}$$

Sinking Funds   Just as the couple in Example 5 was saving for a future purchase, some borrowers, such as municipalities, may have a debt that must be paid in a single large sum on a specified future date. If these borrowers make periodic deposits that will produce that sum on a specified date, we say that they have established a **sinking fund.** If the deposits are all the same size and are made regularly, they form an ordinary annuity whose future value (on a specified date) is the amount of the debt. To find the size of these periodic deposits, we solve for R in the equation for the future value of an annuity:

$$S = R\left[\frac{(1 + i)^n - 1}{i}\right]$$

● **EXAMPLE 6** *Sinking Fund*

A company establishes a sinking fund to discharge a debt of $100,000 due in 5 years by making equal semiannual deposits, the first due in 6 months. If the deposits are placed in an account that pays 6%, compounded semiannually, what is the size of the deposits?

Solution

For this sinking fund, we want to find the payment size, R, given that the future value is $S = \$100,000$, $n = 2(5) = 10$, and $i = 0.06/2 = 0.03$. Thus we have

$$S = R\left[\frac{(1 + i)^n - 1}{i}\right]$$

$$\$100,000 = R\left[\frac{(1 + 0.03)^{10} - 1}{0.03}\right]$$

$$\$100,000 = R(11.463879)$$

$$\frac{\$100,000}{11.463879} = R \quad \text{so} \quad R = \$8723.05$$

Thus the semiannual deposit is \$8723.05.

● **Checkpoint**

1. Suppose that \$500 is deposited at the end of every quarter for 6 years in an account that pays 8%, compounded quarterly.
   (a) What is the total number of payments (periods)?
   (b) What is the interest rate per period?
   (c) What formula is used to find the future value of the annuity?
   (d) Find the future value of the annuity.
2. A sinking fund of \$100,000 is to be established with equal payments at the end of each half-year for 15 years. Find the amount of each payment if money is worth 10%, compounded semiannually.

Annuities Due

Deposits in savings accounts, rent payments, and insurance premiums are examples of **annuities due.** Unlike an ordinary annuity, an annuity due has the periodic payments made at the *beginning* of the period. The *term* of an annuity due is from the first payment to the end of one period after the last payment. Thus an annuity due draws interest for one period more than the ordinary annuity.

We can find the future value of an annuity due by treating each payment as though it were made at the *end* of the preceding period in an ordinary annuity. Then that amount, $Rs_{\overline{n}|i}$, remains in the account for one additional period (see Figure 6.5) and its future value is $Rs_{\overline{n}|i}(1 + i)$.

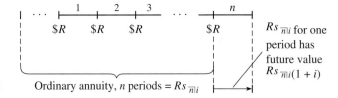

Figure 6.5   Ordinary annuity, n periods $= Rs_{\overline{n}|i}$

Thus the formula for the future value of an annuity due is as follows.

Future Value of an Annuity Due

$$S_{due} = Rs_{\overline{n}|i}(1 + i) = R\left[\frac{(1 + i)^n - 1}{i}\right](1 + i)$$

● **EXAMPLE 7** *Future Value*

Find the future value of an investment if $150 is deposited at the beginning of each month for 9 years and the interest rate is 7.2%, compounded monthly.

Solution

Because deposits are made at the *beginning* of each month, this is an annuity due with $R = \$150, n = 9(12) = 108$, and $i = 0.072/12 = 0.006$.

$$S_{due} = R\left[\frac{(1 + i)^n - 1}{i}\right](1 + i)$$

$$= \$150\left[\frac{(1 + 0.006)^{108} - 1}{0.006}\right](1 + 0.006)$$

$$= \$150(151.3359308)(1.006) = \$22{,}836.59, \text{ to the nearest cent}$$

We can also use the formula for the future value of an annuity due to determine the payment size required to reach an investment goal.

● **EXAMPLE 8** *Required Payment*

Suppose a company wants to have $450,000 after $2\frac{1}{2}$ years to modernize its production equipment. How much of each previous quarter's profits should be deposited at the beginning of the current quarter to reach this goal, if the company's investment earns 6.8%, compounded quarterly?

Solution

We seek the payment size, R, for an annuity due with $S_{due} = \$450{,}000$, $n = (2.5)(4) = 10$, and $i = 0.068/4 = 0.017$.

$$S_{due} = R\left[\frac{(1 + i)^n - 1}{i}\right](1 + i)$$

$$\$450{,}000 = R\left[\frac{(1 + 0.017)^{10} - 1}{0.017}\right](1 + 0.017)$$

$$\$450{,}000 = R(10.80073308)(1.017) = (10.98434554)R$$

$$R = \frac{\$450{,}000}{10.98434554} = \$40{,}967.39, \text{ to the nearest cent}$$

Thus, the company needs to deposit about $40,967 at the beginning of each quarter for the next $2\frac{1}{2}$ years to reach its goal.

● *Checkpoint*

3. Suppose $100 is deposited at the beginning of each month for 3 years in an account that pays 6%, compounded monthly.
   (a) What is the total number of payments (or periods)?
   (b) What is the interest rate per period?
   (c) What formula is used to find the future value of the annuity?
   (d) Find the future value.

● **Checkpoint Solutions**

1. (a) $n = 4(6) = 24$ periods   (b) $i = 0.08/4 = 0.02$ per period

(c) $S = R\left[\dfrac{(1 + i)^n - 1}{i}\right]$ for an ordinary annuity

(d) $S = \$500\left[\dfrac{(1 + 0.02)^{24} - 1}{0.02}\right] = \$15{,}210.93$, to the nearest cent

2. Use $n = 2(15) = 30$, $i = 0.10/2 = 0.05$, and $S = \$100{,}000$ in the formula

$$S = R\left[\dfrac{(1 + i)^n - 1}{i}\right]$$

$$\$100{,}000 = R\left[\dfrac{(1 + 0.05)^{30} - 1}{0.05}\right] = R(66.4388475)$$

$$\dfrac{\$100{,}000}{66.4388475} = R \quad \text{so} \quad R = \$1505.14, \text{ to the nearest cent}$$

3. (a) $n = 3(12) = 36$ periods   (b) $i = 0.06/12 = 0.005$ per period

(c) $S_{due} = R\left[\dfrac{(1 + i)^n - 1}{i}\right](1 + i)$ for an annuity due

(d) $S_{due} = \$100\left[\dfrac{(1 + 0.005)^{36} - 1}{0.005}\right](1 + 0.005) = \$3953.28$, to the nearest cent

---

6.3 Exercises

ORDINARY ANNUITIES

1. The following figure shows a graph that compares the future values, at 8% compounded annually, of an annuity of $1000 per year and one of $1120 per year.
   (a) Decide which graph corresponds to which annuity.
   (b) Verify your conclusion to (a) by finding the value of each annuity and the difference between them at $t = 25$ years.

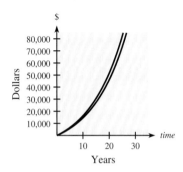

2. The following figure shows a graph that compares the future values, at 9% compounded monthly, of an annuity of $50 per month and one of $60 per month.
   (a) Decide which graph corresponds to which annuity.

(b) Use the graph to estimate (to the nearest 10 months) how long it will be before the larger annuity is $10,000 more than the smaller one.

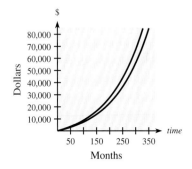

3. Find the future value of an annuity of $1300 paid at the end of each year for 5 years, if interest is earned at a rate of 6%, compounded annually.

4. Find the future value of an annuity of $5000 paid at the end of each year for 10 years, if it earns 9%, compounded annually.

5. Find the future value of an ordinary annuity of $80 paid quarterly for 3 years, if the interest rate is 8%, compounded quarterly.

6. Find the future value of an ordinary annuity of $300 paid quarterly for 5 years, if the interest rate is 12%, compounded quarterly.

7. Parents agree to invest $500 (at 10%, compounded semiannually) for their son on the December 31 or June 30 following each semester that he makes the dean's list during his 4 years in college. If he makes the dean's list in each of the 8 semesters, how much money will his parents have to give him when he graduates?

8. Jake Werkheiser decides to invest $2000 in an IRA each April 15 for the next 10 years. If he makes these investments, and if the certificates pay 12%, compounded annually, how much will he have at the end of the 10 years?

9. How much will have to be invested at the end of each year at 10%, compounded annually, to pay off a debt of $50,000 in 8 years?

10. How much will have to be invested at the end of each year at 12%, compounded annually, to pay off a debt of $30,000 in 6 years?

11. A family wants to have a $200,000 college fund for their children at the end of 20 years. What contribution must be made at the end of each quarter if their investment pays 7.6%, compounded quarterly?

12. What is the amount of the payment that must be put in an account at the end of each quarter to establish an ordinary annuity that has a future value of $50,000 in 14 years, if the investment pays 12%, compounded quarterly?

13. The Weidmans want to save $20,000 in 2 years for a down payment on a house. If they make monthly deposits in an account paying 12%, compounded monthly, what is the size of the payments that are required to meet their goal?

14. A sinking fund is established to discharge a debt of $80,000 in 10 years. If deposits are made at the end of each 6-month period and interest is paid at the rate of 8%, compounded semiannually, what is the amount of each deposit?

15. If $2500 is deposited at the end of each quarter in an account that earns 5% compounded quarterly, after how many quarters will the account contain $80,000?

16. If $4000 is deposited at the end of each half year in an account that earns 6.2% compounded semiannually, how long will it be before the account contains $120,000?

17. When you establish a sinking fund, which interest rate is better? Explain.
    (a) 10%
    (b) 6%

18. If you set up a sinking fund, which interest rate is better? Explain.
    (a) 12% compounded monthly
    (b) 12% compounded annually

19. In this section's Application Preview, we considered the investment strategies of twins and found that starting early and stopping was a significantly better strategy than waiting, in terms of total contributions made as well as total value in the account at retirement. Suppose now that twin 1 invests $2000 at the end of each year for 10 years only (until age 33) in an account that earns 8%, compounded annually. Suppose that twin 2 waits until turning 40 to begin investing. How much must twin 2 put aside at the end of each year for the next 25 years in an account that earns 8% compounded annually in order to have the same amount as twin 1 at the end of these 25 years (when they turn 65)?

20. (a) Patty Stacey deposits $2000 at the end of each of the 5 years she qualifies for an IRA. If she leaves the money that has accumulated in the IRA account for 25 additional years, how much is in her account at the end of the 30-year period? Assume an interest rate of 9%, compounded annually.
    (b) Suppose that Patty's husband delays starting an IRA for the first 10 years he works but then makes $2000 deposits at the end of each of the next 15 years. If the interest rate is 9%, compounded annually, and if he leaves the money in his account for 5 additional years, how much will be in his account at the end of the 30-year period?
    (c) Does Patty or her husband have more IRA money?

ANNUITIES DUE

21. Find the future value of an annuity due of $100 each quarter for $2\frac{1}{2}$ years at 12%, compounded quarterly.

22. Find the future value of an annuity due of $1500 each month for 3 years if the interest rate is 12%, compounded monthly.

23. Find the future value of an annuity due of $200 paid at the beginning of each 6-month period for 8 years if the interest rate is 6%, compounded semiannually.

24. A house is rented for $1800 per quarter, with each quarter's rent payable in advance. If money is worth 8%, compounded quarterly, and the rent is deposited in an account, what is the future value of the rent for one year?

25. How much must be deposited at the beginning of each year in an account that pays 8%, compounded annually, so that the account will contain $24,000 at the end of 5 years?

26. What is the size of the payments that must be deposited at the beginning of each 6-month period in an account that pays 7.8%, compounded semiannually, so that the account will have a future value of $120,000 at the end of 15 years?

MISCELLANEOUS PROBLEMS

In Problems 27–36, (a) state whether the problem relates to an ordinary annuity or an annuity due, then (b) solve the problem.

27. A couple has determined that they need $300,000 to establish an annuity when they retire in 25 years. How much money should they deposit at the end of each month in an investment plan that pays 10%, compounded monthly, so they will have the $300,000 in 25 years?

28. Sam deposits $500 at the end of every 6 months in an account that pays 8%, compounded semiannually. How much will he have at the end of 8 years?

29. Mr. Gordon plans to invest $300 at the end of each month in an account that pays 9%, compounded monthly. After how many months will the account be worth $50,000?

30. For 3 years, $400 is placed in a savings account at the beginning of each 6-month period. If the account pays interest at 10%, compounded semiannually, how much will be in the account at the end of the 3 years?

31. Grandparents plan to open an account on their grandchild's birthday and contribute each month until she goes to college. How much must they contribute at the beginning of each month in an investment that pays 12%, compounded monthly, if they want the balance to be $180,000 at the end of 18 years?

32. How much money should a couple deposit at the end of each month in an investment plan that pays 7.5%, compounded monthly, so they will have $800,000 in 30 years?

33. Jane Adele deposits $500 in an account at the beginning of each 3-month period for 9 years. If the account pays interest at the rate of 8%, compounded quarterly, how much will she have in her account after 9 years?

34. A company establishes a sinking fund to discharge a debt of $75,000 due in 8 years by making equal semiannual deposits, the first due in 6 months. If the investment pays 12%, compounded semiannually, what is the size of the deposits?

35. A sinking fund is established by a working couple so that they will have $20,000 to pay for part of their daughter's education when she enters college. If they make deposits at the end of each 3-month period for 10 years, and if interest is paid at 12%, compounded quarterly, what size deposits must they make?

36. A property owner has several rental units and wants to build more. How much of each month's rental income should be deposited at the beginning of each month in an account that earns 6.6%, compounded monthly, if the goal is to have $100,000 at the end of 4 years?

37. Suppose a recent college graduate's first job allows her to deposit $100 at the end of each month in a savings plan that earns 9%, compounded monthly. This savings plan continues for 8 years before new obligations make it impossible to continue. If the accrued amount remains in the plan for the next 15 years without deposits or withdrawals, how much money will be in the account 23 years after the plan began?

38. Suppose a young couple deposits $1000 at the end of each quarter in an account that earns 7.6%, compounded quarterly, for a period of 8 years. After the 8 years, they start a family and find they can contribute only $200 per quarter. If they leave the money from the first 8 years in the account and continue to contribute $200 at the end of each quarter for the next $18\frac{1}{2}$ years, how much will they have in the account (to help with their child's college expenses)?

39. A small business owner contributes $3000 at the end of each quarter to a retirement account that earns 8% compounded quarterly.
    (a) How long will it be until the account is worth $150,000?
    (b) Suppose when the account reaches $150,000, the business owner increases the contributions to $5000 at the end of each quarter. What will the total value of the account be after 15 more years?

40. A young executive deposits $300 at the end of each month for 8 years and then increases the deposits. If the account earns 7.2%, compounded monthly, how much (to the nearest dollar) should each new deposit be in order to have a total of $400,000 after 25 years?

In Problems 41 and 42, use a spreadsheet or financial program on a calculator or computer.

41. Compare the future value of an ordinary annuity of $100 per month and that of an annuity due of $100 per month if each is invested at 7.2%, compounded monthly. Find the amount in each account at the end of each month for 10 years.
    (a) How much is contributed to each annuity?
    (b) Which account has more money after 10 years? Explain why.

42. Investigate the effect that small differences in payment size can have over time. Consider two ordinary annuities, both of which earn 8% compounded quarterly, one with payments of $100 at the end of each quarter and one with payments of $110 at the end of each quarter. Track the future value of each annuity at the end of each quarter for a period of 25 years.
    (a) How much is contributed to each annuity after 10 years, 20 years, and 25 years?
    (b) What is the difference in the future values after 10 years, 20 years, and 25 years? Why are these amounts more than the differences in the amounts contributed?

Present Values of Annuities

OBJECTIVES

- To compute the present values of ordinary annuities, annuities due, and deferred annuities
- To compute the payments for a specified present value for an ordinary annuity, an annuity due, and a deferred annuity
- To find how long an annuity will last

◗ Application Preview

If you wanted to receive, at retirement, $1000 at the end of each month for 16 years, what lump sum would you need to invest in an annuity that paid 9%, compounded monthly? (See Example 2.) We call this lump sum the **present value** of the annuity. Note that the annuity in this case is an account from which a person receives equal periodic payments (withdrawals).

We have discussed how contributing to an annuity program will result in a sum of money, and we have called that sum the future value of the annuity. Just as the term *annuity* is used to describe an account in which a person makes equal periodic payments (deposits), this term is also used to describe an account from which a person receives equal periodic payments (withdrawals). That is, if you invest a lump sum of money in an account today, so that at regular intervals you will receive a fixed sum of money, you have established an annuity.

Many people who are retiring purchase an annuity to supplement their income. This annuity pays them a fixed sum of money at regular intervals (usually each month). The single sum of money required to purchase an annuity that will provide these payments at regular intervals is the **present value** of the annuity.

Ordinary Annuities

Suppose we wish to invest a lump sum of money (denoted by A_n) in an annuity that earns interest at rate i per period in order to receive (withdraw) payments of size $\$R$ from this account at the end of each of n periods (after which time the account balance will be $\$0$). Recall that receiving payments at the end of each period means that this is an ordinary annuity.

To find a formula for A_n, we can find the present value of each future payment and then add these present values (see Figure 6.6).

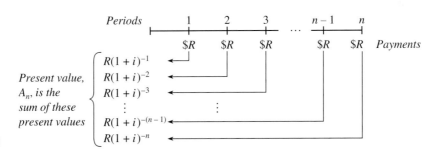

Figure 6.6

Figure 6.6 shows that we can express A_n as follows.

$$A_n = R(1 + i)^{-1} + R(1 + i)^{-2} + R(1 + i)^{-3} + \cdots + R(1 + i)^{-(n-1)} + R(1 + i)^{-n} \quad (1)$$

Multiplying both sides of equation (1) by $(1 + i)$ gives

$$(1 + i)A_n = R + R(1 + i)^{-1} + R(1 + i)^{-2} + \cdots + R(1 + i)^{-(n-2)} + R(1 + i)^{-(n-1)} \quad (2)$$

If we subtract equation (1) from equation (2), we obtain $iA_n = R - R(1 + i)^{-n}$. Then solving for A_n gives

$$A_n = R\left[\frac{1 - (1 + i)^{-n}}{i}\right]$$

Thus we have the following.

Present Value of an Ordinary Annuity

If a payment of $R is to be made at the end of each period for n periods from an account that earns interest at a rate of i per period, then the account is an ordinary annuity, and the present value is

$$A_n = R \cdot a_{\overline{n}|i} = R \cdot \left[\frac{1 - (1 + i)^{-n}}{i}\right]$$

where $a_{\overline{n}|i}$ represents the present value of an ordinary annuity of $1 per period for n periods, with an interest rate of i per period.

The values of $a_{\overline{n}|i}$ can be computed directly with a calculator or found in Table II in the Appendix.

● EXAMPLE 1 *Present Value*

What is the present value of an annuity of $1500 payable at the end of each 6-month period for 2 years if money is worth 8%, compounded semiannually?

Solution
We are given that $R = \$1500$ and $i = 0.08/2 = 0.04$. Because a payment is made twice a year for 2 years, the number of periods is $n = (2)(2) = 4$. Thus,

$$A_n = R \cdot a_{\overline{4}|0.04} = \$1500\left[\frac{1 - (1 + 0.04)^{-4}}{0.04}\right]$$
$$= \$1500(3.629895)$$
$$= \$5444.84, \text{ to the nearest cent}$$

If we had used Table II in the Appendix, we would have seen that $a_{\overline{4}|0.04}$, the present value of an ordinary annuity of $1 per period for 4 periods with an interest rate of $0.04 = 4\%$ per period, is 3.629895, and we would have obtained the same present value, $5444.84.

◗ EXAMPLE 2 *Present Value* (Application Preview)

Find the lump sum that one must invest in an annuity in order to receive $1000 at the end of each month for the next 16 years, if the annuity pays 9%, compounded monthly.

Solution
The sum we seek is the present value of an ordinary annuity, A_n, with $R = \$1000$, $i = 0.09/12 = 0.0075$, and $n = (16)(12) = 192$.

$$A_n = R\left[\frac{1 - (1 + i)^{-n}}{i}\right]$$
$$= \$1000\left[\frac{1 - (1.0075)^{-192}}{0.0075}\right] = \$1000(101.5727689) = \$101,572.77$$

Thus the required lump sum, to the nearest dollar, is $101,573.

It is important to note that all annuities involve both periodic payments and a lump sum of money. It is whether this lump sum is in the present or in the future that distinguishes problems that use formulas for present values of annuities from those that use formulas for future values of annuities. In Example 2, the lump sum was needed now (in the present) to generate the $1000 payments, so we used the present value formula.

Figure 6.7 shows a graph that compares the present value of an annuity of $1000 per year at an interest rate of 6%, compounded annually, with the same annuity at an interest rate of 10%, compounded annually. Note the impact that the higher interest rate has on the present value needed to establish such an annuity for a longer period of time. Specifically, at 6% interest, a present value of more than $12,500 is needed to generate 25 years of $1000 payments, but at an interest rate of 10%, the necessary present value is less than $10,000.

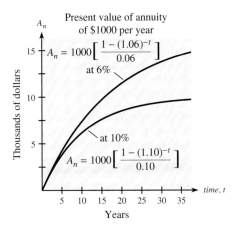

Figure 6.7

Spreadsheet
Note

The analysis shown graphically in Figure 6.7 can also be done with a spreadsheet, as Table 6.5 shows for years 5, 10, 15, 20, and 25.

TABLE 6.5

	A	B	C
1		Present Value giving	R = $1000 for n years
2	Year number	at 6% compounded annually	at 10% compounded annually
3	5	$4212.36	$3790.79
4	10	$7360.09	$6144.57
5	15	$9712.25	$7606.08
6	20	$11469.92	$8513.56
7	25	$12783.36	$9077.04

Our graphical and spreadsheet comparisons in Figure 6.7 and Table 6.5 and those from other sections emphasize the truth of the saying "Time is money." We have seen that, given enough time, relatively small differences in contributions or in interest rates can result in substantial differences in amounts (both present and future values).

● **EXAMPLE 3** *Payments from an Annuity*

Suppose that a couple plans to set up an ordinary annuity with a $100,000 inheritance they received. What is the size of the quarterly payments they will receive for the next 6 years (while their children are in college) if the account pays 7%, compounded quarterly?

Solution

The $100,000 is the amount the couple has now, so it is the present value of an ordinary annuity whose payment size, R, we seek. Using present value $A_n = \$100,000$, $n = 6(4) = 24$, and $i = 0.07/4 = 0.0175$, we solve for R.

$$A_n = R\left[\frac{1 - (1 + i)^{-n}}{i}\right]$$

$$\$100,000 = R\left[\frac{1 - (1 + 0.0175)^{-24}}{0.0175}\right]$$

$$\$100,000 = R(19.46068565)$$

$$R = \frac{\$100,000}{19.46068565} = \$5138.57, \text{ to the nearest cent}$$

● *Checkpoint*

1. Suppose an annuity pays $2000 at the end of each 3-month period for $3\frac{1}{2}$ years and money is worth 4%, compounded quarterly.
   (a) What is the total number of periods?
   (b) What is the interest rate per period?
   (c) What formula is used to find the present value of the annuity?
   (d) Find the present value.
2. An inheritance of $400,000 will provide how much at the end of each year for the next 20 years, if money is worth 7%, compounded annually?

● **EXAMPLE 4** *Number of Payments from an Annuity*

An inheritance of $250,000 is invested at 9%, compounded monthly. If $2500 is withdrawn at the end of each month, how long will it be until the account balance is $0?

Solution

The regular withdrawals form an ordinary annuity with present value $A_n = \$250,000$, payment $R = \$2500$, $i = 0.09/12 = 0.0075$, and $n = $ the number of months.

Use $A_n = R\left[\dfrac{1 - (1 + i)^{-n}}{i}\right]$ and solve for n in $250,000 = 2500\left[\dfrac{1 - (1.0075)^{-n}}{0.0075}\right]$.

This is an exponential equation, and to solve it we isolate $(1.0075)^{-n}$ and then take the natural logarithm of both sides.

$$\frac{250,000(0.0075)}{2500} = 1 - (1.0075)^{-n}$$

$$0.75 = 1 - (1.0075)^{-n}$$

$$(1.0075)^{-n} = 0.25$$

$$\ln\left[(1.0075)^{-n}\right] = \ln(0.25) \quad \Rightarrow \quad -n\left[\ln(1.0075)\right] = \ln(0.25)$$

$$-n = \frac{\ln(0.25)}{\ln(1.0075)} \approx -185.5, \quad \text{so} \quad n \approx 185.5$$

Thus, the account balance will be $0 in 186 months.

Annuities Due

Recall that an annuity due is one in which payments are made at the beginning of each period. This means that the present value of an annuity due of n payments (denoted $A_{(n,due)}$) of R at interest rate i per period can be viewed as an initial payment of R plus the payment program for an ordinary annuity of $n - 1$ payments of R at interest rate i per period (see Figure 6.8).

Present Value of an Annuity Due

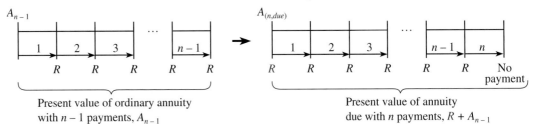

Figure 6.8

Thus, from Figure 6.8, we have

$$A_{(n,due)} = R + A_{n-1} = R + R\left[\frac{1 - (1 + i)^{-(n-1)}}{i}\right]$$

$$= R\left[1 + \frac{1 - (1 + i)^{-(n-1)}}{i}\right] = R\left[\frac{i}{i} + \frac{1 - (1 + i)^{-(n-1)}}{i}\right]$$

$$= R\left[\frac{(1 + i) - (1 + i)^{-(n-1)}}{i}\right] = R(1 + i)\left[\frac{1 - (1 + i)^{-n}}{i}\right]$$

Thus we have the following formula for the present value of an annuity due.

Present Value of an Annuity Due

If a payment of R is to be made at the beginning of each period for n periods from an account that earns interest rate i per period, then the account is an annuity due, and its present value is given by

$$A_{(n,due)} = R\left[\frac{1 - (1 + i)^{-n}}{i}\right](1 + i) = Ra_{\overline{n}|i}(1 + i)$$

where $a_{\overline{n}|i}$ denotes the present value of an ordinary annuity of $1 per period for n periods at interest rate i per period.

● **EXAMPLE 5** *Lottery Prize*

A lottery prize worth $1,200,000 is awarded in payments of $10,000 at the beginning of each month for 10 years. Suppose money is worth 7.8%, compounded monthly. What is the *real* value of the prize?

Solution

The *real* value of this prize is its present value when it is awarded. That is, it is the present value of an annuity due with $R = \$10,000$, $i = 0.078/12 = 0.0065$, and $n = 12(10) = 120$. Thus

$$A_{(120,due)} = \$10,000\left[\frac{1 - (1 + 0.0065)^{-120}}{0.0065}\right](1 + 0.0065)$$

$$= \$10,000(83.1439199)(1.0065) = \$836,843.55, \text{ to the nearest cent.}$$

This means the lottery operator needs this amount to generate the 120 monthly payments of $10,000 each.

● **EXAMPLE 6** *Court Settlement Payments*

Suppose that a court settlement results in a $750,000 award. If this is invested at 9%, compounded semiannually, how much will it provide at the beginning of each half-year for a period of 7 years?

Solution

Because payments are made at the beginning of each half-year, this is an annuity due. We seek the payment size, R, and use the present value $A_{(n,due)} = \$750,000$, $n = 2(7) = 14$, and $i = 0.09/2 = 0.045$.

$$A_{(n,due)} = R\left[\frac{1 - (1 + i)^{-n}}{i}\right](1 + i)$$

$$\$750,000 = R\left[\frac{1 - (1 + 0.045)^{-14}}{0.045}\right](1 + 0.045)$$

$$\$750,000 = R(10.682852)$$

$$R = \frac{\$750,000}{10.682852} = \$70,205.97, \text{ to the nearest cent}$$

● **Checkpoint**

3. What lump sum will be needed to generate payments of $5000 at the beginning of each quarter for a period of 5 years if money is worth 7%, compounded quarterly?

Deferred Annuities

A **deferred annuity** is one in which the first payment is made not at the beginning or end of the first period, but at some later date. An annuity that is deferred for k periods and then has payments of R per period at the end of each of the next n periods is an ordinary deferred annuity and can be illustrated by Figure 6.9.

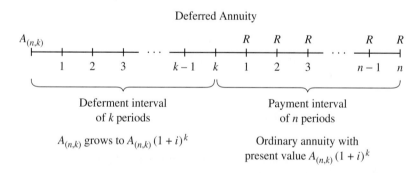

Figure 6.9

We now consider how to find the present value of such a deferred annuity when the interest rate is i per period. If payment is deferred for k periods, then the present value deposited now, denoted by $A_{(n,k)}$, is a compound interest investment for these k periods, and its future value is $A_{(n,k)}(1 + i)^k$. This amount then becomes the present value of the ordinary annuity for the next n periods. From Figure 6.9, we see that we have two equivalent expressions for the amount at the beginning of the first payment period.

$$A_{(n,k)}(1 + i)^k = R\left[\frac{1 - (1 + i)^{-n}}{i}\right]$$

Multiplying both sides by $(1 + i)^{-k}$ gives the following.

Present Value of a
Deferred Annuity

The present value of a deferred annuity of $R per period for n periods, deferred for k periods with interest rate i per period, is given by

$$A_{(n,k)} = R\left[\frac{1 - (1 + i)^{-n}}{i}\right](1 + i)^{-k} = Ra_{\overline{n}|i}(1 + i)^{-k}$$

● **EXAMPLE 7** *Present Value of a Deferred Annuity*

A deferred annuity is purchased that will pay $10,000 per quarter for 15 years after being deferred for 5 years. If money is worth 6% compounded quarterly, what is the present value of this annuity?

Solution
We use R = $10,000, n = 4(15) = 60, k = 4(5) = 20, and i = 0.06/4 = 0.015 in the formula for the present value of a deferred annuity.

$$A_{(60,20)} = R\left[\frac{1 - (1 + i)^{-60}}{i}\right](1 + i)^{-20}$$

$$= \$10,000\left[\frac{1 - (1.015)^{-60}}{0.015}\right](1.015)^{-20}$$

$$= \$10,000(39.38026889)(0.7424704182)$$

$$= \$292,386.85, \text{ to the nearest cent}$$

● **EXAMPLE 8** *Lottery Prize Payments*

Suppose a lottery prize of $50,000 is invested by a couple for future use as their child's college fund. The family plans to use the money as 8 semiannual payments at the end of each 6-month period after payments are deferred for 10 years. How much would each payment be if the money can be invested at 8.6% compounded semiannually?

Solution
We seek the payment R for a deferred annuity with n = 8 payment periods, deferred for k = 2(10) = 20 periods, i = 0.086/2 = 0.043, and $A_{(n,k)}$ = $50,000.

$$A_{(n,k)} = R\left[\frac{1 - (1 + i)^{-n}}{i}\right](1 + i)^{-k}$$

$$\$50,000 = R\left[\frac{1 - (1.043)^{-8}}{0.043}\right](1.043)^{-20}$$

$$\$50,000 = R(6.650118184)(0.4308378316)$$

$$\$50,000 = R(2.865122499)$$

$$R = \frac{\$50,000}{2.865122499} = \$17,451.26, \text{ to the nearest cent}$$

In Example 8, note the effect of the deferral time. The family receives 8($17,451.26) = $139,610.08 from the original $50,000 investment.

● *Checkpoint*

4. Suppose an annuity at 6% compounded semiannually will pay $5000 at the end of each 6-month period for 5 years with the first payment deferred for 10 years.
   (a) What is the number of payment periods and the number of deferral periods?
   (b) What is the interest rate per period?
   (c) What formula is used to find the present value of this annuity?
   (d) Find the present value of this annuity.

● *Checkpoint Solutions*

1. (a) $n = 3.5(4) = 14$
   (b) $i = 0.04/4 = 0.01$
   (c) $A_n = R\left[\dfrac{1 - (1 + i)^{-n}}{i}\right]$ for an ordinary annuity
   (d) $A_n = \$2000\left[\dfrac{1 - (1.01)^{-14}}{0.01}\right] = \$26,007.41$, to the nearest cent

2. We seek R for an ordinary annuity with $A_n = \$400,000$, $n = 20$, and $i = 0.07$.

$$A_n = R\left[\frac{1 - (1 + i)^{-n}}{i}\right]$$

$$\$400,000 = R\left[\frac{1 - (1.07)^{-20}}{0.07}\right] = R(10.59401425)$$

$$R = \frac{\$400,000}{10.59401425} = \$37,757.17, \text{ to the nearest cent}$$

3. With payments at the beginning of each quarter, we seek the present value $A_{(n,due)}$ with $n = 4(5) = 20$, $i = 0.07/4 = 0.0175$, and $R = \$5000$. Hence,

$$A_{(20,due)} = R\left[\frac{1 - (1 + i)^{-n}}{i}\right](1 + i)$$

$$= \$5000\left[\frac{1 - (1 + 0.0175)^{-20}}{0.0175}\right](1 + 0.0175)$$

$$= \$5000(16.752881)(1.0175) = \$85,230.28, \text{ to the nearest cent}$$

4. (a) $n = 5(2) = 10$ payment periods, and $k = 10(2) = 20$ deferral periods
   (b) $0.06/2 = 0.03$
   (c) $A_{(n,k)} = R\left[\dfrac{1 - (1 + i)^{-n}}{i}\right](1 + i)^{-k}$ for a deferred annuity
   (d) $A_{(n,k)} = \$5000\left[\dfrac{1 - (1.03)^{-10}}{0.03}\right](1.03)^{-20} = \$23,614.83$, to the nearest cent

6.4 Exercises

ORDINARY ANNUITIES

1. Find the present value of an annuity of $6000 paid at the end of each 6-month period for 8 years if the interest rate is 8%, compounded semiannually.
2. Find the present value of an annuity that pays $3000 at the end of each 6-month period for 6 years if the interest rate is 6%, compounded semiannually.

3. Suppose a state lottery prize of $5 million is to be paid in 20 payments of $250,000 each at the end of each of the next 20 years. If money is worth 10%, compounded annually, what is the present value of the prize?
4. How much is needed in an account that earns 8.4% compounded monthly in order to withdraw $1000 at the end of each month for 20 years?

5. With a present value of $135,000, what is the size of the withdrawals that can be made at the end of each quarter for the next 10 years if money is worth 6.4%, compounded quarterly?

6. If $88,000 is invested in an annuity that earns 5.8%, compounded quarterly, what payments will it provide at the end of each quarter for the next $5\frac{1}{2}$ years?

7. Dr. Jane Kodiak plans to sell her practice to an HMO. The HMO will pay her $750,000 now or will make a partial payment now and make additional payments at the end of each year for the next 10 years. If money is worth 6.5%, compounded annually, is it better to take $750,000 now or $250,000 now and $70,000 at the end of each year for the next 10 years? Justify your choice.

8. Suppose Becky has her choice of $10,000 at the end of each month for life or a single prize of $1.5 million. She is 35 years old and her life expectancy is 40 more years.
   (a) Find the present value of the annuity if money is worth 7.2%, compounded monthly.
   (b) If she takes the $1.5 million, spends $700,000 of it, and invests the remainder at 7.2% compounded monthly, what monthly annuity will she receive for the next 40 years?

9. As the most successful contestant in the history of *Jeopardy*, Ken Jennings won more than $2.5 million during 2004. Suppose he invested $1.5 million in an ordinary annuity that earned 7.2%, compounded monthly. How much would he receive at the end of each month for the next 20 years?

10. As a result of a November 2004 court settlement, Goldman Sachs agreed to pay $13.4 million to Ford Motor Company, with $10 million of it going to a Ford charitable trust. (*Source: The New York Times,* November 4, 2004) If the trust invested this money at 6.3%, compounded annually, how much could be awarded to worthwhile organizations at the end of each year for the next 20 years?

11. A personal account earmarked as a retirement supplement contains $242,400. Suppose $200,000 is used to establish an annuity that earns 6%, compounded quarterly, and pays $4500 at the end of each quarter. How long will it be until the account balance is $0?

12. A professional athlete invested $2.5 million of a bonus in an account that earns 6.8%, compounded semiannually. If $120,000 is to be withdrawn at the end of each six months, how long will it be until the account balance is $0?

13. The following figure shows a graph that compares the present values of two ordinary annuities of $1000 annually, one at 8% compounded annually and one at 10% compounded annually.

    (a) Determine which graph corresponds to the 8% rate and which to the 10% rate.
    (b) Use the graph to estimate the difference between the present values of these annuities for 25 years.
    (c) Write a sentence that explains this difference.

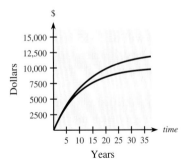

14. The following figure shows a graph that compares the present values of two ordinary annuities of $800 quarterly, one at 6% compounded quarterly and one at 9% compounded quarterly.
    (a) Determine which graph corresponds to the 6% rate and which to the 9% rate.
    (b) Use the graph to estimate the difference between the present values of these annuities for 25 years (100 quarters).
    (c) Write a sentence that explains this difference.

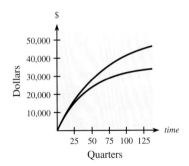

ANNUITIES DUE

15. Explain the difference between an ordinary annuity and an annuity due.

16. Is there any difference between the present values in (a) and (b)? Explain.
    (a) An annuity due that pays $1000 at the beginning of each year for 10 years
    (b) Taking $1000 now and establishing an ordinary annuity that pays $1000 at the end of each year for 9 years

17. Find the present value of an annuity due that pays $3000 at the beginning of each quarter for the next 7 years. Assume that money is worth 5.8%, compounded quarterly.

18. Find the present value of an annuity due that pays $25,000 every 6 months for the next $2\frac{1}{2}$ years if money is worth 6.2%, compounded semiannually.

19. What amount must be set aside now to generate payments of $50,000 at the beginning of each year for the next 12 years if money is worth 5.92%, compounded annually?

20. Suppose an annuity will pay $15,000 at the beginning of each year for the next 7 years. How much money is needed to start this annuity if it earns 7.3%, compounded annually?

21. A year-end bonus of $25,000 will generate how much money at the beginning of each month for the next year, if it can be invested at 6.48%, compounded monthly?

22. A couple inherits $89,000. How much can this generate at the beginning of each month over the next 5 years, if money is worth 6.3%, compounded monthly?

23. Recent sales of some real estate and record profits make it possible for a manufacturer to set aside $800,000 in a fund to be used for modernization and remodeling. How much can be withdrawn from this fund at the beginning of each half-year for the next 3 years if the fund earns 7.7%, compounded semiannually?

24. As a result of a court settlement, an accident victim is awarded $1.2 million. The attorney takes one-third of this amount, another third is used for immediate expenses, and the remaining third is used to set up an annuity. What amount will this annuity pay at the beginning of each quarter for the next 5 years if the annuity earns 7.6%, compounded quarterly?

25. A used piece of rental equipment has $2\frac{1}{2}$ years of useful life remaining. When rented, the equipment brings in $800 per month (paid at the beginning of the month). If the equipment is sold now and money is worth 4.8%, compounded monthly, what must the selling price be to recoup the income that the rental company loses by selling the equipment "early"?

26. A $2.4 million state lottery pays $10,000 at the beginning of each month for 20 years. How much money must the state actually have in hand to set up the payments for this prize if money is worth 6.3%, compounded monthly?

MISCELLANEOUS PROBLEMS FOR ORDINARY ANNUITIES AND ANNUITIES DUE

In Problems 27–32, (a) decide whether the problem relates to an ordinary annuity or an annuity due, then (b) solve the problem.

27. An insurance settlement of $750,000 must replace Trixie Eden's income for the next 40 years. What income will this settlement provide at the end of each month if it is invested in an annuity that earns 8.4%, compounded monthly?

28. A trust will provide $10,000 to a county library at the beginning of each 3-month period for the next $2\frac{1}{2}$ years. If money is worth 7.4%, compounded quarterly, find the amount in the trust when it begins.

29. A company wants to have $40,000 at the beginning of each 6-month period for the next $4\frac{1}{2}$ years. If an annuity is set up for this purpose, how much must be invested now if the annuity earns 6.68%, compounded semiannually?

30. Is it more economical to buy an automobile for $29,000 cash or to pay $8000 down and $3000 at the end of each quarter for 2 years, if money is worth 8% compounded quarterly?

31. Juanita Domingo's parents want to establish a college trust for her. They want to make 16 quarterly withdrawals of $2000, with the first withdrawal 3 months from now. If money is worth 7.2%, compounded quarterly, how much must be deposited now to provide for this trust?

32. A retiree inherits $93,000 and invests it at 6.6%, compounded monthly, in an annuity that provides an amount at the end of each month for the next 12 years. Find the monthly amount.

For each of Problems 33 and 34, answer the following questions.
(a) How much is in the account after the last deposit is made?
(b) How much was deposited?
(c) What is the amount of each withdrawal?
(d) What is the total amount withdrawn?

33. Suppose an individual makes an initial investment of $2500 in an account that earns 7.8%, compounded monthly, and makes additional contributions of $100 at the end of each month for a period of 12 years. After these 12 years, this individual wants to make withdrawals at the end of each month for the next 5 years (so that the account balance will be reduced to $0).

34. Suppose that Nam Banh deposits his $12,500 bonus in an account that earns 8%, compounded quarterly, and makes additional deposits of $500 at the end of each quarter for the next $22\frac{1}{2}$ years, until he retires. To supplement his retirement, Nam wants to make withdrawals at the end of each quarter for the next 12 years (at which time the account balance will be $0).

35. A young couple wants to have a college fund that will pay $30,000 at the end of each half-year for 8 years.
    (a) If they can invest at 8%, compounded semiannually, how much do they need to invest at the end of each 6-month period for the next 18 years in order to begin making their college withdrawals 6 months after their last investment?

(b) Suppose 8 years after beginning the annuity payments, they receive an inheritance of $38,000 that they contribute to the account, and they continue to make their regular payments as found in part (a). How many college withdrawals will they be able to make before the account balance is $0?

36. A recent college graduate begins a savings plan at age 27 by investing $400 at the end of each month in an account that earns 7.5%, compounded monthly.
    (a) If this plan is followed for 10 years, how much should the monthly contributions be for the next 28 years in order to be able to withdraw $10,000 at the end of each month from the account for the next 25 years?
    (b) What is the total amount contributed?
    (c) What is the total amount withdrawn?

DEFERRED ANNUITIES

37. Find the present value of an annuity of $2000 per year at the end of each of 8 years after being deferred for 6 years, if money is worth 7% compounded annually.

38. Find the present value of an annuity of $2000, at the end of each quarter for 5 years after being deferred for 3 years, if money is worth 8% compounded quarterly.

39. The terms of a single parent's will indicate that a child will receive an ordinary annuity of $16,000 per year from age 18 to age 24 (so the child can attend college) and that the balance of the estate goes to a niece. If the parent dies on the child's 14th birthday, how much money must be removed from the estate to purchase the annuity? (Assume an interest rate of 6%, compounded annually.)

40. On his 48th birthday, a man wants to set aside enough money to provide an income of $300 at the end of each month from his 60th birthday to his 65th birthday. If he earns 6%, compounded monthly, how much will this supplemental retirement plan cost him on his 48th birthday?

41. The semiannual tuition payment at a major university is expected to be $30,000 for the 4 years beginning 18 years from now. What lump sum payment should the university accept now, in lieu of tuition payments beginning 18 years, 6 months from now? Assume that money is worth 7%, compounded semiannually, and that tuition is paid at the end of each half-year for 4 years.

42. A grateful alumnus wishes to provide a scholarship of $2000 per year for 5 years to his alma mater, with the first scholarship awarded on his 60th birthday. If money is worth 6%, compounded annually, how much money must he donate on his 50th birthday?

43. Danny Metzger's parents invested $1600 when he was born. This money is to be used for Danny's college education and is to be withdrawn in four equal annual payments beginning when Danny is age 19. Find the amount that will be available each year, if money is worth 6%, compounded annually.

44. Carol Goldsmith received a trust fund inheritance of $10,000 on her 30th birthday. She plans to use the money to supplement her income with 20 quarterly payments beginning on her 60th birthday. If money is worth 7.6% compounded quarterly, how much will each quarterly payment be?

45. Hockey goalie Dominik Hasek, formerly of the NHL's Buffalo Sabres and the Olympic gold medal-winning Czech Republic, received a $4 million signing bonus with his 1998 contract extension with the Sabres (as reported in *USA Today* on March 20, 1998). Suppose Hasek set aside $2.5 million in a fund that earns 9.6%, compounded monthly. If he deferred this amount for 15 years, how much per month would it provide for the 40 years after the deferral period?

46. A couple received a $134,000 inheritance the year they turned 48 and invested it in a fund that earns 7.7% compounded semiannually. If this amount is deferred for 14 years (until they retire), how much will it provide at the end of each half-year for the next 20 years after they retire?

In Problems 47 and 48, use a spreadsheet or financial program on a calculator or computer.

47. Suppose a couple have $100,000 at retirement that they can invest in an ordinary annuity that earns 7.8%, compounded monthly. Track the balance in this annuity account until it reaches $0, if the couple receive the following monthly payments.
    (a) $1000   (b) $2500

48. Suppose you invested $250,000 in an annuity that earned interest compounded monthly. This annuity paid $3000 at the end of each month. Experiment with the following different interest rates (compounded monthly) to see how long it will be, with each rate, until this annuity has an account balance of $0.
    (a) 6.5%   (b) 9%

OBJECTIVES

- To find the regular payments required to amortize a debt
- To find the amount that can be borrowed for a specified payment
- To develop an amortization schedule
- To find the unpaid balance of a loan
- To find the effect of paying an extra amount

Loans and Amortization

◗ Application Preview

Chuckie and Angelica have been married for 4 years and are ready to buy their first home. They have saved $30,000 for a down payment, and their budget can accommodate a monthly mortgage payment of $850.00. They want to know in what price range they can purchase a home if the rate for a 30-year loan is 7.8% compounded monthly. (See Example 3.)

When businesses, individuals, or families (such as Chuckie and Angelica) borrow money to make a major purchase, four questions commonly arise:

1. What will the payments be?
2. How much can be borrowed and still fit a budget?
3. What is the payoff amount of the loan before the final payment is due?
4. What is the total of all payments needed to pay off the loan?

In this section we discuss the most common way that consumers and businesses discharge debts—regular payments of fixed size made on a loan (called **amortization**)—and determine ways to answer the Chuckie and Angelica question (Question 2) and the other three questions posed above.

Amortization

Just as we invest money to earn interest, banks and lending institutions lend money and collect interest for its use. Although your aunt may lend you money with the understanding that you will repay the full amount of the money plus simple interest at the end of a year, financial institutions generally expect you to make partial payments on a regular basis (often monthly). Most consumer loans (for automobiles, appliances, televisions, and the like) are classed as **installment loans.**

Federal law now requires that the full cost of any loan and the **true annual percentage rate (APR)** be disclosed with the loan. Because of this law, loans now are usually paid off by a series of partial payments with interest charged on the unpaid balance at the end of each payment period. In this case, the stated interest rate is the same as the APR.

This type of loan is usually repaid by making all payments (including principal and interest) of equal size. This process of repaying the loan is called **amortization.**

When a bank makes a loan of this type, it is purchasing from the borrower an ordinary annuity that pays a fixed return each payment period. The lump sum the bank gives to the borrower (the principal of the loan) is the present value of the ordinary annuity, and each payment the bank receives from the borrower is a payment from the annuity. Hence, to find the size of these periodic payments, we solve for R in the formula for the present value of an ordinary annuity,

$$A_n = R\left[\frac{1 - (1 + i)^{-n}}{i}\right]$$

which yields the following algebraically equivalent formula.

Amortization Formula

If the debt of $\$A_n$, with interest at a rate of i per period, is amortized by n equal periodic payments (each payment being made at the end of a period), the size of each payment is

$$R = A_n \cdot \left[\frac{i}{1 - (1 + i)^{-n}}\right]$$

● **EXAMPLE 1** *Payments to Amortize a Debt*

A debt of $1000 with interest at 16%, compounded quarterly, is to be amortized by 20 quarterly payments (all the same size) over the next 5 years. What will the size of these payments be?

Solution
The amortization of this loan is an ordinary annuity with present value $1000. Therefore, $A_n = \$1000$, $n = 20$, and $i = 0.16/4 = 0.04$. Thus we have

$$R = A_n\left[\frac{i}{1 - (1 + i)^{-n}}\right] = \$1000\left[\frac{0.04}{1 - (1.04)^{-20}}\right]$$
$$= \$1000(0.07358175) = \$73.58, \text{ to the nearest cent}$$

● **EXAMPLE 2** *Buying a Home*

A man buys a house for $200,000. He makes a $50,000 down payment and agrees to amortize the rest of the debt with quarterly payments over the next 10 years. If the interest on the debt is 12%, compounded quarterly, find (a) the size of the quarterly payments, (b) the total amount of the payments, and (c) the total amount of interest paid.

Solution
(a) We know that $A_n = \$200,000 - \$50,000 = \$150,000$, $n = 4(10) = 40$, and $i = 0.12/4 = 0.03$. Thus, for the quarterly payment, we have

$$R = \$150,000\left[\frac{0.03}{1 - (1.03)^{-40}}\right]$$
$$= (\$150,000)(0.043262378) = \$6489.36, \text{ to the nearest cent}$$

(b) The man made 40 payments of $6489.36, so his payments totaled
$$(40)(\$6489.36) = \$259,574.40$$
plus the $50,000 down payment, or $309,574.40.

(c) Of the $309,574.40 paid, $200,000 was for payment of the house. The remaining $109,574.40 was the total amount of interest paid.

◗ **EXAMPLE 3** *Affordable Home* **(Application Preview)**

Chuckie and Angelica have $30,000 for a down payment, and their budget can accommodate a monthly mortgage payment of $850.00. What is the most expensive home they can buy if they can borrow money for 30 years at 7.8%, compounded monthly?

Solution
We seek the amount that Chuckie and Angelica can borrow, or A_n, knowing that $R = \$850$, $n = 30(12) = 360$, and $i = 0.078/12 = 0.0065$. We can use these values in the amortization formula and solve for A_n (or use the formula for the present value of an ordinary annuity).

$$R = A_n\left[\frac{i}{1 - (1 + i)^{-n}}\right]$$
$$\$850 = A_n\left[\frac{0.0065}{1 - (1.0065)^{-360}}\right]$$
$$A_n = \$850\left[\frac{1 - (1.0065)^{-360}}{0.0065}\right]$$
$$A_n = \$850(138.9138739) = \$118,076.79, \text{ to the nearest cent}$$

Thus, if they borrow $118,077 (to the nearest dollar) and put down $30,000, the most expensive home they can buy would cost $118,077 + $30,000 = $148,077.

● Checkpoint

1. A debt of $25,000 is to be amortized with equal quarterly payments over 6 years, and money is worth 7%, compounded quarterly.
   (a) Find the total number of payments.
   (b) Find the interest rate per period.
   (c) Write the formula used to find the size of each payment.
   (d) Find the amount of each payment.

2. A new college graduate determines that she can afford a car payment of $350 per month. If the auto manufacturer is offering a special 2.1% financing rate, compounded monthly for 4 years, how much can she borrow and still have a $350 monthly payment?

Amortization Schedule

We can construct an **amortization schedule** that summarizes all the information regarding the amortization of a loan.

For example, a loan of $10,000 with interest at 10% could be repaid in 5 equal annual payments of size

$$R = \$10,000\left[\frac{0.10}{1 - (1 + 0.10)^{-5}}\right] = \$10,000(0.263797) = \$2637.97$$

Each time this $2637.97 payment is made, some is used to pay the interest on the unpaid balance, and some is used to reduce the principal. For the first payment, the unpaid balance is $10,000, so the interest payment is 10% of $10,000, or $1000. The remaining $1637.97 is applied to the principal. Hence, after this first payment, the unpaid balance is $10,000 − $1637.97 = $8362.03.

For the second payment of $2637.97, the amount used for interest is 10% of $8362.03, or $836.20; the remainder, $1801.77, is used to reduce the principal.

This information for these two payments and for the remaining payments is summarized in the following amortization schedule.

Period	Payment	Interest	Balance Reduction	Unpaid Balance
				10000.00
1	2637.97	1000.00	1637.97	8362.03
2	2637.97	836.20	1801.77	6560.26
3	2637.97	656.03	1981.94	4578.32
4	2637.97	457.83	2180.14	2398.18
5	2638.00	239.82	2398.18	0.00
Total	13189.88	3189.88	10000.00	

Note that the last payment was increased by 3¢ so that the balance was reduced to $0 at the end of the 5 years. Such an adjustment is normal in amortizing a loan.

Spreadsheet Note

The amortization schedule above involves only a few payments. For loans involving more payments, a spreadsheet program is an excellent tool to develop an amortization schedule. The spreadsheet requires the user to understand the interrelationships among the columns. The accompanying spreadsheet output in Table 6.6 shows the amortization schedule for the first 24 monthly payments on a $100,000 loan amortized for 30 years at 6%, compounded monthly. It is interesting to observe how little the unpaid balance changes during these first 2 years—and, consequently, how much of each payment is devoted to paying interest due rather than to reducing the debt. In fact, the column labeled "Total Interest" shows a running total of how much has been paid toward interest. We see that after 12 payments of $599.55, a total of $7194.60 has been paid, and $5966.59 of this has been interest payments.

TABLE 6.6

	A	B	C	D	E	F
1	Payment Number	Payment Amount	Interest	Balance Reduction	Unpaid Balance	Total Interest
2	0	0	0	0	$100,000.00	$0.00
3	1	$599.55	$500.00	$99.55	$99,900.45	$500.00
4	2	$599.55	$499.50	$100.05	$99,800.40	$999.50
5	3	$599.55	$499.00	$100.55	$99,699.85	$1,498.50
6	4	$599.55	$498.50	$101.05	$99,598.80	$1,997.00
7	5	$599.55	$497.99	$101.56	$99,497.25	$2,495.00
8	6	$599.55	$497.49	$102.06	$99,395.18	$2,992.48
9	7	$599.55	$496.98	$102.57	$99,292.61	$3,489.46
10	8	$599.55	$496.46	$103.09	$99,189.52	$3,985.92
11	9	$599.55	$495.95	$103.60	$99,085.92	$4,481.87
12	10	$599.55	$495.43	$104.12	$98,981.80	$4,977.30
13	11	$599.55	$494.91	$104.64	$98,877.16	$5,472.21
14	12	$599.55	$494.39	$105.16	$98,771.99	$5,966.59
15	13	$599.55	$493.86	$105.69	$98,666.30	$6,460.45
16	14	$599.55	$493.33	$106.22	$98,560.09	$6,953.79
17	15	$599.55	$492.80	$106.75	$98,453.34	$7,446.59
18	16	$599.55	$492.27	$107.28	$98,346.05	$7,938.85
19	17	$599.55	$491.73	$107.82	$98,238.23	$8,430.58
20	18	$599.55	$491.19	$108.36	$98,129.87	$8,921.77
21	19	$599.55	$490.65	$108.90	$98,020.97	$9,412.42
22	20	$599.55	$490.10	$109.45	$97,911.53	$9,902.53
23	21	$599.55	$489.56	$109.99	$97,801.54	$10,392.09
24	22	$599.55	$489.01	$110.54	$97,690.99	$10,881.09
25	23	$599.55	$488.45	$111.10	$97,579.90	$11,369.55
26	24	$599.55	$487.90	$111.65	$97,468.25	$11,857.45

Unpaid Balance of a Loan

Many people who borrow money, such as for a car or a home, do not pay on the loan for its entire term. Rather, they pay off the loan early by making a final lump sum payment. The unpaid balance found in the amortization schedule is the "payoff amount" of the loan and represents the lump sum payment that must be made to complete payment on the loan. When the number of payments is large, we may wish to find the unpaid balance of a loan without constructing an amortization schedule.

Recall that calculations for amortization of a debt are based on the present value formula for an ordinary annuity. Because of this, the **unpaid balance of a loan** (also called the **payoff amount** and the **outstanding principal of the loan**) is the present value needed to generate all the remaining payments.

Unpaid Balance or Payoff Amount of a Loan

For a loan of n payments of R per period at interest rate i per period, the unpaid balance after k payments have been made is the present value of an ordinary annuity with $n - k$ payments. That is, with $n - k$ payments remaining,

$$\text{Unpaid balance} = A_{n-k} = R\left[\frac{1 - (1 + i)^{-(n-k)}}{i}\right]$$

● **EXAMPLE 4** *Unpaid Balance*

In Example 2, we found that the quarterly payment for a loan of $150,000 at 12%, compounded quarterly, for 10 years is $6489.36 (to the nearest cent). Find the unpaid balance immediately after the 15th payment.

Solution

The unpaid balance after the 15th payment is the present value of the annuity with $40 - 15 = 25$ payments remaining. Thus we use $R = \$6489.36$, $i = 0.03$, and $n - k = 25$ in the formula for the unpaid balance of a loan.

$$A_{n-k} = R\left[\frac{1 - (1 + i)^{-(n-k)}}{i}\right] = \$6489.36\left[\frac{1 - (1.03)^{-25}}{0.03}\right]$$

$$= (\$6489.36)\,(17.4131477) = \$113{,}000.18, \text{ to the nearest cent}$$

● *Checkpoint*

3. A 42-month auto loan has monthly payments of $411.35. If the interest rate is 8.1%, compounded monthly, find the unpaid balance immediately after the 24th payment.

Figure 6.10

The curve in Figure 6.10 shows the unpaid balance of a $100,000 loan at 6%, compounded monthly, for 30 years (with monthly payments of $599.55, as in Table 6.6). In the figure, the straight line represents how the unpaid balance would decrease if each payment diminished the debt by the same amount. Note how the curve decreases slowly at first and more rapidly as the unpaid balance nears zero. The reason for this behavior is that when the unpaid balance is large, much of each payment is devoted to interest. Similarly, when the debt decreases, more of each payment goes toward the principal. An interesting question then is what would be the effect of paying an extra amount (directly toward the principal) with each payment.

● **EXAMPLE 5** *Effect of Paying an Extra Amount*

Consider the loan in Table 6.6: $100,000 borrowed at 6% compounded monthly for 30 years, with monthly payments of $599.55. As Table 6.6 shows, the unpaid balance after 24 monthly payments is $97,468.25. Suppose that from this point the borrower decides to pay $650 per month.

(a) How many more payments must be made?
(b) How much would this save over the life of the loan?

Solution

(a) Use $A_n = R\left[\dfrac{1 - (1 + i)^{-n}}{i}\right]$ with $A_n = \$97{,}468.25$, $R = \$650$, $i = 0.06/12 = 0.005$, and solve for n (by first isolating the exponential $(1.005)^{-n}$).

$$97{,}468.25 = 650\left[\frac{1 - (1.005)^{-n}}{0.005}\right] \tag{1}$$

$$\frac{(97{,}468.25)(0.005)}{650} = 1 - (1.005)^{-n}$$

$$(1.005)^{-n} = 1 - \frac{(97{,}468.25)(0.005)}{650}$$

$$\ln\left[(1.005)^{-n}\right] = \ln(0.2502442308)$$

$$-n\left[\ln(1.005)\right] = \ln(0.2502442308)$$

$$n = \frac{\ln(0.2502442308)}{-\ln(1.005)} \approx 277.8$$

(b) For the purposes of determining the savings, assume $n = 277.8$ payments of $650 were made. Thus the total paid with each method is as follows.

Paying Extra: (24 payments of $599.55) + (277.8 payments of $650) = $194,959.20

Paying $599.55:   (360 payments of $599.55) = $215,838.00

Hence by increasing the payments slightly, the borrower saved

$$\$215,838.00 - \$194,959.20 = \$20,878.80$$

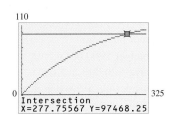

Figure 6.11

Notice in Example 5 that we could have solved equation (1) graphically by finding where $y_1 = 97,468.25$ and $y_2 = 650\left[\dfrac{1 - (1.005)^{-x}}{0.005}\right]$ intersect (see Figure 6.11).

● *Checkpoint Solutions*

1. (a) $n = 6(4) = 24$ payments
   (b) $i = 0.07/4 = 0.0175$ per period
   (c) $R = A_n\left[\dfrac{i}{1 - (1 + i)^{-n}}\right]$
   (d) $R = \$25,000\left[\dfrac{0.0175}{1 - (1.0175)^{-24}}\right] = \$1284.64$, to the nearest cent

2. Use $R = A_n\left[\dfrac{i}{1 - (1 + i)^{-n}}\right]$ with $i = 0.021/12 = 0.00175$, $n = 48$, and $R = \$350$.

$$\$350 = A_n\left[\dfrac{0.00175}{1 - (1.00175)^{-48}}\right] = A_n(0.021738795)$$

$$A_n = \dfrac{\$350}{0.021738795} = \$16,100.25, \text{ to the nearest cent}$$

3. Use $R = \$411.35$, $i = 0.081/12 = 0.00675$, and $n - k = 42 - 24 = 18$.

$$A_{n-k} = R\left[\dfrac{1 - (1 + i)^{-(n-k)}}{i}\right] = \$411.35\left[\dfrac{1 - (1.00675)^{-18}}{0.00675}\right]$$

$$= \$6950.13, \text{ to the nearest cent}$$

6.5 Exercises

1. Answer parts (a) and (b) with respect to loans for 10 years and 25 years. Assume the same interest rate.
   (a) Which loan results in more of each payment being directed toward principal? Explain.
   (b) Which loan results in a lower periodic payment? Explain.

2. When a debt is amortized, which interest rate is better for the borrower, 10% or 6%? Explain.

3. A debt of $8000 is to be amortized with 8 equal semi-annual payments. If the interest rate is 12%, compounded semiannually, what is the size of each payment?

4. A loan of $10,000 is to be amortized with 10 equal quarterly payments. If the interest rate is 6%, compounded quarterly, what is the periodic payment?

5. A recent graduate's student loans total $18,000. If these loans are at 4.2%, compounded quarterly, for 10 years, what are the quarterly payments?

6. For equipment upgrades a business borrowed $400,000 at 8%, compounded semiannually, for 5 years. What are the semiannual payments?

7. A homeowner planning a kitchen remodeling can afford a $200 monthly payment. How much can the home-owner borrow for 5 years at 6%, compounded monthly, and still stay within the budget?

8. AdriAnne and Anna's Auto Repair wants to add a new service bay. How much can they borrow at 5%, compounded quarterly, if the desired quarterly payment is $6000?

In Problems 9–12, develop an amortization schedule for the loan described.

9. $100,000 for 3 years at 9% compounded annually
10. $30,000 for 5 years at 7% compounded annually
11. $20,000 for 1 year at 12% compounded quarterly
12. $50,000 for $2\frac{1}{2}$ years at 10% compounded semiannually
13. A $10,000 loan is to be amortized for 10 years with quarterly payments of $334.27. If the interest rate is 6%, compounded quarterly, what is the unpaid balance immediately after the sixth payment?
14. A debt of $8000 is to be amortized with 8 equal semi-annual payments of $1288.29. If the interest rate is 12%, compounded semiannually, find the unpaid balance immediately after the fifth payment.
15. When Maria Acosta bought a car $2\frac{1}{2}$ years ago, she borrowed $14,000 for 48 months at 8.1% compounded monthly. Her monthly payments are $342.44, but she'd like to pay off the loan early. How much will she owe just after her payment at the $2\frac{1}{2}$-year mark?
16. Six and a half years ago, a small business borrowed $50,000 for 10 years at 9%, compounded semiannually, in order to update some equipment. Now the company would like to pay off this loan. Find the payoff amount just after the company makes the 14th semiannual payment of $3843.81.

Problems 17–20 describe a debt to be amortized. In each problem, find:
(a) the size of each payment,
(b) the total amount paid over the life of the loan, and
(c) the total interest paid over the life of the loan.

17. A man buys a house for $175,000. He makes a $75,000 down payment and amortizes the rest of the debt with semiannual payments over the next 10 years. The interest rate on the debt is 12%, compounded semiannually.
18. Sean Lee purchases $20,000 worth of supplies for his restaurant by making a $3000 down payment and amortizing the rest with quarterly payments over the next 5 years. The interest rate on the debt is 16%, compounded quarterly.
19. John Fare purchased $10,000 worth of equipment by making a $2000 down payment and promising to pay the remainder of the cost in semiannual payments over the next 4 years. The interest rate on the debt is 10%, compounded semiannually.
20. A woman buys an apartment house for $1,250,000 by making a down payment of $250,000 and amortizing the rest of the debt with semiannual payments over the next 10 years. The interest rate on the debt is 8%, compounded semiannually.
21. A man buys a car for $36,000. If the interest rate on the loan is 12%, compounded monthly, and if he wants to make monthly payments of $900 for 36 months, how much must he put down?

22. A woman buys a car for $40,000. If the interest rate on the loan is 12%, compounded monthly, and if she wants to make monthly payments of $700 for 3 years, how much must she have for a down payment?
23. A couple purchasing a home budget $900 per month for their loan payment. If they have $20,000 available for a down payment and are considering a 25-year loan, how much can they spend on the home at each of the following rates?
   (a) 6.9% compounded monthly
   (b) 7.5% compounded monthly
24. A developer wants to buy a certain parcel of land. The developer feels she can afford payments of $22,000 each half-year for the next 7 years. How much can she borrow and hold to this budget at each of the following interest rates?
   (a) 8.9% compounded semiannually
   (b) 7.3% compounded semiannually
25. A couple who borrow $90,000 for 30 years at 7.2%, compounded monthly, must make monthly payments of $610.91.
   (a) Find their unpaid balance after 1 year.
   (b) During that first year, how much interest do they pay?
26. A company that purchases a piece of equipment by borrowing $250,000 for 10 years at 6%, compounded monthly, has monthly payments of $2775.51.
   (a) Find the unpaid balance on this loan after 1 year.
   (b) During that first year, how much interest does the company pay?
27. A recent college graduate buys a new car by borrowing $18,000 at 8.4%, compounded monthly, for 5 years. She decides to pay an extra $15 per payment.
   (a) What is the monthly payment required by the loan, and how much does she decide to pay each month?
   (b) How many payments (that include the extra $15) will she make?
   (c) How much will she save by paying the extra $15?
28. A young couple buying their first home borrow $85,000 for 30 years at 7.2%, compounded monthly, and make payments of $576.97. After 3 years, they are able to make a one-time payment of $2000 along with their 36th payment.
   (a) Find the unpaid balance immediately after they pay the extra $2000 and their 36th payment.
   (b) How many regular payments of $576.97 will amortize the unpaid balance from part (a)?
   (c) How much will the couple save over the life of the loan by paying the extra $2000?

A debt of $100,000 is amortized at 6%, compounded monthly, over 25 years with 300 monthly payments of $644.30 each. Figure 6.12 includes two graphs: one shows the total amount paid (in monthly payments) as a function of time (in months), and the other shows the amount paid toward the principal of the debt as a function of time. Use this figure to complete Problems 29 and 30.

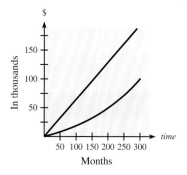

Figure 6.12

29. (a) Correctly label each graph.
    (b) Draw a vertical segment whose length represents the total amount of interest paid on the debt after 250 months.
30. Draw a vertical segment whose length represents the outstanding principal of the debt (or the payoff amount of the loan) after 150 months.
31. What difference does 0.5% make on a loan? To answer this question, find (to the nearest dollar) the monthly payment and total interest paid over the life of the loan for each of the following.
    (a) An auto loan of $15,000 at 8.0% versus 8.5%, compounded monthly, for 4 years.
    (b) A mortgage loan of $80,000 at 6.75% versus 7.25%, compounded monthly, for 30 years.
    (c) In each of these 0.5% differences, what seems to have the greatest effect on the borrower: amount borrowed, interest rate, or duration of the loan? Explain.
32. Some banks now have biweekly mortgages (that is, with payments every other week). Compare a 20-year, $100,000 loan at 8.1% by finding the payment size and the total interest paid over the life of the loan under each of the following conditions.
    (a) Payments are monthly, and the rate is 8.1%, compounded monthly.
    (b) Payments are biweekly, and the rate is 8.1%, compounded biweekly.
33. Many banks charge points on mortgage loans. Each point is the equivalent of a 1% charge on the amount borrowed and is paid before the loan is made as part of

the closing costs of buying a home (closing costs include points, title fees, attorney's fees, assessor's fees, and so on).
(a) If $100,000 is borrowed for 25 years, for each of the following, find the payment size and the total paid over the life of the loan (including points).
    (i) $7\frac{1}{2}$%, compounded monthly, with 0 points
    (ii) $7\frac{1}{4}$%, compounded monthly, with 1 point
    (iii) 7%, compounded monthly, with 2 points
(b) Which loan in (a) has the lowest total cost over the life of the loan?
34. Time-share sales provide an opportunity for vacationers to own a resort condo for 1 week (or more) each year forever. The owners may use their week at their own condo or trade the week and vacation elsewhere. Time-share vacation sales usually require payment in full or financing through the time-share company, and interest rates are usually in the 13% to 18% range. Suppose the cost to buy a 1-week time share in a 3-bedroom condo is $21,833. Also suppose a 10% down payment is required, with the balance financed for 15 years at 16.5%, compounded monthly.
(a) Find the monthly payment.
(b) Determine the total cost over the life of the loan.
(c) Suppose maintenance fees for this condo are $400 per year. Find the annual cost of the condo over the life of the loan. Assume that the annual maintenance fees remain constant.
(d) Use (c) and the 10% down payment to determine the average annual cost for having this vacation condo for 1 week over the life of the loan.
35. During four years of college, Nolan MacGregor's student loans are $4000, $3500, $4400, and $5000 for freshman year through senior year, respectively. Each loan amount gathers interest of 1%, compounded quarterly, while Nolan is in school and 3%, compounded quarterly, during a 6-month grace period after graduation.
(a) What is the loan balance after the grace period? Assume the freshman year loan earns 1% interest for 3/4 year during the first year, then for 3 full years until graduation. Make similar assumptions for the loans for the other years.
(b) After the grace period, the loan is amortized over the next 10 years at 3%, compounded quarterly. Find the quarterly payment.
(c) If Nolan decides to pay an additional $90 per payment, how many payments of this size will amortize the debt?
(d) How much will Nolan save by paying the extra $90 with each payment?

36. Clark and Lana take a 30-year home mortgage of $121,000 at 7.8%, compounded monthly. They make their regular monthly payments for 5 years, then decide to pay $1000 per month.
    (a) Find their regular monthly payment.
    (b) Find the unpaid balance when they begin paying the $1000.
    (c) How many payments of $1000 will it take to pay off the loan?
    (d) How much interest will they save by paying the loan in this way?

Use a spreadsheet or financial program on a calculator or computer to complete Problems 37 and 38.

37. Develop an amortization schedule for a 4-year car loan if $16,700 is borrowed at 8.2%, compounded monthly.
38. Develop an amortization schedule for a 10-year mortgage loan of $80,000 at 7.2%, compounded monthly.

Key Terms and Formulas

Section	Key Terms	Formulas
6.1	Simple interest	$I = Prt$
	Future value of a simple interest investment or loan	$S = P + I$
	Sequence function	
	Arithmetic sequence	$a_n = a_{n-1} + d \quad (n > 1)$
	Common difference	d
	nth term of	$a_n = a_1 + (n - 1)d$
	Sum of first n terms	$s_n = \dfrac{n}{2}(a_1 + a_n)$
6.2	Future value of compound interest investment	
	Periodic compounding	$S = P(1 + i)^n = P\left(1 + \dfrac{r}{m}\right)^{mt}$
	Continuous compounding	$S = Pe^{rt}$
	Annual percentage yield	
	Periodic compounding	$\text{APY} = \left(1 + \dfrac{r}{m}\right)^m - 1$ $= (1 + i)^m - 1$
	Continuous compounding	$\text{APY} = e^r - 1$
	Geometric sequence	$a_n = ra_{n-1} \quad (n > 1)$
	Common ratio	r
	nth term of	$a_n = a_1 r^{n-1}$
	Sum of first n terms	$s_n = \dfrac{a_1(1 - r^n)}{1 - r} \quad (\text{if } r \neq 1)$
6.3	Future values of annuities	
	Ordinary annuity	$S = R\left[\dfrac{(1 + i)^n - 1}{i}\right]$
	Required payment into an ordinary annuity (sinking fund)	
	Annuity due	$S_{due} = R\left[\dfrac{(1 + i)^n - 1}{i}\right](1 + i)$

Section	Key Terms	Formulas
6.4	Present values of annuities	
	Ordinary annuity	$A_n = R\left[\dfrac{1 - (1 + i)^{-n}}{i}\right]$
	Annuity due	$A_{(n,due)} = R\left[\dfrac{1 - (1 + i)^{-n}}{i}\right](1 + i)$
	Deferred annuity	$A_{(n,k)} = R\left[\dfrac{1 - (1 + i)^{-n}}{i}\right](1 + i)^{-k}$
6.5	True annual percentage rate (APR)	
	Amortization (same as present value of an ordinary annuity)	$R = A_n\left[\dfrac{i}{1 - (1 + i)^{-n}}\right]$
	Unpaid balance ($n - k$ payments left) Amortization schedule	$A_{n-k} = R\left[\dfrac{1 - (1 + i)^{-(n-k)}}{i}\right]$

Review Exercises

For miscellaneous financial problems, see Problems 41–56.

Section 6.1

1. Find the first 4 terms of the sequence with nth term

$$a_n = \frac{1}{n^2}$$

2. Identify any arithmetic sequences and find the common differences.
   (a) $12, 7, 2, -3, \ldots$ (b) $1, 3, 6, 10, \ldots$
   (c) $\frac{1}{6}, \frac{1}{3}, \frac{1}{2}, \frac{2}{3}, \ldots$
3. Find the 80th term of the arithmetic sequence with first term -2 and common difference 3.
4. Find the 36th term of the arithmetic sequence with third term 10 and eighth term 25.
5. Find the sum of the first 60 terms of the arithmetic sequence $\frac{1}{3}, \frac{1}{2}, \frac{2}{3}, \ldots$

Section 6.2

6. Identify any geometric sequences and their common ratios.
   (a) $\frac{1}{4}, 2, 16, 128, \ldots$ (b) $16, -12, 9, -\frac{27}{4}, \ldots$
   (c) $4, 16, 36, 64, \ldots$
7. Find the fourth term of the geometric sequence with first term 64 and eighth term $\frac{1}{2}$.

8. Find the sum of the first 16 terms of the geometric sequence $\frac{1}{9}, \frac{1}{3}, 1, \ldots$.

APPLICATIONS

Section 6.1

9. *Simple interest* If $8000 is borrowed at 12% simple interest for 3 years, what is the future value of the loan at the end of the 3 years?

10. *Simple interest* Mary Toy borrowed $2000 from her parents and repaid them $2100 after 9 months. What simple interest rate did she pay?

11. *Simple interest* How much summer earnings must a college student deposit on August 31 in order to have $3000 for tuition and fees on December 31 of the same year, if the investment earns 6% simple interest?

12. *Contest payments* Suppose the winner of a contest receives $10 on the first day of the month, $20 on the second day, $30 on the third day, and so on for a 30-day month. What is the total won?

13. *Salaries* Suppose you are offered two identical jobs: one paying a starting salary of $20,000 with yearly raises of $1000 and one paying a starting salary of $18,000 with yearly raises of $1200. Which job will pay more money over a 10-year period?

Section 6.2

14. *Investments*   An investment is made at 8%, compounded quarterly, for 10 years.
    (a) Find the number of periods.
    (b) Find the interest rate per period.

15. *Future value*   Write the formula for the future value of a compound interest investment, if interest is compounded
    (a) periodically.   (b) continuously.

16. *Compound interest*   Which compounding method, at 6%, earns more money?
    (a) semiannually   (b) monthly

17. *Interest*   If $1000 is invested for 4 years at 8%, compounded quarterly, how much interest will be earned?

18. *Present value*   How much must one invest now in order to have $18,000 in 4 years if the investment earns 5.4%, compounded monthly?

19. *Future value*   What is the future value if $1000 is invested for 6 years at 8%, compounded continuously?

20. *Present value*   A couple received an inheritance and plan to invest some of it for their grandchild's college education. How much must they invest if they would like the fund to have $100,000 after 15 years, and their investment earns 10.31%, compounded continuously?

21. *Investments*   If $15,000 is invested at 6%, compounded quarterly, how long will it be before it grows to $25,000?

22. *Continuous compounding*   (a) If an initial investment of $35,000 grows to $257,000 in 15 years, what annual interest rate, continuously compounded, was earned? (b) What is the annual percentage yield on this investment?

23. *Annual percentage yield*   Find the annual percentage yield equivalent to a nominal rate of 7.2% (a) compounded quarterly and (b) compounded continuously.

24. *Chess legend*   Legend has it that when the king of Persia wanted to reward the inventor of chess with whatever he desired, the inventor asked for one grain of wheat on the first square of the chessboard, with the number of grains doubled on each square thereafter for the remaining 63 squares. Find the number of grains on the 64th square. (The sum of all the grains of wheat on all the squares would cover Alaska more than 3 inches deep with wheat!)

25. *Chess legend*   If, in Problem 24, the king granted the inventor's wish for *half* a chessboard, how many grains would the inventor receive?

Section 6.3

26. *Ordinary annuity*   Find the future value of an ordinary annuity of $800 paid at the end of every 6-month period for 10 years, if it earns interest at 12%, compounded semiannually.

27. *Sinking fund*   How much would have to be invested at the end of each year at 6%, compounded annually, to pay off a debt of $80,000 in 10 years?

28. *Annuity due*   Find the future value of an annuity due of $800 paid at the beginning of every 6-month period for 10 years, if it earns interest at 12%, compounded semiannually.

29. *Payment into an annuity due*   A company wants to have $250,000 available in $4\frac{1}{2}$ years for new construction. How much must be deposited at the beginning of each quarter to reach this goal if the investment earns 10.2%, compounded quarterly?

30. *Time to reach a goal*   If $1200 is deposited at the end of each quarter in an account that earns 7.2%, compounded quarterly, how long will it be until the account is worth $60,000?

Section 6.4

31. *Present value of an ordinary annuity*   What lump sum would have to be invested at 9%, compounded semiannually, to provide an annuity of $10,000 at the end of each half-year for 10 years?

32. *Deferred annuity*   A couple wish to set up an annuity that will provide 6 monthly payments of $3000 while they take an extended cruise. How much of a $15,000 inheritance must be set aside if they plan to leave in 5 years and want the first payment before they leave? Assume that money is worth 7.8%, compounded monthly.

33. *Powerball lottery*   Winners of lotteries receive the jackpot distributed over a period of years, usually 20 or 25 years. The winners of the Powerball lottery on July 29, 1998, elected to take a one-time cash payout rather than receive the $295.7 million jackpot in 25 annual payments beginning on the date the lottery was won.
    (a) How much money would the winners have received at the beginning of each of the 25 years?
    (b) If the value of money was 5.91%, compounded annually, what one-time payout did they receive in lieu of the annual payments?

34. *Payment from an ordinary annuity*   A recent college graduate's gift from her grandparents is $20,000. How much will this provide at the end of each month for the

next 12 months while the graduate travels? Assume that money is worth 6.6%, compounded monthly.

35. *Payment from an annuity due*   An IRA of $250,000 is rolled into an annuity that pays a retired couple at the beginning of each quarter for the next 20 years. If the annuity earns 6.2%, compounded quarterly, how much will the couple receive each quarter?

36. *Duration of an annuity*   A retirement account that earns 6.8%, compounded semiannually, contains $488,000. How long can $40,000 be withdrawn at the end of each half-year until the account balance is $0?

Section 6.5

37. *Amortization*   A debt of $1000 with interest at 12%, compounded monthly, is amortized by 12 monthly payments (of equal size). What is the size of each payment?

38. *Unpaid balance*   A debt of $8000 is amortized with eight semiannual payments of $1288.29 each. If money is worth 12%, compounded semiannually, find the unpaid balance after five payments have been made.

39. *Cash value*   A woman paid $90,000 down for a house and agreed to pay 18 quarterly payments of $1500 each. If money is worth 4%, compounded quarterly, how much would the house have cost if she had paid cash?

40. *Amortization schedule*   Complete the next two lines of the amortization schedule for a $100,000 loan for 30 years at 7.5%, compounded monthly, with monthly payments of $699.22.

Payment Number	Payment Amount	Interest	Balance Reduction	Unpaid Balance
56	$699.22	$594.67	$104.55	$95,042.20

MISCELLANEOUS FINANCIAL PROBLEMS

41. If an initial investment of $2500 grows to $38,000 in 18 years, what annual interest rate, compounded annually, did this investment earn?

42. How much must be deposited at the end of each month in an account that earns 8.4%, compounded monthly, if the goal is to have $40,000 after 10 years?

43. What is the future value if $1000 is invested for 6 years at 8% (a) simple interest and (b) compounded semiannually?

44. Quarterly payments of $500 are deposited in an account that pays 8%, compounded quarterly. How much will have accrued in the account at the end of 4 years if each payment is made at the end of each quarter?

45. If $8000 is invested at 7%, compounded continuously, how long will it be before it grows to $22,000?

46. An investment broker bought some stock at $87.89 per share and sold it after 3 months for $105.34 per share. What was the annual simple interest rate earned on this transaction?

47. Kevin Patrick paid off the loan he took out to buy his car, but once the loan was paid, he continued to deposit $400 on the first of each month in an account that paid 5.4%, compounded monthly. After four years of making these deposits, Kevin was ready to buy a new car. How much did he have in the account?

48. A young couple receive an inheritance of $72,000 that they want to set aside for a college fund for their two children. How much will this provide at the end of each half-year for a period of 9 years if it is deferred for 11 years and can be invested at 7.3%, compounded semiannually?

49. A bank is trying to decide whether to advertise some new 18-month certificates of deposit (CDs) at 6.52%, compounded quarterly, or at 6.48%, compounded continuously. Which rate is a better investment for the consumer who buys such a CD? Which rate is better for the bank?

50. A divorce settlement of $40,000 is paid in $1000 payments at the end of each of 40 months. What is the present value of this settlement if money is worth 12%, compounded monthly?

51. If $8000 is invested at 12%, compounded continuously, for 3 years, what is the total interest earned at the end of the 3 years?

52. A couple borrowed $92,000 to buy a condominium. Their loan was for 25 years and money is worth 6%, compounded monthly.
   (a) Find their monthly payment size.
   (b) Find the total amount they would pay over 25 years.
   (c) Find the total interest they would pay over 25 years.
   (d) Find the unpaid balance after 7 years.

53. Suppose a salesman invests his $12,500 bonus in a fund that earns 10.8%, compounded monthly. Suppose also that he makes contributions of $150 at the end of each month to this fund.
   (a) Find the future value after $12\frac{1}{2}$ years.
   (b) If after the $12\frac{1}{2}$ years, the fund is used to set up an annuity, how much will it pay at the end of each month for the next 10 years?

54. Three years from now, a couple plan to spend 4 months traveling in China, Japan, and Southeast Asia. When they take their trip, they would like to withdraw $5000 at the beginning of each month to cover their expenses for that month. Starting now, how much must they deposit at the beginning of each month for the next 3 years so that the account will provide the money they want while they are traveling? Assume that such an account pays 6.6%, compounded monthly.

55. At age 22, Aruam Sdlonyer receives a $2000 IRA from her parents. At age 30, she decides that at age 67 she'd like to have a retirement fund that would pay $10,000 at the end of each month for 20 years. Suppose all investments earn 8.4%, compounded monthly. How much does Aruam need to deposit at the end of each month from ages 30 to 67 to realize her goal?

56. A company borrows $2.6 million for 15 years at 5.6%, compounded quarterly. After 2 years of regular payments, the company's profits are such that management feels they can increase the quarterly payments to $70,000 each. How long will it take to pay off the loan, and how much interest will be saved?

6 Chapter Test

1. If $8000 is invested at 7%, compounded continuously, how long will it be before it grows to $47,000?

2. How much must you put aside at the end of each half-year if, in 6 years, you want to have $12,000, and the current interest rate is 6.2%, compounded semiannually?

3. To buy a bond that matures to $5000 in 9 months, you must pay $4756.69. What simple interest rate is earned?

4. A debt of $280,000 is amortized with 40 equal semi-annual payments of $14,357.78. If interest is 8.2%, compounded semiannually, find the unpaid balance of the debt after 25 payments have been made.

5. If an investment of $1000 grew to $13,500 in 9 years, what interest rate, compounded annually, did this investment earn?

6. A couple borrowed $97,000 at 7.2%, compounded monthly, for 25 years to purchase a condominium.
   (a) Find their monthly payment.
   (b) Over the 25 years, how much interest will they pay?

7. If you borrow $2500 for 15 months at 4% simple interest, how much money must you repay after the 15 months?

8. Suppose you invest $100 at the end of each month in an account that earns 6.9%, compounded monthly. What is the future value of the account after $5\frac{1}{2}$ years?

9. Find the annual percentage yield (the effective annual rate) for an investment that earns 8.4%, compounded monthly.

10. An accountant wants to withdraw $3000 from an investment at the beginning of each quarter for the next 15 years. How much must be deposited originally if the investment earns 6%, compounded quarterly? Assume that after the 15 years, the balance is zero.

11. If $10,000 is invested at 7%, compounded continuously, what is the value of the investment after 20 years?

12. A collector made a $10,000 down payment for a classic car and agreed to make 18 quarterly payments of $1500 at the end of each quarter. If money is worth 8%, compounded quarterly, how much would the car have cost if the collector had paid cash?

13. What amount must you invest in an account that earns 6.8%, compounded quarterly, if you want to have $9500 after 5 years?

14. If $1500 is deposited at the end of each half-year in a retirement account that earns 8.2%, compounded semi-annually, how long will it be before the account contains $500,000?

15. A couple would like to have $50,000 at the end of $4\frac{1}{2}$ years. They can invest $10,000 now and make additional contributions at the end of each quarter over the $4\frac{1}{2}$ years. If the investment earns 7.7%, compounded quarterly, what size payments will enable them to reach their goal?

16. A woman makes $3000 contributions at the end of each half-year to a retirement account for a period of 8 years. For the next 20 years, she makes no additional contributions and no withdrawals.
   (a) If the account earns 7.5%, compounded semiannually, find the value of the account after the 28 years.
   (b) If this account is used to set up an annuity that pays her an amount at the beginning of each 6-month period for the next 20 years, how much will each payment be?

17. Maxine deposited $80 at the beginning of each month for 15 years in an account that earned 6%, compounded monthly. Find the value of the account after the 15 years.

18. Grandparents plan to establish a college trust for their youngest grandchild. How much is needed now so the trust, which earns 6.4%, compounded quarterly, will provide $4000 at the end of each quarter for 4 years after being deferred for 10 years?

19. The following sequence is arithmetic: 298.8, 293.3, 287.8, 282.3, ...
   (a) If you were not told that it was arithmetic, explain how you could tell that it was.
   (b) Find the 51st term of this sequence.
   (c) Find the sum of the first 51 terms.

20. A 400-milligram dose of heart medicine is taken daily. During each 24-hour period, the body eliminates 40% of this drug (so that 60% remains in the body). Thus

the amount of drug in the bloodstream just after the 31st dose is given by

$$400 + 400(0.6) + 400(0.6)^2 + 400(0.6)^3 + \cdots$$
$$+ 400(0.6)^{29} + 400(0.6)^{30}$$

Find the level of the drug in the bloodstream at this time.

21. If you took a home mortgage loan of $150,000 for 30 years at 8.4%, compounded monthly, your monthly payments would be $1142.76. Suppose after 2 years you had an additional $2000. Would you save more over the life of the loan by paying an extra $2000 with your 24th payment or by paying $1160 per month beginning with the 25th payment? To help you answer this question, complete the following.

(a) Find the unpaid balance after 24 payments (both including and not including the $2000 payment).

(b) Determine how long it takes to pay off the loan with the regular $1142.76 payments and the additional $2000 payment. Then find the total interest paid.

(c) Determine how long it takes to pay off the loan without the additional $2000 payment, but with monthly payments of $1160 from the 25th payment. Then find the total interest paid.

(d) Which payment method costs less?

Extended Applications & Group Projects

I. Mail Solicitation Home Equity Loan: Is This a Good Deal?

Discover® offers a home equity line of credit that lowers the total of the monthly payments on several hypothetical loans by $366.12. (See table below.)

The home equity loan payment is based on "10.49% APR annualized over a 10-year term." This means the loan is amortized at 10.49% with monthly payments.

(a) Using this interest rate and monthly payments for the 10 years, how much money will still be owed at the end of the 10 years?
(b) How much is paid during the 10 years?
(c) How much would the payment have to be to have the debt paid in the 10 years?
(d) How long would it take to pay off the debts under the original plan if interest were 15% on each loan?

| | Current Monthly Payment Example | | Discover Home Equity | |
| | | Monthly | | Monthly |
Loan Type	Loan Amount	Payment	Loan Amount	Payment
Bank cards	$10,000	$186.00	PAID OFF	NONE
Auto loan	$12,500	$320.03	PAID OFF	NONE
Department store cards	$2,500	$78.63	PAID OFF	NONE
Discover home equity line of credit			$25,000	$218.54
Total	$25,000	$584.66	$25,000	$218.54

Source: Discover Loan Center mailer

II. Profit Reinvestment

T. C. Hardware Store wants to construct a new building at a second location. If construction could begin immediately, the cost would be $250,000. At this time, however, the owners do not have the required 20% down payment, so they plan to invest $1000 per month of their profits until they have the necessary amount. They can invest their money in an annuity account that pays 6%, compounded monthly, but they are concerned about the 3% average inflation rate in the construction industry. They would like you to give them some projections about how 3% inflation will affect the time required to accrue the down payment and the building's eventual cost. They would also like to know how their projected profits, after the new building is complete, will affect their schedule for paying off the mortgage loan.

Specifically, the owners would like you to prepare a report that answers the following questions.

1. If the 3% annual inflation rate is accurate, how long will it take to get the down payment? *Hint:* Assume inflation is compounded monthly. Then the required down payment will be 20% of

$$250{,}000\left(1 + \frac{0.03}{12}\right)^n$$

where n is in months. Thus you must find n such that this amount equals the future value of the owners' annuity. The solution to the resulting equation is difficult to obtain by algebraic methods but can be found with a graphing utility.

2. If the 3% annual inflation rate is accurate (compounded monthly), what will T. C. Hardware's projected construction costs be (to the nearest hundred dollars) when it has its down payment?

3. If T. C. Hardware borrows 80% of its construction costs and amortizes that amount at 7.8%, compounded monthly, for 15 years, what will its monthly payment be?

4. Finally, the owners believe that 2 years after this new building is begun (that is, after 2 years of payments on the loan), their profits will increase enough so that they'll be able to make double mortgage payments each month until the loan is paid. Under these assumptions, how long will it take T. C. Hardware to pay off the loan?

III. Purchasing a Home

Buying a home is the biggest single investment or purchase that most individuals make. This project is designed to give you some insight into the home-buying process and the associated costs.

Find all the costs associated with buying a home by making a 20% down payment and borrowing 80% of the cost of the home.

1. Decide how much your home will cost (and how much you'll have to put down and how much to mortgage). Call a realtor to determine the cost of a home in your area (this could be a typical starter home, an average home for the area, or your dream house). Ask the realtor to identify and estimate all the closing costs associated with buying your home. Closing costs include your down payment, attorney's fees, title fees, and so on.

2. Call at least three different banks to determine mortgage loan rates for 15 years, 25 years, and 30 years and the costs associated with obtaining your loan. These associated costs of a loan are paid at closing and include an application fee, an appraisal fee, points, and so on.

3. For each bank and each different loan duration, develop a summary that contains
   (a) an itemization and explanation of all closing costs and the total amount due at closing
   (b) the monthly payment
   (c) the total amount paid over the life of the loan
   (d) the amortization schedule for each loan (use a spreadsheet or financial software package)

4. For each duration (15 years, 25 years, and 30 years), identify which bank gives the best loan rate and explain why you think it is best. Be sure to consider what is best for someone who is likely to remain in the home for 7 years or less compared with someone who is likely to stay 20 years or more.

7

Introduction to Probability

An economist cannot predict exactly how the gross national product will change, a physician cannot determine exactly the cause of lung cancer, and a psychologist cannot determine the exact effect of environment on behavior. But each of these determinations can be made with varying probabilities. Thus, an understanding of the meaning and determination of the probabilities of events occurring is important to success in business, economics, the life sciences, and the social sciences. Probability was initially developed to solve gambling problems, but it is now the basis for solving problems in a wide variety of areas.

The topics and applications discussed in this chapter include the following.

Chapter Warm-up

Prerequisite Problem Type	For Section	Answer	Section for Review
(a) The sum of $x, \frac{1}{4}$, and $\frac{2}{3}$ is 1. What is x? (b) If the sum of $x, \frac{1}{3}$, and $\frac{1}{3}$ is 1, what is x?	7.1	(a) $x = \frac{1}{12}$ (b) $x = \frac{1}{3}$	1.1 Linear equations
If 12 numbers are weighted so that each of the 12 numbers has an equal weight and the sum of the weights is 1, what weight should be assigned to each number?	7.1	$\frac{1}{12}$	1.1 Linear equations
(a) List the set of integers that satisfies $2 \leq x < 6$. (b) List the set of integers that satisfies $4 \leq s \leq 12$.	7.1	(a) {2, 3, 4, 5} (b) {4, 5, 6, 7, 8, 9, 10, 11, 12}	1.1 Linear inequalities
For the sets $E = \{1, 2, 3, 4, 5\}$ and $F = \{2, 4, 6, 8\}$, find (a) $E \cup F$ (b) $E \cap F$	7.1 7.2	 (a) {1, 2, 3, 4, 5, 6, 8} (b) {2, 4}	0.1 Sets
Find the product: $[0.9 \quad 0.1] \begin{bmatrix} 0.8 & 0.2 \\ 0.3 & 0.7 \end{bmatrix}$	7.7	$[0.75 \quad 0.25]$	3.2 Matrix multiplication
Solve the system using matrices: $\begin{cases} 0.5V_1 + 0.4V_2 + 0.3V_3 = V_1 \\ 0.4V_1 + 0.5V_2 + 0.3V_3 = V_2 \\ 0.1V_1 + 0.1V_2 + 0.4V_3 = V_3 \end{cases}$	7.7	$V_1 = 3V_3$ $V_2 = 3V_3$ $V_3 = $ any real number	3.3 Gauss-Jordan elimination

7.1

Probability; Odds

OBJECTIVES

- To compute the probability of the occurrence of an event
- To construct a sample space for a probability experiment
- To compute the odds that an event will occur
- To compute the empirical probability that an event will occur

◗ Application Preview

Before the General Standard Company can place a 3-year warranty on all of the faucets it produces, it needs to find the **probability** that each faucet it produces will function properly (not leak) for 3 years. (See Example 8.) In this section we will use **sample spaces** to find the probability that an event will occur by using assumptions or knowledge of the entire **population** of data under consideration. When knowledge of the entire population is not available, we can compute the **empirical probability** that an event will occur by using data collected about the probability experiment. This method can be used to compute the probability that the General Standard faucets will work properly if information about a large **sample** of the faucets is gathered.

Sample Spaces and Probability

When we toss a coin, it can land in one of two ways, heads or tails. If the coin is a "fair" coin, these two possible **outcomes** have an equal chance of occurring, and we say the outcomes of this **probability experiment** are **equally likely.** If we seek the probability that an experiment has a certain result, that result is called an **event.**

Suppose an experiment can have a total of *n* equally likely outcomes, and that *k* of these outcomes would be considered successes. Then the probability of achieving a success in the experiment is k/n. That is, the probability of a success in an experiment is the number of ways the experiment can result in a success divided by the total number of possible outcomes.

Probability of a Single Event

If an event E can happen in k ways out of a total of n equally likely possibilities, the probability of the occurrence of the event is denoted by

$$\Pr(E) = \frac{\text{Number of successes}}{\text{Number of possible outcomes}} = \frac{k}{n}$$

● EXAMPLE 1 *Probability*

If we draw a ball from a bag containing 4 white balls and 6 black balls, what is the probability of

(a) getting a white ball?   (b) getting a black ball?   (c) not getting a white ball?

Solution

(a) A white ball can occur (be drawn) in 4 ways out of a total of 10 equally likely possibilities. Thus the probability of drawing a white ball is

$$\Pr(W) = \frac{4}{10} = \frac{2}{5}$$

(b) We have 6 chances to succeed at drawing a black ball out of a total of 10 possible outcomes. Thus the probability of getting a black ball is

$$\Pr(B) = \frac{6}{10} = \frac{3}{5}$$

(c) The probability of not getting a white ball is $\Pr(B)$, because not getting a white ball is the same as getting a black ball. Thus $\Pr(\text{not } W) = \Pr(B) = 3/5$.

Suppose a card is selected at random* from a box containing cards numbered 1 through 12. To find the probability that the card drawn has an even number on it, it is necessary to list (at least mentally) all the possible outcomes of the experiment (selecting a number at random). A set that contains all the possible outcomes of an experiment is called a **sample space.** For the experiment of drawing a numbered card, a sample space is

$$S = \{1, 2, 3, 4, 5, 6, 7, 8, 9, 10, 11, 12\}$$

Each element of the sample space is called a **sample point,** and an **event** is a subset of the sample space. Each sample point in the sample space is assigned a nonnegative **probability measure** or **probability weight** such that the sum of the weights in the sample space is 1. The probability of an event is the sum of the weights of the sample points in the event's **subspace** (subset of S).

For example, the experiment "drawing a number at random from the numbers 1 through 12" has the sample space S given above. Because each element of S is equally likely to occur, we assign a probability weight of $1/12$ to each element. The event "drawing an even number" in this experiment has the subspace

$$E = \{2, 4, 6, 8, 10, 12\}$$

Because each of the 6 elements in the subspace has weight $1/12$, the probability of event E is

$$\text{Pr(even)} = \text{Pr}(E) = \frac{1}{12} + \frac{1}{12} + \frac{1}{12} + \frac{1}{12} + \frac{1}{12} + \frac{1}{12} = \frac{6}{12} = \frac{1}{2}$$

Note that using the sample space S for this experiment gives the same result as using the formula $\text{Pr}(E) = k/n$.

Some experiments have an infinite number of outcomes, but we will concern ourselves only with experiments having finite sample spaces. We can usually construct more than one sample space for an experiment. The sample space in which each sample point is equally likely is called an **equiprobable sample space.** Suppose the number of elements in a sample space is $n(S) = n$, with each element representing an equally likely outcome of a probability experiment. Then the weight (probability) assigned to each element is

$$\frac{1}{n(S)} = \frac{1}{n}$$

If an event E that is a subset of an equiprobable sample space contains $n(E) = k$ elements of S, then we can restate the probability of an event E in terms of the number of elements of E.

Probability of an Event

If an event E can occur in $n(E) = k$ ways out of $n(S) = n$ equally likely ways, then

$$\text{Pr}(E) = \frac{n(E)}{n(S)} = \frac{k}{n}$$

● **EXAMPLE 2** *Probability of an Event*

If a number is to be selected at random from the integers 1 through 12, what is the probability that it is

(a) divisible by 3?   (b) even and divisible by 3?   (c) even or divisible by 3?

*Selecting a card at random means every card has an equal chance of being selected.

Solution

The set $S = \{1, 2, 3, 4, 5, 6, 7, 8, 9, 10, 11, 12\}$ is an equiprobable sample space for this experiment (as mentioned previously).

(a) The set $E = \{3, 6, 9, 12\}$ contains the numbers that are divisible by 3. Thus

$$\text{Pr(divisible by 3)} = \frac{n(E)}{n(S)} = \frac{4}{12} = \frac{1}{3}$$

(b) The numbers 1 through 12 that are even *and* divisible by 3 are 6 and 12, so

$$\text{Pr(even and divisible by 3)} = \frac{2}{12} = \frac{1}{6}$$

(c) The numbers that are even *or* divisible by 3 are 2, 3, 4, 6, 8, 9, 10, 12, so

$$\text{Pr(even or divisible by 3)} = \frac{8}{12} = \frac{2}{3}$$

● *Checkpoint*

1. If a coin is drawn from a box containing 4 gold, 10 silver, and 16 copper coins, all the same size, what is the probability of
   (a) getting a gold coin?
   (b) getting a copper coin?
   (c) not getting a copper coin?

If an event E is certain to occur, E contains all of the elements of the sample space, S. Hence the sum of the probability weights of E is the same as that of S, so

$$\text{Pr}(E) = 1 \qquad \text{if } E \text{ is certain to occur}$$

If event E is impossible, $E = \varnothing$, so

$$\text{Pr}(E) = 0 \qquad \text{if } E \text{ is impossible}$$

Because all probability weights of elements of a sample space are nonnegative,

$$0 \leq \text{Pr}(E) \leq 1 \qquad \text{for any event } E$$

● **EXAMPLE 3 *Sample Spaces***

Suppose a coin is tossed 3 times.

(a) Construct an equiprobable sample space for the experiment.
(b) Find the probability of obtaining 0 heads.
(c) Find the probability of obtaining 2 heads.

Solution

(a) Perhaps the most obvious way to record the possibilities for this experiment is to list the number of heads that could result: $\{0, 1, 2, 3\}$. But the probability of obtaining 0 heads is different from the probability of obtaining 2 heads, so this sample space is not equiprobable. A sample space in which each outcome is equally likely is $\{$HHH, HHT, HTH, THH, HTT, THT, TTH, TTT$\}$, where HHT indicates that the first two tosses were heads and the third was a tail.

(b) Because there are 8 equally likely possible outcomes, $n(S) = 8$. Only one of the eight possible outcomes, $E = \{TTT\}$, gives 0 heads, so $n(E) = 1$. Thus

$$\Pr(0 \text{ heads}) = \frac{n(E)}{n(S)} = \frac{1}{8}$$

(c) The event "two heads" is $F = \{HHT, HTH, THH\}$, so

$$\Pr(2 \text{ heads}) = \frac{n(F)}{n(S)} = \frac{3}{8}$$

To find the probability of obtaining a given sum when a pair of dice is rolled, we need to determine the outcomes. However, the sums $\{2, 3, 4, 5, 6, 7, 8, 9, 10, 11, 12\}$ are not equally likely; there is only one way to obtain a sum of 2, but several ways to obtain a sum of 6. If we distinguish between the two dice we are rolling, and we record all the possible outcomes for each die, we see that there are 36 possibilities, each of which is equally likely (see Table 7.1).

TABLE 7.1

First Die	Second Die					
	1	2	3	4	5	6
1	(1, 1)	(1, 2)	(1, 3)	(1, 4)	(1, 5)	(1, 6)
2	(2, 1)	(2, 2)	(2, 3)	(2, 4)	(2, 5)	(2, 6)
3	(3, 1)	(3, 2)	(3, 3)	(3, 4)	(3, 5)	(3, 6)
4	(4, 1)	(4, 2)	(4, 3)	(4, 4)	(4, 5)	(4, 6)
5	(5, 1)	(5, 2)	(5, 3)	(5, 4)	(5, 5)	(5, 6)
6	(6, 1)	(6, 2)	(6, 3)	(6, 4)	(6, 5)	(6, 6)

This list of possible outcomes for finding the sum of two dice is an equiprobable sample space for the experiment. Because the 36 elements in the sample space are equally likely, we can find the probability that a given sum results by determining the number of ways this sum can occur and dividing that number by 36. Thus the probability that sum 6 will occur is 5/36, and the probability that sum 9 will occur is $4/36 = 1/9$. (See the four ways the sum 9 can occur in Table 7.1.) We can see that the probabilities of events can be found more easily if an equiprobable sample space is used to determine the possibilities.

● **EXAMPLE 4** *Rolling Dice*

Use Table 7.1 to find the following probabilities if a pair of distinguishable dice is rolled.

(a) Pr(sum is 5)   (b) Pr(sum is 2)   (c) Pr(sum is 8)

Solution

(a) If E is the event "sum of 5," then $E = \{(4, 1), (3, 2), (2, 3), (1, 4)\}$.

Therefore, $\Pr(\text{sum is } 5) = \dfrac{n(E)}{n(S)} = \dfrac{4}{36} = \dfrac{1}{9}$.

(b) The sum 2 results only from $\{(1, 1)\}$. Thus $\Pr(\text{sum is } 2) = 1/36$.

(c) The sample points that give a sum of 8 are in the subspace

$$F = \{(6, 2), (5, 3), (4, 4), (3, 5), (2, 6)\}$$

Thus $\Pr(\text{sum is } 8) = \dfrac{n(F)}{n(S)} = \dfrac{5}{36}$.

2. A ball is to be drawn from a bag containing balls numbered 1, 2, 3, 4, and 5. To find the probability that a ball with an even number is drawn, we can use the sample space $S_1 = \{\text{even, odd}\}$ or $S_2 = \{1, 2, 3, 4, 5\}$.
   (a) Which of these sample spaces is an equiprobable sample space?
   (b) What is the probability of drawing a ball that is even-numbered?
3. A coin is tossed and a die is rolled.
   (a) Write an equiprobable sample space listing the possible outcomes for this experiment.
   (b) Find the probability of getting a head on the coin and an even number on the die.
   (c) Find the probability of getting a tail on the coin and a number divisible by 3 on the die.

Odds and Probability

We sometimes use **odds** to describe the probability that an event will occur. At the middle of the 2004 football season, vegas.com stated that the probability of the New England Patriots winning the 2005 Super Bowl was 2/7. This means that the ratio of the chance of the team winning to its chance of losing was 2 to 5. We say that its odds were 2 to 5, or 2:5. The odds in favor of an event E occurring and the odds against E occurring are found as follows.

Odds

If the probability that event E occurs is $\Pr(E) \neq 1$, then the odds that E will occur are

$$\text{Odds in favor of } E = \frac{\Pr(E)}{1 - \Pr(E)}$$

The odds that E will not occur, if $\Pr(E) \neq 0$, are

$$\text{Odds against } E = \frac{1 - \Pr(E)}{\Pr(E)}$$

The odds $\dfrac{a}{b}$ are stated as "a to b" or "$a{:}b$."

● **EXAMPLE 5** *Odds*

If the probability of drawing a queen from a deck of playing cards is 1/13, what are the odds

(a) in favor of drawing a queen?
(b) against drawing a queen?

Solution

(a) $\dfrac{1/13}{1 - 1/13} = \dfrac{1/13}{12/13} = \dfrac{1}{12}$

The odds in favor of drawing a queen are 1 to 12, which we write as 1:12.

(b) $\dfrac{12/13}{1/13} = \dfrac{12}{1}$

The odds against drawing a queen are 12 to 1, denoted 12:1.

Odds and Probability

If the odds in favor of an event E occurring are $a{:}b$, the probability that E will occur is

$$\Pr(E) = \frac{a}{a + b}$$

If the odds against E are $b{:}a$, then

$$\Pr(E) = \frac{a}{a + b}$$

● **EXAMPLE 6** *Industrial Safety*

If the odds against dying in an industrial accident are 3992:3, what is the probability of dying in an industrial accident?

Solution
The probability is

$$\frac{3}{3 + 3992} = \frac{3}{3995}$$

Empirical Probability

Up to this point, the probabilities assigned to events were determined *theoretically,* either by assumption (a fair coin has probability 1/2 of obtaining a head in one toss) or by knowing the entire population under consideration (if 1000 people are in a room and 400 of them are males, the probability of picking a person at random and getting a male is $400/1000 = 2/5$). We can also determine probabilities **empirically,** where the assignment of the probability of an event is frequently derived from our experiences or from data that have been gathered. Probabilities developed in this way are called **empirical probabilities.** An empirical probability is formed by conducting an experiment, or by observing a situation a number of times, and noting how many times a certain event occurs.

Empirical Probability of an Event

$$\Pr(\text{Event}) = \frac{\text{Number of observed occurrences of the event}}{\text{Total number of trials or observations}}$$

In some cases empirical probability is the only way to assign a probability to an event. For example, life insurance premiums are based on the probability that an individual will live to a certain age. These probabilities are developed empirically and organized into mortality tables.

In other cases, empirical probabilities may have a theoretical framework. For example, for tossing a coin,

$$\Pr(\text{Head}) = \Pr(\text{Tail}) = \frac{1}{2}$$

gives the theoretical probability. If a coin was actually tossed 1000 times, it might come up heads 517 times and tails 483 times, giving empirical probabilities for these 1000 tosses of

$$\Pr(\text{Head}) = \frac{517}{1000} = 0.517$$

$$\Pr(\text{Tail}) = \frac{483}{1000} = 0.483$$

On the other hand, if the coin was tossed only 4 times, it might come up heads 3 times and tails once, giving empirical probabilities vastly different from the theoretical ones. In general, empirical probabilities can differ from theoretical probabilities, but they are likely to reflect the theoretical probabilities when the number of observations is large. Moreover, when this is not the case, there may be reason to suspect that the theoretical model does not fit the event (for example, that a coin may be biased).

● EXAMPLE 7 *Quality Control*

Experience has shown that 800 of the 500,000 microcomputer diskettes manufactured by a firm are defective. On the basis of this evidence, determine the probability that a diskette picked at random from those manufactured by this firm will be defective.

Solution

The 800 defective diskettes occurred in a very large number of diskettes manufactured, so the ratio of the number of defective diskettes to the total number of diskettes manufactured provides the *empirical* probability of a diskette picked at random being defective.

$$\text{Pr(defective)} = \frac{800}{500,000} = \frac{1}{625}$$

◐ EXAMPLE 8 *Quality Control* **(Application Preview)**

The General Standard Company can use information about a large sample of its faucets to compute the empirical probability that a faucet chosen at random will function properly (not leak) for 3 years. Of a sample of 12,316 faucets produced in 2003, 137 were found to fail within 3 years. Find the empirical probability that any faucet the company produces will work properly for 3 years.

Solution

If 137 of these faucets fail to work properly, then $12,316 - 137 = 12,179$ will work properly. Thus the probability that any faucet selected at random will function properly is

$$\text{Pr(not leak)} = \frac{12,179}{12,316} \approx 0.989$$

● ***Checkpoint Solutions***

1. (a) The number of coins is 30 and the number of gold coins is 4, so Pr(gold) = $4/30 = 2/15$.
   (b) The number of copper coins is 16, so Pr(copper) = $16/30 = 8/15$.
   (c) The number of coins that are not copper is 14, so Pr(not copper) = $14/30 = 7/15$.
2. (a) $S_1 = \{\text{even, odd}\}$ is not an equiprobable sample space for this experiment, because there are not equal numbers of even and odd numbers.
   $S_2 = \{1, 2, 3, 4, 5\}$ is an equiprobable sample space.
   (b) Pr(even) = $2/5$
3. (a) {H1, H2, H3, H4, H5, H6, T1, T2, T3, T4, T5, T6}
   (b) Pr(head and even number) = $3/12 = 1/4$
   (c) Pr(tail and number divisible by 3) = $2/12 = 1/6$

7.1 Exercises

1. One ball is drawn at random from a bag containing 4 red balls and 6 white balls. What is the probability that the ball is
   (a) red?　　(b) green?　　　(c) red or white?

2. One ball is drawn at random from a bag containing 4 red balls and 6 white balls. What is the probability that the ball is
   (a) white?　　(b) white or red?　　(c) red and white?

3. If you draw one card at random from a deck of 12 cards numbered 1 through 12, inclusive, what is the probability that the number you draw is divisible by 4?

4. An ordinary die is tossed. What is the probability of getting a 3 or a 4?

5. A die is rolled. Find the probability of getting a number greater than 0.

6. A die is rolled. What is the probability that
   (a) a 4 will result?
   (b) a 7 will result?
   (c) an odd number will result?

7. An urn contains three red balls numbered 1, 2, 3, four white balls numbered 4, 5, 6, 7, and three black balls numbered 8, 9, 10. A ball is drawn from the urn. What is the probability that
   (a) it is red?
   (b) it is odd-numbered?
   (c) it is red and odd-numbered?
   (d) it is red or odd-numbered?
   (e) it is not black?

8. An urn contains three red balls numbered 1, 2, 3, four white balls numbered 4, 5, 6, 7, and three black balls numbered 8, 9, 10. A ball is drawn from the urn. What is the probability that the ball is
   (a) white?
   (b) white and odd?
   (c) white or even?
   (d) black or white?
   (e) black and white?

9. From a deck of 52 ordinary playing cards, one card is drawn. Find the probability that it is
   (a) a queen.
   (b) a red card.
   (c) a spade.

10. From a deck of 52 ordinary playing cards, one card is drawn. Find the probability that it is
    (a) a red king.　　(b) a king or a black card.

11. Suppose a fair coin is tossed two times. Construct an equiprobable sample space for the experiment, and determine each of the following probabilities:
    (a) Pr(0 heads)　　(b) Pr(1 head)　　(c) Pr(2 heads)

12. Suppose a fair coin is tossed four times. Construct an equiprobable sample space for the experiment, and determine each of the following probabilities:
    (a) Pr(2 heads)　　(b) Pr(3 heads)　　(c) Pr(4 heads)

13. Use Table 7.1 to determine the following probabilities if a distinguishable pair of dice is rolled.
    (a) Pr(sum is 4)
    (b) Pr(sum is 10)
    (c) Pr(sum is 12)

14. (a) When a pair of distinguishable dice is rolled, what sum is most likely to occur?
    (b) When a pair of distinguishable dice is rolled, what is $\Pr(4 \leq S \leq 8)$, where S represents the sum rolled?

15. If a pair of dice, one green and one red, is rolled, what is
    (a) $\Pr(4 \leq S \leq 7)$, where S is the sum rolled?
    (b) $\Pr(8 \leq S \leq 12)$, where S is the sum rolled?

16. If a green die and a red die are rolled, find
    (a) $\Pr(2 \leq S \leq 6)$, where S is the sum rolled on the two dice.
    (b) $\Pr(4 \leq S)$, where S is the sum rolled on the two dice.

17. Suppose a die is tossed 1200 times and a 6 comes up 431 times.
    (a) Find the empirical probability for a 6 to occur.
    (b) On the basis of a comparison of the empirical probability and the theoretical probability, do you think the die is fair or biased?

18. Suppose that a coin is tossed 3000 times and 1800 heads result. What is the empirical probability that a head will occur with this coin? Is there evidence that the coin is a fair coin?

19. If the probability that an event will occur is 2/5, what are the odds
    (a) in favor of the event occurring?
    (b) against the event occurring?

20. If the probability that an event E will not occur is 8/11, what are the odds
    (a) that E will occur?
    (b) that E will not occur?

21. If the odds that a particular horse will win a race are 1:20, what is the probability
    (a) that the horse will win the race?
    (b) that the horse will lose the race?

22. If the odds that an event E will not occur are 23:57, what is the probability
    (a) that event E will occur?
    (b) that event E will not occur?

APPLICATIONS

23. *Drug use* Forty-six percent of marijuana use among youth occurs in the inner cities (*Source:* Partnership for a Drugfree America). If an instance of such marijuana use is chosen at random, what is the probability that the use occurs in an inner city?

24. *Breast cancer* According to the American Cancer Society, 199 of 200 mammograms turn out to be normal. What is the probability that the mammogram of a woman chosen at random will turn out to be normal?

25. *Car maintenance* A car rental firm has 425 cars. Sixty-three of these cars have defective turn signals and 32 have defective tires. What is the probability that one of these cars selected at random
(a) has defective turn signals?
(b) has defective tires?

26. *Testing* An unprepared student must take a 7-question multiple-choice test that has 5 possible answers per question. If the student guesses on the first question, what is the probability that she will answer that question incorrectly?

27. *Voting* The following table gives the average number of voters in each of three political parties during the last 12 years, along with the average number that voted in presidential elections during this period.
(a) For each political party, use these data to find the probability that a person selected at random from the registered voters in the party will vote in the next election.
(b) For which party is the probability highest?

	Political Party		
	Republican	Democratic	Independent
Registered voters	4500	6100	2200
Voted	2835	2501	1122

28. *Voting* In a survey of 350 registered voters in Belair, 123 said that they would vote for the Democratic candidate for mayor. If a registered voter is selected at random, what is the probability that this person will vote for the Democratic candidate?

29. *Sales promotion* In a sales promotion, a clothing store gives its customers a chance to draw a ticket from a box that contains a discount on their next purchase. The box contains 3000 tickets giving a 10% discount, 500 giving a 30% discount, 100 giving a 50% discount, and 1 giving a 100% discount. What is the probability that a given customer will randomly draw a ticket giving
(a) a 100% discount?
(b) a 50% discount?
(c) a discount of less than 50%?

(d) Is a given customer more likely to get a ticket with a 30% discount or with a discount higher than 30%?

30. *Management* A dry cleaning firm has 12 employees: 7 women and 5 men. Three of the women and five of the men are 40 years old or older. The remainder are over 20 years of age and under 40. If a person is chosen at random from this firm, what is the probability that the person is
(a) a woman?
(b) under 40 years of age?
(c) 20 years old?

31. *World population* The following table gives the world population, in millions, in 2001.
(a) What is the probability that a person picked at random is from a developed nation?
(b) What is the probability that a person picked at random is age 20–39?
(c) What is the probability that a person picked at random is age 60–79?

Age	In Developed Nations	In Less Developed Nations	Total
0–19	295	2098	2393
20–39	341	1605	1946
40–59	318	882	1200
60–79	196	349	545
80+	38	35	73

Source: 2002 *World Almanac*

32. *Vigorous activity* The following table gives the percent of U.S. students in grades 9–12 who participated in moderate to vigorous activity during 1999.
(a) What is the probability that a white, non-Hispanic male selected at random participated in moderate to vigorous activity?
(b) If a female in grade 11 is selected at random, what is the probability that she participated in moderate to vigorous activity?
(c) Can the probability that an Hispanic person selected at random participated in moderate to vigorous activity be found by adding the percents for Hispanic males and Hispanic females in the table? How can you be sure of your answer?

	Grade 9	Grade 10	Grade 11	Grade 12
Male	81.6	76.3	70.5	75.6
Female	72.9	61.6	54.4	59.7

	White, Non-Hispanic	Black, Non-Hispanic	Hispanic
Male	78.1	68.1	75.5
Female	65.3	52.5	54.3

Source: Centers for Disease Control, National Center for Chronic Disease Prevention and Health Promotion

33. *Blood types* Human blood is classified by blood type, which indicates the presence or absence of the antigens A, B, and Rh, as follows.

    | A present | Type A |
    | B present | Type B |
    | Both A and B present | Type AB |
    | Neither A nor B present | Type O |

    Each of these types is combined with a + or a − sign to indicate whether the Rh antigen is present or not. Write a sample space containing all possible blood types. Do you think this is an equiprobable sample space?

34. *Genetics* Construct an equiprobable sample space that gives the possible combinations of male and female children in a family with three children.

35. *Blood types* Four percent of the population of the United States has type AB blood (*Source:* aabb.org, 2005). What is the probability that a U.S. resident chosen at random
    (a) will have type AB blood?
    (b) will not have type AB blood?

36. *Blood types* Sixteen percent of the population of the United States has a negative Rh (Rhesus) factor in their blood, and the remainder have a positive Rh factor (*Source:* aabb.org, 2005). What is the probability that a resident of the United States chosen at random
    (a) will have a negative Rh factor?
    (b) will have a positive Rh factor?

37. *Minorities* Among the residents of Los Angeles, 10.0% are classified as of Asian descent (*Source: Census 2000,* U.S. Census Bureau). If an L.A. resident is chosen at random, what is the probability that he or she is
    (a) of Asian descent?   (b) not of Asian descent?

38. *Minorities* Among the population of Houston, 37.4% are of Hispanic descent (*Source: Census 2000,* U.S. Census Bureau). What is the probability that a Houston resident chosen at random is
    (a) of Hispanic descent?
    (b) not of Hispanic descent?

39. *Fraud* A company selling substandard drugs to developing countries sold 2,000,000 capsules with 60,000 of them empty (*Source:* "60 Minutes," January 18, 1998). What is the probability that a person who takes a randomly chosen capsule from this company will get an empty capsule?

40. *Inventory* Forty percent of a company's total output consists of baseballs, 30% consists of softballs, and 10% consists of tennis balls. Its only remaining prod-

uct is handballs. If we place balls in a box in the same ratios as the company's output and select a ball at random from the box, what is the probability that
    (a) the ball is a baseball?
    (b) the ball is a tennis ball?
    (c) the ball is not a softball?
    (d) the ball is a handball?

41. *Lactose intolerance* Lactose intolerance affects about 75% of African, Asian, and Native Americans (*Source:* Jean Carper, "Eat Smart," *USA Weekend*). If a person is selected from this group of people, what is the probability that the person will have lactose intolerance?

42. *Lactose intolerance* Lactose intolerance affects about 20% of non-Hispanic white Americans (*Source:* Jean Carper, "Eat Smart," *USA Weekend*). If a person is selected from this group of people, what is the probability that the person will be lactose-intolerant?

43. *Management* Because of a firm's growth, it is necessary to transfer one of its employees to one of its branch stores. Three of the nine employees are women, and all of the nine employees are equally qualified for the transfer. If the person to be transferred is chosen at random, what is the probability that the transferred person is a woman?

44. *Quality control* A supply of 500 television picture tubes has 6 defective tubes. What is the probability that a tube picked at random from the supply is not defective?

45. *Education* A professor assigns 6 homework projects and then, during the next class, rolls a die to determine which one of the 6 projects to grade. If a student has completed 2 of the assignments, what is the probability that the professor will grade a project that the student has completed?

46. *Marketing* Meow Chow offers a free coin in each bag of cat food. If each 10,000-bag shipment has one gold coin, 2 silver dollars, 5 quarters, 20 dimes, 50 nickels, and 9922 pennies, what is the probability that a bag of Meow Chow cat food selected at random will have a penny in it?

47. *Minorities* The NAACP claimed that in Volusia County, Florida, police stopped dark-skinned drivers on I-95 who were not violating the law much more frequently than they stopped white drivers, and then they conducted searches of the cars for drugs or for large sums of money that could be related to drug sales (*Source: American Journal,* August 30, 1994). If police made routine stops of randomly selected motorists who were not violating a law, and if 22% of the drivers on I-95 passing through this county were dark-skinned, what is the

probability that the police stopped a driver who was not violating a law but who was dark-skinned? If 39% of the motorists stopped for routine checks were dark-skinned people, did the NAACP claim seem reasonable?

48. *Quality control*   A computer store offers used computers free to local middle schools. Of the 54 machines available, four have defective memories, six have defective keys, and the remainder have no defects. If a teacher picks one at random, what is the probability that she will select a defective computer?

49. *Super Bowl*   Near the end of the 2004 season, vegas.com stated the odds as 5 to 8 that the Philadelphia Eagles would win the 2005 Super Bowl. What was the probability that they would win, according to these odds?

50. *World Series*   Shortly after the 2004 World Series, vegas.com stated the odds as 2 to 7 that the New York Yankees would win the 2005 World Series. What was the probability that they would win, according to these odds?

51. *Genetics*   A newly married couple plan to have three children. Assuming the probability of a girl being born equals that of a boy being born, what is the probability that exactly two of the three children born will be girls? (*Hint:* Construct the sample space.)

52. *Management*   A frustrated store manager is asked to make four different yes-no decisions that have no relation to each other. Because he is impatient to leave work, he flips a coin for each decision. If the correct decision in each case was yes, what is the probability that
    (a) all of his decisions were correct?
    (b) none of his decisions was correct?
    (c) half of his decisions were correct?

53. *Genetics*   A couple plan to have two children and want to know the probability of having one boy and one girl.
    (a) Does {0, 1, 2} represent an equiprobable sample space for finding the probability of exactly one boy in two children?
    (b) Construct an equiprobable sample space for this problem.
    (c) What is the probability of getting one boy and one girl in two children?

54. *Genetics*   What is the probability that a couple planning to have two children will have
    (a) at least one child of each sex?
    (b) two children of the same sex?

55. *Testing*   A quiz consists of 8 matching questions with 8 possible matches from which to choose. If a student guesses on the first question, what is the probability that he will get the correct answer?

56. *Testing*   In taking a quiz with 6 multiple-choice questions, a student knows that two of the possible five answers to the first question are not correct. What is the probability of her getting the correct answer to the first question?

57. *Quality control*   On the average, 6 articles out of each 250 produced by a certain machine are defective. What is the probability that an article chosen at random is defective?

58. *Baseball*   If a baseball player has 1150 at bats and 320 hits, what is the probability that he will get a hit his next time at bat?

59. *Traffic safety*   In a certain town, citizens' groups have identified three intersections as potentially dangerous. They collected the following data.

	Intersections		
	A	B	C
Number of vehicles/day	25,500	3890	8580
Number of vehicles in accidents/year	182	51	118

For each intersection, find the probability that a vehicle entering the intersection will be involved in an accident. Which intersection is the most dangerous?

60. *Traffic safety*   A traffic survey showed that 5680 cars entered an intersection of the main street of a city and that 1460 of them turned onto the intersecting street at this intersection.
    (a) What is the empirical probability a car will turn onto the intersecting street?
    (b) What is the probability a car will not turn onto the intersecting street?

61. *Elections*   Prior to an election, a poll of 1200 citizens is taken. Suppose 557 plan to vote for A, 533 plan to vote for B, and the remainder indicate they will not vote.
    (a) If the poll is accurate, find the probability that a citizen chosen at random will vote for A.
    (b) If the poll is accurate, what is the probability that a person picked at random will not vote?

62. *Elections*   Suppose a county has 800,000 registered voters, 480,000 of whom are Democrats and 290,000 of whom are Republicans. If a registered voter is chosen at random, what is the probability that
    (a) he or she is a Democrat?
    (b) he or she is neither a Democrat nor a Republican?

63. *Births*   Suppose that at a certain hospital the births on a given day consisted of 8 girls and 2 boys. Suppose that at the same hospital, the births in a given year consisted of 1220 girls and 1194 boys.
    (a) Find the empirical probabilities for the birth of each sex on the given day.
    (b) Find the empirical probabilities for the given year.
    (c) What are the theoretical probabilities, and which empirical probabilities do you think more accurately reflect reality?

 7.2

OBJECTIVES

- To find the probability of the intersection of two events
- To find the probability of the union of two events
- To find the probability of the complement of an event

Unions and Intersections of Events: One-Trial Experiments

❶ **Application Preview**

Suppose that the directors of the Chatham County School Board consist of 8 Democrats, 3 of whom are female, and 4 Republicans, 2 of whom are female. If a director is chosen at random to discuss an issue on television, we can find the probability that the person will be a Democrat or female by finding the probability of the union of the two events "a Democrat is chosen" and "a female is chosen." (See Example 4.) In this section we will discuss methods of finding the probabilities of the intersection and union of events by using sample spaces and formulas.

Unions, Intersections, and Complements of Events

As we stated in the previous section, events determine subsets of the sample space for a probability experiment. The intersection, union, and complement of events are defined as follows:

If E and F are two events in a sample space S, then the **intersection of E and F** is

$$E \cap F = \{a : a \in E \quad \text{and} \quad a \in F\}$$

the **union of E and F** is

$$E \cup F = \{a : a \in E \quad \text{or} \quad a \in F\}$$

and the **complement of E** is

$$E' = \{a : a \in S \quad \text{and} \quad a \notin E\}$$

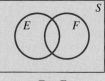 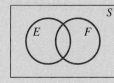 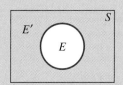

$E \cap F$ $E \cup F$

Using these definitions, we can write the probabilities that certain events will occur as follows:

$$\text{Pr}(E \text{ and } F \text{ both occur}) = \text{Pr}(E \cap F)$$
$$\text{Pr}(E \text{ or } F \text{ occurs}) = \text{Pr}(E \cup F)$$
$$\text{Pr}(E \text{ does } \textbf{not} \text{ occur}) = \text{Pr}(E')$$

In Section 7.1, "Probability; Odds," we investigated probability questions such as these by identifying the sample space elements for each described event. We now apply the properties just stated.

● EXAMPLE 1 *Unions and Intersections of Events*

A card is drawn from a box containing 15 cards numbered 1 to 15. What is the probability that the card is

(a) even and divisible by 3?   (b) even or divisible by 3?   (c) not even?

Solution

The sample space S contains the 15 numbers.

$$S = \{1, 2, 3, 4, 5, 6, 7, 8, 9, 10, 11, 12, 13, 14, 15\}$$

If we let E represent "even-numbered" and D represent "number divisible by 3," we have

$$E = \{2, 4, 6, 8, 10, 12, 14\} \quad \text{and} \quad D = \{3, 6, 9, 12, 15\}$$

These sets are shown in the Venn diagram in Figure 7.1.

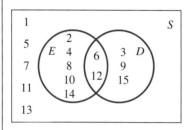

Figure 7.1

(a) The event "even and divisible by 3" is $E \cap D = \{6, 12\}$, so

$$\text{Pr(even and divisible by 3)} = \text{Pr}(E \cap D) = \frac{n(E \cap D)}{n(S)} = \frac{2}{15}$$

(b) The event "even or divisible by 3" is $E \cup D = \{2, 3, 4, 6, 8, 9, 10, 12, 14, 15\}$, so

$$\text{Pr(even or divisible by 3)} = \text{Pr}(E \cup D) = \frac{n(E \cup D)}{n(S)} = \frac{10}{15} = \frac{2}{3}$$

(c) The event "not even," or E', contains all elements of S not in E, so
$E' = \{1, 3, 5, 7, 9, 11, 13, 15\}$, and the probability that the card is not even is

$$\text{Pr(not } E) = \text{Pr}(E') = \frac{n(E')}{n(S)} = \frac{8}{15}$$

● **Checkpoint**

1. If a ball is drawn from a bag containing 4 red balls numbered 1, 2, 3, 4 and 3 white balls numbered 5, 6, 7, what is the probability that the ball is
   (a) red and even?   (b) white and even?

Probability of the Complement

Because the complement of an event contains all the elements of S *except* for those elements in E (see Figure 7.2), the number of elements in E' is $n(S) - n(E)$, so

$$\text{Pr}(E') = \frac{n(S) - n(E)}{n(S)} = 1 - \frac{n(E)}{n(S)} = 1 - \text{Pr}(E)$$

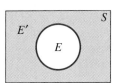

Figure 7.2

$$\Pr(\text{not } E) = \Pr(E') = 1 - \Pr(E)$$

● **EXAMPLE 2** *Diversity*

A dry cleaning firm has 12 employees: 7 women and 5 men. Three of the women and 5 of the men are 40 years old or older. If a person is chosen at random from this firm, what is the probability that the person is under 40 years of age?

Solution
Eight people (3 women and 5 men) are 40 or older, so

$$\Pr(\text{under } 40) = 1 - \Pr(40+) = 1 - \frac{8}{12} = \frac{4}{12} = \frac{1}{3}$$

Inclusion-Exclusion Principle

The number of elements in the set $E \cup F$ is the number of elements in E plus the number of elements in F, *minus* the number of elements in $E \cap F$ [because this number was counted twice in $n(E) + n(F)$]. Thus,

$$n(E \cup F) = n(E) + n(F) - n(E \cap F)$$

Thus,

$$\Pr(E \cup F) = \frac{n(E \cup F)}{n(S)} = \frac{n(E)}{n(S)} + \frac{n(F)}{n(S)} - \frac{n(E \cap F)}{n(S)}$$
$$= \Pr(E) + \Pr(F) - \Pr(E \cap F)$$

This establishes what is called the **inclusion-exclusion principle,** which is summarized as follows.

Inclusion-Exclusion Principle

If E and F are any two events, then the probability that one event or the other will occur, denoted $\Pr(E \text{ or } F)$, is given by

$$\Pr(E \text{ or } F) = \Pr(E \cup F) = \Pr(E) + \Pr(F) - \Pr(E \cap F)$$

Suppose that of 30 people interviewed for a position, 18 had a business degree, 15 had previous experience, and 8 of those with experience also had a business degree. The Venn diagram in Figure 7.3 illustrates this situation.

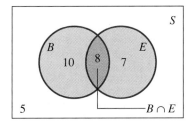

Figure 7.3

Because 8 of the 30 people have both a business degree and experience,

$$\Pr(B \cap E) = \frac{8}{30}$$

We also see that

$$\Pr(B) = \frac{18}{30} \quad \text{and} \quad \Pr(E) = \frac{15}{30}$$

Thus the probability of choosing a person at random who has a business degree or experience is

$$\Pr(B \cup E) = \Pr(B) + \Pr(E) - \Pr(B \cap E)$$
$$= \frac{18}{30} + \frac{15}{30} - \frac{8}{30} = \frac{25}{30} = \frac{5}{6}$$

● **EXAMPLE 3** *Diversity*

A dry cleaning firm has 12 employees: 7 women and 5 men. Three of the women and 5 of the men are older than 40. If a person is chosen at random from this firm, what is the probability that the person is a woman or older than 40?

Solution
There are 7 women and 8 people over 40 years of age with 3 of the women over 40. Thus the probability that the person is a woman or over 40 is given by

$$\Pr(\text{woman or older than 40}) = \Pr(\text{woman}) + \Pr(40+) - \Pr(\text{woman and }40+)$$
$$= \frac{7}{12} + \frac{8}{12} - \frac{3}{12} = \frac{12}{12} = 1$$

The probability is 1, which means that it is certain that a person selected at random will be a woman or over 40. (Note that all the men are over 40.)

◗ **EXAMPLE 4** *Politics* **(Application Preview)**

The directors of the Chatham County School Board consist of 8 Democrats, 3 of whom are female, and 4 Republicans, 2 of whom are female. If a director is chosen at random to discuss an issue on television, what is the probability that the person will be a Democrat or female?

Solution
There are 12 directors, of whom 8 are Democrats and 5 are female. But 3 of the directors are Democrats *and* females and cannot be counted twice. Thus

$$\Pr(D \text{ or } f) = \Pr(D \cup f) = \Pr(D) + \Pr(f) - \Pr(D \text{ and } f)$$
$$= \frac{8}{12} + \frac{5}{12} - \frac{3}{12} = \frac{10}{12} = \frac{5}{6}$$

● *Checkpoint*

2. If a ball is drawn from a bag containing 4 red balls numbered 1, 2, 3, 4 and 3 white balls numbered 5, 6, 7, what is the probability that the ball is
   (a) red or even?
   (b) white or even?

Mutually Exclusive Events

We say that events E and F are **mutually exclusive** if and only if $E \cap F = \emptyset$. Thus

$$\Pr(E \cup F) - \Pr(E) + \Pr(F) - 0 = \Pr(E) + \Pr(F)$$

if and only if E and F are mutually exclusive.

Mutually Exclusive Events

If E and F are **mutually exclusive,** then $\Pr(E \cap F) = 0$, and

$$\Pr(E \text{ or } F) = \Pr(E \cup F) = \Pr(E) + \Pr(F)$$

Mutually exclusive

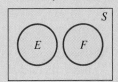

● **EXAMPLE 5** *Mutually Exclusive Events*

Find the probability of obtaining a 6 or a 4 in one roll of a die.

Solution
Rolling a 6 and rolling a 4 on one roll of a die are mutually exclusive events. Let E be the event "rolling a 6" and F be the event "rolling a 4."

$$\Pr(E) = \Pr(\text{rolling } 6) = \frac{1}{6} \quad \text{and} \quad \Pr(F) = \Pr(\text{rolling } 4) = \frac{1}{6}$$

Then

$$\Pr(E \cup F) = \Pr(E) + \Pr(F) = \frac{1}{6} + \frac{1}{6} = \frac{1}{3}$$

● **EXAMPLE 6** *Property Development*

A firm is considering three possible locations for a new factory. The probability that site A will be selected is $1/3$ and the probability that site B will be selected is $1/5$. Only one location will be chosen.

(a) What is the probability that site A or site B will be chosen?
(b) What is the probability that neither site A nor site B will be chosen?

Solution
(a) The two events are mutually exclusive, so

$$\Pr(\text{site A or site B}) = \Pr(\text{site A} \cup \text{site B})$$

$$= \Pr(\text{site A}) + \Pr(\text{site B}) = \frac{1}{3} + \frac{1}{5} = \frac{8}{15}$$

(b) The probability that neither will be chosen is $1 - 8/15 = 7/15$.

The formula for the probability of the union of mutually exclusive events can be extended beyond two events.

$$\Pr(E_1 \cup E_2 \cup \ldots \cup E_n) = \Pr(E_1) + \Pr(E_2) + \cdots + \Pr(E_n)$$

for mutually exclusive events $E_1, E_2, \ldots, E_n$.

● **EXAMPLE 7** *Dice*

Find the probability of rolling two distinct dice and obtaining a sum greater than 9.

Solution

The event "a sum greater than 9" consists of the mutually exclusive events "sum = 10," "sum = 11," and "sum = 12," and these probabilities can be found in Table 7.1 in Section 7.1.

$$\text{Pr(10 or 11 or 12)} = \text{Pr}(10 \cup 11 \cup 12)$$

$$= \text{Pr}(10) + \text{Pr}(11) + \text{Pr}(12) = \frac{3}{36} + \frac{2}{36} + \frac{1}{36} = \frac{6}{36} = \frac{1}{6}$$

● *Checkpoint Solutions*

1. (a) The sample space contains 7 balls: $S = \{R1, R2, R3, R4, W5, W6, W7\}$. The probability that a ball drawn at random is red and even is found by using the subspace that describes that event. Thus

$$\text{Pr(red and even)} = \text{Pr(red} \cap \text{even)} = \frac{2}{7}$$

   (b) The set $\{W6\}$ describes the event "white and even," so

$$\text{Pr(white and even)} = \text{Pr(white} \cap \text{even)} = \frac{1}{7}$$

2. (a) Pr(red or even)
      $= \text{Pr(red} \cup \text{even)} = \text{Pr(red)} + \text{Pr(even)} - \text{Pr(red} \cap \text{even)}$
      $= \frac{4}{7} + \frac{3}{7} - \frac{2}{7} = \frac{5}{7}$
   (b) Pr(white or even)
      $= \text{Pr(white} \cup \text{even)} = \text{Pr(white)} + \text{Pr(even)} - \text{Pr(white} \cap \text{even)}$
      $= \frac{3}{7} + \frac{3}{7} - \frac{1}{7} = \frac{5}{7}$

7.2 Exercises

1. If you draw one card at random from a deck of 12 cards numbered 1 through 12, what is the probability that the number you draw is divisible by 3 and even?

2. If you draw one card at random from a deck of 12 cards numbered 1 through 12, what is the probability that the number you draw is even and divisible by 5?

3. An ordinary die is tossed. What is the probability of getting a number divisible by 3 or an odd number?

4. A card is drawn at random from a deck of 52 playing cards. Find the probability that it is a club or a king.

5. If the probability that event E will occur is $3/5$, what is the probability that E will not occur?

6. If the probability that an event will occur is $7/9$, what is the probability that the event will not occur?

7. A bag contains 5 red balls numbered 1, 2, 3, 4, 5 and 9 white balls numbered 6, 7, 8, 9, 10, 11, 12, 13, 14. If a ball is drawn, what is the probability that
   (a) the ball is red and even-numbered?
   (b) the ball is red or even-numbered?

8. A bag contains 5 red balls numbered 1, 2, 3, 4, 5 and 9 white balls numbered 6, 7, 8, 9, 10, 11, 12, 13, 14. If a ball is drawn, what is the probability that it is
   (a) white and odd-numbered?
   (b) white or odd-numbered?
   (c) white or even-numbered?

9. If you draw one card from a deck of 12 cards, numbered 1 through 12, what is the probability you will get an odd number or a number divisible by 4?

10. Suppose a die is biased (unfair) so that each odd-numbered face has probability $1/4$ of resulting, and each even face has probability $1/12$ of resulting. Find the probability of getting a number greater than 3.

11. If you draw one card from a deck of 12 cards numbered 1 through 12, what is the probability that the card will be odd or divisible by 3?

12. A card is drawn from a deck of 52. What is the probability that it will be an ace, king, or jack?

13. A bag contains 4 white, 7 black, and 6 green balls. What is the probability that a ball drawn at random from the bag is white or green?

14. In a game where only one player can win, the probability that Jack will win is 1/5 and the probability that Bill will win is 1/4. Find the probability that one of them will win.

15. A cube has 2 faces painted red, 2 painted white, and 2 painted blue. What is the probability of getting a red face or a white face in one roll?

16. A cube has 3 faces painted white, 2 faces painted red, and 1 face painted blue. What is the probability that a roll will result in a red or blue face?

17. A ball is drawn from a bag containing 13 red balls numbered 1–13 and 5 white balls numbered 14–18. What is the probability that
    (a) the ball is not even-numbered?
    (b) the ball is red and even-numbered?
    (c) the ball is red or even-numbered?
    (d) the ball is neither red nor even-numbered?

18. A ball is drawn from a bag containing 13 red balls numbered 1–13 and 5 white balls numbered 14–18. What is the probability that
    (a) the ball is not red?
    (b) the ball is white and odd-numbered?
    (c) the ball is white or odd-numbered?

APPLICATIONS

19. *Drug use* Forty-six percent of marijuana use among youth occurs in the inner cities (*Source:* Partnership for a Drugfree America). If an instance of such marijuana use is chosen at random, what is the probability that the use does not occur in an inner city?

20. *Breast cancer* According to the American Cancer Society, 199 of 200 mammograms turn out to be normal. What is the probability that the mammogram of a woman chosen at random will show an abnormality?

21. *Car maintenance* A car rental firm has 425 cars. Sixty-three of these cars have defective turn signals and 32 have defective tires. What is the probability that one of these cars selected at random
    (a) does not have defective turn signals?
    (b) has no defects if no car has 2 defects?

22. *Testing* An unprepared student must take a 7-question multiple-choice test that has 5 possible answers per question. If the student guesses on the first question, what is the probability that she will answer that question incorrectly?

23. *Linguistics* Of 100 students, 24 can speak French, 18 can speak German, and 8 can speak both French and German. If a student is picked at random, what is the probability that he or she can speak French or German?

24. *Management* A company employs 65 people. Eight of the 30 men and 21 of the 35 women work in the business office. What is the probability that an employee picked at random is a woman or works in the business office?

25. *Politics* Sacco and Rosen are among three candidates running for public office, and polls indicate that the probability that Sacco will win is 1/3 and the probability that Rosen will win is 1/2. Only one candidate can win.
    (a) What is the probability that Sacco or Rosen will win?
    (b) What is the probability that neither Sacco nor Rosen will win?

26. *Photography* Rob Lee knows his camera will take a good picture unless the flash is defective or the batteries are dead. The probability of having a defective flash is 0.05, the probability of the batteries being dead is 0.3, and the probability that both these problems occur is 0.01. What is the probability that the picture will be good?

27. *Salaries* The following table gives the percent of employees of the Ace Company in each of three salary brackets, categorized by the sex of the employees. An employee is selected at random.
    (a) What is the probability that the person selected is female and makes less than $30,000?
    (b) What is the probability that the person selected makes at least $50,000?
    (c) What is the probability that the person is male or makes less than $30,000?

	Earns Less Than $30,000	Earns at Least $30,000 and Less Than $50,000	Earns at Least $50,000
Male	25%	18%	5%
Female	35%	14%	3%

28. *Salaries* At the Baker Engineering Company, 40% of the employees are female and 60% of the employees earn less than $40,000. If 30% of the employees are

females who earn less than $40,000, what is the probability that an employee picked at random is female or earns less than $40,000?

29. *World population*   The following table gives the world population, in millions, in 2001.
    (a) What is the probability that a person picked at random is from a developed nation, or that the person is age 40–59?
    (b) What is the probability that a person picked at random is age 20–39, and that he or she is from a less developed nation?
    (c) What is the probability that a person picked at random is age 40 or more?

Age	In Developed Nations	In Less Developed Nations	Total
0–19	295	2098	2393
20–39	341	1605	1946
40–59	318	882	1200
60–79	196	349	545
80+	38	35	73

Source: 2002 *World Almanac*

30. *Causes of death*   In the United States in 1999, there were 725,192 deaths from heart disease, 549,838 from cancer, and 167,366 from stroke (*Source:* National Center for Health Statistics, U.S. Department of Health and Human Services). If a death is chosen at random from the 2,391,399 deaths that occurred in 1999, what is the probability that the cause of that death was from cancer or stroke?

31. *AIDS deaths*   The following table gives the numbers of AIDS deaths in 2000 for people over age 13 in various categories. Use the table to find the probability that a person who died of AIDS in 2000
    (a) was Hispanic.
    (b) was female.
    (c) was female or Hispanic.
    (d) was male or black (non-Hispanic)

	White, non-Hispanic	Black, non-Hispanic	Hispanic	Other	Total
Male	5759	6667	2674	211	15,311
Female	940	3278	763	81	5,062

Source: Health, U.S. 2001, U.S. Department of Health and Human Services

32. *Accidental deaths*   In 2000, the death rate from motor vehicle accidents was 15.6 per 100,000 people and the death rate from falls was 5.9 per 100,000 people (*Source:* Health, U.S. 2001, U.S. Department of Health and Human Services). What is the probability that an accidental death in 2000 was from a motor vehicle accident or from a fall?

33. *Cognitive complexity*   The cognitive complexity of a structure was studied by Scott* using a technique in which a person was asked to specify a number of objects and group them into as many groupings as he or she found meaningful. A person groups 12 objects into three groups in such a way that

> 7 objects are in group A.
> 7 objects are in group B.
> 8 objects are in group C.
> 3 objects are in both group A and group B.
> 5 objects are in both group B and group C.
> 4 objects are in both group A and group C.
> 2 objects are in all three groups.

   (a) What is the probability that an object chosen at random has been placed in group A or group B?
   (b) What is the probability that an object chosen at random has been placed in group B or group C?

34. *Politics*   A group of 100 people contains 60 Democrats and 35 Republicans. If there are 60 women and if 40 of the Democrats are women, what is the probability that a person selected at random is a Democrat or a woman?

35. *Education*   A mathematics class consists of 16 engineering majors, 12 science majors, and 4 liberal arts majors.
    (a) What is the probability that a student selected at random will be a science or liberal arts major?
    (b) What is the probability that a student selected at random will be an engineering or science major?
    (c) Five of the engineering students, 6 of the science majors, and 2 of the liberal arts majors are female. What is the probability that a student selected at random is an engineering major or is a female?

36. *Parts delivery*   Repairing a copy machine requires that two parts be delivered from two suppliers. The probability that part A will be delivered on Thursday is 0.6, and the probability that part B will be delivered on Thursday is 0.8. If the probability that one or the other part will arrive on Thursday is 0.9, what is the probability that both will be delivered on Thursday?

*W. Scott, "Cognitive Complexity and Cognitive Flexibility," *Sociometry* 25 (1962), pp. 405–414.

Military spending **Suppose the following table summarizes the opinions of various groups on the issue of increased military spending. Use this table to calculate probabilities in Problems 37–40.**

Opinion	Whites		Nonwhites		
	Reps.	Dems.	Reps.	Dems.	Total
Favor	300	100	25	10	435
Oppose	100	250	25	190	565
Total	400	350	50	200	1000

37. Find the probability that an individual chosen at random is a Republican or favors increased military spending.

38. Find the probability that an individual chosen at random is a Democrat or opposes increased military spending.

39. Find the probability that an individual chosen at random is white or opposes increased military spending.

40. Find the probability that an individual chosen at random is nonwhite or favors increased military spending.

41. *Drinking age* A survey questioned 1000 people regarding raising the legal drinking age from 18 to 21. Of the 560 who favored raising the age, 390 were female. Of the 440 opposition responses, 160 were female. A person is selected at random.
    (a) What is the probability the person is a female or favors raising the age?
    (b) What is the probability the person is a male or favors raising the age?
    (c) What is the probability the person is a male or opposes raising the age?

42. *Job bids* Three construction companies have bid for a job. Max knows that the two companies he is competing with have probabilities 1/3 and 1/6, respectively, of getting the job. What is the probability that Max will get the job?

43. *Management* A company employs 80 people. Twenty of the 50 men and 18 of the 30 women work in the business office. What is the probability that an employee picked at random is a man or does not work in the business office?

44. *Television* The probability that a wife watches a certain television show is 0.55, that her husband watches it is 0.45, and that both watch the show is 0.30. What is the probability that the husband or the wife watches this show?

45. *Cancer testing* A long-term study has revealed that the prostate-specific antigen (PSA) test for prostate cancer in men is very effective. The study shows that 87% of the men for which the test is positive actually have prostate cancer (*Source: Journal of the American Medical Association,* January 1995). If a man selected at random tests positive for prostate cancer with this test, what is the probability that he does not have prostate cancer?

46. *Cancer testing* The study described in Problem 45 also revealed that 91% of the men for whom the PSA test was negative do not have prostate cancer (*Source: Journal of the American Medical Association,* January 1995). If a man selected at random tests negative for prostate cancer with this test, what is the probability that he does have prostate cancer?

7.3

Conditional Probability: The Product Rule

OBJECTIVES

- To solve probability problems involving conditional probability
- To compute the probability that two or more independent events will occur
- To compute the probability that two or more dependent events will occur

◗ Application Preview

Before prepaid phone cards, AT&T calling cards used each subscriber's home phone number plus a 4-digit personal identification number (PIN). Because criminals could observe users of public phones to record and sell PINs for fraudulent use, AT&T discontinued using home phone numbers and changed to random 14-digit PINs. To see how this reduces the chance of someone obtaining a PIN, we can find the probability that someone can guess a PIN in these two cases. To see which PIN would be safer, we can calculate the probability that someone can guess your PIN if you have a 4-digit PIN and if you have a 14-digit PIN by using the **Product Rule.** (See Example 9.) In this section we will use the Product Rule to find the probability that two events will occur, and we will solve problems involving **conditional probability.**

Conditional Probability   Each probability that we have computed has been relative to the sample space for the experiment. Sometimes information is given that reduces the sample space needed to solve a stated problem. For example, if a die is rolled, the probability that a 5 occurs, given that an odd number is rolled, is denoted by

$$\text{Pr}(5 \text{ rolled} \mid \text{odd number rolled})$$

The knowledge that an odd number occurs reduces the sample space from the numbers on a die, $S = \{1, 2, 3, 4, 5, 6\}$, to the sample space for odd numbers on a die, $S_1 = \{1, 3, 5\}$, so the event $E = $ "a 5 occurs" has probability

$$\Pr(5 \mid \text{odd number}) = \frac{n(5 \text{ and odd number rolled})}{n(\text{odd number rolled})} = \frac{n(E \cap S_1)}{n(S_1)} = \frac{1}{3}$$

To find $\Pr(A \mid B)$, we seek the probability that A occurs, given that B occurs. Figure 7.4 shows the original sample space S with the reduced sample space B shaded. Because all elements must be contained in B, we evaluate $\Pr(A \mid B)$ by dividing the number of elements in $A \cap B$ by the number of elements in B.

Figure 7.4

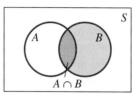

The probability that A occurs, given that B occurs, is denoted by $\Pr(A \mid B)$ and is given by

$$\Pr(A \mid B) = \frac{n(A \cap B)}{n(B)}$$

● **EXAMPLE 1 *Public Opinion***

Suppose the following table summarizes the opinions of various groups on the issue of increased military spending. What is the probability that a person selected at random from this group of people

(a) is nonwhite?
(b) is nonwhite and favors increased military spending?
(c) favors increased military spending, given that the person is nonwhite?

	Whites		Nonwhites		
Opinion	Republicans	Democrats	Republicans	Democrats	Totals
Favor	300	100	25	10	435
Oppose	100	250	25	190	565
Total	400	350	50	200	1000

Solution

(a) Of the 1000 people in the group, $50 + 200 = 250$ are nonwhite, so

$$\Pr(\text{nonwhite}) = \frac{250}{1000} = \frac{1}{4}$$

(b) There are 1000 people in the group and $25 + 10 = 35$ who both are nonwhite and favor increased spending, so

$$\Pr(\text{nonwhite and favor}) = \frac{35}{1000} = \frac{7}{200}$$

(c) There are $50 + 200 = 250$ nonwhite and $25 + 10 = 35$ people who both are nonwhite and favor increased spending, so

$$\Pr(\text{favor} \mid \text{nonwhite}) = \frac{n(\text{favor and nonwhite})}{n(\text{nonwhite})} = \frac{35}{250} = \frac{7}{50}$$

The probability found in Example 1(c) is called **conditional probability.** The conditional probability $\text{Pr}(A \mid B)$ can be found by using the number of elements in $A \cap B$ and in B or by using the probabilities of $A \cap B$ and B. To find the formula by using probabilities, we divide both the numerator and the denominator of the right-hand side of

$$\text{Pr}(A \mid B) = \frac{n(A \cap B)}{n(B)}$$

by $n(S)$, the number of elements in the sample space, which gives

$$\text{Pr}(A \mid B) = \frac{\dfrac{n(A \cap B)}{n(S)}}{\dfrac{n(B)}{n(S)}} = \frac{\text{Pr}(A \cap B)}{\text{Pr}(B)}$$

Conditional Probability

Let A and B be events in the sample space S with $\text{Pr}(B) > 0$. The conditional probability that event A occurs, given that event B occurs, is given by

$$\text{Pr}(A \mid B) = \frac{\text{Pr}(A \cap B)}{\text{Pr}(B)}$$

● **EXAMPLE 2** *Conditional Probability*

A red die and a green die are rolled. What is the probability that the sum rolled on the dice is 6, given that the sum is less than 7?

Solution

Using the conditional probability formula and the sample space in Table 7.1 in Section 7.1, we have

$$\text{Pr}(\text{sum is } 6 \mid \text{sum} < 7) = \frac{\text{Pr}(\text{sum is } 6 \cap \text{sum} < 7)}{\text{Pr}(\text{sum} < 7)}$$

Because

$$\text{Pr}(\text{sum is } 6 \cap \text{sum} < 7) = \text{Pr}(\text{sum is } 6) = \frac{5}{36}$$

the probability is

$$\text{Pr}(\text{sum is } 6 \mid \text{sum} < 7) = \frac{5/36}{1/36 + 2/36 + 3/36 + 4/36 + 5/36}$$

$$= \frac{5/36}{15/36} = \frac{1}{3}$$

● **Checkpoint**

1. Suppose that one ball is drawn from a bag containing 4 red balls numbered 1, 2, 3, 4 and 3 white balls numbered 5, 6, 7. What is the probability that the ball is white, given that it is an even-numbered ball?

The Product Rule

We can solve the formula

$$\text{Pr}(A \mid B) = \frac{\text{Pr}(A \cap B)}{\text{Pr}(B)}$$

for $\Pr(A \cap B)$ by multiplying both sides of the equation by $\Pr(B)$. This gives

$$\Pr(A \cap B) = \Pr(B) \cdot \Pr(A \mid B)$$

Similarly, we can solve

$$\Pr(B \mid A) = \frac{\Pr(B \cap A)}{\Pr(A)} = \frac{\Pr(A \cap B)}{\Pr(A)}$$

for $\Pr(A \cap B)$, getting

$$\Pr(A \cap B) = \Pr(A) \cdot \Pr(B \mid A)$$

Product Rule

> If A and B are probability events, then the probability of the event "A and B" is $\Pr(A \text{ and } B)$, and it can be found by one or the other of these two formulas.
>
> $$\Pr(A \text{ and } B) = \Pr(A \cap B) = \Pr(A) \cdot \Pr(B \mid A)$$
>
> or
>
> $$\Pr(A \text{ and } B) = \Pr(A \cap B) = \Pr(B) \cdot \Pr(A \mid B)$$

● **EXAMPLE 3** *Cards*

Suppose that from a deck of 52 cards, two cards are drawn in succession, without replacement. Find the probability that both cards are kings.

Solution

We seek the probability that the first card will be a king and the probability that the second card will be a king given that the first card was a king. The probability for the second card is denoted

$$\Pr(\text{2nd king} \mid \text{1st king})$$

and it is 3/51 because one king has been removed from the deck on the first draw.

Thus the probability that two cards (drawn without replacement) are both kings can be found as follows.

$$\Pr(\text{1st card is a king}) = \frac{4}{52}$$

$$\Pr(\text{2d card is a king} \mid \text{1st card is a king}) = \frac{3}{51}$$

$$\Pr(\text{2 kings}) = \Pr(\text{1st king}) \cdot \Pr(\text{2d king} \mid \text{1st king})$$

$$= \frac{4}{52} \cdot \frac{3}{51} = \frac{1}{221}$$

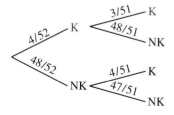

Figure 7.5

We can represent the outcomes in Example 3 with a **probability tree,** as shown in Figure 7.5. The outcome of the first draw is either a king or not a king; it is illustrated by the beginning two branches of the tree. The possible outcomes of the second trial can be treated as branches that emanate from each of the outcomes of the first trial. The tree shows that if a king occurs on the first trial, the outcome of the second trial is either a king or not a king, and the probabilities of these outcomes depend on the fact that a king occurred on the first trial. On the other hand, the probabilities of getting or not getting a king are different if the first draw does not yield a king, so the branches illustrating these possible outcomes have different probabilities.

There is only one "path" through the tree that illustrates getting kings on both draws, and the probability that both draws result in kings is the product of the probabilities on the branches of this path. Note that this gives $\frac{4}{52} \cdot \frac{3}{51} = \frac{1}{221}$, which is the same Product Rule probability found in Example 3.

In general, the connection between the Product Rule and a probability tree is as follows.

Probability Trees

> The probability attached to *each branch* of a probability tree is the conditional probability that the specified event will occur, given that the events on the preceding branches have occurred. When an event can be described by one path through a probability tree, the probability that that event will occur is the *product* of the probabilities on the branches *along the path* that represents the event.

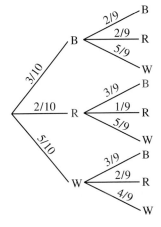

Figure 7.6

● EXAMPLE 4 *Probability Trees*

A box contains 3 black balls, 2 red balls, and 5 white balls. One ball is drawn, it is *not* replaced, and a second ball is drawn. Find the probability that the first ball is red and the second is black.

Solution

Let E_1 be "red ball first" and E_2 be "black ball second." The probability tree for this experiment is shown in Figure 7.6. The path representing the event whose probability we seek is the path through R (red ball) to B (black ball). The probability of this event is the product of the probabilities along this path.

$$\Pr(E_1 \cap E_2) = \Pr(E_1) \cdot \Pr(E_2 \mid E_1) = \frac{2}{10} \cdot \frac{3}{9} = \frac{1}{15}$$

● *Checkpoint*

2. Two balls are drawn, without replacement, from the box described in Example 4.
   (a) Find the probability that both balls are white.
   (b) Find the probability that the first ball is white and the second is red.

● EXAMPLE 5 *Delinquent Accounts*

Polk's Department Store has observed that 80% of its charge accounts have men's names on them and that 16% of the accounts with men's names on them have been delinquent at least once, while 5% of the accounts with women's names on them have been delinquent at least once. What is the probability that an account selected at random

(a) is in a man's name and is delinquent?
(b) is in a woman's name and is delinquent?

Solution

(a) Pr(man's account) = 0.80
    Pr(delinquent | man's) = 0.16
    Pr(man's and delinquent) = Pr(man's $\cap$ delinquent)
    $\qquad\qquad\qquad\qquad\quad$ = Pr(man's) · Pr(delinquent | man's)
    $\qquad\qquad\qquad\qquad\quad$ = (0.80)(0.16) = 0.128

Using the probability tree in Figure 7.7 gives the same probability.

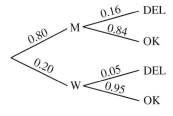

Figure 7.7

(b) Pr(woman's account) = 0.20
Pr(delinquent | woman's) = 0.05

Using the probability tree in Figure 7.7 gives the probability

Pr(woman's and delinquent) = (0.20)(0.05) = 0.01

● EXAMPLE 6 *Manufacturing Inspections*

All products on an assembly line must pass two inspections. It has been determined that the probability that the first inspector will miss a defective item is 0.09. If a defective item gets past the first inspector, the probability that the second inspector will not detect it is 0.01. What is the probability that a defective item will not be rejected by either inspector? (All good items pass both inspections.)

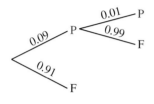

Solution

The second inspector inspects only items passed by the first inspector. The probability that a defective item will pass both inspections is as follows.

$$Pr(\text{defective pass both}) = Pr(\text{pass first} \cap \text{pass second})$$
$$= Pr(\text{pass first}) \cdot Pr(\text{pass second} \mid \text{passed first})$$
$$= (0.09)(0.01) = 0.0009$$

Independent Events If a coin is tossed twice, the sample space is $S = \{HH, HT, TH, TT\}$, so

$$Pr(\text{2d is H} \mid \text{1st is H}) = \frac{Pr(\text{1st is H} \cap \text{2d is H})}{Pr(\text{1st is H})} = \frac{Pr(\text{Both H})}{Pr(\text{1st is H})} = \frac{1/4}{1/2} = \frac{1}{2}$$

We can see that the probability that the second toss of a coin is a head, given that the first toss was a head, was 1/2, which is the same as the probability that the second toss is a head, regardless of what happened on the first toss. This is because the results of the coin tosses are **independent events.** We can define *independent events* as follows.

Independent Events

The events A and B are independent if and only if

$$Pr(A \mid B) = Pr(A) \qquad \text{and} \qquad Pr(B \mid A) = Pr(B)$$

This means that the occurrence or nonoccurrence of one event does not affect the other.

If A and B are independent events, then this definition can be used to simplify the Product Rule:

$$Pr(A \cap B) = Pr(A) \cdot Pr(B \mid A) = Pr(A) \cdot Pr(B)$$

Thus we have the following.

Product Rule for Independent Events

If A and B are independent events, then

$$Pr(A \text{ and } B) = Pr(A \cap B) = Pr(A) \cdot Pr(B)$$

● EXAMPLE 7 *Independent Events*

A die is rolled and a coin is tossed. Find the probability of getting a 4 on the die and a head on the coin.

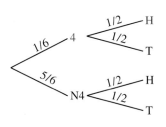

Solution

Let E_1 be "4 on the die" and E_2 be "head on the coin." The events are independent because what occurs on the die does not affect what happens to the coin.

$$\Pr(E_1) = \Pr(4 \text{ on die}) = \frac{1}{6} \quad \text{and} \quad \Pr(E_2) = \Pr(\text{head on coin}) = \frac{1}{2}$$

Then

$$\Pr(E_1 \cap E_2) = \Pr(E_1) \cdot \Pr(E_2) = \frac{1}{6} \cdot \frac{1}{2} = \frac{1}{12}$$

The Product Rules can be expanded to include more than two events, as the next example shows.

● **EXAMPLE 8** *Extending the Product Rules*

A bag contains 3 red marbles, 4 white marbles, and 3 black marbles. Find the probability of getting a red marble on the first draw, a black marble on the second draw, and a white marble on the third draw (a) if the marbles are drawn with replacement, and (b) if the marbles are drawn without replacement.

Solution

(a) The marbles are replaced after each draw, so the events are independent. Let E_1 be "red on first," let E_2 be "black on second," and let E_3 be "white on third." Then

$$\Pr(E_1 \cap E_2 \cap E_3) = \Pr(E_1) \cdot \Pr(E_2) \cdot \Pr(E_3)$$
$$= \frac{3}{10} \cdot \frac{3}{10} \cdot \frac{4}{10} = \frac{36}{1000} = \frac{9}{250}$$

(b) The marbles are not replaced, so the events are dependent.

$$\Pr(E_1 \cap E_2 \cap E_3) = \Pr(E_1) \cdot \Pr(E_2 \mid E_1) \cdot \Pr(E_3 \mid E_1 \text{ and } E_2)$$
$$= \frac{3}{10} \cdot \frac{3}{9} \cdot \frac{4}{8} = \frac{1}{20}$$

◗ **EXAMPLE 9** *PIN Numbers* **(Application Preview)**

Suppose that a person knows your home phone number and has observed the first 3 digits of your PIN. Calculate the probability that the person can guess your PIN if

(a) you have a 4-digit PIN.
(b) you have a 14-digit PIN.

With which PIN would you feel safer?

Solution

(a) Because the person knows the first 3 digits, he or she need guess only the fourth digit, so the probability is

$$1 \cdot 1 \cdot 1 \cdot \frac{1}{10} = \frac{1}{10}$$

(b) The person knows the first 3 digits, so the person must guess the remaining 11 digits. The probability that he or she can do this is

$$1 \cdot 1 \cdot 1 \cdot \frac{1}{10} \cdot \frac{1}{10} \cdot \frac{1}{10} \cdot \frac{1}{10} \cdot \frac{1}{10} \cdot \frac{1}{10} \cdot \frac{1}{10} \cdot \frac{1}{10} \cdot \frac{1}{10} \cdot \frac{1}{10} \cdot \frac{1}{10} = \frac{1}{100,000,000,000}$$

Clearly the 14-digit PIN is much safer than the 4-digit PIN. However, some professional criminals are able to steal even the 14-digit PIN if it is not guarded carefully.

● EXAMPLE 10 *Coin Tossing*

A coin is tossed 4 times. Find the probability that

(a) no heads occur.
(b) at least one head is obtained.

Solution

(a) Each toss is independent, and on each toss, the probability that a tail occurs is $1/2$. Thus the probability of 4 tails in 4 tosses is

$$\text{Pr(TTTT)} = \frac{1}{2} \cdot \frac{1}{2} \cdot \frac{1}{2} \cdot \frac{1}{2} = \frac{1}{16}$$

(b) The probability that at least 1 head will occur can be found by computing

Pr(1 head or 2 heads or 3 heads or 4 heads)

The equiprobable sample space for this probability experiment is

$$S = \{\text{HHHH, HHHT, HHTH, HTHH, THHH, HHTT, HTHT, HTTH,}$$
$$\text{THTH, TTHH, THHT, HTTT, THTT, TTHT, TTTH, TTTT}\}$$

It contains 16 elements, and 15 of these satisfy the conditions of the problem. The only element that does not satisfy the conditions of the problem is the element TTTT, which gives 0 heads. Thus the probability that at least one head occurs is

Pr(at least one head occurs) = Pr(not 0 heads)

$$= 1 - \text{Pr(0 heads)} = 1 - \frac{1}{16} = \frac{15}{16}$$

In general, we can calculate the probability of at least one success in n trials as follows.

Pr(at least 1 success in n trials) $= 1 - $ Pr(0 successes)

● **Checkpoint Solutions**

1. $\text{Pr(white | even-numbered)} = \dfrac{\text{Pr(white and even)}}{\text{Pr(even)}} = \dfrac{1/7}{3/7} = \dfrac{1}{3}$

2. (a) Pr(white on both draws) $= \text{Pr}(W_1 \cap W_2) = \text{Pr}(W_1) \cdot \text{Pr}(W_2 \mid W_1)$

$$= \frac{5}{10} \cdot \frac{4}{9} = \frac{2}{9}$$

(b) Pr(white on first and red on second) $= \text{Pr}(W_1 \cap R_2) = \text{Pr}(W_1) \cdot \text{Pr}(R_2 \mid W_1)$

$$= \frac{5}{10} \cdot \frac{2}{9} = \frac{1}{9}$$

7.3 Exercises

1. A card is drawn from a deck of 52 playing cards. Given that it is a red card, what is the probability that
   (a) it is a heart?   (b) it is a king?

2. A card is drawn from a deck of 52 playing cards. Given that it is an ace, king, queen, or jack, what is the probability that it is a jack?

3. A die has been "loaded" so that the probability of rolling any even number is 2/9 and the probability of rolling any odd number is 1/9. What is the probability of
   (a) rolling a 6, given that an even number is rolled?
   (b) rolling a 3, given that a number divisible by 3 is rolled?

4. A die has been "loaded" so that the probability of rolling any even number is 2/9 and the probability of rolling any odd number is 1/9. What is the probability of
   (a) rolling a 5?
   (b) rolling a 5, given that an even number is rolled?
   (c) rolling a 5, given that an odd number is rolled?

5. A bag contains 9 red balls numbered 1, 2, 3, 4, 5, 6, 7, 8, 9 and 6 white balls numbered 10, 11, 12, 13, 14, 15. One ball is drawn from the bag. What is the probability that the ball is red, given that the ball is even-numbered?

6. If one ball is drawn from the bag in Problem 5, what is the probability that the ball is white, given that the ball is odd-numbered?

7. A bag contains 4 red balls and 6 white balls. Two balls are drawn without replacement.
   (a) What is the probability that the second ball is white, given that the first ball is red?
   (b) What is the probability that the second ball is red, given that the first ball is white?
   (c) Answer (a) if the first ball is replaced before the second is drawn.

8. A fair die is rolled. Find the probability that the result is a 4, given that the result is even.

9. A fair coin is tossed 3 times. Find the probability of
   (a) throwing 3 heads, given that the first toss is a head.
   (b) throwing 3 heads, given that the first two tosses result in heads.

10. A fair coin is tossed 14 times. What is the probability of tossing 14 heads, given that the first 13 tosses are heads?

11. A die is thrown twice. What is the probability that a 3 will result the first time and a 6 the second time?

12. A die is rolled and a coin is tossed. What is the probability that the die shows an even number and the coin toss results in a head?

13. (a) A coin is tossed three times. What is the probability of getting a head on all three tosses?
    (b) A coin is tossed three times. What is the probability of getting at least one tail? [*Hint:* Use the result from part (a)].

14. The probability that Sam will win in a certain game whenever he plays is 2/5. If he plays two games, what is the probability that he will win just the first game?

15. (a) A box contains 3 red balls, 2 white balls, and 5 black balls. Two balls are drawn at random from the box (with replacement of the first before the second is drawn). What is the probability of getting a red ball on the first draw and a white ball on the second?
    (b) Answer the question in part (a) if the first ball is not replaced before the second is drawn.
    (c) Are the events in part (a) or in part (b) independent? Explain.

16. (a) One card is drawn at random from a deck of 52 cards. The first card is replaced, and a second card is drawn. Find the probability that both are hearts.
    (b) Answer the question in part (a) if the first card is not replaced before the second card is drawn.
    (c) Are the events in part (a) or in part (b) independent? Explain.

17. Two balls are drawn from a bag containing 3 white balls and 2 red balls. If the first ball is replaced before the second is drawn, what is the probability that
    (a) both balls are red?   (b) both balls are white?
    (c) the first ball is red and the second is white?
    (d) one of the balls is black?

18. Answer Problem 17 if the first ball is not replaced before the second is drawn.

19. A bag contains 9 nickels, 4 dimes, and 5 quarters. If you draw 3 coins at random from the bag, without replacement, what is the probability that you will get a nickel, a quarter, and a nickel, in that order?

20. A bag contains 6 red balls and 8 green balls. If two balls are drawn together, find the probability that
    (a) both are red.   (b) both are green.

21. A red ball and 4 white balls are in a box. If two balls are drawn, without replacement, what is the probability
    (a) of getting a red ball on the first draw and a white ball on the second?
    (b) of getting 2 white balls?
    (c) of getting 2 red balls?

22. From a deck of 52 playing cards, two cards are drawn, one after the other without replacement. What is the probability that
    (a) the first will be a king and the second will be a jack?
    (b) the first will be a king and the second will be a jack of the same suit?

23. One card is drawn at random from a deck of 52 cards. The first card is not replaced, and a second card is drawn. Find the probability that
    (a) both cards are spades.
    (b) the first card is a heart and the second is a club.

24. Two cards are drawn from a deck of 52 cards. What is the probability that both are aces
    (a) if the first card is replaced before the second is drawn?
    (b) if the cards are drawn without replacement?

25. Two balls are drawn, without replacement, from a bag containing 13 red balls numbered 1–13 and 5 white balls numbered 14–18. What is the probability that
    (a) the second ball is red, given that the first ball is white?
    (b) both balls are even-numbered?
    (c) the first ball is red and even-numbered and the second ball is even-numbered?

26. A red die and a green die are rolled. What is the probability that a 6 results on the green die and that a 4 results on the red die?

APPLICATIONS

Universal health care **The following table gives the results of a 2005 survey of 1000 people regarding the funding of universal health care by employers (employer mandate). Use the table to answer Problems 27–30.**

	Favor	Oppose	No Opinion	Total
Democrat	310	150	60	520
Republican	125	345	10	480
Total	435	495	70	1000

27. What is the probability that a person selected at random from this group will favor universal health care with an employer mandate, given that the person is a Democrat?

28. What is the probability that a person selected at random from this group will oppose universal health care with an employer mandate, if the person is a Republican?

29. If the group surveyed represents the people of the United States, what is the probability that a citizen selected at random will favor universal health care with an employer mandate, given that the person is a Republican?

30. If the group surveyed represents the people of the United States, what is the probability that a citizen selected at random will oppose universal health care with an employer mandate, given that the person is a Democrat?

Military spending **Suppose the following table summarizes the opinions of various groups on the issue of increased military spending. Use this table to calculate the empirical probabilities in Problems 31–38.**

| Opinion | Whites | | Nonwhites | | Total |
	Reps.	Dems.	Reps.	Dems.	
Favor	300	100	25	10	435
Oppose	100	250	25	190	565
Total	400	350	50	200	1000

31. Given that a randomly selected individual is nonwhite, find the probability that he or she opposes increased military spending.

32. Given that a randomly selected individual is a Democrat, find the probability that he or she opposes increased military spending.

33. Given that a randomly selected individual is in favor of increased military spending, find the probability that he or she is a Republican.

34. Given that a randomly selected individual is opposed to increased military spending, find the probability that he or she is a Democrat.

35. Find the probability that a person who favors increased military spending is nonwhite.

36. Find the probability that an individual is white and opposes increased military spending.

37. Find the probability that an individual is a white Republican opposed to increased military spending.

38. Find the probability that an individual is a Democrat and opposes increased military spending.

39. *Golf* On June 16, 1997, two amateur golfers playing together hit back-to-back holes in one (*Source: The Island Packet,* June 19, 1997). According to the National Hole-in-One Association, the probability of an amateur golfer getting a hole-in-one is 1/12,000. If the golfer's shots are independent of each other, what is the probability that two amateur golfers will get back-to-back holes in one?

40. *Lottery* A name is drawn at random from the 96 entrants in a golf tournament. The person whose name is selected draws a ball from a bag containing four balls, one of which is embossed with the Peggos Company logo. If the person draws the ball with the logo, he or she wins a set of Ping irons. If you are one of the entrants in the tournament, what is the probability that you will win the Ping irons?

41. *Blood types* In the pretrial hearing of the O. J. Simpson case, the prosecution stated that Mr. Simpson's blood markers included type A blood, which 33.7% of the population has; blood SD subtype 1, which 79.6% of the population has; and PGM 2+2−, which 1.6% of the population has. If these blood markers are independent, what is the probability that a person selected at random will have the same blood markers as O. J. Simpson?

42. *Education* Fifty-one percent of the U.S. population are female, and 23.6% of the female population have college degrees (*Source:* Bureau of the Census, Current Population Survey: *Educational Attainment in the U.S., March 2000* (2001)). If a U.S. resident is chosen at random, what is the probability that the person is a female with a college degree?

43. *Quality control*   Each computer component that the Peggos Company produces is tested twice before it is shipped. There is a 0.7 probability that a defective component will be so identified by the first test and a 0.8 probability that it will be identified as being defective by the second test. What is the probability that a defective component will not be identified as defective before it is shipped?

44. *Lactose intolerance*   Lactose intolerance affects about 20% of non-Hispanic white Americans, and 75.6% of the residents of the United States are non-Hispanic whites (*Source:* Jean Carper, "Eat Smart," *USA Weekend*). If a U.S. resident is selected at random, what is the probability that the person will be a non-Hispanic white and have lactose intolerance?

45. *Lactose intolerance*   Lactose intolerance affects about 50% of Hispanic Americans, and 9% of the residents of the United States are Hispanic (*Source:* Jean Carper, "Eat Smart," *USA Weekend*). If a U.S. resident is selected at random, what is the probability that the person will be Hispanic and have lactose intolerance?

46. *Quality control*   If 3% of all light bulbs a company manufactures are defective, the probability of any one bulb being defective is 0.03. What is the probability that three bulbs drawn independently from the company's stock will be defective?

47. *Quality control*   To test its shotgun shells, a company fires 5 of them. What is the probability that all 5 will fire properly if 5% of the company's shells are actually defective?

48. *Quality control*   One machine produces 30% of a product for a company. If 10% of the products from this machine are defective and the other machines produce no defective items, what is the probability that an article produced by this company is defective?

49. *Advertising*   A company estimates that 30% of the country has seen its commercial and that if a person sees its commercial, there is a 20% probability that the person will buy its product. What is the probability that a person chosen at random in the country will have seen the commercial and bought the product?

50. *Employment*   Ronald Lee has been told by a company that the probability that he will be offered a job in the quality control department is 0.6 and the probability that he will be asked to be foreman of the department, if he is offered the job, is 0.1. What is the probability that he will be offered the job and asked to be foreman?

51. *Birth control*   Suppose a birth control pill is 99% effective in preventing pregnancy.

(a) What is the probability that none of 100 women using the pill will become pregnant?

(b) What is the probability that at least one woman per 100 users will become pregnant?

52. *Combat*   If a fighter pilot has a 4% chance of being shot down on each mission during a war, is it certain he will be shot down in 25 missions?

53. *Maintenance*   Twenty-three percent of the cars owned by a car rental firm have some defect. What is the probability that of 3 cars selected at random,

(a) none has a defect?

(b) at least one has a defect?

54. *Testing*   An unprepared student must take a 7-question multiple-choice test that has 5 possible answers per question. If the student guesses on every question, what is the probability that

(a) she will answer every question correctly?

(b) she will answer every question incorrectly?

(c) she will answer at least one question correctly?

55. *Testing*   An unprepared student must take a 7-question multiple-choice test that has 5 possible answers per question. If the student can eliminate two of the possible answers on the first three questions, and if she guesses on every question, what is the probability that

(a) she will answer every question correctly?

(b) she will answer every question incorrectly?

(c) she will answer at least one question correctly?

56. *Testing*   What is the probability that a student who guesses on every question of a 10-question true-or-false test will get at least one answer correct?

57. *Racing*   Suppose that the odds in favor of the horse Portia winning a race are 3:8 and that the odds of the horse Trinka winning the same race are 1:10. What is the probability that Portia or Trinka will win the race? What are the odds that one or the other will win?

58. *Racing*   Suppose that the odds that Blackjack will win a race are 1 to 3 and the odds that Snowball will win the same race are 1 to 5. If only one horse can win, what odds should be given that one of these two horses will win?

59. *Birth dates*   Assuming there are 365 different birthdays, find the probability that two people chosen at random will have

(a) different birthdays.   (b) the same birthday.

60. *Birth dates*   Assuming there are 365 different birthdays, find the probability that of three people chosen at random,

(a) no two will have the same birthday.

(b) at least two will have the same birthday.

61. *Birth dates* Assuming there are 365 different birthdays, find the probability that of 20 people chosen randomly,
    (a) no two will have the same birthday.
    (b) at least two will have the same birthday.

62. *Crime* In an actual case,* probability was used to convict a couple of mugging an elderly woman. Shortly after the mugging, a young, white woman with blonde hair worn in a ponytail was seen running from the scene of the crime and entering a yellow car that was driven away by a black man with a beard. A couple matching this description was arrested for the crime. A prosecuting attorney argued that the couple arrested had to be the couple at the scene of the crime because the probability of a second couple matching the description was

very small. He estimated the probabilities of six events as follows:

> Probability of black-white couple: 1/1000
> Probability of black man: 1/3
> Probability of bearded man: 1/10
> Probability of blonde woman: 1/4
> Probability of hair in ponytail: 1/10
> Probability of yellow car: 1/10

He multiplied these probabilities and concluded that the probability that another couple would have these characteristics is 1/12,000,000. On the basis of this circumstantial evidence, the couple was convicted and sent to prison. The conviction was overturned by the state supreme court because the prosecutor made an incorrect assumption. What error do you think he made?

*Time, January 8, 1965, p. 42, and April 26, 1968, p. 41.

7.4 Probability Trees and Bayes' Formula

OBJECTIVES

● To use probability trees to solve problems
● To use Bayes' formula to solve probability problems

◗ Application Preview

Suppose that a test for diagnosing a certain serious illness is successful in detecting the disease in 95% of all infected persons who are tested but that it incorrectly diagnoses 4% of all healthy people as having the serious disease. Suppose also that it incorrectly diagnoses 12% of all people having another minor disease as having the serious disease. If it is known that 2% of the population have the serious disease, 90% are healthy, and 8% have the minor disease, we can use a formula called **Bayes' formula** to find the probability that a person has the serious disease if the test indicates that he or she does. (See Example 3.) Such a **Bayes problem** can be solved with the formula or with the use of probability trees.

Probability Trees

In Section 7.3, "Conditional Probability: The Product Rule," we used probability trees to find probabilities in experiments involving more than one trial. Trees provide a systematic way to analyze probability experiments that have two or more trials or that use multiple paths within the tree. In this section, we will use trees to solve some additional probability problems and to solve Bayes problems. Recall that in a probability tree, we find the probability of an event as follows.

Probability Trees

The probability attached to each branch is the conditional probability that the specified event will occur, given that the events on the preceding branches have occurred.

1. For each event that can be described by a single path through the tree, the probability that that event will occur is the product of the probabilities on the branches along the path that represents the event.
2. If an event can be described by two or more paths through a probability tree, its probability is found by adding the probabilities from the paths.

Consider the following example.

● **EXAMPLE 1** *Probability Trees*

A bag contains 5 red balls, 4 blue balls, and 3 white balls. Two balls are drawn, one after the other, without replacement. Draw a tree representing the experiment and find

(a) Pr(blue on first draw and white on second draw).
(b) Pr(white on both draws).
(c) Pr(drawing a blue ball and a white ball).
(d) Pr(second ball is red).

Solution
The tree is shown in Figure 7.8.

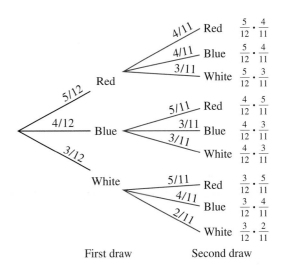

Figure 7.8　　　　　First draw　　　　　Second draw

(a) By multiplying the probabilities along the path that represents blue on the first draw and white on the second draw, we obtain

$$\text{Pr(blue on the first draw and white on second draw)} = \frac{4}{12} \cdot \frac{3}{11} = \frac{1}{11}$$

(b) Multiplying the probabilities along the path that represents white on both draws gives

$$\text{Pr(white on both draws)} = \frac{3}{12} \cdot \frac{2}{11} = \frac{1}{22}$$

(c) This result can occur by drawing a blue ball first and then a white ball *or* by drawing a white ball and then a blue ball. Thus either of two paths leads to this result. Because the results represented by these paths are mutually exclusive, the probability is found by *adding* the probabilities from the two paths. That is, letting B represent a blue ball and W represent a white ball, we have

$$\text{Pr}(B \text{ and } W) = \text{Pr}(B \cap W)$$
$$= \text{Pr}(B \text{ first, then } W) + \text{Pr}(W \text{ first, then } B)$$
$$= \frac{4}{12} \cdot \frac{3}{11} + \frac{3}{12} \cdot \frac{4}{11} = \frac{2}{11}$$

(d) The second ball can be red after the first is red, blue, or white.

$$\text{Pr(2d ball is red)} = \text{Pr}(R \text{ then } R, \text{ or } B \text{ then } R, \text{ or } W \text{ then } R)$$
$$= \text{Pr}(R \text{ then } R) + \text{Pr}(B \text{ then } R) + \text{Pr}(W \text{ then } R)$$
$$= \frac{5}{12} \cdot \frac{4}{11} + \frac{4}{12} \cdot \frac{5}{11} + \frac{3}{12} \cdot \frac{5}{11} = \frac{5}{12}$$

1. Urn I contains 3 gold coins, urn II contains 1 gold coin and 2 silver coins, and urn III contains 1 gold coin and 1 silver coin. If an urn is selected at random and a coin is drawn from the urn, construct a probability tree and find the probability that a gold coin will be drawn.

● EXAMPLE 2 *Quality Control*

A security system is manufactured with a "fail-safe" provision so that it functions properly if any two or more of its three main components, A, B, and C, are functioning properly. The probabilities that components A, B, and C are functioning properly are 0.95, 0.90, and 0.92, respectively. What is the probability that the system functions properly?

Solution

The probability tree in Figure 7.9 shows all of the possibilities for the components of the system. Each stage of the tree corresponds to a component, and G indicates that the component is good; N indicates it is not good. Four paths through the probability tree give at least two good (functioning) components, so there are four mutually exclusive possible outcomes. The probability that the system functions properly is

$$\text{Pr(at least } 2G) = \text{Pr}(GGG \cup GGN \cup GNG \cup NGG)$$

$$= (0.95)(0.90)(0.92) + (0.95)(0.90)(0.08) + (0.95)(0.10)(0.92) + (0.05)(0.90)(0.92)$$

$$= 0.9838$$

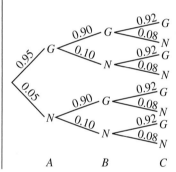

Figure 7.9 A B C

Bayes' Formula

Using a probability tree permits us to answer more difficult questions regarding conditional probability. For example, in the experiment described in Example 1, we can find the probability that the first ball is red, given that the second ball is blue. To find this probability, we recall the formula for conditional probability, which is

$$\text{Pr}(E_2 \mid E_1) = \frac{\text{Pr}(E_1 \cap E_2)}{\text{Pr}(E_1)}$$

so

$$\text{Pr(1st is red} \mid \text{2d is blue)} = \frac{\text{Pr(1st is red and 2d is blue)}}{\text{Pr(2d is blue)}}$$

One path (red-blue) in Figure 7.8 in Example 1 describes "1st is red and 2d is blue" and three paths (red-blue, blue-blue, white-blue) describe "2d is blue," so

$$\text{Pr}(1\text{st is red} \mid 2\text{d is blue}) = \frac{\dfrac{5}{12} \cdot \dfrac{4}{11}}{\dfrac{5}{12} \cdot \dfrac{4}{11} + \dfrac{4}{12} \cdot \dfrac{3}{11} + \dfrac{3}{12} \cdot \dfrac{4}{11}}$$

$$= \frac{5/33}{5/33 + 1/11 + 1/11} = \frac{5}{11}$$

This example illustrates a special type of problem, which we call a **Bayes' problem.** In this type of problem, we know the result of the second stage of a two-stage experiment and wish to find the probability of a specified result in the first stage. In the example above we knew the second ball drawn was blue, and we sought the probability that the first ball was red.

Suppose there are n possible outcomes in the first stage of the experiment, denoted $E_1, E_2, \ldots, E_n$, and m possible outcomes in the second stage, denoted $F_1, F_2, \ldots, F_m$ (see Figure 7.10). Then the probability that event E_1 occurs in the first stage, given that F_1 has occurred in the second stage, is

$$\text{Pr}(E_1 \mid F_1) = \frac{\text{Pr}(E_1 \cap F_1)}{\text{Pr}(F_1)} \tag{1}$$

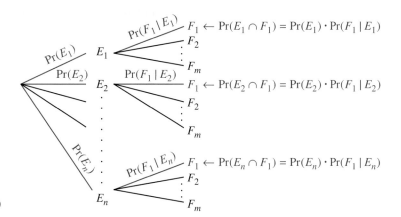

Figure 7.10

By looking at the probability tree in Figure 7.10, we see that

$$\text{Pr}(E_1 \cap F_1) = \text{Pr}(E_1) \cdot \text{Pr}(F_1 \mid E_1)$$

and that

$$\text{Pr}(F_1) = \text{Pr}(E_1 \cap F_1) + \text{Pr}(E_2 \cap F_1) + \cdots + \text{Pr}(E_n \cap F_1)$$

Using these facts, equation (1) becomes

$$\text{Pr}(E_1 \mid F_1) = \frac{\text{Pr}(E_1) \cdot \text{Pr}(F_1 \mid E_1)}{\text{Pr}(E_1 \cap F_1) + \text{Pr}(E_2 \cap F_1) + \cdots + \text{Pr}(E_n \cap F_1)} \tag{2}$$

Using the fact that for any E_i, $\text{Pr}(E_i \text{ and } F_1) = \text{Pr}(E_i) \cdot \text{Pr}(F_1 \mid E_i)$, we can rewrite equation (2) in a form called **Bayes' formula.**

Bayes' Formula

If a probability experiment has n possible outcomes in its first stage given by E_1, $E_2, \ldots, E_n$, and if F_1 is an event in the second stage, then the probability that event E_1 occurs in the first stage, given that F_1 has occurred in the second stage, is

$$\Pr(E_1 | F_1) = \frac{\Pr(E_1) \cdot \Pr(F_1 | E_1)}{\Pr(E_1) \cdot \Pr(F_1 | E_1) + \Pr(E_2) \cdot \Pr(F_1 | E_2) + \cdots + \Pr(E_n) \cdot \Pr(F_1 | E_n)}$$

Note that Bayes problems can be solved either by this formula or by the use of a probability tree. Of the two methods, many students find using the tree easier because it is less abstract than the formula.

Bayes' Formula and Trees

$$\Pr(E_1 | F_1) = \frac{\text{Product of branch probabilities on path leading to } F_1 \text{ through } E_1}{\text{Sum of all branch products on paths leading to } F_1}$$

◗ EXAMPLE 3 *Medical Tests* (Application Preview)

Suppose a test for diagnosing a certain serious disease is successful in detecting the disease in 95% of all persons infected but that it incorrectly diagnoses 4% of all healthy people as having the serious disease. Suppose also that it incorrectly diagnoses 12% of all people having another minor disease as having the serious disease. It is known that 2% of the population have the serious disease, 90% of the population are healthy, and 8% have the minor disease. Find the probability that a person selected at random has the serious disease if the test indicates that he or she does. Use H to represent healthy, M to represent having the minor disease, and D to represent having the serious disease.

Solution

The tree that represents the health condition of a person chosen at random and the results of the test on that person are shown in Figure 7.11. (A test that indicates that a person has the disease is called a *positive* test.)

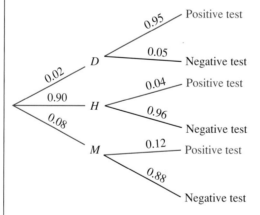

Figure 7.11

We seek $\Pr(D \mid \text{pos. test})$. Using the tree and the conditional probability formula gives

$$\Pr(D \mid \text{pos. test}) = \frac{\Pr(D \cap \text{pos. test})}{\Pr(\text{pos. test})}$$

$$= \frac{(0.02)(0.95)}{(0.90)(0.04) + (0.02)(0.95) + (0.08)(0.12)}$$

$$= \frac{0.0190}{0.0360 + 0.0190 + 0.0096}$$

$$\approx 0.2941$$

Using Bayes' formula directly gives the same result.

$\Pr(D \mid \text{pos. test})$

$$= \frac{\Pr(D) \cdot \Pr(\text{pos. test} \mid D)}{\Pr(H) \cdot \Pr(\text{pos. test} \mid H) + \Pr(D) \cdot \Pr(\text{pos. test} \mid D) + \Pr(M) \cdot \Pr(\text{pos. test} \mid M)}$$

$$= \frac{(0.02)(0.95)}{(0.90)(0.04) + (0.02)(0.95) + (0.08)(0.12)} \approx 0.2941$$

● *Checkpoint*

2. Use the information (and tree) from Example 3 to find the probability that a person chosen at random is healthy if the test is negative (a negative result indicates that the person does not have the disease).

Summary   Table 7.2 gives a summary of probability formulas and when they are used.

TABLE 7.2   Summary of Probability Formulas

		One Trial	Two Trials	More Than Two Trials
Pr(*A* and *B*)	Independent	Sample space	Pr(*A*) · Pr(*B*)	Product of probabilities
	Dependent	Sample space	Pr(*A*) · Pr(*B* | *A*)	Product of conditional probabilities
Pr(*A* or *B*)	Mutually exclusive	Pr(*A*) + Pr(*B*)	Probability tree	Probability tree
	Not mutually exclusive	Pr(*A*) + Pr(*B*) − Pr(*A* and *B*)	Probability tree	Probability tree

● *Checkpoint Solutions*

1. The probability tree for this probability experiment is given below. From the tree we have

$$\Pr(\text{gold}) = \Pr[(\text{I and } G) \text{ or } (\text{II and } G) \text{ or } (\text{III and } G)]$$

$$= \frac{1}{3} \cdot \frac{3}{3} + \frac{1}{3} \cdot \frac{1}{3} + \frac{1}{3} \cdot \frac{1}{2} = \frac{11}{18}$$

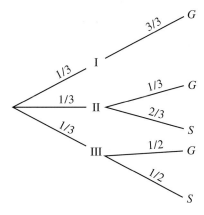

2. The probability that a person is healthy, given that the test is negative, is

$$\Pr(H \mid \text{neg. test}) = \frac{\Pr(H \cap \text{neg. test})}{\Pr(\text{neg. test})}$$

$$= \frac{(0.90)(0.96)}{(0.02)(0.05) + (0.90)(0.96) + (0.08)(0.88)}$$

$$\approx 0.9237$$

7.4 Exercises

In Problems 1–12, use probability trees to find the probabilities of the indicated outcomes.

1. A bag contains 5 coins, 4 of which are fair and 1 that has a head on each side. If a coin is selected from the bag and tossed twice, what is the probability of obtaining 2 heads?

2. Three balls are drawn, without replacement, from a bag containing 4 red balls and 5 white balls. Find the probability that
   (a) three white balls are drawn.
   (b) two white balls and one red ball are drawn.
   (c) the third ball drawn is red.

3. Two cards are drawn, without replacement, from a regular deck of 52 cards. What is the probability that both cards are aces, given that they are the same color?

4. An urn contains 4 red, 5 white, and 6 black balls. One ball is drawn from the urn, it is replaced, and a second ball is drawn.
   (a) What is the probability that both balls are white?
   (b) What is the probability that one ball is white and one is red?
   (c) What is the probability that at least one ball is black?

5. An urn contains 4 red, 5 white, and 6 black balls. One ball is drawn from the urn, it is not replaced, and a second ball is drawn.
   (a) What is the probability that both balls are white?
   (b) What is the probability that one ball is white and one is red?
   (c) What is the probability that at least one ball is black?

6. An urn contains 4 red, 5 white, and 6 black balls. Three balls are drawn, without replacement, from the urn.
   (a) What is the probability that all three balls are red?
   (b) What is the probability that exactly two balls are red?
   (c) What is the probability that at least two balls are red?

7. A bag contains 4 white balls and 6 red balls. Three balls are drawn, without replacement, from the bag.
   (a) What is the probability that all three balls are white?
   (b) What is the probability that exactly one ball is white?
   (c) What is the probability that at least one ball is white?

8. A bag contains 5 coins, of which 4 are fair; the remaining coin has a head on both sides. If a coin is selected

at random from the bag and tossed three times, what is the probability that heads will occur exactly twice?

9. A bag contains 5 coins, of which 4 are fair; the remaining coin has a head on both sides. If a coin is selected at random from the bag and tossed three times, what is the probability that heads will occur at least twice?

10. A bag contains 4 white balls and 7 red balls. If 2 balls are drawn at random from the bag, with the first ball being replaced before the second ball is drawn, what is the probability that
    (a) the first ball is red and the second is white?
    (b) one ball is red and the other is white?

11. A bag contains 4 white balls and 6 red balls. What is the probability that if 2 balls are drawn (with replacement),
    (a) the first ball is red and the second is white?
    (b) both balls are red?
    (c) one ball is red and one is white?
    (d) the first ball is red or the second is white?

12. Two balls are drawn, without replacement, from a bag containing 13 red balls numbered 1–13 and 5 white balls numbered 14–18. What is the probability that
    (a) the second ball is even-numbered, given that the first ball is even-numbered?
    (b) the first ball is red and the second ball is even-numbered?
    (c) the first ball is even-numbered and the second is white?

In Problems 13–16, use (a) a probability tree and (b) Bayes' formula to find the probabilities.

In Problems 13 and 14, each of urns I and II has 5 red balls, 3 white balls, and 2 green balls. Urn III has 1 red ball, 1 white ball, and 8 green balls.

13. An urn is selected at random and a ball is drawn. If the ball is green, find the probability that urn III was selected.

14. An urn is selected at random and a ball is drawn. If the ball is red, what is the probability that urn III was selected?

15. There are 3 urns containing coins. Urn I contains 3 gold coins, urn II contains 1 gold coin and 1 silver coin, and urn III contains 2 silver coins. An urn is selected and a coin is drawn from the urn. If the selected coin is gold, what is the probability that the urn selected was urn I?

16. In Problem 15, what is the probability that the urn selected was urn III if the coin selected was silver?

APPLICATIONS

17. *Lactose intolerance* Lactose intolerance affects about 20% of non-Hispanic white Americans, 50% of Hispanic Americans, and 75% of African, Asian, and Native Americans (*Source:* Jean Carper, "Eat Smart," *USA Weekend*). Seventy-six percent of U.S. residents are non-Hispanic whites, 9% of them are Hispanic, and 15% are African, Asian, or Native American. If a person is selected from this group of people, what is the probability that the person will have lactose intolerance?

18. *Genetics* What is the probability that a couple will have at least two sons if they plan to have 3 children and if the probability of having a son equals the probability of having a daughter?

19. *Marksmanship* Suppose that a marksman hits the bull's-eye 15,000 times in 50,000 shots. If the next 4 shots are independent, find the probability that
(a) the next 4 shots hit the bull's-eye.
(b) two of the next 4 shots hit the bull's-eye.

20. *Majors* In a random survey of students concerning student activities, 30 engineering majors, 25 business majors, 20 science majors, and 15 liberal arts majors were selected. If two students are selected at random, what is the probability of getting
(a) two science majors?
(b) a science major and an engineering major?

21. *Quality control* Suppose a box contains 3 defective transistors and 12 good transistors. If 2 transistors are drawn from the box without replacement, what is the probability that
(a) the first transistor is good and the second transistor is defective?
(b) the first transistor is defective and the second one is good?
(c) one of the transistors drawn is good and one of them is defective?

22. *Education* The probability that an individual without a college education earns more than $30,000 is 0.2, whereas the probability that a person with a B.S. or higher degree earns more than $30,000 is 0.6. The probability that a person chosen at random has a B.S. degree is 0.3. What is the probability that a person has at least a B.S. degree if it is known that he or she earns more than $30,000?

23. *Alcoholism* A small town has 8000 adult males and 6000 adult females. A sociologist conducted a survey and found that 40% of the males and 30% of the females drink heavily. An adult is selected at random from the town.
(a) What is the probability the person is a male?
(b) What is the probability the person drinks heavily?
(c) What is the probability the person is a male or drinks heavily?
(d) What is the probability the person is a male, if it is known that the person drinks heavily?

24. *TV violence* One hundred boys and 100 girls were asked if they had ever been frightened by a television program. Thirty of the boys and 60 of the girls said they had been frightened. One of these children is selected at random.
(a) What is the probability that he or she has been frightened?
(b) What is the probability the child is a girl, given that he or she has been frightened?

25. *Drinking age* A survey questioned 1000 people regarding raising the legal drinking age from 18 to 21. Of the 560 who favored raising the age, 390 were female. Of the 440 opposition responses, 160 were female. If a person selected at random from this group is a man, what is the probability that the person favors raising the drinking age?

26. *Politics* A candidate for public office knows that he has a 60% chance of being elected to the office if he is nominated by his political party and that his party has a 50% chance of winning with another nominee. If the probability that he will be nominated by his party is 0.25, and his party wins the election, what is the probability that he is the winning candidate?

27. *Pregnancy test* A self-administered pregnancy test detects 85% of those who are pregnant but does not detect pregnancy in 15%. It is 90% accurate in indicating women who are not pregnant but indicates 10% of this group as being pregnant. Suppose it is known that 1% of the women in a neighborhood are pregnant. If a woman is chosen at random from those living in this neighborhood, and if the test indicates she is pregnant, what is the probability that she really is?

28. *Drinking age* A survey questioned 1000 people regarding raising the legal drinking age from 18 to 21. Of the 560 who favored raising the age, 390 were female. Of the 440 who expressed opposition, 160 were female.
(a) What is the probability that a person selected at random is a female?
(b) What is the probability that a person selected at random favors raising the age if the person is a woman?

29. *Survey* In a SAPS (Student Activity Participation Study) survey, 30 engineering majors, 25 business majors, 20 science majors, 15 liberal arts majors, and 10 human development majors were selected. Ten of the engineering, 12 of the science, 6 of the human development, 13 of the business, and 8 of the liberal arts majors selected for the study were female.
    (a) What is the probability of selecting a female if one person from this group is selected randomly?
    (b) What is the probability that a student selected randomly from this group is a science major, given that she is female?

30. *Survey* If a student is selected at random from the group in Problem 29, what is the probability the student is
    (a) an engineering major, given that she is female?
    (b) an engineering major, given that he is male?

7.5 Counting: Permutations and Combinations

◗ Application Preview

During a national television advertising campaign, Little Caesar's Pizza stated that for $9.95, you could get 2 medium-sized pizzas, each with any of 0 to 5 toppings chosen from 11 that are available. The commercial asked the question, "How many different pairs of pizzas can you get?" (See Example 8.) Answering this question uses the computation of **combinations,** which are discussed in this section. Other counting techniques, including **permutations,** are also discussed in this section.

Fundamental Counting Principle

Suppose that you decide to dine at a restaurant that offers 3 appetizers, 8 entrees, and 6 desserts. How many different meals can you have? We can answer questions of this type using the **Fundamental Counting Principle.**

Fundamental Counting Principle

If there are $n(A)$ ways in which an event A can occur, and if there are $n(B)$ ways in which a second event B can occur after the first event has occurred, then the two events can occur in $n(A) \cdot n(B)$ ways.

The Fundamental Counting Principle can be extended to any number of events as long as they are independent. Thus the total number of possible meals at the restaurant mentioned above is

$$n(\text{appetizers}) \cdot n(\text{entrees}) \cdot n(\text{desserts}) = 3 \cdot 8 \cdot 6 = 144$$

● EXAMPLE 1 *Head Table Seating*

Suppose a person planning a banquet cannot decide how to seat 6 honored guests at the head table. In how many arrangements can they be seated in the 6 chairs on one side of a table?

Solution
If we think of the 6 chairs as spaces, we can determine how many ways each space can be filled. The planner could place any one of the 6 people in the first space (say the first chair on the left). One of the 5 remaining persons could then be placed in the second space, one of the 4 remaining persons in the third space, and so on, as illustrated below.

$$\underline{6} \quad \underline{5} \quad \underline{4} \quad \underline{3} \quad \underline{2} \quad \underline{1}$$

By the Fundamental Counting Principle, the total number of arrangements that can be made is the product of these numbers. That is, the total number of possible arrangements is

$$6 \cdot 5 \cdot 4 \cdot 3 \cdot 2 \cdot 1 = 720$$

Because special products such as $6 \cdot 5 \cdot 4 \cdot 3 \cdot 2 \cdot 1$ frequently occur in counting theory, we use special notation to denote them. We write 6! (read "6 factorial") to denote $6 \cdot 5 \cdot 4 \cdot 3 \cdot 2 \cdot 1$. Likewise, $4! = 4 \cdot 3 \cdot 2 \cdot 1$.

Factorial

For any positive integer n, we define n **factorial,** written $n!$, as

$$n! = n(n - 1)(n - 2) \cdots 3 \cdot 2 \cdot 1$$

We define $0! = 1$.

● EXAMPLE 2 *Banquet Seating*

Suppose the planner in Example 1 knows that 8 people feel they should be at the head table, but only 6 spaces are available. How many arrangements can be made that place 6 of the 8 people at the head table?

Solution

There are again 6 spaces to fill, but any of 8 people can be placed in the first space, any of 7 people in the second space, and so on. Thus the total number of possible arrangements is

$$8 \cdot 7 \cdot 6 \cdot 5 \cdot 4 \cdot 3 = 20,160$$

● EXAMPLE 3 *License Plates*

Pennsylvania offers a special "Save Wild Animals" license plate, which gives Pennsylvania zoos $15 for each plate sold. These plates have the two letters P and Z, followed by four numbers. When plates like these have all been purchased, additional plates will replace the final digit with a letter other than O. What is the maximum amount that the zoos could receive from the sale of the plates?

Solution

The following numbers represent the number of possible digits or letters in each of the spaces on the license plates as illustrated below.

$$\underline{1} \quad \underline{1} \quad \underline{10} \quad \underline{10} \quad \underline{10} \quad \underline{35}$$

Note that the last space can be a number from 0 to 9 or a letter other than O. The product of these numbers gives the number of license plates that can be made satisfying the given conditions.

$$1 \cdot 1 \cdot 10 \cdot 10 \cdot 10 \cdot 35 = 35,000$$

Thus 35,000 automobiles can be licensed with these plates, so the zoos could receive ($15)(35,000) = $525,000.

● *Checkpoint*

1. If a state permits either a letter or a nonzero digit to be used in each of six places on its license plates, how many different plates can it issue?

Permutations

The number of possible arrangements in Example 2 is called the number of **permutations** of 8 things taken 6 at a time, and it is denoted $_8P_6$. Note that $_8P_6$ gives the first 6 factors of 8!, so we can use factorial notation to write the product.

$$_8P_6 = \frac{8!}{2!} = \frac{8!}{(8-6)!}$$

Permutations

The number of possible distinct arrangements of r objects chosen from a set of n objects is called the number of **permutations** of n objects taken r at a time, and it equals

$$_nP_r = \frac{n!}{(n-r)!}$$

Note that $_nP_n = n!$

● **EXAMPLE 4 Club Officers**

In how many ways can a president, a vice president, a secretary, and a treasurer be selected from an organization with 20 members?

Solution

We seek the number of orders (arrangements) in which 4 people can be selected from a group of 20. This number is

$$_{20}P_4 = \frac{20!}{(20-4)!} = 116,280$$

● *Checkpoint*

2. In a 5-question matching test, how many different answer sheets are possible if no answer can be used twice and there are
   (a) 5 answers available for matching?
   (b) 7 answers available for matching?

Combinations

When we talk about arrangements of people at a table or arrangements of digits on a license plate, order is important. We now consider counting problems in which order is not important. Suppose you are the president of a company and you want to pick 2 secretaries to work for you. If 5 people are qualified, how many different pairs of people can you select? If you select 2 people from the 5 in a specific order, the number of *arrangements* of the 5 people taken 2 at a time would be

$$_5P_2 = 20$$

But $_5P_2$ gives the number of ways you can select *and* order the pairs, and we seek only the number of ways you can select them. If the names of the people are A, B, C, D, and E, then two of the arrangements, AB and BA, would represent only one pair. Because each of the pairs of people can be arranged in 2! = 2 ways, the number of pairs that can be selected without regard to order is

$$\frac{_5P_2}{2!} = \frac{20}{2} = 10$$

To find the number of ways you can select 3 people from the 5 without regard to order, we would first find $_5P_3$ and then divide by the number of ways the 3 people could be ordered (3!). Thus 3 people can be selected from 5 (without regard for order) in $_5P_3/3! = 10$ ways. We say that the number of combinations of 5 things taken 3 at a time is 10.

Combinations

> The number of ways in which r objects can be chosen from a set of n objects without regard to the order of selection is called the number of **combinations** of n objects taken r at a time, and it equals
>
> $$_nC_r = \frac{_nP_r}{r!}$$

Note that the fundamental difference between permutations and combinations is that permutations are used when order is a factor in the selection, and combinations are used when it is not. Because $_nP_r = n!/(n - r)!$, we have

$$_nC_r = \frac{_nP_r}{r!} = \frac{\dfrac{n!}{(n - r)!}}{r!} = \frac{n!}{r!(n - r)!}$$

Combination Formula

> The number of combinations of n objects taken r at a time is also denoted by $\binom{n}{r}$; that is,
>
> $$\binom{n}{r} = {_nC_r} = \frac{n!}{r!(n - r)!}$$

● **EXAMPLE 5** *Auto Options*

An auto dealer is offering any 4 of 6 special options at one price on a specially equipped car being sold. How many different choices of specially equipped cars do you have?

Solution
The order in which you choose the options is not relevant, so we seek the number of combinations of 6 things taken 4 at a time. Thus the number of possible choices is

$$_6C_4 = \frac{6!}{4!(6 - 4)!} = \frac{6!}{4!2!} = 15$$

● **EXAMPLE 6** *Committees*

Suppose you are the junior class president, and you want to pick 4 people to serve on a committee. If 10 people are willing to serve, in how many different ways can you pick your committee?

Solution
Because the order in which the committee members are selected is not important, there are

$$_{10}C_4 = \frac{10!}{4!6!} = 210$$

ways in which the committee could be picked.

● **EXAMPLE 7** *Committees*

Six men and eight women have volunteered to serve on a committee. How many different committees can be formed containing three men and three women?

Solution

Three men can be selected from six in $\binom{6}{3}$ ways, and three women can be selected from eight in $\binom{8}{3}$ ways. Therefore, three men *and* three women can be selected in

$$\binom{6}{3}\binom{8}{3} = \frac{6!}{3!3!} \cdot \frac{8!}{5!3!} = 1120 \text{ ways}$$

● *Checkpoint*

Determine whether each of the following counting problems can be solved by using permutations or combinations, and answer each question.

3. Find the number of ways in which 8 members of a space shuttle crew can be selected from 20 available astronauts.
4. The command structure on a space shuttle flight is determined by the order in which astronauts are selected for a flight. How many different command structures are possible if 8 astronauts are selected from the 20 that are available?
5. If 14 men and 6 women are available for a space shuttle flight, how many crews are possible that have 5 men and 3 women astronauts?

◗ **EXAMPLE 8** *Little Caesar's Pizza Choices* **(Application Preview)**

During a national television advertising campaign, Little Caesar's Pizza stated that for $9.95, you could get 2 medium-sized pizzas, each with any of 0 to 5 toppings chosen from 11 that are available. The commercial asked the question, "How many different pairs of pizzas can you get?" Answer the question, if the first pizza has a thin crust and the second has a thick crust.

Solution

Because you can get 0, 1, 2, 3, 4, or 5 toppings from the 11 choices, the possible choices for the first pizza are

$$_{11}C_0 + {}_{11}C_1 + {}_{11}C_2 + {}_{11}C_3 + {}_{11}C_4 + {}_{11}C_5 = 1 + 11 + 55 + 165 + 330 + 462 = 1024$$

Because the same number of choices is available for the second pizza, the number of choices for two pizzas is

$$1024 \cdot 1024 = 1,048,576$$

Technology Note

Most calculators have the capability to evaluate permutations and combinations.   ■

● *Checkpoint Solutions*

1. Any of 35 items (26 letters and 9 digits) can be used in any of the six places, so the number of different license plates is

$$35 \cdot 35 \cdot 35 \cdot 35 \cdot 35 \cdot 35 = 35^6 = 1,838,265,625$$

2. (a) There are 5 answers available for the first question, 4 for the second, 3 for the third, 2 for the fourth, and 1 for the fifth, so there are

$$5 \cdot 4 \cdot 3 \cdot 2 \cdot 1 = 120$$

possible answer sheets. Note that this is the number of permutations of 5 things taken 5 at a time,

$$_5P_5 = 5! = 120$$

(b) There are 7 answers available for the first question, 6 for the second, 5 for the third, 4 for the fourth, and 3 for the fifth, so there are

$$7 \cdot 6 \cdot 5 \cdot 4 \cdot 3 = 2520$$

possible answer sheets. Note that this is the number of permutations of 7 things taken 5 at a time, $_7P_5 = 2520$.

3. The members of the crew are to be selected without regard to the order of selection, so we seek the number of combinations of 8 people selected from 20. There are

$$_{20}C_8 = 125{,}970$$

possible crews.

4. The order in which the members are selected is important, so we seek the number of permutations of 8 people selected from 20. There are

$$_{20}P_8 = 5{,}079{,}110{,}400$$

possible command structures.

5. As in Problem 3, the order of selection is not relevant, so we seek the number of combinations of 5 men out of 14 *and* 3 women out of 6. There are

$$_{14}C_5 \cdot {_6C_3} = 2002 \cdot 20 = 40{,}040$$

possible crews that have 5 men and 3 women.

7.5 Exercises

PERMUTATIONS

1. Compute $_6P_4$.
2. Compute $_8P_5$.
3. Compute $_{10}P_6$.
4. Compute $_7P_4$.
5. Compute $_5P_0$.
6. Compute $_3P_3$.
7. How many four-digit numbers can be formed from the digits 1, 3, 5, 7, 8, and 9
   (a) if each digit may be used once in each number?
   (b) if each digit may be used repeatedly in each number?
8. How many four-digit numbers can be formed from the digits 1, 3, 5, 7, 8, and 9
   (a) if the numbers must be even and digits are not used repeatedly?
   (b) if the numbers are less than 3000 and digits are not repeated? (*Hint:* In both parts, begin with the digit where there is a restriction on the choices.)
9. Compute $_nP_n$.
10. Compute $_nP_1$.
11. Find $\dfrac{(n+1)!}{n!}$.
12. Find $\dfrac{(2n+2)!}{(2n)!}$.
13. If $(n+1)! = 17n!$, find n.
14. If $(n+1)! = 30(n-1)!$, find n.

COMBINATIONS

15. Compute $_{100}C_{98}$.
16. Compute $_{80}C_{76}$.
17. Compute $_4C_4$.
18. Compute $_3C_1$.
19. Compute $\dbinom{5}{0}$.
20. Compute $\dbinom{n}{0}$.
21. If $_nC_6 = {_nC_4}$, find n.
22. If $_nC_8 = {_nC_7}$, find n.

APPLICATIONS

23. *Toyota Matrix* The 2003 Toyota Matrix was first made available with 2 different engines and with 3 levels of trim, plus either front-wheel or all-wheel drive.
    (a) How many different Toyota Matrix cars (not counting color) were available for purchase?
    (b) If we include color and there were 6 colors available, how many different Toyota Matrix cars were available for purchase?

24. *Ice cream cones* Baskin Robbins ice cream stores offer 31 flavors of ice cream cones in sizes small, medium, and large.
    (a) How many different selections of cones are possible if each cone has one flavor?

(b) If 4 toppings are available to put on a cone, and a cone can be bought without a topping, how many different selections of ice cream cones are available if each cone has one flavor?

25. *License plates* If a state wants each of its license plates to contain 7 different digits, how many different license plates can it make?

26. *License plates* If a state wants each of its license plates to contain any 7 digits, how many different license plates are possible?

27. *Signaling* A sailboat owner received 6 different signal flags with his new sailboat. If the order in which the flags are arranged on the mast determines the signal being sent, how many 3-flag signals can be sent?

28. *Racing* Eight horses are entered in a race. In how many ways can the horses finish? Assume no ties.

29. *Management* Four candidates for manager of a department store are ranked according to the weighted average of several criteria. How many different rankings are possible if no two candidates receive the same rank?

30. *Testing* An examination consists of 12 questions. If 10 questions must be answered, find the number of different orders in which a student can answer the questions.

31. *Molecules* Biologists have identified 4 kinds of small molecules: adenine, cytosine, quanine, and thymine, which link together to form larger molecules in genes. How many 3-molecule chains can be formed if the order of linking is important and each small molecule can occur more than once in a chain?

32. *Families* A member of a family of four is asked to rank all members of the family in the order of their power in the family. How many possible rankings are there?

33. *Management* A department store manager wants to display 6 brands of a product along one shelf of an aisle. In how many ways can he arrange the brands?

34. *Call letters* The call letters for radio stations begin with K or W, followed by 3 additional letters. How many sets of call letters having 4 letters are possible?

35. *Testing* How many ways can a 10-question true-false test be answered?

36. *Testing* How many ways can a 10-question multiple-choice test be answered if each question has 4 possible answers?

37. *Testing* In a 10-part matching question, how many ways can the matches be made if no two parts are used twice and if there are 10 possible matches?

38. *Testing* Answer Problem 37 if there are 10 parts and 15 possible matches.

39. *Committees* In how many ways can a committee of 5 be selected from 10 people willing to serve?

40. *Astronauts* If 8 people are qualified for the next flight to a space station, how many different groups of 3 people can be chosen for the flight?

41. *Sales* A traveling salesperson has 30 products to sell but has room in her sample case for only 20 of the products. If all the products are the same size, in how many ways can she select 20 different products for her case?

42. *Gambling* A poker hand consists of 5 cards dealt from a deck of 52 cards. How many different poker hands are possible?

43. *Psychology* A psychological study of outstanding salespeople is to be made. If 5 salespeople are needed and 12 are qualified, in how many ways can the 5 be selected?

44. *Banking* A company wishes to use the services of 3 different banks in the city. If 15 banks are available, in how many ways can it choose the 3 banks?

45. *Politics* To determine voters' feelings regarding an issue, a candidate asks a sample of people to pick 4 words out of 10 that they feel best describe the issue. How many different groupings of 4 words are possible?

46. *Sociology* A sociologist wants to pick 3 fifth-grade students from each of four schools. If each of the schools has 50 fifth-grade students, how many different groups of 12 students can he select? (Set up in combination symbols; do not work out.)

47. *Committees* In how many ways can a committee consisting of 6 men and 6 women be selected from a group consisting of 20 men and 22 women?

48. *Bridge* In how many ways can a hand consisting of 6 spades, 4 hearts, 2 clubs, and 1 diamond be selected from a deck of 52 cards?

49. *Sins* The Iranian government issued an order requiring minibuses in Tehran to have separate sections to avoid millions of "sins," because every day 370,000 women ride the minibuses (*Source: USA Today*). If each of these women is brushed (intentionally or accidentally) 10 times each day by one or more men, how many sins would be committed daily?

50. *Politics* In how many ways can a committee of 12 members be selected from the 100-member U.S. Senate?

OBJECTIVE

● To use counting techniques to solve probability problems

Permutations, Combinations, and Probability

◑ Application Preview

Suppose that of the 20 prospective jurors for a trial, 12 favor the death penalty, whereas 8 do not, and that 12 are chosen at random from these 20. To find the probability that 7 of the 12 will favor the death penalty, we can use combinations to find the number of ways to choose 12 jurors from 20 prospective jurors and to find the number of ways to choose 7 from the 12 who favor the death penalty and 5 from the 8 who oppose it. (See Example 5.) We frequently use other counting principles to determine the number of ways in which an event can occur, $n(E)$, the total number of outcomes, $n(S)$, and thus the probability that the event E occurs.

As we saw in Section 7.1, "Probability; Odds," if an event E can happen in $n(E)$ ways out of a total of $n(S)$ equally likely possibilities, the probability of the occurrence of the event is

$$\Pr(E) = \frac{n(E)}{n(S)}$$

● **EXAMPLE 1** *Intelligence Tests*

A psychologist claims she can teach four-year-old children to spell three-letter words very quickly. To test her, one of her students was given cards with the letters A, C, D, K, and T on them and told to spell CAT. What is the probability the child will spell CAT by chance?

Solution

The number of different three-letter "words" that can be formed using these 5 cards is $n(S) = {}_5P_3 = 5!/(5 - 3)! = 60$. Because only one of those 60 arrangements will be CAT, $n(E) = 1$, and the probability that the child will be successful by guessing is $1/60$.

● **EXAMPLE 2** *License Plates*

Suppose that license plates for a state each contain six digits. If a license plate from this state is selected at random, what is the probability that it will have six different digits?

Solution

The number of ways in which six different digits can be selected from 0, 1, 2, 3, 4, 5, 6, 7, 8, 9 is $n(E) = n(\text{all digits different}) = {}_{10}P_6$.

The total number of license plates that can be made with six digits will include those with digits repeated, so the number is

$$n(S) = 10^6$$

Thus the probability that a plate chosen at random will have six different digits is

$$\Pr(E) = \frac{n(E)}{n(S)} = \frac{{}_{10}P_6}{10^6} = \frac{10 \cdot 9 \cdot 8 \cdot 7 \cdot 6 \cdot 5}{1,000,000} = 0.1512$$

● *Checkpoint*

1. If three wires (red, black, and white) are randomly attached to a three-way switch (which has 3 poles to which wires can be attached), what is the probability that the wires will be attached at random in the one order that makes the switch work properly?

● **EXAMPLE 3** *Quality Control*

A box of 24 shotgun shells contains 4 shells that will not fire. If 8 shells are selected from the box, what is the probability

(a) that all 8 shells will be good?
(b) that 6 shells will be good and 2 will not fire?

Solution

(a) The 8 shells can be selected from 24 in $n(S) = {}_{24}C_8 = \binom{24}{8}$ ways, and 8 good shells can be selected from the 20 good ones in $n(E) = {}_{20}C_8 = \binom{20}{8}$ ways. Thus the probability that all 8 shells will be good is

$$\Pr(8 \text{ good}) = \frac{n(E)}{n(S)} = \frac{\dfrac{20!}{12!8!}}{\dfrac{24!}{16!8!}} \approx 0.17$$

(b) Six good shells *and* two defective shells can be selected in

$$n(E) = {}_{20}C_6 \cdot {}_4C_2 = \binom{20}{6} \cdot \binom{4}{2}$$

ways, so

$$\Pr(6 \text{ good and 2 defective}) = \frac{n(E)}{n(S)} = \frac{\dbinom{20}{6} \cdot \dbinom{4}{2}}{\dbinom{24}{8}} = \frac{\dfrac{20!}{14!6!} \cdot \dfrac{4!}{2!2!}}{\dfrac{24!}{16!8!}} \approx 0.316$$

● **EXAMPLE 4** *Quality Control*

A manufacturing process for computer chips is such that 5 out of 100 chips are defective. If 10 chips are chosen at random from a box containing 100 newly manufactured chips, what is the probability that none of the chips will be defective?

Solution

The number of ways in which any 10 chips can be chosen from 100 is

$$n(S) = {}_{100}C_{10} = \binom{100}{10} = \frac{100!}{10!90!}$$

The number of ways in which 10 chips can be chosen with none of them being defective is the same as the number of ways 10 good chips can be chosen from the 95 good chips in the box. Thus

$$n(E) = {}_{95}C_{10} = \binom{95}{10} = \frac{95!}{10!85!}$$

Thus the probability that none of the chips chosen will be defective is

$$\Pr(E) = \frac{n(E)}{n(S)} = \frac{\dfrac{95!}{10!85!}}{\dfrac{100!}{10!90!}} \approx 0.58375$$

◗ EXAMPLE 5 Jury Selection (Application Preview)

Suppose that of the 20 prospective jurors for a trial, 12 favor the death penalty and 8 do not. If 12 jurors are chosen at random from these 20, what is the probability that 7 of the jurors will favor the death penalty?

Solution
The number of ways in which 7 jurors of the 12 will favor the death penalty is the number of ways in which 7 favor and 5 do not favor, or

$$n(E) = {}_{12}C_7 \cdot {}_8C_5 = \frac{12!}{7!5!} \cdot \frac{8!}{5!3!}$$

The total number of ways in which 12 jurors can be selected from 20 people is

$$n(S) = {}_{20}C_{12} = \frac{20!}{12!8!}$$

Thus the probability we seek is

$$\Pr(E) = \frac{n(E)}{n(S)} = \frac{\frac{12!}{7!5!} \cdot \frac{8!}{5!3!}}{\frac{20!}{12!8!}} \approx 0.3521$$

● **Checkpoint**

2. If 5 people are chosen from a group that contains 4 men and 6 women, what is the probability that 3 men and 2 women will be chosen?

● **EXAMPLE 6 License Plates**

Pennsylvania auto license plates have three letters followed by four numbers. If Janie's initials are J. J. R. and she lives at 1125 Spring Street, what is the probability that the license plate she draws at random will

(a) have her initials and house number on it?   (b) have her initials on it?

Solution
(a) The total number of possible license plates with three letters followed by four numbers is

$$26 \cdot 26 \cdot 26 \cdot 10 \cdot 10 \cdot 10 \cdot 10 = 175{,}760{,}000$$

Of all these plates, only one will have Janie's initials and house number, so the probability that she will get this plate at random is 1/175,760,000.
(b) The total number of possible license plates is 175,760,000, and the number with Janie's initials and any four numbers on it is $1 \cdot 1 \cdot 1 \cdot 10 \cdot 10 \cdot 10 \cdot 10 = 10{,}000$, so the probability that Janie will get a plate with her initials is 10,000/175,760,000 = 1/17,576.

● **EXAMPLE 7 Auto Keys**

Beginning in 1990, General Motors began to use a theft deterrent key on some of its cars. The key has six parts, with three patterns for each part, plus an electronic chip containing a code from 1 to 15. What is the probability that one of these keys selected at random will start a GM car requiring a key of this type?

Solution

Each of the six parts has three patterns, and the chip can have any of 15 codes, so there are

$$3 \cdot 3 \cdot 3 \cdot 3 \cdot 3 \cdot 3 \cdot 15 = 10,935$$

possible keys. Because there is only one key that will start the car, the probability that a key selected at random will start a given car is 1/10,935.

● *Checkpoint Solutions*

1. To find the total number of orders in which the wires can be attached, we note that the first pole can have any of the 3 wires attached to it, the second pole can have any of 2, and the third pole can have only the remaining wire, so there are $3 \cdot 2 \cdot 1 = 6$ possible different orders. Only one of these orders permits the switch to work properly, so the probability is 1/6.

2. Five people can be chosen from 10 people in $_{10}C_5 = 252$ ways. Three men and 2 women can be chosen in $_4C_3 \cdot {_6}C_2 = 4 \cdot 15 = 60$ ways. Thus the probability is $60/252 = 5/21$.

7.6 Exercises

APPLICATIONS

1. *Education* If a child is given cards with A, C, D, G, O, and T on them, what is the probability he or she could spell DOG by guessing the correct arrangement of 3 cards from the 6?

2. *Racing* Eight horses in a race wear numbers 1, 2, 3, 4, 5, 6, 7, and 8. What is the probability that the first three horses to finish the race will be numbered 1, 2, and 3, respectively?

3. *Politics* A senator asks his constituents to rank five issues in order of importance.
   (a) How many rankings are possible?
   (b) What is the probability that one reply chosen at random will rank the issues in alphabetical order?

4. *Photography* Two men and a woman are lined up to have their picture taken. If they are arranged at random, what is the probability that
   (a) the woman will be on the left in the picture?
   (b) the woman will be in the middle in the picture?

5. *License plates* Suppose that all license plates in a state have three letters and three digits. If a plate is chosen at random, what is the probability that all three letters and all three numbers on the plate will be different?

6. *Spelling*
   (a) In how many different orders can the letters R, A, N, D, O, M be written?
   (b) If the letters R, A, N, D, O, M are placed in a random order, what is the probability that they will spell RANDOM?

7. *ATMs*
   (a) An automatic teller machine requires that each customer enter a four-digit personal identification number (PIN) when he or she inserts a bank card. If a person finds a bank card and guesses at a PIN to use the card fraudulantly, what is the probability that the person will succeed in one attempt?
   (b) If the person knows that the PIN will not have any digit repeated, what is the probability the person will succeed in guessing in one attempt?

8. *Keys* Keys for older General Motors cars had six parts, with three patterns for each part.
   (a) How many different key designs are possible for these cars?
   (b) If you find an older GM key and own an older GM car, what is the probability that it will fit your trunk?

9. *Telephones* If the first digit of a seven-digit telephone number cannot be a 0 or a 1, what is the probability that a number chosen at random will have all seven digits the same?

10. *License plates* If a license plate has seven different digits, what is the probability that it contains a zero?

11. *Rewards* As a reward for a record year, the Ace Software Company is randomly selecting 4 people from its 500 employees for a free trip to Hawaii, but it will not pay for a traveling companion. If John and Jill are married and both are employees, what is the probability that they will both win?

12. *Mutual funds*   The retirement plan for a company allows employees to invest in 10 different mutual funds. If Sam selected 4 of these funds at random and 6 of the 10 grew by at least 10% over the last year, what is the probability that 3 of Sam's 4 funds grew by at least 10% last year?

13. *Testing*   In a 10-question matching test with 10 possible answers to match and no answer used more than once, what is the probability of guessing and getting every answer correct?

14. *Testing*   In a 10-question matching test with 15 possible answers to match and no answer used more than once, what is the probability of guessing and getting every answer correct?

15. *Quality control*   A box of 12 transistors has 3 defective ones. If 2 transistors are drawn from the box together, what is the probability
    (a) that both transistors are defective?
    (b) that neither transistor is defective?
    (c) that one transistor is defective?

16. *Quality control*   Suppose that 6 transistors are drawn at random from a box containing 18 good transistors and 2 defective transistors. What is the probability that
    (a) all 6 of the transistors are good?
    (b) exactly 4 of the transistors are good?
    (c) exactly 2 of the transistors are good?

17. *Quality control*   A retailer purchases 100 of a new brand of DVD player, of which 2 are defective. The purchase agreement says that if he tests 5 chosen at random and finds 1 or more defective, he receives all the DVD players free of charge. What is the probability that he will not have to pay for the DVD players?

18. *Quality control*   A box containing 500 central processing units (CPUs) for microcomputers has 20 defective CPUs. If 10 CPUs are selected together from the box, what is the probability that
    (a) all 10 are defective?
    (b) none is defective?
    (c) half of them are defective?

19. *Banking*   To see whether a bank has enough minority construction company loans, a social agency selects 30 loans to construction companies at random and finds that 2 of them are loans to minority companies. If the bank's claim that 10 of every 100 of its loans to construction companies are minority loans is true, what is the probability that 2 loans out of 30 are minority loans? Leave your answer with combination symbols.

20. *Diversity*   A high school principal must select 12 girls at random from the freshman class to serve as hostesses

at the junior-senior prom. There are 200 freshman girls, including 20 from minorities, and the principal would like at least one minority girl to have this honor. If he selects the girls at random, what is the probability that
    (a) he will select exactly one minority girl?
    (b) he will select no minority girls?
    (c) he will select at least one minority girl?

21. *Diversity*   Suppose that an employer plans to hire four people from a group of nine equally qualified people, of whom three are minority candidates. If the employer does not know which candidates are minority candidates, and if she selects her employees at random, what is the probability that
    (a) no minority candidates are hired?
    (b) all three minority candidates are hired?
    (c) one minority candidate is hired?

22. *Lottery*   Four men and three women are semifinalists in a lottery. From this group, three finalists are to be selected by a drawing. What is the probability that all three finalists will be men?

23. *Management*   Suppose that a children's basketball coach knows the best 6 players to use on his team, but pressure from parents to give everyone a chance to start in a game causes him to pick the starting team by choosing 5 players at random from the 10 team members. What is the probability that this will give him a team with 5 of his 6 best players?

24. *Sales*   A car dealer has 12 different cars that he would like to display, but he has room to display only 5.
    (a) In how many ways can he pick 5 cars to display?
    (b) Suppose 8 of the cars are the same color, with the remaining 4 having distinct colors. If the dealer tells a salesperson to display any 5 cars, what is the probability that all 5 cars will be the same color?

25. *Diversity*   Suppose that two openings on an appellate court bench are to be filled from current municipal court judges. The municipal court judges consist of 23 men and 4 women. Find the probability
    (a) that both appointees are men.
    (b) that one man and one woman are appointed.
    (c) that at least one woman is appointed.

26. *Politics*   Suppose that four of the eight students running for class officers (president, vice president, secretary, and treasurer) have grade-point averages (GPAs) above 3.0. If the officers are selected at random, what is the probability that all four officers will be students with GPAs above 3.0?

27. *Management*   Suppose that an indecisive company owner has selected the three top officers of his company

at random but claims that they earned their jobs because of ability.

(a) What is the probability that the most able person is at the top?

(b) What is the probability that the top three officers are ranked according to their ability?

28. *Quality control* Suppose that 10 computer chips are drawn from a box containing 12 good chips and 4 defective chips. What is the probability that

(a) exactly 4 of the chips are defective?

(b) all 10 of the chips are good?

(c) exactly 8 of the chips are good?

29. *Lotteries* In New York State there is a state lottery game called Pick 10. The state chooses 20 numbers from 80 numbers, and each player chooses 10 numbers from the same 80 numbers. If all 10 of a player's numbers are among the 20 numbers that the state picked, then that player is a "big winner." Find the probability of being a big winner.

30. *Diversity* A task force studying sex discrimination wishes to establish a subcommittee. If the subcommittee is to consist of 11 members chosen from a group of 23 men and 19 women, find the probability that the subcommittee consists of

(a) 6 men and 5 women.   (b) all women.

31. *Insurance* An insurance company receives 25 claims in a certain month. Company policy mandates that a random selection of 5 of these claims be thoroughly investigated. If 2 of the 25 claims are fraudulent, find the probability that

(a) both fraudulent claims are thoroughly investigated.

(b) neither fraudulent claim is thoroughly investigated.

32. *Evaluation* Employees of a firm receive annual reviews. In a certain department, 4 employees received excellent ratings, 15 received good ratings, and 1 received a marginal rating. If 3 employees in this department are randomly selected to complete a form for an internal study of the firm, find the probability that

(a) all 3 selected were rated excellent.

(b) one from each category was selected.

33. *Poker* What is the probability of being dealt a poker hand of 5 cards containing

(a) 5 spades?   (b) 5 cards of the same suit?

34. *Bridge* A standard deck of cards contains 13 spades, 13 hearts, 13 diamonds, and 13 clubs. A bridge hand contains 13 cards selected at random from the 52-card deck. Find the probability that a hand contains

(a) 13 hearts.   (b) exactly 8 spades.

35. *Bridge* A standard deck of cards contains 13 spades, 13 hearts, 13 diamonds, and 13 clubs. A bridge hand contains 13 cards selected at random from the 52-card deck. Find the probability that a hand contains 4 spades, 4 hearts, 4 diamonds, and 1 club.

36. *Bridge* Find the probability of being dealt a hand containing 1 spade, 6 hearts, 4 diamonds, and 2 clubs. Assume that a normal deck of 52 cards is used.

7.7

Markov Chains

OBJECTIVES

- To use transition matrices to find specific stages of Markov chains
- To write transition matrices for Markov chain problems
- To find steady-state vectors for certain Markov chain problems

◖ Application Preview

Suppose a department store has determined that a woman who uses a credit card in the current month will use it again next month with probability 0.8. That is,

Pr(woman will use card next month | she used it this month) = 0.8

and

Pr(woman will not use card next month | she used it this month) = 0.2

Furthermore, suppose the store has determined that the probability that a woman who does not use a credit card this month will not use it next month is 0.7. That is,

Pr(woman will not use card next month | she did not use it this month) = 0.7

and

Pr(woman will use card next month | she did not use it this month) = 0.3

To find the probability that the woman will use the credit card in some later month, we use **Markov chains** and put the conditional probabilities in a transition matrix (see Example 1). Markov chains are named for the famous Russian mathematician A. A. Markov, who developed this theory in 1906. They are useful in the analysis of price movements, consumer behavior, laboratory animal behavior, and many other processes in business, the social sciences, and the life sciences.

Markov Chains   Suppose that a sequence of experiments with the same set of outcomes is performed, with the probabilities of the outcomes in one experiment depending on the outcomes of the previous experiment. In this case the sequence of experiments forms a **Markov chain.** The outcomes of the experiments are called **states,** and the probability of moving from one state to another is called the **transition probability.** These probabilities can sometimes be organized into a **transition matrix.**

> ◗ **EXAMPLE 1** *Credit Card Use* **(Application Preview)**
>
> Create the transition matrix that contains the conditional probabilities from the problem described in the Application Preview.
>
> **Solution**
> If the card is used this month, the probabilities that the woman will or will not use the card next month can be represented by the probability tree

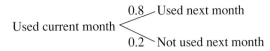

$$\text{Used current month} \begin{cases} 0.8 \; \text{Used next month} \\ 0.2 \; \text{Not used next month} \end{cases}$$

We can enter these numbers in the first row of the transition matrix:

$$\begin{array}{cc} & \textit{Next month} \\ & \begin{matrix} \text{Card} & \text{Card} \\ \text{used} & \text{not used} \end{matrix} \\ \textit{Current month} \; \begin{matrix} \text{Card used} \\ \text{Card not used} \end{matrix} & \begin{bmatrix} 0.8 & 0.2 \\ & \end{bmatrix} \end{array}$$

If the card is not used this month, the probabilities that the woman will or will not use the card next month can be represented by the probability tree

$$\text{Not used current month} \begin{cases} 0.7 \; \text{Not used next month} \\ 0.3 \; \text{Used next month} \end{cases}$$

Putting these conditional probabilities in the second row completes the transition matrix:

$$\begin{array}{cc} & \textit{Next month} \\ & \begin{matrix} \text{Card} & \text{Card} \\ \text{used} & \text{not used} \end{matrix} \\ \textit{Current month} \\ P = \begin{matrix} \text{Card used} \\ \text{Card not used} \end{matrix} & \begin{bmatrix} 0.8 & 0.2 \\ 0.3 & 0.7 \end{bmatrix} \end{array}$$

In general, a transition matrix is characterized by the following.

1. The *ij*-entry in the matrix is Pr(moving from state *i* to state *j*). Note that this is really the probability of moving to state *j* given that you are in state *i*.
2. The entries are nonnegative.
3. The sum of the entries in each row equals 1.
4. The matrix is square.

Now suppose the department store data show that during the first month a credit card is received, the probabilities for each state are as follows:

$$\text{Pr(a woman used a credit card)} = 0.9$$

$$\text{Pr(a woman did not use a credit card)} = 0.1$$

These probabilities can be placed in the row matrix

$$A = [0.9 \quad 0.1]$$

which is called the **initial-probability vector** for the problem. This vector gives the probabilities for each state at the outset of the trials. In general, a probability vector is defined as follows.

Probability Vector

> For a vector to be a probability vector, the sum of its entries must be 1 and each of its entries must be nonnegative.

Note that if a woman used her credit card in the initial month, the probability is 0.8 that she will use it the next month, and if she didn't use the card, the probability is 0.3 that she will use it the next month. Thus,

$$\text{Pr(will use 2nd month)} = \text{Pr(used 1st month)} \cdot \text{Pr(will use 2nd | used 1st)}$$
$$+ \text{Pr(not used 1st month)} \cdot \text{Pr(will use 2nd | not used 1st)}$$
$$= (0.9)(0.8) + (0.1)(0.3)$$
$$= 0.72 + 0.03 = 0.75$$

Note that this is one element of the product matrix AP. Similarly, the second element of the product matrix is the probability that a woman will not use the card in the second month.

$$AP = [0.9 \quad 0.1] \begin{bmatrix} 0.8 & 0.2 \\ 0.3 & 0.7 \end{bmatrix} = [0.75 \quad 0.25]$$

$$\underset{\substack{\text{Initial} \\ \text{month}}}{\quad} \underset{\substack{\text{Transition} \\ \text{matrix}}}{\quad} \underset{\substack{\text{Second} \\ \text{month}}}{\quad}$$

Thus

$$\text{Pr(a woman will use a credit card in 2nd month)} = 0.75$$

and

$$\text{Pr(a woman will not use a credit card in 2nd month)} = 0.25$$

To find the probabilities for each state for the third month, we multiply the second-month probability vector times the transition matrix.

$$[0.75 \quad 0.25] \begin{bmatrix} 0.8 & 0.2 \\ 0.3 & 0.7 \end{bmatrix} = [0.675 \quad 0.325]$$

$$\underset{\substack{\text{Second} \\ \text{month}}}{\quad} \underset{\substack{\text{Transition} \\ \text{matrix}}}{\quad} \underset{\substack{\text{Third} \\ \text{month}}}{\quad}$$

Continuing in this manner, we can find the probabilities for later months.

● **EXAMPLE 2** *Politics*

Suppose that in a certain city the Democratic, Republican, and Consumer parties always nominate candidates for mayor. The probability of winning in any election depends on the party in power and is given by the following transition matrix P.

	Party in office next term		
Party in office now	D	R	C
Democrat	0.5	0.4	0.1
$P =$ Republican	0.4	0.5	0.1
Consumer	0.3	0.3	0.4

Suppose the probability of winning the next election is given by [0.4 0.4 0.2]; that is, Pr(D) = 0.4, Pr(R) = 0.4, Pr(C) = 0.2. Find the probability of each party winning the *fourth* election (the third election after this one).

Solution

The probability of each party winning the second election is found using the probabilities given for the first election.

$$[0.4 \quad 0.4 \quad 0.2] \begin{bmatrix} 0.5 & 0.4 & 0.1 \\ 0.4 & 0.5 & 0.1 \\ 0.3 & 0.3 & 0.4 \end{bmatrix} = [0.42 \quad 0.42 \quad 0.16]$$

Using the probabilities for the second election, we can find the probabilities for the third election.

$$[0.42 \quad 0.42 \quad 0.16] \begin{bmatrix} 0.5 & 0.4 & 0.1 \\ 0.4 & 0.5 & 0.1 \\ 0.3 & 0.3 & 0.4 \end{bmatrix} = [0.426 \quad 0.426 \quad 0.148]$$

Using the probabilities for the third election, we can find the probabilities for the fourth election.

$$[0.426 \quad 0.426 \quad 0.148] \begin{bmatrix} 0.5 & 0.4 & 0.1 \\ 0.4 & 0.5 & 0.1 \\ 0.3 & 0.3 & 0.4 \end{bmatrix} = [0.4278 \quad 0.4278 \quad 0.1444]$$

Steady-State Vectors

We can perform matrix multiplication with a graphing calculator, or we can use a computer spreadsheet to perform repeated multiplications. Continuing the multiplication process with the vectors in Example 2 eventually gives a vector with the probabilities

$$[0.42857 \quad 0.42857 \quad 0.14286]$$

to five decimal places, and repeated multiplications will not change the probabilities, so the system appears to be stabilizing. This vector is the vector $[V_1 \quad V_2 \quad V_3]$ such that

$$[V_1 \quad V_2 \quad V_3] \cdot P = [V_1 \quad V_2 \quad V_3]$$

The exact probability vector for which this is true in Example 2 is $\left[\frac{3}{7} \quad \frac{3}{7} \quad \frac{1}{7}\right]$, because

$$\left[\tfrac{3}{7} \quad \tfrac{3}{7} \quad \tfrac{1}{7}\right] \begin{bmatrix} 0.5 & 0.4 & 0.1 \\ 0.4 & 0.5 & 0.1 \\ 0.3 & 0.3 & 0.4 \end{bmatrix} = \left[\tfrac{3}{7} \quad \tfrac{3}{7} \quad \tfrac{1}{7}\right]$$

This vector is called the **fixed-probability vector** for the given transition matrix. Because this vector is where the process eventually stabilizes, it is also called the **steady-state vector.**

Rather than trying to guess this steady-state vector by repeated multiplication, we can solve the equation

$$[V_1 \quad V_2 \quad V_3] \begin{bmatrix} 0.5 & 0.4 & 0.1 \\ 0.4 & 0.5 & 0.1 \\ 0.3 & 0.3 & 0.4 \end{bmatrix} = [V_1 \quad V_2 \quad V_3]$$

for V_1, V_2, and V_3 as follows:

$$[V_1 \quad V_2 \quad V_3] \begin{bmatrix} 0.5 & 0.4 & 0.1 \\ 0.4 & 0.5 & 0.1 \\ 0.3 & 0.3 & 0.4 \end{bmatrix} = [V_1 \quad V_2 \quad V_3]$$

$$0.5V_1 + 0.4V_2 + 0.3V_3 = V_1$$
$$0.4V_1 + 0.5V_2 + 0.3V_3 = V_2$$
$$0.1V_1 + 0.1V_2 + 0.4V_3 = V_3$$

Rewriting with zeros on the right and solving by Gauss-Jordan elimination give

$$\begin{bmatrix} -0.5 & 0.4 & 0.3 & | & 0 \\ 0.4 & -0.5 & 0.3 & | & 0 \\ 0.1 & 0.1 & -0.6 & | & 0 \end{bmatrix} \xrightarrow[\text{Switch } R_1 \,\&\, R_3]{10 \text{ (each row)}} \begin{bmatrix} 1 & 1 & -6 & | & 0 \\ 4 & -5 & 3 & | & 0 \\ -5 & 4 & 3 & | & 0 \end{bmatrix}$$

$$\xrightarrow[5R_1 + R_3 \to R_3]{-4R_1 + R_2 \to R_2} \begin{bmatrix} 1 & 1 & -6 & | & 0 \\ 0 & -9 & 27 & | & 0 \\ 0 & 9 & -27 & | & 0 \end{bmatrix} \xrightarrow[\substack{-R_2 + R_1 \to R_1 \\ -9R_2 + R_3 \to R_3}]{\substack{-\frac{1}{9}R_2 \to R_2 \\ \text{then}}} \begin{bmatrix} 1 & 0 & -3 & | & 0 \\ 0 & 1 & -3 & | & 0 \\ 0 & 0 & 0 & | & 0 \end{bmatrix}$$

so

$$V_1 = 3V_3 \qquad V_2 = 3V_3 \qquad V_3 = \text{any real number}$$

Any vector that satisfies these conditions is a solution of the system. Because $[V_1 \quad V_2 \quad V_3]$ is a probability vector, we have $V_1 + V_2 + V_3 = 1$. Substituting gives $3V_3 + 3V_3 + V_3 = 1$, so $V_3 = \frac{1}{7}$, and the steady-state vector is

$$\begin{bmatrix} \dfrac{3}{7} & \dfrac{3}{7} & \dfrac{1}{7} \end{bmatrix}$$

Any transition matrix P such that some power of P, P^m, contains only positive entries is called a **regular transition matrix,** and it can be shown that each Markov chain using a regular transition matrix will have a steady-state vector.

Note that a regular transition matrix P may contain zeros; the only requirement is that some power of P contain all positive entries. In this text we will limit ourselves to regular transition matrices.

Furthermore, if a steady-state vector exists, then the effect of the initial probabilities diminishes as the number of steps in the process increases. This means that the probabilities for the elections would approach

$$\begin{bmatrix} \dfrac{3}{7} & \dfrac{3}{7} & \dfrac{1}{7} \end{bmatrix}$$

regardless of the probabilities for the initial election. For example, even if a Democrat were certain to be elected in the initial election (that is, the initial vector is $[1 \quad 0 \quad 0]$), the probabilities for the fifth election are $[0.4292 \quad 0.4291 \quad 0.1417]$, which approximates the steady-state vector. (See Problem 17 of the exercise set for this section.)

● *Checkpoint*   1. Use the transition matrix $\begin{bmatrix} 0.1 & 0.3 & 0.6 \\ 0.3 & 0.5 & 0.2 \\ 0.1 & 0.1 & 0.8 \end{bmatrix}$ and the initial probability vector

$[0.2 \quad 0.3 \quad 0.5]$ to
(a) find the probability vector for the third stage (two steps after the initial stage).
(b) find the steady-state vector associated with the matrix.

● **EXAMPLE 3** *Educational Attainment*

Suppose that in a certain city the probabilities that a woman with less than a high school education has a daughter with less than a high school education, with a high school degree, and with education beyond high school are 0.2, 0.6, and 0.2, respectively. The probabilities that a woman with a high school degree has a daughter with less than a high school education, with a high school degree, and with education beyond high school are 0.1, 0.5, and 0.4, respectively. The probabilities that a woman with education beyond high school has a daughter with less than a high school education, with a high school degree, and with education beyond high school are 0.1, 0.1, and 0.8, respectively. The population of women in the city is now 60% with less than a high school education, 30% with a high school degree, and 10% with education beyond high school.

(a) What is the transition matrix for this information?
(b) What is the probable distribution of women according to educational level two generations from now?
(c) What is the steady-state vector for this information?

Solution
(a) The transition matrix is

	Daughter's educational level		
Mother's	Less than	H.S.	More than
educational level	H.S.	degree	H.S.
Less than H.S.	0.2	0.6	0.2
H.S. degree	0.1	0.5	0.4
More than H.S.	0.1	0.1	0.8

(b) The probable distribution after one generation is

$$[0.6 \quad 0.3 \quad 0.1] \begin{bmatrix} 0.2 & 0.6 & 0.2 \\ 0.1 & 0.5 & 0.4 \\ 0.1 & 0.1 & 0.8 \end{bmatrix} = [0.16 \quad 0.52 \quad 0.32]$$

The probable distribution after two generations is

$$[0.16 \quad 0.52 \quad 0.32] \begin{bmatrix} 0.2 & 0.6 & 0.2 \\ 0.1 & 0.5 & 0.4 \\ 0.1 & 0.1 & 0.8 \end{bmatrix} = [0.116 \quad 0.388 \quad 0.496]$$

(c) The steady-state vector is found by solving

$$[V_1 \quad V_2 \quad V_3] \begin{bmatrix} 0.2 & 0.6 & 0.2 \\ 0.1 & 0.5 & 0.4 \\ 0.1 & 0.1 & 0.8 \end{bmatrix} = [V_1 \quad V_2 \quad V_3]$$

$$\begin{cases} 0.2V_1 + 0.1V_2 + 0.1V_3 = V_1 \\ 0.6V_1 + 0.5V_2 + 0.1V_3 = V_2 \\ 0.2V_1 + 0.4V_2 + 0.8V_3 = V_3 \end{cases} \text{ or } \begin{cases} -0.8V_1 + 0.1V_2 + 0.1V_3 = 0 \\ 0.6V_1 - 0.5V_2 + 0.1V_3 = 0 \\ 0.2V_1 + 0.4V_2 - 0.2V_3 = 0 \end{cases}$$

$$\begin{bmatrix} -0.8 & 0.1 & 0.1 & | & 0 \\ 0.6 & -0.5 & 0.1 & | & 0 \\ 0.2 & 0.4 & -0.2 & | & 0 \end{bmatrix} \rightarrow \begin{bmatrix} 1 & 2 & -1 & | & 0 \\ 0 & -17 & 7 & | & 0 \\ 0 & 17 & -7 & | & 0 \end{bmatrix}$$

$$\rightarrow \begin{bmatrix} 1 & 0 & -\frac{3}{17} & | & 0 \\ 0 & 1 & -\frac{7}{17} & | & 0 \\ 0 & 0 & 0 & | & 0 \end{bmatrix}$$

Thus $V_1 = \frac{3}{17}V_3$ and $V_2 = \frac{7}{17}V_3$. The probability vector that satisfies these conditions and $V_1 + V_2 + V_3 = 1$ must satisfy

$$\frac{3}{17}V_3 + \frac{7}{17}V_3 + V_3 = 1$$

Thus

$$\frac{27}{17}V_3 = 1, \quad \text{so} \quad V_3 = \frac{17}{27}$$

and the steady-state vector is

$$\begin{bmatrix} \frac{3}{27} & \frac{7}{27} & \frac{17}{27} \end{bmatrix}$$

● **Checkpoint Solutions**

1. (a) The second-stage probability vector is

$$[0.2 \quad 0.3 \quad 0.5] \begin{bmatrix} 0.1 & 0.3 & 0.6 \\ 0.3 & 0.5 & 0.2 \\ 0.1 & 0.1 & 0.8 \end{bmatrix} = [0.16 \quad 0.26 \quad 0.58]$$

and that for the third stage is

$$[0.16 \quad 0.26 \quad 0.58] \begin{bmatrix} 0.1 & 0.3 & 0.6 \\ 0.3 & 0.5 & 0.2 \\ 0.1 & 0.1 & 0.8 \end{bmatrix} = [0.152 \quad 0.236 \quad 0.612]$$

(b) To find the steady-state vector, we solve

$$[V_1 \quad V_2 \quad V_3] \begin{bmatrix} 0.1 & 0.3 & 0.6 \\ 0.3 & 0.5 & 0.2 \\ 0.1 & 0.1 & 0.8 \end{bmatrix} = [V_1 \quad V_2 \quad V_3]$$

Multiplying and combining like terms give the following system of equations.

$$\begin{cases} -0.9V_1 + 0.3V_2 + 0.1V_3 = 0 \\ 0.3V_1 - 0.5V_2 + 0.1V_3 = 0 \\ 0.6V_1 + 0.2V_2 - 0.2V_3 = 0 \end{cases}$$

We solve this by using the augmented matrix

$$\begin{bmatrix} -0.9 & 0.3 & 0.1 & | & 0 \\ 0.3 & -0.5 & 0.1 & | & 0 \\ 0.6 & 0.2 & -0.2 & | & 0 \end{bmatrix}$$

The solution gives the steady-state vector $\begin{bmatrix} \frac{1}{7} & \frac{3}{14} & \frac{9}{14} \end{bmatrix}$.

7.7 Exercises

Can the vectors in Problems 1–4 be probability vectors? If not, why?

1. $\begin{bmatrix} \frac{1}{4} & \frac{3}{4} \end{bmatrix}$

2. $\begin{bmatrix} \frac{5}{6} & \frac{2}{3} \end{bmatrix}$

3. $\begin{bmatrix} \frac{1}{4} & \frac{3}{5} & \frac{1}{6} \end{bmatrix}$

4. $\begin{bmatrix} \frac{2}{3} & \frac{1}{12} & \frac{1}{4} \end{bmatrix}$

Can the matrices in Problems 5–8 be transition matrices? If not, why?

5. $\begin{bmatrix} 0.3 & 0.5 & 0.2 \\ 0.1 & 0.6 & 0.3 \end{bmatrix}$

6. $\begin{bmatrix} 0.6 & 0.2 & 0.2 \\ 0.1 & 0.5 & 0.4 \\ 0.3 & 0.5 & 0.1 \end{bmatrix}$

7. $\begin{bmatrix} 0.1 & 0.3 & 0.6 \\ 0.5 & 0.2 & 0.3 \\ 0.1 & 0.7 & 0.2 \end{bmatrix}$

8. $\begin{bmatrix} 0.6 & 0.5 & -0.1 \\ 0.2 & 0.6 & 0.2 \\ 0.3 & 0.3 & 0.4 \end{bmatrix}$

In Problems 9–12, use the transition matrix and the initial-probability vector to find the probability vector for the third stage (two after the initial stage) of the Markov chain.

$$A = \begin{bmatrix} 0.1 & 0.9 \\ 0.3 & 0.7 \end{bmatrix} \qquad B = \begin{bmatrix} 0.5 & 0.5 \\ 0.9 & 0.1 \end{bmatrix}$$

$$C = \begin{bmatrix} 0.5 & 0.3 & 0.2 \\ 0.3 & 0.5 & 0.2 \\ 0.1 & 0.1 & 0.8 \end{bmatrix} \qquad D = \begin{bmatrix} 0.8 & 0.1 & 0.1 \\ 0.2 & 0.6 & 0.2 \\ 0.3 & 0.3 & 0.4 \end{bmatrix}$$

9. Transition matrix A, initial-probability vector [0.2 0.8].
10. Transition matrix B, initial-probability vector [0.6 0.4].
11. Transition matrix C, initial-probability vector [0.1 0.3 0.6].
12. Transition matrix D, initial-probability vector [0.9 0 0.1].

In Problems 13–16, find the steady-state vector associated with the given transition matrix from Problems 9–12.

13. *A*   14. *B*   15. *C*   16. *D*

APPLICATIONS

Elections   Use the following information for Problems 17 and 18. In a certain city, the Democratic, Republican, and Consumer parties always nominate candidates for mayor. The probability of winning in any election depends on the party in power and is given by the following transition matrix.

Party in office now	Party in office next term		
	D	**R**	**C**
Democrat	0.5	0.4	0.1
Republican	0.4	0.5	0.1
Consumer	0.3	0.3	0.4

17. Using the given transition matrix and assuming the initial-probability vector is [1 0 0], find the probability vectors for the next four steps of the Markov chain. (This initial-probability vector indicates that a Democrat is certain to win the initial election.)
18. Using the given transition matrix and assuming that a Republican is certain to win the initial election, find the probability vectors for the next four steps of the Markov chain.

Church attendance   Use the following information for Problems 19–22. The probability that daughters of a mother who attends church regularly will also attend church regularly is 0.8, whereas the probability that daughters of a mother who does not attend regularly will attend regularly is 0.3.

19. What is the transition matrix for this information?
20. If a woman attends church regularly and has one daughter who in turn has one daughter, what is the probability that the granddaughter attends church regularly?
21. If a woman does not attend church, what is the probability that her granddaughter attends church regularly?
22. What is the steady-state vector for this information?

Car selection   Use the following information for Problems 23–28. A man owns an Audi, a Ford, and a VW. He drives every day and never drives the same car two days in a row. These are the probabilities that he drives each of the other cars the next day:

Pr(Ford after Audi) = 0.7   Pr(VW after Audi) = 0.3
Pr(Audi after Ford) = 0.6   Pr(VW after Ford) = 0.4
Pr(Audi after VW) = 0.8   Pr(Ford after VW) = 0.2

23. Write the transition matrix for his selection of a car.
24. If he drove the Ford today, what are the probabilities for the cars that he will drive 2 days from today?
25. If he drove the Ford today, what are the probabilities for the cars that he will drive 4 days from today?
26. If he drove the Audi today, what are the probabilities for the cars that he will drive 4 days from now?
27. What is the steady-state vector for this problem?
28. Would the probabilities for which car he drives 100 days from now depend on whether he drove a Ford or an Audi today?

29. *Population demographics*   Suppose a government study estimated that the probability of successive generations of a rural family remaining in a rural area was 0.7 and the probability of successive generations of an urban family remaining in an urban area was 0.9. Assuming a Markov chain applies to these facts, find the steady-state vector.

30. *Psychology*   A psychologist found that a laboratory mouse placed in a T-maze will go down the same branch with probability 0.8 if there was food there on the previous trial and that it will go down the same branch with probability 0.6 if there was no food there on the previous trial. Find the steady-state vector for this experiment. If the mouse takes the branches with equal probability when no reward is found, does it appear that the mouse "learns" to choose the food path?

31. *Charitable contributions*   The local community-service funding organization in a certain county has classified

the population into those who did not give the previous year, those who gave less than their "fair share," and those who met or exceeded their "fair share." Suppose the organization developed the following transition matrix for these groups.

		This year	
Last year	N	L	F
No contribution	0.7	0.2	0.1
Less than fair share	0.1	0.6	0.3
Fair share	0	0.1	0.9

Find the steady-state vector for this county's contribution habits.

Advertising   Use the following information for Problems 32 and 33. A local business A has two competitors, B and C. No customer patronizes more than one of these businesses at the same time. Initially the probabilities that a customer patronizes A, B, or C are 0.2, 0.6, and 0.2, respectively. Suppose A initiates an advertising campaign to improve its business and finds the following transition matrix to describe the effect.

Last week's customers went to		This week's customers go to	
	A	B	C
A	0.7	0.2	0.1
B	0.4	0.4	0.2
C	0.4	0.4	0.2

32. If A runs the advertising campaign for 3 weeks, find the probability of a customer patronizing each business.

33. Find the steady-state vector for this market—that is, the long-range share of the market that each business can expect if the transition matrix holds.

Genetics   Use the following information for Problems 34–36. For species that reproduce sexually, characteristics are determined by a gene from each parent. Suppose that for a certain trait there are two possible genes available from each parent: a dominant gene D and a recessive gene r. Then the different gene combinations (called *genotypes*) for the offspring are DD, Dr, rD, and rr, where Dr and rD produce the same trait. Suppose further that these genotypes are the states of a Markov chain with the following transition matrix.

Parent genotype		Offspring genotype	
	DD	Dr	rr
DD	0.7	0.3	0
Dr	0.35	0.5	0.15
rr	0	0.7	0.3

34. If the initial occurrence of those genotypes in the population is 0.5 for DD, 0.4 for Dr, and 0.1 for rr, find the distribution (probability) of each type after the first and second generations.

35. Find the steady-state vector for this genotype.

36. To which generation(s) does the steady-state vector correspond? This is called the **Hardy-Weinberg model.**

Key Terms and Formulas

Section	Key Terms	Formulas
7.1	Equally likely events	$\Pr(E) = k/n$
	Sample space	
	Certain event	$\Pr(E) = 1$
	Impossible event	$\Pr(E) = 0$
	Odds	
	Empirical probability	
7.2	Intersection of events	$\Pr(E \text{ and } F) = \Pr(E \cap F)$
	Union of events	$\Pr(E \text{ or } F) = \Pr(E \cup F)$
		$= \Pr(E) + \Pr(F) - \Pr(E \cap F)$
	Complement	$\Pr(\text{not } E) = \Pr(E') = 1 - \Pr(E)$
	Mutually exclusive events	$\Pr(E \text{ or } F) = \Pr(E) + \Pr(F)$

Section	Key Terms	Formulas
7.3	Conditional probability	$\Pr(E \mid F) = \dfrac{\Pr(E \cap F)}{\Pr(F)}$
	Product Rule	$\Pr(E \cap F) = \Pr(F) \cdot \Pr(E \mid F)$
	Independent events	$\Pr(A \mid B) = \Pr(A)$ and
		$\Pr(A \cap B) = \Pr(A) \cdot \Pr(B)$
7.4	Probability tree	
	Bayes' formula and trees	$\Pr(E_1 \mid F_1) = \dfrac{\text{Branch product on path leading through } E_1 \text{ to } F_1}{\text{Sum of all branch products on paths leading to } F_1}$
7.5	Fundamental Counting Principle	$n(A) \cdot n(B)$
	n-factorial	$n! = n(n-1)\cdots(3)(2)(1)$ and $0! = 1$
	Permutations	$_nP_r = \dfrac{n!}{(n-r)!}$
	Combinations	$_nC_r = \dfrac{n!}{r!(n-r)!}$
7.6	Probability using permutations and combinations	
7.7	Markov chains	
	Vectors	
	Initial-probability vector	
	Fixed-probability or steady-state vector	

Review Exercises

Section 7.1

1. If one ball is drawn from a bag containing nine balls numbered 1 through 9, what is the probability that the ball's number is
   (a) odd?
   (b) divisible by 3?
   (c) odd and divisible by 3?

2. Suppose one ball is drawn from a bag containing nine red balls numbered 1 through 9 and three white balls numbered 10, 11, 12. What is the probability that
   (a) the ball is red?
   (b) the ball is odd-numbered?
   (c) the ball is white and even-numbered?
   (d) the ball is white or odd-numbered?

3. If the probability that an event E occurs is 3/7, what are the odds that
   (a) E will occur?   (b) E will not occur?

4. Suppose that a fair coin is tossed two times. Construct an equiprobable sample space for the experiment and determine each of the following probabilities.
   (a) Pr(0 heads)   (b) Pr(1 head)   (c) Pr(2 heads)

5. Suppose that a fair coin is tossed three times. Construct an equiprobable sample space for the experiment and determine each of the following probabilities.
   (a) Pr(2 heads)   (b) Pr(3 heads)   (c) Pr(1 head)

Section 7.2

6. A card is drawn at random from an ordinary deck of 52 playing cards. What is the probability that it is a queen or a jack?

7. A deck of 52 cards is shuffled. A card is drawn, it is replaced, the pack is again shuffled, and a second card is drawn. What is the probability that each card drawn is an ace, king, queen, or jack?

8. If the probability of winning a game is 1/4 and there can be no ties, what is the probability of losing the game?

9. A card is drawn from a deck of 52 playing cards. What is the probability that it is an ace or a 10?

10. A card is drawn at random from a deck of playing cards. What is the probability that it is a king or a red card?

11. A bag contains 4 red balls numbered 1, 2, 3, 4 and 5 white balls numbered 5, 6, 7, 8, 9. A ball is drawn. What is the probability that the ball
    (a) is red and even-numbered?
    (b) is red or even-numbered?
    (c) is white or odd-numbered?

Section 7.3

12. A bag contains 4 red balls and 3 black balls. Two balls are drawn at random from the bag without replacement. Find the probability that both balls are red.

13. A box contains 2 red balls and 3 black balls. Two balls are drawn from the box without replacement. Find the probability that the second ball is red, given that the first ball is black.

14. A bag contains 8 white balls, 5 red balls, and 7 black balls. If three balls are drawn at random from the bag, with replacement, what is the probability that the first two balls are red and the third is black?

15. A bag contains 8 white balls, 5 red balls, and 7 black balls. If three balls are drawn at random from the bag, without replacement, what is the probability that the first two balls are red and the third is black?

Section 7.4

16. A bag contains 3 red, 5 black, and 8 white balls. Three balls are drawn, without replacement. What is the probability that one of each color is drawn?

17. An urn contains 4 red and 6 white balls. One ball is drawn, it is not replaced, and a second ball is drawn. What is the probability that one ball is white and one is red?

18. Urn I contains 3 red and 4 white balls and urn II contains 5 red and 2 white balls. An urn is selected and a ball is drawn.
    (a) What is the probability that urn I is selected and a red ball is drawn?
    (b) What is the probability that a red ball is selected?
    (c) If a red ball is selected, what is the probability that urn I was selected?

19. Bag A contains 3 red, 6 black, and 5 white balls, and bag B contains 4 red, 5 black, and 7 white balls. A bag is selected at random and a ball is drawn. If the ball is white, what is the probability that bag B was selected?

Section 7.5

20. Compute $_6P_2$.   21. Compute $_7C_3$.

22. How many 3-letter sets of initials are possible?

23. How many combinations of 8 things taken 5 at a time are possible?

Section 7.7

24. Explain why each given matrix is not a transition matrix for a Markov chain.

    (a) $\begin{bmatrix} 0.2 & 0.7 & 0.1 \\ 0.3 & 0.2 & 0.5 \end{bmatrix}$ (b) $\begin{bmatrix} 0.4 & 0.7 \\ 0.6 & 0.3 \end{bmatrix}$

25. For the initial probability vector $\begin{bmatrix} 0.1 & 0.9 \end{bmatrix}$ and the transition matrix A, find the probability vectors for each of the next two stages.

    $$A = \begin{bmatrix} 0.4 & 0.6 \\ 0.8 & 0.2 \end{bmatrix}$$

26. Find the steady state vector associated with the transition matrix.

    $$\begin{bmatrix} 0.6 & 0.4 \\ 0.1 & 0.9 \end{bmatrix}$$

APPLICATIONS

Section 7.1

27. *Senior citizens*   In a certain city, 30,000 citizens out of 80,000 are over 50 years of age. What is the probability that a citizen selected at random will be 50 years old or younger?

28. *World Series*   Suppose the odds that the Chicago Cubs will win the 2010 World Series are 1 to 6. What is the associated probability that the Cubs will win this series?

Section 7.2

29. *United Nations*   Of 100 job applicants to the United Nations, 30 speak French, 40 speak German, and 12 speak both French and German. If an applicant is chosen at random, what is the probability that the applicant speaks French or German?

Productivity   **Suppose that in a study of leadership style versus industrial productivity, the following data were obtained. Use these empirical data to answer Problems 30–32.**

| Productivity | Leadership Style | | | |
	Demo-cratic	Authori-tarian	Laissez-faire	Total
Low	40	15	40	95
Medium	25	75	10	110
High	25	30	20	75
Total	90	120	70	280

30. Find the probability that a person chosen at random has a democratic style and medium productivity.

31. Find the probability that an individual chosen at random has high productivity or an authoritarian style.

Section 7.3

32. Find the probability that an individual chosen at random has an authoritarian style, given that he or she has medium productivity.

33. *Quality control* A product must pass an initial inspection, where the probability that it will be rejected is 0.2. If it passes this inspection, it must also pass a second inspection where the probability that it will be rejected is 0.1. What is the probability that it will pass both inspections?

Section 7.4

34. *Color blindness* Sixty men out of 1000 and 3 women out of 1000 are color blind. A person is picked at random from a group containing 10 men and 10 women.
    (a) What is the probability that the person is color blind?
    (b) What is the probability that the person is a man if the person is color blind?

35. *Purchasing* A regional survey found that 70% of all families who indicated an intention to buy a new car bought a new car within 3 months, that 10% of families who did not indicate an intention to buy a new car bought one within 3 months, and that 22% indicated an intention to buy a new car. If a family chosen at random bought a car, find the probability that the family had not previously indicated an intention to buy a car.

Section 7.5

36. *Management* A personnel director ranks 4 applicants for a job. How many rankings are possible?

37. *Management* An organization wants to select a president, vice president, secretary, and treasurer. If 8 people are willing to serve and each of them is eligible for any of the offices, in how many different ways can the offices be filled?

38. *Utilities* A utility company sends teams of 4 people each to perform repairs. If it has 12 qualified people, how many different ways can people be assigned to a team?

39. *Committees* An organization wants to select a committee of 4 members from a group of 8 eligible members. How many different committees are possible?

40. *Juries* A jury can be deadlocked if one person disagrees with the rest. There are 12 ways in which a jury can be deadlocked if one person disagrees, because any one of the 12 jurors could disagree.

    (a) In how many ways can a jury be deadlocked if 2 people disagree?
    (b) If 3 people disagree?

41. *License plates* Because Atlanta was the site of the 1996 Olympics, the state of Georgia released vanity license plates containing the Olympic symbol, and a large number of these plates were sold. Because of the symbol, each plate could contain at most 5 digits and/or letters. How many possible plates could have been produced using from 1 to 5 digits or letters?

42. *Blood types* In a book describing the mass murder of 18 people in northern California, a policewoman was quoted as stating that there are blood groups O, A, B, and AB, positive and negative, blood types M, N, MN, and P, and 8 Rh blood types, so there are 288 unique groups (*Source:* Joseph Harrington and Robert Berger, *Eye of Evil,* St. Martin's Press, 1993). Comment on this conclusion.

Section 7.6

43. *Scheduling* A college registrar adds 4 new courses to the list of offerings for the spring semester. If he added the course names in random order at the end of the list, what is the probability that these 4 courses are listed in alphabetical order?

44. *Lottery* Pennsylvania's Daily Number pays 500 to 1 to people who play the winning 3-digit number exactly as it is drawn. However, players can "box" a number so they can win $80 from a $1 bet if the 3 digits they pick come out in any order. What is the probability that a "boxed" number with 3 different digits will be a winner?

45. *Lottery* If a person "boxes" (see Problem 44) a 4-digit number with 4 different digits in Pennsylvania's Pick Four Lottery, what is the probability that the person will win?

46. *Quality control* A supplier has 200 compact discs, of which 10% are known to be defective. If a music store purchases 10 of these discs, what is the probability that
    (a) none of the discs is defective?
    (b) 2 of the discs are defective?

47. *Stocks* Mr. Way must sell stocks from 3 of the 5 companies whose stocks he owns so that he can send his children to college. If he chooses the companies at random, what is the probability that the 3 companies will be the 3 with the best future earnings?

48. *Quality control* A sample of 6 fuses is drawn from a lot containing 10 good fuses and 2 defective fuses. Find the probability that the number of defective fuses is
    (a) exactly 1.   (b) at least 1.

Section 7.7

Income levels Use the following information for Problems 49 and 50. Suppose people in a certain community are classified as being low-income (L), middle-income (M), or high-income (H). Suppose further that the probabilities for children being in a given state depend on which state their parents were in, according to the following Markov chain transition matrix.

	Income level of children		
Income level of parents	L	M	H
L	0.15	0.6	0.25
M	0.15	0.35	0.5
H	0	0.2	0.8

49. If the initial distribution probabilities of families in a certain population are $[0.7 \quad 0.2 \quad 0.1]$, find the distribution for the next three generations.

50. Find the expected long-term probability distribution of families by income level.

7 Chapter Test

A bag contains three white balls numbered 1, 2, 3 and four black balls numbered 4, 5, 6, 7. If one ball is drawn at random, find the probability of each event described in Problems 1–3.

1. (a) The ball is odd-numbered.
   (b) The ball is white.
2. (a) The ball is black and even-numbered.
   (b) The ball is white or even-numbered.
3. (a) The ball is red.
   (b) The ball is numbered less than 8.

A bag contains three white balls numbered 1, 2, 3 and four black balls numbered 4, 5, 6, 7. Two balls are drawn without replacement. Find the probability of each event described in Problems 4–9.

4. The sum of the numbers is 7.
5. Both balls are white.
6. (a) The first ball is white, and the second is black.
   (b) A white ball and a black ball are drawn.
7. Both balls are odd-numbered.
8. The second ball is white. (Use a tree.)
9. The first ball is black, given that the second ball is white.
10. A cat hits the letters on a computer keyboard three times. What is the probability that the word RAT appears?
11. In a certain region of the country with a large number of cars, 45% of the cars are from Asian manufacturers, 35% are American, and 20% are European. If three cars are chosen at random, what is the probability that two American cars are chosen?
12. In a group of 20 people, 15 are right-handed, 4 are left-handed, and 1 is ambidextrous. What is the probability that a person selected at random is
   (a) left-handed?   (b) ambidextrous?

13. In a group of 20 people, 15 are right-handed, 4 are left-handed, and 1 is ambidextrous. If 2 people are selected at random, without replacement, use a tree to find the following probabilities.
   (a) Both are left-handed.
   (b) One is right-handed, and one is left-handed.
   (c) Two are right-handed.
   (d) Two are ambidextrous.
14. The state of Arizona has a "6 of 42" game in which 6 balls are drawn without replacement from 42 balls numbered 1–42. The order in which the balls are drawn does not matter.
   (a) How many different drawing results are possible?
   (b) What is the probability that a person holding one ticket will win (match all the numbers drawn)?
15. The multistate "5 of 50" game draws 5 balls without replacement from 50 balls numbered 1–50.
   (a) How many different drawing results are possible if order does not matter?
   (b) What is the probability that a person holding one ticket will win (match all the numbers drawn)?
16. Computer chips come from two suppliers, with 80% coming from supplier 1 and 20% coming from supplier 2. Six percent of the chips from supplier 1 are defective, and 8% of the chips from supplier 2 are defective. If a chip is chosen at random, what is the probability that it is defective?
17. A placement test is given by a university to predict student success in a calculus course. On average, 70% of students who take the test pass it, and 87% of those who pass the test also pass the course, whereas 8% of those who fail the test pass the course.

(a) What is the probability that a student taking the placement test and the calculus course will pass the course?

(b) If a student passed the course, what is the probability that he or she passed the test?

18. Lactose intolerance affects about 20% of non-Hispanic white Americans and 50% of white Hispanic Americans. Nine percent of Americans are Hispanic whites and 75.6% are non-Hispanic whites (*Source:* Jean Carper, "Eat Smart," *USA Weekend*). If a white American is chosen at random and is lactose-intolerant, what is the probability that he or she is Hispanic?

19. A car rental firm has 350 cars. Seventy of the cars have only defective windshield wipers and 25 have only defective tail lights. Two hundred of the cars have no defects; the remainder have other defects. What is the probability that a car chosen at random

(a) has defective windshield wipers?

(b) has defective tail lights?

(c) does not have defective tail lights?

20. NACO Body Shops has found that 14% of the cars on the road need to be painted and that 3% need major body work and painting. What is the probability that a car selected at random needs major body work if it is known that it needs to be painted?

21. Garage door openers have 10 on-off switches on the opener in the garage and on the remote control used to open the door. The door opens when the remote control is pressed and all the switches on both the opener and the remote agree.

(a) How many different sequences of the on-off switches are possible on the door opener?

(b) If the switches on the remote control are set at random, what is the probability that it will open a given garage door?

(c) Remote controls and the openers in the garage for a given brand of door opener have the controls and openers matched in one of three sequences. If an owner does not change the sequence of switches on her or his remote control and opener after purchase, what is the probability that a different new remote control purchased from this company will open that owner's door?

(d) What should the owner of a new garage door opener do to protect the contents of his or her garage (and home)?

22. A publishing company has determined that a new edition of an existing mathematics textbook will be re-adopted by 80% of its current users and will be adopted by 7% of the users of other texts if the text is not changed radically. To determine whether it should change the book radically to attract more sales, it uses Markov chains. Assume that the text in question currently has 25% of its possible market.

(a) Create the transition matrix for this chain.

(b) Find the probability vector for the text three editions later and, from that, determine the percent of the market for that future edition.

(c) Find the steady-state vector for this text to determine what percent of its market this text will have if this policy is continued.

Extended Applications & Group Projects

I. Phone Numbers

Primarily because of the rapid growth in the use of cellular phones, the number of available phone numbers had to be increased. This was accomplished by increasing the number of available area codes in the United States. Investigate how this was done by completing the following, remembering that an area code cannot begin with a 0 or a 1.

1. Consult a phone book that was printed before 1995 and observe the restrictions on the middle digit of area codes at that time.
2. How many area codes were possible in the United States prior to 1995?
3. Consult a current phone book to see how the restrictions on area codes have been changed.
4. How many area codes are now possible because of the change? How many times more phone numbers are now possible because of the change?
5. What further changes could be made to the area code (other than adding a fourth digit) that would increase the available phone numbers? How many times more numbers would be available?
6. After the area code, every number mentioned on television dramas has 555 as its first three digits. How many phone numbers are available in each area code of TV Land?
7. Find how many area codes Los Angeles has, then determine how many different phone numbers are available in TV Land's Los Angeles.

II. Competition in the Telecommunications Industry

Of the 100,000 persons in a particular city who have cellular telephones, 85,000 use Cingular, 10,000 use Verizon, and 5000 use Sprint. A research firm has surveyed cell users to determine the level of customer satisfaction and to learn what company each expects to be using in a year's time. The results are depicted below.

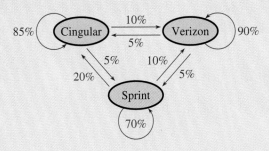

Although Cingular clearly dominates this city's cellular service now, the situation could change in the future.

1. Calculate the percent of the cell users who will be using each company in a year's time.
2. What will the long-term end result be if this pattern of change continues year after year?
3. You should have found, in your answer to Question 1, that the number of Sprint customers increases in a year's time. Explain how this can happen when Sprint loses 30% of its customers and gains only 5% of Cingular's and Verizon's customers.
4. Verizon managers have been presented with two proposals for increasing their market share, and each proposal has the same cost. It is estimated that plan A would increase their customer retention rate from 90% to 95% through an enhanced customer services department. Plan B, an advertising campaign targeting Cingular customers, is intended to increase the percent of people switching from Cingular to Verizon from 10% to 11%. If the company has sufficient funds to implement one plan only, which one would you advise it to approve and why?
5. Suppose that because of budget constraints, Verizon implements neither plan. A year later it faces the same choice. What should it do then?

8

Further Topics in Probability; Data Description

The probability that the birth of three children will result in 2 boys and a girl can be found efficiently by treating it as a **binomial experiment.** In this chapter, we discuss how a set of data can be described with **descriptive statistics,** including mode, median, mean, and standard deviation. Descriptive statistics are used by businesses to summarize data about advertising effectiveness, production costs, and profit. We continue our discussion of probability by considering **binomial probability distributions, discrete probability distributions,** and **normal probability distributions.** Social and behavioral scientists collect data about carefully selected **samples** and use probability distributions to reach conclusions about the populations from which the samples were drawn.

The topics and applications discussed in this chapter include the following.

Sections	Applications
8.1 Binomial Probability Experiments	*Gambling, birth defects, quality control*
8.2 Data Description	*Test scores, payrolls, basketball, air quality*
8.3 Discrete Probability Distributions Measures of dispersion The binomial distribution	*Expected value, raffles, insurance, gambling, eye surgery*
8.4 Normal Probability Distribution	*IQ scores, mean height, demand*

Chapter Warm-up

Prerequisite Problem Type	For Section	Answer	Section for Review
If a fair coin is tossed 3 times, find Pr(2 heads and 1 tail).	8.1	$\frac{3}{8}$	7.4 Probability trees
Evaluate: (a) $\binom{4}{3}$ (b) $\binom{5}{2}$ (c) $\binom{24}{0}$	8.1 8.3	(a) 4 (b) 10 (c) 1	7.5 Counting: permutations and combinations
Expand: (a) $(p + q)^2$ (b) $(p + q)^3$	8.3	(a) $p^2 + 2pq + q^2$ (b) $p^3 + 3p^2q + 3pq^2 + q^3$	0.5 Special products
Graph all real numbers satisfying: (a) $100 \leq x \leq 115$ (b) $-1 \leq z \leq 2$	8.4	(a) (b)	1.1 Linear inequalities
If 68% of the IQ scores of adults lie between 85 and 115, what is the probability that an adult chosen at random will have an IQ score between 85 and 115?	8.4	0.68	7.1 Probability

OBJECTIVE

● To solve probability problems related to binomial experiments

Binomial Probability Experiments

❍ Application Preview

It has been determined with probability 0.01 that any child whose mother lives in a specific area near a chemical dump will be born with a birth defect. To find the probability that 2 of 10 children whose mothers live in the area will be born with birth defects, we could use a probability tree that lists all the possibilities and then select the paths that give 2 children with birth defects. However, it is easier to find the probability by using a formula developed in this section. (See Example 2.) This formula can be used when the probability experiment is a **binomial probability experiment.**

Suppose you have a coin that is biased so that when the coin is tossed the probability of getting a head is 2/3 and the probability of getting a tail is 1/3. What is the probability of getting 2 heads in 3 tosses of this coin?

Suppose the question were "What is the probability that the first two tosses will be heads and the third will be tails?" In that case, the answer would be

$$\Pr(HHT) = \frac{2}{3} \cdot \frac{2}{3} \cdot \frac{1}{3} = \frac{4}{27}$$

However, the original question did not specify the order, so we must consider other orders that will give us 2 heads and 1 tail. We can use a tree to find all the possibilities (see Figure 8.1). We can see that there are 3 paths through the tree that correspond to 2 heads and 1 tail in 3 tosses. Because the probability for each successful path is 4/27,

$$\Pr(2H \text{ and } 1T) = 3 \cdot \frac{4}{27} = \frac{4}{9}$$

In this problem we can find the probability of 2 heads in 3 tosses by finding the probability for any one path of the tree (such as HHT) and then multiplying that probability by the number of paths that result in 2 heads. Notice that the 3 paths in the tree that result in 2 heads correspond to the different orders in which 2 heads and 1 tail can occur.

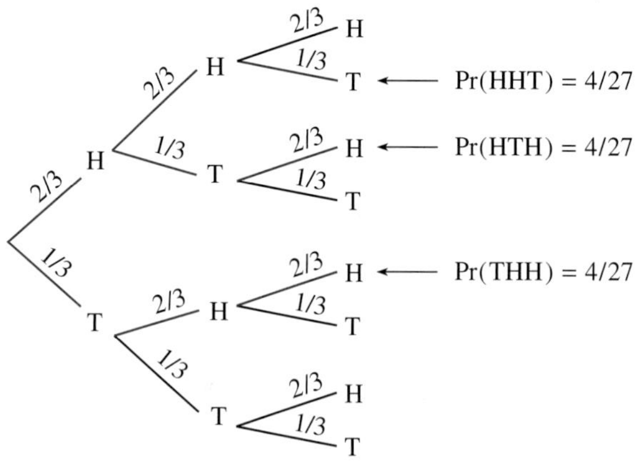

Figure 8.1

This is the number of ways that we can pick 2 positions out of 3 for the heads to occur, which is $_3C_2 = \binom{3}{2}$. Thus the probability of 2 heads resulting from 3 tosses is

$$\Pr(2H \text{ and } 1T) = \binom{3}{2}\frac{2}{3} \cdot \frac{2}{3} \cdot \frac{1}{3} = \binom{3}{2}\left(\frac{2}{3}\right)^2\left(\frac{1}{3}\right)^1 = \frac{4}{9}$$

The experiment discussed above (tossing a coin 3 times) is an example of a general class of experiments called **binomial probability experiments.** These experiments are also called **Bernoulli experiments,** or **Bernoulli trials,** after the 18th-century mathematician Jacob Bernoulli.

A **binomial probability experiment** satisfies the following properties:

1. There are n repeated trials of the experiment.
2. Each trial results in one of two outcomes, with one denoted success (S) and the other failure (F).
3. The trials are independent, with the probability of success, p, the same for every trial. The probability of failure is $q = 1 - p$ for every trial.
4. The probability of x successes in n trials is

$$\Pr(x) = \binom{n}{x}p^x q^{n-x}$$

$\Pr(x)$ is called a **binomial probability.**

● **EXAMPLE 1** *Binomial Exponents*

A die is rolled 4 times and the number of times a 6 results is recorded.

(a) Is the experiment a binomial experiment?
(b) What is the probability that three 6's will result?

Solution

(a) Yes. There are 4 trials, resulting in success (a 6) or failure (not a 6). The result of each roll is independent of previous rolls; the probability of success on each roll is $1/6$.
(b) There are $n = 4$ trials with $\Pr(\text{rolling a 6}) = p = 1/6$ and $\Pr(\text{not rolling a 6}) = q = 5/6$, and we seek the probability of $x = 3$ successes (because we want three 6's). Thus

$$\Pr(\text{three 6's}) = \binom{4}{3}\left(\frac{1}{6}\right)^3\left(\frac{5}{6}\right)^1 = \frac{4!}{1!3!}\left(\frac{1}{216}\right)\left(\frac{5}{6}\right) = \frac{5}{324}$$

● *Checkpoint*

Determine whether the following experiments satisfy the conditions for a binomial probability experiment.

1. A coin is tossed 8 times and the number of times a tail occurs is recorded.
2. A die is rolled twice and the sum that occurs is recorded.
3. A ball is drawn from a bag containing 4 white and 7 red balls and then replaced. This is done a total of 6 times, and the number of times a red ball is drawn is recorded.

◗ **EXAMPLE 2** *Birth Defects* **(Application Preview)**

It has been found that the probability is 0.01 that any child whose mother lives in an area near a chemical dump will be born with a birth defect. Suppose 10 children whose mothers live in the area are born in a given month.

(a) What is the probability that 2 of the 10 will be born with birth defects?
(b) What is the probability that at least 2 of them will have birth defects?

Solution

(a) Note that we may consider a child having a birth defect as a success for this problem even though it is certainly not true in reality. Each of the 10 births is independent, with probability of success $p = 0.01$ and probability of failure $q = 0.99$. Therefore, the experiment is a binomial experiment, and

$$Pr(2) = \binom{10}{2}(0.01)^2(0.99)^8 = 0.00415$$

(b) The probability of at least 2 of the children having birth defects is

$$1 - [Pr(0) + Pr(1)] = 1 - \left[\binom{10}{0}(0.01)^0(0.99)^{10} + \binom{10}{1}(0.01)^1(0.99)^9\right]$$
$$= 1 - [0.90438 + 0.09135] = 0.00427$$

We interpret this to mean that there is a 0.427% likelihood that at least 2 of these children will have birth defects.

● **EXAMPLE 3** *Quality Control*

A manufacturer of motorcycle parts guarantees that a box of 24 parts will contain at most 1 defective part. If the records show that the manufacturer's machines produce 1% defective parts, what is the probability that a box of parts will satisfy the guarantee?

Solution

The problem is a binomial experiment problem with $n = 24$ and (considering getting a defective part as a success) $p = 0.01$. Then the probability that the manufacturer will have no more than 1 defective part in a box is

$$Pr(x \le 1) = Pr(0) + Pr(1) = \binom{24}{0}(0.01)^0(0.99)^{24} + \binom{24}{1}(0.01)^1(0.99)^{23}$$
$$= 1(1)(0.7857) + 24(0.01)(0.7936) = 0.9762$$

We interpret this to mean that there is a 97.62% likelihood that there will be no more than 1 defective part in a box of 24 parts.

● **Checkpoint**

4. A bag contains 5 red balls and 3 white balls. If 3 balls are drawn, with replacement, from the bag, what is the probability that
   (a) two balls will be red?   (b) at least two balls will be red?

● **Checkpoint Solutions**

1. A tail is a success on each trial, and all of the conditions of a binomial probability distribution are satisfied.
2. Each trial in this experiment does not result in a success or failure, so this is not a binomial probability experiment.
3. A red ball is a success on each trial, and each condition of a binomial distribution is satisfied.
4. (a) $Pr(2 \text{ red}) = \binom{3}{2}\left(\frac{5}{8}\right)^2\left(\frac{3}{8}\right)$

    $= \dfrac{225}{512}$

   (b) $Pr(\text{at least 2 red}) = \binom{3}{2}\left(\frac{5}{8}\right)^2\left(\frac{3}{8}\right) + \binom{3}{3}\left(\frac{5}{8}\right)^3\left(\frac{3}{8}\right)^0$

    $= \dfrac{225}{512} + \dfrac{125}{512} = \dfrac{175}{256}$

8.1 Exercises

1. If the probability of success on each trial of an experiment is 0.3, what is the probability of 4 successes in 6 independent trials?

2. If the probability of success on each trial of an experiment is 0.4, what is the probability of 5 successes in 8 trials?

3. Suppose a fair die is rolled 18 times.
   (a) What is the probability p that a 4 will occur each time the die is rolled?
   (b) What is the probability q that a 4 will not occur each time the die is rolled?
   (c) What is n for this experiment?
   (d) What is the probability that a 4 will occur 6 times in the 18 rolls?

4. Suppose a fair coin is tossed 12 times.
   (a) What is the probability p that a head will occur each time the coin is tossed?
   (b) What is the probability q that a head will not occur each time the coin is tossed?
   (c) What is n for this experiment?
   (d) What is the probability that a head will occur 6 times in the 12 tosses?

5. Suppose a fair coin is tossed 6 times. What is the probability that
   (a) 6 heads will occur?
   (b) 3 heads will occur?
   (c) 2 heads will occur?

6. If a fair die is rolled 3 times, what is the probability that
   (a) a 5 will result 2 times out of the 3 rolls?
   (b) an odd number will result 2 times?
   (c) a number divisible by 3 will occur 3 times?

7. Suppose a fair die is rolled 12 times. What is the probability that 5 "aces" (1's) will occur?

8. In Problem 7, what is the probability that no aces will occur?

9. A bag contains 6 red balls and 4 black balls. We draw 5 balls, with each one replaced before the next is drawn.
   (a) What is the probability that 2 balls drawn will be red?
   (b) What is the probability that 5 black balls will be drawn?
   (c) What is the probability that at least 3 black balls will be drawn?

10. A bag contains 6 red balls and 4 black balls. If we draw 5 balls, with each one replaced before the next is drawn, what is the probability that at least 2 balls drawn will be red?

11. Suppose a pair of dice is thrown 4 times. What is the probability that a sum of 9 occurs exactly 2 times?

12. If a pair of dice is thrown 4 times, find the probability that a sum of 7 occurs at least 3 times.

13. Suppose the probability that a marksman will hit a target each time he shoots is 0.85. If he fires 10 shots at a target, what is the probability he will hit it 8 times?

14. The probability that a sharpshooter will hit a target each time he shoots is 0.98. What is the probability that he will hit the target 10 times in 10 shots?

APPLICATIONS

15. *Genetics* A family has 4 children. If the probability that each child is a girl is 0.5, what is the probability that
    (a) half of the children are girls?
    (b) all of the children are girls?

16. *Testing* A multiple-choice test has 10 questions and 5 choices for each question. If a student is totally unprepared and guesses on each question, what is the probability that
    (a) she will answer every question correctly?
    (b) she will answer half of the questions correctly?

17. *Management* The manager of a store buys portable radios in lots of 12. Suppose that, on the average, 2 out of each group of 12 are defective. The manager randomly selects 4 radios out of the group to test.
    (a) What is the probability that he will find 2 defective radios?
    (b) What is the probability that he will find no defective radios?

18. *Quality control* A hospital buys thermometers in lots of 1000. On the average, one out of 1000 is defective. If 10 are selected from one lot, what is the probability that none is defective?

19. *Genetics* The probability that a certain couple will have a blue-eyed child is 1/4, and they have 4 children. What is the probability that
    (a) 1 of their children has blue eyes?
    (b) 2 of their children have blue eyes?
    (c) none of their children has blue eyes?

20. *Genetics* If the probability that a certain couple will have a blue-eyed child is 1/2, and they have 4 children, what is the probability that
    (a) none of the children has blue eyes?
    (b) at least 1 child has blue eyes?

21. *Health care* Suppose that 10% of the patients who have a certain disease die from it. If 5 patients have the disease, what is the probability that
    (a) exactly 2 patients will die from it?
    (b) no patients will die from it?
    (c) no more than 2 patients will die from it?

22. *Employee benefits* In a certain school district, 3% of the faculty use none of their sick days in a school year. Find the probability that 5 faculty members selected at random used no sick days in a given year.

23. *Genetics* If the ratio of boys born to girls born is 105 to 100, and if 6 children are born in a certain hospital in a day, what is the probability that 4 of them are boys?

24. *Biology* It has been determined empirically that the probability that a given cell will survive for a given period of time is 0.4. Find the probability that 3 out of 6 of these cells will survive for this period of time.

25. *Insurance* If records indicate that 4 houses out of 1000 are expected to be damaged by fire in any year, what is the probability that a woman who owns 10 houses will have fire damage in 2 of them in a year?

26. *Psychology* Suppose 4 rats are placed in a T-maze in which they must turn right or left. If each rat makes a choice by chance, what is the probability that 2 of the rats will turn to the right?

27. *Sports* A baseball player has a lifetime batting average of 0.300. If he comes to bat 5 times in a given game, what is the probability that
    (a) he will get 3 hits?
    (b) he will get more than 3 hits?

28. *Quality control* If the probability of an automobile part being defective is 0.02, find the probability of getting exactly 3 defective parts in a sample of 12.

29. *Quality control* A company produces shotgun shells in batches of 300. A sample of 10 is tested from each batch, and if more than one defect is found, the entire batch is tested. If 1% of the shells are actually defective,
    (a) what is the probability of 0 defective shells in the sample?
    (b) what is the probability of 1 defective shell?
    (c) what is the probability of more than 1 defective shell?

30. *Quality control* A baseball pitching machine throws a bad pitch with probability 0.1. What is the probability that the machine will throw 5 or fewer bad pitches out of 20 pitches? (Set up only.)

31. *Testing* A quiz consists of 10 multiple-choice questions with 5 choices for each question. Suppose a student is sure of the first 5 answers and has each of the last 5 questions narrowed to 3 of the possible 5 choices. If the student guesses among the narrowed choices on the last 5 questions, find
    (a) the probability of passing the quiz (getting at least 60%).
    (b) the probability of getting at least a B (at least 80%).

32. *Demographics* In a certain community, 10% of the population is Jewish. A study shows that of 7 social service agencies, 4 have board presidents who are Jewish. Find the probability that this could happen by chance.

33. *Suicide rates* Suppose the probability of suicide among a certain age group is 0.003. If a randomly selected group of 100 Native Americans within this age group had no suicides, find the probability of this occurring by chance.

8.2 Data Description

OBJECTIVES

- To set up frequency tables and construct frequency histograms for sets of data
- To find the mode of a set of scores (numbers)
- To find the median of a set of scores
- To find the mean of a set of scores
- To find the range of a set of data
- To find the variance and standard deviation of a set of data

◗ Application Preview

In modern business, a vast amount of data is collected for use in making decisions about the production, distribution, and sale of merchandise. Businesses also collect and summarize data about advertising effectiveness, production costs, sales, wages, and profits. Behavioral scientists attempt to reach conclusions about general behavioral characteristics by studying the characteristics of small samples of people. For example, election predictions are based on careful sampling of the votes; correct predictions are frequently announced on television even though only 5% of the vote has been counted. Life scientists can use statistical methods with laboratory animals to detect substances that may be dangerous to humans. The following table gives the numbers of days per year that San Diego, California failed to meet acceptable air quality standards for the years 1990–1999. We will see in this section how to display these data with a histogram and how to summarize them by finding the mean and standard deviation of the data. (See Example 10.)

Year	1990	1991	1992	1993	1994	1995	1996	1997	1998	1999
Unacceptable days	96	67	66	58	46	48	31	14	33	16

Source: U.S. Environmental Protection Agency, Office of Air Quality Planning and Standards

Histograms   A set of data taken from some source of data for the purpose of learning about the source is called a **sample;** the source of the data is called a **population.** The branch of statistics that deals with the collection, organization, summarization, and presentation of sample data is called **data description** or **descriptive statistics.** The descriptive statistics methods discussed in this section include graphical representation of data, measures of central tendency, and measures of dispersion.

The first step in statistical work is the collection of data. Data are regarded as statistics data if the numbers collected have some relationship. The heights of all female first-year college students are statistical data because a definite relationship exists among the collected numbers.

A *graph* frequently is used to provide a picture of the statistical data that have been gathered and to show relationships among various quantities. The advantage of the graph for showing data is that a person can quickly get an idea of the relationships that exist. The disadvantage is that the data can be shown only approximately on a graph. One type of graph that is used frequently in statistical work is the **frequency histogram.** A histogram is really a bar graph. It is constructed by putting the scores (numbers collected) along the horizontal axis and the frequencies with which they occur along the vertical axis. A frequency table is usually set up to prepare the data for the histogram.

● **EXAMPLE 1** *Frequency Histograms*

Construct a frequency histogram for the following scores: 38, 37, 36, 40, 35, 40, 38, 37, 36, 37, 39, 38.

Solution
We first construct the frequency table and then use the table to construct the histogram shown in Figure 8.2.

Score	Frequency
35	1
36	2
37	3
38	3
39	1
40	2

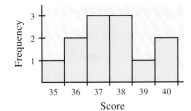

Figure 8.2

Graphing calculators, spreadsheets, and computer programs can be used to plot the histogram of a set of data. The width of each bar of the histogram can be determined by the calculator or adjusted by the user when the histogram is plotted.

● **EXAMPLE 2** *Frequency Histograms*

Construct a histogram from the following frequency table.

Interval	Frequency
0–5	0
6–10	2
11–15	5
16–20	1
21–25	3

Solution

The histogram is shown in Figure 8.3.

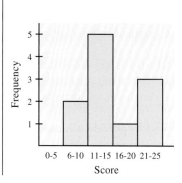

Figure 8.3

Mode   A set of data can be described by listing all the scores or by drawing a frequency histogram. But we often wish to describe a set of scores by giving the *average score* or the typical score in the set. There are three types of measures that are called *averages,* or measures of central tendency. They are the **mode,** the **median,** and the **mean.**

    To determine what value represents the most "typical" score in a set of scores, we use the **mode** of the scores.

Mode

> The **mode** of a set of scores is the score that occurs most frequently. The mode of a set of scores may not be unique.

That is, the mode is the most popular score—the one that is most likely to occur. The mode can be readily determined from a frequency table or frequency histogram because it is determined according to the frequencies of the scores. The score associated with the highest bar on a histogram is the mode.

● **EXAMPLE 3** *Mode*

(a) Find the mode of the following scores: 10, 4, 3, 6, 4, 2, 3, 4, 5, 6, 8, 10, 2, 1, 4, 3.
(b) Determine the mode of the scores shown in the histogram in Figure 8.4.

Solution

(a) Arranging the scores in order gives: 1, 2, 2, 3, 3, 3, 4, 4, 4, 4, 5, 6, 6, 8, 10, 10. The mode is 4, because it occurs more frequently than any other score.
(b) The most frequent score, as the histogram reveals, is 2.

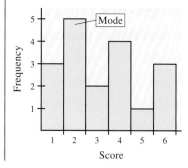

Figure 8.4

● **EXAMPLE 4** *Mode*

Find the mode of the following scores: 4, 3, 2, 4, 6, 5, 5, 7, 6, 5, 7, 3, 1, 7, 2.

Solution
Arranging the scores in order gives: 1, 2, 2, 3, 3, 4, 4, 5, 5, 5, 6, 6, 7, 7, 7. Both 5 and 7 occur three times. Thus they are both modes. This set of data is said to be *bimodal* because it has two modes.

Median   Although the mode of a set of scores tells us the most frequent score, it does not always represent the typical performance, or central tendency, of the set of scores. For example, the scores 2, 3, 3, 3, 5, 8, 9, 10, 13 have a mode of 3. But 3 is not a good measure of central tendency for this set of scores, for it is nowhere near the middle of the distribution.
    One value that gives a better measure of central tendency is the **median.**

Median   The **median** is the score or point above and below which an equal number of the scores lie when the scores are arranged in ascending or descending order. If there is an odd number of scores, we find the median by ranking the scores from smallest to largest and picking the middle score.
    If the number of scores is even, there will be no middle score when the scores are ranked, so the *median* is the point that is the average *of the two middle scores.*

● **EXAMPLE 5** *Median*

Find the median of each of the following sets of scores.

(a) 2, 3, 16, 5, 15, 38, 18, 17, 12
(b) 3, 2, 6, 8, 12, 4, 3, 2, 1, 6

Solution
(a) We rank the scores from smallest to largest: 2, 3, 5, 12, 15, 16, 17, 18, 38. The median is the middle score, which is 15. Note that there are 4 scores above and 4 scores below 15.
(b) We rank the scores from smallest to largest: 1, 2, 2, 3, 3, 4, 6, 6, 8, 12. The two scores that lie in the middle of the distribution are 3 and 4. Thus the median is a point that is midway between 3 and 4; that is, it is the arithmetic average of 3 and 4. The median is

$$\frac{3+4}{2} = \frac{7}{2} = 3.5$$

In this case we say the median is a point because there is no *score* that is 3.5.

The median is a measure of central tendency that is easy to find and that is not influenced by extreme scores. On a histogram, the area to the left of the median is equal to the area to the right of the median. The scores in the histogram shown in Figure 8.5 are 1, 1, 1, 2, 2, 2, 2, 2, 3, 3, 4, 4, 4, 4, 5, 6, 6, 6, and the median is 3. Note that the area of the histogram to the left of 3 equals the area to the right of 3.

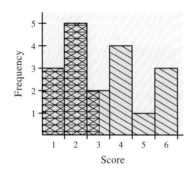

Figure 8.5

Some calculators have keys or functions for finding the median of a set of data. Such calculators, or computers, are valuable when the set of data is large.

Mean The median is the most easily interpreted measure of central tendency, and it is the best indicator of central tendency when the set of scores contains a few extreme values. However, the most frequently used measure of central tendency is the **mean.**

Mean

The symbol $\bar{x}$ is used to represent the **mean** of a set of scores in a sample. Thus, if n is the number of scores, the formula for the mean is

$$\bar{x} = \frac{\Sigma x}{n} = \frac{\text{sum of scores}}{\text{number of scores}}$$

The mean is used most often as a measure of central tendency because it is more useful than the median in the general applications of statistics.

● **EXAMPLE 6** *Mean*

Find the mean of the following sample of numbers: 12, 8, 7, 10, 6, 14, 7, 6, 12, 9.

Solution

$$\bar{x} = \frac{\Sigma x}{10} = \frac{12 + 8 + 7 + 10 + 6 + 14 + 7 + 6 + 12 + 9}{10} = \frac{91}{10}$$

so $\bar{x} = 9.1$.
   Note that the mean need not be one of the numbers (scores) given.

If the data are given in a frequency table or a frequency histogram, we can use a more efficient formula for finding the mean.

Mean of Grouped Data

If a set of n scores is grouped into k classes with x representing the score in a given class and $f(x)$ representing the number of scores in that class, then $n = \Sigma f(x)$ and the mean of the data is given by

$$\bar{x} = \frac{\Sigma(x \cdot f(x))}{\Sigma f(x)}$$

If we are given data in this frequency table,

x	1	2	3	4
f	3	1	4	2

we can find the mean $\bar{x}$ by using the following abbreviated formula

$$\bar{x} = \frac{\Sigma(x \cdot f)}{\Sigma f}$$

where each value of x is used once.

The mean for the frequency table above is

$$\bar{x} = \frac{1(3) + 2(1) + 3(4) + 4(2)}{3 + 1 + 4 + 2} = 2.5$$

● **EXAMPLE 7** *Test Scores*

Find the mean of the following sample of test scores for a math class.

Scores	Class Marks	Frequencies
40–49	44.5	2
50–59	54.5	0
60–69	64.5	6
70–79	74.5	12
80–89	84.5	8
90–99	94.5	2

Solution

For the purpose of finding the mean of interval data, we assume that all scores within an interval are represented by the class mark (midpoint of the interval). Thus

$$\bar{x} = \frac{(44.5)(2) + (54.5)(0) + (64.5)(6) + (74.5)(12) + (84.5)(8) + (94.5)(2)}{30}$$

$$= 74.5$$

If a set of data is represented by a histogram, the mean of the data is at the point where the histogram could be balanced. The mean of the data in Figure 8.6 is

$$\frac{1(3) + 2(6) + 3(2) + 4(4) + 5(1) + 6(2)}{3 + 6 + 2 + 4 + 1 + 2} = \frac{54}{18} = 3$$

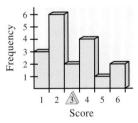

Figure 8.6

Any of the three measures of central tendency (mode, median, mean) may be referred to as an average. The measure of central tendency that should be used with data depends on the purpose for which the data were collected. The mode is the number that occurs with the greatest frequency. It is used if we want the most popular value. For example, the mode salary for a company is the salary that is received by most of the employees of the company. The median salary for the company is the salary that falls in the middle of the distribution.

The mean salary is the arithmetic average of the salaries. It is quite possible that the mode, median, and mean salaries for a company will be different. Figure 8.7 illustrates how the mode, median, and mean salaries might look for a company.

Payroll of Ace Cap Company

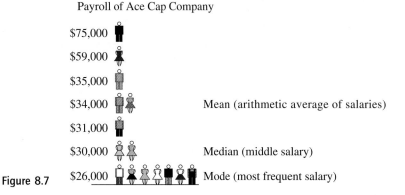

$75,000

$59,000

$35,000

$34,000   Mean (arithmetic average of salaries)

$31,000

$30,000   Median (middle salary)

Figure 8.7 $26,000   Mode (most frequent salary)

Which of the three measures represents the "average" salary? To say the average salary is $34,000 (the mean) helps conceal two things: the bosses' (or owners') very large salaries ($75,000 and $59,000), and the laborers' very small salaries ($26,000). The mode tells us more: The most common salary in this company is $26,000. The median gives us more information about these salaries than any other single figure. It tells us that if the salaries are ranked from lowest to highest, then the salary that lies in the middle is $30,000.

Because any of the three measures may be referred to as the average of a set of data, we must be careful to avoid being misled. The owner of the Ace Cap Company will probably state that the average salary for his company is $34,000 because it makes it appear that he pays high salaries. The local union president will likely claim that the mode ($26,000) is the "average" salary. Advertising agencies may also make use of an "average" that will make their product appear in the best light. We run the risk of being misled if we do not know which average is being cited.

One reason we may be misled concerning these three measures is that they frequently fall very close to each other, as we shall see in more detail later. Also, since the mean is the most frequently used, we tend to associate it with the word *average.*

● *Checkpoint*

Consider the data 10, 12, 12, 8, 30, 12, 8, 10, 5, 23.
1. Find the mode.
2. Find the mean.
3. Find the median.

Range

Although the mean of a set of data is useful in locating the center of the distribution of the data, it doesn't tell us as much about the distribution as we might think at first. For example, a basketball team with five 6-ft players is quite different from a team with one 6-ft, two 5-ft, and two 7-ft players. The distributions of the heights of these teams differ not in the mean height but in how the heights *vary* from the mean.

One measure of how a distribution varies is the **range** of the distribution.

Range

> The **range** of a set of numbers is the difference between the largest and smallest numbers in the set.

Consider the following set of heights of players on a basketball team, in inches: 69, 70, 75, 69, 73, 78, 74, 73, 78, 71. The range of this set of heights is $78 - 69 = 9$ inches. Note that this range is determined by only two numbers and does not give any information about how the other heights vary.

Variance and Standard Deviation

If we calculated the deviation of each score from the mean, $(x - \bar{x})$, and then attempted to average the deviations, we would see that the deviations add to 0. In an effort to find a meaningful measure of dispersion about the mean, statisticians developed the **variance,** which squares the deviations (to make them all positive) and then averages the squared deviations.

Frequently, we do not have all the data for a population and must use a sample of these data. To estimate the variance of a population from a sample, statisticians compensate for the fact that there is usually less variability in a sample than in the population itself by summing the squared deviations and dividing them by $n - 1$ rather than averaging them.

To get a measure comparable to the original deviations before they were squared, the **standard deviation,** which is the square root of the variance, was introduced. The formulas for the variance and standard deviation of sample data follow.

Variance and Standard Deviation

$$\text{Sample Variance} \quad s^2 = \frac{\Sigma(x - \bar{x})^2}{n - 1}$$

$$\text{Sample Standard Deviation} \quad s = \sqrt{\frac{\Sigma(x - \bar{x})^2}{n - 1}}$$

The standard deviation of a sample is a measure of the concentration of the scores about their mean. The smaller the standard deviation, the closer the scores lie to the mean. For example, the histograms in Figure 8.8 describe two sets of data with the same means and the same ranges. From looking at the histograms, we see that more of the data are concentrated about the mean $\bar{x} = 4$ in Figure 8.8(b) than in Figure 8.8(a). In the following example we will see that the standard deviation is smaller for the data in Figure 8.8(b) than in Figure 8.8(a).

● **EXAMPLE 8** *Standard Deviation*

Figure 8.8(a) is the histogram for the sample data 1, 1, 1, 3, 3, 4, 4, 5, 6, 6, 7, 7. Figure 8.8(b) is the histogram for the data 1, 2, 3, 3, 4, 4, 4, 4, 5, 5, 6, 7. Both sets of data have a range of 6 and a mean of 4. Find the standard deviations of these samples.

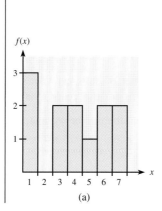
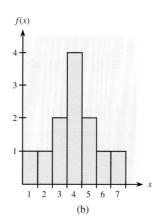

Figure 8.8
(a)   (b)

Solution

For the sample data in Figure 8.8(a), we have

$$\Sigma(x - \bar{x})^2 = (1 - 4)^2 + (1 - 4)^2 + (1 - 4)^2 + (3 - 4)^2 + (3 - 4)^2 + (4 - 4)^2$$
$$+ (4 - 4)^2 + (5 - 4)^2 + (6 - 4)^2 + (6 - 4)^2 + (7 - 4)^2 + (7 - 4)^2$$
$$= (-3)^2 + (-3)^2 + (-3)^2 + (-1)^2 + (-1)^2 + 0^2 + 0^2 + 1^2 + 2^2 + 2^2$$
$$+ 3^2 + 3^2$$
$$= 56$$

$$s^2 = \frac{\Sigma(x - \bar{x})^2}{n - 1} = \frac{56}{11} \approx 5.0909$$

$$s = \sqrt{s^2} = \sqrt{5.0909} \approx 2.26$$

For the sample data in Figure 8.8(b), we have

$$\Sigma(x - \bar{x})^2 = (1 - 4)^2 + (2 - 4)^2 + (3 - 4)^2 + (3 - 4)^2 + (4 - 4)^2 + (4 - 4)^2$$
$$+ (4 - 4)^2 + (4 - 4)^2 + (5 - 4)^2 + (5 - 4)^2 + (6 - 4)^2 + (7 - 4)^2$$
$$= (-3)^2 + (-2)^2 + (-1)^2 + (-1)^2 + 0^2 + 0^2 + 0^2 + 0^2 + 1^2 + 1^2$$
$$+ 2^2 + 3^2$$
$$= 30$$

$$s^2 = \frac{30}{11} \approx 2.7273$$

$$s = \sqrt{\frac{30}{11}} \approx 1.65$$

Calculator   Scientific calculators and graphing calculators have keys or built-in functions that compute
Note   the mean and standard deviation of a set of data. The ease of using calculators to find the
standard deviation makes it the preferred method.   ■

 ● **EXAMPLE 9** *Basketball*

The heights of the basketball team members mentioned earlier are 69, 70, 75, 69, 73, 78, 74, 73, 78, and 71 inches. Find the mean and standard deviation of this sample.

Solution

By using the formula or a calculator, we find that the mean is 73 and the sample standard deviation is

$$s = 3.333$$

 ◗ **EXAMPLE 10** *Air Quality* **(Application Preview)**

The following table gives the number of days per year that San Diego, California failed to meet acceptable air quality standards for the years 1990–1999.

(a) Construct a bar graph to represent the number of unacceptable days in San Diego per year.
(b) Find the mean number of unacceptable days per year for the 10-year period.

(c) What is the standard deviation of the data?

Year	1990	1991	1992	1993	1994	1995	1996	1997	1998	1999
Unacceptable days	96	67	66	58	46	48	31	14	33	16

Source: U.S. Environmental Protection Agency, Office of Air Quality Planning and Standards

Solution

(a) The bar graph displaying the number of unacceptable days per year is shown in Figure 8.9.

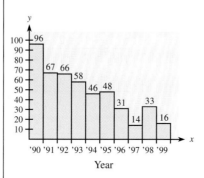

Figure 8.9

(b) The mean number of unacceptable days per year over this 10-year period is

$$\bar{x} = \frac{\Sigma x}{n} = \frac{475}{10} = 47.5$$

(c) The standard deviation of the number of unacceptable days over this period is

$$s = 25.35$$

This shows that the air quality fluctuated considerably over this period of time.

● *Checkpoint*

4. Find the standard deviation of the data:

$$10,\ 12,\ 12,\ 8,\ 30,\ 12,\ 8,\ 10,\ 5,\ 23$$

● *Checkpoint Solutions*

1. The mode (most frequent) score is 12.
2. The mean is

$$\frac{10 + 12 + 12 + 8 + 30 + 12 + 8 + 10 + 5 + 23}{10} = 13$$

3. Arranging the 10 scores from smallest to largest gives 5, 8, 8, 10, 10, 12, 12, 12, 23, 30. Thus the median is the average of the fifth and sixth scores.

$$\frac{10 + 12}{2} = 11$$

4. Using technology, the standard deviation is found to be

$$s = 7.63$$

8.2 Exercises

Construct frequency histograms for the data given in the frequency tables in Problems 1–6.

1.
Score	Frequency
12	2
13	3
14	4
15	3
16	1

2.
Score	Frequency
22	1
23	4
24	3
25	2
26	3

3.
x	3	4	5	6	7	8	9	10	11	12
$f(x)$	6	0	5	3	4	3	2	1	0	4

4.
x	5	6	8	9	11	12
$f(x)$	3	4	2	7	3	2

5.
Interval	Frequency
1–4	1
5–8	0
9–12	2
13–16	3
17–20	1

6.
Interval	Frequency
10–14	2
15–19	4
20–24	3
25–29	2
30–34	3

Construct frequency histograms for the data in Problems 7 and 8.

7. 3, 2, 5, 6, 3, 2, 6, 5, 4, 2, 1, 6
8. 5, 4, 6, 5, 4, 6, 3, 6, 5, 4, 1, 7

Find the modes of the sets of scores in Problems 9–11.

9. 3, 4, 3, 2, 2, 3, 5, 7, 6, 2, 3
10. 5, 8, 10, 12, 5, 4, 6, 3, 5
11. 14, 17, 13, 16, 15, 12, 13, 12, 13

Find the medians of the sets of scores in Problems 12–15.

12. 1, 3, 6, 7, 5
13. 2, 1, 3, 4, 2, 1, 2
14. 4, 7, 9, 18, 36, 14, 12
15. 1, 0, 2, 1, 1, 0, 2, 14, 37

Find the modes, medians, and means of the scores in Problems 16–22.

16. 5, 7, 7, 7, 9, 12, 4
17. 3, 2, 1, 6, 8, 12, 14, 2
18. 5, 8, 7, 6, 1, 1, 31
19. 14, 17, 20, 31, 17, 42
20. 3.2, 3.2, 3.5, 3.7, 3.4
21. 2.8, 6.4, 5.3, 5.3, 6.8
22. 1.14, 2.28, 7.58, 6.32, 5.17
23. Use class marks to find the mean, mode, and median of the data in Problem 5.
24. Use class marks to find the mean, mode, and median of the data in Problem 6.

In Problems 25–28, find the range of the set of numbers given.

25. 3, 5, 7, 8, 2, 11, 6, 5
26. 5, 1, 3, 1.4, 6.3, 8
27. −1, 2, 4, 3, 6, 11, −3, 4, 2
28. 2, 3, 3, 3, 3, 3, 9

Find the mean, variance, and standard deviation of each of the sets of sample data in Problems 29–32.

29. 5, 7, 1, 3, 0, 8, 6, 2
30. 7, 13, 5, 11, 8, 10, 9
31. 11, 12, 13, 14, 15, 16, 17
32. 4, 3, 6, 7, 8, 9, 12

In Problems 33–36, find the mean and standard deviation of the sample data in each table.

33.
x	1	2	3	4	5
$f(x)$	3	1	4	2	1

34.
x	3	4	5	6	8
$f(x)$	1	3	4	1	3

35.
x	3	4	5	6	7	8	9	10	11	12
$f(x)$	6	0	5	3	4	3	2	1	0	4

36.
x	5	6	7	8	9	10	11	12	13	14	15
$f(x)$	3	2	1	0	2	5	3	1	2	4	6

APPLICATIONS

37. *Unemployment rates* The following table gives the U.S. unemployment rates for civilian workers for selected years from 1960 to 2004. (a) Display these data with a bar graph and (b) summarize the data by finding the mean and standard deviation of the unemployment rate.

Year	1960	1970	1980	1990	2000	2004
Unemployment rate	5.5	4.9	7.1	5.6	4.0	5.6

Source: Bureau of Labor Statistics, U.S. Department of Labor

38. *Scholarship awards* The amounts of money awarded (in millions of dollars) in South Carolina's LIFE Scholarships for the years 2001 through 2004 are shown in the following table.
    (a) Construct a histogram of these data.
    (b) Find the mean amount awarded during this period.

Year	2001	2002	2003	2004
$ millions	46.4	54.4	106.5	119.2

Source: www.che.sc.gov

39. *U.S. farms* The following table gives the size of the average farm (in acres) for selected years from 1940 to 2000. (a) Display these data with a bar graph and (b) summarize the data by finding the mean of the farm sizes.

Year	1940	1950	1960	1970	1980	1990	2000
Farm size	174	213	297	374	426	460	434

Source: U.S. Department of Agriculture

40. *Product testing*   A taxi company tests a new brand of tires by putting a set of four on each of three taxis. The following table indicates the numbers of miles the tires lasted. Display these data with a histogram and find the mean number of miles the tires lasted.

Number of Miles	Number of Tires
30,000	4
32,000	2
34,000	3
36,000	2
38,000	1

Real estate   **Use the following information for Problems 41–43. Suppose you live in a neighborhood with a few expensive homes and many modest homes.**

41. If you wanted to impress people with the neighborhood where you lived, which measure would you give as the "average" property value?

42. Which "average" would you cite to the property tax committee if you wanted to convince them that property values aren't very high in your neighborhood?

43. What measure of central tendency would give the most representative "average" property value for your neighborhood? Explain.

44. *Salaries*   A survey revealed that of University of North Carolina alumni entering employment in 1984, those with a major in geography had the highest mean starting salary of all majors. Discuss this conclusion, using the fact that Michael Jordan was in this group and majored in geography.

45. *Voting*   The table below gives the 2004 votes in favor of the measure "Amend the state constitution to recognize marriage only between a man and a woman" in states that had it on the ballot.
    (a) Find the mean of the state percents favoring this measure.
    (b) Find the standard deviation of the percents.

State	% Favoring	State	% Favoring
Arkansas	75	North Dakota	73
Georgia	76	Ohio	62
Kentucky	75	Oklahoma	76
Michigan	59	Oregon	57
Mississippi	86	Utah	66
Montana	66		

Source: The New York Times, 11/04/04

46. *Malpractice*   Four states had 2004 ballot measures about imposing limits on damage awards or attorneys' fees in medical malpractice cases. Find the mean and the standard deviation of the numbers of voters per state in favor of this measure, using the data below.

State	Number in Favor
Florida	4,435,179
Nevada	465,321
Oregon	744,267
Wyoming	15,628

Source: The New York Times, 11/04/04

47. *U.S. households*   The percents of total households that have married couples for the years 1981 to 2000 are given in the table below. Construct a bar graph of these data with the average percent for each 5-year period.

Year	Percent	Year	Percent
1981	60	1991	56
1982	59	1992	55
1983	59	1993	55
1984	59	1994	55
1985	58	1995	54
1986	58	1996	54
1987	58	1997	53
1988	57	1998	53
1989	56	1999	53
1990	56	2000	52

Source: U.S. Bureau of the Census, U.S. Department of Commerce

48. *Gasoline prices*   The following table gives the average retail prices for unleaded regular gasoline in U.S. cities for the years 1980–2004. Construct a bar graph showing the average price for each 5-year period. What happened in 2005?

Year	Average Price	Year	Average Price
1980	124.5	1993	110.8
1981	137.8	1994	111.2
1982	129.6	1995	114.7
1983	124.1	1996	123.1
1984	121.2	1997	123.4
1985	120.2	1998	105.9
1986	92.7	1999	116.5
1987	94.8	2000	151.0
1988	94.6	2001	146.1
1989	102.1	2002	135.8
1990	116.4	2003	149.7
1991	114.0	2004	196.9
1992	112.7		

Source: U.S. Department of Energy

49. *Birth weights* The birth weights (in kilograms) of a sample of 160 children are given in the following table. Find the mean and standard deviation of the weights.

Weight (kg)	Frequency	Weight (kg)	Frequency
2.0	4	3.5	26
2.3	12	3.8	20
2.6	20	4.1	16
2.9	26	4.4	10
3.2	20	4.7	6

50. *Salaries* S & S Printing has 15 employees and 5 job classifications. Positions and wages for these employees are given in the table.

Job	Number	Weekly Salary
Supervisor	1	$1200
Printers	8	600
Camera-Darkroom	3	750
Secretaries	2	550
Delivery	1	820

(a) What is the mean salary per job classification?
(b) What is the mean salary per person?

51. *Salaries* Suppose a company has 10 employees, 1 earning $160,000, 1 earning $120,000, 2 earning $60,000, 1 earning $40,000, and 5 earning $32,000.
(a) What is the mean salary for the company?
(b) What is the median salary?
(c) What is the mode of the salaries?

52. *Sales* A new car with a $19,000 list price can be bought for different prices from different dealers. In one city the car can be bought for $18,200 from 2 dealers, for $18,000 from 1 dealer, for $17,800 from 3 dealers, for $17,600 from 2 dealers, and for $17,500 from 2 dealers. What are the mean and standard deviation of this sample of car prices?

53. *Insurance rates* The following table gives the monthly insurance rates for a $100,000 life insurance policy for smokers aged 35–49.
(a) Find the mean monthly insurance rate for these ages.
(b) Find the standard deviation of the monthly insurance rates for these ages.

Age (years)	Monthly Insurance Rate (dollars)	Age (years)	Monthly Insurance Rate (dollars)
35	17.32	43	23.71
36	17.67	44	25.11
37	18.02	45	26.60
38	18.46	46	28.00
39	19.07	47	29.40
40	19.95	48	30.80
41	21.00	49	32.55
42	22.22		

Source: American General Life Insurance Company

54. *Educational funding* Data on the amounts of federal on-budget funds for research programs at universities and related institutions for selected years appear in the table below.
(a) Find the mean of the federal on-budget funds for research programs for this period.
(b) Find the standard deviation of the federal on-budget funds for research programs for this period.
(c) Does the standard deviation indicate that there is little variation over the years?

Year	1965	1970	1975	1980	1985	1990	1995	2000
Federal funds (billions of dollars)	1.82	2.28	3.42	5.80	8.84	12.61	15.68	21.02

Source: U.S. Department of Education, National Center for Educational Statistics

55. *Treasury bonds* The average percent yields of long-term Treasury bonds from 1986 to 2001 are shown in the following table. Use a calculator or spreadsheet to compute the mean yield over
(a) 1986–1993
(b) 1994–2001
(c) 1986–2001
(d) Find the standard deviation over 1986–2001.

Year	T-Bond Yield (%)	Year	T-Bond Yield (%)
1986	7.37	1994	7.87
1987	9.12	1995	6.06
1988	9.01	1996	6.55
1989	7.90	1997	5.99
1990	8.24	1998	5.06
1991	7.70	1999	6.35
1992	7.44	2000	5.49
1993	6.25	2001	5.67

Source: U.S. Department of the Treasury

56. *Private school enrollments*   The following table gives the private school enrollments as percents of total school enrollment (actual and projected) for selected school years from 1899–1900 to 2009–2010. Use a calculator or spreadsheet to find the mean and standard deviation of the percents per year of private school enrollments.

School Year	% Private	School Year	% Private
1899–1900	8.7	1959–1960	16.1
1909–1910	8.7	1969–1970	12.1
1919–1920	7.9	1979–1980	12.0
1929–1930	10.3	1989–1990	11.4
1939–1940	10.3	1999–2000	11.4
1949–1950	13.5	2009–2010	11.1

Source: U.S. Department of Education

57. *Executions*   The data in the following table give the total numbers of prisoners executed in the United States from 1984 to 2004.

(a) Construct a histogram with x as the years (in groups of 3) and with $f(x)$ as the number of prisoners executed.

(b) Find the mean number of executions per year.

(c) Find the standard deviation of the number of executions.

(d) Does the standard deviation indicate that there is little variation over the years?

Year	Total	Year	Total
1984	21	1995	56
1985	18	1996	45
1986	18	1997	74
1987	25	1998	68
1988	11	1999	98
1989	16	2000	85
1990	23	2001	66
1991	14	2002	71
1992	31	2003	65
1993	38	2004	59
1994	31		

Source: www.clarkprosecutor.org

8.3

OBJECTIVES

- To identify random variables
- To verify that a table or formula describes a discrete probability distribution
- To compute the mean and expected value of a discrete probability distribution
- To make decisions by using expected value
- To find the variance and standard deviation of a discrete probability distribution
- To find the mean and standard deviation of a binomial distribution
- To graph a binomial distribution
- To expand a binomial to a power, using the binomial formula

Discrete Probability Distributions; The Binomial Distribution

❶ Application Preview

The T. J. Cooper Insurance Company insures 100,000 automobiles. Company records over a 5-year period give the following payouts and their probabilities during each 6-month period.

Payment	$100,000	$50,000	$25,000	$5000	$1000
Probability	0.0001	0.001	0.002	0.008	0.02

The company can determine how much to charge each driver by finding the **expected value** of the **discrete probability distribution** determined by this table. (See Example 4.)

A special discrete probability distribution, the **binomial probability distribution,** can be used to find the expected value for binomial experiments. For example, clinical studies have shown that 5% of photorefractive keratectomy procedures are unsuccessful in correcting nearsightedness. Because each eye operation is independent, the number of failures follows the binomial distribution, and we can easily find the expected number of unsuccessful operations per 1000 operations. (See Example 7.)

Discrete Probability Distributions

If a player rolls a die and receives $1 for each dot on the face she rolls, the amount of money won on one roll can be represented by the variable x. If the die is rolled once, the following table gives the possible outcomes of the experiment and their probabilities.

x	1	2	3	4	5	6
Pr(x)	1/6	1/6	1/6	1/6	1/6	1/6

TABLE 8.1

x	Pr(x)
0	1/16
1	1/4
2	3/8
3	1/4
4	1/16

Note that we have used the variable x to denote the possible numerical outcomes of this experiment. Because x results from a probability experiment, we call x a **random variable,** and because there are a finite number of possible values of x, we say that x is a discrete random variable. Whenever all possible values of a random variable can be listed (or counted), the random variable is a **discrete random variable.**

If x represents the possible number of heads that can occur in the toss of four coins, we can use the sample space {0, 1, 2, 3, 4} to list the possible numerical values of x. Table 8.1 shows the values that the random variable of the coin-tossing experiment can assume and how the probability is distributed over these values.

Discrete Probability Distribution

A table, graph, or formula that assigns to each value of a discrete random variable x a probability Pr(x) describes a **discrete probability distribution** if the following two conditions hold.

(i) $0 \leq \text{Pr}(x) \leq 1$, for any value of x.
(ii) The sum of all the probabilities is 1. We use Σ to denote "the sum of" and write $\Sigma \text{Pr}(x) = 1$, where the sum is taken over all values of x.

● **EXAMPLE 1** *Discrete Probability Distribution*

An experiment consists of selecting a ball from a bag containing 15 balls, each with a number 1 through 5. If the probability of selecting a ball with the number x on it is Pr(x) = $x/15$, where x is an integer and $1 \leq x \leq 5$, verify that $f(x) = \text{Pr}(x) = x/15$ describes a discrete probability distribution for the random variable x.

Solution
For each integer $x(1 \leq x \leq 5)$, Pr(x) satisfies $0 \leq \text{Pr}(x) \leq 1$.

$$\sum \text{Pr}(x) = \text{Pr}(1) + \text{Pr}(2) + \text{Pr}(3) + \text{Pr}(4) + \text{Pr}(5)$$
$$= \frac{1}{15} + \frac{2}{15} + \frac{3}{15} + \frac{4}{15} + \frac{5}{15} = 1$$

Hence Pr(x) = $x/15$ describes a discrete probability distribution.

Probability Density Histograms

We can visualize the possible values of a discrete random variable and their associated probabilities by constructing a graph. This graph, called a **probability density histogram,** is designed so that centered over each value of the discrete random variable x along the horizontal axis is a bar having width equal to 1 unit and height (and thus, area) equal to Pr(x). Figure 8.10 gives the probability density histogram for the experiment described in Example 1.

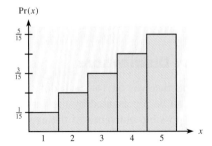

Figure 8.10

Mean and Expected Value

We found the **mean** of a set of n numbers grouped into k classes by using the formula

$$\frac{\sum [x \cdot f(x)]}{n}, \text{ or with } \sum \left[x \cdot \frac{f(x)}{n} \right]$$

We can also find the **theoretical mean** of a probability distribution by using the values x of the random variable, and the probabilities that those values will occur. This mean is also called the **expected value** of x because it gives the average outcome of a probability experiment.

Mean and Expected Value

> If x is a discrete random variable with values $x_1, x_2, \ldots, x_n$, then the **mean, μ,** of the distribution is the **expected value of x,** denoted by **$E(x)$,** and is given by
>
> $$\mu = E(x) = \sum [x \Pr(x)] = x_1 \Pr(x_1) + x_2 \Pr(x_2) + \cdots + x_n \Pr(x_n)$$

Many decisions in business and science are made on the basis of what the outcomes of specific decisions will be, so expected value is very important in decision making and planning.

● **EXAMPLE 2** *Testing*

A five-question multiple choice test has 5 possible answers to each question. If a student guesses the answer to each question, the possible numbers of answers she could get correct and the probability of getting each number are given in the table. What is the expected number of answers she will get correct?

x	$\Pr(x)$
0	1024/3125
1	256/625
2	128/625
3	32/625
4	4/625
5	1/3125

Solution
The expected value of this distribution is

$$E(x) = 0 \cdot \frac{1024}{3125} + 1 \cdot \frac{256}{625} + 2 \cdot \frac{128}{625} + 3 \cdot \frac{32}{625} + 4 \cdot \frac{4}{625} + 5 \cdot \frac{1}{3125} = 1$$

This means that a student can expect to get 1 answer correct if she guesses at every answer.

● *Checkpoint*

1. Consider a game in which you roll a die and receive $1 for each spot that occurs.
   (a) What are the expected winnings from this game?
   (b) If you paid $4 to play this game, how much would you lose, on average, each time you played the game?
2. An experiment has the following possible outcomes for the random variable x: 1, 2, 3, 4, 6, 7, 8, 9, 10. The probability that the value x occurs is $x/50$. Find the expected value of x for the experiment.

● **EXAMPLE 3** *Raffle*

A fire company sells chances to win a new car, with each ticket costing $5. If the car is worth $18,000 and the company sells 6000 tickets, what is the expected value for a person buying one ticket?

Solution
For a person buying one ticket, there are two possible outcomes from the drawing: winning or losing. The probability of winning is 1/6000, and the amount won would be

$18,000 − $5 (the cost of the ticket). The probability of losing is 5999/6000, and the amount lost is written as a winning of −$5. Thus the expected value is

$$17{,}995\left(\frac{1}{6000}\right) + (-5)\left(\frac{5999}{6000}\right) = \frac{17{,}995}{6000} - \frac{29{,}995}{6000} = -2$$

Thus, on the average, a person can expect to lose $2 on every ticket. It is not possible to lose exactly $2 on a ticket—the person will either win $18,000 or lose $5; the expected value means that the fire company will make $2 on each ticket it sells if it sells all 6000.

◑ EXAMPLE 4 *Insurance (Application Preview)*

The T. J. Cooper Insurance Company insures 100,000 cars. Company records over a 5-year period indicate that during each 6-month period it will pay out the following amounts for accidents.

$100,000 with probability 0.0001	$5000 with probability 0.008
$50,000 with probability 0.001	$1000 with probability 0.02
$25,000 with probability 0.002	$0 with probability 0.9689

How could the company use the expected payout per car for each 6-month period to help determine its rates?

Solution
The expected value of the company's payments is

$$\$100{,}000(0.0001) + \$50{,}000(0.001) + \$25{,}000(0.002) + \$5000(0.008)$$
$$+ \$1000(0.02) = \$10 + \$50 + \$50 + \$40 + \$20 = \$170$$

Thus the average premium the company would charge each driver per car for a 6-month period would be

$$\$170 + \text{operating costs} + \text{profit} + \text{reserve for bad years}$$

Measures of Dispersion

Figure 8.11 shows two probability histograms that have the same mean (expected value). However, these histograms look very different. This is because the histograms differ in the **dispersion,** or **variation,** of the values of the random variables.

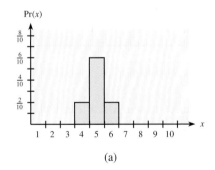

$$\mu = E(x) = 4\left(\frac{2}{10}\right) + 5\left(\frac{6}{10}\right) + 6\left(\frac{2}{10}\right)$$
$$= \frac{8 + 30 + 12}{10} = 5$$

Figure 8.11

(a)

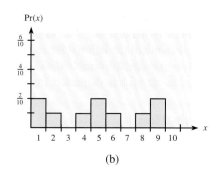

$$\mu = E(x) = 1\left(\frac{2}{10}\right) + 2\left(\frac{1}{10}\right) + 4\left(\frac{1}{10}\right)$$

$$+ 5\left(\frac{2}{10}\right) + 6\left(\frac{1}{10}\right) + 8\left(\frac{1}{10}\right) + 9\left(\frac{2}{10}\right)$$

$$= \frac{2 + 2 + 4 + 10 + 6 + 8 + 18}{10} = 5$$

Figure 8.11 (cont'd) (b)

Thus we need a way to indicate how distributions with the same mean, such as the ones above, differ. As with sample data, we can find the **variance** of a probability distribution by finding the mean of the *squares* of these deviations. The square root of this measure, called the **standard deviation,** is very useful in describing how the values of x are concentrated about the mean of the probability distribution.

Variance and Standard Deviation

If x is a discrete random variable with values $x_1, x_2, \ldots, x_n$ and mean μ, then the variance of the distribution of x is

$$\sigma^2 = \sum (x - \mu)^2 \text{Pr}(x)$$

The standard deviation is

$$\sigma = \sqrt{\sigma^2} = \sqrt{\sum (x - \mu)^2 \text{Pr}(x)}$$

● **EXAMPLE 5** *Standard Deviation of a Distribution*

Find the standard deviations of the distributions described in (a) Figure 8.11(a) and (b) Figure 8.11(b).

Solution

(a) $\sigma^2 = \sum (x - \mu)^2 \text{Pr}(x)$

$$= (4 - 5)^2 \cdot \frac{2}{10} + (5 - 5)^2 \cdot \frac{6}{10} + (6 - 5)^2 \cdot \frac{2}{10}$$

$$= 1 \cdot \frac{2}{10} + 0 \cdot \frac{6}{10} + 1 \cdot \frac{2}{10}$$

$$= 0.4$$

$$\sigma = \sqrt{0.4} \approx 0.632$$

(b) $\sigma^2 = (1 - 5)^2 \cdot \frac{2}{10} + (2 - 5)^2 \cdot \frac{1}{10} + (4 - 5)^2 \cdot \frac{1}{10} + (5 - 5)^2 \cdot \frac{2}{10}$

$$+ (6 - 5)^2 \cdot \frac{1}{10} + (8 - 5)^2 \cdot \frac{1}{10} + (9 - 5)^2 \cdot \frac{2}{10}$$

$$= 16 \cdot \frac{2}{10} + 9 \cdot \frac{1}{10} + 1 \cdot \frac{1}{10} + 0 \cdot \frac{2}{10} + 1 \cdot \frac{1}{10} + 9 \cdot \frac{1}{10} + 16 \cdot \frac{2}{10}$$

$$= 8.4$$

$$\sigma = \sqrt{8.4} \approx 2.898$$

As we can see by referring to Figure 8.11 and to the standard deviations calculated in Example 5, smaller values of σ indicate that the values of x are clustered nearer μ (as is true with sample mean and standard deviation). When σ is large, the values of x are more widely dispersed from μ. As we saw for sample data in Section 8.2, a large standard deviation s indicates that the data are widely dispersed from the sample mean $\bar{x}$.

The Binomial Distribution

A special discrete probability distribution is the **binomial probability distribution,** which describes all possible outcomes of a binomial experiment with their probabilities. Table 8.2 gives the binomial probability distribution for the experiment of tossing a fair coin 6 times, with x equal to the number of heads.

TABLE 8.2

x	$\Pr(x)$	x	$\Pr(x)$
0	$\binom{6}{0}\left(\frac{1}{2}\right)^0\left(\frac{1}{2}\right)^6 = \frac{1}{64}$	4	$\binom{6}{4}\left(\frac{1}{2}\right)^4\left(\frac{1}{2}\right)^2 = \frac{15}{64}$
1	$\binom{6}{1}\left(\frac{1}{2}\right)^1\left(\frac{1}{2}\right)^5 = \frac{6}{64}$	5	$\binom{6}{5}\left(\frac{1}{2}\right)^5\left(\frac{1}{2}\right)^1 = \frac{6}{64}$
2	$\binom{6}{2}\left(\frac{1}{2}\right)^2\left(\frac{1}{2}\right)^4 = \frac{15}{64}$	6	$\binom{6}{6}\left(\frac{1}{2}\right)^6\left(\frac{1}{2}\right)^0 = \frac{1}{64}$
3	$\binom{6}{3}\left(\frac{1}{2}\right)^3\left(\frac{1}{2}\right)^3 = \frac{20}{64}$		

Binomial Probability Distribution

If x is a variable that assumes the values $0, 1, 2, \ldots, r, \ldots, n$ with probabilities $\binom{n}{0}p^0q^n, \binom{n}{1}p^1q^{n-1}, \binom{n}{2}p^2q^{n-2}, \ldots, \binom{n}{r}p^rq^{n-r}, \ldots, \binom{n}{n}p^nq^0$, respectively, where p is the probability of success and $q = 1 - p$, then x is called a **binomial variable.**

The values of x and their corresponding probabilities described above form the **binomial probability distribution.**

Recall that $\binom{n}{r} = {}_nC_r$ can be evaluated easily on a calculator.

● *Checkpoint*

3. Show that the values of x and the associated probabilities in Table 8.2 satisfy the conditions for a discrete probability distribution.

Mean and Standard Deviation

The expected value of the number of successes (heads) for the coin toss experiment above is given by

$$E(x) = 0 \cdot \frac{1}{64} + 1 \cdot \frac{6}{64} + 2 \cdot \frac{15}{64} + 3 \cdot \frac{20}{64} + 4 \cdot \frac{15}{64} + 5 \cdot \frac{6}{64} + 6 \cdot \frac{1}{64} = \frac{192}{64} = 3$$

This expected number of successes seems reasonable. If the probability of success is $1/2$, we would expect to succeed on half of the 6 trials. We could also use the formula for expected value to see that the expected number of heads in 100 tosses is $100(1/2) = 50$. For any binomial distribution the expected number of successes is given by np, where n is the number of trials and p is the probability of success. Because the expected value of any probability distribution is defined as the mean of that distribution, this implies that we can find the expected value, or mean, of a binomial distribution with an easier formula.

Mean of a Binomial Distribution

The theoretical mean of any binomial distribution is

$$\mu = np$$

where n is the number of trials in the corresponding binomial experiment and p is the probability of success on each trial.

A simple formula can also be developed for the standard deviation of a binomial distribution.

Standard Deviation of a Binomial Distribution

The standard deviation of a binomial distribution is

$$\sigma = \sqrt{npq}$$

where n is the number of trials, p is the probability of success on each trial, and $q = 1 - p$.

The standard deviation of the binomial distribution corresponding to the number of heads resulting when a coin is tossed 16 times is

$$\sigma = \sqrt{16 \cdot \frac{1}{2} \cdot \frac{1}{2}} = \sqrt{4} = 2$$

● **EXAMPLE 6** *Nursing Homes*

According to the *New England Journal of Medicine,* there is a 43 percent chance that a person 65 years or older will enter a nursing home sometime in his or her lifetime. Of 100 people in this age group, what is the expected number of people who will eventually enter a nursing home, and what is the standard deviation?

Solution

The number of people who will eventually enter a nursing home follows a binomial distribution, with the probability that any one person in this group will enter a home equal to $p = 0.43$ and the probability that he or she will not equal to $q = 0.57$. Thus the expected number out of 100 that will eventually enter a nursing home is

$$E(x) = np = 100(0.43) = 43$$

The standard deviation of this distribution is

$$\sigma = \sqrt{npq} = \sqrt{100 \cdot 0.43 \cdot 0.57} \approx 4.95$$

◐ **EXAMPLE 7** *Eye Surgery* **(Application Preview)**

Clinical studies have shown that 5 percent of patients operated on for nearsightedness with photorefractive keratectomy (PRK) still need to wear glasses. Find the expected number of unsuccessful operations (patients still wearing glasses) per 1000 operations.

Solution

Each operation is independent, so the number of failures follows the binomial distribution. The probability of an unsuccessful operation is $p = 0.05$ and $n = 1000$. Thus the expected number of unsuccessful operations in 1000 operations is

$$E(x) = \mu = 1000(0.05) = 50$$

● *Checkpoint*

4. A fair cube is colored so that 4 faces are green and 2 are red. The cube is tossed 9 times. If success is defined as a red face being up, find

(a) the mean number of red faces and

(b) the standard deviation of the number of red faces that occur.

Binomial Formula

The binomial probability distribution is closely related to the powers of a binomial. For example, if a binomial experiment has 3 trials with probability of success p on each trial, then the binomial probability distribution is given in Table 8.3, with the values of x written from 3 to 0. Compare this with the expansion of $(p + q)^3$, which we introduced in Chapter 0, "Algebraic Concepts."

$$(p + q)^3 = p^3 + 3p^2q + 3pq^2 + q^3$$

TABLE 8.3

x	$Pr(x)$
3	$\binom{3}{3}p^3q^0 = p^3$
2	$\binom{3}{2}p^2q = 3p^2q$
1	$\binom{3}{1}pq^2 = 3pq^2$
0	$\binom{3}{0}p^0q^3 = q^3$

The formula we can use to expand a binomial $(a + b)$ to any positive integer power n is as follows.

Binomial Formula

$$(a + b)^n = \binom{n}{n}a^n + \binom{n}{n-1}a^{n-1}b + \binom{n}{n-2}a^{n-2}b^2 + \cdots$$
$$+ \binom{n}{2}a^2b^{n-2} + \binom{n}{1}ab^{n-1} + \binom{n}{0}b^n$$

● **EXAMPLE 8 *Binomial Formula***

Expand $(x + y)^4$.

Solution

$$(x + y)^4 = \binom{4}{4}x^4 + \binom{4}{3}x^3y + \binom{4}{2}x^2y^2 + \binom{4}{1}xy^3 + \binom{4}{0}y^4$$
$$= x^4 + 4x^3y + 6x^2y^2 + 4xy^3 + y^4$$

● *Checkpoint Solutions*

1. (a) $E(x) = \$1\left(\dfrac{1}{6}\right) + \$2\left(\dfrac{1}{6}\right) + \$3\left(\dfrac{1}{6}\right) + \$4\left(\dfrac{1}{6}\right) + \$5\left(\dfrac{1}{6}\right) + \$6\left(\dfrac{1}{6}\right)$

$= \$3.50$

(b) $\$4 - \$3.50 = \$.50$. You would lose $\$0.50$, on average, from each play.

2. The expected value of x for this experiment is

$$E(x) = 1 \cdot \frac{1}{50} + 2 \cdot \frac{2}{50} + 3 \cdot \frac{3}{50} + 4 \cdot \frac{4}{50} + 6 \cdot \frac{6}{50} + 7 \cdot \frac{7}{50} + 8 \cdot \frac{8}{50}$$
$$+ 9 \cdot \frac{9}{50} + 10 \cdot \frac{10}{50} = 7.2$$

3. Each value of the random variable x has a probability, with $0 \le \Pr(x) \le 1$, and with $\Sigma \Pr(x) = 1$. Thus the conditions of a probability distribution are satisfied.

4. $\Pr(\text{red face}) = \dfrac{1}{3} = p$, so $q = \dfrac{2}{3}$ and $n = 9$.

   (a) $\mu = np = 9\left(\dfrac{1}{3}\right) = 3$ (b) $\sigma = \sqrt{npq} = \sqrt{9\left(\dfrac{1}{3}\right)\left(\dfrac{2}{3}\right)} = \sqrt{2}$

8.3 Exercises

Determine whether each table in Problems 1–4 describes a discrete probability distribution. Explain.

1.

x	$\Pr(x)$
1	$-1/5$
2	$1/4$
3	$5/10$
4	$9/20$

2.

x	$\Pr(x)$
0	$-4/5$
1	$2/5$
2	$3/5$
3	$4/5$

3.

x	$\Pr(x)$
-1	$1/4$
1	$3/8$
2	$1/4$
4	$1/8$

4.

x	$\Pr(x)$
0	$1/3$
4	$1/6$
8	$1/2$
12	0

Determine whether each formula in Problems 5–8 describes a discrete probability distribution.

5. $\Pr(x) = \dfrac{x}{21}; x = 0, 1, 2, 3, 4, 5, 6$

6. $\Pr(x) = \dfrac{x}{10}; x = 1, 2, 3, 4$

7. $\Pr(x) = \dfrac{10 - x}{10}; x = 1, 2, 3, 4, 5$

8. $\Pr(x) = \dfrac{x + 1}{15}; x = 1, 2, 3, 4$

Each of the tables in Problems 9–12 defines a discrete probability distribution. Find the expected value of each distribution.

9.

x	0	1	2	3
$\Pr(x)$	$1/8$	$1/4$	$1/4$	$3/8$

10.

x	1	2	3	4
$\Pr(x)$	$1/10$	$1/5$	$3/10$	$2/5$

11.

x	4	5	6	7
$\Pr(x)$	$1/3$	$1/3$	$1/3$	0

12.

x	1	2	3	4
$\Pr(x)$	$1/15$	$4/15$	$1/5$	$7/15$

In Problems 13–16, find the mean, variance, and standard deviation for each probability distribution.

13.

x	$\Pr(x)$
0	$1/4$
1	$1/4$
2	$1/8$
3	$3/8$

14.

x	$\Pr(x)$
3	$1/4$
4	$1/4$
5	$1/4$
6	$1/4$

15. $\Pr(x) = \dfrac{x}{21}; x = 0, 1, 2, 3, 4, 5, 6$

16. $\Pr(x) = \dfrac{x}{10}; x = 2, 3, 5$

17. Suppose an experiment has five possible outcomes for x: 0, 1, 2, 3, 4. The probability that each of these outcomes occurs is $x/10$. What is the expected value of x for the experiment?

18. Suppose an experiment has six possible outcomes for x: 0, 1, 2, 3, 4, 5. The probability that each of these outcomes occurs is $x/15$. What is the expected value of x for the experiment?

19. Five slips of paper containing the numbers 0, 1, 2, 3, 4 are placed in a hat. If the experiment consists of drawing one number, and if the experiment is repeated a large number of times, what is the expected value of the number drawn?

20. An experiment consists of rolling a die once. If the experiment is repeated a large number of times, what is the expected value of the number rolled?

21. A die is rolled 3 times, and success is rolling a 5.
    (a) Construct the binomial distribution that describes this experiment.
    (b) Find the mean of this distribution.
    (c) Find the standard deviation of this distribution.

22. A die is rolled 3 times. Success is rolling a number divisible by 3.
    (a) Construct the binomial distribution that describes this experiment.
    (b) Find the mean of this distribution.
    (c) Find the standard deviation of this distribution.

23. A variable x has a binomial distribution with probability of success 0.7 for each trial. For a total of 60 trials, what are
    (a) the mean and
    (b) the standard deviation of the distribution?

24. A variable x has a binomial distribution with probability of success equal to 0.8 for each trial with a total of 40 trials. What are
    (a) the mean and
    (b) the standard deviation of the distribution?

25. A coin is "loaded" so that the probability of tossing a head is 3/5. If it is tossed 50 times, what are
    (a) the mean and
    (b) the standard deviation of the number of heads that occur?

26. Suppose a pair of dice is thrown 1200 times. How many times would we expect a sum of 7 to occur?

27. Suppose a pair of dice is thrown 900 times. How many times would we expect a sum of 6 to occur?

28. Suppose that a die is "loaded" so that the probability of getting each even number is 1/4 and the probability of getting each odd number is 1/12. If it is rolled 240 times, how many times would we expect to get a 3?

29. Expand $(a + b)^6$.

30. Expand $(x + y)^5$.

31. Expand $(x + h)^4$.

32. Write the general expression for $(x + h)^n$.

APPLICATIONS

33. *Animal relocation* In studying endangered species, scientists have found that when animals are relocated, it takes x years without offspring before the first young are born, where x and the probability of x are given below. What is the expected number of years before the first young are born?

x	0	1	2	3	4
Pr(x)	0.04	0.35	0.38	0.18	0.05

34. *Raffle* A charity sells 1000 raffle tickets for $1 each. There is one grand prize of $250, two prizes of $75 each, and five prizes of $10 each. If x is the net winnings for someone who buys a ticket, then the following table gives the probability distribution for x. Find the expected winnings of someone who buys a ticket.

x	Pr(x)	x	Pr(x)
249	0.001	9	0.005
74	0.002	−1	0.992

35. *Campaigning* A candidate must decide whether he should spend his time and money on TV commercials or making personal appearances. His staff determines that by using TV he can reach 100,000 people with probability 0.01, 50,000 people with probability 0.47, and 25,000 people with probability 0.52; by making personal appearances he can reach 80,000 people with probability 0.02, 50,000 people with probability 0.47, and 20,000 people with probability 0.51. In the following tables, x represents the number of people reached by each choice. In each case, find the expected value of x to decide which method will reach more people.

TV Commercials		Personal Appearances	
x	Pr(x)	x	Pr(x)
100,000	0.01	80,000	0.02
50,000	0.47	50,000	0.47
25,000	0.52	20,000	0.51

36. *Genetics* If the probability that a newborn child is a male is 1/2, what is the expected number of male children in a family having 4 children?

37. *Raffle* Living Arrangements for the Developmentally Disabled (LADD), Inc., a nonprofit organization, sells chances for a $40,000 Corvette at $100 per ticket. It sells 1500 tickets and offers four prizes, summarized in the table that follows. What are the expected winnings (or loss) for each ticket?

Prizes	Amount
First	$40,000
Second	5,000
Third	2,500
Fourth	1,500

Source: Automobile, February 2002

38. *Lottery* A charity sells raffle tickets for $1 each. First prize is $500, second prize is $100, third prize is $10. If you bought one of the 1000 tickets sold, what are your expected winnings?

39. *Gambling* Suppose a student is offered a chance to draw a card from an ordinary deck of 52 playing cards and win $15 for an ace, $10 for a king, and $1 for a queen. If $4 must be paid to play the game, what is the expected winnings every time the game is played by the student?

40. *Revenue* The Rent-to-Own Company estimates that 35% of its rentals result in the sale of the product, with an average revenue of 100%, 56% of the rentals are returned in good condition, with an average revenue of 35% on these rentals, and the remainder of the rentals are stolen or returned in poor condition, giving a loss

of 15% on these rentals. If these estimates are accurate, what is the expected percent of revenue for this company?

41. *Sales* A young man plans to sell umbrellas at the city's Easter Parade. He knows that he can sell 180 umbrellas at $5 each if it rains hard, he can sell 50 if it rains lightly, and he can sell 10 if it doesn't rain at all. Past records show it rains hard 25% of the time on Easter, rains lightly 20% of the time, and does not rain at all 55% of the time. If he can buy 0, 100, or 200 umbrellas at $2 each and return the unsold ones for $1 each, how many should he buy?

42. *Advertising* A candidate must decide whether he should spend his advertising dollars on radio commercials or on telemarketing. His staff determines that by using radio he can reach 80,000 people with probability 0.01, reach 40,000 people with probability 0.47, and reach 25,000 with probability 0.52; by using telemarketing he can reach 70,000 people with probability 0.04, reach 50,000 people with probability 0.38, and reach 30,000 with probability 0.58. Which method will reach more people?

43. *Insurance* A car owner must decide whether she should take out a $100-deductible collision policy in addition to her liability insurance policy. Records show that each year, in her area, 8% of the drivers have an accident that is their fault or for which no fault is assigned, and that the average cost of repairs for these types of accidents is $1000. If the $100-deductible collision policy costs $100 per year, would she save money in the long run by buying the insurance or "taking the chance"? (*Hint:* Find the expected values if she has the policy and if she doesn't have the policy and compare them.)

44. *Altimeters* Suppose a company manufactures altimeters so that they are accurate at an average height of 1000 feet. Is the accuracy of this single measurement sufficient to ensure the safety of the passengers and crew of planes using these altimeters? Explain.

45. *Machining* A machinist makes pipes with an average diameter of 2 inches. If he machines 100 pipes with an average diameter of 2 inches, does this mean that all of them are usable? (A pipe is usable if its diameter is within 0.01 inch of 2 inches.) Explain.

46. *Budgeting* Suppose a youth has a part-time job in an ice cream shop. He receives $40 if he is called to work a full day and $20 if he is called to work a half-day. Over the past year he has been called to work a full day an average of 8 days per month and for a half-day an

average of 14 days per month. How much can he expect to earn *per day* during a 30-day month?

47. *Health care* Suppose that 10% of the patients who have a certain disease will die from it.
    (a) If 100 people have the disease, how many would we expect to die from it?
    (b) What is the standard deviation of the number of deaths that could occur?

48. *Cancer research* Suppose it has been determined that the probability that a rat injected with cancerous cells will live is 0.6. If 35 rats are injected, how many would be expected to die?

49. *Voting* A candidate claims that 60% of the voters in his district will vote for him.
    (a) If his district contains 100,000 voters, how many votes does he expect to get from his district?
    (b) What is the standard deviation of the number of these votes?

50. *Genetics* In a family with 2 children, the probability that both children will be boys is 1/4.
    (a) If 1200 families with two children are selected at random, how many of the families would we expect to have 2 boys?
    (b) What is the standard deviation of the number of families with 2 boys?

51. *Voting* If the number of votes the candidate in Problem 49 gets is 2 standard deviations below what he expected, how many votes will he get (approximately)?

52. *Genetics* Suppose that of the 1200 families in Problem 50, 315 families have two boys. How many standard deviations above the mean is this number (315) of families?

53. *Testing* A multiple-choice test has 20 questions and 5 choices for each question. If a student is totally unprepared and guesses on each question, the mean is the number of questions she can expect to answer correctly.
    (a) How many questions can she expect to answer correctly?
    (b) What is the standard deviation of the distribution?

54. *Birth control* Suppose a birth control pill is 99% effective in preventing pregnancy. What number of women would be expected to get pregnant out of 100 women using this pill?

55. *Quality control* A certain calculator circuit board is manufactured in lots of 200. If 1% of the boards are defective, find the mean and standard deviation of the number of defects in each lot.

56. *CPA exam*   Forty-eight percent of accountants taking a CPA exam fail the first time. If 1000 candidates take the exam for the first time, what is the expected number that will pass?

57. *Seed germination*   A certain type of corn seed has an 85% germination rate. If 2000 seeds were planted, how many seeds would be expected not to germinate?

58. *Quality control*   The probability that a manufacturer produces a defective medical thermometer is 0.001. What is the expected number of defective thermometers in a shipment of 3000?

8.4

Normal Probability Distribution

◗ Application Preview

Discrete probability distributions are important, but they do not apply to many kinds of measurements. For example, the weights of people, the heights of trees, and the IQ scores of college students cannot be measured with whole numbers because each of them can assume any one of an infinite number of values on a measuring scale, and the values cannot be listed or counted, as is possible for a discrete random variable. The measurements mentioned above follow a special **continuous probability distribution** called the **normal distribution.**

IQ scores follow the normal distribution with a mean of 100 and a standard deviation of 15. The graph of this normal distribution is a bell-shaped curve that is symmetric about the mean of 100. We can use this information to find the probability that a person picked at random will have an IQ between 85 and 115. (See Example 1.)

The normal distribution is perhaps the most important probability distribution, because so many measurements that occur in nature follow this particular distribution.

The **normal distribution** has the following properties.

1. Its graph is a bell-shaped curve like that in Figure 8.12.* The graph is called the **normal curve.** It approaches but never touches the horizontal axis as it extends in both directions.

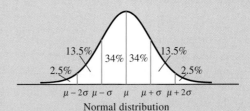

Figure 8.12   Normal distribution

2. The curve is symmetric about a vertical center line that passes through the value that is the mean, the median, *and* the mode of the distribution. That is, the mean, median, and mode are the same for a normal distribution. (That is why they are all called average.)
3. A normal distribution is completely determined when its mean μ and its standard deviation σ are known.
4. Approximately 68% of all scores lie within 1 standard deviation of the mean. Approximately 95% of all scores lie within 2 standard deviations of the mean. More than 99.5% of all scores will lie within 3 standard deviations of the mean.

*Percents shown are approximate.

◗ EXAMPLE 1 *IQ Scores* (Application Preview)

IQ scores follow a normal distribution with mean 100 and standard deviation 15.

(a) What percent of the scores will be
   (i) between 100 and 115?
   (ii) between 85 and 115?
   (iii) between 70 and 130?
   (iv) greater than 130?
(b) Find the probability that a person picked at random has an IQ score
   (i) between 85 and 115.
   (ii) greater than 130.

Solution

(a) Because the mean is 100 and the standard deviation is 15, IQs of 85 and 115 are 1 standard deviation from 100, and IQs of 70 and 130 are 2 standard deviations from 100. The approximate percents associated with these values are shown on the graph of a normal distribution in Figure 8.13. We can use this graph to answer the questions.
   (i) 34%   (ii) 68%   (iii) 95%   (iv) 2.5%

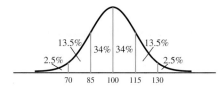

Figure 8.13

(b) The total area under the normal curve is 1. The area under the curve from value x_1 to value x_2 represents the percent of the scores that lie between x_1 and x_2. Thus the percent of the scores between x_1 and x_2 equals the *area* under the curve from x_1 to x_2 and both *represent the probability* that a score chosen at random will lie between x_1 and x_2. Thus, (i) the probability that a person chosen at random has an IQ between 85 and 115 is 0.68, and (ii) the probability that the IQ is greater than 130 is 0.025.

● *Checkpoint*

1. The mean of a normal distribution is 25 and the standard deviation is 3.
   (a) What percent of the scores will be between 22 and 28?
   (b) What is the probability that a score chosen at random will be between 22 and 28?

We have seen that 34% of the scores lie between 100 and 115 in the normal distribution graph in Figure 8.13. As we have mentioned, this also means that the area under the curve from $x = 100$ to $x = 115$ is 0.34 and that the probability that a score chosen at random from this normal population lies between 100 and 115 is 0.34. We can write this as

$$\Pr(100 \leq x \leq 115) = 0.34$$

The mean and standard deviation of the IQ scores are given to be 100 and 15, respectively, so a score of 115 is 15 points above the mean, which is 1 standard deviation above the mean.

z-Scores

Because the probability of obtaining a score from a normal distribution is always related to how many standard deviations the score is away from the mean, it is desirable to convert all scores from a normal distribution to **standard scores,** or *z*-**scores.** The *z*-score for any score x is found by determining how many standard deviations x is from the mean μ. The formula for converting scores to *z*-scores follows.

z-Scores

If σ is the standard deviation of the population data, then the number of standard deviations that x is from the mean μ is given by the z-score

$$z = \frac{x - \mu}{\sigma}$$

This formula enables us to convert scores from any normal distribution to a distribution of z-scores with no units of measurement.

The distribution of z-scores will always be a normal distribution with mean 0 and standard deviation 1. This distribution is called the **standard normal distribution.** Figure 8.14 shows the graph of the standard normal distribution, with approximate percents shown. The total area under the curve is 1, with 0.5 on either side of the mean 0.

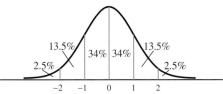

Figure 8.14   Standard normal distribution

By comparing Figure 8.14 with Figure 8.12, we see that each unit from 0 in the standard normal distribution corresponds to 1 standard deviation from the mean of the normal distribution.

We can use a table to determine more accurately the area under the standard normal curve between two z-scores. Table III in the Appendix gives the area under the normal curve from $z = 0$ to $z = z_0$, for values of z_0 from 0 to 3. As with the normal curve, the area under the curve from $z = 0$ to $z = z_0$ is the probability that a z-score lies between 0 and z_0.

● **EXAMPLE 2 *Standard Normal Distribution***

(a) Find the area under the standard normal curve from $z = 0$ to $z = 1.50$.
(b) Find $\Pr(0 \leq z \leq 1.50)$.

Solution

(a) Looking in the column headed by z in Table III in the Appendix, we see 1.50. Across from 1.50 in the column headed by A is 0.4332. Thus the area under the standard normal curve between $z = 0$ and $z = 1.50$ is $A_{1.50} = 0.4332$. (See Figure 8.15.)

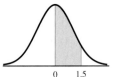

Figure 8.15

(b) The area under the curve from 0 to 1.50 equals the probability that z lies between 0 and 1.50. Thus

$$\Pr(0 \leq z \leq 1.50) = A_{1.50} = 0.4332$$

The following facts, which are direct results of the symmetry of the normal curve about μ, are useful in calculating probabilities using Table III in the Appendix.

1. $\Pr(-z_0 \leq z \leq 0) = \Pr(0 \leq z \leq z_0) = A_{z_0}$
2. $\Pr(z \geq 0) = \Pr(z \leq 0) = 0.5$

● **EXAMPLE 3** *Probabilities with z-Scores*

Find the following probabilities for the random variable z with standard normal distribution.

(a) $\Pr(-1 \leq z \leq 0)$ (b) $\Pr(-1 \leq z \leq 1.5)$ (c) $\Pr(1 \leq z \leq 1.5)$
(d) $\Pr(z > 2)$ (e) $\Pr(z < 1.35)$

Solution

(a) $\Pr(-1 \leq z \leq 0) = \Pr(0 \leq z \leq 1) = A_1 = 0.3413$

(b) We find $\Pr(-1 \leq z \leq 1.5)$ by using A_1 and $A_{1.5}$ as follows:

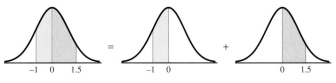

$$
\begin{aligned}
\Pr(-1 \leq z \leq 1.5) &= \Pr(-1 \leq z \leq 0) &+ \Pr(0 \leq z \leq 1.5) \\
&= A_1 &+ A_{1.5} \\
&= 0.3413 &+ 0.4332 \\
&= 0.7745
\end{aligned}
$$

(c) We find $\Pr(1 \leq z \leq 1.5)$ by using A_1 and $A_{1.5}$ as follows:

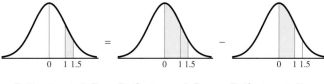

$$
\begin{aligned}
\Pr(1 \leq z \leq 1.5) &= \Pr(0 \leq z \leq 1.5) &- \Pr(0 \leq z \leq 1) \\
&= A_{1.5} &- A_1 \\
&= 0.4332 &- 0.3413 \\
&= 0.0919
\end{aligned}
$$

(d) We find $\Pr(z > 2)$ by using 0.5 and A_2 as follows:

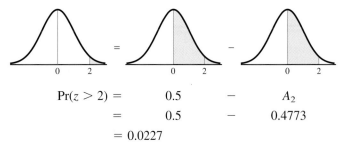

$$
\begin{aligned}
\Pr(z > 2) &= 0.5 &- A_2 \\
&= 0.5 &- 0.4773 \\
&= 0.0227
\end{aligned}
$$

(e) We find $\Pr(z < 1.35)$ by using 0.5 and $A_{1.35}$ as follows:

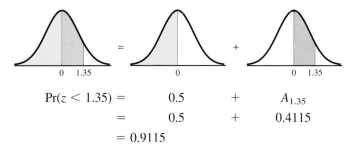

$$
\begin{aligned}
\Pr(z < 1.35) &= 0.5 &+ A_{1.35} \\
&= 0.5 &+ 0.4115 \\
&= 0.9115
\end{aligned}
$$

● *Checkpoint*   2. For the standard normal distribution, find the following probabilities.
(a) $\Pr(0 \leq z \leq 2.5)$
(b) $\Pr(z > 2.5)$
(c) $\Pr(z \leq 2.5)$

If a population follows a normal distribution, but with mean and/or standard deviation different from 0 and 1, respectively, we can convert the scores to **standard scores,** or **z-scores.** Recall that the z-score for any score x is found by determining how many standard deviations x is from the mean μ by using the formula

$$z = \frac{x - \mu}{\sigma}$$

Thus

$$\Pr(a \leq x \leq b) = \Pr(z_a \leq z \leq z_b)$$

where

$$z_a = \frac{a - \mu}{\sigma} \qquad \text{and} \qquad z_b = \frac{b - \mu}{\sigma}$$

and we can use Table III in the Appendix, the table of standard scores (z-scores), to find these probabilities. For example, the normal distribution of IQ scores has mean $\mu = 100$ and standard deviation $\sigma = 15$. Thus the z-score for $x = 115$ is

$$z = \frac{115 - 100}{15} = 1$$

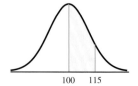

A z-score of 1 indicates that 115 is 1 standard deviation above the mean (see Figure 8.16), and

$$\Pr(100 \leq x \leq 115) = \Pr(0 \leq z \leq 1) = 0.3413$$

Figure 8.16

● **EXAMPLE 4 *Height***

If the mean height of a population of students is $\mu = 68$ in. with standard deviation $\sigma = 3$ in., what is the probability that a person chosen at random from the population will be between

(a) 68 and 74 in. tall?   (b) 65 and 74 in. tall?

Solution
(a) To find $\Pr(68 \leq x \leq 74)$, we convert 68 and 74 to z-scores.

$$\text{For 68: } z = \frac{68 - 68}{3} = 0 \qquad \text{For 74: } z = \frac{74 - 68}{3} = 2$$

Thus

$$\Pr(68 \leq x \leq 74) = \Pr(0 \leq z \leq 2) = A_2 = 0.4773$$

(b) The z-score for 65 is

$$z = \frac{65 - 68}{3} = -1$$

and the z-score for 74 is 2. Thus

$$\Pr(65 \leq x \leq 74) = \Pr(-1 \leq z \leq 2)$$
$$= A_1 + A_2$$
$$= 0.3413 + 0.4773 = 0.8186$$

Checkpoint

3. For the normal distribution with mean 70 and standard deviation 5, find the following probabilities.
    (a) $\Pr(70 \le x \le 77)$
    (b) $\Pr(68 \le x \le 73)$
    (c) $\Pr(x > 75)$

● **EXAMPLE 5 *Demand***

Recessionary pressures have forced an electrical supplies distributor to consider reducing its inventory. It has determined that, over the past year, the daily demand for 48-in. fluorescent light fixtures is normally distributed with a mean of 432 and a standard deviation of 86. If the distributor decides to reduce its daily inventory to 500 units, what percent of the time will its inventory be insufficient to meet the demand? Will this cost the distributor business?

Solution

The distribution is normal, with $\mu = 432$ and $\sigma = 86$. The probability that more than 500 units will be demanded is

$$\Pr(x > 500) = \Pr\left(z > \frac{500 - 432}{86}\right) = \Pr(z > 0.79)$$

$$= 0.5 - A_{0.79} = 0.5 - 0.2852 = 0.2148$$

The distributor will have an insufficient inventory 21.48% of the time. This will probably result in a loss of customers, so the distributor should not decrease the inventory to 500 units per day.

● **EXAMPLE 6 *Graphs of Normal Distributions***

The function

$$f(x) = \frac{1}{\sigma\sqrt{2\pi}}e^{-(x-\mu)^2/(2\sigma^2)}$$

gives the equation for a normal curve with mean μ and standard deviation σ.
(a) Graph the following normal curves to see how different means yield different distributions.

$$\mu = 0, \sigma = 1, y = \frac{1}{\sqrt{2\pi}}e^{-x^2/2} \quad \text{and}$$

$$\mu = 3, \sigma = 1, y = \frac{1}{\sqrt{2\pi}}e^{-(x-3)^2/2}$$

If the standard deviations are the same, how does increasing the size of the mean affect the graph of the distribution?
(b) To see how normal distributions differ for different standard deviations, graph the normal distributions with

$$\mu = 0, \sigma = 3 \quad \text{and}$$

$$\mu = 0, \sigma = 2$$

What happens to the height of the curve as the standard deviation decreases? Do the data appear to be clustered nearer the mean when the standard deviation is smaller?

Solution

(a) The graphs are shown in Figure 8.17. The graphs are the same size but are centered over different means. A different mean causes the graph to shift to the right or left.

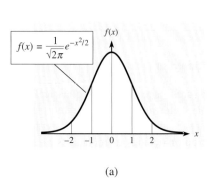

Figure 8.17

(a)   (b)

(b) The graphs are shown in Figure 8.18. The height of the graph will be larger if the standard deviation decreases. Yes, the data will be clustered closer to the mean when the standard deviation is smaller.

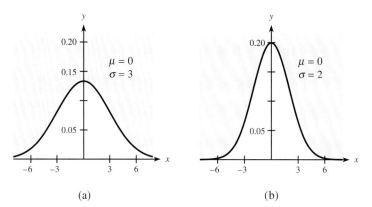

Figure 8.18

(a)   (b)

● **Checkpoint Solutions**

1. (a) These scores lie within 1 standard deviation of the mean, 25, so 68% lie in this range.
   (b) The percent of the scores in this range is 68%, so the probability that a score chosen at random from this normal distribution will be in this range is 0.68.

2. (a) $\Pr(0 \leq z \leq 2.5) = A_{2.5} = 0.4938$
   (b) $\Pr(z > 2.5) = 0.5 - A_{2.5} = 0.5 - 0.4938 = 0.0062$
   (c) $\Pr(z \leq 2.5) = 0.5 + A_{2.5} = 0.5 + 0.4938 = 0.9938$

3. (a) $\Pr(70 \leq x \leq 77) = \Pr(0 \leq z \leq 1.4) = A_{1.4} = 0.4192$
   (b) $\Pr(68 \leq x \leq 73) = \Pr(-0.4 \leq z \leq 0.6) = A_{0.4} + A_{0.6}$
   $$= 0.1554 + 0.2258 = 0.3812$$
   (c) $\Pr(x > 75) = \Pr(z > 1) = 0.5 - A_1 = 0.5 - 0.3413 = 0.1587$

8.4 Exercises

Use Table III in the Appendix to find the probability that a z-score from the standard normal distribution will lie within each of the intervals in Problems 1–14.

1. $0 \leq z \leq 1.8$
2. $0 \leq z \leq 2.4$
3. $-1.8 \leq z \leq 0$
4. $-3 \leq z \leq 0$
5. $-1.5 \leq z \leq 2.1$
6. $-1.25 \leq z \leq 3$
7. $-1.9 \leq z \leq -1.1$
8. $-2.45 \leq z \leq -1.45$
9. $2.1 \leq z \leq 3.0$
10. $1.85 \leq z \leq 2.85$
11. $z > 2$
12. $z > 1.5$
13. $z < 1.2$
14. $z > -1.6$

Suppose a population of scores x is normally distributed with $\mu = 20$ and $\sigma = 5$. In Problems 15–18, use the standard normal distribution to find the probabilities indicated.

15. $\Pr(20 \le x \le 22.5)$ 16. $\Pr(20 \le x \le 21.25)$
17. $\Pr(13.75 \le x \le 20)$ 18. $\Pr(18.5 \le x \le 20)$

Suppose a population of scores x is normally distributed with $\mu = 50$ and $\sigma = 10$. In Problems 19–22, use the standard normal distribution to find the probabilities indicated.

19. $\Pr(45 \le x \le 55)$ 20. $\Pr(42 \le x \le 58)$
21. $\Pr(35 \le x \le 60)$ 22. $\Pr(32 \le x \le 68)$

Suppose a population of scores x is normally distributed with $\mu = 110$ and $\sigma = 12$. In Problems 23–26, use the standard normal distribution to find the probabilities indicated.

23. $\Pr(x < 134)$ 24. $\Pr(x < 88)$
25. $\Pr(x > 128)$ 26. $\Pr(x > 96)$

APPLICATIONS

27. *Growth*   The Fish Commission states that the lengths of all fish in Spring Run are normally distributed, with mean $\mu = 15$ cm and standard deviation $\sigma = 4$ cm. What is the probability that a fish caught in Spring Run will be between
(a) 15 and 19 cm long?   (b) 10 and 15 cm long?

28. *Butterfat*   A quart of Parker's milk contains a mean of 39 grams of butterfat, with a standard deviation of 2 grams. If the butterfat is normally distributed, find the probability that a quart of this brand of milk chosen at random will contain between
(a) 39 and 43 grams of butterfat.
(b) 36 and 39 grams of butterfat.

29. *Growth*   The heights of a certain species of plant are normally distributed, with mean $\mu = 20$ cm and standard deviation $\sigma = 4$ cm. What is the probability that a plant chosen at random will be between 10 and 30 cm tall?

30. *Mating calls*   The durations of the mating calls of a population of tree toads are normally distributed, with mean 189 milliseconds (msec) and standard deviation 32 msec. What proportion of these calls would be expected to last between
(a) 157 and 221 msec?   (b) 205 and 253 msec?

31. *Growth*   The mean weight of a group of boys is 160 lb, with a standard deviation of 15 lb. If the weights are normally distributed, find the probability that one of the boys picked at random from the group weighs
(a) between 160 and 181 lb.
(b) more than 190 lb.
(c) between 181 and 190 lb.
(d) between 130 and 181 lb.

32. *Testing*   The Scholastic Aptitude Test (SAT) scores in mathematics at a certain high school are normally distributed, with a mean of 500 and a standard deviation of 100. What is the probability that an individual chosen at random has a score
(a) greater than 700?   (b) less than 300?
(c) between 550 and 600?

33. *Mileage*   A certain model of automobile has its gas mileage (in miles per gallon, or mpg) normally distributed, with a mean of 28 mpg and a standard deviation of 4 mpg. Find the probability that a car selected at random has gas mileage
(a) less than 22 mpg.   (b) greater than 30 mpg.
(c) between 26 and 30 mpg.

34. *Quality control*   A machine precision-cuts tubing such that $\mu = 40.00$ inches and $\sigma = 0.03$ inch. If the lengths are normally distributed, find the probability that a piece of tubing selected at random measures
(a) greater than 40.05 inches.
(b) less than 40.045 inches.
(c) between 39.95 and 40.05 inches.

35. *Blood pressure*   Systolic blood pressure for a group of women is normally distributed, with a mean of 120 and a standard deviation of 12. Find the probability that a woman selected at random has blood pressure
(a) greater than 140.   (b) less than 110.
(c) between 110 and 130.

36. *Industrial waste*   In a certain state, the daily amounts of industrial waste are normally distributed, with a mean of 8000 tons and a standard deviation of 2000 tons. Find the probability that on a randomly selected day the amount of industrial waste is
(a) greater than 11,000 tons.
(b) less than 6000 tons.
(c) between 9000 and 11,000 tons.

37. *Reaction time*   Reaction time is normally distributed, with a mean of 0.7 second and a standard deviation of 0.1 second. Find the probability that an individual selected at random has a reaction time
(a) greater than 0.9 second.
(b) less than 0.6 second.
(c) between 0.6 and 0.9 second.

38. *Inventory*   Suppose that the electrical supplies distributor in Example 5 wants to reduce its daily inventory but wants to limit to 5% the number of days that it will have an insufficient inventory of 48-in. fluorescent light bulbs to meet daily demand. At what level should it maintain its daily inventory? (Recall that the daily demand over the past year had a mean of 432 and a standard deviation of 86; test values of 550, 575, and 600 units.)

Key Terms and Formulas

Section	Key Terms	Formulas
8.1	Binomial probability experiment	$\Pr(x) = \binom{n}{x} p^x q^{n-x}$
8.2	Descriptive statistics Frequency histogram Mode Median	
	Mean Range	$\bar{x} = \dfrac{\Sigma x}{n}$
	Sample variance	$s^2 = \dfrac{\Sigma(x - \bar{x})^2}{n - 1}$
	Sample standard deviation	$s = \sqrt{\dfrac{\Sigma(x - \bar{x})^2}{n - 1}}$
8.3	Discrete probability distribution Random variable Probability density histogram Expected value	$E(x) = \sum[x\Pr(x)]$
	Mean	$\mu = E(x) = \sum[xPr(x)]$
	Variance	$\sigma^2 = \sum(x - \mu)^2\Pr(x)$

	Standard deviation
	Binomial probability distribution
	  Mean
	  Standard deviation
		$\sigma = \sqrt{\sigma^2}$
		
		$\mu = np$
		$\sigma = \sqrt{npq}$

	Binomial formula	$(a + b)^n = \binom{n}{n}a^n + \binom{n}{n-1}a^{n-1}b$ $\qquad + \cdots + \binom{n}{1}ab^{n-1} + \binom{n}{0}b^n$
8.4	Normal distribution Standard normal distribution	
	z-scores	$z = \dfrac{x - \mu}{\sigma}$

Review Exercises

Section 8.1

1. If the probability of success on each trial of an experiment is 0.4, what is the probability of 5 successes in 7 trials?

2. A bag contains 5 black balls and 7 white balls. If we draw 4 balls, with each one replaced before the next is drawn, what is the probability that
   (a) 2 balls are black?
   (b) at least 2 balls are black?

3. If a die is rolled 4 times, what is the probability that a number greater than 4 is rolled at least 2 times?

Section 8.2

Consider the data in the following frequency table, and use these data in Problems 4–7.

Score	Frequency
1	4
2	6
3	8
4	3
5	5

4. Construct a frequency histogram for the data.
5. What is the mode of the data?
6. What is the mean of the data?
7. What is the median of the data?
8. Construct a frequency histogram for the following set of test scores: 14, 16, 15, 14, 17, 16, 12, 12, 13, 14.
9. What is the median of the scores in Problem 8?
10. What is the mode of the scores in Problem 8?
11. What is the mean of the scores in Problem 8?
12. Find the mean, variance, and standard deviation of the following data: 4, 3, 4, 6, 8, 0, 2.
13. Find the mean, variance, and standard deviation of the following data: 3, 2, 1, 5, 1, 4, 0, 2, 1, 1.

Section 8.3

14. For the following probability distribution, find the expected value $E(x)$.

x	$Pr(x)$
1	0.2
2	0.3
3	0.4
4	0.1

Determine whether each function or table in Problems 15–18 represents a discrete probability distribution.

15. $Pr(x) = x/15, x = 1, 2, 3, 4, 5$
16. $Pr(x) = x/55, x = 1, 2, 3, 4, 5, 6, 7, 8, 9$

17.

x	$Pr(x)$
3	1/6
4	1/3
5	1/3
6	1/6

18.

x	$Pr(x)$
2	1/2
3	1/4
4	−1/4
5	1/2

19. For the following probability distribution, find the expected value $E(x)$.

x	1	2	3	4
$Pr(x)$	0.4	0.3	0.2	0.1

20. For the discrete probability distribution described by $Pr(x) = x/16, x = 1, 2, 3, 4, 6$, find the
    (a) mean.
    (b) variance.
    (c) standard deviation.

21. For the probability distribution described by the table, find the
    (a) mean.
    (b) variance.
    (c) standard deviation.

x	$Pr(x)$
1	1/12
2	1/6
3	1/3
4	5/12

22. A coin has been altered so that the probability that a head will occur is 2/3. If the coin is tossed 6 times, give the mean and standard deviation of the distribution of the number of heads.
23. Suppose a pair of dice is thrown 18 times. How many times would we expect a sum of 7 to occur?
24. Expand $(x + y)^5$.

Section 8.4

25. What is the area under the standard normal curve between $z = -1.6$ and $z = 1.9$?
26. If z is a standard normal score, find $Pr(-1 \le z \le -0.5)$.
27. Find $Pr(1.23 \le z \le 2.55)$.
28. If a variable x is normally distributed, with $\mu = 25$ and $\sigma = 5$, find $Pr(25 \le x \le 30)$.
29. If a variable x is normally distributed, with $\mu = 25$ and $\sigma = 5$, find $Pr(20 \le x \le 30)$.
30. If a variable x is normally distributed, with $\mu = 25$ and $\sigma = 5$, find $Pr(30 \le x \le 35)$.

APPLICATIONS

Section 8.1

31. *Genetics* Suppose the probability that a certain couple will have a blond child is 1/4. If they have 6 children, what is the probability that 2 of them will be blond?

32. *Sampling* Suppose 70% of a population opposes a proposal and a sample of size 5 is drawn from the population. What is the probability that the majority of the sample will favor the proposal?

33. *Crime* Suppose 20% of the population are victims of crime. Out of 5 people selected at random, find the probability that 2 are crime victims.

34. *Disease* One person in 100,000 develops a certain disease. Calculate
    (a) Pr(exactly 1 person in 100,000 has the disease).
    (b) Pr(at least 1 person in 100,000 has the disease).

Section 8.2

Farm families **The distribution of farm families (as a percent of the population) in a 50-county survey is given in the table. Use these data in Problems 35–37.**

Percent	Number of Counties
10–19	5
20–29	16
30–39	25
40–49	3
50–59	1

35. Make a frequency histogram for these data.
36. Find the mean percent of farm families in these 50 counties.
37. Find the standard deviation of the percent of farm families in these 50 counties.

Section 8.3

38. *Cancer testing* The probability that a man with prostate cancer will test positive with the prostate specific antigen (PSA) test is 0.91. If 500 men with prostate cancer are tested, how many would we expect to test positive?

39. *Fraud* A company selling substandard drugs to developing countries sold 2,000,000 capsules with 60,000 of them empty (*Source: "*60 Minutes," Jan. 18, 1998). If a person gets 100 randomly chosen capsules from this company, what is the expected number of empty capsules that this person will get?

40. *Racing* Suppose you paid $2 for a bet on a horse and as a result you will win $100 if your horse wins a given race. Find the expected value of the bet if the odds that the horse will win are 3 to 12.

41. *Lotteries* A state lottery pays $500 to anyone who selects the correct 3-digit number from a ran-

x	Pr(x)
-1	0.999
499	0.001

dom drawing. If it costs $1 to play, then the probability distribution for the amount won is given in the table. What are the expected winnings of a person who plays?

42. *Crime* Refer to Problem 33.
    (a) What is the expected number of people in this sample that are crime victims?
    (b) Find the probability that exactly the expected number of people in this sample are victims of crime.

Section 8.4

43. *Testing* Suppose the mean SAT score for students admitted to a university is 1000, with a standard deviation of 200. Suppose that a student is selected at random. If the scores are normally distributed, find the probability that the student's SAT score is
    (a) between 1000 and 1400.
    (b) between 1200 and 1400.
    (c) greater than 1400.

44. *Net worth* The mean net worth of the residents of Sun City, a retirement community, is $611,000 (*Source: The Island Packet,* August 22, 1998). If their net worth is normally distributed with a standard deviation of $96,000, what percent of the residents have net worths between $700,000 and $800,000?

8 Chapter Test

1. If a coin is "loaded" so that the probability that a head will occur on each toss is 1/3, what is the probability that
   (a) 3 heads will result from 5 tosses of the coin?
   (b) at least 3 heads will occur in 5 tosses of the coin?
2. Suppose the coin in Problem 1 is tossed 12 times,
   (a) What is the expected number of heads?
   (b) Find the mean, variance, and standard deviation of the distribution of heads.
3. If x is a random variable, what properties must Pr(x) satisfy in order to be a discrete probability distribution function?
4. Find the expected value of the variable x for the probability distribution given in the table.

x	3	4	5	6	7	8
Pr(x)	0.2	0.3	0.1	0.1	0.2	0.1

5. For the probability distribution described by the table below, find the mean, variance, and standard deviation.

x	10	12	15	18	20	25
Pr(x)	0.1	0.3	0.1	0.2	0.1	0.2

6. For the sample data in the table below, find the mean, median, and mode.

Score	Frequency
20	5
21	7
22	3
23	4
24	2

7. If the variable x is normally distributed, with a mean of 16 and a standard deviation of 6, find the following probabilities.
   (a) Pr($14 \le x \le 22$)
   (b) Pr($x \le 22$)
   (c) Pr($22 \le x \le 24$)
8. For the normal distribution with mean 70 and standard deviation 12, find:
   (a) Pr($73 \le x \le 97$)
   (b) Pr($65 \le x \le 84$)
   (c) Pr($x > 84$)

The following table gives the ages of householders in primary families. Use this table in Problems 9 and 10.

Age of Householder (years)	Number (thousands)
15–24	3,079
25–34	14,082
35–44	18,274
45–54	13,746
55–64	8,895
65+	11,236

Source: U.S. Bureau of the Census

9. Construct a histogram of the data.
10. Use class marks and the given number (frequency) in each class to find the mean and standard deviation of the ages of the householders. Use 72.5 as the class mark for the 65+ class.
11. The following table gives the 2000 population of the United States.
    (a) Use the class marks and the given number (frequency) in each class to find the mean age.
    (b) In third world countries where the birth rate is higher and the life span is shorter, which group, under 30, 30–59, or 60 and over, would be most likely to be the largest? How would this affect a third world country's mean age relative to that of the United States?

Age	Class Mark	Number (millions)
Under 15	7	60.3
15–29	22	58.7
30–44	37	65.7
45–59	52	51.6
60–74	67	29.2
Over 75	82	16.7

Source: U.S. Bureau of the Census

12. The following table gives selected 2002 year-end data for cellular phone use in 45 countries worldwide.
    (a) Use class marks and the number of countries (frequency) to find the mean number of cellular telephone subscribers per 100 of population.
    (b) If these data were reexamined in 2010, how would you expect the mean found in (a) might have changed? Explain.

Number of Subscribers (per 100 pop.)	Number of Countries
90–99	4
80–89	11
70–79	11
60–69	9
50–59	5
40–49	5

Source: International Telecommunications Union

13. Suppose that 2% of the computer chips shipped to an assembly plant are defective.
    (a) What is the probability that 10 of 100 chips chosen from the shipment will be defective?
    (b) What is the expected number of defective chips in the shipment if it contains 1500 chips?

The following table gives the percents of women who become pregnant during 1 year of use with different types of contraceptives. The "Typical Use" column includes failures from incorrect or inconsistent use. Use this table in Problems 14–16.

Failure Rates Over 1 Year		
Contraceptive Type	Correct and Consistent Use	Typical Use
Spermicides	6%	21%
Diaphragm	6	18
Male latex condom	3	12
Unprotected	N/A	85

Source: R. A. Hatcher, J. Trussel, F. Stewart, et al., *Contraceptive Technology,* 1994, Irvington Pub.

14. If a group of 30 women use diaphragms correctly and consistently, how many of these 30 women could be expected to become pregnant in a year?
15. If a group of 30 women use diaphragms typically, how many of these 30 women could be expected to become pregnant in a year?
16. If spermicides and male latex condoms are used correctly by a group of 30 couples, how many of the 30 women could be expected to become pregnant in a year? Assume that these uses are independent events.
17. A new light bulb has an expected life of 18,620 hours, with a standard deviation of 4012 hours. If the hours of life of these bulbs are normally distributed, find the probability that one of these bulbs chosen at random will last
    (a) less than 10,000 hours.
    (b) more than 24,000 hours.
    (c) between 15,000 and 21,000 hours.

Extended Applications & Group Projects

I. Lotteries

One of the reasons the July 29, 1998 Powerball Lottery (sponsored by 10 states and the District of Columbia) was so large ($295.7 million) is that the odds of winning are so small and the pot continues to grow until someone wins. To win the Powerball Jackpot, a player's 5 game balls and 1 red powerball must match those chosen. On November 2, 1997, the game was changed to lower the probability of winning, as shown in the following table.

Type of Game	Old Powerball	New Powerball
Game	5-of-45	5-of-49
Powerballs	1-of-45	1-of-42

Source: Lottery America

1. Use the information in the table to find the probability of winning the old 5-of-45 game and that of winning the new 5-of-49 game. Note that the order in which the first five numbers occur does not matter, and the Powerball is not used in this game.

2. Use the information in the table to find the probability of winning the Powerball Jackpot in the old game and in the new game. Note that the order in which the first five numbers occur does not matter. Approximate odds are given at http://www.powerball.com/pbprizesNodds.shtm. Use these approximations to check your results for the new game.

   The minimum payoff for the Jackpot has increased from $5 million to $10 million, and the following table gives the payoffs for each game.

Payoffs		
Match	Old Game Payout	New Game Payout
5 + Powerball	Jackpot	Jackpot
5	$100,000	$100,000
4 + Powerball	5,000	5,000
4	100	100
3 + Powerball	100	100
3	5	7
2 + Powerball	5	7
1 + Powerball	2	4
Powerball	1	3

Source: Lottery America

3. Assume that a share of the Jackpot is $10 million, and find the expected winnings from one ticket in the old game and in the new game. The sponsors of the Powerball game claim that making the probability lower improves the game. How do they justify that claim?

II. Statistics in Medical Research; Hypothesis Testing

Bering Research Laboratories is involved in a national study organized by the National Centers for Disease Control to determine whether a new birth control pill affects the blood pressures of women who are using it. A random sample of 100 women using the pill was selected, and the participants' blood pressures were measured. The mean blood pressure was $\bar{x} = 112.5$. If the blood pressures of all women are normally distributed, with mean $\mu = 110$ and standard deviation $\sigma = 12$, is the mean of this sample sufficiently far from the mean of the population of all women to indicate that the birth control pill affects blood pressure?

To answer this question, researchers at Bering Labs first need to know how the means of samples drawn from a normal population are distributed. As a result of the **Central Limit Theorem,** if all possible samples of size n are drawn from a normal population that has mean μ and standard deviation σ, then the distribution of the means of these samples will also be normally distributed with mean μ but with standard deviation given by

$$\sigma_{\bar{x}} = \frac{\sigma}{\sqrt{n}}$$

To conclude whether women using this birth control pill have blood pressures that are significantly different from those of all women in the population, researchers must decide whether the mean $\bar{x}$ of their sample is so far from the population mean μ that it is unlikely that the sample was chosen from the population.

In this statistical test, the researchers will assume that the sample has been drawn from the population of women with mean $\mu = 110$ unless the blood pressures collected in the sample are so different from those of the population that it is not reasonable to make this assumption. In statistics, we can say that the assumption is unreasonable if the probability that the sample mean, $\bar{x} = 112.5$, could be drawn from this population, with mean 110, is less than 0.05. If this probability is less than 0.05, we say that the sample is drawn from a population with a mean different from 110, which strongly suggests that the women who have taken the birth control pill have, on average, blood pressures different from those of the women who are not using the pill. That is, there is statistically significant evidence that the pill affects the blood pressures of women.

Because 95% of all normally distributed data points lie within 2 standard deviations of the mean, researchers at Bering Labs need to determine whether the sample mean $\bar{x} = 112.5$ is more than 2 standard deviations from $\mu = 110$. To determine how many standard deviations $\bar{x} = 112.5$ is from $\mu = 110$, they compute the z-score for $\bar{x}$ with the formula

$$z_{\bar{x}} = \frac{\bar{x} - \mu}{\sigma/\sqrt{n}}$$

Use this to determine if there is sufficient evidence that this birth control pill affects blood pressure.

9

Derivatives

If a firm receives $30,000 in revenue during a 30-day month, its average revenue per day is $30,000/30 = $1000. This does not necessarily mean the actual revenue was $1000 on any one day, just that the average was $1000 per day. Similarly, if a person drove 50 miles in one hour, the average velocity was 50 miles per hour, but the driver could have received a speeding ticket for traveling 70 miles per hour.

The smaller the time interval, the nearer the average velocity will be to the instantaneous velocity (the speedometer reading). Similarly, changes in revenue over a smaller number of units can give information about the instantaneous rate of change of revenue. The mathematical bridge from average rates of change to instantaneous rates of change is the **limit.**

This chapter is concerned with *limits* and *rates of change*. We will see that the *derivative* of a function can be used to determine instantaneous rates of change.

The topics and applications discussed in this chapter include the following.

Chapter Warm-up

Prerequisite Problem Type	For Section	Answer	Section for Review
If $f(x) = \dfrac{x^2 - x - 6}{x + 2}$, then find (a) $f(-3)$ (b) $f(-2.5)$ (c) $f(-2.1)$ (d) $f(-2)$	9.1–9.9	(a) -6 (b) -5.5 (c) -5.1 (d) undefined	1.2 Function notation
Factor: (a) $x^2 - x - 6$ (b) $x^2 - 4$ (c) $x^2 + 3x + 2$	9.1 9.7	(a) $(x + 2)(x - 3)$ (b) $(x - 2)(x + 2)$ (c) $(x + 1)(x + 2)$	0.6 Factoring
Write as a power: (a) $\sqrt{t}$ (b) $\dfrac{1}{x}$ (c) $\dfrac{1}{\sqrt[3]{x^2 + 1}}$	9.4–9.8	(a) $t^{1/2}$ (b) x^{-1} (c) $(x^2 + 1)^{-1/3}$	0.3, 0.4 Exponents and radicals
Simplify: (a) $\dfrac{4(x + h)^2 - 4x^2}{h}$, if $h \neq 0$ (b) $(2x^3 + 3x + 1)(2x) + (x^2 + 4)(6x^2 + 3)$ (c) $\dfrac{x(3x^2) - x^3(1)}{x^2}$, if $x \neq 0$	9.3 9.5 9.7	(a) $8x + 4h$ (b) $10x^4 + 33x^2 + 2x + 12$ (c) $2x$	0.5 Simplifying algebraic expressions
Simplify: (a) $\dfrac{x^2 - x - 6}{x + 2}$ if $x \neq -2$ (b) $\dfrac{x^2 - 4}{x - 2}$ if $x \neq 2$	9.1 9.7	(a) $x - 3$ (b) $x + 2$	0.7 Simplifying fractions
If $f(x) = 3x^2 + 2x$, find $\dfrac{f(x + h) - f(x)}{h}$.	9.3	$6x + 3h + 2$	1.2 Function notation
Find the slope of the line passing through $(1, 2)$ and $(2, 4)$.	9.3	2	1.3 Slopes
Write the equation of the line passing through $(1, 5)$ with slope 8.	9.3 9.4 9.6	$y = 8x - 3$	1.3 Point-slope equation of a line

Limits

OBJECTIVES

- To use graphs and numerical tables to find limits of functions, when they exist
- To find limits of polynomial functions
- To find limits of rational functions

◗ **Application Preview**

Although everyone recognizes the value of eliminating any and all particulate pollution from smokestack emissions of factories, company owners are concerned about the cost of removing this pollution. Suppose that USA Steel has shown that the cost C of removing p percent of the particulate pollution from the emissions at one of its plants is

$$C = C(p) = \frac{7300p}{100 - p}$$

To investigate the cost of removing as much of the pollution as possible, we can evaluate the **limit** as p (the percent) approaches 100 from values less than 100. (See Example 6.) Using a limit is important in this case, because this function is undefined at $p = 100$ (it is impossible to remove 100% of the pollution).

In various applications we have seen the importance of the slope of a line as a rate of change. In particular, the slope of a linear total cost, total revenue, or profit function for a product tells us the marginals or rates of change of these functions. When these functions are not linear, how do we define marginals (and slope)?

We can get an idea about how to extend the notion of slope (and rate of change) to functions that are not linear. Observe that for many curves, if we take a very close (or "zoom-in") view near a point, the curve appears straight. See Figure 9.1. We can think of the slope of the "straight" line as the slope of the curve. The mathematical process used to obtain this "zoom-in" view is the process of taking limits.

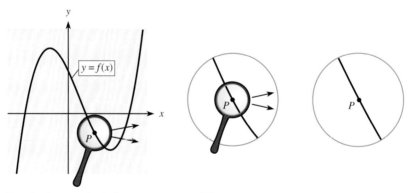

Figure 9.1   Zooming in near point P, the curve appears straight.

Notion of a Limit   We have used the notation $f(c)$ to indicate the value of a function $f(x)$ at $x = c$. If we need to discuss a value that $f(x)$ approaches as x approaches c, we use the idea of a **limit.** For example, if

$$f(x) = \frac{x^2 - x - 6}{x + 2}$$

then we know that $x = -2$ is not in the domain of $f(x)$, so $f(-2)$ does not exist even though $f(x)$ exists for every value of $x \neq -2$. Figure 9.2 shows the graph of $y = f(x)$ with an open

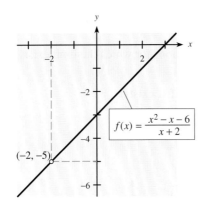

Figure 9.2

TABLE 9.1

Left of −2	
x	$f(x) = \dfrac{x^2 - x - 6}{x + 2}$
−3.000	−6.000
−2.500	−5.500
−2.100	−5.100
−2.010	−5.010
−2.001	−5.001

Right of −2	
x	$f(x) = \dfrac{x^2 - x - 6}{x + 2}$
−1.000	−4.000
−1.500	−4.500
−1.900	−4.900
−1.990	−4.990
−1.999	−4.999

circle where $x = -2$. The open circle indicates that $f(-2)$ does not exist but shows that points near $x = -2$ have functional values that lie on the line on either side of the open circle. Even though $f(-2)$ is not defined, the figure shows that as x approaches -2 from either side of -2, the graph approaches the open circle at $(-2, -5)$ and the values of $f(x)$ approach -5. Thus -5 is the limit of $f(x)$ as x approaches -2, and we write

$$\lim_{x \to -2} f(x) = -5, \quad \text{or} \quad f(x) \to -5 \quad \text{as} \quad x \to -2$$

This conclusion is fairly obvious from the graph, but it is not so obvious from the equation for $f(x)$.

We can use the values near $x = -2$ in Table 9.1 to help verify that $f(x) \to -5$ as $x \to -2$. Note that to the left of -2, the values of x increase from -3.000 to -2.001 in small increments, and in the corresponding column for $f(x)$, the values of the function $f(x)$ increase from -6.000 to -5.001. To the right of -2, the values of x decrease from -1.000 to -1.999 while the corresponding values of $f(x)$ decrease from -4.000 to -4.999. Hence, Table 9.1 and Figure 9.2 indicate that the value of $f(x)$ approaches -5 as x approaches -2 from both sides of $x = -2$.

From our discussion of the graph in Figure 9.2 and Table 9.1, we see that as x approaches -2 from either side of -2, the limit of the function is the value L that the function approaches. This limit L is not necessarily the value of the function at $x = -2$. This leads to our intuitive definition of *limit*.

Limit

Let $f(x)$ be a function defined on an open interval containing c, except perhaps at c. Then

$$\lim_{x \to c} f(x) = L$$

is read "the limit of $f(x)$ as x approaches c equals L." The number L exists if we can make values of $f(x)$ as close to L as we desire by choosing values of x sufficiently close to c. When the values of $f(x)$ do not approach a single finite value L as x approaches c, we say the limit does not exist.

As the definition states, a limit as $x \to c$ can exist only if the function approaches a single finite value as x approaches c from both the left and right of c.

● **EXAMPLE 1** *Limits*

Figure 9.3 shows three functions for which the limit exists as *x* approaches 2. Use this figure to find the following.

(a) $\lim_{x \to 2} f(x)$ and $f(2)$ (if it exists)

(b) $\lim_{x \to 2} g(x)$ and $g(2)$ (if it exists)

(c) $\lim_{x \to 2} h(x)$ and $h(2)$ (if it exists)

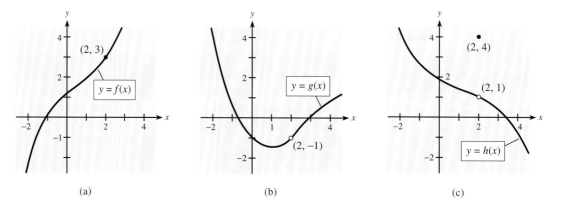

Figure 9.3   (a)   (b)   (c)

Solution

(a) From the graph in Figure 9.3(a), we see that as *x* approaches 2 from both the left and the right, the graph approaches the point (2, 3). Thus *f(x)* approaches the single value 3. That is,

$$\lim_{x \to 2} f(x) = 3$$

The value of *f(2)* is the *y*-coordinate of the point on the graph at *x* = 2. Thus *f(2)* = 3.

(b) Figure 9.3(b) shows that as *x* approaches 2 from both the left and the right, the graph approaches the open circle at (2, −1). Thus

$$\lim_{x \to 2} g(x) = -1$$

The figure also shows that at *x* = 2 there is no point on the graph. Thus *g(2)* is undefined.

(c) Figure 9.3(c) shows that

$$\lim_{x \to 2} h(x) = 1$$

The figure also shows that at *x* = 2 there is a point on the graph at (2, 4). Thus *h(2)* = 4, and we see that $\lim_{x \to 2} h(x) \neq h(2)$.

As Example 1 shows, the limit of the function as *x* approaches *c* may or may not be the same as the value of the function at *x* = *c*.

In Example 1 we saw that the limit as *x* approaches 2 meant the limit as *x* approaches 2 from both the left and the right. We can also consider limits only from the left or only from the right; these are called one-sided limits.

One-Sided Limits

> **Limit from the Right:** $\lim_{x \to c^+} f(x) = L$
>
> means the values of $f(x)$ approach the value L as $x \to c$ but $x > c$.
>
> **Limit from the Left:** $\lim_{x \to c^-} f(x) = M$
>
> means the values of $f(x)$ approach the value M as $x \to c$ but $x < c$.

Note that when one or both one-sided limits fail to exist, then the limit does not exist. Also, when the one-sided limits differ, such as if $L \neq M$ above, then the values of $f(x)$ do not approach a *single* value as x approaches c, and $\lim_{x \to c} f(x)$ does not exist. So far, the examples have illustrated limits that existed and for which the two one-sided limits were the same. In the next example, we use one-sided limits and consider cases where a limit does not exist.

● **EXAMPLE 2** *One-Sided Limits*

Using the functions graphed in Figure 9.4, determine why the limit as $x \to 2$ does not exist for

(a) $f(x)$
(b) $g(x)$
(c) $h(x)$

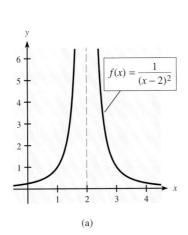
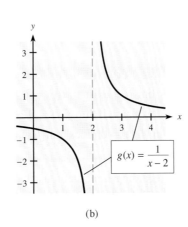
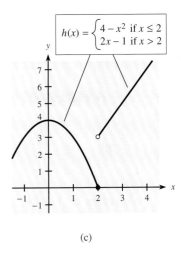

$$h(x) = \begin{cases} 4 - x^2 & \text{if } x \leq 2 \\ 2x - 1 & \text{if } x > 2 \end{cases}$$

$$f(x) = \frac{1}{(x-2)^2}$$

$$g(x) = \frac{1}{x - 2}$$

Figure 9.4   (a)   (b)   (c)

Solution

(a) As $x \to 2$ from the left side and the right side of $x = 2$, $f(x)$ increases without bound, which we denote by saying that $f(x)$ approaches $+\infty$ as $x \to 2$. In this case, $\lim_{x \to 2} f(x)$ does not exist [denoted by $\lim_{x \to 2} f(x)$ DNE] because $f(x)$ does not approach a finite value as $x \to 2$. In this case, we write

$$f(x) \to +\infty \text{ as } x \to 2$$

The graph has a vertical asymptote at $x = 2$.

(b) As $x \to 2$ from the left, $g(x)$ approaches $-\infty$, and as $x \to 2$ from the right, $g(x)$ approaches $+\infty$, so $g(x)$ does not approach a finite value as $x \to 2$. Therefore, the limit does not exist. The graph of $y = g(x)$ has a vertical asymptote at $x = 2$.

In this case we summarize by writing

$$\lim_{x \to 2^-} g(x) \text{ DNE} \quad \text{or} \quad g(x) \to -\infty \text{ as } x \to 2^-$$

$$\lim_{x \to 2^+} g(x) \text{ DNE} \quad \text{or} \quad g(x) \to +\infty \text{ as } x \to 2^+$$

and

$$\lim_{x \to 2} g(x) \text{ DNE}$$

(c) As $x \to 2$ from the left, the graph approaches the point at (2, 0), so $\lim_{x \to 2^-} h(x) = 0$. As $x \to 2$ from the right, the graph approaches the open circle at (2, 3), so $\lim_{x \to 2^+} h(x) = 3$. Because these one-sided limits differ, $\lim_{x \to 2} h(x)$ does not exist.

Examples 1 and 2 illustrate the following two important facts regarding limits.

1. The limit of a function as x approaches c is independent of the value of the function at c. When $\lim_{x \to c} f(x)$ exists, the value of the function at c may be (i) the same as the limit, (ii) undefined, or (iii) defined but different from the limit (see Figure 9.3 and Example 1).
2. The limit is said to exist only if the following conditions are satisfied:
   (a) The limit L is a finite value (real number).
   (b) The limit as x approaches c from the left equals the limit as x approaches c from the right. That is, we must have

   $$\lim_{x \to c^-} f(x) = \lim_{x \to c^+} f(x)$$

   Figure 9.4 and Example 2 illustrate cases where $\lim_{x \to c} f(x)$ does not exist.

● **Checkpoint**

1. Can $\lim_{x \to c^-} f(x)$ exist if $f(c)$ is undefined?
2. Does $\lim_{x \to c} f(x)$ exist if $f(c) = 0$?
3. Does $f(c) = 1$ if $\lim_{x \to c} f(x) = 1$?
4. If $\lim_{x \to c^-} f(x) = 0$, does $\lim_{x \to c} f(x)$ exist?

Properties of Limits, Algebraic Evaluation

Fact 1 regarding limits tells us that the value of the limit of a function as $x \to c$ will not always be the same as the value of the function at $x = c$. However, there are many functions for which the limit and the functional value agree [see Figure 9.3(a)], and for these functions we can easily evaluate limits. The following properties of limits allow us to identify certain classes or types of functions for which $\lim_{x \to c} f(x)$ equals $f(c)$.

Properties of Limits

If k is a constant, $\lim_{x \to c} f(x) = L$, and $\lim_{x \to c} g(x) = M$, then the following are true.

I. $\lim_{x \to c} k = k$

II. $\lim_{x \to c} x = c$

III. $\lim_{x \to c} [f(x) \pm g(x)] = L \pm M$

IV. $\lim_{x \to c} [f(x) \cdot g(x)] = LM$

V. $\lim_{x \to c} \dfrac{f(x)}{g(x)} = \dfrac{L}{M}$ if $M \neq 0$

VI. $\lim_{x \to c} \sqrt[n]{f(x)} = \sqrt[n]{\lim_{x \to c} f(x)} = \sqrt[n]{L}$, provided that $L > 0$ when n is even.

If f is a polynomial function, then Properties I–IV imply that $\lim\limits_{x \to c} f(x)$ can be found by evaluating $f(c)$. Moreover, if h is a rational function whose denominator is not zero at $x = c$, then Property V implies that $\lim\limits_{x \to c} h(x)$ can be found by evaluating $h(c)$. The following summarizes these observations and recalls the definitions of polynomial and rational functions.

Function	Definition	Limit
Polynomial function	The function $$f(x) = a_n x^n + a_{n-1} x^{n-1} + \cdots + a_1 x + a_0,$$ where $a_n \neq 0$ and n is a positive integer, is called a **polynomial function** of degree n.	$\lim\limits_{x \to c} f(x) = f(c)$ for all values c (by Properties I–IV)
Rational function	The function $$h(x) = \frac{f(x)}{g(x)}$$ where both $f(x)$ and $g(x)$ are polynomial functions, is called a **rational function.**	$\lim\limits_{x \to c} h(x) = \lim\limits_{x \to c} \dfrac{f(x)}{g(x)} = \dfrac{f(c)}{g(c)}$ when $g(c) \neq 0$ (by Property V)

● **EXAMPLE 3** *Limits*

Find the following limits, if they exist.

(a) $\lim\limits_{x \to -1} (x^3 - 2x)$

(b) $\lim\limits_{x \to 4} \dfrac{x^2 - 4x}{x - 2}$

Solution

(a) Note that $f(x) = x^3 - 2x$ is a polynomial, so

$$\lim_{x \to -1} f(x) = f(-1) = (-1)^3 - 2(-1) = 1$$

Figure 9.5(a) shows the graph of $f(x) = x^3 - 2x$.

(b) Note that this limit has the form

$$\lim_{x \to c} \frac{f(x)}{g(x)}$$

where $f(x)$ and $g(x)$ are polynomials and $g(c) \neq 0$. Therefore, we have

$$\lim_{x \to 4} \frac{x^2 - 4x}{x - 2} = \frac{4^2 - 4(4)}{4 - 2} = \frac{0}{2} = 0$$

Figure 9.5(b) shows the graph of $g(x) = \dfrac{x^2 - 4x}{x - 2}$.

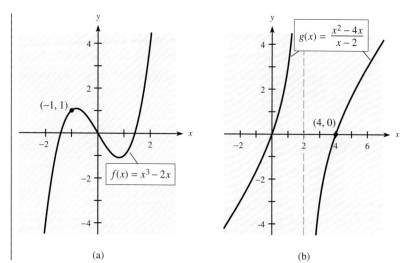

Figure 9.5   (a)   (b)

We have seen that we can use Property V to find the limit of a rational function $f(x)/g(x)$ as long as the denominator is *not* zero. If the limit of the denominator of $f(x)/g(x)$ *is* zero, then there are two possible cases.

I. Both $\lim_{x \to c} g(x) = 0$ and $\lim_{x \to c} f(x) = 0$, or

II. $\lim_{x \to c} g(x) = 0$ and $\lim_{x \to c} f(x) \neq 0$.

In case I we say that $f(x)/g(x)$ has the form 0/0 at $x = c$. We call this the **0/0 indeterminate form;** the limit cannot be evaluated until $x - c$ is factored from both $f(x)$ and $g(x)$ and the fraction is reduced. Example 4 will illustrate this case.

In case II, the limit has the form $a/0$, where a is a constant, $a \neq 0$. This expression is undefined, and the limit does not exist. Example 5 will illustrate this case.

● **EXAMPLE 4 0/0 Indeterminate Form**

Evaluate the following limits, if they exist.

(a) $\lim_{x \to 2} \dfrac{x^2 - 4}{x - 2}$

(b) $\lim_{x \to 1} \dfrac{x^2 - 3x + 2}{x^2 - 1}$

Solution

(a) We cannot find the limit by using Property V because the denominator is zero at $x = 2$. The numerator is also zero at $x = 2$, so the expression

$$\frac{x^2 - 4}{x - 2}$$

has the 0/0 indeterminate form at $x = 2$. Thus we can factor $x - 2$ from both the numerator and the denominator and reduce the fraction. (We can divide by $x - 2$ because $x - 2 \neq 0$ while $x \to 2$.)

$$\lim_{x \to 2} \frac{x^2 - 4}{x - 2} = \lim_{x \to 2} \frac{(x - 2)(x + 2)}{x - 2} = \lim_{x \to 2} (x + 2) = 4$$

Figure 9.6(a) shows the graph of $f(x) = (x^2 - 4)/(x - 2)$. Note the open circle at (2, 4).

(b) By substituting 1 for x in $(x^2 - 3x + 2)/(x^2 - 1)$, we see that the expression has the 0/0 indeterminate form at $x = 1$, so $x - 1$ is a factor of both the numerator and the denominator. (We can then reduce the fraction because $x - 1 \neq 0$ while $x \to 1$.)

$$\lim_{x \to 1} \frac{x^2 - 3x + 2}{x^2 - 1} = \lim_{x \to 1} \frac{(x - 1)(x - 2)}{(x - 1)(x + 1)}$$

$$= \lim_{x \to 1} \frac{x - 2}{x + 1}$$

$$= \frac{1 - 2}{1 + 1} = \frac{-1}{2} \qquad \text{(by Property V)}$$

Figure 9.6(b) shows the graph of $g(x) = (x^2 - 3x + 2)/(x^2 - 1)$. Note the open circle at $(1, -\frac{1}{2})$.

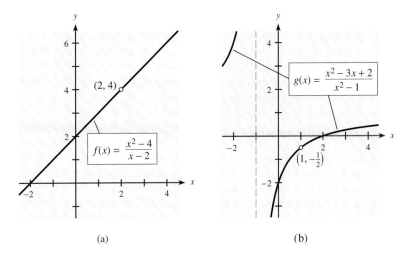

Figure 9.6

(a)   (b)

Note that although both problems in Example 4 had the 0/0 indeterminate form, they had different answers.

● EXAMPLE 5 *Limit with a/0 Form*

Find $\displaystyle\lim_{x \to 1} \frac{x^2 + 3x + 2}{x - 1}$, if it exists.

Solution

Substituting 1 for x in the function results in 6/0, so this limit has the form $a/0$, with $a \neq 0$, and is like case II discussed previously. Hence the limit does not exist. Because the numerator is not zero when $x = 1$, we know that $x - 1$ is *not* a factor of the numerator, and we cannot divide numerator and denominator as we did in Example 4. Table 9.2 confirms that this limit does not exist, because the values of the expression are unbounded near $x = 1$.

TABLE 9.2

Left of $x = 1$		Right of $x = 1$	
x	$\dfrac{x^2 + 3x + 2}{x - 1}$	x	$\dfrac{x^2 + 3x + 2}{x - 1}$
0	-2	2	12
0.5	-7.5	1.5	17.5
0.7	-15.3	1.2	35.2
0.9	-55.1	1.1	65.1
0.99	-595.01	1.01	605.01
0.999	$-5,995.001$	1.001	6,005.001
0.9999	$-59,995.0001$	1.0001	60,005.0001
$\displaystyle\lim_{x \to 1^-} \dfrac{x^2 + 3x + 2}{x - 1}$ DNE		$\displaystyle\lim_{x \to 1^+} \dfrac{x^2 + 3x + 2}{x - 1}$ DNE	
$(f(x) \to -\infty$ as $x \to 1^-)$		$(f(x) \to +\infty$ as $x \to 1^+)$	

The left-hand and right-hand limits do not exist. Thus $\displaystyle\lim_{x \to 1} \dfrac{x^2 + 3x + 2}{x - 1}$ does not exist.

In Example 5, even though the left-hand and right-hand limits do not exist (see Table 9.2), knowledge that the functional values are unbounded (that is, that they become infinite) is helpful in graphing. The graph is shown in Figure 9.7. We see that $x = 1$ is a vertical asymptote.

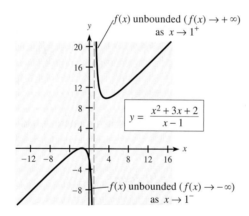

Figure 9.7

The results of Examples 4 and 5 can be summarized as follows.

Rational Functions: Evaluating Limits of the Form $\displaystyle\lim_{x \to c} \dfrac{f(x)}{g(x)}$ where $\displaystyle\lim_{x \to c} g(x) = 0$

Type I. If $\displaystyle\lim_{x \to c} f(x) = 0$ and $\displaystyle\lim_{x \to c} g(x) = 0$, then the fractional expression has the **0/0 indeterminate form** at $x = c$. We can factor $x - c$ from $f(x)$ and $g(x)$, reduce the fraction, and then find the limit of the resulting expression, if it exists.

Type II. If $\displaystyle\lim_{x \to c} f(x) \neq 0$ and $\displaystyle\lim_{x \to c} g(x) = 0$, then $\displaystyle\lim_{x \to c} \dfrac{f(x)}{g(x)}$ does not exist. In this case, the values of $f(x)/g(x)$ are unbounded near $x = c$; the line $x = c$ is a vertical asymptote.

◑ EXAMPLE 6 *Cost-Benefit (Application Preview)*

USA Steel has shown that the cost C of removing p percent of the particulate pollution from the smokestack emissions at one of its plants is

$$C = C(p) = \frac{7300p}{100 - p}$$

To investigate the cost of removing as much of the pollution as possible, find:

(a) the cost of removing 50% of the pollution.
(b) the cost of removing 90% of the pollution.
(c) the cost of removing 99% of the pollution.
(d) the cost of removing 100% of the pollution.

Solution

(a) The cost of removing 50% of the pollution is $7300 because

$$C(50) = \frac{7300(50)}{100 - 50} = \frac{365,000}{50} = 7300$$

(b) The cost of removing 90% of the pollution is $65,700 because

$$C(90) = \frac{7300(90)}{100 - 90} = \frac{657,000}{10} = 65,700$$

(c) The cost of removing 99% of the pollution is $722,700 because

$$C(99) = \frac{7300(99)}{100 - 99} = \frac{722,700}{1} = 722,700$$

(d) The cost of removing 100% of the pollution is undefined because the denominator of the function is 0 when $p = 100$. To see what the cost approaches as p approaches 100 from values smaller than 100, we evaluate $\lim\limits_{x \to 100^-} \frac{7300p}{100 - p}$. This limit has the Type II form for rational functions. Thus $\frac{7300}{100 - p} \to +\infty$ as $x \to 100^-$, which means that as the amount of pollution that is removed approaches 100%, the cost increases without bound. (That is, it is impossible to remove 100% of the pollution.)

● **Checkpoint**

5. Evaluate the following limits (if they exist).

(a) $\lim\limits_{x \to -3} \dfrac{2x^2 + 5x - 3}{x^2 - 9}$ (b) $\lim\limits_{x \to 5} \dfrac{x^2 - 3x - 3}{x^2 - 8x + 1}$ (c) $\lim\limits_{x \to -3/4} \dfrac{4x}{4x + 3}$

In Problems 6–9, assume that f, g, and h are polynomials.

6. Does $\lim\limits_{x \to c} f(x) = f(c)$?

7. Does $\lim\limits_{x \to c} \dfrac{g(x)}{h(x)} = \dfrac{g(c)}{h(c)}$?

8. If $g(c) = 0$ and $h(c) = 0$, can we be certain that

(a) $\lim\limits_{x \to c} \dfrac{g(x)}{h(x)} = 0$?   (b) $\lim\limits_{x \to c} \dfrac{g(x)}{h(x)}$ exists?

9. If $g(c) \neq 0$ and $h(c) = 0$, what can be said about $\lim\limits_{x \to c} \dfrac{g(x)}{h(x)}$ and $\lim\limits_{x \to c} \dfrac{h(x)}{g(x)}$?

Limits of Piecewise Defined Functions

As we noted in Section 2.4, "Special Functions and Their Graphs," many applications are modeled by piecewise defined functions. To see how we evaluate a limit involving a piecewise defined function, consider the following example.

● **EXAMPLE 7** *Limit of a Piecewise Defined Function*

Find $\lim_{x \to 1^-} f(x)$, $\lim_{x \to 1^+} f(x)$, and $\lim_{x \to 1} f(x)$, if they exist, for

$$f(x) = \begin{cases} x^2 + 1 \text{ for } x \le 1 \\ x\ \ + 2 \text{ for } x > 1 \end{cases}$$

Solution

Because $f(x)$ is defined by $x^2 + 1$ when $x < 1$,

$$\lim_{x \to 1^-} f(x) = \lim_{x \to 1^-} (x^2 + 1) = 2$$

Because $f(x)$ is defined by $x + 2$ when $x > 1$,

$$\lim_{x \to 1^+} f(x) = \lim_{x \to 1^+} (x + 2) = 3$$

And because

$$2 = \lim_{x \to 1^-} f(x) \ne \lim_{x \to 1^+} f(x) = 3$$

$\lim_{x \to 1} f(x)$ does not exist.

Table 9.3 and Figure 9.8 show these results numerically and graphically.

TABLE 9.3

Left of 1	
x	$f(x) = x^2 + 1$
0.1	1.01
0.9	1.81
0.99	1.98
0.999	1.998
0.9999	1.9998

Right of 1	
x	$f(x) = x + 2$
1.2	3.2
1.01	3.01
1.001	3.001
1.0001	3.0001
1.00001	3.00001

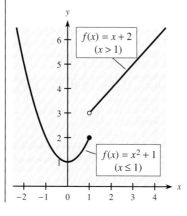

Figure 9.8

Calculator Note

We have used graphical, numerical, and algebraic methods to understand and evaluate limits. Graphing calculators can be especially effective when we are exploring limits graphically or numerically. ■

● **EXAMPLE 8** *Limits: Graphically, Numerically, and Algebraically*

Consider the following limits.

(a) $\lim_{x \to 5} \dfrac{x^2 + 2x - 35}{x^2 - 6x + 5}$ (b) $\lim_{x \to -1} \dfrac{2x}{x + 1}$

Investigate each limit by using the following methods.

(i) Graphically: Graph the function with a graphing utility and trace near the limiting *x*-value.

(ii) Numerically: Use the table feature of a graphing utility to evaluate the function very close to the limiting *x*-value.

(iii) Algebraically: Use properties of limits and algebraic techniques.

Solution

(a) $\lim_{x \to 5} \dfrac{x^2 + 2x - 35}{x^2 - 6x + 5}$

  (i) Figure 9.9(a) shows the graph of $y = (x^2 + 2x - 35)/(x^2 - 6x + 5)$. Tracing near $x = 5$ shows y-values getting close to 3.

  (ii) Figure 9.9(b) shows a table for $y_1 = (x^2 + 2x - 35)/(x^2 - 6x + 5)$ with x-values approaching 5 from both sides (note that the function is undefined at $x = 5$). Again, the y-values approach 3.

Both (i) and (ii) strongly suggest $\lim_{x \to 5} \dfrac{x^2 + 2x - 35}{x^2 - 6x + 5} = 3$.

  (iii) Algebraic evaluation of this limit confirms what the graph and the table suggest.

$$\lim_{x \to 5} \frac{x^2 + 2x - 35}{x^2 - 6x + 5} = \lim_{x \to 5} \frac{(x + 7)(x - 5)}{(x - 1)(x - 5)} = \lim_{x \to 5} \frac{x + 7}{x - 1} = \frac{12}{4} = 3$$

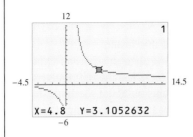

(a)

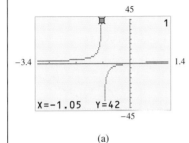

(b)

Figure 9.9

(b) $\lim_{x \to -1} \dfrac{2x}{x + 1}$

  (i) Figure 9.10(a) shows the graph of $y = 2x/(x + 1)$; it indicates a break in the graph near $x = -1$. Evaluation confirms that the break occurs at $x = -1$ and also suggests that the function becomes unbounded near $x = -1$. In addition, we can see that as x approaches -1 from opposite sides, the function is headed in different directions. All this suggests that the limit does not exist.

  (ii) Figure 9.10(b) shows a graphing calculator table of values for $y_1 = 2x/(x + 1)$ and with x-values approaching $x = -1$. The table reinforces our preliminary conclusions from the graph that the limit does not exist, because the function is unbounded near $x = -1$.

  (iii) Algebraically we see that this limit has the form $-2/0$. Thus $\lim_{x \to -1} \dfrac{2x}{x + 1}$ DNE.

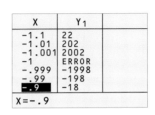

Figure 9.10

(a)

(b)

We could also use the graphing and table features of spreadsheets to explore limits.

● *Checkpoint Solutions*

1. Yes. For example, Figure 9.2 and Table 9.1 show that this is possible for
$f(x) = \dfrac{x^2 - x - 6}{x + 2}$. Remember that $\lim\limits_{x \to c} f(x)$ does not depend on $f(c)$.

2. Not necessarily. Figure 9.4(c) shows the graph of $y = h(x)$ with $h(2) = 0$, but $\lim\limits_{x \to 2} h(x)$ does not exist.

3. Not necessarily. Figure 9.3(c) shows the graph of $y = h(x)$ with $\lim\limits_{x \to 2} h(x) = 1$ but $h(2) = 4$.

4. Not necessarily. For example, Figure 9.4(c) shows the graph of $y = h(x)$ with $\lim\limits_{x \to 2^-} h(x) = 0$, but with $\lim\limits_{x \to 2^+} h(x) = 2$, so the limit doesn't exist. Recall that if $\lim\limits_{x \to c^-} f(x) = \lim\limits_{x \to c^+} f(x) = L$, then $\lim\limits_{x \to c} f(x) = L$.

5. (a) $\lim\limits_{x \to -3} \dfrac{2x^2 + 5x - 3}{x^2 - 9} = \lim\limits_{x \to -3} \dfrac{(2x - 1)(x + 3)}{(x + 3)(x - 3)} = \lim\limits_{x \to -3} \dfrac{2x - 1}{x - 3} = \dfrac{-7}{-6} = \dfrac{7}{6}$

   (b) $\lim\limits_{x \to 5} \dfrac{x^2 - 3x - 3}{x^2 - 8x + 1} = \dfrac{7}{-14} = -\dfrac{1}{2}$

   (c) Substituting $x = -3/4$ gives $-3/0$, so $\lim\limits_{x \to -3/4} \dfrac{4x}{4x + 3}$ does not exist.

6. Yes, Properties I–IV yield this result.

7. Not necessarily. If $h(c) \neq 0$, then this is true. Otherwise, it is not true.

8. For both (a) and (b), $g(x)/h(x)$ has the $0/0$ indeterminate form at $x = c$. In this case we can make no general conclusion about the limit. It is possible for the limit to exist (and be zero or nonzero) or not to exist. Consider the following $0/0$ indeterminate forms.

   (i) $\lim\limits_{x \to 0} \dfrac{x^2}{x} = \lim\limits_{x \to 0} x = 0$ (ii) $\lim\limits_{x \to 0} \dfrac{x(x + 1)}{x} = \lim\limits_{x \to 0} (x + 1) = 1$

   (iii) $\lim\limits_{x \to 0} \dfrac{x}{x^2} = \lim\limits_{x \to 0} \dfrac{1}{x}$, which does not exist

9. $\lim\limits_{x \to c} \dfrac{g(x)}{h(x)}$ does not exist and $\lim\limits_{x \to c} \dfrac{h(x)}{g(x)} = 0$

9.1 Exercises

In Problems 1–6, a graph of $y = f(x)$ is shown and a c-value is given. For each problem, use the graph to find the following, whenever they exist.

(a) $\lim\limits_{x \to c} f(x)$ and   (b) $f(c)$

1. $c = 4$

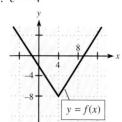

2. $c = 6$

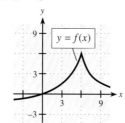

3. $c = 20$

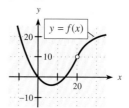

4. $c = -10$

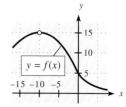

5. $c = -8$

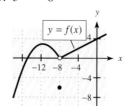

6. $c = -2$

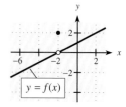

In Problems 7–10, use the graph of $y = f(x)$ and the given c-value to find the following, whenever they exist.

(a) $\lim\limits_{x \to c^-} f(x)$ (b) $\lim\limits_{x \to c^+} f(x)$

(c) $\lim\limits_{x \to c} f(x)$ (d) $f(c)$

7. $c = -10$

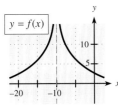

8. $c = 2$

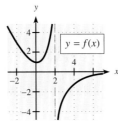

9. $c = -4\frac{1}{2}$

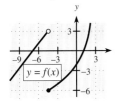

10. $c = 2$

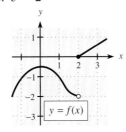

In Problems 11–14, complete each table and predict the limit, if it exists.

11. $f(x) = \dfrac{2 - x - x^2}{x - 1}$

$\lim\limits_{x \to 1} f(x) = ?$

x	$f(x)$
0.9	
0.99	
0.999	
↓	↓
1	?
↑	↑
1.001	
1.01	
1.1	

12. $f(x) = \dfrac{2x + 1}{\frac{1}{4} - x^2}$

$\lim\limits_{x \to -0.5} f(x) = ?$

x	$f(x)$
-0.51	
-0.501	
-0.5001	
↓	↓
-0.5	?
↑	↑
-0.4999	
-0.499	
-0.49	

13. $f(x) = \begin{cases} 5x - 1 & \text{for } x < 1 \\ 8 - 2x - x^2 & \text{for } x \geq 1 \end{cases}$

$\lim\limits_{x \to 1} f(x) = ?$

x	$f(x)$
0.9	
0.99	
0.999	
↓	↓
1	?
↑	↑
1.001	
1.01	
1.1	

14. $f(x) = \begin{cases} 4 - x^2 & \text{for } x \leq -2 \\ x^2 + 2x & \text{for } x > -2 \end{cases}$

$\lim\limits_{x \to -2} f(x) = ?$

x	$f(x)$
-2.1	
-2.01	
-2.001	
↓	↓
-2	?
↑	↑
-1.999	
-1.99	

In Problems 15–36, use properties of limits and algebraic methods to find the limits, if they exist.

15. $\lim\limits_{x \to -35} (34 + x)$

16. $\lim\limits_{x \to 80} (82 - x)$

17. $\lim\limits_{x \to -1} (4x^3 - 2x^2 + 2)$

18. $\lim\limits_{x \to 3} (2x^3 - 12x^2 + 5x + 3)$

19. $\lim\limits_{x \to -1/2} \dfrac{4x - 2}{4x^2 + 1}$

20. $\lim\limits_{x \to -1/3} \dfrac{1 - 3x}{9x^2 + 1}$

21. $\lim\limits_{x \to 3} \dfrac{x^2 - 9}{x - 3}$

22. $\lim\limits_{x \to -4} \dfrac{x^2 - 16}{x + 4}$

23. $\lim\limits_{x \to 7} \dfrac{x^2 - 8x + 7}{x^2 - 6x - 7}$

24. $\lim\limits_{x \to -5} \dfrac{x^2 + 8x + 15}{x^2 + 5x}$

25. $\lim\limits_{x \to -2} \dfrac{x^2 + 4x + 4}{x^2 + 3x + 2}$

26. $\lim\limits_{x \to 10} \dfrac{x^2 - 8x - 20}{x^2 - 11x + 10}$

27. $\lim\limits_{x \to 3} f(x)$, where $f(x) = \begin{cases} 10 - 2x & \text{for } x < 3 \\ x^2 - x & \text{for } x \geq 3 \end{cases}$

28. $\lim\limits_{x \to 5} f(x)$, where $f(x) = \begin{cases} 7x - 10 & \text{for } x < 5 \\ 25 & \text{for } x \geq 5 \end{cases}$

29. $\lim\limits_{x \to -1} f(x)$, where $f(x) = \begin{cases} x^2 + \dfrac{4}{x} & \text{for } x \leq -1 \\ 3x^3 - x - 1 & \text{for } x > -1 \end{cases}$

30. $\lim\limits_{x \to 2} f(x)$, where $f(x) = \begin{cases} \dfrac{x^3 - 4}{x - 3} & \text{for } x \leq 2 \\ \dfrac{3 - x^2}{x} & \text{for } x > 2 \end{cases}$

31. $\lim\limits_{x \to 2} \dfrac{x^2 + 6x + 9}{x - 2}$

32. $\lim\limits_{x \to 5} \dfrac{x^2 - 6x + 8}{x - 5}$

33. $\lim\limits_{x \to -1} \dfrac{x^2 + 5x + 6}{x + 1}$

34. $\lim\limits_{x \to 3} \dfrac{x^2 + 2x - 3}{x - 3}$

35. $\lim\limits_{h \to 0} \dfrac{(x + h)^3 - x^3}{h}$

36. $\lim\limits_{h \to 0} \dfrac{2(x + h)^2 - 2x^2}{h}$

In Problems 37–40, graph each function with a graphing utility and use it to predict the limit. Check your work either by using the table feature of the graphing utility or by finding the limit algebraically.

37. $\lim\limits_{x \to 10} \dfrac{x^2 - 19x + 90}{3x^2 - 30x}$

38. $\lim\limits_{x \to -3} \dfrac{x^4 + 3x^3}{2x^4 - 18x^2}$

39. $\lim\limits_{x \to -1} \dfrac{x^3 - x}{x^2 + 2x + 1}$ 40. $\lim\limits_{x \to 5} \dfrac{x^2 - 7x + 10}{x^2 - 10x + 25}$

 In Problems 41–44, use the table feature of a graphing utility to predict each limit. Check your work by using either a graphical or an algebraic approach.

41. $\lim\limits_{x \to -2} \dfrac{x^4 - 4x^2}{x^2 + 8x + 12}$ 42. $\lim\limits_{x \to -4} \dfrac{x^3 + 4x^2}{2x^2 + 7x - 4}$

43. $\lim\limits_{x \to 4} f(x)$, where $f(x) = \begin{cases} 12 - \dfrac{3}{4}x & \text{for } x \le 4 \\ x^2 - 7 & \text{for } x > 4 \end{cases}$

44. $\lim\limits_{x \to 7} f(x)$, where $f(x) = \begin{cases} 2 + x - x^2 & \text{for } x \le 7 \\ 13 - 9x & \text{for } x > 7 \end{cases}$

45. Use values 0.1, 0.01, 0.001, 0.0001, and 0.00001 with your calculator to approximate

$$\lim\limits_{a \to 0} (1 + a)^{1/a}$$

to three decimal places. This limit equals the special number e that is discussed in Section 5.1, "Exponential Functions," and Section 6.2, "Compound Interest; Geometric Sequences."

46. If $\lim\limits_{x \to 2}[f(x) + g(x)] = 5$ and $\lim\limits_{x \to 2} g(x) = 11$, find
    (a) $\lim\limits_{x \to 2} f(x)$
    (b) $\lim\limits_{x \to 2}\{[f(x)]^2 - [g(x)]^2\}$
    (c) $\lim\limits_{x \to 2} \dfrac{3g(x)}{f(x) - g(x)}$

47. If $\lim\limits_{x \to 3} f(x) = 4$ and $\lim\limits_{x \to 3} g(x) = -2$, find
    (a) $\lim\limits_{x \to 3}[f(x) + g(x)]$ (b) $\lim\limits_{x \to 3}[f(x) - g(x)]$
    (c) $\lim\limits_{x \to 3}[f(x) \cdot g(x)]$ (d) $\lim\limits_{x \to 3} \dfrac{g(x)}{f(x)}$

48. (a) If $\lim\limits_{x \to 2^+} f(x) = 5$, $\lim\limits_{x \to 2^-} f(x) = 5$, and $f(2) = 0$, find $\lim\limits_{x \to 2} f(x)$, if it exists. Explain your conclusions.
    (b) If $\lim\limits_{x \to 0^+} f(x) = 3$, $\lim\limits_{x \to 0^-} f(x) = 0$, and $f(0) = 0$, find $\lim\limits_{x \to 0} f(x)$, if it exists. Explain your conclusions.

APPLICATIONS

49. *Revenue*   The total revenue for a product is given by

$$R(x) = 1600x - x^2$$

where x is the number of units sold. What is $\lim\limits_{x \to 100} R(x)$?

50. *Profit*   If the profit function for a product is given by

$$P(x) = 92x - x^2 - 1760$$

find $\lim\limits_{x \to 40} P(x)$.

51. *Sales and training*   The average monthly sales volume (in thousands of dollars) for a firm depends on the number of hours x of training of its sales staff, according to

$$S(x) = \dfrac{4}{x} + 30 + \dfrac{x}{4}, \quad 4 \le x \le 100$$

(a) Find $\lim\limits_{x \to 4^+} S(x)$.   (b) Find $\lim\limits_{x \to 100^-} S(x)$.

52. *Sales and training*   During the first 4 months of employment, the monthly sales S (in thousands of dollars) for a new salesperson depend on the number of hours x of training, as follows:

$$S = S(x) = \dfrac{9}{x} + 10 + \dfrac{x}{4}, \quad x \ge 4$$

(a) Find $\lim\limits_{x \to 4^+} S(x)$.   (b) Find $\lim\limits_{x \to 10} S(x)$.

53. *Advertising and sales*   Suppose that the daily sales S (in dollars) t days after the end of an advertising campaign are

$$S = S(t) = 400 + \dfrac{2400}{t + 1}$$

(a) Find $S(0)$.   (b) Find $\lim\limits_{t \to 7} S(t)$.
(c) Find $\lim\limits_{t \to 14} S(t)$.

54. *Advertising and sales*   Sales y (in thousands of dollars) are related to advertising expenses x (in thousands of dollars) according to

$$y = y(x) = \dfrac{200x}{x + 10}, \quad x \ge 0$$

(a) Find $\lim\limits_{x \to 10} y(x)$.   (b) Find $\lim\limits_{x \to 0^+} y(x)$.

55. *Productivity*   During an 8-hour shift, the rate of change of productivity (in units per hour) of children's phonographs assembled after t hours on the job is

$$r(t) = \dfrac{128(t^2 + 6t)}{(t^2 + 6t + 18)^2}, \quad 0 \le t \le 8$$

(a) Find $\lim\limits_{t \to 4} r(t)$.   (b) Find $\lim\limits_{t \to 8^-} r(t)$.
(c) Is the rate of productivity higher near the lunch break (at $t = 4$) or near quitting time (at $t = 8$)?

56. *Revenue*   If the revenue for a product is $R(x) = 100x - 0.1x^2$, and the average revenue per unit is

$$\bar{R}(x) = \dfrac{R(x)}{x}, \quad x > 0$$

find (a) $\lim\limits_{x \to 100} \dfrac{R(x)}{x}$ and (b) $\lim\limits_{x \to 0^+} \dfrac{R(x)}{x}$.

57. *Cost-benefit*   Suppose that the cost C of obtaining water that contains p percent impurities is given by

$$C(p) = \dfrac{120{,}000}{p} - 1200$$

(a) Find $\lim\limits_{p \to 100^-} C(p)$, if it exists. Interpret this result.
(b) Find $\lim\limits_{p \to 0^+} C(p)$, if it exists.
(c) Is complete purity possible? Explain.

58. *Cost-benefit* Suppose that the cost C of removing p percent of the particulate pollution from the smokestacks of an industrial plant is given by

$$C(p) = \frac{730{,}000}{100 - p} - 7300$$

(a) Find $\lim\limits_{p \to 80} C(p)$.
(b) Find $\lim\limits_{p \to 100^-} C(p)$, if it exists.
(c) Can 100% of the particulate pollution be removed? Explain.

59. *Federal income tax* Use the following tax rate schedule for single taxpayers, and create a table of values that could be used to find the following limits, if they exist. Let x represent the amount of taxable income, and let $T(x)$ represent the tax due.

(a) $\lim\limits_{x \to 29{,}050^-} T(x)$
(b) $\lim\limits_{x \to 29{,}050^+} T(x)$
(c) $\lim\limits_{x \to 29{,}050} T(x)$

Schedule X—Use if your filing status is **Single**

If your taxable income is:		The tax is:	of the amount over—
Over—	But not over—		
$0	$7,150	- - - - - - 10%	$0
7,150	29,050	$715.00 + 15%	7,150
29,050	70,350	4,000.00 + 25%	29,050
70,350	146,750	14,325.00 + 28%	70,350
146,750	319,100	35,717.00 + 33%	146,750
319,100	- - - - -	92,592.50 + 35%	319,100

Source: Internal Revenue Service, 2004, Form 1040 Instructions

60. *Parking costs* The Ace Parking Garage charges $5.00 for parking for 2 hours or less, and $1.50 for each extra hour or part of an hour after the 2-hour minimum. The parking charges for the first 5 hours could be written as a function of the time as follows:

$$f(t) = \begin{cases} \$5.00 & \text{if } 0 < t \leq 2 \\ \$6.50 & \text{if } 2 < t \leq 3 \\ \$8.00 & \text{if } 3 < t \leq 4 \\ \$9.50 & \text{if } 4 < t \leq 5 \end{cases}$$

(a) Find $\lim\limits_{t \to 1} f(t)$, if it exists.
(b) Find $\lim\limits_{t \to 2} f(t)$, if it exists.

61. *Municipal water rates* The Corner Water Corp. of Shippenville, Pennsylvania has the following rates per 1000 gallons of water used.

Usage (x)	Cost per 1000 Gallons (C(x))
First 10,000 gallons	$7.98
Next 110,000 gallons	6.78
Over 120,000 gallons	5.43

If Corner Water has a monthly service fee of $3.59, write a function $C = C(x)$ that models the charges (where x is thousands of gallons) and find $\lim\limits_{x \to 10} C(x)$ (that is, as usage approaches 10,000 gallons).

62. *Phone card charges* A certain calling card costs 3.7 cents per minute to make a call. However, when the card is used at a pay phone there is a 10-minute charge for the first minute of the call, and then the regular charge thereafter. If $C = C(t)$ is the charge from a pay phone for a call lasting t minutes, create a table of charges for calls lasting close to 1 minute and use it to find the following limits, if they exist.

(a) $\lim\limits_{t \to 1^-} C(t)$ (b) $\lim\limits_{t \to 1^+} C(t)$ (c) $\lim\limits_{t \to 1} C(t)$

Dow Jones Industrial Average **The graph in the following figure shows the Dow Jones Industrial Average (DJIA) at 1-minute intervals for Monday, October 5, 2004. Use the graph for Problems 63 and 64, with t as the time of day and $D(t)$ as the DJIA at time t.**

63. Estimate $\lim\limits_{t \to 9:30AM^+} D(t)$, if it exists. Explain what this limit corresponds to.

64. Estimate $\lim\limits_{t \to 4:00PM^-} D(t)$, if it exists. Explain what this limit corresponds to.

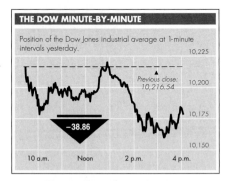

Source: Bloomberg Financial Markets, *The New York Times*, October 6, 2004. Copyright © 2004. The New York Times Co. Reprinted by permission.

Farm workers The percent of U.S. workers in farm occupations during certain years is shown in the table.

Year	Percent	Year	Percent	Year	Percent
1820	71.8	1930	21.2	1980	2.7
1850	63.7	1940	17.4	1985	2.8
1870	53	1950	11.6	1990	2.4
1900	37.5	1960	6.1	1994	2.5
1920	27	1970	3.6	2002	2.5

Source: Bureau of Labor Statistics

Assume that the percent of U.S. workers in farm occupations can be modeled with the function

$$f(t) = 1000 \cdot \frac{-7.4812t + 1560.2}{1.2882t^2 - 122.18t + 21{,}483}$$

where *t* is the number of years past 1800. (A graph of *f(t)* along with the data in the table is shown in the accompanying figure.) Use the table and equation in Problems 65 and 66.

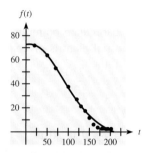

65. (a) Find $\lim\limits_{t \to 210} f(t)$, if it exists.
    (b) What does this limit predict?
    (c) Is the equation accurate as $t \to 210$? Explain.

66. (a) Find $\lim\limits_{t \to 140} f(t)$, if it exists.
    (b) What does this limit predict?
    (c) Is the equation accurate as $t \to 140$? Explain.

9.2

Continuous Functions; Limits at Infinity

OBJECTIVES

- To determine whether a function is continuous or discontinuous at a point
- To determine where a function is discontinuous
- To find limits at infinity and horizontal asymptotes

◗ Application Preview

Suppose that a friend of yours and her husband have a taxable income of $117,250, and she tells you that she doesn't want to make any more money because that would put them in a higher tax bracket. She makes this statement because the tax rate schedule for married taxpayers filing a joint return (shown in the table) appears to have a jump in taxes for taxable income at $117,250.

Schedule Y-1—If your filing status is
Married filing jointly or Qualifying widow(er)

If your taxable income is: Over—	But not over—	The tax is:	of the amount over—
$0	$14,300	- - - - - - 10%	$0
14,300	58,100	$1,430.00 + 15%	14,300
58,100	117,250	8,000.00 + 25%	58,100
117,250	178,650	22,787.50 + 28%	117,250
178,650	319,100	39,979.50 + 33%	178,650
319,100	- - - - -	86,328.00 + 35%	319,100

Source: Internal Revenue Service, 2004, Form 1040 Instructions

To see whether the couple's taxes would jump to some higher level, we will write the function that gives income tax for married taxpayers as a function of taxable income and show that the function is **continuous** (see Example 3). That is, we will see that the tax paid does not jump at $117,250 even though the tax on income above $117,250 is collected at a higher rate. In this section, we will show how to determine whether a function is continuous, and we will investigate some different types of discontinuous functions.

We have found that $f(c)$ is the same as the limit as $x \to c$ for any polynomial function $f(x)$ and any real number c. Any function for which this special property holds is called a **continuous function.** The graphs of such functions can be drawn without lifting the pencil from the paper, and graphs of others may have holes, vertical asymptotes, or jumps that make it impossible to draw them without lifting the pencil. Even though a function may not be continuous everywhere, it is likely to have some points where the limit of the function as $x \to c$ is the same as $f(c)$. In general, we define continuity of a function at the value $x = c$ as follows.

Continuity at a Point

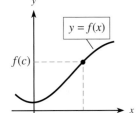

The function f is **continuous at $x = c$** if *all* of the following conditions are satisfied.

1. $f(c)$ exists 2. $\lim\limits_{x \to c} f(x)$ exists 3. $\lim\limits_{x \to c} f(x) = f(c)$

The figure at the left illustrates these three conditions.
If one or more of the conditions above do not hold, we say the function is **discontinuous at $x = c$.**

Figure 9.11 shows graphs of some functions that are discontinuous at $x = 2$.

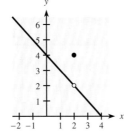

(a) $f(x) = \dfrac{1}{x - 2}$

$\lim\limits_{x \to 2} f(x)$ and $f(2)$ do not exist.

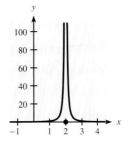

(b) $f(x) = \dfrac{x^3 - 2x^2 - x + 2}{x - 2}$

$f(2)$ does not exist.

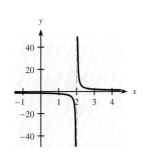

(c) $f(x) = \begin{cases} 4 - x & \text{if } x \neq 2 \\ 4 & \text{if } x = 2 \end{cases}$

$\lim\limits_{x \to 2} f(x) = 2 \neq 4 = f(2)$

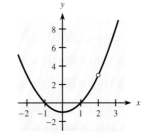

(d) $f(x) = \begin{cases} 1/(x - 2)^2 & \text{if } x \neq 2 \\ 0 & \text{if } x = 2 \end{cases}$

$\lim\limits_{x \to 2} f(x)$ does not exist.

Figure 9.11

In the previous section, we saw that if f is a polynomial function, then $\lim\limits_{x \to c} f(x) = f(c)$ for every real number c, and also that $\lim\limits_{x \to c} h(x) = h(c)$ if $h(x) = \dfrac{f(x)}{g(x)}$ is a rational function and $g(c) \neq 0$. Thus, by definition, we have the following.

Every polynomial function is continuous for all real numbers.

Every rational function is continuous at all values of x except those that make the denominator 0.

● EXAMPLE 1 *Discontinuous Functions*

For what values of x, if any, are the following functions continuous?

(a) $h(x) = \dfrac{3x + 2}{4x - 6}$

(b) $f(x) = \dfrac{x^2 - x - 2}{x^2 - 4}$

Solution

(a) This is a rational function, so it is continuous for all values of x except for those that make the denominator, $4x - 6$, equal to 0. Because $4x - 6 = 0$ at $x = 3/2$, $h(x)$ is continuous for all real numbers except $x = 3/2$. Figure 9.12(a) shows a vertical asymptote at $x = 3/2$.

(b) This is a rational function, so it is continuous everywhere except where the denominator is 0. To find the zeros of the denominator, we factor $x^2 - 4$.

$$f(x) = \frac{x^2 - x - 2}{x^2 - 4} = \frac{x^2 - x - 2}{(x - 2)(x + 2)}$$

Because the denominator is 0 for $x = 2$ and for $x = -2$, $f(2)$ and $f(-2)$ do not exist (recall that division by 0 is undefined). Thus the function is discontinuous at $x = 2$ and $x = -2$. The graph of this function (see Figure 9.12(b)) shows a hole at $x = 2$ and a vertical asymptote at $x = -2$.

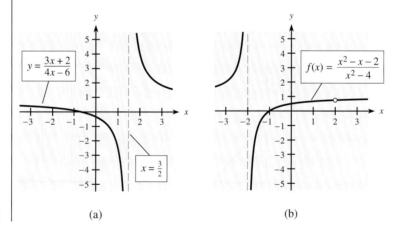

Figure 9.12   (a)   (b)

● **Checkpoint**

1. Find any x-values where the following functions are discontinuous.

(a) $f(x) = x^3 - 3x + 1$ (b) $g(x) = \dfrac{x^3 - 1}{(x - 1)(x + 2)}$

If the pieces of a piecewise defined function are polynomials, the only values of x where the function might be discontinuous are those at which the definition of the function changes.

● **EXAMPLE 2** *Piecewise Defined Functions*

Determine the values of x, if any, for which the following functions are discontinuous.

(a) $g(x) = \begin{cases} (x + 2)^3 + 1 & \text{if } x \le -1 \\ 3 & \text{if } x > -1 \end{cases}$ (b) $f(x) = \begin{cases} 4 - x^2 & \text{if } x < 2 \\ x - 2 & \text{if } x \ge 2 \end{cases}$

Solution

(a) $g(x)$ is a piecewise defined function in which each part is a polynomial. Thus, to see whether a discontinuity exists, we need only check the value of x for which the definition of the function changes—that is, at $x = -1$. Note that $x = -1$ satisfies $x \le -1$, so $g(-1) = (-1 + 2)^3 + 1 = 2$. Because $g(x)$ is defined differently for $x < -1$ and $x > -1$, we use left- and right-hand limits. For $x \to -1^-$, we know that $x < -1$, so $g(x) = (x + 2)^3 + 1$:

$$\lim_{x \to -1^-} g(x) = \lim_{x \to -1^-} [(x + 2)^3 + 1] = (-1 + 2)^3 + 1 = 2$$

Similarly, for $x \to -1^+$, we know that $x > -1$, so $g(x) = 3$:

$$\lim_{x \to -1^+} g(x) = \lim_{x \to -1^+} 3 = 3$$

Because the left- and right-hand limits differ, $\lim_{x \to -1} g(x)$ does not exist, so $g(x)$ is discontinuous at $x = -1$. This result is confirmed by examining the graph of g, shown in Figure 9.13.

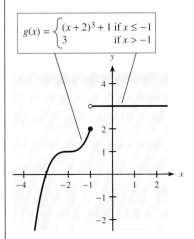

$$g(x) = \begin{cases} (x + 2)^3 + 1 & \text{if } x \le -1 \\ 3 & \text{if } x > -1 \end{cases}$$

Figure 9.13

(b) As with $g(x)$, $f(x)$ is continuous everywhere except perhaps at $x = 2$, where the definition of $f(x)$ changes. Because $x = 2$ satisfies $x \ge 2$, $f(2) = 2 - 2 = 0$. The left- and right-hand limits are

$$\lim_{x \to 2^-} f(x) = \lim_{x \to 2^-} (4 - x^2) = 4 - 2^2 = 0$$

and

$$\lim_{x \to 2^+} f(x) = \lim_{x \to 2^+} (x - 2) = 2 - 2 = 0$$

Because the right- and left-hand limits are equal, we conclude that $\lim_{x \to 2} f(x) = 0$. The limit is equal to the functional value

$$\lim_{x \to 2} f(x) = f(2)$$

so we conclude that f is continuous at $x = 2$ and thus f is continuous for all values of x. This result is confirmed by the graph of f, shown in Figure 9.14.

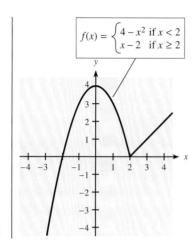

$$f(x) = \begin{cases} 4 - x^2 & \text{if } x < 2 \\ x - 2 & \text{if } x \geq 2 \end{cases}$$

Figure 9.14

◑ EXAMPLE 3 *Taxes **(Application Preview)***

The tax rate schedule for married taxpayers filing a joint return (shown in the table) appears to have a jump in taxes for taxable income at $117,250.

Schedule Y-1—If your filing status is
Married filing jointly or **Qualifying widow(er)**

If your taxable income is:		The tax is:	of the amount over—
Over—	But not over—		
$0	$14,300	------ 10%	$0
14,300	58,100	$1,430.00 + 15%	14,300
58,100	117,250	8,000.00 + 25%	58,100
117,250	178,650	22,787.50 + 28%	117,250
178,650	319,100	39,979.50 + 33%	178,650
319,100	-----	86,328.00 + 35%	319,100

Source: Internal Revenue Service, 2004, Form 1040 Instructions

(a) Use the table and write the function that gives income tax for married taxpayers as a function of taxable income, x.
(b) Is the function in part (a) continuous at $x = 117,250$?
(c) A married friend of yours and her husband have a taxable income of $117,250, and she tells you that she doesn't want to make any more money because doing so would put her in a higher tax bracket. What would you tell her to do if she is offered a raise?

Solution
(a) The function that gives the tax due for married taxpayers is

$$T(x) = \begin{cases} 0.10x & \text{if} & 0 \leq x \leq 14,300 \\ 1430 + 0.15(x - 14,300) & \text{if} & 14,300 < x \leq 58,100 \\ 8000 + 0.25(x - 58,100) & \text{if} & 58,100 < x \leq 117,250 \\ 22,787.50 + 0.28(x - 117,250) & \text{if} & 117,250 < x \leq 178,650 \\ 39,979.50 + 0.33(x - 178,650) & \text{if} & 178,650 < x \leq 319,100 \\ 86,328 + 0.35(x - 319,100) & \text{if} & x > 319,100 \end{cases}$$

(b) We examine the three conditions for continuity at $x = 117,250$.
  (i) $T(117,250) = 22,787.50$, so $T(117,250)$ exists.
  (ii) Because the function is piecewise defined near 117,250, we evaluate $\lim\limits_{x \to 117,250} T(x)$ by evaluating one-sided limits:

From the left: $\lim\limits_{x \to 117,250^-} T(x) = \lim\limits_{x \to 117,250^-}[8000 + 0.25(x - 58,100)] = 22,787.50$

From the right: $\lim\limits_{x \to 117,250^+} T(x) = \lim\limits_{x \to 117,250^+}[22,787.50 + 0.28(x - 117,250)] = 22,787.50$

Because these one-sided limits agree, the limit exists and is

$$\lim\limits_{x \to 117,250} T(x) = 22,787.50$$

  (iii) Because $\lim\limits_{x \to 117,250} T(x) = T(117,250) = 22,787.50$, the function is continuous at 117,250.

(c) If your friend earned more than \$117,250, she and her husband would pay taxes at a higher rate on the money earned *above* the \$117,250, but it would not increase the tax rate on any income *up to* \$117,250. Thus she should take any raise that's offered.

● *Checkpoint*

2. If $f(x)$ and $g(x)$ are polynomials, $h(x) = \begin{cases} f(x) & \text{if } x \le a \\ g(x) & \text{if } x > a \end{cases}$ is continuous everywhere except perhaps at _____.

Limits at Infinity

We noted in Section 2.4, "Special Functions and Their Graphs," that the graph of $y = 1/x$ has a vertical asymptote at $x = 0$ (shown in Figure 9.15(a)). By graphing $y = 1/x$ and evaluating the function for very large x values we can see that $y = 1/x$ never becomes negative for positive x-values regardless of how large the x-value is. Although no value of x makes $1/x$ equal to 0, it is easy to see that $1/x$ approaches 0 as x gets very large. This is denoted by

$$\lim\limits_{x \to +\infty} \frac{1}{x} = 0$$

and means that the line $y = 0$ (the x-axis) is a horizontal asymptote for $y = 1/x$. We also see that $y = 1/x$ approaches 0 as x decreases without bound, and we denote this by

$$\lim\limits_{x \to -\infty} \frac{1}{x} = 0$$

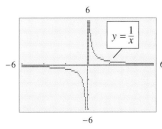

(a)

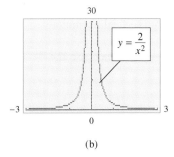

(b)

Figure 9.15

These limits for $f(x) = 1/x$ can also be established with numerical tables.

x	$f(x) = 1/x$	x	$f(x) = 1/x$
100	0.01	-100	-0.01
100,000	0.00001	$-100,000$	-0.00001
100,000,000	0.00000001	$-100,000,000$	-0.00000001
$\downarrow$	$\downarrow$	$\downarrow$	$\downarrow$
$+\infty$	0	$-\infty$	0

$$\lim\limits_{x \to +\infty} \frac{1}{x} = 0 \qquad\qquad \lim\limits_{x \to -\infty} \frac{1}{x} = 0$$

We can use the graph of $y = 2/x^2$ in Figure 9.15(b) to see that the x-axis ($y = 0$) is a horizontal asymptote and that

$$\lim\limits_{x \to +\infty} \frac{2}{x^2} = 0 \qquad \text{and} \qquad \lim\limits_{x \to -\infty} \frac{2}{x^2} = 0$$

By using graphs and/or tables of values, we can generalize the results for the functions shown in Figure 9.15 and conclude the following.

Limits at Infinity

If c is any constant, then

1. $\displaystyle\lim_{x\to+\infty} c = c$ and $\displaystyle\lim_{x\to-\infty} c = c$.

2. $\displaystyle\lim_{x\to+\infty} \frac{c}{x^p} = 0$, where $p > 0$.

3. $\displaystyle\lim_{x\to-\infty} \frac{c}{x^n} = 0$, where $n > 0$ is any integer.

In order to use these properties for finding the limits of rational functions as x approaches $+\infty$ or $-\infty$, we first divide each term of the numerator and denominator by the highest power of x present and then determine the limit of the resulting expression.

● **EXAMPLE 4** *Limits at Infinity*

Find each of the following limits, if they exist.

(a) $\displaystyle\lim_{x\to+\infty} \frac{2x-1}{x+2}$ (b) $\displaystyle\lim_{x\to-\infty} \frac{x^2+3}{1-x}$

Solution

(a) The highest power of x present is x^1, so we divide each term in the numerator and denominator by x and then use the properties for limits at infinity.

$$\lim_{x\to+\infty} \frac{2x-1}{x+2} = \lim_{x\to+\infty} \frac{\dfrac{2x}{x} - \dfrac{1}{x}}{\dfrac{x}{x} + \dfrac{2}{x}} = \lim_{x\to+\infty} \frac{2 - \dfrac{1}{x}}{1 + \dfrac{2}{x}}$$

$$= \frac{2-0}{1+0} = 2 \quad \text{(by Properties 1 and 2)}$$

Figure 9.16(a) shows the graph of this function with the y-coordinates of the graph approaching 2 as x approaches $+\infty$ and as x approaches $-\infty$. That is, $y = 2$ is a horizontal asymptote. Note also that there is a discontinuity (vertical asymptote) where $x = -2$.

(b) We divide each term in the numerator and denominator by x^2 and then use the properties.

$$\lim_{x\to-\infty} \frac{x^2+3}{1-x} = \lim_{x\to-\infty} \frac{\dfrac{x^2}{x^2} + \dfrac{3}{x^2}}{\dfrac{1}{x^2} - \dfrac{x}{x^2}} = \lim_{x\to-\infty} \frac{1 + \dfrac{3}{x^2}}{\dfrac{1}{x^2} - \dfrac{1}{x}}$$

This limit does not exist because the numerator approaches 1 and the denominator approaches 0 through positive values. Thus

$$\frac{x^2+3}{1-x} \to +\infty \text{ as } x \to -\infty$$

The graph of this function, shown in Figure 9.16(b), has y-coordinates that increase without bound as x approaches $-\infty$ and that decrease without bound as x approaches $+\infty$. (There is no horizontal asymptote.) Note also that there is a vertical asymptote at $x = 1$.

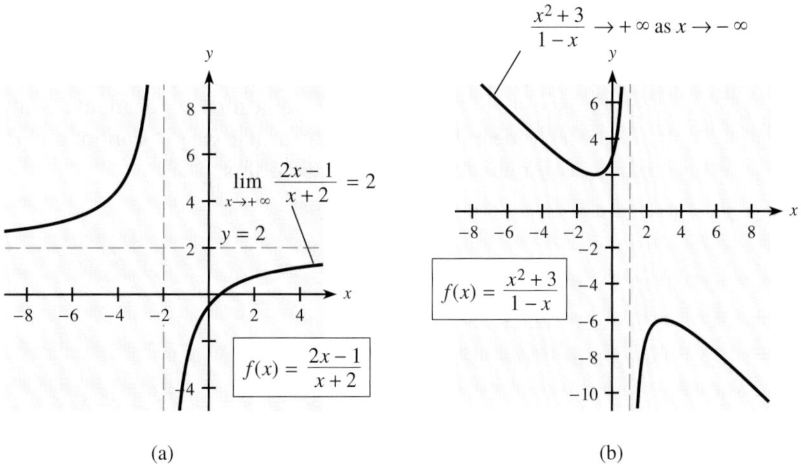

$$\frac{x^2 + 3}{1 - x} \to +\infty \text{ as } x \to -\infty$$

$$\lim_{x \to \infty} \frac{2x - 1}{x + 2} = 2$$

$$y = 2$$

$$f(x) = \frac{2x - 1}{x + 2}$$

$$f(x) = \frac{x^2 + 3}{1 - x}$$

Figure 9.16

(a)

(b)

In our work with limits at infinity, we have mentioned horizontal asymptotes several times. The connection between these concepts follows.

Limits at Infinity and Horizontal Asymptotes

If $\lim_{x \to \infty} f(x) = b$ or $\lim_{x \to -\infty} f(x) = b$, where b is a constant, then the line $y = b$ is a horizontal asymptote for the graph of $y = f(x)$. Otherwise, $y = f(x)$ has no horizontal asymptotes.

● **Checkpoint**

3. (a) Evaluate $\lim_{x \to +\infty} \dfrac{x^2 - 4}{2x^2 - 7}$.

(b) What does part (a) say about horizontal asymptotes for $f(x) = (x^2 - 4)/(2x^2 - 7)$?

Calculator Note

We can use the graphing and table features of a graphing calculator to help locate and investigate discontinuities and limits at infinity (horizontal asymptotes). A graphing calculator can be used to focus our attention on a possible discontinuity and to support or suggest appropriate algebraic calculations. ■

● **EXAMPLE 5** *Limits with Technology*

Use a graphing utility to investigate the continuity of the following functions.

(a) $f(x) = \dfrac{x^2 + 1}{x + 1}$ (b) $g(x) = \dfrac{x^2 - 2x - 3}{x^2 - 1}$

(c) $h(x) = \dfrac{|x + 1|}{x + 1}$ (d) $k(x) = \begin{cases} \dfrac{-x^2}{2} - 2x & \text{if } x \le -1 \\ \dfrac{x}{2} + 2 & \text{if } x > -1 \end{cases}$

Solution

(a) Figure 9.17(a) shows that $f(x)$ has a discontinuity (vertical asymptote) near $x = -1$. Because $f(-1)$ DNE, we know that $f(x)$ is not continuous at $x = -1$.

(b) Figure 9.17(b) shows that $g(x)$ is discontinuous (vertical asymptote) near $x = 1$, and this looks like the only discontinuity. However, the denominator of $g(x)$ is zero at $x = 1$ and $x = -1$, so $g(x)$ must have discontinuities at both of these x-values. Evaluating or using the table feature confirms that $x = -1$ is a discontinuity (a hole, or missing point). The figure also shows a horizontal asymptote; evaluation of $\lim_{x \to \infty} g(x)$ confirms this is the line $y = 1$.

(c) Figure 9.17(c) shows a discontinuity (jump) at $x = -1$. We also see that $h(-1)$ DNE, which confirms the observations from the graph.

(d) The graph in Figure 9.17(d) appears to be continuous. The only "suspicious" x-value is $x = -1$, where the formula for $k(x)$ changes. Evaluating $k(-1)$ and examining a table near $x = -1$ indicates that $k(x)$ is continuous there. Algebraic evaluations of the two one-sided limits confirm this.

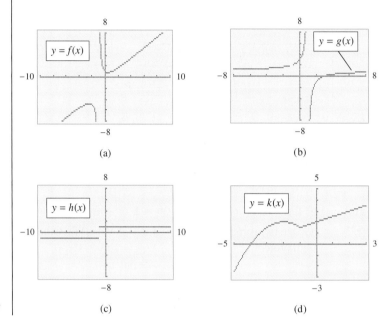

(a)

(b)

Figure 9.17

(c)

(d)

Summary

The following information is useful in discussing continuity of functions.

A. A polynomial function is continuous everywhere.

B. A rational function is a function of the form $\dfrac{f(x)}{g(x)}$, where $f(x)$ and $g(x)$ are polynomials.

1. If $g(x) \neq 0$ at any value of x, the function is continuous everywhere.
2. If $g(c) = 0$, the function is discontinuous at $x = c$.
   (a) If $g(c) = 0$ and $f(c) \neq 0$, then there is a vertical asymptote at $x = c$.
   (b) If $g(c) = 0$ and $\lim\limits_{x \to c} \dfrac{f(x)}{g(x)} = L$, then the graph has a missing point at (c, L).

C. A piecewise defined function *may* have a discontinuity at any x-value where the function changes its formula. One-sided limits must be used to see whether the limit exists.

The following steps are useful when we are evaluating limits at infinity for a rational function $f(x) = p(x)/q(x)$.

1. Divide both $p(x)$ and $q(x)$ by the highest power of x found in either polynomial.
2. Use the properties of limits at infinity to complete the evaluation.

● *Checkpoint Solutions*

1. (a) This is a polynomial function, so it is continuous at all values of x (discontinuous at none).
   (b) This is a rational function. It is discontinuous at $x = 1$ and $x = -2$ because these values make its denominator 0.
2. $x = a$.

3. (a) $\lim\limits_{x \to +\infty} \dfrac{x^2 - 4}{2x^2 - 7} = \lim\limits_{x \to +\infty} \dfrac{1 - \dfrac{4}{x^2}}{2 - \dfrac{7}{x^2}} = \dfrac{1 - 0}{2 - 0} = \dfrac{1}{2}$

(b) The line $y = 1/2$ is a horizontal asymptote.

9.2 Exercises

Problems 1 and 2 refer to the following figure. For each given x-value, use the figure to determine whether the function is continuous or discontinuous at that x-value. If the function is discontinuous, state which of the three conditions that define continuity is not satisfied.

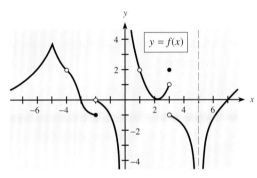

1. (a) $x = -5$ (b) $x = 1$ (c) $x = 3$ (d) $x = 0$
2. (a) $x = 2$ (b) $x = -4$ (c) $x = -2$ (d) $x = 5$

In Problems 3–8, determine whether each function is continuous or discontinuous at the given x-value. Examine the three conditions in the definition of continuity.

3. $f(x) = \dfrac{x^2 - 4}{x - 2}$, $x = -2$

4. $y = \dfrac{x^2 - 9}{x + 3}$, $x = 3$

5. $y = \dfrac{x^2 - 9}{x + 3}$, $x = -3$

6. $f(x) = \dfrac{x^2 - 4}{x - 2}$, $x = 2$

7. $f(x) = \begin{cases} x - 3 & \text{if } x \le 2 \\ 4x - 7 & \text{if } x > 2 \end{cases}$, $x = 2$

8. $f(x) = \begin{cases} x^2 + 1 & \text{if } x \le 1 \\ 2x^2 - 1 & \text{if } x > 1 \end{cases}$, $x = 1$

In Problems 9–16, determine whether the given function is continuous. If it is not, identify where it is discontinuous and which condition fails to hold. You can verify your conclusions by graphing each function with a graphing utility, if one is available.

9. $f(x) = 4x^2 - 1$

10. $y = 5x^2 - 2x$

11. $g(x) = \dfrac{4x^2 + 3x + 2}{x + 2}$

12. $y = \dfrac{4x^2 + 4x + 1}{x + 1/2}$

13. $y = \dfrac{x}{x^2 + 1}$

14. $y = \dfrac{2x - 1}{x^2 + 3}$

15. $f(x) = \begin{cases} 3 & \text{if } x \le 1 \\ x^2 + 2 & \text{if } x > 1 \end{cases}$

16. $f(x) = \begin{cases} x^3 + 1 & \text{if } x \le 1 \\ 2 & \text{if } x > 1 \end{cases}$

In Problems 17–20, use the trace and table features of a graphing utility to investigate whether each of the following functions has any discontinuities.

17. $y = \dfrac{x^2 - 5x - 6}{x + 1}$

18. $y = \dfrac{x^2 - 5x + 4}{x - 4}$

19. $f(x) = \begin{cases} x - 4 & \text{if } x \le 3 \\ x^2 - 8 & \text{if } x > 3 \end{cases}$

20. $f(x) = \begin{cases} x^2 + 4 & \text{if } x \ne 1 \\ 8 & \text{if } x = 1 \end{cases}$

Each of Problems 21–24 contains a function and its graph. For each problem, answer parts (a) and (b).
(a) Use the graph to determine, as well as you can,
   (i) vertical asymptotes, (ii) $\lim\limits_{x \to +\infty} f(x)$,
   (iii) $\lim\limits_{x \to -\infty} f(x)$, (iv) horizontal asymptotes.
(b) Check your conclusions in (a) by using the functions to determine items (i)–(iv) analytically.

21. $f(x) = \dfrac{8}{x + 2}$

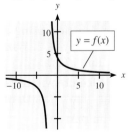

22. $f(x) = \dfrac{x - 3}{x - 2}$

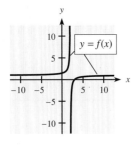

23. $f(x) = \dfrac{2(x + 1)^3(x + 5)}{(x - 3)^2(x + 2)^2}$

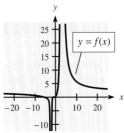

24. $f(x) = \dfrac{4x^2}{x^2 - 4x + 4}$

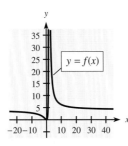

In Problems 25–32, complete (a) and (b).
(a) Use analytic methods to evaluate each limit.
(b) What does the result from (a) tell you about horizontal asymptotes?
You can verify your conclusions by graphing the functions with a graphing utility, if one is available.

25. $\lim\limits_{x \to +\infty} \dfrac{3}{x + 1}$

26. $\lim\limits_{x \to -\infty} \dfrac{4}{x^2 - 2x}$

27. $\lim\limits_{x \to +\infty} \dfrac{x^3 - 1}{x^3 + 4}$

28. $\lim\limits_{x \to -\infty} \dfrac{3x^2 + 2}{x^2 - 4}$

29. $\lim\limits_{x \to -\infty} \dfrac{5x^3 - 4x}{3x^3 - 2}$

30. $\lim\limits_{x \to +\infty} \dfrac{4x^2 + 5x}{x^2 - 4x}$

31. $\lim\limits_{x \to +\infty} \dfrac{3x^2 + 5x}{6x + 1}$

32. $\lim\limits_{x \to -\infty} \dfrac{5x^3 - 8}{4x^2 + 5x}$

In Problems 33 and 34, use a graphing utility to complete (a) and (b).
(a) Graph each function using a window with $0 \le x \le 300$ and $-2 \le y \le 2$. What does the graph indicate about $\lim\limits_{x \to +\infty} f(x)$?
(b) Use the table feature with x-values larger than 10,000 to investigate $\lim\limits_{x \to +\infty} f(x)$. Does the table support your conclusions in part (a)?

33. $f(x) = \dfrac{x^2 - 4}{3 + 2x^2}$

34. $f(x) = \dfrac{5x^3 - 7x}{1 - 3x^3}$

In Problems 35 and 36, complete (a)–(c). Use analytic methods to find (a) any points of discontinuity and (b) limits as $x \to +\infty$ and $x \to -\infty$. (c) Then explain why, for these functions, a graphing utility is better as a support tool for the analytic methods than as the primary tool for investigation.

35. $f(x) = \dfrac{1000x - 1000}{x + 1000}$

36. $f(x) = \dfrac{3000x}{4350 - 2x}$

For Problems 37 and 38, let
$$f(x) = \frac{a_n x^n + a_{n-1}x^{n-1} + \cdots + a_1 x + a_0}{b_m x^m + b_{m-1}x^{m-1} + \cdots + b_1 x + b_0}$$
be a rational function.

37. If $m = n$, show that $\lim\limits_{x \to \infty} f(x) = \dfrac{a_n}{b_n}$, and hence that $y = \dfrac{a_n}{b_n}$ is a horizontal asymptote.

38. (a) If $m > n$, show that $\lim\limits_{x \to \infty} f(x) = 0$, and hence that $y = 0$ is a horizontal asymptote.
(b) If $m < n$, find $\lim\limits_{x \to \infty} f(x)$. What does this say about horizontal asymptotes?

APPLICATIONS

39. *Sales volume* Suppose that the weekly sales volume (in thousands of units) for a product is given by
$$y = \frac{32}{(p + 8)^{2/5}}$$
where p is the price in dollars per unit. Is this function continuous
(a) for all values of p? (b) at $p = 24$?
(c) for all $p \ge 0$?
(d) What is the domain for this application?

40. *Worker productivity* Suppose that the average number of minutes M that it takes a new employee to assemble one unit of a product is given by
$$M = \frac{40 + 30t}{2t + 1}$$
where t is the number of days on the job. Is this function continuous
(a) for all values of t? (b) at $t = 14$?
(c) for all $t \ge 0$?
(d) What is the domain for this application?

41. *Demand* Suppose that the demand for a product is defined by the equation
$$p = \frac{200,000}{(q + 1)^2}$$
where p is the price and q is the quantity demanded.
(a) Is this function discontinuous at any value of q? What value?

(b) Because q represents quantity, we know that $q \geq 0$. Is this function continuous for $q \geq 0$?

42. *Advertising and sales* The sales volume y (in thousands of dollars) is related to advertising expenditures x (in thousands of dollars) according to

$$y = \frac{200x}{x + 10}$$

(a) Is this function discontinuous at any points?
(b) Advertising expenditures x must be nonnegative. Is this function continuous for these values of x?

43. *Annuities* If an annuity makes an infinite series of equal payments at the end of the interest periods, it is called a **perpetuity.** If a lump sum investment of A_n is needed to result in n periodic payments of R when the interest rate per period is i, then

$$A_n = R\left[\frac{1 - (1 + i)^{-n}}{i}\right]$$

(a) Evaluate $\lim_{n \to \infty} A_n$ to find a formula for the lump sum payment for a perpetuity.
(b) Find the lump sum investment needed to make payments of $100 per month in perpetuity if interest is 12%, compounded monthly.

44. *Response to adrenalin* Experimental evidence suggests that the response y of the body to the concentration x of injected adrenalin is given by

$$y = \frac{x}{a + bx}$$

where a and b are experimental constants.
(a) Is this function continuous for all x?
(b) On the basis of your conclusion in (a) and the fact that in reality $x \geq 0$ and $y \geq 0$, must a and b be both positive, be both negative, or have opposite signs?

45. *Cost-benefit* Suppose that the cost C of removing p percent of the impurities from the waste water in a manufacturing process is given by

$$C(p) = \frac{9800p}{101 - p}$$

Is this function continuous for all those p-values for which the problem makes sense?

46. *Pollution* Suppose that the cost C of removing p percent of the particulate pollution from the exhaust gases at an industrial site is given by

$$C(p) = \frac{8100p}{100 - p}$$

Describe any discontinuities for $C(p)$. Explain what each discontinuity means.

47. *Pollution* The percent p of particulate pollution that can be removed from the smokestacks of an industrial plant by spending C dollars is given by

$$p = \frac{100C}{7300 + C}$$

Find the percent of the pollution that could be removed if spending C were allowed to increase without bound. Can 100% of the pollution be removed? Explain.

48. *Cost-benefit* The percent p of impurities that can be removed from the waste water of a manufacturing process at a cost of C dollars is given by

$$p = \frac{100C}{8100 + C}$$

Find the percent of the impurities that could be removed if cost were no object (that is, if cost were allowed to increase without bound). Can 100% of the impurities be removed? Explain.

49. *Federal income tax* The tax owed by a married couple filing jointly and their tax rates can be found in the following tax rate schedule.

Schedule Y-1—If your filing status is
Married filing jointly or Qualifying widow(er)

If your taxable income is:		The tax is:	of the amount over—
Over—	But not over—		
$0	$14,300	------ 10%	$0
14,300	58,100	$1,430.00 + 15%	14,300
58,100	117,250	8,000.00 + 25%	58,100
117,250	178,650	22,787.50 + 28%	117,250
178,650	319,100	39,979.50 + 33%	178,650
319,100	-----	86,328.00 + 35%	319,100

Source: Internal Revenue Service, 2004, Form 1040 Instructions

From this schedule, the tax rate $R(x)$ is a function of taxable income x, as follows.

$$R(x) = \begin{cases} 0.10 & \text{if} & 0 \leq x \leq 14,300 \\ 0.15 & \text{if} & 14,300 < x \leq 58,100 \\ 0.25 & \text{if} & 58,100 < x \leq 117,250 \\ 0.28 & \text{if} & 117,250 < x \leq 178,650 \\ 0.33 & \text{if} & 178,650 < x \leq 319,100 \\ 0.35 & \text{if} & x > 319,100 \end{cases}$$

Identify any discontinuities in $R(x)$.

50. *Calories and temperature* Suppose that the number of calories of heat required to raise 1 gram of water (or ice) from −40°C to x°C is given by

$$f(x) = \begin{cases} \frac{1}{2}x + 20 & \text{if } -40 \leq x < 0 \\ x + 100 & \text{if } 0 \leq x \end{cases}$$

(a) What can be said about the continuity of the function $f(x)$?

(b) What happens to water at 0°C that accounts for the behavior of the function at 0°C?

51. *Electrical usage costs* The monthly charge in dollars for x kilowatt-hours (kWh) of electricity used by a residential consumer of Excelsior Electric Membership Corporation from November through June is given by the function

$$C(x) = \begin{cases} 10 + 0.094x & \text{if} & 0 \leq x \leq 100 \\ 19.40 + 0.075(x - 100) & \text{if} & 100 < x \leq 500 \\ 49.40 + 0.05(x - 500) & \text{if} & x > 500 \end{cases}$$

(a) What is the monthly charge if 1100 kWh of electricity is consumed in a month?

(b) Find $\lim\limits_{x \to 100} C(x)$ and $\lim\limits_{x \to 500} C(x)$, if the limits exist.

(c) Is C continuous at $x = 100$ and at $x = 500$?

52. *Postage costs* First-class postage is 37 cents for the first ounce or part of an ounce that a letter weighs and is an additional 23 cents for each additional ounce or part of an ounce above 1 ounce. Use the table or graph of the postage function, $f(x)$, to determine the following.

(a) $\lim\limits_{x \to 2.5} f(x)$

(b) $f(2.5)$

(c) Is $f(x)$ continuous at 2.5?

(d) $\lim\limits_{x \to 4} f(x)$

(e) $f(4)$

(f) Is $f(x)$ continuous at 4?

Weight x	Postage f(x)
$0 < x \leq 1$	$0.37
$1 < x \leq 2$	0.60
$2 < x \leq 3$	0.83
$3 < x \leq 4$	1.06
$4 < x \leq 5$	1.29

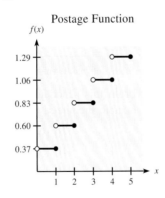

Postage Function

53. **Modeling** *Public debt of the United States* The interest paid on the public debt of the United States of America as a percent of federal expenditures for selected years is shown in the following table.

Year	Interest Paid as a Percent of Federal Expenditures	Year	Interest Paid as a Percent of Federal Expenditures
1930	0	1975	9.8
1940	10.5	1980	12.7
1950	13.4	1985	18.9
1955	9.4	1990	21.1
1960	10.0	1995	22.0
1965	9.6	2000	20.3
1970	9.9	2003	14.7

Source: Bureau of Public Debt, Department of the Treasury

If t is the number of years past 1900, use the table to complete the following.

(a) Use the data in the table to find a fourth-degree function $d(t)$ that models the percent of federal expenditures devoted to payment of interest on the public debt.

(b) Use $d(t)$ to predict the percent of federal expenditures devoted to payment of interest in 2009.

(c) Calculate $\lim\limits_{t \to +\infty} d(t)$.

(d) Can $d(t)$ be used to predict the percent of federal expenditures devoted to payment of interest on the public debt for large values of t? Explain.

(e) For what years can you guarantee that $d(t)$ cannot be used to predict the percent of federal expenditures devoted to payment of interest on the public debt? Explain.

54. *Students per computer* By using data from Quality Education Data Inc., Denver, CO, the number of students per computer in U.S. public schools (1983–2002) can be modeled with the function

$$f(x) = \frac{375.5 - 14.9x}{x + 0.02}$$

where x is the number of years past the school year ending in 1981.

(a) Is this function continuous for school years from 1981 onward?

(b) Find the long-range projection of this model by finding $\lim\limits_{x \to \infty} f(x)$. Explain what this tells us about the validity of the model.

9.3

OBJECTIVES

- To define and find average rates of change
- To define the derivative as a rate of change
- To use the definition of derivative to find derivatives of functions
- To use derivatives to find slopes of tangents to curves

Average and Instantaneous Rates of Change: The Derivative

◐ Application Preview

In Chapter 1, "Linear Equations and Functions," we studied linear revenue functions and defined the marginal revenue for a product as the rate of change of the revenue function. For linear revenue functions, this rate is also the slope of the line that is the graph of the revenue function. In this section, we will define **marginal revenue** as the rate of change of the revenue function, even when the revenue function is not linear.

Thus, if an oil company's revenue (in thousands of dollars) is given by

$$R = 100x - x^2, \quad x \geq 0$$

where x is the number of thousands of barrels of oil sold per day, we can find and interpret the marginal revenue when 20,000 barrels are sold (see Example 4).

We will discuss the relationship between the marginal revenue at a given point and the slope of the line tangent to the revenue function at that point. We will see how the derivative of the revenue function can be used to find both the slope of this tangent line and the marginal revenue.

Average Rates of Change

For linear functions, we have seen that the slope of the line measures the average rate of change of the function and can be found from any two points on the line. However, for a function that is not linear, the slope between different pairs of points no longer always gives the same number, but it can be interpreted as an average rate of change. We use this connection between average rates of change and slopes for linear functions to define the average rate of change for any function.

Average Rate of Change

The **average rate of change** of a function $y = f(x)$ from $x = a$ to $x = b$ is defined by

$$\text{Average rate of change} = \frac{f(b) - f(a)}{b - a}$$

The figure shows that this average rate is the same as the slope of the segment joining the points $(a, f(a))$ and $(b, f(b))$.

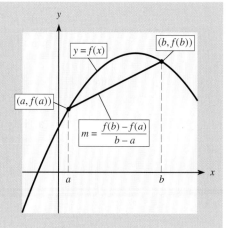

● EXAMPLE 1 *Total Cost*

Suppose a company's total cost in dollars to produce x units of its product is given by

$$C(x) = 0.01x^2 + 25x + 1500$$

Find the average rate of change of total cost for (a) the first 100 units produced (from $x = 0$ to $x = 100$) and (b) the second 100 units produced.

Solution
(a) The average rate of change of total cost from $x = 0$ to $x = 100$ units is

$$\frac{C(100) - C(0)}{100 - 0} = \frac{(0.01(100)^2 + 25(100) + 1500) - (1500)}{100}$$

$$= \frac{4100 - 1500}{100} = \frac{2600}{100} = 26 \text{ dollars per unit}$$

(b) The average rate of change of total cost from $x = 100$ to $x = 200$ units is

$$\frac{C(200) - C(100)}{200 - 100} = \frac{(0.01(200)^2 + 25(200) + 1500) - (4100)}{100}$$

$$= \frac{6900 - 4100}{100} = \frac{2800}{100} = 28 \text{ dollars per unit}$$

● **EXAMPLE 2** *Elderly in the Work Force*

Figure 9.18 shows the percents of elderly men and of elderly women in the work force in selected census years from 1890 to 2000. For the years from 1950 to 2000, find and interpret the average rate of change of the percent of (a) elderly men in the work force and (b) elderly women in the work force. (c) What caused these trends?

Elderly in the Labor Force, 1890-2000
(labor force participation rate; figs. for 1910 not available)

Figure 9.18   *Source:* Bureau of the Census, U.S. Department of Commerce

Solution
(a) From 1950 to 2000, the annual average rate of change in the percent of elderly men in the work force is

$$\frac{\text{Change in men's percent}}{\text{Change in years}} = \frac{18.6 - 41.4}{2000 - 1950} = \frac{-22.8}{50} = -0.456 \text{ percent per year}$$

This means that from 1950 to 2000, *on average,* the percent of elderly men in the work force dropped by 0.456% per year.
(b) Similarly, the average rate of change for women is

$$\frac{\text{Change in women's percent}}{\text{Change in years}} = \frac{10.0 - 7.8}{2000 - 1950} = \frac{2.2}{50} = 0.044 \text{ percent per year}$$

In like manner, this means that from 1950 to 2000, *on average,* the percent of elderly women in the work force increased by 0.044% each year.
(c) In general, from 1950 to 1990, people have been retiring earlier, but the number of women in the work force has increased dramatically.

Instantaneous Rates of Change: Velocity

Another common rate of change is velocity. For instance, if we travel 200 miles in our car over a 4-hour period, we know that we averaged 50 mph. However, during that trip there may have been times when we were traveling on an Interstate at faster than 50 mph and times when we were stopped at a traffic light. Thus, for the trip we have not only an average velocity but also instantaneous velocities (or instantaneous speeds as displayed on the speedometer). Let's see how average velocity can lead us to instantaneous velocity.

Suppose a ball is thrown straight upward at 64 feet per second from a spot 96 feet above ground level. The equation that describes the height y of the ball after x seconds is

$$y = f(x) = 96 + 64x - 16x^2$$

Figure 9.19 shows the graph of this function for $0 \leq x \leq 5$. The average velocity of the ball over a given time interval is the change in the height divided by the length of time that has passed. Table 9.4 shows some average velocities over time intervals beginning at $x = 1$.

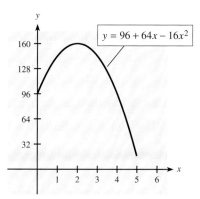

$$y = 96 + 64x - 16x^2$$

Figure 9.19

TABLE 9.4 Average Velocities

Time (seconds)			Height (feet)			Average Velocity (ft/sec)
Beginning	Ending	Change (Δx)	Beginning	Ending	Change (Δy)	($\Delta y/\Delta x$)
1	2	1	144	160	16	16/1 = 16
1	1.5	0.5	144	156	12	12/0.5 = 24
1	1.1	0.1	144	147.04	3.04	3.04/0.1 = 30.4
1	1.01	0.01	144	144.3184	0.3184	0.3184/0.01 = 31.84

In Table 9.4, the smaller the time interval, the more closely the average velocity approximates the instantaneous velocity at $x = 1$. Thus the instantaneous velocity at $x = 1$ is closer to 31.84 ft/s than to 30.4 ft/s.

If we represent the change in time by h, then the average velocity from $x = 1$ to $x = 1 + h$ approaches the instantaneous velocity at $x = 1$ as h approaches 0. (Note that h can be positive or negative.) This is illustrated in the following example.

● **EXAMPLE 3** *Velocity*

Suppose a ball is thrown straight upward so that its height $f(x)$ (in feet) is given by the equation

$$f(x) = 96 + 64x - 16x^2$$

where x is time (in seconds).

(a) Find the average velocity from $x = 1$ to $x = 1 + h$.
(b) Find the instantaneous velocity at $x = 1$.

Solution

(a) Let h represent the change in x (time) from 1 to $1 + h$. Then the corresponding change in $f(x)$ (height) is

$$f(1 + h) - f(1) = [96 + 64(1 + h) - 16(1 + h)^2] - [96 + 64 - 16]$$
$$= 96 + 64 + 64h - 16(1 + 2h + h^2) - 144$$
$$= 16 + 64h - 16 - 32h - 16h^2$$
$$= 32h - 16h^2$$

The average velocity V_{av} is the change in height divided by the change in time.

$$V_{av} = \frac{f(1 + h) - f(1)}{h}$$
$$= \frac{32h - 16h^2}{h}$$
$$= 32 - 16h$$

(b) The instantaneous velocity V is the limit of the average velocity as h approaches 0.

$$V = \lim_{h \to 0} V_{av} = \lim_{h \to 0} (32 - 16h)$$
$$= 32 \text{ ft/s}$$

Note that average velocity is found over a time interval. Instantaneous velocity is usually called **velocity,** and it can be found at any time x, as follows.

Velocity

Suppose that an object moving in a straight line has its position y at time x given by $y = f(x)$. Then the **velocity** of the object at time x is

$$V = \lim_{h \to 0} \frac{f(x + h) - f(x)}{h}$$

provided that this limit exists.

The instantaneous rate of change of any function (commonly called *rate of change*) can be found in the same way we find velocity. The function that gives this instantaneous rate of change of a function f is called the **derivative** of f.

Derivative

If f is a function defined by $y = f(x)$, then the **derivative** of $f(x)$ at any value x, denoted $f'(x)$, is

$$f'(x) = \lim_{h \to 0} \frac{f(x + h) - f(x)}{h}$$

if this limit exists. If $f'(c)$ exists, we say that f is **differentiable** at c.

The following procedure illustrates how to find the derivative of a function $y = f(x)$ at any value x.

Derivative Using the Definition

Procedure	Example
To find the derivative of $y = f(x)$ at any value x:	Find the derivative of $f(x) = 4x^2$.
1. Let h represent the change in x from x to $x + h$.	1. The change in x from x to $x + h$ is h.
2. The corresponding change in $y = f(x)$ is $$f(x + h) - f(x)$$	2. The change in $f(x)$ is $$\begin{aligned} f(x + h) - f(x) &= 4(x + h)^2 - 4x^2 \\ &= 4(x^2 + 2xh + h^2) - 4x^2 \\ &= 4x^2 + 8xh + 4h^2 - 4x^2 \\ &= 8xh + 4h^2 \end{aligned}$$
3. Form the difference quotient $\dfrac{f(x + h) - f(x)}{h}$ and simplify.	3. $\dfrac{f(x + h) - f(x)}{h} = \dfrac{8xh + 4h^2}{h}$ $= 8x + 4h$
4. Find $\displaystyle\lim_{h \to 0} \dfrac{f(x + h) - f(x)}{h}$ to determine $f'(x)$, the derivative of $f(x)$.	4. $f'(x) = \displaystyle\lim_{h \to 0} \dfrac{f(x + h) - f(x)}{h}$ $f'(x) = \displaystyle\lim_{h \to 0} (8x + 4h) = 8x$

Note that in the example above, we could have found the derivative of the function $f(x) = 4x^2$ at a particular value of x, say $x = 3$, by evaluating the derivative formula at that value:

$$f'(x) = 8x \quad \text{so} \quad f'(3) = 8(3) = 24$$

In addition to $f'(x)$, the derivative at any point x may be denoted by

$$\frac{dy}{dx}, \quad y', \quad \frac{d}{dx} f(x), \quad D_x y, \quad \text{or} \quad D_x f(x)$$

We can, of course, use variables other than x and y to represent functions and their derivatives. For example, we can represent the derivative of the function defined by $p = 2q^2 - 1$ by dp/dq.

● **Checkpoint**

1. Find the average rate of change of $f(x) = 30 - x - x^2$ over $[1, 4]$.
2. For the function $y = f(x) = x^2 - x + 1$, find

   (a) $f(x + h) - f(x)$

   (b) $\dfrac{f(x + h) - f(x)}{h}$

   (c) $f'(x) = \displaystyle\lim_{h \to 0} \dfrac{f(x + h) - f(x)}{h}$

   (d) $f'(2)$

In Section 1.6, "Applications of Functions in Business and Economics," we defined the **marginal revenue** for a product as the rate of change of the total revenue function for the product. If the total revenue function for a product is not linear, we define the marginal revenue for the product as the instantaneous rate of change, or the derivative, of the revenue function.

Marginal Revenue

Suppose that the total revenue function for a product is given by $R = R(x)$, where x is the number of units sold. Then the **marginal revenue** at x units is

$$\overline{MR} = R'(x) = \lim_{h \to 0} \frac{R(x + h) - R(x)}{h}$$

provided that the limit exists.

Note that the marginal revenue (derivative of the revenue function) can be found by using the steps in the Procedure/Example table on the preceding page. These steps can also be combined, as they are in Example 4.

❍ EXAMPLE 4 *Revenue (Application Preview)*

Suppose that an oil company's revenue (in thousands of dollars) is given by the equation

$$R = R(x) = 100x - x^2, \quad x \geq 0$$

where x is the number of thousands of barrels of oil sold each day.

(a) Find the function that gives the marginal revenue at any value of x.
(b) Find the marginal revenue when 20,000 barrels are sold (that is, at $x = 20$).

Solution

(a) The marginal revenue function is the derivative of $R(x)$.

$$R'(x) = \lim_{h \to 0} \frac{R(x + h) - R(x)}{h}$$

$$= \lim_{h \to 0} \frac{\left[100(x + h) - (x + h)^2\right] - (100x - x^2)}{h}$$

$$= \lim_{h \to 0} \frac{100x + 100h - (x^2 + 2xh + h^2) - 100x + x^2}{h}$$

$$= \lim_{h \to 0} \frac{100h - 2xh - h^2}{h} = \lim_{h \to 0} (100 - 2x - h) = 100 - 2x$$

Thus, the marginal revenue function is $\overline{MR} = R'(x) = 100 - 2x$.

(b) The function found in (a) gives the marginal revenue at *any* value of x. To find the marginal revenue when 20 units are sold, we evaluate $R'(20)$.

$$R'(20) = 100 - 2(20) = 60$$

Hence the marginal revenue at $x = 20$ is $60,000 per thousand barrels of oil. Because the marginal revenue is used to approximate the revenue from the sale of one additional unit, we interpret $R'(20) = 60$ to mean that the expected revenue from the sale of the next thousand barrels (after 20,000) will be approximately $60,000. [*Note:* The actual revenue from this sale is $R(21) - R(20) = 1659 - 1600 = 59$ (thousand dollars).]

Tangent to a Curve

As mentioned earlier, the rate of change of revenue (the marginal revenue) for a linear revenue function is given by the slope of the line. In fact, the slope of the revenue curve gives us the marginal revenue even if the revenue function is not linear. We will show that the slope of the graph of a function at any point is the same as the derivative at that point. In order to show this, we must define the slope of a curve at a point on the curve. We will define the slope of a curve at a point as the slope of the line tangent to the curve at the point.

In geometry, a **tangent** to a circle is defined as a line that has one point in common with the circle. (See Figure 9.20(a).) This definition does not apply to all curves, as Figure 9.20(b) shows. Many lines can be drawn through the point *A* that touch the curve only at *A*. One of the lines, line *l*, looks like it is tangent to the curve.

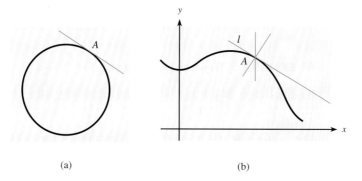

Figure 9.20 (a) (b)

We can use **secant lines** (lines that intersect the curve at two points) to determine the tangent to a curve at a point. In Figure 9.21, we have a set of secant lines s_1, s_2, s_3, and s_4 that pass through a point *A* on the curve and points Q_1, Q_2, Q_3, and Q_4 on the curve near *A*. (For points and secant lines to the left of point *A*, there would be a similar figure and discussion.) The line *l* represents the tangent line to the curve at point *A*. We can get a secant line as close as we wish to the tangent line *l* by choosing a "second point" *Q* sufficiently close to point *A*.

As we choose points on the curve closer and closer to *A* (from both sides of *A*), the limiting position of the secant lines that pass through *A* is the **tangent line** to the curve at point *A*, and the slopes of those secant lines approach the slope of the tangent line at *A*. Thus we can find the slope of the tangent line by finding the slope of a secant line and taking the limit of this slope as the "second point" *Q* approaches *A*. To find the slope of the tangent to the graph of $y = f(x)$ at $A(x_1, f(x_1))$, we first draw a secant line from point *A* to a second point $Q(x_1 + h, f(x_1 + h))$ on the curve (see Figure 9.22).

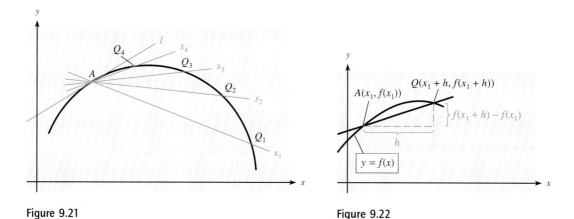

Figure 9.21   **Figure 9.22**

The slope of this secant line is

$$m_{AQ} = \frac{f(x_1 + h) - f(x_1)}{h}$$

As *Q* approaches *A*, we see that the difference between the *x*-coordinates of these two points decreases, so *h* approaches 0. Thus the slope of the tangent is given by the following.

Slope of the Tangent

The **slope of the tangent** to the graph of $y = f(x)$ at point $A(x_1, f(x_1))$ is

$$m = \lim_{h \to 0} \frac{f(x_1 + h) - f(x_1)}{h}$$

if this limit exists. That is, $m = f'(x_1)$, the derivative at $x = x_1$.

● **EXAMPLE 5** *Slope of the Tangent*

Find the slope of $y = f(x) = x^2$ at the point $A(2, 4)$.

Solution
The formula for the slope of the tangent to $y = f(x)$ at $(2, 4)$ is

$$m = f'(2) = \lim_{h \to 0} \frac{f(2 + h) - f(2)}{h}$$

Thus for $f(x) = x^2$, we have

$$m = f'(2) = \lim_{h \to 0} \frac{(2 + h)^2 - 2^2}{h}$$

Taking the limit immediately would result in both the numerator and the denominator approaching 0. To avoid this, we simplify the fraction before taking the limit.

$$m = \lim_{h \to 0} \frac{4 + 4h + h^2 - 4}{h} = \lim_{h \to 0} \frac{4h + h^2}{h} = \lim_{h \to 0} (4 + h) = 4$$

Thus the slope of the tangent to $y = x^2$ at $(2, 4)$ is 4 (see Figure 9.23).

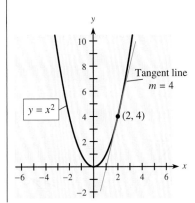

Figure 9.23

The statement "the slope of the tangent to the curve at $(2, 4)$ is 4" is frequently simplified to the statement "the slope of the curve at $(2, 4)$ is 4." Knowledge that the slope is a positive number on an interval tells us that the function is increasing on that interval, which means that a point moving along the graph of the function rises as it moves to the right on that interval. If the derivative (and thus the slope) is negative on an interval, the curve is decreasing on the interval; that is, a point moving along the graph falls as it moves to the right on that interval.

● **EXAMPLE 6** *Tangent Line*

Given $y = f(x) = 3x^2 + 2x + 11$, find

(a) the derivative of $f(x)$ at any point $(x, f(x))$.
(b) the slope of the curve at $(1, 16)$.
(c) the equation of the line tangent to $y = 3x^2 + 2x + 11$ at $(1, 16)$.

Solution
(a) The derivative of $f(x)$ at any value x is denoted by $f'(x)$ and is

$$y' = f'(x) = \lim_{h \to 0} \frac{f(x + h) - f(x)}{h}$$

$$= \lim_{h \to 0} \frac{[3(x + h)^2 + 2(x + h) + 11] - (3x^2 + 2x + 11)}{h}$$

$$= \lim_{h \to 0} \frac{3(x^2 + 2xh + h^2) + 2x + 2h + 11 - 3x^2 - 2x - 11}{h}$$

$$= \lim_{h \to 0} \frac{6xh + 3h^2 + 2h}{h}$$

$$= \lim_{h \to 0} (6x + 3h + 2)$$

$$= 6x + 2$$

(b) The derivative is $f'(x) = 6x + 2$, so the slope of the tangent to the curve at $(1, 16)$ is $f'(1) = 6(1) + 2 = 8$.
(c) The equation of the tangent line uses the given point $(1, 16)$ and the slope $m = 8$. Using $y - y_1 = m(x - x_1)$ gives $y - 16 = 8(x - 1)$, or $y = 8x + 8$.

The discussion in this section indicates that the derivative of a function has several interpretations.

Interpretations of the Derivative

For a given function, each of the following means "find the **derivative.**"

1. Find the **velocity** of an object moving in a straight line.
2. Find the **instantaneous rate of change** of a function.
3. Find the **marginal revenue** for a given revenue function.
4. Find the **slope of the tangent** to the graph of a function.

That is, all the terms printed in boldface are mathematically the same, and the answers to questions about any one of them give information about the others. For example, if we know the slope of the tangent to the graph of a revenue function at a point, then we know the marginal revenue at that point.

Calculator Note

Note in Figure 9.23 that near the point of tangency at $(2, 4)$, the tangent line and the function look coincident. In fact, if we graphed both with a graphing calculator and repeatedly zoomed in near the point $(2, 4)$, the two graphs would eventually appear as one. Try this for yourself. Thus the derivative of $f(x)$ at the point where $x = a$ can be approximated by finding the slope between $(a, f(a))$ and a second point that is nearby. ∎

In addition, we know that the slope of the tangent to $f(x)$ at $x = a$ is defined by

$$f'(a) = \lim_{h \to 0} \frac{f(a + h) - f(a)}{h}$$

Hence we could also estimate $f'(a)$—that is, the slope of the tangent at $x = a$—by evaluating

$$\frac{f(a + h) - f(a)}{h} \quad \text{when } h \approx 0 \text{ and } h \neq 0$$

● **EXAMPLE 7 *Approximating the Slope of the Tangent Line***

(a) Let $f(x) = 2x^3 - 6x^2 + 2x - 5$. Use $\dfrac{f(a + h) - f(a)}{h}$ and two values of h to make estimates of the slope of the tangent to $f(x)$ at $x = 3$ on opposite sides of $x = 3$.

(b) Use the following table of values of x and $g(x)$ to estimate $g'(3)$.

x	1	1.9	2.7	2.9	2.999	3	3.002	3.1	4	5
$g(x)$	1.6	4.3	11.4	10.8	10.513	10.5	10.474	10.18	6	−5

Solution

The table feature of a graphing utility can facilitate the following calculations.

(a) We can use $h = 0.0001$ and $h = -0.0001$ as follows:

$$\text{With } h = 0.0001: \quad f'(3) \approx \frac{f(3 + 0.0001) - f(3)}{0.0001}$$

$$= \frac{f(3.0001) - f(3)}{0.0001} = 20.0012 \approx 20$$

$$\text{With } h = -0.0001: \quad f'(3) \approx \frac{f(3 + (-0.0001)) - f(3)}{-0.0001}$$

$$= \frac{f(2.9999) - f(3)}{-0.0001} = 19.9988 \approx 20$$

(b) We use the given table and measure the slope between $(3, 10.5)$ and another point that is nearby (the closer, the better). Using $(2.999, 10.513)$, we obtain

$$g'(3) \approx \frac{y_2 - y_1}{x_2 - x_1} = \frac{10.5 - 10.513}{3 - 2.999} = \frac{-0.013}{0.001} = -13$$

Calculator Note
Most graphing calculators have a feature called the **numerical derivative** (usually denoted by nDer or nDeriv) that can approximate the derivative of a function at a point. On most calculators this feature uses a calculation similar to our method in part (a) of Example 7 and produces the same estimate. The numerical derivative of $f(x) = 2x^3 - 6x^2 + 2x - 5$ with respect to x at $x = 3$ can be found as follows on many graphing calculators:

$$\text{nDeriv}(2x^3 - 6x^2 + 2x - 5, x, 3) = 20 \qquad ■$$

Differentiability and Continuity

So far we have talked about how the derivative is defined, what it represents, and how to find it. However, there are functions for which derivatives do not exist at every value of x. Figure 9.24 shows some common cases where $f'(c)$ does not exist but where $f'(x)$ exists for all other values of x. These cases occur where there is a discontinuity, a corner, or a vertical tangent line.

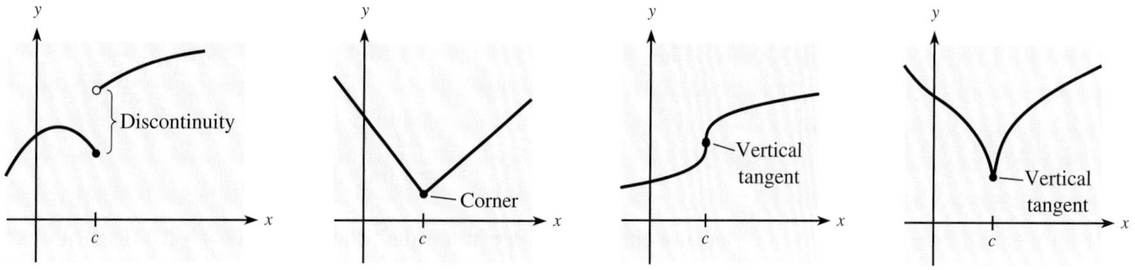

(a) Not differentiable
at $x = c$

(b) Not differentiable
at $x = c$

(c) Not differentiable
at $x = c$

(d) Not differentiable
at $x = c$

Figure 9.24

From Figure 9.24 we see that a function may be continuous at $x = c$ even though $f'(c)$ does not exist. Thus continuity does not imply differentiability at a point. However, differentiability does imply continuity.

Differentiability Implies Continuity   If a function f is differentiable at $x = c$, then f is continuous at $x = c$.

● **EXAMPLE 8** *Water Usage Costs*

The monthly charge for water in a small town is given by

$$y = f(x) = \begin{cases} 18 & \text{if } 0 \leq x \leq 20 \\ 0.1x + 16 & \text{if } x > 20 \end{cases}$$

(a) Is this function continuous at $x = 20$?
(b) Is this function differentiable at $x = 20$?

Solution
(a) We must check the three properties for continuity.

1. $f(x) = 18$ for $x \leq 20$ so $f(20) = 18$
2. $\left.\begin{array}{l} \lim\limits_{x \to 20^-} f(x) = \lim\limits_{x \to 20^-} 18 = 18 \\ \lim\limits_{x \to 20^+} f(x) = \lim\limits_{x \to 20^+} (0.1x + 16) = 18 \end{array}\right\} \Rightarrow \lim\limits_{x \to 20} f(x) = 18$
3. $\lim\limits_{x \to 20} f(x) = f(20)$

Thus $f(x)$ is continuous at $x = 20$.

(b) Because the function is defined differently on either side of $x = 20$, we need to test to see whether $f'(20)$ exists by evaluating both

(i) $\lim\limits_{h \to 0^-} \dfrac{f(20 + h) - f(20)}{h}$ and   (ii) $\lim\limits_{h \to 0^+} \dfrac{f(20 + h) - f(20)}{h}$

and determining whether they are equal.

(i) $\lim\limits_{h \to 0^-} \dfrac{f(20 + h) - f(20)}{h} = \lim\limits_{h \to 0^-} \dfrac{18 - 18}{h}$

$= \lim\limits_{h \to 0^-} 0 = 0$

(ii) $\displaystyle\lim_{h\to 0^+}\frac{f(20+h)-f(20)}{h}=\lim_{h\to 0^+}\frac{[0.1(20+h)+16]-18}{h}$

$$=\lim_{h\to 0^+}\frac{0.1h}{h}$$

$$=\lim_{h\to 0^+}0.1=0.1$$

Because these limits are not equal, the derivative $f'(20)$ does not exist.

● **Checkpoint**

3. Which of the following are given by $f'(c)$?
   (a) The slope of the tangent when $x=c$
   (b) The y-coordinate of the point where $x=c$
   (c) The instantaneous rate of change of $f(x)$ at $x=c$
   (d) The marginal revenue at $x=c$, if $f(x)$ is the revenue function
4. Must a graph that has no discontinuity, corner, or cusp at $x=c$ be differentiable at $x=c$?

Calculator Note

We can use a graphing calculator to explore the relationship between secant lines and tangent lines. For example, if the point (a, b) lies on the graph of $y=x^2$, then the equation of the secant line to $y=x^2$ from $(1, 1)$ to (a, b) has the equation

$$y-1=\frac{b-1}{a-1}(x-1),\quad\text{or}\quad y=\frac{b-1}{a-1}(x-1)+1$$

Figure 9.25 illustrates the secant lines for three different choices for the point (a, b).

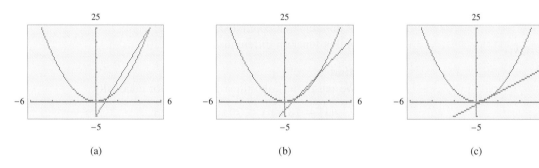

(a)   (b)   (c)

Figure 9.25

We see that as the point (a, b) moves closer to $(1, 1)$, the secant line looks more like the tangent line to $y=x^2$ at $(1, 1)$. Furthermore, (a, b) approaches $(1, 1)$ as $a\to 1$, and the slope of the secant approaches the following limit.

$$\lim_{a\to 1}\frac{b-1}{a-1}=\lim_{a\to 1}\frac{a^2-1}{a-1}=\lim_{a\to 1}(a+1)=2$$

This limit, 2, is the slope of the tangent line at $(1, 1)$. That is, the derivative of $y=x^2$ at $(1, 1)$ is 2. [Note that a graphing utility's calculation of the numerical derivative of $f(x)=x^2$ with respect to x at $x=1$ gives $f'(1)=2$.] ■

● Checkpoint Solutions

1. $\dfrac{f(4) - f(1)}{4 - 1} = \dfrac{10 - 28}{3} = \dfrac{-18}{3} = -6$

2. (a) $f(x + h) - f(x) = \left[(x + h)^2 - (x + h) + 1\right] - (x^2 - x + 1)$

   $= x^2 + 2xh + h^2 - x - h + 1 - x^2 + x - 1$

   $= 2xh + h^2 - h$

   (b) $\dfrac{f(x + h) - f(x)}{h} = \dfrac{2xh + h^2 - h}{h}$

   $= 2x + h - 1$

   (c) $f'(x) = \lim_{h \to 0} \dfrac{f(x + h) - f(x)}{h} = \lim_{h \to 0} (2x + h - 1)$

   $= 2x - 1$

   (d) $f'(x) = 2x - 1$, so $f'(2) = 3$.

3. Parts (a), (c), and (d) are given by $f'(c)$. The y-coordinate where $x = c$ is given by $f(c)$.

4. No. Figure 9.24(c) shows such an example.

9.3 Exercises

In Problems 1–4, for each given function find the average rate of change over each specified interval.

1. $f(x) = x^2 + x - 12$ over (a) $[0, 5]$ and (b) $[-3, 10]$

2. $f(x) = 6 - x - x^2$ over (a) $[-1, 2]$ and (b) $[1, 10]$

3. For $f(x)$ given by the table, over (a) $[2, 5]$ and (b) $[3.8, 4]$

x	0	2	2.5	3	3.8	4	5
$f(x)$	14	20	22	19	17	16	30

4. For $f(x)$ given in the table, over (a) $[3, 3.5]$ and (b) $[2, 6]$

x	1	2	3	3.5	3.7	6
$f(x)$	40	25	18	15	18	38

5. Given $f(x) = 2x - x^2$, find the average rate of change of $f(x)$ over each of the following pairs of intervals.
   (a) $[2.9, 3]$ and $[2.99, 3]$ (b) $[3, 3.1]$ and $[3, 3.01]$
   (c) What do the calculations in parts (a) and (b) suggest the instantaneous rate of change of $f(x)$ at $x = 3$ might be?

6. Given $f(x) = x^2 + 3x + 7$, find the average rate of change of $f(x)$ over each of the following pairs of intervals.
   (a) $[1.9, 2]$ and $[1.99, 2]$
   (b) $[2, 2.1]$ and $[2, 2.01]$
   (c) What do the calculations in parts (a) and (b) suggest the instantaneous rate of change of $f(x)$ at $x = 2$ might be?

7. In the Procedure/Example table in this section we were given $f(x) = 4x^2$ and found $f'(x) = 8x$. Find
   (a) the instantaneous rate of change of $f(x)$ at $x = 4$.
   (b) the slope of the tangent to the graph of $y = f(x)$ at $x = 4$.
   (c) the point on the graph of $y = f(x)$ at $x = 4$.

8. In Example 6 in this section we were given $f(x) = 3x^2 + 2x + 11$ and found $f'(x) = 6x + 2$. Find
   (a) the instantaneous rate of change of $f(x)$ at $x = 6$.
   (b) the slope of the tangent to the graph of $y = f(x)$ at $x = 6$.
   (c) the point on the graph of $y = f(x)$ at $x = 6$.

9. Let $f(x) = 2x^2 - x$.
   (a) Use the definition of derivative and the Procedure/Example table in this section to verify that $f'(x) = 4x - 1$.
   (b) Find the instantaneous rate of change of $f(x)$ at $x = -1$.
   (c) Find the slope of the tangent to the graph of $y = f(x)$ at $x = -1$.
   (d) Find the point on the graph of $y = f(x)$ at $x = -1$.

10. Let $f(x) = 9 - \dfrac{1}{2}x^2$.
    (a) Use the definition of derivative and the Procedure/Example table in this section to verify that $f'(x) = -x$.
    (b) Find the instantaneous rate of change of $f(x)$ at $x = 2$.

(c) Find the slope of the tangent to the graph of $y = f(x)$ at $x = 2$.

(d) Find the point on the graph of $y = f(x)$ at $x = 2$.

In Problems 11–14, the tangent line to the graph of $f(x)$ at $x = 1$ is shown. On the tangent line, P is the point of tangency and A is another point on the line.

(a) Find the coordinates of the points P and A.

(b) Use the coordinates of P and A to find the slope of the tangent line.

(c) Find $f'(1)$.

(d) Find the instantaneous rate of change of $f(x)$ at P.

11.

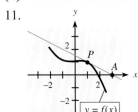

12.

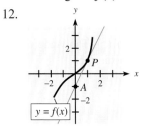

13.

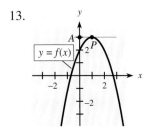

14.

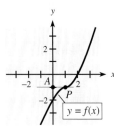

For each function in Problems 15–18, find

(a) the derivative, by using the definition.

(b) the instantaneous rate of change of the function at any value and at the given value.

(c) the slope of the tangent at the given value.

15. $f(x) = 4x^2 - 2x + 1$; $x = -3$

16. $f(x) = 16x^2 - 4x + 2$; $x = 1$

17. $p(q) = q^2 + 4q + 1$; $q = 5$

18. $p(q) = 2q^2 - 4q + 5$; $q = 2$

For each function in Problems 19–22, approximate $f'(a)$ in the following ways.

(a) Use the numerical derivative feature of a graphing utility.

(b) Use $\dfrac{f(a + h) - f(a)}{h}$ with $h = 0.0001$.

(c) Graph the function on a graphing utility. Then zoom in near the point until the graph appears straight, pick two points, and find the slope of the line you see.

19. $f'(2)$ for $f(x) = 3x^4 - 7x - 5$

20. $f'(-1)$ for $f(x) = 2x^3 - 11x + 9$

21. $f'(4)$ for $f(x) = (2x - 1)^3$

22. $f'(3)$ for $(x) = \dfrac{3x + 1}{2x - 5}$

In Problems 23 and 24, use the given tables to approximate $f'(a)$ as accurately as you can.

23.

x	12.0	12.99	13	13.1	$a = 13$
$f(x)$	1.41	17.42	17.11	22.84	

24.

x	-7.4	-7.50	-7.51	-7	$a = -7.5$
$f(x)$	22.12	22.351	22.38	24.12	

In the figures given in Problems 25 and 26, at each point A and B draw an approximate tangent line and then use it to complete parts (a) and (b).

(a) Is $f'(x)$ greater at point A or at point B? Explain.

(b) Estimate $f'(x)$ at point B.

25.

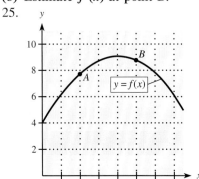

26.

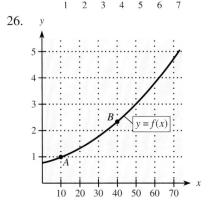

In Problems 27 and 28, a point (a, b) on the graph of $y = f(x)$ is given, and the equation of the line tangent to the graph of $f(x)$ at (a, b) is given. In each case, find $f'(a)$ and $f(a)$.

27. $(-3, -9)$; $5x - 2y = 3$

28. $(-1, 6)$; $x + 10y = 59$

29. If the instantaneous rate of change of $f(x)$ at $(1, -1)$ is 3, write the equation of the line tangent to the graph of $f(x)$ at $x = 1$.

30. If the instantaneous rate of change of $g(x)$ at $(-1, -2)$ is $1/2$, write the equation of the line tangent to the graph of $g(x)$ at $x = -1$.

Because the derivative of a function represents both the slope of the tangent to the curve and the instantaneous rate of change of the function, it is possible to use information about one to gain information about the other. In Problems 31 and 32, use the graph of the function $y = f(x)$ given in Figure 9.26.

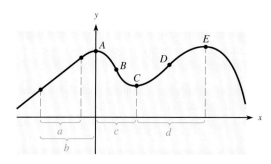

Figure 9.26

31. (a) Over what interval(s) (*a*) through (*d*) is the rate of change of $f(x)$ positive?
    (b) Over what interval(s) (*a*) through (*d*) is the rate of change of $f(x)$ negative?
    (c) At what point(s) *A* through *E* is the rate of change of $f(x)$ equal to zero?
32. (a) At what point(s) *A* through *E* does the rate of change of $f(x)$ change from positive to negative?
    (b) At what point(s) *A* through *E* does the rate of change of $f(x)$ change from negative to positive?
33. Given the graph of $y = f(x)$ in Figure 9.27, determine for which *x*-values *A*, *B*, *C*, *D*, or *E* the function is
    (a) continuous.   (b) differentiable.
34. Given the graph of $y = f(x)$ in Figure 9.27, determine for which *x*-values *F*, *G*, *H*, *I*, or *J* the function is
    (a) continuous.   (b) differentiable.

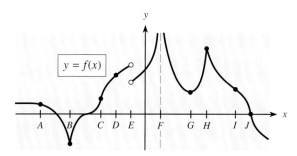

Figure 9.27

In Problems 35–38, (a) find the slope of the tangent to the graph of $f(x)$ at any point, (b) find the slope of the tangent at the given *x*-value, (c) write the equation of the line tangent to the graph of $f(x)$ at the given point, and (d) graph both $f(x)$ and its tangent line (use a graphing utility if one is available).

35. (a) $f(x) = x^2 + x$
    (b) $x = 2$
    (c) $(2, 6)$
36. (a) $f(x) = x^2 + 3x$
    (b) $x = -1$
    (c) $(-1, -2)$
37. (a) $f(x) = x^3 + 3$
    (b) $x = 1$
    (c) $(1, 4)$
38. (a) $f(x) = 5x^3 + 2$
    (b) $x = -1$
    (c) $(-1, -3)$

APPLICATIONS

39. *Total cost*   Suppose total cost in dollars from the production of *x* printers is given by
$$C(x) = 0.0001x^3 + 0.005x^2 + 28x + 3000$$
    Find the average rate of change of total cost when production changes
    (a) from 100 to 300 printers.
    (b) from 300 to 600 printers.
    (c) Interpret the results from parts (a) and (b).

40. *Average velocity*   If an object is thrown upward at 64 ft/s from a height of 20 feet, its height *S* after *x* seconds is given by
$$S(x) = 20 + 64x - 16x^2$$
    What is the average velocity in the
    (a) first 2 seconds after it is thrown?
    (b) next 2 seconds?

41. *Demand*   If the demand for a product is given by
$$D(p) = \frac{1000}{\sqrt{p}} - 1$$
    what is the average rate of change of demand when *p* increases from
    (a) 1 to 25?
    (b) 25 to 100?

42. *Revenue*   If the total revenue function for a blender is
$$R(x) = 36x - 0.01x^2$$
    where *x* is the number of units sold, what is the average rate of change in revenue $R(x)$ as *x* increases from 10 to 20 units?

43. *Total cost*   Suppose the figure shows the total cost graph for a company. Arrange the average rates of change of total cost from *A* to *B*, *B* to *C*, and *A* to *C* from smallest to greatest, and explain your choice.

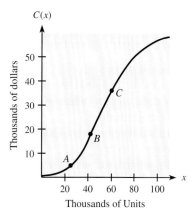

44. *Students per computer* The following figure shows the number of students per computer in U.S. public schools for the school years that ended in 1984 through 2002.
    (a) Use the figure to find the average rate of change in the number of students per computer from 1990 to 2000. Interpret your result.
    (b) From the figure, determine for what two consecutive school years the average rate of change of the number of students per computer is closest to zero.

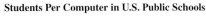

Students Per Computer in U.S. Public Schools

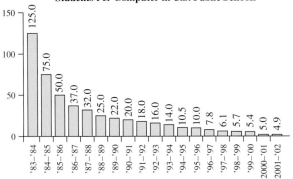

Source: Quality Education Data, Inc., Denver, Co. Reprinted by permission.

45. *Marginal revenue* Say the revenue function for a stereo system is

$$R(x) = 300x - x^2 \quad \text{dollars}$$

where x denotes the number of units sold.
    (a) What is the function that gives marginal revenue?
    (b) What is the marginal revenue if 50 units are sold, and what does it mean?
    (c) What is the marginal revenue if 200 units are sold, and what does it mean?
    (d) What is the marginal revenue if 150 units are sold, and what does it mean?
    (e) As the number of units sold passes through 150, what happens to revenue?

46. *Marginal revenue* Suppose the total revenue function for a blender is

$$R(x) = 36x - 0.01x^2 \quad \text{dollars}$$

where x is the number of units sold.
    (a) What function gives the marginal revenue?
    (b) What is the marginal revenue when 600 units are sold, and what does it mean?
    (c) What is the marginal revenue when 2000 units are sold, and what does it mean?
    (d) What is the marginal revenue when 1800 units are sold, and what does it mean?

47. *Labor force and output* The monthly output at the Olek Carpet Mill is

$$Q(x) = 15,000 + 2x^2 \quad \text{units}, \quad (40 \leq x \leq 60)$$

where x is the number of workers employed at the mill. If there are currently 50 workers, find the instantaneous rate of change of monthly output with respect to the number of workers. That is, find $Q'(50)$.

48. *Consumer expenditure* Suppose that the demand for x units of a product is

$$x = 10,000 - 100p$$

where p dollars is the price per unit. Then the consumer expenditure for the product is

$$E(p) = px = p(10,000 - 100p)$$
$$= 10,000p - 100p^2$$

What is the instantaneous rate of change of consumer expenditure with respect to price at
    (a) any price p?   (b) $p = 5$?   (c) $p = 20$?

In Problems 49–52, find derivatives with the numerical derivative feature of a graphing utility.

49. *Profit* Suppose that the profit function for the monthly sales of a car by a dealership is

$$P(x) = 500x - x^2 - 100$$

where x is the number of cars sold. What is the instantaneous rate of change of profit when
    (a) 200 cars are sold? Explain its meaning.
    (b) 300 cars are sold? Explain its meaning.

50. *Profit* If the total revenue function for a toy is

$$R(x) = 2x$$

and the total cost function is

$$C(x) = 100 + 0.2x^2 + x$$

what is the instantaneous rate of change of profit if 10 units are produced and sold? Explain its meaning.

51. *Heat index* The highest recorded temperature in the state of Alaska was 100°F and occurred on June 27, 1915, at Fort Yukon. The *heat index* is the apparent temperature of the air at a given temperature and humidity level. If x denotes the relative humidity (in percent), then the heat index (in degrees Fahrenheit) for an air temperature of 100°F can be approximated by the function

$$f(x) = 0.009x^2 + 0.139x + 91.875$$

    (a) At what rate is the heat index changing when the humidity is 50%?
    (b) Write a sentence that explains the meaning of your answer in part (a).

52. *Receptivity* In learning theory, receptivity is defined as the ability of students to understand a complex concept. Receptivity is highest when the topic is introduced and tends to decrease as time passes in a lecture. Suppose

that the receptivity of a group of students in a mathematics class is given by

$$g(t) = -0.2t^2 + 3.1t + 32$$

where t is minutes after the lecture begins.

(a) At what rate is receptivity changing 10 minutes after the lecture begins?

(b) Write a sentence that explains the meaning of your answer in part (a).

53. *Marginal revenue*   Suppose the graph shows a manufacturer's total revenue, in thousands of dollars, from the sale of x cellular telephones to dealers.

(a) Is the marginal revenue greater at 300 cell phones or at 700? Explain.

(b) Use part (a) to decide whether the sale of the 301st cell phone or the 701st brings in more revenue. Explain.

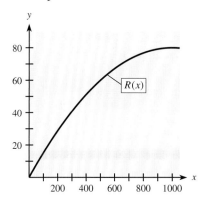

54. *Social Security beneficiaries*   The graph shows a model for the number of millions of Social Security beneficiaries (actual to 2000 and projected beyond 2000). The model was developed with data from the 2000 Social Security Trustees Report.

(a) Was the instantaneous rate of change of the number of beneficiaries with respect to the year greater in 1960 or in 1980? Justify your answer.

(b) Is the instantaneous rate of change of the number of beneficiaries projected to be greater in 2000 or in 2030? Justify your answer.

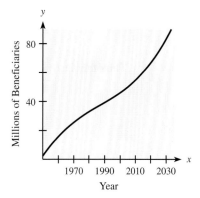

9.4 Derivative Formulas

◗ Application Preview

For more than 30 years, U.S. total personal income has experienced steady growth. With Bureau of Economic Analysis, U.S. Department of Commerce data for selected years from 1975 to 2002, U.S. total personal income I, in billions of current dollars, can be modeled by

$$I = I(t) = 5.33t^2 + 81.5t + 822$$

where t is the number of years past 1970. We can find the rate of growth of total U.S. personal income in 2007 by using the derivative $I'(t)$ of the total personal income function. (See Example 8.)

As we discussed in the previous section, the derivative of a function can be used to find the rate of change of the function. In this section we will develop formulas that will make it easier to find certain derivatives.

Derivative of $f(x) = x^n$

We can use the definition of derivative to show the following:

$$\text{If } f(x) = x^2, \text{ then } f'(x) = 2x.$$
$$\text{If } f(x) = x^3, \text{ then } f'(x) = 3x^2.$$
$$\text{If } f(x) = x^4, \text{ then } f'(x) = 4x^3.$$
$$\text{If } f(x) = x^5, \text{ then } f'(x) = 5x^4.$$

Do you recognize a pattern that could be used to find the derivative of $f(x) = x^6$? What is the derivative of $f(x) = x^n$? If you said that the derivative of $f(x) = x^6$ is $f'(x) = 6x^5$ and the derivative of $f(x) = x^n$ is $f'(x) = nx^{n-1}$, you're right. We can use the definition of derivative to show this. If n is a positive integer, then

$$f'(x) = \lim_{h \to 0} \frac{f(x+h) - f(x)}{h}$$

$$= \lim_{h \to 0} \frac{(x+h)^n - x^n}{h}$$

Because we are assuming that n is a positive integer, we can use the binomial formula, developed in Section 8.3, to expand $(x + h)^n$. This formula can be stated as follows:

$$(x+h)^n = x^n + nx^{n-1}h + \frac{n(n-1)}{1 \cdot 2}x^{n-2}h^2 + \cdots + h^n$$

Thus the derivative formula yields

$$f'(x) = \lim_{h \to 0} \frac{\left[x^n + nx^{n-1}h + \dfrac{n(n-1)}{1 \cdot 2}x^{n-2}h^2 + \cdots + h^n \right] - x^n}{h}$$

$$= \lim_{h \to 0} \left[nx^{n-1} + \frac{n(n-1)}{1 \cdot 2}x^{n-2}h + \cdots + h^{n-1} \right]$$

Now, each term after nx^{n-1} contains h as a factor, so all terms except nx^{n-1} will approach 0 as $h \to 0$. Thus

$$f'(x) = nx^{n-1}$$

Even though we proved this derivative rule only for the case when n is a positive integer, the rule applies for any real number n.

Powers of x Rule

If $f(x) = x^n$, where n is a real number, then $f'(x) = nx^{n-1}$.

● **EXAMPLE 1** *Powers of x Rule*

Find the derivatives of the following functions.

(a) $g(x) = x^6$ (b) $f(x) = x^{-2}$ (c) $y = x^{14}$ (d) $y = x^{1/3}$

Solution
(a) If $g(x) = x^6$, then $g'(x) = 6x^{6-1} = 6x^5$.
(b) The Powers of x Rule applies for all real values. Thus for $f(x) = x^{-2}$, we have

$$f'(x) = -2x^{-2-1} = -2x^{-3} = \frac{-2}{x^3}$$

(c) If $y = x^{14}$, then $dy/dx = 14x^{14-1} = 14x^{13}$.
(d) The Powers of x Rule applies to $y = x^{1/3}$.

$$\frac{dy}{dx} = \frac{1}{3}x^{1/3-1} = \frac{1}{3}x^{-2/3} = \frac{1}{3x^{2/3}}$$

In Example 1 we took the derivative with respect to x of *both sides* of each equation. We denote the operation "take the derivative with respect to x" by $\dfrac{d}{dx}$. Thus for $y = x^{14}$, in (c), we can use this notation on both sides of the equation.

$$\frac{d}{dx}(y) = \frac{d}{dx}(x^{14}) \quad \text{gives} \quad \frac{dy}{dx} = 14x^{13}$$

Similarly, we can use this notation to indicate the derivative of an expression.

$$\frac{d}{dx}(x^{-2}) = -2x^{-3}$$

The differentiation rules are stated and proved for the independent variable x, but they also apply to other independent variables. The following examples illustrate differentiation involving variables other than x and y.

● **EXAMPLE 2** *Derivatives*

Find the derivatives of the following functions.

(a) $u(s) = s^8$ (b) $p = q^{2/3}$ (c) $C(t) = \sqrt{t}$ (d) $s = \dfrac{1}{\sqrt{t}}$

Solution

(a) If $u(s) = s^8$, then $u'(s) = 8s^{8-1} = 8s^7$.

(b) If $p = q^{2/3}$, then

$$\frac{dp}{dq} = \frac{2}{3}q^{2/3-1} = \frac{2}{3}q^{-1/3} = \frac{2}{3q^{1/3}}$$

(c) Writing $\sqrt{t}$ in its equivalent form, $t^{1/2}$, permits us to use the derivative formula.

$$C'(t) = \frac{1}{2}t^{1/2-1} = \frac{1}{2}t^{-1/2}$$

Writing the derivative in radical form gives

$$C'(t) = \frac{1}{2} \cdot \frac{1}{t^{1/2}} = \frac{1}{2\sqrt{t}}$$

(d) Writing $1/\sqrt{t}$ as a power of t gives

$$s = \frac{1}{t^{1/2}} = t^{-1/2}$$

so

$$\frac{ds}{dt} = -\frac{1}{2}t^{-1/2-1} = -\frac{1}{2}t^{-3/2}$$

Writing the derivative in a form similar to that of the original function gives

$$\frac{ds}{dt} = -\frac{1}{2} \cdot \frac{1}{t^{3/2}} = -\frac{1}{2\sqrt{t^3}}$$

● **EXAMPLE 3** *Tangent Line*

Write the equation of the tangent line to the graph of $y = x^3$ at $x = 1$.

Solution

Writing the equation of the tangent line to $y = x^3$ at $x = 1$ involves three steps.

1. Evaluate the function to find the point of tangency.
   At $x = 1$: $y = (1)^3 = 1$, so the point is $(1, 1)$
2. Evaluate the derivative to find the slope of the tangent.
   At any point: $m_{\tan} = y' = 3x^2$
   At $x = 1$: $m_{\tan} = y'|_{x=1} = 3(1^2) = 3$
3. Use $y - y_1 = m(x - x_1)$ with the point $(1, 1)$ and slope $m = 3$.
   $y - 1 = 3(x - 1) \Rightarrow y = 3x - 3 + 1 \Rightarrow y = 3x - 2$

Figure 9.28 shows the graph of $y = x^3$ and the tangent line at $x = 1$.

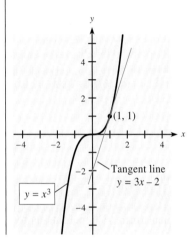

Figure 9.28

Derivative of a Constant

A function of the form $y = f(x) = c$, where c is a constant, is called a **constant function.** We can show that the derivative of a constant function is 0, as follows.

$$f'(x) = \lim_{h \to 0} \frac{f(x + h) - f(x)}{h} = \lim_{h \to 0} \frac{c - c}{h} = \lim_{h \to 0} 0 = 0$$

We can state this rule formally.

Constant Function Rule

If $f(x) = c$, where c is a constant, then $f'(x) = 0$.

● **EXAMPLE 4** *Derivative of a Constant*

Find the derivative of the function defined by $y = 4$.

Solution

Because 4 is a constant, $\dfrac{dy}{dx} = 0$.

Recall that the function defined by $y = 4$ has a horizontal line as its graph. Thus the slope of the line (and the derivative of the function) is 0.

Derivative of
y = c · u(x)

We now can take derivatives of constant functions and powers of x. But we do not yet have a rule for taking derivatives of functions of the form $f(x) = 4x^5$ or $g(t) = \frac{1}{2}t^2$. The following rule provides a method for handling functions of this type.

Coefficient Rule

If $f(x) = c \cdot u(x)$, where c is a constant and $u(x)$ is a differentiable function of x, then $f'(x) = c \cdot u'(x)$.

The formula above says that the derivative of a constant times a function is the constant times the derivative of the function.

Using Properties of Limits II and IV (from Section 9.1, "Limits"), we can show

$$\lim_{h \to 0} c \cdot g(h) = c \cdot \lim_{h \to 0} g(h)$$

We can use this result to verify the Coefficient Rule. If $f(x) = c \cdot u(x)$, then

$$f'(x) = \lim_{h \to 0} \frac{f(x + h) - f(x)}{h} = \lim_{h \to 0} \frac{c \cdot u(x + h) - c \cdot u(x)}{h}$$

$$= \lim_{h \to 0} c \cdot \left[\frac{u(x + h) - u(x)}{h} \right] = c \cdot \lim_{h \to 0} \frac{u(x + h) - u(x)}{h}$$

so $f'(x) = c \cdot u'(x)$.

● **EXAMPLE 5** *Coefficient Rule for Derivatives*

Find the derivatives of the following functions.

(a) $f(x) = 4x^5$ (b) $g(t) = \frac{1}{2}t^2$ (c) $p = \dfrac{5}{\sqrt{q}}$

Solution
(a) $f'(x) = 4(5x^4) = 20x^4$
(b) $g'(t) = \frac{1}{2}(2t) = t$
(c) $p = \dfrac{5}{\sqrt{q}} = 5q^{-1/2}$, so

$$\frac{dp}{dq} = 5\left(-\frac{1}{2}q^{-3/2}\right) = -\frac{5}{2\sqrt{q^3}}$$

● **EXAMPLE 6** *World Tourism*

Over the past 20 years, world tourism has grown into one of the world's major industries. From 1990 to 2002, the receipts from world tourism y, in billions of dollars, can be modeled by the function

$$y = 113.54x^{0.52584}$$

where x is the number of years past 1985 (*Source:* World Tourism Organization).

(a) Name the function that models the rate of change of the receipts from world tourism.
(b) Find the function from part (a).
(c) Find the rate of change in world tourism in 2007.

Solution

(a) The rate of change of world tourism receipts is modeled by the derivative.

(b) $\dfrac{dy}{dx} = 113.54(0.52584x^{0.52584 - 1}) \approx 59.704x^{-0.47416}$

(c) $\dfrac{dy}{dx}\bigg|_{x=22} = 59.704(22^{-0.47416}) \approx 13.787$

Thus, in 2007, the model predicts that world tourism will change by about \$13.787 billion per year.

Derivatives of Sums and Differences

In Example 6 of Section 9.3, "Average and Instantaneous Rates of Change: The Derivative," we found the derivative of $f(x) = 3x^2 + 2x + 11$ to be $f'(x) = 6x + 2$. This result, along with the results of several of the derivatives calculated in the exercises for that section, suggest that we can find the derivative of a function by finding the derivatives of its terms and combining them. The following rules state this formally.

Sum Rule

If $f(x) = u(x) + v(x)$, where u and v are differentiable functions of x, then $f'(x) = u'(x) + v'(x)$.

We can prove this rule as follows. If $f(x) = u(x) + v(x)$, then

$$f'(x) = \lim_{h \to 0} \frac{f(x + h) - f(x)}{h}$$

$$= \lim_{h \to 0} \frac{[u(x + h) + v(x + h)] - [u(x) + v(x)]}{h}$$

$$= \lim_{h \to 0} \left[\frac{u(x + h) - u(x)}{h} + \frac{v(x + h) - v(x)}{h} \right]$$

$$= \lim_{h \to 0} \frac{u(x + h) - u(x)}{h} + \lim_{h \to 0} \frac{v(x + h) - v(x)}{h}$$

$$= u'(x) + v'(x)$$

A similar rule applies to the difference of functions.

Difference Rule

If $f(x) = u(x) - v(x)$, where u and v are differentiable functions of x, then $f'(x) = u'(x) - v'(x)$.

● **EXAMPLE 7 Sum and Difference Rules**

Find the derivatives of the following functions.

(a) $y = 3x + 5$ (b) $f(x) = 4x^3 - 2x^2 + 5x - 3$

(c) $p = \frac{1}{3}q^3 + 2q^2 - 3$ (d) $u(x) = 5x^4 + x^{1/3}$

(e) $y = 4x^3 + \sqrt{x}$ (f) $s = 5t^6 - \dfrac{1}{t^2}$

Solution

(a) $y' = 3 \cdot 1 + 0 = 3$

(b) The rules regarding the derivatives of sums and differences of two functions also apply if more than two functions are involved. We may think of the functions that are added and subtracted as terms of the function f. Then it would be correct to say that we may take the derivative of a function term by term. Thus,

$$f'(x) = 4(3x^2) - 2(2x) + 5(1) - 0 = 12x^2 - 4x + 5$$

(c) $\dfrac{dp}{dq} = \frac{1}{3}(3q^2) + 2(2q) - 0 = q^2 + 4q$

(d) $u'(x) = 5(4x^3) + \frac{1}{3}x^{-2/3} = 20x^3 + \dfrac{1}{3x^{2/3}}$

(e) We may write the function as $y = 4x^3 + x^{1/2}$, so

$$y' = 4(3x^2) + \frac{1}{2}x^{-1/2} = 12x^2 + \dfrac{1}{2x^{1/2}}$$

$$= 12x^2 + \dfrac{1}{2\sqrt{x}}$$

(f) We may write $s = 5t^6 - 1/t^2$ as $s = 5t^6 - t^{-2}$, so

$$\dfrac{ds}{dt} = 5(6t^5) - (-2t^{-3}) = 30t^5 + 2t^{-3}$$

$$= 30t^5 + \dfrac{2}{t^3}$$

Each derivative in Example 7 has been *simplified*. This means that the final form of the derivative contains no negative exponents and the use of radicals or fractional exponents matches the original problem.

Also, in part (a) of Example 7, we saw that the derivative of $y = 3x + 5$ is 3. Because the slope of a line is the same at all points on the line, it is reasonable that the derivative of a linear equation is a constant. In particular, the slope of the graph of the equation $y = mx + b$ is m at all points on its graph because the derivative of $y = mx + b$ is $y' = f'(x) = m$.

◗ EXAMPLE 8 *Personal Income* **(Application Preview)**

Suppose that t is the number of years past 1970 and that

$$I = I(t) = 5.33t^2 + 81.5t + 822$$

models the U.S. total personal income in billions of current dollars. For 2007, find the model's prediction for

(a) the U.S. total personal income.
(b) the rate of change of U.S. total personal income.

Solution
(a) For 2007, $t = 37$ and

$$I(37) = 5.33(37^2) + 81.5(37) + 822 = \$11{,}134.27 \text{ billion (current dollars)}$$

(b) The rate of change of U.S. total personal income is given by

$$I'(t) = 10.66t + 81.5$$

The predicted rate for 2007 is

$$I'(37) = \$475.92 \text{ billion (current dollars) per year}$$

This predicts that U.S. total personal income will change by about \$475.92 billion from 2007 to 2008.

● **EXAMPLE 9** *Slope of the Tangent*

Find the slope of the tangent to $f(x) = \frac{1}{2}x^2 + 5x$ at each of the following.

(a) $x = 2$ (b) $x = -5$

Solution
The derivative of $f(x) = \frac{1}{2}x^2 + 5x$ is $f'(x) = x + 5$.

(a) At $x = 2$, the slope is $f'(2) = 2 + 5 = 7$.
(b) At $x = -5$, the slope is $f'(-5) = -5 + 5 = 0$. Thus when $x = -5$, the tangent to the curve is a horizontal line.

● **Checkpoint**

1. True or false: The derivative of a constant times a function is equal to the constant times the derivative of the function.
2. True or false: The derivative of the sum of two functions is equal to the sum of the derivatives of the two functions.
3. True or false: The derivative of the difference of two functions is equal to the difference of the derivatives of the two functions.
4. Does the Coefficient Rule apply to $f(x) = x^n/c$, where c is a constant? Explain.
5. Find the derivative of each of the following functions.

   (a) $f(x) = x^{10} - 10x + 5$ (b) $s = \dfrac{1}{t^5} - 10^7 + 1$

6. Find the slope of the line tangent to $f(x) = x^3 - 4x^2 + 1$ at $x = -1$.

● **EXAMPLE 10** *Horizontal Tangents*

Find all points on the graph of $f(x) = x^3 + 3x^2 - 45x + 4$ where the tangent line is horizontal.

Solution
A horizontal line has slope equal to 0. Thus, to find the desired points, we solve $f'(x) = 0$.

$$f'(x) = 3x^2 + 6x - 45$$

We solve $3x^2 + 6x - 45 = 0$ as follows:

$$3x^2 + 6x - 45 = 0$$
$$3(x^2 + 2x - 15) = 0$$
$$3(x + 5)(x - 3) = 0$$

Solving $3(x + 5)(x - 3) = 0$ gives $x = -5$ and $x = 3$. The y-coordinates for these x-values come from $f(x)$. The desired points are $(-5, f(-5)) = (-5, 179)$ and $(3, f(3)) = (3, -77)$. Figure 9.29 shows the graph of $y = f(x)$ with these points and the tangent lines at them indicated.

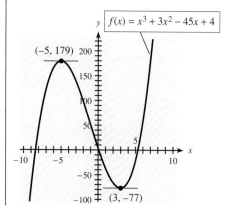

Figure 9.29

Marginal Revenue   The marginal revenue $R'(x)$ is used to estimate the change in revenue caused by the sale of one additional unit.

● **EXAMPLE 11 *Revenue***

Suppose that a manufacturer of a product knows that because of the demand for this product, his revenue is given by

$$R(x) = 1500x - 0.02x^2, \qquad 0 \le x \le 1000$$

where x is the number of units sold and $R(x)$ is in dollars.

(a) Find the marginal revenue at $x = 500$.
(b) Find the change in revenue caused by the increase in sales from 500 to 501 units.
(c) Find the difference between the marginal revenue found in (a) and the change in revenue found in (b).

Solution
(a) The marginal revenue for any value of x is

$$R'(x) = 1500 - 0.04x$$

The marginal revenue at $x = 500$ is

$$R'(500) = 1500 - 20 = 1480 \text{ (dollars per unit)}$$

We can interpret this to mean that the approximate revenue from the sale of the 501st unit will be \$1480.
(b) The revenue at $x = 500$ is $R(500) = 745{,}000$, and the revenue at $x = 501$ is $R(501) = 746{,}479.98$, so the change in revenue is

$$R(501) - R(500) = 746{,}479.98 - 745{,}000 = 1479.98 \text{ (dollars)}$$

(c) The difference is $1480 - 1479.98 = 0.02$. Thus we see that the marginal revenue at $x = 500$ is a good estimate of the revenue from the 501st unit.

Calculator Note

We have mentioned that graphing calculators have a numerical derivative feature that can be used to estimate the derivative of a function at a specific value of x. This feature can also be used to check the derivative of a function that has been computed with a formula. We graph both the derivative calculated with a formula and the numerical derivative. If the two graphs lie on top of one another, the computed derivative agrees with the numerical derivative. Figure 9.30 illustrates this idea for the derivative of $f(x) = \frac{1}{3}x^3 - 2x^2 + 4$. Figure 9.30(a) shows $f'(x) = x^2 - 4x$ as y_1 and the calculator's numerical derivative of $f(x)$ as y_2. Figure 9.30(b) shows the graphs of both y_1 and y_2 (the graphs are coincident).

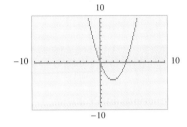

Figure 9.30   (a)   (b)

● **EXAMPLE 12** *Comparing f(x) and f'(x)*

(a) Graph $f(x) = x^3 - 3x + 3$ and its derivative $f'(x)$ on the same set of axes so that all values of x that make $f'(x) = 0$ are in the x-range.
(b) Investigate the graph of $y = f(x)$ near values of x where $f'(x) = 0$. Does the graph of $y = f(x)$ appear to turn at values where $f'(x) = 0$?
(c) Compare the interval of x values where $f'(x) < 0$ with the interval where the graph of $y = f(x)$ is decreasing from left to right.
(d) What is the relationship between the intervals where $f'(x) > 0$ and where the graph of $y = f(x)$ is increasing from left to right?

Solution

(a) The graphs of $f(x) = x^3 - 3x + 3$ and $f'(x) = 3x^2 - 3$ are shown in Figure 9.31.
(b) The values where $f'(x) = 0$ are the x-intercepts, $x = -1$ and $x = 1$. The graph of $y = x^3 - 3x + 3$ appears to turn at both these values.
(c) $f'(x) < 0$ where the graph of $y = f'(x)$ is below the x-axis, for $-1 < x < 1$. The graph of $y = f(x)$ appears to be decreasing on this interval.
(d) They appear to be the same intervals.

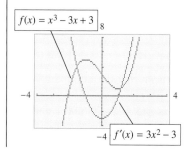

Figure 9.31

● ***Checkpoint Solutions***

1. True, by the Coefficient Rule.
2. True, by the Sum Rule.
3. True, by the Difference Rule.
4. Yes, $f(x) = x^n/c = (1/c)x^n$, so the coefficient is $(1/c)$.
5. (a) $f'(x) = 10x^9 - 10$
   (b) Note that $s = t^{-5} - 10^7 + 1$ and that 10^7 and 1 are constants.

$$\frac{ds}{dt} = -5t^{-6} = \frac{-5}{t^6}$$

6. The slope of the tangent at $x = -1$ is $f'(-1)$.

$$f'(x) = 3x^2 - 8x, \qquad f'(-1) = 3(-1)^2 - 8(-1) = 11$$

9.4 Exercises

Find the derivatives of the functions in Problems 1–10.

1. $y = 4$
2. $f(s) = 6$
3. $y = x$
4. $s = t^2$
5. $f(x) = 2x^3 - x^5$
6. $f(x) = 3x^4 - x^9$
7. $y = 6x^4 - 5x^2 + x - 2$
8. $y = 3x^5 - 5x^3 - 8x + 8$
9. $g(x) = 10x^9 - 5x^5 + 7x^3 + 5x - 6$
10. $h(x) = 12x^{20} + 8x^{10} - 2x^7 + 17x - 9$

In Problems 11–14, at the indicated points, find
(a) the slope of the tangent to the curve, and
(b) the instantaneous rate of change of the function.

11. $y = 4x^2 + 3x, \quad x = 2$
12. $C(x) = 3x^2 - 5, \quad (3, 22)$
13. $P(x) = x^2 - 4x, \quad (2, -4)$
14. $R(x) = 16x + x^2, \quad x = 1$

In Problems 15–22, find the derivative of each function.

15. $y = x^{-5} + x^{-8} - 3$ 16. $y = x^{-1} - x^{-2} + 13$
17. $y = 3x^{11/3} - 2x^{7/4} - x^{1/2} + 8$
18. $y = 5x^{8/5} - 3x^{5/6} + x^{1/3} + 5$
19. $f(x) = 5x^{-4/5} + 2x^{-4/3}$
20. $f(x) = 6x^{-8/3} - x^{-2/3}$
21. $g(x) = \dfrac{3}{x^4} + \dfrac{2}{x^5} + 5\sqrt[3]{x}$

22. $h(x) = \dfrac{7}{x^7} - \dfrac{3}{x^3} + 8\sqrt{x}$

In Problems 23–26, write the equation of the tangent line to each curve at the indicated point. As a check, graph both the function and the tangent line.

23. $y = x^3 - 3x^2 + 5 \quad$ at $x = 1$
24. $y = x^4 - 4x^3 - 2 \quad$ at $x = 2$

25. $f(x) = 4x^2 - \dfrac{1}{x} \quad$ at $x = -\dfrac{1}{2}$

26. $f(x) = \dfrac{x^3}{3} - \dfrac{3}{x^3} \quad$ at $x = -1$

In Problems 27–30, find the coordinates of points where the graph of $f(x)$ has horizontal tangents. As a check, graph $f(x)$ and see whether the points you found look as though they have horizontal tangents.

27. $f(x) = -x^3 + 9x^2 - 15x + 6$

28. $f(x) = \dfrac{1}{3}x^3 - 3x^2 - 16x + 8$

29. $f(x) = x^4 - 4x^3 + 9$
30. $f(x) = 3x^5 - 5x^3 + 2$

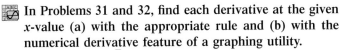 In Problems 31 and 32, find each derivative at the given x-value (a) with the appropriate rule and (b) with the numerical derivative feature of a graphing utility.

31. $y = 5 - 2\sqrt{x} \quad$ at $x = 4$
32. $y = 1 + 3x^{2/3} \quad$ at $x = -8$

In Problems 33–36, complete the following.
(a) Calculate the derivative of each function with the appropriate formula.
(b) Check your result from part (a) by graphing your calculated derivative and the numerical derivative of the given function with respect to x evaluated at x.

33. $f(x) = 2x^3 + 5x - \pi^4 + 8$
34. $f(x) = 3x^2 - 8x + 2^5 - 20$
35. $h(x) = \dfrac{10}{x^3} - \dfrac{10}{\sqrt[5]{x^2}} + x^2 + 1$

36. $g(x) = \dfrac{5}{x^{10}} + \dfrac{4}{\sqrt[4]{x^3}} + x^5 - 4$

The tangent line to a curve at a point closely approximates the curve near the point. In fact, for x-values close enough to the point of tangency, the function and its tangent line are virtually indistinguishable. Problems 37 and 38 explore this relationship. Use each given function and the indicated point to complete the following.
(a) Write the equation of the tangent line to the curve at the indicated point.
(b) Use a graphing utility to graph both the function and its tangent line. Be sure your graph shows the point of tangency.
(c) Repeatedly zoom in on the point of tangency until the function and the tangent line cannot be distinguished. Identify the x- and y-ranges in this window.

37. $f(x) = 3x^2 + 2x \quad$ at $x = 1$
38. $f(x) = 4x - x^2 \quad$ at $x = 5$

For each function in Problems 39–42, do the following.
(a) Find $f'(x)$.
(b) Graph both $f(x)$ and $f'(x)$ with a graphing utility.
(c) Use the graph of $f'(x)$ to identify x-values where $f'(x) = 0$, $f'(x) > 0$, and $f'(x) < 0$.
(d) Use the graph of $f(x)$ to identify x-values where $f(x)$ has a maximum or minimum point, where the graph of $f(x)$ is rising, and where the graph of $f(x)$ is falling.

39. $f(x) = 8 - 2x - x^2$
40. $f(x) = x^2 + 4x - 12$
41. $f(x) = x^3 - 12x - 5$

42. $f(x) = 7 - 3x^2 - \dfrac{x^3}{3}$

APPLICATIONS

43. *Revenue*   Suppose that a wholesaler expects that his monthly revenue, in dollars, for small television sets will be

$$R(x) = 100x - 0.1x^2, \quad 0 \le x \le 800$$

where x is the number of units sold. Find his marginal revenue and interpret it when the quantity sold is
(a) $x = 300$
(b) $x = 600$

44. Revenue The total revenue, in dollars, for a commodity is described by the function

$$R = 300x - 0.02x^2$$

(a) What is the marginal revenue when 40 units are sold?
(b) Interpret your answer to part (a).

45. Workers and output The number of units of weekly output of a certain product is

$$Q(x) = 200x + 6x^2$$

where x is the number of workers on the assembly line. There are presently 60 workers on the line.
(a) Find $Q'(x)$ and estimate the change in the weekly output caused by the addition of one worker.
(b) Calculate $Q(61) - Q(60)$ to see the actual change in the weekly output.

46. Capital investment and output The monthly output of a certain product is

$$Q(x) = 800x^{5/2}$$

where x is the capital investment in millions of dollars. Find dQ/dx, which can be used to estimate the effect on the output if an additional capital investment of $1 million is made.

47. Demand The demand for q units of a product depends on the price p (in dollars) according to

$$q = \frac{1000}{\sqrt{p}} - 1, \qquad \text{for } p > 0$$

Find and explain the meaning of the instantaneous rate of change of demand with respect to price when the price is
(a) $25   (b) $100

48. Demand Suppose that the demand for a product depends on the price p according to

$$D(p) = \frac{50{,}000}{p^2} - \frac{1}{2}, \qquad p > 0$$

where p is in dollars. Find and explain the meaning of the instantaneous rate of change of demand with respect to price when
(a) $p = 50$ (b) $p = 100$

49. Cost and average cost Suppose that the total cost function, in dollars, for the production of x units of a product is given by

$$C(x) = 4000 + 55x + 0.1x^2$$

Then the average cost of producing x items is

$$\overline{C(x)} = \frac{\text{total cost}}{x} = \frac{4000}{x} + 55 + 0.1x$$

(a) Find the instantaneous rate of change of average cost with respect to the number of units produced, at any level of production.
(b) Find the level of production at which this rate of change equals zero.
(c) At the value found in part (b), find the instantaneous rate of change of cost and find the average cost. What do you notice?

50. Cost and average cost Suppose that the total cost function, in dollars, for a certain commodity is given by

$$C(x) = 40{,}500 + 190x + 0.2x^2$$

where x is the number of units produced.
(a) Find the instantaneous rate of change of the average cost

$$\overline{C} = \frac{40{,}500}{x} + 190 + 0.2x$$

for any level of production.
(b) Find the level of production where this rate of change equals zero.
(c) At the value found in part (b), find the instantaneous rate of change of cost and find the average cost. What do you notice?

51. Cost-benefit Suppose that for a certain city the cost C, in dollars, of obtaining drinking water that contains p percent impurities (by volume) is given by

$$C = \frac{120{,}000}{p} - 1200$$

(a) Find the rate of change of cost with respect to p when impurities account for 1% (by volume).
(b) Write a sentence that explains the meaning of your answer in part (a).

52. Cost-benefit Suppose that the cost C, in dollars, of processing the exhaust gases at an industrial site to ensure that only p percent of the particulate pollution escapes is given by

$$C(p) = \frac{8100(100 - p)}{p}$$

(a) Find the rate of change of cost C with respect to the percent of particulate pollution that escapes when $p = 2$ (percent).
(b) Write a sentence interpreting your answer to part (a).

53. Wind chill One form of the formula that meteorologists use to calculate wind chill temperature (WC) is

$$WC = 48.064 + 0.474t - 0.020ts - 1.85s$$
$$+ 0.304t\sqrt{s} - 27.74\sqrt{s}$$

where s is the wind speed in mph and t is the actual air temperature in degrees Fahrenheit. Suppose temperature is constant at 15°.
(a) Express wind chill WC as a function of wind speed s.
(b) Find the rate of change of wind chill with respect to wind speed when the wind speed is 25 mph.
(c) Interpret your answer to part (b).

54. *Allometric relationships—crabs* For fiddler crabs, data gathered by Thompson* show that the allometric relationship between the weight C of the claw and the weight W of the body is given by

$$C = 0.11W^{1.54}$$

Find the function that gives the rate of change of claw weight with respect to body weight.

55. **Modeling** *U.S. cellular subscriberships* The following table shows the number of U.S. cellular subscriberships (in millions) from 1985 to 2002.

Year	Subscriberships	Year	Subscriberships
1985	0.340	1994	24.134
1986	0.682	1995	33.786
1987	1.231	1996	44.043
1988	2.069	1997	55.312
1989	3.509	1998	69.209
1990	5.283	1999	86.047
1991	7.557	2000	109.478
1992	11.033	2001	128.375
1993	16.009	2002	140.767

Source: The CTIA Semi-Annual Wireless Industry Survey

(a) Model these data with a power function $S(x)$, where S is the number of U.S. cellular subscriberships (in millions) and x is the number of years past 1980.
(b) For the period from 1992 to 2002, use the data to find the average rate of change of U.S. cellular subscriberships.
(c) Find and interpret the instantaneous rate of change of the modeling function $S(x)$ for the year 2002.

56. **Modeling** *U.S. poverty threshold* The table shows the yearly income poverty threshold for a single person for selected years from 1990 to 2002.

Year	Income	Year	Income
1990	$6652	2000	$8794
1995	7763	2001	9039
1998	8316	2002	9182
1999	8501		

Source: U.S. Bureau of the Census

*d'Arcy Thompson, *On Growth and Form* (Cambridge, England: Cambridge University Press, 1961).

(a) Model these data with a power function $p(t)$, where p is the poverty threshold income for a single person and t is the number of years past 1980.
(b) Find the function that models the rate of change of the poverty threshold income for a single person.
(c) Use the model in (b) to find and interpret the rate of change of the poverty threshold income in 2002.

57. **Modeling** *U.S. population* The table gives the U.S. population to the nearest million (actual or projected) for selected years.

Year	Population (in millions)	Year	Population (in millions)
1960	181	1995	263
1965	194	1998	271
1970	205	2000	281
1975	216	2003	294
1980	228	2025	358
1985	238	2050	409
1990	250		

Source: U.S. Bureau of the Census

(a) Find a cubic function $P(t)$ that models these data, where P is the U.S. population in millions and t is the number of years past 1960.
(b) Find the function that models the instantaneous rate of change of the U.S. population.
(c) Find and interpret the instantaneous rates of change in 2000 and 2025.

58. **Modeling** *Population below poverty level* The table below shows the number of millions of people in the United States who lived below the poverty level for selected years between 1960 and 2002.

Year	Persons Living Below the Poverty Level (millions)	Year	Persons Living Below the Poverty Level (millions)
1960	39.9	1993	39.3
1965	33.2	1996	36.5
1970	25.4	1997	35.6
1975	25.9	1998	34.5
1980	29.3	1999	32.3
1986	32.4	2000	31.1
1990	33.6	2002	34.6

Source: Bureau of the Census, U.S. Department of Commerce

(a) Model these data with a cubic function $p = p(t)$, where p is the number of millions of people and t represents the number of years since 1960.
(b) Find the rate of growth in the number of persons living below the poverty level in 2002.
(c) Interpret your answer to part (b).

9.5

The Product Rule and the Quotient Rule

OBJECTIVES

- To use the Product Rule to find the derivatives of certain functions
- To use the Quotient Rule to find the derivatives of certain functions

◗ **Application Preview**

When medicine is administered, reaction (measured in change of blood pressure or temperature) can be modeled by

$$R = m^2\left(\frac{c}{2} - \frac{m}{3}\right)$$

where c is a positive constant and m is the amount of medicine absorbed into the blood.* The rate of change of R with respect to m is the sensitivity of the body to medicine. To find an expression for sensitivity as a function of m, we calculate dR/dm. We can find this derivative with the **Product Rule** for derivatives. See Example 6.

Product Rule

We have simple formulas for finding the derivatives of the sums and differences of functions. But we are not so lucky with products. The derivative of a product is *not* the product of the derivatives. To see this, we consider the function $f(x) = x \cdot x$. Because this function is $f(x) = x^2$, its derivative is $f'(x) = 2x$. But the product of the derivatives of x and x would give $1 \cdot 1 = 1 \neq 2x$. Thus we need a different formula to find the derivative of a product. This formula is given by the Product Rule.

Product Rule

> If $f(x) = u(x) \cdot v(x)$, where u and v are differentiable functions of x, then
>
> $$f'(x) = u(x) \cdot v'(x) + v(x) \cdot u'(x)$$

Thus the derivative of a product of two functions is the first function times the derivative of the second plus the second function times the derivative of the first.

We can prove the Product Rule as follows. If $f(x) = u(x) \cdot v(x)$, then

$$\lim_{h \to 0} \frac{f(x + h) - f(x)}{h} = \lim_{h \to 0} \frac{u(x + h) \cdot v(x + h) - u(x) \cdot v(x)}{h}$$

Subtracting and adding $u(x + h) \cdot v(x)$ in the numerator gives

$$f'(x) = \lim_{h \to 0} \frac{u(x + h) \cdot v(x + h) - u(x + h) \cdot v(x) + u(x + h) \cdot v(x) - u(x) \cdot v(x)}{h}$$

$$= \lim_{h \to 0}\left(u(x + h)\left[\frac{v(x + h) - v(x)}{h}\right] + v(x)\left[\frac{u(x + h) - u(x)}{h}\right]\right)$$

Properties III and IV of limits (from Section 9.1, "Limits") give

$$f'(x) = \lim_{h \to 0} u(x + h) \cdot \lim_{h \to 0} \frac{v(x + h) - v(x)}{h} + \lim_{h \to 0} v(x) \cdot \lim_{h \to 0} \frac{u(x + h) - u(x)}{h}$$

Because u is differentiable and hence continuous, it follows that $\lim_{h \to 0} u(x + h) = u(x)$, so we have the formula we seek:

$$f'(x) = u(x) \cdot v'(x) + v(x) \cdot u'(x)$$

*Source: R. M. Thrall et al., *Some Mathematical Models in Biology,* U.S. Department of Commerce, 1967.

● **EXAMPLE 1** *Product Rule*

Find dy/dx if $y = (2x^3 + 3x + 1)(x^2 + 4)$.

Solution
Using the Product Rule with $u(x) = 2x^3 + 3x + 1$ and $v(x) = x^2 + 4$, we have

$$\frac{dy}{dx} = (2x^3 + 3x + 1)(2x) + (x^2 + 4)(6x^2 + 3)$$

$$= 4x^4 + 6x^2 + 2x + 6x^4 + 3x^2 + 24x^2 + 12$$

$$= 10x^4 + 33x^2 + 2x + 12$$

We could, of course, avoid using the Product Rule by multiplying the two factors before taking the derivative. But multiplying the factors first may involve more work than using the Product Rule.

● **EXAMPLE 2** *Slope of a Tangent Line*

Given $f(x) = (4x^3 + 5x^2 - 6x + 5)(x^3 - 4x^2 + 1)$, find the slope of the tangent to the graph of $y = f(x)$ at $x = 1$.

Solution

$$f'(x) = (4x^3 + 5x^2 - 6x + 5)(3x^2 - 8x) + (x^3 - 4x^2 + 1)(12x^2 + 10x - 6)$$

If we substitute $x = 1$ into $f'(x)$, we find that the slope of the curve at $x = 1$ is $f'(1) = 8(-5) + (-2)(16) = -72$.

Quotient Rule   For a quotient of two functions, we might be tempted to take the derivative of the numerator divided by the derivative of the denominator; but this is incorrect. With the example $f(x) = x^3/x$ (which equals x^2 if $x \neq 0$), this approach would give $3x^2/1 = 3x^2$ as the derivative, rather than $2x$. Thus, finding the derivative of a function that is the quotient of two functions requires a new formula.

Quotient Rule   If $f(x) = \dfrac{u(x)}{v(x)}$, where u and v are differentiable functions of x, with $v(x) \neq 0$, then

$$f'(x) = \frac{v(x) \cdot u'(x) - u(x) \cdot v'(x)}{[v(x)]^2}$$

The preceding formula says that the derivative of a quotient is the denominator times the derivative of the numerator minus the numerator times the derivative of the denominator, all divided by the square of the denominator.

To see that this rule is reasonable, again consider the function $f(x) = x^3/x$, $x \neq 0$. Using the Quotient Rule, with $u(x) = x^3$ and $v(x) = x$, we get

$$f'(x) = \frac{x(3x^2) - x^3(1)}{x^2} = \frac{3x^3 - x^3}{x^2} = \frac{2x^3}{x^2} = 2x$$

We see that $f'(x) = 2x$ is the correct derivative. The proof of the Quotient Rule is left for the student in Problem 37 of the exercises in this section.

● **EXAMPLE 3** *Quotient Rule*

(a) If $f(x) = \dfrac{x^2 - 4x}{x + 5}$, find $f'(x)$.

(b) If $f(x) = \dfrac{x^3 - 3x^2 + 2}{x^2 - 4}$, find the instantaneous rate of change of $f(x)$ at $x = 3$.

Solution

(a) Using the Quotient Rule with $u(x) = x^2 - 4x$ and $v(x) = x + 5$, we get

$$f'(x) = \frac{(x + 5)(2x - 4) - (x^2 - 4x)(1)}{(x + 5)^2}$$

$$= \frac{2x^2 + 6x - 20 - x^2 + 4x}{(x + 5)^2}$$

$$= \frac{x^2 + 10x - 20}{(x + 5)^2}$$

(b) We evaluate $f'(x)$ at $x = 3$ to find the desired rate of change. Using the Quotient Rule with $u(x) = x^3 - 3x^2 + 2$ and $v(x) = x^2 - 4$, we get

$$f'(x) = \frac{(x^2 - 4)(3x^2 - 6x) - (x^3 - 3x^2 + 2)(2x)}{(x^2 - 4)^2}$$

$$= \frac{(3x^4 - 6x^3 - 12x^2 + 24x) - (2x^4 - 6x^3 + 4x)}{(x^2 - 4)^2}$$

$$= \frac{x^4 - 12x^2 + 20x}{(x^2 - 4)^2}$$

Thus, the instantaneous rate of change at $x = 3$ is

$$f'(3) = \frac{3^4 - 12(3^2) + 20(3)}{(3^2 - 4)^2} = \frac{33}{25} = 1.32$$

● **EXAMPLE 4** *Quotient Rule*

Use the Quotient Rule to find the derivative of $y = 1/x^3$.

Solution

Letting $u(x) = 1$ and $v(x) = x^3$, we get

$$y' = \frac{x^3(0) - 1(3x^2)}{(x^3)^2} = -\frac{3x^2}{x^6} = -\frac{3}{x^4}$$

Note that we could have found the derivative more easily by rewriting y.

$$y = 1/x^3 = x^{-3} \qquad \text{gives} \qquad y' = -3x^{-4} = -\frac{3}{x^4}$$

Recall that we proved the Powers of x Rule for positive integer powers and assumed that it was true for all real number powers. In Problem 38 of the exercises in this section, you will be asked to use the Quotient Rule to show that the Powers of x Rule applies to negative integers.

It is not necessary to use the Quotient Rule when the denominator of the function in question contains only a constant. For example, the function $y = (x^3 - 3x)/3$ can be written $y = \frac{1}{3}(x^3 - 3x)$, so the derivative is $y' = \frac{1}{3}(3x^2 - 3) = x^2 - 1$.

● Checkpoint

1. True or false: The derivative of the product of two functions is equal to the product of the derivatives of the two functions.
2. True or false: The derivative of the quotient of two functions is equal to the quotient of the derivatives of the two functions.
3. Find $f'(x)$ for each of the following.
    (a) $f(x) = (x^{12} + 8x^5 - 7)(10x^7 - 4x + 19)$ Do not simplify.
    (b) $f(x) = \dfrac{2x^4 + 3}{3x^4 + 2}$ Simplify.
4. If $y = \frac{4}{3}(x^2 + 3x - 4)$, does finding y' require the Product Rule? Explain.
5. If $y = f(x)/c$, where c is a constant, does finding y' require the Quotient Rule? Explain.

● EXAMPLE 5 *Marginal Revenue*

Suppose that the revenue function for a product is given by

$$R(x) = 10x + \frac{100x}{3x + 5}$$

where x is the number of units sold and R is in dollars.

(a) Find the marginal revenue function.
(b) Find the marginal revenue when $x = 15$.

Solution

(a) We must use the Quotient Rule to find the marginal revenue (the derivative).

$$\overline{MR} = R'(x) = 10 + \frac{(3x + 5)(100) - 100x(3)}{(3x + 5)^2}$$

$$= 10 + \frac{300x + 500 - 300x}{(3x + 5)^2} = 10 + \frac{500}{(3x + 5)^2}$$

(b) The marginal revenue when $x = 15$ is $R'(15)$.

$$R'(15) = 10 + \frac{500}{[(3)(15) + 5]^2} = 10 + \frac{500}{(50)^2}$$

$$= 10 + \frac{500}{2500} = 10.20 \text{ (dollars per unit)}$$

Recall that $R'(15)$ estimates the revenue from the sale of the 16th item.

◐ EXAMPLE 6 *Sensitivity to a Drug* **(Application Preview)**

When medicine is administered, reaction (measured in change of blood pressure or temperature) can be modeled by

$$R = m^2\left(\frac{c}{2} - \frac{m}{3}\right)$$

where c is a positive constant and m is the amount of medicine absorbed into the blood. The rate of change of R with respect to m is the sensitivity of the body to medicine. Find an expression for sensitivity s as a function of m.

Solution

The sensitivity is the rate of change of R with respect to m, or the derivative

$$\frac{dR}{dm} = m^2\left(0 - \frac{1}{3}\right) + \left(\frac{c}{2} - \frac{1}{3}m\right)(2m)$$

$$= -\frac{1}{3}m^2 + mc - \frac{2}{3}m^2 = mc - m^2$$

so sensitivity is given by

$$s = \frac{dR}{dm} = mc - m^2$$

● EXAMPLE 7 *Derivatives and Graphs*

(a) Graph $f(x) = \dfrac{8x}{x^2 + 4}$ and its derivative on the same set of axes over an interval that contains the x-values where $f'(x) = 0$.

(b) Determine the values of x where $f'(x) = 0$ and the intervals where $f'(x) > 0$ and where $f'(x) < 0$.

(c) What is the relationship between the derivative $f'(x)$ being positive or negative and the graph of $y = f(x)$ increasing (rising) or decreasing (falling)?

Solution

We find the derivative of $f(x)$ by using the Quotient Rule.

$$f'(x) = \frac{(x^2 + 4)(8) - (8x)(2x)}{(x^2 + 4)^2}$$

$$= \frac{32 - 8x^2}{(x^2 + 4)^2}$$

(a) The graphs of $f(x) = \dfrac{8x}{x^2 + 4}$ and $f'(x) = \dfrac{32 - 8x^2}{(x^2 + 4)^2}$ are shown in Figure 9.32.

(b) We can use ZERO or another solution feature to find that $f'(x) = 0$ at $x = -2$ and $x = 2$. In the interval $-2 < x < 2$, the graph of $y = f'(x)$ is above the x-axis, so $f'(x)$ is positive for $-2 < x < 2$. In the intervals $x < -2$ and $x > 2$, the graph of $y = f'(x)$ is below the x-axis, so $f'(x)$ is negative there.

(c) The graph of $y = f(x)$ appears to turn at $x = -2$, and at $x = 2$, where $f'(x) = 0$. The graph of $y = f(x)$ is increasing where $f'(x) > 0$ and decreasing where $f'(x) < 0$.

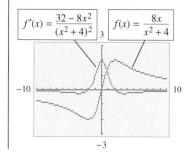

Figure 9.32

● ***Checkpoint Solutions***

1. False. The derivative of a product is equal to the first function times the derivative of the second plus the second function times the derivative of the first. That is,

$$\frac{d}{dx}(fg) = f \cdot \frac{dg}{dx} + g \cdot \frac{df}{dx}$$

2. False. The derivative of a quotient is equal to the denominator times the derivative of the numerator minus the numerator times the derivative of the denominator, all divided by the square of the denominator. That is,

$$\frac{d}{dx}\left(\frac{f}{g}\right) = \frac{g \cdot f' - f \cdot g'}{g^2}$$

3. (a) $f'(x) = (x^{12} + 8x^5 - 7)(70x^6 - 4) + (10x^7 - 4x + 19)(12x^{11} + 40x^4)$

   (b) $f'(x) = \dfrac{(3x^4 + 2)(8x^3) - (2x^4 + 3)(12x^3)}{(3x^4 + 2)^2}$

   $= \dfrac{24x^7 + 16x^3 - 24x^7 - 36x^3}{(3x^4 + 2)^2} = \dfrac{-20x^3}{(3x^4 + 2)^2}$

4. No; y' can be found with the Coefficient Rule:

$$y' = \frac{4}{3}(2x + 3)$$

5. No; y' can be found with the Coefficient Rule:

$$y' = \left(\frac{1}{c}\right)f'(x)$$

9.5 Exercises

1. Find y' if $y = (x + 3)(x^2 - 2x)$.

2. Find $\dfrac{ds}{dt}$ if $s = (t^4 + 1)(t^3 - 1)$.

3. If $f(x) = (x^{12} + 3x^4 + 4)(4x^3 - 1)$, find $f'(x)$.

4. If $y = (3x^7 + 4)(8x^6 - 6x^4 - 9)$, find $\dfrac{dy}{dx}$.

In Problems 5–8, find the derivative, but do not simplify your answer.

5. $y = (7x^6 - 5x^4 + 2x^2 - 1)(4x^9 + 3x^7 - 5x^2 + 3x)$

6. $y = (9x^9 - 7x^7 - 6x)(3x^5 - 4x^4 + 3x^3 - 8)$

7. $y = (x^2 + x + 1)(\sqrt[3]{x} - 2\sqrt{x} + 5)$

8. $y = (\sqrt[5]{x} - 2\sqrt[4]{x} + 1)(x^3 - 5x - 7)$

**In Problems 9 and 10, at each indicated point find
(a) the slope of the tangent line, and
(b) the instantaneous rate of change of the function.**

9. $y = (x^2 + 1)(x^3 - 4x)$ at $(-2, 0)$

10. $y = (x^3 - 3)(x^2 - 4x + 1)$ at $(2, -15)$

In Problems 11–20, find the indicated derivatives and simplify.

11. $\dfrac{dp}{dq}$ for $p = \dfrac{q^2 + 1}{q - 2}$

12. $C'(x)$ for $C(x) = \dfrac{2x^3}{3x^4 + 2}$

13. $\dfrac{dy}{dx}$ for $y = \dfrac{1 - 2x^2}{x^4 - 2x^2 + 5}$

14. $\dfrac{ds}{dt}$ for $s = \dfrac{t^3 - 4}{t^3 - 2t^2 - t - 5}$

15. $\dfrac{dz}{dx}$ for $z = x^2 + \dfrac{x^2}{1 - x - 2x^2}$

16. $\dfrac{dy}{dx}$ for $y = 200x - \dfrac{100x}{3x + 1}$

17. $\dfrac{dp}{dq}$ for $p = \dfrac{3\sqrt[3]{q}}{1 - q}$

18. $\dfrac{dy}{dx}$ for $y = \dfrac{2\sqrt{x} - 1}{1 - 4\sqrt{x^3}}$

19. y' for $y = \dfrac{x(x^2 + 4)}{x - 2}$

20. $f'(x)$ for $f(x) = \dfrac{(x + 1)(x - 2)}{x^2 + 1}$

In Problems 21 and 22, at the indicated point for each function, find
(a) the slope of the tangent line, and
(b) the instantaneous rate of change of the function.

21. $y = \dfrac{x^2 + 1}{x + 3}$ at $(2, 1)$

22. $y = \dfrac{x^2 - 4x}{x^2 + 2x}$ at $\left(2, -\dfrac{1}{2}\right)$

In Problems 23–26, write the equation of the tangent line to the graph of the function at the indicated point. Check the reasonableness of your answer by graphing both the function and the tangent line.

23. $y = (9x^2 - 6x + 1)(1 + 2x)$ at $x = 1$

24. $y = (4x^2 + 4x + 1)(7 - 2x)$ at $x = 0$

25. $y = \dfrac{3x^4 - 2x - 1}{4 - x^2}$ at $x = 1$

26. $y = \dfrac{x^2 - 4x}{2x - x^3}$ at $x = 2$

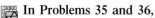 In Problems 27–30, use the numerical derivative feature of a graphing utility to find the derivative of each function at the given x-value.

27. $y = \left(4\sqrt{x} + \dfrac{3}{x}\right)\left(3\sqrt[3]{x} - \dfrac{5}{x^2} - 25\right)$ at $x = 1$

28. $y = (3\sqrt[4]{x^5} + \sqrt[5]{x^4} - 1)\left(\dfrac{2}{x^3} - \dfrac{1}{\sqrt{x}}\right)$ at $x = 1$

29. $f(x) = \dfrac{4x - 4}{3x^{2/3}}$ at $x = 1$

30. $f(x) = \dfrac{3\sqrt[3]{x} + 1}{x + 2}$ at $x = -1$

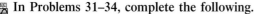 In Problems 31–34, complete the following.
(a) Find the derivative of each function, and check your work by graphing both your calculated derivative and the numerical derivative of the function.
(b) Use your graph of the derivative to find points where the original function has horizontal tangent lines.
(c) Use a graphing utility to graph the function and indicate the points found in part (b) on the graph.

31. $f(x) = (x^2 + 4x + 4)(x - 7)$
32. $f(x) = (x^2 - 14x + 49)(2x + 1)$

33. $y = \dfrac{x^2}{x - 2}$ 34. $y = \dfrac{x^2 - 7}{4 - x}$

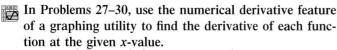 In Problems 35 and 36,
(a) find $f'(x)$.
(b) graph both $f(x)$ and $f'(x)$ with a graphing utility.
(c) identify the x-values where $f'(x) = 0$, $f'(x) > 0$, and $f'(x) < 0$.
(d) identify x-values where $f(x)$ has a maximum point or a minimum point, where $f(x)$ is increasing, and where $f(x)$ is decreasing.

35. $f(x) = \dfrac{10x^2}{x^2 + 1}$ 36. $f(x) = \dfrac{8 - x^2}{x^2 + 4}$

37. Prove the Quotient Rule for differentiation. (*Hint:* Add $[-u(x) \cdot v(x) + u(x) \cdot v(x)]$ to the expanded numerator and use steps similar to those used to prove the Product Rule.)

38. Use the Quotient Rule to show that the Powers of x Rule applies to negative integer powers. That is, show that $(d/dx)x^n = nx^{n-1}$ when $n = -k$, $k > 0$, by finding the derivative of $f(x) = 1/(x^k)$.

APPLICATIONS

39. *Cost-benefit*   If the cost C (in dollars) of removing p percent of the particulate pollution from the exhaust gases at an industrial site is given by

$$C(p) = \frac{8100p}{100 - p}$$

find the rate of change of C with respect to p.

40. *Cost-benefit*   If the cost C (in dollars) of removing p percent of the impurities from the waste water in a manufacturing process is given by

$$C(p) = \frac{9800p}{101 - p}$$

find the rate of change of C with respect to p.

41. *Revenue*   Suppose the revenue (in dollars) from the sale of x units of a product is given by

$$R(x) = \frac{60x^2 + 74x}{2x + 2}$$

Find the marginal revenue when 49 units are sold. Interpret your result.

42. *Revenue*   The revenue (in dollars) from the sale of x units of a product is given by

$$R(x) = \frac{3000}{2x + 2} + 80x - 1500$$

Find the marginal revenue when 149 units are sold. Interpret your result.

43. *Revenue*   A travel agency will plan a group tour for groups of size 25 or larger. If the group contains exactly 25 people, the cost is \$300 per person. If each person's cost is reduced by \$10 for each additional person above the 25, then the revenue is given by

$$R(x) = (25 + x)(300 - 10x)$$

where x is the number of additional people above 25. Find the marginal revenue if the group contains 30 people. Interpret your result.

44. *Revenue* McRobert's TV Shop sells 200 sets per month at a price of $400 per unit. Market research indicates that the shop can sell one additional set for each $1 it reduces the price, and in this case the total revenue is

$$R(x) = (200 + x)(400 - x)$$

where x is the number of additional sets beyond the 200. If the shop sells a total of 250 sets, find the marginal revenue. Interpret your result.

45. *Response to a drug* The reaction R to an injection of a drug is related to the dosage x (in milligrams) according to

$$R(x) = x^2\left(500 - \frac{x}{3}\right)$$

where 1000 mg is the maximum dosage. If the rate of reaction with respect to the dosage defines the sensitivity to the drug, find the sensitivity.

46. *Nerve response* The number of action potentials produced by a nerve, t seconds after a stimulus, is given by

$$N(t) = 25t + \frac{4}{t^2 + 2} - 2$$

Find the rate at which the action potentials are produced by the nerve.

47. *Test reliability* If a test having reliability r is lengthened by a factor n, the reliability of the new test is given by

$$R = \frac{nr}{1 + (n - 1)r}, \quad 0 < r \le 1$$

Find the rate at which R changes with respect to n.

48. *Advertising and sales* The sales of a product s (in thousands of dollars) are related to advertising expenses (in thousands of dollars) by

$$s = \frac{200x}{x + 10}$$

Find and interpret the meaning of the rate of change of sales with respect to advertising expenses when
(a) $x = 10$ (b) $x = 20$

49. *Candidate recognition* Suppose that the proportion P of voters who recognize a candidate's name t months after the start of the campaign is given by

$$P(t) = \frac{13t}{t^2 + 100} + 0.18$$

(a) Find the rate of change of P when $t = 6$, and explain its meaning.
(b) Find the rate of change of P when $t = 12$, and explain its meaning.
(c) One month prior to the election, is it better for $P'(t)$ to be positive or negative? Explain.

50. *Endangered species population* It is determined that a wildlife refuge can support a group of up to 120 of a certain endangered species. If 75 are introduced onto the refuge and their population after t years is given by

$$p(t) = 75\left(1 + \frac{4t}{t^2 + 16}\right)$$

find the rate of population growth after t years. Find the rate after each of the first 7 years.

51. *Wind chill* According to the National Climatic Data Center, during 1991, the lowest temperature recorded in Indianapolis, Indiana, was 0°F. If x is the wind speed in miles per hour and $x \ge 5$, then the wind chill (in degrees Fahrenheit) for an air temperature of 0°F can be approximated by the function

$$f(x) = \frac{289.173 - 58.5731x}{x + 1}$$

(a) At what rate is the wind chill changing when the wind speed is 20 mph?
(b) Explain the meaning of your answer to part (a).

52. *Response to injected adrenalin* Experimental evidence has shown that the concentration of injected adrenaline x is related to the response y of a muscle according to the equation

$$y = \frac{x}{a + bx}$$

where a and b are constants. Find the rate of change of response with respect to the concentration.

53. *Social Security beneficiaries* The following table gives the number of millions of Social Security beneficiaries for selected years from 1950 to 2000, with projections through 2030.

Year	Number of Beneficiaries (millions)	Year	Number of Beneficiaries (millions)
1950	2.9	2000	44.8
1960	14.3	2010	53.3 (projected)
1970	25.2	2020	68.8 (projected)
1980	35.1	2030	82.7 (projected)
1990	39.5		

Source: 2000 Social Security Trustees Report

With $B(t)$ representing the number of beneficiaries (in millions) t years past 1950, these data can be modeled by the function

$$B(t) = (0.01t + 3)(0.02383t^2 - 9.79t + 3097.19) - 9289.371$$

(a) Find the function that gives the instantaneous rate of change of the number of beneficiaries.
(b) Find and interpret the instantaneous rate of change in 2010.
(c) Use the data to determine which of the average rates of change (from 2000 to 2010, from 2010 to 2020, or from 2000 to 2020) best approximates the instantaneous rate from part (b).

54. *International phone traffic* The following table shows the billions of minutes of international phone traffic from 1990 to 2003. If M represents the minutes of international phone traffic (in billions) and t is the number of years past 1985, then the data can be modeled with the function

$$M(t) = (2t + 5)(-0.0168t^2 + 0.7641t - 6.98) + 88.44$$

(a) Find the function that gives the annual instantaneous rate of change of the minutes of international phone traffic.
(b) Find and interpret the instantaneous rate of change of the minutes of international phone traffic in 2003.

Year	International Phone Minutes (in billions)	Year	International Phone Minutes (in billions)
1990	33	1997	79
1991	38	1998	89
1992	43	1999	100
1993	49	2000	118
1994	57	2001	127
1995	63	2002	135
1996	71	2003	140

Source: International Telecommunications Union

 55. *Farm workers* The percent of U.S. workers in farm occupations during certain years is shown in the table below.

Year	Percent of All Workers in Farm Occupations	Year	Percent of All Workers in Farm Occupations
1820	71.8	1960	6.1
1850	63.7	1970	3.6
1870	53	1980	2.7
1900	37.5	1985	2.8
1920	27	1990	2.4
1930	21.2	1994	2.5
1940	17.4	2002	2.5
1950	11.6		

Source: Bureau of Labor Statistics

Assume that the percent of U.S. workers in farm occupations can be modeled with the function

$$f(t) = 1000 \cdot \frac{-7.4812t + 1560.2}{1.2882t^2 - 122.18t + 21{,}483}$$

where t is the number of years past 1800.
(a) Find the function that models the instantaneous rate of change of the percent of U.S. workers in farm occupations.
(b) Use the model in part (a) to find the instantaneous rates of change in 1870 and in 2000.
(c) Interpret each of the rates of change in part (b).

56. *Students per computer* The number of students per computer in public schools for selected school years from the 1983–84 through the 2001–2002 school years is given in the following table.

School Year	Students per Computer	School Year	Students per Computer
1983–84	125	1993–94	14
1985–86	50	1995–96	10
1987–88	32	1997–98	6.1
1989–90	22	1999–2000	5.4
1991–92	18	2001–2002	4.9

Source: Quality Education Data, Inc., Denver, CO

Assume the data can be modeled with the function

$$y = \frac{375.5 - 14.9x}{x + 0.02}$$

where y is the number of students per computer and x is the number of years past the school year ending in 1981.
(a) Find the function that gives the instantaneous rate of change of this model.
(b) Find and interpret the instantaneous rate of change of the students per computer for the school year ending in 2001.
(c) Use the data to find the average rate of change of the number of students per computer from the school year ending in 2000 to the school year ending in 2002. How well does this approximate the instantaneous rate found in part (b)?

9.6 — The Chain Rule and the Power Rule

OBJECTIVES

● To use the Chain Rule to differentiate functions
● To use the Power Rule to differentiate functions

◗ Application Preview

The demand x for a product is given by

$$x = \frac{98}{\sqrt{2p + 1}} - 1$$

where p is the price per unit. To find how fast demand is changing when price is \$24, we take the derivative of x with respect to p. If we write this function with a power rather than a radical, it has the form

$$x = 98(2p + 1)^{-1/2} - 1$$

The formulas learned so far cannot be used to find this derivative. We use a new formula, the **Power Rule,** to find this derivative. (See Example 7.) In this section we will discuss the **Chain Rule** and the Power Rule, which is one of the results of the Chain Rule, and we will use these formulas to solve applied problems.

Composite Functions

Recall from Section 1.2, "Functions," that if f and g are functions, then the composite functions g of f (denoted $g \circ f$) and f of g (denoted $f \circ g$) are defined as follows:

$$(g \circ f)(x) = g(f(x)) \quad \text{and} \quad (f \circ g)(x) = f(g(x))$$

● **EXAMPLE 1** *Composite Function*

If $f(x) = 3x^2$ and $g(x) = 2x - 1$, find $F(x) = f(g(x))$.

Solution
Substituting $g(x) = 2x - 1$ for x in $f(x)$ gives

$$f(g(x)) = f(2x - 1) = 3(2x - 1)^2$$

Thus $F(x) = 3(2x - 1)^2$.

Chain Rule

We could find the derivative of the function $F(x) = 3(2x - 1)^2$ by multiplying out the expression $3(2x - 1)^2$. Then

$$F(x) = 3(4x^2 - 4x + 1) = 12x^2 - 12x + 3$$

so $F'(x) = 24x - 12$. But we can also use a very powerful rule, called the **Chain Rule,** to find derivatives of composite functions. If we write the composite function $y = f(g(x))$ in the form $y = f(u)$, where $u = g(x)$, we state the Chain Rule as follows.

Chain Rule

If f and g are differentiable functions with $y = f(u)$, and $u = g(x)$, then y is a differentiable function of x, and

$$\frac{dy}{dx} = f'(u) \cdot g'(x)$$

or equivalently

$$\frac{dy}{dx} = \frac{dy}{du} \cdot \frac{du}{dx}$$

Note that dy/du represents the derivative of $y = f(u)$ *with respect to u* and du/dx represents the derivative of $u = g(x)$ *with respect to x.* For example, if $y = 3(2x - 1)^2$, then the outside function, f, is the squaring function, and the inside function, g, is $2x - 1$, so we may write $y = f(u) = 3u^2$, where $u = g(x) = 2x - 1$. Then the derivative is

$$\frac{dy}{dx} = \frac{dy}{du} \cdot \frac{du}{dx} = 6u \cdot 2 = 12u$$

To write this derivative in terms of x, we substitute $2x - 1$ for u. Thus

$$\frac{dy}{dx} = 12(2x - 1) = 24x - 12$$

Note that we get the same result by using the Chain Rule as we did by multiplying out $f(x) = 3(2x - 1)^2$. The Chain Rule is important because it is not always possible to rewrite the function as a polynomial. Consider the following example.

● **EXAMPLE 2 Chain Rule**

If $y = \sqrt{x^2 - 1}$, find $\dfrac{dy}{dx}$.

Solution

If we write this function as $y = f(u) = \sqrt{u} = u^{1/2}$, when $u = x^2 - 1$, we can find the derivative.

$$\frac{dy}{dx} = \frac{dy}{du} \cdot \frac{du}{dx} = \frac{1}{2} \cdot u^{-1/2} \cdot 2x = u^{-1/2} \cdot x = \frac{1}{\sqrt{u}} \cdot x = \frac{x}{\sqrt{u}}$$

To write this derivative in terms of x alone, we substitute $x^2 - 1$ for u. Then

$$\frac{dy}{dx} = \frac{x}{\sqrt{x^2 - 1}}$$

Note that we could not find the derivative of a function like that of Example 2 by the methods learned previously.

● **EXAMPLE 3 Chain Rule**

If $y = \dfrac{1}{x^2 + 3x + 1}$, find $\dfrac{dy}{dx}$.

Solution

If we let $u = x^2 + 3x + 1$, we can write $y = f(u) = \dfrac{1}{u}$, or $y = u^{-1}$. Then

$$\frac{dy}{dx} = \frac{dy}{du} \cdot \frac{du}{dx} = -u^{-2}(2x + 3) = \frac{-2x - 3}{u^2}$$

Substituting for u gives

$$\frac{dy}{dx} = \frac{-2x - 3}{(x^2 + 3x + 1)^2}$$

Note in Example 3 that we also could have found the derivative with the Quotient Rule and obtained the same result.

$$\frac{dy}{dx} = \frac{(x^2 + 3x + 1)(0) - (1)(2x + 3)}{(x^2 + 3x + 1)^2}$$

$$= \frac{-(2x + 3)}{(x^2 + 3x + 1)^2} = \frac{-2x - 3}{(x^2 + 3x + 1)^2}$$

● **EXAMPLE 4** *Allometric Relationships*

The relationship between the length L (in meters) and weight W (in kilograms) of a species of fish in the Pacific Ocean is given by $W = 10.375L^3$. The rate of growth in length is given by $\frac{dL}{dt} = 0.36 - 0.18L$, where t is measured in years.

(a) Determine a formula for the rate of growth in weight $\frac{dW}{dt}$ in terms of L.

(b) If a fish weighs 30 kilograms, approximate its rate of growth in weight using the formula found in (a).

Solution

(a) The rate of change uses the Chain Rule, as follows:

$$\frac{dW}{dt} = \frac{dW}{dL} \cdot \frac{dL}{dt} = 31.125L^2(0.36 - 0.18L) = 11.205L^2 - 5.6025L^3$$

(b) From $W = 10.375L^3$ and $W = 30$ kg, we can find L by solving

$$30 = 10.375L^3$$

$$\frac{30}{10.375} = L^3 \quad \text{so} \quad L = \sqrt[3]{\frac{30}{10.375}} \approx 1.4247 \text{ m}$$

Hence, the rate of growth in weight is

$$\frac{dW}{dt} = 11.205(1.4247)^2 - 5.6025(1.4247)^3 = 6.542 \text{ kilograms/year}$$

Power Rule

The Chain Rule is very useful and will be extremely important with functions that we will study later. A special case of the Chain Rule, called the **Power Rule,** is useful for the algebraic functions we have studied so far, composite functions where the outside function is a power.

Power Rule

If $y = u^n$, where u is a differentiable function of x, then

$$\frac{dy}{dx} = nu^{n-1} \cdot \frac{du}{dx}$$

● **EXAMPLE 5** *Power Rule*

(a) If $y = (x^2 - 4x)^6$, find $\frac{dy}{dx}$. (b) If $p = \frac{4}{3q^2 + 1}$, find $\frac{dp}{dq}$.

Solution

(a) This has the form $y = u^n = u^6$, with $u = x^2 - 4x$. Thus, by the Power Rule,

$$\frac{dy}{dx} = nu^{n-1} \cdot \frac{du}{dx} = 6u^5(2x - 4)$$

Substituting $x^2 - 4x$ for u gives

$$\frac{dy}{dx} = 6(x^2 - 4x)^5(2x - 4) = (12x - 24)(x^2 - 4x)^5$$

(b) We can use the Power Rule to find dp/dq if we write the equation in the form

$$p = 4(3q^2 + 1)^{-1}$$

Then

$$\frac{dp}{dq} = 4[-1(3q^2 + 1)^{-2}(6q)] = \frac{-24q}{(3q^2 + 1)^2}$$

The derivative of the function in Example 5(b) can also be found by using the Quotient Rule, but the Power Rule provides a more efficient method.

● **EXAMPLE 6** *Power Rule with a Radical*

Find the derivatives of (a) $y = 3\sqrt[3]{x^2 - 3x + 1}$ and (b) $g(x) = \dfrac{1}{\sqrt{(x^2 + 1)^3}}$.

Solution

(a) Because $y = 3(x^2 - 3x + 1)^{1/3}$, we can make use of the Power Rule with $u = x^2 - 3x + 1$.

$$y' = 3\left(nu^{n-1}\frac{du}{dx}\right) = 3\left[\frac{1}{3}u^{-2/3}(2x - 3)\right]$$

$$= (x^2 - 3x + 1)^{-2/3}(2x - 3)$$

$$= \frac{2x - 3}{(x^2 - 3x + 1)^{2/3}}$$

(b) Writing $g(x)$ as a power gives

$$g(x) = (x^2 + 1)^{-3/2}$$

Then

$$g'(x) = -\frac{3}{2}(x^2 + 1)^{-5/2}(2x) = -3x \cdot \frac{1}{(x^2 + 1)^{5/2}} = \frac{-3x}{\sqrt{(x^2 + 1)^5}}$$

● *Checkpoint*

1. (a) If $f(x) = (3x^4 + 1)^{10}$, does $f'(x) = 10(3x^4 + 1)^9$?

   (b) If $f(x) = (2x + 1)^5$, does $f'(x) = 10(2x + 1)^4$?

   (c) If $f(x) = \dfrac{[u(x)]^n}{c}$, where c is a constant, does $f'(x) = \dfrac{n[u(x)]^{n-1} \cdot u'(x)}{c}$?

2. (a) If $f(x) = \dfrac{12}{2x^2 - 1}$, find $f'(x)$ by using the Power Rule (not the Quotient Rule).

   (b) If $f(x) = \dfrac{\sqrt{x^3 - 1}}{3}$, find $f'(x)$ by using the Power Rule (not the Quotient Rule).

⚬ EXAMPLE 7 Demand (Application Preview)

The demand for x hundred units of a product is given by

$$x = 98(2p + 1)^{-1/2} - 1$$

where p is the price per unit in dollars. Find the rate of change of the demand with respect to price when $p = 24$.

Solution

The rate of change of demand with respect to price is

$$\frac{dx}{dp} = 98\left[-\frac{1}{2}(2p + 1)^{-3/2}(2) \right] = -98(2p + 1)^{-3/2}$$

When $p = 24$, the rate of change is

$$\frac{dx}{dp}\bigg|_{p=24} = -98(48 + 1)^{-3/2} = -98 \cdot \frac{1}{49^{3/2}}$$

$$= -98 \cdot \frac{1}{343} = -\frac{2}{7}$$

This means that when the price is \$24, demand is changing at the rate of $-2/7$ hundred units per dollar, or if the price changes by \$1, demand will change by about $-200/7$ units.

● **Checkpoint Solutions**

1. (a) No, $f'(x) = 10(3x^4 + 1)^9(12x^3)$.   (b) Yes   (c) Yes

2. (a) $f(x) = 12(2x^2 - 1)^{-1}$

$$f'(x) = -12(2x^2 - 1)^{-2}(4x) = \frac{-48x}{(2x^2 - 1)^2}$$

   (b) $f(x) = \frac{1}{3}(x^3 - 1)^{1/2}$

$$f'(x) = \frac{1}{6}(x^3 - 1)^{-1/2}(3x^2) = \frac{x^2}{2\sqrt{x^3 - 1}}$$

9.6 Exercises

In Problems 1–4, find $\dfrac{dy}{du}, \dfrac{du}{dx}$, and $\dfrac{dy}{dx}$.

1. $y = u^3$ and $u = x^2 + 1$
2. $y = u^4$ and $u = x^2 + 4x$
3. $y = u^4$ and $u = 4x^2 - x + 8$
4. $y = u^{10}$ and $u = x^2 + 5x$

Differentiate the functions in Problems 5–20.

5. $f(x) = (2x^4 - 5)^{25}$ 6. $g(x) = (3 - 2x)^{10}$
7. $h(x) = \frac{2}{3}(x^6 + 3x^2 - 11)^8$
8. $k(x) = \frac{5}{7}(2x^3 - x + 6)^{14}$
9. $g(x) = (x^2 + 4x)^{-2}$ 10. $p = (q^3 + 1)^{-5}$
11. $f(x) = \dfrac{1}{(x^2 + 2)^3}$ 12. $g(x) = \dfrac{1}{4x^3 + 1}$

13. $g(x) = \dfrac{1}{(2x^3 + 3x + 5)^{3/4}}$

14. $y = \dfrac{1}{(3x^3 + 4x + 1)^{3/2}}$

15. $y = \sqrt{x^2 + 4x + 5}$ 16. $y = \sqrt{x^2 + 3x}$

17. $y = \dfrac{8(x^2 - 3)^5}{5}$ 18. $y = \dfrac{5\sqrt{1 - x^3}}{6}$

19. $y = \dfrac{(3x + 1)^5 - 3x}{7}$

20. $y = \dfrac{\sqrt{2x - 1} - \sqrt{x}}{2}$

At the indicated point, for each function in Problems 21–24, find
(a) the slope of the tangent line, and
(b) the instantaneous rate of change of the function.
A graphing utility's numerical derivative feature can be used to check your work.

21. $y = (x^3 + 2x)^4$ at $x = 2$
22. $y = \sqrt{5x^2 + 2x}$ at $x = 1$
23. $y = \sqrt{x^3 + 1}$ at $(2, 3)$
24. $y = (4x^3 - 5x + 1)^3$ at $(1, 0)$

In Problems 25–28, write the equation of the line tangent to the graph of each function at the indicated point. As a check, graph both the function and the tangent line you found to see whether it looks correct.

25. $y = (x^2 - 3x + 3)^3$ at $(1, 1)$
26. $y = (x^2 + 1)^3$ at $(2, 125)$
27. $y = \sqrt{3x^2 - 2}$ at $x = 3$
28. $y = \left(\dfrac{1}{x^3 - x}\right)^3$ at $x = 2$

 In Problems 29 and 30, complete the following for each function.
(a) Find $f'(x)$.
(b) Check your result in part (a) by graphing both it and the numerical derivative of the function.
(c) Find x-values for which the slope of the tangent is 0.
(d) Find points (x, y) where the slope of the tangent is 0.
(e) Use a graphing utility to graph the function and locate the points found in part (d).

29. $f(x) = (x^2 - 4)^3 + 12$
30. $f(x) = 10 - (x^2 - 2x - 8)^2$

 In Problems 31 and 32, do the following for each function $f(x)$.
(a) Find $f'(x)$.
(b) Graph both $f(x)$ and $f'(x)$ with a graphing utility.
(c) Determine x-values where $f'(x) = 0$, $f'(x) > 0$, $f'(x) < 0$.
(d) Determine x-values for which $f(x)$ has a maximum or minimum point, where the graph is increasing, and where it is decreasing.

31. $f(x) = 12 - 3(1 - x^2)^{4/3}$
32. $f(x) = 3 + \dfrac{1}{16}(x^2 - 4x)^4$

In Problems 33 and 34, find the derivative of each function.

33. (a) $y = \dfrac{2x^3}{3}$ (b) $y = \dfrac{2}{3x^3}$
(c) $y = \dfrac{(2x)^3}{3}$ (d) $y = \dfrac{2}{(3x)^3}$

34. (a) $y = \dfrac{3}{(5x)^5}$ (b) $y = \dfrac{3x^5}{5}$
(c) $y = \dfrac{3}{5x^5}$ (d) $y = \dfrac{(3x)^5}{5}$

APPLICATIONS

35. *Ballistics*   Ballistics experts are able to identify the weapon that fired a certain bullet by studying the markings on the bullet. Tests are conducted by firing into a bale of paper. If the distance s, in inches, that the bullet travels into the paper is given by

$$s = 27 - (3 - 10t)^3$$

for $0 \le t \le 0.3$ second, find the velocity of the bullet one-tenth of a second after it hits the paper.

36. *Population of microorganisms*   Suppose that the population of a certain microorganism at time t (in minutes) is given by

$$P = 1000 - 1000(t + 10)^{-1}$$

Find the rate of change of population.

37. *Revenue*   The revenue from the sale of a product is

$$R = 1500x + 3000(2x + 3)^{-1} - 1000 \quad \text{dollars}$$

where x is the number of units sold. Find the marginal revenue when 100 units are sold. Interpret your result.

38. *Revenue*   The revenue from the sale of x units of a product is

$$R = 15(3x + 1)^{-1} + 50x - 15 \quad \text{dollars}$$

Find the marginal revenue when 40 units are sold. Interpret your result.

39. *Pricing and sales*   Suppose that the weekly sales volume y (in thousands of units sold) depends on the price per unit of the product according to

$$y = 32(3p + 1)^{-2/5}, \quad p > 0$$

where p is in dollars.
(a) What is the rate of change in sales volume when the price is \$21?
(b) Interpret your answer to part (a).

40. *Pricing and sales*   A chain of auto service stations has found that its monthly sales volume y (in thousands of dollars) is related to the price p (in dollars) of an oil change according to

$$y = \dfrac{90}{\sqrt{p + 5}}, \quad p > 10$$

(a) What is the rate of change of sales volume when the price is $20?

(b) Interpret your answer to part (a).

41. *Demand* Suppose that the demand for q units of a product priced at p per unit is described by

$$p = \frac{200,000}{(q + 1)^2}$$

(a) What is the rate of change of price with respect to the quantity demanded when $q = 49$?

(b) Interpret your answer to part (a).

Stimulus-response **The relation between the magnitude of a sensation y and the magnitude of the stimulus x is given by**

$$y = k(x - x_0)^n$$

where k is a constant, x_0 is the threshold of effective stimulus, and n depends on the type of stimulus. Find the rate of change of sensation with respect to the amount of stimulus for each of Problems 42–44.

42. For the stimulus of visual brightness $y = k(x - x_0)^{1/3}$

43. For the stimulus of warmth $y = k(x - x_0)^{8/5}$

44. For the stimulus of electrical stimulation $y = k(x - x_0)^{7/2}$

45. *Demand* If the demand for q units of a product priced at p per unit is described by the equation

$$p = \frac{100}{\sqrt{2q + 1}}$$

find the rate of change of p with respect to q.

46. *Advertising and sales* The daily sales S (in thousands of dollars) attributed to an advertising campaign are given by

$$S = 1 + \frac{3}{t + 3} - \frac{18}{(t + 3)^2}$$

where t is the number of weeks the campaign runs. What is the rate of change of sales at

(a) $t = 8$?   (b) $t = 10$?

(c) Should the campaign be continued after the 10th week? Explain.

47. *Body-heat loss* The description of body-heat loss due to convection involves a coefficient of convection, K_c, which depends on wind velocity according to the following equation.

$$K_c = 4\sqrt{4v + 1}$$

Find the rate of change of the coefficient with respect to the wind velocity.

48. *Data entry speed* The data entry speed (in entries per minute) of a data clerk trainee is

$$S = 10\sqrt{0.8x + 4}, \quad 0 \le x \le 100$$

where x is the number of hours of training he has had. What is the rate at which his speed is changing and what does this rate mean when he has had

(a) 15 hours of training?

(b) 40 hours of training?

49. *Investments* If an IRA is a variable-rate investment for 20 years at rate r percent per year, compounded monthly, then the future value S that accumulates from an initial investment of $1000 is

$$S = 1000\left[1 + \frac{0.01r}{12}\right]^{240}$$

What is the rate of change of S with respect to r and what does it tell us if the interest rate is (a) 6%? (b) 12%?

50. *Concentration of body substances* The concentration C of a substance in the body depends on the quantity of the substance Q and the volume V through which it is distributed. For a static substance, this is given by

$$C = \frac{Q}{V}$$

For a situation like that in the kidneys, where the fluids are moving, the concentration is the ratio of the rate of change of quantity with respect to time and the rate of change of volume with respect to time.

(a) Formulate the equation for concentration of a moving substance.

(b) Show that this is equal to the rate of change of quantity with respect to volume.

51. *Public debt of the United States* The interest paid on the public debt of the United States of America, as a percent of federal expenditures for selected years, is shown in the table.

    Assume that the percent of federal expenditures devoted to payment of interest can be modeled with the function

$$d(t) = -0.197(0.1t + 3)^4 + 5.32(0.1t + 3)^3$$
$$- 51.3(0.1t + 3)^2 + 209.6(0.1t + 3) - 295$$

where t is the number of years past 1930.

(a) Use this model to determine and interpret the instantaneous rate of change of the percent of federal expenditures devoted to payment of interest on the public debt in 1980 and 2000.
(b) Use the data to find an average rate of change that approximates the instantaneous rate in 1980.

Year	Interest Paid as a Percent of Federal Expenditures	Year	Interest Paid as a Percent of Federal Expenditures
1930	0	1975	9.8
1940	10.5	1980	12.7
1950	13.4	1985	18.9
1955	9.4	1990	21.1
1960	10.0	1995	22.0
1965	9.6	2000	20.3
1970	9.9	2003	14.7

Source: Bureau of Public Debt, Department of the Treasury

52. *Budget deficit or surplus*   The table gives the yearly U.S. federal budget deficit (as a negative value) or surplus (as a positive value) in billions of dollars from 1990 to 2004.

Year	Deficit or Surplus	Year	Deficit or Surplus
1990	−221.2	1998	70.0
1991	−269.4	1999	124.4
1992	−290.4	2000	237.0
1993	−255.0	2001	127.0
1994	−203.1	2002	−159.0
1995	−164.0	2003	−374.0
1996	−107.5	2004	−445.0
1997	−22.0		

Source: Budget of the United States Government

Assume the federal budget deficit (or surplus) can be modeled with the function

$$D(t) = -1975(0.1t - 1)^3 + 3357.4(0.1t - 1)^2 - 1064.68(0.1t - 1) - 205.453$$

where D is in billions of dollars and t is the number of years past 1980.
(a) Use the model to find and interpret the instantaneous rate of change of the U.S. federal deficit (or surplus) in 1998 and 2003.
(b) Use the data to find an average rate of change that approximates the instantaneous rate in 1998.

53. *Persons living below the poverty level*   The following table shows the number of millions of people in the United States who lived below the poverty level for selected years between 1960 and 2002.

Year	Persons Living Below the Poverty Level (millions)	Year	Persons Living Below the Poverty Level (millions)
1960	39.9	1993	39.3
1965	33.2	1996	36.5
1970	25.4	1997	35.6
1975	25.9	1998	34.5
1980	29.3	1999	32.3
1986	32.4	2000	31.1
1990	33.6	2002	34.6

Source: Bureau of the Census, U.S. Department of Commerce

Assume that the number of persons below the poverty level (in millions) can be modeled by

$$p(t) = -0.001905(t + 60)^3 + 0.4783(t + 60)^2 - 39.37(t + 60) + 1092$$

where t is the number of years past 1960.
(a) Use the model to find and interpret the instantaneous rate of change of the number of persons below the poverty level in 1970 and in 1998.
(b) Use the data to find the average rate of change from 1975 to 1990.
(c) For which year in the table (between 1975 and 1990) is the instantaneous rate closest to the average rate found in (b)?

54. *Gross domestic product*   The following table gives the U.S. gross domestic product (GDP) in billions of dollars for selected years.

Year	GDP	Year	GDP
1940	837	1980	3746
1945	1559	1985	4207
1950	1328	1990	4853
1955	1700	1995	5439
1960	1934	2000	9817
1965	2373	2001	10,128
1970	2847	2002	10,487
1975	3173	2003	11,004

Source: U.S. Bureau of Economic Analysis

Assume that the GDP can be modeled with the function

$$f(t) = 0.1156(t - 20)^3 - 7.637(t - 20)^2 + 186.46(t - 20) + 597.97$$

where f is in billions of dollars and t is the number of years past 1920.
(a) Using only the data in the table, how could you approximate the instantaneous rate of change of the GDP for 2003? Find this approximation.
(b) Use the model to find and interpret the instantaneous rate of change of the GDP in 1960 and 2003.

Using Derivative Formulas

OBJECTIVE

● To use derivative formulas separately and in combination with each other

◗ Application Preview

Suppose the weekly revenue function for a product is given by

$$R(x) = \frac{36,000,000x}{(2x + 500)^2}$$

where $R(x)$ is the dollars of revenue from the sale of x units. We can find marginal revenue by finding the derivative of the revenue function. This revenue function contains both a quotient and a power, so its derivative is found by using both the Quotient Rule and the Power Rule. But first we must decide the order in which to apply these formulas. (See Example 4.)

We have used the Power Rule to find the derivatives of functions like

$$y = (x^3 - 3x^2 + x + 1)^5$$

but we have not found the derivatives of functions like

$$y = [(x^2 + 1)(x^3 + x + 1)]^5$$

This function is different because the function u (which is raised to the fifth power) is the product of two functions, $(x^2 + 1)$ and $(x^3 + x + 1)$. The equation is of the form $y = u^5$, where $u = (x^2 + 1)(x^3 + x + 1)$. This means that the Product Rule should be used to find du/dx. Then

$$\frac{dy}{dx} = 5u^4 \cdot \frac{du}{dx}$$

$$= 5[(x^2 + 1)(x^3 + x + 1)]^4[(x^2 + 1)(3x^2 + 1) + (x^3 + x + 1)(2x)]$$

$$= 5[(x^2 + 1)(x^3 + x + 1)]^4(5x^4 + 6x^2 + 2x + 1)$$

$$= (25x^4 + 30x^2 + 10x + 5)[(x^2 + 1)(x^3 + x + 1)]^4$$

A different type of problem involving the Power Rule and the Product Rule is finding the derivative of $y = (x^2 + 1)^5(x^3 + x + 1)$. We may think of y as the *product* of two functions, one of which is a power. Thus the fundamental formula we should use is the Product Rule. The two functions are $u(x) = (x^2 + 1)^5$ and $v(x) = x^3 + x + 1$. The Product Rule gives

$$\frac{dy}{dx} = u(x) \cdot v'(x) + v(x) \cdot u'(x)$$

$$= (x^2 + 1)^5(3x^2 + 1) + (x^3 + x + 1)[5(x^2 + 1)^4(2x)]$$

Note that the Power Rule was used to find $u'(x)$, since $u(x) = (x^2 + 1)^5$.

We can simplify dy/dx by factoring $(x^2 + 1)^4$ from both terms:

$$\frac{dy}{dx} = (x^2 + 1)^4[(x^2 + 1)(3x^2 + 1) + (x^3 + x + 1) \cdot 5 \cdot 2x]$$

$$= (x^2 + 1)^4(13x^4 + 14x^2 + 10x + 1)$$

● **EXAMPLE 1 Power of a Quotient**

If $y = \left(\dfrac{x^2}{x - 1}\right)^5$, find y'.

Solution

We again have an equation of the form $y = u^n$, but this time u is a quotient. Thus we will need the Quotient Rule to find du/dx.

$$y' = nu^{n-1} \cdot \frac{du}{dx} = 5u^4 \frac{(x-1) \cdot 2x - x^2 \cdot 1}{(x-1)^2}$$

Substituting for u and simplifying gives

$$y' = 5\left(\frac{x^2}{x-1}\right)^4 \cdot \frac{2x^2 - 2x - x^2}{(x-1)^2}$$

$$= \frac{5x^8(x^2 - 2x)}{(x-1)^6} = \frac{5x^{10} - 10x^9}{(x-1)^6}$$

● **EXAMPLE 2 *Quotient of Two Powers***

Find $f'(x)$ if $f(x) = \dfrac{(x-1)^2}{(x^4+3)^3}$.

Solution

This function is the quotient of two functions, $(x-1)^2$ and $(x^4+3)^3$, so we must use the Quotient Rule to find the derivative of $f(x)$, but taking the derivatives of $(x-1)^2$ and $(x^4+3)^3$ will require the Power Rule.

$$f'(x) = \frac{\left[v(x) \cdot u'(x) - u(x) \cdot v'(x)\right]}{[v(x)]^2}$$

$$= \frac{(x^4+3)^3[2(x-1)(1)] - (x-1)^2[3(x^4+3)^2 4x^3]}{[(x^4+3)^3]^2}$$

$$= \frac{2(x^4+3)^3(x-1) - 12x^3(x-1)^2(x^4+3)^2}{(x^4+3)^6}$$

We see that 2, $(x^4+3)^2$, and $(x-1)$ are all factors in both terms of the numerator, so we can factor them from both terms and reduce the fraction.

$$f'(x) = \frac{2(x^4+3)^2(x-1)[(x^4+3) - 6x^3(x-1)]}{(x^4+3)^6}$$

$$= \frac{2(x-1)(-5x^4 + 6x^3 + 3)}{(x^4+3)^4}$$

● **EXAMPLE 3 *Product with a Power***

Find $f'(x)$ if $f(x) = (x^2 - 1)\sqrt{3 - x^2}$.

Solution

The function is the product of two functions, $x^2 - 1$ and $\sqrt{3 - x^2}$. Therefore, we will use the Product Rule to find the derivative of $f(x)$, but the derivative of $\sqrt{3 - x^2} = (3 - x^2)^{1/2}$ will require the Power Rule.

$$f'(x) = u(x) \cdot v'(x) + v(x) \cdot u'(x)$$

$$= (x^2 - 1)\left[\frac{1}{2}(3 - x^2)^{-1/2}(-2x)\right] + (3 - x^2)^{1/2}(2x)$$

$$= (x^2 - 1)[-x(3 - x^2)^{-1/2}] + (3 - x^2)^{1/2}(2x)$$

$$= \frac{-x^3 + x}{(3 - x^2)^{1/2}} + 2x(3 - x^2)^{1/2}$$

We can combine these terms over the common denominator $(3 - x^2)^{1/2}$ as follows:

$$f'(x) = \frac{-x^3 + x}{(3 - x^2)^{1/2}} + \frac{2x(3 - x^2)^1}{(3 - x^2)^{1/2}} = \frac{-x^3 + x + 6x - 2x^3}{(3 - x^2)^{1/2}}$$

$$= \frac{-3x^3 + 7x}{(3 - x^2)^{1/2}}$$

We should note that in Example 3 we could have written $f'(x)$ in the form

$$f'(x) = (-x^3 + x)(3 - x^2)^{-1/2} + 2x(3 - x^2)^{1/2}$$

Now the factor $(3 - x^2)$, to different powers, is contained in both terms of the expression. Thus we can factor $(3 - x^2)^{-1/2}$ from both terms. (We choose the $-1/2$ power because it is the smaller of the two powers.) Dividing $(3 - x^2)^{-1/2}$ into the first term gives $(-x^3 + x)$, and dividing it into the second term gives $2x(3 - x^2)^1$. Why? Thus we have

$$f'(x) = (3 - x^2)^{-1/2}[(-x^3 + x) + 2x(3 - x^2)]$$

$$= \frac{-3x^3 + 7x}{(3 - x^2)^{1/2}}$$

which agrees with our previous answer.

● **Checkpoint**

1. If a function has the form $y = [u(x)]^n \cdot v(x)$, where n is a constant, we begin to find the derivative by using the _____ Rule and then use the _____ Rule to find the derivative of $[u(x)]^n$.
2. If a function has the form $y = [u(x)/v(x)]^n$, where n is a constant, we begin to find the derivative by using the _____ Rule and then use the _____ Rule.
3. Find the derivative of each of the following and simplify.

   (a) $f(x) = 3x^4(2x^4 + 7)^5$ (b) $g(x) = \dfrac{(4x + 3)^7}{2x - 9}$

● **EXAMPLE 4** *Revenue* **(Application Preview)**

Suppose that the weekly revenue function for a product is given by

$$R(x) = \frac{36{,}000{,}000x}{(2x + 500)^2}$$

where $R(x)$ is the dollars of revenue from the sale of x units.

(a) Find the marginal revenue function.
(b) Find the marginal revenue when 50 units are sold.

Solution
(a) $\overline{MR} = R'(x)$

$$= \frac{(2x + 500)^2(36{,}000{,}000) - 36{,}000{,}000x[2(2x + 500)^1(2)]}{(2x + 500)^4}$$

$$= \frac{36{,}000{,}000(2x + 500)(2x + 500 - 4x)}{(2x + 500)^4}$$

$$= \frac{36{,}000{,}000(500 - 2x)}{(2x + 500)^3}$$

(b) $\overline{MR}(50) = R'(50) = \dfrac{36{,}000{,}000(500 - 100)}{(100 + 500)^3}$

$= \dfrac{36{,}000{,}000(400)}{(600)^3}$

$= \dfrac{200}{3} \approx 66.67$

The marginal revenue is \$66.67 when 50 units are sold. That is, the predicted revenue from the sale of the 51st unit is approximately \$66.67.

It may be helpful to review the formulas needed to find the derivatives of various types of functions. Table 9.5 presents examples of different types of functions and the formulas needed to find their derivatives.

TABLE 9.5   Derivative Formulas Summary

Examples	Formulas
$f(x) = 14$	If $f(x) = c$, then $f'(x) = 0$.
$y = x^4$	If $f(x) = x^n$, then $f'(x) = nx^{n-1}$.
$g(x) = 5x^3$	If $g(x) = cf(x)$, then $g'(x) = cf'(x)$.
$y = 3x^2 + 4x$	If $f(x) = u(x) + v(x)$, then $f'(x) = u'(x) + v'(x)$.
$y = (x^2 - 2)(x + 4)$	If $f(x) = u(x) \cdot v(x)$, then $f'(x) = u(x) \cdot v'(x) + v(x) \cdot u'(x)$.
$f(x) = \dfrac{x^3}{x^2 + 1}$	If $f(x) = \dfrac{u(x)}{v(x)}$ then $f'(x) = \dfrac{v(x) \cdot u'(x) - u(x) \cdot v'(x)}{[v(x)]^2}$.
$y = (x^3 - 4x)^{10}$	If $y = u^n$ and $u = g(x)$, then $\dfrac{dy}{dx} = nu^{n-1} \cdot \dfrac{du}{dx}$.
$y = \left(\dfrac{x-1}{x^2+3}\right)^3$	Power Rule, then Quotient Rule to find $\dfrac{du}{dx}$ where $u = \dfrac{x-1}{x^2+3}$.
$y = (x + 1)\sqrt{x^3 + 1}$	Product Rule, then Power Rule to find $v'(x)$, where $v(x) = \sqrt{x^3 + 1}$.
$y = \dfrac{(x^2 - 3)^4}{x + 1}$	Quotient Rule, then Power Rule to find the derivative of the numerator.

● *Checkpoint Solutions*

1. Product, Power
2. Power, Quotient
3. (a) $f'(x) = 3x^4[5(2x^4 + 7)^4(8x^3)] + (2x^4 + 7)^5(12x^3)$

   $= 120x^7(2x^4 + 7)^4 + 12x^3(2x^4 + 7)^5$

   $= 12x^3(2x^4 + 7)^4[10x^4 + (2x^4 + 7)] = 12x^3(12x^4 + 7)(2x^4 + 7)^4$

   (b) $g'(x) = \dfrac{(2x - 9)[7(4x + 3)^6(4)] - (4x + 3)^7(2)}{(2x - 9)^2}$

   $= \dfrac{2(4x + 3)^6[14(2x - 9) - (4x + 3)]}{(2x - 9)^2} = \dfrac{2(24x - 129)(4x + 3)^6}{(2x - 9)^2}$

9.7 Exercises

Find the derivatives of the functions in Problems 1–32. Simplify and express the answer using positive exponents only.

1. $f(x) = \pi^4$

2. $f(x) = \dfrac{1}{4}$

3. $g(x) = \dfrac{4}{x^4}$

4. $y = \dfrac{x^4}{4}$

5. $g(x) = 5x^3 + \dfrac{4}{x}$

6. $y = 3x^2 + 4\sqrt{x}$

7. $y = (x^2 - 2)(x + 4)$

8. $y = (x^3 - 5x^2 + 1)(x^3 - 3)$

9. $f(x) = \dfrac{x^3 + 1}{x^2}$

10. $y = \dfrac{1 + x^2 - x^4}{1 + x^4}$

11. $y = \dfrac{(x^3 - 4x)^{10}}{10}$

12. $y = \dfrac{5}{2}(3x^4 - 6x^2 + 2)^5$

13. $y = \dfrac{5}{3}x^3(4x^5 - 5)^3$

14. $y = 3x^4(2x^5 + 1)^7$

15. $y = (x - 1)^2(x^2 + 1)$

16. $f(x) = (5x^3 + 1)(x^4 + 5x)^2$

17. $y = \dfrac{(x^2 - 4)^3}{x^2 + 1}$

18. $y = \dfrac{(x^2 - 3)^4}{x}$

19. $p = [(q + 1)(q^3 - 3)]^3$

20. $y = [(4 - x^2)(x^2 + 5x)]^4$

21. $R(x) = [x^2(x^2 + 3x)]^4$

22. $c(x) = [x^3(x^2 + 1)]^{-3}$

23. $y = \left(\dfrac{2x - 1}{x^2 + x}\right)^4$

24. $y = \left(\dfrac{5 - x^2}{x^4}\right)^3$

25. $g(x) = (8x^4 + 3)^2(x^3 - 4x)^3$

26. $y = (3x^3 - 4x)^3(4x^2 - 8)^2$

27. $f(x) = \dfrac{\sqrt[3]{x^2 + 5}}{4 - x^2}$

28. $g(x) = \dfrac{\sqrt[3]{2x - 1}}{2x + 1}$

29. $y = x^2\sqrt[4]{4x - 3}$

30. $y = 3x\sqrt[3]{4x^4 + 3}$

31. $c(x) = 2x\sqrt{x^3 + 1}$

32. $R(x) = x\sqrt[3]{3x^3 + 2}$

In Problems 33 and 34, find the derivative of each function.

33. (a) $F_1(x) = \dfrac{3(x^4 + 1)^5}{5}$

    (b) $F_2(x) = \dfrac{3}{5(x^4 + 1)^5}$

    (c) $F_3(x) = \dfrac{(3x^4 + 1)^5}{5}$

    (d) $F_4(x) = \dfrac{3}{(5x^4 + 1)^5}$

34. (a) $G_1(x) = \dfrac{2(x^3 - 5)^3}{3}$

    (b) $G_2(x) = \dfrac{(2x^3 - 5)^3}{3}$

    (c) $G_3(x) = \dfrac{2}{3(x^3 - 5)^3}$

    (d) $G_4(x) = \dfrac{2}{(3x^3 - 5)^3}$

APPLICATIONS

35. *Physical output* The total physical output P of workers is a function of the number of workers, x. The function $P = f(x)$ is called the physical productivity function. Suppose that the physical productivity of x construction workers is given by

$$P = 10(3x + 1)^3 - 10$$

Find the marginal physical productivity, dP/dx.

36. *Revenue* Suppose that the revenue function for a certain product is given by

$$R(x) = 15(2x + 1)^{-1} + 30x - 15$$

where x is in thousands of units and R is in thousands of dollars.

(a) Find the marginal revenue when 2000 units are sold.

(b) How is revenue changing when 2000 units are sold?

37. *Revenue* Suppose that the revenue in dollars from the sale of x campers is given by

$$R(x) = 60{,}000x + 40{,}000(10 + x)^{-1} - 4000$$

(a) Find the marginal revenue when 10 units are sold.

(b) How is revenue changing when 10 units are sold?

38. *Production* Suppose that the production of x items of a new line of products is given by

$$x = 200[(t + 10) - 400(t + 40)^{-1}]$$

where t is the number of weeks the line has been in production. Find the rate of production, dx/dt.

39. *National consumption* If the national consumption function is given by

$$C(y) = 2(y + 1)^{1/2} + 0.4y + 4$$

find the marginal propensity to consume, dC/dy.

40. *Demand* Suppose that the demand function for q units of an appliance priced at $\$p$ per unit is given by

$$p = \dfrac{400(q + 1)}{(q + 2)^2}$$

Find the rate of change of price with respect to the number of appliances.

41. *Volume* When squares of side x inches are cut from the corners of a 12-inch-square piece of cardboard, an open-top box can be formed by folding up the sides. The volume of this box is given by

$$V = x(12 - 2x)^2$$

Find the rate of change of volume with respect to the size of the squares.

42. *Advertising and sales* Suppose that sales (in thousands of dollars) are directly related to an advertising campaign according to

$$S = 1 + \frac{3t - 9}{(t + 3)^2}$$

where t is the number of weeks of the campaign.
(a) Find the rate of change of sales after 3 weeks.
(b) Interpret the result in part (a).

43. *Advertising and sales* An inferior product with an extensive advertising campaign does well when it is released, but sales decline as people discontinue use of the product. If the sales S (in thousands of dollars) after t weeks are given by

$$S(t) = \frac{200t}{(t + 1)^2}, \quad t \geq 0$$

what is the rate of change of sales when $t = 9$? Interpret your result.

44. *Advertising and sales* An excellent film with a very small advertising budget must depend largely on word-of-mouth advertising. If attendance at the film after t weeks is given by

$$A = \frac{100t}{(t + 10)^2}$$

what is the rate of change in attendance and what does it mean when (a) $t = 10$? (b) $t = 20$?

45. *Farm workers* The percent of U.S. workers in farm occupations during certain years is shown in the table.

Year	Percent	Year	Percent
1820	71.8	1960	6.1
1850	63.7	1970	3.6
1870	53	1980	2.7
1900	37.5	1985	2.8
1920	27	1990	2.4
1930	21.2	1994	2.5
1940	17.4	2002	2.5
1950	11.6		

Source: Bureau of Labor Statistics

Assume that the percent of U.S. workers in farm occupations can be modeled with the function

$$f(t) = \frac{1000[-7.4812(t + 20) + 1560.2]}{1.2882(t + 20)^2 - 122.18(t + 20) + 21{,}483}$$

where t is the number of years past 1820.
(a) Find the function that models the instantaneous rate of change of the percent of U.S. workers in farm occupations.
(b) Use the result of part (a) to find the instantaneous rates of change in 1850 and in 1950.
(c) Interpret the two rates of change in part (b).

9.8

Higher-Order Derivatives

OBJECTIVE

● To find second derivatives and higher-order derivatives of certain functions

◗ Application Preview

Since DVD players were introduced, their popularity has increased enormously. Figure 9.33(a) shows a graph of DVD player sales revenue (actual and projected from Alexander & Associates, 2001), in billions of dollars, as a function of the number of years past 1999. (*Source: Newsweek,* August 20, 2001) Note that revenue is always increasing and that the rate of change of revenue

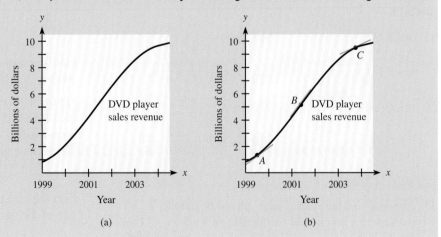

Figure 9.33   (a)   (b)

(as seen from tangent lines to the graph) is always positive. However, the tangent lines shown in Figure 9.33(b) indicate that the rate of change of revenue is greater at *B* than at either *A* or *C*.

Furthermore, the rate of change of revenue (the slopes of tangents) increases from *A* to *B* and then decreases from *B* to *C*. To learn how the rate of change of revenue is changing, we are interested in finding the derivative of the rate of change of revenue—that is, the derivative of the derivative of the revenue. (See Example 4.) This is called the **second derivative.** In this section we will discuss second and higher-order derivatives.

Second Derivatives

Because the derivative of a function is itself a function, we can take a derivative of the derivative. The derivative of a first derivative is called a **second derivative.** We can find the second derivative of a function *f* by differentiating it twice. If *f'* represents the first derivative of a function, then *f''* represents the second derivative of that function.

● **EXAMPLE 1** *Second Derivative*

(a) Find the second derivative of $y = x^4 - 3x^2 + x^{-2}$.
(b) If $f(x) = 3x^3 - 4x^2 + 5$, find $f''(x)$.

Solution
(a) The first derivative is $y' = 4x^3 - 6x - 2x^{-3}$.
The second derivative, which we may denote by y'', is

$$y'' = 12x^2 - 6 + 6x^{-4}$$

(b) The first derivative is $f'(x) = 9x^2 - 8x$.
The second derivative is $f''(x) = 18x - 8$.

It is also common to use $\dfrac{d^2y}{dx^2}$ and $\dfrac{d^2f(x)}{dx^2}$ to denote the second derivative of a function.

● **EXAMPLE 2** *Second Derivative*

If $y = \sqrt{2x - 1}$, find d^2y/dx^2.

Solution
The first derivative is

$$\frac{dy}{dx} = \frac{1}{2}(2x - 1)^{-1/2}(2) = (2x - 1)^{-1/2}$$

The second derivative is

$$\frac{d^2y}{dx^2} = -\frac{1}{2}(2x - 1)^{-3/2}(2) = -(2x - 1)^{-3/2}$$

$$= \frac{-1}{(2x - 1)^{3/2}} = \frac{-1}{\sqrt{(2x - 1)^3}}$$

Higher-Order Derivatives

We can also find third, fourth, fifth, and higher derivatives, continuing indefinitely. The third, fourth, and fifth derivatives of a function *f* are denoted by f''', $f^{(4)}$, and $f^{(5)}$, respectively. Other notations for the third and fourth derivatives include

$$y''' = \frac{d^3y}{dx^3} = \frac{d^3f(x)}{dx^3}, \qquad y^{(4)} = \frac{d^4y}{dx^4} = \frac{d^4f(x)}{dx^4}$$

● **EXAMPLE 3** *Higher-Order Derivatives*

Find the first four derivatives of $f(x) = 4x^3 + 5x^2 + 3$.

Solution

$$f'(x) = 12x^2 + 10x, \qquad f''(x) = 24x + 10, \qquad f'''(x) = 24, \qquad f^{(4)}(x) = 0$$

Just as the first derivative, $f'(x)$, can be used to determine the rate of change of a function $f(x)$, the second derivative, $f''(x)$, can be used to determine the rate of change of $f'(x)$.

◗ **EXAMPLE 4** *DVD Sales Revenue* **(Application Preview)**

Alexander & Associates' actual and projected revenue (in billions of dollars) from the sale of DVD players can be modeled by

$$R(t) = -0.094t^3 + 0.680t^2 + 0.715t + 0.826$$

where t is the number of years past 1999. (*Source: Newsweek,* August 20, 2001.)

(a) Find the instantaneous rate of change of the DVD player sales revenue function.
(b) Find the instantaneous rate of change of the function found in part (a).
(c) Find where the function in part (b) equals zero. Then explain how the rate of change of revenue is changing for one t-value before and one after this value.

Solution
(a) The instantaneous rate of change of $R(t)$ is

$$R'(t) = -0.282t^2 + 1.360t + 0.715$$

(b) The instantaneous rate of change of $R'(t)$ is

$$R''(t) = -0.564t + 1.360$$

(c)
$$0 = -0.564t + 1.360$$
$$0.564t = 1.360$$
$$t = \frac{1.360}{0.564} \approx 2.411$$

For $t = 2$, which is less than 2.411,

$$R''(2) = -0.564(2) + 1.360 = 0.232$$

This means that for $t = 2$, the rate of change of revenue is changing at the rate of 0.232 billion dollars per year, per year. Thus for $t = 2$ the rate of change of revenue is increasing, so revenue is increasing at an increasing rate. See Figure 9.33(a) earlier in this section. For $t = 3$,

$$R''(3) = -0.564(3) + 1.360 = -0.332$$

This means that the rate of change of revenue is changing at the rate of -0.332 billion dollars per year, per year. Thus for $t = 3$ the rate of change of revenue is decreasing, so revenue is (still) increasing but at a decreasing (slower) rate. See Figure 9.33(a).

● **EXAMPLE 5** *Rate of Change of a Derivative*

Let $f(x) = 3x^4 + 6x^3 - 3x^2 + 4$.

(a) How fast is $f(x)$ changing at (1, 10)?
(b) How fast is $f'(x)$ changing at (1, 10)?
(c) Is $f'(x)$ increasing or decreasing at (1, 10)?

Solution

(a) Because $f'(x) = 12x^3 + 18x^2 - 6x$, we have

$$f'(1) = 12 + 18 - 6 = 24$$

Thus the rate of change of $f(x)$ at $(1, 10)$ is 24 (y units per x unit).

(b) Because $f''(x) = 36x^2 + 36x - 6$, we have

$$f''(1) = 66$$

Thus the rate of change of $f'(x)$ at $(1, 10)$ is 66 (y units per x unit per x unit).

(c) Because $f''(1) = 66 > 0, f'(x)$ is increasing at $(1, 10)$.

● **EXAMPLE 6** *Acceleration*

Suppose that a particle travels according to the equation

$$s = 100t - 16t^2 + 200$$

where s is the distance in feet and t is the time in seconds. Then ds/dt is the velocity, and $d^2s/dt^2 = dv/dt$ is the acceleration of the particle. Find the acceleration.

Solution

The velocity is $v = ds/dt = 100 - 32t$ feet per second, and the acceleration is

$$\frac{dv}{dt} = \frac{d^2s}{dt^2} = -32 \text{ (ft/sec)/sec} = -32 \text{ ft/sec}^2$$

● *Checkpoint*

Suppose that the vertical distance a particle travels is given by

$$s = 4x^3 - 12x^2 + 6$$

where s is in feet and x is in seconds.

1. Find the function that describes the velocity of this particle.
2. Find the function that describes the acceleration of this particle.
3. Is the acceleration always positive?
4. When does the *velocity* of this particle increase?

Calculator
Note

We can use the numerical derivative feature of a graphing calculator to find the second derivative of a function at a point. ■

  ● **EXAMPLE 7** *Second Derivative*

Find $f''(2)$ if $f(x) = \sqrt{x^3 - 1}$.

Solution

We need the derivative of the derivative function, evaluated at $x = 2$. Figure 9.34 shows how the numerical derivative feature of a graphing calculator can be used to obtain this result.

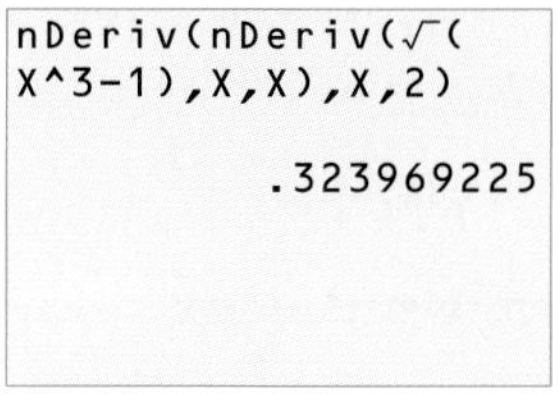

```
nDeriv(nDeriv(√(
X^3-1),X,X),X,2)
               .323969225
```

Figure 9.34

Thus Figure 9.34 shows that $f''(2) = 0.323969225 \approx 0.32397$. We can check this result by calculating $f''(x)$ with formulas.

$$f'(x) = \frac{1}{2}(x^3 - 1)^{-\frac{1}{2}}(3x^2)$$

$$f''(x) = \frac{1}{2}(x^3 - 1)^{-\frac{1}{2}}(6x) + (3x^2)\left[-\frac{1}{4}(x^3 - 1)^{-\frac{3}{2}}(3x^2)\right]$$

$$f''(2) = 0.3239695483 \approx 0.32397$$

Thus we see that the numerical derivative approximation is quite accurate.

● **EXAMPLE 8 Second Derivatives and Graphs**

(a) Given $f(x) = x^4 - 12x^2 + 2$, graph $f(x)$ and its second derivative on the same set of axes over an interval that contains all x-values where $f''(x) = 0$.
(b) When the graph of $y = f(x)$ is opening downward, is $f''(x) > 0, f''(x) < 0$, or $f''(x) = 0$?
(c) When the graph of $y = f(x)$ is opening upward, is $f''(x) > 0, f''(x) < 0$, or $f''(x) = 0$?

Solution

(a) $f'(x) = 4x^3 - 24x$ and $f''(x) = 12x^2 - 24 = 12(x^2 - 2)$. Because $f''(x) = 0$ at $x = -\sqrt{2}$ and at $x = \sqrt{2}$, we use an x-range that contains $x = -\sqrt{2}$ and $\sqrt{2}$. The graph of $f(x) = x^4 - 12x^2 + 2$ and its second derivative, $f''(x) = 12x^2 - 24$, are shown in Figure 9.35.
(b) The graph of $y = f(x)$ appears to be opening downward on the same interval for which $f''(x) < 0$.
(c) The graph appears to be opening upward on the same intervals for which $f''(x) > 0$.

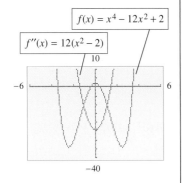

$f(x) = x^4 - 12x^2 + 2$

$f''(x) = 12(x^2 - 2)$

Figure 9.35

● *Checkpoint Solutions*

1. The velocity is described by $s'(x) = 12x^2 - 24x$.
2. The acceleration is described by $s''(x) = 24x - 24$.
3. No; the acceleration is positive when $s''(x) > 0$—that is, when $24x - 24 > 0$. It is zero when $24x - 24 = 0$ and negative when $24x - 24 < 0$. Thus acceleration is negative when $x < 1$ second, zero when $x = 1$ second, and positive when $x > 1$ second.
4. The velocity increases when the acceleration is positive. Thus the velocity is increasing after 1 second.

9.8 Exercises

In Problems 1–6, find the second derivative.

1. $f(x) = 2x^{10} - 18x^5 - 12x^3 + 4$
2. $y = 6x^5 - 3x^4 + 12x^2$
3. $g(x) = x^3 - \dfrac{1}{x}$ 4. $h(x) = x^2 - \dfrac{1}{x^2}$
5. $y = x^3 - \sqrt{x}$ 6. $y = 3x^2 - \sqrt[3]{x^2}$

In Problems 7–12, find the third derivative.

7. $y = x^5 - 16x^3 + 12$ 8. $y = 6x^3 - 12x^2 + 6x$
9. $f(x) = 2x^9 - 6x^6$ 10. $f(x) = 3x^5 - x^6$
11. $y = 1/x$ 12. $y = 1/x^2$

In Problems 13–24, find the indicated derivative.

13. If $y = x^5 - x^{1/2}$, find $\dfrac{d^2y}{dx^2}$.

14. If $y = x^4 + x^{1/3}$, find $\dfrac{d^2y}{dx^2}$.

15. If $f(x) = \sqrt{x + 1}$, find $f'''(x)$.

16. If $f(x) = \sqrt{x - 5}$, find $f'''(x)$.

17. Find $\dfrac{d^4y}{dx^4}$ if $y = 4x^3 - 16x$.

18. Find $y^{(4)}$ if $y = x^6 - 15x^3$.

19. Find $f^{(4)}(x)$ if $f(x) = \sqrt{x}$.

20. Find $f^{(4)}(x)$ if $f(x) = 1/x$.

21. Find $y^{(4)}$ if $y' = \sqrt{4x - 1}$.

22. Find $y^{(5)}$ if $\dfrac{d^2y}{dx^2} = \sqrt[3]{3x + 2}$.

23. Find $f^{(6)}(x)$ if $f^{(4)}(x) = x(x + 1)^{-1}$.

24. Find $f^{(3)}(x)$ if $f'(x) = \dfrac{x^2}{x^2 + 1}$.

25. If $f(x) = 16x^2 - x^3$, what is the rate of change of $f'(x)$ at $(1, 15)$?

26. If $y = 36x^2 - 6x^3 + x$, what is the rate of change of y' at $(1, 31)$?

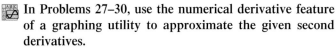 **In Problems 27–30, use the numerical derivative feature of a graphing utility to approximate the given second derivatives.**

27. $f''(3)$ for $f(x) = x^3 - \dfrac{27}{x}$

28. $f''(-1)$ for $f(x) = \dfrac{x^2}{4} - \dfrac{4}{x^2}$

29. $f''(21)$ for $f(x) = \sqrt{x^2 + 4}$

30. $f''(3)$ for $f(x) = \dfrac{1}{\sqrt{x^2 + 7}}$

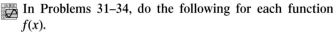 **In Problems 31–34, do the following for each function $f(x)$.**

(a) Find $f'(x)$ and $f''(x)$.

(b) Graph $f(x), f'(x),$ and $f''(x)$ with a graphing utility.

(c) Identify x-values where $f''(x) = 0, f''(x) > 0,$ and $f''(x) < 0$.

(d) Identify x-values where $f'(x)$ has a maximum point or a minimum point, where $f'(x)$ is increasing, and where $f'(x)$ is decreasing.

(e) When $f(x)$ has a maximum point, is $f''(x) > 0$ or $f''(x) < 0$?

(f) When $f(x)$ has a minimum point, is $f''(x) > 0$ or $f''(x) < 0$?

31. $f(x) = x^3 - 3x^2 + 5$ 32. $f(x) = 2 + 3x - x^3$

33. $f(x) = -\dfrac{1}{3}x^3 - x^2 + 3x + 7$

34. $f(x) = \dfrac{1}{3}x^3 - \dfrac{3x^2}{2} - 4x + 10$

APPLICATIONS

35. *Acceleration* A particle travels as a function of time according to the formula

$$s = 100 + 10t + 0.01t^3$$

where s is in meters and t is in seconds. Find the acceleration of the particle when $t = 2$.

36. *Acceleration* If the formula describing the distance s (in feet) an object travels as a function of time (in seconds) is

$$s = 100 + 160t - 16t^2$$

what is the acceleration of the object when $t = 4$?

37. *Revenue* The revenue (in dollars) from the sale of x units of a certain product can be described by

$$R(x) = 100x - 0.01x^2$$

Find the instantaneous rate of change of the marginal revenue.

38. *Revenue* Suppose that the revenue (in dollars) from the sale of a product is given by

$$R = 70x + 0.5x^2 - 0.001x^3$$

where x is the number of units sold. How fast is the marginal revenue $\overline{MR}$ changing when $x = 100$?

39. *Sensitivity* When medicine is administered, reaction (measured in change of blood pressure or temperature) can be modeled by

$$R = m^2\left(\frac{c}{2} - \frac{m}{3}\right)$$

where c is a positive constant and m is the amount of medicine absorbed into the blood (*Source:* R. M. Thrall et al., *Some Mathematical Models in Biology*, U.S. Department of Commerce, 1967). The sensitivity to the medication is defined to be the rate of change of reaction R with respect to the amount of medicine m absorbed in the blood.

(a) Find the sensitivity.

(b) Find the instantaneous rate of change of sensitivity with respect to the amount of medicine absorbed in the blood.

(c) Which order derivative of reaction gives the rate of change of sensitivity?

40. *Photosynthesis* The amount of photosynthesis that takes place in a certain plant depends on the intensity of light x according to the equation

$$f(x) = 145x^2 - 30x^3$$

(a) Find the rate of change of photosynthesis with respect to the intensity.

(b) What is the rate of change when $x = 1$? when $x = 3$?

(c) How fast is the rate found in part (a) changing when $x = 1$? when $x = 3$?

41. *Revenue* The revenue (in thousands of dollars) from the sale of a product is

$$R = 15x + 30(4x + 1)^{-1} - 30$$

where x is the number of units sold.
(a) At what rate is the marginal revenue $\overline{MR}$ changing when the number of units being sold is 25?
(b) Interpret your result in part (a).

42. *Advertising and sales* The sales of a product S (in thousands of dollars) are given by

$$S = \frac{600x}{x + 40}$$

where x is the advertising expenditure (in thousands of dollars).
(a) Find the rate of change of sales with respect to advertising expenditure.
(b) Use the second derivative to find how this rate is changing at $x = 20$.
(c) Interpret your result in part (b).

43. *Advertising and sales* The daily sales S (in thousands of dollars) that are attributed to an advertising campaign are given by

$$S = 1 + \frac{3}{t + 3} - \frac{18}{(t + 3)^2}$$

where t is the number of weeks the campaign runs.
(a) Find the rate of change of sales at any time t.
(b) Use the second derivative to find how this rate is changing at $t = 15$.
(c) Interpret your result in part (b).

44. *Advertising and sales* A product with a large advertising budget has its sales S (in millions of dollars) given by

$$S = \frac{500}{t + 2} - \frac{1000}{(t + 2)^2}$$

where t is the number of months the product has been on the market.
(a) Find the rate of change of sales at any time t.
(b) What is the rate of change of sales at $t = 2$?
(c) Use the second derivative to find how this rate is changing at $t = 2$.
(d) Interpret your results from parts (b) and (c).

45. *Persons living below the poverty level* The table below shows the number of millions of people in the United States who lived below the poverty level for selected years between 1960 and 2002.

Year	Persons Living Below the Poverty Level (millions)	Year	Persons Living Below the Poverty Level (millions)
1960	39.9	1993	39.3
1965	33.2	1996	36.5
1970	25.4	1997	35.6
1975	25.9	1998	34.5
1980	29.3	1999	32.3
1986	32.4	2000	31.1
1990	33.6	2002	34.6

Source: Bureau of the Census, U.S. Department of Commerce

Assume that the number of millions of people in the United States who lived below the poverty level can be modeled by the function

$$p(t) = -0.0019t^3 + 0.1355t^2 - 2.542t + 40.62$$

where t is the number of years past 1960.
(a) Find the function that models the instantaneous rate of change of $p(t)$.
(b) Use the second derivative to find how this rate is changing in 1980 and 1998.
(c) Interpret the meanings of $p'(20)$ and $p''(20)$. Write a sentence that explains each.

46. *U.S. cellular subscriberships* The following table shows the numbers of U.S. cellular subscriberships (in millions) from 1985 to 2002.

Year	Subscriberships	Year	Subscriberships
1985	0.340	1994	24.134
1986	0.682	1995	33.786
1987	1.231	1996	44.043
1988	2.069	1997	55.312
1989	3.509	1998	69.209
1990	5.283	1999	86.047
1991	7.557	2000	109.478
1992	11.033	2001	128.375
1993	16.009	2002	140.767

Source: The CTIA Semi-Annual Wireless Industry Survey

Assume that the number of U.S. cellular subscriberships, in millions, can be modeled by the function

$$S(t) = 0.0003488t^{4.2077}$$

where t is the number of years past 1980.
(a) Find the function that models the instantaneous rate of change of the U.S. cellular subscriberships.
(b) Use the second derivative to determine how fast the rate found in (a) was changing in 1990 and 2002.
(c) Write sentences that explain the meanings of $S'(20)$ and $S''(20)$.

47. **Modeling** *U.S. poverty threshold* The table shows the yearly income poverty thresholds for a single person for selected years from 1990 to 2002.

Year	Income	Year	Income
1990	$6652	2000	$8794
1995	7763	2001	9039
1998	8316	2002	9182
1999	8501		

Source: U.S. Bureau of the Census

(a) Model these data with a power function $p(t)$, where p is the poverty threshold income for a single person and t is the number of years past 1980.

(b) Find the function that models the rate of change of the poverty threshold income for a single person.

(c) Use the second derivative to determine how fast this rate was changing in 1995 and 2002.

(d) Write sentences that explain the meanings of $p'(22)$ and $p''(22)$.

 48. **Modeling** *U.S. Population* The table gives the U.S. population to the nearest million (actual or projected) for selected years.

Year	Population (in millions)	Year	Population (in millions)
1960	181	1995	263
1965	194	1998	271
1970	205	2000	281
1975	216	2003	294
1980	228	2025	358
1985	238	2050	409
1990	250		

Source: U.S. Bureau of the Census

(a) Find a cubic function $P(t)$ that models these data, where P is the U.S. population in millions and t is the number of years past 1960.

(b) Find the function that models the instantaneous rate of change of the U.S. population.

(c) Use the second derivative to determine how fast this rate was changing in 1990 and 2020.

(d) Write sentences that explain the meanings of $P'(30)$ and $P''(30)$.

9.9

Applications of Derivatives in Business and Economics

OBJECTIVES

- To find the marginal cost and marginal revenue at different levels of production
- To find the marginal profit function, given information about cost and revenue

◗ Application Preview

If the total cost in dollars to produce x kitchen blenders is given by

$$C(x) = 0.001x^3 - 0.3x^2 + 32x + 2500$$

what is the marginal cost function and what does it tell us about how costs are changing when $x = 80$ and $x = 200$ blenders are produced? See Example 3.

In Section 1.6, "Applications of Functions in Business and Economics," we defined the marginals for linear total cost, total revenue, and profit functions as the rate of change or slope of the respective function. In Section 9.3, we extended the notion of marginal revenue to nonlinear total revenue functions by defining marginal revenue as the derivative of total revenue. In this section, for any total cost or profit function we can extend the notion of marginal to nonlinear functions.

Marginal Revenue

As we saw in Section 9.3, "Average and Instantaneous Rates of Change: The Derivative," the instantaneous rate of change (the derivative) of the revenue function is the marginal revenue.

Marginal Revenue

If $R = R(x)$ is the total revenue function for a commodity, then the **marginal revenue function** is $\overline{MR} = R'(x)$.

Recall that if the demand function for a product in a monopoly market is $p = f(x)$, then the total revenue from the sale of x units is

$$R(x) = px = f(x) \cdot x$$

● **EXAMPLE 1** *Revenue and Marginal Revenue*

If the demand for a product in a monopoly market is given by

$$p = 16 - 0.02x$$

where x is the number of units and p is the price per unit, (a) find the total revenue function, and (b) find the marginal revenue for this product at $x = 40$.

Solution

(a) The total revenue function is

$$R(x) = px = (16 - 0.02x)x$$
$$= 16x - 0.02x^2$$

(b) The marginal revenue function is

$$\overline{MR} = R'(x) = 16 - 0.04x$$

At $x = 40, R'(40) = 16 - 1.6 = 14.40$ dollars per unit. Thus the 41st item sold will increase the total revenue by approximately $14.40.

   The marginal revenue is an approximation or estimate of the revenue gained from the sale of 1 additional unit. We have used marginal revenue in Example 1 to find that the revenue from the sale of the 41st item will be approximately $14.40. The actual increase in revenue from the sale of the 41st item is

$$R(41) - R(40) = 622.38 - 608 = \$14.38$$

Marginal revenue (and other marginals) can be used to predict for more than one additional unit. For instance, in Example 1, $\overline{MR}(40) = \$14.40$ per unit means that the expected or approximate revenue for the 41st through the 45th items sold would be 5($14.40) = $72.00. The actual revenue for these 5 items is $R(45) - R(40) = \$679.50 - \$608 = \$71.50$.

● **EXAMPLE 2** *Maximum Revenue*

Use the graphs in Figure 9.36 to determine the x-value where the revenue function has its maximum. What is happening to the marginal revenue at and near this x-value?

Solution

Figure 9.36(a) shows that the total revenue function has a maximum value at $x = 400$. After that, the total revenue function decreases. This means that the total revenue will be reduced each time a unit is sold if more than 400 are produced and sold. The graph of the marginal revenue function in Figure 9.36(b) shows that the marginal revenue is positive to the left

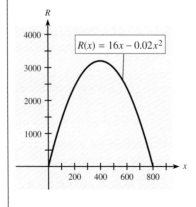

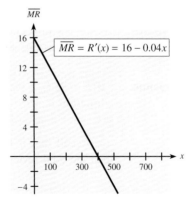

Figure 9.36   (a) Total revenue function   (b) Marginal revenue function

of 400. This indicates that the rate at which the total revenue is changing is positive until 400 units are sold; thus the total revenue is increasing. Then, at 400 units, the rate of change is 0. After 400 units are sold, the marginal revenue is negative, which indicates that the total revenue is now decreasing. It is clear from looking at either graph that 400 units should be produced and sold to maximize the total revenue function $R(x)$. That is, the *total revenue* function has its maximum at $x = 400$.

Marginal Cost   Next we define the marginal cost function.

Marginal Cost

> If $C = C(x)$ is a total cost function for a commodity, then its derivative, $\overline{MC} = C'(x)$, is the **marginal cost function.**

Notice that the linear total cost function with equation

$$C(x) = 300 + 6x \qquad \text{(in dollars)}$$

has marginal cost $6 per unit because its slope is 6. Taking the derivative of $C(x)$ also gives

$$\overline{MC} = C'(x) = 6$$

which verifies that the marginal cost is $6 per unit at all levels of production.

◐ EXAMPLE 3 *Marginal Cost* **(Application Preview)**

Suppose the daily total cost in dollars for a certain factory to produce x kitchen blenders is given by

$$C(x) = 0.001x^3 - 0.3x^2 + 32x + 2500$$

(a) Find the marginal cost function for these blenders.
(b) Find and interpret the marginal cost when $x = 80$ and $x = 200$.

Solution
(a) The marginal cost function is the derivative of $C(x)$.

$$\overline{MC} = C'(x) = 0.003x^2 - 0.6x + 32$$

(b) $C'(80) = 0.003(80)^2 - 0.6(80) + 32 = 3.2$ dollars per unit

  $C'(200) = 0.003(200)^2 - 0.6(200) + 32 = 32$ dollars per unit

These values for the marginal cost can be used to *estimate* the amount that total cost would change if production were increased by one blender. Thus,

$$C'(80) = 3.2$$

means total cost would increase by *about* $3.20 if an 81st blender were produced. Note that

$$C(81) - C(80) = 3.141 \approx \$3.14$$

is the actual increase in total cost for an 81st blender.

   Also, $C'(200) = 32$ means that total cost would increase by *about* $32 if a 201st blender were produced.

   Whenever $C'(x)$ is positive it means that an additional unit produced adds to or increases the total cost. In addition, the value of the derivative (or marginal cost) measures how fast $C(x)$ is increasing. Thus, our calculations above indicate that $C(x)$ is increasing faster at $x = 200$ than at $x = 80$. The graphs of the total cost function and the marginal cost function in Figure 9.37(a) and (b) also illustrate these facts.

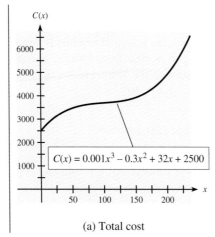

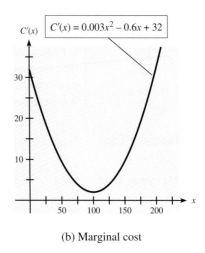

Figure 9.37   (a) Total cost   (b) Marginal cost

The graphs of many marginal cost functions tend to be U-shaped; they eventually will rise, even though there may be an initial interval where they decrease. Looking at the marginal cost graph in Figure 9.37(b), we see that marginal cost reaches its minimum near $x = 100$. We can also see this in Figure 9.37(a) by noting that tangent lines drawn to the total cost graph would have slopes that decrease until about $x = 100$ and then would increase.

Because producing more units can never reduce the total cost of production, the following properties are valid for total cost functions.

1. The total cost can never be negative. If there are fixed costs, the cost of producing 0 units is positive; otherwise, the cost of producing 0 units is 0.
2. The total cost function is always increasing; the more units produced, the higher the total cost. Thus the marginal cost is always positive.
3. There may be limitations on the units produced, such as those imposed by plant space.

● **EXAMPLE 4** *Total and Marginal Cost*

Suppose the graph in Figure 9.38 shows the monthly total cost for producing a product.

(a) Estimate the total cost of producing 400 items.
(b) Estimate the cost of producing the 401st item.
(c) Will producing the 151st item cost more or less than producing the 401st item?

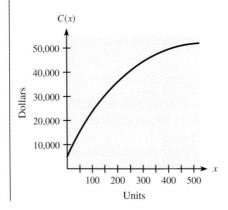

Figure 9.38

Solution

(a) The total cost of producing 400 items is the height of the total cost graph when $x = 400$, or about $50,000.

(b) The approximate cost of the 401st item is the marginal cost when $x = 400$, or the slope of the tangent line drawn to the graph at $x = 400$. Figure 9.39 shows the total cost graph with a tangent line at $x = 400$.

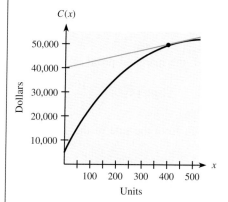

Figure 9.39

Note that the tangent line passes through the point $(0, 40,000)$, so we can find the slope by using the points $(0, 40,000)$ and $(400, 50,000)$.

$$m = \frac{y_2 - y_1}{x_2 - x_1} = \frac{50,000 - 40,000}{400 - 0} = \frac{10,000}{400} = 25$$

Thus the marginal cost at $x = 400$ is 25 dollars per item, so the approximate cost of the 401st item is $25.

(c) From Figure 9.39 we can see that if a tangent line to the graph were drawn where $x = 150$, it would be steeper than the one at $x = 400$. Because the slope of the tangent is the marginal cost and the marginal cost predicts the cost of the next item, this means that it would cost more to produce the 151st item than to produce the 401st.

● **Checkpoint**

Suppose the total cost function for a commodity is $C(x) = 0.01x^3 - 0.9x^2 + 33x + 3000$.
1. Find the marginal cost function.
2. What is the marginal cost if $x = 50$ units are produced?
3. Use marginal cost to estimate the cost of producing the 51st unit.
4. Calculate $C(51) - C(50)$ to find the actual cost of producing the 51st unit.
5. True or false: For products that have linear cost functions, the actual cost of producing the $(x + 1)$st unit is equal to the marginal cost at x.

Marginal Profit

As with marginal cost and marginal revenue, the derivative of a profit function for a commodity will give us the marginal profit function for the commodity.

Marginal Profit

If $P = P(x)$ is the profit function for a commodity, then the **marginal profit function** is $\overline{MP} = P'(x)$.

● **EXAMPLE 5** *Marginal Profit*

If the total profit, in thousands of dollars, for a product is given by $P(x) = 20\sqrt{x + 1} - 2x$, what is the marginal profit at a production level of 15 units?

Solution
The marginal profit function is

$$\overline{MP} = P'(x) = 20 \cdot \frac{1}{2}(x + 1)^{-1/2} - 2 = \frac{10}{\sqrt{x + 1}} - 2$$

If 15 units are produced, the marginal profit is

$$P'(15) = \frac{10}{\sqrt{15 + 1}} - 2 = \frac{1}{2}$$

This means that the profit from the sale of the 16th unit is approximately $\frac{1}{2}$ (thousand dollars), or $500.

In a **competitive market,** each firm is so small that its actions in the market cannot affect the price of the product. The price of the product is determined in the market by the intersection of the market demand curve (from all consumers) and the market supply curve (from all firms that supply this product). The firm can sell as little or as much as it desires at the given market price, which it cannot change.

Therefore, a firm in a competitive market has a total revenue function given by $R(x) = px$, where p is the market equilibrium price for the product and x is the quantity sold.

● **EXAMPLE 6** *Profit in a Competitive Market*

A firm in a competitive market must sell its product for $200 per unit. The cost per unit (per month) is $80 + x$, where x represents the number of units sold per month. Find the marginal profit function.

Solution
If the cost per unit is $80 + x$, then the total cost of x units is given by the equation $C(x) = (80 + x)x = 80x + x^2$. The revenue per unit is $200, so the total revenue is given by $R(x) = 200x$. Thus the profit function is

$$P(x) = R(x) - C(x) = 200x - (80x + x^2), \quad \text{or} \quad P(x) = 120x - x^2$$

The marginal profit is $P'(x) = 120 - 2x$.

The marginal profit in Example 6 is not always positive, so producing and selling a certain number of items will maximize profit. Note that the marginal profit will be negative (that is, profit will decrease) if more than 60 items per month are produced. We will discuss methods of maximizing total revenue and profit, and of minimizing average cost, in the next chapter.

● *Checkpoint*   If the total profit function for a product is $P(x) = 20\sqrt{x + 1} - 2x$, then the marginal profit is

$$P'(x) = \frac{10}{\sqrt{x + 1}} - 2 \quad \text{and} \quad P''(x) = \frac{-5}{\sqrt{(x + 1)^3}}$$

6. Is $P''(x) < 0$ for all values of $x \geq 0$?
7. Is the marginal profit decreasing for all $x \geq 0$?

● **EXAMPLE 7** *Marginal Profit*

Figure 9.40 shows graphs of a company's total revenue and total cost functions.

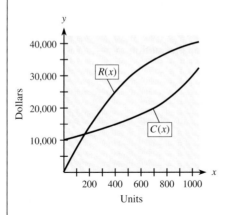

Figure 9.40

(a) If 100 units are being produced and sold, will producing the 101st item increase or decrease the profit?
(b) If 300 units are being produced and sold, will producing the 301st item increase or decrease the profit?
(c) If 1000 units are being produced and sold, will producing the 1001st item increase or decrease the profit?

Solution
At a given x-value, the slope of the tangent line to each function gives the marginal at that x-value.

(a) At $x = 100$, we can see that the graph of $R(x)$ is steeper than the graph of $C(x)$. Thus, the tangent line to $R(x)$ will be steeper than the tangent line to $C(x)$. Hence $\overline{MR}(100) > \overline{MC}(100)$, which means that the revenue from the 101st item will exceed the cost. Therefore, profit will increase when the 101st item is produced. Note that at $x = 100$, total costs are greater than total revenue, so the company is losing money but should still sell the 101st item because it will reduce the amount of loss.
(b) At $x = 300$, we can see that the tangent line to $R(x)$ again will be steeper than the tangent line to $C(x)$. Hence $\overline{MR}(300) > \overline{MC}(300)$, which means that the revenue from the 301st item will exceed the cost. Therefore, profit will increase when the 301st item is produced.
(c) At $x = 1000$, we can see that the tangent line to $C(x)$ will be steeper than the tangent line to $R(x)$. Hence $\overline{MC}(1000) > \overline{MR}(1000)$, which means that the cost of the 1001st item will exceed the revenue. Therefore, profit will decrease when the 1001st item is produced.

● **EXAMPLE 8** *Profit and Marginal Profit*

In Example 5, we found that the profit (in thousands of dollars) for a company's products is given by $P(x) = 20\sqrt{x + 1} - 2x$ and its marginal profit is given by $P'(x) = \dfrac{10}{\sqrt{x + 1}} - 2.$

(a) Use the graphs of $P(x)$ and $P'(x)$ to determine the relationship between the two functions.
(b) When is the marginal profit 0? What is happening to profit at this level of production?

Solution

(a) By comparing the graphs of the two functions (shown in Figure 9.41), we see that for $x > 0$, profit $P(x)$ is increasing over the interval where the marginal profit $P'(x)$ is positive, and profit is decreasing over the interval where the marginal profit $P'(x)$ is negative.

(b) By using ZERO or SOLVER, or by using algebra, we see that $P'(x) = 0$ when $x = 24$. This level of production ($x = 24$) is where profit is maximized, at 52 (thousand dollars).

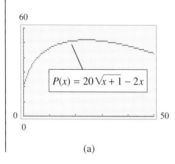

Figure 9.41

(a)   (b)

● *Checkpoint Solutions*

1. $\overline{MC} = C'(x) = 0.03x^2 - 1.8x + 33$
2. $C'(50) = 18$
3. $C'(50) = 18$, so it will cost approximately $18 to produce the 51st unit.
4. $C(51) - C(50) = 3668.61 - 3650 = 18.61$
5. True   6. Yes   7. Yes, because $P''(x) < 0$ for $x \geq 0$.

9.9 Exercises

MARGINAL REVENUE, COST, AND PROFIT

In Problems 1–8, total revenue is in dollars and x is the number of units.

1. (a) If the total revenue function for a product is $R(x) = 4x$, what is the marginal revenue function for that product?
   (b) What does this marginal revenue function tell us?

2. If the total revenue function for a product is $R(x) = 32x$, what is the marginal revenue for the product? What does this mean?

3. Suppose that the total revenue function for a commodity is $R = 36x - 0.01x^2$.
   (a) Find $R(100)$ and tell what it represents.
   (b) Find the marginal revenue function.
   (c) Find the marginal revenue at $x = 100$, and tell what it predicts about the sale of the next unit and the next 3 units.
   (d) Find $R(101) - R(100)$ and explain what this value represents.

4. Suppose that the total revenue function for a commodity is $R(x) = 25x - 0.05x^2$.
   (a) Find $R(50)$ and tell what it represents.
   (b) Find the marginal revenue function.
   (c) Find the marginal revenue at $x = 50$, and tell what it predicts about the sale of the next unit and the next 2 units.
   (d) Find $R(51) - R(50)$ and explain what this value represents.

5. Suppose that demand for local cable TV service is given by

$$p = 80 - 0.4x$$

where p is the monthly price in dollars and x is the number of subscribers (in hundreds).
   (a) Find the total revenue as a function of the number of subscribers.
   (b) Find the number of subscribers when the company charges $50 per month for cable service. Then find the total revenue for $p = \$50$.
   (c) How could the company attract more subscribers?
   (d) Find and interpret the marginal revenue when the price is $50 per month. What does this suggest about the monthly charge to subscribers?

6. Suppose that in a monopoly market, the demand function for a product is given by

$$p = 160 - 0.1x$$

where x is the number of units and p is the price in dollars.
   (a) Find the total revenue from the sale of 500 units.
   (b) Find and interpret the marginal revenue at 500 units.
   (c) Is more revenue expected from the 501st unit sold or from the 701st? Explain.

7. (a) Graph the marginal revenue function from Problem 3.
   (b) At what value of x will total revenue be maximized for Problem 3.
   (c) What is the maximum revenue?

8. (a) Graph the marginal revenue function from Problem 4.
   (b) Determine the number of units that must be sold to maximize total revenue.
   (c) What is the maximum revenue?

In Problems 9–16, cost is in dollars and x is the number of units. Find the marginal cost functions for the given cost functions.

9. $C(x) = 40 + 8x$ 10. $C(x) = 200 + 16x$
11. $C(x) = 500 + 13x + x^2$
12. $C(x) = 300 + 10x + \frac{1}{100}x^2$
13. $C = x^3 - 6x^2 + 24x + 10$
14. $C = 0.1x^3 - 1.5x^2 + 9x + 15$
15. $C = 400 + 27x + x^3$ 16. $C(x) = 50 + 48x + x^3$
17. Suppose that the cost function for a commodity is

$$C(x) = 40 + x^2 \quad \text{dollars}$$

   (a) Find the marginal cost at $x = 5$ units and tell what this predicts about the cost of producing 1 additional unit.
   (b) Calculate $C(6) - C(5)$ to find the actual cost of producing 1 additional unit.

18. Suppose that the cost function for a commodity is

$$C(x) = 300 + 6x + \tfrac{1}{20}x^2 \quad \text{dollars}$$

   (a) Find the marginal cost at $x = 8$ units and tell what this predicts about the cost of producing 1 additional unit.
   (b) Calculate $C(9) - C(8)$ to find the actual cost of producing 1 additional unit.

19. If the cost function for a commodity is

$$C(x) = x^3 - 4x^2 + 30x + 20 \quad \text{dollars}$$

find the marginal cost at $x = 4$ units and tell what this predicts about the cost of producing 1 additional unit and 3 additional units.

20. If the cost function for a commodity is

$$C(x) = \tfrac{1}{90}x^3 + 4x^2 + 4x + 10 \quad \text{dollars}$$

find the marginal cost at $x = 3$ units and tell what this predicts about the cost of producing 1 additional unit and 2 additional units.

21. If the cost function for a commodity is

$$C(x) = 300 + 4x + x^2$$

graph the marginal cost function.

22. If the cost function for a commodity is

$$C(x) = x^3 - 12x^2 + 63x + 15$$

graph the marginal cost function.

In each of Problems 23 and 24, the graph of a company's total cost function is shown. For each problem, use the graph to answer the following questions.
(a) Will the 101st item or the 501st item cost more to produce? Explain.
(b) Does this total cost function represent a manufacturing process that is getting more efficient or less efficient? Explain.

23.

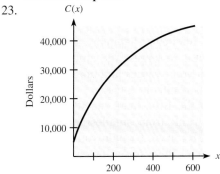

24.

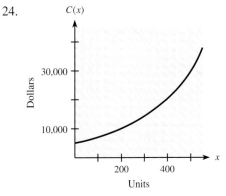

In Problems 25–28, cost, revenue, and profit are in dollars and x is the number of units.

25. If the total profit function is $P(x) = 5x - 25$, find the marginal profit. What does this mean?
26. If the total profit function is $P(x) = 16x - 32$, find the marginal profit. What does this mean?
27. Suppose that the total revenue function for a product is $R(x) = 32x$ and that the total cost function is $C(x) = 200 + 2x + x^2$.

(a) Find the profit from the production and sale of 20 units.
(b) Find the marginal profit function.
(c) Find $\overline{MP}$ at $x = 20$ and explain what it predicts.
(d) Find $P(21) - P(20)$ and explain what this value represents.

28. Suppose that the total revenue function is given by

$$R(x) = 46x$$

and that the total cost function is given by

$$C(x) = 100 + 30x + \tfrac{1}{10}x^2$$

(a) Find $P(100)$.
(b) Find the marginal profit function.
(c) Find $\overline{MP}$ at $x = 100$ and explain what it predicts.
(d) Find $P(101) - P(100)$ and explain what this value represents.

In each of Problems 29 and 30, the graphs of a company's total revenue function and total cost function are shown. For each problem, use the graph to answer the following questions.
(a) From the sale of 100 items, 400 items, and 700 items, rank from smallest to largest the amount of profit received. Explain your choices and note whether any of these scenarios results in a loss.
(b) From the sale of the 101st item, the 401st item, and the 701st item, rank from smallest to largest the amount of profit received. Explain your choices, and note whether any of these scenarios results in a loss.

29.
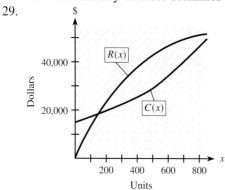

30.
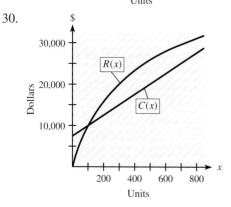

In each of Problems 31 and 32, the graph of a company's profit function is shown. For each problem, use the graph to answer the following questions about points A, B, and C.
(a) Rank from smallest to largest the amounts of profit received at these points. Explain your choices, and note whether any point results in a loss.
(b) Rank from smallest to largest the marginal profit at these points. Explain your choices, and note whether any marginal is negative and what this means.

31.
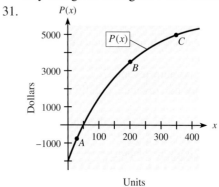

32.
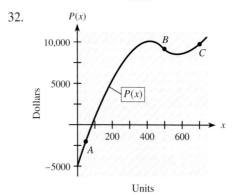

33. (a) Graph the marginal profit function for the profit function $P(x) = 30x - x^2 - 200$, where $P(x)$ is in thousands of dollars and x is the number of units.
   (b) What level of production and sales will give a 0 marginal profit?
   (c) At what level of production and sales will profit be at a maximum?
   (d) What is the maximum profit?

34. (a) Graph the marginal profit function for the profit function $P(x) = 16x - 0.1x^2 - 100$, where $P(x)$ is in dollars and x is the number of units.
   (b) What level of production and sales will give a 0 marginal profit?
   (c) At what level of production and sales will profit be at a maximum?
   (d) What is the maximum profit?

35. The price of a product in a competitive market is $300. If the cost per unit of producing the product is $160 + x$ dollars, where x is the number of units produced per

month, how many units should the firm produce and sell to maximize its profit?

36. The cost per unit of producing a product is $60 + 2x$ dollars, where x represents the number of units produced per week. If the equilibrium price determined by a competitive market is $220, how many units should the firm produce and sell each week to maximize its profit?

37. If the daily cost per unit of producing a product by the Ace Company is $10 + 2x$ dollars, and if the price on the competitive market is $50, what is the maximum daily profit the Ace Company can expect on this product?

38. The Mary Ellen Candy Company produces chocolate Easter bunnies at a cost per unit of $0.10 + 0.01x$ dollars, where x is the number produced. If the price on the competitive market for a bunny this size is $2.50, how many should the company produce to maximize its profit?

Key Terms and Formulas

Section	Key Terms	Formulas
9.1	Limit One-sided limits Properties of limits 0/0 indeterminate form	
9.2	Continuous function Vertical asymptote Horizontal asymptote Limit at infinity	
9.3	Average rate of change of f over $[a, b]$	$\dfrac{f(b) - f(a)}{b - a}$
	Average velocity Velocity Instantaneous rate of change	
	Derivative	$f'(x) = \lim\limits_{h \to 0} \dfrac{f(x + h) - f(x)}{h}$
	Marginal revenue Tangent line Secant line Slope of a curve Differentiability and continuity	$\overline{MR} = R'(x)$
9.4	Powers of x Rule	$\dfrac{d(x^n)}{dx} = nx^{n-1}$
	Constant Function Rule	$\dfrac{d(c)}{dx} = 0 \quad$ for constant c
	Coefficient Rule	$\dfrac{d}{dx}[c \cdot f(x)] = c \cdot f'(x)$
	Sum Rule	$\dfrac{d}{dx}[u + v] = \dfrac{du}{dx} + \dfrac{dv}{dx}$
	Difference Rule	$\dfrac{d}{dx}[u - v] = \dfrac{du}{dx} - \dfrac{dv}{dx}$

Section	Key Terms	Formulas
9.5	Product Rule	$\dfrac{d}{dx}[uv] = uv' + vu'$
	Quotient Rule	$\dfrac{d}{dx}\left(\dfrac{u}{v}\right) = \dfrac{vu' - uv'}{v^2}$
9.6	Chain Rule	$\dfrac{dy}{dx} = \dfrac{dy}{du} \cdot \dfrac{du}{dx}$
	Power Rule	$\dfrac{d}{dx}(u^n) = nu^{n-1}\dfrac{du}{dx}$
9.8	Second derivative; third derivative; higher-order derivatives	
9.9	Marginal revenue function	$\overline{MR} = R'(x)$
	Marginal cost function	$\overline{MC} = C'(x)$
	Marginal profit function	$\overline{MP} = P'(x)$

Review Exercises

Section 9.1

In Problems 1–6, use the graph of $y = f(x)$ in Figure 9.42 to find the functional values and limits, if they exist.

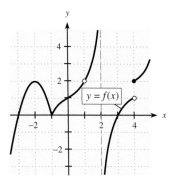

Figure 9.42

1. (a) $f(-2)$ (b) $\lim\limits_{x \to -2} f(x)$
2. (a) $f(-1)$ (b) $\lim\limits_{x \to -1} f(x)$
3. (a) $f(4)$ (b) $\lim\limits_{x \to 4^-} f(x)$
4. (a) $\lim\limits_{x \to 4^+} f(x)$ (b) $\lim\limits_{x \to 4} f(x)$
5. (a) $f(1)$ (b) $\lim\limits_{x \to 1} f(x)$
6. (a) $f(2)$ (b) $\lim\limits_{x \to 2} f(x)$

In Problems 7–20, find each limit, if it exists.

7. $\lim\limits_{x \to 4}(3x^2 + x + 3)$
8. $\lim\limits_{x \to 4} \dfrac{x^2 - 16}{x + 4}$
9. $\lim\limits_{x \to -1} \dfrac{x^2 - 1}{x + 1}$
10. $\lim\limits_{x \to 3} \dfrac{x^2 - 9}{x - 3}$

11. $\lim\limits_{x \to 2} \dfrac{4x^3 - 8x^2}{4x^3 - 16x}$
12. $\lim\limits_{x \to -\frac{1}{2}} \dfrac{x^2 - \frac{1}{4}}{6x^2 + x - 1}$
13. $\lim\limits_{x \to 3} \dfrac{x^2 - 16}{x - 3}$
14. $\lim\limits_{x \to -3} \dfrac{x^2 - 9}{x - 3}$
15. $\lim\limits_{x \to 1} \dfrac{x^2 - 9}{x - 3}$
16. $\lim\limits_{x \to 2} \dfrac{x^2 - 8}{x - 2}$
17. $\lim\limits_{x \to 1} f(x)$ where $f(x) = \begin{cases} 4 - x^2 & \text{if } x < 1 \\ 4 & \text{if } x = 1 \\ 2x + 1 & \text{if } x > 1 \end{cases}$
18. $\lim\limits_{x \to -2} f(x)$ where $f(x) = \begin{cases} x^3 - x & \text{if } x < -2 \\ 2 - x^2 & \text{if } x \geq -2 \end{cases}$
19. $\lim\limits_{h \to 0} \dfrac{3(x + h)^2 - 3x^2}{h}$
20. $\lim\limits_{h \to 0} \dfrac{[(x + h) - 2(x + h)^2] - (x - 2x^2)}{h}$

 In Problems 21 and 22, use tables to investigate each limit. Check your result analytically or graphically.

21. $\lim\limits_{x \to 2} \dfrac{x^2 + 10x - 24}{x^2 - 5x + 6}$
22. $\lim\limits_{x \to -\frac{1}{2}} \dfrac{x^2 + \frac{1}{6}x - \frac{1}{6}}{x^2 + \frac{5}{6}x + \frac{1}{6}}$

Section 9.2

Use the graph of $y = f(x)$ in Figure 9.42 to answer the questions in Problems 23 and 24.

23. Is $f(x)$ continuous at
    (a) $x = -1$?   (b) $x = 1$?

24. Is $f(x)$ continuous at
  (a) $x = -2$? (b) $x = 2$?

In Problems 25–30, suppose that

$$f(x) = \begin{cases} x^2 + 1 & \text{if } x \le 0 \\ x & \text{if } 0 < x < 1 \\ 2x^2 - 1 & \text{if } x \ge 1 \end{cases}$$

25. What is $\lim_{x \to -1} f(x)$?
26. What is $\lim_{x \to 0} f(x)$, if it exists?
27. What is $\lim_{x \to 1} f(x)$, if it exists?
28. Is $f(x)$ continuous at $x = 0$?
29. Is $f(x)$ continuous at $x = 1$?
30. Is $f(x)$ continuous at $x = -1$?

For the functions in Problems 31–34, determine which are continuous. Identify discontinuities for those that are not continuous.

31. $y = \dfrac{x^2 + 25}{x - 5}$
32. $y = \dfrac{x^2 - 3x + 2}{x - 2}$
33. $f(x) = \begin{cases} x + 2 & \text{if } x \le 2 \\ 5x - 6 & \text{if } x > 2 \end{cases}$
34. $y = \begin{cases} x^4 - 3 & \text{if } x \le 1 \\ 2x - 3 & \text{if } x > 1 \end{cases}$

In Problems 35 and 36, use the graphs to find (a) the points of discontinuity, (b) $\lim_{x \to +\infty} f(x)$, and (c) $\lim_{x \to -\infty} f(x)$.

35.
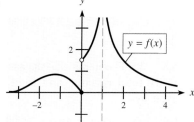

36.
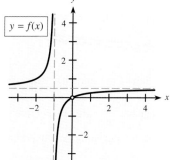

In Problems 37 and 38, evaluate the limits, if they exist.

37. $\lim_{x \to -\infty} \dfrac{2x^2}{1 - x^2}$
38. $\lim_{x \to +\infty} \dfrac{3x^{2/3}}{x + 1}$

Section 9.3

39. Find the average rate of change of
$$f(x) = 2x^4 - 3x + 7 \text{ over } [-1, 2].$$

In Problems 40 and 41, decide whether the statements are true or false.

40. $\lim_{h \to 0} \dfrac{f(x + h) - f(x)}{h}$ gives the formula for the slope of the tangent and the instantaneous rate of change of $f(x)$ at any value of x.
41. $\lim_{h \to 0} \dfrac{f(c + h) - f(c)}{h}$ gives the equation of the tangent line to $f(x)$ at $x = c$.
42. Use the definition of derivative to find $f'(x)$ for $f(x) = 3x^2 + 2x - 1$.
43. Use the definition of derivative to find $f'(x)$ if $f(x) = x - x^2$.

Use the graph of $y = f(x)$ in Figure 9.42 to answer the questions in Problems 44–46.

44. Explain which is greater: the average rate of change of f over $[-3, 0]$ or over $[-1, 0]$.
45. Is $f(x)$ differentiable at
  (a) $x = -1$? (b) $x = 1$?
46. Is $f(x)$ differentiable at
  (a) $x = -2$? (b) $x = 2$?

47. Let $f(x) = \dfrac{\sqrt[3]{4x}}{(3x^2 - 10)^2}$. Approximate $f'(2)$
  (a) by using the numerical derivative feature of a graphing utility, and
  (b) by evaluating $\dfrac{f(2 + h) - f(2)}{h}$ with $h = 0.0001$.

48. Use the given table of values for $g(x)$ to
  (a) find the average rate of change of $g(x)$ over $[2, 5]$.
  (b) approximate $g'(4)$ as accurately as possible.

x	2	2.3	3.1	4	4.3	5
$g(x)$	13.2	12.1	9.7	12.2	14.3	18.1

Use the following graph of $f(x)$ to complete Problems 49 and 50.

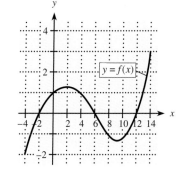

49. Estimate $f'(4)$.
50. Rank the following from smallest to largest and explain.
    $A: f'(2)$ $B: f'(6)$
    C: the average rate of change of $f(x)$ over $[2, 10]$

Section 9.4

51. If $c = 4x^5 - 6x^3$, find c'.
52. If $f(x) = 4x^2 - 1$, find $f'(x)$.
53. If $p = 3q + \sqrt{7}$, find dp/dq.
54. If $y = \sqrt{x}$, find y'.
55. If $f(z) = \sqrt[3]{z^4}$, find $f'(z)$.
56. If $v(x) = 4/\sqrt[3]{x}$, find $v'(x)$.
57. If $y = \dfrac{1}{x} - \dfrac{1}{\sqrt{x}}$, find y'.
58. If $f(x) = \dfrac{3}{2x^2} - \sqrt[3]{x} + 4^5$, find $f'(x)$.
59. Write the equation of the line tangent to the graph of $y = 3x^5 - 6$ at $x = 1$.
60. Write the equation of the line tangent to the curve $y = 3x^3 - 2x$ at the point where $x = 2$.

In Problems 61 and 62, (a) find all x-values where the slope of the tangent equals zero, (b) find points (x, y) where the slope of the tangent equals zero, and (c) use a graphing utility to graph the function and label the points found in (b).
61. $f(x) = x^3 - 3x^2 + 1$ 62. $f(x) = x^6 - 6x^4 + 8$

Section 9.5

63. If $f(x) = (3x - 1)(x^2 - 4x)$, find $f'(x)$.
64. Find y' if $y = (x^2 + 1)(3x^3 + 1)$.
65. If $p = \dfrac{2q - 1}{q^2}$, find $\dfrac{dp}{dq}$.
66. Find $\dfrac{ds}{dt}$ if $s = \dfrac{\sqrt{t}}{(3t + 1)}$.
67. Find $\dfrac{dy}{dx}$ for $y = \sqrt{x}(3x + 2)$.
68. Find $\dfrac{dC}{dx}$ for $C = \dfrac{5x^4 - 2x^2 + 1}{x^3 + 1}$.

Section 9.6

69. If $y = (x^3 - 4x^2)^3$, find y'.
70. If $y = (5x^6 + 6x^4 + 5)^6$, find y'.
71. If $y = (2x^4 - 9)^9$, find $\dfrac{dy}{dx}$.
72. Find $g'(x)$ if $g(x) = \dfrac{1}{\sqrt{x^3 - 4x}}$.

Section 9.7

73. Find $f'(x)$ if $f(x) = x^2(2x^4 + 5)^8$.
74. Find S' if $S = \dfrac{(3x + 1)^2}{x^2 - 4}$.
75. Find $\dfrac{dy}{dx}$ if $y = [(3x + 1)(2x^3 - 1)]^{12}$.
76. Find y' if $y = \left(\dfrac{x + 1}{1 - x^2}\right)^3$.
77. Find y' if $y = x\sqrt{x^2 - 4}$.
78. Find $\dfrac{dy}{dx}$ if $y = \dfrac{x}{\sqrt[3]{3x - 1}}$.

Section 9.8

In Problems 79 and 80, find the second derivatives.
79. $y = \sqrt{x} - x^2$ 80. $y = x^4 - \dfrac{1}{x}$

In Problems 81 and 82, find the fifth derivatives.
81. $y = (2x + 1)^4$
82. $y = \dfrac{(1 - x)^6}{24}$
83. If $\dfrac{dy}{dx} = \sqrt{x^2 - 4}$, find $\dfrac{d^3y}{dx^3}$.
84. If $\dfrac{d^2y}{dx^2} = \dfrac{x}{x^2 + 1}$, find $\dfrac{d^4y}{dx^4}$.

APPLICATIONS

Sections 9.1 and 9.2

Cost, revenue, and profit In Problems 85–88, assume that a company's monthly total revenue and total cost (both in dollars) are given by

$$R(x) = 140x - 0.01x^2 \quad \text{and} \quad C(x) = 60x + 70,000$$

where x is the number of units. (Let $P(x)$ denote the profit function.)
85. Find (a) $\lim\limits_{x \to 4000} R(x)$, (b) $\lim\limits_{x \to 4000} C(x)$, and (c) $\lim\limits_{x \to 4000} P(x)$.
86. Find and interpret (a) $\lim\limits_{x \to 0^+} C(x)$ and (b) $\lim\limits_{x \to 1000} P(x)$.
87. If $\overline{R}(x) = \dfrac{R(x)}{x}$ and $\overline{C}(x) = \dfrac{C(x)}{x}$ are, respectively, the company's average revenue per unit and average cost per unit, find
    (a) $\lim\limits_{x \to 0^+} \overline{R}(x)$ (b) $\lim\limits_{x \to 0^+} \overline{C}(x)$
88. Evaluate and explain the meanings of
    (a) $\lim\limits_{x \to \infty} C(x)$ (b) $\lim\limits_{x \to \infty} \overline{C}(x) = \lim\limits_{x \to \infty} \dfrac{C(x)}{x}$

Section 9.3

Elderly in the work force **Problems 89 and 90 use the following graph, which shows the percent of elderly men and women in the work force for selected census years from 1890 to 2000.**

89. For the period from 1900 to 2000, find and interpret the annual average rate of change of
    (a) elderly men in the work force and
    (b) elderly women in the work force.

90. (a) During what decade was there an increase in the percent of elderly men in the work force? Find and interpret the annual average rate of change for elderly men in that decade.
    (b) During what decade was there the largest increase in the percent of elderly women in the work force? Find and interpret the average rate of change for elderly women in that decade.

Elderly in the Labor Force, 1890-2000
(labor force participation rate; figs. for 1910 not available)

	Men	Women
1890	68.3	7.6
1900	63.1	8.3
1920	55.6	7.3
1930	54.0	7.3
1940	41.8	6.1
1950	41.4	7.8
1960	30.5	10.3
1970	24.8	10.0
1980	19.3	8.2
1990	17.6	8.4
2000	18.6	10.0

Source: Bureau of the Census, U.S. Department of Commerce

Section 9.4

91. *Demand*   Suppose that the demand for x units of a product is given by $x = (100/p) - 1$, where p is the price per unit of the product. Find and interpret the rate of change of demand with respect to price if the price is
    (a) \$10   (b) \$20

92. *Paved roads and streets*   By using data from the U.S. Department of Transportation, the percent of paved roads and streets in the United States can be modeled by

    $$P(t) = 6.61t^{0.559}$$

    where t is the number of years past 1940. Find and interpret $P(65)$ and $P'(65)$.

93. *Revenue*   The following graph shows the revenue function for a commodity. Will the $(A + 1)$st item sold or the $(B + 1)$st item sold produce more revenue? Explain.

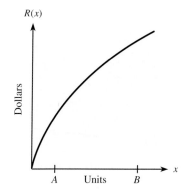

94. *Revenue*   In a 100-unit apartment building, when the price charged per apartment rental is $(830 + 30x)$ dollars, then the number of apartments rented is $100 - x$ and the total revenue for the building is

    $$R(x) = (830 + 30x)(100 - x)$$

    where x is the number of \$30 rent increases (and also the resulting number of unrented apartments). Find the marginal revenue when $x = 10$. Does this tell you that the rent should be raised (causing more vacancies) or lowered? Explain.

95. *Productivity*   Suppose the productivity of a worker (in units per hour) after x hours of training and time on the job is given by

    $$P(x) = 3 + \frac{70x^2}{x^2 + 1000}$$

    (a) Find and interpret $P(20)$.
    (b) Find and interpret $P'(20)$.

Section 9.6

96. *Demand*   The demand q for a product at price p is given by

    $$q = 10,000 - 50\sqrt{0.02p^2 + 500}$$

    Find the rate of change of demand with respect to price.

97. *Supply*   The number of units x of a product that is supplied at price p is given by

$$x = \sqrt{p - 1}, \quad p \geq 1$$

If the price p is \$10, what is the rate of change of the supply with respect to the price, and what does it tell us?

Section 9.8

98. *Acceleration*   Suppose an object moves so that its distance to a sensor, in feet, is given by

$$s(t) = 16 + 140t + 8\sqrt{t}$$

where t is the time in seconds. Find the acceleration at time $t = 4$ seconds.

99. *Profit*   Suppose a company's profit (in dollars) is given by

$$P(x) = 70x - 0.1x^2 - 5500$$

where x is the number of units. Find and interpret $P'(300)$ and $P''(300)$.

Section 9.9

In Problems 100–107, cost, revenue, and profit are in dollars and x is the number of units.

100. *Cost*   If the cost function for a particular good is $C(x) = 3x^2 + 6x + 600$, what is the
   (a) marginal cost function?
   (b) marginal cost if 30 units are produced?
   (c) interpretation of your answer in part (b)?

101. *Cost*   If the total cost function for a commodity is $C(x) = 400 + 5x + x^3$, what is the marginal cost when 4 units are produced, and what does it mean?

102. *Revenue*   The total revenue function for a commodity is $R = 40x - 0.02x^2$, with x representing the number of units.
   (a) Find the marginal revenue function.
   (b) At what level of production will marginal revenue be 0?

103. *Profit*   If the total revenue function for a product is given by $R(x) = 60x$ and the total cost function is given by $C = 200 + 10x + 0.1x^2$, what is the marginal profit at $x = 10$? What does the marginal profit at $x = 10$ predict?

104. *Revenue*   The total revenue function for a commodity is given by $R = 80x - 0.04x^2$.
   (a) Find the marginal revenue function.
   (b) What is the marginal revenue at $x = 100$?
   (c) Interpret your answer in part (b).

105. *Revenue*   If the revenue function for a product is

$$R(x) = \frac{60x^2}{2x + 1}$$

find the marginal revenue.

106. *Profit*   A firm has monthly costs given by

$$C = 45{,}000 + 100x + x^3$$

where x is the number of units produced per month. The firm can sell its product in a competitive market for \$4600 per unit. Find the marginal profit.

107. *Profit*   A small business has weekly costs of

$$C = 100 + 30x + \frac{x^2}{10}$$

where x is the number of units produced each week. The competitive market price for this business's product is \$46 per unit. Find the marginal profit.

108. *Cost, revenue, and profit*   The following graph shows the total revenue and total cost functions for a company. Use the graph to decide (and justify) at which of points A, B, and C
   (a) the cost of the next item will be least.
   (b) the profit will be greatest.
   (c) the profit from the sale of the next item will be greatest.
   (d) the next item sold will reduce the profit.

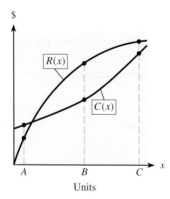

9 Chapter Test

1. Evaluate the following limits, if they exist. Use algebraic methods.
   (a) $\lim\limits_{x \to -2} \dfrac{4x - x^2}{4x - 8}$
   (b) $\lim\limits_{x \to \infty} \dfrac{8x^2 - 4x + 1}{2 + x - 5x^2}$
   (c) $\lim\limits_{x \to 7} \dfrac{x^2 - 5x - 14}{x^2 - 6x - 7}$
   (d) $\lim\limits_{x \to -5} \dfrac{5x - 25}{x + 5}$

2. (a) Write the limit definition for $f'(x)$.
   (b) Use the definition from (a) to find $f'(x)$ for $f(x) = 3x^2 - x + 9$.

3. Let $f(x) = \dfrac{4x}{x^2 - 8x}$. Identify all x-values where $f(x)$ is *not* continuous.

4. Use derivative formulas to find the derivative of each of the following. Simplify, except for (d).
   (a) $B = 0.523W - 5176$
   (b) $p = 9t^{10} - 6t^7 - 17t + 23$
   (c) $y = \dfrac{3x^3}{2x^7 + 11}$
   (d) $f(x) = (3x^5 - 2x + 3)(4x^{10} + 10x^4 - 17)$
   (e) $g(x) = \frac{3}{4}(2x^5 + 7x^3 - 5)^{12}$
   (f) $y = (x^2 + 3)(2x + 5)^6$
   (g) $f(x) = 12\sqrt{x} - \dfrac{10}{x^2} + 17$

5. Find $\dfrac{d^3y}{dx^3}$ for $y = x^3 - x^{-3}$.

6. Let $f(x) = x^3 - 3x^2 - 24x - 10$.
   (a) Write the equation of the line tangent to the graph of $y = f(x)$ at $x = -1$.
   (b) Find all points (both x- and y-coordinates) where $f'(x) = 0$.

7. Find the average rate of change of $f(x) = 4 - x - 2x^2$ over $[1, 6]$.

8. Use the given tables to evaluate the following limits, if they exist.
   (a) $\lim\limits_{x \to 5} f(x)$ (b) $\lim\limits_{x \to 5} g(x)$ (c) $\lim\limits_{x \to 5^-} g(x)$

x	4.99	4.999	→ 5 ←	5.001	5.01
$f(x)$	2.01	2.001	→ ? ←	1.999	1.99

x	4.99	4.999	→ 5 ←	5.001	5.01
$g(x)$	−3.99	−3.999	→ ? ←	6.999	6.99

9. Use the definition of continuity to investigate whether $g(x)$ is continuous at $x = -2$. Show your work.
   $$g(x) = \begin{cases} 6 - x & \text{if } x \le -2 \\ x^3 & \text{if } x > -2 \end{cases}$$

In Problems 10 and 11, suppose a company has its total cost for a product given by $C(x) = 200x + 10{,}000$ dollars and its total revenue given by $R(x) = 250x - 0.01x^2$ dollars, where x is the number of units produced and sold.

10. (a) Find the marginal revenue function.
    (b) Find $R(72)$ and $R'(72)$ and tell what each represents or predicts.

11. (a) Form the profit function for this product.
    (b) Find the marginal profit function.
    (c) Find the marginal profit when $x = 1000$, and then write a sentence that interprets this result.

12. Suppose that $f(x)$ is a differentiable function. Use the table of values to approximate $f'(3)$ as accurately as possible.

x	2	2.5	2.999	3	3.01	3.1
$f(x)$	0	18.4	44.896	45	46.05	56.18

13. Use the graph to perform the evaluations (a)–(f) and to answer (g)–(i). If no value exists, so indicate.
    (a) $f(1)$
    (b) $\lim\limits_{x \to 6} f(x)$
    (c) $\lim\limits_{x \to 3^-} f(x)$
    (d) $\lim\limits_{x \to -4} f(x)$
    (e) $\lim\limits_{x \to -\infty} f(x)$
    (f) Estimate $f'(4)$
    (g) Find all x-values where $f'(x)$ does not exist.
    (h) Find all x-values where $f(x)$ is not continuous.
    (i) Rank from smallest to largest: $f'(-2), f'(2)$, and the average rate of change of $f(x)$ over $[-2, 2]$.

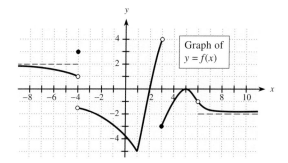

Graph of $y = f(x)$

14. Given that the line $y = \frac{2}{3}x - 8$ is tangent to the graph of $y = f(x)$ at $x = 6$, find
    (a) $f'(6)$
    (b) $f(6)$
    (c) the instantaneous rate of change of $f(x)$ with respect to x at $x = 6$

15. The following graph shows the total revenue and total cost functions for a company. Use the graph to decide (and justify) at which of points A, B, and C
    (a) profit is the greatest.
    (b) there is a loss.
    (c) producing and selling another item will increase profit.
    (d) the next item sold will decrease profit.

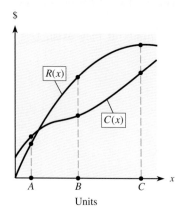

Extended Applications & Group Projects

I. Marginal Return to Sales

A tire manufacturer studying the effectiveness of television advertising and other promotions on sales of its GRIPPER-brand tires attempted to fit data it had gathered to the equation

$$S = a_0 + a_1x + a_2x^2 + b_1y$$

where S is sales revenue in millions of dollars, x is millions of dollars spent on television advertising, y is millions of dollars spent on other promotions, and a_0, a_1, a_2, and b_1 are constants. The data, gathered in two different regions of the country where expenditures for other promotions were kept constant (at B_1 and B_2), resulted in the following quadratic equations relating TV advertising and sales.

$$\text{Region 1:} \quad S_1 = 30 + 20x - 0.4x^2 + B_1$$
$$\text{Region 2:} \quad S_2 = 20 + 36x - 1.3x^2 + B_2$$

The company wants to know how to make the best use of its advertising dollars in the regions and whether the current allocation could be improved. Advise management about current advertising effectiveness, allocation of additional expenditures, and reallocation of current advertising expenditures by answering the following questions.

1. In the analysis of sales and advertising, **marginal return to sales** is usually used, and it is given by dS_1/dx for Region 1 and dS_2/dx for Region 2.
   (a) Find $\dfrac{dS_1}{dx}$ and $\dfrac{dS_2}{dx}$.
   (b) If \$10 million is being spent on TV advertising in each region, what is the marginal return to sales in each region?
2. Which region would benefit more from additional advertising expenditure, if \$10 million is currently being spent in each region?
3. If any additional money is made available for advertising, in which region should it be spent?
4. How could money already being spent be reallocated to produce more sales revenue?

II. Tangent Lines and Optimization in Business and Economics

In business and economics, common questions of interest include how to maximize profit or revenue, how to minimize average costs, and how to maximize average productivity. For questions such as these, the description of how to obtain the desired maximum or minimum represents an optimal (or best possible) solution to the problem. And the process of finding an optimal solution is called optimization. Answering these questions often involves tangent lines. In this project, we examine how tangent lines can be used to minimize average costs.

Suppose that Wittage, Inc. manufactures paper shredders for home and office use and that its weekly total costs (in dollars) for x shredders are given by

$$C(x) = 0.03x^2 + 12.75x + 6075$$

The average cost per unit for x units (denoted by $\overline{C}(x)$) is the total cost divided by the number of units. Thus, the average cost function for Wittage, Inc. is

$$\overline{C}(x) = \frac{C(x)}{x} = \frac{0.03x^2 + 12.75x + 6075}{x} = \frac{0.03x^2}{x} + \frac{12.75x}{x} + \frac{6075}{x}$$

$$\overline{C}(x) = 0.03x + 12.75 + \frac{6075}{x}$$

Wittage, Inc. would like to know how the company can use marginal costs to gain information about minimizing average costs. To investigate this relationship, answer the following questions.

1. Find Wittage's marginal cost function, then complete the following table.

x	100	200	300	400	500	600
$\overline{C}(x)$						
$\overline{MC}(x)$						

2. (a) For what x-values in the table is $\overline{C}$ getting smaller?
   (b) For these x-values, is $\overline{MC} < \overline{C}, \overline{MC} > \overline{C}$, or $\overline{MC} = \overline{C}$?

3. (a) For what x-values in the table is $\overline{C}$ getting larger?
   (b) For these x-values, is $\overline{MC} < \overline{C}, \overline{MC} > \overline{C}$, or $\overline{MC} = \overline{C}$?

4. Between what pair of consecutive x-values in the table will $\overline{C}$ reach its minimum?

5. Interpret (a) $\overline{C}(200)$ and $\overline{MC}(200)$ and (b) $\overline{C}(400)$ and $\overline{MC}(400)$.

6. On the basis of what $\overline{C}$ and $\overline{MC}$ tell us, and the work so far, complete the following statements by filling the blank with *increase, decrease* or *stay the same*. Explain your reasoning.

   (a) $\overline{MC} < \overline{C}$ means that if Wittage makes one more unit, then $\overline{C}$ will _____.
   (b) $\overline{MC} > \overline{C}$ means that if Wittage makes one more unit, then $\overline{C}$ will _____.

7. (a) On the basis of your answers to question 6, how will $\overline{C}$ and $\overline{MC}$ be related when $\overline{C}$ is minimized? Write a careful statement.

   (b) Use your idea from part (a) to find x where $\overline{C}$ is minimized. Then check your calculations by graphing $\overline{C}$ and locating the minimum from the graph.

Next, let's examine graphically the relationship that emerged in question 7 to see whether it is true in general. Figure 9.43 shows a typical total cost function (Wittage's total cost function has a similar shape) and an arbitrary point on the graph.

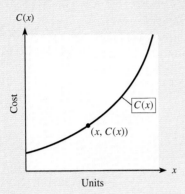

Figure 9.43

8. (a) Measure the slope of the line joining the point $(x, C(x))$ and the origin.
   (b) How is this slope related to the function $\overline{C}$ at $(x, C(x))$?
9. Figure 9.44(a) shows a typical total cost function with several points labeled.
   (a) Explain how you can tell at a glance that $\overline{C}(x)$ is decreasing as the level of production moves from A to B to C to D.
   (b) Explain how you can tell at a glance that $\overline{C}(x)$ is increasing as the level of production moves from Q to R to S to T.

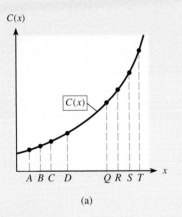
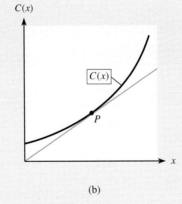

(a)   (b)

Figure 9.44

10. In Figure 9.44(b), a line from $(0, 0)$ is drawn tangent to $C(x)$ and the point of tangency is labeled P. At point P in Figure 9.44(b), what represents
    (a) the average cost, $\overline{C}(x)$?
    (b) the marginal cost, $\overline{MC}(x)$?
    (c) Explain how Figure 9.44 confirms that $\overline{C}(x)$ is minimized when $\overline{MC}(x) = \overline{C}(x)$.

On the basis of the numerical approach for Wittage's total cost function and the general graphical approach, what recommendation would you give Wittage about determining the number of units that will minimize average costs (even if the total cost function changes)?

10

Applications of Derivatives

The derivative can be used to determine where a function has a "turning point" on its graph, so that we can determine where the graph reaches its highest or lowest point within a particular interval. These points are called the **relative maxima** and **relative minima**, respectively, and are useful in sketching the graph of the function. The techniques for finding these points are also useful in solving applied problems, such as finding the maximum profit, the minimum average cost, and the maximum productivity. The second derivative can be used to find **points of inflection** of the graph of a function and to find the point of diminishing returns in certain applications.

The topics and applications discussed in this chapter include the following.

Chapter Warm-up

Prerequisite Problem Type	For Section	Answer	Section for Review
Write $\frac{1}{3}(x^2 - 1)^{-2/3}(2x)$ with positive exponents.	10.2	$\dfrac{2x}{3(x^2 - 1)^{2/3}}$	0.3, 0.4 Exponents and radicals
Factor: (a) $x^3 - x^2 - 6x$ (b) $8000 - 80x - 3x^2$	10.1 10.2 10.3	 (a) $x(x - 3)(x + 2)$ (b) $(40 - x)(200 + 3x)$	0.6 Factoring
(a) For what values of x is $\dfrac{2}{3\sqrt[3]{x + 2}}$ undefined? (b) For what values of x is $\frac{1}{3}(x^2 - 1)^{-2/3}(2x)$ undefined?	10.1 10.2 10.5	(a) $x = -2$ (b) $x = -1, x = 1$	1.2 Domains of functions
If $f(x) = \frac{1}{3}x^3 - x^2 - 3x + 2$, and $f'(x) = x^2 - 2x - 3$, (a) find $f(-1)$.　(b) find $f'(-2)$.	10.1	 (a) $\dfrac{11}{3}$　(b) 5	1.2 Function notation
(a) Solve $0 = x^2 - 2x - 3$. (b) If $f'(x) = 3x^2 - 3$, what values of x make $f'(x) = 0$?	10.1–10.5	(a) $x = -1, x = 3$ (b) $x = -1, x = 1$	2.1 Solving quadratic equations
Does $\displaystyle\lim_{x \to -2} \dfrac{2x - 4}{3x + 6}$ exist?	10.5	No; unbounded	9.1 Limits
(a) Find $f''(x)$ if $f(x) = x^3 - 4x^2 + 3$. (b) Find $P''(x)$ if $P(x) = 48x - 1.2x^2$.	10.2	(a) $f''(x) = 6x - 8$ (b) $P''(x) = -2.4$	9.8 Higher-order derivatives
Find the derivatives: (a) $y = \dfrac{1}{3}x^3 - x^2 - 3x + 2$ (b) $f = x + 2\left(\dfrac{80{,}000}{x}\right)$ (c) $p(t) = 1 + \dfrac{4t}{t^2 + 16}$ (d) $y = (x + 2)^{2/3}$ (e) $y = \sqrt[3]{x^2 - 1}$	10.1 10.2 10.3 10.4	 (a) $y' = x^2 - 2x - 3$ (b) $f' = 1 - \dfrac{160{,}000}{x^2}$ (c) $p'(t) = \dfrac{64 - 4t^2}{(t^2 + 16)^2}$ (d) $y' = \dfrac{2}{3(x + 2)^{1/3}}$ (e) $y' = \dfrac{2x}{3(x^2 - 1)^{2/3}}$	9.4, 9.5, 9.6 Derivatives

Relative Maxima and Minima: Curve Sketching

- To find relative maxima and minima and horizontal points of inflection of functions
- To sketch graphs of functions by using information about maxima, minima, and horizontal points of inflection

◗ Application Preview

When a company initiates an advertising campaign, there is typically a surge in weekly sales. As the effect of the campaign lessens, sales attributable to it usually decrease. For example, suppose a company models its weekly sales revenue during an advertising campaign by

$$S = \frac{100t}{t^2 + 100}$$

where t is the number of weeks since the beginning of the campaign. The company would like to determine accurately when the revenue function is increasing, when it is decreasing, and when sales revenue is maximized. (See Example 5.)

In this section we will use the derivative of a function to decide whether the function is increasing or decreasing on an interval and to find where the function has relative maximum points and relative minimum points. We will use the information about derivatives of functions to graph the functions and to solve applied problems.

Recall that we can find the maximum value or the minimum value of a quadratic function by finding the vertex of its graph. But for functions of higher degree, the special features of their graphs may be harder to find accurately, even when a graphing utility is used. In addition to intercepts and asymptotes, we can use the first derivative as an aid in graphing. The first derivative identifies the "turning points" of the graph, which help us determine the general shape of the graph and choose a viewing window that includes the interesting points of the graph.

In Figure 10.1(a) we see that the graph of $y = \frac{1}{3}x^3 - x^2 - 3x + 2$ has two "turning points," at $\left(-1, \frac{11}{3}\right)$ and $(3, -7)$. The curve has a relative maximum at $\left(-1, \frac{11}{3}\right)$ because this point is higher than any other point "near" it on the curve; the curve has a relative minimum at $(3, -7)$ because this point is lower than any other point "near" it on the curve. A formal definition follows.

Relative Maxima and Minima

The point $(x_1, f(x_1))$ is a **relative maximum point** of the function f if there is an interval around x_1 on which $f(x_1) \geq f(x)$ for all x in the interval. In this case, we say the relative maximum *occurs* at $x = x_1$ and the relative maximum is $f(x_1)$.

The point $(x_2, f(x_2))$ is a **relative minimum point** of the function f if there is an interval around x_2 on which $f(x_2) \leq f(x)$ for all x in the interval. In this case, we say the relative minimum *occurs* at $x = x_2$ and the relative minimum is $f(x_2)$.

In order to determine whether a turning point of a function is a maximum point or a minimum point, it is frequently helpful to know what the graph of the function does in intervals on either side of the turning point. We say a function is **increasing** on an interval if the values of the function increase as the x-values increase (that is, if the graph rises as we move from left to right on the interval). Similarly, a function is **decreasing** on an interval if the values of the function decrease as the x-values increase (that is, if the graph falls as we move from left to right on the interval).

We have seen that if the slope of a line is positive, then the linear function is increasing and its graph is rising. Similarly, if $f(x)$ is differentiable over an interval and if each tangent line to the curve over that interval has positive slope, then the curve is rising over the interval and the function is increasing. Because the derivative of the function gives the slope of the tangent to the curve, we see that if $f'(x) > 0$ on an interval, then $f(x)$ is increasing on that interval. A similar conclusion can be reached when the derivative is negative on the interval.

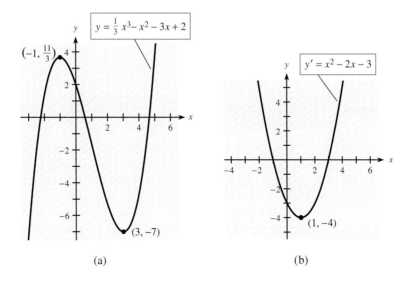

Figure 10.1   (a)   (b)

Increasing and Decreasing Functions

If f is a function that is differentiable on an interval (a, b), then

   if $f'(x) > 0$ for all x in (a, b), f is increasing on (a, b).

   if $f'(x) < 0$ for all x in (a, b), f is decreasing on (a, b).

Figure 10.1(a) shows the graph of a function, and Figure 10.1(b) shows the graph of its derivative. The figures show that the graph of $y = f(x)$ is increasing for the same x-values for which the graph of $y' = f'(x)$ is above the x-axis (when $f'(x) > 0$). Similarly, the graph of $y = f(x)$ is decreasing for the same x-values ($-1 < x < 3$) for which the graph of $y' = f'(x)$ is below the x-axis (when $f'(x) < 0$).

The derivative $f'(x)$ can change signs only at values of x at which $f'(x) = 0$ or $f'(x)$ is undefined. We call these values of x **critical values.** The point corresponding to a critical value for x is a **critical point.*** Because a curve changes from increasing to decreasing at a relative maximum (see Figure 10.1(a)), we have the following fact.

Relative Maximum

If f has a relative maximum at $x = x_0$, then $f'(x_0) = 0$ or $f'(x_0)$ is undefined.

Figure 10.2 shows a function with two relative maxima, one at $x = x_1$ and the second at $x = x_3$. At $x = x_1$ the derivative is 0, and at $x = x_3$ the derivative does not exist. In Figure 10.2, we also see that a relative minimum occurs at $x = x_2$ and $f'(x_2) = 0$. As Figure 10.2 shows, the function changes from decreasing to increasing at a relative minimum. Thus we have the following fact.

---

*There may be some critical values at which both $f'(x)$ and $f(x)$ are undefined. Critical points do not occur at these values, but studying the derivative on either side of such values may be of interest.

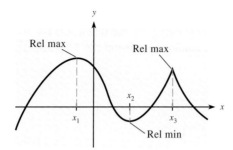

Figure 10.2

Relative Minimum   If f has a relative minimum at $x = x_0$, then $f'(x_0) = 0$ or $f'(x_0)$ is undefined.

Thus we can find relative maxima and minima for a curve by finding values of x for which the function has critical points. The behavior of the derivative to the left and right of (and near) these points will tell us whether they are relative maxima, relative minima, or neither.

Because the critical values are the only values at which the graph can have turning points, the derivative cannot change sign anywhere except at a critical value. Thus, in an interval between two critical values, the sign of the derivative at any value in the interval will be the sign of the derivative at all values in the interval.

Using the critical values of $f(x)$ and the sign of $f'(x)$ between those critical values, we can create a **sign diagram for $f'(x)$.** The sign diagram for the graph in Figure 10.2 is shown in Figure 10.3. This sign diagram was created from the graph of f, but it is also possible to predict the shape of a graph from a sign diagram.

Direction of graph of $f(x)$:

Signs and values of $f'(x)$: $+ + + 0 - - - 0 + + + * - - -$

x-axis with critical values:

x_1 x_2 x_3

Figure 10.3   *means $f'(x_3)$ is undefined.

Suppose that the point (x_1, y_1) is a critical point. If $f'(x)$ is positive to the left of, and near, this critical point, and if $f'(x)$ is negative to the right of, and near, this critical point, then the curve is increasing to the left of the point and decreasing to the right. This means that a relative maximum occurs at the point. If we drew tangent lines to the curve on the left and right of this critical point, they would fit on the curve in one of the two ways shown in Figure 10.4.

Similarly, suppose that the point (x_2, y_2) is a critical point, $f'(x)$ is negative to the left of and near (x_2, y_2), and $f'(x)$ is positive to the right of and near this critical point. Then the curve is decreasing to the left of the point and increasing to the right, and a relative minimum occurs at the point. If we drew tangent lines to the curve to the left of and to the right of this critical point, they would fit on the curve in one of the two ways shown in Figure 10.5.

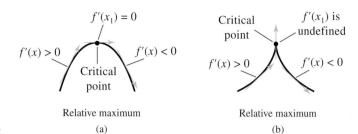

Figure 10.4

Relative maximum
(a)

Relative maximum
(b)

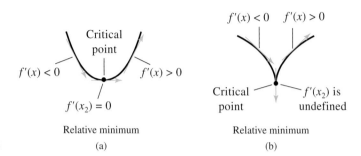

Figure 10.5

Relative minimum
(a)

Relative minimum
(b)

● **EXAMPLE 1** *Relative Maximum*

Show that the graph of $f(x) = 3x^2 - 2x^3$ has a relative maximum at $x = 1$.

Solution
We need to show that

$$f'(1) = 0 \text{ or } f'(1) \text{ is undefined.}$$
$$f'(x) > 0 \text{ to the left of and near } x = 1.$$
$$f'(x) < 0 \text{ to the right of and near } x = 1.$$

Because $f'(x) = 6x - 6x^2 = 6x(1 - x)$ is 0 only at $x = 0$ and at $x = 1$, we can test (evaluate) the derivative at any value in the interval $(0, 1)$ to see what the curve is doing to the left of $x = 1$, and we can test the derivative at any value to the right of $x = 1$ to see what the curve is doing for $x > 1$.

$$f'\left(\frac{1}{2}\right) = 3 - \frac{3}{2} = \frac{3}{2} > 0 \Rightarrow \text{increasing to left of } x = 1$$

$$f'(1) = 0 \Rightarrow \text{horizontal tangent at } x = 1$$

$$f'\left(\frac{3}{2}\right) = 9 - \frac{27}{2} = -\frac{9}{2} < 0 \Rightarrow \text{decreasing to right of } x = 1$$

A partial sign diagram for $f'(x)$ is given at the right.
The sign diagram shows that the graph of the function has a relative maximum at $x = 1$.

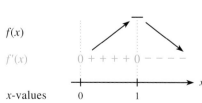

$f(x)$

$f'(x)$ $0 + + + + 0 - - - -$

x-values 0 1

The preceding discussion suggests the following procedure for finding relative maxima and minima of a function.

First-Derivative Test

Procedure	Example

To find relative maxima and minima of a function:

Find the relative maxima and minima of $f(x) = \frac{1}{3}x^3 - x^2 - 3x + 2$.

1. Find the first derivative of the function.

1. $f'(x) = x^2 - 2x - 3$

2. Set the derivative equal to 0, and solve for values of x that satisfy $f'(x) = 0$. These are called **critical values.** Values that make $f'(x)$ undefined are also critical values.

2. $0 = x^2 - 2x - 3 = (x + 1)(x - 3)$ has solutions $x = -1, x = 3$. No values of x make $f'(x) = x^2 - 2x - 3$ undefined. Critical values are -1 and 3.

3. Substitute the critical values into the *original function* to find the **critical points.**

3. $f(-1) = \frac{11}{3}$ $f(3) = -7$
The critical points are $\left(-1, \frac{11}{3}\right)$ and $(3, -7)$.

4. Evaluate $f'(x)$ at a value of x to the left and one to the right of each critical point to develop a sign diagram.

   (a) If $f'(x) > 0$ to the left and $f'(x) < 0$ to the right of the critical value, the critical point is a relative maximum point.
   (b) If $f'(x) < 0$ to the left and $f'(x) > 0$ to the right of the critical value, the critical point is a relative minimum point.

4. $f'(-2) = 5 > 0$ and $f'(0) = -3 < 0$
Thus $(-1, 11/3)$ is a relative maximum point.
$f'(2) = -3 < 0$ and $f'(4) = 5 > 0$
Thus $(3, -7)$ is a relative minimum point.
The sign diagram for $f'(x)$ is

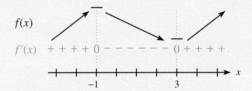

5. Use the information from the sign diagram and selected points to sketch the graph.

5. The information from this sign diagram is shown in Figure 10.6(a). Plotting additional points gives the graph of the function, which is shown in Figure 10.6(b).

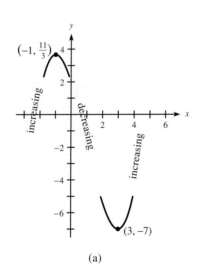

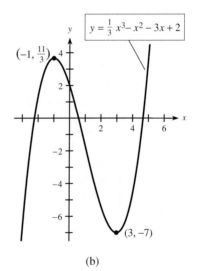

Figure 10.6   (a)   (b)

Because the critical values are the only x-values at which the graph can have turning points, we can test to the left and right of each critical value by testing to the left of the smallest critical value, then testing a value *between* each two successive critical values, and then testing to the right of the largest critical value. The following example illustrates this procedure.

● **EXAMPLE 2** *Maxima and Minima*

Find the relative maxima and minima of $f(x) = \frac{1}{4}x^4 - \frac{1}{3}x^3 - 3x^2 + 8$, and sketch its graph.

Solution
1. $f'(x) = x^3 - x^2 - 6x$
2. Setting $f'(x) = 0$ gives $0 = x^3 - x^2 - 6x$. Solving for x gives

$$0 = x(x - 3)(x + 2)$$

$x = 0$	$x - 3 = 0$	$x + 2 = 0$
	$x = 3$	$x = -2$

   Thus the critical values are $x = 0$, $x = 3$, and $x = -2$.
3. Substituting the critical values into the original function gives the critical points:

$$f(-2) = \tfrac{8}{3}, \quad \text{so} \left(-2, \tfrac{8}{3}\right) \text{ is a critical point.}$$
$$f(0) = 8, \quad \text{so } (0, 8) \text{ is a critical point.}$$
$$f(3) = -\tfrac{31}{4}, \quad \text{so} \left(3, -\tfrac{31}{4}\right) \text{ is a critical point.}$$

4. Testing $f'(x)$ to the left of the smallest critical value, then between the critical values, and then to the right of the largest critical value will give the sign diagram. Evaluating $f'(x)$ at the test values $x = -3$, $x = -1$, $x = 1$, and $x = 4$ gives the signs to determine relative maxima and minima.

The sign diagram
for $f'(x)$ is

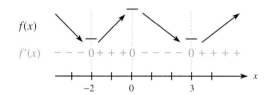

Thus we have

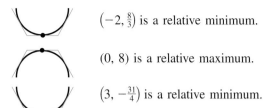

$\left(-2, \tfrac{8}{3}\right)$ is a relative minimum.

$(0, 8)$ is a relative maximum.

$\left(3, -\tfrac{31}{4}\right)$ is a relative minimum.

5. Figure 10.7(a) shows the graph of the function near the critical points, and Figure 10.7(b) shows the graph of the function.

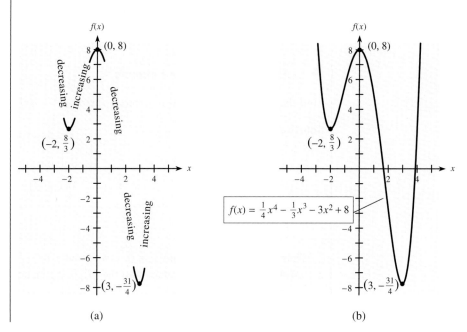

Figure 10.7   (a)   (b)

Note that we substitute the critical values into the *original function f(x)* to find the *y*-values of the critical points, but we test for relative maxima and minima by substituting values near the critical values into the *derivative of the function, f'(x)*.

Only four values were needed to test three critical points in Example 2. This method will work *only if* the critical values are tested in order from smallest to largest.

If the first derivative of *f* is 0 at x_0 but does not change from positive to negative or from negative to positive as *x* passes through x_0, then the critical point at x_0 is neither a relative maximum nor a relative minimum. In this case we say that *f* has a **horizontal point of inflection** (abbreviated HPI) at x_0.

● **EXAMPLE 3 *Maxima, Minima, and Horizontal Points of Inflection***

Find the relative maxima, relative minima, and horizontal points of inflection of $h(x) = \frac{1}{4}x^4 - \frac{2}{3}x^3 - 2x^2 + 8x + 4$, and sketch its graph.

Solution
1. $h'(x) = x^3 - 2x^2 - 4x + 8$
2. $0 = x^3 - 2x^2 - 4x + 8$ or $0 = x^2(x - 2) - 4(x - 2)$. Therefore, we have $0 = (x - 2)(x^2 - 4)$. Thus $x = -2$ and $x = 2$ are solutions.
3. The critical points are $\left(-2, -\frac{32}{3}\right)$ and $\left(2, \frac{32}{3}\right)$.

4. Using test values, such as $x = -3$, $x = 0$, and $x = 3$, gives the sign diagram for $h'(x)$.

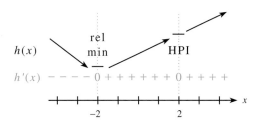

5. Figure 10.8(a) shows the graph of the function near the critical points, and Figure 10.8(b) shows the graph of the function.

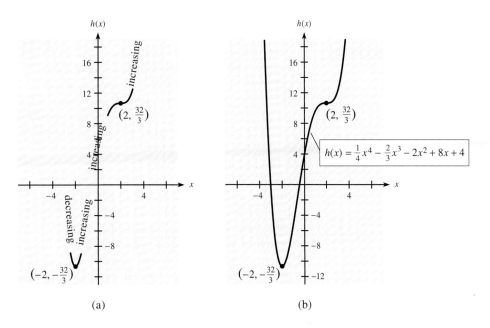

Figure 10.8

(a)   (b)

● **EXAMPLE 4** *Undefined Derivatives*

Find the relative maxima and minima (if any) of the graph of $y = (x + 2)^{2/3}$.

Solution

1. $y' = f'(x) = \dfrac{2}{3}(x + 2)^{-1/3} = \dfrac{2}{3\sqrt[3]{x + 2}}$

2. $0 = \dfrac{2}{3\sqrt[3]{x + 2}}$ has no solutions; $f'(x)$ is undefined at $x = -2$.

3. $f(-2) = 0$, so the critical point is $(-2, 0)$.
4. The sign diagram for $f'(x)$ is

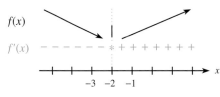

Thus a relative minimum occurs at $(-2, 0)$.

*means $f'(-2)$ is undefined.

5. Figure 10.9(a) shows the graph of the function near the critical point, and Figure 10.9(b) shows the graph.

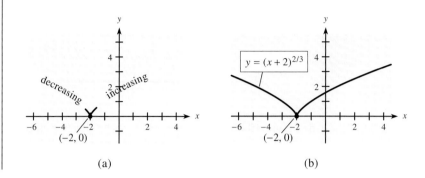

Figure 10.9   (a)   (b)

● **Checkpoint**

1. The x-values of critical points are found where $f'(x)$ is _____ or _____.
2. Decide whether the following are true or false.
   (a) If $f'(1) = 7$, then $f(x)$ is increasing at $x = 1$.
   (b) If $f'(-2) = 0$, then a relative maximum or a relative minimum occurs at $x = -2$.
   (c) If $f'(-3) = 0$ and $f'(x)$ changes from positive on the left to negative on the right of $x = -3$, then a relative minimum occurs at $x = -3$.
3. If $f(x) = 7 + 3x - x^3$, then $f'(x) = 3 - 3x^2$. Use these functions to decide whether the following are true or false.
   (a) The only critical value is $x = 1$.
   (b) The critical points are $(1, 0)$ and $(-1, 0)$.
4. If $f'(x)$ has the following partial sign diagram, make a "stick-figure" sketch of $f(x)$ and label where any maxima and minima occur. Assume that $f(x)$ is defined for all real numbers.

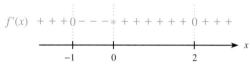

$*$means $f'(0)$ is undefined.

◑ EXAMPLE 5 *Advertising (Application Preview)*

The weekly sales S of a product during an advertising campaign are given by

$$S = \frac{100t}{t^2 + 100}, \qquad 0 \le t \le 20$$

where t is the number of weeks since the beginning of the campaign and S is in thousands of dollars.

(a) Over what interval are sales increasing? decreasing?
(b) What are the maximum weekly sales?
(c) Sketch the graph for $0 \le t \le 20$.

Solution
(a) To find where S is increasing, we first find $S'(t)$

$$S'(t) = \frac{(t^2 + 100)100 - (100t)2t}{(t^2 + 100)^2}$$

$$= \frac{10,000 - 100t^2}{(t^2 + 100)^2}$$

We see that $S'(t) = 0$ when $10,000 - 100t^2 = 0$, or

$$100(100 - t^2) = 0$$
$$100(10 + t)(10 - t) = 0$$
$$t = -10 \quad \text{or} \quad t = 10$$

Because $S'(t)$ is never undefined ($t^2 + 100 \neq 0$ for any real t) and because $0 \le t \le 20$, our only critical value is $t = 10$. Testing $S'(t)$ to the left and right of $t = 10$ gives the sign diagram.

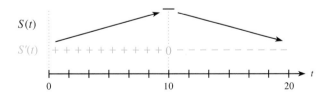

Hence, S is increasing on the interval [0, 10) and decreasing on the interval (10, 20].

(b) Because S is increasing to the left of $t = 10$ and S is decreasing to the right of $t = 10$, the maximum value of S occurs at $t = 10$ and is

$$S = S(10) = \frac{100(10)}{10^2 + 100} = \frac{1000}{200} = 5 \, (\text{thousand dollars})$$

(c) Plotting some additional points gives the graph; see Figure 10.10.

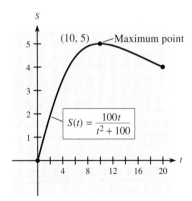

Figure 10.10

Calculator Note With a graphing calculator, choosing an appropriate viewing window is the key to understanding the graph of a function. We saw that the derivative can be used to determine the critical values of a function and hence the interesting points of its graph. Therefore, the derivative can be used to determine the viewing window that provides an accurate representation of the graph. The derivative can also be used to discover graphical behavior that might be overlooked in a graph with a standard window. ■

● **EXAMPLE 6** *Critical Points and Viewing Windows*

Find the critical values of $f(x) = 0.0001x^3 + 0.003x^2 - 3.6x + 5$. Use them to determine an appropriate viewing window. Then sketch the graph.

Solution

Suppose that we first graph this function using a standard viewing window. The graph of this function, for $-10 \leq x \leq 10$, is shown in Figure 10.11(a). The graph looks like a line in this window, but the function is not linear. We could explore the function by tracing or zooming, but using the critical points is more helpful in graphing the function. We begin by finding $f'(x)$.

$$f'(x) = 0.0003x^2 + 0.006x - 3.6$$

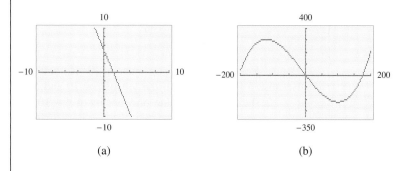

Figure 10.11   (a)   (b)

Solve $f'(x) = 0$ to find the critical values.

$$0 = 0.0003x^2 + 0.006x - 3.6$$
$$0 = 0.0003(x^2 + 20x - 12{,}000)$$
$$0 = 0.0003(x + 120)(x - 100)$$
$$x = -120 \quad \text{or} \quad x = 100$$

We choose a window that includes $x = -120$ and $x = 100$, and use VALUE to find $f(-120) = 307.4$ and $f(100) = -225$. We then graph $y = f(x)$ in a window that includes $y = -225$ and $y = 307.4$ (see Figure 10.11(b)). On this graph we can see that $(-120, 307.4)$ is a relative maximum and that $(100, -225)$ is a relative minimum.

We can also use a table of values of $f'(x)$ on a graphing utility or Excel to create a sign diagram that will identify the relative maxima and minima. Figure 10.12(a) shows values of $y_1 = f(x)$ and $y_2 = f'(x)$ for different values of x at and near the critical values. Using the x- and y_2-values gives the sign diagram in Figure 10.12(b), which shows that $(-120, 307.4)$ is a relative maximum and that $(100, -225)$ is a relative minimum.

X	Y₁	Y₂
-150	275	2.25
-120	307.4	0
0	5	-3.6
100	-225	0
150	-130	4.05
X=0		

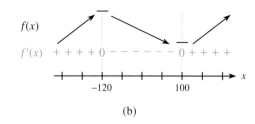

Figure 10.12   (a)   (b)

Modeling

● **EXAMPLE 7** *Marijuana Use*

The table below gives the percents of high school seniors who used marijuana in selected years from 1975 to 2000.

(a) Using x as the number of years past 1970, develop an equation that models the data.
(b) During what years does the model indicate that the maximum and minimum use occurred?

Year	Percent Using Marijuana	Year	Percent Using Marijuana
1975	47.3	1990	27.0
1977	56.4	1992	21.9
1980	60.3	1994	30.7
1982	58.7	1996	35.8
1985	54.2	1998	49.1
1987	36.3	2000	48.8

Source: National Institute on Drug Abuse

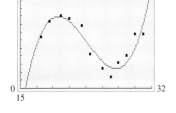

Figure 10.13

Solution

An equation that models the data is $y = 0.0235x^3 - 1.153x^2 + 15.37x - 2.490$. (See Figure 10.13.)

To find the critical values, we solve $y' = f'(x) = 0$, where $f'(x) = 0.0705x^2 - 2.306x + 15.37$. We can solve

$$0 = 0.0705x^2 - 2.306x + 15.37$$

with the quadratic formula or with a graphing utility. The two solutions are approximately

$$x = 9.322 \quad \text{and} \quad x = 23.39$$

The sign diagram shows that $x = 9.322$ gives a relative maximum (during 1980) and that $x = 23.39$ gives a relative minimum (during 1994).

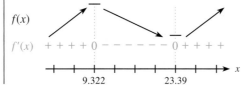

Spreadsheet
Note

Excel can also be used to develop and graph a model for data (as in Example 7). Also, the Solver feature of Excel can be used to solve optimization problems from calculus. Details are given in the *Excel Guide* that accompanies this text. ■

● *Checkpoint Solutions*

1. $f'(x) = 0$ or $f'(x)$ is undefined.
2. (a) True. $f(x)$ is increasing when $f'(x) > 0$.
   (b) False. There may be a horizontal point of inflection at $x = -2$. (See Figure 10.8 earlier in this section.)
   (c) False. A relative maximum occurs at $x = -3$.

3. (a) False. Critical values are solutions to $3 - 3x^2 = 0$, or $x = 1$ and $x = -1$.
   (b) False. *y*-coordinates of critical points come from $f(x) = 7 + 3x - x^3$. Thus critical points are $(1, 9)$ and $(-1, 5)$.

4.

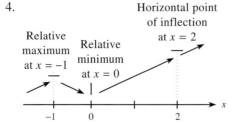

10.1 Exercises

In Problems 1 and 2, use the indicated points on the graph of $y = f(x)$ to identify points at which $f(x)$ has (a) a relative maximum, (b) a relative minimum, and (c) a horizontal point of inflection.

1.

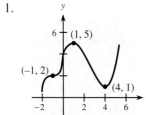

2.

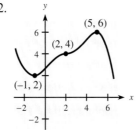

3. Use the graph of $y = f(x)$ in Problem 1 to identify at which of the indicated points the derivative $f'(x)$ (a) changes from positive to negative, (b) changes from negative to positive, and (c) does not change sign.

4. Use the graph of $y = f(x)$ in Problem 2 to identify at which of the indicated points the derivative $f'(x)$ (a) changes from positive to negative, (b) changes from negative to positive, and (c) does not change sign.

In Problems 5 and 6, use the sign diagram for $f'(x)$ to determine (a) the critical values of $f(x)$, (b) intervals on which $f(x)$ increases, (c) intervals on which $f(x)$ decreases, (d) *x*-values at which relative maxima occur, and (e) *x*-values at which relative minima occur.

5. $f'(x)$ $- - - 0 + + + + + 0 - - -$

   3   7

6. $f'(x)$ $+ + + + 0 + + + + + + + 0 - - - -$

   -5 8

In Problems 7 and 8, find the critical values of the function.

7. $y = 2x^3 - 12x^2 + 6$ 8. $y = x^3 - 3x^2 + 6x + 1$

In Problems 9 and 10, make a sign diagram for the function and determine the relative maxima and minima. (See Problems 7 and 8.)

9. $y = 2x^3 - 12x^2 + 6$ 10. $y = x^3 - 3x^2 + 6x + 1$

For each function and graph in Problems 11–14:
(a) Estimate the coordinates of the relative maxima, relative minima, or horizontal points of inflection by observing the graph.
(b) Use $y' = f'(x)$ to find the critical values.
(c) Find the critical points.
(d) Do the results in part (c) confirm your estimates in part (a)?

11. $y = x^3 - 3x + 4$ 12. $y = x - \frac{1}{3}x^3$

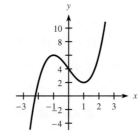 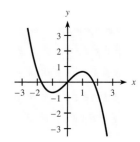

13. $y = x^3 + 3x^2 + 3x - 2$ 14. $y = x^3 - 6x^2 + 12x + 1$

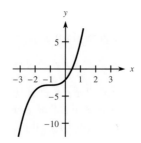 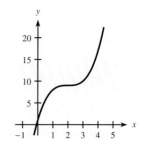

For each function in Problems 15–20:
(a) Find $y' = f'(x)$.
(b) Find the critical values.
(c) Find the critical points.
(d) Find intervals of x-values where the function is increasing and where it is decreasing.
(e) Classify the critical points as relative maxima, relative minima, or horizontal points of inflection. In each case, you may check your conclusions with a graphing utility.

15. $y = \frac{1}{2}x^2 - x$

16. $y = x^2 + 4x$

17. $y = \dfrac{x^3}{3} + \dfrac{x^2}{2} - 2x + 1$

18. $y = \dfrac{x^4}{4} - \dfrac{x^3}{3} - 2$

19. $y = x^{2/3}$

20. $y = -(x - 3)^{2/3}$

For each function and graph in Problems 21–24:
(a) Use the graph to identify x-values for which $y' > 0$, $y' < 0$, $y' = 0$, and y' does not exist.
(b) Use the derivative to check your conclusions.

21. $y = 6 - x - x^2$

22. $y = \frac{1}{2}x^2 - 4x + 1$

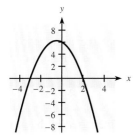

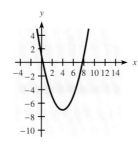

23. $y = 6 + x^3 - \frac{1}{15}x^5$

24. $y = x^4 - 2x^2 - 1$

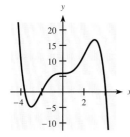

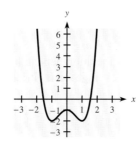

For each function in Problems 25–30, find the relative maxima, relative minima, horizontal points of inflection, and sketch the graph. You may check your graph with a graphing utility.

25. $y = \frac{1}{3}x^3 - x^2 + x + 1$

26. $y = \frac{1}{4}x^4 - \frac{2}{3}x^3 + \frac{1}{2}x^2 - 2$

27. $y = \frac{1}{3}x^3 + x^2 - 24x + 20$

28. $C(x) = x^3 - \frac{3}{2}x^2 - 18x + 5$

29. $y = 3x^5 - 5x^3 + 1$

30. $y = \frac{1}{6}x^6 - x^4 + 7$

In Problems 31–36, both a function and its derivative are given. Use them to find critical values, critical points, intervals on which the function is increasing and decreasing, relative maxima, relative minima, and horizontal points of inflection; sketch the graph of each function.

31. $y = (x^2 - 2x)^2$ $\qquad \dfrac{dy}{dx} = 4x(x - 1)(x - 2)$

32. $f(x) = (x^2 - 4)^2$ $\qquad f'(x) = 4x(x + 2)(x - 2)$

33. $y = \dfrac{x^3(x - 5)^2}{27}$ $\qquad \dfrac{dy}{dx} = \dfrac{5x^2(x - 3)(x - 5)}{27}$

34. $y = \dfrac{x^2(x - 5)^3}{27}$ $\qquad \dfrac{dy}{dx} = \dfrac{5x(x - 2)(x - 5)^2}{27}$

35. $f(x) = x^{2/3}(x - 5)$ $\qquad f'(x) = \dfrac{5(x - 2)}{3x^{1/3}}$

36. $f(x) = x - 3x^{2/3}$ $\qquad f'(x) = \dfrac{x^{1/3} - 2}{x^{1/3}}$

In Problems 37–42, use the derivative to locate critical points and determine a viewing window that shows all features of the graph. Use a graphing utility to sketch a complete graph.

37. $f(x) = x^3 - 225x^2 + 15{,}000x - 12{,}000$

38. $f(x) = x^3 - 15x^2 - 16{,}800x + 80{,}000$

39. $f(x) = x^4 - 160x^3 + 7200x^2 - 40{,}000$

40. $f(x) = x^4 - 240x^3 + 16{,}200x^2 - 60{,}000$

41. $y = 7.5x^4 - x^3 + 2$

42. $y = 2 - x^3 - 7.5x^4$

In each of Problems 43–46, a graph of $f'(x)$ is given. Use the graph to determine the critical values of $f(x)$, where $f(x)$ is increasing, where it is decreasing, and where it has relative maxima, relative minima, and horizontal points of inflection. In each case sketch a possible graph for $f(x)$ that passes through (0, 0).

43. $f'(x) = x^2 - x - 2$

44. $f'(x) = 4x - x^2$

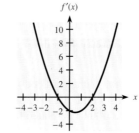

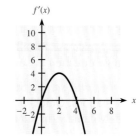

45. $f'(x) = x^3 - 3x^2$

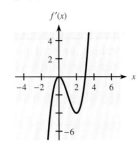

46. $f'(x) = x(x - 2)^2$

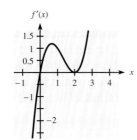

In Problems 47 and 48, two graphs are given. One is the graph of f and the other is the graph of f'. Decide which is which and explain your reasoning.

47.

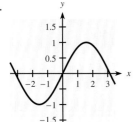
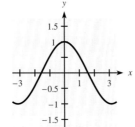

48.

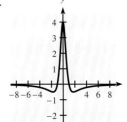
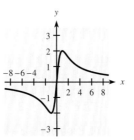

APPLICATIONS

49. *Advertising and sales* Suppose that the daily sales (in dollars) t days after the end of an advertising campaign are given by

$$S = 1000 + \frac{400}{t + 1}, \qquad t \geq 0$$

Does S increase for all $t \geq 0$, decrease for all $t \geq 0$, or change direction at some point?

50. *Pricing and sales* Suppose that a chain of auto service stations, Quick-Oil, Inc., has found that its monthly sales volume y (in thousands of dollars) is related to the price p (in dollars) of an oil change by

$$y = \frac{90}{\sqrt{p + 5}}, \qquad p > 10$$

Is y increasing or decreasing for all values of $p > 10$?

51. *Productivity* A time study showed that, on average, the productivity of a worker after t hours on the job can be modeled by

$$P(t) = 27t + 6t^2 - t^3, \qquad 0 \leq t \leq 8$$

where P is the number of units produced per hour.
(a) Find the critical values of this function.
(b) Which critical value makes sense in this model?
(c) For what values of t is P increasing?
(d) Graph the function for $0 \leq t \leq 8$.

52. *Production* Analysis of daily output of a factory shows that, on average, the number of units per hour y produced after t hours of production is

$$y = 70t + \frac{1}{2}t^2 - t^3, \qquad 0 \leq t \leq 8$$

(a) Find the critical values of this function.
(b) Which critical values make sense in this particular problem?
(c) For which values of t, for $0 \leq t \leq 8$, is y increasing?
(d) Graph this function.

53. *Production costs* Suppose that the average cost, in dollars, of producing a shipment of a certain product is

$$\overline{C} = 5000x + \frac{125,000}{x}, \qquad x > 0$$

where x is the number of machines used in the production process.
(a) Find the critical values of this function.
(b) Over what interval does the average cost decrease?
(c) Over what interval does the average cost increase?

54. *Average costs* Suppose the average costs of a mining operation depend on the number of machines used, and average costs, in dollars, are given by

$$\overline{C}(x) = 2900x + \frac{1,278,900}{x}, \qquad x > 0$$

where x is the number of machines used.
(a) Find the critical values of $\overline{C}(x)$ that lie in the domain of the problem.
(b) Over what interval in the domain do average costs decrease?
(c) Over what interval in the domain do average costs increase?
(d) How many machines give minimum average costs?
(e) What is the minimum average cost?

55. *Marginal revenue* Suppose the weekly marginal revenue function for selling x units of a product is given by the graph in the figure.

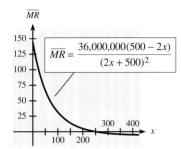

$$\overline{MR} = \frac{36,000,000(500 - 2x)}{(2x + 500)^2}$$

(a) At each of $x = 150$, $x = 250$, and $x = 350$, what is happening to revenue?
(b) Over what interval is revenue increasing?
(c) How many units must be sold to maximize revenue?

56. *Earnings* Suppose that the rate of change $f'(x)$ of the average annual earnings of new car salespersons is shown in the figure.

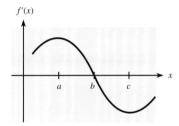

(a) If a, b, and c represent certain years, what is happening to $f(x)$, the average annual earnings of the salespersons, at a, b, and c?
(b) Over what interval (involving a, b, or c) is there an increase in $f(x)$, the average annual earnings of the salespersons?

57. *Revenue* The weekly revenue of a certain recently released film is given by

$$R(t) = \frac{50t}{t^2 + 36}, \qquad t \geq 0$$

where R is in millions of dollars and t is in weeks.
(a) Find the critical values.
(b) For how many weeks will weekly revenue increase?

58. *Medication* Suppose that the concentration C of a medication in the bloodstream t hours after an injection is given by

$$C(t) = \frac{0.2t}{t^2 + 1}$$

(a) Determine the number of hours before C attains its maximum.
(b) Find the maximum concentration.

59. *Candidate recognition* Suppose that the proportion P of voters who recognize a candidate's name t months after the start of the campaign is given by

$$P(t) = \frac{13t}{t^2 + 100} + 0.18$$

(a) How many months after the start of the campaign is recognition at its maximum?
(b) To have greatest recognition on November 1, when should a campaign be launched?

60. *Medication* The number of milligrams x of a medication in the bloodstream t hours after a dose is taken can be modeled by

$$x(t) = \frac{2000t}{t^2 + 16}$$

(a) For what t-values is x increasing?
(b) Find the t-value at which x is maximum.
(c) Find the maximum value of x.

61. *Federal budget* The graph below shows the U.S. federal deficits and surpluses for the years 1989–2005. Suppose they are given by the function

$$f(x) = -2.0x^3 + 35.3x^2 - 106.0x - 205.4$$

where $x = 0$ represents 1990, and where $f(x)$ is in billions and is positive for surpluses.
(a) Find the year in which this model indicates the maximum surplus. Does this agree with the data?
(b) Discuss the events that may have caused deficits to occur after 2001.

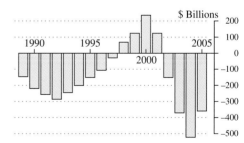

Source: http://www.kowaldesign.com/budget/

62. *Consumer prices* With data for selected years from 1920 to 2002, the consumer price index for urban consumers (CPI-U) can be modeled by the function

$$y = 0.000121x^3 + 0.0266x^2 - 1.424x + 29.74$$

where x is the number of years past 1915 and y is the consumer price index in year $1915 + x$. (*Source:* U.S. Bureau of the Census) Use this model to find the year in which the CPI-U reached its minimum. What was happening in the United States around this time that helps explain why the minimum occurred in this year?

63. **Modeling** *Homicide rate* The numbers of homicides per 100,000 people for selected years from 1950 to 2000 are given in the following table.

Year	Homicides per 100,000	Year	Homicides per 100,000
1950	4.6	1985	7.9
1955	4.1	1990	9.4
1960	5.1	1995	8.2
1965	5.1	1997	6.8
1970	7.9	1998	6.3
1975	9.6	1999	5.7
1980	10.2	2000	5.6

Source: FBI, Uniform Crime Statistics

The scatter plot of these data uses $x = 0$ in 1950 and is shown in the figure.
(a) Does the scatter plot of the data indicate that a quadratic, a cubic, or a quartic function is the best fit for the data?
(b) Use technology to find the equation that models the data and graph the data and the model on the same axes.
(c) Find the year in which the maximum number of homicides occurs, according to this model.

64. **Modeling** *Poverty* The table shows the numbers of millions of people in the United States who lived below the poverty level for selected years from 1960 to 2002.

(a) Find a cubic model that approximately fits the data, using x as years past 1960.
(b) Find the year when poverty was at a minimum, and the year when it was at a maximum, according to the model.

Year	Persons Living Below the Poverty Level (millions)	Year	Persons Living Below the Poverty Level (millions)
1960	39.9	1993	39.3
1965	33.2	1994	38.1
1970	25.4	1995	36.4
1975	25.9	1996	36.5
1980	29.3	1997	35.6
1986	32.4	1998	34.5
1989	31.5	1999	32.3
1990	33.6	2000	31.1
1991	35.7	2002	34.6
1992	38.0		

Source: Bureau of the Census, U.S. Department of Commerce

65. **Modeling** *Congressional grants* Congress earmarks millions of dollars each year for academic projects. These awards are often the result of Congressional connections. The following table gives the money earmarked by Congress for academic projects, in millions of dollars, for the years 1990–2000.
(a) Use a scatter plot to determine the degree of the function that is the best fit for the data, with x equal to the number of years past 1990.
(b) Use technology to find a function that models the millions of dollars in earmarked grants as a function of the number of years past 1990.
(c) Find the relative maximum on the graph of the model.
(d) Interpret the relative maximum found in part (c).
(e) Compare the model's function value found in (c) with its function values for 1990 and 2000. Which of these is largest?

Year	$ million	Year	$ million
1990	230	1996	640
1991	450	1997	410
1992	640	1998	470
1993	680	1999	800
1994	610	2000	1040
1995	600		

Source: Associated Press

Concavity: Points of Inflection

- To find points of inflection of graphs of functions
- To use the second-derivative test to graph functions

◗ Application Preview

Suppose that a retailer wishes to sell his store and uses the graph in Figure 10.14 to show how profits have increased since he opened the store and the potential for profit in the future. Can we conclude that profits will continue to grow, or should we be concerned about future earnings?

Note that although profits are still increasing in 2007, they seem to be increasing more slowly than in previous years. Indeed, they appear to have been growing at a decreasing rate since about 2005. We can analyze this situation more carefully by identifying the point at which profit begins to grow more slowly, that is, the *point of diminishing returns.* (See Example 3.)

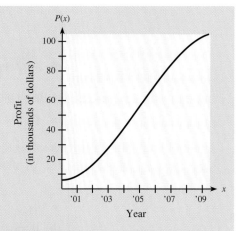

Figure 10.14

Just as we used the first derivative to determine whether a curve was increasing or decreasing on a given interval, we can use the second derivative to determine whether the curve is concave up or concave down on an interval.

A curve is said to be **concave up** on an interval (a, b) if at each point on the interval the curve is above its tangent at the point (Figure 10.15(a)). If the curve is below all its tangents on a given interval, it is **concave down** on the interval (Figure 10.15(b)).

Looking at Figure 10.15(a), we see that the *slopes* of the tangent lines increase over the interval where the graph is concave up. Because $f'(x)$ gives the slopes of those tangents, it follows that $f'(x)$ is increasing over the interval where $f(x)$ is concave up. However, we know that $f'(x)$ is increasing when its derivative, $f''(x)$, is positive. That is, if the second derivative is positive, the curve is concave up.

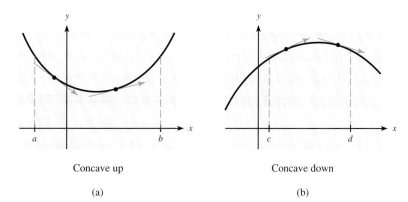

Concave up

Concave down

Figure 10.15

(a)

(b)

Similarly, if the second derivative of a function is negative over an interval, the slopes of the tangents to the graph decrease over that interval. This happens when the tangent lines are above the graph, as in Figure 10.15(b), so the graph must be concave down on this interval.

Thus we see that the second derivative can be used to determine the concavity of a curve.

Concave Up and
Concave Down

Assume that the first and second derivatives of function f exist. If $f''(x) > 0$ on an interval I, the graph of f is **concave up** on the interval. If $f''(x) < 0$ on an interval I, then the graph of f is **concave down** on I. We also say that the graph of $y = f(x)$ is concave up at $(a, f(a))$ if $f''(a) > 0$ and that the graph is concave down at $(b, f(b))$ if $f''(b) < 0$.

● **EXAMPLE 1** *Concavity at a Point*

Is the graph of $f(x) = x^3 - 4x^2 + 3$ concave up or down at the point

(a) $(1, 0)$?   (b) $(2, -5)$?

Solution
(a) We must find $f''(x)$ before we can answer this question.

$$f'(x) = 3x^2 - 8x \qquad f''(x) = 6x - 8$$

Then $f''(1) = 6(1) - 8 = -2$, so the graph is concave down at $(1, 0)$.
(b) Because $f''(2) = 6(2) - 8 = 4$, the graph is concave up at $(2, -5)$. The graph of $f(x) = x^3 - 4x^2 + 3$ is shown in Figure 10.16(a).

Points of Inflection

Looking at the graph of $y = x^3 - 4x^2 + 3$ (Figure 10.16(a)), we see that the curve is concave down on the left and concave up on the right. Thus it has changed from concave down to concave up, and from Example 1 we would expect the concavity to change somewhere between $x = 1$ and $x = 2$. Figure 10.16(b) shows the graph of $y'' = f''(x) = 6x - 8$, and we can see that $y'' = 0$ when $x = \frac{4}{3}$ and that $y'' < 0$ for $x < \frac{4}{3}$ and $y'' > 0$ for $x > \frac{4}{3}$. Thus the second derivative changes sign at $x = \frac{4}{3}$, so the concavity of the graph of $y = f(x)$ changes at $x = \frac{4}{3}$, $y = -\frac{47}{27}$. The point where concavity changes is called a **point of inflection.**

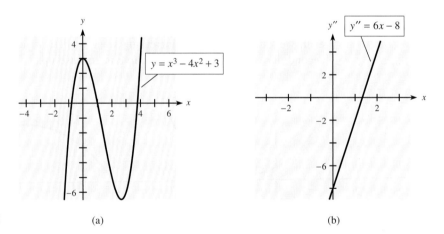

Figure 10.16   (a)   (b)

Point of Inflection

A point (x_0, y_0) on the graph of a function f is called a **point of inflection** if the curve is concave up on one side of the point and concave down on the other side. The second derivative at this point, $f''(x_0)$, will be 0 or undefined.

In general, we can find points of inflection and information about concavity as follows.

Finding Points of Inflection and Concavity

Procedure	Example
To find the point(s) of inflection of a curve and intervals where it is concave up and where it is concave down:	Find the points of inflection and concavity of the graph of $y = \dfrac{x^4}{2} - x^3 + 5$.
1. Find the second derivative of the function.	1. $y' = f'(x) = 2x^3 - 3x^2$ $y'' = f''(x) = 6x^2 - 6x$
2. Set the second derivative equal to 0, and solve for x. Potential points of inflection occur at these values of x or at values of x where $f(x)$ is defined and $f''(x)$ is undefined.	2. $0 = 6x^2 - 6x = 6x(x - 1)$ has solutions $x = 0, x = 1$. $f''(x)$ is defined everywhere.
3. Find the potential points of inflection.	3. $(0, 5)$ and $\left(1, \frac{9}{2}\right)$ are potential points of inflection.
4. If the second derivative has opposite signs on the two sides of one of these values of x, a point of inflection occurs. The curve is concave up where $f''(x) > 0$ and concave down where $f''(x) < 0$. The changes in the sign of $f''(x)$ correspond to changes in concavity and occur at points of inflection.	4. A **sign diagram for** $f''(x)$ is 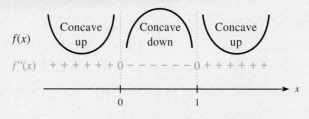 $(0, 5)$ and $\left(1, \frac{9}{2}\right)$ are points of inflection. See the graph in Figure 10.17.

The graph of $y = \frac{1}{2}x^4 - x^3 + 5$ is shown in Figure 10.17. Note the points of inflection at $(0, 5)$ and $\left(1, \frac{9}{2}\right)$. The point of inflection at $(0, 5)$ is a horizontal point of inflection because $f'(x)$ is also 0 at $x = 0$.

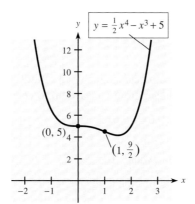

Figure 10.17

● **EXAMPLE 2** *Water Purity*

Suppose that a real estate developer wishes to remove pollution from a small lake so that she can sell lakefront homes on a "crystal clear" lake. The graph in Figure 10.18 shows the relation between dollars spent on cleaning the lake and the purity of the water. The point of inflection on the graph is called the **point of diminishing returns** on her investment because it is where the *rate* of return on her investment changes from increasing to decreasing. Show that the rate of change in the purity of the lake, $f'(x)$, is maximized at this point, $x = c$. Assume that $f(c)$, $f'(c)$, and $f''(c)$ are defined.

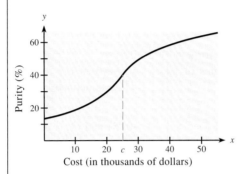

Figure 10.18

Solution

Because $x = c$ is a point of inflection for $f(x)$, we know that the concavity must change at $x = c$. From the figure we see the following.

$$x < c: \quad f(x) \text{ is concave up, so } f''(x) > 0.$$
$$f''(x) > 0 \text{ means that } f'(x) \text{ is increasing.}$$
$$x > c: \quad f(x) \text{ is concave down, so } f''(x) < 0.$$
$$f''(x) < 0 \text{ means that } f'(x) \text{ is decreasing.}$$

Thus $f'(x)$ has $f'(c)$ as its relative maximum.

● **EXAMPLE 3** *Diminishing Returns* **(Application Preview)**

Suppose the annual profit for a store (in thousands of dollars) is given by

$$P(x) = -0.2x^3 + 3x^2 + 6$$

where x is the number of years past 2000. If this model is accurate, find the point of diminishing returns for the profit.

Solution

The point of diminishing returns occurs at the point of inflection. Thus we seek the point where the graph of this function changes from concave up to concave down, if such a point exists.

$$P'(x) = -0.6x^2 + 6x$$
$$P''(x) = -1.2x + 6$$
$$P''(x) = 0 \quad \text{when} \quad 0 = -1.2x + 6 \quad \text{or} \quad \text{when } x = 5$$

Thus $x = 5$ is a possible point of inflection. We test $P''(x)$ to the left and the right of $x = 5$.

$$P''(4) = 1.2 > 0 \Rightarrow \text{concave up to the left of } x = 5$$
$$P''(6) = -1.2 < 0 \Rightarrow \text{concave down to the right of } x = 5$$

Thus (5, 56) is the point of inflection for the graph, and the point of diminishing returns for the profit is when $x = 5$ (in the year 2005) and is $P(5) = 56$ thousand dollars. Figure 10.19

shows the graphs of $P(x)$, $P'(x)$, and $P''(x)$. At $x = 5$, we see that the point of diminishing returns on the graph of $P(x)$ corresponds to the maximum point of the graph of $P'(x)$ and the zero (or x-intercept) of the graph of $P''(x)$.

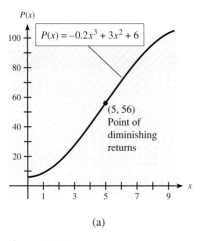

(a)

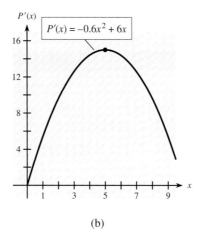

(b)

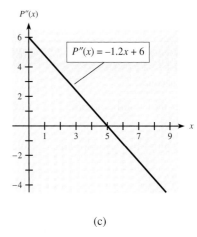

(c)

Figure 10.19

• ***Checkpoint***

1. If $f''(x) > 0$, then $f(x)$ is concave _____.
2. At what value of x does the graph $y = \frac{1}{3}x^3 - 2x^2 + 2x$ have a point of inflection?
3. On the graph below, locate any points of inflection (approximately) and label where the curve satisfies $f''(x) > 0$ and $f''(x) < 0$.

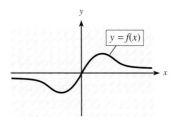

4. Determine whether the following is true or false. If $f''(0) = 0$, then $f(x)$ has a point of inflection at $x = 0$.

Second-Derivative Test

We can use information about points of inflection and concavity to help sketch graphs. For example, if we know that the curve is concave up at a critical point where $f'(x) = 0$, then the point must be a relative minimum because the tangent to the curve is horizontal at the critical point, and only a point at the bottom of a "concave up" curve could have a horizontal tangent (see Figure 10.20(a)).

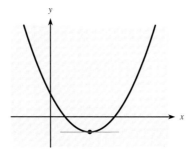

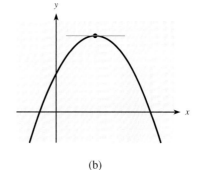

Figure 10.20   (a)   (b)

On the other hand, if the curve is concave down at a critical point where $f'(x) = 0$, then the point is a relative maximum (see Figure 10.20(b)).

Thus we can use the **second-derivative test** to determine whether a critical point where $f'(x) = 0$ is a relative maximum or minimum.

Second-Derivative Test

Procedure	Example
To find relative maxima and minima of a function:	Find the relative maxima and minima of $y = f(x) = \frac{1}{3}x^3 - x^2 - 3x + 2$.
1. Find the critical values of the function.	1. $f'(x) = x^2 - 2x - 3$ $0 = (x - 3)(x + 1)$ has solutions $x = -1$ and $x = 3$. No values of x make $f'(x)$ undefined.
2. Substitute the critical values into $f(x)$ to find the critical points.	2. $f(-1) = \frac{11}{3}$ $f(3) = -7$ The critical points are $\left(-1, \frac{11}{3}\right)$ and $(3, -7)$.
3. Evaluate $f''(x)$ at each critical value for which $f'(x) = 0$. (a) If $f''(x_0) < 0$, a relative maximum occurs at x_0. (b) If $f''(x_0) > 0$, a relative minimum occurs at x_0. (c) If $f''(x_0) = 0$, or $f''(x_0)$ is undefined, the second-derivative test fails; use the first-derivative test.	3. $f''(x) = 2x - 2$ $f''(-1) = 2(-1) - 2 = -4 < 0$, so $\left(-1, \frac{11}{3}\right)$ is a relative maximum point. $f''(3) = 2(3) - 2 = 4 > 0$, so $(3, -7)$ is a relative minimum point. (The graph is shown in Figure 10.21.)

In the foregoing Procedure/Example, we found the relative maxima and minima for the function $f(x) = \frac{1}{3}x^3 - x^2 - 3x + 2$ using the second-derivative test, and in the Procedure/Example in Section 10.1, we found the relative maxima and minima for this function with the first-derivative test. Because both of these tests are methods for classifying critical values, the conclusions were the same in both cases. The second-derivative test has one advantage and two disadvantages when compared with the first-derivative test.

Advantage of the Second-Derivative Test	Disadvantages of the Second-Derivative Test
It is quick and easy for many functions.	The second derivative is difficult to find for some functions. The second-derivative test sometimes fails to give results.

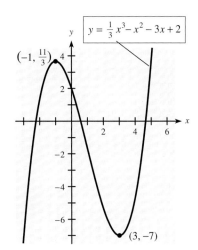

$$y = \frac{1}{3}x^3 - x^2 - 3x + 2$$

$\left(-1, \frac{11}{3}\right)$

$(3, -7)$

Figure 10.21

● **EXAMPLE 4** *Maxima, Minima, and Points of Inflection*

Find the relative maxima and minima and points of inflection of $y = 3x^4 - 4x^3 - 2$.

Solution

$$y' = f'(x) = 12x^3 - 12x^2$$

Solving $0 = 12x^3 - 12x^2 = 12x^2(x - 1)$ gives $x = 1$ and $x = 0$. Thus the critical points are $(1, -3)$ and $(0, -2)$.

$$y'' = f''(x) = 36x^2 - 24x$$

$$f''(1) = 12 > 0 \Rightarrow (1, -3) \text{ is a relative minimum point.}$$

$$f''(0) = 0 \Rightarrow \text{the second-derivative test fails.}$$

Because the second-derivative test fails, we must use the first-derivative test at the critical point $(0, -2)$.

$$\left.\begin{array}{l} f'(-1) = -24 < 0 \\ f'\left(\frac{1}{2}\right) = -\frac{3}{2} < 0 \end{array}\right\} \Rightarrow (0, -2) \text{ is a horizontal point of inflection.}$$

We look for points of inflection by setting $f''(x) = 0$ and solving for x. We find that $0 = 36x^2 - 24x$ has solutions $x = 0$ and $x = \frac{2}{3}$.

$$\left.\begin{array}{l} f''(-1) = 60 > 0 \\ f''\left(\frac{1}{2}\right) = -3 < 0 \end{array}\right\} \Rightarrow (0, -2) \text{ is a point of inflection.}$$

Thus we see again that $(0, -2)$ is a horizontal point of inflection. This is a special point, where the curve changes concavity *and* has a horizontal tangent (see Figure 10.22). Testing for concavity on either side of $x = \frac{2}{3}$ gives

$$\left.\begin{array}{l} f''\left(\frac{1}{2}\right) = -3 < 0 \\ f''(1) = 12 > 0 \end{array}\right\} \Rightarrow \left(\frac{2}{3}, -\frac{70}{27}\right) \text{ is a point of inflection.}$$

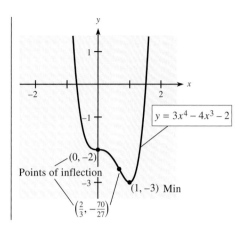

$$y = 3x^4 - 4x^3 - 2$$

(0, -2)
Points of inflection
(1, -3) Min
$\left(\frac{2}{3}, -\frac{70}{27}\right)$

Figure 10.22

Technology We can use a graphing calculator or Excel to explore the relationships among f, f', and f'',
Note as we did in the previous section for f and f'. ■

● **EXAMPLE 5** *Concavity from the Graph of y = f(x)*

Figure 10.23 shows the graph of $f(x) = \frac{1}{6}(2x^3 - 3x^2 - 12x + 12)$.

(a) From the graph, identify points where $f''(x) = 0$.
(b) From the graph, observe intervals where $f''(x) > 0$ and where $f''(x) < 0$.
(c) Check the conclusions from (a) and (b) by calculating $f''(x)$ and graphing it.

Solution

(a) From Figure 10.23, we can make an initial estimate of the x-value of the point of inflection. It appears to be near $x = \frac{1}{2}$, so we expect $f''(x) = 0$ at (or very near) $x = \frac{1}{2}$.

(b) We see that the graph is concave downward (so $f''(x) < 0$) to the left of the point of inflection. That is, $f''(x) < 0$ when $x < \frac{1}{2}$. Similarly, $f''(x) > 0$ when $x > \frac{1}{2}$.

(c) $f(x) = \frac{1}{6}(2x^3 - 3x^2 - 12x + 12)$
$f'(x) = \frac{1}{6}(6x^2 - 6x - 12) = x^2 - x - 2$
$f''(x) = 2x - 1$
Thus $f''(x) = 0$ when $x = \frac{1}{2}$.
Figure 10.24 shows the graph of $f''(x) = 2x - 1$. We see that the graph crosses the x-axis ($f''(x) = 0$) when $x = \frac{1}{2}$, is below the x-axis ($f''(x) < 0$) when $x < \frac{1}{2}$, and is above the x-axis ($f''(x) > 0$) when $x > \frac{1}{2}$. This verifies our conclusions from (a) and (b).

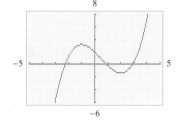

Figure 10.23

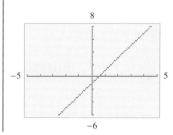

Figure 10.24

Spreadsheet We can investigate the function in Example 5 by using Excel to find the outputs of $f(x) =$
Note $\frac{1}{6}(2x^3 - 3x^2 - 12x + 12), f'(x),$ and $f''(x)$ for values of x at and near the critical points.

	A	B	C	D
	x	f(x)	f'(x)	f''(x)
1	x	f(x)	f'(x)	f''(x)
2	-2	1.3333	4	-5
3	-1	3.1667	0	-3
4	0	2	-2	-1
5	1/2	0.9167	-2.25	0
6	1	-0.1667	-2	1
7	2	-1.3333	0	3
8	3	0.5	4	5

By looking at column C, we see that $f'(x)$ changes from positive to negative as x passes
through the value $x = -1$ and that the derivative changes from negative to positive as x
passes through the value $x = 2$. Thus the graph has a relative maximum at $x = -1$ and a
relative minimum at $x = 2$. By looking at column D, we can also see that $f''(x)$ changes
from negative to positive as x passes through the value $x = 1/2$, so the graph has a point
of inflection at $x = 1/2$. ■

● **EXAMPLE 6** *Concavity from the Graph of* y = f′(x)

Figure 10.25 shows the graph of $f'(x) = -x^2 - 2x$. Use the graph of $f'(x)$ to do the
following.

(a) Find intervals on which $f(x)$ is concave downward and concave upward.
(b) Find x-values at which $f(x)$ has a point of inflection.
(c) Check the conclusions from (a) and (b) by finding $f''(x)$ and graphing it.
(d) For $f(x) = \frac{1}{3}(9 - x^3 - 3x^2)$, calculate $f'(x)$ to verify that this could be $f(x)$.

Solution

(a) Concavity for $f(x)$ can be found from the sign of $f''(x)$. Because $f''(x)$ is the first
derivative of $f'(x)$, wherever the graph of $f'(x)$ is increasing, it follows that
$f''(x) > 0$. Thus $f''(x) > 0$ and $f(x)$ is concave upward when $x < -1$. Similarly,
$f''(x) < 0$, and $f(x)$ is concave downward, when $f'(x)$ is decreasing—that is, when
$x > -1$.
(b) From (a) we know that $f''(x)$ changes sign at $x = -1$, so $f(x)$ has a point of inflection at $x = -1$. Note that $f'(x)$ has its maximum at the x-value where $f(x)$ has a
point of inflection. In fact, points of inflection for $f(x)$ will correspond to relative
extrema for $f'(x)$.
(c) For $f'(x) = -x^2 - 2x$, we have $f''(x) = -2x - 2$. Figure 10.26 shows the graph of
$y = f''(x)$ and verifies our conclusions from (a) and (b).
(d) If $f(x) = \frac{1}{3}(9 - x^3 - 3x^2)$, then $f'(x) = \frac{1}{3}(-3x^2 - 6x) = -x^2 - 2x$. Figure 10.27
shows the graph of $f(x) = \frac{1}{3}(9 - x^3 - 3x^2)$. Note that the point of inflection and the
concavity correspond to what we discovered in (a) and (b).

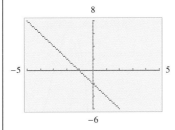

Figure 10.26

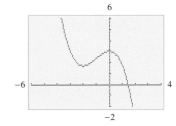

Figure 10.27

Figure 10.25

The relationships among $f(x)$, $f'(x)$, and $f''(x)$ that we explored in Example 6 can be summarized as follows.

$f(x)$	Concave Upward	Concave Downward	Point of Inflection	
$f'(x)$	increasing	decreasing	maximum	minimum
$f''(x)$	positive $(+)$	negative $(-)$	$(+)$ to $(-)$	$(-)$ to $(+)$

● **Checkpoint Solutions**

1. Up
2. The second derivative of $y = \frac{1}{3}x^3 - 2x^2 + 2x$ is $y'' = 2x - 4$, so $y'' = 0$ at $x = 2$. Because $y'' < 0$ for $x < 2$ and $y'' > 0$ for $x > 2$, a point of inflection occurs at $x = 2$.
3. Points of inflection at A, B, and C
   $f''(x) < 0$ to the left of A and between B and C
   $f''(x) > 0$ between A and B and to the right of C

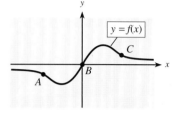

4. (a) False. For example, if $f(x) = x^4$, then $f'(x) = 4x^3$ and $f''(x) = 12x^2$. Note that $f''(0) = 0$, but $f''(x)$ does not change sign from $x < 0$ to $x > 0$.

10.2 Exercises

In Problems 1–4, determine whether each function is concave up or concave down at the indicated points.

1. $f(x) = x^3 - 3x^2 + 1$ at (a) $x = -2$ (b) $x = 3$
2. $f(x) = x^3 + 6x - 4$ at (a) $x = -5$ (b) $x = 7$
3. $y = 2x^3 + 4x - 8$ at (a) $x = -1$ (b) $x = 4$
4. $y = 4x^3 - 3x^2 + 2$ at (a) $x = 0$ (b) $x = 5$

In Problems 5–10, use the indicated x-values on the graph of $y = f(x)$ to find the following.

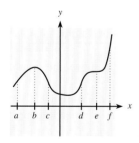

5. Find intervals over which the graph is concave down.
6. Find intervals over which the graph is concave up.
7. Find intervals where $f''(x) > 0$.
8. Find intervals where $f''(x) < 0$.
9. Find the x-coordinates of three points of inflection.
10. Find the x-coordinate of a horizontal point of inflection.

In Problems 11–14, a function and its graph are given. Use the second derivative to determine intervals on which the function is concave up, to determine intervals on which it is concave down, and to locate points of inflection. Check these results against the graph shown.

11. $f(x) = x^3 - 6x^2 + 5x + 6$

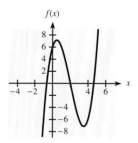

12. $y = x^3 - 9x^2$

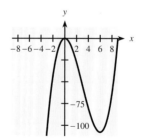

13. $f(x) = \frac{1}{4}x^4 + \frac{1}{2}x^3 - 3x^2 + 3$

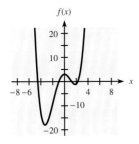

14. $y = 2x^4 - 6x^2 + 4$

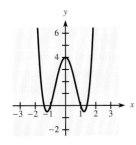

Find the relative maxima, relative minima, and points of inflection, and sketch the graphs of the functions, in Problems 15–20.

15. $y = x^2 - 4x + 2$

16. $y = x^3 - x^2$

17. $y = \frac{1}{3}x^3 - 2x^2 + 3x + 2$

18. $y = x^3 - 3x^2 + 6$

19. $y = x^4 - 16x^2$

20. $y = x^4 - 8x^3 + 16x^2$

In Problems 21–24, a function and its first and second derivatives are given. Use these to find critical values, relative maxima, relative minima, and points of inflection; sketch the graph of each function.

21. $f(x) = 3x^5 - 20x^3$
$f'(x) = 15x^2(x - 2)(x + 2)$
$f''(x) = 60x(x^2 - 2)$

22. $f(x) = x^5 - 5x^4$
$f'(x) = 5x^3(x - 4)$
$f''(x) = 20x^2(x - 3)$

23. $y = x^{1/3}(x - 4)$
$y' = \dfrac{4(x - 1)}{3x^{2/3}}$
$y'' = \dfrac{4(x + 2)}{9x^{5/3}}$

24. $y = x^{4/3}(x - 7)$
$y' = \dfrac{7x^{1/3}(x - 4)}{3}$
$y'' = \dfrac{28(x - 1)}{9x^{2/3}}$

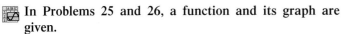 In Problems 25 and 26, a function and its graph are given.

(a) From the graph, estimate where $f''(x) > 0$, where $f''(x) < 0$, and where $f''(x) = 0$.

(b) Use (a) to decide where $f'(x)$ has its relative maxima and relative minima.

(c) Verify your results in parts (a) and (b) by finding $f'(x)$ and $f''(x)$ and then graphing each with a graphing utility.

25. $f(x) = -\frac{1}{3}x^3 + x^2 + 8x - 12$

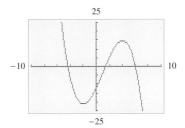

26. $f(x) = \frac{1}{3}x^3 + 2x^2 - 12x - 20$

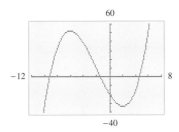

In Problems 27 and 28, $f'(x)$ and its graph are given. Use the graph of $f'(x)$ to determine the following.

(a) Where is the graph of $f(x)$ concave up and where is it concave down?

(b) Where does $f(x)$ have any points of inflection?

(c) Find $f''(x)$ and graph it. Then use that graph to check your conclusions from parts (a) and (b).

(d) Sketch a possible graph of $f(x)$.

27. $f'(x) = 4x - x^2$

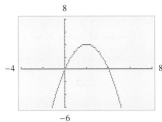

28. $f'(x) = x^2 - x - 2$

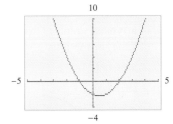

In Problems 29 and 30, use the graph shown in Figure 10.28 and identify points from A through I that satisfy the given conditions.

29. (a) $f'(x) > 0$ and $f''(x) > 0$
(b) $f'(x) < 0$ and $f''(x) < 0$
(c) $f'(x) = 0$ and $f''(x) > 0$
(d) $f'(x) > 0$ and $f''(x) = 0$
(e) $f'(x) = 0$ and $f''(x) = 0$

30. (a) $f'(x) > 0$ and $f''(x) < 0$
    (b) $f'(x) < 0$ and $f''(x) > 0$
    (c) $f'(x) = 0$ and $f''(x) < 0$
    (d) $f'(x) < 0$ and $f''(x) = 0$

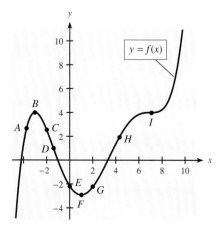

Figure 10.28

In Problems 31 and 32, a graph is given. Tell where $f(x)$ is concave up, where it is concave down, and where it has points of inflection on the interval $-2 < x < 2$, if the given graph is the graph of
(a) $f(x)$ (b) $f'(x)$ (c) $f''(x)$

31. 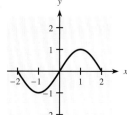   32.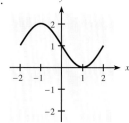

APPLICATIONS

33. *Productivity—diminishing returns*   The figure below is a typical graph of worker productivity as a function of time on the job.

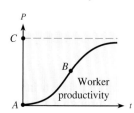

(a) If P represents the productivity and t represents the time, write a mathematical symbol that represents the rate of change of productivity with respect to time.
(b) Which of A, B, and C is the critical point for the rate of change found in part (a)? This point actually corresponds to the point at which the rate of

production is maximized, or the point for maximum worker efficiency. In economics, this is called the point of diminishing returns.
(c) Which of A, B, and C corresponds to the upper limit of production?

34. *Population growth*   The following figure shows the growth of a population as a function of time.

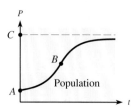

(a) If P represents the population and t represents the time, write a mathematical symbol that represents the rate of change (growth rate) of the population with respect to time.
(b) Which of A, B, and C corresponds to the point at which the growth *rate* attains its maximum?
(c) Which of A, B, and C corresponds to the upper limit of population?

35. *Advertising and sales*   The figure below shows the daily sales volume S as a function of time t since an ad campaign began.

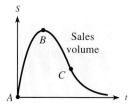

(a) Which of A, B, and C is the point of inflection for the graph?
(b) On which side of C is $d^2S/dt^2 > 0$?
(c) Does the *rate of change* of sales volume attain its minimum at C?

36. *Oxygen purity*   The figure below shows the oxygen level P (for purity) in a lake t months after an oil spill.

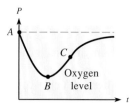

(a) Which of A, B, and C is the point of inflection for the graph?
(b) On which side of C is $d^2P/dt^2 < 0$?
(c) Does the *rate of change* of purity attain its maximum at C?

37. *Production* Suppose that the total number of units produced by a worker in *t* hours of an 8-hour shift can be modeled by the production function $P(t)$:

$$P(t) = 27t + 12t^2 - t^3$$

(a) Find the number of hours before production is maximized.
(b) Find the number of hours before the rate of production is maximized. That is, find the point of diminishing returns.

38. *Poiseuille's law—velocity of blood* According to Poiseuille's law, the speed *S* of blood through an artery of radius *r* at a distance *x* from the artery wall is given by

$$S = k[r^2 - (r - x)^2]$$

where *k* is a constant. Find the distance *x* that maximizes the speed.

39. *Advertising and sales—diminishing returns* Suppose that a company's daily sales volume attributed to an advertising campaign is given by

$$S(t) = \frac{3}{t + 3} - \frac{18}{(t + 3)^2} + 1$$

(a) Find how long it will be before sales volume is maximized.
(b) Find how long it will be before the rate of change of sales volume is minimized. That is, find the point of diminishing returns.

40. *Oxygen purity—diminishing returns* Suppose that the oxygen level *P* (for purity) in a body of water *t* months after an oil spill is given by

$$P(t) = 500\left[1 - \frac{4}{t + 4} + \frac{16}{(t + 4)^2}\right]$$

(a) Find how long it will be before the oxygen level reaches its minimum.
(b) Find how long it will be before the rate of change of *P* is maximized. That is, find the point of diminishing returns.

41. *International phone traffic* The following table gives the minutes of international phone traffic (in billions) for the years from 1990 to 2003. The equation that models these data is

$$y = -0.03354x^3 + 1.4436x^2 - 10.134x + 53.554$$

where *x* is the number of years past 1985. Use the model to find the year when the second derivative predicts that the rate of change of international phone traffic begins to decrease.

Year	International Phone Minutes (in billions)	Year	International Phone Minutes (in billions)
1990	33	1997	79
1991	38	1998	89
1992	43	1999	100
1993	49	2000	118
1994	57	2001	127
1995	63	2002	135
1996	71	2003	140

Source: International Telecommunications Union

42. *Consumer prices* The data in the table can be modeled by the function

$$c(x) = -0.0023x^3 + 0.3662x^2 - 5.086x + 74.74$$

where $x = 0$ represents 1945 and $c(x)$ is the consumer price index (CPI) in year $1945 + x$. During what year does the model predict that the rate of change of the CPI reached its maximum?

Year	CPI	Year	CPI
1945	53.9	1980	248.8
1950	72.1	1985	322.2
1955	80.2	1990	391.4
1960	88.7	1995	456.5
1965	94.5	2000	515.8
1970	116.3	2003	549.2
1975	161.2		

Source: Bureau of Labor Statistics

43. **Modeling** *Foreign-born population* The figure gives the percent of the U.S. population that was foreign-born in census years from 1900 to 2000.
(a) Find the cubic function that is the best fit for the data. Use $x = 0$ to represent 1900.
(b) Find the critical point of the graph of the cubic model for $x > 50$.
(c) Interpret this point in terms of the percent of foreign-born people in the U.S. population.

Percent of Population That Was Foreign-Born, 1900–2000

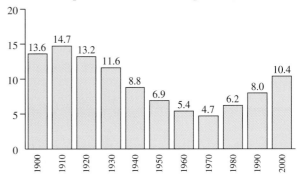

Source: Bureau of the Census, U.S. Department of Commerce

44. **Modeling** *Elderly men in the workforce* The table below gives the percent of men 65 years or older in the work force for selected years from 1890 to 2002.
    (a) With $x = 0$ representing 1890, find the cubic function that models these data.
    (b) Use the model to determine when the rate of change of the percent of elderly men in the work force reached its minimum.
    (c) On the graph of this model, to what does the result in (b) correspond?

Year	Percent	Year	Percent
1890	68.3	1960	30.5
1900	63.1	1970	24.8
1920	55.6	1980	19.3
1930	54.0	1990	17.6
1940	41.8	2000	18.6
1950	41.4	2002	18.0

Source: U.S. Bureau of the Census

45. **Modeling** *Gross national product* The following table gives the U.S. gross national product (GNP) for the years 1913–1922.

(a) Using $x = 0$ to represent 1912, find the function that models the data.
(b) During what year in this time interval does the model predict a minimum GNP?
(c) During what year in this time interval does the model predict a maximum GNP?

Year	Gross National Product (GNP)
1913	39.6
1914	38.6
1915	40.0
1916	48.3
1917	60.4
1918	76.4
1919	84.0
1920	91.5
1921	69.6
1922	74.1

Source: National Debt in Perspective, Oscar Falconi, Wholesale Nutrition (Internet)

10.3 Optimization in Business and Economics

OBJECTIVES

- To find absolute maxima and minima
- To maximize revenue, given the total revenue function
- To minimize the average cost, given the total cost function
- To find the maximum profit from total cost and total revenue functions, or from a profit function

◗ Application Preview

Suppose a travel agency charges $300 per person for a certain tour when the tour group has 25 people, but will reduce the price by $10 per person for each additional person above the 25. In order to determine the group size that will maximize the revenue, the agency must create a model for the revenue. With such a model, the techniques of calculus can be useful in determining the maximum revenue. (See Example 2.)

Most companies (such as the travel agency) are interested in obtaining the greatest possible revenue or profit. Similarly, manufacturers of products are concerned about producing their products for the lowest possible average cost per unit. Therefore, rather than just finding the relative maxima or relative minima of a function, we will consider where the **absolute maximum** or **absolute minimum** of a function occurs in a given interval. This requires evaluating the function at the endpoints of the given interval as well as finding the relative extrema.

As their name implies, **absolute extrema** are the functional values that are the largest or smallest values over the entire domain of the function (or over the interval of interest).

Absolute Extrema

The value $f(a)$ is the **absolute maximum** of f if $f(a) \geq f(x)$ for all x in the domain of f (or over the interval of interest).

The value $f(b)$ is the **absolute minimum** of f if $f(b) \leq f(x)$ for all x in the domain of f (or over the interval of interest).

In this section we will discuss how to find the absolute extrema of a function and then use these techniques to solve applications involving revenue, cost, and profit.

Let us begin by considering the graph of $y = (x - 1)^2$, shown in Figure 10.29(a). This graph has a relative minimum at (1, 0). Note that the relative minimum is the lowest point on the graph. In this case, the point (1, 0) is an **absolute minimum point,** and 0 is the absolute minimum for the function. Similarly, when there is a point that is the highest point on the graph over the domain of the function, we call the point an **absolute maximum point** of the graph of the function.

In Figure 10.29(a), we see that there is no relative maximum. However, if the domain of the function is restricted to the interval $[\frac{1}{2}, 2]$, then we get the graph shown in Figure 10.29(b). In this case, there is an absolute maximum of 1 at the point (2, 1) and the absolute minimum of 0 is still at (1, 0).

If the domain of $y = (x - 1)^2$ is restricted to the interval [2, 3], the resulting graph is that shown in Figure 10.29(c). In this case, the absolute minimum is 1 and occurs at the point (2, 1), and the absolute maximum is 4 and occurs at (3, 4).

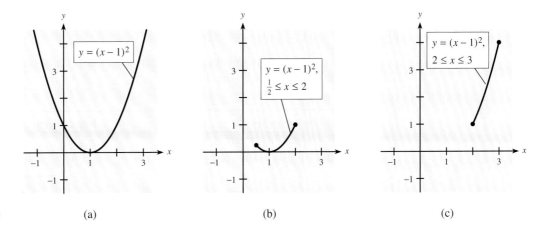

Figure 10.29   (a)   (b)   (c)

As the preceding discussion indicates, if the domain of a continuous function is limited to a closed interval, the absolute maximum or minimum may occur at an endpoint of the domain. In testing functions with limited domains for absolute maxima and minima, we must compare the function values at the endpoints of the domain with the function values at the critical values found by taking derivatives. In the management, life, and social sciences, a limited domain occurs very often, because many quantities are required to be positive, or at least nonnegative.

Maximizing Revenue

Because the marginal revenue is the first derivative of the total revenue, it should be obvious that the total revenue function will have a critical point at the point where the marginal revenue equals 0. With the total revenue function $R(x) = 16x - 0.02x^2$, the point where $R'(x) = 0$ is clearly a maximum because $R(x)$ is a parabola that opens downward. But the domain may be limited, the revenue function may not always be a parabola, or the critical point may not always be a maximum, so it is important to verify where the maximum value occurs.

● EXAMPLE 1 *Revenue*

The total revenue in dollars for a firm is given by

$$R(x) = 8000x - 40x^2 - x^3$$

where x is the number of units sold per day. If only 50 units can be sold per day, find the number of units that must be sold to maximize revenue. Find the maximum revenue.

Solution

This revenue function is limited to x in the interval [0, 50]. Thus, the maximum revenue will occur at a critical value in this interval or at an endpoint. $R'(x) = 8000 - 80x - 3x^2$, so we must solve $8000 - 80x - 3x^2 = 0$ for x.

$$(40 - x)(200 + 3x) = 0$$
$$40 - x = 0 \quad 200 + 3x = 0$$

so

$$x = 40 \quad \text{or} \quad x = -\frac{200}{3}$$

We reject the negative value of x and then use either the second-derivative test or the first-derivative test with a sign diagram.

Second-Derivative Test	First-Derivative Test
$R''(x) = -80 - 6x$, $R''(40) = -320 < 0$, so 40 is a relative maximum.	$R(x)$ ↗ ─── ↘ $R'(x)$ + + + + + + + + 0 − − − − − − →x 40

These tests show that a relative maximum occurs at $x = 40$, giving revenue $R(40) = \$192,000$. Checking the endpoints $x = 0$ and $x = 50$ gives $R(0) = \$0$ and $R(50) = \$175,000$. Thus $R = \$192,000$ at $x = 40$ is the (absolute) maximum revenue.

● *Checkpoint*

1. True or false: If $R(x)$ is the revenue function, we find all possible points where $R(x)$ could be maximized by solving $\overline{MR} = 0$ for x.

◑ EXAMPLE 2 *Revenue* **(Application Preview)**

A travel agency will plan tours for groups of 25 or larger. If the group contains exactly 25 people, the price is $300 per person. However, the price per person is reduced by $10 for each additional person above the 25. What size group will produce the largest revenue for the agency?

Solution

The total revenue is

$$R = (\text{number of people})(\text{price per person})$$

The table below shows how the revenue is changed by increases in the size of the group.

Number in Group	Price per Person	Revenue
25	300	7500
25 + 1	300 − 10	7540
25 + 2	300 − 20	7560
⋮	⋮	⋮
25 + x	300 − 10x	(25 + x)(300 − 10x)

Thus when x is the number of people added to the 25, the total revenue will be a function of x,

$$R = R(x) = (25 + x)(300 - 10x)$$

or

$$R(x) = 7500 + 50x - 10x^2$$

This is a quadratic function, so its graph is a parabola that is concave down. A maximum will occur at its vertex, where $R'(x) = 0$.

$R'(x) = 50 - 20x$, and the solution to $0 = 50 - 20x$ is $x = 2.5$. Thus adding 2.5 people to the group should maximize the total revenue. But we cannot add half a person, so we will test the total revenue function for 27 people and 28 people. This will determine the optimal group size because $R(x)$ is concave down for all x.

For $x = 2$ (giving 27 people) we get $R(2) = 7500 + 50(2) - 10(2)^2 = 7560$. For $x = 3$ (giving 28 people) we get $R(3) = 7500 + 50(3) - 10(3)^2 = 7560$. Note that both 27 and 28 people give the same total revenue and that this revenue is greater than the revenue for 25 people. Thus the revenue is maximized at either 27 or 28 people in the group. (See Figure 10.30.)

Figure 10.30

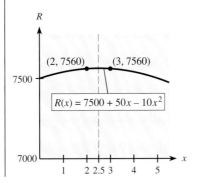

Minimizing Average Cost

Because the total cost function is always increasing for $x \geq 0$, the number of units that will make the total cost a minimum is always $x = 0$ units, which gives an absolute minimum. However, it is more useful to find the number of units that will make the average cost per unit a minimum.

Average Cost

If the total cost is represented by $C = C(x)$, then the **average cost per unit** is

$$\overline{C} = \frac{C(x)}{x}$$

For example, if $C = 3x^2 + 4x + 2$ is the total cost function for a commodity, the **average cost function** is

$$\overline{C} = \frac{3x^2 + 4x + 2}{x} = 3x + 4 + \frac{2}{x}$$

Note that the average cost per unit is undefined if no units are produced.

● **Checkpoint**   2. If $C(x) = \dfrac{x^2}{20} + 10x + 2500$, form $\overline{C}(x)$, the average cost function.

We can use derivatives to find the minimum of the average cost function, as the following example shows.

● **EXAMPLE 3 *Average Cost***

If the total cost function for a commodity is given by $C = \frac{1}{4}x^2 + 4x + 100$ dollars, where x represents the number of units produced, producing how many units will result in a minimum *average cost* per unit? Find the minimum average cost.

Solution
The average cost function is given by

$$\overline{C} = \frac{\frac{1}{4}x^2 + 4x + 100}{x} = \frac{1}{4}x + 4 + \frac{100}{x}$$

Then

$$\overline{C}' = \overline{C}'(x) = \frac{1}{4} - \frac{100}{x^2}$$

Setting $\overline{C}' = 0$ gives

$$0 = \frac{1}{4} - \frac{100}{x^2}$$

$$0 = x^2 - 400, \quad \text{or} \quad x = \pm 20$$

Because the quantity produced must be positive, 20 units should minimize the average cost per unit. We show that it is an absolute minimum by using the second derivative.

$$\overline{C}''(x) = \frac{200}{x^3} \quad \text{so} \quad \overline{C}''(x) > 0 \quad \text{when } x > 0$$

Thus the minimum average cost per unit occurs if 20 units are produced. The graph of the average cost per unit is shown in Figure 10.31. The minimum average cost per unit is $\overline{C}(20) = \$14$.

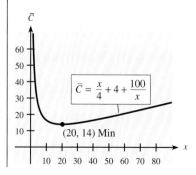

Figure 10.31

The graph of the cost function for the commodity in Example 3 is shown in Figure 10.32, along with several lines that join (0, 0) to a point of the form $(x, C(x))$ on the total cost graph. Note that the slope of each of these lines has the form

$$\frac{C(x) - 0}{x - 0} = \frac{C(x)}{x}$$

so the slope of each line is the *average cost* for the given number of units at the point on $C(x)$. Note that the line from the origin to the point on the curve where $x = 20$ is tangent to the total cost curve. Hence, when $x = 20$, the slope of this line represents both the derivative of the cost function (the *marginal cost*) and the *average cost*. All lines from the origin to points with x-values larger than 20 or smaller than 20 are steeper (and therefore have greater slopes) than the line to the point where $x = 20$. Thus the minimum average cost occurs where the average cost equals the marginal cost. You will be asked to show this analytically in Problems 21 and 22 of the 10.3 Exercises.

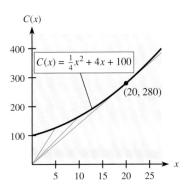

Figure 10.32

Maximizing Profit

In the previous chapter, we defined the marginal profit function as the derivative of the profit function. That is,

$$\overline{MP} = P'(x)$$

In this chapter we have seen how to use the derivative to find maxima and minima for various functions. Now we can apply those same techniques in the context of **profit maximization.** We can use marginal profit to maximize profit functions.

If there is a physical limitation on the number of units that can be produced in a given period of time, then the endpoints of the interval caused by these limitations should also be checked.

● **EXAMPLE 4 *Profit***

Suppose that the production capacity for a certain commodity cannot exceed 30. If the total profit for this commodity is

$$P(x) = 4x^3 - 210x^2 + 3600x - 200 \quad \text{dollars}$$

where x is the number of units sold, find the number of items that will maximize profit.

Solution
The restrictions on capacity mean that $P(x)$ is restricted by $0 \le x \le 30$. The marginal profit function is

$$P'(x) = 12x^2 - 420x + 3600$$

Setting $P'(x)$ equal to 0, we get

$$0 = 12(x - 15)(x - 20)$$

so $P'(x) = 0$ at $x = 15$ *and* $x = 20$. Testing to the right and left of these values (i.e., using the first-derivative test), we get

$$
\left.
\begin{array}{l}
P'(0) = 3600 > 0 \\
P'(18) = -72 < 0 \\
P'(25) = 600 > 0
\end{array}
\right\}
\left.
\begin{array}{l}
\Rightarrow \text{relative maximum at } x = 15 \\
\Rightarrow \text{relative minimum at } x = 20
\end{array}
\right.
$$

Thus, at (15, 20,050) the total profit function has a *relative* maximum, but we must check the endpoints (0 and 30) before deciding whether it is the absolute maximum.

$$P(0) = -200 \quad \text{and} \quad P(30) = 26{,}800$$

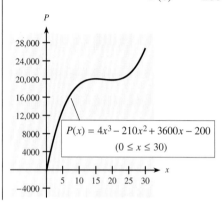

Thus the absolute maximum profit is $26,800, and it occurs at the endpoint, $x = 30$. Figure 10.33 shows the graph of the profit function.

$$P(x) = 4x^3 - 210x^2 + 3600x - 200$$
$$(0 \le x \le 30)$$

Figure 10.33

In a **monopolistic market,** the seller who has a monopoly can control the price by regulating the supply of the product. The seller controls the supply, so he or she can force the price higher by limiting supply.

If the demand function for the product is $p = f(x)$, total revenue for the sale of x units is $R(x) = px = f(x) \cdot x$. Note that the price p is fixed by the market in a competitive market but varies with output for the monopolist.

If $\overline{C} = \overline{C}(x)$ represents the average cost per unit sold, then the total cost for the x units sold is $C = \overline{C} \cdot x = \overline{C}x$. Because we have both total cost and total revenue as a function of the quantity, x, we can maximize the profit function, $P(x) = px - \overline{C}x$, where p represents the demand function $p = f(x)$ and $\overline{C}$ represents the average cost function $\overline{C} = \overline{C}(x)$.

● **EXAMPLE 5** *Profit in a Monopoly Market*

The price of a product in dollars is related to the number of units x demanded daily by

$$p = 168 - 0.2x$$

A monopolist finds that the daily average cost for this product is

$$\overline{C} = 120 + x \quad \text{dollars}$$

(a) How many units must be sold daily to maximize profit?
(b) What is the selling price at this "optimal" level of production?
(c) What is the maximum possible daily profit?

Solution
(a) The total revenue function for the product is

$$R(x) = px = (168 - 0.2x)x = 168x - 0.2x^2$$

and the total cost function is

$$C(x) = \overline{C} \cdot x = (120 + x)x = 120x + x^2$$

Thus the profit function is

$$P(x) = R(x) - C(x) = 168x - 0.2x^2 - (120x + x^2) = 48x - 1.2x^2$$

Then $P'(x) = 48 - 2.4x$, so $P'(x) = 0$ when $x = 20$. We see that $P''(20) = -2.4$, so by the second-derivative test, $P(x)$ has a maximum at $x = 20$. That is, selling 20 units will maximize profit.

(b) The selling price is determined by $p = 168 - 0.2x$, so the price that will result from supplying 20 units per day is $p = 168 - 0.2(20) = 164$. That is, the "optimal" selling price is $164 per unit.

(c) The profit at $x = 20$ is $P(20) = 48(20) - 1.2(20)^2 = 960 - 480 = 480$. Thus the maximum possible profit is $480 per day.

In a **competitive market,** each firm is so small that its actions in the market cannot affect the price of the product. The price of the product is determined in the market by the intersection of the market demand curve (from all consumers) and market supply curve (from all firms that supply this product). The firm can sell as little or as much as it desires at the market equilibrium price.

Therefore, a firm in a competitive market has a total revenue function given by $R(x) = px$, where p is the market equilibrium price for the product and x is the quantity sold.

● **EXAMPLE 6 *Profit in a Competitive Market***

A firm in a competitive market must sell its product for $200 per unit. The average cost per unit (per month) is $\overline{C} = 80 + x$, where x is the number of units sold per month. How many units should be sold to maximize profit?

Solution
If the average cost per unit is $\overline{C} = 80 + x$, then the total cost of x units is given by $C(x) = (80 + x)x = 80x + x^2$. The revenue per unit is $200, so the total revenue is given by $R(x) = 200x$. Thus the profit function is

$$P(x) = R(x) - C(x) = 200x - (80x + x^2), \quad \text{or} \quad P(x) = 120x - x^2$$

Then $P'(x) = 120 - 2x$. Setting $P'(x) = 0$ and solving for x gives $x = 60$. Because $P''(60) = -2$, the profit is maximized when the firm sells 60 units per month.

● *Checkpoint*

3. (a) If $p = 5000 - x$ gives the demand function in a monopoly market, find $R(x)$, if it is possible with this information.
   (b) If $p = 5000 - x$ gives the demand function in a competitive market, find $R(x)$, if it is possible with this information.
4. If $R(x) = 400x - 0.25x^2$ and $C(x) = 150x + 0.25x^2 + 8500$ are the total revenue and total cost (both in dollars) for x units of a product, find the number of units that gives maximum profit and find the maximum profit.

Calculator Note

As we have seen, graphing calculators can be used to locate maximum values. In addition, if it is difficult to determine critical values algebraically, we may be able to find them graphically. ▪

● **EXAMPLE 7** *Maximizing Profit*

Suppose total revenue and total costs for a company are given by

$$R(x) = 3000 - \frac{3000}{x + 1}$$

$$C(x) = 500 + 12x + x^2$$

where x is in thousands of units and revenue and costs are in thousands of dollars. Graph $P(x)$ and $P'(x)$ to determine the number of units that yields maximum profit and the amount of the maximum profit.

Solution

$$P(x) = R(x) - C(x)$$

$$= 3000 - \frac{3000}{x + 1} - (500 + 12x + x^2)$$

$$= 2500 - \frac{3000}{x + 1} - 12x - x^2$$

$$P'(x) = \frac{3000}{(x + 1)^2} - 12 - 2x$$

Finding the critical values by solving $P'(x) = 0$ is very difficult in this case (try it). That is why we are using a graphing approach. Figure 10.34(a) shows the graph of $P(x)$ and Figure 10.34(b) shows the graph of $P'(x)$. These figures indicate that the maximum profit occurs near $x = 10$.

By adjusting the viewing window for $P'(x)$, we obtain the graph in Figure 10.35. This shows that $P'(x) = 0$ when $x = 9$ (or when 9000 units are sold). The maximum profit is

$$P(9) = 2500 - 300 - 108 - 81 = 2011 \text{ (thousands of dollars)}$$

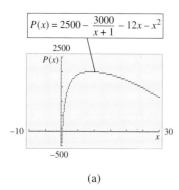

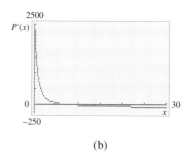

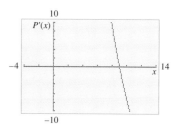

(a)   (b)

Figure 10.34   Figure 10.35

● **Checkpoint Solutions**

1. False. $\overline{MR} = R'(x)$, but there may also be critical points where $R'(x)$ is undefined, or $R(x)$ may be maximized at endpoints of a restricted domain.

2. $\overline{C}(x) = \dfrac{C(x)}{x} = \dfrac{x^2/20 + 10x + 2500}{x} = \dfrac{x}{20} + 10 + \dfrac{2500}{x}$

3. (a) $R(x) = p \cdot x = (5000 - x)x = 5000x - x^2$

   (b) In a competitive market, $R(x) = p \cdot x$, where p is the constant equilibrium price. Thus we need to know the supply function and find the equilibrium price before we can form $R(x)$.

4. $P(x) = R(x) - C(x) = 250x - 0.5x^2 - 8500$ and $P'(x) = 250 - x$. $P'(x) = 0$ when $x = 250$ and $P''(x) = -1$ for all x. Hence the second-derivative test shows that the (absolute) maximum profit occurs when $x = 250$ and is $P(250) = \$22,750$.

10.3 Exercises

In Problems 1–4, find the absolute maxima and minima for $f(x)$ on the interval $[a, b]$.
1. $f(x) = x^3 - 2x^2 - 4x + 2, [-1, 3]$
2. $f(x) = x^3 - 3x + 3, [-3, 1.5]$
3. $f(x) = x^3 + x^2 - x + 1, [-2, 0]$
4. $f(x) = x^3 - x^2 - x, [-0.5, 2]$

MAXIMIZING REVENUE

5. (a) If the total revenue function for a radio is $R = 36x - 0.01x^2$, then sale of how many units, x, will maximize the total revenue in dollars? Find the maximum revenue.
   (b) Find the maximum revenue if production is limited to at most 1500 radios.
6. (a) If the total revenue function for a blender is $R(x) = 25x - 0.05x^2$, sale of how many units, x, will provide the maximum total revenue in dollars? Find the maximum revenue.
   (b) Find the maximum revenue if production is limited to at most 200 blenders.
7. If the total revenue function for a computer is $R(x) = 2000x - 20x^2 - x^3$, find the level of sales, x, that maximizes revenue and find the maximum revenue in dollars.
8. A firm has total revenues given by

$$R(x) = 2800x - 8x^2 - x^3 \quad \text{dollars}$$

   for x units of a product. Find the maximum revenue from sales of that product.
9. An agency charges $10 per person for a trip to a concert if 30 people travel in a group. But for each person above the 30, the charge will be reduced by $0.20. How many people will maximize the total revenue for the agency if the trip is limited to at most 50 people?
10. A company handles an apartment building with 50 units. Experience has shown that if the rent for each of the units is $360 per month, all the units will be filled, but 1 unit will become vacant for each $10 increase in the monthly rate. What rent should be charged to maximize the total revenue from the building if the upper limit on the rent is $450 per month?
11. A cable TV company has 1000 customers paying $20 each month. If each $1 reduction in price attracts 100 new customers, find the price that yields maximum revenue. Find the maximum revenue.
12. If club members charge $5 admission to a classic car show, 1000 people will attend, and for each $1 increase in price, 100 fewer people will attend. What price will give the maximum revenue for the show? Find the maximum revenue.
13. The function $\overline{R}(x) = R(x)/x$ defines the average revenue for selling x units. For

$$R(x) = 2000x + 20x^2 - x^3$$

   (a) find the maximum average revenue.
   (b) show that $\overline{R}(x)$ attains its maximum at an x-value where $\overline{R}(x) = \overline{MR}$.
14. For the revenue function given by

$$R(x) = 2800x + 8x^2 - x^3$$

   (a) find the maximum average revenue.
   (b) show that $\overline{R}(x)$ attains its maximum at an x-value where $\overline{R}(x) = \overline{MR}$.

MINIMIZING AVERAGE COST

15. If the total cost function for a lamp is $C(x) = 25 + 13x + x^2$ dollars, producing how many units, x, will result in a minimum average cost per unit? Find the minimum average cost.
16. If the total cost function for a product is $C(x) = 300 + 10x + 0.03x^2$ dollars, producing how many units, x, will result in a minimum average cost per unit? Find the minimum average cost.
17. If the total cost function for a product is $C(x) = 100 + x^2$ dollars, producing how many units, x, will result in a minimum average cost per unit? Find the minimum average cost.
18. If the total cost function for a product is $C(x) = 250 + 6x + 0.1x^2$ dollars, producing how many units, x, will minimize the average cost? Find the minimum average cost.
19. If the total cost function for a product is $C(x) = (x + 4)^3$ dollars, where x represents the number of hundreds of units produced, producing how many units will minimize average cost? Find the minimum average cost.

20. If the total cost function for a product is $C(x) = (x + 5)^3$ dollars, where x represents the number of hundreds of units produced, producing how many units will minimize average cost? Find the minimum average cost.

21. For the cost function $C(x) = 25 + 13x + x^2$, show that average costs are minimized at the x-value where

$$\overline{C}(x) = \overline{MC}$$

22. For the cost function $C(x) = 300 + 10x + 0.03x^2$, show that average costs are minimized at the x-value where

$$\overline{C}(x) = \overline{MC}$$

The graphs in Problems 23 and 24 show total cost functions. For each problem:
(a) Explain how to use the total cost graph to determine the level of production at which average cost is minimized.
(b) Determine that level of production.

23.

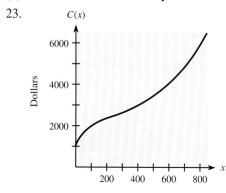

24.

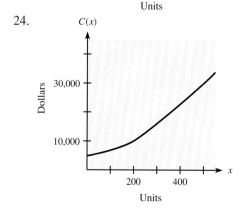

MAXIMIZING PROFIT

25. If the profit function for a product is $P(x) = 5600x + 85x^2 - x^3 - 200,000$ dollars, selling how many items, x, will produce a maximum profit? Find the maximum profit.

26. If the profit function for a commodity is $P = 6400x - 18x^2 - \frac{1}{3}x^3 - 40,000$ dollars, selling how many units, x, will result in a maximum profit? Find the maximum profit.

27. A manufacturer estimates that its product can be produced at a total cost of $C(x) = 45,000 + 100x + x^3$

dollars. If the manufacturer's total revenue from the sale of x units is $R(x) = 4600x$ dollars, determine the level of production x that will maximize the profit. Find the maximum profit.

28. A product can be produced at a total cost $C(x) = 800 + 100x^2 + x^3$ dollars, where x is the number produced. If the total revenue is given by $R(x) = 60,000x - 50x^2$ dollars, determine the level of production, x, that will maximize the profit. Find the maximum profit.

29. A firm can produce only 1000 units per month. The monthly total cost is given by $C(x) = 300 + 200x$ dollars, where x is the number produced. If the total revenue is given by $R(x) = 250x - \frac{1}{100}x^2$ dollars, how many items, x, should the firm produce for maximum profit? Find the maximum profit.

30. A firm can produce 100 units per week. If its total cost function is $C = 500 + 1500x$ dollars and its total revenue function is $R = 1600x - x^2$ dollars, how many units, x, should it produce to maximize its profit? Find the maximum profit.

31. *Marginal revenue and marginal cost* The following figure shows the graph of a quadratic revenue function and a linear cost function.
    (a) At which of the four x-values shown is the distance between the revenue and the cost greatest?
    (b) At which of the four x-values shown is the profit largest?
    (c) At which of the four x-values shown is the slope of the tangent to the revenue curve equal to the slope of the cost line?
    (d) What is the relationship between marginal cost and marginal revenue when profit is at its maximum value?

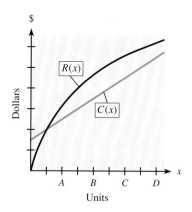

32. *Marginal revenue and marginal cost* The following figure shows the graph of revenue function $y = R(x)$ and cost function $y = C(x)$.
    (a) At which of the four x-values shown is the profit largest?
    (b) At which of the four x-values shown is the slope of the tangent to the revenue curve equal to the slope of the tangent to the cost curve?

(c) What is the relationship between marginal cost and marginal revenue when profit is at its maximum value?

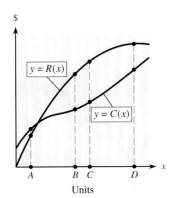

33. A company handles an apartment building with 50 units. Experience has shown that if the rent for each of the units is $720 per month, all of the units will be filled, but 1 unit will become vacant for each $20 increase in this monthly rate. If the monthly cost of maintaining the apartment building is $12 per rented unit, what rent should be charged per month to maximize the profit?

34. A travel agency will plan a tour for groups of size 25 or larger. If the group contains exactly 25 people, the cost is $500 per person. However, each person's cost is reduced by $10 for each additional person above the 25. If the travel agency incurs a cost of $125 per person for the tour, what size group will give the agency the maximum profit?

35. A firm has monthly average costs, in dollars, given by

$$\overline{C} = \frac{45,000}{x} + 100 + x$$

where x is the number of units produced per month. The firm can sell its product in a competitive market for $1600 per unit. If production is limited to 600 units per month, find the number of units that gives maximum profit, and find the maximum profit.

36. A small business has weekly average costs, in dollars, of

$$\overline{C} = \frac{100}{x} + 30 + \frac{x}{10}$$

where x is the number of units produced each week. The competitive market price for this business's product is $46 per unit. If production is limited to 150 units per week, find the level of production that yields maximum profit, and find the maximum profit.

37. The weekly demand function for x units of a product sold by only one firm is $p = 600 - \frac{1}{2}x$ dollars, and the average cost of production and sale is $\overline{C} = 300 + 2x$ dollars.
    (a) Find the quantity that will maximize profit.
    (b) Find the selling price at this optimal quantity.
    (c) What is the maximum profit?

38. The monthly demand function for x units of a product sold by a monopoly is $p = 8000 - x$ dollars, and its average cost is $\overline{C} = 4000 + 5x$ dollars.

(a) Determine the quantity that will maximize profit.
(b) Determine the selling price at the optimal quantity.
(c) Determine the maximum profit.

39. The monthly demand function for a product sold by a monopoly is $p = 1960 - \frac{1}{3}x^2$ dollars, and the average cost is $\overline{C} = 1000 + 2x + x^2$ dollars. Production is limited to 1000 units and x is in hundreds of units.
    (a) Find the quantity that will give maximum profit.
    (b) Find the maximum profit.

40. The monthly demand function for x units of a product sold by a monopoly is $p = 5900 - \frac{1}{2}x^2$ dollars, and its average cost is $\overline{C} = 3020 + 2x$ dollars. If production is limited to 100 units, find the number of units that maximizes profit. Will the maximum profit result in a profit or loss?

41. An industry with a monopoly on a product has its average weekly costs, in dollars, given by

$$\overline{C} = \frac{10,000}{x} + 60 - 0.03x + 0.00001x^2$$

The weekly demand for x units of the product is given by $p = 120 - 0.015x$ dollars. Find the price the industry should set and the number of units it should produce to obtain maximum profit. Find the maximum profit.

42. A large corporation with monopolistic control in the marketplace has its average daily costs, in dollars, given by

$$\overline{C} = \frac{800}{x} + 100x + x^2$$

The daily demand for x units of its product is given by $p = 60,000 - 50x$ dollars. Find the quantity that gives maximum profit, and find the maximum profit. What selling price should the corporation set for its product?

MISCELLANEOUS APPLICATIONS

43. **Modeling** *Social Security beneficiaries* The numbers of millions of Social Security beneficiaries for selected years and projected into the future are given in the following table.
    (a) Find the cubic function that models these data, with x equal to the number of years past 1950.
    (b) Find the point of inflection of the graph of this function for $x > 0$.
    (c) Graph this function and discuss what the point of inflection indicates.

Year	Number of Beneficiaries (millions)	Year	Number of Beneficiaries (millions)
1950	2.9	2002	44.8
1960	14.3	2010	53.3
1970	25.2	2020	68.8
1980	35.1	2030	82.7
1990	39.5		

Source: 2000 Social Security Trustees Report

44. **Modeling** *Federal budget* The following table gives the U.S. federal budget deficit (as a negative value) or surplus (as a positive value) in billions of dollars for selected years.
    (a) Find a cubic function that models these data, with $x = 0$ representing 1990.
    (b) For the years 1990 to 2005, use the model to determine when the absolute maximum (a surplus) and absolute minimum (a deficit) occur and determine their amounts.

Year	Deficit or Surplus	Year	Deficit or Surplus
1990	−221.2	1998	70.0
1991	−269.4	1999	124.4
1992	−290.4	2000	237.0
1993	−255.0	2001	127.0
1994	−203.1	2002	−159.0
1995	−164.0	2003	−374.0
1996	−107.5	2004	−445.0
1997	−22.0		

Source: Budget of the U.S. Government

45. *Dow Jones Industrial Average* The figure shows the Dow Jones Industrial Average from summer 2000 through mid-January 2002.
    (a) Approximate when during 2001 the Dow reached its absolute maximum for that year.
    (b) When do you think the Dow reached its absolute minimum for this period? What happened to trigger this?

STOCKS Dow Jones Industrial Average

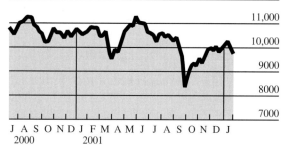

Source: The Wall Street Journal, January 17, 2002. Copyright © 2002 by Dow Jones & Co. Reprinted by permission of Dow Jones & Co. via Copyright Clearance Center.

46. *U.S. trade deficit* The following figure shows the U.S. trade deficit, in billions of dollars, from late 1990 to late 1993.
    (a) Approximately when during this period did the trade deficit reach its absolute minimum, and what was the minimum?
    (b) Approximately when during this period did the 12-month moving average reach its absolute minimum, and what was the minimum?

(c) Approximately when did the trade deficit reach its absolute maximum, and what was the maximum?
(d) Approximately when did the 12-month moving average reach its absolute maximum, and what was the maximum?

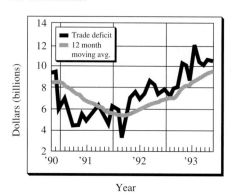

Source: The Wall Street Journal, December 17, 1993. Copyright © 1993 by Dow Jones & Co. Reprinted by permission of Dow Jones & Co. via Copyright Clearance Center.

47. *Social Security support* The graph shows the number of workers, $W = f(t)$, still in the work force per Social Security beneficiary (historically and projected into the future) as a function of time t, in calendar years with $1950 \leq t \leq 2050$. Use the graph to answer the following.

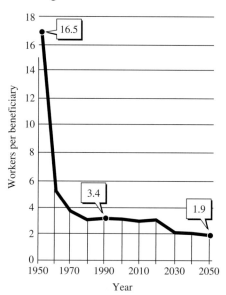

Source: Social Security Administration

(a) What is the absolute maximum of $f(t)$?
(b) What is the absolute minimum of $f(t)$?
(c) Does this graph suggest that Social Security taxes will rise or will fall in the early 21st century? Explain.

10.4

Applications of Maxima and Minima

OBJECTIVE

- To apply the procedures for finding maxima and minima to solve problems from the management, life, and social sciences

◗ **Application Preview**

Suppose that a company needs 1,000,000 items during a year and that preparation costs are $800 for each production run. Suppose further that it costs the company $6 to produce each item and $1 to store each item for up to a year. Find the number of units that should be produced in each production run so that the total costs of production and storage are minimized. This question is answered using an **inventory cost model** (see Example 4).

This inventory-cost determination is a typical example of the kinds of questions and important business applications that require the use of the derivative for finding maxima and minima. As managers, workers, or consumers, we may be interested in such things as maximum revenue, maximum profit, minimum cost, maximum medical dosage, maximum utilization of resources, and so on.

If we have functions that model cost, revenue, or population growth, we can apply the methods of this chapter to find the maxima and minima of those functions.

● **EXAMPLE 1** *Company Growth*

Suppose that a new company begins production in 2006 with eight employees and the growth of the company over the next 10 years is predicted by

$$N = N(t) = 8\left(1 + \frac{160t}{t^2 + 16}\right), \qquad 0 \le t \le 10$$

where N is the number of employees t years after 2006.

Determine in what year the number of employees will be maximized and the maximum number of employees.

Solution

This function will have a relative maximum when $N'(t) = 0$.

$$N'(t) = 8\left[\frac{(t^2 + 16)(160) - (160t)(2t)}{(t^2 + 16)^2}\right]$$

$$= 8\left[\frac{160t^2 + 2560 - 320t^2}{(t^2 + 16)^2}\right]$$

$$= 8\left[\frac{2560 - 160t^2}{(t^2 + 16)^2}\right]$$

Because $N'(t) = 0$ when its numerator is 0 (note that this denominator is never 0), we must solve

$$2560 - 160t^2 = 0$$
$$160(4 + t)(4 - t) = 0$$

so

$$t = -4 \quad \text{or} \quad t = 4$$

We are interested only in positive t-values, so we test $t = 4$.

$$\left.\begin{array}{l} N'(0) = 8\left[\dfrac{2560}{256}\right] > 0 \\[2mm] N'(10) = 8\left[\dfrac{-13{,}440}{(116)^2}\right] < 0 \end{array}\right\} \Rightarrow \textit{relative maximum}$$

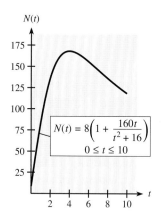

Figure 10.36

The relative maximum is

$$N(4) = 8\left(1 + \frac{640}{32}\right) = 168$$

At $t = 0$, the number of employees is $N(0) = 8$, and it increases to $N(4) = 168$. After $t = 4$ (in 2010), $N(t)$ decreases to $N(10) = 118$ (approximately), so $N(4) = 168$ is the maximum number of employees. Figure 10.36 verifies these conclusions.

Sometimes we must develop the function we need from the statement of the problem. In this case, it is important to understand what is to be maximized or minimized and to express that quantity as a function of *one* variable.

● **EXAMPLE 2** *Minimizing Cost*

A farmer needs to enclose a rectangular pasture containing 1,600,000 square feet. Suppose that along the road adjoining his property he wants to use a more expensive fence and that he needs no fence on one side perpendicular to the road because a river bounds his property on that side. If the fence costs $15 per foot along the road and $10 per foot along the two remaining sides that must be fenced, what dimensions of his rectangular field will minimize his cost? (See Figure 10.37.)

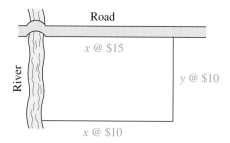

Figure 10.37

Solution

In Figure 10.37, x represents the length of the pasture along the road (and parallel to the road) and y represents the width. The cost function for the fence used is

$$C = 15x + 10y + 10x = 25x + 10y$$

We cannot use a derivative to find where C is minimized unless we write C as a function of x or y only. Because the area of the rectangular field must be 1,600,000 square feet, we have

$$A = xy = 1,600,000$$

Solving for y in terms of x and substituting give C as a function of x.

$$y = \frac{1,600,000}{x}$$

$$C = 25x + 10\left(\frac{1,600,000}{x}\right) = 25x + \frac{16,000,000}{x}$$

The derivative of C with respect to x is

$$C'(x) = 25 - \frac{16,000,000}{x^2}$$

and we find the relative minimum of C as follows:

$$0 = 25 - \frac{16,000,000}{x^2}$$

$$0 = 25x^2 - 16,000,000$$

$$25x^2 = 16,000,000$$

$$x^2 = 640,000 \quad \Rightarrow \quad x = \pm 800$$

We use $x = 800$ feet

Testing to see if $x = 800$ gives the minimum cost, we find

$$C''(x) = \frac{32,000,000}{x^3}$$

$C''(x) > 0$ for $x > 0$, so $C(x)$ is concave up for all positive x. Thus $x = 800$ gives the absolute minimum, and $C(800) = 40,000$ is the minimum cost. The other dimension of the rectangular field is $y = 1,600,000/800 = 2000$ feet. Figure 10.38 verifies that $C(x)$ reaches its minimum (of 40,000) at $x = 800$.

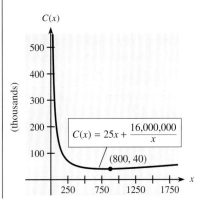

Figure 10.38

● EXAMPLE 3 *Postal Restrictions*

Postal restrictions limit the size of packages sent through the mail. If the restrictions are that the length plus the girth may not exceed 108 in., find the volume of the largest box with square cross section that can be mailed.

Solution

Let l equal the length of the box, and let s equal a side of the square end. See Figure 10.39(a). The volume we seek to maximize is given by

$$V = s^2 l$$

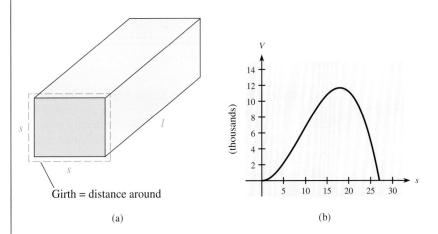

Figure 10.39   (a)   (b)

We can use the restriction that girth plus length equals 108,

$$4s + l = 108$$

to express V as a function of s or l. Because $l = 108 - 4s$, the equation for V becomes

$$V = s^2(108 - 4s) = 108s^2 - 4s^3$$

Thus we can use dV/ds to find the critical values.

$$\frac{dV}{ds} = 216s - 12s^2$$

$$0 = s(216 - 12s)$$

The critical values are $s = 0$, $s = \frac{216}{12} = 18$. The critical value $s = 0$ will not maximize the volume, for in this case, $V = 0$. Testing to the left and right of $s = 18$ gives

$$V'(17) > 0 \quad \text{and} \quad V'(19) < 0$$

Thus $s = 18$ in. and $l = 108 - 4(18) = 36$ in. yield a maximum volume of 11,664 cubic inches. Once again we can verify our results graphically. Figure 10.39(b) on the preceding page shows that $V = s^2(108 - 4s)$ achieves its maximum when $s = 18$.

● ***Checkpoint***

Suppose we want to find the minimum value of $C = 5x + 2y$ and we know that x and y must be positive and that $xy = 1000$.
1. What equation do we differentiate to solve this problem?
2. Find the critical values.
3. Find the minimum value of C.

Consider the **inventory cost model** in which x items are produced in each run and items are removed from inventory at a fixed constant rate. Then the number of units in storage changes with time and is illustrated in Figure 10.40. To see how these inventory cost models work, consider the following example.

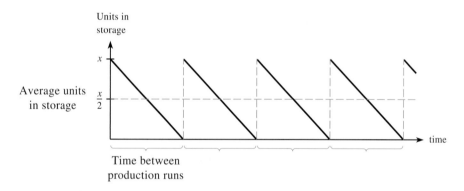

Figure 10.40

Units in storage

Average units in storage $\frac{x}{2}$

x

Time between production runs

time

❍ EXAMPLE 4 *Production Costs (Application Preview)*

Suppose that a company needs 1,000,000 items during a year and that preparation costs are $800 for each production run. Suppose further that it costs the company $6 to produce each item and $1 to store an item for up to a year. If each production run consists of x items, find x so that the total costs of production and storage are minimized.

Solution
The total production costs are given by

$$\begin{pmatrix} \text{No. of} \\ \text{runs} \end{pmatrix}\begin{pmatrix} \text{cost} \\ \text{per run} \end{pmatrix} + \begin{pmatrix} \text{no. of} \\ \text{items} \end{pmatrix}\begin{pmatrix} \text{cost} \\ \text{per item} \end{pmatrix} = \left(\frac{1,000,000}{x}\right)(\$800) + (1,000,000)(\$6)$$

The total storage costs are

$$\begin{pmatrix} \text{Average} \\ \text{no. stored} \end{pmatrix}\begin{pmatrix} \text{storage cost} \\ \text{per item} \end{pmatrix} = \left(\frac{x}{2}\right)(\$1)$$

Thus the total costs of production and storage are

$$C = \left(\frac{1,000,000}{x}\right)(800) + 6,000,000 + \frac{x}{2}$$

We wish to find x so that C is minimized.

$$C' = \frac{-800,000,000}{x^2} + \frac{1}{2}$$

If $x > 0$, critical values occur when $C' = 0$.

$$0 = \frac{-800,000,000}{x^2} + \frac{1}{2}$$

$$\frac{800,000,000}{x^2} = \frac{1}{2}$$

$$1,600,000,000 = x^2$$

$$x = \pm 40,000$$

Because x must be positive, we test $x = 40,000$ with the second derivative.

$$C''(x) = \frac{1,600,000,000}{x^3}, \quad \text{so} \quad C''(40,000) > 0$$

Note that $x = 40,000$ yields an absolute minimum value of C, because $C'' > 0$ for all $x > 0$. That is, production runs of 40,000 items yield minimum total costs for production and storage.

Calculator Note  Problems of the types we've studied in this section could also be solved (at least approximately) with a graphing calculator. With this approach, our first goal is still to express the quantity to be maximized or minimized as a function of one variable. Then that function can be graphed, and the (at least approximate) optimal value can be obtained from the graph. ■

● EXAMPLE 5 *Property Development*

A developer of a campground has to pay for utility line installation to the community center from a transformer on the street at the corner of her property. Because of local restrictions, the lines must be underground on her property. Suppose that the costs are $50 per meter along the street and $100 per meter underground. How far from the transformer should the line enter the property to minimize installation costs? See Figure 10.41(a).

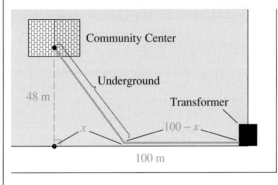

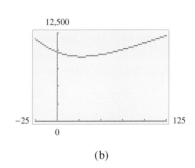

Figure 10.41   (a)   (b)

Solution

If the developer had the cable placed underground from the community center perpendicular to the street and then to the transformer, the cost would be

$$\$100(48) + \$50(100) = \$9800$$

It may be possible to save some money by placing the cable on a diagonal to the street. By using the Pythagorean Theorem, we find that the length of the underground cable that meets the street x meters closer to the transformer is

$$\sqrt{48^2 + x^2} = \sqrt{2304 + x^2} \quad \text{meters}$$

Thus the cost C of installation is given by

$$C = 100\sqrt{2304 + x^2} + 50(100 - x) \quad \text{dollars}$$

Figure 10.41(b) on the preceding page shows the graph of this function over the possible interval for x ($0 \le x \le 100$). Because any extrema must occur in this interval, we can find the minimum by using the MIN command on a calculator or computer. The minimum cost is \$9156.92, when $x = 27.7$ meters (that is, when the cable meets the street 72.3 meters from the transformer). We can use the derivative of C to verify that $x = 27.7$ gives the minimum cost. By finding the derivative and graphing it, we can see that the only relative extremum occurs at $x = 27.7$. (Try this.) Because the cost is lower at $x = 27.7$ than at either endpoint of the interval, the lowest possible cost occurs there.

● **Checkpoint Solutions**

1. We must differentiate C, but first C must be expressed as a function of one variable: $xy = 1000$ means that $y = 1000/x$. If we substitute $1000/x$ for y in $C = 5x + 2y$, we get

$$C(x) = 5x + 2\left(\frac{1000}{x}\right) = 5x + \frac{2000}{x}$$

Find $C'(x)$ to solve the problem.

2. $C'(x) = 5 - 2000/x^2$, so $C'(x) = 0$ when

$$5 - \frac{2000}{x^2} = 0, \quad \text{or} \quad x = \pm 20$$

Because x must be positive, the only critical value is $x = 20$.

3. $C''(x) = 4000/x^3$, so $C''(x) > 0$ for all $x > 0$. Thus $x = 20$ yields the minimum value. Also, when $x = 20$, we have $y = 50$, so the minimum value of C is $C = 5(20) + 2(50) = 200$.

10.4 Exercises

APPLICATIONS

1. *Return to sales* The manufacturer of GRIPPER tires modeled its return to sales from television advertising expenditures in two regions, as follows:

$$\text{Region 1:} \quad S_1 = 30 + 20x_1 - 0.4x_1{}^2$$
$$\text{Region 2:} \quad S_2 = 20 + 36x_2 - 1.3x_2{}^2$$

where S_1 and S_2 are the sales revenue in millions of dollars and x_1 and x_2 are millions of dollars of expenditures for television advertising.

(a) What advertising expenditures would maximize sales revenue in each district?
(b) How much money will be needed to maximize sales revenue in both districts?

2. *Projectiles* A ball thrown into the air from a building 100 ft high travels along a path described by

$$y = \frac{-x^2}{110} + x + 100$$

where y is its height in feet and x is the horizontal distance from the building in feet. What is the maximum height the ball will reach?

3. *Profit*   The profit from a grove of orange trees is given by $x(200 - x)$ dollars, where x is the number of orange trees per acre. How many trees per acre will maximize the profit?

4. *Reaction rates*   The velocity v of an autocatalytic reaction can be represented by the equation

$$v = x(a - x)$$

where a is the amount of material originally present and x is the amount that has been decomposed at any given time. Find the maximum velocity of the reaction.

5. *Productivity*   Analysis of daily output of a factory during an 8-hour shift shows that the hourly number of units y produced after t hours of production is

$$y = 70t + \tfrac{1}{2}t^2 - t^3, \qquad 0 \le t \le 8$$

(a) After how many hours will the hourly number of units be maximized?
(b) What is the maximum hourly output?

6. *Productivity*   A time study showed that, on average, the productivity of a worker after t hours on the job can be modeled by

$$P = 27t + 6t^2 - t^3, \qquad 0 \le t \le 8$$

where P is the number of units produced per hour. After how many hours will productivity be maximized? What is the maximum productivity?

7. *Consumer expenditure*   Suppose that the demand x (in units) for a product is $x = 10{,}000 - 100p$, where p dollars is the market price per unit. Then the consumer expenditure for the product is

$$E = px = 10{,}000p - 100p^2$$

For what market price will expenditure be greatest?

8. *Production costs*   Suppose that the monthly cost in dollars of mining a certain ore is related to the number of pieces of equipment used, according to

$$C = 25{,}000x + \frac{870{,}000}{x}, \qquad x > 0$$

where x is the number of pieces of equipment used. Using how many pieces of equipment will minimize the cost?

Medication   For Problems 9 and 10, consider that when medicine is administered, reaction (measured in change of blood pressure or temperature) can be modeled by

$$R = m^2\left(\frac{c}{2} - \frac{m}{3}\right)$$

where c is a positive constant and m is the amount of medicine absorbed into the blood (*Source:* R. M. Thrall et al., *Some Mathematical Models in Biology*, U.S. Department of Commerce, 1967).

9. Find the amount of medicine that is being absorbed into the blood when the reaction is maximum.

10. The rate of change of reaction R with respect to the amount of medicine m is defined to be the sensitivity.
(a) Find the sensitivity, S.
(b) Find the amount of medicine that is being absorbed into the blood when the sensitivity is maximum.

11. *Advertising and sales*   An inferior product with a large advertising budget sells well when it is introduced, but sales fall as people discontinue use of the product. Suppose that the weekly sales S are given by

$$S = \frac{200t}{(t + 1)^2}, \qquad t \ge 0$$

where S is in millions of dollars and t is in weeks. After how many weeks will sales be maximized?

12. *Revenue*   A newly released film has its weekly revenue given by

$$R(t) = \frac{50t}{t^2 + 36}, \qquad t \ge 0$$

where R is in millions of dollars and t is in weeks.
(a) After how many weeks will the weekly revenue be maximized?
(b) What is the maximum weekly revenue?

13. *News impact*   Suppose that the percent p (as a decimal) of people who could correctly identify two of eight defendants in a drug case t days after their trial began is given by

$$p(t) = \frac{6.4t}{t^2 + 64} + 0.05$$

Find the number of days before the percent is maximized, and find the maximum percent.

14. *Candidate recognition*   Suppose that in an election year the proportion p of voters who recognize a certain candidate's name t months after the campaign started is given by

$$p(t) = \frac{7.2t}{t^2 + 36} + 0.2$$

After how many months is the proportion maximized?

15. *Minimum fence*   Two equal rectangular lots are enclosed by fencing the perimeter of a rectangular lot and then putting a fence across its middle. If each lot is to contain 1200 square feet, what is the minimum amount of fence needed to enclose the lots (include the fence across the middle)?

16. *Minimum fence*   The running yard for a dog kennel must contain at least 900 square feet. If a 20-foot side of the kennel is used as part of one side of a rectangular yard with 900 square feet, what dimensions will require the least amount of fencing?

17. *Minimum cost*   A rectangular field with one side along a river is to be fenced. Suppose that no fence is needed along the river, the fence on the side opposite the river costs \$20 per foot, and the fence on the other sides costs \$5 per foot. If the field must contain 45,000 square feet, what dimensions will minimize costs?

18. *Minimum cost*   From a tract of land a developer plans to fence a rectangular region and then divide it into two identical rectangular lots by putting a fence down the middle. Suppose that the fence for the outside boundary costs \$5 per foot and the fence for the middle costs \$2 per foot. If each lot contains 13,500 square feet, find the dimensions of each lot that yield the minimum cost for the fence.

19. *Optimization at a fixed cost*   A rectangular area is to be enclosed and divided into thirds. The family has \$800 to spend for the fencing material. The outside fence costs \$10 per running foot installed, and the dividers cost \$20 per running foot installed. What are the dimensions that will maximize the area enclosed? (The answer contains a fraction.)

20. *Minimum cost*   A kennel of 640 square feet is to be constructed as shown. The cost is \$4 per running foot for the sides and \$1 per running foot for the ends and dividers. What are the dimensions of the kennel that will minimize the cost?

21. *Minimum cost*   The base of a rectangular box is to be twice as long as it is wide. The volume of the box

is 256 cubic inches. The material for the top costs \$0.10 per square inch and the material for the sides and bottom costs \$0.05 per square inch. Find the dimensions that will make the cost a minimum.

22. *Velocity of air during a cough*   According to B. F. Visser, the velocity v of air in the trachea during a cough is related to the radius r of the trachea according to

$$v = ar^2(r_0 - r)$$

where a is a constant and r_0 is the radius of the trachea in a relaxed state. Find the radius r that produces the maximum velocity of air in the trachea during a cough.

23. *Inventory cost model*   Suppose that a company needs 1,500,000 items during a year and that preparation for each production run costs \$600. Suppose also that it costs \$15 to produce each item and \$2 per year to store an item. Use the inventory cost model to find the number of items in each production run so that the total costs of production and storage are minimized.

24. *Inventory cost model*   Suppose that a company needs 60,000 items during a year and that preparation for each production run costs \$400. Suppose further that it costs \$4 to produce each item and \$0.75 to store an item for one year. Use the inventory cost model to find the number of items in each production run that will minimize the total costs of production and storage.

25. *Inventory cost model*   A company needs 150,000 items per year. It costs the company \$360 to prepare a production run of these items and \$7 to produce each item. If it also costs the company \$0.75 per year for each item stored, find the number of items that should be produced in each run so that total costs of production and storage are minimized.

26. *Inventory cost model*   A company needs 450,000 items per year. Production costs are \$500 to prepare for a production run and \$10 for each item produced. Inventory costs are \$2 per item per year. Find the number of items that should be produced in each run so that the total costs of production and storage are minimized.

27. *Volume*   A rectangular box with a square base is to be formed from a square piece of metal with 12-inch sides. If a square piece with side x is cut from each corner of the metal and the sides are folded up to form an open box, the volume of the box is $V = (12 - 2x)^2 x$. What value of x will maximize the volume of the box?

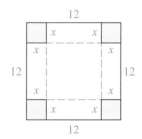

28. *Volume*
    (a) A square piece of cardboard 36 centimeters on a side is to be formed into a rectangular box by cutting squares with length x from each corner and folding up the sides. What is the maximum volume possible for the box?
    (b) Show that if the piece of cardboard is k centimeters on each side, cutting squares of size $k/6$ and folding up the sides gives the maximum volume.

29. *Revenue* The owner of an orange grove must decide when to pick one variety of oranges. She can sell them for $8 a bushel if she sells them now, with each tree yielding an average of 5 bushels. The yield increases by half a bushel per week for the next 5 weeks, but the price per bushel decreases by $0.50 per bushel each week. When should the oranges be picked for maximum return?

30. *Minimum material*
    (a) A box with an open top and a square base is to be constructed to contain 4000 cubic inches. Find the dimensions that will require the minimum amount of material to construct the box.
    (b) A box with an open top and a square base is to be constructed to contain k cubic inches. Show that the minimum amount of material is used to construct the box when each side of the base is

    $$x = (2k)^{1/3} \text{ and the height is } y = \left(\frac{k}{4}\right)^{1/3}$$

31. *Minimum cost* A printer has a contract to print 100,000 posters for a political candidate. He can run the posters by using any number of plates from 1 to 30 on his press. If he uses x metal plates, they will produce x copies of the poster with each impression of the press. The metal plates cost $2.00 to prepare, and it costs $12.50 per hour to run the press. If the press can make 1000 impressions per hour, how many metal plates should the printer make to minimize costs?

32. *Shortest time* A vacationer on an island 8 miles offshore from a point that is 48 miles from town must travel to town occasionally. (See the figure.) The vacationer has a boat capable of traveling 30 mph and can go by auto along the coast at 55 mph. At what point should the car be left to minimize the time it takes to get to town?

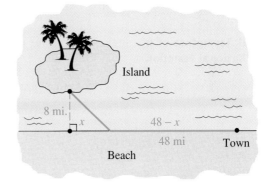

33. *DVD players* The sales revenue for DVD players is given by

    $$R(t) = -0.094t^3 + 0.680t^2 + 0.7175t + 0.826$$

    in billions of dollars, where t is the number of years past 1999. Use a graphing utility to find the year when sales revenue is maximized.

10.5 Rational Functions: More Curve Sketching

OBJECTIVES

- To locate horizontal asymptotes
- To locate vertical asymptotes
- To sketch graphs of functions that have vertical and/or horizontal asymptotes

◗ Application Preview

Suppose that the total cost of producing a shipment of a product is

$$C(x) = 5000x + \frac{125,000}{x}, \quad x > 0$$

where x is the number of machines used in the production process. To find the number of machines that will minimize the total cost, we find the minimum value of this rational function. (See Example 3.) The graph of this function contains a vertical asymptote at $x = 0$. We will discuss graphs and applications involving asymptotes in this section.

The procedures for using the first-derivative test and the second-derivative test are given in previous sections, but none of the graphs discussed in those sections contains vertical asymptotes or horizontal asymptotes. In this section, we consider how to use information about asymptotes along with the first and second derivatives, and we present a unified approach to curve sketching.

Asymptotes

In Section 2.4, "Special Functions and Their Graphs," we first discussed asymptotes and saw that they are important features of the graphs that have them. Then, in our discussion of limits in Sections 9.1 and 9.2, we discovered the relationship between certain limits and asymptotes. The formal defination of vertical asymptotes uses limits.

Vertical Asymptote

> The line $x = x_0$ is a **vertical asymptote** of the graph of $y = f(x)$ if the values of $f(x)$ approach $+\infty$ or $-\infty$ as x approaches x_0 (from the left or the right).

From our work with limits, recall that a vertical asymptote will occur on the graph of a function at an x-value at which the denominator (but not the numerator) of the function is equal to zero. These observations allow us to determine where vertical asymptotes occur.

Vertical Asymptote of a Rational Function

> The graph of the rational function
>
> $$h(x) = \frac{f(x)}{g(x)}$$
>
> has a vertical asymptote at $x = c$ if $g(c) = 0$ and $f(c) \neq 0$.

Because a horizontal asymptote tells us the behavior of the values of the function (y-coordinates) when x increases or decreases without bound, we use limits at infinity to determine the existence of horizontal asymptotes.

Horizontal Asymptote

> The graph of a rational function $y = f(x)$ will have a **horizontal asymptote** at $y = b$, for a constant b, if
>
> $$\lim_{x \to +\infty} f(x) = b \quad \text{or} \quad \lim_{x \to -\infty} f(x) = b$$
>
> Otherwise, the graph has no horizontal asymptote.

For a rational function f, $\lim_{x \to +\infty} f(x) = b$ if and only if $\lim_{x \to -\infty} f(x) = b$, so we only need to find one of these limits to locate a horizontal asymptote. In Problems 37 and 38 in the 9.2 Exercises, the following statements regarding horizontal asymptotes of the graphs of rational functions were proved.

Horizontal Asymptotes of Rational Functions

Consider the rational function $y = \dfrac{f(x)}{g(x)} = \dfrac{a_n x^n + \cdots + a_1 x + a_0}{b_m x^m + \cdots + b_1 x + b_0}$.

1. If $n < m$ (that is, if the degree of the numerator is less than that of the denominator), a horizontal asymptote occurs at $y = 0$ (the x-axis).

2. If $n = m$ (that is, if the degree of the numerator equals that of the denominator), a horizontal asymptote occurs at $y = \dfrac{a_n}{b_m}$ (the ratio of the leading coefficients).

3. If $n > m$ (that is, if the degree of the numerator is greater than that of the denominator), there is no horizontal asymptote.

● **EXAMPLE 1** *Vertical and Horizontal Asymptotes*

Find any vertical and horizontal asymptotes for

(a) $f(x) = \dfrac{2x - 1}{x + 2}$

(b) $f(x) = \dfrac{x^2 + 3}{1 - x}$

Solution

(a) The denominator of this function is 0 at $x = -2$, and because this value does not make the numerator 0, there is a vertical asymptote at $x = -2$. Because the function is rational, with the degree of the numerator equal to that of the denominator and with the ratio of the leading coefficients equal to 2, the graph of the function has a horizontal asymptote at $y = 2$. The graph is shown in Figure 10.42(a).

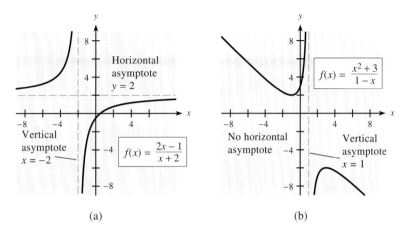

Figure 10.42

(a)   (b)

(b) At $x = 1$, the denominator of $f(x)$ is 0 and the numerator is not, so a vertical asymptote occurs at $x = 1$. The function is rational with the degree of the numerator greater than that of the denominator, so there is no horizontal asymptote. The graph is shown in Figure 10.42(b).

More Curve Sketching

We now extend our first- and second-derivative techniques of curve sketching to include functions that have asymptotes.

In general, the following steps are helpful when we sketch the graph of a function.

1. Determine the domain of the function. The domain may be restricted by the nature of the problem or by the equation.
2. Look for vertical asymptotes, especially if the function is a rational function.
3. Look for horizontal asymptotes, especially if the function is a rational function.
4. Find the relative maxima and minima by using the first-derivative test or the second-derivative test.
5. Use the second derivative to find the points of inflection if this derivative is easily found.
6. Use other information (intercepts, for example) and plot additional points to complete the sketch of the graph.

● **EXAMPLE 2 *Graphing with Asymptotes***

Sketch the graph of the function $f(x) = \dfrac{x^2}{(x + 1)^2}$.

Solution

1. The domain is the set of all real numbers except $x = -1$.
2. Because $x = -1$ makes the denominator 0 and does not make the numerator 0, there is a vertical asymptote at $x = -1$.
3. Because $\dfrac{x^2}{(x + 1)^2} = \dfrac{x^2}{x^2 + 2x + 1}$

   the function is rational with the degree of the numerator equal to that of the denominator and with the ratio of the leading coefficients equal to 1. Hence, the graph of the function has a horizontal asymptote at $y = 1$.
4. To find any maxima and minima, we first find $f'(x)$.

$$f'(x) = \frac{(x + 1)^2(2x) - x^2[2(x + 1)]}{(x + 1)^4}$$

$$= \frac{2x(x + 1)[(x + 1) - x]}{(x + 1)^4}$$

$$= \frac{2x}{(x + 1)^3}$$

Thus $f'(x) = 0$ when $x = 0$ (and $y = 0$), and $f'(x)$ is undefined at $x = -1$ (where the vertical asymptote occurs). Testing $f'(x)$ on either side of $x = 0$ and $x = -1$ gives the following sign diagram. The sign diagram for f' shows that the critical point $(0, 0)$ is a relative minimum and shows how the graph approaches the vertical asymptote at $x = -1$.

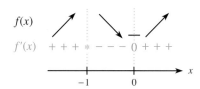

*$x = -1$ is a vertical asymptote.

5. The second derivative is

$$f''(x) = \frac{(x + 1)^3(2) - 2x[3(x + 1)^2]}{(x + 1)^6}$$

Factoring $(x + 1)^2$ from the numerator and simplifying give

$$f''(x) = \frac{2 - 4x}{(x + 1)^4}$$

We can see that $f''(0) = 2 > 0$, so the second-derivative test also shows that $(0, 0)$ is a relative minimum. We see that $f''(x) = 0$ when $x = \frac{1}{2}$. Checking $f''(x)$ between

$x = -1$ (where it is undefined) and $x = \frac{1}{2}$ shows that the graph is concave up on this interval. Note also that $f''(x) < 0$ for $x > \frac{1}{2}$, so the point $\left(\frac{1}{2}, \frac{1}{9}\right)$ is a point of inflection. Also see the sign diagram for $f''(x)$.

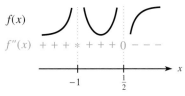

*$x = -1$ is a vertical asymptote.

6. To see how the graph approaches the horizontal asymptote, we check $f(x)$ for large values of $|x|$.

$$f(-100) = \frac{(-100)^2}{(-99)^2} = \frac{10{,}000}{9{,}801} > 1, \quad f(100) = \frac{100^2}{101^2} = \frac{10{,}000}{10{,}201} < 1$$

Thus the graph has the characteristics shown in Figure 10.43(a). The graph is shown in Figure 10.43(b).

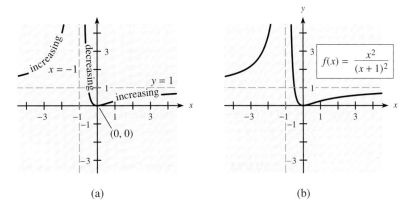

Figure 10.43 (a) (b)

When we wish to learn about a function $f(x)$ or sketch its graph, it is important to understand what information we obtain from $f(x)$, from $f'(x)$, and from $f''(x)$. The following summary may be helpful.

Summary

Source	Information Provided
$f(x)$	y-coordinates; horizontal asymptotes, vertical asymptotes; domain restrictions
$f'(x)$	Increasing [$f'(x) > 0$]; decreasing [$f'(x) < 0$]; critical points [$f'(x) = 0$ *or* $f'(x)$ undefined]; sign-diagram tests for maxima and minima
$f''(x)$	Concave up [$f''(x) > 0$]; concave down [$f''(x) < 0$]; possible points of inflection [$f''(x) = 0$ or $f''(x)$ undefined]; sign-diagram tests for points of inflection; second-derivative test for maxima and minima

● **Checkpoint**

1. Let $f(x) = \dfrac{2x + 10}{x - 1}$ and decide whether the following are true or false.

   (a) $f(x)$ has a vertical asymptote at $x = 1$.

   (b) $f(x)$ has $y = 2$ as its horizontal asymptote.

2. Let $f(x) = \dfrac{x^3 - 16}{x} + 1$; then $f'(x) = \dfrac{2x^3 + 16}{x^2}$ and $f''(x) = \dfrac{2x^3 - 32}{x^3}$. Use these to determine whether the following are true or false.

   (a) There are no asymptotes.

   (b) $f'(x) = 0$ when $x = -2$

   (c) A partial sign diagram for $f'(x)$ is

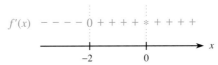

   *means $f'(0)$ is undefined.

   (d) There is a relative minimum at $x = -2$.

   (e) A partial sign diagram for $f''(x)$ is

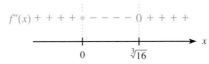

   *means $f''(0)$ is undefined.

   (f) There are points of inflection at $x = 0$ and $x = \sqrt[3]{16}$.

◆ EXAMPLE 3 *Production Costs (Application Preview)*

Suppose that the total cost of producing a shipment of a certain product is

$$C(x) = 5000x + \frac{125{,}000}{x}, \qquad x > 0$$

where x is the number of machines used in the production process.

(a) Graph this total cost function, and identify any asymptotes.

(b) How many machines should be used to minimize the total cost?

Solution

(a) Writing this function with all terms over a common denominator gives

$$C(x) = 5000x + \frac{125{,}000}{x} = \frac{5000x^2 + 125{,}000}{x}$$

The domain of $C(x)$ does not include 0, and $C \to +\infty$ as $x \to 0^+$, so there is a vertical asymptote at $x = 0$. Thus the cost increases without bound as the number of machines used in the process approaches zero. The function decreases initially, but eventually increases, and (because the numerator has a higher degree than the denominator) there is no horizontal asymptote (see Figure 10.44).

(b) Finding the derivative of $C(x)$ gives

$$C'(x) = 5000 - \frac{125{,}000}{x^2} = \frac{5000x^2 - 125{,}000}{x^2}$$

Setting $C'(x) = 0$ and solving for x gives the critical values of x.

$$0 = \frac{5000(x + 5)(x - 5)}{x^2}$$

$$x = 5 \quad \text{or} \quad x = -5$$

Because $C''(x) = 250,000x^{-3} = \dfrac{250,000}{x^3}$ is positive for all $x > 0$, using 5 machines minimizes the cost at $C(5) = 50,000$ (see Figure 10.44).

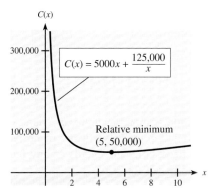

$$C(x) = 5000x + \frac{125,000}{x}$$

Relative minimum
(5, 50,000)

Figure 10.44

Calculator Note

If a graphing calculator is not available, the procedures previously outlined in this section are necessary to generate a complete and accurate graph. With a graphing calculator, the graph of a function is easily generated as long as the viewing window dimensions are appropriate. Although a graphing calculator may reveal the existence of asymptotes, it cannot always precisely locate them. Also, we sometimes need information provided by derivatives to obtain a window that shows all features of a graph. ∎

● **EXAMPLE 4 *Horizontal and Vertical Asymptotes***

Figure 10.45 shows the graph of $f(x) = \dfrac{71x^2}{28(3 - 2x^2)}$.

(a) Determine whether the function has horizontal or vertical asymptotes, and estimate where they occur.
(b) Check your conclusions to part (a) analytically.
(c) Discuss which method is more accurate.

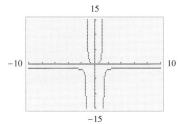

Figure 10.45

Solution

(a) The graph appears to have a horizontal asymptote somewhere between $y = -1$ and $y = -2$, perhaps near $y = -1.5$. Also, there are two vertical asymptotes located approximately at $x = 1.25$ and $x = -1.25$.

(b) The function

$$f(x) = \frac{71x^2}{28(3 - 2x^2)} = \frac{71x^2}{84 - 56x^2}$$

is a rational function with the degree of the numerator equal to that of the denominator and with the ratio of the leading coefficients equal to $-71/56$. Thus the graph of the function has a horizontal asymptote at $y = -71/56$.

Vertical asymptotes occur at x-values where $28(3 - 2x^2) = 0$, or

$$3 - 2x^2 = 0$$
$$3 = 2x^2$$
$$\frac{3}{2} = x^2$$
$$\pm\sqrt{\frac{3}{2}} = x \quad \text{or} \quad x \approx \pm 1.225$$

(c) The analytic method is more accurate, of course, because asymptotes reveal extreme behavior of the function either vertically or horizontally. An accurate graph shows all features, but not necessarily the details of any feature. Despite this, our estimates from the graph were not too bad.

Even with a graphing calculator, sometimes analytic methods are needed to determine an appropriate viewing window.

● **EXAMPLE 5   *Graphing with Technology***

The standard viewing window of the graph of $f(x) = \dfrac{x + 10}{x^2 + 300}$ appears blank (check and see). Find any asymptotes, maxima, and minima, and determine an appropriate viewing window. Sketch the graph.

Solution
Because $x^2 + 300 = 0$ has no real solution, there are no vertical asymptotes. The function is rational with the degree of the numerator less than that of the denominator, so the horizontal asymptote is $y = 0$, which is the x-axis.

We then find an appropriate viewing window by locating the critical points.

$$f'(x) = \frac{(x^2 + 300)(1) - (x + 10)(2x)}{(x^2 + 300)^2}$$
$$= \frac{x^2 + 300 - 2x^2 - 20x}{(x^2 + 300)^2} = \frac{300 - 20x - x^2}{(x^2 + 300)^2}$$

$f'(x) = 0$ when the numerator is zero. Thus

$$300 - 20x - x^2 = 0$$
$$0 = x^2 + 20x - 300$$
$$0 = (x + 30)(x - 10)$$
$$x + 30 = 0 \qquad x - 10 = 0$$
$$x = -30 \qquad x = 10$$

The critical points are $x = -30$, $y = -\frac{1}{60} \approx$ -0.01666667 and $x = 10$, $y = \frac{1}{20} = 0.05$. A sign diagram for $f'(x)$ is shown at the right.

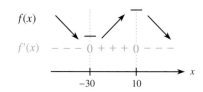

Without using the information above, a graphing utility may not give a useful graph. An x-range that includes -30 and 10 is needed. Because $y = 0$ is a horizontal asymptote, these relative extrema are absolute, and the y-range must be quite small for the shape of the graph to be seen clearly. Figure 10.46 shows the graph.

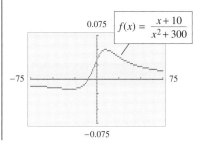

Figure 10.46

 ● **EXAMPLE 6** *Profit*

A profit function for a product is given by

$$P(x) = \frac{-x^2 + 16x - 4}{4x^2 + 16}, \qquad \text{for } x \geq 0$$

where x is in thousands of units and $P(x)$ is in billions of dollars. Because of fixed costs, profit is negative when fewer than 255 units are produced and sold. Will a loss occur at any other level of production and sales?

Solution

Looking at the graph of this function over the range $0 \leq x \leq 10$ (see Figure 10.47(a)), it is not clear whether the graph will eventually cross the x-axis. To see whether a loss ever occurs and to see what profit is approached as the number of units produced and sold becomes large, we consider the behavior of $P(x)$ as $x \to \infty$ (i.e., whether $P(x)$ has a horizontal asymptote).

Note that $P(x)$ is a rational function, and from the highest-power terms in the numerator and denominator we see that $y = -\frac{1}{4}$ is a horizontal asymptote. Thus a loss of $\$\frac{1}{4}$ billion is approached as the number of units increases without bound. Figure 10.47(b) shows that the graph does cross the x-axis, where $-x^2 + 16x - 4 = 0$. Using ZERO, SOLVER, the Quadratic Formula, or a program gives $P(x) = 0$ at $x \approx 0.254$ and $x \approx 15.7459$. Thus if $15,746$ units or more are produced and sold, the profit is negative.

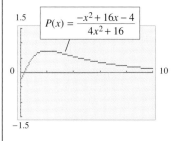

(a)

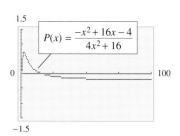

(b)

Figure 10.47

● **Checkpoint Solutions**

1. (a) True, $x = 1$ makes the denominator of $f(x)$ equal to zero, whereas the numerator is nonzero.

   (b) True, the function is rational with the degree of the numerator equal to that of the denominator and with the ratio of the leading coefficients equal to 2.

2. (a) False. There are no horizontal asymptotes, but $x = 0$ is a vertical asymptote.

   (b) True

   (c) True

   (d) True. The relative minimum point is $(-2, f(-2)) = (-2, 13)$.

   (e) True

   (f) False. There is a point of inflection only at $(\sqrt[3]{16}, 1)$. At $x = 0$ the vertical asymptote occurs, so there is no point on the graph and hence no point of inflection.

10.5 Exercises

In Problems 1–4, a function and its graph are given. Use the graph to find each of the following, if they exist. Then confirm your results analytically.

(a) vertical asymptotes

(b) $\lim\limits_{x \to \infty} f(x)$

(c) horizontal asymptotes

(d) $\lim\limits_{x \to -\infty} f(x)$

1. $f(x) = \dfrac{x - 4}{x - 2}$

2. $f(x) = \dfrac{8}{x + 2}$

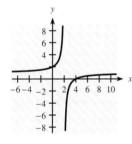

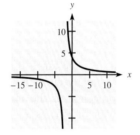

3. $f(x) = \dfrac{3(x^4 + 2x^3 + 6x^2 + 2x + 5)}{(x^2 - 4)^2}$

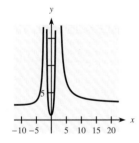

4. $f(x) = \dfrac{x^2}{(x - 2)^2}$

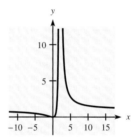

In Problems 5–10, find any horizontal and vertical asymptotes for each function.

5. $y = \dfrac{2x}{x - 3}$

6. $y = \dfrac{3x - 1}{x + 5}$

7. $y = \dfrac{x + 1}{x^2 - 4}$

8. $y = \dfrac{4x}{9 - x^2}$

9. $y = \dfrac{3x^3 - 6}{x^2 + 4}$

10. $y = \dfrac{6x^3}{4x^2 + 9}$

For each function in Problems 11–18, find any horizontal and vertical asymptotes, and use information from the first derivative to sketch the graph.

11. $f(x) = \dfrac{2x + 2}{x - 3}$

12. $f(x) = \dfrac{5x - 15}{x + 2}$

13. $y = \dfrac{x^2 + 4}{x}$

14. $y = \dfrac{x^2 + 4}{x^2}$

15. $y = \dfrac{27x^2}{(x + 1)^3}$

16. $y = \left(\dfrac{x + 2}{x - 3}\right)^2$

17. $f(x) = \dfrac{16x}{x^2 + 1}$

18. $f(x) = \dfrac{4x^2}{x^4 + 1}$

In Problems 19–24, a function and its first and second derivatives are given. Use these to find any horizontal and vertical asymptotes, critical points, relative maxima, relative minima, and points of inflection. Then sketch the graph of each function.

19. $y = \dfrac{x}{(x-1)^2}$

$y' = -\dfrac{x+1}{(x-1)^3}$

$y'' = \dfrac{2x+4}{(x-1)^4}$

20. $y = \dfrac{(x-1)^2}{x^2}$

$y' = \dfrac{2(x-1)}{x^3}$

$y'' = \dfrac{6-4x}{x^4}$

21. $y = x + \dfrac{3}{\sqrt[3]{x-3}}$

$y' = 1 - \dfrac{1}{(x-3)^{4/3}}$

$y'' = \dfrac{4}{3(x-3)^{7/3}}$

22. $y = 3\sqrt[3]{x} + \dfrac{1}{x}$

$y' = \dfrac{x^{4/3} - 1}{x^2}$

$y'' = \dfrac{6 - 2x^{4/3}}{3x^3}$

23. $f(x) = \dfrac{9(x-2)^{2/3}}{x^2}$

$f'(x) = \dfrac{12(3-x)}{x^3(x-2)^{1/3}}$

$f''(x) = \dfrac{4(7x^2 - 42x + 54)}{x^4(x-2)^{4/3}}$

24. $f(x) = \dfrac{3x^{2/3}}{x+1}$

$f'(x) = \dfrac{2-x}{x^{1/3}(x+1)^2}$

$f''(x) = \dfrac{2(2x^2 - 8x - 1)}{3x^{4/3}(x+1)^3}$

In Problems 25–28, a function and its graph are given.
(a) Use the graph to estimate the locations of any horizontal or vertical asymptotes.
(b) Use the function to determine precisely the locations of any asymptotes.

25. $f(x) = \dfrac{9x}{17 - 4x}$

26. $f(x) = \dfrac{5 - 13x}{3x + 20}$

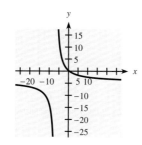

27. $f(x) = \dfrac{20x^2 + 98}{9x^2 - 49}$

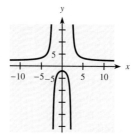

28. $f(x) = \dfrac{15x^2 - x}{7x^2 - 35}$

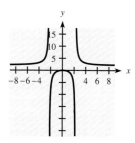

For each function in Problems 29–34, complete the following steps.
(a) Use a graphing utility to graph the function in the standard viewing window.
(b) Analytically determine the location of any asymptotes and extrema.
(c) Graph the function in a viewing window that shows all features of the graph. State the ranges for *x*-values and *y*-values for your viewing window.

29. $f(x) = \dfrac{x + 25}{x^2 + 1400}$

30. $f(x) = \dfrac{x - 50}{x^2 + 1100}$

31. $f(x) = \dfrac{100(9 - x^2)}{x^2 + 100}$

32. $f(x) = \dfrac{200x^2}{x^2 + 100}$

33. $f(x) = \dfrac{1000x - 4000}{x^2 - 10x - 2000}$

34. $f(x) = \dfrac{900x + 5400}{x^2 - 30x - 1800}$

APPLICATIONS

35. *Cost-benefit* The percent p of particulate pollution that can be removed from the smokestacks of an industrial plant by spending C dollars is given by

$$p = \dfrac{100C}{7300 + C}$$

(a) Find any C-values at which the rate of change of p with respect to C does not exist. Make sure that these make sense in the problem.
(b) Find C-values for which p is increasing.
(c) If there is a horizontal asymptote, find it.
(d) Can 100% of the pollution be removed?

36. *Cost-benefit*   The percent p of impurities that can be removed from the waste water of a manufacturing process at a cost of C dollars is given by

$$p = \frac{100C}{8100 + C}$$

(a) Find any C-values at which the rate of change of p with respect to C does not exist. Make sure that these make sense in the problem.
(b) Find C-values for which p is increasing.
(c) Find any horizontal asymptotes.
(d) Can 100% of the pollution be removed?

37. *Revenue*   A recently released film has its weekly revenue given by

$$R(t) = \frac{50t}{t^2 + 36}, \qquad t \geq 0$$

where $R(t)$ is in millions of dollars and t is in weeks.
(a) Graph $R(t)$.
(b) When will revenue be maximized?
(c) Suppose that if revenue decreases for 4 consecutive weeks, the film will be removed from theaters and will be released as a video 12 weeks later. When will the video come out?

38. *Minimizing average cost*   If the total daily cost, in dollars, of producing plastic rafts for swimming pools is given by

$$C(x) = 500 + 8x + 0.05x^2$$

where x is the number of rafts produced per day, then the average cost per raft produced is given by $\overline{C}(x) = C(x)/x$, for $x > 0$.
(a) Graph this function.
(b) Discuss what happens to the average cost as the number of rafts decreases, approaching 0.
(c) Find the level of production that minimizes average cost.

39. *Wind chill*   If x is the wind speed in miles per hour and is greater than or equal to 5, then the wind chill (in degrees Fahrenheit) for an air temperature of 0°F can be approximated by the function

$$f(x) = \frac{289.173 - 58.5731x}{x + 1}, \qquad x \geq 5$$

(a) Ignoring the restriction $x \geq 5$, does $f(x)$ have a vertical asymptote? If so, what is it?
(b) Does $f(x)$ have a vertical asymptote within its domain?

(c) Does $f(x)$ have a horizontal asymptote? If so, what is it?
(d) In the context of wind chill, does $\lim_{x \to \infty} f(x)$ have a physical interpretation? If so, what is it, and is it meaningful?

40. *Profit*   An entrepreneur starts new companies and sells them when their growth is maximized. Suppose that the annual profit for a new company is given by

$$P(x) = 22 - \frac{1}{2}x - \frac{18}{x + 1}$$

where P is in thousands of dollars and x is the number of years after the company is formed. If she wants to sell the company before profits begin to decline, after how many years should she sell it?

41. *Productivity*   The figure is a typical graph of worker productivity per hour P as a function of time t on the job.
(a) What is the horizontal asymptote?
(b) What is $\lim_{x \to \infty} P(t)$?
(c) What is the horizontal asymptote for $P'(t)$?
(d) What is $\lim_{x \to \infty} P'(t)$?

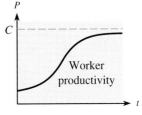

42. *Sales volume*   The figure shows a typical curve that gives the volume of sales S as a function of time t after an ad campaign.
(a) What is the horizontal asymptote?
(b) What is $\lim_{t \to \infty} S(t)$?
(c) What is the horizontal asymptote for $S'(t)$?
(d) What is $\lim_{t \to \infty} S'(t)$?

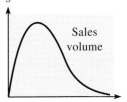

Farm workers   **The percent of U.S. workers in farm occupations during certain years is shown in the table.**

Year	Percent	Year	Percent	Year	Percent
1820	71.8	1930	21.2	1980	2.7
1850	63.7	1940	17.4	1985	2.8
1870	53.0	1950	11.6	1990	2.4
1900	37.5	1960	6.1	1994	2.5
1920	27.0	1970	3.6	2002	2.5

Source: Bureau of Labor Statistics, U.S. Department of Labor

Assume that the percent of U.S. workers in farm occupations can be modeled by the function

$$f(t) = 1000 \cdot \frac{-7.4812t + 1560.2}{1.2882t^2 - 122.18t + 21,483}$$

where t is the number of years past 1800. Use this model in Problems 43 and 44.

43. (a) Find $\lim_{t \to \infty} f(t)$.

(b) Interpret your answer to part (a).

(c) Does $f(t)$ have any vertical asymptotes within its domain $t \geq 0$?

44. (a) Use a graphing utility to graph $f(t)$ for $t = 0$ to $t = 220$.

(b) From the graph, identify t-values where the model is inappropriate, and explain why it is inappropriate.

45. *Barometric pressure* The figure shows a barograph readout of the barometric pressure as recorded by Georgia Southern University's meteorological equipment. The figure shows a tremendous drop in barometric pressure on Saturday morning, March 13, 1993.

(a) If $B(t)$ is barometric pressure expressed as a function of time, as shown in the figure, does $B(t)$ have

a vertical asymptote sometime after 8 A.M. on Saturday, March 13, 1993? Explain why or why not.

(b) Consult your library or some other resource to find out what happened in Georgia (and in the eastern United States) on March 13, 1993, to cause such a dramatic drop in barometric pressure.

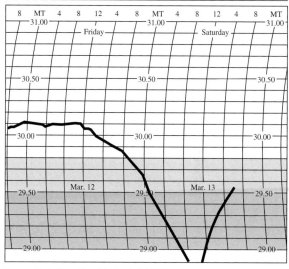

Source: *Statesboro Herald*, March 14, 1993.

Key Terms and Formulas

Section	Key Terms	Formulas
10.1	Relative maxima and minima	
	Increasing	$f'(x) > 0$
	Decreasing	$f'(x) < 0$
	Critical points	$f'(x) = 0$ or $f'(x)$ undefined
	Sign diagram for $f'(x)$	
	First-derivative test	
	Horizontal point of inflection	
10.2	Concave up	$f''(x) > 0$
	Concave down	$f''(x) < 0$
	Point of inflection	May occur where $f''(x) = 0$ or $f''(x)$ undefined
	Sign diagram for $f''(x)$	
	Second-derivative test	

Section	Key Terms	Formulas
10.3	Absolute extrema Average cost Profit maximization Competitive market Monopolistic market	$\bar{C}(x) = C(x)/x$ $R(x) = p \cdot x$ where p = equilibrium price $R(x) = p \cdot x$ where $p = f(x)$ is the demand function
10.4	Inventory cost models	
10.5	Asymptotes Horizontal: $y = b$ Vertical: $x = c$ For rational function $y = f(x)/g(x)$	$f(x) \to b$ as $x \to \infty$ or $x \to -\infty$ y unbounded near $x = c$ Vertical: $x = c$ if $g(c) = 0$ and $f(c) \neq 0$ Horizontal: Key on highest-power terms of $f(x)$ and $g(x)$

Review Exercises

Section 10.1

In Problems 1–4, find all critical points and determine whether they are relative maxima, relative minima, or horizontal points of inflection.

1. $y = -x^2$
2. $p = q^2 - 4q - 5$
3. $f(x) = 1 - 3x + 3x^2 - x^3$
4. $f(x) = \dfrac{3x}{x^2 + 1}$

In Problems 5–10:
(a) Find all critical values, including those at which $f'(x)$ is undefined.
(b) Find the relative maxima and minima, if any exist.
(c) Find the horizontal points of inflection, if any exist.
(d) Sketch the graph.

5. $y = x^3 + x^2 - x - 1$
6. $f(x) = 4x^3 - x^4$
7. $f(x) = x^3 - \dfrac{15}{2}x^2 - 18x + \dfrac{3}{2}$
8. $y = 5x^7 - 7x^5 - 1$
9. $y = x^{2/3} - 1$
10. $y = x^{2/3}(x - 4)^2$

Section 10.2

11. Is the graph of $y = x^4 - 3x^3 + 2x - 1$ concave up or concave down at $x = 2$?
12. Find intervals on which the graph of $y = x^4 - 2x^3 - 12x^2 + 6$ is concave up and intervals on which it is concave down, and find points of inflection.
13. Find the relative maxima, relative minima, and points of inflection of the graph of $y = x^3 - 3x^2 - 9x + 10$.

In Problems 14 and 15, find any relative maxima, relative minima, and points of inflection, and sketch each graph.

14. $y = x^3 - 12x$
15. $y = 2 + 5x^3 - 3x^5$

Section 10.3

16. Given $R = 280x - x^2$, find the absolute maximum and minimum for R when (a) $0 \le x \le 200$ and (b) $0 \le x \le 100$.

17. Given $y = 6400x - 18x^2 - \dfrac{x^3}{3}$, find the absolute maximum and minimum for y when (a) $0 \le x \le 50$ and (b) $0 \le x \le 100$.

Section 10.5

In Problems 18 and 19, use the graphs to find the following items.
(a) vertical asymptotes
(b) horizontal asymptotes
(c) $\lim\limits_{x \to +\infty} f(x)$
(d) $\lim\limits_{x \to -\infty} f(x)$

18.

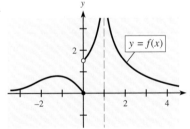

19.

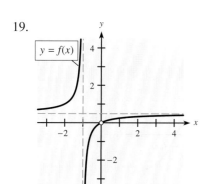

In Problems 20 and 21, find any horizontal asymptotes and any vertical asymptotes.

20. $y = \dfrac{3x + 2}{2x - 4}$　　　　21. $y = \dfrac{x^2}{1 - x^2}$

In Problems 22–24:
(a) Find any horizontal and vertical asymptotes.
(b) Find any relative maxima and minima.
(c) Sketch each graph.

22. $y = \dfrac{3x}{x + 2}$　　　　23. $y = \dfrac{8(x - 2)}{x^2}$

24. $y = \dfrac{x^2}{x - 1}$

Sections 10.1 and 10.2

In Problems 25 and 26, a function and its graph are given.
(a) Use the graph to determine (estimate) x-values where $f'(x) > 0$, where $f'(x) < 0$, and where $f'(x) = 0$.
(b) Use the graph to determine x-values where $f''(x) > 0$, where $f''(x) < 0$, and where $f''(x) = 0$.
(c) Check your conclusions to (a) by finding $f'(x)$ and graphing it with a graphing utility.
(d) Check your conclusions to (b) by finding $f''(x)$ and graphing it with a graphing utility.

25. $f(x) = x^3 - 4x^2 + 4x$

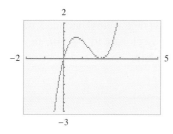

26. $f(x) = 0.0025x^4 + 0.02x^3 - 0.48x^2 + 0.08x + 4$

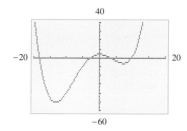

In Problems 27 and 28, $f'(x)$ and its graph are given.
(a) Use the graph of $f'(x)$ to determine (estimate) where the graph of $f(x)$ is increasing, where it is decreasing, and where it has relative extrema.
(b) Use the graph of $f'(x)$ to determine where $f''(x) > 0$, where $f''(x) < 0$, and where $f''(x) = 0$.
(c) Verify that the given $f(x)$ has $f'(x)$ as its derivative, and graph $f(x)$ to check your conclusions in (a).
(d) Calculate $f''(x)$ and graph it to check your conclusions in (b).

27. $f'(x) = x^2 + 4x - 5$　$\left(\text{for } f(x) = \dfrac{x^3}{3} + 2x^2 - 5x \right)$

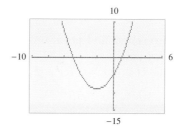

28. $f'(x) = 6x^2 - x^3$　$\left(\text{for } f(x) = 2x^3 - \dfrac{x^4}{4} \right)$

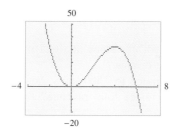

In Problems 29 and 30, $f''(x)$ and its graph are given.
(a) Use the graph to determine (estimate) where the graph of $f(x)$ is concave up, where it is concave down, and where it has points of inflection.
(b) Verify that the given $f(x)$ has $f''(x)$ as its second derivative, and graph $f(x)$ to check your conclusions in (a).

29. $f''(x) = 4 - x$　$\left(\text{for } f(x) = 2x^2 - \dfrac{x^3}{6} \right)$

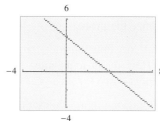

30. $f''(x) = 6 - x - x^2$ $\left(\text{for } f(x) = 3x^2 - \dfrac{x^3}{6} - \dfrac{x^4}{12}\right)$

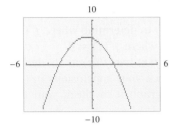

APPLICATIONS

Sections 10.1–10.3

In Problems 31–36, cost, revenue, and profit are in dollars and x is the number of units.

31. *Cost*   Suppose the total cost function for a product is

$$C(x) = 3x^2 + 15x + 75$$

How many units will minimize the average cost? Find the minimum average cost.

32. *Revenue*   Suppose the total revenue function for a product is given by

$$R(x) = 32x - 0.01x^2$$

   (a) How many units will maximize the total revenue? Find the maximum revenue.
   (b) If production is limited to 1500 units, how many units will maximize the total revenue? Find the maximum revenue.

33. *Profit*   Suppose the profit function for a product is

$$P(x) = 1080x + 9.6x^2 - 0.1x^3 - 50{,}000$$

Find the maximum profit.

34. *Profit*   How many units (x) will maximize profit if $R(x) = 46x - 0.01x^2$ and $C(x) = 0.05x^2 + 10x + 1100$?

35. *Profit*   A product can be produced at a total cost of $C(x) = 800 + 4x$, where x is the number produced and is limited to at most 150 units. If the total revenue is given by $R(x) = 80x - \frac{1}{4}x^2$, determine the level of production that will maximize the profit.

36. *Average cost*   The total cost function for a product is $C = 2x^2 + 54x + 98$. How many units must be produced to minimize average cost?

37. *Marginal profit*   The following figure shows the graph of a marginal profit function for a company. At what level of sales will profit be maximized? Explain.

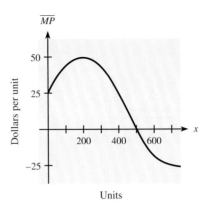

38. *Productivity—diminishing returns*   Suppose the productivity P of an individual worker (in number of items produced per hour) is a function of the number of hours of training t according to

$$P(t) = 5 + \frac{95t^2}{t^2 + 2700}$$

Find the number of hours of training at which the rate of change of productivity is maximized. (That is, find the point of diminishing returns.)

39. *Output*   The figure shows a typical graph of output y (in thousands of dollars) as a function of capital investment I (also in thousands of dollars).
   (a) Is the point of diminishing returns closest to the point at which $I = 20$, $I = 60$, or $I = 120$? Explain.
   (b) The average output per dollar of capital investment is defined as the total output divided by the amount of capital investment; that is,

$$\text{Average output} = \frac{f(I)}{I}$$

   Calculate the slope of a line from $(0, 0)$ to an arbitrary point $(I, f(I))$ on the output graph. How is this slope related to the average output?
   (c) Is the maximum average output attained when the capital investment is closest to $I = 40$, to $I = 70$, or to $I = 140$? Explain.

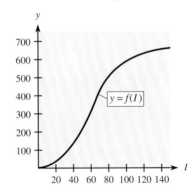

40. *Revenue*   MMR II Extreme Bike Shop sells 54 basic-style mountain bikes per month at a price of $385 each. Market research indicates that MMR II could sell 10 more of these bikes if the price were $25 lower. At what selling price will MMR II maximize the revenue from these bikes?

41. *Profit*   If in Problem 40 the mountain bikes cost the shop $200 each, at what selling price will MMR II's profit be a maximum?

42. *Profit*   Suppose that for a product in a competitive market, the demand function is $p = 1200 - 2x$ and the supply function is $p = 200 + 2x$, where x is the number of units and p is in dollars. A firm's average cost function for this product is

$$\overline{C}(x) = \frac{12,000}{x} + 50 + x$$

Find the maximum profit. (*Hint:* First find the equilibrium price.)

43. *Profit*   The monthly demand function for x units of a product sold at $\$p$ per unit by a monopoly is $p = 800 - x$, and its average cost is $\overline{C} = 200 + x$.
(a) Determine the quantity that will maximize profit.
(b) Find the selling price at the optimal quantity.

44. *Profit*   Suppose that in a monopolistic market, the demand function for a commodity is

$$p = 7000 - 10x - \frac{x^2}{3}$$

where x is the number of units and p is in dollars. If a company's average cost function for this commodity is

$$\overline{C}(x) = \frac{40,000}{x} + 600 + 8x$$

find the maximum profit.

Section 10.4

45. *Reaction to a drug*   The reaction R to an injection of a drug is related to the dose x (in milligrams) according to

$$R(x) = x^2 \left(500 - \frac{x}{3} \right)$$

Find the dose that yields the maximum reaction.

46. *Productivity*   The number of parts produced per hour by a worker is given by

$$N = 4 + 3t^2 - t^3$$

where t is the number of hours on the job without a break. If the worker starts at 8 A.M., when will she be at maximum productivity during the morning?

47. *Population*   Population estimates show that the equation $P = 300 + 10t - t^2$ represents the size of the graduating class of a high school, where t represents the number of years after 2005, $0 \le t \le 10$. What will be the largest graduating class in the next 10 years?

48. *Night brightness*   Suppose that an observatory is to be built between cities A and B, which are 30 miles apart. For the best viewing, the observatory should be located where the night brightness from these cities is minimum. If the night brightness of city A is 8 times that of city B, then the night brightness b between the two cities and x miles from A is given by

$$b = \frac{8k}{x^2} + \frac{k}{(30 - x)^2}$$

where k is a constant. Find the best location for the observatory; that is, find x that minimizes b.

49. *Product design*   A playpen manufacturer wants to make a rectangular enclosure with maximum play area. To remain competitive, he wants the perimeter of the base to be only 16 feet. What dimensions should the playpen have?

50. *Printing design*   A printed page is to contain 56 square inches and have a $\frac{3}{4}$-inch margin at the bottom and 1-inch margins at the top and on both sides. Find the dimensions that minimize the size of the page (and hence the costs for paper).

51. *Drug sensitivity*   The reaction R to an injection of a drug is related to the dose x, in milligrams, according to

$$R(x) = x^2 \left(500 - \frac{x}{3} \right)$$

The sensitivity to the drug is defined by dR/dx. Find the dose that maximizes sensitivity.

52. *Federal tax per capita*   For the years from 1983 to 2003, the federal tax per capita T (in dollars) can be modeled by the function

$$T(x) = -0.848x^3 + 33.17x^2 - 151.4x + 2920$$

where x is the number of years past 1980. (*Source:* Tax Foundation)
(a) When does the rate of change of federal tax per capita reach its maximum?
(b) On a graph of $T(x)$, what feature occurs at the x-value found in (a)?

53. *Inventory cost model* A company needs to produce 288,000 items per year. Production costs are $1500 to prepare for a production run and $30 for each item produced. Inventory costs are $1.50 per year for each item stored. Find the number of items that should be produced in each run so that the total costs of production and storage are minimum.

Section 10.5

54. *Average cost* Suppose the total cost of producing x units of a product is given by

$$C(x) = 4500 + 120x + 0.05x^2 \quad \text{dollars}$$

(a) Find any asymptotes of the average cost function $\overline{C}(x) = C(x)/x$.
(b) Graph the average cost function.

55. *Market share* Suppose a company's percent share of the market (actual and projected) for a new product t quarters after its introduction is given by

$$M(t) = \frac{3.8t^2 + 3}{0.1t^2 + 1}$$

(a) Find the company's market share when the product is introduced.
(b) Find any horizontal asymptote of the graph of $M(t)$, and write a sentence that explains the meaning of this asymptote.

---

10 Chapter Test

Find the local maxima, local minima, points of inflection, and asymptotes, if they exist, for each of the functions in Problems 1–3. Graph each function.

1. $f(x) = x^3 + 6x^2 + 9x + 3$
2. $y = 4x^3 - x^4 - 10$
3. $y = \dfrac{x^2 - 3x + 6}{x - 2}$

In Problems 4–6, use the function $y = 3x^5 - 5x^3 + 2$.

4. Over what intervals is the graph of this function concave up?
5. Find the points of inflection of this function.
6. Find the relative maxima and minima of this function.
7. Find the absolute maximum and minimum for $f(x) = 2x^3 - 15x^2 + 3$ on the interval $[-2, 8]$.
8. Find all horizontal and vertical asymptotes of the function $f(x) = \dfrac{200x - 500}{x + 300}$.
9. Use the following graph of $y = f(x)$ and the indicated points to complete the following chart. Enter $+$, $-$, or 0, according to whether f, f', and f'' are positive, negative, or zero at each point.

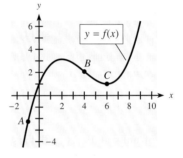

Point	f	f'	f''
A			
B			
C			

10. Use the figure to complete the following.
    (a) $\displaystyle\lim_{x \to -\infty} f(x) = \,?$
    (b) What is the vertical asymptote?
    (c) Find any horizontal asymptotes.

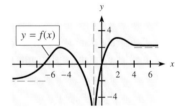

11. If $f(6) = 10, f'(6) = 0$, and $f''(6) = -3$, what can we conclude about the point on the graph of $y = f(x)$ where $x = 6$? Explain.

12. If x represents the number of years past 1900, the equation that models the number of Catholics per priest in the United States is

$$y = 0.19x^2 - 16.59x + 1038.29$$

(*Source:* Dr. Robert G. Kennedy, University of St. Thomas)
   (a) During which year does this model indicate that the number of Catholics per priest was a minimum?
   (b) What does this mean about the number of priests relative to the number of Catholics in the United States?

13. The revenue function for a product is $R(x) = 164x$ dollars and the cost function for the product is

$$C(x) = 0.01x^2 + 20x + 300 \quad \text{dollars}$$

where x is the number of units produced and sold.
   (a) How many units of the product should be sold to obtain maximum profit?
   (b) What is the maximum possible profit?

14. The cost of producing x units of a product is given by

$$C(x) = 100 + 20x + 0.01x^2 \quad \text{dollars}$$

How many units should be produced to minimize average cost?

15. A firm sells 100 TV sets per month at $300 each, but market research indicates that it can sell 1 more set per month for each $2 reduction of the price. At what price will the revenue be maximized?

16. An open-top box is made by cutting squares from the corners of a piece of tin and folding up the sides. If the piece of tin was originally 20 centimeters on a side, how long should the sides of the removed squares be to maximize the resulting volume?

17. A company estimates that it will need 784,000 items during the coming year. It costs $420 to manufacture each item, $2500 to prepare for each production run, and $5 per year for each item stored. How many units should be in each production run so that the total costs of production and storage are minimized?

18. **Modeling**   The table below gives the national debt, in trillions of dollars, for the years 1990–2003.
   (a) Use x as the number of years past 1990 to create a model using these data.
   (b) During what year does the model indicate that the rate of change of the national debt reaches its minimum?
   (c) What feature of the graph of the function found in (a) is the result found in (b)?

Year	National Debt ($ trillions)
1990	3.2
1991	3.6
1992	4.0
1993	4.4
1994	4.6
1995	4.9
1996	5.2
1997	5.4
1998	5.5
1999	5.7
2000	5.7
2001	5.8
2002	6.2
2003	6.8

Source: Bureau of Public Debt, U.S. Department of Treasury

Extended Applications & Group Projects

I. Production Management

Metal Containers, Inc. is reviewing the way it submits bids on U.S. Army contracts. The army often requests open-top boxes, with square bases and of specified volumes. The army also specifies the materials for the boxes, and the base is usually made of a different material than the sides. The box is put together by riveting a bracket at each of the eight corners. For Metal Containers, the total cost of producing a box is the sum of the cost of the materials for the box and the labor costs associated with affixing each bracket.

Instead of estimating each job separately, the company wants to develop an overall approach that will allow it to cost out proposals more easily. To accomplish this, company managers need you to devise a formula for the total cost of producing each box and determine the dimensions that allow a box of specified volume to be produced at minimum cost. Use the following notation to help you solve this problem.

Cost of the material for the base = A per square unit

Cost of the material for the sides = B per square unit

Cost of each bracket = C

Cost to affix each bracket = D

Length of the sides of the base = x

Height of the box = h

Volume specified by the army = V

1. Write an expression for the company's total cost in terms of these quantities.
2. At the time an order is received for boxes of a specified volume, the costs of the materials and labor will be fixed and only the dimensions will vary. Find a formula for each dimension of the box so that the total cost is a minimum.
3. The army requests bids on boxes of 48 cubic feet with base material costing the container company $12 per square foot and side material costing $8 per square foot. Each bracket costs $5, and the associated labor cost is $1 per bracket. Use your formulas to find the dimensions of the box that meet the army's requirements at a minimum cost. What is this cost?

Metal Containers asks you to determine how best to order the brackets it uses on its boxes. You are able to obtain the following information: The company uses approximately 100,000 brackets a year, and the purchase price of each is $5. It buys the same number of brackets (say, n) each time it places an order with the supplier, and it costs $60 to process each order. Metal Containers also has additional costs associated with storing, insuring, and financing its inventory of brackets. These carrying costs amount to 15% of the average value of inventory annually. The brackets are used steadily and deliveries are made just as inventory reaches zero, so that inventory fluctuates between zero and n brackets.

4. If the total annual cost associated with the bracket supply is the sum of the annual purchasing cost and the annual carrying costs, what order size n would minimize the total cost?
5. In the general case of the bracket-ordering problem, the order size n that minimizes the total cost of the bracket supply is called the economic order quantity, or EOQ. Use the following notations to determine a general formula for the EOQ.

Fixed cost per order = F

Unit cost = C

Quantity purchased per year = P

Carrying cost (as a decimal rate) = r

II. Room Pricing in the Off-Season (Modeling)

The data in the table below, from a survey of resort hotels with comparable rates on Hilton Head Island, show that room occupancy during the off-season (November through February) is related to the price charged for a basic room.

Price per Day	Occupancy Rate, %
$ 69	53
89	47
95	46
99	45
109	40
129	32

The goal is to use these data to help answer the following questions.

A. What price per day will maximize the daily off-season revenue for a typical hotel in this group if it has 200 rooms available?
B. Suppose that for this typical hotel the daily cost is $4592 plus $30 per occupied room. What price will maximize the profit for this hotel in the off-season?

The price per day that will maximize the off-season profit for this typical hotel applies to this group of hotels. To find the room price per day that will maximize the daily revenue and the room price per day that will maximize the profit for this hotel (and thus the group of hotels) in the off-season, complete the following.

1. Multiply each occupancy rate by 200 to get the hypothetical room occupancy. Create the revenue data points that compare the price with the revenue, R, which is equal to price times the room occupancy.
2. Use technology to create an equation that models the revenue, R, as a function of the price per day, x.
3. Use maximization techniques to find the price that these hotels should charge to maximize the daily revenue.
4. Use technology to get the occupancy as a function of the price, and use the occupancy function to create a daily cost function.
5. Form the profit function.
6. Use maximization techniques to find the price that will maximize the profit.

11 Derivatives Continued

In this chapter we will develop derivative formulas for logarithmic and exponential functions, focusing primarily on base *e* exponentials and logarithms. We will apply logarithmic and exponential functions and use their derivatives to solve maximization and minimization problems in the management and life sciences.

We will also develop methods for finding the derivative of one variable with respect to another even when the first variable is not a function of the other. This method is called **implicit differentiation.** We will use implicit differentiation with respect to time to solve problems involving rates of change of two or more variables. These problems are called **related rates** problems.

The topics and applications discussed in this chapter include the following.

Chapter Warm-up

Prerequisite Problem Type	For Section	Answer	Section for Review
(a) Simplify: $\dfrac{1}{(3x)^{1/2}} \cdot \dfrac{1}{2}(3x)^{-1/2} \cdot 3$ (b) Write with a positive exponent: $\sqrt{x^2 - 1}$	11.1	(a) $\dfrac{1}{2x}$ (b) $(x^2 - 1)^{1/2}$	0.4 Rational exponents
(a) Vertical lines have _____ slopes. (b) Horizontal lines have _____ slopes.	11.3	(a) Undefined (b) 0	1.3 Slopes
Write the equation of the line passing through $(-2, -2)$ with slope 5.	11.3	$y = 5x + 8$	1.3 Equations of lines
Solve: $x^2 + y^2 - 9 = 0$, for y.	11.3	$y = \pm\sqrt{9 - x^2}$	2.1 Quadratic equations
(a) Write $\log_a(x + h) - \log_a x$ as an expression involving one logarithm. (b) Does $\ln x^4 = 4 \ln x$? (c) Does $\dfrac{x}{h}\log_a\left(\dfrac{x+h}{x}\right) = \log_a\left(1 + \dfrac{h}{x}\right)^{x/h}$? (d) Expand $\ln(xy)$ to separate x and y. (e) If $y = a^x$, then $x =$ _____. (f) Simplify $\ln e^x$.	11.1 11.2	(a) $\log_a\left(\dfrac{x+h}{x}\right)$ (b) Yes (c) Yes (d) $\ln x + \ln y$ (e) $\log_a y$ (f) x	5.2 Logarithms
Find the derivative of (a) $y = x^2 - 2x - 2$ (b) $T(q) = 400q - \dfrac{4}{3}q^2$ (c) $y = \sqrt{9 - x^2}$	11.1–11.5	 (a) $y' = 2x - 2$ (b) $T'(q) = 400 - \dfrac{8}{3}q$ (c) $y' = -x(9 - x^2)^{-1/2}$	9.4–9.6 Derivatives
If $\dfrac{dy}{dx} = \dfrac{-1}{2\sqrt{x}}$, find the slope of the tangent to $y = f(x)$ at $x = 4$.	11.3	Slope $= -\dfrac{1}{4}$	9.3–9.7 Derivatives

11.1 Derivatives of Logarithmic Functions

◑ Application Preview

The table below shows the expected life spans at birth for people born in certain years in the United States. Assume that for these years, the expected life span at birth can be modeled by the function $l(x) = 11.406 + 14.202 \ln x$, where x is the number of years past 1900. The graph of this function is shown in Figure 11.1. If we wanted to use this model to find the rate of change of life span with respect to the number of years past 1900, we would need the derivative of this function and hence the derivative of the logarithmic function $\ln x$. (See Example 5.)

Year	Life Span (years)	Year	Life Span (years)
1920	54.1	1990	75.4
1930	59.7	1992	75.5
1940	62.9	1994	75.7
1950	68.2	1996	76.1
1960	69.7	1998	76.7
1970	70.8	1999	76.7
1975	72.6	2000	76.9
1980	73.7	2001	77.2
1985	74.7		

Source: National Center for Health Statistics

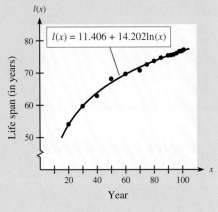

Figure 11.1

Logarithmic Functions

In Chapter 5, "Exponential and Logarithmic Functions," we defined the logarithmic function $y = \log_a x$ as follows:

Logarithmic Function

For $a > 0$ and $a \neq 1$, the **logarithmic function**

$$y = \log_a x \quad \text{(logarithmic form)}$$

has domain $x > 0$, base a, and is defined by

$$a^y = x \quad \text{(exponential form)}$$

The a is called the **base** in both $\log_a x = y$ and $a^y = x$, and y is the *logarithm* in $\log_a x = y$ and the *exponent* in $a^y = x$. Thus **a logarithm is an exponent.** Although logarithmic functions can have any base a, where $a > 0$ and $a \neq 1$, most problems in calculus and many of the applications to the management, life, and social sciences involve logarithms with base e, called **natural logarithms.** In this section we'll see why this base is so important. Recall that the natural logarithmic function ($y = \log_e (x)$) is written $y = \ln x$; see Figure 11.2 for the graph.

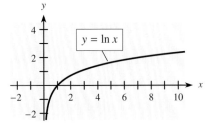

Figure 11.2

Derivative of y = ln x   From Figure 11.2 we see that for $x > 0$, the graph of $y = \ln x$ is always increasing, so the slope of the tangent line to any point must be positive. This means that the derivative of $y = \ln x$ is always positive. Figure 11.3 shows the graph of $y = \ln x$ with tangent lines drawn at several points and with their slopes indicated.

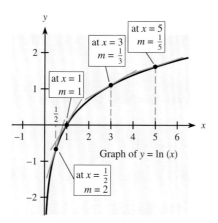

Figure 11.3

Note in Figure 11.3 that at each point where a tangent line is drawn, the slope of the tangent line is the reciprocal of the *x*-coordinate. In fact, this is true for every point on $y = \ln x$, so the slope of the tangent at any point is given by $1/x$. Thus we have the following:

Derivative of y = ln x

If $y = \ln x$, then $\dfrac{dy}{dx} = \dfrac{1}{x}$.

This formula for the derivative of $y = \ln x$ is proved at the end of this section.

● **EXAMPLE 1 *Derivatives Involving Logarithms***

(a) If $y = x^3 + 3 \ln x$, find dy/dx.
(b) If $y = x^2 \ln x$, find y'.

Solution

(a) $\dfrac{dy}{dx} = 3x^2 + 3\left(\dfrac{1}{x}\right) = 3x^2 + \dfrac{3}{x}$

(b) By the Product Rule,

$$y' = x^2 \cdot \dfrac{1}{x} + (\ln x)(2x) = x + 2x \ln x$$

We can use the Chain Rule to find the formula for the derivative of $y = \ln u$, where $u = f(x)$.

Derivatives of Natural Logarithmic Functions

If $y = \ln u$, where u is a differentiable function of x, then

$$\dfrac{dy}{dx} = \dfrac{1}{u} \cdot \dfrac{du}{dx}$$

● **EXAMPLE 2** *Derivative y = ln (u)*

Find $f'(x)$ for each of the following.

(a) $f(x) = \ln(x^2)$ (b) $f(x) = \ln(x^4 - 3x + 7)$

Solution

(a) $f'(x) = \dfrac{1}{x^2}(2x) = \dfrac{2x}{x^2} = \dfrac{2}{x}$ (b) $f'(x) = \dfrac{1}{x^4 - 3x + 7}(4x^3 - 3) = \dfrac{4x^3 - 3}{x^4 - 3x + 7}$

● **EXAMPLE 3** *Derivatives Involving Logarithms*

(a) Find $f'(x)$ if $f(x) = \frac{1}{3}\ln(2x^6 - 3x + 2)$.

(b) Find $g'(x)$ if $g(x) = \dfrac{\ln(2x + 1)}{2x + 1}$.

Solution

(a) $f'(x)$ is $\frac{1}{3}$ of the derivative of $\ln(2x^6 - 3x + 2)$.

$$f'(x) = \frac{1}{3} \cdot \frac{1}{2x^6 - 3x + 2}(12x^5 - 3)$$

$$= \frac{4x^5 - 1}{2x^6 - 3x + 2}$$

(b) We begin with the Quotient Rule.

$$g'(x) = \frac{(2x + 1)\dfrac{1}{2x + 1}(2) - [\ln(2x + 1)]2}{(2x + 1)^2}$$

$$= \frac{2 - 2\ln(2x + 1)}{(2x + 1)^2}$$

● *Checkpoint*

1. If $y = \ln(3x^2 + 2)$, find y'.
2. If $y = \ln x^6$, find y'.

Using Properties of Logarithms

A logarithmic function of products, quotients, or powers, such as $y = \ln[x(x^5 - 2)^{10}]$, can be rewritten with properties of logarithms so that finding the derivative is much easier. The properties of logarithms, which were introduced in Section 5.2, are stated here for logarithms with an arbitrary base a (with $a > 0$ and $a \neq 1$) and for natural logarithms.

Properties of Logarithms

Let M, N, p, and a be real numbers with $M > 0$, $N > 0$, $a > 0$, and $a \neq 1$.

Base a Logarithms	Natural Logarithms
I. $\log_a(a^x) = x$ (for any real x)	I. $\ln(e^x) = x$
II. $a^{\log_a(x)} = x$ (for $x > 0$)	II. $e^{\ln x} = x$
III. $\log_a(MN) = \log_a M + \log_a N$	III. $\ln(MN) = \ln M + \ln N$
IV. $\log_a\left(\dfrac{M}{N}\right) = \log_a M - \log_a N$	IV. $\ln\left(\dfrac{M}{N}\right) = \ln M - \ln N$
V. $\log_a(M^p) = p\log_a M$	V. $\ln(M^p) = p\ln M$

For example, to find the derivative of $f(x) = \ln \sqrt[3]{2x^6 - 3x + 2}$, it is easier to rewrite this as follows:

$$f(x) = \ln\left[(2x^6 - 3x + 2)^{1/3}\right] = \tfrac{1}{3} \ln (2x^6 - 3x + 2)$$

Then take the derivative, as in Example 3(a) above.

● **EXAMPLE 4** *Logarithm Properties and Derivatives*

(a) Use logarithm properties to find dy/dx when

$$y = \ln\left[x(x^5 - 2)^{10}\right]$$

(b) Find the derivative of

$$f(x) = \ln\left(\frac{\sqrt[3]{3x + 5}}{x^2 + 11}\right)^4$$

Solution

(a) We use logarithm Properties III and V to rewrite the function.

$$y = \ln x + \ln (x^5 - 2)^{10} \qquad \text{Property III}$$
$$y = \ln x + 10 \ln (x^5 - 2) \qquad \text{Property V}$$

We now take the derivative.

$$\frac{dy}{dx} = \frac{1}{x} + 10 \cdot \frac{1}{x^5 - 2} \cdot 5x^4$$

$$= \frac{1}{x} + \frac{50x^4}{x^5 - 2}$$

(b) Again we begin by using logarithm properties.

$$f(x) = \ln\left(\frac{\sqrt[3]{3x + 5}}{x^2 + 11}\right)^4 = 4 \ln\left(\frac{\sqrt[3]{3x + 5}}{x^2 + 11}\right) \qquad \text{Property V}$$

$$f(x) = 4\left[\tfrac{1}{3} \ln (3x + 5) - \ln (x^2 + 11)\right] \qquad \text{Properties IV and V}$$

We now take the derivative.

$$f'(x) = 4\left[\frac{1}{3} \cdot \frac{1}{3x + 5} \cdot 3 - \frac{1}{x^2 + 11} \cdot 2x\right]$$

$$= 4\left[\frac{1}{3x + 5} - \frac{2x}{x^2 + 11}\right] = \frac{4}{3x + 5} - \frac{8x}{x^2 + 11}$$

● **Checkpoint**

3. If $y = \ln \sqrt[3]{x^2 + 1}$, find y'.

4. Find $f'(x)$ for $f(x) = \ln\left(\frac{2x^4}{(5x + 7)^5}\right)$.

◑ **EXAMPLE 5** *Life Span* **(Application Preview)**

Assume that the average life span (in years) for people born from 1920 to 2001 can be modeled by

$$l(x) = 11.406 + 14.202 \ln x$$

where x is the number of years past 1900.

(a) Find the function that models the rate of change of life span.
(b) Does $l(x)$ have a maximum value for $x > 0$?
(c) Evaluate $\lim\limits_{x \to \infty} l'(x)$.
(d) What do (b) and (c) tell us about the average life span?

Solution
(a) The rate of change of life span is given by the derivative.

$$l'(x) = 0 + 14.202\left(\frac{1}{x}\right) = \frac{14.202}{x}$$

(b) For $x > 0$, we see that $l'(x) > 0$. Hence $l(x)$ is increasing for all values of $x > 0$, so $l(x)$ never achieves a maximum value. That is, there is no maximum life span.
(c) $\lim\limits_{x \to \infty} l'(x) = \lim\limits_{x \to \infty} \dfrac{14.202}{x} = 0$
(d) If this model is accurate, life span will continue to increase, but at an ever slower rate.

● **EXAMPLE 6** *Cost*

Suppose the cost function for x units of a product is given by

$$C(x) = 18{,}250 + 615 \ln (4x + 10)$$

where $C(x)$ is in dollars. Find the marginal cost when 100 units are produced, and explain what it means.

Solution
Marginal cost is given by $C'(x)$.

$$\overline{MC} = C'(x) = 615\left(\frac{1}{4x + 10}\right)(4) = \frac{2460}{4x + 10}$$

$$\overline{MC}(100) = \frac{2460}{4(100) + 10} = \frac{2460}{410} = 6$$

When 100 units are produced, the marginal cost is 6. This means that the approximate cost of producing the 101st item is $6.

Derivative of
$y = \log_a (x)$

So far we have found derivatives of natural logarithmic functions. If we have a logarithmic function with a base other than e, then we can use the **change-of-base formula,** which was introduced in Section 5.2, to express it as a natural logarithm:

$$\log_a x = \frac{\ln x}{\ln a}$$

We can apply this change-of-base formula to find the derivative of a logarithm with any base, as the following example illustrates.

● **EXAMPLE 7** *Derivative of $y = \log_a (u)$*

If $y = \log_4 (x^3 + 1)$, find dy/dx.

Solution

By using the change-of-base formula, we have

$$y = \log_4 (x^3 + 1) = \frac{\ln (x^3 + 1)}{\ln 4} = \frac{1}{\ln 4} \cdot \ln (x^3 + 1)$$

Thus

$$\frac{dy}{dx} = \frac{1}{\ln 4} \cdot \frac{1}{x^3 + 1} \cdot 3x^2 = \frac{3x^2}{(x^3 + 1) \ln 4}$$

Note that this formula means that logarithms with bases other than e will have the *constant* $1/\ln a$ as a factor in their derivatives (as Example 7 had $1/\ln 4$ as a factor). This means that derivatives involving natural logarithms have a simpler form, and we see why base e logarithms are used more frequently in calculus.

● **EXAMPLE 8** *Critical Values*

Let $f(x) = x \ln x - x$. Use the graph of the derivative of $f(x)$ for $x > 0$ to answer the following questions.

(a) At what value a does the graph of $f'(x)$ cross the x-axis (that is, where is $f'(x) = 0$)?
(b) What value a is a critical value for $y = f(x)$?
(c) Does $f(x)$ have a relative maximum or a relative minimum at $x = a$?

Solution

$f'(x) = x \cdot \dfrac{1}{x} + \ln x - 1 = \ln x$. The graph of $f'(x) = \ln x$ is shown in Figure 11.4.

(a) The graph crosses the x-axis at $x = 1$, so $f'(a) = 0$ if $a = 1$.
(b) $a = 1$ is a critical value of $f(x)$.
(c) Because $f'(x)$ is negative for $x < 1$, $f(x)$ is decreasing for $x < 1$.
   Because $f'(x)$ is positive for $x > 1$, $f(x)$ is increasing for $x > 1$.
   Therefore, $f(x)$ has a relative minimum at $x = 1$.
   The graph of $f(x) = x \ln x - x$ is shown in Figure 11.5. It has a relative minimum at $x = 1$.

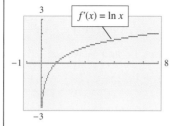

Figure 11.4

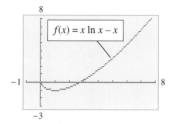

Figure 11.5

Proof that $\dfrac{d}{dx}(\ln x) = \dfrac{1}{x}$

For completeness, we now include the formal proof that if $y = \ln x$, then $dy/dx = 1/x$.

$$\frac{dy}{dx} = \lim_{h \to 0} \frac{f(x+h) - f(x)}{h}$$

$$= \lim_{h \to 0} \frac{\ln(x+h) - \ln x}{h}$$

$$= \lim_{h \to 0} \frac{\ln\left(\dfrac{x+h}{x}\right)}{h} \qquad \text{Property IV}$$

$$= \lim_{h \to 0} \frac{x}{x} \cdot \frac{1}{h} \ln\left(\frac{x+h}{x}\right) \qquad \text{Introduce } \frac{x}{x}.$$

$$= \lim_{h \to 0} \frac{1}{x} \cdot \frac{x}{h} \ln\left(1 + \frac{h}{x}\right)$$

$$= \lim_{h \to 0} \frac{1}{x} \ln\left(1 + \frac{h}{x}\right)^{x/h} \qquad \text{Property V}$$

$$= \frac{1}{x} \lim_{h \to 0} \left[\ln\left(1 + \frac{h}{x}\right)^{x/h} \right]^{*}$$

$$= \frac{1}{x} \ln \left[\lim_{h \to 0} \left(1 + \frac{h}{x}\right)^{x/h} \right]$$

If we let $a = \dfrac{h}{x}$, then $h \to 0$ means $a \to 0$, and we have

$$\frac{dy}{dx} = \frac{1}{x} \ln \left[\lim_{a \to 0} (1 + a)^{1/a} \right]$$

In Problem 45 in the 9.1 Exercises, we saw that $\lim\limits_{a \to 0} (1 + a)^{1/a} = e$. Hence,

$$\frac{dy}{dx} = \frac{1}{x} \ln e = \frac{1}{x}$$

● **Checkpoint Solutions**

1. $y' = \dfrac{6x}{3x^2 + 2}$

2. $y' = \dfrac{1}{x^6} \cdot 6x^5 = \dfrac{6}{x}$

3. By first using logarithm Property V, we get $y = \ln(x^2 + 1)^{1/3} = \frac{1}{3} \ln(x^2 + 1)$. Then

$$y' = \frac{1}{3} \cdot \frac{2x}{x^2 + 1} = \frac{2x}{3(x^2 + 1)}$$

4. By using logarithm Properties IV and V, we get

$$f(x) = \ln(2x^4) - \ln\left[(5x + 7)^5\right] = \ln(2x^4) - 5\ln(5x + 7)$$

Thus

$$f'(x) = \frac{1}{2x^4} \cdot 8x^3 - 5 \cdot \frac{1}{5x + 7} \cdot 5 = \frac{4}{x} - \frac{25}{5x + 7}$$

---

*The next step uses a new limit property for continuous composite functions. In particular, $\lim\limits_{x \to c} \ln(f(x)) = \ln\left[\lim\limits_{x \to c} f(x)\right]$ when $\lim\limits_{x \to c} f(x)$ exists and is positive.

11.1 Exercises

Find the derivatives of the functions in Problems 1–10.

1. $f(x) = 4 \ln x$
2. $y = 3 \ln x$
3. $y = \ln 8x$
4. $y = \ln 5x$
5. $y = \ln x^4$
6. $f(x) = \ln x^3$
7. $f(x) = \ln (4x + 9)$
8. $y = \ln (6x + 1)$
9. $y = \ln (2x^2 - x) + 3x$
10. $y = \ln (8x^3 - 2x) - 2x$
11. Find dp/dq if $p = \ln (q^2 + 1)$.
12. Find $\dfrac{ds}{dq}$ if $s = \ln \left(\dfrac{q^2}{4} + 1 \right)$.

In each of Problems 13–18, find the derivative of the function in (a). Then find the derivative of the function in (b) or show that the function in (b) is the same function as that in (a).

13. (a) $y = \ln x - \ln (x - 1)$
    (b) $y = \ln \dfrac{x}{x - 1}$
14. (a) $y = \ln (x - 1) + \ln (2x + 1)$
    (b) $y = \ln [(x - 1)(2x + 1)]$
15. (a) $y = \frac{1}{3} \ln (x^2 - 1)$
    (b) $y = \ln \sqrt[3]{x^2 - 1}$
16. (a) $y = 3 \ln (x^4 - 1)$
    (b) $y = \ln (x^4 - 1)^3$
17. (a) $y = \ln (4x - 1) - 3 \ln x$
    (b) $y = \ln \left(\dfrac{4x - 1}{x^3} \right)$
18. (a) $y = 3 \ln x - \ln (x + 1)$
    (b) $y = \ln \left(\dfrac{x^3}{x + 1} \right)$
19. Find $\dfrac{dp}{dq}$ if $p = \ln \left(\dfrac{q^2 - 1}{q} \right)$.
20. Find $\dfrac{ds}{dt}$ if $s = \ln [t^3(t^2 - 1)]$.
21. Find $\dfrac{dy}{dt}$ if $y = \ln \left(\dfrac{t^2 + 3}{\sqrt{1 - t}} \right)$.
22. Find $\dfrac{dy}{dx}$ if $y = \ln \left(\dfrac{3x + 2}{x^2 - 5} \right)^{1/4}$.
23. Find $\dfrac{dy}{dx}$ if $y = \ln (x^3 \sqrt{x + 1})$.
24. Find $\dfrac{dy}{dx}$ if $y = \ln (x^2(x^4 - x + 1))$.

In Problems 25–38, find y'.

25. $y = x - \ln x$
26. $y = x^2 \ln (2x + 3)$
27. $y = \dfrac{\ln x}{x}$
28. $y = \dfrac{1 + \ln x}{x^2}$
29. $y = \ln (x^4 + 3)^2$
30. $y = \ln (3x + 1)^{1/2}$
31. $y = (\ln x)^4$
32. $y = (\ln x)^{-1}$
33. $y = [\ln (x^4 + 3)]^2$
34. $y = \sqrt{\ln (3x + 1)}$
35. $y = \log_4 x$
36. $y = \log_5 x$
37. $y = \log_6 (x^4 - 4x^3 + 1)$
38. $y = \log_2 (1 - x - x^2)$

In Problems 39–42, find the relative maxima and relative minima, and sketch the graph with a graphing utility to check your results.

39. $y = x \ln x$
40. $y = x^2 \ln x$
41. $y = x^2 - 8 \ln x$
42. $y = \ln x - x$

APPLICATIONS

43. *Marginal cost* Suppose that the total cost (in dollars) for a product is given by

    $$C(x) = 1500 + 200 \ln (2x + 1)$$

    where x is the number of units produced.
    (a) Find the marginal cost function.
    (b) Find the marginal cost when 200 units are produced, and interpret your result.
    (c) Total cost functions always increase because producing more items costs more. What then must be true of the marginal cost function? Does it apply in this problem?

44. *Investing* The number of years t that it takes for an investment to double is a function of the interest rate r, compounded continuously, according to

    $$t = \frac{\ln 2}{r}$$

    (a) At what rate $\dfrac{dt}{dr}$ is the required time changing with respect to r if $r = 10\%$, compounded continuously?
    (b) What happens to the rate of change if r is (i) very large or (ii) very small?

45. *Marginal revenue* The total revenue, in dollars, from the sale of x units of a product is given by

    $$R(x) = \frac{2500x}{\ln (10x + 10)}$$

    (a) Find the marginal revenue function.
    (b) Find the marginal revenue when 100 units are sold, and interpret your result.

46. *Supply* Suppose that the supply of x units of a product at price p dollars per unit is given by

    $$p = 10 + 50 \ln (3x + 1)$$

    (a) Find the rate of change of supply price with respect to the number of units supplied.
    (b) Find the rate of change of supply price when the number of units is 33.
    (c) Approximate the price increase associated with the number of units supplied changing from 33 to 34.

47. *Demand*   The demand function for a product is given by $p = 4000/\ln (x + 10)$, where p is the price per unit in dollars when x units are demanded.
    (a) Find the rate of change of price with respect to the number of units sold when 40 units are sold.
    (b) Find the rate of change of price with respect to the number of units sold when 90 units are sold.
    (c) Find the second derivative to see whether the rate at which the price is changing at 40 units is increasing or decreasing.

48. *pH level*   The pH of a solution is given by

    $$pH = -\log [H^+]$$

    where $[H^+]$ is the concentration of hydrogen ions (in gram atoms per liter). What is the rate of change of pH with respect to $[H^+]$?

49. *Drug concentration*   Concentration (in mg/ml) in the bloodstream of a certain drug is related to the time t (in minutes) after an injection and can be calculated using y in the equation

    $$y = A \ln (t) - Bt + C$$

    where A, B, and C are positive constants. In terms of A and B, find t at which y (and hence the drug concentration) reaches its maximum.

50. *Decibels*   The loudness of sound (L, measured in decibels) perceived by the human ear depends on intensity levels (I) according to

    $$L = 10 \log (I/I_0)$$

    where I_0 is the standard threshold of audibility. If $x = I/I_0$, then using the change-of-base formula, we get

    $$L = \frac{10 \ln (x)}{\ln 10}$$

    At what rate is the loudness changing with respect to x when the intensity is 100 times the standard threshold of audibility (that is, when $x = 100$)?

51. *Richter scale*   The Richter scale reading, R, used for measuring the magnitude of an earthquake with intensity I is determined by

    $$R = \frac{\ln (I/I_0)}{\ln 10}$$

    where I_0 is a standard minimum threshold of intensity. If $I_0 = 1$, what is the rate of change of the Richter scale reading with respect to intensity?

52. *Women in the work force*   Between the years 1960 and 2002, the percent of women in the work force can be modeled by

    $$w(x) = 2.552 + 14.569 \ln x$$

where x is the number of years past 1950 (*Source:* U.S. Bureau of Labor Statistics). If this model is accurate beyond 2002, at what rate will the percent be changing in 2010?

53. **Modeling**   *Poverty threshold*   The table below gives the average poverty thresholds for individuals for 1987–2002.
    (a) Use a logarithmic equation to model these data, with x equal to the number of years past 1980.
    (b) Use this model to predict the rate at which the poverty threshold will be growing in 2010.

Year	Poverty Threshold Income	Year	Poverty Threshold Income
1987	$5778	1995	$7763
1988	6022	1996	7995
1989	6310	1997	8183
1990	6652	1998	8316
1991	6932	1999	8501
1992	7143	2000	8794
1993	7363	2001	9039
1994	7547	2002	9182

Source: U.S. Bureau of the Census

54. **Modeling**   *U.S. households with cable TV*   The following data give the percent of U.S. households with cable TV from 1980 to 2002.
    (a) Let $x = 0$ represent 1970 and find a logarithmic function $C(x)$ that models these data.
    (b) Find the function that models the rate of change of the percent of households with cable TV.
    (c) Find and interpret $C(40)$ and $C'(40)$.
    (d) The data show a decline in this percent from 2001 to 2002. What may have caused this decline? Does this mean that the model may not be valid past 2002 and that the results in part (c) may be inaccurate?

Year	Percent	Year	Percent
1980	22.6	1992	61.5
1981	28.3	1993	62.5
1982	35.0	1994	63.4
1983	40.5	1995	65.7
1984	43.7	1996	66.7
1985	46.2	1997	67.3
1986	48.1	1998	67.4
1987	50.5	1999	68.0
1988	53.8	2000	68.0
1989	57.1	2001	69.2
1990	59.0	2002	68.9
1991	60.6		

Source: Nielson Media Research

11.2 Derivatives of Exponential Functions

◗ Application Preview

We saw in Chapter 6, "Mathematics of Finance," that the amount that accrues when $100 is invested at 8%, compounded continuously, is

$$S(t) = 100e^{0.08t}$$

where t is the number of years. If we want to find the rate at which the money in this account is growing at the end of 1 year, then we need to find the derivative of this exponential function. (See Example 5.)

Derivatives of $y = e^x$

Just as base e logarithms are most convenient, we begin by focusing on base e exponentials. Figure 11.6(a) shows the graph of $f(x) = e^x$, and Figure 11.6(b) shows the same graph with tangent lines drawn to several points.

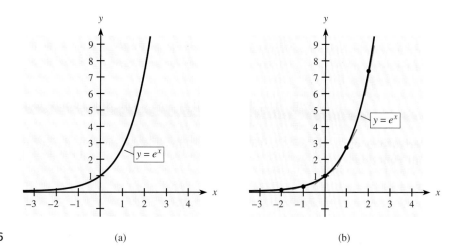

Figure 11.6   (a)   (b)

Note in Figure 11.6(b) that when $x < 0$, tangent lines have slopes near 0, much like the y-coordinates of $f(x) = e^x$. Furthermore, as x increases to $x = 0$ and for $x > 0$, the slopes of the tangents increase just as the values of the function do. This suggests that the function that gives the slope of the tangent to the graph of $f(x) = e^x$ (that is, the derivative) is similar to the function itself. In fact, the derivative of $f(x) = e^x$ is exactly the function itself, which we prove as follows:

From Property I of logarithms, we know that

$$\ln e^x = x$$

Taking the derivative, with respect to x, of both sides of this equation, we have

$$\frac{d}{dx}(\ln e^x) = \frac{d}{dx}(x)$$

Using the Chain Rule for logarithms gives

$$\frac{1}{e^x} \cdot \frac{d}{dx}(e^x) = 1$$

and solving for $\dfrac{d}{dx}(e^x)$ yields

$$\frac{d}{dx}(e^x) = e^x$$

Thus we can conclude the following.

Derivative of $y = e^x$

> If $y = e^x$, then $\dfrac{dy}{dx} = e^x$.

● **EXAMPLE 1** *Derivatives of an Exponential Function*

If $y = 3e^x + 4x - 11$, find $\dfrac{dy}{dx}$.

Solution

$$\frac{dy}{dx} = 3e^x + 4$$

Derivative of $y = e^u$

As with logarithmic functions, the Chain Rule permits us to expand our derivative formulas.

Derivatives of Exponential Functions

> If $y = e^u$, where u is a differentiable function of x, then
>
> $$\frac{dy}{dx} = e^u \cdot \frac{du}{dx}$$

● **EXAMPLE 2** *Derivative of $y = e^u$*

(a) If $f(x) = e^{4x^3}$, find $f'(x)$.
(b) If $s = 3te^{3t^2 + 5t}$, find ds/dt.

Solution
(a) $f'(x) = e^{4x^3} \cdot (12x^2) = 12x^2\, e^{4x^3}$
(b) We begin with the Product Rule.

$$\frac{ds}{dt} = 3t \cdot e^{3t^2 + 5t}(6t + 5) + e^{3t^2 + 5t} \cdot 3$$

$$= (18t^2 + 15t)e^{3t^2 + 5t} + 3e^{3t^2 + 5t}$$

$$= 3e^{3t^2 + 5t}(6t^2 + 5t + 1)$$

● **EXAMPLE 3** *Quotient with an Exponential*

If $u = w/e^{3w}$, find u'.

Solution

The function is a quotient, with the denominator equal to e^{3w}. Using the Quotient Rule gives

$$u' = \frac{(e^{3w})(1) - (w)(e^{3w} \cdot 3)}{(e^{3w})^2}$$

$$= \frac{e^{3w} - 3we^{3w}}{e^{6w}}$$

$$= \frac{1 - 3w}{e^{3w}}$$

● **EXAMPLE 4** *Derivatives and Logarithmic Properties*

If $y = e^{\ln x^2}$, find y'.

Solution

$$y' = e^{\ln x^2} \cdot \frac{1}{x^2} \cdot 2x = \frac{2}{x} e^{\ln x^2}$$

By Property II of logarithms (see the previous section), $e^{\ln u} = u$, and we can simplify the derivative to

$$y' = \frac{2}{x} \cdot x^2 = 2x$$

Note that if we had used this property *before* taking the derivative, we would have had

$$y = e^{\ln x^2} = x^2$$

Then the derivative is $y' = 2x$.

● **Checkpoint**   1. If $y = 2e^{4x}$, find y'.   2. If $y = e^{x^2 + 6x}$, find y'.   3. If $s = te^{t^2}$, find ds/dt.

◗ **EXAMPLE 5** *Future Value* **(Application Preview)**

When $100 is invested at 8% compounded continuously, the amount that accrues after t years, which is called the future value, is $S(t) = 100e^{0.08t}$. At what rate is the money in this account growing

(a) at the end of 1 year?   (b) at the end of 10 years?

Solution

The rate of growth of the money is given by

$$S'(t) = 100e^{0.08t}(0.08) = 8e^{0.08t}$$

(a) The rate of growth of the money at the end of 1 year is

$$S'(1) = 8e^{0.08} \approx 8.666$$

Thus the future value will change by about $8.67 during the next year.

(b) The rate of growth of the money at the end of 10 years is

$$S'(10) = 8e^{0.08(10)} \approx 17.804$$

Thus the future value will change by about $17.80 during the next year.

● **EXAMPLE 6** *Revenue*

North Forty, Inc. is a wilderness camping equipment manufacturer. The revenue function for its best-selling backpack, the Sierra, can be modeled by the function

$$R(x) = 25xe^{(1-0.01x)}$$

where $R(x)$ is the revenue in thousands of dollars from the sale of x thousand Sierra backpacks. Find the marginal revenue when 75,000 packs are sold, and explain what it means.

Solution

The marginal revenue function is given by $R'(x)$, and to find this derivative we use the Product Rule.

$$R'(x) = \overline{MR} = 25x[e^{(1-0.01x)} \cdot (-0.01)] + e^{(1-0.01x)}(25)$$
$$\overline{MR} = 25e^{(1-0.01x)}(1 - 0.01x)$$

To find the marginal revenue when 75,000 packs are sold, we use $x = 75$.

$$\overline{MR}(75) = 25e^{(1-0.75)}(1 - 0.75) \approx 8.025$$

This means that the sale of one (thousand) more Sierra backpacks will yield approximately $8.025 (thousand) in additional revenue.

● **Checkpoint**

4. If the sales of a product are given by $S = 1000e^{-0.2x}$, where x is the number of days after the end of an advertising campaign, what is the rate of decline in sales 20 days after the end of the campaign?

Derivative of $y = a^u$

In a manner similar to that used to find the derivative of $y = e^x$, we can develop a formula for the derivative of $y = a^x$ for any base $a > 0$ and $a \neq 1$.

Derivative of $y = a^u$

If $y = a^x$, with $a > 0, a \neq 1$, then

$$\frac{dy}{dx} = a^x \ln a$$

If $y = a^u$, with $a > 0, a \neq 1$, where u is a differentiable function of x, then

$$\frac{dy}{dx} = a^u \frac{du}{dx} \ln a$$

● **EXAMPLE 7** *Derivatives of $y = a^u$*

(a) If $y = 4^x$, find dy/dx.
(b) If $y = 5^{x^2+x}$, find y'.

Solution

(a) $\dfrac{dy}{dx} = 4^x \ln 4$

(b) $y' = 5^{x^2+x}(2x + 1) \ln 5$

Calculator Note

We can make use of a graphing calculator to study the behavior of an exponential function and its derivative. ∎

● **EXAMPLE 8** *Technology and Critical Values*

For the function $y = e^x - 3x^2$, complete the following.

(a) Approximate the critical values of the function to four decimal places.
(b) Determine whether relative maxima or relative minima occur at the critical values.

Solution

(a) The derivative is $y' = e^x - 6x$. Using the built-in features of a graphing utility, we find that $y' = 0$ at $x \approx 0.2045$ (see Figure 11.7(a)) and at $x \approx 2.8331$.
(b) From the graph of $y' = e^x - 6x$ in Figure 11.7(a), we can observe where $y' > 0$ and where $y' < 0$. From this we can make a sign diagram to determine relative maxima and relative minima.

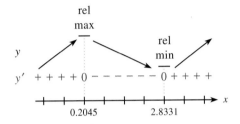

The graph of $y = e^x - 3x^2$ in Figure 11.7(b) shows that the relative maximum point is $(0.2045, 1.1015)$ and that the relative minimum point is $(2.8331, -7.0813)$.

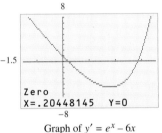

Graph of $y' = e^x - 6x$

(a)

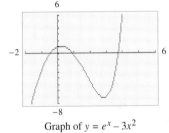

Graph of $y = e^x - 3x^2$

(b)

Figure 11.7

● *Checkpoint Solutions*

1. $y' = 2e^{4x}(4) = 8e^{4x}$
2. $y' = (2x + 6)e^{x^2 + 6x}$
3. By the Product Rule, $\dfrac{ds}{dt} = e^{t^2}(1) + t[e^{t^2}(2t)] = e^{t^2} + 2t^2 e^{t^2}$.
4. The rate of decline is given by dS/dx.

$$\frac{dS}{dx} = 1000e^{-0.2x}(-0.2) = -200e^{-0.2x}$$

$$\frac{dS}{dx}\bigg|_{x=20} = -200e^{(-0.2)(20)}$$

$$= -200e^{-4}$$

$$\approx -3.663 \text{ sales/day}$$

11.2 Exercises

Find the derivatives of the functions in Problems 1–32.

1. $y = 5e^x - x$
2. $y = x^2 - 3e^x$
3. $f(x) = e^x - x^e$
4. $f(x) = 4e^x - \ln x$
5. $y = e^{x^3}$
6. $y = e^{x^2 - 1}$
7. $y = 6e^{3x^2}$
8. $y = 1 - 2e^{-x^3}$
9. $y = 2e^{(x^2 + 1)^3}$
10. $y = e^{\sqrt{x^2 - 9}}$
11. $y = e^{\ln x^3}$
12. $y = e^3 + e^{\ln x}$
13. $y = e^{-1/x}$
14. $y = 2e^{\sqrt{x}}$
15. $y = e^{-1/x^2} + e^{-x^2}$
16. $y = \dfrac{2}{e^{2x}} + \dfrac{e^{2x}}{2}$
17. $s = t^2 e^t$
18. $p = 4qe^{q^3}$
19. $y = e^{x^4} - (e^x)^4$
20. $y = 4(e^x)^3 - 4e^{x^3}$
21. $y = \ln(e^{4x} + 2)$
22. $y = \ln(e^{2x} + 1)$
23. $y = e^{-3x} \ln(2x)$
24. $y = e^{2x^2} \ln(4x)$
25. $y = \dfrac{1 + e^{5x}}{e^{3x}}$
26. $y = \dfrac{x}{1 + e^{2x}}$
27. $y = (e^{3x} + 4)^{10}$
28. $y = \dfrac{e^x - e^{-x}}{e^x + e^{-x}}$
29. $y = 6^x$
30. $y = 3^x$
31. $y = 4^{x^2}$
32. $y = 5^{x-1}$

33. (a) What is the slope of the line tangent to $y = xe^{-x}$ at $x = 1$?
    (b) Write the equation of the line tangent to the graph of $y = xe^{-x}$ at $x = 1$.

34. (a) What is the slope of the line tangent to $y = e^{-x}/(1 + e^{-x})$ at $x = 0$?
    (b) Write the equation of the line tangent to the graph of $y = e^{-x}/(1 + e^{-x})$ at $x = 0$.

 35. The equation for the standard normal probability distribution is

$$y = \frac{1}{\sqrt{2\pi}} e^{-z^2/2}$$

(a) At what value of z will the curve be at its highest point?
(b) Graph this function with a graphing utility to verify your answer.

 36. (a) Find the mode of the normal distribution* given by

$$y = \frac{1}{\sqrt{2\pi}} e^{-(x - 10)^2/2}$$

(b) What is the mean of this normal distribution?
(c) Use a graphing utility to verify your answer.

 In Problems 37–40, find any relative maxima and minima. Use a graphing utility to check your results.

37. $y = \dfrac{e^x}{x}$
38. $y = \dfrac{x}{e^x}$

*The mode occurs at the highest point on normal curves and equals the mean.

39. $y = x - e^x$
40. $y = \dfrac{x^2}{e^x}$

APPLICATIONS

41. *Future value* If \$P is invested for n years at 10% compounded continuously, the future value that results after n years is given by the function

$$S = Pe^{0.1n}$$

(a) At what rate is the future value growing at any time (for any nonnegative n)?
(b) At what rate is the future value growing after 1 year ($n = 1$)?
(c) Is the rate of growth of the future value after 1 year greater than 10%? Why?

42. *Future value* The future value that accrues when \$700 is invested at 9%, compounded continuously, is

$$S(t) = 700e^{0.09t}$$

where t is the number of years.
(a) At what rate is the money in this account growing when $t = 4$?
(b) At what rate is it growing when $t = 10$?

43. *Sales decay* After the end of an advertising campaign, the sales of a product are given by

$$S = 100{,}000e^{-0.5t}$$

where S is weekly sales in dollars and t is the number of weeks since the end of the campaign.
(a) Find the rate of change of S (that is, the rate of *sales decay*).
(b) Give a reason from looking at the function and another reason from looking at the derivative that explain how you know sales are decreasing.

44. *Sales decay* The sales decay for a product is given by

$$S = 50{,}000e^{-0.8t}$$

where S is the daily sales in dollars and t is the number of days since the end of a promotional campaign. Find the rate of sales decay.

45. *Marginal cost* Suppose that the total cost in dollars of producing x units of a product is given by

$$C(x) = 10{,}000 + 20xe^{x/600}$$

Find the marginal cost when 600 units are produced.

46. *Marginal revenue*   Suppose that the revenue in dollars from the sale of x units of a product is given by

$$R(x) = 1000xe^{-x/50}$$

Find the marginal revenue function.

47. *Drugs in a bloodstream*   The percent concentration y of a certain drug in the bloodstream at any time t (in hours) is given by

$$y = 100(1 - e^{-0.462t})$$

(a) What function gives the instantaneous rate of change of the concentration of the drug in the bloodstream?
(b) Find the rate of change of the concentration after 1 hour. Give your answer to three decimal places.

48. *Radioactive decay*   The amount of the radioactive isotope thorium-234 present at time t in years is given by

$$Q(t) = 100e^{-0.02828t}$$

(a) What function describes how rapidly the isotope is decaying?
(b) Find the rate of radioactive decay of the isotope.

49. *Spread of disease*   Suppose that the spread of a disease through the student body at an isolated college campus can be modeled by

$$y = \frac{10,000}{1 + 9999e^{-0.99t}}$$

where y is the total number affected at time t (in days). Find the rate of change of y.

50. *Spread of a rumor*   The number of people $N(t)$ in a community who are reached by a particular rumor at time t (in days) is given by

$$N(t) = \frac{50,500}{1 + 100e^{-0.7t}}$$

Find the rate of change of $N(t)$.

51. *Pollution*   Pollution levels in Lake Erie have been modeled by the equation

$$x = 0.05 + 0.18e^{-0.38t}$$

where x is the volume of pollutants (in cubic kilometers) and t is the time (in years). (Adapted from R. H. Rainey, *Science* 155 (1967), 1242–1243.) What is the rate of change of x with respect to time?

52. *Drug concentration*   Suppose the concentration $C(t)$, in mg/ml, of a drug in the bloodstream t minutes after an injection is given by

$$C(t) = 20te^{-0.04t}$$

(a) Find the instantaneous rate of change of the concentration after 10 minutes.
(b) Find the maximum concentration and when it occurs.

53. *World population*   Using figures from the U.S. Bureau of the Census, International Data Base, world population can be considered to be growing according to

$$P(t) = \frac{10.94}{1 + 3.97e^{-0.029t}}$$

where $t = 0$ represents 1945 and $P(t)$ is in billions.
(a) Find the rate of change of $P(t)$.
(b) Find and interpret $P'(100)$.
(c) Use the numerical derivative feature of a graphing calculator to find whether the rate of change of population growth is predicted to be increasing or decreasing in 2040.

54. *Blood pressure*   Medical research has shown that between heartbeats, the pressure in the aorta of a normal adult is a function of time in seconds and can be modeled by the equation

$$P = 95e^{-0.491t}$$

(a) Use the derivative to find the rate at which the pressure is changing at any time t.
(b) Use the derivative to find the rate at which the pressure is changing after 0.1 second.
(c) Is the pressure increasing or decreasing?

55. *Richter scale*   The intensity of an earthquake is related to the Richter scale reading R by

$$\frac{I}{I_0} = 10^R$$

where I_0 is a standard minimum intensity. If $I_0 = 1$, what is the rate of change of the intensity I with respect to the Richter scale reading?

56. *Decibel readings*   The intensity level of sound, I, is given by

$$\frac{I}{I_0} = 10^{L/10}$$

where L is the decibel reading and I_0 is the standard threshold of audibility. If $I/I_0 = y$, at what rate is y changing with respect to L when $L = 20$?

57. *National health care*   Using data from the U.S. Department of Health and Human Services, the national health expenditure H can be modeled by

$$H = 29.57e^{0.097t}$$

where t is the number of years past 1960 and H is in billions of dollars. If this model is accurate, at what rate will health care expenditures change in 2008?

58. *Wireless subscribers*   By using data from the C.T.I.A. Semi-Annual Wireless Survey, the number of wireless subscribers (in millions) can be modeled by

$$N = 12.58e^{0.2096t}$$

where t is the number of years past 1990. If this model is accurate, find and interpret the rate of change of the number of wireless subscribers in 2010.

59. *U.S. debt* For selected years from 1900 to 2003, the national debt d, in billions of dollars, can be modeled by

$$d = 1.68e^{0.0828t}$$

where t is the number of years past 1900 (*Source: Bureau of Public Debt, U.S. Treasury*).
(a) What function describes how fast the national debt is changing?
(b) Find the instantaneous rate of change of the national debt model $d(t)$ in 1940 and 2000.
(c) The Gramm-Rudman Balanced Budget Act (passed in 1985) was designed to limit the growth of the national debt and hence change the function that modeled it. For the years 1999, 2000, and 2001, the debt (in billions of dollars) was 5656.3, 5674.2, and 5807.5, respectively. Find the average annual rates of change for 1999 to 2000 and 2000 to 2001. When these average rates are compared with the instantaneous rate for 2000 found from the model in part (b), does it seem that the model may need to be changed? Explain.
(d) What event disrupted the government's effort to balance the budget?

60. *Mutual funds* For selected years from 1978 to 2000, the number of mutual funds N, excluding money market funds, can be modeled by

$$N = 276.1e^{0.1351t}$$

where t is the number of years past 1975 (*Source: Investment Company Institute, Mutual Fund Fact Book, 2001*).
(a) What function describes how rapidly the number of mutual funds is changing?
(b) Find and interpret the rate of change of the number of mutual funds in 2000.

61. *Purchasing power* The following table gives the purchasing power of $1 based on consumer prices for 1970–2002. Using these data with $x = 0$ representing 1960, the dollar's purchasing power, P, can be modeled by

$$P = 3.655e^{-0.049x}$$

(a) Use the model to find the rate of decay of the purchasing power of $1 in 2002.
(b) Interpret the result in part (a).
(c) Use the data to find the average rate of change from 2001 to 2002, which approximates the rate of change in 2002.

Year	Purchasing Power of $1	Year	Purchasing Power of $1	Year	Purchasing Power of $1
1970	2.574	1981	1.098	1992	0.713
1971	2.466	1982	1.035	1993	0.692
1972	2.391	1983	1.003	1994	0.675
1973	2.251	1984	0.961	1995	0.656
1974	2.029	1985	0.928	1996	0.637
1975	1.859	1986	0.913	1997	0.623
1976	1.757	1987	0.880	1998	0.613
1977	1.649	1988	0.846	1999	0.600
1978	1.532	1989	0.807	2000	0.581
1979	1.380	1990	0.766	2001	0.565
1980	1.215	1991	0.734	2002	0.556

Source: U.S. Bureau of Labor Statistics

62. *Personal income* Total personal income in the U.S. (in billions of dollars) for selected years from 1960 to 2002 is given in the following table.

Year	1960	1970	1980	1990	2000	2002
Personal income	411.7	836.1	2285.7	4791.6	8406.6	8929.1

Source: Bureau of Economic Analysis, U.S. Department of Commerce

These data can be modeled by

$$I = 434.24e^{0.075x}$$

where x is the number of years past 1960.
(a) If this model is accurate, find the rate of change of the total U.S. personal income in 2002.
(b) Use the data to find the average rate of change from 2000 to 2002, which can be used to approximate the rate of change in 2002.
(c) Interpret your answers from parts (a) and (b).

63. **Modeling** *Federal tax per capita* The following table shows the U.S. federal tax per capita for selected years from 1960 to 2002.
(a) Use $x = 0$ to represent 1950 and model these data with an exponential function.
(b) If this model remains accurate, find and interpret the rate of change of U.S. federal tax per capita in 2010.

Year	Tax per Capita	Year	Tax per Capita
1960	$507.97	1990	$4222.50
1965	$588.95	1995	$5216.70
1970	$955.31	1999	$6957.79
1975	$1375.84	2000	$7595.24
1980	$2275.66	2001	$7448.90
1985	$3098.97	2002	$6967.40

Source: Internal Revenue Service, 2002 *Data Book* (2003)

64. **Modeling** *Consumer price index* The consumer price index (CPI) is calculated by finding the total price of various items that have been averaged according to a prescribed formula. The following table gives the consumer price indexes of all urban consumers (CPI-U) for selected years from 1940 to 2002.
    (a) With x representing years past 1900, find an exponential equation that models these data.
    (b) Use your model to predict the rate of growth in this price index in 2010.
    (c) For these data, name and find the model for another type of function that involves exponential functions.
    (d) With the model from (c), use the numerical derivative feature of your calculator to find the rate of change in 2010.

Year	Consumer Price Index	Year	Consumer Price Index
1940	14.0	1980	82.4
1950	24.1	1990	130.7
1960	29.6	2000	172.2
1970	38.8	2002	179.9

Source: U.S. Bureau of the Census

65. **Modeling** *Prison population* The following table shows the prison population in the United States for selected years from 1980 to 2002.
    (a) Find a logistic model for these data. Use $x = 0$ to represent 1980.
    (b) If the function from part (a) is $P(x)$, find and interpret $P(20)$ and $P'(20)$.

Year	Population
1980	329,821
1985	502,507
1990	773,919
1995	1,125,874
1998	1,300,573
2000	1,381,892
2002	1,440,655

Source: Bureau of Justice Statistics Bulletin, *Prisoners in 2002* (2003).

11.3

Implicit Differentiation

OBJECTIVES

- To find derivatives by using implicit differentiation
- To find slopes of tangents by using implicit differentiation

◗ Application Preview

In the retail electronics industry, suppose the monthly demand for Precision, Inc. stereo headphones is given by

$$p = \frac{10,000}{(x + 1)^2}$$

where p is the price in dollars per set of headphones and x is demand in hundreds of sets of headphones. If we want to find the rate of change of the quantity demanded with respect to the price, then we need to find dx/dp. (See Example 9.) Although we can solve this equation for x so that dx/dp can be found, the resulting equation does not define x as a function of p. In this case, and in other cases where we cannot solve equations for the variables we need, we can find derivatives with a technique called **implicit differentiation.**

Up to this point, functions involving x and y have been written in the form $y = f(x)$, defining y as an *explicit function* of x. Taking the derivative of both sides of the equation $y = f(x)$ gives a derivative of the form

$$\frac{dy}{dx} = f'(x)$$

For example, $y - x^2 = 0$ could be written as $y = x^2$, and then the derivative is $\frac{dy}{dx} = 2x$.

However, not all equations involving x and y can be written in the form $y = f(x)$, and we need a new technique for taking their derivatives. For example, solving $y^2 = x$ for y gives

$y = \pm\sqrt{x}$ so that y is not a function of x. We can write $y = \sqrt{x}$ and $y = -\sqrt{x}$, but then finding the derivative $\dfrac{dy}{dx}$ at a point on the graph of $y^2 = x$ would require determining which of these functions applies before taking the derivative. Alternatively, we can *imply* that y is a function of x and use a technique called **implicit differentiation.** When y is an implied function of x, we find $\dfrac{dy}{dx}$ by differentiating both sides of the equation with respect to x and then algebraically solving for $\dfrac{dy}{dx}$.

● **EXAMPLE 1 *Implicit Differentiation***

(a) Use implicit differentiation to find $\dfrac{dy}{dx}$ for $y^2 = x$.

(b) Find the slopes of the tangents to the graph of $y^2 = x$ at the points $(4, 2)$ and $(4, -2)$.

Solution

(a) First take the derivative of both sides of the equation with respect to x.

$$\frac{d}{dx}(y^2) = \frac{d}{dx}(x)$$

Because y is an implied, or implicit, function of x, we can think of $y^2 = x$ as meaning $[u(x)]^2 = x$ for some function u. We use the Chain Rule to take the derivative of y^2 in the same way we would for $[u(x)]^2$. Thus

$$\frac{d}{dx}(y^2) = \frac{d}{dx}(x) \quad \text{gives} \quad 2y\frac{dy}{dx} = 1$$

Solving for $\dfrac{dy}{dx}$ gives $\dfrac{dy}{dx} = \dfrac{1}{2y}$.

(b) To find the slopes of the tangents at the points $(4, 2)$ and $(4, -2)$, we use the coordinates of the points to evaluate $\dfrac{dy}{dx}$ at those points.

$$\frac{dy}{dx}\bigg|_{(4,\,2)} = \frac{1}{2(2)} = \frac{1}{4} \qquad \text{and} \qquad \frac{dy}{dx}\bigg|_{(4,\,-2)} = \frac{1}{2(-2)} = \frac{-1}{4}$$

Figure 11.8 shows the graph of $y^2 = x$ with tangent lines drawn at $(4, 2)$ and $(4, -2)$.

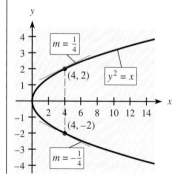

Figure 11.8

As noted previously, solving $y^2 = x$ for y gives the two functions $y = \sqrt{x} = x^{1/2}$ and $y = -\sqrt{x} = -x^{1/2}$; see Figure 11.9(a) and (b).

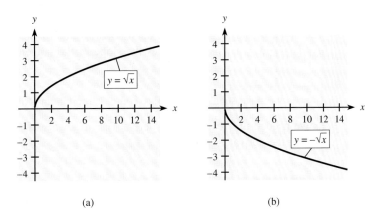

Figure 11.9   (a)   (b)

Let us now compare the results obtained in Example 1 with the results from the derivatives of the two functions $y = \sqrt{x}$ and $y = -\sqrt{x}$. The derivatives of these two functions are as follows:

$$\text{For } y = x^{1/2}, \text{ then } \frac{dy}{dx} = \frac{1}{2}x^{-1/2} = \frac{1}{2\sqrt{x}}$$

$$\text{For } y = -x^{1/2}, \text{ then } \frac{dy}{dx} = -\frac{1}{2}x^{-1/2} = \frac{-1}{2\sqrt{x}}$$

We cannot find the slope of a tangent line at $x = 4$ unless we also know the y-coordinate and hence which function and which derivative to use.

$$\text{At } (4, 2) \text{ use } y = \sqrt{x} \quad \text{so} \quad \frac{dy}{dx}\bigg|_{(4, 2)} = \frac{1}{2\sqrt{4}} = \frac{1}{4}$$

$$\text{At } (4, -2) \text{ use } y = -\sqrt{x} \quad \text{so} \quad \frac{dy}{dx}\bigg|_{(4, -2)} = \frac{-1}{2\sqrt{4}} = \frac{-1}{4}$$

These are the same results that we obtained directly after implicit differentiation. In this example, we could easily solve for y in terms of x and work the problem two different ways. However, sometimes solving explicitly for y is difficult or impossible. For these functions, implicit differentiation is necessary and extends our ability to find derivatives.

● **EXAMPLE 2** *Slope of a Tangent*

Find the slope of the tangent to the graph of $x^2 + y^2 - 9 = 0$ at $(\sqrt{5}, 2)$.

Solution
We find the derivative dy/dx from $x^2 + y^2 - 9 = 0$ by taking the derivative term by term on both sides of the equation.

$$\frac{d}{dx}(x^2) + \frac{d}{dx}(y^2) + \frac{d}{dx}(-9) = \frac{d}{dx}(0)$$

$$2x + 2y \cdot \frac{dy}{dx} + 0 = 0$$

Solving for dy/dx gives

$$\frac{dy}{dx} = -\frac{2x}{2y} = -\frac{x}{y}$$

The slope of the tangent to the curve is the value of the derivative at $(\sqrt{5}, 2)$. Evaluating $dy/dx = -x/y$ at this point gives the slope of the tangent as $-\sqrt{5}/2$.

As we have seen in Examples 1 and 2, the Chain Rule yields $\frac{d}{dx}(y^n) = ny^{n-1}\frac{dy}{dx}$. Just as we needed this result to find derivatives of powers of y, the next example discusses how to find the derivative of xy.

● **EXAMPLE 3** *Equation of a Tangent Line*

Write the equation of the tangent to the graph of $x^3 + xy + 4 = 0$ at the point $(2, -6)$.

Solution
Note that the second term is the *product* of x and y. Because we are assuming that y is a function of x, and because x is a function of x, we must use the Product Rule to find $\frac{d}{dx}(xy)$.

$$\frac{d}{dx}(xy) = x \cdot 1\frac{dy}{dx} + y \cdot 1 = x\frac{dy}{dx} + y$$

Thus we have

$$\frac{d}{dx}(x^3) + \frac{d}{dx}(xy) + \frac{d}{dx}(4) = \frac{d}{dx}(0)$$

$$3x^2 + \left(x\frac{dy}{dx} + y\right) + 0 = 0$$

Solving for dy/dx gives

$$\frac{dy}{dx} = \frac{-3x^2 - y}{x}$$

The slope of the tangent to the curve at $x = 2, y = -6$ is

$$m = \frac{-3(2)^2 - (-6)}{2} = -3$$

The equation of the tangent line is

$$y - (-6) = -3[x - (2)], \quad \text{or} \quad y = -3x$$

A graphing utility can be used to graph the function of Example 3 and the line that is tangent to the curve at $(2, -6)$. To graph the equation, we solve the equation for y, getting

$$y = \frac{-x^3 - 4}{x}$$

The graph of the equation and the line that is tangent to the curve at $(2, -6)$ are shown in Figure 11.10.

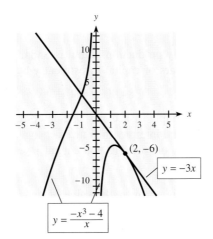

Figure 11.10

● **EXAMPLE 4** *Implicit Differentiation*

Find dy/dx if $x^4 + 5xy^4 = 2y^2 + x - 1$.

Solution
By viewing $5xy^4$ as the product $(5x)(y^4)$ and differentiating implicitly, we get

$$\frac{d}{dx}(x^4) + \frac{d}{dx}(5xy^4) = \frac{d}{dx}(2y^2) + \frac{d}{dx}(x) - \frac{d}{dx}(1)$$

$$4x^3 + 5x\left(4y^3\frac{dy}{dx}\right) + y^4(5) = 4y\frac{dy}{dx} + 1$$

To solve for $\dfrac{dy}{dx}$, we first rewrite the equation with terms containing $\dfrac{dy}{dx}$ on one side and the other terms on the other side.

$$20xy^3\frac{dy}{dx} - 4y\frac{dy}{dx} = 1 - 4x^3 - 5y^4$$

Now $\dfrac{dy}{dx}$ is a factor of one side. To complete the solution, we factor out $\dfrac{dy}{dx}$ and divide both sides by its coefficient.

$$(20xy^3 - 4y)\frac{dy}{dx} = 1 - 4x^3 - 5y^4$$

$$\frac{dy}{dx} = \frac{1 - 4x^3 - 5y^4}{20xy^3 - 4y}$$

● **Checkpoint**

1. Find the following:

   (a) $\dfrac{d}{dx}(x^3)$ (b) $\dfrac{d}{dx}(y^4)$ (c) $\dfrac{d}{dx}(x^2y^5)$

2. Find $\dfrac{dy}{dx}$ for $x^3 + y^4 = x^2y^5$.

● **EXAMPLE 5** *Production*

Suppose a company's weekly production output is $384,000 and that this output is related to hours of labor x and dollars of capital investment y by

$$384{,}000 = 30x^{1/3}\,y^{2/3}$$

(This relationship is an example of a Cobb-Douglas production function, studied in more detail in Chapter 14.) Find and interpret the rate of change of capital investment with respect to labor hours when labor hours are 512 and capital investment is $64,000.

Solution

The desired rate of change is given by the value of dy/dx when $x = 512$ and $y = 64{,}000$. Taking the derivative implicitly gives

$$\frac{d}{dx}(384{,}000) = \frac{d}{dx}(30x^{1/3}\,y^{2/3})$$

$$0 = 30x^{1/3}\left(\frac{2}{3}y^{-1/3}\frac{dy}{dx}\right) + y^{2/3}(10x^{-2/3})$$

$$0 = \frac{20x^{1/3}}{y^{1/3}}\cdot\frac{dy}{dx} + \frac{10y^{2/3}}{x^{2/3}}$$

$$\frac{-20x^{1/3}}{y^{1/3}}\cdot\frac{dy}{dx} = \frac{10y^{2/3}}{x^{2/3}}$$

Multiplying both sides by $\dfrac{-y^{1/3}}{20x^{1/3}}$ gives

$$\frac{dy}{dx} = \left(\frac{10y^{2/3}}{x^{2/3}}\right)\left(\frac{-y^{1/3}}{20x^{1/3}}\right) = \frac{-y}{2x}$$

When $x = 512$ and $y = 64{,}000$, we obtain

$$\frac{dy}{dx} = \frac{-64{,}000}{2(512)} = -62.5$$

This means that when labor hours are 512 and capital investment is $64,000, if labor hours change by 1 hour, then capital investment could decrease by about $62.50.

● **EXAMPLE 6** *Horizontal and Vertical Tangents*

At what point(s) does $x^2 + 4y^2 - 2x + 4y - 2 = 0$ have a horizontal tangent? At what point(s) does it have a vertical tangent?

Solution

First we find the derivative implicitly, with y' representing $\dfrac{dy}{dx}$.

$$2x + 8y \cdot y' - 2 + 4y' - 0 = 0$$

We isolate the y' terms, factor out y', and solve for y'.

$$8yy' + 4y' = 2 - 2x$$

$$(8y + 4)y' = 2 - 2x$$

$$y' = \frac{2 - 2x}{8y + 4} = \frac{1 - x}{4y + 2}$$

Horizontal tangents will occur where $y' = 0$; that is, where $1 - x = 0$, or $x = 1$. We can now find the corresponding y-value(s) by substituting 1 for x in the original equation and solving.

$$1 + 4y^2 - 2 + 4y - 2 = 0$$
$$4y^2 + 4y - 3 = 0$$
$$(2y - 1)(2y + 3) = 0$$
$$y = \frac{1}{2} \quad \text{or} \quad y = -\frac{3}{2}$$

Thus horizontal tangents occur at $\left(1, \frac{1}{2}\right)$ and $\left(1, -\frac{3}{2}\right)$; see Figure 11.11.

Figure 11.11

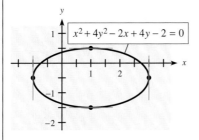

Vertical tangents will occur where the derivative is undefined; that is, where $4y + 2 = 0$, or $y = -\frac{1}{2}$. To find the corresponding x-value(s), we substitute $-\frac{1}{2}$ in the equation for y and solve for x.

$$x^2 + 4\left(-\frac{1}{2}\right)^2 - 2x + 4\left(-\frac{1}{2}\right) - 2 = 0$$
$$x^2 - 2x - 3 = 0$$
$$(x - 3)(x + 1) = 0$$
$$x = 3 \quad \text{or} \quad x = -1$$

Thus vertical tangents occur at $\left(3, -\frac{1}{2}\right)$ and $\left(-1, -\frac{1}{2}\right)$; see Figure 11.11.

● **EXAMPLE 7** *Implicit Derivatives and Logarithms*

If $\ln xy = 6$, find dy/dx.

Solution
Using the properties of logarithms, we have

$$\ln x + \ln y = 6$$

which leads to the implicit derivative

$$\frac{1}{x} + \frac{1}{y}\frac{dy}{dx} = 0$$

Solving gives

$$\frac{1}{y}\frac{dy}{dx} = -\frac{1}{x}$$
$$\frac{dy}{dx} = -\frac{y}{x}$$

● **EXAMPLE 8** *Implicit Derivatives and Exponentials*

Find dy/dx if $4x^2 + e^{xy} = 6y$.

Solution

We take the derivative of both sides and obtain

$$8x + e^{xy} \cdot \frac{d}{dx}(xy) = 6\frac{dy}{dx}$$

$$8x + e^{xy}\left(x\frac{dy}{dx} + y\right) = 6\frac{dy}{dx}$$

$$8x + xe^{xy}\frac{dy}{dx} + ye^{xy} = 6\frac{dy}{dx}$$

$$8x + ye^{xy} = 6\frac{dy}{dx} - xe^{xy}\frac{dy}{dx}$$

$$8x + ye^{xy} = (6 - xe^{xy})\frac{dy}{dx}$$

$$\frac{8x + ye^{xy}}{6 - xe^{xy}} = \frac{dy}{dx}$$

◗ **EXAMPLE 9** *Demand* **(Application Preview)**

Suppose the demand for Precision, Inc. stereo headphones is given by

$$p = \frac{10,000}{(x + 1)^2}$$

where p is the price per set in dollars and x is in hundreds of headphone sets demanded. Find the rate of change of demand with respect to price when 19 (hundred) sets are demanded.

Solution

The rate of change of demand with respect to price is dx/dp. Using implicit differentiation, we get the following.

$$\frac{d}{dp}(p) = \frac{d}{dp}\left[\frac{10,000}{(x + 1)^2}\right] = \frac{d}{dp}\left[10,000(x + 1)^{-2}\right]$$

$$1 = 10,000\left[-2(x + 1)^{-3}\frac{dx}{dp}\right]$$

$$1 = \frac{-20,000}{(x + 1)^3} \cdot \frac{dx}{dp}$$

$$\frac{(x + 1)^3}{-20,000} = \frac{dx}{dp}$$

When 19 (hundred) headphone sets are demanded we use $x = 19$, and the rate of change of demand with respect to price is

$$\left.\frac{dx}{dp}\right|_{x=19} = \frac{(19 + 1)^3}{-20,000} = \frac{8000}{-20,000} = -0.4$$

This result means that when 19 (hundred) headphone sets are demanded, if the price per set is increased by \$1, then the expected change in demand is a decrease of 0.4 hundred, or 40, headphone sets.

Calculator Note To graph a function with a graphing calculator, we need to write y as an *explicit* function of x (such as $y = \sqrt{4 - x^2}$). If an equation defines y as an *implicit* function of x, we have to solve for y in terms of x before we can use the graphing calculator. Sometimes we cannot solve for y, and other times, such as for

$$x^{2/3} + y^{2/3} = 8^{2/3}$$

y cannot be written as a single function of x. If this equation is solved for y and a graphing calculator is used to graph that function, the resulting graph usually shows only the portion of the graph that lies in Quadrants I and II, and sometimes only the part in Quadrant I. The complete graph is shown in Figure 11.12. Thus, for the graph of an implicitly defined function, a graphing calculator must be used carefully (and sometimes cannot be used at all).

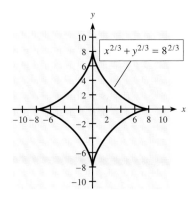

Figure 11.12

● ***Checkpoint Solutions***

1. (a) $\dfrac{d}{dx}(x^3) = 3x^2$ (b) $\dfrac{d}{dx}(y^4) = 4y^3\dfrac{dy}{dx}$

   (c) $\dfrac{d}{dx}(x^2y^5) = x^2\left(5y^4\dfrac{dy}{dx}\right) + y^5(2x)$ (by the Product Rule)

2. For $x^3 + y^4 = x^2y^5$, we can use the answers to Question 1 to obtain

$$3x^2 + 4y^3\frac{dy}{dx} = x^2\left(5y^4\frac{dy}{dx}\right) + y^5(2x)$$

$$3x^2 + 4y^3\frac{dy}{dx} = 5x^2y^4\frac{dy}{dx} + 2xy^5$$

$$3x^2 - 2xy^5 = (5x^2y^4 - 4y^3)\frac{dy}{dx}$$

$$\frac{3x^2 - 2xy^5}{5x^2y^4 - 4y^3} = \frac{dy}{dx}$$

11.3 Exercises

In Problems 1–6, find dy/dx at the given point without first solving for y.

1. $x^2 - 4y - 17 = 0$ at $(1, -4)$
2. $3x^2 - 10y + 400 = 0$ at $(10, 70)$
3. $xy^2 = 8$ at $(2, 2)$
4. $e^y = x$ at $(1, 0)$
5. $x^2 + 3xy - 4 = 0$ at $(1, 1)$
6. $x^2 + 5xy + 4 = 0$ at $(1, -1)$

Find *dy/dx* for the functions in Problems 7–10.

7. $x^2 + 2y^2 - 4 = 0$
8. $x + y^2 - 4y + 6 = 0$
9. $x^2 + 4x + y^2 - 3y + 1 = 0$
10. $x^2 - 5x + y^3 - 3y - 3 = 0$
11. If $x^2 + y^2 = 4$, find y'.
12. If $p^2 + 4p - q = 4$, find dp/dq.
13. If $xy^2 - y^2 = 1$, find y'.
14. If $p^2 - q = 4$, find dp/dq.
15. If $p^2q = 4p - 2$, find dp/dq.
16. If $x^2 - 3y^4 = 2x^5 + 7y^3 - 5$, find dy/dx.
17. If $3x^5 - 5y^3 = 5x^2 + 3y^5$, find dy/dx.
18. If $x^2 + 3x^2y^4 = y + 8$, find dy/dx.
19. If $x^4 + 2x^3y^2 = x - y^3$, find dy/dx.
20. If $(x + y)^2 = 5x^4y^3$, find dy/dx.
21. Find dy/dx for $x^4 + 3x^3y^2 - 2y^5 = (2x + 3y)^2$.
22. Find y' for $2x + 2y = \sqrt{x^2 + y^2}$.

For Problems 23–26, find the slope of the tangent to the curve.

23. $x^2 + 4x + y^2 + 2y - 4 = 0$ at $(1, -1)$
24. $x^2 - 4x + 2y^2 - 4 = 0$ at $(2, 2)$
25. $x^2 + 2xy + 3 = 0$ at $(-1, 2)$
26. $y + x^2 = 4$ at $(0, 4)$
27. Write the equation of the line tangent to the curve $x^2 - 2y^2 + 4 = 0$ at $(2, 2)$.
28. Write the equation of the line tangent to the curve $x^2 + y^2 + 2x - 3 = 0$ at $(-1, 2)$.
29. Write the equation of the line tangent to the curve $4x^2 + 3y^2 - 4y - 3 = 0$ at $(-1, 1)$.
30. Write the equation of the line tangent to the curve $xy + y^2 = 0$ at $(3, 0)$.
31. If $\ln x = y^2$, find dy/dx.
32. If $\ln (x + y) = y^2$, find dy/dx.
33. If $y^2 \ln x = 4$, find dy/dx.
34. If $\ln xy = 2$, find dy/dx.
35. Find the slope of the tangent to the curve $x^2 + \ln y = 4$ at the point $(2, 1)$.
36. Write the equation of the line tangent to the curve $x \ln y + 2xy = 2$ at the point $(1, 1)$.
37. If $xe^y = 6$, find dy/dx.
38. If $x + e^{xy} = 10$, find dy/dx.
39. If $e^{xy} = 4$, find dy/dx.
40. If $x - xe^y = 3$, find dy/dx.
41. If $ye^x - y = 3$, find dy/dx.
42. If $x^2y = e^{x+y}$, find dy/dx.
43. Find the slope of the line tangent to the graph of $ye^x = y^2 + x - 2$ at $(0, 2)$.
44. Find the slope of the line tangent to the curve $xe^y = 3x^2 + y - 24$ at $(3, 0)$.
45. Write the equation of the line tangent to the curve $xe^y = 2y + 3$ at $(3, 0)$.
46. Write the equation of the line tangent to the curve $ye^x = 2y + 1$ at $(0, -1)$.

47. At what points does the curve defined by $x^2 + 4y^2 - 4x - 4 = 0$ have
    (a) horizontal tangents?
    (b) vertical tangents?
48. At what points does the curve defined by $x^2 + 4y^2 - 4 = 0$ have
    (a) horizontal tangents?
    (b) vertical tangents?
49. In Problem 11, the derivative y' was found to be

$$y' = \frac{-x}{y}$$

when $x^2 + y^2 = 4$.
    (a) Take the implicit derivative of the equation for y' to show that

$$y'' = \frac{-y + xy'}{y^2}$$

    (b) Substitute $-x/y$ for y' in the expression for y'' in part (a) and simplify to show that

$$y'' = -\frac{(x^2 + y^2)}{y^3}$$

    (c) Does $y'' = -4/y^3$? Why or why not?
50. (a) Find y' implicitly for $x^3 - y^3 = 8$.
    (b) Then, by taking derivatives implicitly, use part (a) to show that

$$y'' = \frac{2x(y - xy')}{y^3}$$

    (c) Substitute x^2/y^2 for y' in the expression for y'' and simplify to show that

$$y'' = \frac{2x(y^3 - x^3)}{y^5}$$

    (d) Does $y'' = -16x/y^5$? Why or why not?
51. Find y'' for $\sqrt{x} + \sqrt{y} = 1$ and simplify.
52. Find y'' for $\dfrac{1}{x} - \dfrac{1}{y} = 1$.

⬚ **In Problems 53 and 54, find the maximum and minimum values of *y*. Use a graphing utility to verify your conclusion.**

53. $x^2 + y^2 - 9 = 0$
54. $4x^2 + y^2 - 8x = 0$

APPLICATIONS

55. *Advertising and sales*   Suppose that a company's sales volume y (in thousands of units) is related to its advertising expenditures x (in thousands of dollars) according to

$$xy - 20x + 10y = 0$$

Find the rate of change of sales volume with respect to advertising expenditures when $x = 10$ (thousand dollars).

56. *Insect control*   Suppose that the number of mosquitoes N (in thousands) in a certain swampy area near a community is related to the number of pounds of insecticide x sprayed on the nesting areas according to

$$Nx - 10x + N = 300$$

Find the rate of change of N with respect to x when 49 pounds of insecticide is used.

57. *Production*   Suppose that a company can produce 12,000 units when the number of hours of skilled labor y and unskilled labor x satisfy

$$384 = (x + 1)^{3/4}(y + 2)^{1/3}$$

Find the rate of change of skilled-labor hours with respect to unskilled-labor hours when $x = 255$ and $y = 214$. This can be used to approximate the change in skilled-labor hours required to maintain the same production level when unskilled-labor hours are increased by 1 hour.

58. *Production*   Suppose that production of 10,000 units of a certain agricultural crop is related to the number of hours of labor x and the number of acres of the crop y according to

$$300x + 30,000y = 11xy - 0.0002x^2 - 5y$$

Find the rate of change of the number of hours with respect to the number of acres.

59. *Demand*   If the demand function for q units of a product at $\$p$ per unit is given by

$$p(q + 1)^2 = 200,000$$

find the rate of change of quantity with respect to price when $p = \$80$. Interpret this result.

60. *Demand*   If the demand function for q units of a commodity at $\$p$ per unit is given by

$$p^2(2q + 1) = 100,000$$

find the rate of change of quantity with respect to price when $p = \$50$. Interpret this result.

61. *Radioactive decay*   The number of grams of radium, y, that will remain after t years if 100 grams existed originally can be found by using the equation

$$-0.000436t = \ln\left(\frac{y}{100}\right)$$

Use implicit differentiation to find the rate of change of y with respect to t—that is, the rate at which the radium will decay.

62. *Disease control*   Suppose the proportion P of people affected by a certain disease is described by

$$\ln\left(\frac{P}{1 - P}\right) = 0.5t$$

where t is the time in months. Find dP/dt, the rate at which P grows.

63. *Temperature-humidity index*   The temperature-humidity index (THI) is given by

$$\text{THI} = t - 0.55(1 - h)(t - 58)$$

where t is the air temperature in degrees Fahrenheit and h is the relative humidity. If the THI remains constant, find the rate of change of humidity with respect to temperature if the temperature is 70°F (*Source:* "Temperature-Humidity Indices," *UMAP Journal*, Fall 1989).

11.4   Related Rates

OBJECTIVE

● To use implicit differentiation to solve problems that involve related rates

◗ Application Preview

According to Poiseuille's law, the flow of blood F is related to the radius r of the vessel according to

$$F = kr^4$$

where k is a constant. When the radius of a blood vessel is reduced, such as by cholesterol deposits, the flow of blood is also restricted. Drugs can be administered that increase the radius of the blood vessel and, hence, the flow of blood. The rate of change of the blood flow and the rate of change of the radius of the blood vessel are time rates of change that are related to each other, so they are called **related rates**. We can use these related rates to find the **percent rate of change** in the blood flow that corresponds to the percent rate of change in the radius of the blood vessel caused by the drug. (See Example 2.)

Related Rates

We have seen that the derivative represents the instantaneous rate of change of one variable with respect to another. When the derivative is taken with respect to time, it represents the rate at which that variable is changing with respect to time (or the velocity). For example, if distance x is measured in miles and time t in hours, then dx/dt is measured in miles per hour and indicates how fast x is changing. Similarly, if V represents the volume (in cubic feet) of water in a swimming pool and t is time (in minutes), then dV/dt is measured in cubic feet per minute (ft^3/min) and might measure the rate at which the pool is being filled with water or being emptied.

Sometimes, two (or more) quantities that depend on time are also related to each other. For example, the height of a tree h (in feet) is related to the radius r (in inches) of its trunk, and this relationship can be modeled by

$$h = kr^{2/3}$$

where k is a constant.* Of course, both h and r are also related to time, so the rates of change dh/dt and dr/dt are related to each other. Thus they are called **related rates.**

The specific relationship between dh/dt and dr/dt can be found by differentiating $h = kr^{2/3}$ implicitly with respect to time t.

● **EXAMPLE 1** *Tree Height and Trunk Radius*

Suppose that for a certain type of tree, the height of the tree (in feet) is related to the radius of its trunk (in inches) by

$$h = 15r^{2/3}$$

Suppose that the rate of change of r is $\frac{3}{4}$ inch per year. Find how fast the height is changing when the radius is 8 inches.

Solution

To find how the rates dh/dt and dr/dt are related, we differentiate $h = 15r^{2/3}$ implicitly with respect to time t.

$$\frac{dh}{dt} = 10r^{-1/3}\frac{dr}{dt}$$

Using $r = 8$ inches and $dr/dt = \frac{3}{4}$ inch per year gives

$$\frac{dh}{dt} = 10(8)^{-1/3}(3/4) = \frac{15}{4} = 3\tfrac{3}{4} \text{ feet per year}$$

Percent Rates of Change

The work in Example 1 shows how to obtain related rates, but the different units (feet per year and inches per year) may be somewhat difficult to interpret. For this reason, many applications in the life sciences deal with **percent rates of change.** The percent rate of change of a quantity is the rate of change of the quantity divided by the quantity.

◖ **EXAMPLE 2** *Blood Flow (Application Preview)*

Poiseuille's law expresses the flow of blood F as a function of the radius r of the vessel according to

$$F = kr^4$$

where k is a constant. When the radius of a blood vessel is restricted, such as by cholesterol deposits, drugs can be administered that will increase the radius of the blood vessel (and hence the blood flow). Find the percent rate of change of the flow of blood that corresponds to the percent rate of change of the radius of a blood vessel caused by the drug.

*T. McMahon, "Size and Shape in Biology," *Science* 179 (1979): 1201.

Solution

We seek the percent rate of change of flow, $(dF/dt)/F$, that results from a given percent rate of change of the radius $(dr/dt)/r$. We first find the related rates of change by differentiating

$$F = kr^4$$

implicitly with respect to time.

$$\frac{dF}{dt} = k\left(4r^3\frac{dr}{dt}\right)$$

Then the percent rate of change of flow can be found by dividing both sides of the equation by F.

$$\frac{\frac{dF}{dt}}{F} = \frac{4kr^3\frac{dr}{dt}}{F}$$

If we replace F on the right side of the equation with kr^4 and reduce, we get

$$\frac{\frac{dF}{dt}}{F} = \frac{4kr^3\frac{dr}{dt}}{kr^4} = 4\left(\frac{\frac{dr}{dt}}{r}\right)$$

Thus we see that the percent rate of change of the flow of blood is 4 times the corresponding percent rate of change of the radius of the blood vessel. This means that a drug that would cause a 12% increase in the radius of a blood vessel at a certain time would produce a corresponding 48% increase in blood flow through that vessel at that time.

Solving Related-Rates Problems

In the examples above, the equation relating the time-dependent variables has been given. For some problems, the original equation relating the variables must first be developed from the statement of the problem. These problems can be solved with the aid of the following procedure.

Solving Related-Rates Problems

Procedure	Example
To solve related-rates problems:	Sand falls at a rate of 5 ft³/min on a conical pile, with the diameter always equal to the height of the pile. At what rate is the height increasing when the pile is 10 ft high?
1. Use geometric and/or physical conditions to write an equation that relates the time-dependent variables.	1. The conical pile has its volume given by $$V = \frac{1}{3}\pi r^2 h$$
2. Substitute into the equation values or relationships that are true *at all times*.	2. The radius $r = \frac{1}{2}h$ at all times, so $$V = \frac{1}{3}\pi\left[\frac{1}{4}h^2\right]h = \frac{\pi}{12}h^3$$
3. Differentiate both sides of the equation implicitly with respect to time. This equation is valid for all times.	3. $\dfrac{dV}{dt} = \dfrac{\pi}{12}\left[3h^2\dfrac{dh}{dt}\right] = \dfrac{\pi}{4}h^2\dfrac{dh}{dt}$

(continued)

Procedure	Example

Solving Related-Rates Problems *(continued)*

Procedure	Example
4. Substitute the values that are known at the instant specified, and solve the equation.	4. $\dfrac{dV}{dt} = 5$ at all times, so when $h = 10$, $$5 = \frac{\pi}{4}(10^2)\frac{dh}{dt}$$
5. Solve for the specified quantity at the given time.	5. $\dfrac{dh}{dt} = \dfrac{20}{100\pi} = \dfrac{1}{5\pi}$ (ft/min)

Note that you should *not* substitute numerical values for any quantity that varies with time until after the derivative is taken. If values are substituted before the derivative is taken, that quantity will have the constant value resulting from the substitution and hence will have a derivative equal to zero.

● **EXAMPLE 3** *Hot Air Balloon*

A hot air balloon has a velocity of 50 ft/min and is flying at a constant height of 500 ft. An observer on the ground is watching the balloon approach. How fast is the distance between the balloon and the observer changing when the balloon is 1000 ft from the observer?

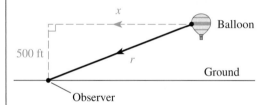

Figure 11.13

Solution

If we let r be the distance between the balloon and the observer and let x be the horizontal distance from the balloon to a point directly above the observer, then we see that these quantities are related by the equation

$$x^2 + 500^2 = r^2 \qquad \text{(See Figure 11.13)}$$

Because the distance x is decreasing, we know that dx/dt must be negative. Thus we are given that $dx/dt = -50$ at all times, and we need to find dr/dt when $r = 1000$. Taking the derivative with respect to t of both sides of the equation $x^2 + 500^2 = r^2$ gives

$$2x\frac{dx}{dt} + 0 = 2r\frac{dr}{dt}$$

Using $dx/dt = -50$ and $r = 1000$, we get

$$2x(-50) = 2000\frac{dr}{dt}$$

$$\frac{dr}{dt} = \frac{-100x}{2000} = \frac{-x}{20}$$

Using $r = 1000$ in $x^2 + 500^2 = r^2$ gives $x^2 = 750{,}000$. Thus $x = 500\sqrt{3}$, and

$$\frac{dr}{dt} = \frac{-500\sqrt{3}}{20} = -25\sqrt{3} \approx -43.3 \text{ ft/min}$$

The distance is decreasing at 43.3 ft/min.

● *Checkpoint*

1. If V represents volume, write a mathematical symbol that represents "the rate of change of volume with respect to time."
2. (a) Differentiate $x^2 + 64 = y^2$ implicitly with respect to time.
   (b) Suppose that we know that y is increasing at 2 units per minute. Use (a) to find the rate of change of x at the instant when $x = 6$ and $y = 10$.
3. True or false: In solving a related-rates problem, we substitute all numerical values into the equation before we take derivatives.

● **EXAMPLE 4** *Spread of an Oil Slick*

Suppose that oil is spreading in a circular pattern from a leak at an offshore rig. If the rate at which the radius of the oil slick is growing is 1 ft/min, at what rate is the area of the oil slick growing when the radius is 600 ft?

Solution

The area of the circular oil slick is given by

$$A = \pi r^2$$

where r is the radius. The rate at which the area is changing is

$$\frac{dA}{dt} = 2\pi r \frac{dr}{dt}$$

Using $r = 600$ ft and $dr/dt = 1$ ft/min gives

$$\frac{dA}{dt} = 2\pi(600 \text{ ft})(1 \text{ ft/min}) = 1200\pi \text{ ft}^2/\text{min}$$

Thus when the radius of the oil slick is 600 ft, the area is growing at the rate of 1200π ft²/min, or approximately 3770 ft²/min.

● *Checkpoint Solutions*

1. dV/dt
2. (a) $2x \dfrac{dx}{dt} + 0 = 2y \dfrac{dy}{dt}$

   $x \dfrac{dx}{dt} = y \dfrac{dy}{dt}$

   (b) Use $\dfrac{dy}{dt} = 2$, $x = 6$, and $y = 10$ in part (a) to obtain

   $6\dfrac{dx}{dt} = 10(2)$ or $\dfrac{dx}{dt} = \dfrac{20}{6} = \dfrac{10}{3}$
3. False. The numerical values for any quantity that is varying with time should not be substituted until after the derivative is taken.

11.4 Exercises

In Problems 1–4, find dy/dt using the given values.
1. $y = x^3 - 3x$ for $x = 2$, $dx/dt = 4$
2. $y = 3x^3 + 5x^2 - x$ for $x = 4$, $dx/dt = 3$
3. $xy = 4$ for $x = 8$, $dx/dt = -2$
4. $xy = x + 3$ for $x = 3$, $dx/dt = -1$

In Problems 5–8, assume that x and y are differentiable functions of t. In each case, find dx/dt given that $x = 5$, $y = 12$, and $dy/dt = 2$.
5. $x^2 + y^2 = 169$ 6. $y^2 - x^2 = 119$
7. $y^2 = 2xy + 24$ 8. $x^2(y - 6) = 12y + 6$

9. If $x^2 + y^2 = z^2$, find dy/dt when $x = 3$, $y = 4$, $dx/dt = 10$, and $dz/dt = 2$.

10. If $s = 2\pi r(r + h)$, find dr/dt when $r = 2$, $h = 8$, $dh/dt = 3$, and $ds/dt = 10\pi$.

11. A point is moving along the graph of the equation $y = -4x^2$. At what rate is y changing when $x = 5$ and is changing at a rate of 2 units/sec?

12. A point is moving along the graph of the equation $y = 5x^3 - 2x$. At what rate is y changing when $x = 4$ and is changing at a rate of 3 units/sec?

13. The radius of a circle is increasing at a rate of 2 ft/min. At what rate is its area changing when the radius is 3 ft? (Recall that for a circle, $A = \pi r^2$.)

14. The area of a circle is changing at a rate of 1 in.²/sec. At what rate is its radius changing when the radius is 2 in.?

15. The volume of a cube is increasing at a rate of 64 in.³/sec. At what rate is the length of each edge of the cube changing when the edges are 6 in. long? (Recall that for a cube, $V = x^3$.)

16. The lengths of the edges of a cube are increasing at a rate of 8 ft/min. At what rate is the surface area changing when the edges are 24 ft long? (Recall that for a cube, $S = 6x^2$.)

APPLICATIONS

17. *Profit*   Suppose that the daily profit (in dollars) from the production and sale of x units of a product is given by

$$P = 180x - \frac{x^2}{1000} - 2000$$

At what rate per day is the profit changing when the number of units produced and sold is 100 and is increasing at a rate of 10 units per day?

18. *Profit*   Suppose that the monthly revenue and cost (in dollars) for x units of a product are

$$R = 400x - \frac{x^2}{20} \quad \text{and} \quad C = 5000 + 70x$$

At what rate per month is the profit changing if the number of units produced and sold is 100 and is increasing at a rate of 10 units per month?

19. *Demand*   Suppose that the price p (in dollars) of a product is given by the demand function

$$p = \frac{1000 - 10x}{400 - x}$$

where x represents the quantity demanded. If the daily demand is *decreasing* at a rate of 20 units per day, at what rate is the price changing when the demand is 20 units?

20. *Supply*   The supply function for a product is given by $p = 40 + 100\sqrt{2x + 9}$, where x is the number of units supplied and p is the price in dollars. If the price is increasing at a rate of $1 per month, at what rate is the supply changing when $x = 20$?

21. *Capital investment and production*   Suppose that for a particular product, the number of units x produced per month depends on the number of thousands of dollars y invested, with $x = 30y + 20y^2$. At what rate will production increase if $10,000 is invested and if the investment capital is increasing at a rate of $1000 per month?

22. *Boyle's law*   Boyle's law for enclosed gases states that at a constant temperature, the pressure is related to the volume by the equation

$$P = \frac{k}{V}$$

where k is a constant. If the volume is increasing at a rate of 5 cubic inches per hour, at what rate is the pressure changing when the volume is 30 cubic inches and $k = 2$ inch-pounds?

Tumor growth   **For Problems 23 and 24, suppose that a tumor in a person's body has a spherical shape and that treatment is causing the radius of the tumor to decrease at a rate of 1 millimeter per month.**

23. At what rate is the volume decreasing when the radius is 3 millimeters? (Recall that $V = \frac{4}{3}\pi r^3$.)

24. At what rate is the surface area of the tumor decreasing when the radius is 3 mm? (Recall that for a sphere, $S = 4\pi r^2$.)

25. *Allometric relationships—fish*   For many species of fish, the allometric relationship between the weight W and the length L is approximately $W = kL^3$, where k is a constant. Find the percent rate of change of the weight as a corresponding percent rate of change of the length.

26. *Blood flow*   The resistance R of a blood vessel to the flow of blood is a function of the radius r of the blood vessel and is given by

$$R = \frac{k}{r^4}$$

where k is a constant. Find the percent rate of change of the resistance of a blood vessel in terms of the percent rate of change in the radius of the blood vessel.

27. *Allometric relationships—crabs* For fiddler crabs, data gathered by Thompson* show that the allometric relationship between the weight C of the claw and the weight W of the body is given by

$$C = 0.11W^{1.54}$$

Find the percent rate of change of the claw weight in terms of the percent rate of change of the body weight for fiddler crabs.

28. *Body weight and surface area* For human beings, the surface area S of the body is related to the body's weight W according to

$$S = kW^{2/3}$$

where k is a constant. Find the percent rate of change of the body's surface area in terms of the percent rate of change of the body's weight.

29. *Cell growth* A bacterial cell has a spherical shape. If the volume of the cell is increasing at a rate of 4 cubic micrometers per day, at what rate is the radius of the cell increasing when it is 2 micrometers? (Recall that for a sphere, $V = \frac{4}{3}\pi r^3$.)

30. *Water purification* Assume that water is being purified by causing it to flow through a conical filter that has a height of 15 inches and a radius of 5 inches. If the depth of the water is decreasing at a rate of 1 inch per minute when the depth is 6 inches, at what rate is the volume of water flowing out of the filter at this instant?

31. *Volume and radius* Suppose that air is being pumped into a spherical balloon at a rate of 5 in.^{3}/min. At what rate is the radius of the balloon increasing when the radius is 5 in.?

32. *Boat docking* Suppose that a boat is being pulled toward a dock by a winch that is 5 ft above the level of the boat deck. If the winch is pulling the cable at a rate of 3 ft/min, at what rate is the boat approaching the dock when it is 12 ft from the dock? Use the figure below.

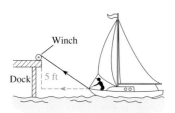

33. *Ladder safety* A 30-ft ladder is leaning against a wall. If the bottom is pulled away from the wall at a rate of 1 ft/sec, at what rate is the top of the ladder sliding down the wall when the bottom is 18 ft from the wall?

34. *Flight* A kite is 30 ft high and is moving horizontally at a rate of 10 ft/min. If the kite string is taut, at what rate is the string being played out when 50 ft of string is out?

35. *Flight* A plane is flying at a constant altitude of 1 mile and a speed of 300 mph. If it is flying toward an observer on the ground, how fast is the plane approaching the observer when it is 5 miles from the observer?

36. *Distance* Two boats leave the same port at the same time, with boat A traveling north at 15 knots (nautical miles per hour) and boat B traveling east at 20 knots. How fast is the distance between them changing when boat A is 30 nautical miles from port?

37. *Distance* Two cars are approaching an intersection on roads that are perpendicular to each other. Car A is north of the intersection and traveling south at 40 mph. Car B is east of the intersection and traveling west at 55 mph. How fast is the distance between the cars changing when car A is 15 miles from the intersection and car B is 8 miles from the intersection?

38. *Water depth* Water is flowing into a barrel in the shape of a right circular cylinder at the rate of 200 in.^{3}/min. If the radius of the barrel is 18 in., at what rate is the depth of the water changing when the water is 30 in. deep?

39. *Water depth* Suppose that water is being pumped into a rectangular swimming pool of uniform depth at 10 ft^3/hr. If the pool is 10 ft wide and 25 ft long, at what rate is the water rising when it is 4 ft deep?

*d'Arcy Thompson, *On Growth and Form* (Cambridge, England: Cambridge University Press, 1961).

Applications in Business and Economics

◗ Application Preview

Suppose that the demand for a product is given by

$$p = \frac{1000}{(q + 1)^2}$$

We can measure how sensitive the demand for this product is to price changes by finding the **elasticity of demand.** (See Example 2.) This elasticity can be used to measure the effects that price changes have on total revenue. We also consider **taxation in a competitive market,** which examines how a tax levied on goods shifts market equilibrium. We will find the tax per unit that, despite changes in market equilibrium, maximizes tax revenue.

Elasticity of Demand

We know from the law of demand that consumers will respond to changes in prices; if prices increase, the quantities demanded will decrease. But the degree of responsiveness of the consumers to price changes will vary widely for different products. For example, a price increase in insulin will not greatly decrease the demand for it by diabetics, but a price increase in clothes may cause consumers to buy less and wear their old clothes longer. When the response to price changes is considerable, we say the demand is *elastic*. When price changes cause relatively small changes in demand for a product, the demand is said to be *inelastic* for that product.

Economists measure the **elasticity of demand** on an interval by dividing the percent change of demand by the percent change of price. We may write this as

$$E_d = -\frac{\text{change in quantity demanded}}{\text{original quantity demanded}} \div \frac{\text{change in price}}{\text{original price}}$$

or

$$E_d = -\frac{\Delta q}{q} \div \frac{\Delta p}{p}$$

The demand curve usually has a negative slope, so we have introduced a negative sign into the formula to give us a positive elasticity.

We can write the equation for elasticity as

$$E_d = -\frac{p}{q} \cdot \frac{\Delta q}{\Delta p}$$

If q is a function of p, then we can write

$$\frac{\Delta q}{\Delta p} = \frac{f(p + \Delta p) - f(p)}{\Delta p}$$

and the limit as Δp approaches 0 gives the **point elasticity of demand:**

$$\eta = \lim_{\Delta p \to 0} \left(-\frac{p}{q} \cdot \frac{\Delta q}{\Delta p} \right) = -\frac{p}{q} \cdot \frac{dq}{dp}$$

Elasticity

The **elasticity of demand** at the point (q_A, p_A) is

$$\eta = -\frac{p}{q} \cdot \frac{dq}{dp}\bigg|_{(q_A, p_A)}$$

● **EXAMPLE 1** *Elasticity*

Find the elasticity of the demand function $p + 5q = 100$ when

(a) the price is \$40,   (b) the price is \$60, and   (c) the price is \$50.

Solution
Solving the demand function for q gives $q = 20 - \frac{1}{5}p$. Then $dq/dp = -\frac{1}{5}$ and

$$\eta = -\frac{p}{q}\left(-\frac{1}{5}\right)$$

(a) When $p = 40$, $q = 12$ and $\eta = -\frac{p}{q}\left(-\frac{1}{5}\right)\bigg|_{(12,\,40)} = -\frac{40}{12}\left(-\frac{1}{5}\right) = \frac{2}{3}$.

(b) When $p = 60$, $q = 8$ and $\eta = -\frac{p}{q}\left(-\frac{1}{5}\right)\bigg|_{(8,\,60)} = -\frac{60}{8}\left(-\frac{1}{5}\right) = \frac{3}{2}$.

(c) When $p = 50$, $q = 10$ and $\eta = -\frac{p}{q}\left(-\frac{1}{5}\right)\bigg|_{(10,\,50)} = -\frac{50}{10}\left(-\frac{1}{5}\right) = 1$.

Note that in Example 1 the demand equation was $p + 5q = 100$, so the demand "curve" is a straight line, with slope $m = -5$. But the elasticity was $\eta = \frac{2}{3}$ at (12, 40), $\eta = \frac{3}{2}$ at (8, 60), and $\eta = 1$ at (10, 50). This illustrates that the elasticity of demand may be different at different points on the demand curve, even though the slope of the demand "curve" is constant. (See Figure 11.14.)

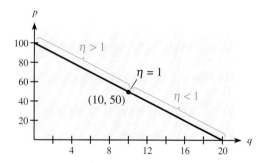

Figure 11.14

This example shows that the elasticity of demand is more than just the slope of the demand curve, which is the rate at which the demand is changing. Recall that the elasticity measures the consumers' degree of responsiveness to a price change.

Economists use η to measure how responsive demand is to price at different points on the demand curve for a product.

Elasticity of Demand

■ If $\eta > 1$, the demand is **elastic,** and the percent decrease in demand is greater than the corresponding percent increase in price.

■ If $\eta < 1$, the demand is **inelastic,** and the percent decrease in demand is less than the corresponding percent increase in price.

■ If $\eta = 1$, the demand is **unitary elastic,** and the percent decrease in demand is approximately equal to the corresponding percent increase in price.

We can also use implicit differentiation to find dq/dp in evaluating the point elasticity of demand.

❍ EXAMPLE 2 *Elasticity (Application Preview)*

The demand for a certain product is given by

$$p = \frac{1000}{(q + 1)^2}$$

where p is the price per unit in dollars and q is demand in units of the product. Find the elasticity of demand with respect to price when $q = 19$.

Solution

To find the elasticity, we need to find dq/dp. Using implicit differentiation, we get the following:

$$\frac{d}{dp}(p) = \frac{d}{dp}\left[1000(q + 1)^{-2}\right]$$

$$1 = 1000\left[-2(q + 1)^{-3}\frac{dq}{dp}\right]$$

$$1 = \frac{-2000}{(q + 1)^3}\frac{dq}{dp}$$

$$\frac{(q + 1)^3}{-2000} = \frac{dq}{dp}$$

When $q = 19$, we have $p = 1000/(19 + 1)^2 = 1000/400 = 5/2$ and

$$\frac{dq}{dp}\bigg|_{(q=19)} = \frac{(19 + 1)^3}{-2000} = \frac{8000}{-2000} = -4$$

The elasticity of demand when $q = 19$ is

$$\eta = \frac{-p}{q} \cdot \frac{dq}{dp} = -\frac{(5/2)}{19} \cdot (-4) = \frac{10}{19} < 1$$

Thus the demand for this product is inelastic.

Elasticity and Revenue

Elasticity is related to revenue in a special way. We can see how by computing the derivative of the revenue function

$$R = pq$$

with respect to p.

$$\frac{dR}{dp} = p \cdot \frac{dq}{dp} + q \cdot 1$$

$$= \frac{q}{q} \cdot p \cdot \frac{dq}{dp} + q = q \cdot \frac{p}{q} \cdot \frac{dq}{dp} + q$$

$$= q(-\eta) + q$$

$$= q(1 - \eta)$$

From this we can summarize the relationship of elasticity and revenue.

Elasticity and Revenue

The rate of change of revenue R with respect to price p is related to elasticity in the following way.

- Elastic $(\eta > 1)$ means $\dfrac{dR}{dp} < 0.$ $\begin{cases} \text{Hence if price increases, revenue decreases,} \\ \text{and if price decreases, revenue increases.} \end{cases}$

- Inelastic $(\eta < 1)$ means $\dfrac{dR}{dp} > 0.$ $\begin{cases} \text{Hence if price increases, revenue increases,} \\ \text{and if price decreases, revenue decreases.} \end{cases}$

- Unitary elastic $(\eta = 1)$ means $\dfrac{dR}{dp} = 0.$ Hence an increase or decrease in price will not change revenue. Revenue is optimized at this point.

● **EXAMPLE 3** *Elasticity and Revenue*

The demand for a product is given by

$$p = 10\sqrt{100 - q}, \qquad 0 \le q \le 100$$

(a) Find the point at which demand is of unitary elasticity, and find intervals in which the demand is inelastic and in which it is elastic.
(b) Find q where revenue is increasing, where it is decreasing, and where it is maximized.
(c) Use a graphing utility to show the graph of the revenue function $R = pq$, with $0 \le q \le 100$, and confirm the results from part (b).

Solution

The elasticity is

$$\eta = \frac{-p}{q} \cdot \frac{dq}{dp} = -\frac{10\sqrt{100 - q}}{q} \cdot \frac{dq}{dp}$$

Finding dq/dp implicitly, we have

$$1 = 10 \left[\frac{1}{2}(100 - q)^{-1/2}\left(-\frac{dq}{dp}\right) \right] = \frac{-5}{\sqrt{100 - q}} \cdot \frac{dq}{dp}$$

so

$$\frac{dq}{dp} = \frac{-\sqrt{100 - q}}{5}$$

Thus

$$\eta = -\frac{10\sqrt{100 - q}}{q}\left[\frac{-\sqrt{100 - q}}{5} \right] = \frac{200 - 2q}{q}$$

(a) Unitary elasticity occurs where $\eta = 1$.

$$1 = \frac{200 - 2q}{q}$$
$$q = 200 - 2q$$
$$3q = 200$$
$$q = 66\tfrac{2}{3}$$

so unitary elasticity occurs when $66\frac{2}{3}$ units are sold, at a price of \$57.74. For values of q between 0 and $66\frac{2}{3}$, $\eta > 1$ and demand is elastic. For values of q between $66\frac{2}{3}$ and 100, $\eta < 1$ and demand is inelastic.

(b) When q increases over $0 < q < 66\frac{2}{3}$, p decreases, so $\eta > 1$ means R increases. Similarly, when q increases over $66\frac{2}{3} < q < 100$, p decreases, so $\eta < 1$ means R decreases. Revenue is maximized where $\eta = 1$, at $q = 66\frac{2}{3}$ and $p = 57.74$.

(c) The graph of this revenue function,

$$R = 10q\sqrt{100 - q}$$

is shown in Figure 11.15 and confirms our conclusions from (b).

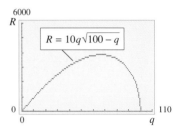

Figure 11.15

● *Checkpoint*

1. Write the formula for point elasticity, η.
2. (a) If $\eta > 1$, the demand is called _____.
   (b) If $\eta < 1$, the demand is called _____.
   (c) If $\eta = 1$, the demand is called _____.
3. Find the elasticity of demand for $q = \dfrac{100}{p} - 1$ when $p = 10$ and $q = 9$.

Taxation in a Competitive Market

Many taxes imposed by governments are "hidden." That is, the tax is levied on goods produced, and the producers must pay the tax. Of course, the tax becomes a cost to the producers, and they pass that cost on to the consumer in the form of higher prices for goods.

Suppose the government imposes a tax of t dollars on each unit produced and sold by producers. If we are in pure competition in which the consumers' demand depends only on price, the *demand function* will not change. The tax will change the supply function, of course, because at each level of output q, the firm will want to charge a price higher by the amount of the tax.

The graphs of the market demand function, the original market supply function, and the market supply function after taxes are shown in Figure 11.16. Because the tax added to each item is constant, the graph of the supply function is t units above the original supply function. If $p = f(q)$ defines the original supply function, then $p = f(q) + t$ defines the supply function after taxation.

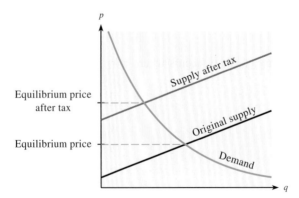

Figure 11.16

Note in this case that after the taxes are imposed, *no* items are supplied at the price that was the equilibrium price before taxation. After the taxes are imposed, the consumers simply have to pay more for the product. Because taxation does not change the demand curve, the quantity purchased at market equilibrium will be less than it was before taxation. Thus governments planning taxes should recognize that they will not collect taxes on the original equilibrium quantity. They will collect on the *new* equilibrium quantity, a quantity reduced by their taxation. Thus a large tax on each item may reduce the quantity demanded at the new market equilibrium so much that very little revenue results from the tax!

If the tax revenue is represented by $T = tq$, where t is the tax per unit and q is the equilibrium quantity of the supply and demand functions after taxation, we can use the following procedure for maximizing the total tax revenue in a competitive market.

Maximizing Total Tax Revenue

Procedure	Example
To find the tax per item (under pure competition) that will maximize total tax revenue:	Suppose the demand and supply functions are given by $p = 600 - q$ and $p = 200 + \frac{1}{3}q$, respectively, where $\$p$ is the price per unit and q is the number of units. Find the tax t that will maximize the total tax revenue T.
1. Write the supply function after taxation.	1. $p = 200 + \dfrac{1}{3}q + t$
2. Set the demand function and the new supply function equal, and solve for t in terms of q.	2. $600 - q = 200 + \dfrac{1}{3}q + t$ $400 - \dfrac{4}{3}q = t$
3. Form the total tax revenue function, $T = tq$, by multiplying the expression for t by q, and then take its derivative with respect to q.	3. $T = tq = 400q - \dfrac{4}{3}q^2$ $T'(q) = \dfrac{dT}{dq} = 400 - \dfrac{8}{3}q$
4. Set $T' = 0$, and solve for q. This is the q that should maximize T. Use the second-derivative test to verify it.	4. $0 = 400 - \dfrac{8}{3}q$ $q = 150$ $T''(q) = -\dfrac{8}{3}$. Thus T is maximized at $q = 150$.
5. Substitute the value of q in the equation for t (in step 2). This is the value of t that will maximize T.	5. $t = 400 - \dfrac{4}{3}(150) = 200$ A tax of \$200 per item will maximize the total tax revenue. The total tax revenue for the period would be $\$200 \cdot (150) = \$30,000$.

Note that in the example just given, if a tax of \$300 were imposed, the total tax revenue the government would receive would be

$$(\$300)(75) = \$22{,}500$$

This means that consumers would spend $75 more for each item, suppliers would sell 75 fewer items, and the government would lose $7500 in tax revenue. Thus everyone would suffer if the tax rate were raised to $300.

● *Checkpoint*   4. For problems involving taxation in a competitive market, if supply is $p = f(q)$ and demand is $p = g(q)$, is the tax t added to $f(q)$ or to $g(q)$?

● **EXAMPLE 4** *Maximizing Tax Revenue*

The demand and supply functions for a product are $p = 900 - 20q - \frac{1}{3}q^2$ and $p = 200 + 10q$, respectively, where p is in dollars and q is the number of units. Find the tax per unit that will maximize the tax revenue T.

Solution

After taxation, the supply function is $p = 200 + 10q + t$, where t is the tax per unit. The demand function will meet the new supply function where

$$900 - 20q - \frac{1}{3}q^2 = 200 + 10q + t$$

so

$$t = 700 - 30q - \frac{1}{3}q^2$$

Then the total tax T is $T = tq = 700q - 30q^2 - \frac{1}{3}q^3$, and we maximize T as follows:

$$T'(q) = 700 - 60q - q^2$$
$$0 = -(q + 70)(q - 10)$$
$$q = 10 \quad \text{or} \quad q = -70$$

Because $q = -70$ is meaningless, we test $q = 10$.

$$\left. \begin{array}{l} T(0) = 0 \\ T'(0) = 700 > 0 \\ T'(q) < 0 \text{ for all } q > 10 \end{array} \right\} \Rightarrow \text{ absolute maximum at } q = 10$$

The maximum possible tax revenue is

$$T(10) \approx \$3666.67$$

The tax per unit that maximizes T is

$$t = 700 - 30(10) - \frac{1}{3}(10)^2 \approx \$366.67$$

An infamous example of a tax increase that resulted in decreased tax revenue and economic disaster is the "luxury tax" that went into effect in 1991. This was a 10% excise tax on the sale of expensive jewelry, furs, airplanes, certain expensive boats, and automobiles costing over $30,000. The Congressional Joint Tax Committee had estimated that the luxury tax would raise $6 million from airplanes alone, but it raised only $53,000 while it destroyed the small-airplane market (one company lost $130 million and 480 jobs in a single year). It also capsized the boat market. The luxury tax was repealed at the end of 1993 (except for automobiles).*

*Fortune, Sept. 6, 1993; *Motor Trend,* December 1993.

● **Checkpoint Solutions**

1. $\eta = \dfrac{-p}{q} \cdot \dfrac{dq}{dp}$

2. (a) elastic   (b) inelastic   (c) unitary elastic

3. $\dfrac{dq}{dp} = \dfrac{-100}{p^2}$ and $\dfrac{dq}{dp} = -1$ when $p = 10, q = 9$

   $\eta = \dfrac{-10}{9}(-1) = \dfrac{10}{9}$(elastic)

4. Tax t is added to supply: $p = f(q) + t$.

11.5 Exercises

ELASTICITY OF DEMAND

In Problems 1–8, p is in dollars and q is the number of units.

1. (a) Find the elasticity of the demand function $p + 4q = 80$ at $(10, 40)$.
   (b) How will a price increase affect total revenue?

2. (a) Find the elasticity of the demand function $2p + 3q = 150$ at the price $p = 15$.
   (b) How will a price increase affect total revenue?

3. (a) Find the elasticity of the demand function $p^2 + 2p + q = 49$ at $p = 6$.
   (b) How will a price increase affect total revenue?

4. (a) Find the elasticity of the demand function $pq = 81$ at $p = 3$.
   (b) How will a price increase affect total revenue?

5. Suppose that the demand for a product is given by $pq + p = 5000$.
   (a) Find the elasticity when $p = \$50$ and $q = 99$.
   (b) Tell what type of elasticity this is: unitary, elastic, or inelastic.
   (c) How would revenue be affected by a price increase?

6. Suppose that the demand for a product is given by $2p^2q = 10{,}000 + 9000p^2$.
   (a) Find the elasticity when $p = \$50$ and $q = 4502$.
   (b) Tell what type of elasticity this is: unitary, elastic, or inelastic.
   (c) How would revenue be affected by a price increase?

7. Suppose that the demand for a product is given by $pq + p + 100q = 50{,}000$.
   (a) Find the elasticity when $p = \$401$.
   (b) Tell what type of elasticity this is.
   (c) How would a price increase affect revenue?

8. Suppose that the demand for a product is given by
   $$(p + 1)\sqrt{q + 1} = 1000$$
   (a) Find the elasticity when $p = \$39$.

   (b) Tell what type of elasticity this is.
   (c) How would a price increase affect revenue?

9. Suppose the demand function for a product is given by
   $$p = \frac{1}{2}\ln\left(\frac{5000 - q}{q + 1}\right)$$
   where p is in hundreds of dollars and q is the number of tons.
   (a) What is the elasticity of demand when the quantity demanded is 2 tons and the price is \$371?
   (b) Is the demand elastic or inelastic?

10. Suppose the weekly demand function for a product is
    $$q = \frac{5000}{1 + e^{2p}} - 1$$
    where p is the price in thousands of dollars and q is the number of units demanded. What is the elasticity of demand when the price is \$1000 and the quantity demanded is 595?

In Problems 11 and 12, the demand functions for specialty steel products are given, where p is in dollars and q is the number of units. For both problems:

(a) Find the elasticity of demand as a function of the quantity demanded, q.

(b) Find the point at which the demand is of unitary elasticity and find intervals in which the demand is inelastic and in which it is elastic.

(c) Use information about elasticity in part (b) to decide where the revenue is increasing, where it is decreasing, and where it is maximized.

(d) Graph the revenue function $R = pq$, and use it to find where revenue is maximized. Is it at the same quantity as that determined in part (c)?

11. $p = 120\sqrt[3]{125 - q}$

12. $p = 30\sqrt{49 - q}$

TAXATION IN A COMPETITIVE MARKET

In Problems 13–22, p is the price per unit in dollars and q is the number of units.

13. If the weekly demand function is $p = 30 - q$ and the supply function before taxation is $p = 6 + 2q$, what tax per item will maximize the total tax revenue?

14. If the demand function for a fixed period of time is given by $p = 38 - 2q$ and the supply function before taxation is $p = 8 + 3q$, what tax per item will maximize the total tax revenue?

15. If the demand and supply functions for a product are $p = 800 - 2q$ and $p = 100 + 0.5q$, respectively, find the tax per unit t that will maximize the tax revenue T.

16. If the demand and supply functions for a product are $p = 2100 - 3q$ and $p = 300 + 1.5q$, respectively, find the tax per unit t that will maximize the tax revenue T.

17. If the weekly demand function is $p = 200 - 2q^2$ and the supply function before taxation is $p = 20 + 3q$, what tax per item will maximize the total tax revenue?

18. If the monthly demand function is $p = 7230 - 5q^2$ and the supply function before taxation is $p = 30 + 30q^2$, what tax per item will maximize the total revenue?

19. Suppose the weekly demand for a product is given by $p + 2q = 840$ and the weekly supply before taxation is given by $p = 0.02q^2 + 0.55q + 7.4$. Find the tax per item that produces maximum tax revenue. Find the tax revenue.

20. If the daily demand for a product is given by the function $p + q = 1000$ and the daily supply before taxation is $p = q^2/30 + 2.5q + 920$, find the tax per item that maximizes tax revenue. Find the tax revenue.

21. If the demand and supply functions for a product are $p = 2100 - 10q - 0.5q^2$ and $p = 300 + 5q + 0.5q^2$, respectively, find the tax per unit t that will maximize the tax revenue T.

22. If the demand and supply functions for a product are $p = 5000 - 20q - 0.7q^2$ and $p = 500 + 10q + 0.3q^2$, respectively, find the tax per unit t that will maximize the tax revenue T.

Key Terms and Formulas

Section	Key Terms	Formulas
11.1	Logarithmic function	$y = \log_a x$, defined by $x = a^y$
	Natural logarithm	$\ln x = \log_e x$; $y = \ln (x)$ means $e^y = x$
	Logarithmic Properties I–V for natural logarithms	$\ln e^x = x$; $e^{\ln x} = x$
		$\ln (MN) = \ln M + \ln N$
		$\ln (M/N) = \ln M - \ln N$
		$\ln (M^p) = p (\ln M)$
	Change-of-base formula	$\log_a x = \dfrac{\ln x}{\ln a}$
	Derivatives of logarithmic functions	$\dfrac{d}{dx} (\ln x) = \dfrac{1}{x}$
		$\dfrac{d}{dx} (\ln u) = \dfrac{1}{u} \cdot \dfrac{du}{dx}$
11.2	Derivatives of exponential functions	$\dfrac{d}{dx}(e^x) = e^x$
		$\dfrac{d}{dx} e^u = e^u \dfrac{du}{dx}$
		$\dfrac{d}{dx} a^u = a^u \dfrac{du}{dx} \ln a$
11.3	Implicit differentiation	$\dfrac{d}{dx}(y^n) = ny^{n-1} \dfrac{dy}{dx}$
11.4	Related rates Percent rates of change	Differentiate implicitly with respect to time, t

Section	Key Terms	Formulas
11.5	Elasticity of demand	$\eta = \dfrac{-p}{q} \cdot \dfrac{dq}{dp}$
	Elastic	$\eta > 1$
	Inelastic	$\eta < 1$
	Unitary elastic	$\eta = 1$
	Taxation in competitive market	
	Supply function after taxation	$p = f(q) + t$

Review Exercises

Sections 11.1 and 11.2

In Problems 1–12, find the indicated derivative.

1. If $y = e^{3x^2 - x}$, find dy/dx.
2. If $y = \ln e^{x^2}$, find y'.
3. If $p = \ln\left(\dfrac{q}{q^2 - 1}\right)$, find $\dfrac{dp}{dq}$.
4. If $y = xe^{x^2}$, find dy/dx.
5. If $f(x) = 5e^{2x} - 40e^{-0.1x} + 11$, find $f'(x)$.
6. If $g(x) = (2e^{3x+1} - 5)^3$, find $g'(x)$.
7. If $y = \ln(3x^4 + 7x^2 - 12)$, find dy/dx.
8. If $s = \dfrac{3}{4}\ln(x^{12} - 2x^4 + 5)$, find ds/dx.
9. If $y = 3^{3x-4}$, find dy/dx.
10. If $y = 1 + \log_8(x^{10})$, find dy/dx.
11. If $y = \dfrac{\ln x}{x}$, find $\dfrac{dy}{dx}$.
12. If $y = \dfrac{1 + e^{-x}}{1 - e^{-x}}$, find $\dfrac{dy}{dx}$.
13. Write the equation of the line tangent to $y = 4e^{x^3}$ at $x = 1$.
14. Write the equation of the line tangent to $y = x \ln x$ at $x = 1$.

Section 11.3

In Problems 15–20, find the indicated derivative.

15. If $y \ln x = 5y^2 + 11$, find dy/dx.
16. Find dy/dx for $e^{xy} = y$.
17. Find dy/dx for $y^2 = 4x - 1$.
18. Find dy/dx for $x^2 + 3y^2 + 2x - 3y + 2 = 0$.
19. Find dy/dx for $3x^2 + 2x^3y^2 - y^5 = 7$.
20. Find the second derivative y'' if $x^2 + y^2 = 1$.
21. Find the slope of the tangent to the curve $x^2 + 4x - 3y^2 + 6 = 0$ at $(3, 3)$.
22. Find the points where tangents to the graph of the equation in Problem 21 are horizontal.

Section 11.4

23. Suppose $3x^2 - 2y^3 = 10y$, where x and y are differentiable functions of t. If $dx/dt = 2$, find dy/dt when $x = 10$ and $y = 5$.
24. A right triangle with legs of lengths x and y has its area given by

$$A = \frac{1}{2}xy$$

If the rate of change of x is 2 units per minute and the rate of change of y is 5 units per minute, find the rate of change of the area when $x = 4$ and $y = 1$.

APPLICATIONS

Section 11.1

25. *Deforestation* One of the major causes of rain forest deforestation is agricultural and residential development. The number of hectares destroyed in a particular year t can be modeled by

$$y = -3.91435 + 2.62196 \ln t$$

where $t = 0$ represents 1950.
(a) Find $y'(t)$.
(b) Find and interpret $y(50)$ and $y'(50)$.

Section 11.2

26. *Compound interest* If the future value of $1000 invested for n years at 12%, compounded continuously, is given by

$$S = 1000e^{0.12n} \quad \text{dollars}$$

find the rate at which the future value is growing after 1 year.

27. *Compound interest*
    (a) In Problem 26, find the rate of growth of the future value after 2 years.
    (b) How much faster is the future value growing at the end of 2 years than after 1 year?

28. *Radioactive decay* A breeder reactor converts stable uranium-238 into the isotope plutonium-239. The decay of this isotope is given by

$$A(t) = A_0 e^{-0.00002876t}$$

where $A(t)$ is the amount of isotope at time t, in years, and A_0 is the original amount. This isotope has a half-life of 24,101 years (that is, half of it will decay away in 24,101 years).
    (a) At what rate is $A(t)$ decaying at this point in time?
    (b) At what rate is $A(t)$ decaying after 1 year?
    (c) Is the rate of decay at its half-life greater or less than after 1 year?

29. *Marginal cost* The average cost of producing x units of a product is $\overline{C} = 600e^{x/600}$ dollars per unit. What is the marginal cost when 600 units are produced?

30. *Inflation* The impact of inflation on a $20,000 pension can be measured by the purchasing power P of $20,000 after t years. For an inflation rate of 5% per year, compounded annually, P is given by

$$P = 20,000e^{-0.0495t}$$

At what rate is purchasing power changing when $t = 10$? (*Source: Viewpoints, VALIC*)

Section 11.4

31. *Evaporation* A spherical droplet of water evaporates at a rate of 1 mm³/min. Find the rate of change of the radius when the droplet has a radius of 2.5 mm.

32. *Worker safety* A sign is being lowered over the side of a building at the rate of 2 ft/min. A worker handling a guide line is 7 ft away from a spot directly below the sign. How fast is the worker taking in the guide line at the instant the sign is 25 ft from the worker's hands? See the following figure.

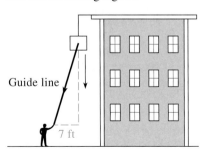

33. *Environment* Suppose that in a study of water birds, the relationship between the area A of wetlands (in square miles) and the number of different species S of birds found in the area was determined to be

$$S = kA^{1/3}$$

where k is constant. Find the percent rate of change of the number of species in terms of the percent rate of change of the area.

Section 11.5

34. *Taxes* Can increasing the tax per unit sold actually lead to a decrease in tax revenues?

35. *Taxes* Suppose the demand and supply functions for a product are

$$p = 2800 - 8q - \frac{q^2}{3} \quad \text{and} \quad p = 400 + 2q$$

respectively, where p is in dollars and q is the number of units. Find the tax per unit t that will maximize the tax revenue T, and find the maximum tax revenue.

36. *Taxes* Suppose the supply and demand functions for a product are

$$p = 40 + 20q \quad \text{and} \quad p = \frac{5000}{q + 1}$$

respectively, where p is in dollars and q is the number of units. Find the tax t that maximizes the tax revenue T, and find the maximum tax revenue.

37. *Elasticity* A demand function is given by

$$pq = 27$$

where p is in dollars and q is the number of units.
    (a) Find the elasticity of demand at (9, 3).
    (b) How will a price increase affect total revenue?

38. *Elasticity* Suppose the demand for a product is given by

$$p^2(2q + 1) = 10,000$$

where p is in dollars and q is the number of units. Find the elasticity of demand when $p = \$20$.

39. *Elasticity* Suppose the weekly demand function for a product is given by

$$p = 100e^{-0.1q}$$

where p is the price in dollars and q is the number of tons demanded. What is the elasticity of demand when the price is $36.79 and the quantity demanded is 10?

40. *Revenue* A product has the demand function

$$p = 100 - 0.5q$$

where p is in dollars and q is the number of units.

(a) Find the elasticity $\eta(q)$ as a function of q, and graph the function

$$f(q) = \eta(q)$$

(b) Find where $f(q) = 1$, which gives the quantity for which the product has unitary elasticity.

(c) The revenue function for this product is

$$R(q) = pq = (100 - 0.5q)q$$

Graph $R(q)$ and find the q-value for which the maximum revenue occurs.

(d) What is the relationship between elasticity and maximum revenue?

11 Chapter Test

In Problems 1–8, find the derivative of each function.

1. $y = 5e^{x^3} + x^2$
2. $y = 4 \ln (x^3 + 1)$
3. $y = \ln (x^4 + 1)^3$
4. $f(x) = 10(3^{2x})$
5. $S = te^{t^4}$
6. $y = \dfrac{e^{x^3 + 1}}{x}$
7. $y = \dfrac{\ln x}{x}$
8. $g(x) = 2 \log_5 (4x + 7)$
9. Find y' if $3x^4 + 2y^2 + 10 = 0$.
10. Let $x^2 + y^2 = 100$.

    If $\dfrac{dx}{dt} = 2$, find $\dfrac{dy}{dt}$ when $x = 6$ and $y = 8$.

11. Find y' if $xe^y = 10y$.
12. Suppose the weekly revenue and weekly cost (both in dollars) for a product are given by $R(x) = 300x - 0.001x^2$ and $C(x) = 4000 + 30x$, respectively, where x is the number of units produced and sold. Find the rate at which profit is changing with respect to time when the number of units produced and sold is 50 and is increasing at a rate of 5 units per week.
13. Suppose the demand for a product is $p^2 + 3p + q = 1500$, where p is in dollars and q is the number of units. Find the elasticity of demand at $p = 30$. If the price is raised to \$31, does revenue increase or decrease?
14. Suppose the demand function for a product is given by $(p + 1)q^2 = 10,000$, where p is the price and q is the quantity. Find the rate of change of quantity with respect to price when $p = \$99$.
15. The sales of a product are given by $S = 80,000e^{-0.4t}$, where S is the daily sales and t is the number of days after the end of an advertising campaign. Find the rate of sales decay 10 days after the end of the ad campaign.
16. The following table gives the average daily shares traded (in millions) on the New York Stock Exchange (NYSE) for selected years from 1970 to 2002. If the equation that models these data is

    $$y = 0.4843e^{0.1513(x - 50)}$$

where $x = 0$ represents 1900, what does the model give for the rate of growth of the NYSE average daily shares traded in (a) 1990 and (b) 2010?

Year	Daily Average	Year	Daily Average
1970	11.6	2000	1041.6
1980	44.9	2001	1240.0
1990	156.8	2002	1441.0
1995	346.1		

Source: New York Stock Exchange, *Fact Book* 2002

17. Suppose the demand and supply functions for a product are $p = 1100 - 5q$ and $p = 20 + 0.4q$, respectively, where p is in dollars and q is the number of units. Find the tax per unit t that will maximize the tax revenue $T = tq$.
18. **Modeling** The following table shows various years from 1950 to 2000 and, for those years, the percent of U.S. roads and streets that were paved.
    (a) Find a logarithmic equation $y = f(x)$ that models these data. Use $x = 0$ to represent 1940.
    (b) Find the function that models the rate of change of the percent of paved roads and streets.
    (c) Find and interpret $y(70)$ and $y'(70)$.

Year	1950	1960	1970	1980	1990	1995	1999	2000
Percent	23.5	34.7	44.5	53.7	58.4	60.8	62.4	63.4

Source: U.S. Department of Transportation, *Highway Statistics Summary to 1995* and *Highway Statistics, 2000.*

19. By using data from the U.S. Bureau of Labor Statistics for the years 1970–2002, the purchasing power P of a 1983 dollar can be modeled by the function

    $$P(t) = 3.655(0.9522)^t$$

where t is the number of years past 1960.

Use the model to predict the rate of change of the purchasing power of the 1983 dollar in the year 2010.

Extended Applications & Group Projects

I. Inflation

Hollingsworth Pharmaceuticals specializes in manufacturing generic medicines. Recently it developed an antibiotic with outstanding profit potential. The new antibiotic's total costs, sales, and sales growth, as well as projected inflation, are described as follows.

Total monthly costs, in dollars, to produce x units (1 unit is 100 capsules):

$$C(x) = \begin{cases} 15,000 + 10x & 0 \leq x \leq 11,000 \\ 15,000 + 10x + 0.001(x - 11,000)^2 & x \geq 11,000 \end{cases}$$

Sales: 10,000 units per month and growing at 1.25% per month, compounded continuously
Selling price: $17 per unit
Inflation: Approximately 0.25% per month, compounded continuously, affecting both total costs and selling price

Company owners are pleased with the sales growth but are concerned about the projected increase in variable costs when production levels exceed 11,000 units per month. The consensus is that improvements eventually can be made that will reduce costs at higher production levels, thus altering the current cost function model. To plan properly for these changes. Hollingsworth Pharmaceuticals would like you to determine when the company's profits will begin to decrease. To help you determine this, answer the following.

1. If inflation is assumed to be compounded continuously, the selling price and total costs must be multiplied by the factor $e^{0.0025t}$. In addition, if sales growth is assumed to be compounded continuously, then sales must be multiplied by a factor of the form e^{rt}, where r is the monthly sales growth rate (expressed as a decimal) and t is time in months. Use these factors to write each of the following as a function of time t:
   (a) selling price p per unit (including inflation)
   (b) number of units x sold per month (including sales growth)
   (c) total revenue (Recall that $R = px$.)
2. Determine how many months it will be before monthly sales exceed 11,000 units.
3. If you restrict your attention to total costs when $x \geq 11,000$, then, after expanding and collecting like terms, $C(x)$ can be written as follows:

$$C(x) = 136,000 - 12x + 0.001x^2 \quad \text{for } x \geq 11,000$$

   Use this form for $C(x)$ with your result from Question 1(b) and with the inflationary factor $e^{0.0025t}$ to express these total costs as a function of time.
4. Form the profit function that would be used when monthly sales exceed 11,000 units by using the total revenue function from Question 1(c) and the total cost function from Question 3. This profit function should be a function of time t.
5. Find how long it will be before the profit is maximized. You may have to solve $P'(t) = 0$ by using a graphing calculator or computer to find the t-intercept of the graph of $P'(t)$. In addition, because $P'(t)$ has large numerical coefficients, you may want to divide both sides of $P'(t) = 0$ by 1000 before solving or graphing.

II. Knowledge Workers (Modeling)

In January 1997, *Working Woman* made the following points about the U.S. economy and the place of women in the economy.

■ The telecommunications industry employs more people than the auto and auto parts industries combined.

■ More Americans make semiconductors than make construction equipment.

■ Almost twice as many Americans make surgical and medical instruments as make plumbing and heating products.

■ The ratio of male to female knowledge workers (engineers, scientists, technicians, professionals, and senior managers) was 3 to 2 in 1983. The following table, which gives the numbers (in millions) of male and female knowledge workers from 1983 to 1997, shows how that ratio is changing.

Year	Female Knowledge Workers (millions)	Male Knowledge Workers (millions)	Year	Female Knowledge Workers (millions)	Male Knowledge Workers (millions)
1983	11.0	15.4	1991	16.1	18.4
1984	11.6	15.9	1992	16.7	18.66
1985	12.3	16.3	1993	17.3	18.7
1986	12.9	16.7	1994	18.0	19.0
1987	13.6	16.8	1995	18.5	19.8
1988	14.3	17.6	1996	19.0	19.6
1989	15.3	18.1	1997	19.5	19.8
1990	15.9	18.6			

Source: Working Woman, January 1997

To compare how the growth in the number of female knowledge workers compares with that of male knowledge workers, do the following:

1. Find a logarithmic equation (with $x = 0$ representing 1980) that models the number of females, and find a logarithmic equation (with $x = 0$ representing 1980) that models the number of males.
2. Use the models found in (1) to find the rate of growth with respect to time of (a) the number of female knowledge workers and (b) the number of male knowledge workers.
3. Compare the two rates of growth in the year 2000 and determine which rate is larger.
4. If these models indicate that it is possible for the number of females to equal the number of males, during what year do they indicate that this will occur?
5. Research this topic on the web to determine whether the trends of your models are supported by current data. Be sure to cite your sources and include recent data.

12

Indefinite Integrals

When we know the derivative of a function, it is often useful to determine the function itself. For example, accountants can use linear regression to translate information about marginal cost into a linear equation defining (approximately) the marginal cost function, and then use the process of **antidifferentiation** (or **integration**) as part of finding the (approximate) total cost function. We can also use integration to find total revenue functions from marginal revenue functions, to optimize profit from information about marginal cost and marginal revenue, and to find national consumption from information about marginal propensity to consume.

Integration can also be used in the social and life sciences to predict growth or decay from expressions giving rates of change. For example, we can find equations for population size from the rate of change of growth, we can write equations for the number of radioactive atoms remaining in a substance if we know the rate of the decay of the substance, and we can determine the volume of blood flow from information about rate of flow.

The topics and applications discussed in this chapter include the following.

Sections	Applications
12.1 The Indefinite Integral	*Revenue, population growth*
12.2 The Power Rule	*Revenue, productivity*
12.3 Integrals Involving Exponential and Logarithmic Functions	*Real estate inflation, population growth*
12.4 Applications of the Indefinite Integral in Business and Economics Total cost and profit National consumption and savings	*Cost, revenue, maximum profit, national consumption and savings*
12.5 Differential Equations Solution of differential equations Separable differential equations Applications of differential equations	*Investment, carbon-14 dating, drug in an organ*

Chapter Warm-up

Prerequisite Problem Type	For Section	Answer	Section for Review
Write as a power: (a) $\sqrt{x}$ (b) $\sqrt{x^2 - 9}$	12.1–12.4	(a) $x^{1/2}$ (b) $(x^2 - 9)^{1/2}$	0.4 Radicals
Expand $(x^2 + 4)^2$.	12.2	$x^4 + 8x^2 + 16$	0.5 Special powers
Divide $x^4 - 2x^3 + 4x^2 - 7x - 1$ by $x^2 - 2x$.	12.3	$x^2 + 4 + \dfrac{x - 1}{x^2 - 2x}$	0.5 Division
Find the derivative of (a) $f(x) = 2x^{1/2}$ (b) $u = x^3 - 3x$	12.1 12.2 12.3	(a) $f'(x) = x^{-1/2}$ (b) $u' = 3x^2 - 3$	9.4 Derivatives
If $y = \dfrac{(x^2 + 4)^6}{6}$, what is y'?	12.2	$(x^2 + 4)^5 2x$	9.6 Derivatives
(a) If $y = \ln u$, what is y'? (b) If $y = e^u$, what is y'?	12.3	(a) $y' = \dfrac{1}{u} \cdot u'$ (b) $y' = e^u \cdot u'$	11.1, 11.2 Derivatives
Solve for y: $\ln y = kt + C$	12.5	$y = e^{kt + C}$	5.2 Logarithmic functions
Solve for k: $0.5 = e^{5600k}$	12.5	$k \approx -0.00012378$	5.3 Exponential equations

The Indefinite Integral

- To find certain indefinite integrals

◗ Application Preview

In our study of the theory of the firm, we have worked with total cost, total revenue, and profit functions and have found their marginal functions. In practice, it is often easier for a company to measure marginal cost, revenue, and profit and use these data to form marginal functions from which it can find total cost, revenue, and profit functions. For example, Jarus Technologies manufactures computer motherboards, and the company's sales records show that the marginal revenue (in dollars per unit) for its motherboards is given by

$$\overline{MR} = 300 - 0.2x$$

where x is the number of units sold. If we want to use this function to find the total revenue function for Jarus Technologies' motherboards, we need to find $R(x)$ from $\overline{MR} = R'(x)$. (See Example 7.) In this situation, we need to be able to reverse the process of differentiation. This reverse process is called **antidifferentiation,** or **integration.**

Indefinite Integrals

We have studied procedures for and applications of finding derivatives of a given function. We now turn our attention to reversing this process of differentiation. When we know the derivative of a function, the process of finding the function itself is called **antidifferentiation.** For example, if the derivative of a function is $2x$, we know that the function could be $f(x) = x^2$ because $\dfrac{d}{dx}(x^2) = 2x$. But the function could also be $f(x) = x^2 + 4$ because $\dfrac{d}{dx}(x^2 + 4) = 2x$. It is clear that any function of the form $f(x) = x^2 + C$, where C is an arbitrary constant, will have $f'(x) = 2x$ as its derivative. Thus we say that the **general antiderivative** of $f'(x) = 2x$ is $f(x) = x^2 + C$, where C is an arbitrary constant.

The process of finding an antiderivative is also called **integration.** The function that results when integration takes place is called an **indefinite integral** or, more simply, an **integral.** We can denote the indefinite integral (that is, the general antiderivative) of a function $f(x)$ by $\int f(x)\,dx$. Thus we can write $\int 2x\,dx$ to indicate the general antiderivative of the function $f(x) = 2x$. The expression is read as "the integral of $2x$ with respect to x." In this case, $2x$ is called the **integrand.** The **integral sign,** $\int$, indicates the process of integration, and the dx indicates that the integral is to be taken with respect to x. Because the antiderivative of $2x$ is $x^2 + C$, we can write

$$\int 2x\,dx = x^2 + C$$

● EXAMPLE 1 *Antidifferentiation*

If $f'(x) = 3x^2$, what is $f(x)$?

Solution
The derivative of the function $f(x) = x^3$ is $f'(x) = 3x^2$. But other functions also have this derivative. They will all be of the form $f(x) = x^3 + C$, where C is a constant. Thus we write

$$\int 3x^2\,dx = x^3 + C$$

● **EXAMPLE 2** *Integration*

If $f'(x) = x^3$, what is $f(x)$?

Solution

We know that $\dfrac{d}{dx}(x^4) = 4x^3$, so the derivative of $f(x) = \frac{1}{4}x^4$ is $f'(x) = x^3$. Thus

$$f(x) = \int f'(x)\,dx = \int x^3\,dx = \tfrac{1}{4}x^4 + C$$

Powers of x Formula   It is easily seen that

$$\int x^4\,dx = \frac{x^5}{5} + C \quad \text{because} \quad \frac{d}{dx}\!\left(\frac{x^5}{5} + C\right) = x^4$$

$$\int x^5\,dx = \frac{x^6}{6} + C \quad \text{because} \quad \frac{d}{dx}\!\left(\frac{x^6}{6} + C\right) = x^5$$

In general, we have the following:

Powers of x Formula

$$\int x^n\,dx = \frac{x^{n+1}}{n+1} + C \quad \text{(for } n \neq -1)$$

In the Powers of x Formula, we see that $n \neq -1$ is essential, because if $n = -1$, then the denominator $n + 1 = 0$. We will discuss the case when $n = -1$ later. (Can you think what function has $1/x$ as its derivative?)

In addition, we can see that this Powers of x Formula applies for any $n \neq -1$ by noting that

$$\frac{d}{dx}\!\left(\frac{x^{n+1}}{n+1} + C\right) = \frac{d}{dx}\!\left(\frac{1}{n+1}x^{n+1} + C\right) = \frac{n+1}{n+1}x^n = x^n$$

● **EXAMPLE 3** *Powers of x Formula*

Evaluate $\int x^{-1/2}\,dx$.

Solution

Using the formula, we get

$$\int x^{-1/2}\,dx = \frac{x^{-1/2+1}}{-1/2+1} + C = \frac{x^{1/2}}{1/2} + C = 2x^{1/2} + C$$

We can check by noting that the derivative of $2x^{1/2} + C$ is $x^{-1/2}$. ✔

Note that the indefinite integral in Example 3 is a function (actually a number of functions, one for each value of C). Graphs of several members of this family of functions are shown in Figure 12.1. Note that at any given x-value, the tangent line to each curve would have the same slope, indicating that all family members have the same derivative.

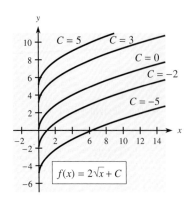

Figure 12.1

● **EXAMPLE 4** *Powers of x Formula*

Find (a) $\int \sqrt[3]{x}\, dx$ and (b) $\int \dfrac{1}{x^2}\, dx$.

Solution

(a) $\displaystyle\int \sqrt[3]{x}\, dx = \int x^{1/3}\, dx = \frac{x^{4/3}}{4/3} + C$

$\qquad = \dfrac{3}{4} x^{4/3} + C = \dfrac{3}{4} \sqrt[3]{x^4} + C$

(b) We write the power of x in the numerator so that the integral has the form in the formula above.

$$\int \frac{1}{x^2}\, dx = \int x^{-2}\, dx = \frac{x^{-2+1}}{-2+1} + C = \frac{x^{-1}}{-1} + C = \frac{-1}{x} + C$$

Other Formulas and Properties

Other formulas will be useful in evaluating integrals. The following table shows how some new integration formulas result from differentiation formulas.

Integration Formulas	
Derivative	Resulting Integral
$\dfrac{d}{dx}(x) = 1$	$\displaystyle\int 1\, dx = \int dx = x + C$
$\dfrac{d}{dx}[c \cdot u(x)] = c \cdot \dfrac{d}{dx} u(x)$	$\displaystyle\int c\, u(x)\, dx = c \int u(x)\, dx$
$\dfrac{d}{dx}[u(x) \pm v(x)] = \dfrac{d}{dx} u(x) \pm \dfrac{d}{dx} v(x)$	$\displaystyle\int [u(x) \pm v(x)]\, dx = \int u(x)\, dx \pm \int v(x)\, dx$

The formulas above indicate that we can integrate functions term by term just as we were able to take derivatives term by term.

● **EXAMPLE 5** *Using Integration Formulas*

Evaluate: (a) $\int 4\,dx$ (b) $\int 8x^5\,dx$ (c) $\int (x^3 + 4x)\,dx$

Solution

(a) $\int 4\,dx = 4\int dx = 4(x + C_1) = 4x + C$

    (Because C_1 is an unknown constant, we can write $4C_1$ as the unknown constant C.)

(b) $\displaystyle \int 8x^5\,dx = 8\int x^5\,dx = 8\left(\frac{x^6}{6} + C_1\right) = \frac{4x^6}{3} + C$

(c) $\int (x^3 + 4x)\,dx = \int x^3\,dx + \int 4x\,dx$

$$= \left(\frac{x^4}{4} + C_1\right) + \left(4 \cdot \frac{x^2}{2} + C_2\right)$$

$$= \frac{x^4}{4} + 2x^2 + C_1 + C_2$$

$$= \frac{x^4}{4} + 2x^2 + C$$

Note that we need only one constant because the sum of C_1 and C_2 is just a new constant.

● **EXAMPLE 6** *Integral of a Polynomial*

Evaluate $\int (x^2 - 4)^2\,dx$.

Solution

We expand $(x^2 - 4)^2$ so that the integrand is in a form that fits the basic integration formulas.

$$\int (x^2 - 4)^2\,dx = \int (x^4 - 8x^2 + 16)\,dx = \frac{x^5}{5} - \frac{8x^3}{3} + 16x + C$$

● **Checkpoint**

1. True or false:
   (a) $\int (4x^3 - 2x)\,dx = \int 4x^3\,dx - \int 2x\,dx$
   $$= (x^4 + C) - (x^2 + C) = x^4 - x^2$$
   (b) $\displaystyle \int \frac{1}{3x^2}\,dx = \frac{1}{3(x^3/3)} + C = \frac{1}{x^3} + C$

2. Evaluate $\int (2x^3 + x^{-1/2} - 4x^{-5})\,dx$.

◗ **EXAMPLE 7** *Revenue* **(Application Preview)**

Sales records at Jarus Technologies show that the rate of change of the revenue (that is, the marginal revenue) in dollars per unit for a motherboard is $\overline{MR} = 300 - 0.2x$, where x represents the quantity sold. Find the total revenue function for the product. Then find the total revenue from the sale of 1000 motherboards.

Solution

We know that the marginal revenue can be found by differentiating the total revenue function. That is,

$$R'(x) = 300 - 0.2x$$

Thus integrating the marginal revenue function gives the total revenue function.

$$R(x) = \int (300 - 0.2x) \, dx = 300x - 0.1x^2 + K^*$$

We can use the fact that there is no revenue when no units are sold to evaluate K. Setting $x = 0$ and $R = 0$ gives $0 = 300(0) - 0.1(0)^2 + K$, so $K = 0$. Thus the total revenue function is

$$R(x) = 300x - 0.1x^2$$

The total revenue from the sale of 1000 motherboards is

$$R(1000) = 300(1000) - 0.1(1000^2) = \$200,000$$

We can check that the $R(x)$ we found in Example 7 is correct by verifying that $R'(x) = 300 - 0.2x$ and $R(0) = 0$. Also, graphs can help us check the reasonableness of our result. Figure 12.2 shows the graphs of $\overline{MR} = 300 - 0.2x$ and of the $R(x)$ we found. Note that $R(x)$ passes through the origin, indicating $R(0) = 0$. Also, reading both graphs from left to right, we see that $R(x)$ increases when $\overline{MR} > 0$, attains its maximum when $\overline{MR} = 0$, and decreases when $\overline{MR} < 0$.

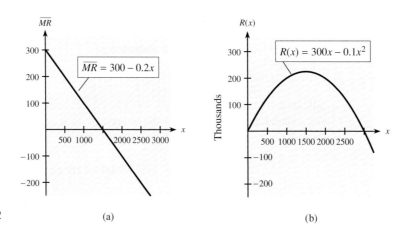

Figure 12.2   (a)   (b)

Calculator Note    We mentioned in Chapter 9, "Derivatives," that graphing calculators have a numerical derivative feature that can be used to check graphically the derivative of a function that has been calculated with a formula. We can also use the numerical integration feature on graphing calculators to check our integration (if we assume temporarily that the constant of integration is 0). We do this by graphing the integral calculated with a formula and the numerical integral from the graphing calculator on the same set of axes. If the graphs lie on top of one another, the integrals agree. Figure 12.3 illustrates this for the function $f(x) = 3x^2 - 2x + 1$. Its integral, with the constant of integration set equal to 0, is shown as $y_1 = x^3 - x^2 + x$ in Figure 12.3(a), and the graphs of the two integrals are shown in Figure 12.3(b). Of course, it is often easier to use the derivative to check integration.

---

*Here we are using K rather than C to represent the constant of integration to avoid confusion between the constant C and the cost function $C = C(x)$.

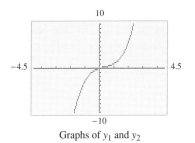

```
Y1=X^3-X^2+X
Y2=fnInt(3X^2-2X
+1,X,0,X)
Y3=
Y4=
Y5=
Y6=
Y7=
```

Graphs of y_1 and y_2

Figure 12.3   (a)   (b)

● **Checkpoint Solutions**

1. (a) False, $\int (4x^3 - 2x)\, dx = \int 4x^3\, dx - \int 2x\, dx = (x^4 + C_1) - (x^2 + C_2)$
$$= x^4 - x^2 + C$$

   (b) False, $\displaystyle\int \frac{1}{3x^2}\, dx = \int \frac{1}{3} \cdot \frac{1}{x^2}\, dx = \frac{1}{3}\int x^{-2}\, dx$
$$= \frac{1}{3} \cdot \frac{x^{-1}}{-1} + C = \frac{-1}{3x} + C$$

2. $\displaystyle\int (2x^3 + x^{-1/2} - 4x^{-5})\, dx = \frac{2x^4}{4} + \frac{x^{1/2}}{1/2} - \frac{4x^{-4}}{-4} + C$
$$= \frac{x^4}{2} + 2x^{1/2} + x^{-4} + C$$

12.1 Exercises

1. If $f'(x) = 4x^3$, what is $f(x)$?
2. If $f'(x) = 5x^4$, what is $f(x)$?
3. If $f'(x) = x^6$, what is $f(x)$?
4. If $g'(x) = x^4$, what is $g(x)$?

Evaluate the integrals in Problems 5–26. Check your answers by differentiating.

5. $\int x^7\, dx$ 6. $\int x^5\, dx$

7. $\int 8x^5\, dx$ 8. $\int 16x^9\, dx$

9. $\int (3^3 + x^{13})\, dx$ 10. $\int (5^2 + x^{10})\, dx$

11. $\int (3 - x^{3/2})\, dx$ 12. $\int (8 + x^{2/3})\, dx$

13. $\int (x^4 - 9x^2 + 3)\, dx$ 14. $\int (3x^2 - 4x - 4)\, dx$

15. $\int (2 + 2\sqrt{x})\, dx$ 16. $\int (17 + \sqrt{x^3})\, dx$

17. $\int 6\sqrt[4]{x}\, dx$ 18. $\int 3\sqrt[3]{x^2}\, dx$

19. $\displaystyle\int \frac{5}{x^4}\, dx$ 20. $\displaystyle\int \frac{6}{x^5}\, dx$

21. $\displaystyle\int \frac{dx}{2\sqrt[3]{x^2}}$ 22. $\displaystyle\int \frac{2\, dx}{5\sqrt{x^3}}$

23. $\displaystyle\int \left(x^3 - 4 + \frac{5}{x^6}\right) dx$

24. $\displaystyle\int \left(x^3 - 7 - \frac{3}{x^4}\right) dx$

25. $\displaystyle\int \left(x^9 - \frac{1}{x^3} + \frac{2}{\sqrt[3]{x}}\right) dx$

26. $\displaystyle\int \left(3x^8 + \frac{4}{x^8} - \frac{5}{\sqrt[5]{x}}\right) dx$

In Problems 27–30, use algebra to rewrite the integrands; then integrate and simplify.

27. $\int (4x^2 - 1)^2 x^3\, dx$ 28. $\int (x^3 + 1)^2 x\, dx$

29. $\displaystyle\int \frac{x + 1}{x^3}\, dx$ 30. $\displaystyle\int \frac{x - 3}{\sqrt{x}}\, dx$

In Problems 31 and 32, find the antiderivatives and graph the resulting family members that correspond to $C = 0$, $C = 4$, $C = -4$, $C = 8$, and $C = -8$.

31. $\int (2x + 3)\, dx$

32. $\int (4 - x)\, dx$

33. If $\int f(x)\, dx = 2x^9 - 7x^5 + C$, find $f(x)$.

34. If $\int g(x)\, dx = 11x^{10} - 4x^3 + C$, find $g(x)$.

In each of Problems 35–38, a family of functions is given and graphs of some members of the family are shown. Write the indefinite integral that gives the family.

35. $F(x) = 5x - \dfrac{x^2}{4} + C$

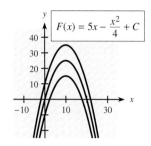

36. $F(x) = \dfrac{x^2}{2} + 3x + C$

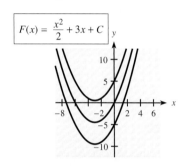

37. $F(x) = x^3 - 3x^2 + C$

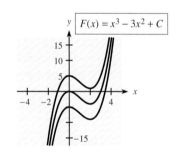

38. $F(x) = 12x - x^3 + C$

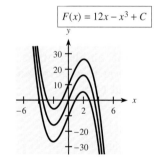

APPLICATIONS

39. *Revenue* If the marginal revenue (in dollars per unit) for a month for a commodity is $\overline{MR} = -0.4x + 30$, find the total revenue function.

40. *Revenue* If the marginal revenue (in dollars per unit) for a month for a commodity is $\overline{MR} = -0.05x + 25$, find the total revenue function.

41. *Revenue* If the marginal revenue (in dollars per unit) for a month is given by $\overline{MR} = -0.3x + 450$, what is the total revenue from the production and sale of 50 units?

42. *Revenue* If the marginal revenue (in dollars per unit) for a month is given by $\overline{MR} = -0.006x + 36$, find the total revenue from the sale of 75 units.

43. *Stimulus-response* Suppose that when a sense organ receives a stimulus at time t, the total number of action potentials is $P(t)$. If the rate at which action potentials are produced is $t^3 + 4t^2 + 6$, and if there are 0 action potentials when $t = 0$, find the formula for $P(t)$.

44. *Projectiles* Suppose that a particle has been shot into the air in such a way that the rate at which its height is changing is $v = 320 - 32t$, in feet per second, and suppose that it is 1600 feet high when $t = 10$ seconds. Write the equation that describes the height of the particle at any time t.

45. *Pollution* A factory is dumping pollutants into a river at a rate given by $dx/dt = t^{3/4}/600$ tons per week, where t is the time in weeks since the dumping began and x is the number of tons of pollutants.
    (a) Find the equation for total tons of pollutants dumped.
    (b) How many tons were dumped during the first year?

46. *Population growth* The rate of growth of the population of a city is predicted to be

$$\frac{dp}{dt} = 1000t^{1.08}$$

where p is the population at time t, and t is measured in years from the present. Suppose that the current population is 100,000. What is the predicted
    (a) rate of growth 5 years from the present?
    (b) population 5 years from the present?

47. *Average cost* The DeWitt Company has found that the rate of change of its average cost for a product is

$$\overline{C}'(x) = \frac{1}{4} - \frac{100}{x^2}$$

where x is the number of units and cost is in dollars. The average cost of producing 20 units is $40.00.
    (a) Find the average cost function for the product.
    (b) Find the average cost of 100 units of the product.

48. *Oil leakage* An oil tanker hits a reef and begins to leak. The efforts of the workers repairing the leak cause the rate at which the oil is leaking to decrease. The oil was leaking at a rate of 31 barrels per hour at the end

of the first hour after the accident, and the rate is decreasing at a rate of one barrel per hour.
(a) What function describes the rate of loss?
(b) How many barrels of oil will leak in the first 6 hours?
(c) When will the oil leak be stopped? How much will have leaked altogether?

49. *Social Security Trust Fund* With fewer workers and more early retirements, future income from payroll taxes used to fund Social Security benefits won't keep pace with scheduled benefits. In fact, the Social Security Trust Fund balance B has been changing at a rate given by

$$\frac{dB}{dt} = -2.1136t + 8.2593$$

billion dollars per year, where t is the number of years past 2000 (*Source:* Social Security Administration).
(a) How long until the trust fund balance begins to decrease?
(b) If the trust fund balance in 2000 was \$74.07 billion, find the function that models the trust fund balance.
(c) How long until the balance is predicted to be zero?

50. *Internet use* The rate of change of the number of millions of U.S. Internet users, y, can be modeled by

$$\frac{dy}{dt} = -2.486t + 30.41$$

where $t = 0$ represents 1995 (*Source:* Jupiter Media Metrix).
(a) When does this model predict the number of Internet users will begin to decrease?
(b) If there were 56.8 million Internet users in 1997, find the function that models the number of Internet users.

51. *Mutual funds* Household assets invested in mutual funds, A, in billions of dollars, have been changing at a rate given by

$$\frac{dA}{dt} = 160.869t^{0.5307}$$

where t is the number of years past 1990 (*Source:* Federal Reserve, *Time,* April 3, 2000).
(a) Are both the amount of assets and the rate of change of assets increasing? Explain.
(b) If there were \$1234.5 billion in household assets invested in 1995, find the function that models the total amount of household assets invested in mutual funds.

52. *Tax load* According to data from the Bureau of Economic Analysis, U.S. Department of Commerce, the percent of personal income, p, used to pay personal taxes has been changing at a rate given by

$$\frac{dp}{dt} = 0.07250t - 0.12312$$

percent per year, where t is the number of years past 1990.
(a) According to this model, did the percent ever decrease during the 1990s? Explain.
(b) In 1991, 12.6% of personal income was used to pay personal taxes. Use this information to find the function that models the percent of personal income used to pay personal taxes.
(c) When does the model predict that 25% of personal income will be used to pay personal taxes?

53. *National health care* National expenditures for health care H (measured in billions of dollars) have risen dramatically since 1960 when the total was \$26.7 billion. According to data from the U.S. Department of Health and Human Services, Office of National Health Statistics, the rate of change of expenditures can be modeled by

$$\frac{dH}{dt} = 0.0126t^2 + 1.532t - 5.08$$

billions of dollars per year, where $t = 0$ represents 1960.
(a) Find the function that models the national expenditures for health care.
(b) Use the model from part (a) to predict the national health care expenditures for 2010.

54. *Consumer price index—medical care* The consumer price index (CPI) measures how prices have changed for consumers. With 1982 as a reference of 100, a year with CPI = 150 indicates that medical care costs in that year were 1.5 times as high as in 1982. Using Bureau of Labor Statistics data for selected years from 1980 to 2002, it is found that the CPI for urban medical care, M, has been changing at a rate given by

$$\frac{dM}{dt} = -0.01668t^2 + 0.4452t + 7.242$$

points per year, where t is the number of years past 1980.
(a) Medical care costs have been steadily increasing since 1980. According to the model, when did they begin to increase at a slower rate? That is, when did the rate begin to decrease?
(b) Find the function that models the CPI for medical care if the CPI was 75 in 1980.
(c) When does the function from part (b) predict that medical care costs will be 3.5 times what they were in 1982?

OBJECTIVE

● To evaluate integrals of the form $\int u^n \cdot u'\, dx = \int u^n\, du$ if $n \neq -1$

The Power Rule

◗ Application Preview

In the previous section, we saw that total revenue could be found by integrating marginal revenue. That is,

$$R(x) = \int \overline{MR}\, dx$$

For example, if the marginal revenue for a product is given by

$$\overline{MR} = \frac{600}{\sqrt{3x+1}} + 2$$

then

$$R(x) = \int \left[\frac{600}{\sqrt{3x+1}} + 2 \right] dx$$

To evaluate this integral, however, we need a more general formula than the Powers of *x* Formula. (See Example 9.)

In this section, we will extend the Powers of *x* Formula to a rule for powers of a function of *x*.

Differentials

Our goal in this section is to extend the Powers of *x* Formula,

$$\int x^n\, dx = \frac{x^{n+1}}{n+1} + C \quad (n \neq -1)$$

to powers of a function of *x*. In order to do this, we must understand the symbol *dx*.

Recall from Section 9.3, "Average and Instantaneous Rates of Change: The Derivative," that the derivative of $y = f(x)$ with respect to *x* can be denoted by *dy/dx*. As we will see, there are advantages to using *dy* and *dx* as separate quantities whose ratio *dy/dx* equals $f'(x)$.

Differentials

If $y = f(x)$ is a differentiable function with derivative $dy/dx = f'(x)$, then the **differential of *x*** is *dx*, and the **differential of *y*** is *dy*, where

$$dy = f'(x)\, dx$$

Although differentials are useful in certain approximation problems, we are interested in the differential notation at this time.

● EXAMPLE 1 *Differential*

Find (a) *dy* if $y = x^3 - 4x^2 + 5$ and (b) *du* if $u = 5x^4 + 11$.

Solution
(a) $dy = f'(x)\, dx = (3x^2 - 8x)\, dx$
(b) If the dependent variable in a function is *u*, then $du = u'(x)\, dx$.

$$du = u'(x)\, dx = 20x^3\, dx$$

The Power Rule

In terms of our goal of extending the Powers of *x* Formula, we would suspect that if *x* is replaced by a function of *x*, then *dx* should be replaced by the differential of that function. Let's see whether this is true.

Recall that if $y = [u(x)]^n$, the derivative of y is

$$\frac{dy}{dx} = n[u(x)]^{n-1} \cdot u'(x)$$

Using this formula for derivatives, we can see that

$$\int n[u(x)]^{n-1} \cdot u'(x)\, dx = [u(x)]^n + C$$

It is easy to see that this formula is equivalent to the following formula, which is called the **Power Rule for Integration.**

Power Rule for Integration

$$\int [u(x)]^n \cdot u'(x)\, dx = \frac{[u(x)]^{n+1}}{n+1} + C, \qquad n \neq -1$$

Using the fact that

$$du = u'(x)\, dx \quad \text{or} \quad du = u'\, dx$$

we can write the Power Rule in the following alternative form.

Power Rule (Alternative Form)

If $u = u(x)$, then

$$\int u^n\, du = \frac{u^{n+1}}{n+1} + C, \qquad n \neq -1$$

Note that this formula has the same form as the formula

$$\int x^n\, dx = \frac{x^{n+1}}{n+1} + C, \qquad n \neq -1$$

with the *function u substituted for x and du substituted for dx.*

● **EXAMPLE 2 Power Rule**

Evaluate $\int (3x^2 + 4)^5 \cdot 6x\, dx$.

Solution
To use the Power Rule, we must be sure that we have the function $u(x)$, its derivative $u'(x)$, and n.

$$u = 3x^2 + 4, \qquad n = 5$$
$$u' = 6x$$

All required parts are present, so the integral is of the form

$$\int (3x^2 + 4)^5\, 6x\, dx = \int u^5 \cdot u'\, dx = \int u^5\, du$$

$$= \frac{u^6}{6} + C = \frac{(3x^2 + 4)^6}{6} + C$$

We can check the integration by noting that the derivative of

$$\frac{(3x^2 + 4)^6}{6} + C \quad \text{is} \quad (3x^2 + 4)^5 \cdot 6x$$

● **EXAMPLE 3**   *Power Rule*

Evaluate $\int \sqrt{2x + 3} \cdot 2 \, dx$.

Solution

If we let $u = 2x + 3$, then $u' = 2$, and so we have

$$\int \sqrt{2x + 3} \cdot 2 \, dx = \int \sqrt{u} \, u' \, dx = \int \sqrt{u} \, du$$

$$= \int u^{1/2} \, du = \frac{u^{3/2}}{3/2} + C$$

Because $u = 2x + 3$, we have

$$\int \sqrt{2x + 3} \cdot 2 \, dx = \frac{2}{3}(2x + 3)^{3/2} + C$$

Check: The derivative of $\frac{2}{3}(2x + 3)^{3/2} + C$ is $(2x + 3)^{1/2} \cdot 2$. ✔

Some members of the family of functions given by

$$\int \sqrt{2x + 3} \cdot 2 \, dx = \frac{2}{3}(2x + 3)^{3/2} + C$$

are shown in Figure 12.4. Note from the graphs that the domain of each function is $x \geq -3/2$. This is because $2x + 3$ must be nonnegative so that $(2x + 3)^{3/2} = (\sqrt{2x + 3})^3$ is a real number.

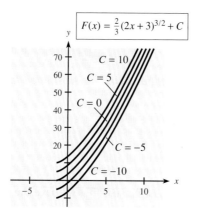

Figure 12.4

● **EXAMPLE 4** *Power Rule*

Evaluate $\int x^3(5x^4 + 11)^9 \, dx$.

Solution

If we let $u = 5x^4 + 11$, then $u' = 20x^3$. Thus we do not have an integral of the form $\int u^n \cdot u' \, dx$, as we had in Example 2 and Example 3; the factor 20 is not in the integrand. To get the integrand in the correct form, we can multiply by 20 and divide it out as follows:

$$\int x^3(5x^4 + 11)^9 \, dx = \int (5x^4 + 11)^9 \cdot x^3 \, dx = \int (5x^4 + 11)^9 \left(\frac{1}{20}\right)(20x^3) \, dx$$

Because $\frac{1}{20}$ is a constant factor, we can factor it outside the integral sign, getting

$$\int (5x^4 + 11)^9 \cdot x^3 \, dx = \frac{1}{20} \int (5x^4 + 11)^9 (20x^3) \, dx$$

$$= \frac{1}{20} \int u^9 \cdot u' \, dx = \frac{1}{20} \cdot \frac{u^{10}}{10} + C$$

$$= \frac{1}{200} (5x^4 + 11)^{10} + C$$

● **EXAMPLE 5 *Power Rule***

Evaluate $\int 5x^2 \sqrt{x^3 - 4} \, dx$.

Solution
If we let $u = x^3 - 4$, then $u' = 3x^2$. Thus we need the factor 3, rather than 5, in the integrand. If we first reorder the factors and then multiply by the constant factor 3 (and divide it out), we have

$$\int \sqrt{x^3 - 4} \cdot 5x^2 \, dx = \int \sqrt{x^3 - 4} \cdot \frac{5}{3} (3x^2) \, dx$$

$$= \frac{5}{3} \int (x^3 - 4)^{1/2} \cdot 3x^2 \, dx$$

This integral is of the form $\frac{5}{3} \int u^{1/2} \cdot u' \, dx$, resulting in

$$\frac{5}{3} \cdot \frac{u^{3/2}}{3/2} + C = \frac{5}{3} \cdot \frac{(x^3 - 4)^{3/2}}{3/2} + C = \frac{10}{9} (x^3 - 4)^{3/2} + C$$

Note that we can factor a constant outside the integral sign to obtain the integrand in the form we seek, but if the integral requires the introduction of a variable to obtain the form $u^n \cdot u' \, dx$, we *cannot* use this form and must try something else.

● **EXAMPLE 6 *Power Rule Fails***

Evaluate $\int (x^2 + 4)^2 \, dx$.

Solution
If we let $u = x^2 + 4$, then $u' = 2x$. Because we would have to introduce a variable to get u' in the integral, we cannot solve this problem by using the Power Rule. We must find another method. We can evaluate this integral by squaring and then integrating term by term.

$$\int (x^2 + 4)^2 \, dx = \int (x^4 + 8x^2 + 16) \, dx$$

$$= \frac{x^5}{5} + \frac{8x^3}{3} + 16x + C$$

Note that if we tried to introduce the factor $2x$ into the integral in Example 6, we would get

$$\int (x^2 + 4)^2 \, dx = \int (x^2 + 4)^2 \cdot \frac{1}{2x} (2x) \, dx$$

Although it is tempting to factor $1/2x$ outside the integral and use the Power Rule, this leads to an "answer" that does not check. That is, the derivative of the "answer" is not the integrand. (Try it and see.) To emphasize again, we can introduce only *a constant factor* to get an integral in the proper form.

● **EXAMPLE 7** *Power Rule*

Evaluate $\int (2x^2 - 4x)^2(x - 1)\, dx$.

Solution

If we want to treat this as an integral of the form $\int u^n u'\, dx$, we will have to let $u = 2x^2 - 4x$. Then u' will be $4x - 4$. Multiplying and dividing by 4 will give us this form, as follows.

$$\int (2x^2 - 4x)^2(x - 1)\, dx = \int (2x^2 - 4x)^2 \cdot \frac{1}{4} \cdot 4(x - 1)\, dx$$

$$= \frac{1}{4}\int (2x^2 - 4x)^2(4x - 4)\, dx$$

$$= \frac{1}{4}\int u^2 u'\, dx = \frac{1}{4} \cdot \frac{u^3}{3} + C$$

$$= \frac{1}{4}\frac{(2x^2 - 4x)^3}{3} + C$$

$$= \frac{1}{12}(2x^2 - 4x)^3 + C$$

● **EXAMPLE 8** *Power Rule*

Evaluate $\int \dfrac{x^2 - 1}{(x^3 - 3x)^3}\, dx$.

Solution

This integral can be treated as $\int u^{-3} u'\, dx$ if we let $u = x^3 - 3x$.

$$\int \frac{x^2 - 1}{(x^3 - 3x)^3}\, dx = \int (x^3 - 3x)^{-3}(x^2 - 1)\, dx$$

Then $u' = 3x^2 - 3$ and we can multiply and divide by 3 to get the form we need.

$$= \int (x^3 - 3x)^{-3} \cdot \frac{1}{3} \cdot 3(x^2 - 1)\, dx$$

$$= \frac{1}{3}\int (x^3 - 3x)^{-3}(3x^2 - 3)\, dx$$

$$= \frac{1}{3}\left[\frac{(x^3 - 3x)^{-2}}{-2}\right] + C$$

$$= \frac{-1}{6(x^3 - 3x)^2} + C$$

1. Which of the following can be evaluated with the Power Rule?
   (a) $\int (4x^2 + 1)^{10}(8x \, dx)$
   (b) $\int (4x^2 + 1)^{10}(x \, dx)$
   (c) $\int (4x^2 + 1)^{10}(8 \, dx)$
   (d) $\int (4x^2 + 1)^{10} \, dx$

2. Which of the following is equal to $\int (2x^3 + 5)^{-2}(6x^2 \, dx)$?
   (a) $\dfrac{[(2x^4)/4 + 5x]^{-1}}{-1} \cdot \dfrac{6x^3}{3} + C$
   (b) $\dfrac{(2x^3 + 5)^{-1}}{-1} \cdot \dfrac{6x^3}{3} + C$
   (c) $\dfrac{(2x^3 + 5)^{-1}}{-1} + C$

3. True or false: Constants can be factored outside the integral sign.

4. Evaluate:
   (a) $\int (x^3 + 9)^5(3x^2 \, dx)$
   (b) $\int (x^3 + 9)^{15}(x^2 \, dx)$
   (c) $\int (x^3 + 9)^2(x \, dx)$

● **EXAMPLE 9** *Revenue* **(Application Preview)**

Suppose that the marginal revenue for a product is given by

$$\overline{MR} = \frac{600}{\sqrt{3x + 1}} + 2$$

Find the total revenue function.

Solution

$$R(x) = \int \overline{MR} \, dx = \int \left[\frac{600}{(3x + 1)^{1/2}} + 2 \right] dx$$

$$= \int 600(3x + 1)^{-1/2} \, dx + \int 2 \, dx$$

$$= 600 \left(\frac{1}{3} \right) \int (3x + 1)^{-1/2}(3 \, dx) + 2 \int dx$$

$$= 200 \frac{(3x + 1)^{1/2}}{1/2} + 2x + K$$

$$= 400 \sqrt{3x + 1} + 2x + K$$

We know that $R(0) = 0$, so we have

$$0 = 400\sqrt{1} + 0 + K \quad \text{or} \quad K = -400$$

Thus the total revenue function is

$$R(x) = 400\sqrt{3x + 1} + 2x - 400$$

Note in Example 9 that even though $R(0) = 0$, the constant of integration K was *not* 0. This is because $x = 0$ does not necessarily mean that $u(x)$ will also be 0.

Summary   The formulas we have stated and used in this and the previous section are all the result of "reversing" derivative formulas.

Integration Formula		**Derivative Formula**
$\int dx = x + C$	because	$\dfrac{d}{dx}(x + C) = 1$
$\int x^n\, dx = \dfrac{x^{n+1}}{n+1} + C \quad (n \ne -1)$	because	$\dfrac{d}{dx}\left(\dfrac{x^{n+1}}{n+1} + C\right) = x^n$
$\int [u(x) + v(x)]\, dx = \int u(x)\, dx + \int v(x)\, dx$	because	$\dfrac{d}{dx}[u(x) + v(x)] = \dfrac{du}{dx} + \dfrac{dv}{dx}$
$\int [u(x) - v(x)]\, dx = \int u(x)\, dx - \int v(x)\, dx$	because	$\dfrac{d}{dx}[u(x) - v(x)] = \dfrac{du}{dx} - \dfrac{dv}{dx}$
$\int cf(x)\, dx = c\int f(x)\, dx$	because	$\dfrac{d}{dx}[cf(x)] = c\left[\dfrac{df}{dx}\right]$
$\int u^n u'\, dx = \dfrac{u^{n+1}}{n+1} + C \quad (n \ne -1)$	because	$\dfrac{d}{dx}\left(\dfrac{u^{n+1}}{n+1} + C\right) = u^n \cdot u'$

Note that there are no integration formulas that correspond to "reversing" the derivative formulas for a product or for a quotient. This means that functions that may be easy to differentiate can be quite difficult (or even impossible) to integrate. Hence, in general, integration is a more difficult process than differentiation. In fact, some functions whose derivatives can be readily found cannot be integrated, such as

$$f(x) = \sqrt{x^3 + 1} \quad \text{and} \quad g(x) = \frac{2x^2}{x^4 + 1}$$

In the next section, we will add to this list of basic integration formulas by "reversing" the derivative formulas for exponential and logarithmic functions.

● *Checkpoint Solutions*

1. Expressions (a) and (b) can be evaluated with the Power Rule. For expression (a), if we let $u = 4x^2 + 1$, then $u' = 8x$ so that the integral becomes

$$\int u^{10} u'\, dx = \int u^{10}\, du$$

For expression (b), we let $u = 4x^2 + 1$ again. Then $u' = 8x$ and the integral becomes

$$\int (4x^2 + 1)^{10} \cdot x\, dx = \int (4x^2 + 1)^{10} \cdot \frac{1}{8}(8x)\, dx = \frac{1}{8}\int u^{10} u'\, dx = \frac{1}{8}\int u^{10}\, du$$

Expressions (c) and (d) do not fit the format of the Power Rule, because neither integral has an x with the dx outside the power (so they need to be multiplied out before integrating).

2. $u = 2x^3 + 5$, so $u' = 6x^2$

$$\int (2x^3 + 5)^{-2}(6x^2)\, dx = \int u^{-2}u'\, dx = \int u^{-2}\, du = u^{-1}/(-1) + C$$
$$= -u^{-1} + C = -(2x^3 + 5)^{-1} + C$$

Thus expression (c) is the correct choice.

3. True

4. (a) $\int (x^3 + 9)^5(3x^2\, dx) = \int u^5 u'\, dx = \dfrac{u^6}{6} + C = \dfrac{(x^3 + 9)^6}{6} + C$

(b) $\displaystyle\int (x^3 + 9)^{15}(x^2\,dx) = \frac{1}{3}\int (x^3 + 9)^{15}(3x^2\,dx)$

$$= \frac{1}{3}\cdot\frac{(x^3 + 9)^{16}}{16} + C = \frac{(x^3 + 9)^{16}}{48} + C$$

(c) The Power Rule does not fit, so we expand the integrand.

$$\int (x^3 + 9)^2(x\,dx) = \int (x^6 + 18x^3 + 81)x\,dx$$

$$= \int (x^7 + 18x^4 + 81x)\,dx$$

$$= \frac{x^8}{8} + \frac{18x^5}{5} + \frac{81x^2}{2} + C$$

12.2 Exercises

In Problems 1 and 2, find *du*.

1. $u = 2x^5 + 9$ 2. $u = 3x^4 - 4x^3$

Evaluate the integrals in Problems 3–32. Check your results by differentiation.

3. $\int (x^2 + 3)^3 2x\,dx$

4. $\int (3x^3 + 1)^4 9x^2\,dx$

5. $\int (15x^2 + 10)^4(30x)\,dx$

6. $\int (8x^4 + 5)^3(32x^3)\,dx$

7. $\int (3x - x^3)^2(3 - 3x^2)\,dx$

8. $\int (4x^2 - 3x)^4(8x - 3)\,dx$

9. $\int (x^2 + 5)^3 x\,dx$

10. $\int (3x^2 - 4)^6 x\,dx$

11. $\int 7(4x - 1)^6\,dx$

12. $\int 3(5 - x)^{-3}\,dx$

13. $\int (x^2 + 1)^{-3} x\,dx$

14. $\int (3x^4 + 7)^{-4}(5x^3\,dx)$

15. $\int (x - 1)(x^2 - 2x + 5)^4\,dx$

16. $\int (2x^3 - x)(x^4 - x^2)^6\,dx$

17. $\int 2(x^3 - 1)(x^4 - 4x + 3)^{-5}\,dx$

18. $\int 3(x^5 - 2x)(x^6 - 6x^2 + 7)^{-2}\,dx$

19. $\int 7x^3\sqrt{x^4 + 6}\,dx$

20. $\int 3x\sqrt{5 - x^2}\,dx$

21. $\int (x^3 + 1)^2(3x\,dx)$

22. $\int (x^2 - 5)^2(2x^2\,dx)$

23. $\int (3x^2 - 1)^2(8x^2\,dx)$

24. $\int (2x^4 + 3)^2(8x\,dx)$

25. $\int \sqrt{x^3 - 3x}\,(x^2 - 1)\,dx$

26. $\int \sqrt[3]{x^2 + 2x}\,(x + 1)\,dx$

27. $\displaystyle\int \frac{3x^4\,dx}{(2x^5 - 5)^4}$

28. $\displaystyle\int \frac{5x^3\,dx}{(x^4 - 8)^3}$

29. $\displaystyle\int \frac{x^3 - 1}{(x^4 - 4x)^3}\,dx$

30. $\displaystyle\int \frac{3x^5 - 2x^3}{(x^6 - x^4)^5}\,dx$

31. $\displaystyle\int \frac{x^2 - 4x}{\sqrt{x^3 - 6x^2 + 2}}\,dx$

32. $\displaystyle\int \frac{x^2 + 1}{\sqrt{x^3 + 3x + 10}}\,dx$

33. If $\int f(x)\,dx = (7x - 13)^{10} + C$, find $f(x)$.

34. If $\int g(x)\,dx = (5x^2 + 2)^6 + C$, find $g(x)$.

In Problems 35 and 36, (a) evaluate each integral and (b) graph the members of the solution family for $C = -5$, $C = 0$, and $C = 5$.

35. $\int x(x^2 - 1)^3\,dx$

36. $\int (3x - 11)^{1/3}\,dx$

Each of Problems 37 and 38 has the form $\int f(x)\,dx$.

(a) Evaluate each integral to obtain a family of functions.

(b) Find and graph the family member that passes through the point (0, 2). Call that function $F(x)$.

(c) Find any x-values where $f(x)$ is not defined but $F(x)$ is.

(d) At the x-values found in part (c), what kind of tangent line does $F(x)$ have?

37. $\displaystyle\int \frac{3\,dx}{(2x - 1)^{3/5}}$

38. $\displaystyle\int \frac{x^2\,dx}{(x^3 - 1)^{1/3}}$

In each of Problems 39 and 40, a family of functions is given, together with the graphs of some functions in the family. Write the indefinite integral that gives the family.

39. $F(x) = (x^2 - 1)^{4/3} + C$

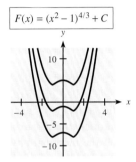

$F(x) = (x^2 - 1)^{4/3} + C$

40. $F(x) = 54(4x^2 + 9)^{-1} + C$

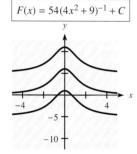

In parts (a)–(c) of Problems 41 and 42, three integrals are given. Integrate those that can be done by the methods studied so far. Additionally, as part (d), give your own example of an integral that looks as though it might use the Power Rule but that cannot be integrated by using methods studied so far.

41. (a) $\displaystyle\int \frac{7x^3\ dx}{(x^3 + 4)^2}$ (b) $\displaystyle\int \frac{7x^2\ dx}{(x^3 + 4)^2}$

   (c) $\displaystyle\int \frac{\sqrt{x^2 + 4}}{x}\ dx$

42. (a) $\displaystyle\int \frac{(2x^5 + 1)^{7/2}}{3x^4}\ dx$ (b) $\displaystyle\int 10x(2x^5 + 1)^{7/2}\ dx$

   (c) $\displaystyle\int \frac{5x^3}{(2x^4 + 1)^{7/2}}\ dx$

APPLICATIONS

43. *Revenue*   Suppose that the marginal revenue for a product is given by

$$\overline{MR} = \frac{-30}{(2x + 1)^2} + 30$$

where x is the number of units and revenue is in dollars. Find the total revenue.

44. *Revenue*   The marginal revenue for a new calculator is given by

$$\overline{MR} = 60{,}000 - \frac{40{,}000}{(10 + x)^2}$$

where x represents hundreds of calculators and revenue is in dollars. Find the total revenue function for these calculators.

45. *Physical productivity*   The total physical output of a number of machines or workers is called *physical productivity* and is a function of the number of machines or workers. If $P = f(x)$ is the productivity, dP/dx is the marginal physical productivity. If the marginal physical productivity for bricklayers is $dP/dx = 90(x + 1)^2$, where P is the number of bricks laid per day and x is the number of bricklayers, find the physical productivity of 4 bricklayers. (*Note:* $P = 0$ when $x = 0$.)

46. *Production*   The rate of production of a new line of products is given by

$$\frac{dx}{dt} = 200\left[1 + \frac{400}{(t + 40)^2}\right]$$

where x is the number of items and t is the number of weeks the product has been in production.
   (a) Assuming that $x = 0$ when $t = 0$, find the total number of items produced as a function of time t.
   (b) How many items were produced in the fifth week?

47. *Data entry speed*   The rate of change in data entry speed of the average student is $ds/dx = 5(x + 1)^{-1/2}$, where x is the number of lessons the student has had and s is in entries per minute.
   (a) Find the data entry speed as a function of the number of lessons if the average student can complete 10 entries per minute with no lessons ($x = 0$).
   (b) How many entries per minute can the average student complete after 24 lessons?

48. *Productivity*   Because a new employee must learn an assigned task, production will increase with time. Suppose that for the average new employee, the rate of performance is given by

$$\frac{dN}{dt} = \frac{1}{2\sqrt{t + 1}}$$

where N is the number of units completed t hours after beginning a new task. If 2 units are completed after 3 hours, how many units are completed after 8 hours?

49. *Film attendance*   An excellent film with a very small advertising budget must depend largely on word-of-mouth advertising. In this case, the rate at which weekly attendance might grow can be given by

$$\frac{dA}{dt} = \frac{-100}{(t + 10)^2} + \frac{2000}{(t + 10)^3}$$

where t is the time in weeks since release and A is attendance in millions.
   (a) Find the function that describes weekly attendance at this film.
   (b) Find the attendance at this film in the tenth week.

50. *Product quality and advertising*   An inferior product with a large advertising budget does well when it is introduced, but sales decline as people discontinue use of the product. Suppose that the rate of weekly sales is given by

$$S'(t) = \frac{400}{(t + 1)^3} - \frac{200}{(t + 1)^2}$$

where S is sales in millions and t is time in weeks.
   (a) Find the function that describes the weekly sales.
   (b) Find the sales for the first week and the ninth week.

51. *Demographics*   Because of job outsourcing, a western Pennsylvania town predicts that its public school population will decrease at the rate

$$\frac{dN}{dx} = \frac{-300}{\sqrt{x + 9}}$$

where x is the number of years and N is the total school population. If the present population ($x = 0$) is 8000, what population size is expected in 7 years?

52. *Franchise growth*   A new fast-food firm predicts that the number of franchises for its products will grow at the rate

$$\frac{dn}{dt} = 9\sqrt{t + 1}$$

where t is the number of years, $0 \leq t \leq 10$. If there is one franchise ($n = 1$) at present ($t = 0$), how many franchises are predicted for 8 years from now?

53. *Poverty line*   Suppose the rate of change of the number of people (in millions) in the United States who lived below the poverty level can be modeled by

$$\frac{dp}{dt} = -0.005715(t + 60)^2 + 0.9566(t + 60) - 39.37$$

where t is the number of years past 1960.

Year	Persons Below the Poverty Level (millions)	Year	Persons Below the Poverty Level (millions)
1960	39.9	1993	39.3
1965	33.2	1996	36.5
1970	25.4	1997	35.6
1975	25.9	1998	34.5
1980	29.3	1999	32.3
1986	32.4	2000	31.1
1989	31.5	2002	34.6
1990	33.6		

Source: Bureau of the Census, U.S. Department of Commerce

(a) Use integration and the data point for 1990 to find the function that models the number of people, in millions, in the United States who lived below the poverty level. Graph this function.
(b) The data in the table show the numbers of people, in millions, in the United States who lived below the poverty level for selected years. Graph the function from part (a) with the data in the table with $t = 0$ representing 1960.
(c) How well does the model fit the data?

54. *U.S. public debt*   Suppose the rate of change of the interest paid on the public debt of the United States as a percent of federal expenditures can be modeled by

$$\frac{dD}{dt} = -0.0788(0.1t + 3)^3 + 1.596(0.1t + 3)^2$$
$$-10.26(0.1t + 3) + 20.96$$

where t is the number of years past 1930.

Year	Interest Paid as a Percent of Federal Expenditures	Year	Interest Paid as a Percent of Federal Expenditures
1930	0.0	1975	9.8
1940	10.5	1980	12.7
1950	13.4	1985	18.9
1955	9.4	1990	21.1
1960	10.0	1995	22.0
1965	9.6	2000	20.3
1970	9.9	2003	14.7

Source: Bureau of Public Debt, U.S. Department of the Treasury

(a) Use integration and the data point for 1965 to find the function $D(t)$ that models the interest paid on the public debt of the United States as a percent of federal expenditures. Graph this function.
(b) The data in the table give the interest paid on the public debt of the United States as a percent of federal expenditures for selected years. Graph the function from part (a) with the data in the table with $t = 0$ representing 1930.
(c) How well does the model fit the data?

55. *Budget deficit*   Suppose the rate of change of the federal budget deficit (or surplus) can be modeled by the function

$$D'(t) = -592.5(0.1t - 1)^2$$
$$+ 671.48(0.1t - 1) - 106.47$$

where D is in billions of dollars and t is the number of years past 1980.

(a) Use integration and the data point for 1990 to find the function $D(t)$ that models the federal budget deficit (or surplus). Graph this function.
(b) The data in the table give the yearly U.S. federal budget deficit (as a negative value) or surplus (as a positive value) in billions of dollars from 1990 to 2004. Graph these data (with $t = 0$ representing 1980) with the function found in part (a).
(c) Comment on the model's fit to the data.

Year	Deficit or Surplus	Year	Deficit of Surplus
1990	−221.2	1998	70.0
1991	−269.4	1999	124.4
1992	−290.4	2000	237.0
1993	−255.0	2001	127.0
1994	−203.1	2002	−159.0
1995	−164.0	2003	−374.0
1996	−107.5	2004	−445.0
1997	−22.0		

Source: Budget of the United States Government

12.3 Integrals Involving Exponential and Logarithmic Functions

OBJECTIVES

● To evaluate integrals of the form $\int e^u u'\, dx$ or, equivalently, $\int e^u\, du$

● To evaluate integrals of the form $\int \dfrac{u'}{u}\, dx$ or, equivalently, $\int \dfrac{1}{u}\, du$

◗ **Application Preview**

The rate of growth of the market value of a home has typically exceeded the inflation rate. Suppose, for example, that the real estate market has an average annual inflation rate of 8%. Then the rate of change of the value of a house that cost $100,000 can be modeled by

$$\frac{dV}{dt} = 7.7e^{0.077t}$$

where V is the value of the home in thousands of dollars and t is the time in years since the home was purchased. To find the market value of such a home 10 years after it was purchased, we would first have to integrate dV/dt. That is, we must be able to integrate an exponential. (See Example 3.)

In this section, we consider integration formulas that result in natural logarithms and formulas for integrating exponentials.

Integrals Involving Exponential Functions

We know that

$$\frac{d}{dx}(e^x) = e^x \qquad \text{and} \qquad \frac{d}{dx}(e^u) = e^u \cdot u'$$

The corresponding integrals are given by the following.

Exponential Formula

If u is a function of x,

$$\int e^u \cdot u'\, dx = \int e^u\, du = e^u + C$$

In particular, $\displaystyle\int e^x\, dx = e^x + C.$

● **EXAMPLE 1 Integral of an Exponential**

Evaluate $\int 5e^x\, dx$.

Solution
$\int 5e^x\, dx = 5 \int e^x\, dx = 5e^x + C$

● **EXAMPLE 2 Integral of $e^u\, du$**

Evaluate: (a) $\int 2xe^{x^2}\, dx$ (b) $\displaystyle\int \frac{x^2\, dx}{e^{x^3}}$

Solution
(a) Letting $u = x^2$ implies that $u' = 2x$, and the integral is of the form $\int e^u \cdot u'\, dx$. Thus

$$\int 2xe^{x^2}\, dx = \int e^{x^2}(2x)\, dx = \int e^u \cdot u'\, dx = e^u + C = e^{x^2} + C$$

(b) In order to use $\int e^u \cdot u' \, dx$, we write the exponential in the numerator. Thus

$$\int \frac{x^2 \, dx}{e^{x^3}} = \int e^{-x^3}(x^2 \, dx)$$

This is *almost* of the form $\int e^u \cdot u' \, dx$. Letting $u = -x^3$ gives $u' = -3x^2$. Thus

$$\int e^{-x^3}(x^2 \, dx) = -\frac{1}{3}\int e^{-x^3}(-3x^2 \, dx) = -\frac{1}{3}e^{-x^3} + C = \frac{-1}{3e^{x^3}} + C$$

● **Checkpoint**

1. True or false:

(a) $\int e^{x^2}(2x \, dx) = e^{x^2} \cdot x^2 + C$ (b) $\int e^{-3x} \, dx = -\frac{1}{3}e^{-3x} + C$

(c) $\int \frac{dx}{e^{3x}} = \frac{1}{3}\left(\frac{1}{e^{3x}}\right) + C$ (d) $\int e^{3x+1}(3 \, dx) = \frac{e^{3x+2}}{3x+2} + C$

◑ EXAMPLE 3 *Real Estate Inflation* **(Application Preview)**

Suppose the rate of change of the value of a house that cost \$100,000 can be modeled by

$$\frac{dV}{dt} = 7.7e^{0.077t}$$

where V is the market value of the home in thousands of dollars and t is the time in years since the home was purchased.

(a) Find the function that expresses the value V in terms of t.
(b) Find the predicted value after 10 years.

Solution

(a) $V = \displaystyle\int \frac{dV}{dt} \, dt = \int 7.7e^{0.077t} \, dt$

$$V = 7.7 \int e^{0.077t}\left(\frac{1}{0.077}\right)(0.077 \, dt)$$

$$V = 7.7\left(\frac{1}{0.077}\right)\int e^{0.077t}(0.077 \, dt)$$

$$V = 100e^{0.077t} + C$$

Using $V = 100$ (thousand) when $t = 0$, we have

$$100 = 100 + C$$
$$0 = C$$

Thus we have the value as a function of time given by

$$V = 100e^{0.077t}$$

(b) The value after 10 years is found by using $t = 10$.

$$V = 100e^{0.077(10)} = 100e^{0.77} \approx 215.98$$

Thus, after 10 years, the predicted value of the home is \$215,980.

● **EXAMPLE 4** *Graphs of Functions and Integrals*

Figure 12.5 shows the graphs of $g(x) = 5e^{-x^2}$ and $h(x) = -10xe^{-x^2}$. One of these functions is $f(x)$ and the other is $\int f(x)\,dx$ with $C = 0$.

(a) Decide which of $g(x)$ and $h(x)$ is $f(x)$ and which is $\int f(x)\,dx$.
(b) How can the graph of $f(x)$ be used to locate and classify the extrema of $\int f(x)\,dx$?
(c) What feature of the graph of $f(x)$ occurs at the same x-values as the inflection points of the graph of $\int f(x)\,dx$?

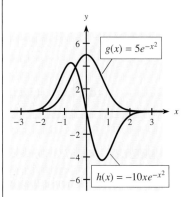

Figure 12.5

Solution
(a) The graph of $h(x)$ looks like the graph of $g'(x)$ because $h(x) > 0$ where $g(x)$ is increasing, $h(x) < 0$ where $g(x)$ is decreasing, and $h(x) = 0$ where $g(x)$ has its maximum. However, if $h(x) = g'(x)$, then, equivalently,

$$\int h(x)\,dx = \int g'(x)\,dx$$
$$= g(x) + C$$

so $h(x) = f(x)$ and $g(x) = \int f(x)\,dx$. We can verify this by noting that

$$\int -10xe^{-x^2}\,dx = 5\int e^{-x^2}(-2x\,dx)$$
$$= 5e^{-x^2} + C$$

(b) We know that $f(x)$ is the derivative of $\int f(x)\,dx$, so, as we saw in (a), the x-intercepts of $f(x)$ locate the critical values and extrema of $\int f(x)\,dx$.
(c) The first derivative of $\int f(x)\,dx$ is $f(x)$, and its second derivative is $f'(x)$. Hence the inflection points of $\int f(x)\,dx$ occur where $f'(x) = 0$. But $f(x)$ has its extrema where $f'(x) = 0$. Thus the extrema of $f(x)$ occur at the same x-values as the inflection points of $\int f(x)\,dx$.

Integrals Involving Logarithmic Functions

Recall that the Power Rule for integrals applies only if $n \neq -1$. That is,

$$\int u^n u'\,dx = \frac{u^{n+1}}{n+1} + C \qquad \text{if } n \neq -1$$

The following formula applies when $n = -1$.

Logarithmic Formula

If u is a function of x, then

$$\int u^{-1} u' \, dx = \int \frac{u'}{u} \, dx = \int \frac{1}{u} \, du = \ln |u| + C$$

In particular, $\int \frac{1}{x} \, dx = \ln |x| + C.$

We use the absolute value of u in the integral because the logarithm is defined only when the quantity is positive. This logarithmic formula is a direct result of the fact that

$$\frac{d}{dx} (\ln |u|) = \frac{1}{u} \cdot u'$$

We can see this result by considering the following.

For $u > 0$: $\quad \dfrac{d}{dx} (\ln |u|) = \dfrac{d}{dx} (\ln u) = \dfrac{1}{u} \cdot u'$

For $u < 0$: $\quad \dfrac{d}{dx} (\ln |u|) = \dfrac{d}{dx} [\ln (-u)] = \dfrac{1}{(-u)} \cdot (-u') = \dfrac{1}{u} \cdot u'$

In addition to this verification, we can graphically illustrate the need for the absolute value sign. Figure 12.6(a) shows that $f(x) = 1/x$ is defined for $x \neq 0$, and from Figures 12.6(b) and 12.6(c), we see that $F(x) = \int 1/x \, dx = \ln |x|$ is also defined for $x \neq 0$, but $y = \ln x$ is defined only for $x > 0$.

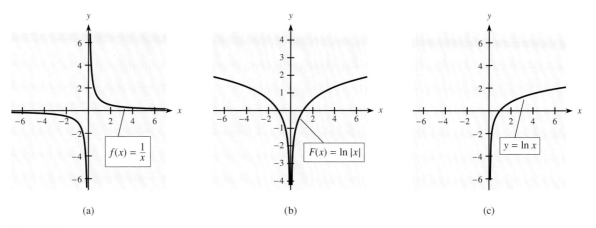

(a)   (b)   (c)

Figure 12.6

● **EXAMPLE 5** *Integrals Resulting in Logarithmic Functions*

Evaluate $\displaystyle\int \frac{4}{4x + 8} \, dx.$

Solution
This integral is of the form

$$\int \frac{u'}{u} \, dx = \ln |u| + C$$

with $u = 4x + 8$ and $u' = 4$. Thus

$$\int \frac{4}{4x + 8} \, dx = \ln |4x + 8| + C$$

Figure 12.7 shows several members of the family

$$F(x) = \int \frac{4 \, dx}{4x + 8} = \ln |4x + 8| + C$$

We can choose different values for C and use a graphing utility to graph families of curves such as those in Figure 12.7.

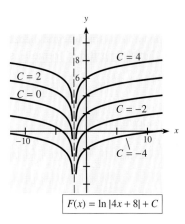

Figure 12.7

$$F(x) = \ln |4x + 8| + C$$

● **EXAMPLE 6 *Integral of du/u***

Evaluate $\int \dfrac{x - 3}{x^2 - 6x + 1} \, dx$.

Solution

This integral is of the form $\int (u'/u) \, dx$, *almost.* If we let $u = x^2 - 6x + 1$, then $u' = 2x - 6$. If we multiply (and divide) the numerator by 2, we get

$$\int \frac{x - 3}{x^2 - 6x + 1} \, dx = \frac{1}{2} \int \frac{2(x - 3)}{x^2 - 6x + 1} \, dx$$

$$= \frac{1}{2} \int \frac{2x - 6}{x^2 - 6x + 1} \, dx$$

$$= \frac{1}{2} \int \frac{u'}{u} \, dx = \frac{1}{2} \ln |u| + C$$

$$= \frac{1}{2} \ln |x^2 - 6x + 1| + C$$

● **EXAMPLE 7 *Population Growth***

Because the world contains only about 10 billion acres of arable land, world population is limited. Suppose that world population is limited to 40 billion people and that the rate of population growth per year is given by

$$\frac{dP}{dt} = k(40 - P)$$

where P is the population in billions at time t and k is a positive constant. Then the relationship between the year and the population during that year is given by the integral

$$t = \frac{1}{k} \int \frac{1}{40 - P} \, dP$$

(a) Evaluate this integral to find the relationship.
(b) Use properties of logarithms and exponential functions to write P as a function of t.

Solution

(a) $t = \dfrac{1}{k} \displaystyle\int \dfrac{1}{40 - P}\, dP = -\dfrac{1}{k} \displaystyle\int \dfrac{-dP}{40 - P} = -\dfrac{1}{k} \ln |40 - P| + C_1$

(b) $t = -\dfrac{1}{k} \ln |40 - P| + C_1$

Solving this equation for P requires converting to exponential form.

$$-k(t - C_1) = \ln |40 - P|$$
$$e^{C_1 k - kt} = 40 - P$$
$$e^{C_1 k} \cdot e^{-kt} = 40 - P$$

Because $e^{C_1 k}$ is an unknown constant, we replace it with C and solve for P.

$$P = 40 - Ce^{-kt}$$

If an integral contains a fraction in which the degree of the numerator is equal to or greater than that of the denominator, we should divide the denominator into the numerator as a first step.

● **EXAMPLE 8 Integrals Requiring Division**

Evaluate $\displaystyle\int \dfrac{x^4 - 2x^3 + 4x^2 - 7x - 1}{x^2 - 2x}\, dx$.

Solution
Because the numerator is of higher degree than the denominator, we begin by dividing $x^2 - 2x$ into the numerator.

$$
\begin{array}{r}
x^2 + 4 \\
x^2 - 2x \overline{)\, x^4 - 2x^3 + 4x^2 - 7x - 1\,} \\
\underline{x^4 - 2x^3 } \\
4x^2 - 7x - 1 \\
\underline{4x^2 - 8x } \\
x - 1
\end{array}
$$

Thus

$$
\int \frac{x^4 - 2x^3 + 4x^2 - 7x - 1}{x^2 - 2x}\, dx = \int \left(x^2 + 4 + \frac{x - 1}{x^2 - 2x} \right) dx
$$

$$
= \int (x^2 + 4)\, dx + \frac{1}{2} \int \frac{2(x - 1)\, dx}{x^2 - 2x}
$$

$$
= \frac{x^3}{3} + 4x + \frac{1}{2} \ln |x^2 - 2x| + C
$$

● **Checkpoint**

2. True or false:

(a) $\int \dfrac{3x^2\,dx}{x^3 + 4} = \ln|x^3 + 4| + C$ (b) $\int \dfrac{2x\,dx}{\sqrt{x^2 + 1}} = \ln|\sqrt{x^2 + 1}| + C$

(c) $\int \dfrac{2}{x}\,dx = 2\ln|x| + C$ (d) $\int \dfrac{x}{x + 1}\,dx = x\int \dfrac{1}{x + 1}\,dx = x\ln|x + 1| + C$

(e) To evaluate $\int \dfrac{4x}{4x + 1}\,dx$, our first step is to divide $4x + 1$ into $4x$.

3. (a) Divide $4x + 1$ into $4x$.   (b) Evaluate $\int \dfrac{4x}{4x + 1}\,dx$.

● **Checkpoint Solutions**

1. (a) False. The correct solution is $e^{x^2} + C$ (see Example 2a).
   (b) True
   (c) False; $\displaystyle\int \dfrac{dx}{e^{3x}} = \int e^{-3x}\,dx = -\dfrac{1}{3}e^{-3x} + C = \dfrac{-1}{3e^{3x}} + C$
   (d) False; $\displaystyle\int e^{3x+1}(3\,dx) = e^{3x+1} + C$

2. (a) True
   (b) False; $\displaystyle\int \dfrac{2x\,dx}{\sqrt{x^2 + 1}} = \int (x^2 + 1)^{-1/2}(2x\,dx)$
   $$= \dfrac{(x^2 + 1)^{1/2}}{1/2} + C = 2(x^2 + 1)^{1/2} + C$$
   (c) True
   (d) False. We cannot factor the variable x outside the integral sign.
   (e) True

3. (a) $4x + 1\overline{)4x}$ so $\dfrac{4x}{4x + 1} = 1 - \dfrac{1}{4x + 1}$
       $\underline{4x + 1}$
       -1

   (b) $\displaystyle\int \dfrac{4x\,dx}{4x + 1} = \int\left(1 - \dfrac{1}{4x + 1}\right)dx = x - \dfrac{1}{4}\ln|4x + 1| + C$

12.3 Exercises

Evaluate the integrals in Problems 1–32.

1. $\int 3e^{3x}\,dx$

2. $\int 4e^{4x}\,dx$

3. $\int e^{-x}\,dx$

4. $\int e^{2x}\,dx$

5. $\int 1000e^{0.1x}\,dx$

6. $\int 1600e^{0.4x}\,dx$

7. $\int 840e^{-0.7x}\,dx$

8. $\int 250e^{-0.5x}\,dx$

9. $\int x^3 e^{3x^4}\,dx$

10. $\int xe^{2x^2}\,dx$

11. $\int \dfrac{3}{e^{2x}}\,dx$

12. $\int \dfrac{4}{e^{1-2x}}\,dx$

13. $\int \dfrac{x^5}{e^{2 - 3x^6}}\,dx$

14. $\int \dfrac{x^3}{e^{4x^4}}\,dx$

15. $\int\left(e^{4x} - \dfrac{3}{e^{x/2}}\right)dx$

16. $\int\left(xe^{3x^2} - \dfrac{5}{e^{x/3}}\right)dx$

17. $\int \dfrac{3x^2}{x^3 + 4}\,dx$

18. $\int \dfrac{8x^7}{x^8 - 1}\,dx$

19. $\int \dfrac{dz}{4z + 1}$

20. $\int \dfrac{y}{y^2 + 1}\,dy$

21. $\int \dfrac{x^3}{x^4 + 1}\,dx$

22. $\int \dfrac{x^2}{x^3 - 9}\,dx$

23. $\int \dfrac{4x}{x^2 - 4}\,dx$

24. $\int \dfrac{5x^2}{x^3 - 1}\,dx$

25. $\int \dfrac{3x^2 - 2}{x^3 - 2x}\,dx$

26. $\int \dfrac{4x^3 + 2x}{x^4 + x^2}\,dx$

27. $\int \dfrac{z^2 + 1}{z^3 + 3z + 17}\,dz$

28. $\int \dfrac{(x + 2)\,dx}{x^2 + 4x - 9}$

29. $\int \dfrac{x^3 - x^2 + 1}{x - 1}\,dx$ 30. $\int \dfrac{2x^3 + x^2 + 2x + 3}{2x + 1}\,dx$

31. $\int \dfrac{x^2 + x + 3}{x^2 + 3}\,dx$ 32. $\int \dfrac{x^4 - 2x^2 + x}{x^2 - 2}\,dx$

In Problems 33 and 34, graphs of two functions labeled *g(x)* and *h(x)* are given. Decide which is the graph of *f(x)* and which is one member of the family ∫ *f(x) dx*. Check your conclusions by evaluating the integral.

33.
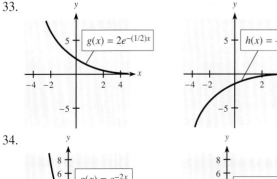

34.
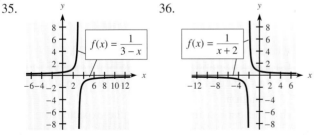

In Problems 35 and 36, a function *f(x)* and its graph are given. Find the family *F(x)* = ∫ *f(x) dx* and graph the member that satisfies *F*(0) = 0.

35.   36.
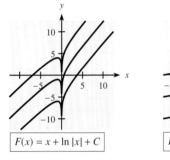
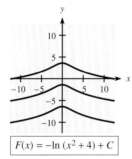

In Problems 37–40, a family of functions is given and graphs of some members are shown. Find the function *f(x)* such that the family is given by ∫ *f(x) dx*.

37. $F(x) = x + \ln|x| + C$ 38. $F(x) = -\ln(x^2 + 4) + C$

39. $F(x) = 5xe^{-x} + C$ 40. $F(x) = e^{0.4x} + e^{-0.4x} + C$
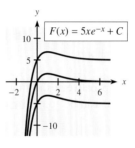
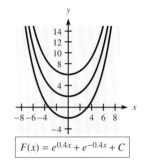

In parts (a)–(d) of Problems 41 and 42, integrate those that can be done by the methods studied so far.

41. (a) $\int xe^{x^3}\,dx$ (b) $\int \dfrac{x + 2}{x^2 + 2x + 7}\,dx$

(c) $\int \dfrac{x^2 + 2x}{x^3 + 3x^2 + 7}\,dx$ (d) $\int 5x^3e^{2x^4}\,dx$

42. (a) $\int \dfrac{3x - 1}{x^3 - x + 2}\,dx$ (b) $\int \dfrac{3x - 1}{6x^2 - 4x + 9}\,dx$

(c) $\int 5\sqrt{x}\,e^{\sqrt{x}}\,dx$ (d) $\int 6xe^{-x^2/8}\,dx$

APPLICATIONS

43. *Revenue* Suppose that the marginal revenue from the sale of *x* units of a product is $\overline{MR} = R'(x) = 6e^{0.01x}$. What is the revenue in dollars from the sale of 100 units of the product?

44. *Concentration of a drug* Suppose that the rate at which the concentration of a drug in the blood changes with respect to time *t* is given by

$$C'(t) = \dfrac{c}{b - a}(be^{-bt} - ae^{-at}), \qquad t \ge 0$$

where *a*, *b*, and *c* are constants depending on the drug administered, with *b > a*. Assuming that *C(t)* = 0 when *t* = 0, find the formula for the concentration of the drug in the blood at any time *t*.

45. *Radioactive decay* The rate of disintegration of a radioactive substance can be described by

$$\dfrac{dn}{dt} = n_0(-K)e^{-Kt}$$

where n_0 is the number of radioactive atoms present when time *t* is 0, and *K* is a positive constant that depends on the substance involved. Using the fact that the constant of integration is 0, integrate *dn/dt* to find the number of atoms *n* that are still radioactive after time *t*.

46. *Radioactive decay* Radioactive substances decay at a rate that is proportional to the amount present. Thus, if the amount present is x, the decay rate is

$$\frac{dx}{dt} = kx \quad (t \text{ in hours})$$

This means that the relationship between the time and the amount of substance present can be found by evaluating the integral

$$t = \int \frac{dx}{kx}$$

(a) Evaluate the integral to find the relationship.
(b) Use properties of logarithms and exponential functions to write x as a function of t.

47. *Memorization* The rate of vocabulary memorization of the average student in a foreign language course is given by

$$\frac{dv}{dt} = \frac{40}{t+1}$$

where t is the number of continuous hours of study, $0 < t \le 4$, and v is the number of words. How many words would the average student memorize in 3 hours?

48. *Population growth* The rate of growth of world population can be modeled by

$$\frac{dn}{dt} = N_0(1+r)^t \ln(1+r), \quad r < 1$$

where t is the time in years from the present and N_0 and r are constants. What function describes world population if the present population is N_0? Use the formula $\int (a^u \ln a)u'\, dx = a^u + C$.

49. *Compound interest* If $\$P$ is invested for n years at 10% compounded continuously, the rate at which the future value is growing is

$$\frac{dS}{dn} = 0.1Pe^{0.1n}$$

(a) What function describes the future value at the end of n years?
(b) In how many years will the future value double?

50. *Temperature changes* When an object is moved from one environment to another, its temperature T changes at a rate given by

$$\frac{dT}{dt} = kCe^{kt}$$

where t is the time in the new environment (in hours), C is the temperature difference (old – new) between the two environments, and k is a constant. If the temperature

of the object (and the old environment) is 70°F, and $C = -10$°F, what function describes the temperature T of the object t hours after it is moved?

51. *Blood pressure in the aorta* The rate at which blood pressure decreases in the aorta of a normal adult after a heartbeat is

$$\frac{dp}{dt} = -46.645e^{-0.491t}$$

where t is time in seconds.
(a) What function describes the blood pressure in the aorta if $p = 95$ when $t = 0$?
(b) What is the blood pressure 0.1 second after a heartbeat?

52. *Sales and advertising* A store finds that its sales decline after the end of an advertising campaign, with its daily sales for the period declining at the rate $S'(t) = -147.78e^{-0.2t}, 0 \le t \le 100$, where t is the number of days since the end of the campaign. Suppose that $S = 7389$ units when $t = 0$.
(a) Find the function that describes the number of daily sales t days after the end of the campaign.
(b) Find the total number of sales 10 days after the end of the advertising campaign.

53. *Life expectancy* Suppose the rate of change of the expected life span l at birth of people born in the United States can be modeled by

$$\frac{dl}{dt} = \frac{14.202}{t+20}$$

where t is the number of years past 1920.
(a) Use integration and the data point for 1975 to find the function that models the life span. Graph this function.
(b) The data in the table give the expected life spans for people born in various years. Graph the function from part (a) with the data, with $t = 0$ representing 1920.
(c) How well does the model fit the data?

Year	Life Span (years)	Year	Life Span (years)
1920	54.1	1990	75.4
1930	59.7	1992	75.5
1940	62.9	1994	75.7
1950	68.2	1996	76.1
1960	69.7	1998	76.7
1970	70.8	1999	76.7
1975	72.6	2000	76.9
1980	73.7	2001	77.2
1985	74.7		

Source: National Center for Health Statistics

54. *U.S. households with cable TV* Suppose the rate of change of the percent P of U.S. households with cable TV can be modeled by

$$\frac{dP}{dt} = \frac{39.48}{t + 5}$$

where t is the number of years past 1975.
(a) Use integration and the data point for 1990 to find the function $P(t)$ that models the percent of U.S. households with cable TV.
(b) How well does the model from part (a) fit the data in the following table? Explain.
(c) If the model remains valid, use it to predict the percent of U.S. households with cable TV in 2020.

Year	Percent	Year	Percent
1980	22.6	1991	60.6
1981	28.3	1992	61.5
1982	35.0	1993	62.5
1983	40.5	1994	63.4
1984	43.7	1995	65.7
1985	46.2	1996	66.7
1986	48.1	1997	67.3
1987	50.5	1998	67.4
1988	53.8	1999	68.0
1989	57.1	2000	67.8
1990	59.0	2001	69.2

Source: Nielson Media Research

 55. *Consumer price index* The U.S. consumer price index (CPI) measures the purchasing power of \$1, relative to the cost of goods in 1983. Thus a CPI (represented by P) of 0.75 in some year means that in that year \$1 purchased as much as \$0.75 did in 1983. Suppose the rate of change of the CPI can be modeled by

$$\frac{dP}{dt} = -0.0548e^{-0.0343t}$$

where t is the number of years past 1970.

(a) Does the model for the rate reflect the fact that the data in the table show that the CPI is decreasing? Explain.
(b) Use integration and the data point for 2002 to find the function that models the CPI.
(c) Find and interpret $P(45)$ and $P'(45)$.

Year	CPI	Year	CPI
1980	1.215	1993	0.692
1983	1.003	1995	0.656
1985	0.928	1997	0.623
1987	0.880	2000	0.581
1990	0.766	2002	0.556

Source: U.S. Bureau of Labor Statistics

 56. *Personal income* Suppose the rate of change of total personal income I in the United States (in billions of dollars) can be modeled by

$$\frac{dI}{dt} = 32.232e^{0.0747t}$$

where t is the number of years past 1960.
(a) Is the average rate of change of total personal income from 1999 to 2000 found from the data below a better estimate for the instantaneous rate for 1999 or for 2000?
(b) Use the data point from 1960 to find the function that models $I(t)$.
(c) Find and interpret $I(50)$ and $I'(50)$.

Year	1960	1970	1980	1990	1999	2000	2002
Personal income	409.4	831.4	2258.5	4673.8	7777.3	8319.2	8929.1

Source: Bureau of Economic Analysis, U.S. Department of Commerce

---

OBJECTIVES

● To use integration to find total cost functions from information involving marginal cost
● To optimize profit, given information regarding marginal cost and marginal revenue
● To use integration to find national consumption functions from information about marginal propensity to consume and marginal propensity to save

12.4 Applications of the Indefinite Integral in Business and Economics

◗ Application Preview

If we know that the consumption of a nation is \$9 billion when income is \$0 and the marginal propensity to save is 0.25, we can easily find the marginal propensity to consume and use integration to find the national consumption function. (See Example 6.)

In this section, we also use integration to derive total cost and profit functions from the marginal cost and marginal revenue functions. One of the reasons for the marginal approach in economics is that firms can observe marginal changes in real life. If they know the marginal cost and the total cost when a given quantity is sold, they can develop their total cost function.

Total Cost and Profit

We know that the marginal cost for a commodity is the derivative of the total cost function—that is, $\overline{MC} = C'(x)$, where $C(x)$ is the total cost function. Thus if we have the marginal cost function, we can integrate (or "reverse" the process of differentiation) to find the total cost. That is, $C(x) = \int \overline{MC}\, dx$.

If, for example, the marginal cost is $\overline{MC} = 4x + 3$, the total cost is given by

$$C(x) = \int \overline{MC}\, dx$$

$$= \int (4x + 3)\, dx$$

$$= 2x^2 + 3x + K$$

where K represents the constant of integration. Now, we know that the total revenue is 0 if no items are produced, but the total cost may not be 0 if nothing is produced. The fixed costs accrue whether goods are produced or not. Thus the value of the constant of integration depends on the fixed costs FC of production.

Thus we cannot determine the total cost function from the marginal cost unless additional information is available to help us determine the fixed costs.

● EXAMPLE 1 *Total Cost*

Suppose the marginal cost function for a month for a certain product is $\overline{MC} = 3x + 50$, where x is the number of units and cost is in dollars. If the fixed costs related to the product amount to \$100 per month, find the total cost function for the month.

Solution
The total cost function is

$$C(x) = \int (3x + 50)\, dx$$

$$= \frac{3x^2}{2} + 50x + K$$

But the constant of integration K is found by using the fact that $C(0) = FC = 100$. Thus

$$3(0)^2 + 50(0) + K = 100, \quad \text{so } K = 100$$

and the total cost for the month is given by

$$C(x) = \frac{3x^2}{2} + 50x + 100$$

● EXAMPLE 2 *Cost*

Suppose monthly records show that the rate of change of the cost (that is, the marginal cost) for a product is $\overline{MC} = 3(2x + 25)^{1/2}$, where x is the number of units and cost is in dollars. If the fixed costs for the month are \$11,125, what would be the total cost of producing 300 items per month?

Solution

We can integrate the marginal cost to find the total cost function.

$$C(x) = \int \overline{MC}\, dx = \int 3(2x + 25)^{1/2}\, dx$$

$$= 3 \cdot \left(\frac{1}{2}\right) \int (2x + 25)^{1/2}(2\, dx)$$

$$= \left(\frac{3}{2}\right) \frac{(2x + 25)^{3/2}}{3/2} + K$$

$$= (2x + 25)^{3/2} + K$$

We can find K by using the fact that fixed costs are $11,125.

$$C(0) = 11{,}125 = (25)^{3/2} + K$$

$$11{,}125 = 125 + K, \quad \text{or} \quad K = 11{,}000$$

Thus the total cost function is

$$C(x) = (2x + 25)^{3/2} + 11{,}000$$

and the cost of producing 300 items per month is

$$C(300) = (625)^{3/2} + 11{,}000$$

$$= 26{,}625 \quad \text{(dollars)}$$

It can be shown that the profit is usually maximized when $\overline{MR} = \overline{MC}$. To see that this does not always give us a maximum *positive* profit, consider the following facts concerning the manufacture of widgets over the period of a month.

1. The marginal revenue is $\overline{MR} = 400 - 30x$.
2. The marginal cost is $\overline{MC} = 20x + 50$.
3. When 5 widgets are produced and sold, the total cost is $1750. The profit *should* be maximized when $\overline{MR} = \overline{MC}$, or when $400 - 30x = 20x + 50$. Solving for x gives $x = 7$. To see whether our profit is maximized when 7 units are produced and sold, let us examine the profit function.

The profit function is given by $P(x) = R(x) - C(x)$, where

$$R(x) = \int \overline{MR}\, dx \quad \text{and} \quad C(x) = \int \overline{MC}\, dx$$

Integrating, we get

$$R(x) = \int (400 - 30x)\, dx = 400x - 15x^2 + K$$

but $R(0) = 0$ gives $K = 0$ for this total revenue function, so

$$R(x) = 400x - 15x^2$$

The total cost function is

$$C(x) = \int (20x + 50)\, dx = 10x^2 + 50x + K$$

The value of fixed cost can be determined by using the fact that 5 widgets cost $1750. This tells us that $C(5) = 1750 = 250 + 250 + K$, so $K = 1250$.

Thus the total cost is $C(x) = 10x^2 + 50x + 1250$. Thus, the profit is

$$P(x) = R(x) - C(x) = (400x - 15x^2) - (10x^2 + 50x + 1250)$$

Simplifying gives

$$P(x) = 350x - 25x^2 - 1250$$

We have found that $\overline{MR} = \overline{MC}$ if $x = 7$, and the graph of $P(x)$ is a parabola that opens downward, so profit is maximized at $x = 7$. But if $x = 7$, profit is

$$P(7) = 2450 - 1225 - 1250 = -25$$

That is, the production and sale of 7 items result in a loss of $25.

The preceding discussion indicates that, although setting $\overline{MR} = \overline{MC}$ may optimize profit, it does not indicate the level of profit or loss, as forming the profit function does.

If the widget firm is in a competitive market, and its optimal level of production results in a loss, it has two options. It can continue to produce at the optimal level in the short run until it can lower or eliminate its fixed costs, even though it is losing money; or it can take a larger loss (its fixed cost) by stopping production. Producing 7 units causes a loss of $25 per month, and ceasing production results in a loss of $1250 (the fixed cost) per month. If this firm and many others like it cease production, the supply will be reduced, causing an eventual increase in price. The firm can resume production when the price increase indicates that it can make a profit.

● EXAMPLE 3 *Maximum Profit*

Given that $\overline{MR} = 200 - 4x$, $\overline{MC} = 50 + 2x$, and the total cost of producing 10 Wagbats is $700, at what level should the Wagbat firm hold production in order to maximize the profits?

Solution
Setting $\overline{MR} = \overline{MC}$, we can solve for the production level that maximizes profit.

$$200 - 4x = 50 + 2x$$
$$150 = 6x$$
$$25 = x$$

The level of production that should optimize profit is 25 units. To see whether 25 units maximizes profits or minimizes the losses (in the short run), we must find the total revenue and total cost functions.

$$R(x) = \int (200 - 4x)\, dx = 200x - 2x^2 + K$$
$$= 200x - 2x^2, \quad \text{because } K = 0$$
$$C(x) = \int (50 + 2x)\, dx = 50x + x^2 + K$$

We find K by noting that $C(x) = 700$ when $x = 10$.

$$700 = 50(10) + (10)^2 + K$$

so $K = 100$.

Thus the cost is given by $C = C(x) = 50x + x^2 + 100$. At $x = 25$, $R = R(25) = 200(25) - 2(25)^2 = \$3750$ and $C = C(25) = 50(25) + (25)^2 + 100 = \$1975$.

We see that the total revenue is greater than the total cost, so production should be held at 25 units, which results in a maximum profit.

Calculator
Note

If it is difficult to solve $\overline{MC} = \overline{MR}$ analytically, we could use a graphing calculator to solve this equation by finding the point of intersection of the graphs of $\overline{MC}$ and $\overline{MR}$. We may also be able to integrate $\overline{MC}$ and $\overline{MR}$ to find the functions $C(x)$ and $R(x)$ and then use a graphing calculator to graph them. From the graphs of $C(x)$ and $R(x)$ we can learn about these functions—and hence about profit. ∎

● **EXAMPLE 4** *Cost, Revenue, and Profit*

Suppose that $\overline{MC} = 1.01(x + 190)^{0.01}$ and $\overline{MR} = (1/\sqrt{2x + 1}) + 2$, where x is the number of thousands of units and both revenue and cost are in thousands of dollars. Suppose further that fixed costs are \$100,236 and that production is limited to at most 180 thousand units.

(a) Determine $C(x)$ and $R(x)$ and graph them to determine whether a profit can be made.
(b) Estimate the level of production that yields maximum profit, and find the maximum profit.

Solution

(a) $C(x) = \displaystyle\int \overline{MC} \, dx = \int 1.01(x + 190)^{0.01} \, dx$

$$= 1.01\frac{(x + 190)^{1.01}}{1.01} + K$$

When we say that fixed costs equal \$100,236, we mean $C(0) = 100.236$.

$$100.236 = C(0) = (190)^{1.01} + K$$
$$100.236 = 200.236 + K$$
$$-100 = K$$

Thus $C(x) = (x + 190)^{1.01} - 100$.

$$R(x) = \int \overline{MR} \, dx = \int \left[(2x + 1)^{-1/2} + 2\right] dx$$

$$= \frac{1}{2}\int (2x + 1)^{-1/2}(2 \, dx) + \int 2 \, dx$$

$$= \frac{\frac{1}{2}(2x + 1)^{1/2}}{1/2} + 2x + K$$

$R(0) = 0$ means

$$0 = R(0) = (1)^{1/2} + 0 + K, \quad \text{or} \quad K = -1$$

Thus $R(x) = (2x + 1)^{1/2} + 2x - 1$.

The graphs of $C(x)$ and $R(x)$ are shown in Figure 12.8 on the next page. (The x-range is chosen to include the production range from 0 to 180 (thousand) units. The y-range is chosen to extend beyond fixed costs of about 100 thousand dollars.)

From the figure we see that a profit can be made as long as the number of units sold exceeds about 95 (thousand). We could locate this value more precisely by using INTERSECT.

(b) From the graph we also see that $R(x) - C(x) = P(x)$ is at its maximum at the right edge of the graph. Because production is limited to at most 180 thousand units, profit will be maximized when $x = 180$ and the maximum profit is

$$P(180) = R(180) - C(180)$$
$$= [(361)^{1/2} + 360 - 1] - [(370)^{1.01} - 100]$$
$$\approx 85.46 \quad \text{(thousand dollars)}$$

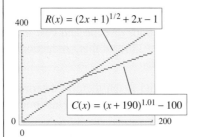

Figure 12.8

● *Checkpoint*

1. True or false:
   (a) If $C(x) = \int \overline{MC}\, dx$, then the constant of integration equals the fixed costs.
   (b) If $R(x) = \int \overline{MR}\, dx$, then the constant of integration equals 0.
2. Find $C(x)$ if $\overline{MC} = \dfrac{100}{\sqrt{x+1}}$ and fixed costs are $8000.

National Consumption and Savings

The consumption function is one of the basic ingredients in a larger discussion of how an economy can have persistent high unemployment or persistent high inflation. This study is often called **Keynesian analysis,** after its founder John Maynard Keynes (pronounced "canes").

If C represents national consumption (in billions of dollars), then a **national consumption function** has the form $C = f(y)$, where y is disposable national income (also in billions of dollars). The **marginal propensity to consume** is the derivative of the national consumption function with respect to y, or $dC/dy = f'(y)$. For example, suppose that

$$C = f(y) = 0.8y + 6$$

is a national consumption function; then the marginal propensity to consume is $f'(y) = 0.8$.

If we know the marginal propensity to consume, we can integrate with respect to y to find national consumption:

$$C = \int f'(y)\, dy = f(y) + K$$

We can find the unique national consumption function if we have additional information to help us determine the value of K, the constant of integration.

● **EXAMPLE 5** *National Consumption*

If consumption is $6 billion when disposable income is 0, and if the marginal propensity to consume is $dC/dy = 0.3 + 0.4/\sqrt{y}$ (in billions of dollars), find the national consumption function.

Solution

If

$$\frac{dC}{dy} = 0.3 + \frac{0.4}{\sqrt{y}}$$

then

$$C = \int \left(0.3 + \frac{0.4}{\sqrt{y}} \right) dy$$

$$= \int (0.3 + 0.4y^{-1/2}) \, dy$$

$$= 0.3y + 0.4\frac{y^{1/2}}{1/2} + K = 0.3y + 0.8y^{1/2} + K$$

Now, if $C = 6$ when $y = 0$, then $6 = 0.3(0) + 0.8\sqrt{0} + K$. Thus the constant of integration is $K = 6$, and the consumption function is

$$C = 0.3y + 0.8\sqrt{y} + 6 \qquad \text{(billions of dollars)}$$

If S represents national savings, we can assume that the disposable national income is given by $y = C + S$, or $S = y - C$. Then the **marginal propensity to save** is $dS/dy = 1 - dC/dy$.

◗ EXAMPLE 6 Consumption and Savings (Application Preview)

If the consumption is $9 billion when income is 0, and if the marginal propensity to save is 0.25, find the consumption function.

Solution
If $dS/dy = 0.25$, then $0.25 = 1 - dC/dy$, or $dC/dy = 0.75$. Thus

$$C = \int 0.75 \, dy = 0.75y + K$$

If $C = 9$ when $y = 0$, then $9 = 0.75(0) + K$, or $K = 9$. Then the consumption function is $C = 0.75y + 9$ (billions of dollars).

● **Checkpoint**

3. If the marginal propensity to save is

$$\frac{dS}{dy} = 0.7 - \frac{0.4}{\sqrt{y}}$$

find the marginal propensity to consume.
4. Find the national consumption function if the marginal propensity to consume is

$$\frac{dC}{dy} = \frac{1}{\sqrt{y+4}} + 0.2$$

and national consumption is $6.8 billion when disposable income is 0.

● **Checkpoint Solutions**

1. (a) False. $C(0)$ equals the fixed costs. It may or may not be the constant of integration. (See Examples 1 and 2.)

   (b) False. We use $R(0) = 0$ to determine the constant of integration, but it may be nonzero. See Example 4.

2. $C(x) = 100 \int (x+1)^{-1/2}\, dx = 100 \left[\dfrac{(x+1)^{1/2}}{1/2} \right] + K$

   $C(x) = 200\sqrt{x+1} + K$

   When $x = 0,\ C(x) = 8000$, so

   $$8000 = 200\sqrt{1} + K$$
   $$7800 = K$$
   $$C(x) = 200\sqrt{x+1} + 7800$$

3. $\dfrac{dC}{dy} = 1 - \dfrac{dS}{dy} = 1 - \left(0.7 - \dfrac{0.4}{\sqrt{y}} \right) = 0.3 + \dfrac{0.4}{\sqrt{y}}$

4. $C(y) = \displaystyle\int \left(\dfrac{1}{\sqrt{y+4}} + 0.2 \right) dy$

   $= \int \left[(y+4)^{-1/2} + 0.2 \right] dy$

   $= \dfrac{(y+4)^{1/2}}{1/2} + 0.2y + K$

   $C(y) = 2\sqrt{y+4} + 0.2y + K$

   Using $C(0) = 6.8$ gives $6.8 = 2\sqrt{4} + 0 + K$, or $K = 2.8$.

   Thus $C(y) = 2\sqrt{y+4} + 0.2y + 2.8$.

12.4 Exercises

TOTAL COST AND PROFIT

In Problems 1–12, cost, revenue, and profit are in dollars and x is the number of units.

1. If the monthly marginal cost for a product is $\overline{MC} = 2x + 100$, with fixed costs amounting to $200, find the total cost function for the month.

2. If the monthly marginal cost for a product is $\overline{MC} = x + 30$, and the related fixed costs are $50, find the total cost function for the month.

3. If the marginal cost for a product is $\overline{MC} = 4x + 2$, and the production of 10 units results in a total cost of $300, find the total cost function.

4. If the marginal cost for a product is $\overline{MC} = 3x + 50$, and the total cost of producing 20 units is $2000, what will be the total cost function?

5. If the marginal cost for a product is $\overline{MC} = 4x + 40$, and the total cost of producing 25 units is $3000, what will be the cost of producing 30 units?

6. If the marginal cost for producing a product is $\overline{MC} = 5x + 10$, with a fixed cost of $800, what will be the cost of producing 20 units?

7. A firm knows that its marginal cost for a product is $\overline{MC} = 3x + 20$, that its marginal revenue is $\overline{MR} = 44 - 5x$, and that the cost of production and sale of 80 units is $11,400.
   (a) Find the optimal level of production.
   (b) Find the profit function.
   (c) Find the profit or loss at the optimal level.

8. A certain firm's marginal cost for a product is $\overline{MC} = 6x + 60$, its marginal revenue is $\overline{MR} = 180 - 2x$, and its total cost of production of 10 items is $1000.
   (a) Find the optimal level of production.
   (b) Find the profit function.
   (c) Find the profit or loss at the optimal level of production.
   (d) Should production be continued for the short run?
   (e) Should production be continued for the long run?

9. Suppose that the marginal revenue for a product is $\overline{MR} = 900$ and the marginal cost is $\overline{MC} = 30\sqrt{x+4}$, with a fixed cost of $1000.
   (a) Find the profit or loss from the production and sale of 5 units.
   (b) How many units will result in a maximum profit?

10. Suppose that the marginal cost for a product is $\overline{MC} = 60\sqrt{x+1}$ and its fixed cost is $340.00. If the marginal revenue for the product is $\overline{MR} = 80x$, find the profit or loss from production and sale of
    (a) 3 units.   (b) 8 units.

11. The average cost of a product changes at the rate

$$\overline{C}'(x) = -6x^{-2} + 1/6$$

and the average cost of 6 units is $10.00.
(a) Find the average cost function.
(b) Find the average cost of 12 units.

12. The average cost of a product changes at the rate

$$\overline{C}'(x) = \frac{-10}{x^2} + \frac{1}{10}$$

and the average cost of 10 units is $20.00.
(a) Find the average cost function.
(b) Find the average cost of 20 units.

 13. Suppose that marginal cost for a certain product is given by $\overline{MC} = 1.05(x + 180)^{0.05}$ and marginal revenue is given by $\overline{MR} = (1/\sqrt{0.5x + 4}) + 2.8$, where x is in thousands of units and both revenue and cost are in thousands of dollars. Fixed costs are $200,000 and production is limited to at most 200 thousand units.
(a) Find $C(x)$ and $R(x)$.
(b) Graph $C(x)$ and $R(x)$ to determine whether a profit can be made.
(c) Determine the level of production that yields maximum profit, and find the maximum profit (or minimum loss).

14. Suppose that the marginal cost for a certain product is given by $\overline{MC} = 1.02(x + 200)^{0.02}$ and marginal revenue is given by $\overline{MR} = (2/\sqrt{4x + 1}) + 1.75$, where x is in thousands of units and revenue and cost are in thousands of dollars. Suppose further that fixed costs are $150,000 and production is limited to at most 200 thousand units.
(a) Find $C(x)$ and $R(x)$.
(b) Graph $C(x)$ and $R(x)$ to determine whether a profit can be made.
(c) Determine what level of production yields maximum profit, and find the maximum profit (or minimum loss).

NATIONAL CONSUMPTION AND SAVINGS

15. If consumption is $7 billion when disposable income is 0, and if the marginal propensity to consume is 0.80, find the national consumption function (in billions of dollars).

16. If national consumption is $9 billion when income is 0, and if the marginal propensity to consume is 0.30, what is consumption when disposable income is $20 billion?

17. If consumption is $8 billion when income is 0, and if the marginal propensity to consume is

$$\frac{dC}{dy} = 0.3 + \frac{0.2}{\sqrt{y}} \qquad \text{(in billions of dollars)}$$

find the national consumption function.

18. If consumption is $5 billion when disposable income is 0, and if the marginal propensity to consume is

$$\frac{dC}{dy} = 0.4 + \frac{0.3}{\sqrt{y}} \qquad \text{(in billions of dollars)}$$

find the national consumption function.

19. If consumption is $6 billion when disposable income is 0, and if the marginal propensity to consume is

$$\frac{dC}{dy} = \frac{1}{\sqrt{y + 1}} + 0.4 \qquad \text{(in billions of dollars)}$$

find the national consumption function.

20. If consumption is $5.8 billion when disposable income is 0, and if the marginal propensity to consume is

$$\frac{dC}{dy} = \frac{1}{\sqrt{2y + 9}} + 0.8 \qquad \text{(in billions of dollars)}$$

find the national consumption function.

21. Suppose that the marginal propensity to consume is

$$\frac{dC}{dy} = 0.7 - e^{-2y} \qquad \text{(in billions of dollars)}$$

and that consumption is $5.65 billion when disposable income is 0. Find the national consumption function.

22. Suppose that the marginal propensity to consume is

$$\frac{dC}{dy} = 0.04 + \frac{\ln (y + 1)}{y + 1} \qquad \text{(in billions of dollars)}$$

and that consumption is $6.04 billion when disposable income is 0. Find the national consumption function.

23. Suppose that the marginal propensity to save is

$$\frac{dS}{dy} = 0.15 \qquad \text{(in billions of dollars)}$$

and that consumption is $5.15 billion when disposable income is 0. Find the national consumption function.

24. Suppose that the marginal propensity to save is

$$\frac{dS}{dy} = 0.22 \qquad \text{(in billions of dollars)}$$

and that consumption is $8.6 billion when disposable income is 0. Find the national consumption function.

25. Suppose that the marginal propensity to save is

$$\frac{dS}{dy} = 0.2 - \frac{1}{\sqrt{3y + 7}} \qquad \text{(in billions of dollars)}$$

and that consumption is $6 billion when disposable income is 0. Find the national consumption function.

26. If consumption is $3 billion when disposable income is 0, and if the marginal propensity to save is

$$\frac{dS}{dy} = 0.2 + e^{-1.5y} \qquad \text{(in billions of dollars)}$$

find the national consumption function.

12.5

OBJECTIVES

- To show that a function is the solution to a differential equation
- To use integration to find the general solution to a differential equation
- To find particular solutions to differential equations using given conditions
- To solve separable differential equations
- To solve applied problems involving separable differential equations

Differential Equations

◗ **Application Preview**

Carbon-14 dating, used to determine the age of fossils, is based on three facts. First, the half-life of carbon-14 is 5600 years. Second, the amount of carbon-14 in any living organism is essentially constant. Third, when an organism dies, the rate of change of carbon-14 in the organism is proportional to the amount present. If y represents the amount of carbon-14 present in the organism, then we can express the rate of change of carbon-14 by the **differential equation**

$$\frac{dy}{dt} = ky$$

where k is a constant and t is time in years. In this section, we study methods that allow us to find a function y that satisfies this differential equation, and then we use that function to date a fossil. (See Example 6.)

Recall that we introduced the derivative as an instantaneous rate of change and denoted the instantaneous rate of change of y with respect to time as dy/dt. For many growth or decay processes, such as carbon-14 decay, the rate of change of the amount of a substance with respect to time is proportional to the amount present. As we noted above, this can be represented by the equation

$$\frac{dy}{dt} = ky \qquad (k = \text{constant})$$

An equation of this type, where y is an unknown function of x or t, is called a **differential equation** because it contains derivatives (or differentials). In this section, we restrict ourselves to differential equations where the highest derivative present in the equation is the first derivative. These differential equations are called **first-order differential equations.** Examples are

$$f'(x) = \frac{1}{x+1}, \qquad \frac{dy}{dt} = 2t, \quad \text{and} \quad x\,dy = (y+1)\,dx$$

Solution of Differential Equations

The solution to a differential equation is a function [say $y = f(x)$] that, when used in the differential equation, results in an identity.

● **EXAMPLE 1** *Differential Equation*

Show that $y = 4e^{-5t}$ is a solution to $dy/dt + 5y = 0$.

Solution
We must show that substituting $y = 4e^{-5t}$ into the equation $dy/dt + 5y = 0$ results in an identity:

$$\frac{d}{dt}(4e^{-5t}) + 5(4e^{-5t}) = 0$$

$$-20e^{-5t} + 20e^{-5t} = 0$$

$$0 = 0$$

Thus $y = 4e^{-5t}$ is a solution.

Now that we know what it means for a function to be a solution to a differential equation, let us consider how to find solutions.

The most elementary differential equations are of the form

$$\frac{dy}{dx} = f(x)$$

where $f(x)$ is a continuous function. These equations are elementary to solve because the solutions are found by integration:

$$y = \int f(x)\, dx$$

● **EXAMPLE 2** *Solving a Differential Equation*

Find the solution of

$$f'(x) = \frac{1}{x+1}$$

Solution
The solution is

$$f(x) = \int f'(x)\, dx = \int \frac{1}{x+1}\, dx = \ln|x+1| + C$$

The solution in Example 2, $f(x) = \ln|x+1| + C$, is called the **general solution** because every solution to the equation has this form, and different values of C give different **particular solutions.** Figure 12.9 shows the graphs of several members of the family of solutions to this differential equation. (We cannot, of course, show all of them.)

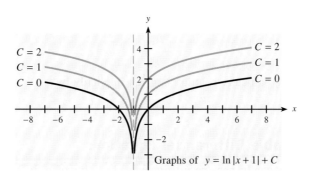

Figure 12.9

We can find a particular solution to a differential equation when we know that the solution must satisfy additional conditions, such as **initial conditions** or **boundary conditions.** For instance, to find the particular solution to

$$f'(x) = \frac{1}{x+1} \qquad \text{with the condition that} \qquad f(-2) = 2$$

we use $f(-2) = 2$ in the general solution, $f(x) = \ln|x+1| + C$.

$$2 = f(-2) = \ln|-2+1| + C$$
$$2 = \ln|-1| + C \quad \text{so} \quad C = 2$$

Thus the particular solution is

$$f(x) = \ln |x + 1| + 2$$

and is shown in Figure 12.9 with $C = 2$.

We frequently denote the value of the solution function $y = f(t)$ at the initial time $t = 0$ as $y(0)$ instead of $f(0)$.

● *Checkpoint*   1. Given $f'(x) = 2x - [1/(x + 1)], f(0) = 4$,
   (a) find the general solution to the differential equation.
   (b) find the particular solution that satisfies $f(0) = 4$.

Just as we can find the differential of both sides of an equation, we can find the solution to a differential equation of the form

$$G(y) \, dy = f(x) \, dx$$

by integrating both sides.

● **EXAMPLE 3 *Differential Equation with a Boundary Condition***

Solve $3y^2 \, dy = 2x \, dx$, if $y(1) = 2$.

Solution
We find the general solution by integrating both sides.

$$\int 3y^2 \, dy = \int 2x \, dx$$

$$y^3 + C_1 = x^2 + C_2$$

$$y^3 = x^2 + C, \qquad \text{where } C = C_2 - C_1$$

By using $y(1) = 2$, we can find C.

$$2^3 = 1^2 + C$$

$$7 = C$$

Thus the particular solution is given implicitly by

$$y^3 = x^2 + 7$$

Separable Differential Equations

It is frequently necessary to change the form of a differential equation before it can be solved by integrating both sides.

For example, the equation

$$\frac{dy}{dx} = y^2$$

cannot be solved by simply integrating both sides of the equation with respect to x because we cannot evaluate $\int y^2 \, dx$.

However, we can multiply both sides of $dy/dx = y^2$ by dx/y^2 to obtain an equation that has all terms containing y on one side of the equation and all terms containing x on the other side. That is, we obtain

$$\frac{dy}{y^2} = dx$$

Separable Differential Equations

When a differential equation can be equivalently expressed in the form

$$g(y)\, dy = f(x)\, dx$$

we say that the equation is **separable.**

The solution of a separable differential equation is obtained by integrating both sides of the equation after the variables have been separated.

● **EXAMPLE 4** *Separable Differential Equation*

Solve the differential equation

$$(x^2y + x^2)\, dy = x^3\, dx$$

Solution

To write the equation in separable form, we first factor x^2 from the left side and divide both sides by it.

$$x^2(y + 1)\, dy = x^3\, dx$$

$$(y + 1)\, dy = \frac{x^3}{x^2}\, dx$$

The equation is now separated, so we integrate both sides.

$$\int (y + 1)\, dy = \int x\, dx$$

$$\frac{y^2}{2} + y + C_1 = \frac{x^2}{2} + C_2$$

This equation, as well as the equation

$$y^2 + 2y - x^2 = C, \qquad \text{where} \qquad C = 2(C_2 - C_1)$$

gives the solution implicitly.

Note that we need not write both C_1 and C_2 when we integrate, because it is always possible to combine the two constants into one.

● **EXAMPLE 5** *Separable Differential Equation*

Solve the differential equation

$$\frac{dy}{dt} = ky \qquad (k = \text{constant})$$

Solution

To solve the equation, we write it in separated form and integrate both sides:

$$\frac{dy}{y} = k\, dt \quad \Rightarrow \quad \int \frac{dy}{y} = \int k\, dt \quad \Rightarrow \quad \ln |y| = kt + C_1$$

Assuming that $y > 0$ and writing this equation in exponential form gives

$$y = e^{kt + C_1}$$

$$y = e^{kt} \cdot e^{C_1} = Ce^{kt}, \qquad \text{where } C = e^{C_1}$$

This solution,

$$y = Ce^{kt}$$

is the general solution to the differential equation $dy/dt = ky$ because all solutions have this form, with different values of C giving different particular solutions. The case of $y < 0$ is covered by values of $C < 0$.

● *Checkpoint*

2. True or false:
   (a) The general solution to $dy = (x/y) \, dx$ can be found from

$$\int y \, dy = \int x \, dx$$

   (b) The first step in solving $dy/dx = -2xy^2$ is to separate it.
   (c) The equation $dy/dx = -2xy^2$ separates as $y^2 \, dy = -2x \, dx$.
3. Suppose that $(xy + x)(dy/dx) = x^2y + y$.
   (a) Separate this equation.   (b) Find the general solution.

In many applied problems that can be modeled with differential equations, we know conditions that allow us to obtain a particular solution.

Applications of Differential Equations

We now consider two applications that can be modeled by differential equations. These are radioactive decay and one-container mixture problems (as a model for drugs in an organ).

◑ EXAMPLE 6 *Carbon-14 Dating* *(Application Preview)*

When an organism dies, the rate of change of the amount of carbon-14 present is proportional to the amount present and is represented by the differential equation

$$\frac{dy}{dt} = ky$$

where y is the amount present, k is a constant, and t is time in years. If we denote the initial amount of carbon-14 in an organism as y_0, then $y = y_0$ represents the amount present at time $t = 0$ (when the organism dies). Suppose that anthropologists discover a fossil that contains 1% of the initial amount of carbon-14. Find the age of the fossil. (Recall that the half-life of carbon-14 is 5600 years.)

Solution
We must find a particular solution to

$$\frac{dy}{dt} = ky$$

subject to the fact that when $t = 0$, $y = y_0$, and we must determine the value of k on the basis of the half-life of carbon-14 ($t = 5600$ years, $y = \frac{1}{2}y_0$ units.) From Example 5, we know that the general solution to the differential equation $dy/dt = ky$ is $y = Ce^{kt}$. Using $y = y_0$ when $t = 0$, we obtain $y_0 = C$, so the equation becomes $y = y_0e^{kt}$. Using $t = 5600$ and $y = \frac{1}{2}y_0$ in this equation gives

$$\frac{1}{2}y_0 = y_0e^{5600k} \qquad \text{or} \qquad 0.5 = e^{5600k}$$

Rewriting this equation in logarithmic form and then solving for k, we get

$$\ln(0.5) = 5600k$$

$$-0.69315 = 5600k$$

$$-0.00012378 = k$$

Thus the equation we seek is

$$y = y_0 e^{-0.00012378t}$$

Using the fact that $y = 0.01y_0$ when the fossil was discovered, we can find its age t by solving

$$0.01y_0 = y_0 e^{-0.00012378t} \quad \text{or} \quad 0.01 = e^{-0.00012378t}$$

Rewriting this in logarithmic form and then solving give

$$\ln(0.01) = -0.00012378t$$

$$-4.6051702 = -0.00012378t$$

$$37{,}204 = t$$

Thus the fossil is approximately 37,200 years old.

The differential equation $dy/dt = ky$, which describes the decay of radioactive substances in Example 6, also models the rate of growth of an investment that is compounded continuously and the rate of decay of purchasing power due to inflation.

Another application of differential equations comes from a group of applications called *one-container mixture problems*. In problems of this type, there is a substance whose amount in a container is changing with time, and the goal is to determine the amount of the substance at any time t. The differential equations that model these problems are of the following form:

$$\begin{bmatrix} \text{Rate of change} \\ \text{of the amount} \\ \text{of the substance} \end{bmatrix} = \begin{bmatrix} \text{Rate at which} \\ \text{the substance} \\ \text{enters the container} \end{bmatrix} - \begin{bmatrix} \text{Rate at which} \\ \text{the substance} \\ \text{leaves the container} \end{bmatrix}$$

We consider this application as it applies to the amount of a drug in an organ.

● EXAMPLE 7 *Drug in an Organ*

A liquid carries a drug into an organ of volume 300 cc at a rate of 5 cc/s, and the liquid leaves the organ at the same rate. If the concentration of the drug in the entering liquid is 0.1 g/cc, and if x represents the amount of drug in the organ at any time t, then using the fact that the rate of change of the amount of the drug in the organ, dx/dt, equals the rate at which the drug enters minus the rate at which it leaves, we have

$$\frac{dx}{dt} = \left(\frac{5\text{ cc}}{\text{s}}\right)\left(\frac{0.1\text{ g}}{\text{cc}}\right) - \left(\frac{5\text{ cc}}{\text{s}}\right)\left(\frac{x\text{ g}}{300\text{ cc}}\right)$$

or

$$\frac{dx}{dt} = 0.5 - \frac{x}{60} = \frac{30}{60} - \frac{x}{60} = \frac{30 - x}{60}, \quad \text{in g/s}$$

Find the amount of the drug in the organ as a function of time t.

Solution

Multiplying both sides of the equation $\dfrac{dx}{dt} = \dfrac{30 - x}{60}$ by $\dfrac{dt}{(30 - x)}$ gives

$$\frac{dx}{30 - x} = \frac{1}{60} dt$$

The equation is now separated, so we can integrate both sides.

$$\int \frac{dx}{30 - x} = \int \frac{1}{60} dt$$

$$-\ln (30 - x) = \frac{1}{60}t + C_1 \qquad (30 - x > 0)$$

$$\ln (30 - x) = -\frac{1}{60}t - C_1$$

Rewriting this in exponential form gives

$$30 - x = e^{-t/60 - C_1} = e^{-t/60} \cdot e^{-C_1}$$

Letting $C = e^{-C_1}$ yields

$$30 - x = Ce^{-t/60}$$

so

$$x = 30 - Ce^{-t/60}$$

and we have the desired function.

● *Checkpoint Solutions*

1. (a) $f(x) = \displaystyle\int\left[2x - \frac{1}{x + 1}\right]dx = x^2 - \ln |x + 1| + C$

   (b) If $f(0) = 4$, then $4 = 0^2 - \ln |1| + C$, so $C = 4$.
   Thus $f(x) = x^2 - \ln |x + 1| + 4$ is the particular solution.

2. (a) True, and the solution is

$$\int y \, dy = \int x \, dx \qquad \Rightarrow \qquad \frac{y^2}{2} = \frac{x^2}{2} + C$$

   (b) True
   (c) False. It separates as $dy/y^2 = -2x \, dx$, and the solution is

$$\int \frac{-dy}{y^2} = \int 2x \, dx \qquad \Rightarrow \qquad y^{-1} = x^2 + C$$

$$\frac{1}{y} = x^2 + C$$

$$y = \frac{1}{x^2 + C}$$

3. (a) $x(y + 1)\dfrac{dy}{dx} = (x^2 + 1)y$

$$\frac{y + 1}{y} dy = \frac{x^2 + 1}{x} dx$$

$$\left(1 + \frac{1}{y}\right)dy = \left(x + \frac{1}{x}\right)dx$$

   (b) $\displaystyle\int\left(1 + \frac{1}{y}\right)dy = \int\left(x + \frac{1}{x}\right)dx$

$$y + \ln |y| = \frac{x^2}{2} + \ln |x| + C$$

   This is the general solution.

12.5 Exercises

In Problems 1–4, show that the given function is a solution to the differential equation.

1. $y = x^2$; $4y - 2xy' = 0$
2. $y = x^3$; $3y - xy' = 0$
3. $y = 3x^2 + 1$; $2y \, dx - x \, dy = 2 \, dx$
4. $y = 4x^3 + 2$; $3y \, dx - x \, dy = 6 \, dx$

In Problems 5–10, use integration to find the general solution to each differential equation.

5. $dy = xe^{x^2 + 1} \, dx$
6. $dy = x^2 e^{x^3 - 1} \, dx$
7. $2y \, dy = 4x \, dx$
8. $4y \, dy = 4x^3 \, dx$
9. $3y^2 \, dy = (2x - 1) \, dx$
10. $4y^3 \, dy = (3x^2 + 2x) \, dx$

In Problems 11–14, find the particular solution.

11. $y' = e^{x-3}$; $y(0) = 2$
12. $y' = e^{2x+1}$; $y(0) = e$
13. $dy = \left(\dfrac{1}{x} - x\right) dx$; $y(1) = 0$
14. $dy = \left(x^2 - \dfrac{1}{x+1}\right) dx$; $y(0) = \dfrac{1}{3}$

In Problems 15–28, find the general solution to the given differential equation.

15. $\dfrac{dy}{dx} = \dfrac{x^2}{y}$ 16. $y^3 \, dx = \dfrac{dy}{x^3}$
17. $dx = x^3 y \, dy$ 18. $dy = x^2 y^3 \, dx$
19. $dx = (x^2 y^2 + x^2) \, dy$ 20. $dy = (x^2 y^3 + xy^3) \, dx$
21. $y^2 \, dx = x \, dy$ 22. $y \, dx = x \, dy$
23. $\dfrac{dy}{dx} = \dfrac{x}{y}$ 24. $\dfrac{dy}{dx} = \dfrac{x^2 + x}{y + 1}$
25. $(x + 1)\dfrac{dy}{dx} = y$ 26. $x^2 y \dfrac{dy}{dx} = y^2 + 1$
27. $e^{2x} y \, dy = (y + 1) \, dx$
28. $e^{4x}(y + 1) \, dx + e^{2x} y \, dy = 0$

In Problems 29–36, find the particular solution to each differential equation.

29. $\dfrac{dy}{dx} = \dfrac{x^2}{y^3}$,   when $x = 1, y = 1$
30. $\dfrac{dy}{dx} = \dfrac{x + 1}{xy}$,   when $x = 1, y = 3$
31. $2y^2 \, dx = 3x^2 \, dy$,   when $x = 2, y = -1$
32. $(x + 1) \, dy = y^2 \, dx$,   when $x = 0, y = 2$
33. $x^2 e^{2y} \, dy = (x^3 + 1) \, dx$,   when $x = 1, y = 0$
34. $y' = \dfrac{1}{xy}$,   when $x = 1, y = 3$

35. $2xy \dfrac{dy}{dx} = y^2 + 1$,   when $x = 1, y = 2$
36. $xe^y \, dx = (x + 1) \, dy$,   when $x = 0, y = 0$

APPLICATIONS

37. *Allometric growth*   If x and y are measurements of certain parts of an organism, then the rate of change of y with respect to x is proportional to the ratio of y to x. That is, these measurements satisfy

$$\frac{dy}{dx} = k\frac{y}{x}$$

which is referred to as an allometric law of growth. Solve this differential equation.

38. *Bimolecular chemical reactions*   A bimolecular chemical reaction is one in which two chemicals react to form another substance. Suppose that one molecule of each of the two chemicals react to form two molecules of a new substance. If x represents the number of molecules of the new substance at time t, then the rate of change of x is proportional to the product of the numbers of molecules of the original chemicals available to be converted. That is, if each of the chemicals initially contained A molecules, then

$$\frac{dx}{dt} = k(A - x)^2$$

If 40% of the initial amount A is converted after 1 hour, how long will it be before 90% is converted?

Compound interest   **In Problems 39 and 40, use the following information.**

When interest is compounded continuously, the rate of change of the amount x of the investment is proportional to the amount present. In this case, the proportionality constant is the annual interest rate r (as a decimal); that is,

$$\frac{dx}{dt} = rx$$

39. (a) If \$10,000 is invested at 6%, compounded continuously, find an equation for the future value of the investment as a function of time t in years.
    (b) What is the future value of the investment after 1 year? after 5 years?
    (c) How long will it take for the investment to double?

40. (a) If \$2000 is invested at 8%, compounded continuously, find an equation for the future value of the investment as a function of time t, in years.
    (b) How long will it take for the investment to double?
    (c) What will be the future value of this investment after 35 years?

41. *Investing* When the interest on an investment is compounded continuously, the investment grows at a rate that is proportional to the amount in the account, so that if the amount present is P, then

$$\frac{dP}{dt} = kP$$

where P is in dollars, t is in years, and k is a constant. If \$100,000 is invested (when $t = 0$) and the amount in the account after 15 years is \$211,700, find the function that gives the value of the investment as a function of t. What is the interest rate on this investment?

42. *Investing* When the interest on an investment is compounded continuously, the investment grows at a rate that is proportional to the amount in the account. If \$20,000 is invested (when $t = 0$) and the amount in the account after 22 years is \$280,264, find the function that gives the value of the investment as a function of t. What is the interest rate on this investment?

43. *Bacterial growth* Suppose that the growth of a certain population of bacteria satisfies

$$\frac{dy}{dt} = ky$$

where y is the number of organisms and t is the number of hours. If initially there are 10,000 organisms and the number triples after 2 hours, how long will it be before the population reaches 100 times the original population?

44. *Bacterial growth* Suppose that, for a certain population of bacteria, growth occurs according to

$$\frac{dy}{dt} = ky \quad (t \text{ in hours})$$

If the doubling rate depends on temperature, find how long it takes for the number of bacteria to reach 50 times the original number at each given temperature in (a) and (b).
    (a) At 90°F, the number doubles after 30 minutes ($\frac{1}{2}$ hour).
    (b) At 40°F, the number doubles after 3 hours.

45. *Sales and pricing* Suppose that in a certain, company, the relationship between the price per unit p of its product and the weekly sales volume y, in thousands of

dollars, is given by

$$\frac{dy}{dp} = -\frac{2}{5}\left(\frac{y}{p + 8}\right)$$

Solve this differential equation if $y = 8$ when $p = \$24$.

46. *Sales and pricing* Suppose that a chain of auto service stations, Quick-Oil, Inc., has found that the relationship between its price p for an oil change and its monthly sales volume y, in thousands of dollars, is

$$\frac{dy}{dp} = -\frac{1}{2}\left(\frac{y}{p + 5}\right)$$

Solve this differential equation if $y = 18$ when $p = \$20$.

47. *Half-life* A breeder reactor converts uranium-238 into an isotope of plutonium-239 at a rate proportional to the amount present at any time. After 10 years, 0.03% of the radioactivity has dissipated (that is, 0.9997 of the initial amount remains). Suppose that initially there is 100 pounds of this substance. Find the half-life.

48. *Radioactive decay* A certain radioactive substance has a half-life of 50 hours. Find how long it will take for 90% of the radioactivity to be dissipated if the amount of material x satisfies

$$\frac{dx}{dt} = kx \quad (t \text{ in hours})$$

49. *Drug in an organ* Suppose that a liquid carries a drug into a 100-cc organ at a rate of 5 cc/s and leaves the organ at the same rate. Suppose that the concentration of the drug entering is 0.06 g/cc. If initially there is no drug in the organ, find the amount of drug in the organ as a function of time t.

50. *Drug in an organ* Suppose that a liquid carries a drug into a 250-cc organ at a rate of 10 cc/s and leaves the organ at the same rate. Suppose that the concentration of the drug entering is 0.15 g/cc. Find the amount of drug in the organ as a function of time t if initially there is none in the organ.

51. *Drug in an organ* Suppose that a liquid carries a drug with concentration 0.1 g/cc into a 200-cc organ at a rate of 5 cc/s and leaves the organ at the same rate. If initially there is 10 g of the drug in the organ, find the amount of drug in the organ as a function of time t.

52. *Drug in an organ* Suppose that a liquid carries a drug with concentration 0.05 g/cc into a 150-cc organ at a rate of 6 cc/s and leaves at the same rate. If initially there is 1.5 g of drug in the organ, find the amount of drug in the organ as a function of time t.

53. *Tumor volume*   Let V denote the volume of a tumor, and suppose that the growth rate of the tumor satisfies

$$\frac{dV}{dt} = 0.2Ve^{-0.1t}$$

If the initial volume of the tumor is 1.86 units, find an equation for V as a function of t.

54. *Gompertz curves*   The differential equation

$$\frac{dx}{dt} = x(a - b \ln x)$$

where x represents the number of objects at time t, and a and b are constants, is the model for Gompertz curves. Recall from Section 5.3, "Solutions of Exponential Equations," that Gompertz curves can be used to study growth or decline of populations, organizations, and revenue from sales of a product, as well as forecast equipment maintenance costs. Solve the differential equation to obtain the Gompertz curve formula

$$x = e^{a/b}e^{-ce^{-bt}}$$

55. *Cell growth*   If V is the volume of a spherical cell, then in certain cell growth and for some fetal growth models, the rate of change of V is given by

$$\frac{dV}{dt} = kV^{2/3}$$

where k is a constant depending on the organism. If $V = 0$ when $t = 0$, find V as a function of t.

56. *Atmospheric pressure*   The rate of change of atmospheric pressure P with respect to the altitude above sea level h is proportional to the pressure. That is,

$$\frac{dP}{dh} = kP$$

Suppose that the pressure at sea level is denoted by P_0, and at 18,000 ft the pressure is half what it is at sea level. Find the pressure, as a percent of P_0, at 25,000 ft.

57. *Newton's law of cooling*   Newton's law of cooling (and warming) states that the rate of change of temperature $u = u(t)$ of an object is proportional to the temperature difference between the object and its surroundings, where T is the constant temperature of the surroundings. That is,

$$\frac{du}{dt} = k(u - T)$$

Suppose an object at 0°C is placed in a room where the temperature is 20°C. If the temperature of the object is 8°C after 1 hour, how long will it take for the object to reach 18°C?

58. *Newton's law of cooling*   Newton's law of cooling can be used to estimate time of death. (Actually the estimate may be quite rough, because cooling does not begin until metabolic processes have ceased.) Suppose a corpse is discovered at noon in a 70°F room, and at that time the body temperature is 96.1°F. If at 1:00 P.M. the body temperature is 94.6°F, use Newton's law of cooling to estimate the time of death.

59. *Fossil-fuel emissions*   The amount of carbon in the atmosphere has been estimated to have increased at a rate of about 3.15% per year from 1860 until 1983, except during the Great Depression and the world wars.* This increase is due to carbon emissions from fossil-fuel burning and can be modeled by the differential equation

$$\frac{dE}{dt} = 0.0315E$$

where t is the number of years since 1860 and E is fossil-fuel emissions in gigatons per year.
(a) Solve this differential equation, and find a particular solution that satisfies $E(0) = 0.13$.
(b) Graph your solution and compare it with the graph below, which shows actual emissions since 1860.

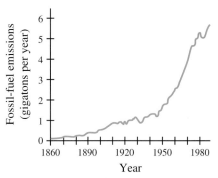

Source: American Scientist, Vol. 79 (July–August 1990): 313

60. *Mutual funds*   The number of mutual funds N, excluding money market funds, can be modeled by

$$\frac{dN}{dt} = 0.125N, \quad N(10) = 1071$$

where t is the number of years past 1975 (*Source: Mutual Fund Fact Book,* 2003).
(a) Find the particular solution to this differential equation.
(b) What does the model predict as the number of mutual funds (excluding money market funds) in 2000?

*American Scientist, Vol. 79 (July–August 1990): 313.

61. *Impact of inflation*   The impact of a 5% inflation rate on a \$50,000-per-year pension can be severe. If P represents the purchasing power (in dollars) of a \$50,000 pension, then the effect of a 5% inflation rate can be modeled by the differential equation

$$\frac{dP}{dt} = -0.05P, \quad P(0) = 50,000$$

where t is in years.
(a) Find the particular solution to this differential equation.
(b) Find the purchasing power after 30 years.

62. *Students per computer*   The following figure shows the numbers of students per computer in public schools for school years 1983–1984 to 2001–2002. If S represents the students per computer, then the rate of change of these data can be modeled by the differential equation

$$\frac{dS}{dt} = -0.1657S; \quad S(16) = 10$$

where t is the number of school years past the 1979–1980 school year.
(a) Find the particular solution to this differential equation.
(b) Check the model you found in part (a) against the graph in the figure. Is the model more accurate for school data near 1983–1984 or near 1999–2000?

Students Per Computer in U.S. Public Schools

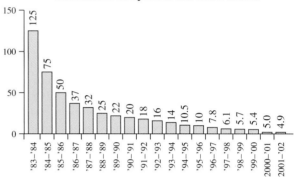

Source: Quality Education Data, Inc., Denver, CO. Reprinted by permission.

Key Terms and Formulas

Section	Key Terms	Formulas
12.1	General antiderivative of $f'(x)$	$f(x) + C$
	Integral	$\int f(x)\, dx$
	Powers of x Formula	$\int x^n\, dx = \dfrac{x^{n+1}}{n+1} + C \quad (n \neq -1)$
	Integration Formulas	$\int dx = x + C$
		$\int cu(x)\, dx = c \int u(x)\, dx \quad (c = \text{a constant})$
		$\int \left[u(x) \pm v(x) \right] dx = \int u(x)\, dx \pm \int v(x)\, dx$
12.2	Power Rule	$\int \left[u(x) \right]^n u'(x)\, dx = \dfrac{\left[u(x) \right]^{n+1}}{n+1} + C \quad (n \neq -1)$
12.3	Exponential Formula	$\int e^u u'\, dx = \int e^u\, du = e^u + C$
	Logarithmic Formula	$\int \dfrac{u'}{u}\, dx = \int \dfrac{1}{u}\, du = \ln |u| + C$

Section	Key Terms	Formulas
12.4	Total cost	$C(x) = \int \overline{MC}\, dx$
	Total revenue	$R(x) = \int \overline{MR}\, dx$
	Profit	
	Marginal propensity to consume	$\dfrac{dC}{dy}$
	Marginal propensity to save	$\dfrac{dS}{dy} = 1 - \dfrac{dC}{dy}$
	National consumption	$C = \int f'(y)\, dy = \int \dfrac{dC}{dy}\, dy$
12.5	Differential equations	
	Solutions	
	General	
	Particular	
	First order	$\dfrac{dy}{dx} = f(x) \Rightarrow y = \int f(x)\, dx$
	Separable	$g(y)\, dy = f(x)\, dx \Rightarrow \int g(y)\, dy = \int f(x)\, dx$
	Radioactive decay	$\dfrac{dy}{dt} = ky$
	Drug in an organ	Rate = (rate in) − (rate out)

Review Exercises

Sections 12.1–12.3

Evaluate the integrals in Problems 1–26.

1. $\int x^6\, dx$

2. $\int x^{1/2}\, dx$

3. $\int (12x^3 - 3x^2 + 4x + 5)\, dx$

4. $\int 7(x^2 - 1)^2\, dx$

5. $\int 7x(x^2 - 1)^2\, dx$

6. $\int (x^3 - 3x^2)^5(x^2 - 2x)\, dx$

7. $\int (x^3 + 4)^2 3x\, dx$

8. $\int 5x^2(3x^3 + 7)^6\, dx$

9. $\int \dfrac{x^2}{x^3 + 1}\, dx$

10. $\int \dfrac{x^2}{(x^3 + 1)^2}\, dx$

11. $\int \dfrac{x^2\, dx}{\sqrt[3]{x^3 - 4}}$

12. $\int \dfrac{x^2\, dx}{x^3 - 4}$

13. $\int \dfrac{x^3 + 1}{x^2}\, dx$

14. $\int \dfrac{x^3 - 3x + 1}{x - 1}\, dx$

15. $\int y^2 e^{y^3}\, dy$

16. $\int (x - 1)^2\, dx$

17. $\int \dfrac{3x^2}{2x^3 - 7}\, dx$

18. $\int \dfrac{5\, dx}{e^{4x}}$

19. $\int (x^3 - e^{3x})\, dx$

20. $\int xe^{1 + x^2}\, dx$

21. $\int \dfrac{6x^7}{(5x^8 + 7)^3}\, dx$

22. $\int \dfrac{7x^3}{\sqrt{1 - x^4}}\, dx$

23. $\int \left(\dfrac{e^{2x}}{2} + \dfrac{2}{e^{2x}} \right) dx$

24. $\int \left[x - \dfrac{1}{(x + 1)^2} \right] dx$

25. (a) $\int (x^2 - 1)^4 x\, dx$ (b) $\int (x^2 - 1)^{10} x\, dx$
   (c) $\int (x^2 - 1)^7 3x\, dx$ (d) $\int (x^2 - 1)^{-2/3} x\, dx$

26. (a) $\int \dfrac{2x\, dx}{x^2 - 1}$ (b) $\int \dfrac{2x\, dx}{(x^2 - 1)^2}$
   (c) $\int \dfrac{3x\, dx}{\sqrt{x^2 - 1}}$ (d) $\int \dfrac{3x\, dx}{x^2 - 1}$

Section 12.5

In Problems 27–32, find the general solution to each differential equation.

27. $\dfrac{dy}{dt} = 4.6e^{-0.05t}$

28. $dy = (64 + 76x - 36x^2)\, dx$

29. $\dfrac{dy}{dx} = \dfrac{4x}{y - 3}$

30. $t\, dy = \dfrac{dt}{y + 1}$

31. $\dfrac{dy}{dx} = \dfrac{x}{e^y}$

32. $\dfrac{dy}{dt} = \dfrac{4y}{t}$

In Problems 33 and 34, find the particular solution to each differential equation.

33. $y' = \dfrac{x^2}{y+1}$, $\quad y(0) = 4$

34. $y' = \dfrac{2x}{1+2y}$, $\quad y(2) = 0$

APPLICATIONS

Section 12.1

35. *Revenue* If the marginal revenue for a month for a product is $\overline{MR} = 6x + 12$, find the total revenue, in dollars, from the sale of $x = 4$ units of the product.

36. *Productivity* Suppose that the rate of change of production of the average worker at a factory is given by

$$\frac{dp}{dt} = 27 + 24t - 3t^2, \qquad 0 \le t \le 8$$

where p is the number of units the worker produces in t hours. How many units will the average worker produce in an 8-hour shift? (Assume that $p = 0$ when $t = 0$.)

Section 12.2

37. *Oxygen levels in water* The rate of change of the oxygen level (in mmol/l) per month in a body of water after an oil spill is given by

$$P'(t) = 400\left[\frac{5}{(t+5)^2} - \frac{50}{(t+5)^3}\right]$$

where t is the number of months after the spill. What function gives the oxygen level P at any time t if $P = 400$ mmol/l when $t = 0$?

38. *Bacterial growth* A population of bacteria grows at the rate

$$\frac{dp}{dt} = \frac{100{,}000}{(t+100)^2}$$

where p is the population and t is time. If the population is 1000 when $t = 1$, write the equation that gives the size of the population at any time t.

Section 12.3

39. *Market share* The rate of change of the market share (as a percent) a firm expects for a new product is

$$\frac{dy}{dt} = 2.4e^{-0.04t}$$

where t is the number of months after the product is introduced.
(a) Write the equation that gives the expected market share y at any time t. (Note that $y = 0$ when $t = 0$.)
(b) What market share does the firm expect after 1 year?

40. *Revenue* If the marginal revenue for a product is $\overline{MR} = \dfrac{800}{x+1}$, find the total revenue function.

Section 12.4

41. *Cost* The marginal cost for a product is $\overline{MC} = 6x + 4$ dollars per unit, and the cost of producing 100 items is $31,400.
(a) Find the fixed costs.
(b) Find the total cost function.

42. *Profit* Suppose a product has a daily marginal revenue $\overline{MR} = 46$ and a daily marginal cost $\overline{MC} = 30 + \frac{1}{5}x$, both in dollars per unit. If the daily fixed cost is $200, how many units will give maximum profit and what is the maximum profit?

43. *National consumption* If consumption is $8.5 billion when disposable income is 0, and if the marginal propensity to consume is

$$\frac{dC}{dy} = \frac{1}{\sqrt{2y+16}} + 0.6 \quad \text{(in billions of dollars)}$$

find the national consumption function.

44. *National consumption* Suppose that the marginal propensity to save is

$$\frac{dS}{dy} = 0.2 - 0.1e^{-2y} \quad \text{(in billions of dollars)}$$

and consumption is $7.8 billion when disposable income is 0. Find the national consumption function.

Section 12.5

45. *Allometric growth* For many species of fish, the length L and weight W of a fish are related by

$$\frac{dW}{dL} = \frac{3W}{L}$$

The general solution to this differential equation expresses the allometric relationship between the length and weight of a fish. Find the general solution.

46. *Investment* When the interest on an investment is compounded continuously, the investment grows at a

rate that is proportional to the amount in the account, so that if the amount present is P, then

$$\frac{dP}{dt} = kP \qquad \text{where } P \text{ is in dollars, } t \text{ is in years, and } k \text{ is a constant}$$

(a) Solve this differential equation to find the relationship.
(b) Use properties of logarithms and exponential functions to write P as a function of t.
(c) If \$50,000 is invested (when $t = 0$) and the amount in the account after 10 years is \$135,914, find the function that gives the value of the investment as a function of t.
(d) In part (c), what does the value of k represent?

47. *Fossil dating* Radioactive beryllium is sometimes used to date fossils found in deep-sea sediment. The amount of radioactive material x satisfies

$$\frac{dx}{dt} = kx$$

Suppose that 10 units of beryllium are present in a living organism and that the half-life of beryllium is 4.6 million years. Find the age of a fossil if 20% of the original radioactivity is present when the fossil is discovered.

48. *Drug in an organ* Suppose that a liquid carries a drug into a 120-cc organ at a rate of 4 cc/s and leaves the organ at the same rate. If initially there is no drug in the organ and if the concentration of drug in the liquid is 3 g/cc, find the amount of drug in the organ as a function of time.

49. *Chemical mixture* A 300-gal tank initially contains a solution with 100 lb of a chemical. A mixture containing 2 lb/gal of the chemical enters the tank at 3 gal/min, and the well-stirred mixture leaves at the same rate. Find an equation that gives the amount of the chemical in the tank as a function of time. How long will it be before there is 500 lb of chemical in the tank?

12 Chapter Test

Evaluate the integrals in Problems 1–8.

1. $\displaystyle\int (6x^2 + 8x - 7)\, dx$

2. $\displaystyle\int \left(4 + \sqrt{x} - \frac{1}{x^2}\right) dx$

3. $\displaystyle\int 5x^2(4x^3 - 7)^9\, dx$

4. $\displaystyle\int (3x^2 - 6x + 1)^{-3}(2x - 2)\, dx$

5. $\displaystyle\int \frac{s^3}{2s^4 - 5}\, ds$

6. $\displaystyle\int 100e^{-0.01x}\, dx$

7. $\displaystyle\int 5y^3 e^{2y^4 - 1}\, dy$

8. $\displaystyle\int \left(e^x + \frac{5}{x} - 1\right) dx$

9. Evaluate $\displaystyle\int \frac{x^2}{x + 1}\, dx$. Use long division.

10. If $\displaystyle\int f(x)\, dx = 2x^3 - x + 5e^x + C$, find $f(x)$.

In Problems 11 and 12, find the particular solution to each differential equation.

11. $y' = 4x^3 + 3x^2$, if $y(0) = 4$

12. $\dfrac{dy}{dx} = e^{4x}$, if $y(0) = 2$

13. Find the general solution of the separable differential equation $\dfrac{dy}{dx} = x^3 y^2$.

14. Suppose the rate of growth of the population of a city is predicted to be

$$\frac{dp}{dt} = 2000t^{1.04}$$

where p is the population and t is the number of years past 2005. If the population in the year 2005 is 50,000, what is the predicted population in the year 2015?

15. Suppose that the marginal cost for x units of a product is $\overline{MC} = 4x + 50$, the marginal revenue is $\overline{MR} = 500$, and the cost of the production and sale of 10 units is \$1000. What is the profit function for this product?

16. Suppose the marginal propensity to save is given by

$$\frac{dS}{dy} = 0.22 - \frac{0.25}{\sqrt{0.5y + 1}} \quad \text{(in billions of dollars)}$$

and national consumption is \$6.6 billion when disposable income is \$0. Find the national consumption function.

17. A certain radioactive material has a half-life of 100 days. If the amount of material present, x, satisfies $\dfrac{dx}{dt} = kx$, where t is in days, how long will it take for 90% of the radioactivity to dissipate?

Extended Applications & Group Projects

I. Employee Production Rate

The manager of a plant has been instructed to hire and train additional employees to manufacture a new product. She must hire a sufficient number of new employees so that within 30 days they will be producing 2500 units of the product each day.

Because a new employee must learn an assigned task, production will increase with training. Suppose that research on similar projects indicates that production increases with training according to the learning curve, so that for the average employee, the rate of production per day is given by

$$\frac{dN}{dt} = be^{-at}$$

where N is the number of units produced per day after t days of training and a and b are constants that depend on the project. Because of experience with a similar project, the manager expects the rate for this project to be

$$\frac{dN}{dt} = 2.5e^{-0.05t}$$

The manager tested her training program with 5 employees and learned that the average employee could produce 11 units per day after 5 days of training. On the basis of this information, she must decide how many employees to hire and begin to train so that a month from now they will be producing 2500 units of the product per day. She estimates that it will take her 10 days to hire the employees, and thus she will have 15 days remaining to train them. She also expects a 10% attrition rate during this period.

How many employees would you advise the plant manager to hire? Check your advice by answering the following questions.

1. Use the expected rate of production and the results of the manager's test to find the function relating N and t—that is, $N = N(t)$.
2. Find the number of units the average employee can produce after 15 days of training. How many such employees would be needed to maintain a production rate of 2500 units per day?
3. Explain how you would revise this last result to account for the expected 10% attrition rate. How many new employees should the manager hire?

II. Supply and Demand

If p is the price in dollars of a given commodity at time t, then we can think of price as a function of time. Similarly, the number of units demanded by consumers q_d at any time, and the number of units supplied by producers q_s at any time, may also be considered as functions of time as well as functions of price.

Both the quantity demanded and the quantity supplied depend not only on the price at the time, but also on the direction and rate of change that consumers and producers ascribe to prices. For example, even when prices are high, if consumers feel that prices are rising, the demand may rise. Similarly, if prices are low but producers feel they may go lower, the supply may rise.

If we assume that prices are determined in the marketplace by supply and demand, then the equilibrium price is the one we seek.

Suppose the supply and demand functions for a certain commodity in a competitive market are given, in hundreds of units, by

$$q_s = 30 + p + 5\frac{dp}{dt}$$

$$q_d = 51 - 2p + 4\frac{dp}{dt}$$

where dp/dt denotes the rate of change of the price with respect to time. If, at $t = 0$, the market equilibrium price is \$12, we can express the market equilibrium price as a function of time.

Our goals are as follows.

A. To express the market equilibrium price as a function of time.
B. To determine whether there is price stability in the marketplace for this item (that is, to determine whether the equilibrium price approaches a constant over time).

To achieve these goals, do the following.

1. Set the expressions for q_s and q_d equal to each other.
2. Solve this equation for $\dfrac{dp}{dt}$.
3. Write this equation in the form $f(p)\,dp = g(t)\,dt$.
4. Integrate both sides of this separated differential equation.
5. Solve the resulting equation for p in terms of t.
6. Use the fact that $p = 12$ when $t = 0$ to find C, the constant of integration, and write the market equilibrium price p as a function of time t.
7. Find $\lim\limits_{t \to \infty} p$, which gives the price we can expect this product to approach. If this limit is finite, then for this item there is price stability in the marketplace. If $p \to +\infty$ as $t \to +\infty$, then price will continue to increase until economic conditions change.

13

Definite Integrals: Techniques of Integration

We saw some applications of the indefinite integral in Chapter 12. In this chapter we define the definite integral and discuss a theorem and techniques that are useful in evaluating or approximating it. We will also see how it can be used in many interesting applications such as consumer's and producer's surplus and total value, present value, and future value of continuous income streams. Improper integrals can be used to find the capital value of a continuous income stream.

The topics and applications discussed in this chapter include the following.

Chapter Warm-up

Prerequisite Problem Type	For Section	Answer	Section for Review
Simplify: $\dfrac{1}{n^3}\left[\dfrac{n(n+1)(2n+1)}{6} - \dfrac{2n(n+1)}{2} + n\right]$	13.1	$\dfrac{2n^2 - 3n + 1}{6n^2}$	0.7 Fractions
(a) If $F(x) = \dfrac{x^4}{4} + 4x + C$, what is $F(4) - F(2)$? (b) If $F(x) = -\dfrac{1}{9}\ln\left[\dfrac{9 + \sqrt{81 - 9x^2}}{3x}\right]$, what is $F(3) - F(2)$?	13.2 13.5	(a) 68 (b) $\dfrac{1}{9}\ln\left(\dfrac{3 + \sqrt{5}}{2}\right)$	1.2 Function notation
Find the limit: (a) $\displaystyle\lim_{n \to +\infty} \dfrac{n^2 + n}{2n^2}$ (b) $\displaystyle\lim_{n \to +\infty} \dfrac{2n^2 - 3n + 1}{6n^2}$ (c) $\displaystyle\lim_{b \to \infty}\left(1 - \dfrac{1}{b}\right)$ (d) $\displaystyle\lim_{b \to \infty}\left(\dfrac{-100,000}{e^{0.10b}} + 100,000\right)$	13.1 13.7	(a) $\frac{1}{2}$ (b) $\frac{1}{3}$ (c) 1 (d) 100,000	9.2 Limits at infinity
Find the derivative of $y = \ln x$.	13.6	$\dfrac{1}{x}$	11.1 Derivatives of logarithmic functions
Integrate: (a) $\int (x^3 + 4)\, dx$ (b) $\int x\sqrt{x^2 - 9}\, dx$ (c) $\int e^{2x}\, dx$	13.2–13.7	(a) $\dfrac{x^4}{4} + 4x + C$ (b) $\frac{1}{3}(x^2 - 9)^{3/2} + C$ (c) $\frac{1}{2}e^{2x} + C$	12.1, 12.2, 12.3 Integration

13.1 Area Under a Curve

◗ Application Preview

One way to find the accumulated production (such as the production of ore from a mine) over a period of time is to graph the rate of production as a function of time and find the area under the resulting curve over a specified time interval. For example, if a coal mine produces at a rate of 30 tons per day, the production over 10 days ($30 \cdot 10 = 300$) could be represented by the area under the line $y = 30$ between $x = 0$ and $x = 10$ (see Figure 13.1).

Using area to determine the accumulated production is very useful when the rate-of-production function varies at different points in time. For example, if the rate of production (in tons per day) is represented by

$$y = 100e^{-0.1x}$$

where x represents the number of days, then the area under the curve (and above the x-axis) from $x = 0$ to $x = 10$ represents the total production over the 10-day period (see Figure 13.2(a) and Example 1). In order to determine the accumulated production and to solve other types of problems, we need a method for finding areas under curves. This is the goal of this section.

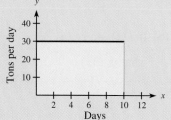

Figure 13.1

Estimating the Area Under a Curve

To estimate the accumulated production for the example in the Application Preview, we approximate the area under the graph of the production rate function. We can find a rough approximation of the area under this curve by fitting two rectangles to the curve as shown in Figure 13.2(b). The area of the first rectangle is $5 \cdot 100 = 500$ square units, and the area of the second rectangle is $(10 - 5)[100e^{-0.1(5)}] \approx 5(60.65) = 303.25$ square units, so this rough approximation is 803.25 square units or 803.25 tons of ore. This approximation is clearly larger than the exact area under the curve. Why?

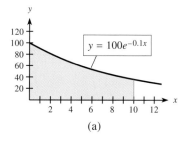
(a)

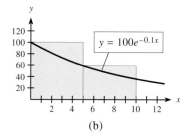
(b)

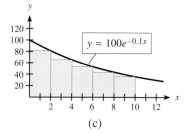
(c)

Figure 13.2

◗ EXAMPLE 1 *Ore Production* **(Application Preview)**

Find another, more accurate, approximation of the tons of ore produced by approximating the area under the curve in Figure 13.2(a). Fit five rectangles with equal bases inside the area under the curve $y = 100e^{-0.1x}$, and use them to approximate the area under the curve from $x = 0$ to $x = 10$ (see Figure 13.2(c)).

Solution

Each of the five rectangles has base 2, and the height of each rectangle is the value of the function at the right-hand endpoint of the interval forming its base. Thus the areas of the rectangles are as follows:

Rectangle	Base	Height	Area = Base × Height
1	2	$100e^{-0.1(2)} \approx 81.87$	$2(81.87) = 163.74$
2	2	$100e^{-0.1(4)} \approx 67.03$	$2(67.03) = 134.06$
3	2	$100e^{-0.1(6)} \approx 54.88$	$2(54.88) = 109.76$
4	2	$100e^{-0.1(8)} \approx 44.93$	$2(44.93) = 89.86$
5	2	$100e^{-0.1(10)} \approx 36.79$	$2(36.79) = 73.58$

The area under the curve is approximately equal to

$$163.74 + 134.06 + 109.76 + 89.86 + 73.58 = 571$$

So approximately 571 tons of ore are produced in the 10-day period. The area is actually 632.12, to two decimal places (or 632.12 tons of ore), so the approximation 571 is smaller than the actual area but is much better than the one we obtained with just two rectangles. In general, if we use bases of equal width, the approximation of the area under a curve improves when more rectangles are used.

Suppose that we wish to find the area between the curve $y = 2x$ and the x-axis from $x = 0$ to $x = 1$ (see Figure 13.3). As we saw in Example 1, one way to approximate this area is to use the areas of rectangles whose bases are on the x-axis and whose heights are the vertical distances from points on their bases to the curve. We can divide the interval $[0, 1]$ into n equal subintervals and use them as the bases of n rectangles whose heights are determined by the curve (see Figure 13.4). The width of each of these rectangles is $1/n$. Using the function value at the right-hand endpoint of each subinterval as the height of the rectangle, we get n rectangles as shown in Figure 13.4. Because part of each rectangle lies above the curve, the sum of the areas of the rectangles will overestimate the area.

Then, with $y = f(x) = 2x$ and subinterval width $1/n$, the areas of the rectangles are as follows.

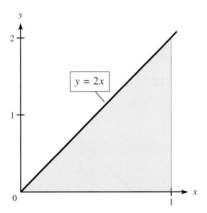

Figure 13.3

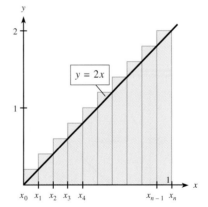

Figure 13.4

Rectangle	Base	Right Endpoint	Height	Area = Base × Height
1	$\dfrac{1}{n}$	$x_1 = \dfrac{1}{n}$	$f(x_1) = 2\left(\dfrac{1}{n}\right)$	$\dfrac{1}{n} \cdot \dfrac{2}{n} = \dfrac{2}{n^2}$
2	$\dfrac{1}{n}$	$x_2 = \dfrac{2}{n}$	$f(x_2) = 2\left(\dfrac{2}{n}\right)$	$\dfrac{1}{n} \cdot \dfrac{4}{n} = \dfrac{4}{n^2}$
3	$\dfrac{1}{n}$	$x_3 = \dfrac{3}{n}$	$f(x_3) = 2\left(\dfrac{3}{n}\right)$	$\dfrac{1}{n} \cdot \dfrac{6}{n} = \dfrac{6}{n^2}$
⋮				
i	$\dfrac{1}{n}$	$x_i = \dfrac{i}{n}$	$f(x_i) = 2\left(\dfrac{i}{n}\right)$	$\dfrac{1}{n} \cdot \dfrac{2i}{n} = \dfrac{2i}{n^2}$
⋮				
n	$\dfrac{1}{n}$	$x_n = \dfrac{n}{n}$	$f(x_n) = 2\left(\dfrac{n}{n}\right)$	$\dfrac{1}{n} \cdot \dfrac{2n}{n} = \dfrac{2n}{n^2}$

Note that $2i/n^2$ gives the area of the *i*th rectangle for *any* value of *i*. Thus for any value of *n*, this area can be approximated by the sum

$$A \approx \frac{2}{n^2} + \frac{4}{n^2} + \frac{6}{n^2} + \cdots + \frac{2i}{n^2} + \cdots + \frac{2n}{n^2}$$

In particular, we have the following approximations of this area for specific values of *n* (the number of rectangles).

$$n = 5: \qquad A \approx \frac{2}{25} + \frac{4}{25} + \frac{6}{25} + \frac{8}{25} + \frac{10}{25} = \frac{30}{25} = 1.20$$

$$n = 10: \qquad A \approx \frac{2}{100} + \frac{4}{100} + \frac{6}{100} + \cdots + \frac{20}{100} = \frac{110}{100} = 1.10$$

$$n = 100: \qquad A \approx \frac{2}{10,000} + \frac{4}{10,000} + \frac{6}{10,000} + \cdots + \frac{200}{10,000}$$

$$= \frac{10,100}{10,000} = 1.01$$

Figure 13.5 shows the rectangles associated with each of these approximations ($n = 5, 10$, and 100) to the area under $y = 2x$ from $x = 0$ to $x = 1$. For larger *n*, the rectangles closely approximate the area under the curve.

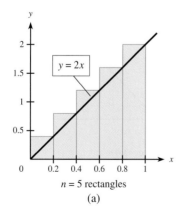

$n = 5$ rectangles

(a)

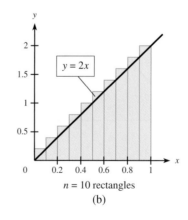

$n = 10$ rectangles

(b)

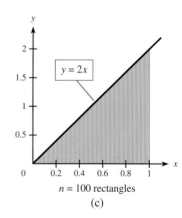

$n = 100$ rectangles

(c)

Figure 13.5

We can find this sum for any n more easily if we observe that the common denominator is n^2 and that the numerator is twice the sum of the first n terms of an arithmetic sequence with first term 1 and last term n. As you may recall from Section 6.1, "Simple Interest; Sequences," the first n terms of this arithmetic sequence add to $n(n + 1)/2$. Thus the area is approximated by

$$A \approx \frac{2(1 + 2 + 3 + \cdots + n)}{n^2} = \frac{2[n(n + 1)/2]}{n^2} = \frac{n + 1}{n}$$

Using this formula, we see the following.

$$n = 5: \qquad A \approx \frac{5 + 1}{5} = \frac{6}{5} = 1.20$$

$$n = 10: \qquad A \approx \frac{10 + 1}{10} = \frac{11}{10} = 1.10$$

$$n = 100: \qquad A \approx \frac{100 + 1}{100} = \frac{101}{100} = 1.01$$

As Figure 13.5 indicated, as n gets larger, the number of rectangles increases, the area of each rectangle decreases, and the approximation becomes more accurate. If we let n increase without bound, the approximation approaches the exact area.

$$A = \lim_{n \to +\infty} \frac{n + 1}{n} = \lim_{n \to +\infty} \left(1 + \frac{1}{n}\right) = 1$$

We can see that this area is correct, for we are computing the area of a triangle with base 1 and height 2. The formula for the area of a triangle gives

$$A = \frac{1}{2}bh = \frac{1}{2} \cdot 1 \cdot 2 = 1$$

Summation Notation

A special notation exists that uses the Greek letter Σ (sigma) to express the sum of numbers or expressions. (We used sigma notation informally in Chapter 8, "Further Topics in Probability; Data Description.") We may indicate the sum of the n numbers $a_1, a_2, a_3, a_4, \ldots, a_n$ by

$$\sum_{i=1}^{n} a_i = a_1 + a_2 + a_3 + \cdots + a_n$$

This may be read as "The sum of a_i as i goes from 1 to n." The subscript i in a_i is replaced first by 1, then by 2, then by 3, . . . , until it reaches the value above the sigma. The i is called the **index of summation,** and it starts with the lower limit, 1, and ends with the upper limit, n. For example, if $x_1 = 2, x_2 = 3, x_3 = -1$, and $x_4 = -2$, then

$$\sum_{i=1}^{4} x_i = x_1 + x_2 + x_3 + x_4 = 2 + 3 + (-1) + (-2) = 2$$

The area of the triangle under $y = 2x$ that we discussed above was approximated by

$$A \approx \frac{2}{n^2} + \frac{4}{n^2} + \frac{6}{n^2} + \cdots + \frac{2i}{n^2} + \cdots + \frac{2n}{n^2}$$

Using **sigma notation,** we can write this sum as

$$A \approx \sum_{i=1}^{n} \left(\frac{2i}{n^2}\right)$$

Sigma notation allows us to represent the sums of the areas of the rectangles in an abbreviated fashion. Some formulas that simplify computations involving sums follow.

Sum Formulas

I. $\sum_{i=1}^{n} 1 = n$

II. $\sum_{i=1}^{n} cx_i = c \sum_{i=1}^{n} x_i$ $(c = \text{constant})$

III. $\sum_{i=1}^{n} (x_i + y_i) = \sum_{i=1}^{n} x_i + \sum_{i=1}^{n} y_i$

IV. $\sum_{i=1}^{n} i = \dfrac{n(n + 1)}{2}$

V. $\sum_{i=1}^{n} i^2 = \dfrac{n(n + 1)(2n + 1)}{6}$

We have found that the area of the triangle discussed above was approximated by

$$A \approx \sum_{i=1}^{n} \frac{2i}{n^2}$$

We can use Formulas II and IV to simplify this sum as follows.

$$\sum_{i=1}^{n} \frac{2i}{n^2} = \frac{2}{n^2} \sum_{i=1}^{n} i = \frac{2}{n^2} \left[\frac{n(n + 1)}{2} \right] = \frac{n + 1}{n}$$

Note that this is the same formula we obtained previously using other methods.

We can also use these sum formulas to evaluate a particular sum. For example,

$$\sum_{i=1}^{100} (2i^2 - 3) = \sum_{i=1}^{100} 2i^2 - \sum_{i=1}^{100} 3(1) \qquad \text{Formula III}$$

$$= 2 \sum_{i=1}^{100} i^2 - 3 \sum_{i=1}^{100} 1 \qquad \text{Formula II}$$

$$= 2 \left[\frac{100(101)(201)}{6} \right] - 3(100) \qquad \text{Formulas I and V with } n = 100$$

$$= 676{,}400$$

Areas and Summation Notation

The following example shows that we can find the area by evaluating the function at the left-hand endpoints of the subintervals.

● EXAMPLE 2 *Area Under a Curve*

Use rectangles to find the area under $y = x^2$ (and above the x-axis) from $x = 0$ to $x = 1$.

Solution

We again divide the interval $[0, 1]$ into n equal subintervals of length $1/n$. If we evaluate the function at the left-hand endpoints of these subintervals to determine the heights of the rectangles, the sum of the areas of the rectangles will underestimate the area (see Figure 13.6). Thus we have the information shown in the following table:

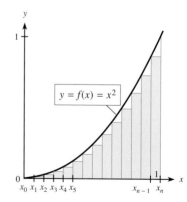

Figure 13.6

Rectangle	Base	Left Endpoint	Height	Area = Base × Height
1	$\dfrac{1}{n}$	$x_0 = 0$	$f(x_0) = 0$	$\dfrac{1}{n} \cdot 0 = 0$
2	$\dfrac{1}{n}$	$x_1 = \dfrac{1}{n}$	$f(x_1) = \dfrac{1}{n^2}$	$\dfrac{1}{n} \cdot \dfrac{1}{n^2} = \dfrac{1}{n^3}$
3	$\dfrac{1}{n}$	$x_2 = \dfrac{2}{n}$	$f(x_2) = \dfrac{4}{n^2}$	$\dfrac{1}{n} \cdot \dfrac{4}{n^2} = \dfrac{4}{n^3}$
4	$\dfrac{1}{n}$	$x_3 = \dfrac{3}{n}$	$f(x_3) = \dfrac{9}{n^2}$	$\dfrac{1}{n} \cdot \dfrac{9}{n^2} = \dfrac{9}{n^3}$
⋮				
i	$\dfrac{1}{n}$	$x_{i-1} = \dfrac{i-1}{n}$	$\dfrac{(i-1)^2}{n^2}$	$\dfrac{(i-1)^2}{n^3}$
⋮				
n	$\dfrac{1}{n}$	$x_{n-1} = \dfrac{n-1}{n}$	$\dfrac{(n-1)^2}{n^2}$	$\dfrac{(n-1)^2}{n^3}$

Thus: Area $= A \approx 0 + \dfrac{1}{n^3} + \dfrac{4}{n^3} + \dfrac{9}{n^3} + \cdots + \dfrac{(i-1)^2}{n^3} + \cdots + \dfrac{(n-1)^2}{n^3}$.

Note that $(i-1)^2/n^3 = (i^2 - 2i + 1)/n^3$ gives the area of the ith rectangle for *any* value of i. The sum of these areas may be written as

$$S = \sum_{i=1}^{n} \frac{i^2 - 2i + 1}{n^3} = \frac{1}{n^3}\left(\sum_{i=1}^{n} i^2 - 2\sum_{i=1}^{n} i + \sum_{i=1}^{n} 1 \right) \qquad \text{Formulas II and III}$$

$$= \frac{1}{n^3}\left[\frac{n(n+1)(2n+1)}{6} - \frac{2n(n+1)}{2} + n \right] \qquad \text{Formulas V, IV, and I}$$

$$= \frac{2n^3 + 3n^2 + n}{6n^3} - \frac{n^2 + n}{n^3} + \frac{n}{n^3} = \frac{2n^2 - 3n + 1}{6n^2}$$

We can use this formula to find the approximate area (value of S) for different values of n.

$$\text{If } n = 10: \qquad \text{Area} \approx S(10) = \frac{200 - 30 + 1}{600} = 0.285$$

$$\text{If } n = 100: \qquad \text{Area} \approx S(100) = \frac{20{,}000 - 300 + 1}{60{,}000} = 0.328$$

Figure 13.7 shows the rectangles associated with each of these approximations.

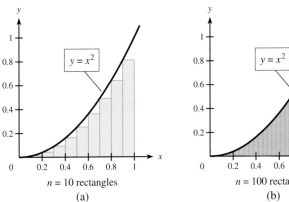
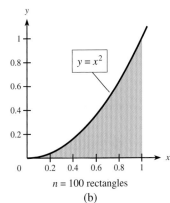

n = 10 rectangles
(a)

n = 100 rectangles
(b)

Figure 13.7

As Figure 13.7 shows, the larger the value of n, the better the value of the sum approximates the exact area under the curve. If we let n increase without bound, we find the exact area.

$$A = \lim_{n \to \infty} \left(\frac{2n^2 - 3n + 1}{6n^2} \right) = \lim_{n \to \infty} \left(\frac{2 - \frac{3}{n} + \frac{1}{n^2}}{6} \right) = \frac{1}{3}$$

Note that the approximations with $n = 10$ and $n = 100$ are less than $\frac{1}{3}$. This is because all the rectangles are *under* the curve (see Figure 13.7).

Thus we see that we can determine the area under a curve $y = f(x)$ from $x = a$ to $x = b$ by dividing the interval $[a, b]$ into n equal subintervals of width $(b - a)/n$ and evaluating

$$A = \lim_{n \to \infty} S_R = \lim_{n \to \infty} \sum_{i=1}^{n} f(x_i) \left(\frac{b - a}{n} \right) \qquad \text{(using right-hand endpoints)}$$

or

$$A = \lim_{n \to \infty} S_L = \lim_{n \to \infty} \sum_{i=1}^{n} f(x_{i-1}) \left(\frac{b - a}{n} \right) \qquad \text{(using left-hand endpoints)}$$

● **Checkpoint**

1. For the interval $[0, 2]$, determine whether the following are true or false.

   (a) For 4 subintervals, each subinterval has width $\frac{1}{2}$.

   (b) For 200 subintervals, each subinterval has width $\frac{1}{100}$.

   (c) For n subintervals, each subinterval has width $\frac{2}{n}$.

   (d) For n subintervals, $x_0 = 0$, $x_1 = \frac{2}{n}$, $x_2 = 2\left(\frac{2}{n}\right)$, ..., $x_i = i\left(\frac{2}{n}\right)$, ..., $x_n = 2$.

2. Find the area under $y = f(x) = 3x - x^2$ from $x = 0$ to $x = 2$ using right-hand endpoints (see the figure). To accomplish this, use $\frac{b - a}{n} = \frac{2}{n}$, $x_i = \frac{2i}{n}$, and $f(x) = 3x - x^2$; find and simplify the following.

   (a) $f(x_i)$ (b) $f(x_i)\left(\frac{b - a}{n} \right)$

   (c) $S_R = \sum_{i=1}^{n} f(x_i) \left(\frac{b - a}{n} \right)$ (d) $A = \lim_{n \to \infty} S_R = \lim_{n \to \infty} \sum_{i=1}^{n} f(x_i) \left(\frac{b - a}{n} \right)$

(figure at left)

$y = 3x - x^2$

Technology
Note

Graphing calculators and spreadsheets can be used to approximate the area under a curve. These technologies are especially useful when summations approximating the area do not simplify easily. A calculator or computer program can be used to compute sums, or a spreadsheet such as Excel can be used to find the sums. Consider the following example. ■

● **EXAMPLE 3 *Estimating Areas with Technology***

To approximate the area under the graph of $y = \sqrt{x}$ on the interval $[0, 4]$, do the following.

(a) Use a calculator or computer to find S_L and S_R for $n = 8$ on the interval $[0, 4]$.
(b) Predict the area under the curve in the graph by evaluating S_L and S_R for larger values of n.

Solution

Figure 13.8(a) shows the graph of $y = \sqrt{x}$ with 8 rectangles whose heights are determined by evaluating the function at the left-hand endpoint of each interval (the first of these rectangles has height 0). Figure 13.8(b) shows the same graph with 8 rectangles whose heights are determined at the right-hand endpoints.

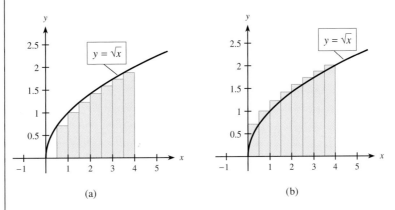

Figure 13.8

(a)   (b)

(a) The formula for the sum of the areas of n rectangles over the interval $[0, 4]$, using left-hand endpoints, is

$$S_L = \sum_{i=1}^{n} \sqrt{\frac{4(i-1)}{n}} \frac{4}{n} = \frac{8}{n^{3/2}} \sum_{i=1}^{n} \sqrt{i-1}$$

and the formula for the sum of the areas of n rectangles over the interval $[0, 4]$, using right-hand endpoints, is

$$S_R = \sum_{i=1}^{n} \sqrt{\frac{4(i)}{n}} \frac{4}{n} = \frac{8}{n^{3/2}} \sum_{i=1}^{n} \sqrt{i}$$

We have no formulas to write either of these summations in a simpler form, but we can find the sums by hand if n is small and we can use calculators, computer programs, or spreadsheets to find the sums for large n. For $n = 8$, the sums are

$$S_L = \frac{1}{\sqrt{8}} \sum_{i=1}^{8} \sqrt{i-1} \qquad \text{and} \qquad S_R = \frac{1}{\sqrt{8}} \sum_{i=1}^{8} \sqrt{i}$$

The following Excel output shows the sum for $n = 8$ rectangles.

	A	B	C	D
1	Area for	n=8		
2		i	SL	SR
3		1	0	.35355
4		2	.35355	.5
5		3	.5	.61237
6		4	.61237	.70711
7		5	.70711	.79057
8		6	.79057	.86603
9		7	.86603	.93541
10		8	.93541	1
11	Total		4.76504	5.76504

As line 11 shows, $S_L = 4.76504$ and $S_R = 5.76504$ for $n = 8$.

(b) Larger values of n give better approximations of the area under the curve (try some). For example, $n = 1000$ gives $S_L = 5.3293$ and $S_R = 5.3373$, and because the area under the curve is between these values, the area is approximately 5.33, to two decimal places. By using values of n larger than 1000, we can get even better approximations of the area.

● **Checkpoint Solutions**

1. All parts are true.

2. (a) $f(x_i) = f\left(\dfrac{2i}{n}\right) = 3\left(\dfrac{2i}{n}\right) - \left(\dfrac{2i}{n}\right)^2 = \dfrac{6i}{n} - \dfrac{4i^2}{n^2}$

(b) $f(x_i)\dfrac{b-a}{n} = \left(\dfrac{6i}{n} - \dfrac{4i^2}{n^2}\right)\left(\dfrac{2}{n}\right) = \dfrac{12i}{n^2} - \dfrac{8i^2}{n^3}$

(c) $S_R = \displaystyle\sum_{i=1}^{n} f(x_i)\dfrac{b-a}{n} = \sum_{i=1}^{n}\left(\dfrac{12i}{n^2} - \dfrac{8i^2}{n^3}\right)$

$= \displaystyle\sum_{i=1}^{n}\dfrac{12i}{n^2} - \sum_{i=1}^{n}\dfrac{8i^2}{n^3}$

$= \dfrac{12}{n^2}\displaystyle\sum_{i=1}^{n} i - \dfrac{8}{n^3}\sum_{i=1}^{n} i^2$

$= \dfrac{12}{n^2}\left[\dfrac{n(n+1)}{2}\right] - \dfrac{8}{n^3}\left[\dfrac{n(n+1)(2n+1)}{6}\right]$

$= \dfrac{6(n+1)}{n} - \dfrac{4(n+1)(2n+1)}{3n^2}$

(d) $A = \displaystyle\lim_{n\to\infty}\sum_{i=1}^{n} f(x_i)\dfrac{b-a}{n} = \lim_{n\to\infty}\left(\dfrac{6n+6}{n} - \dfrac{8n^2+12n+4}{3n^2}\right) = 6 - \dfrac{8}{3} = \dfrac{10}{3}$

13.1 Exercises

In Problems 1–4, approximate the area under each curve over the specified interval by using the indicated number of subintervals (or rectangles) and evaluating the function at the *right-hand* endpoints of the subintervals. (See Example 1.)

1. $f(x) = 4x - x^2$ from $x = 0$ to $x = 2$; 2 subintervals
2. $f(x) = x^3$ from $x = 0$ to $x = 3$; 3 subintervals
3. $f(x) = 9 - x^2$ from $x = 1$ to $x = 3$; 4 subintervals
4. $f(x) = x^2 + x + 1$ from $x = -1$ to $x = 1$; 4 subintervals

In Problems 5–8, approximate the area under each curve by evaluating the function at the *left-hand* endpoints of the subintervals.

5. $f(x) = 4x - x^2$ from $x = 0$ to $x = 2$; 2 subintervals
6. $f(x) = x^3$ from $x = 0$ to $x = 3$; 3 subintervals
7. $f(x) = 9 - x^2$ from $x = 1$ to $x = 3$; 4 subintervals
8. $f(x) = x^2 + x + 1$ from $x = -1$ to $x = 1$; 4 subintervals

When the area under $f(x) = x^2 + x$ from $x = 0$ to $x = 2$ is approximated, the formulas for the sum of n rectangles using *left-hand* endpoints and *right-hand* endpoints are

Left-hand endpoints: $\quad S_L = \dfrac{14}{3} - \dfrac{6}{n} + \dfrac{4}{3n^2}$

Right-hand endpoints: $\quad S_R = \dfrac{14n^2 + 18n + 4}{3n^2}$

Use these formulas to answer Problems 9–13.

9. Find $S_L(10)$ and $S_R(10)$.
10. Find $S_L(100)$ and $S_R(100)$.
11. Find $\displaystyle\lim_{n\to\infty} S_L$ and $\lim_{n\to\infty} S_R$.
12. Compare the right-hand and left-hand values by finding $S_R - S_L$ for $n = 10$, for $n = 100$, and as $n \to \infty$. (Use Problems 9–11.)
13. Because $f(x) = x^2 + x$ is increasing over the interval from $x = 0$ to $x = 2$, function values at the right-hand endpoints are maximum values for each subinterval, and function values at the left-hand endpoints are minimum values for each subinterval. How would the

approximate area using $n = 10$ and *any* other point within each subinterval compare with $S_L(10)$ and $S_R(10)$? What would happen to the area result as $n \to \infty$ if any other point in each subinterval were used?

In Problems 14–19, find the value of each sum.

14. $\sum_{k=1}^{3} x_k$, if $x_1 = 1, x_2 = 3, x_3 = -1, x_4 = 5$

15. $\sum_{i=1}^{4} x_i$, if $x_1 = 3, x_2 = -1, x_3 = 3, x_4 = -2$

16. $\sum_{i=3}^{5} (i^2 + 1)$ 17. $\sum_{j=2}^{5} (j^2 - 3)$

18. $\sum_{i=4}^{7} \left(\dfrac{i - 3}{i^2} \right)$ 19. $\sum_{j=0}^{4} (j^2 - 4j + 1)$

In Problems 20–25, use the sum formulas I–V to express each of the following without the summation symbol. In Problems 20–23, find the numerical value.

20. $\sum_{k=1}^{50} 1$ 21. $\sum_{j=1}^{60} 3$

22. $\sum_{k=1}^{50} (6k^2 + 5)$ 23. $\sum_{k=1}^{30} (k^2 + 4k)$

24. $\sum_{i=1}^{n} \left(1 - \dfrac{i^2}{n^2} \right)\left(\dfrac{2}{n} \right)$ 25. $\sum_{i=1}^{n} \left(1 - \dfrac{2i}{n} + \dfrac{i^2}{n^2} \right)\left(\dfrac{3}{n} \right)$

Use the function $y = 2x$ from $x = 0$ to $x = 1$ and n equal subintervals with the function evaluated at the *left-hand* endpoint of each subinterval for Problems 26 and 27.

26. What is the area of the
    (a) first rectangle?   (b) second rectangle?
    (c) *i*th rectangle?

27. (a) Find a formula for the sum of the areas of the n rectangles (call this S). Then find
    (b) $S(10)$.   (c) $S(100)$.
    (d) $S(1000)$.   (e) $\lim_{n \to \infty} S$.

28. How do your answers to Problems 27(a)–(e) compare with the corresponding calculations in the discussion (after Example 1) of the area under $y = 2x$ using *right-hand* endpoints?

For Problems 29(a)–(e), use the function $y = x^2$ from $x = 0$ to $x = 1$ and n equal subintervals with the function evaluated at the *right-hand* endpoints.

29. (a) Find a formula for the sum of the areas of the n rectangles (call this S). Then find
    (b) $S(10)$.   (c) $S(100)$.
    (d) $S(1000)$.   (e) $\lim_{n \to \infty} S$.

30. How do your answers to Problems 29(a)–(e) compare with the corresponding calculations in Example 2?

31. Use rectangles to find the area between $y = x^2 - 6x + 8$ and the x-axis from $x = 0$ to $x = 2$. Divide the interval $[0, 2]$ into n equal subintervals so that each subinterval has length $2/n$.

32. Use rectangles to find the area between $y = 4x - x^2$ and the x-axis from $x = 0$ to $x = 4$. Divide the interval

$[0, 4]$ into n equal subintervals so that each subinterval has length $4/n$.

APPLICATIONS

33. *Telecommunications* Spending by big telecommunications companies for the years 1999–2002 is shown in the figure. Estimate the total amount spent from 1999 to 2002 by using rectangles with heights evaluated at the left-hand endpoints.

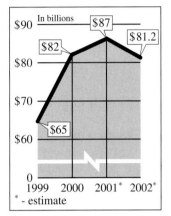

34. *Oil imports* Crude oil is imported continuously by the United States. The following table and graph show the rate of U.S. oil imports (in billions of barrels per year). (*Source:* Energy Information Administration)
    (a) Use $n = 5$ equal subdivisions and left-hand endpoints to estimate the area under the graph from 1999 to 2004.
    (b) What does this area represent in terms of U.S. oil imports?

Year	Rate
1999	3.187
2000	3.311
2001	3.405
2002	3.336
2003	3.435
2004	3.375

35. *Speed trials* The graph in the figure gives the times that it takes a Porsche 911 to reach speeds from 0 mph to 100 mph, in increments of 10 mph, with a curve connecting them. The area under this curve from $t = 0$ seconds to $t = 14$ seconds represents the total amount of distance traveled over the 14-second period. Count the squares under the curve to estimate this distance. This

car will travel 1/4 mile in 14 seconds, to a speed of 100.2 mph. Is your estimate close to this result? (Be careful with time units.)

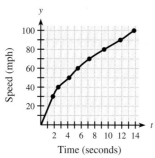

Source: Motor Trend

36. *Speed trials* The graph in the figure gives the times that it takes a Mitsubishi Eclipse GSX to reach speeds from 0 mph to 100 mph, in increments of 10 mph, with a curve connecting them. The area under this curve from $t = 0$ seconds to $t = 21.1$ seconds represents the total amount of distance traveled over the 21.1-second period. Count the squares under the curve to estimate this distance. This car will travel 1/4 mile in 15.4 seconds, to a speed of 89.0 mph, so your estimate should be more than 1/4 mile. Is it? (Be careful with time units.)

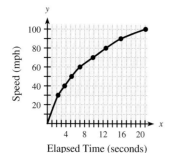

Source: Road & Track

37. *Pollution monitoring* Suppose the presence of phosphates in certain waste products dumped into a lake promotes the growth of algae. Rampant growth of algae affects the oxygen supply in the water, so an environmental group wishes to estimate the area of algae growth. The group measures the length across the algae growth (see the figure) and obtains the following data (in feet).

x	Length	x	Length
0	0	50	27
10	15	60	24
20	18	70	23
30	18	80	0
40	30		

Use 8 rectangles with bases of 10 feet and lengths measured at the left-hand endpoints to approximate the area of the algae growth.

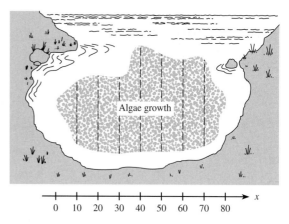

38. *Drug levels in the blood* The manufacturer of a medicine wants to test how a new 300-milligram capsule is released into the bloodstream. After a volunteer is given a capsule, blood samples are drawn every half-hour, and the number of milligrams of the drug in the bloodstream is calculated. The results obtained follow.

Time t (hr)	$N(t)$ (mg)	Time t (hr)	$N(t)$ (mg)
0	0	2.0	178.3
0.5	247.3	2.5	113.9
1.0	270.0	3.0	56.2
1.5	236.4	3.5	19.3

Use 7 rectangles, each with height $N(t)$ at the left endpoint and with width 0.5 hr, to estimate the area under the graph representing these data. Divide this area by 3.5 hr to estimate the average drug level over this time period.

39. *Wireless communications* Spending for U.S. wireless communications services (in billions of dollars per year) can be modeled by

$$S(t) = 11.23t + 6.205$$

where $t = 0$ represents 1995. (*Source:* The Strategis Group; Donaldson, Lufkin & Jenrette; Cellular Telecommunications and Internet Association; Telecommunications Industry Association) Use $n = 10$ equal subdivisions with right-hand endpoints to approximate the area under the graph of $S(t)$ between $t = 5$ and $t = 10$. What does this area represent?

40. *World tourism* Worldwide spending for tourism (in billions of dollars per year) can be modeled by

$$T(t) = -54.846 + 192.7 \ln(t)$$

where $t = 0$ represents 1985. (*Source:* World Tourism Organization) Use $n = 10$ equal subdivisions with right-hand endpoints to approximate the area under the graph of $T(t)$ between $t = 15$ and $t = 20$. What does this area represent?

The Definite Integral: The Fundamental Theorem of Calculus

13.2

OBJECTIVES

- To evaluate definite integrals using the Fundamental Theorem of Calculus
- To use definite integrals to find the area under a curve

◗ Application Preview

Suppose that money flows continuously into a slot machine at a casino and grows at a rate given by $A'(t) = 100e^{0.1t}$, where t is the time in hours and $0 \leq t \leq 10$. Then the **definite integral**

$$\int_0^{10} 100e^{0.1t}\, dt$$

gives the total amount of money that accumulates over the 10-hour period, if no money is paid out. (See Example 7.)

In the previous section, we used the sums of areas of rectangles to approximate the areas under curves. In this section, we will see how such sums are related to the definite integral and how to evaluate definite integrals. In addition, we will see how definite integrals can be used to solve several types of applied problems.

Riemann Sums and the Definite Integral

In the previous section, we saw that we could determine the area under a curve using equal subintervals and the function values at either the left-hand endpoints or the right-hand endpoints of the subintervals. In fact, we can use subintervals that are not of equal length, and we can use any point within each subinterval to determine the height of each rectangle. Suppose that we wish to find the area above the x-axis and under the curve $y = f(x)$ over a closed interval $[a, b]$. We can divide the interval into n subintervals (not necessarily equal), with the endpoints of these intervals at $x_0 = a, x_1, x_2 \ldots, x_n = b$. We now choose a point (*any* point) in each subinterval, and denote the points $x_1^*, x_2^*, \ldots, x_i^*, \ldots, x_n^*$. Then the ith rectangle (for any i) has height $f(x_i^*)$ and width $x_i - x_{i-1}$, so its area is $f(x_i^*)(x_i - x_{i-1})$. Then the sum of the areas of the n rectangles is

$$S = \sum_{i=1}^{n} f(x_i^*)(x_i - x_{i-1}) = \sum_{i=1}^{n} f(x_i^*)\,\Delta x_i, \qquad \text{where } \Delta x_i = x_i - x_{i-1}$$

Because the points in the subinterval may be chosen anywhere in the subinterval, we cannot be sure whether the rectangles will underestimate or overestimate the area under the curve. But increasing the number of subintervals (increasing n) and making sure that every interval becomes smaller (just increasing n will not guarantee this if the subintervals are unequal) will in the long run improve the estimation. Thus for any subdivision of $[a, b]$ and any x_i^*, the area is given by

$$A = \lim_{\substack{\max \Delta x_i \to 0 \\ (n \to \infty)}} \sum_{i=1}^{n} f(x_i^*)\,\Delta x_i, \qquad \text{provided that this limit exists}$$

The preceding discussion takes place in the context of finding the area under a curve. However, if f is any function (not necessarily nonnegative) defined on $[a, b]$, then for each subdivision of $[a, b]$ and each choice of x_i^*, we define the sum

$$S = \sum_{i=1}^{n} f(x_i^*)\Delta x_i$$

as the **Riemann sum** of f for the subdivision of $[a, b]$. In addition, the limit of the Riemann sum (as max $\Delta x_i \to 0$) has other important applications and is called the **definite integral** of $f(x)$ over interval $[a, b]$.

Definite Integral

If f is a function on the interval $[a, b]$, then, for any subdivision of $[a, b]$ and any choice of x_i^* in the ith subinterval, the *definite integral* of f from a to b is

$$\int_a^b f(x)\, dx = \lim_{\substack{\max \Delta x_i \to 0 \\ (n \to \infty)}} \sum_{i=1}^n f(x_i^*)\, \Delta x_i$$

If f is a continuous function, and $\Delta x_i \to 0$ as $n \to \infty$, then the limit exists and we say that f is integrable on $[a, b]$.

Note that for some intervals, values of f may be negative. In this case, the product $f(x_i^*)\Delta x_i$ will be negative and can be thought of geometrically as a "signed area." (Remember that area is a positive number.) Thus a definite integral can be thought of geometrically as the sum of signed areas, just as a derivative can be thought of geometrically as the slope of a tangent line. In the case where $f(x)$ is positive for all x from a to b, the definite integral equals the area between the graph of $y = f(x)$ and the x-axis.

Fundamental Theorem of Calculus

The obvious question is how this definite integral is related to the indefinite integral (the antiderivative) discussed in Chapter 12. The connection between these two concepts is the most important result in calculus, because it connects derivatives, indefinite integrals, and definite integrals.

To help see the connection, consider the marginal revenue function

$$R'(x) = 300 - 0.2x$$

and the revenue function

$$R(x) = \int (300 - 0.2x)\, dx = 300x - 0.1x^2$$

that is the indefinite integral of the marginal revenue function. (See Figure 13.9(a).)

Using this revenue function, we can find the revenue from the sale of 1000 units to be

$$R(1000) = 300(1000) - 0.1(1000)^2 = 200{,}000 \text{ dollars}$$

and the revenue from the sale of 500 units to be

$$R(500) = 300(500) - 0.1(500)^2 = 125{,}000 \text{ dollars}$$

Thus the additional revenue received from the sale of the 500 additional units is

$$200{,}000 - 125{,}000 = 75{,}000 \text{ dollars}$$

If we used the definition of the definite integral (or geometry) to find the area under the graph of the marginal revenue function from $x = 500$ to $x = 1000$, we would find that the area is 75,000 (see Figure 13.9(b)). Note in Figure 13.9(b) that the shaded area is given by

Area = Area (Δ with base from 500 to 1500) − Area (Δ with base from 1000 to 1500)

Recall that the area of a triangle is $\frac{1}{2}$(base)(height) so we have

$$\text{Area} = \tfrac{1}{2}(1000)(200) - \tfrac{1}{2}(500)(100) = 75{,}000$$

Thus we can also find this additional revenue when sales are increased from 500 to 1000 by evaluating the *definite integral*

$$\int_{500}^{1000} (300 - 0.2x)\, dx$$

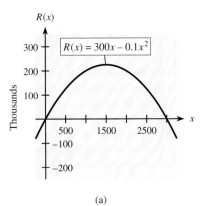

Figure 13.9   (a)   (b)

In general, the definite integral

$$\int_a^b f(x)\,dx$$

can be used to find the change in the function $F(x)$ when x changes from a to b, where $f(x)$ is the derivative of $F(x)$. This result is the **Fundamental Theorem of Calculus.**

**Fundamental Theorem
of Calculus**

Let f be a continuous function on the closed interval $[a, b]$; then the definite integral of f exists on this interval, and

$$\int_a^b f(x)\,dx = F(b) - F(a)$$

where F is any function such that $F'(x) = f(x)$ for all x in $[a, b]$.

Stated differently, this theorem says that if the function F is an indefinite integral of a function f that is continuous on the interval $[a, b]$, then

$$\int_a^b f(x)\,dx = F(b) - F(a)$$

Thus, we apply the Fundamental Theorem of Calculus by using the following two steps:

1. Integration of $f(x)$: $\displaystyle\int_a^b f(x)\,dx = F(x)\Big|_a^b$

2. Evaluation of $F(x)$: $\displaystyle F(x)\Big|_a^b = F(b) - F(a)$

● **EXAMPLE 1 *Definite Integral***

Evaluate $\displaystyle\int_2^4 (x^3 + 4)\,dx$.

Solution

1. $\displaystyle\int_2^4 (x^3 + 4)\,dx = \frac{x^4}{4} + 4x + C\,\Big|_2^4$

2. $\displaystyle\qquad = \left[\frac{(4)^4}{4} + 4(4) + C\right] - \left[\frac{(2)^4}{4} + 4(2) + C\right]$

$\displaystyle\qquad = (64 + 16 + C) - (4 + 8 + C)$

$\displaystyle\qquad = 68 \qquad$ (Note that the C's subtract out.)

Note that the Fundamental Theorem states that F can be *any* indefinite integral of f, so we need not add the constant of integration to the integral.

● **EXAMPLE 2** *Fundamental Theorem*

Evaluate $\displaystyle\int_1^3 (3x^2 + 6x)\, dx$.

Solution

$$\int_1^3 (3x^2 + 6x)\, dx = x^3 + 3x^2 \Big|_1^3$$

$$= (3^3 + 3 \cdot 3^2) - (1^3 + 3 \cdot 1^2)$$

$$= 54 - 4 = 50$$

Properties   The properties of definite integrals given below follow from properties of summations.

1. $\displaystyle\int_a^b [f(x) \pm g(x)]\, dx = \int_a^b f(x)\, dx \pm \int_a^b g(x)\, dx$

2. $\displaystyle\int_a^b kf(x)\, dx = k \int_a^b f(x)\, dx,$ where k is a constant

The following example uses both of these properties.

● **EXAMPLE 3** *Definite Integral*

Evaluate $\displaystyle\int_3^5 (\sqrt{x^2 - 9} + 2)x\, dx$.

Solution

$$\int_3^5 (\sqrt{x^2 - 9} + 2)x\, dx = \int_3^5 \sqrt{x^2 - 9}(x\, dx) + \int_3^5 2x\, dx$$

$$= \frac{1}{2}\int_3^5 (x^2 - 9)^{1/2}(2x\, dx) + \int_3^5 2x\, dx$$

$$= \frac{1}{2}\left[\frac{2}{3}(x^2 - 9)^{3/2}\right]\Big|_3^5 + x^2\Big|_3^5$$

$$= \frac{1}{3}[(16)^{3/2} - (0)^{3/2}] + (25 - 9)$$

$$= \frac{64}{3} + 16 = \frac{64}{3} + \frac{48}{3} = \frac{112}{3}$$

In the integral $\int_a^b f(x)\, dx$, we call a the *lower limit* and b the *upper limit* of integration. Although we developed the definite integral with the assumption that the lower limit was less than the upper limit, the following properties permit us to evaluate the definite integral even when that is not the case.

3. $\displaystyle\int_a^a f(x)\, dx = 0$

4. If f is integrable on $[a, b]$, then

$$\int_b^a f(x)\, dx = -\int_a^b f(x)\, dx$$

The following examples illustrate these properties.

● **EXAMPLE 4** *Properties of Definite Integrals*

Evaluate $\displaystyle\int_4^4 x^2\, dx$.

Solution

Note that because the limits of integration are the same, it is not necessary to integrate to see that the value of the integral is 0. As a check, we use the Fundamental Theorem of Calculus.

$$\int_4^4 x^2\, dx = \frac{x^3}{3}\Big|_4^4 = \frac{4^3}{3} - \frac{4^3}{3} = 0$$

● **EXAMPLE 5** *Properties of Definite Integrals*

Compare $\displaystyle\int_2^4 3x^2\, dx$ and $\displaystyle\int_4^2 3x^2\, dx$.

Solution

$$\int_2^4 3x^2\, dx = x^3\Big|_2^4 = 4^3 - 2^3 = 56$$

$$\int_4^2 3x^2\, dx = x^3\Big|_4^2 = 2^3 - 4^3 = -56$$

Thus, as Property 4 guarantees,

$$\int_4^2 3x^2\, dx = -\int_2^4 3x^2\, dx$$

Another property of definite integrals is called the additive property.

5. If f is continuous on some interval containing a, b, and c,* then

$$\int_a^c f(x)\, dx + \int_c^b f(x)\, dx = \int_a^b f(x)\, dx$$

● **EXAMPLE 6** *Properties of Definite Integrals*

Show that $\displaystyle\int_2^3 4x\, dx + \int_3^5 4x\, dx = \int_2^5 4x\, dx$.

*Note that c need not be between a and b.

Solution

$$\int_2^3 4x \, dx = 2x^2 \Big|_2^3 = 18 - 8 = 10$$

$$\int_3^5 4x \, dx = 2x^2 \Big|_3^5 = 50 - 18 = 32$$

$$\int_2^5 4x \, dx = 2x^2 \Big|_2^5 = 50 - 8 = 42$$

Thus

$$\int_2^3 4x \, dx + \int_3^5 4x \, dx = \int_2^5 4x \, dx$$

The Definite Integral and Areas

Let us now return to area problems, to see the relationship between the definite integral and the area under a curve. By the formula for the area of a triangle or by using rectangles and the limit definition of area, the area under the curve (line) $y = x$ from $x = 0$ to $x = 1$ can be shown to be $\frac{1}{2}$ (see Figure 13.10(a)). Using the definite integral to find the area gives

$$A = \int_0^1 x \, dx = \frac{x^2}{2} \Big|_0^1 = \frac{1}{2} - 0 = \frac{1}{2}$$

In the previous section, we used rectangles to find that the area under $y = x^2$ from $x = 0$ to $x = 1$ was $\frac{1}{3}$ (see Figure 13.10(b)). Using the definite integral, we get

$$A = \int_0^1 x^2 \, dx = \frac{x^3}{3} \Big|_0^1 = \frac{1}{3} - 0 = \frac{1}{3}$$

which agrees with the answer obtained previously.

However, not every definite integral represents the area between the curve and the *x*-axis over an interval. For example,

$$\int_0^2 (x - 2) \, dx = \frac{x^2}{2} - 2x \Big|_0^2 = (2 - 4) - (0) = -2$$

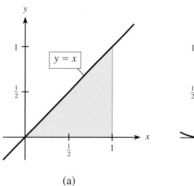
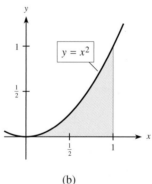
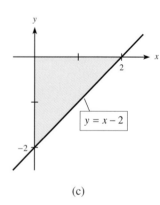

Figure 13.10   (a)   (b)   (c)

This would indicate that the area between the curve and the *x*-axis is negative, but area must be positive. A look at the graph of $y = x - 2$ (see Figure 13.10(c)) shows us what is happening. The region bounded by $y = x - 2$ and the *x*-axis between $x = 0$ and $x = 2$ is a triangle whose base is 2 and whose height is 2, so its area is $\frac{1}{2}bh = \frac{1}{2}(2)(2) = 2$. The integral has value -2 because $y = x - 2$ lies below the x-axis from $x = 0$ to $x = 2$, and the function values over the interval $[0, 2]$ are negative. Thus the value of the definite integral over *this* interval does not represent the area between the curve and the x-axis, but rather represents the "signed" area as mentioned previously.

In general, the definite integral will give the area under the curve and above the x-axis only when $f(x) \geq 0$ for all x in $[a, b]$.

Area Under a Curve

If f is a continuous function on $[a, b]$ and $f(x) \geq 0$ on $[a, b]$, then the exact area between $y = f(x)$ and the x-axis from $x = a$ to $x = b$ is given by

$$\text{Area (shaded)} = \int_a^b f(x)\, dx$$

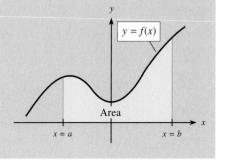

Note also that if $f(x) \leq 0$ for all x in $[a, b]$, then

$$\int_a^b f(x)\, dx = -\text{Area (between } f(x) \text{ and the } x\text{-axis)}$$

● **Checkpoint**

1. True or false:
   (a) For any integral, we can omit the constant of integration (the $+C$).
   (b) $-\int_{-1}^{3} f(x)\, dx = \int_{3}^{-1} f(x)\, dx$, if f is integrable on $[-1, 3]$.

   (c) The area between $f(x)$ and the x-axis on the interval $[a, b]$ is given by $\int_a^b f(x)\, dx$.

2. Evaluate:
   (a) $\int_0^3 (x^2 + 1)\, dx$ (b) $\int_0^3 (x^2 + 1)^4 x\, dx$

If the rate of growth of some function with respect to time t is $f'(t)$, then the total growth of the function during the period from $t = 0$ to $t = k$ can be found by evaluating the definite integral

$$\int_0^k f'(t)\, dt = f(t) \Big|_0^k = f(k) - f(0)$$

For nonnegative rates of growth, this definite integral (and thus growth) is the same as the area under the graph of $f'(t)$ from $t = 0$ to $t = k$.

◗ **EXAMPLE 7** *Income Stream* **(Application Preview)**

Suppose that money flows continuously into a slot machine at a casino and grows at a rate given by

$$A'(t) = 100e^{0.1t}$$

where t is time in hours and $0 \leq t \leq 10$. Find the total amount that accumulates in the machine during the 10-hour period, if no money is paid out.

Solution

The total amount is given by

$$A = \int_0^{10} 100 e^{0.1t} \, dt = \frac{100}{0.1} \int_0^{10} e^{0.1t} (0.1) \, dt$$

$$= 1000 e^{0.1t} \Big|_0^{10}$$

$$= 1000 e - 1000$$

$$\approx 1718.28 \quad \text{(dollars)}$$

In Section 8.4, "Normal Probability Distribution," we stated that the total area under the normal curve is 1 and that the area under the curve from value x_1 to value x_2 represents the probability that a score chosen at random will lie between x_1 and x_2.

The normal distribution is an example of a **continuous probability distribution** because the values of the random variable are considered over intervals rather than at discrete values. The statements above relating probability and area under the graph apply to other continuous probability distributions determined by **probability density functions.** In fact, if x is a continuous random variable with probability density function $f(x)$, then the probability that x is between a and b is

$$\Pr(a \leq x \leq b) = \int_a^b f(x) \, dx$$

● **EXAMPLE 8** *Product Life*

Suppose the probability density function for the life of a computer component is $f(x) = 0.10 e^{-0.10x}$, where $x \geq 0$ is the number of years the component is in use. Find the probability that the component will last between 3 and 5 years.

Solution

The probability that the component will last between 3 and 5 years is the area under the graph of the function between $x = 3$ and $x = 5$. The probability is given by

$$\Pr(3 \leq x \leq 5) = \int_3^5 0.10 e^{-0.10x} \, dx = -e^{-0.10x} \Big|_3^5$$

$$= -e^{-0.5} + e^{-0.3}$$

$$\approx -0.6065 + 0.7408$$

$$= 0.1343$$

Calculator Note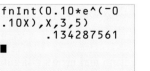

```
fnInt(0.10*e^(-0
.10X),X,3,5)
       .134287561
■
```

Figure 13.11

Most graphing calculators have a numerical integration feature that can be used to get very accurate approximations of definite integrals. This feature can be used to evaluate definite integrals directly or to check those done with the Fundamental Theorem of Calculus. Figure 13.11 shows the numerical integration feature applied to the integral in Example 8. Note that when this answer is rounded to four decimal places, the results agree.

Antiderivatives of some functions cannot be found (an example of this is $\int e^{-x^2} \, dx$). In such cases, finding the approximate numerical value of a definite integral with technology is extremely useful in solving applied problems. ■

1. (a) False. We can omit the constant of integration $(+C)$ only for definite integrals.

   (b) True

   (c) False. Only if $f(x) \geq 0$ on $[a, b]$ is this true.

2. (a) $\displaystyle\int_0^3 (x^2 + 1) \, dx = \frac{x^3}{3} + x \Big|_0^3 = \left(\frac{27}{3} + 3\right) - (0 + 0) = 12$

   (b) $\displaystyle\int_0^3 (x^2 + 1)^4 x \, dx = \frac{1}{2} \int_0^3 (x^2 + 1)^4 (2x \, dx)$

   $$= \frac{1}{2} \cdot \frac{(x^2 + 1)^5}{5} \Big|_0^3$$

   $$= \frac{1}{10}[(3^2 + 1)^5 - (1)^5] = \frac{1}{10}(10^5 - 1) = 9999.9$$

13.2 Exercises

Evaluate the definite integrals in Problems 1–26.

1. $\displaystyle\int_0^3 4x \, dx$

2. $\displaystyle\int_0^1 8x \, dx$

3. $\displaystyle\int_2^4 dx$

4. $\displaystyle\int_1^5 2 \, dy$

5. $\displaystyle\int_2^4 x^3 \, dx$

6. $\displaystyle\int_0^5 x^2 \, dx$

7. $\displaystyle\int_0^5 4\sqrt[3]{x^2} \, dx$

8. $\displaystyle\int_2^4 3\sqrt{x} \, dx$

9. $\displaystyle\int_2^4 (4x^3 - 6x^2 - 5x) \, dx$

10. $\displaystyle\int_0^2 (x^4 - 5x^3 + 2x) \, dx$

11. $\displaystyle\int_3^4 (x - 4)^9 \, dx$

12. $\displaystyle\int_{-1}^0 (x + 2)^{13} \, dx$

13. $\displaystyle\int_2^4 (x^2 + 2)^3 x \, dx$

14. $\displaystyle\int_0^3 (2x - x^2)^4 (1 - x) \, dx$

15. $\displaystyle\int_{-1}^2 (x^3 - 3x^2)^3 (x^2 - 2x) \, dx$

16. $\displaystyle\int_0^4 (3x^2 - 2)^4 x \, dx$

17. $\displaystyle\int_0^4 \sqrt{4x + 9} \, dx$

18. $\displaystyle\int_0^2 \sqrt[3]{2x^3 - 8} \, x^2 \, dx$

19. $\displaystyle\int_1^3 \frac{3}{y^2} \, dy$

20. $\displaystyle\int_1^2 \frac{5}{z^3} \, dz$

21. $\displaystyle\int_0^1 e^{3x} \, dx$

22. $\displaystyle\int_0^2 e^{4x} \, dx$

23. $\displaystyle\int_1^e \frac{4}{z} \, dz$

24. $\displaystyle\int_1^e 3y^{-1} \, dy$

25. $\displaystyle\int_0^2 8x^2 e^{-x^3} \, dx$

26. $\displaystyle\int_0^1 \frac{3x^3 \, dx}{4x^4 + 9}$

In Problems 27–30, evaluate each integral (a) with the Fundamental Theorem of Calculus and (b) with a graphing utility (as a check).

27. $\displaystyle\int_3^6 \frac{x}{3x^2 + 4} \, dx$

28. $\displaystyle\int_0^2 \frac{x}{x^2 + 4} \, dx$

29. $\displaystyle\int_1^2 \frac{x^2 + 3}{x} \, dx$

30. $\displaystyle\int_1^4 \frac{4\sqrt{x} + 5}{\sqrt{x}} \, dx$

31. In the figure, which of the shaded regions (A, B, C, or D) has the area given by

   (a) $\displaystyle\int_a^b f(x) \, dx$?

   (b) $\displaystyle-\int_a^b f(x) \, dx$?

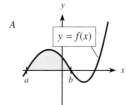

A

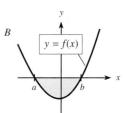

B

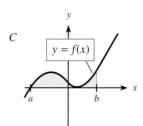

C

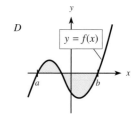

D

32. For which of the following functions $f(x)$ does

$$\int_0^2 f(x)\, dx$$

give the area between the graph of $f(x)$ and the x-axis from $x = 0$ to $x = 2$?

(a) $f(x) = x^2 + 1$ (b) $f(x) = -x^2$

(c) $f(x) = x - 1$

In Problems 33–36, (a) write the integral that describes the area of the shaded region and (b) find the area.

33.

 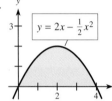
$y = 2x - \frac{1}{2}x^2$

34.

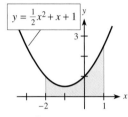
$y = \frac{1}{2}x^2 + x + 1$

35.

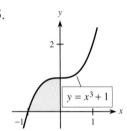
$y = x^3 + 1$

36.

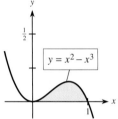
$y = x^2 - x^3$

37. Find the area between the curve $y = -x^2 + 3x - 2$ and the x-axis from $x = 1$ to $x = 2$.

38. Find the area between the curve $y = x^2 + 3x + 2$ and the x-axis from $x = -1$ to $x = 3$.

39. Find the area between the curve $y = xe^{x^2}$ and the x-axis from $x = 1$ to $x = 3$.

40. Find the area between the curve $y = e^{-x}$ and the x-axis from $x = -1$ to $x = 1$.

In Problems 41 and 42, use the figures to decide which of $\int_0^a f(x)\, dx$ or $\int_0^a g(x)\, dx$ is larger or if they are equal. Explain your choices.

41.

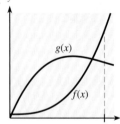
$g(x)$
$f(x)$

42.

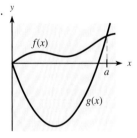
$f(x)$
$g(x)$

In Problems 43–48, use properties of definite integrals.

43. How does $\displaystyle\int_{-1}^{-3} x\sqrt{x^2 + 1}\, dx$ compare with

$$\int_{-3}^{-1} x\sqrt{x^2 + 1}\, dx?$$

44. If $\displaystyle\int_{-1}^{0} x^3\, dx = -\frac{1}{4}$ and $\displaystyle\int_{0}^{1} x^3\, dx = \frac{1}{4}$, what does $\displaystyle\int_{-1}^{1} x^3\, dx$ equal?

45. If $\displaystyle\int_{1}^{2} (2x - x^2)\, dx = \frac{2}{3}$ and $\displaystyle\int_{2}^{4} (2x - x^2)\, dx = -\frac{20}{3}$, what does $\displaystyle\int_{1}^{4} (x^2 - 2x)\, dx$ equal?

46. If $\displaystyle\int_{1}^{2} (2x - x^2)\, dx = \frac{2}{3}$, what does $\displaystyle\int_{1}^{2} 6(2x - x^2)\, dx$ equal?

47. Evaluate $\displaystyle\int_{4}^{4} \sqrt{x^2 - 2}\, dx$.

48. Evaluate $\displaystyle\int_{2}^{2} (x^3 + 4x)^{-6}\, dx$.

APPLICATIONS

49. *Depreciation* The rate of depreciation of a building is given by $D'(t) = 3000(20 - t)$ dollars per year, $0 \le t \le 20$; see the following figure.

(a) Use the graph to find the total depreciation of the building over the first 10 years ($t = 0$ to $t = 10$).

(b) Use the definite integral to find the total depreciation over the first 10 years.

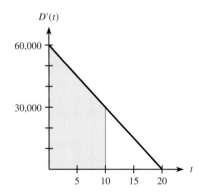

50. *Depreciation* The rate of depreciation of a building is given by $D'(t) = 3000(20 - t)$ dollars per year, $0 \le t \le 20$; see the figure in Problem 49.

(a) Use the graph to find the total depreciation of the building over the first 20 years.

(b) Use the definite integral to find the total depreciation over the first 20 years.

(c) Use the graph to find the total depreciation between 10 years and 20 years and check it with a definite integral.

51. *Sales and advertising*   A store finds that its sales revenue changes at a rate given by

$$S'(t) = -3t^2 + 300t \qquad \text{dollars per day}$$

where t is the number of days after an advertising campaign ends and $0 \le t \le 30$.
   (a) Find the total sales for the first week after the campaign ends ($t = 0$ to $t = 7$).
   (b) Find the total sales for the second week after the campaign ends ($t = 7$ to $t = 14$).

52. *Health care costs*   The annual health care costs in the United States (in billions of dollars) for selected years are given in the table below. The equation

$$y = 1.0275x^2 - 9.0483x + 36.077$$

models the annual health care costs, y (in billions of dollars per year), as a function of the number of years past 1960, x. Use a definite integral and this model to find the total cost of health care over the period 1960–2001.

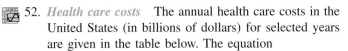

Year	1960	1970	1980	1990
Cost	26.7	73.1	245.8	695.6

Year	1995	1999	2000	2001
Cost	987.0	1210.7	1310.0	1425.5

Source: U.S. Department of Health and Human Services

53. *Total income*   The income from an oil change service chain can be considered as flowing continuously at an annual rate given by

$$f(t) = 10{,}000e^{0.02t} \qquad \text{(dollars/year)}$$

Find the total income for this chain over the first 2 years (from $t = 0$ to $t = 2$).

54. *Total income*   Suppose that a vending machine service company models its income by assuming that money flows continuously into the machines, with the annual rate of flow given by

$$f(t) = 120e^{0.01t}$$

in thousands of dollars per year. Find the total income from the machines over the first 3 years.

55. *Telecommunications revenue*   Total market revenue for worldwide telecommunications (in billions of dollars per year) can be modeled by

$$R(t) = 72.11t + 81.16$$

where $t = 0$ represents 1985. (*Source:* International Telecommunications Union) Evaluate $\int_{15}^{25} R(t)\, dt$ and tell what it represents.

56. *Wireless communications*   Spending for U.S. wireless communications services (in billions of dollars per year) can be modeled by

$$S(t) = 11.23t + 6.205$$

where $t = 0$ represents 1995. (*Source:* The Strategis Group; Donaldson, Lufkin & Jenrette; Cellular Telecommunications and Internet Association; Telecommunications Industry Association) Evaluate $\int_{10}^{15} S(t)\, dt$ and tell what it represents.

Velocity of blood   **In Problems 57 and 58, the velocity of blood through a vessel is given by** $v = K(R^2 - r^2)$**, where** K **is the (constant) maximum velocity of the blood,** R **is the (constant) radius of the vessel, and** r **is the distance of the particular corpuscle from the center of the vessel. The rate of flow can be found by measuring the volume of blood that flows past a point in a given time period. This volume,** V**, is given by**

$$V = \int_0^R v(2\pi r\, dr)$$

57. If $R = 0.30$ cm and $v = (0.30 - 3.33r^2)$ cm/s, find the volume.

58. Develop a general formula for V by evaluating

$$V = \int_0^R v(2\pi r\, dr)$$

using $v = K(R^2 - r^2)$.

Production   **In Problems 59 and 60, the rate of production of a new line of products is given by**

$$\frac{dx}{dt} = 200\left[1 + \frac{400}{(t + 40)^2}\right]$$

where x is the number of items produced and t is the number of weeks the products have been in production.
59. How many units were produced in the first 5 weeks?
60. How many units were produced in the sixth week?

61. *Testing*   The time t (in minutes) needed to read an article appearing on a foreign-language placement test is given by the probability density function

$$f(t) = 0.012t^2 - 0.0012t^3, \qquad 0 \le t \le 10$$

For a test taker chosen at random, find the probability that this person takes 8 minutes or more to read the article.

62. *Response time*   In a small city the response time t (in minutes) of the fire company is given by the probability density function

$$f(t) = \frac{60t^2 - 4t^3}{16{,}875}, \qquad 0 \le t \le 15$$

For a fire chosen at random, find the probability that the response time is 10 minutes or less.

63. *Customer service*   The duration t (in minutes) of customer service calls received by a certain company is given by the probability density function

$$f(t) = 0.3e^{-0.3t}, \qquad t \geq 0$$

Find the probability that a call selected at random lasts
(a) 3 minutes or less.   (b) between 5 and 10 minutes.

64. *Product life*   The useful life of a car battery t (in years) is given by the probability density function

$$f(t) = 0.2e^{-0.2t}, \qquad t \geq 0$$

Find the probability that a battery chosen at random lasts
(a) 2 years or less.   (b) between 4 and 6 years.

65. **Modeling**   *Pollution*   Carbon monoxide (CO) emissions (in millions of short tons per year) for selected years from 1970 to 2001 are given in the table.
(a) Fit a 4th-degree polynomial model, $E(t)$, to these data. Use $t = 0$ to represent 1970.
(b) Evaluate $\int_{20}^{30} E(t)\, dt$ and tell what it represents.

Year	CO	Year	CO
1970	204.043	1995	126.777
1975	188.398	1998	115.382
1980	185.407	1999	117.229
1985	176.884	2000	123.568
1990	154.189	2001	120.759

Source: Environmental Protection Agency

66. **Modeling**   *Dodge Viper acceleration*   Table 13.1(a) shows the time in seconds that a Dodge Viper GTS requires to reach various speeds up to 100 mph. Table 13.1(b) shows the same data, but with speeds in miles per second.
(a) Fit a power model to the data in Table 13.1(b).
(b) Use a definite integral from 0 to 9.2 of the function you found in (a) to find the distance traveled by the Viper as it went from 0 mph to 100 mph in 9.2 seconds.

TABLE 13.1

(a)		(b)	
Time (seconds)	Speed (mph)	Time (seconds)	Speed (mi/s)
1.7	30	1.7	0.00833
2.4	40	2.4	0.01111
3.2	50	3.2	0.01389
4.1	60	4.1	0.01667
5.8	70	5.8	0.01944
6.2	80	6.2	0.02222
7.8	90	7.8	0.02500
9.2	100	9.2	0.02778

Source: Motor Trend

13.3

Area Between Two Curves

OBJECTIVES

● To find the area between two curves
● To find the average value of a function

◗ Application Preview

In economics, the **Lorenz curve** is used to represent the inequality of income distribution among different groups in the population of a country. The curve is constructed by plotting the cumulative percent of families at or below a given income level and the cumulative percent of total personal income received by these families. For example, the table shows the coordinates of some points on the Lorenz curve $y = L(x)$ that divide the income (for the United States in 2001) into 5 equal income levels (quintiles). The point (0.40, 0.122) is on the Lorenz curve because the families with incomes in the bottom 40% of the country received 12.2% of the total income in 2001. The graph of the Lorenz curve $y = L(x)$ is shown in Figure 13.12.

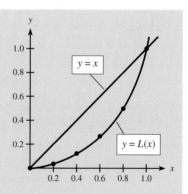

Figure 13.12

U.S. Income Distribution for 2001 (Points on the Lorenz Curve)	
x, Cumulative Proportion of Families Below Income Level	*y* = *L*(*x*), Cumulative Proportion of Total Income
0	0
0.20	0.035
0.40	0.122
0.60	0.268
0.80	0.498
1	1

Source: U.S. Bureau of the Census

Equality of income would result if each family received an equal proportion of the total income, so that the bottom 20% would receive 20% of the total income, the bottom 40% would receive 40%, and so on. The Lorenz curve representing this would have the equation *y* = *x*.

The inequality of income distribution is measured by the **Gini coefficient** of income, which measures how far the Lorenz curve falls below *y* = *x*. It is defined as

$$\frac{\text{Area between } y = x \text{ and } y = L(x)}{\text{Area below } y = x}$$

Because the area of the triangle below *y* = *x* and above the *x*-axis from *x* = 0 to *x* = 1 is 1/2, the Gini coefficient of income is

$$\frac{\text{Area between } y = x \text{ and } y = L(x)}{1/2} = 2 \cdot [\text{area between } y = x \text{ and } y = L(x)]$$

In this section we will use the definite integral to find the area between two curves. We will use the area between two curves to find the Gini coefficient of income (see Example 4) and to find average cost, average revenue, average profit, and average inventory.

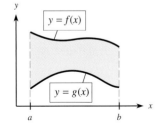

Figure 13.13

We have used the definite integral to find the area of the region between a curve and the *x*-axis over an interval where the curve lies above the *x*-axis. We can easily extend this technique to finding the area between two curves over an interval where one curve lies above the other. (See Figure 13.13.)

Suppose that the graphs of both *y* = *f*(*x*) and *y* = *g*(*x*) lie above the *x*-axis and that the graph of *y* = *f*(*x*) lies above *y* = *g*(*x*) throughout the interval from *x* = *a* to *x* = *b*; that is, *f*(*x*) ≥ *g*(*x*) on [*a*, *b*].

Then $\int_a^b f(x)\,dx$ gives the area between the graph of *y* = *f*(*x*) and the *x*-axis (see Figure 13.14(a)), and $\int_a^b g(x)\,dx$ gives the area between the graph of *y* = *g*(*x*) and the *x*-axis (see Figure 13.14(b)). As Figure 13.14(c) shows, the area of the region between the graphs of *y* = *f*(*x*) and *y* = *g*(*x*) is the difference of these two areas. That is,

$$\text{Area between the curves} = \int_a^b f(x)\,dx - \int_a^b g(x)\,dx$$

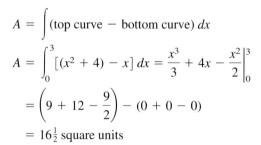

Figure 13.14   (a)   (b)   (c)

Although Figure 13.14(c) shows the graphs of both $y = f(x)$ and $y = g(x)$ lying above the x-axis, this difference of their integrals will always give the area between their graphs if both functions are continuous and if $f(x) \geq g(x)$ on the interval $[a, b]$. Using the fact that

$$\int_a^b f(x)\, dx - \int_a^b g(x)\, dx = \int_a^b [f(x) - g(x)]\, dx$$

we have the following result.

Area Between Two Curves

> If f and g are continuous functions on $[a, b]$ and if $f(x) \geq g(x)$ on $[a, b]$, then the area of the region bounded by $y = f(x)$, $y = g(x)$, $x = a$, and $x = b$ is
>
> $$A = \int_a^b [f(x) - g(x)]\, dx$$

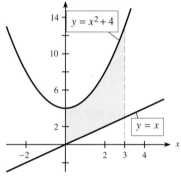

Figure 13.15

● **EXAMPLE 1 *Area Between Two Curves***

Find the area of the region bounded by $y = x^2 + 4$, $y = x$, $x = 0$, and $x = 3$.

Solution
We first sketch the graphs of the functions. The graph of the region is shown in Figure 13.15. Because $y = x^2 + 4$ lies above $y = x$ in the interval from $x = 0$ to $x = 3$, the area is

$$A = \int (\text{top curve} - \text{bottom curve})\, dx$$

$$A = \int_0^3 [(x^2 + 4) - x]\, dx = \left. \frac{x^3}{3} + 4x - \frac{x^2}{2} \right|_0^3$$

$$= \left(9 + 12 - \frac{9}{2} \right) - (0 + 0 - 0)$$

$$= 16\tfrac{1}{2} \text{ square units}$$

We are sometimes asked to find the area enclosed by two curves. In this case, we find the points of intersection of the curves to determine a and b.

● **EXAMPLE 2 *Area Enclosed by Two Curves***

Find the area enclosed by $y = x^2$ and $y = 2x + 3$.

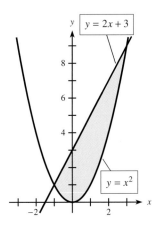

Figure 13.16

Solution

We first find a and b by finding the x-coordinates of the points of intersection of the graphs. Setting the y-values equal gives

$$x^2 = 2x + 3$$
$$x^2 - 2x - 3 = 0$$
$$(x - 3)(x + 1) = 0$$
$$x = 3, \quad x = -1$$

Thus $a = -1$ and $b = 3$.

We next sketch the graphs of these functions on the same set of axes. Because the graphs do not intersect on the interval $(-1, 3)$, we can determine which function is larger on this interval by evaluating $2x + 3$ and x^2 at any value c where $-1 < c < 3$. Figure 13.16 shows the region between the graphs, with $2x + 3 \geq x^2$ from $x = -1$ to $x = 3$. The area of the enclosed region is

$$A = \int_{-1}^{3} [(2x + 3) - x^2] \, dx$$
$$= x^2 + 3x - \frac{x^3}{3} \Big|_{-1}^{3}$$
$$= (9 + 9 - 9) - \left(1 - 3 + \frac{1}{3}\right)$$
$$= 10\frac{2}{3} \text{ square units}$$

Some graphs enclose two or more regions because they have more than two points of intersection.

● **EXAMPLE 3** *A Region with Two Sections*

Find the area of the region enclosed by the graphs of

$$y = f(x) = x^3 - x^2$$

and

$$y = g(x) = 2x$$

Solution

To find the points of intersection of the graphs, we set the y-values equal and solve for x.

$$x^3 - x^2 = 2x$$
$$x^3 - x^2 - 2x = 0$$
$$x(x - 2)(x + 1) = 0$$
$$x = 0, \ x = 2, \ x = -1$$

Graphing these functions between $x = -1$ and $x = 2$, we see that for any x-value in the interval $(-1, 0)$, $f(x) \geq g(x)$, so $f(x) \geq g(x)$ for the region enclosed by the curves from $x = -1$ to $x = 0$. But evaluating the functions for any x-value in the interval $(0, 2)$ shows that $f(x) \leq g(x)$ for the region enclosed by the curves from $x = 0$ to $x = 2$. See Figure 13.17.

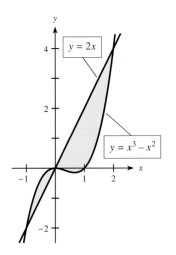

Figure 13.17

Thus we need one integral to find the area of the region from $x = -1$ to $x = 0$ and a second integral to find the area from $x = 0$ to $x = 2$. The area is found by summing these two integrals.

$$A = \int_{-1}^{0} [(x^3 - x^2) - (2x)]\, dx + \int_{0}^{2} [(2x) - (x^3 - x^2)]\, dx$$

$$= \int_{-1}^{0} (x^3 - x^2 - 2x)\, dx + \int_{0}^{2} (2x - x^3 + x^2)\, dx$$

$$= \left(\frac{x^4}{4} - \frac{x^3}{3} - x^2 \right)\Big|_{-1}^{0} + \left(x^2 - \frac{x^4}{4} + \frac{x^3}{3} \right)\Big|_{0}^{2}$$

$$= \left[(0) - \left(\frac{1}{4} - \frac{-1}{3} - 1 \right) \right] + \left[\left(4 - \frac{16}{4} + \frac{8}{3} \right) - (0) \right] = \frac{37}{12}$$

Thus the area between the curves is $\frac{37}{12}$ square units.

Calculator Note Most graphing calculators have a numerical integration feature, and some can perform both symbolic and numerical integration.

A graphing calculator could be used to find the area enclosed by the graphs of $y = f(x) = x^3 - x^2$ and $y = g(x) = 2x$, found in Example 3. By using SOLVER or INTERSECT, we can find that the curves intersect at $x = -1$, $x = 0$, and $x = 2$. From the graph, shown in Figure 13.17, we see that the curves enclose two regions, with $f(x) > g(x)$ for $-1 < x < 0$ and $g(x) > f(x)$ for $0 < x < 2$. Thus the area is given by

$$\int_{-1}^{0} [(x^3 - x^2) - (2x)]\, dx + \int_{0}^{2} [(2x) - (x^3 - x^2)]\, dx$$

By using the numerical integration feature of a graphing calculator, we can find the integral (and the area) to be $0.4166666667 + 2.666666667 = 3.083333333$. This is the same (to nine decimal places) as the decimal representation of $37/12$, found in Example 3. ■

● **Checkpoint**

1. True or false:
   (a) Over the interval $[a, b]$, the area between the continuous functions $f(x)$ and $g(x)$ is

   $$\int_{a}^{b} [f(x) - g(x)]\, dx$$

   (b) If $f(x) \geq g(x)$ and the area between $f(x)$ and $g(x)$ is given by

   $$\int_{a}^{b} [f(x) - g(x)]\, dx$$

   then $x = a$ and $x = b$ represent the left and right boundaries, respectively, of the region.
   (c) To find points of intersection of $f(x)$ and $g(x)$, solve $f(x) = g(x)$.

2. Consider the functions $f(x) = x^2 + 3x - 9$ and $g(x) = \frac{1}{4}x^2$.

   (a) Find the points of intersection of $f(x)$ and $g(x)$.
   (b) Determine which function is greater than the other between the points found in part (a).
   (c) Set up the integral used to find the area between the curves in the interval between the points found in part (a).
   (d) Find the area.

◗ EXAMPLE 4 *Income Distribution (Application Preview)*

The inequality of income distribution is measured by the Gini coefficient of income, which is defined as

$$\frac{\text{Area between } y = x \text{ and } y = L(x)}{\text{Area below } y = x} = \frac{\displaystyle\int_0^1 [x - L(x)]\, dx}{1/2}$$

$$= 2 \int_0^1 [x - L(x)]\, dx$$

The function $y = L(x) = x^{2.52}$ models the 2001 income distribution data in the Application Preview.

(a) Use this $L(x)$ to find the Gini coefficient of income for 2001.
(b) If the Census Bureau Gini coefficient of income for 1991 is 0.428, during which year is the distribution of income more nearly equal?

Solution
(a) The Gini coefficient of income for 2001 is

$$2 \int_0^1 [x - L(x)]\, dx = 2 \int_0^1 (x - x^{2.52})\, dx$$

$$= 2 \left[\frac{x^2}{2} - \frac{x^{3.52}}{3.52} \right]\Bigg|_0^1 = 2 \left[\frac{1}{2} - \frac{1}{3.52} \right] \approx 0.432$$

(b) Absolute equality of income would occur if the Gini coefficient of income were 0; and smaller coefficients indicate more nearly equal incomes. Thus the distribution of income was more nearly equal in 1991 than in 2001.

Average Value   If the graph of $y = f(x)$ lies on or above the x-axis from $x = a$ to $x = b$, then the area between the graph and the x-axis is

$$A = \int_a^b f(x)\, dx \qquad \text{(See Figure 13.18(a))}$$

The area A is also the area of a rectangle with base equal to $b - a$ and height equal to the *average value* (or average height) of the function $y = f(x)$ (see Figure 13.18(b)).

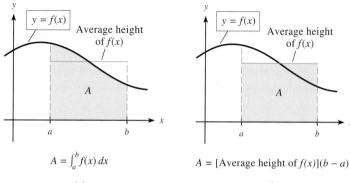

$$A = \int_a^b f(x)\, dx \qquad\qquad A = [\text{Average height of } f(x)](b - a)$$

Figure 13.18   (a)   (b)

Thus the average value of the function in Figure 13.18 is

$$\frac{A}{b-a} = \frac{1}{b-a}\int_a^b f(x)\,dx$$

Even if $f(x) \leq 0$ on all or part of the interval $[a, b]$, we can find the average value by using the integral. Thus we have the following.

Average Value

> The average value of a continuous function $y = f(x)$ over the interval $[a, b]$ is
>
> $$\text{Average value} = \frac{1}{b-a}\int_a^b f(x)\,dx$$

● **EXAMPLE 5** *Average Cost*

Suppose that the cost in dollars for a product is given by $C(x) = 400 + x + 0.3x^2$, where x is the number of units.

(a) What is the average value of $C(x)$ for 10 to 20 units?
(b) Find the average cost per unit if 40 units are produced.

Solution
(a) The average value of $C(x)$ from $x = 10$ to $x = 20$ is

$$\frac{1}{20-10}\int_{10}^{20}(400 + x + 0.3x^2)\,dx = \frac{1}{10}\left(400x + \frac{x^2}{2} + 0.1x^3\right)\Big|_{10}^{20}$$

$$= \frac{1}{10}\big[(8000 + 200 + 800) - (4000 + 50 + 100)\big]$$

$$= 485 \quad \text{(dollars)}$$

Thus for any number of units between $x = 10$ and $x = 20$, the *average total cost* for that number of units is $485.
(b) The average cost per unit if 40 units are produced is the average cost function evaluated at $x = 40$. The average cost function is

$$\overline{C}(x) = \frac{C(x)}{x} = \frac{400}{x} + 1 + 0.3x$$

Thus the *average cost per unit* if 40 units are produced is

$$\overline{C}(40) = \frac{400}{40} + 1 + 0.3(40) = 23 \quad \text{(dollars per unit)}$$

● *Checkpoint*   3. Find the average value of $f(x) = x^2 - 4$ over $[-1, 3]$.

● **EXAMPLE 6** *Average Value of a Function*

Consider the functions $f(x) = x^2 - 4$ and $g(x) = x^3 - 4x$. For each function, do the following.

(a) Graph the function on the interval $[-2, 2]$.
(b) On the graph, "eyeball" the average value (height) of each function on $[-2, 2]$.
(c) Compute the average value of the function over the interval $[-2, 2]$.

Solution

For $f(x) = x^2 - 4$:

(a) The graph of $f(x) = x^2 - 4$ is shown in Figure 13.19(a).
(b) The average height of $f(x)$ over $[-2, 2]$ appears to be near -2.
(c) The average value of $f(x)$ over the interval is given by

$$\frac{1}{2 - (-2)} \int_{-2}^{2} (x^2 - 4)\, dx = \frac{1}{4}\left[\frac{x^3}{3} - 4x\right]_{-2}^{2}$$

$$= \left(\frac{8}{12} - 2\right) - \left(-\frac{8}{12} + 2\right)$$

$$= \frac{4}{3} - 4 = -\frac{8}{3} = -2\frac{2}{3}$$

For $g(x) = x^3 - 4x$:

(a) The graph of $g(x) = x^3 - 4x$ is shown in Figure 13.19(b).
(b) The average height of $g(x)$ over $[-2, 2]$ appears to be approximately 0.
(c) The average value of $g(x)$ is given by

$$\frac{1}{2 - (-2)} \int_{-2}^{2} (x^3 - 4x)\, dx = \frac{1}{4}\left[\frac{x^4}{4} - \frac{4x^2}{2}\right]_{-2}^{2}$$

$$= \left(\frac{16}{16} - \frac{16}{8}\right) - \left(\frac{16}{16} - \frac{16}{8}\right)$$

$$= -1 + 1 = 0$$

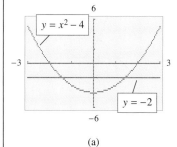

Figure 13.19   (a)   (b)

● **Checkpoint Solutions**

1. (a) False. This is true only if $f(x) \geq g(x)$ over $[a, b]$.
   (b) True   (c) True

2. (a) Solve $f(x) = g(x)$, or $x^2 + 3x - 9 = \frac{1}{4}x^2$.

$$\frac{3}{4}x^2 + 3x - 9 = 0$$

$$3x^2 + 12x - 36 = 0$$

$$x^2 + 4x - 12 = 0$$

$$(x + 6)(x - 2) = 0$$

$$x = -6, x = 2$$

The points of intersection are $(-6, 9)$ and $(2, 1)$.
   (b) Evaluating $g(x)$ and $f(x)$ at any point in the interval $(-6, 2)$ shows that $g(x) > f(x)$, so $g(x) \geq f(x)$ on $[-6, 2]$.

(c) $A = \int_{-6}^{2} \left[\frac{1}{4}x^2 - (x^2 + 3x - 9) \right] dx$

(d) $A = \frac{x^3}{12} - \frac{x^3}{3} - \frac{3x^2}{2} + 9x \Big|_{-6}^{2} = 64$ square units

3. $\dfrac{1}{3 - (-1)} \displaystyle\int_{-1}^{3} (x^2 - 4)\, dx = \dfrac{1}{4}\left(\dfrac{x^3}{3} - 4x\right)\Big|_{-1}^{3}$

$= \dfrac{1}{4}\left[(9 - 12) - \left(-\dfrac{1}{3} + 4\right)\right]$

$= \dfrac{1}{4}\left(\dfrac{-20}{3}\right) = -\dfrac{5}{3}$

13.3 Exercises

For each shaded region in Problems 1–6, (a) form the integral that represents the area of the shaded region and (b) find the area of the region.

For each shaded region in Problems 7–12, (a) find the points of intersection of the curves, (b) form the integral that represents the area of the shaded region, and (c) find the area of the shaded region.

1.
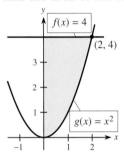
$f(x) = 4$
$(2, 4)$
$g(x) = x^2$

2.
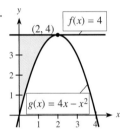
$f(x) = 4$
$(2, 4)$
$g(x) = 4x - x^2$

7.
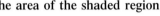
$y = x + 2$
$y = x^2$

8.
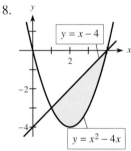
$y = x - 4$
$y = x^2 - 4x$

3.
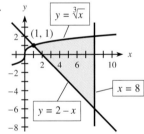
$y = \sqrt[3]{x}$
$(1, 1)$
$x = 8$
$y = 2 - x$

4.
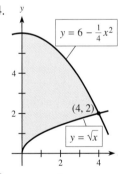
$y = 6 - \frac{1}{4}x^2$
$(4, 2)$
$y = \sqrt{x}$

9.
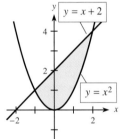
$y = x - x^2$
$y = x^2 - 4x$

10.
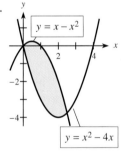
$y = 6 - x^2$
$y = x$

5.
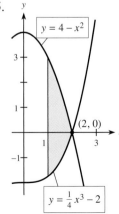
$y = 4 - x^2$
$(2, 0)$
$y = \frac{1}{4}x^3 - 2$

6.
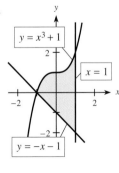
$y = x^3 + 1$
$x = 1$
$y = -x - 1$

11.
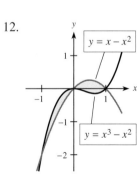
$y = x^3 - 2x$
$y = 2x$

12.
$y = x - x^2$
$y = x^3 - x^2$

In Problems 13–26, equations are given whose graphs enclose a region. In each problem, find the area of the region.

13. $f(x) = x^2 + 2$; $g(x) = -x^2$; $x = 0$; $x = 2$
14. $f(x) = x^2$; $g(x) = -\frac{1}{10}(10 + x)$; $x = 0$; $x = 3$
15. $y = x^3 - 1$; $y = x - 1$; to the right of the y-axis
16. $y = x^2 - 2x + 1$; $y = x^2 - 5x + 4$; $x = 2$
17. $y = \frac{1}{2}x^2$; $y = x^2 - 2x$
18. $y = x^2$; $y = 4x - x^2$
19. $h(x) = x^2$; $k(x) = \sqrt{x}$
20. $g(x) = 1 - x^2$; $h(x) = x^2 + x$
21. $f(x) = x^3$; $g(x) = x^2 + 2x$
22. $f(x) = x^3$; $g(x) = 2x - x^2$
23. $f(x) = \dfrac{3}{x}$; $g(x) = 4 - x$
24. $f(x) = \dfrac{6}{x}$; $g(x) = -x - 5$
25. $y = \sqrt{x + 3}$; $x = -3$; $y = 2$
26. $y = \sqrt{4 - x}$; $x = 4$; $y = 3$

In Problems 27–32, find the average value of each function over the given interval.

27. $f(x) = 9 - x^2$ over $[0, 3]$
28. $f(x) = 2x - x^2$ over $[0, 2]$
29. $f(x) = x^3 - x$ over $[-1, 1]$
30. $f(x) = \frac{1}{2}x^3 + 1$ over $[-2, 0]$
31. $f(x) = \sqrt{x} - 2$ over $[1, 4]$
32. $f(x) = \sqrt[3]{x}$ over $[-8, -1]$

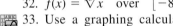 33. Use a graphing calculator or computer to find the area between the curves $y = f(x) = x^3 - 4x$ and $y = g(x) = x^2 - 4$.

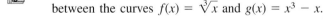 34. Use a graphing calculator or computer to find the area between the curves $f(x) = \sqrt[3]{x}$ and $g(x) = x^3 - x$.

APPLICATIONS

35. *Average profit* For the product whose total cost and total revenue are shown in the figure, represent total revenue by $R(x)$ and total cost by $C(x)$ and write an integral that gives the average profit for the product over the interval from x_0 to x_1.

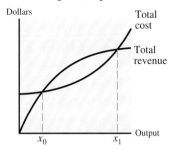

36. *Sales and advertising* The following figure shows the sales growth rates under different levels of distribution and advertising from a to b. Set up an integral to determine the extra sales growth if $4 million is used in advertising rather than $2 million.

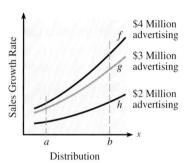

Distribution

37. *Cost* The cost of producing x units of a certain item is $C(x) = x^2 + 400x + 2000$.
    (a) Use $C(x)$ to find the average cost of producing 1000 units.
    (b) Find the average value of the cost function $C(x)$ over the interval from 0 to 1000.

38. *Inventory management* The figure shows how an inventory of a product is depleted each quarter of a given year. What is the average inventory per month for the first 3 months for this product? (Assume that the graph is a line joining (0, 1300) and (3, 100).)

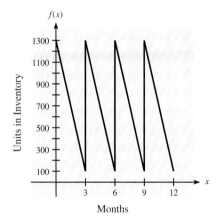

Months

39. *Sales and advertising* The number of daily sales of a product was found to be given by
$$S = 100xe^{-x^2} + 100$$
x days after the start of an advertising campaign for this product.
    (a) Find the average daily sales during the first 20 days of the campaign—that is, from $x = 0$ to $x = 20$.
    (b) If no new advertising campaign is begun, what is the average number of sales per day for the next 10 days (from $x = 20$ to $x = 30$)?

40. *Demand* The demand function for a certain product is given by
$$p = 500 + \frac{1000}{q + 1}$$
where p is the price and q is the number of units demanded. Find the average price as demand ranges from 49 to 99 units.

41. *Tax burden*   The U.S. federal tax burden per capita (in dollars) for selected years from 1960 to 2002 can be modeled by

$$T(t) = 257.8e^{0.067t}$$

where $t = 0$ represents 1950. (*Source:* Internal Revenue Service, *2002 Data Book,* 2003) If the model remains valid, find the predicted average federal tax burden per capita from 2000 to 2010.

42. *Total income*   Suppose that the income from a slot machine in a casino flows continuously at a rate of

$$f(t) = 100e^{0.1t}$$

where t is the time in hours since the casino opened. Then the total income during the first 10 hours is given by

$$\int_0^{10} 100e^{0.1t}\, dt$$

Find the average income over the first 10 hours.

43. *Drug levels in the blood*   A drug manufacturer has developed a time-release capsule with the number of milligrams of the drug in the bloodstream given by

$$S = 30x^{18/7} - 240x^{11/7} + 480x^{4/7}$$

where x is in hours and $0 \le x \le 4$. Find the average number of milligrams of the drug in the bloodstream for the first 4 hours after a capsule is taken.

44. *Income distribution*   The Lorenz curves for the income distribution in the United States in 1950 and in 1970 are given below. Find the Gini coefficient of income for both years and compare the distributions of income for these years.

$$1950: y = x^{2.1521}$$
$$1970: y = x^{2.2024}$$

45. *Income distribution*   U.S. political analysts tend to believe that the tax policies of Republican administrations tend to favor the wealthy and those of Democratic administrations tend to favor the poor. The Lorenz curves for income distribution in 1988 (Reagan's final year) and 2000 (Clinton's final year) are given below. Find the Gini coefficient of income for each year and determine in which administration's final year the income distribution was more nearly equal. Does this support the conventional wisdom regarding the tax policies of Republicans versus Democrats?

$$1988: y = x^{2.3521} \qquad 2000: y = x^{2.4870}$$

46. *Income distribution*   In an effort to make the distribution of income more nearly equal, the government of a country passes a tax law that changes the Lorenz curve from $y = 0.99x^{2.1}$ for one year to $y = 0.32x^2 + 0.68x$ for the next year. Find the Gini coefficient of income for both years and compare the distributions of income before and after the tax law was passed. Interpret the result.

47. *Income distribution*   Data from the U.S. Bureau of the Census yields the Lorenz curves given below for income distribution among blacks and among Asians in the United States for 2003. Find the Gini coefficient of income for each group and determine in which group the income is more nearly equally distributed.

$$\text{Blacks: } y = x^{2.5938}$$
$$\text{Asians: } y = x^{2.5070}$$

48. *Income distribution*   Suppose the Gini coefficient of income for a certain country is 2/5. If the Lorenz curve for this country is

$$L(x) - \frac{1}{3}x + \frac{2}{3}x^p$$

find the value of p.

OBJECTIVES

13.4

Applications of Definite Integrals in Business and Economics

- To use definite integrals to find total income, present value, and future value of continuous income streams
- To use definite integrals to find the consumer's surplus
- To use definite integrals to find the producer's surplus

◗ **Application Preview**

Suppose the oil pumped from a well is considered as a continuous income stream with annual rate of flow (in thousands of dollars per year) at time t years given by

$$f(t) = 600e^{-0.2(t+5)}$$

The company can use a definite integral involving $f(t)$ to estimate the well's present value over the next 10 years (see Example 2).

The definite integral can be used in a number of applications in business and economics. In addition to the present value, the definite integral can be used to find the total income over a fixed number of years from a **continuous income stream.**

Definite integrals also can be used to determine the savings realized in the marketplace by some consumers (called **consumer's surplus**) and some producers (called **producer's surplus**).

Continuous Income Streams

An oil company's profits depend on the amount of oil that can be pumped from a well. Thus we can consider a pump at an oil field as producing a **continuous stream of income** for the owner. Because both the pump and the oil field "wear out" with time, the continuous stream of income is a function of time. Suppose $f(t)$ is the (annual) *rate* of flow of income from this pump; then we can find the total income from the rate of income by using integration. In particular, the total income for k years is given by

$$\text{Total income} = \int_0^k f(t)\, dt$$

● **EXAMPLE 1** *Oil Revenue*

A small oil company considers the continuous pumping of oil from a well as a continuous income stream with its annual rate of flow at time t given by

$$f(t) = 600e^{-0.2t}$$

in thousands of dollars per year. Find an estimate of the total income from this well over the next 10 years.

Solution

$$\text{Total income} = \int_0^{10} f(t)\, dt = \int_0^{10} 600e^{-0.2t}\, dt$$

$$= \frac{600}{-0.2} e^{-0.2t}\Big|_0^{10}$$

$$\approx 2594 \quad \text{(to the nearest integer)}$$

Thus the total income is approximately $2,594,000.

In addition to the total income from a continuous income stream, the **present value** of the stream is also important. The present value is the value today of a continuous income stream that will be providing income in the future. The present value is useful in deciding when to replace machinery or what new equipment to select.

To find the present value of a continuous stream of income with rate of flow $f(t)$, we first graph the function $f(t)$ and divide the time interval from 0 to k into n subintervals of width Δt_i, $i = 1$ to n.

The total amount of income is the area under this curve between $t = 0$ and $t = k$. We can approximate the amount of income in each subinterval by finding the area of the rectangle in that subinterval. (See Figure 13.20.)

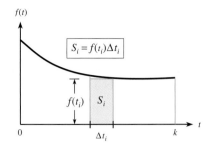

f(t)

$S_i = f(t_i)\Delta t_i$

$f(t_i)$ S_i

Figure 13.20   0 Δt_i k   t

We have shown that the future value S that accrues if \$$P$ is invested for t years at an annual rate r, compounded continuously, is

$$S = Pe^{rt}$$

Thus the present value of the investment that yields the single payment of \$$S$ after t years is

$$P = \frac{S}{e^{rt}} = Se^{-rt}$$

The contribution to S in the ith subinterval is $S_i = f(t_i)\,\Delta t_i$, and the present value of this amount is

$$P_i = f(t_i)\,\Delta t_i e^{-rt_i}$$

Thus the total present value of S can be approximated by

$$\sum_{i=1}^{n} f(t_i)\,\Delta t_i e^{-rt_i}$$

This approximation improves as $\Delta t_i \to 0$ with the present value given by

$$\lim_{\Delta t_i \to 0} \sum_{i=1}^{n} f(t_i)\,\Delta t_i e^{-rt_i}$$

This limit gives the present value as a definite integral.

Present Value of a Continuous Income Stream

If $f(t)$ is the rate of continuous income flow earning interest at rate r, compounded continuously, then the present value of the continuous income stream is

$$\text{Present value} = \int_0^k f(t)e^{-rt}\,dt$$

where $t = 0$ to $t = k$ is the time interval.

◐ EXAMPLE 2 Present Value (Application Preview)

Suppose that the oil company in Example 1 is planning to sell one of its wells because of its remote location. Suppose further that the company wants to use the present value of this well over the next 10 years to help establish its selling price. If the company determines that the annual rate of flow is

$$f(t) = 600e^{-0.2(t+5)}$$

in thousands of dollars per year, and if money is worth 10%, compounded continuously, find this present value.

Solution

$$\text{Present value} = \int_0^{10} f(t)e^{-rt}\, dt$$

$$= \int_0^{10} 600e^{-0.2(t+5)}e^{-0.1t}\, dt$$

$$= \int_0^{10} 600e^{-0.3t-1}\, dt$$

If $u = -0.3t - 1$, then $du = -0.3\, dt$ and this has the form

$$\int 600e^u\left(\frac{1}{-0.3}du\right) = \frac{600}{-0.3}e^{-0.3t-1}\Big|_0^{10}$$

$$= -2000(e^{-4} - e^{-1})$$

$$\approx 699 \quad \text{(to the nearest integer)}$$

Thus the present value is \$699,000.

Recall that the future value of a continuously compounded investment at rate r after k years is Pe^{rk}, where P is the amount invested (or the present value). Thus, for a continuous income stream, the future value is found as follows.

**Future Value of a
Continuous Income Stream**

If $f(t)$ is the rate of continuous income flow for k years earning interest at rate r, compounded continuously, then the future value of the continuous income stream is

$$FV = e^{rk}\int_0^k f(t)e^{-rt}\, dt$$

● **EXAMPLE 3** *Future Value*

If the rate of flow of income from an asset is $1000e^{0.02t}$, in millions of dollars per year, and if the income is invested at 6% compounded continuously, find the future value of the asset 4 years from now.

Solution
The future value is given by

$$FV = e^{rk}\int_0^k f(t)e^{-rt}\, dt$$

$$= e^{(0.06)4}\int_0^4 1000e^{0.02t}e^{-0.06t}\, dt = e^{0.24}\int_0^4 1000e^{-0.04t}\, dt$$

$$= e^{0.24}[-25{,}000e^{-0.04t}]_0^4 = -25{,}000e^{0.24}[e^{-0.16} - 1]$$

$$\approx 4699.05 \quad \text{(millions of dollars)}$$

● *Checkpoint*

1. Suppose that a continuous income stream has an annual rate of flow given by $f(t) = 5000e^{-0.01t}$, and suppose that money is worth 7% compounded continuously. Create the integral used to find
   (a) the total income for the next 5 years.
   (b) the present value for the next 5 years.
   (c) the future value 5 years from now.

Consumer's Surplus

Suppose that the demand for a product is given by $p = f(x)$ and that the supply of the product is described by $p = g(x)$. The price p_1 where the graphs of these functions intersect is the **equilibrium price** (see Figure 13.21(a)). As the demand curve shows, some consumers (but not all) would be willing to pay more than $\$p_1$ for the product.

For example, some consumers would be willing to buy x_3 units if the price were $\$p_3$. Those consumers willing to pay more than $\$p_1$ are benefiting from the lower price. The total gain for all those consumers willing to pay more than $\$p_1$ is called the **consumer's surplus,** and under proper assumptions the area of the shaded region in Figure 13.21(a) represents this consumer's surplus.

Looking at Figure 13.21(b), we see that if the demand curve has equation $p = f(x)$, the consumer's surplus is given by the area between $f(x)$ and the x-axis from 0 to x_1, *minus* the area of the rectangle denoted *TR*:

$$CS = \int_0^{x_1} f(x)\,dx - p_1 x_1$$

Note that with equilibrium price p_1 and equilibrium quantity x_1, the product $p_1 x_1$ is the area of the rectangle that represents the total dollars spent by consumers and received as revenue by producers (see Figure 13.21(b)).

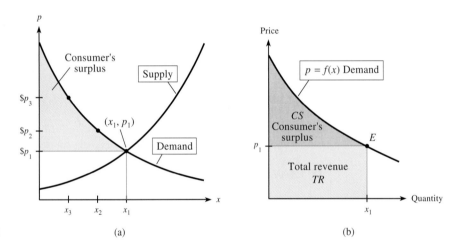

Figure 13.21   (a)   (b)

● **EXAMPLE 4** *Consumer's Surplus*

The demand function for x units of a product is $p = 1020/(x + 1)$ dollars. If the equilibrium price is \$20, what is the consumer's surplus?

Solution

We must first find the quantity that will be purchased at this price. Letting $p = 20$ and solving for x, we get

$$20 = \frac{1020}{x + 1}$$

$$20(x + 1) = 1020$$

$$20x + 20 = 1020 \qquad \text{or} \qquad 20x = 1000$$

$$x = 50$$

Thus the equilibrium point is (50, 20). The consumer's surplus is given by

$$CS = \int_0^{x_1} f(x)\,dx - p_1 x_1 = \int_0^{50} \frac{1020}{x+1}\,dx - 20 \cdot 50$$

$$= 1020 \ln |x+1| \Big|_0^{50} - 1000$$

$$= 1020(\ln 50 - \ln 1) - 1000$$

$$\approx 1020(3.912 - 0) - 1000$$

$$= 3990.24 - 1000$$

$$= 2990.24$$

The consumer's surplus is $2990.24.

● **EXAMPLE 5** *Consumer's Surplus*

A product's demand function is $p = \sqrt{49 - 6x}$ and its supply function is $p = x + 1$, where p is in dollars and x is the number of units. Find the equilibrium point and the consumer's surplus there.

Solution

We can determine the equilibrium point by solving the two equations simultaneously.

$$\sqrt{49 - 6x} = x + 1$$

$$49 - 6x = (x + 1)^2$$

$$0 = x^2 + 8x - 48$$

$$0 = (x + 12)(x - 4)$$

$$x = 4 \quad \text{or} \quad x = -12$$

Thus the equilibrium quantity is 4 and the equilibrium price is $5 (because $x = -12$ is not a solution). The graphs of the supply and demand functions are shown in Figure 13.22.

The consumer's surplus is given by

$$CS = \int_0^4 f(x)\,dx - p_1 x_1$$

$$= \int_0^4 \sqrt{49 - 6x}\,dx - 5 \cdot 4$$

$$= -\frac{1}{6} \int_0^4 \sqrt{49 - 6x}\,(-6\,dx) - 20$$

$$= -\frac{1}{9}(49 - 6x)^{3/2} \Big|_0^4 - 20$$

$$= -\frac{1}{9}\left[(25)^{3/2} - (49)^{3/2}\right] - 20$$

$$= -\frac{1}{9}(125 - 343) - 20$$

$$\approx 24.22 - 20 = 4.22$$

The consumer's surplus is $4.22.

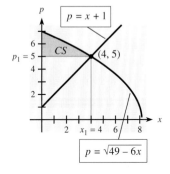

Figure 13.22

● EXAMPLE 6 *Monopoly Market*

Suppose a monopoly has its total cost (in dollars) for a product given by $C(x) = 60 + 2x^2$. Suppose also that demand is given by $p = 30 - x$, where p is in dollars and x is the number of units. Find the consumer's surplus at the point where the monopoly has maximum profit.

Solution

We must first find the point where the profit function is maximized. Because the demand for x units is $p = 30 - x$, the total revenue is

$$R(x) = (30 - x)x = 30x - x^2$$

Thus the profit function is

$$P(x) = R(x) - C(x)$$
$$P(x) = 30x - x^2 - (60 + 2x^2)$$
$$P(x) = 30x - 60 - 3x^2$$

Then $P'(x) = 30 - 6x$. So $0 = 30 - 6x$ has the solution $x = 5$.

Because $P''(5) = -6 < 0$, the profit for the monopolist is maximized when $x = 5$ units are sold at price $p = 30 - x = 25$ dollars per unit.

The consumer's surplus at $x = 5$, $p = 25$ is given by

$$CS = \int_0^5 f(x)\, dx - 5 \cdot 25$$

where $f(x)$ is the demand function.

$$CS = \int_0^5 (30 - x)\, dx - 125$$

$$= 30x - \left. \frac{x^2}{2} \right|_0^5 - 125$$

$$= \left(150 - \frac{25}{2} \right) - 125$$

$$= \frac{25}{2} = 12.50$$

The consumer's surplus is $12.50.

Producer's Surplus

When a product is sold at the equilibrium price, some producers will also benefit, for they would have sold the product at a lower price. The area between the line $p = p_1$ and the supply curve (from $x = 0$ to $x = x_1$) gives the producer's surplus (see Figure 13.23).

If the supply function is $p = g(x)$, the **producer's surplus** is given by the area between the graph of $p = g(x)$ and the x-axis from 0 to x_1 *subtracted from* the area of the rectangle $0x_1Ep_1$.

$$PS = p_1 x_1 - \int_0^{x_1} g(x)\, dx$$

Note that p_1x_1 represents the total revenue at the equilibrium point.

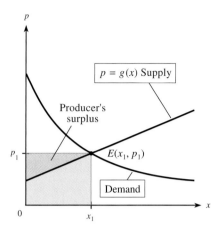

Figure 13.23

● **EXAMPLE 7** *Producer's Surplus*

Suppose that the supply function for x units of a product is $p = x^2 + x$ dollars. If the equilibrium price is $20, what is the producer's surplus?

Solution

Because $p = 20$, we can find the equilibrium quantity x as follows:

$$20 = x^2 + x$$
$$0 = x^2 + x - 20$$
$$0 = (x + 5)(x - 4)$$
$$x = -5, \quad x = 4$$

The equilibrium point is $x = 4$, $p = 20$. The producer's surplus is given by

$$PS = 20 \cdot 4 - \int_0^4 (x^2 + x)\, dx$$

$$= 80 - \left(\frac{x^3}{3} + \frac{x^2}{2}\right)\Big|_0^4$$

$$= 80 - \left(\frac{64}{3} + 8\right)$$

$$\approx 50.67$$

The producer's surplus is $50.67.
See Figure 13.24.

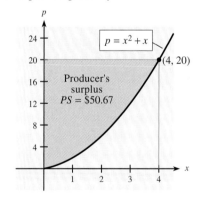

Figure 13.24

● **EXAMPLE 8** *Producer's Surplus*

The demand function for a product is $p = \sqrt{49 - 6x}$ and the supply function is $p = x + 1$. Find the producer's surplus.

Solution

We found the equilibrium point for these functions to be (4, 5) in Example 5 (see Figure 13.22 earlier in this section). The producer's surplus is

$$PS = 5 \cdot 4 - \int_0^4 (x + 1) \, dx = 20 - \left(\frac{x^2}{2} + x \right) \Big|_0^4$$

$$= 20 - (8 + 4) = 8$$

The producer's surplus is $8.

● **Checkpoint**

2. Suppose that for a certain product, the supply function is $p = f(x)$, the demand function is $p = g(x)$, and the equilibrium point is (x_1, p_1). Decide whether the following are true or false.

   (a) $CS = \int_0^{x_1} f(x) \, dx - p_1 x_1$ (b) $PS = \int_0^{x_1} f(x) \, dx - p_1 x_1$

3. If demand is $p = \dfrac{100}{x + 1}$, supply is $p = x + 1$, and the market equilibrium is (9, 10), create the integral used to find the

   (a) consumer's surplus.   (b) producer's surplus.

● **EXAMPLE 9 Consumer's Surplus and Producer's Surplus**

Suppose that for x units of a certain product, the demand function is $p = 200e^{-0.01x}$ dollars and the supply function is $p = \sqrt{200x + 49}$ dollars.

(a) Use a calculator or computer to find the market equilibrium point.
(b) Find the consumer's surplus.
(c) Find the producer's surplus.

Solution

(a) Solving $200e^{-0.01x} = \sqrt{200x + 49}$ requires solving

$$40{,}000e^{-0.02x} = 200x + 49$$

which is very difficult using algebraic techniques. Using SOLVER or INTERSECT gives $x = 60$, to the nearest unit, with a price of $109.76. (See Figure 13.25(a).)

(b) The consumer's surplus, shown in Figure 13.25(b), is

$$\int_0^{60} 200e^{-0.01x} \, dx - 109.76(60) = [-20{,}000e^{-0.01x}]_0^{60} - 6585.60$$

$$= -20{,}000e^{-0.6} + 20{,}000 - 6585.60$$

$$\approx 2438.17 \text{ dollars}$$

(c) The producer's surplus, shown in Figure 13.25(b), is

$$60(109.76) - \int_0^{60} \sqrt{200x + 49} \, dx$$

$$= 6585.60 - \frac{1}{200} \left[\frac{(200x + 49)^{3/2}}{3/2} \right]_0^{60}$$

$$= 6585.60 - \frac{1}{300} [(12{,}049^{3/2} - 49^{3/2})]$$

$$\approx 2178.10 \text{ dollars}$$

Note that we also could have evaluated these definite integrals with the numerical integration feature of a graphing utility, and we would have obtained the same results.

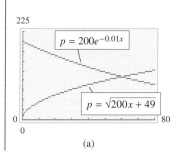

Figure 13.25 (a)   (b)

● **Checkpoint Solutions**

1. (a) $\int_0^5 5000e^{-0.01t}\, dt$

(b) $\int_0^5 (5000e^{-0.01t})(e^{-0.07t})\, dt = \int_0^5 5000e^{-0.08t}\, dt$

(c) $e^{(0.07)(5)} \int_0^5 (5000e^{-0.01t})(e^{-0.07t})\, dt = e^{0.35} \int_0^5 5000e^{-0.08t}\, dt$

2. (a) False. Consumer's surplus uses the demand function, so

$$CS = \int_0^{x_1} g(x)\, dx - p_1 x_1$$

(b) False. Producer's surplus uses the supply function, but the formula is

$$PS = p_1 x_1 - \int_0^{x_1} f(x)\, dx$$

3. (a) $CS = \int_0^9 \dfrac{100}{x+1}\, dx - 90$ (b) $PS = 90 - \int_0^9 (x+1)\, dx$

13.4 Exercises

CONTINUOUS INCOME STREAMS

1. Find the total income over the next 10 years from a continuous income stream that has an annual rate of flow at time t given by $f(t) = 12,000$ (dollars per year).

2. Find the total income over the next 8 years from a continuous income stream with an annual rate of flow at time t given by $f(t) = 8500$ (dollars per year).

3. Suppose that a steel company views the production of its continuous caster as a continuous income stream with a monthly rate of flow at time t given by

$$f(t) = 24,000e^{0.03t} \quad \text{(dollars per month)}$$

Find the total income from this caster in the first year.

4. Suppose that the Quick-Fix Car Service franchise finds that the income generated by its stores can be modeled by assuming that the income is a continuous stream with a monthly rate of flow at time t given by

$$f(t) = 10,000e^{0.02t} \quad \text{(dollars per month)}$$

Find the total income from a Quick-Fix store for the first 2 years of operation.

5. A small brewery considers the output of its bottling machine as a continuous income stream with an annual rate of flow at time t given by

$$f(t) = 80e^{-0.1t}$$

in thousands of dollars per year. Find the income from this stream for the next 10 years.

6. A company that services a number of vending machines considers its income as a continuous stream with an annual rate of flow at time t given by

$$f(t) = 120e^{-0.4t}$$

in thousands of dollars per year. Find the income from this stream over the next 5 years.

7. A franchise models the profit from its store as a continuous income stream with a monthly rate of flow at time t given by

$$f(t) = 3000e^{0.004t} \quad \text{(dollars per month)}$$

When a new store opens, its manager is judged against the model, with special emphasis on the second half of the first year. Find the total profit for the second 6-month period ($t = 6$ to $t = 12$).

8. The Quick-Fix Car Service franchise has a continuous income stream with a monthly rate of flow modeled by $f(t) = 10,000e^{0.02t}$ (dollars per month). Find the total income for years 2 through 5.

9. A continuous income stream has an annual rate of flow at time t given by

$$f(t) = 12,000e^{0.04t} \quad \text{(dollars per year)}$$

If money is worth 8% compounded continuously, find the present value of this stream for the next 8 years.

10. A continuous income stream has an annual rate of flow at time t given by

$$f(t) = 9000e^{0.12t} \quad \text{(dollars per year)}$$

Find the present value of this income stream for the next 10 years, if money is worth 6% compounded continuously.

11. The income from an established chain of laundromats is a continuous stream with its annual rate of flow at time t given by $f(t) = 63,000$ (dollars per year). If money is worth 7% compounded continuously, find the present value and future value of this chain over the next 5 years.

12. The profit from an insurance agency can be considered as a continuous income stream with an annual rate of flow at time t given by $f(t) = 84,000$ (dollars per year). Find the present value and future value of this agency over the next 12 years, if money is worth 8% compounded continuously.

13. Suppose that a printing firm considers the production of its presses as a continuous income stream. If the annual rate of flow at time t is given by

$$f(t) = 97.5e^{-0.2(t+3)}$$

in thousands of dollars per year, and if money is worth 6% compounded continuously, find the present value and future value of the presses over the next 10 years.

14. Suppose that a vending machine company is considering selling some of its machines. Suppose further that the income from these particular machines is a contin-

uous stream with an annual rate of flow at time t given by

$$f(t) = 12e^{-0.4(t+3)}$$

in thousands of dollars per year. Find the present value and future value of the machines over the next 5 years if money is worth 10% compounded continuously.

15. A 58-year-old couple are considering opening a business of their own. They will either purchase an established Gift and Card Shoppe or open a new Video Rental Palace. The Gift Shoppe has a continuous income stream with an annual rate of flow at time t given by

$$G(t) = 30,000 \quad \text{(dollars per year)}$$

and the Video Palace has a continuous income stream with a projected annual rate of flow at time t given by

$$V(t) = 21,600e^{0.08t} \quad \text{(dollars per year)}$$

The initial investment is the same for both businesses, and money is worth 10% compounded continuously. Find the present value of each business over the next 7 years (until the couple reach age 65) to see which is the better buy.

16. If the couple in Problem 15 plan to keep the business until age 70 (for the next 12 years), find each present value to see which business is the better buy in this case.

CONSUMER'S SURPLUS

In Problems 17–26, p and C are in dollars and x is the number of units.

17. The demand function for a product is $p = 34 - x^2$. If the equilibrium price is \$9, what is the consumer's surplus?

18. The demand function for a product is $p = 100 - 4x$. If the equilibrium price is \$40, what is the consumer's surplus?

19. The demand function for a product is $p = 200/(x + 2)$. If the equilibrium quantity is 8 units, what is the consumer's surplus?

20. The demand function for a certain product is $p = 100/(1 + 2x)$. If the equilibrium quantity is 12 units, what is the consumer's surplus?

21. The demand function for a certain product is $p = 81 - x^2$ and the supply function is $p = x^2 + 4x + 11$. Find the equilibrium point and the consumer's surplus there.

22. The demand function for a product is $p = 49 - x^2$ and the supply function is $p = 4x + 4$. Find the equilibrium point and the consumer's surplus there.

23. If the demand function for a product is $p = 12/(x + 1)$ and the supply function is $p = 1 + 0.2x$, find the consumer's surplus under pure competition.

24. If the demand function for a good is $p = 110 - x^2$ and the supply function for it is $p = 2 - \frac{6}{5}x + \frac{1}{5}x^2$, find the consumer's surplus under pure competition.

25. A monopoly has a total cost function $C = 1000 + 120x + 6x^2$ for its product, which has demand function $p = 360 - 3x - 2x^2$. Find the consumer's surplus at the point where the monopoly has maximum profit.

26. A monopoly has a total cost function $C = 500 + 2x^2 + 10x$ for its product, which has demand function $p = -\frac{1}{3}x^2 - 2x + 30$. Find the consumer's surplus at the point where the monopoly has maximum profit.

PRODUCER'S SURPLUS

In Problems 27–36, p is in dollars and x is the number of units.

27. Suppose that the supply function for a good is $p = 4x^2 + 2x + 2$. If the equilibrium price is $422, what is the producer's surplus there?

28. Suppose that the supply function for a good is $p = 0.1x^2 + 3x + 20$. If the equilibrium price is $36, what is the producer's surplus there?

29. If the supply function for a commodity is $p = 10e^{x/3}$, what is the producer's surplus when 15 units are sold?

30. If the supply function for a commodity is $p = 40 + 100(x + 1)^2$, what is the producer's surplus at $x = 20$?

31. Find the producer's surplus for a product if its demand function is $p = 81 - x^2$ and its supply function is $p = x^2 + 4x + 11$.

32. Find the producer's surplus for a product if its demand function is $p = 49 - x^2$ and its supply function is $p = 4x + 4$.

33. Find the producer's surplus for a product with demand function $p = 12/(x + 1)$ and supply function $p = 1 + 0.2x$.

34. Find the producer's surplus for a product with demand function $p = 110 - x^2$ and supply function $p = 2 - \frac{6}{5}x + \frac{1}{5}x^2$.

35. The demand function for a certain product is $p = 144 - 2x^2$ and the supply function is $p = x^2 + 33x + 48$. Find the producer's surplus at the equilibrium point.

36. The demand function for a product is $p = 280 - 4x - x^2$ and the supply function for it is $p = 160 + 4x + x^2$. Find the producer's surplus at the equilibrium point.

13.5

Using Tables of Integrals

OBJECTIVE

● To use tables of integrals to evaluate certain integrals

◑ **Application Preview**

With data from the International Telecommunications Union for 1990 through 2003, the rate of change of total market revenue for worldwide telecommunications (in billions of dollars per year) can be modeled by

$$\frac{dR}{dt} = 27.534(1.0869^t)$$

where $t = 0$ represents 1985 and $R = \$508$ billion in 1990. Finding the function for total market revenue or the predicted revenue for 2010 (when $t = 25$) requires evaluating

$$\int 27.534(1.0869^t)\, dt$$

Evaluating this integral is made easier by using a formula such as those given in Table 13.2 on the next page. (See Example 5.)

The formulas in this table, and others listed in other resources, such as the Chemical Rubber Company's *Standard Mathematical Tables*, extend the number of integrals that can be evaluated. Using the formulas is not quite as easy as it may sound because finding the correct formula and using it properly may present problems. The examples in this section illustrate how some of these formulas are used.

TABLE 13.2 Integration Formulas

1. $\displaystyle\int u^n\, du = \frac{u^{n+1}}{n+1} + C, \quad \text{for } n \neq -1$

2. $\displaystyle\int \frac{du}{u} = \int u^{-1}\, du = \ln|u| + C$

3. $\displaystyle\int a^u\, du = a^u \log_a e + C = \frac{a^u}{\ln a} + C$

4. $\displaystyle\int e^u\, du = e^u + C$

5. $\displaystyle\int \frac{du}{a^2 - u^2} = \frac{1}{2a} \ln\left|\frac{a+u}{a-u}\right| + C$

6. $\displaystyle\int \sqrt{u^2 + a^2}\, du = \frac{1}{2}\left(u\sqrt{u^2+a^2} + a^2 \ln|u + \sqrt{u^2+a^2}|\right) + C$

7. $\displaystyle\int \sqrt{u^2 - a^2}\, du = \frac{1}{2}\left(u\sqrt{u^2-a^2} - a^2 \ln|u + \sqrt{u^2-a^2}|\right) + C$

8. $\displaystyle\int \frac{du}{\sqrt{u^2 + a^2}} = \ln|u + \sqrt{u^2+a^2}| + C$

9. $\displaystyle\int \frac{du}{u\sqrt{a^2 - u^2}} = -\frac{1}{a} \ln\left|\frac{a + \sqrt{a^2 - u^2}}{u}\right| + C$

10. $\displaystyle\int \frac{du}{\sqrt{u^2 - a^2}} = \ln|u + \sqrt{u^2-a^2}| + C$

11. $\displaystyle\int \frac{du}{u\sqrt{a^2 + u^2}} = -\frac{1}{a} \ln\left|\frac{a + \sqrt{a^2 + u^2}}{u}\right| + C$

12. $\displaystyle\int \frac{u\, du}{au + b} = \frac{u}{a} - \frac{b}{a^2} \ln|au + b| + C$

13. $\displaystyle\int \frac{du}{u(au + b)} = \frac{1}{b} \ln\left|\frac{u}{au + b}\right| + C$

14. $\displaystyle\int \ln u\, du = u(\ln u - 1) + C$

15. $\displaystyle\int \frac{u\, du}{(au + b)^2} = \frac{1}{a^2}\left(\ln|au + b| + \frac{b}{au + b}\right) + C$

16. $\displaystyle\int u\sqrt{au + b}\, du = \frac{2(3au - 2b)(au + b)^{3/2}}{15a^2} + C$

● **EXAMPLE 1** *Using Formulas*

Evaluate $\displaystyle\int \frac{dx}{\sqrt{x^2 + 4}}$.

Solution

We must find a formula in Table 13.2 that is of the same form as this integral. We see that formula 8 has the desired form, *if* we let $u = x$ and $a = 2$. Thus

$$\int \frac{dx}{\sqrt{x^2 + 4}} = \ln|x + \sqrt{x^2 + 4}| + C$$

● **EXAMPLE 2** *Fitting a Formula*

Evaluate $\displaystyle\int_1^2 \frac{dx}{x^2 + 2x}$.

Solution

There does not appear to be any formula having exactly the same form as our integral. But if we rewrite our integral as

$$\int_1^2 \frac{dx}{x(x + 2)}$$

we see that formula 13 will work. Letting $u = x$, $a = 1$, and $b = 2$, we get

$$\int_1^2 \frac{dx}{x(x + 2)} = \frac{1}{2} \ln \left| \frac{x}{x + 2} \right| \Big|_1^2 = \frac{1}{2} \ln \left| \frac{2}{4} \right| - \frac{1}{2} \ln \left| \frac{1}{3} \right|$$

$$= \frac{1}{2} \left(\ln \frac{1}{2} - \ln \frac{1}{3} \right)$$

$$= \frac{1}{2} \ln \frac{3}{2} \approx 0.2027$$

Although the formulas in Table 13.2 are given in terms of the variable u, they may be used with any variable.

● **EXAMPLE 3** *Using Formulas*

Evaluate (a) $\displaystyle\int \frac{dq}{9 - q^2}$ and (b) $\displaystyle\int \ln (2x + 1) \, dx$.

Solution

(a) The formula that applies in this case is formula 5, with $a = 3$ and $u = q$. Then

$$\int \frac{dq}{9 - q^2} = \frac{1}{2 \cdot 3} \ln \left| \frac{3 + q}{3 - q} \right| + C = \frac{1}{6} \ln \left| \frac{3 + q}{3 - q} \right| + C$$

(b) This integral has the form of formula 14, with $u = 2x + 1$. But if $u = 2x + 1$, du must be represented by the differential of $2x + 1$ (that is, $2 \, dx$). Thus

$$\int \ln (2x + 1) \, dx = \frac{1}{2} \int \ln (2x + 1)(2 \, dx)$$

$$= \frac{1}{2} \int \ln (u) \, du = \frac{1}{2} u [\ln (u) - 1] + C$$

$$= \frac{1}{2} (2x + 1) [\ln (2x + 1) - 1] + C$$

● *Checkpoint*

1. Can both $\displaystyle\int \frac{dx}{\sqrt{x^2 - 4}}$ and $-\displaystyle\int \frac{dx}{\sqrt{4 - x^2}}$ be evaluated with formula 10 in Table 13.2?

2. Determine the formula used to evaluate $\displaystyle\int \frac{3x}{4x - 5} \, dx$, and show how the formula would be applied.

3. True or false: In order for us to use a formula, the given integral must correspond exactly to the formula, including du.

4. True or false: $\displaystyle\int \frac{dx}{x^2(3x^2 - 7)}$ can be evaluated with formula 13.

5. True or false: $\displaystyle\int \frac{dx}{(6x + 1)^2}$ can be evaluated with either formula 1 or formula 15.

6. True or false: $\int \sqrt{x^2 + 4}\, dx$ can be evaluated with formula 1, formula 6, or formula 16.

● **EXAMPLE 4 *Fitting a Formula***

Evaluate $\displaystyle\int_1^2 \frac{dx}{x\sqrt{81 - 9x^2}}$.

Solution
This integral is similar to that of formula 9 in Table 13.2. Letting $a = 9$, letting $u = 3x$, and multiplying the numerator and denominator by 3 gives the proper form.

$$\int_1^2 \frac{dx}{x\sqrt{81 - 9x^2}} = \int_1^2 \frac{3\,dx}{3x\sqrt{81 - 9x^2}}$$

$$= -\frac{1}{9} \ln \left| \frac{9 + \sqrt{81 - 9x^2}}{3x} \right| \Big|_1^2$$

$$= \left[-\frac{1}{9} \ln \left(\frac{9 + \sqrt{45}}{6} \right) \right] - \left[-\frac{1}{9} \ln \left(\frac{9 + \sqrt{72}}{3} \right) \right]$$

$$\approx 0.088924836$$

Remember that the formulas given in Table 13.2 represent only a very small sample of all possible integration formulas. Additional formulas may be found in books of mathematical tables.

◗ **EXAMPLE 5 *Worldwide Telecommunications Revenue (Application Preview)***

With data from 1990 through 2003, the rate of change of total market revenue for worldwide telecommunications (in billions of dollars per year) can be modeled by

$$\frac{dR}{dt} = 27.534(1.0869^t)$$

where $t = 0$ represents 1985 and $R = \$508$ billion in 1990. (*Source:* International Telecommunications Union)

(a) Find the function that models the total market revenue for worldwide telecommunications.
(b) Find the predicted total market revenue for 2010.

Solution
(a) To find $R(t)$ we integrate dR/dt by using formula 3, with $a = 1.0869$ and $u = t$.

$$R(t) = \int 27.534(1.0869^t)\, dt = 27.534 \int (1.0869^t)\, dt = 27.534\frac{(1.0869^t)}{\ln(1.0869)} + C$$

$$R(t) \approx 330(1.0869^t) + C$$

We use the fact that $R = \$508$ billion in 1990 (when $t = 5$) to find the value of C.

$$508 = 330(1.0869^5) + C \Rightarrow 508 = 500.6 + C \Rightarrow C = 7.4$$

Thus $R(t) = 330(1.0869^t) + 7.4$.

(b) The predicted revenue in 2010 (when $t = 25$) is

$$R(25) = 330(1.0869^{25}) + 7.4 \approx 2657 \text{ billion dollars}$$

Calculator A B C
Note

Numerical integration with a graphing calculator is especially useful in evaluating definite integrals when the formulas for the integrals are difficult to use or are not available. For example, evaluating the definite integral in Example 4 above with the numerical integration feature of a graphing calculator gives 0.08892484. The decimal approximation of the answer found in Example 4 is 0.088924836, so the numerical approximation of the answer agrees for the first eight decimal places. ■

● **Checkpoint Solutions**

1. No. Although $\displaystyle\int \frac{dx}{\sqrt{x^2 - 4}}$ can be evaluated with formula 10 from Table 13.2, $\displaystyle -\int \frac{dx}{\sqrt{4 - x^2}}$ cannot, because $\sqrt{4 - x^2}$ cannot be rewritten in the form $\sqrt{u^2 - a^2}$ as is needed for this formula to be used.

2. Use formula 12 with $u = x$, $a = 4$, $b = -5$, and $du = dx$.

$$\int \frac{3x}{4x - 5}\, dx = 3 \int \frac{x\, dx}{4x - 5} = 3 \int \frac{u\, du}{au + b} = 3\left(\frac{u}{a} - \frac{b}{a^2} \ln |au + b| + C \right)$$

$$= \frac{3x}{4} + \frac{15}{16} \ln |4x - 5| + C$$

3. True. An exact correspondence with the formula and du is necessary.

4. False. With $u = x^2$, we must have $du = 2x\, dx$. In this problem there is no x with dx, so the problem cannot correspond to formula 13.

5. False. The integral can be evaluated with formula 1, but not with formula 15. The correspondence is $u = 6x + 1$, $du = 6\, dx$, and $n = -2$.

6. False. The integral can be evaluated only with formula 6, with $u = x$, $du = dx$, and $a = 2$, but not with formula 1 or formula 16.

13.5 Exercises

Evaluate the integrals in Problems 1–32. Identify the formula used.

1. $\displaystyle\int \frac{dx}{16 - x^2}$

2. $\displaystyle\int \frac{dx}{x(3x + 5)}$

3. $\displaystyle\int_1^4 \frac{dx}{x\sqrt{9 + x^2}}$

4. $\displaystyle\int \frac{dx}{x\sqrt{9 - x^2}}$

5. $\displaystyle\int \ln w\, dw$

6. $\displaystyle\int 4(3^x)\, dx$

7. $\displaystyle\int_0^2 \frac{q\, dq}{6q + 9}$

8. $\displaystyle\int_1^5 \frac{dq}{q\sqrt{25 + q^2}}$

9. $\displaystyle\int \frac{dv}{v(3v + 8)}$

10. $\displaystyle\int_0^3 \sqrt{x^2 + 16}\, dx$

11. $\displaystyle\int_5^7 \sqrt{x^2 - 25}\, dx$

12. $\displaystyle\int \frac{x\, dx}{(3x + 2)^2}$

13. $\displaystyle\int w\sqrt{4w + 5}\, dw$

14. $\displaystyle\int \frac{dy}{\sqrt{9 + y^2}}$

15. $\displaystyle\int x\,5^{x^2}\,dx$

16. $\displaystyle\int \sqrt{9x^2 + 4}\,dx$

17. $\displaystyle\int_0^3 x\sqrt{x^2 + 4}\,dx$

18. $\displaystyle\int x\sqrt{x^4 - 36}\,dx$

19. $\displaystyle\int \frac{5\,dx}{x\sqrt{4 - 9x^2}}$

20. $\displaystyle\int x\,e^{x^2}\,dx$

21. $\displaystyle\int \frac{dx}{\sqrt{9x^2 - 4}}$

22. $\displaystyle\int \frac{dx}{25 - 4x^2}$

23. $\displaystyle\int \frac{3x\,dx}{(2x - 5)^2}$

24. $\displaystyle\int_0^1 \frac{x\,dx}{6 - 5x}$

25. $\displaystyle\int \frac{dx}{\sqrt{(3x + 1)^2 + 1}}$

26. $\displaystyle\int \frac{dx}{9 - (2x + 3)^2}$

27. $\displaystyle\int_0^3 x\sqrt{(x^2 + 1)^2 + 9}\,dx$

28. $\displaystyle\int_1^e x\ln x^2\,dx$

29. $\displaystyle\int \frac{x\,dx}{7 - 3x^2}$

30. $\displaystyle\int_0^1 \frac{e^x}{1 + e^x}\,dx$

31. $\displaystyle\int \frac{dx}{\sqrt{4x^2 + 7}}$

32. $\displaystyle\int e^{2x}\sqrt{3e^x + 1}\,dx$

 Use formulas or numerical integration with a graphing calculator or computer to evaluate the definite integrals in Problems 33–36.

33. $\displaystyle\int_2^3 \frac{e^{\sqrt{x-1}}}{\sqrt{x - 1}}\,dx$

34. $\displaystyle\int_2^4 \frac{3x}{\sqrt{x^4 - 9}}\,dx$

35. $\displaystyle\int_0^1 \frac{x^3\,dx}{(4x^2 + 5)^2}$

36. $\displaystyle\int_0^1 (e^x + 1)^3 e^x\,dx$

APPLICATIONS

37. *Producer's surplus* If the supply function for x units of a commodity is $p = 40 + 100\ln(x + 1)^2$ dollars, what is the producer's surplus at $x = 20$?

38. *Consumer's surplus* If the demand function for wheat is $p = \dfrac{1500}{\sqrt{x^2 + 1}} + 4$ dollars, where x is the number of hundreds of bushels of wheat, what is the consumer's surplus at $x = 7$, $p = 216.13$?

39. *Cost* (a) If the marginal cost for x units of a good is $\overline{MC} = \sqrt{x^2 + 9}$ (dollars per unit) and if the fixed cost is $300, what is the total cost function of the good?
(b) What is the total cost of producing 4 units of this good?

40. *Consumer's surplus* Suppose that the demand function for an appliance is
$$p = \frac{400q + 400}{(q + 2)^2}$$
where q is the number of units and p is in dollars. What is the consumer's surplus if the equilibrium price is $19 and the equilibrium quantity is 18?

41. *Income streams* Suppose that when a new oil well is opened, its production is viewed as a continuous income stream with monthly rate of flow
$$f(t) = 10\ln(t + 1) - 0.1t$$
where t is time in months and $f(t)$ is in thousands of dollars per month. Find the total income over the next 10 years (120 months).

42. *Spread of disease* An isolated community of 1000 people susceptible to a certain disease is exposed when one member returns carrying the disease. If x represents the number infected with the disease at time t (in days), then the rate of change of x is proportional to the product of the number infected, x, and the number still susceptible, $1000 - x$. That is,
$$\frac{dx}{dt} = kx(1000 - x) \quad\text{or}\quad \frac{dx}{x(1000 - x)} = k\,dt$$
(a) If $k = 0.001$, integrate both sides to solve this differential equation.
(b) Find how long it will be before half the population of the community is affected.
(c) Find the rate of new cases, dx/dt, every other day for the first 13 days.

Integration by Parts

- To evaluate integrals using the method of integration by parts

◗ Application Preview

If the value of oil produced by a piece of oil extraction equipment is considered a continuous income stream with an annual rate of flow (in dollars per year) at time t in years given by

$$f(t) = 300{,}000 - 2500t, \qquad 0 \le t \le 10$$

and if money can be invested at 8%, compounded continuously, then the present value of the piece of equipment is

$$\int_0^{10} (300{,}000 - 2500t)e^{-0.08t}\, dt$$

$$= 300{,}000 \int_0^{10} e^{-0.08t}\, dt - 2500 \int_0^{10} te^{-0.08t}\, dt$$

The first integral can be evaluated with the formula for the integral of $e^u\, du$. Evaluating the second integral can be done by using **integration by parts,** which is a special technique that involves rewriting an integral in a form that can be evaluated. (See Example 6.)

Integration by parts is an integration technique that uses a formula that follows from the Product Rule for derivatives (actually differentials) as follows:

$$\frac{d}{dx}(uv) = u\frac{dv}{dx} + v\frac{du}{dx} \quad \text{so} \quad d(uv) = u\, dv + v\, du$$

Rearranging the differential form and integrating both sides give the following.

$$u\, dv = d(uv) - v\, du$$
$$\int u\, dv = \int d(uv) - \int v\, du$$
$$\int u\, dv = uv - \int v\, du$$

Integration by Parts

$$\int u\, dv = uv - \int v\, du$$

Integration by parts is very useful if the integral we seek to evaluate can be treated as the product of one function, u, and the differential dv of a second function, so that the two integrals $\int dv$ and $\int v\, du$ can be found. Let us consider an example using this method.

● EXAMPLE 1 *Integration by Parts*

Evaluate $\int xe^x\, dx$.

Solution

We cannot evaluate this integral using methods we have learned. But we can "split" the integrand into two parts, setting one part equal to u and the other part equal to dv. This "split" must be done in such a way that $\int dv$ and $\int v\, du$ can be evaluated. Letting $u = x$ and letting $dv = e^x\, dx$ are possible choices. If we make these choices, we have

$$u = x \qquad dv = e^x\, dx$$
$$du = 1\, dx \qquad v = \int e^x\, dx = e^x$$

Then

$$\int xe^x \, dx = uv - \int v \, du$$

$$= xe^x - \int e^x \, dx$$

$$= xe^x - e^x + C$$

We see that choosing $u = x$ and $dv = e^x \, dx$ worked in evaluating $\int xe^x \, dx$ in Example 1. If we had chosen $u = e^x$ and $dv = x \, dx$, the results would not have been so successful.

How can we select u and dv to make integration by parts work? As a general guideline, we first identify the types of functions occurring in the problem in the order

Logarithm,   Polynomial (or Power of x),   Radical,   Exponential*

Thus, in Example 1, we had x and e^x, a polynomial and an exponential. Then we choose u to equal the function whose type occurs first on the list; hence in Example 1 we chose $u = x$. Then dv equals the rest of the integrand (and always includes dx) so that $u \, dv$ equals the original integrand. A helpful way to remember the order of the function types that help us choose u is the sentence

"Lazy People Rarely Excel."

in which the first letters, LPRE, coordinate with the order and types of functions. Consider the following examples.

● EXAMPLE 2 *Integration by Parts*

Evaluate $\int x \ln x \, dx$.

Solution
The integral contains a logarithm ($\ln x$) and a polynomial (x). Thus, let $u = \ln x$ and $dv = x \, dx$. Then

$$du = \frac{1}{x} \, dx \qquad \text{and} \qquad v = \frac{x^2}{2}$$

so

$$\int x \ln x \, dx = u \cdot v - \int v \, du$$

$$= (\ln x) \frac{x^2}{2} - \int \frac{x^2}{2} \cdot \frac{1}{x} \, dx$$

$$= \frac{x^2}{2} \ln x - \int \frac{x}{2} \, dx$$

$$= \frac{x^2}{2} \ln x - \frac{x^2}{4} + C$$

Note that letting $dv = \ln x \, dx$ is contrary to our guidelines and would lead to great difficulty in evaluating $\int dv$ and $\int v \, du$.

---

*This order is related to the ease with which the function types can be integrated.

● **EXAMPLE 3** *Integration by Parts*

Evaluate $\int \ln x^2 \, dx$.

Solution

The only function in this problem is a logarithm, the first function type on our list for choosing u. Thus

$$u = \ln x^2 \qquad\qquad dv = dx$$

$$du = \frac{2x}{x^2} \, dx = \frac{2}{x} \, dx \qquad v = x$$

Then

$$\int \ln x^2 \, dx = x \ln x^2 - \int x \cdot \frac{2}{x} \, dx$$

$$= x \ln x^2 - 2x + C$$

Note that if we write $\ln x^2$ as $2 \ln x$, we can also evaluate this integral using formula 14 in Table 13.2, so integration by parts would not be needed.

● **Checkpoint**

1. True or false: In evaluating $\int u \, dv$ by parts,
   (a) the parts u and dv are selected and the parts du and v are calculated.
   (b) the differential (often dx) is always chosen as part of dv.
   (c) the parts du and v are found from u and dv as follows:

$$du = u' \, dx \qquad \text{and} \qquad v = \int dv$$

   (d) For $\int \dfrac{3x}{e^{2x}} \, dx$, we could choose $u = 3x$ and $dv = e^{2x} \, dx$.

2. For $\int \dfrac{\ln x}{x^4} \, dx$,
   (a) identify u and dv.
   (b) find du and v.
   (c) complete the evaluation of the integral.

Sometimes it is necessary to repeat integration by parts to complete the evaluation. When this occurs, at each use of integration by parts it is important to choose u and dv consistently according to our guidelines.

● **EXAMPLE 4** *Repeated Integration by Parts*

Evaluate $\int x^2 e^{2x} \, dx$.

Solution

Let $u = x^2$ and $dv = e^{2x} \, dx$, so $du = 2x \, dx$ and $v = \frac{1}{2}e^{2x}$. Then

$$\int x^2 e^{2x} \, dx = \frac{1}{2}x^2 e^{2x} - \int xe^{2x} \, dx$$

We cannot evaluate $\int xe^{2x}\,dx$ directly, but this new integral is simpler than the original, and a second integration by parts will be successful. Letting $u = x$ and $dv = e^{2x}\,dx$ gives $du = dx$ and $v = \frac{1}{2}e^{2x}$. Thus

$$\int x^2 e^{2x}\,dx = \frac{1}{2}x^2 e^{2x} - \left(\frac{1}{2}xe^{2x} - \int \frac{1}{2}e^{2x}\,dx\right)$$

$$= \frac{1}{2}x^2 e^{2x} - \frac{1}{2}xe^{2x} + \frac{1}{4}e^{2x} + C$$

$$= \frac{1}{4}e^{2x}(2x^2 - 2x + 1) + C$$

The most obvious choices for u and dv are not always the correct ones, as the following example shows. Integration by parts may still involve some trial and error.

● **EXAMPLE 5** *A Tricky Integration by Parts*

Evaluate $\int x^3 \sqrt{x^2 + 1}\,dx$.

Solution
Here we have a polynomial (x^3) and a radical. However, if we choose $u = x^3$, then the resulting $dv = \sqrt{x^2 + 1}\,dx$ cannot be integrated easily. But we can use $\sqrt{x^2 + 1}$ as part of dv, and we can evaluate $\int dv$ if we let $dv = x\sqrt{x^2 + 1}\,dx$. Then

$$u = x^2 \qquad dv = (x^2 + 1)^{1/2}x\,dx$$

$$du = 2x\,dx \qquad v = \int (x^2 + 1)^{1/2}(x\,dx) = \frac{1}{2}\int (x^2 + 1)^{1/2}(2x\,dx)$$

$$v = \frac{1}{2}\frac{(x^2 + 1)^{3/2}}{3/2} = \frac{1}{3}(x^2 + 1)^{3/2}$$

Then $\displaystyle\int x^3 \sqrt{x^2 + 1}\,dx = \frac{x^2}{3}(x^2 + 1)^{3/2} - \int \frac{1}{3}(x^2 + 1)^{3/2}(2x\,dx)$

$$= \frac{x^2}{3}(x^2 + 1)^{3/2} - \frac{1}{3}\frac{(x^2 + 1)^{5/2}}{5/2} + C$$

$$= \frac{x^2}{3}(x^2 + 1)^{3/2} - \frac{2}{15}(x^2 + 1)^{5/2} + C$$

◑ **EXAMPLE 6** *Income Stream* **(Application Preview)**

Suppose that the value of oil produced by a piece of oil extraction equipment is considered a continuous income stream with an annual rate of flow (in dollars per year) at time t in years given by

$$f(t) = 300{,}000 - 2500t, \qquad 0 \le t \le 10$$

and that money is worth 8%, compounded continuously. Find the present value of the piece of equipment.

Solution

The present value of the piece of equipment is given by

$$\int_0^{10} (300,000 - 2500t)e^{-0.08t}\, dt$$

$$= 300,000 \int_0^{10} e^{-0.08t}\, dt - 2500 \int_0^{10} te^{-0.08t}\, dt$$

$$= \frac{300,000}{-0.08} e^{-0.08t}\Big|_0^{10} - 2500 \int_0^{10} te^{-0.08t}\, dt$$

The value of the first integral is

$$\frac{300,000}{-0.08} e^{-0.08t}\Big|_0^{10} = \frac{300,000}{-0.08} e^{-0.8} - \frac{300,000}{-0.08}$$

$$\approx -1,684,983.615 + 3,750,000$$

$$= 2,065,016.385$$

The second of these integrals can be evaluated by using integration by parts, with $u = t$ and $dv = e^{-0.08t}\, dt$. Then $du = 1\, dt$ and $v = \dfrac{e^{-0.08t}}{-0.08}$, and this integral is

$$-2500 \int_0^{10} te^{-0.08t}\, dt = -2500\, \frac{te^{-0.08t}}{-0.08}\Big|_0^{10} + 2500 \int_0^{10} \frac{e^{-0.08t}}{-0.08}\, dt$$

$$= \frac{2500}{0.08} te^{-0.08t}\Big|_0^{10} + \frac{2500}{0.0064} e^{-0.08t}\Big|_0^{10}$$

$$= \frac{2500}{0.08} 10e^{-0.8} + \frac{2500}{0.0064} e^{-0.8} - \frac{2500}{0.0064}$$

$$\approx -74,690.572$$

Thus the sum of the integrals is

$$2,065,016.385 + (-74,690.572) = 1,990,325.813$$

so the present value of this piece of equipment is $1,990,325.81.

One further note about integration by parts. It can be very useful on certain types of problems, but not on all types. Don't attempt to use integration by parts when easier methods are available.

Calculator
Note

Using the numerical integration feature of a graphing calculator to evaluate the integral in Example 6 gives the present value of $1,990,325.80, so this answer is the same as that found in Example 6, to the nearest dollar. ■

● *Checkpoint Solutions*

1. (a) True   (b) True   (c) True
   (d) False. The *product* of u and dv must equal the original integrand. Rewrite as

$$\int \frac{3x}{e^{2x}}\, dx = \int 3xe^{-2x}\, dx$$

and then choose $u = 3x$ and $dv = e^{-2x}\, dx$.

2. First we rewrite:

$$\int \frac{\ln x}{x^4}\,dx = \int \ln x\,(x^{-4}\,dx)$$

Then the integral contains a logarithm and a power of x.

(a) $u = \ln x$ and $dv = x^{-4}\,dx$

(b) $du = \dfrac{1}{x}\,dx$ and $v = \displaystyle\int x^{-4}\,dx = \dfrac{x^{-3}}{-3}$

(c) $\displaystyle\int \frac{\ln x}{x^4}\,dx = uv - \int v\,du$

$$= (\ln x)\left(\frac{x^{-3}}{-3}\right) - \int \frac{x^{-3}}{-3}\cdot\frac{1}{x}\,dx$$

$$= \frac{-\ln x}{3x^3} + \frac{1}{3}\int x^{-4}\,dx$$

$$= -\frac{\ln x}{3x^3} + \frac{1}{3}\left(\frac{x^{-3}}{-3}\right) + C$$

$$= -\frac{\ln x}{3x^3} - \frac{1}{9x^3} + C$$

13.6 Exercises

In Problems 1–16, use integration by parts to evaluate the integral.

1. $\displaystyle\int xe^{2x}\,dx$

2. $\displaystyle\int xe^{-x}\,dx$

3. $\displaystyle\int x^2 \ln x\,dx$

4. $\displaystyle\int x^3 \ln x\,dx$

5. $\displaystyle\int_4^6 q\sqrt{q-4}\,dq$

6. $\displaystyle\int_0^1 y(1-y)^{3/2}\,dy$

7. $\displaystyle\int \frac{\ln x}{x^2}\,dx$

8. $\displaystyle\int \frac{\ln(x-1)}{\sqrt{x-1}}\,dx$

9. $\displaystyle\int_1^e \ln x\,dx$

10. $\displaystyle\int \frac{x}{\sqrt{x-3}}\,dx$

11. $\displaystyle\int x \ln(2x-3)\,dx$

12. $\displaystyle\int x \ln(4x)\,dx$

13. $\displaystyle\int q^3\sqrt{q^2-3}\,dq$

14. $\displaystyle\int \frac{x^3}{\sqrt{9-x^2}}\,dx$

15. $\displaystyle\int_0^4 x^3\sqrt{x^2+9}\,dx$

16. $\displaystyle\int \sqrt{x}\,\ln x\,dx$

In Problems 17–24, use integration by parts to evaluate the integral. Note that evaluation may require integration by parts more than once.

17. $\displaystyle\int x^2 e^{-x}\,dx$

18. $\displaystyle\int_0^1 x^2 e^x\,dx$

19. $\displaystyle\int_0^2 x^3 e^{x^2}\,dx$

20. $\displaystyle\int x^3 e^x\,dx$

21. $\displaystyle\int x^3 \ln^2 x\,dx$

22. $\displaystyle\int \frac{x^2}{\sqrt{x-3}}\,dx$

23. $\displaystyle\int e^{2x}\sqrt{e^x+1}\,dx$

24. $\displaystyle\int_1^2 (\ln x)^2\,dx$

In Problems 25–30, match each of the integrals with the formula or method (I–IV) that should be used to evaluate it. Then evaluate the integral.

I. Integration by parts

II. $\displaystyle\int e^u\,du$

III. $\displaystyle\int \frac{du}{u}$

IV. $\displaystyle\int u^n\,du$

25. $\displaystyle\int xe^{x^2}\,dx$

26. $\displaystyle\int \frac{x}{9-4x^2}\,dx$

27. $\displaystyle\int e^x\sqrt{e^x+1}\,dx$

28. $\displaystyle\int 4x^2 e^{x^3}\,dx$

29. $\displaystyle\int_0^4 \frac{t}{e^t}\,dt$

30. $\displaystyle\int x^2\sqrt{x-1}\,dx$

APPLICATIONS

31. *Producer's surplus*   If the supply function for x units of a commodity is $p = 30 + 50 \ln (2x + 1)^2$ dollars, what is the producer's surplus at $x = 30$?

32. *Cost*   If the marginal cost function for x units of a product is $\overline{MC} = 1 + 3 \ln (x + 1)$ dollars per unit, and if the fixed cost is \$100, find the total cost function.

33. *Present value*   Suppose that a machine's production can be considered as a continuous income stream with annual rate of flow at time t given by

$$f(t) = 10{,}000 - 500t \quad \text{(dollars per year)}$$

If money is worth 10%, compounded continuously, find the present value of the machine over the next 5 years.

34. *Present value*   Suppose that the production of a machine used to mine coal is considered as a continuous income stream with annual rate of flow at time t given by

$$f(t) = 280{,}000 - 14{,}000t \quad \text{(dollars per year)}$$

If money is worth 7%, compounded continuously, find the present value of this machine over the next 8 years.

35. *Income distribution*   Suppose the Lorenz curve for the distribution of income of a certain country is given by

$$y = xe^{x-1}$$

Find the Gini coefficient of income.

36. *Income streams*   Suppose the income from an Internet access business is a continuous income stream with annual rate of flow given by

$$f(t) = 100te^{-0.1t}$$

in thousands of dollars per year. Find the total income over the next 10 years.

37. *World tourism*   With data for 1990 to 2002, the total receipts for world tourism (in billions of dollars per year) can be modeled by the function

$$R(t) = -54.846 + 192.7 \ln t$$

where $t = 0$ represents 1985. (*Source:* World Tourism Organization) Find the predicted total receipts for world tourism for the decade from 2005 to 2015.

38. *Life span*   With data for selected years from 1920 to 2001, the life span of individuals in the United States can be modeled by

$$L(x) = 11.307 + 14.229 \ln x$$

where $x = 0$ represents 1900. (*Source:* National Center for Health Statistics) Find the predicted average life span from 2000 to 2015.

13.7   Improper Integrals and Their Applications

OBJECTIVES

● To evaluate improper integrals
● To apply improper integrals to continuous income streams and to probability density functions

Application Preview

We saw in Section 13.4, "Applications of Definite Integrals in Business and Economics," that the present value of a continuous income stream over a fixed number of years can be found by using a definite integral. When this notion is extended to an infinite time interval, the result is called the **capital value** of the income stream and is given by

$$\text{Capital value} = \int_0^\infty f(t)e^{-rt}\, dt$$

where $f(t)$ is the annual rate of flow at time t, and r is the annual interest rate, compounded continuously. This is called an **improper integral.** For example, the capital value of a trust that provides $f(t) = \$10{,}000$ per year indefinitely (when interest is 10% compounded continuously) is found by evaluating an improper integral. (See Example 2.)

Some applications of calculus to statistics or business (such as capital value) involve definite integrals over intervals of infinite length (improper integrals). The area of a region that extends infinitely to the left or right along the x-axis (see Figure 13.26) could be described by an improper integral.

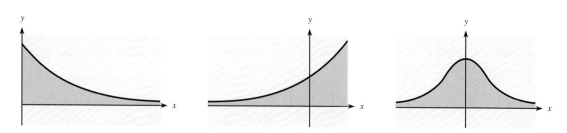

Figure 13.26

To see how to find such an area and hence evaluate an improper integral, let us consider how to find the area between the curve $y = 1/x^2$ and the x-axis to the right of $x = 1$.

To find the area under this curve from $x = 1$ to $x = b$, where b is any number greater than 1 (see Figure 13.27), we evaluate

$$A = \int_1^b \frac{1}{x^2}\,dx = \frac{-1}{x}\bigg|_1^b = \frac{-1}{b} - \left(\frac{-1}{1}\right) = 1 - \frac{1}{b}$$

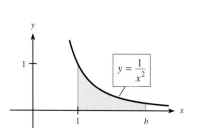

Figure 13.27

Note that the larger b is, the closer the area is to 1. If $b = 100$, $A = 0.99$; if $b = 1000$, $A = 0.999$; and if $b = 1,000,000$, $A = 0.999999$.

We can represent the area of the region under $1/x^2$ to the right of 1 using the notation

$$\lim_{b\to\infty} \int_1^b \frac{1}{x^2}\,dx = \lim_{b\to\infty}\left(1 - \frac{1}{b}\right)$$

where $\lim_{b\to\infty}$ represents the limit as b gets larger without bound. Note that $\frac{1}{b} \to 0$ as $b \to \infty$, so

$$\lim_{b\to\infty}\left(1 - \frac{1}{b}\right) = 1 - 0 = 1$$

Thus the area under the curve $y = 1/x^2$ to the right of $x = 1$ is 1.

In general, we define the area under a curve $y = f(x)$ to the right of $x = a$, with $f(x) \ge 0$, to be

$$\text{Area} = \lim_{b\to\infty}(\text{area from } a \text{ to } b) = \lim_{b\to\infty}\int_a^b f(x)\,dx$$

This motivates the definition that follows.

Improper Integral

$$\int_a^\infty f(x)\,dx = \lim_{b\to\infty}\int_a^b f(x)\,dx$$

If the limit defining the improper integral is a unique finite number, we say that the integral *converges;* otherwise, we say that the integral *diverges.*

● **EXAMPLE 1** *Improper Integrals*

Evaluate the following improper integrals, if they converge.

(a) $\displaystyle\int_1^\infty \frac{1}{x^3}\,dx$ (b) $\displaystyle\int_1^\infty \frac{1}{x}\,dx$

Solution

(a) $\displaystyle\int_1^\infty \frac{1}{x^3}\,dx = \lim_{b\to\infty}\int_1^b \frac{1}{x^3}\,dx = \lim_{b\to\infty}\left[\frac{x^{-2}}{-2}\right]_1^b$

$\displaystyle\qquad\qquad = \lim_{b\to\infty}\left[\frac{-1}{2b^2} - \left(\frac{-1}{2(1)^2}\right)\right]$

$\displaystyle\qquad\qquad = \lim_{b\to\infty}\left(\frac{-1}{2b^2} + \frac{1}{2}\right)$

Now, as $b\to\infty$, $\dfrac{-1}{2b^2}\to 0$, so the limit, and the integral, converge to $\frac{1}{2}$. That is,

$$\int_1^\infty \frac{1}{x^3}\,dx = \frac{1}{2}$$

(b) $\displaystyle\int_1^\infty \frac{1}{x}\,dx = \lim_{b\to\infty}\int_1^b \frac{1}{x}\,dx = \lim_{b\to\infty}\Big[\ln|x|\Big]_1^b$

$\displaystyle\qquad\qquad = \lim_{b\to\infty}(\ln b - \ln 1)$

Now $\ln b$ increases without bound as $b\to\infty$, so the limit, and the integral, diverge. We write this as

$$\int_1^\infty \frac{1}{x}\,dx = \infty$$

From Example 1 we can conclude that the area under the curve $y = 1/x^3$ to the right of $x = 1$ is $\frac{1}{2}$, whereas the corresponding area under the curve $y = 1/x$ is infinite. (We have already seen that the corresponding area under $y = 1/x^2$ is 1.)

As Figure 13.28 shows, the graphs of $y = 1/x^2$ and $y = 1/x$ look similar, but the graph of $y = 1/x^2$ gets "close" to the x-axis much more rapidly than the graph of $y = 1/x$. The area under $y = 1/x$ does not converge to a finite number because as $x \to \infty$ the graph of $1/x$ does not approach the x-axis rapidly enough.

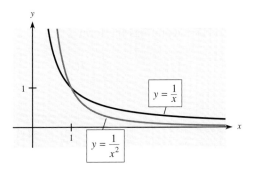

Figure 13.28

◗ EXAMPLE 2 *Capital Value* **(Application Preview)**

Suppose that an organization wants to establish a trust fund that will provide a continuous income stream with an annual rate of flow at time t given by $f(t) = 10{,}000$ dollars per year. If the interest rate remains at 10% compounded continuously, find the capital value of the fund.

Solution

The capital value of the fund is given by

$$\int_0^\infty f(t)e^{-rt}\,dt$$

where $f(t)$ is the annual rate of flow at time t, and r is the annual interest rate, compounded continuously.

$$\int_0^\infty 10{,}000e^{-0.10t}\,dt = \lim_{b\to\infty}\int_0^b 10{,}000e^{-0.10t}\,dt$$

$$= \lim_{b\to\infty}\left[-100{,}000e^{-0.10t}\right]_0^b$$

$$= \lim_{b\to\infty}\left(\frac{-100{,}000}{e^{0.10b}} + 100{,}000\right)$$

$$= 100{,}000$$

Thus the capital value of the fund is $100,000.

Another term for a fund such as the one in Example 2 is a **perpetuity.** Usually the rate of flow of a perpetuity is a constant. If the rate of flow is a constant A, it can be shown that the capital value is given by A/r (see Problem 37 in the 13.7 Exercises).

● **Checkpoint**

1. True or false:

    (a) $\displaystyle\lim_{b\to+\infty}\frac{1}{b^p} = 0$ if $p > 0$ (b) If $p > 0$, $b^p \to \infty$ as $b \to \infty$

    (c) $\displaystyle\lim_{b\to+\infty} e^{-pb} = 0$ if $p > 0$

2. Evaluate the following (if they exist).

    (a) $\displaystyle\int_1^\infty \frac{1}{x^{4/3}}\,dx = \lim_{b\to\infty}\int_1^b x^{-4/3}\,dx$ (b) $\displaystyle\int_0^\infty \frac{dx}{\sqrt{x+1}}$

A second improper integral has the form

$$\int_{-\infty}^b f(x)\,dx$$

and is defined by

$$\int_{-\infty}^b f(x)\,dx = \lim_{a\to\infty}\int_{-a}^b f(x)\,dx$$

The integral converges if the limit is finite. In addition, the improper integral

$$\int_{-\infty}^\infty f(x)\,dx$$

is defined by

$$\int_{-\infty}^\infty f(x)\,dx = \lim_{a\to\infty}\int_{-a}^c f(x)\,dx + \lim_{b\to\infty}\int_c^b f(x)\,dx$$

for any finite constant c. (Often, c is chosen to be 0.) If both limits are finite, the improper integral converges; otherwise, it diverges.

● **EXAMPLE 3** *Improper Integrals*

Evaluate the following integrals.

(a) $\displaystyle\int_{-\infty}^{4} e^{3x}\, dx$ (b) $\displaystyle\int_{-\infty}^{\infty} \frac{x^3}{(x^4 + 3)^2}\, dx$

Solution

(a) $\displaystyle\int_{-\infty}^{4} e^{3x}\, dx = \lim_{a \to \infty} \int_{-a}^{4} e^{3x}\, dx$

$$= \lim_{a \to \infty} \left[\left(\frac{1}{3}\right) e^{3x} \right]_{-a}^{4}$$

$$= \lim_{a \to \infty} \left[\left(\frac{1}{3}\right) e^{12} - \left(\frac{1}{3}\right) e^{-3a} \right]$$

$$= \lim_{a \to \infty} \left[\left(\frac{1}{3}\right) e^{12} - \left(\frac{1}{3}\right)\left(\frac{1}{e^{3a}}\right) \right]$$

$$= \frac{1}{3} e^{12} \qquad \text{(because } 1/e^{3a} \to 0 \text{ as } a \to \infty)$$

(b) $\displaystyle\int_{-\infty}^{\infty} \frac{x^3}{(x^4 + 3)^2}\, dx = \lim_{a \to \infty} \int_{-a}^{0} \frac{x^3}{(x^4 + 3)^2}\, dx + \lim_{b \to \infty} \int_{0}^{b} \frac{x^3}{(x^4 + 3)^2}\, dx$

$$= \lim_{a \to \infty} \left[\frac{1}{4} \frac{(x^4 + 3)^{-1}}{-1} \right]_{-a}^{0} + \lim_{b \to \infty} \left[\frac{1}{4} \frac{(x^4 + 3)^{-1}}{-1} \right]_{0}^{b}$$

$$= \lim_{a \to \infty} \left[-\frac{1}{4}\left(\frac{1}{3} - \frac{1}{a^4 + 3} \right) \right] + \lim_{b \to \infty} \left[-\frac{1}{4}\left(\frac{1}{b^4 + 3} - \frac{1}{3} \right) \right]$$

$$= -\frac{1}{12} + 0 + 0 + \frac{1}{12} = 0$$

$$\left(\text{since } \lim_{a \to \infty} \frac{1}{a^4 + 3} = 0 \quad \text{and} \quad \lim_{b \to \infty} \frac{1}{b^4 + 3} = 0 \right)$$

In Section 13.2, "The Definite Integral: The Fundamental Theorem of Calculus," we calculated the probability that a computer component will last between 3 and 5 years when the probability density function for the life span is $f(x) = 0.10e^{-0.10x}$, where x is the number of years, $x \geq 0$. The probability that the component will last more than 3 years is given by the improper integral

$$\int_{3}^{\infty} 0.10e^{-0.10x}\, dx$$

which gives

$$\lim_{b \to \infty} \int_{3}^{b} 0.10e^{-0.10x}\, dx = \lim_{b \to \infty} \left[-e^{-0.10x} \right]_{3}^{b}$$

$$= \lim_{b \to \infty} \left(-e^{-0.10b} + e^{-0.3} \right)$$

$$= e^{-0.3} \approx 0.7408$$

We noted in Chapter 8, "Further Topics in Probability; Data Description," that the sum of the probabilities for a probability distribution (a probability density function) equals 1. In particular, we stated that the area under the normal probability curve is 1.

Probability Density Function

In general, if $f(x) \geq 0$ for all x, then f is a probability density function for a continuous random variable if and only if

$$\int_{-\infty}^{\infty} f(x)\, dx = 1$$

The complete probability density function for the life span of the computer component mentioned above is

$$f(x) = \begin{cases} 0.10e^{-0.10x} & \text{if } x \geq 0 \\ 0 & \text{if } x < 0 \end{cases}$$

We can verify that

$$\int_{-\infty}^{\infty} f(x)\, dx = 1$$

for this function.

$$\int_{-\infty}^{\infty} f(x)\, dx = \int_{-\infty}^{0} f(x)\, dx + \int_{0}^{\infty} f(x)\, dx = \int_{-\infty}^{0} 0\, dx + \int_{0}^{\infty} 0.10e^{-0.10x}\, dx$$

$$= 0 + \lim_{b \to \infty} \int_{0}^{b} 0.10e^{-0.10x}\, dx$$

$$= \lim_{b \to \infty} \left[-e^{-0.10x} \right]_{0}^{b}$$

$$= \lim_{b \to \infty} \left[-e^{-0.10b} + 1 \right]$$

$$= 1$$

In Section 8.3, "Discrete Probability Distributions; The Binomial Distribution," we found the expected value (mean) of a discrete probability distribution using the formula

$$E(x) = \sum x \Pr(x)$$

For continuous probability distributions, such as the normal probability distribution, the expected value, or mean, can be found by evaluating the improper integral

$$\int_{-\infty}^{\infty} xf(x)\, dx$$

Mean (Expected Value)

If x is a continuous random variable with probability density function f, then the mean of the probability distribution is

$$\mu = \int_{-\infty}^{\infty} xf(x)\, dx$$

The normal distribution density function, in standard form, is

$$f(x) = \frac{1}{\sqrt{2\pi}} e^{-x^2/2}$$

so the mean of the normal probability distribution is given by

$$\mu = \int_{-\infty}^{\infty} x\left(\frac{1}{\sqrt{2\pi}}e^{-x^2/2}\right)dx$$

$$= \lim_{a \to \infty}\int_{-a}^{0}\frac{1}{\sqrt{2\pi}}xe^{-x^2/2}\,dx + \lim_{b \to \infty}\int_{0}^{b}\frac{1}{\sqrt{2\pi}}xe^{-x^2/2}\,dx$$

$$= \lim_{a \to \infty}\frac{1}{\sqrt{2\pi}}\left[-e^{-x^2/2}\right]_{-a}^{0} + \lim_{b \to \infty}\frac{1}{\sqrt{2\pi}}\left[-e^{-x^2/2}\right]_{0}^{b}$$

$$= \frac{1}{\sqrt{2\pi}}(-1 + 0) + \frac{1}{\sqrt{2\pi}}(0 + 1) = 0$$

This verifies the statement in Chapter 8 that the mean of the standard normal distribution is 0.

  ● **EXAMPLE 4** *Area of a Region*

Find the area of the region below the graph of $f(x) = \dfrac{\ln x}{x^2}$ and above the *x*-axis.

This can be done by evaluating the integral over the interval where $f(x)$ is above the *x*-axis. It is not possible to find two points where $f(x) = 0$, so we will use graphs to find the interval and to evaluate the integral.

(a) Graph $y = f(x)$ to find the interval over which to integrate.
(b) Evaluate the integral of $f(x)$ over this interval.

Solution

(a) The graph of $y = f(x)$ is shown in Figure 13.29(a) on the next page. Using ZERO, TABLE, or TRACE shows that $f(1) = 0$ and that $f(x)$ approaches 0 as *x* approaches $+\infty$. Thus the area is found by evaluating

$$\int_{1}^{+\infty} f(x)\,dx = \int_{1}^{+\infty}\frac{\ln x}{x^2}\,dx = \lim_{b \to +\infty}\int_{1}^{b}\frac{\ln x}{x^2}\,dx$$

(b) Evaluating $\displaystyle\int_{1}^{b}\frac{\ln x}{x^2}\,dx$ requires the use of integration by parts, with

$$u = \ln x \quad\text{and}\quad dv = x^{-2}\,dx$$

Thus

$$du = \frac{1}{x}\,dx \quad\text{and}\quad v = \frac{x^{-1}}{-1}$$

$$\int_{1}^{b}\frac{\ln x}{x^2}\,dx = (\ln x)\left(\frac{x^{-1}}{-1}\right)\Bigg|_{1}^{b} - \int_{1}^{b}(-x^{-1})\frac{1}{x}\,dx = -\frac{\ln x}{x}\Bigg|_{1}^{b} + \frac{x^{-1}}{-1}\Bigg|_{1}^{b}$$

$$= \left(-\frac{\ln b}{b} + \frac{\ln 1}{1}\right) - \left(\frac{1}{b} - 1\right) = 1 - \frac{1}{b} - \frac{\ln b}{b}$$

Thus, $\displaystyle\int_{1}^{\infty} f(x)\,dx = \lim_{b \to \infty}\left(1 - \frac{1}{b} - \frac{\ln b}{b}\right) = \lim_{b \to \infty}1 - \lim_{b \to \infty}\frac{1}{b} - \lim_{b \to \infty}\frac{\ln b}{b}.$

We have not developed a method to evaluate $\displaystyle\lim_{b \to \infty}\left(\frac{\ln b}{b}\right)$ or, equivalently, $\displaystyle\lim_{x \to \infty}\left(\frac{\ln x}{x}\right)$.

But we can use TRACE or TABLE to see that the graph of $y = 1 - \dfrac{1}{x} - \dfrac{\ln x}{x}$ approaches 1 as *x* approaches $+\infty$ (see Figure 13.29(b)).

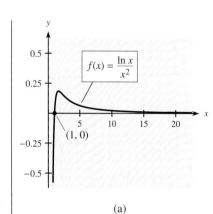
(a)

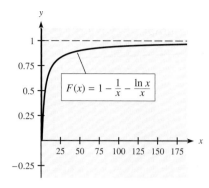
(b)

Figure 13.29

Thus $\lim_{b \to \infty} \left(1 - \dfrac{1}{b} - \dfrac{\ln b}{b} \right) = 1$, and the area under $y = f(x)$ is

$$\int_1^\infty f(x)\, dx = \lim_{b \to \infty} \left(1 - \dfrac{1}{b} - \dfrac{\ln b}{b} \right) = 1.$$

● **Checkpoint Solutions**

1. (a) True   (b) True   (c) True

2. (a) $\lim_{b \to \infty} \int_1^b x^{-4/3}\, dx = \lim_{b \to \infty} \left. \dfrac{x^{-1/3}}{-1/3} \right|_1^b = \lim_{b \to \infty} \left(\dfrac{-3}{b^{1/3}} - \dfrac{-3}{1} \right) = 0 + 3 = 3$

   (b) $\lim_{b \to \infty} \int_0^b (x + 1)^{-1/2}\, dx = \lim_{b \to \infty} \left. \dfrac{(x + 1)^{1/2}}{1/2} \right|_0^b = \lim_{b \to \infty} \left. 2\sqrt{x + 1} \right|_0^b$

   $\qquad\qquad\qquad = \lim_{b \to \infty} (2\sqrt{b + 1} - 2)$

   The limit approaches ∞, so the integral diverges.

13.7 Exercises

In Problems 1–20, evaluate the improper integrals that converge.

1. $\displaystyle\int_1^\infty \dfrac{dx}{x^6}$

2. $\displaystyle\int_1^\infty \dfrac{1}{x^4}\, dx$

3. $\displaystyle\int_1^\infty \dfrac{dt}{t^{3/2}}$

4. $\displaystyle\int_5^\infty \dfrac{dx}{(x - 1)^3}$

5. $\displaystyle\int_1^\infty e^{-x}\, dx$

6. $\displaystyle\int_0^\infty x^2 e^{-x^3}\, dx$

7. $\displaystyle\int_1^\infty \dfrac{dt}{t^{1/3}}$

8. $\displaystyle\int_1^\infty \dfrac{1}{\sqrt{x}}\, dx$

9. $\displaystyle\int_0^\infty e^{3x}\, dx$

10. $\displaystyle\int_1^\infty x e^{x^2}\, dx$

11. $\displaystyle\int_{-\infty}^{-1} \dfrac{10}{x^2}\, dx$

12. $\displaystyle\int_{-\infty}^{-2} \dfrac{x}{\sqrt{x^2 - 1}}\, dx$

13. $\displaystyle\int_{-\infty}^{0} x^2 e^{-x^3}\, dx$

14. $\displaystyle\int_{-\infty}^{0} \dfrac{x}{(x^2 + 1)^2}\, dx$

15. $\displaystyle\int_{-\infty}^{-1} \dfrac{6}{x}\, dx$

16. $\displaystyle\int_{-\infty}^{-2} \dfrac{3x}{x^2 + 1}\, dx$

17. $\displaystyle\int_{-\infty}^{\infty} \dfrac{2x}{x^2 + 1}\, dx$

18. $\displaystyle\int_{-\infty}^{\infty} \dfrac{x}{(x^2 + 1)^2}\, dx$

19. $\displaystyle\int_{-\infty}^{\infty} x^3 e^{-x^4}\, dx$

20. $\displaystyle\int_{-\infty}^{\infty} x^4 e^{-x^5}\, dx$

21. For what value of c does $\displaystyle\int_0^\infty \dfrac{c}{e^{0.5t}}\, dt = 1$?

22. For what value of c does $\displaystyle\int_{10}^\infty \dfrac{c}{x^3}\, dx = 1$?

In Problems 23–26, find the area, if it exists, of the region under the graph of $y = f(x)$ and to the right of $x = 1$.

23. $f(x) = \dfrac{x}{e^{x^2}}$

24. $f(x) = \dfrac{1}{\sqrt[5]{x^3}}$

25. $f(x) = \dfrac{1}{\sqrt[3]{x^5}}$

26. $f(x) = \dfrac{1}{x\sqrt{x}}$

27. Show that the function

$$f(x) = \begin{cases} \dfrac{200}{x^3} & \text{if } x \ge 10 \\ 0 & \text{otherwise} \end{cases}$$

is a probability density function.

28. Show that

$$f(x) = \begin{cases} 3e^{-3t} & \text{if } t \ge 0 \\ 0 & \text{if } t < 0 \end{cases}$$

is a probability density function.

29. For what value of c is the function

$$f(x) = \begin{cases} c/x^2 & \text{if } x \ge 1 \\ 0 & \text{otherwise} \end{cases}$$

a probability density function?

30. For what value of c is the function

$$f(x) = \begin{cases} c/x^3 & \text{if } x \ge 100 \\ 0 & \text{otherwise} \end{cases}$$

a probability density function?

31. If

$$f(x) = \begin{cases} ce^{-x/4} & x \ge 0 \\ 0 & x < 0 \end{cases}$$

is a probability density function, what must be the value of c?

32. If

$$f(x) = \begin{cases} ce^{-kx} & \text{if } x \ge 0 \\ 0 & \text{if } x < 0 \end{cases}$$

is a probability density function, what must be the value of c?

33. Find the mean of the probability distribution if the probability density function is

$$f(x) = \begin{cases} \dfrac{200}{x^3} & \text{if } x \ge 10 \\ 0 & \text{otherwise} \end{cases}$$

34. Find the mean of the probability distribution if the probability density function is

$$f(x) = \begin{cases} c/x^3 & \text{if } x \ge 100 \\ 0 & \text{otherwise} \end{cases}$$

35. Find the area below the graph of $y = f(x)$ and above the x-axis for $f(x) = 24xe^{-3x}$. Use the graph of $y = f(x)$ to find the interval for which $f(x) \ge 0$ and the graph of the integral of $f(x)$ over this interval to find the area.

36. Find the area below the graph of $y = f(x)$ and above the x-axis for $f(x) = x^2e^{-x}$ and $x \ge 0$. Use the graph of the integral of $f(x)$ over this interval to find the area.

APPLICATIONS

37. *Capital value* Suppose that a continuous income stream has an annual rate of flow at time t given by $f(t) = A$, where A is a constant. If the interest rate is r (as a decimal, $r > 0$), compounded continuously, show that the capital value of the stream is A/r.

38. *Capital value* Suppose that a donor wishes to provide a cash gift to a hospital that will generate a continuous income stream with an annual rate of flow at time t given by $f(t) = \$20,000$ per year. If the annual interest rate is 12% compounded continuously, find the capital value of this perpetuity.

39. *Capital value* Suppose that a business provides a continuous income stream with an annual rate of flow at time t given by $f(t) = 120e^{0.04t}$ in thousands of dollars per year. If the interest rate is 9% compounded continuously, find the capital value of the business.

40. *Capital value* Suppose that the output of the machinery in a factory can be considered as a continuous income stream with annual rate of flow at time t given by $f(t) = 450e^{-0.09t}$ in thousands of dollars per year. If the annual interest rate is 6% compounded continuously, find the capital value of the machinery.

41. *Capital value* A business has a continuous income stream with an annual rate of flow at time t given by $f(t) = 56,000e^{0.02t}$ (dollars per year). If the interest rate is 10% compounded continuously, find the capital value of the business.

42. *Capital value* Suppose that a business provides a continuous income stream with an annual rate of flow at time t given by $f(t) = 10,800e^{0.06t}$ (dollars per year). If money is worth 12% compounded continuously, find the capital value of the business.

43. *Repair time* In a manufacturing process involving several machines, the average down time t (in hours) for a machine that needs repair has the probability density function

$$f(t) = 0.5e^{-0.5t} \qquad t \ge 0$$

Find the probability that a failed machine's down time is
(a) 2 hours or more.   (b) 8 hours or more.

44. *Customer service*   The duration t (in minutes) of customer service calls received by a certain company is given by the probability density function

$$f(t) = 0.3e^{-0.3t} \qquad t \geq 0$$

Find the probability that a call selected at random lasts
(a) 4 minutes or more.
(b) 10 minutes or more.

45. *Quality control*   The probability density function for the life span of an electronics part is $f(t) = 0.08e^{-0.08t}$, where t is the number of months in service. Find the probability that any given part of this type lasts longer than 24 months.

46. *Warranties*   A transmission repair firm that wants to offer a lifetime warranty on its repairs has determined that the probability density function for transmission failure after repair is $f(t) = 0.3e^{-0.3t}$, where t is the

number of months after repair. What is the probability that a transmission chosen at random will last
(a) 3 months or less?
(b) more than 3 months?

47. *Radioactive waste*   Suppose that the rate at which a nuclear power plant produces radioactive waste is proportional to the number of years it has been operating, according to $f(t) = 500t$ in pounds per year. Suppose also that the waste decays exponentially at a rate of 3% per year. Then the amount of radioactive waste that will accumulate in b years is given by

$$\int_0^b 500te^{-0.03(b-t)}\,dt$$

(a) Evaluate this integral.
(b) How much waste will accumulate in the long run? Take the limit as $b \to \infty$ in (a).

OBJECTIVES

● To approximate definite integrals by using the Trapezoidal Rule
● To approximate definite integrals by using Simpson's Rule

Numerical Integration Methods: The Trapezoidal Rule and Simpson's Rule

◐ Application Preview

A pharmaceutical company tests the body's assimilation of a new drug by administering a 200-milligram dose and collecting the following data from blood samples (t is time in hours, and $R(t)$ gives the assimilation of the drug in milligrams per hour).

t	0	0.5	1.0	1.5	2.0	2.5	3.0
$R(t)$	0.0	15.3	32.3	51.0	74.8	102.0	130.9

The company would like to find the amount of drug assimilated in 3 hours, which is given by

$$\int_0^3 R(t)\,dt$$

In this section we develop straightforward and quite accurate methods for evaluating integrals such as this, even when, as with $R(t)$, only function data, rather than a formula, are given. (See Example 4.)

We have studied several techniques for integration and have even used tables to evaluate some integrals. Yet some functions that arise in practical problems cannot be integrated by using any formula. For any function $f(x) \geq 0$ on an interval $[a, b]$, however, we have seen that a definite integral can be viewed as an area and that we can usually approximate the area and hence the integral (see Figure 13.30). One such approximation method uses rectangles, as we saw when we defined the definite integral. In this section, we consider two other **numerical integration methods** to approximate a definite integral: the **Trapezoidal Rule** and **Simpson's Rule**.

Trapezoidal Rule   To develop the Trapezoidal Rule formula, we assume that $f(x) \geq 0$ on $[a, b]$ and subdivide the interval $[a, b]$ into n equal pieces, each of length $(b - a)/n = h$. Then, within each subdivision, we can approximate the area by using a trapezoid. As shown in Figure 13.31, we can use the formula for the area of a trapezoid to approximate the area of the first subdivision. Continuing in this way for each trapezoid, we have

$$\int_a^b f(x)\,dx \approx A_1 + A_2 + A_3 + \cdots + A_{n-1} + A_n$$

$$= \left[\frac{f(x_0) + f(x_1)}{2}\right]h + \left[\frac{f(x_1) + f(x_2)}{2}\right]h + \left[\frac{f(x_2) + f(x_3)}{2}\right]h + \cdots + \left[\frac{f(x_{n-1}) + f(x_n)}{2}\right]h$$

$$= \frac{h}{2}\left[f(x_0) + f(x_1) + f(x_1) + f(x_2) + f(x_2) + \cdots + f(x_{n-1}) + f(x_{n-1}) + f(x_n)\right]$$

This can be simplified to obtain the Trapezoidal Rule.

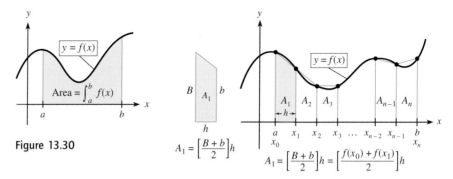

Figure 13.30

Figure 13.31

Trapezoidal Rule   If f is continuous on $[a, b]$, then

$$\int_a^b f(x)\,dx \approx \frac{h}{2}\left[f(x_0) + 2f(x_1) + 2f(x_2) + \cdots + 2f(x_{n-1}) + f(x_n)\right]$$

where $h = \dfrac{b - a}{n}$ and n is the number of subdivisions of $[a, b]$.

Despite the fact that we used areas to develop the Trapezoidal Rule, we can use this rule to evaluate definite integrals even if $f(x) < 0$ on all or part of $[a, b]$.

● **EXAMPLE 1** *Trapezoidal Rule*

Use the Trapezoidal Rule to approximate $\displaystyle\int_1^3 \frac{1}{x}\,dx$ with

(a) $n = 4$.
(b) $n = 8$.

Solution

First, we note that this integral can be evaluated directly:

$$\int_1^3 \frac{1}{x}\, dx = \ln |x| \Big|_1^3 = \ln 3 - \ln 1 = \ln 3 \approx 1.099$$

(a) The interval [1, 3] must be divided into 4 equal subintervals of width $h = \frac{3-1}{4} = \frac{2}{4} = \frac{1}{2}$ as follows:

$$
\begin{array}{ccccc}
1 & 1.5 & 2 & 2.5 & 3 \\
x_0 & x_1 & x_2 & x_3 & x_4
\end{array}
$$

Thus, from the Trapezoidal Rule, we have

$$\int_1^3 \frac{1}{x}\, dx \approx \frac{h}{2}[f(x_0) + 2f(x_1) + 2f(x_2) + 2f(x_3) + f(x_4)]$$

$$= \frac{1/2}{2}[f(1) + 2f(1.5) + 2f(2) + 2f(2.5) + f(3)]$$

$$= \frac{1}{4}\left[1 + 2\left(\frac{1}{1.5}\right) + 2\left(\frac{1}{2}\right) + 2\left(\frac{1}{2.5}\right) + \frac{1}{3}\right]$$

$$\approx 0.25(1 + 1.333 + 1 + 0.8 + 0.3333)$$

$$\approx 1.117$$

(b) In this case, the interval [1, 3] is divided into 8 equal subintervals of width $h = \frac{3-1}{8} = \frac{2}{8} = \frac{1}{4}$ as follows.

$$
\begin{array}{ccccccccc}
1 & 1.25 & 1.5 & 1.75 & 2 & 2.25 & 2.5 & 2.75 & 3 \\
x_0 & x_1 & x_2 & x_3 & x_4 & x_5 & x_6 & x_7 & x_8
\end{array}
$$

Thus, from the Trapezoidal Rule, we have

$$\int_1^3 \frac{1}{x}\, dx \approx \frac{h}{2}[f(x_0) + 2f(x_1) + 2f(x_2) + 2f(x_3) + 2f(x_4) + 2f(x_5)$$

$$+ 2f(x_6) + 2f(x_7) + f(x_8)]$$

$$= \frac{1/4}{2}\left[\frac{1}{1} + 2\left(\frac{1}{1.25}\right) + 2\left(\frac{1}{1.5}\right) + 2\left(\frac{1}{1.75}\right) + 2\left(\frac{1}{2}\right) + 2\left(\frac{1}{2.25}\right)\right.$$

$$\left. + 2\left(\frac{1}{2.5}\right) + 2\left(\frac{1}{2.75}\right) + \frac{1}{3}\right]$$

$$\approx 1.103$$

In Example 1, because we know that the value of the integral is $\ln 3 \approx 1.099$, we can measure the accuracy of each approximation. We can see that a larger value of n (namely, $n = 8$) produced a more accurate approximation to $\ln 3$. In general, larger values of n produce more accurate approximations, but they also make computations more difficult.

Technology Note The TABLE feature of a graphing calculator can be used to find the values of $f(x_0)$, $f(x_1)$, $f(x_2)$, etc. needed for the Trapezoidal Rule. For instance, in Example 1, for $n = 8$ enter $y_1 = 1/x$. Then set (a) the TABLE feature to start at $x = 1$ (the lower limit on the integral), (b) the ΔTbl $= \dfrac{b-a}{n} = \dfrac{2}{8} = 0.25$, and (c) the TABLE to work automatically. See Figure 13.32.

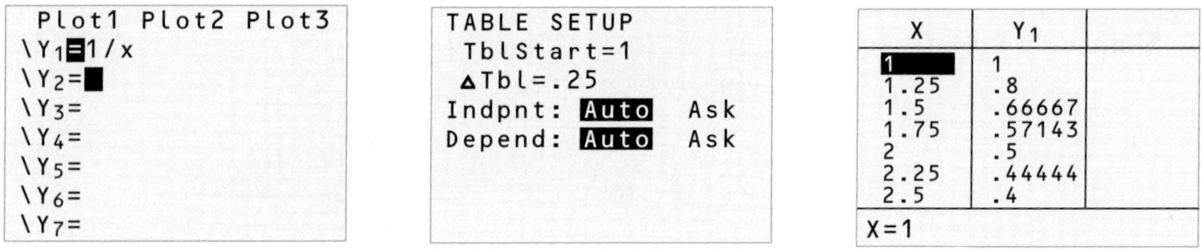

Plot1 Plot2 Plot3
\Y₁◻1/x
\Y₂=■
\Y₃=
\Y₄=
\Y₅=
\Y₆=
\Y₇=

```
TABLE SETUP
 TblStart=1
 ▲Tbl=.25
 Indpnt: Auto  Ask
 Depend: Auto  Ask
```

X	Y₁	
1	1	
1.25	.8	
1.5	.66667	
1.75	.57143	
2	.5	
2.25	.44444	
2.5	.4	
X=1		

Figure 13.32   (a)   (b)   (c)

Note in Figure 13.32 that the TABLE gives values of y_1 only through $x = 2.5$, but scrolling past the last entry gives values for $x = 2.75, 3.0,$ and so on. Thus for the Trapezoidal Rule, we get

$$\int_1^3 \frac{1}{x}\, dx \approx \frac{0.25}{2}\left[\begin{array}{l} 1(1) + 2(0.8) + 2(0.66667) + 2(0.57143) + 2(0.5) \\ \qquad + 2(0.44444) + 2(0.4) + 2(0.36364) + 1(0.33333) \end{array}\right]$$

$$\approx 1.103$$

just as we did in Example 1.

An Excel spreadsheet also can be very useful in finding the numerical integral with the Trapezoidal Rule. A discussion of the use of Excel in finding numerical integrals is given in the *Excel Guide* that accompanies this text.   ■

Because the exact value of an integral is rarely available when an approximation is used, it is important to have some way to judge the accuracy of an answer. The following formula, which we state without proof, can be used to bound the error that results from using the Trapezoidal Rule.

Trapezoidal Rule Error

The error E in using the Trapezoidal Rule to approximate $\int_a^b f(x)\, dx$ satisfies

$$|E| \le \frac{(b-a)^3}{12n^2}\left[\max_{a \le x \le b} |f''(x)|\right]$$

where n is the number of subdivisions of $[a, b]$.

For a numerical method to be worthwhile, there must be some way of assessing its accuracy. Hence, this formula is important. We leave its application, however, to more advanced courses.

Simpson's Rule   The Trapezoidal Rule was developed by using a line segment to approximate the function over each subinterval and then using the areas under the line segments to approximate the area under the curve. Another numerical method, Simpson's Rule, uses a parabola to approximate the function over each pair of subintervals (see Figure 13.33) and then uses the areas under the parabolas to approximate the area under the curve. Because Simpson's Rule is based on pairs of subintervals, n must be even.

Simpson's Rule (*n* Even)

If $f(x)$ is continuous on $[a, b]$, and if $[a, b]$ is divided into an *even* number n of equal subdivisions, then

$$\int_a^b f(x)\, dx \approx \frac{h}{3}\left[f(x_0) + 4f(x_1) + 2f(x_2) + 4f(x_3) + \cdots + 2f(x_{n-2}) + 4f(x_{n-1}) + f(x_n)\right]$$

where $h = \dfrac{b-a}{n}$.

We leave the derivation of Simpson's Rule to more advanced courses.

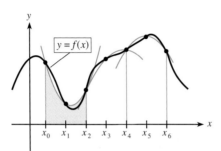

Figure 13.33

● **EXAMPLE 2** *Simpson's Rule*

Use Simpson's Rule with $n = 4$ to approximate $\displaystyle\int_1^3 \frac{1}{x}\, dx$.

Solution

Because $n = 4$ is even, Simpson's Rule can be used, and the interval is divided into four subintervals of length $h = \frac{3-1}{4} = \frac{1}{2}$ as follows:

$$
\begin{array}{ccccc}
1 & 1.5 & 2 & 2.5 & 3 \\
x_0 & x_1 & x_2 & x_3 & x_4
\end{array}
$$

$$
\begin{aligned}
\int_1^3 \frac{1}{x}\, dx &\approx \frac{1/2}{3}\left[f(1) + 4f(1.5) + 2f(2) + 4f(2.5) + f(3)\right] \\
&= \frac{1}{6}\left[\frac{1}{1} + 4\left(\frac{1}{1.5}\right) + 2\left(\frac{1}{2}\right) + 4\left(\frac{1}{2.5}\right) + \frac{1}{3}\right] \\
&= 1.100
\end{aligned}
$$

Note that the result of Example 2 is better than both the $n = 4$ and the $n = 8$ Trapezoidal Rule approximations done in Example 1. In general, Simpson's Rule is more accurate than the Trapezoidal Rule for a given number of subdivisions. We can determine the accuracy of Simpson's Rule approximations by using the following formula.

Simpson's Rule Error Formula

The error E in using Simpson's Rule to approximate $\displaystyle\int_a^b f(x)\, dx$ satisfies

$$|E| \leq \frac{(b-a)^5}{180n^4}\left[\max_{a \leq x \leq b} |f^{(4)}(x)|\right]$$

where n is the number of subdivisions of $[a, b]$.

The presence of the factor $180n^4$ in the denominator indicates that the error will often be quite small for even a modest value of n. Although Simpson's Rule often leads to more accurate results than the Trapezoidal Rule for a fixed choice of n, the Trapezoidal Rule is sometimes used because its error is more easily determined than that of Simpson's Rule or, more importantly, because the number of subdivisions is odd.

● **Checkpoint**

1. Suppose [1, 4] is divided into 6 equal subintervals.
   (a) Find the width h of each subinterval.
   (b) Find the subdivision points.
2. True or false:
   (a) When the Trapezoidal Rule is used, the number of subdivisions, n, must be even.
   (b) When Simpson's Rule is used, the number of subdivisions, n, must be even.
3. We can show that $\int_0^2 4x^3\, dx = 16$. Use $n = 4$ subdivisions to find
   (a) the Trapezoidal Rule approximation of this integral.
   (b) the Simpson's Rule approximation of this integral.

As mentioned previously for the Trapezoidal Rule, the TABLE feature of a graphing calculator can also be used with Simpson's Rule.

● **EXAMPLE 3** *Simpson's Rule*

Use Simpson's Rule with $n = 10$ subdivisions to approximate

$$\int_1^6 (\ln (x))^2\, dx$$

Solution

With $y_1 = (\ln (x))^2$ and $\dfrac{b - a}{n} = \dfrac{6 - 1}{10} = 0.5 = \Delta$ Tbl, we can use a graphing calculator to find the following table of values.

x	y_1	x	y_1
1	0	4	1.9218
1.5	0.1644	4.5	2.2622
2	0.48045	5	2.5903
2.5	0.83959	5.5	2.9062
3	1.2069	6	3.2104
3.5	1.5694		

Thus Simpson's Rule yields

$$\int_1^6 (\ln (x))^2\, dx$$

$$\approx \frac{(6 - 1)/10}{3}\left[\begin{array}{l} 1(0) + 4(0.1644) + 2(0.48045) + 4(0.83959) + 2(1.2069) + 4(1.5694) \\ \quad + 2(1.9218) + 4(2.2622) + 2(2.5903) + 4(2.9062) + 1(3.2104) \end{array}\right]$$

$$\approx 7.7627$$

An Excel spreadsheet also can be very useful with Simpson's Rule.

One advantage that both the Trapezoidal Rule and Simpson's Rule offer is that they may be used when only function values at the subdivision points are known and the function formula itself is not known. This can be especially useful in applied problems.

> ○ **EXAMPLE 4** *Pharmaceutical Testing* **(Application Preview)**
>
> A pharmaceutical company tests the body's ability to assimilate a drug. The test is done by administering a 200-milligram dose and then, every half-hour, monitoring the rate of assimilation. The following table gives the data; t is time in hours, and $R(t)$ is the rate of assimilation in milligrams per hour.
>
> | t | 0 | 0.5 | 1.0 | 1.5 | 2.0 | 2.5 | 3.0 |
> |------|-----|------|------|------|------|-------|-------|
> | $R(t)$ | 0.0 | 15.3 | 32.3 | 51.0 | 74.8 | 102.0 | 130.9 |
>
> To find the total amount of the drug (in milligrams) that is assimilated in the first 3 hours, the company must find
>
> $$\int_0^3 R(t)\, dt$$
>
> Use Simpson's Rule to approximate this definite integral.
>
> **Solution**
> The values of t correspond to the endpoints of the subintervals, and the values of $R(t)$ correspond to function values at those endpoints. From the table we see that $h = \frac{1}{2}$ and $n = 6$ (even); thus Simpson's Rule is applied as follows:
>
> $$\int_0^3 R(t)\, dt \approx \frac{h}{3}\left[R(t_0) + 4R(t_1) + 2R(t_2) + 4R(t_3) + 2R(t_4) + 4R(t_5) + R(t_6)\right]$$
>
> $$= \frac{1}{6}\left[0 + 4(15.3) + 2(32.3) + 4(51.0) + 2(74.8) + 4(102.0) + 130.9\right]$$
>
> $$\approx 169.7 \text{ mg}$$

Thus at the end of 3 hours, the body has assimilated approximately 169.7 mg of the 200-milligram dose.

Note in Example 4 that the Trapezoidal Rule could also be used, because it relies only on the values of $R(t)$ at the subdivision endpoints.

In practice, these kinds of approximations are usually done with a computer, where the computations can be done quickly, even for large values of n. In addition, numerical methods (and hence computer programs) exist that approximate the errors, even in cases such as Example 4 where the function is not known. We leave any further discussion of error approximation formulas and additional numerical techniques for a more advanced course.

● **Checkpoint Solutions**

1. (a) $h = \dfrac{b - a}{n} = \dfrac{4 - 1}{6} = \dfrac{1}{2} = 0.5$

   (b) $x_0 = 1, x_1 = 1.5, x_2 = 2, x_3 = 2.5, x_4 = 3, x_5 = 3.5, x_6 = 4$

2. (a) False. With the Trapezoidal Rule, n can be even or odd.

   (b) True

3. (a) $\dfrac{0.5}{2}[0 + 2(0.5) + 2(4) + 2(13.5) + 32] = 17$

   (b) $\dfrac{0.5}{3}[0 + 4(0.5) + 2(4) + 4(13.5) + 32] = 16$

13.8 Exercises

For each interval $[a, b]$ and value of n given in Problems 1–6, find h and the values of $x_0, x_1, \ldots x_n$.

1. $[0, 2]$ $n = 4$
2. $[0, 4]$ $n = 8$
3. $[1, 4]$ $n = 6$
4. $[2, 5]$ $n = 9$
5. $[-1, 4]$ $n = 5$
6. $[-1, 2]$ $n = 6$

For each integral in Problems 7–12, do the following.
(a) Approximate its value by using the Trapezoidal Rule.
(b) Approximate its value by using Simpson's Rule.
(c) Find its exact value by integration.
(d) State which approximation is more accurate. (Round each result to two decimal places.)

7. $\int_0^3 x^2 \, dx; \quad n = 6$
8. $\int_0^1 x^3 \, dx; \quad n = 4$
9. $\int_1^2 \frac{1}{x^2} \, dx; \quad n = 4$
10. $\int_1^4 \frac{1}{x} \, dx; \quad n = 6$
11. $\int_0^4 x^{1/2} \, dx; \quad n = 8$
12. $\int_0^2 x^{3/2} \, dx; \quad n = 8$

In Problems 13–18, approximate each integral by
(a) the Trapezoidal Rule.
(b) Simpson's Rule.
Use $n = 4$ and round answers to three decimal places.

13. $\int_0^2 \sqrt{x^3 + 1} \, dx$
14. $\int_0^2 \frac{dx}{\sqrt{4x^3 + 1}}$
15. $\int_0^1 e^{-x^2} \, dx$
16. $\int_0^1 e^{x^2} \, dx$
17. $\int_1^5 \ln (x^2 - x + 1) \, dx$
18. $\int_1^5 \ln (x^2 + x + 2) \, dx$

Use the table of values given in each of Problems 19–22 to approximate $\int_a^b f(x) \, dx$. Use Simpson's Rule whenever n is even; otherwise, use the Trapezoidal Rule. Round answers to one decimal place.

19. Find $\int_1^4 f(x) \, dx.$

x	$f(x)$
1	1
1.6	2.2
2.2	1.8
2.8	2.9
3.4	4.6
4.0	2.1

20. Find $\int_1^2 f(x) \, dx.$

x	$f(x)$
1	1
1.2	0.5
1.4	0.3
1.6	0.1
1.8	0.8
2.0	0.1

21. Find $\int_{1.2}^{3.6} f(x) \, dx.$

x	$f(x)$
1.2	6.1
1.6	4.8
2.0	3.1
2.4	2.0
2.8	2.8
3.2	5.6
3.6	9.7

22. Find $\int_0^{1.8} f(x) \, dx.$

x	$f(x)$
0	8.8
0.3	4.6
0.6	1.5
0.9	0
1.2	0.7
1.5	2.8
1.8	7.6

APPLICATIONS

In Problems 23–30, round all calculations to two decimal places.

23. *Total income* Suppose that the production from an assembly line can be considered as a continuous income stream with annual rate of flow given by

$$f(t) = 100 \frac{e^{0.1t}}{t + 1} \quad \text{(in thousands of dollars per year)}$$

Use Simpson's Rule with $n = 4$ to approximate the total income over the first 2 years, given by

$$\text{Total income} = \int_0^2 100 \frac{e^{0.1t}}{t + 1} \, dt$$

24. *Present value* Suppose that the rate of flow of a continuous income stream is given by $f(t) = 500t$ (in thousands of dollars per year). If money is worth 7% compounded continuously, then the present value of this stream over the next 5 years is given by

$$\text{Present value} = \int_0^5 500t \, e^{-0.07t} \, dt$$

Use the Trapezoidal Rule with $n = 5$ to approximate this present value.

25. *Cost* Suppose that a company's total cost (in dollars) of producing x items is given by $C(x) = (x^2 + 1)^{3/2} + 1000$. Use the Trapezoidal Rule with $n = 3$ to approximate the average cost for the production of $x = 30$ to $x = 33$ items.

26. *Demand* Suppose that the demand for q units of a certain product at $\$p$ per unit is given by

$$p = 850 + \frac{100}{q^2 + 1}$$

Use Simpson's Rule with $n = 6$ to approximate the average price as demand ranges from 3 to 9 items.

Supply and demand Use the following supply and demand schedules in Problems 27 and 28, with p in dollars and x as the number of units.

Supply Schedule		Demand Schedule	
x	p	x	p
0	120	0	2400
10	260	10	1500
20	380	20	1200
30	450	30	950
40	540	40	800
50	630	50	730
60	680	60	680
70	720	70	640

27. Use Simpson's Rule to approximate the producer's surplus at market equilibrium. Note that market equilibrium can be found from the tables.

28. Use Simpson's Rule to approximate the consumer's surplus at market equilibrium.

29. *Production* Suppose that the rate of production of a product (in units per week) is measured at the end of each of the first 5 weeks after start-up, and the following data are obtained.

Weeks t	Rate $R(t)$	Weeks t	Rate $R(t)$
0	250.0	3	243.3
1	247.6	4	241.3
2	245.4	5	239.5

Approximate the total number of units produced in the first 5 weeks.

30. *Drug levels in the blood* The manufacturer of a medicine wants to test how a new 300-milligram capsule is released into the bloodstream. After a volunteer is given a capsule, blood samples are drawn every half-hour, and the number of milligrams of the drug in the bloodstream is calculated. The results obtained are as follows.

Time t (hr)	$N(t)$ (mg)	Time t (hr)	$N(t)$ (mg)
0	0	2.0	178.3
0.5	247.3	2.5	113.9
1.0	270	3.0	56.2
1.5	236.4	3.5	19.3

Approximate the *average* number of milligrams in the bloodstream during the first $3\frac{1}{2}$ hours.

Income distribution The following income distribution data define points on Lorenz curves, where x represents the fraction of the U.S. population and $L(x)$ the cumulative percent of income held by fraction x. In Problems 31 and 32, use these data and a numerical method to evaluate $2 \int_0^1 [x - L(x)] \, dx$, and hence to find the Gini coefficient of income.

		x				
	0.0	0.2	0.4	0.6	0.8	1.0
$L(x)$ for Blacks						
2003	0	2.9	11.1	25.8	49.8	100
1990	0	3.1	11.1	26.0	51.1	100
$L(x)$ for Asians						
2003	0	2.7	11.7	27.6	52.1	100
2002	0	3.3	12.4	27.7	51.3	100

Source: U.S. Bureau of the Census

31. Find the Gini coefficient of income for black households in 1990 and 2003. In which year was income for Blacks more equally distributed?

32. Find the Gini coefficient of income for Asian households in 2002 and 2003. In which year was income for Asians more equally distributed?

33. *Pollution monitoring* Suppose that the presence of phosphates in certain waste products dumped into a lake promotes the growth of algae. Rampant growth of algae affects the oxygen supply in the water, so an environmental group wishes to estimate the area of algae growth. Group members measure across the algae growth (see Figure 13.34) and obtain the following data (in feet).

x	Width w	x	Width w
0	0	50	27
10	15	60	24
20	18	70	23
30	18	80	0
40	30		

Use Simpson's Rule to approximate the area of the algae growth.

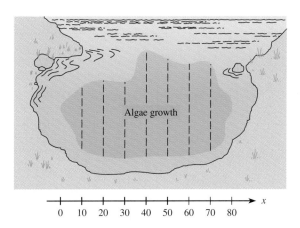

Figure 13.34

34. *Development costs*   A land developer is planning to dig a small lake and build a group of homes around it. To estimate the cost of the project, the area of the lake must be calculated from the proposed measurements (in feet) given in Figure 13.35 and in the following table.

x	Width $w(x)$
0	0
100	300
200	200
300	400
400	0

Use Simpson's Rule to approximate the area of the lake.

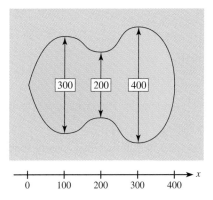

Figure 13.35

Key Terms and Formulas

Section	Key Terms	Formulas
13.1	Sigma notation	$\displaystyle\sum_{i=1}^{n} 1 = n; \quad \sum_{i=1}^{n} i = \frac{n(n+1)}{2}$ $\displaystyle\sum_{i=1}^{n} i^2 = \frac{n(n+1)(2n+1)}{6}$
	Area	
	Right-hand endpoints	$\displaystyle\lim_{n\to\infty} \sum_{i=1}^{n} f(x_i)\frac{b-a}{n}$
	Left-hand endpoints	$\displaystyle\lim_{n\to\infty} \sum_{i=1}^{n} f(x_{i-1})\frac{b-a}{n}$
13.2	Riemann sum	$\displaystyle\sum_{i=1}^{n} f(x_i^*)\Delta x_i$
	Definite integral	$\displaystyle\int_a^b f(x)\,dx = \lim_{\substack{\max\Delta x_i\to 0 \\ (n\to\infty)}} \sum_{i=1}^{n} f(x_i^*)\,\Delta x_i$
	Fundamental Theorem of Calculus	$\displaystyle\int_a^b f(x)\,dx = F(b) - F(a)$, where $F'(x) = f(x)$

Section	Key Terms	Formulas
13.2	Definite integral properties	$\displaystyle\int_a^a f(x)\,dx = 0$
		$\displaystyle\int_b^a f(x)\,dx = -\int_a^b f(x)\,dx$
		$\displaystyle\int_a^b [f(x) \pm g(x)]\,dx = \int_a^b f(x)\,dx \pm \int_a^b g(x)\,dx$
		$\displaystyle\int_a^b kf(x)\,dx = k\int_a^b f(x)\,dx$
		$\displaystyle\int_a^c f(x)\,dx + \int_c^b f(x)\,dx = \int_a^b f(x)\,dx$
	Area under $f(x)$, where $f(x) \geq 0$	$\displaystyle A = \int_a^b f(x)\,dx$
	$f(x)$ is a probability density function	$\displaystyle \Pr(a \leq x \leq b) = \int_a^b f(x)\,dx$
13.3	Area between $f(x)$ and $g(x)$, where $f(x) \geq g(x)$	$\displaystyle A = \int_a^b [f(x) - g(x)]\,dx$
	Average value of f over $[a, b]$	$\displaystyle \frac{1}{b-a}\int_a^b f(x)\,dx$
	Lorenz curve	
	Gini coefficient	$\displaystyle 2\int_0^1 [x - L(x)]\,dx$
13.4	Continuous income streams	
	Total income	$\displaystyle\int_0^k f(t)\,dt$ (for k years)
	Present value	$\displaystyle\int_0^k f(t)e^{-rt}\,dt$, where r is the interest rate
	Future value	$\displaystyle e^{rk}\int_0^k f(t)e^{-rt}\,dt$
	Consumer's surplus [demand is $f(x)$]	$\displaystyle CS = \int_0^{x_1} f(x)\,dx - p_1 x_1$
	Producer's surplus [supply is $g(x)$]	$\displaystyle PS = p_1 x_1 - \int_0^{x_1} g(x)\,dx$
13.5	Integration by formulas	See Table 13.2.
13.6	Integration by parts	$\displaystyle\int u\,dv = uv - \int v\,du$

Section	Key Terms	Formulas
13.7	Improper integrals	$\displaystyle\int_a^\infty f(x)\,dx = \lim_{b\to\infty}\int_a^b f(x)\,dx$
		$\displaystyle\int_{-\infty}^b f(x)\,dx = \lim_{a\to\infty}\int_{-a}^b f(x)\,dx$
		$\displaystyle\int_{-\infty}^\infty f(x)\,dx = \int_{-\infty}^c f(x)\,dx + \int_c^\infty f(x)\,dx$
	Capital value of a continuous income stream	$\displaystyle\int_0^\infty f(t)e^{-rt}\,dt$
	Probability density function, $f(x)$	$f(x) \geq 0$ and $\displaystyle\int_{-\infty}^\infty f(x)\,dx = 1$
	Mean	$\displaystyle\mu = \int_{-\infty}^\infty xf(x)\,dx$
13.8	Trapezoidal Rule for $\displaystyle\int_a^b f(x)\,dx$	$\displaystyle\approx \frac{h}{2}\left[f(x_0) + 2f(x_1) + \cdots + 2f(x_{n-1}) + f(x_n)\right]$ where $h = \dfrac{b-a}{n}$
	Error formula	$\displaystyle|E| \leq \frac{(b-a)^3}{12n^2}\left[\max_{a\leq x\leq b}|f''(x)|\right]$
	Simpson's Rule for $\displaystyle\int_a^b f(x)\,dx$	$\displaystyle\approx \frac{h}{3}\left[f(x_0) + 4f(x_1) + 2f(x_2) + 4f(x_3)\right.$ $\left. + \cdots + 2f(x_{n-2}) + 4f(x_{n-1}) + f(x_n)\right],$ where n is even and $h = \dfrac{b-a}{n}$
	Error formula	$\displaystyle|E| \leq \frac{(b-a)^5}{180n^4}\left[\max_{a\leq x\leq b}|f^{(4)}(x)|\right]$

Review Exercises

Section 13.1

1. Calculate $\displaystyle\sum_{k=1}^8 (k^2 + 1)$.

2. Use formulas to simplify

$$\sum_{i=1}^n \frac{3i}{n^3}$$

3. Use 6 subintervals of the same size to approximate the area under the graph of $y = 3x^2$ from $x = 0$ to $x = 1$. Use the right-hand endpoints of the subintervals to find the heights of the rectangles.

4. Use rectangles to find the area under the graph of $y = 3x^2$ from $x = 0$ to $x = 1$. Use n equal subintervals.

Section 13.2

5. Use a definite integral to find the area under the graph of $y = 3x^2$ from $x = 0$ to $x = 1$.

6. Find the area between the graph of $y = x^3 - 4x + 5$ and the x-axis from $x = 1$ to $x = 3$.

Evaluate the integrals in Problems 7–18.

7. $\displaystyle\int_1^4 4\sqrt{x^3}\,dx$

8. $\displaystyle\int_{-3}^2 (x^3 - 3x^2 + 4x + 2)\,dx$

9. $\displaystyle\int_0^5 (x^3 + 4x)\,dx$

10. $\displaystyle\int_{-1}^3 (3x + 4)^{-2}\,dx$

11. $\displaystyle\int_{-3}^{-1} (x + 1)\,dx$

12. $\displaystyle\int_2^3 \frac{x^2}{2x^3 - 7}\,dx$

13. $\int_{-1}^{2} (x^2 + x)\, dx$

14. $\int_{1}^{4} \left(\frac{1}{x} + \sqrt{x}\right) dx$

15. $\int_{0}^{4} (2x + 1)^{1/2}\, dx$

16. $\int_{0}^{1} \frac{x}{x^2 + 1}\, dx$

17. $\int_{0}^{1} e^{-2x}\, dx$

18. $\int_{0}^{1} xe^{x^2}\, dx$

Section 13.3

Find the area between the curves in Problems 19–22.

19. $y = x^2 - 3x + 2$ and $y = x^2 + 4$ from $x = 0$ to $x = 5$
20. $y = x^2$ and $y = 4x + 5$
21. $y = x^3$ and $y = x$ from $x = -1$ to $x = 0$
22. $y = x^3 - 1$ and $y = x - 1$

Section 13.5

Evaluate the integrals in Problems 23–26, using the formulas in Table 13.2.

23. $\int \sqrt{x^2 - 4}\, dx$

24. $\int_{0}^{1} 3^x\, dx$

25. $\int x \ln x^2\, dx$

26. $\int \frac{dx}{x(3x + 2)}$

Section 13.6

In Problems 27–30, use integration by parts to evaluate.

27. $\int x^5 \ln x\, dx$

28. $\int xe^{-2x}\, dx$

29. $\int \frac{x\, dx}{\sqrt{x + 5}}$

30. $\int_{1}^{e} \ln x\, dx$

Section 13.7

Evaluate the improper integrals in Problems 31–34.

31. $\int_{1}^{\infty} \frac{1}{x}\, dx$

32. $\int_{-\infty}^{-1} \frac{200}{x^3}\, dx$

33. $\int_{0}^{\infty} 5e^{-3x}\, dx$

34. $\int_{-\infty}^{0} \frac{x}{(x^2 + 1)^2}\, dx$

Section 13.8

35. Evaluate $\int_{1}^{3} \frac{2}{x^3}\, dx$

   (a) exactly.
   (b) by using the Trapezoidal Rule with $n = 4$ (to three decimal places).
   (c) by using Simpson's Rule with $n = 4$ (to three decimal places).

36. Use the Trapezoidal Rule with $n = 5$ to approximate

$$\int_{0}^{1} \frac{4\, dx}{x^2 + 1}$$

   Round your answer to three decimal places.

37. Use the table that follows to approximate

$$\int_{1}^{2.2} f(x)\, dx$$

   by using Simpson's Rule. Round your answer to 1 decimal place.

x	$f(x)$
1	0
1.3	2.8
1.6	5.1
1.9	4.2
2.2	0.6

38. Suppose that a definite integral is to be approximated and it is found that to achieve a specified accuracy, n must satisfy $n \geq 4.8$. What is the smallest n that can be used, if
   (a) the Trapezoidal Rule is used?
   (b) Simpson's Rule is used?

APPLICATIONS

Section 13.2

39. *Maintenance* Maintenance costs for buildings increase as the buildings age. If the rate of increase in maintenance costs for a building is

$$M'(t) = \frac{14{,}000}{\sqrt{t + 16}}$$

   where M is in dollars and t is time in years, $0 \leq t \leq 15$, find the total maintenance cost for the first 9 years ($t = 0$ to $t = 9$).

40. *Quality control* Suppose the probability density function for the life expectancy of a "disposable" telephone is

$$f(x) = \begin{cases} 1.4e^{-1.4x} & x \geq 0 \\ 0 & x < 0 \end{cases}$$

   Find the probability that the telephone lasts 2 years or less.

Section 13.3

41. *Savings* The future value of $1000 invested in a savings account at 10%, compounded continuously, is $S = 1000e^{0.1t}$, where t is in years. Find the average amount in the savings account during the first 5 years.

42. *Income streams*   Suppose the total income in dollars from a video machine is given by

$$I = 50e^{0.2t}, \quad 0 \le t \le 4, t \text{ in hours}$$

Find the average income over this 4-hour period.

43. *Income distribution*   In 1969, after "The Great Society" initiatives of the Johnson administration, the Lorenz curve for the U.S. income distribution was $L(x) = x^{2.1936}$. In 2000, after the stock market's historic 10-year growth, the Lorenz curve for the U.S. income distribution was $L(x) = x^{2.4870}$. Find the Gini coefficient of income for both years, and determine in which year income was more equally distributed.

Section 13.4

44. *Consumer's surplus*   The demand function for a product under pure competition is $p = \sqrt{64 - 4x}$, and the supply function is $p = x - 1$, where x is the number of units and p is in dollars.
    (a) Find the market equilibrium.
    (b) Find the consumer's surplus at market equilibrium.

45. *Producer's surplus*   Find the producer's surplus at market equilibrium for Problem 44.

46. *Income streams*   Find the total income over the next 10 years from a continuous income stream that has an annual flow rate at time t given by $f(t) = 125e^{0.05t}$ in thousands of dollars per year.

47. *Income streams*   Suppose that a machine's production is considered a continuous income stream with an annual rate of flow at time t given by $f(t) = 150e^{-0.2t}$ in thousands of dollars per year. Money is worth 8%, compounded continuously.
    (a) Find the present value of the machine's production over the next 5 years.
    (b) Find the future value of the production 5 years from now.

Section 13.5

48. *Average cost*   Suppose the cost function for x units of a product is given by $C(x) = \sqrt{40,000 + x^2}$ dollars. Find the average cost over the first 150 units.

49. *Producer's surplus*   Suppose the supply function for x units of a certain lamp is given by

$$p = 0.02x + 50.01 - \frac{10}{\sqrt{x^2 + 1}}$$

where p is in dollars. Find the producer's surplus if the equilibrium price is $70 and the equilibrium quantity is 1000.

Section 13.6

50. *Income streams*   Suppose the present value of a continuous income stream over the next 5 years is given by

$$P = 9000 \int_0^5 te^{-0.08t} \, dt, \quad P \text{ in dollars, } t \text{ in years}$$

Find the present value.

51. *Cost*   If the marginal cost for x units of a product is $\overline{MC} = 3 + 60(x + 1) \ln (x + 1)$ dollars per unit and if the fixed cost is $2000, find the total cost function.

Section 13.7

52. *Quality control*   Find the probability that a telephone lasts more than 1 year if the probability density function for its life expectancy is given by

$$f(x) = \begin{cases} 1.4e^{-1.4x} & x \ge 0 \\ 0 & x < 0 \end{cases}$$

53. *Capital value*   Find the capital value of a business if its income is considered a continuous income stream with annual rate of flow given by

$$f(t) = 120e^{0.03t}$$

in thousands of dollars per year, and the current interest rate is 6% compounded continuously.

Section 13.8

54. *Total income*   Suppose that a continuous income stream has an annual rate of flow $f(t) = 100e^{-0.01t^2}$ (in thousands of dollars per year). Use Simpson's Rule with $n = 4$ to approximate the total income from this stream over the next 2 years.

55. *Revenue*   A company has the following data from the sale of its product.

x	$\overline{MR}$
0	0
2	480
4	720
6	720
8	480
10	0

If x represents hundreds of units and revenue R is in hundreds of dollars, approximate the total revenue from the sale of 1000 units by approximating

$$\int_0^{10} \overline{MR} \, dx$$

with the Trapezoidal Rule.

13 Chapter Test

1. Use left-hand endpoints and $n = 4$ subdivisions to approximate the area under $f(x) = \sqrt{4 - x^2}$ on the interval $[0, 2]$.

2. Consider $f(x) = 5 - 2x$ from $x = 0$ to $x = 1$ with n equal subdivisions.
   (a) If $f(x)$ is evaluated at right-hand endpoints, find a formula for the sum, S, of the areas of the n rectangles.
   (b) Find $\lim\limits_{n \to \infty} S$.

3. Express the area in Quadrant I under $y = 12 + 4x - x^2$ (shaded in the figure) as an integral. Then evaluate the integral to find the area.

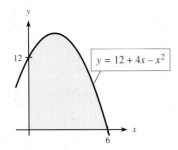

$y = 12 + 4x - x^2$

4. Evaluate the following definite integrals. (Do not approximate with technology.)
   (a) $\displaystyle\int_0^4 (9 - 4x)\, dx$ (b) $\displaystyle\int_0^3 x(8x^2 + 9)^{-1/2}\, dx$
   (c) $\displaystyle\int_1^4 \frac{5}{4x - 1}\, dx$ (d) $\displaystyle\int_1^\infty \frac{7}{x^2}\, dx$
   (e) $\displaystyle\int_4^4 \sqrt{x^3 + 10}\, dx$ (f) $\displaystyle\int_0^1 5x^2 e^{2x^3}\, dx$

5. Use integration by parts to evaluate the following.
   (a) $\displaystyle\int 3xe^x\, dx$ (b) $\displaystyle\int x \ln(2x)\, dx$

6. If $\displaystyle\int_1^4 f(x)\, dx = 3$ and $\displaystyle\int_3^4 f(x)\, dx = 7$, find $\displaystyle\int_1^3 2f(x)\, dx$.

7. Use Table 13.2 in Section 13.5 to evaluate each of the following.
   (a) $\displaystyle\int \ln(2x)\, dx$ (b) $\displaystyle\int x\sqrt{3x - 7}\, dx$

8. Use the numerical integration feature of a graphing utility to approximate $\displaystyle\int_1^4 \sqrt{x^3 + 10}\, dx$.

9. Suppose the supply function for a product is $p = 40 + 0.001x^2$ and the demand function is $p = 120 - 0.2x$, where x is the number of units and p is the price in dollars. If the market equilibrium price is $80, find (a) the consumer's surplus and (b) the producer's surplus.

10. Suppose a continuous income stream has an annual rate of flow $f(t) = 85e^{-0.01t}$, in thousands of dollars per year, and the current interest rate is 7% compounded continuously.
    (a) Find the total income over the next 12 years.
    (b) Find the present value over the next 12 years.
    (c) Find the capital value of the stream.

11. Find the area between $y = 2x + 4$ and $y = x^2 - x$.

12. The figure shows typical supply and demand curves. On the figure, sketch and shade the region whose area represents the consumer's surplus.

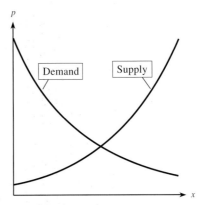

13. In an effort to make the distribution of income more nearly equal, the government of a country passes a tax law that changes the Lorenz curve from $y = 0.998x^{2.6}$ for one year to $y = 0.57x^2 + 0.43x$ for the next year. Find the Gini coefficient of income for each year and determine whether the distribution of income is more or less equitable after the tax law is passed. Interpret the result.

14. With data from the U.S. Department of Energy, the number of thousands of barrels of oil per day imported from the Persian Gulf each year from 1985 to 2000 can be modeled by the function

$$f(t) = 3.722t^3 - 145.3t^2 + 1836.8t - 5735.2$$

where t is the number of years past 1980 (*Monthly Energy Review*, April 2001).

(a) Find the average number of barrels imported per day from 1985 to 2000 (that is, from $t = 5$ to $t = 20$).

(b) Find the average number of barrels imported per day during the 1990s (from $t = 10$ to $t = 20$).

15. Use the Trapezoidal Rule to approximate

$$\int_1^3 (x \ln x)\, dx$$

with $n = 4$ subintervals.

16. To measure the amount of pollution in the Clarion River, it is necessary to find the cross-sectional area of the river at a point downriver from the source of the pollution. The distance across the river is 240 feet, and the following table gives depth measurements, $D(x)$, of the river at various distances x from one bank of the river. Use these measurements to estimate the cross-sectional area with Simpson's Rule.

x	0	40	80	120	160	200	240
$D(x)$	0	22	35	48	32	24	0

Extended Applications & Group Projects

I. Retirement Planning

A 52-year-old client asks an accountant how to plan for his future retirement at age 62. He expects income from Social Security in the amount of $16,000 per year and a retirement pension of $30,000 per year from his employer. He wants to make monthly contributions to an investment plan that pays 8%, compounded monthly, for 10 years so that he will have a total income of $62,000 per year for 30 years. What will the size of the monthly contributions have to be to accomplish this goal, if it is assumed that money will be worth 8%, compounded continuously, throughout the period after he is 62?

To help you answer this question, complete the following.

1. How much money must the client withdraw annually from his investment plan during his retirement so that his total income goal is met?
2. How much money S must the client's account contain when he is 62 so that it will generate this annual amount for 30 years? (*Hint: S* can be considered the present value over 30 years of a continuous income stream with the amount you found in Question 1 as its annual rate of flow.)
3. The monthly contribution R that would, after 10 years, amount to the present value S found in Question 2 can be obtained from the formula

$$R = S\left[\frac{i}{(1 + i)^n - 1}\right]$$

where i represents the monthly interest rate and n the number of months. Find the client's monthly contribution, R.

II. Purchasing Electrical Power (Modeling)

In order to plan its purchases of electrical power from suppliers over the next 5 years, the PAC Electric Company needs to model its load data (demand for power by its customers) and use this model to predict future loads. The company pays for the electrical power each month on the basis of the peak load (demand) at any point during the month. The table gives, for the years 1987–2005, the load in megawatts (that is, in millions of watts) for the month when the maximum load occurred and the load in megawatts for the month when the minimum load occurred. The maximum loads occurred in summer, and the minimum loads occurred in spring or fall.

Year	Maximum Monthly Load	Minimum Monthly Load
1987	40.9367	19.4689
1988	45.7127	22.1504
1989	48.0460	25.3670
1990	56.1712	28.7254
1991	55.5793	31.0460
1992	62.4285	31.3838
1993	76.6536	34.8426
1994	73.8214	38.4544
1995	74.8844	40.6080
1996	83.0590	47.3621
1997	88.3914	45.8393
1998	88.7704	48.7956
1999	94.2620	48.3313
2000	105.1596	52.7710
2001	95.8301	54.4757
2002	97.8854	55.2210
2003	102.8912	55.1360
2004	109.5541	57.2162
2005	111.2516	58.3216

The company wishes to predict the average monthly load over the next 5 years so that it can plan its future monthly purchases. To assist the company, proceed as follows.

1. (a) Using the years and the maximum monthly load given for each year, graph the data, with x representing the number of years past 1987 and y representing the maximum load in megawatts.
   (b) Find the equation that best fits the data, using both a quadratic model and a cubic model.
   (c) Graph the data and both of these models from 1987 to 2007 (that is, from $x = 0$ to $x = 20$).
2. Do the two models appear to fit the data equally well in the interval 1987–2007? Which model appears to be a better predictor for the next decade?
3. Use the quadratic model to predict the maximum monthly load in the year 2010. How can this value be used by the company? Should this number be used to plan monthly power purchases for each month in 2010?
4. To create a "typical" monthly load function:
   (a) Create a table with the year as the independent variable and the average of the maximum and minimum monthly loads as the dependent variable.
   (b) Find the quadratic model that best fits these data points, using $x = 0$ to represent 1987.
5. Use a definite integral with the typical monthly load function to predict the average monthly load over the years 2007–2012.
6. What factors in addition to the average monthly load should be considered when the company plans future purchases of power?

14

Functions of Two or More Variables

In this chapter we will extend our study of functions of two or more variables. We will use these concepts to solve problems in the management, life, and social sciences. In particular, we will discuss joint cost functions, utility functions that describe the customer satisfaction derived from the consumption of two products, Cobb-Douglas production functions, and wind chill temperatures as a function of air temperature and wind speed.

We will use derivatives with respect to one of two variables (called **partial derivatives**) to find marginal cost, marginal productivity, marginal utility, marginal demand, and other rates of change. We will use partial derivatives to find maxima and minima of functions of two variables, and we will use Lagrange multipliers to optimize functions of two variables subject to a condition that constrains these variables. These skills are used to maximize profit, production, and utility and to minimize cost subject to constraints.

The topics and applications discussed in this chapter include the following.

Sections	Applications
14.1 Functions of Two or More Variables	*Utility, Cobb-Douglas production functions*
14.2 Partial Differentiation	*Marginal cost, marginal sales*
14.3 Applications of Functions of Two Variables in Business and Economics Joint cost and marginal cost Production functions Demand functions	*Joint cost, marginal cost, crop harvesting, marginal productivity, marginal demand*
14.4 Maxima and Minima Linear regression	*Maximum profit, inventory*
14.5 Maxima and Minima of Functions Subject to Constraints: Lagrange Multipliers	*Maximum utility, maximum production, marginal productivity of money*

Chapter Warm-up

Prerequisite Problem Type	For Section	Answer	Section for Review
If $y = f(x)$, x is the independent variable and y is the _____ variable.	14.1	Dependent	1.2 Functions
What is the domain of $f(x) = \dfrac{3x}{x-1}$?	14.1	All reals except $x = 1$	1.2 Domains
If $C(x) = 5 + 5x$, what is $C(0.20)$?	14.1	6	1.2 Function notation
(a) Solve for x and y: $\begin{cases} 0 = 50 - 2x - 2y \\ 0 = 60 - 2x - 4y \end{cases}$ (b) Solve for x and y: $\begin{cases} x = 2y \\ x + y - 9 = 0 \end{cases}$	14.4 14.5	(a) $x = 20$, $y = 5$ (b) $x = 6$, $y = 3$	1.5 Systems of equations
If $z = 4x^2 + 5x^3 - 7$, what is $\dfrac{dz}{dx}$?	14.2 14.3 14.4 14.5	$\dfrac{dz}{dx} = 8x + 15x^2$	9.4 Derivatives
If $f(x) = (x^2 - 1)^2$, what is $f'(x)$?	14.2	$f'(x) = 4x(x^2 - 1)$	9.6 Derivatives
If $z = 10y - \ln y$, what is $\dfrac{dz}{dy}$?	14.2	$\dfrac{dz}{dy} = 10 - \dfrac{1}{y}$	11.1 Derivatives of logarithmic functions
If $z = 5x^2 + e^x$, what is $\dfrac{dz}{dx}$?	14.2	$\dfrac{dz}{dx} = 10x + e^x$	11.2 Derivatives of exponential functions
Find the slope of the tangent to $y = 4x^3 - 4e^x$ at $(0, -4)$.	14.2	-4	9.3, 9.4, 11.2 Derivatives

Functions of Two or More Variables

OBJECTIVES

- To find the domain of a function of two or more variables
- To evaluate a function of two or more variables given values for the independent variables

◗ Application Preview

The relations we have studied up to this point have been limited to two variables, with one of the variables assumed to be a function of the other. But there are many instances where one variable may depend on two or more other variables. For example, a company's quantity of output or production Q (measured in units or dollars) can be modeled according to the equation

$$Q = AK^{\alpha}L^{1-\alpha}$$

where A is a constant, K is the company's capital investment, L is the size of the labor force (in work-hours), and α is a constant with $0 < \alpha < 1$. Functions of this type are called **Cobb-Douglas production functions,** and they are frequently used in economics. For example, suppose the Cobb-Douglas production function for a company is given by

$$Q = 4K^{0.4}L^{0.6}$$

where Q is thousands of dollars of production value, K is hundreds of dollars of capital investment, and L is hours of labor. We could use this function to determine the production value for a given amount of capital investment and available work-hours of labor. We could also find how production is affected by changes in capital investment or available work-hours. (See Example 5.)

In addition, the demand function for a commodity frequently depends on the price of the commodity, available income, and prices of competing goods. Other examples from economics will be presented later in this chapter.

We write $z = f(x, y)$ to state that z is a function of both x and y. The variables x and y are called the **independent variables** and z is called the **dependent variable.** Thus the function f associates with each pair of possible values for the independent variables (x and y) exactly one value of the dependent variable (z).

The equation $z = x^2 - xy$ defines z as a function of x and y. We can denote this by writing $z = f(x, y) = x^2 - xy$. The domain of the function is the set of all ordered pairs (of real numbers), and the range is the set of all real numbers.

● **EXAMPLE 1 *Domain***

Give the domain of the function

$$g(x, y) = \frac{x^2 - 3y}{x - y}$$

Solution
The domain of the function is the set of ordered pairs that do not give a 0 denominator. That is, the domain is the set of all ordered pairs where the first and second elements are not equal (that is, where $x \neq y$).

● *Checkpoint*

1. Find the domain of the function

$$f(x, y) = \frac{2}{\sqrt{x^2 - y^2}}$$

We graph the function $z = f(x, y)$ by using three dimensions. We can construct a three-dimensional coordinate space by drawing three mutually perpendicular axes, as in Figure 14.1. By setting up scales of measurement along the three axes from the origin of each axis, we can determine the three coordinates (x, y, z) for any point P. The point shown in Figure 14.1 is $+2$ units in the x-direction, $+3$ units in the y-direction, and $+4$ units in the z-direction, so the coordinates of the point are $(2, 3, 4)$.

The pairs of axes determine the three **coordinate planes;** the xy-plane, the yz-plane, and the xz-plane. The planes divide the space into eight **octants.** The point $P(2, 3, 4)$ is in the first octant.

If we are given a function $z = f(x, y)$, we can find the z-value corresponding to $x = a$ and $y = b$ by evaluating $f(a, b)$.

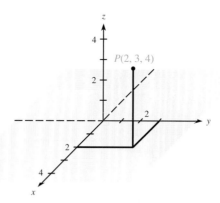

Figure 14.1

● **EXAMPLE 2** *Function Values*

If $z = f(x, y) = x^2 - 4xy + xy^3$, find the following.

(a) $f(1, 2)$　　(b) $f(2, 5)$　　(c) $f(-1, 3)$

Solution
(a) $f(1, 2) = 1^2 - 4(1)(2) + (1)(2)^3 = 1$
(b) $f(2, 5) = 2^2 - 4(2)(5) + (2)(5)^3 = 214$
(c) $f(-1, 3) = (-1)^2 - 4(-1)(3) + (-1)(3)^3 = -14$

● *Checkpoint*

2. If $f(x, y, z) = x^2 + 2y - z$, find $f(2, 3, 4)$.

● **EXAMPLE 3** *Cost*

A small furniture company's cost (in dollars) to manufacture 1 unit of several different all-wood items is given by

$$C(x, y) = 5 + 5x + 22y$$

where x represents the number of board-feet of material used and y represents the number of work-hours of labor required for assembly and finishing. A certain bookcase uses 20 board-feet of material and requires 2.5 work-hours for assembly and finishing. Find the cost of manufacturing this bookcase.

Solution
The cost is

$$C(20, 2.5) = 5 + 5(20) + 22(2.5) = 160 \quad \text{dollars}$$

For a given function $z = f(x, y)$, we can construct a table of values by assigning values to x and y and finding the corresponding values of z. To each pair of values for x and y there corresponds a unique value of z, and thus a unique point in a three-dimensional coordinate system. From a table of values such as this, a finite number of points can be plotted. All points that satisfy the equation form a "surface" in space. Because z is a function of x and y, lines parallel to the z-axis will intersect such a surface at at most one point. The graph of the equation $z = 4 - x^2 - y^2$ is the surface shown in Figure 14.2(a). The portion of the surface above the xy-plane resembles a bullet and is called a **paraboloid.** Some points on the surface are given in Table 14.1.

TABLE 14.1

x	y	z
−2	0	0
0	−2	0
−1	0	3
0	−1	3
−1	−1	2
0	0	4
1	0	3
0	1	3
1	1	2
2	0	0
0	2	0

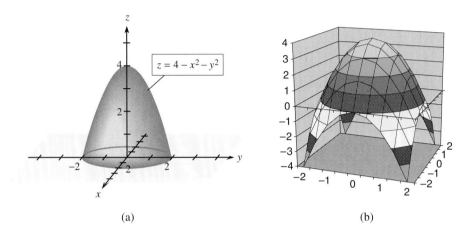

Figure 14.2 (a) (b)

Technology Note We can create the graphs of functions of two variables with Excel in a manner similar to that used to graph functions of one variable. Figure 14.2(b) shows the Excel graph of $z = 4 - x^2 - y^2$. ∎

In practical applications of functions of two variables, we will have little need to construct the graphs of the surfaces. For this reason, we will not discuss methods of sketching the graphs. Although you will not be asked to sketch graphs of these surfaces, the fact that the graphs do *exist* will be used in studying relative maxima and minima of functions of two variables.

The properties of functions of one variable can be extended to functions of two variables. The precise definition of continuity for functions of two variables is technical and may be found in more advanced books. We will limit our study to functions that are continuous and have continuous derivatives in the domain of interest to us. We may think of continuous functions as functions whose graphs consist of surfaces without "holes" or "breaks" in them.

Let the function $U = f(x, y)$ represent the **utility** (that is, satisfaction) derived by a consumer from the consumption of two goods, X and Y, where x and y represent the amounts of X and Y, respectively. Because we will assume that the utility function is continuous, a given level of utility can be derived from an infinite number of combinations of x and y. The graph of all points (x, y) that give the same utility is called an **indifference curve.** A set of indifference curves corresponding to different levels of utility is called an **indifference map** (see Figure 14.3).

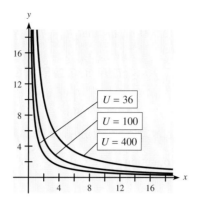

Figure 14.3

● **EXAMPLE 4** *Utility*

Suppose that the utility function for two goods, X and Y, is $U = x^2y^2$ and a consumer purchases 10 units of X and 2 units of Y.

(a) What is the level of utility U for these two products?
(b) If the consumer purchases 5 units of X, how many units of Y must be purchased to retain the same level of utility?
(c) Graph the indifference curve for this level of utility.
(d) Graph the indifference curves for this utility function if $U = 100$ and if $U = 36$.

Solution

(a) If $x = 10$ and $y = 2$ satisfy the utility function, then the level of utility is
$U = 10^2 \cdot 2^2 = 400$.
(b) If x is 5, y must satisfy $400 = 5^2y^2$, so $y = \pm 4$, and 4 units of Y must be purchased.
(c) The indifference curve for $U = 400$ is $400 = x^2y^2$. The graph for positive x and y is shown in Figure 14.3.
(d) The indifference map in Figure 14.3 contains these indifference curves.

A graphing utility can be used to graph each indifference curve in the indifference map shown in Figure 14.3. To graph the indifference curve for a given value of U, we must recognize that y will be positive and solve for y to express it as a function of x. Try it for $U = 100$. Does your graph agree with the one shown in Figure 14.3?

Sometimes functions of two variables are studied by fixing a value for one variable and graphing the resulting function of a single variable. We'll do this in Section 14.3 with production functions.

● **EXAMPLE 5** *Production* **(Application Preview)**

Suppose a company has the Cobb-Douglas production function.

$$Q = 4K^{0.4}L^{0.6}$$

where Q is thousands of dollars of production value, K is hundreds of dollars of capital investment per week, and L is work-hours of labor per week.

(a) If current capital investment is $72,900 per week and work-hours are 3072 per week, find the current weekly production value.
(b) If weekly capital investment is increased to $97,200 and new employees are hired so that there are 4096 total weekly work-hours, find the percent increase in the production value.

Solution

(a) Capital investment of $72,900 means that $K = 729$. We use this value and $L = 3072$ in the production function.

$$Q = 4(729)^{0.4}(3072)^{0.6} = 6912$$

Thus the weekly production value is $6,912,000.

(b) In this case we use $K = 972$ and $L = 4096$.

$$Q = 4(972)^{0.4}(4096)^{0.6} = 9216$$

This is an increase in production value of $9216 - 6912 = 2304$, which is equivalent to a weekly increase of

$$\frac{2304}{6912} = 0.33\tfrac{1}{3} = 33\tfrac{1}{3}\%$$

● *Checkpoint Solutions*

1. The domain is the set of ordered pairs of real numbers where $x^2 - y^2 > 0$ or $x^2 > y^2$; that is, where $|x| > |y|$.
2. $f(2, 3, 4) = 4 + 6 - 4 = 6$

14.1 Exercises

Give the domain of each function in Problems 1–8.

1. $z = x^2 + y^2$

2. $z = 4x - 3y$

3. $z = \dfrac{4x - 3}{y}$

4. $z = \dfrac{x + y^2}{\sqrt{x}}$

5. $z = \dfrac{4x^3y - x}{2x - y}$

6. $z = \sqrt{x - y}$

7. $q = \sqrt{p_1} + 3p_2$

8. $q = \dfrac{p_1 + p_2}{\sqrt{p_1}}$

In Problems 9–22, evaluate the functions at the given values of the independent variables.

9. $z = x^3 + 4xy + y^2$; $x = 1, y = -1$

10. $z = 4x^2 - 3xy^3$; $x = 2, y = 2$

11. $z = \dfrac{x - y}{x + y}$; $x = 4, y = -1$

12. $z = \dfrac{x^2 + xy}{x - y}$; $x = 3, y = 2$

13. $C(x_1, x_2) = 600 + 4x_1 + 6x_2$; $x_1 = 400, x_2 = 50$

14. $C(x_1, x_2) = 500 + 5x_1 + 7x_2$;   find $C(200, 300)$.

15. $q_1(p_1, p_2) = \dfrac{p_1 + 4p_2}{p_1 - p_2}$;   find $q_1(40, 35)$.

16. $q_1(p_1, p_2) = \dfrac{5p_1 - p_2}{p_1 + 3p_2}$;   find $q_1(50, 10)$.

17. $z(x, y) = xe^{x+y}$;   find $z(3, -3)$.

18. $f(x, y) = ye^{2x} + y^2$;   find $f(0, 7)$.

19. $f(x, y) = \dfrac{\ln (xy)}{x^2 + y^2}$;   find $f(-3, -4)$.

20. $z(x, y) = x \ln y - y \ln x$;   find $z(1, 1)$.

21. $w = \dfrac{x^2 + 4yz}{xyz}$ at $(1, 3, 1)$

22. $u = f(w, x, y, z) = \dfrac{wx - yz^2}{xy - wz}$ at $(2, 3, 1, -1)$

APPLICATIONS

23. *Investment*   The future value S of an investment earning 6% compounded continuously is a function of the principal P and the length of time t that the principal has been invested. It is given by

$$S = f(P, t) = Pe^{0.06t}$$

Find $f(2000, 20)$, and interpret your answer.

24. *Amortization*   If $100,000 is borrowed to purchase a home, then the monthly payment R is a function of the interest rate i (expressed as a percent) and the number of years n before the mortgage is paid. It is given by

$$R = f(i, n) = 100,000 \left[\frac{0.01(i/12)}{1 - (1 + 0.01(i/12))^{-12n}} \right]$$

Find $f(7.25, 30)$ and interpret your answer.

25. *Wilson's lot size formula* In economics, the most economical quantity Q of goods (TVs, dresses, gallons of paint, etc.) for a store to order is given by Wilson's lot size formula

$$Q = f(K, M, h) = \sqrt{2KM/h}$$

where K is the cost of placing the order, M is the number of items sold per week, and h is the weekly holding cost for each item (the cost of storage space, utilities, taxes, security, etc.). Find $f(200, 625, 1)$ and interpret your answer.

26. *Gas law* Suppose that a gas satisfies the universal gas law, $V = nRT/P$, with n equal to 10 moles of the gas and R, the universal gas constant, equal to 0.082054. What is V if $T = 10$ K (kelvins, the units in which temperature is measured on the Kelvin scale) and $P = 1$ atmosphere?

Temperature-humidity models **There are different models for measuring the effects of high temperature and humidity. Two of these are the Summer Simmer Index (S) and the Apparent Temperature (A),* and they are given by**

$$S = 1.98T - 1.09(1 - H)(T - 58) - 56.9$$

$$A = 2.70 + 0.885T - 78.7H + 1.20TH$$

where T is the air temperature (in degrees Fahrenheit) and H is the relative humidity (expressed as a decimal). Use these models in Problems 27 and 28.

27. At the Dallas–Fort Worth Airport, the average daily temperatures and humidities for July are

Maximum: 97.8°F with 44% humidity

Minimum: 74.7°F with 80% humidity**

Calculate the Summer Simmer Index S and the Apparent Temperature A for both the average daily maximum and the average daily minimum temperature.

28. In Orlando, Florida, the following represent the average daily temperatures and humidities for August.

Maximum: 91.6°F with 60% humidity

Minimum: 73.4°F with 92% humidity**

Calculate the Summer Simmer Index S and the Apparent Temperature A for both the average daily maximum and the average daily minimum temperature.

*W. Bosch and L. G. Cobb, "Temperature-Humidity Indices," UMAP Unit 691, *The UMAP Journal*, 10, no. 3 (Fall 1989): 237–256.
**James Ruffner and Frank Bair (eds.), *Weather of U.S. Cities*, Gale Research Co., Detroit, MI, 1987.

29. *Mortgage* The following tables show that a monthly mortgage payment, R, is a function of the amount financed, A, in thousands of dollars; the duration of the loan, n, in years; and the annual interest rate, r, as a percent. If $R = f(A, n, r)$, use the tables to find the following, and then write a sentence of explanation for each.

(a) $f(90, 20, 8)$ (b) $f(160, 15, 9)$

**8% Annual Percent Rate
Monthly Payments (Principal and Interest)**

Amount Financed	10 Years	15 Years	20 Years	25 Years	30 Years
$50,000	$606.64	$477.83	$418.22	$385.91	$366.88
60,000	727.97	573.39	501.86	463.09	440.26
70,000	849.29	668.96	585.51	540.27	513.64
80,000	970.62	764.52	669.15	617.45	587.01
90,000	1091.95	860.09	752.80	694.63	660.39
100,000	1213.28	955.65	836.44	771.82	733.76
120,000	1455.94	1146.78	1003.72	926.18	880.52
140,000	1698.58	1337.92	1171.02	1080.54	1027.28
160,000	1941.24	1529.04	1338.30	1234.90	1174.02
180,000	2183.90	1720.18	1505.60	1389.26	1320.78
200,000	2426.56	1911.30	1672.88	1543.64	1467.52

**9% Annual Percent Rate
Monthly Payments (Principal and Interest)**

Amount Financed	10 Years	15 Years	20 Years	25 Years	30 Years
$50,000	$633.38	$507.13	$449.86	$419.60	$402.31
60,000	760.05	608.56	539.84	503.52	482.77
70,000	886.73	709.99	629.81	587.44	563.24
80,000	1013.41	811.41	719.78	671.36	643.70
90,000	1140.08	912.84	809.75	755.28	724.16
100,000	1266.76	1014.27	899.73	839.20	804.62
120,000	1520.10	1217.12	1079.68	1007.04	965.54
140,000	1773.46	1419.96	1259.62	1174.88	1126.48
160,000	2026.82	1622.82	1439.56	1342.72	1287.40
180,000	2280.16	1825.68	1619.50	1510.56	1448.32
200,000	2533.52	2028.54	1799.46	1678.40	1609.24

Source: The Mortgage Money Guide, Federal Trade Commission

30. *Wind chill* Wind and cold temperatures combine to make the air temperature feel colder than it actually is. This combination is reported as wind chill. The following table shows that wind chill temperatures, WC, are a function of wind speed, s, and air temperature, t. If $WC = f(s, t)$, use the table to find the following, and then write a sentence of explanation for each.

(a) $f(25, 5)$ (b) $f(15, -15)$

	Air Temperature (°F)						
	35	25	15	5	−5	−15	−25
5	33	21	12	0	−10	−21	−31
15	16	2	−11	−25	−38	−51	−65
25	8	−7	−22	−36	−51	−66	−81
35	4	−12	−27	−43	−58	−74	−89
45	2	−14	−30	−46	−62	−78	−93

Wind Speed (mph)

Source: National Weather Service, NOAA, U.S. Department of Commerce

 31. *Utility*   Suppose that the utility function for two goods X and Y is given by $U = xy^2$, and a consumer purchases 9 units of X and 6 units of Y.
   (a) If the consumer purchases 9 units of Y, how many units of X must be purchased to retain the same level of utility?
   (b) If the consumer purchases 81 units of X, how many units of Y must be purchased to retain the same level of utility?
   (c) Graph the indifference curve for the utility level found in parts (a) and (b). Use the graph to confirm your answers to parts (a) and (b).

 32. *Utility*   Suppose that an indifference curve for two goods, X and Y, has the equation $xy = 400$.
   (a) If 25 units of X are purchased, how many units of Y must be purchased to remain on this indifference curve?
   (b) Graph this indifference curve and confirm your results in part (a).

 33. *Production*   Suppose that a company's production for Q units of its product is given by the Cobb-Douglas production function

$$Q = 30K^{1/4}L^{3/4}$$

where K is dollars of capital investment and L is labor hours.
   (a) Find Q if $K = \$10,000$ and $L = 625$ hours.
   (b) Show that if *both* K and L are doubled, then the output is doubled.
   (c) If capital investment is held at $10,000, graph Q as a function of L.

 34. *Production*   Suppose that a company's production for Q units of its product is given by the Cobb-Douglas production function

$$Q = 70K^{2/3}L^{1/3}$$

where K is dollars of capital investment and L is labor hours.
   (a) Find Q if $K = \$64,000$ and $L = 512$ hours.
   (b) Show that if both K and L are halved, then Q is also halved.
   (c) If capital investment is held at $64,000, graph Q as a function of L.

35. *Production*   Suppose that the number of units of a good produced, z, is given by $z = 20xy$, where x is the number of machines working properly and y is the average number of work-hours per machine. Find the production for a week in which
   (a) 12 machines are working properly and the average number of work-hours per machine is 30.
   (b) 10 machines are working properly and the average number of work-hours per machine is 25.

36. *Profit*   The Kirk Kelly Kandy Company makes two kinds of candy, Kisses and Kreams. The profit, in dollars, for the company is given by

$$P(x, y) = 10x + 6.4y - 0.001x^2 - 0.025y^2$$

where x is the number of pounds of Kisses sold per week and y is the number of pounds of Kreams. What is the company's profit if it sells
   (a) 20 pounds of Kisses and 10 pounds of Kreams?
   (b) 100 pounds of Kisses and 16 pounds of Kreams?
   (c) 10,000 pounds of Kisses and 256 pounds of Kreams?

37. *Epidemic*   The cost per day to society of an epidemic is

$$C(x, y) = 20x + 200y$$

where C is in dollars, x is the number of people infected on a given day, and y is the number of people who die on a given day. If 14,000 people are infected and 20 people die on a given day, what is the cost to society?

38. *Pesticide*   An area of land is to be sprayed with two brands of pesticide: x liters of brand 1 and y liters of brand 2. If the number of thousands of insects killed is given by

$$f(x, y) = 10,000 - 6500e^{-0.01x} - 3500e^{-0.02y}$$

how many insects would be killed if 80 liters of brand 1 and 120 liters of brand 2 were used?

14.2

Partial Differentiation

OBJECTIVES

- To find partial derivatives of functions of two or more variables
- To evaluate partial derivatives of functions of two or more variables at given points
- To use partial derivatives to find slopes of tangents to surfaces
- To find and evaluate second- and higher-order partial derivatives of functions of two variables.

◗ Application Preview

Suppose that a company's sales are related to its television advertising by

$$s = 20{,}000 + 10nt + 20n^2$$

where n is the number of commercials per day and t is the length of the commercials in seconds. To find the instantaneous rate of change of sales with respect to the number of commercials per day if the company is running ten 30-second commercials, we find the **partial derivative** of s with respect to n. (See Example 7.)

In this section we find partial derivatives of functions of two or more variables and use these derivatives to find rates of change and slopes of tangents to surfaces. We will also find second- and higher-order partial derivatives of functions of two variables.

First-Order Partial Derivatives

If $z = f(x, y)$, we find the partial derivative of z with respect to x (denoted $\partial z/\partial x$) by treating the variable y as a constant and taking the derivative of $z = f(x, y)$ with respect to x. We can also take the derivative of z with respect to y by holding the variable x constant and taking the derivative of $z = f(x, y)$ with respect to y. We denote this derivative as $\partial z/\partial y$. Note that dz/dx represents the derivative of a function of one variable, x, and that $\partial z/\partial x$ represents the partial derivative of a function of two or more variables. Notations used to represent the partial derivative of $z = f(x, y)$ with respect to x are

$$\frac{\partial z}{\partial x}, \quad \frac{\partial f}{\partial x}, \quad \frac{\partial}{\partial x} f(x, y), \quad f_x(x, y), \quad f_x, \quad \text{and} \quad z_x$$

and notations used to represent the partial derivative of $z = f(x, y)$ with respect to y are

$$\frac{\partial z}{\partial y}, \quad \frac{\partial f}{\partial y}, \quad \frac{\partial}{\partial y} f(x, y), \quad f_y(x, y), \quad f_y, \quad \text{and} \quad z_y$$

● **EXAMPLE 1** *Partial Derivatives*

If $z = 4x^2 + 5x^2y^2 + 6y^3 - 7$, find $\partial z/\partial x$ and $\partial z/\partial y$.

Solution
To find $\partial z/\partial x$, y is held constant so that the term $6y^3$ is constant, and its derivative is 0; the term $5x^2y^2$ is viewed as the constant $(5y^2)$ times (x^2), so its derivative is $(5y^2)(2x) = 10xy^2$. Thus

$$\frac{\partial z}{\partial x} = 8x + 10xy^2$$

Similarly for $\partial z/\partial y$, the term $4x^2$ is constant and the partial derivative of $5x^2y^2$ is the constant $5x^2$ times the derivative of y^2, so its derivative is $(5x^2)(2y) = 10x^2y$. Thus

$$\frac{\partial z}{\partial y} = 10x^2y + 18y^2$$

● **EXAMPLE 2** *Marginal Cost*

Suppose that the total cost of manufacturing a product is given by

$$C(x, y) = 5 + 5x + 2y$$

where x represents the cost of 1 ounce of materials and y represents the labor cost in dollars per hour.

(a) Find the rate at which the total cost changes with respect to the material cost x. This is the marginal cost of the product with respect to the material cost.
(b) Find the rate at which the total cost changes with respect to the labor cost y. This is the marginal cost of the product with respect to the labor cost.

Solution
(a) The rate at which the total cost changes with respect to the material cost x is the partial derivative found by treating the y-variable as a constant and taking the derivative of both sides of $C(x, y) = 5 + 5x + 2y$ with respect to x.

$$\frac{\partial C}{\partial x} = 0 + 5 + 0, \quad \text{or} \quad \frac{\partial C}{\partial x} = 5$$

The partial derivative $\partial C/\partial x = 5$ tells us that a change of \$1 in the cost of materials will cause an increase of \$5 in total costs, *if* labor costs remain constant.
(b) The rate at which the total cost changes with respect to the labor cost y is the partial derivative found by treating the x-variable as a constant and taking the derivative of both sides of $C(x, y) = 5 + 5x + 2y$ with respect to y.

$$\frac{\partial C}{\partial y} = 0 + 0 + 2, \quad \text{or} \quad \frac{\partial C}{\partial y} = 2$$

The partial derivative $\partial C/\partial y = 2$ tells us that a change of \$1 in the cost of labor will cause an increase of \$2 in total costs, *if* material costs remain constant.

● **EXAMPLE 3** *Partial Derivatives*

If $z = x^2 y + e^x - \ln y$, find z_x and z_y.

Solution

$$z_x = \frac{\partial z}{\partial x} = 2xy + e^x$$

$$z_y = \frac{\partial z}{\partial y} = x^2 - \frac{1}{y}$$

● **EXAMPLE 4** *Power Rule*

If $f(x, y) = (x^2 - y^2)^2$, find the following.

(a) f_x
(b) f_y

Solution
(a) $f_x = 2(x^2 - y^2)2x = 4x^3 - 4xy^2$
(b) $f_y = 2(x^2 - y^2)(-2y) = -4x^2 y + 4y^3$

● *Checkpoint*

1. If $z = 100x + 10xy - y^2$, find the following.

   (a) z_x

   (b) $\dfrac{\partial z}{\partial y}$

● **EXAMPLE 5** *Quotient Rule*

If $q = \dfrac{p_1 p_2 + 2p_1}{p_1 p_2 - 2p_2}$, find $\partial q / \partial p_1$.

Solution

$$
\begin{aligned}
\frac{\partial q}{\partial p_1} &= \frac{(p_1 p_2 - 2p_2)(p_2 + 2) - (p_1 p_2 + 2p_1)p_2}{(p_1 p_2 - 2p_2)^2} \\
&= \frac{p_1 p_2^2 + 2p_1 p_2 - 2p_2^2 - 4p_2 - p_1 p_2^2 - 2p_1 p_2}{(p_1 p_2 - 2p_2)^2} \\
&= \frac{-2p_2^2 - 4p_2}{p_2^2(p_1 - 2)^2} \\
&= \frac{-2p_2(p_2 + 2)}{p_2^2(p_1 - 2)^2} \\
&= \frac{-2(p_2 + 2)}{p_2(p_1 - 2)^2}
\end{aligned}
$$

We may evaluate partial derivatives by substituting values for x and y in the same way we did with derivatives of functions of one variable. For example, if $\partial z / \partial x = 2x - xy$, the value of the partial derivative with respect to x at $x = 2, y = 3$ is

$$
\left. \frac{\partial z}{\partial x} \right|_{(2,\, 3)} = 2(2) - 2(3) = -2
$$

Other notations used to denote evaluation of partial derivatives with respect to x at (a, b) are

$$
\frac{\partial}{\partial x} f(a, b) \quad \text{and} \quad f_x(a, b)
$$

We denote the evaluation of partial derivatives with respect to y at (a, b) by

$$
\left. \frac{\partial z}{\partial y} \right|_{(a,\, b)}, \quad \frac{\partial}{\partial y} f(a, b), \quad \text{or} \quad f_y(a, b)
$$

● **EXAMPLE 6** *Partial Derivative at a Point*

Find the partial derivative of $f(x, y) = x^2 + 3xy + 4$ with respect to x at the point $(1, 2, 11)$.

Solution

$$
\begin{aligned}
f_x(x, y) &= 2x + 3y \\
f_x(1, 2) &= 2(1) + 3(2) = 8
\end{aligned}
$$

● *Checkpoint*

2. If $g(x, y) = 4x^2 - 3xy + 10y^2$, find the following.

   (a) $\dfrac{\partial g}{\partial x}(1, 3)$ (b) $g_y(4, 2)$

◐ **EXAMPLE 7** *Marginal Sales (Application Preview)*

Suppose that a company's sales are related to its television advertising by

$$s = 20{,}000 + 10nt + 20n^2$$

where n is the number of commercials per day and t is the length of the commercials in seconds. Find the partial derivative of s with respect to n, and use the result to find the instantaneous rate of change of sales with respect to the number of commercials per day, if the company is currently running ten 30-second commercials per day.

Solution
The partial derivative of s with respect to n is $\partial s/\partial n = 10t + 40n$. At $n = 10$ and $t = 30$, the rate of change in sales is

$$\left.\frac{\partial s}{\partial n}\right|_{\substack{n=10\\t=30}} = 10(30) + 40(10) = 700$$

Thus, increasing the number of 30-second commercials by 1 would result in approximately 700 additional sales. This is the marginal sales with respect to the number of commercials per day at $n = 10, t = 30$.

We have seen that the partial derivative $\partial z/\partial x$ is found by holding y constant and taking the derivative of z with respect to x and that the partial derivative $\partial z/\partial y$ is found by holding x constant and taking the derivative of z with respect to y. We now give formal definitions of these partial derivatives.

Partial Derivatives

The partial derivative of $z = f(x, y)$ with respect to x at the point (x, y) is

$$\frac{\partial z}{\partial x} = \frac{\partial}{\partial x} f(x, y) = \lim_{h \to 0} \frac{f(x + h, y) - f(x, y)}{h}$$

provided this limit exists.

The partial derivative of $z = f(x, y)$ with respect to y at the point (x, y) is

$$\frac{\partial z}{\partial y} = \frac{\partial}{\partial y} f(x, y) = \lim_{h \to 0} \frac{f(x, y + h) - f(x, y)}{h}$$

provided this limit exists.

We have already stated that the graph of $z = f(x, y)$ is a surface in three dimensions. The partial derivative with respect to x of such a function may be thought of as the slope of the tangent to the surface at a point (x, y, z) on the surface in the *positive direction of the x-axis.* That is, if a plane parallel to the xz-plane cuts the surface, passing through the point (x_0, y_0, z_0), the line in the plane that is tangent to the surface will have a slope equal to $\partial z/\partial x$ evaluated at the point. Thus

$$\left.\frac{\partial z}{\partial x}\right|_{(x_0,\, y_0)}$$

represents the slope of the tangent to the surface at (x_0, y_0, z_0), in the positive direction of the x-axis (see Figure 14.4).

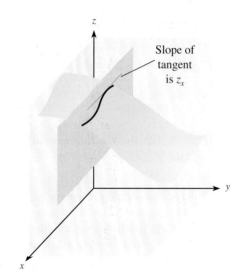

Figure 14.4

Similarly,

$$\left.\frac{\partial z}{\partial y}\right|_{(x_0,\,y_0)} = \frac{\partial}{\partial y} f(x_0, y_0)$$

represents the slope of the tangent to the surface at (x_0, y_0, z_0) in the positive direction of the y-axis (see Figure 14.5).

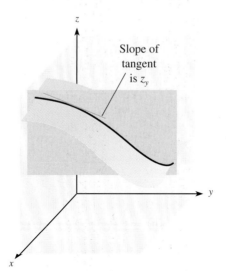

Figure 14.5

● **EXAMPLE 8** *Slopes of Tangents*

Let $z = 4x^3 - 4e^x + 4y^2$ and let P be the point $(0, 2, 12)$. Find the slope of the tangent to z at the point P in the positive direction of (a) the x-axis and (b) the y-axis.

Solution

(a) The slope of z at P in the positive x-direction is given by $\dfrac{\partial z}{\partial x}$, evaluated at P.

$$\frac{\partial z}{\partial x} = 12x^2 - 4e^x \quad \text{and} \quad \left.\frac{\partial z}{\partial x}\right|_{(0,\,2)} = 12(0)^2 - 4e^0 = -4$$

This tells us that z *decreases* approximately 4 units for an increase of 1 unit in x at this point.

(b) The slope of z at P in the positive y-direction is given by $\dfrac{\partial z}{\partial y}$, evaluated at P.

$$\frac{\partial z}{\partial y} = 8y \quad \text{and} \quad \left.\frac{\partial z}{\partial y}\right|_{(0,\,2)} = 8(2) = 16$$

Thus, at the point P (0, 2, 12), the function *increases* approximately 16 units in the z-value for a unit increase in y.

Up to this point, we have considered derivatives of functions of two variables. We can easily extend the concept to functions of three or more variables. We can find the partial derivative with respect to any one independent variable by taking the derivative of the function with respect to that variable while holding all other independent variables constant.

● **EXAMPLE 9** *Functions of Four Variables*

If $u = f(w, x, y, z) = 3x^2y + w^3 - 4xyz$, find the following.

(a) $\dfrac{\partial u}{\partial w}$ (b) $\dfrac{\partial u}{\partial x}$ (c) $\dfrac{\partial u}{\partial y}$ (d) $\dfrac{\partial u}{\partial z}$

Solution

(a) $\dfrac{\partial u}{\partial w} = 3w^2$ (b) $\dfrac{\partial u}{\partial x} = 6xy - 4yz$

(c) $\dfrac{\partial u}{\partial y} = 3x^2 - 4xz$ (d) $\dfrac{\partial u}{\partial z} = -4xy$

● **EXAMPLE 10** *Functions of Three Variables*

If $C = 4x_1 + 2x_1{}^2 + 3x_2 - x_1x_2 + x_3{}^2$, find the following.

(a) $\dfrac{\partial C}{\partial x_1}$ (b) $\dfrac{\partial C}{\partial x_2}$ (c) $\dfrac{\partial C}{\partial x_3}$

Solution

(a) $\dfrac{\partial C}{\partial x_1} = 4 + 4x_1 - x_2$ (b) $\dfrac{\partial C}{\partial x_2} = 3 - x_1$ (c) $\dfrac{\partial C}{\partial x_3} = 2x_3$

● *Checkpoint*

3. If $f(w, x, y, z) = 8xy^2 + 4yz - xw^2$, find

  (a) $\dfrac{\partial f}{\partial x}$ (b) $\dfrac{\partial f}{\partial w}$ (c) $\dfrac{\partial f}{\partial y}(1, 2, 1, 3)$ (d) $\dfrac{\partial f}{\partial z}(0, 2, 1, 3)$

Higher-Order Partial Derivatives

Just as we have taken derivatives of derivatives to obtain higher-order derivatives of functions of one variable, we may also take partial derivatives of partial derivatives to obtain higher-order partial derivatives of a function of more than one variable. If $z = f(x, y)$, then the partial derivative functions z_x and z_y are called *first partials*. Partial derivatives of z_x and z_y are called *second partials*, so $z = f(x, y)$ has *four* **second partial derivatives.** The notations for these second partial derivatives follow.

Second Partial Derivatives

$$z_{xx} = \frac{\partial^2 z}{\partial x^2} = \frac{\partial}{\partial x}\left(\frac{\partial z}{\partial x}\right):$$ both derivatives taken with respect to x

$$z_{yy} = \frac{\partial^2 z}{\partial y^2} = \frac{\partial}{\partial y}\left(\frac{\partial z}{\partial y}\right):$$ both derivatives taken with respect to y

$$z_{xy} = \frac{\partial^2 z}{\partial y\,\partial x} = \frac{\partial}{\partial y}\left(\frac{\partial z}{\partial x}\right):$$ first derivative taken with respect to x, second with respect to y

$$z_{yx} = \frac{\partial^2 z}{\partial x\,\partial y} = \frac{\partial}{\partial x}\left(\frac{\partial z}{\partial y}\right):$$ first derivative taken with respect to y, second with respect to x

● **EXAMPLE 11** *Second Partial Derivatives*

If $z = x^3 y - 3xy^2 + 4$, find each of the second partial derivatives of the function.

Solution
Because $z_x = 3x^2 y - 3y^2$ and $z_y = x^3 - 6xy$,

$$z_{xx} = \frac{\partial}{\partial x}(3x^2 y - 3y^2) = 6xy$$

$$z_{xy} = \frac{\partial}{\partial y}(3x^2 y - 3y^2) = 3x^2 - 6y$$

$$z_{yy} = \frac{\partial}{\partial y}(x^3 - 6xy) = -6x$$

$$z_{yx} = \frac{\partial}{\partial x}(x^3 - 6xy) = 3x^2 - 6y$$

Note that z_{xy} and z_{yx} are equal for the function in Example 11. This will always occur if the derivatives of the function are continuous.

$z_{xy} = z_{yx}$

If the second partial derivatives z_{xy} and z_{yx} of a function $z = f(x, y)$ are continuous at a point, they are equal there.

● **EXAMPLE 12** *Second Partial Derivatives*

Find each of the second partial derivatives of $z = x^2 y + e^{xy}$.

Solution
Because $z_x = 2xy + e^{xy} \cdot y = 2xy + ye^{xy}$,

$$z_{xx} = 2y + e^{xy} \cdot y^2 = 2y + y^2 e^{xy}$$
$$z_{xy} = 2x + (e^{xy} \cdot 1 + ye^{xy} \cdot x)$$
$$= 2x + e^{xy} + xye^{xy}$$

Because $z_y = x^2 + e^{xy} \cdot x = x^2 + xe^{xy}$,

$$z_{yx} = 2x + (e^{xy} \cdot 1 + xe^{xy} \cdot y)$$
$$= 2x + e^{xy} + xye^{xy}$$
$$z_{yy} = 0 + xe^{xy} \cdot x = x^2 e^{xy}$$

● **Checkpoint**

4. If $z = 4x^3y^4 + 4xy$, find the following.

   (a) z_{xx} (b) z_{yy} (c) z_{xy} (d) z_{yx}

5. If $z = x^2 + 4e^{xy}$, find z_{xy}.

We can find partial derivatives of orders higher than the second. For example, we can find the third-order partial derivatives z_{xyx} and z_{xyy} for the function in Example 11 from the second derivative $z_{xy} = 3x^2 - 6y$.

$$z_{xyx} = 6x$$

$$z_{xyy} = -6$$

● **EXAMPLE 13 *Third Partial Derivatives***

If $y = 4y \ln x + e^{xy}$, find z_{xyy}.

Solution

$$z_x = 4y \cdot \frac{1}{x} + e^{xy} \cdot y$$

$$z_{xy} = 4 \cdot \frac{1}{x} \cdot 1 + e^{xy} \cdot 1 + y \cdot e^{xy} \cdot x = \frac{4}{x} + e^{xy} + xye^{xy}$$

$$z_{xyy} = 0 + e^{xy} \cdot x + xy \cdot e^{xy} \cdot x + e^{xy} \cdot x = x^2ye^{xy} + 2xe^{xy}$$

● **Checkpoint Solutions**

1. (a) $z_x = 100 + 10y$

   (b) $\dfrac{\partial z}{\partial y} = 10x - 2y$

2. (a) $\dfrac{\partial g}{\partial x} = 8x - 3y$ and $\dfrac{\partial g}{\partial x}(1, 3) = 8(1) - 3(3) = -1$

   (b) $g_y = -3x + 20y$ and $g_y(4, 2) = -3(4) + 20(2) = 28$

3. (a) $\dfrac{\partial f}{\partial x} = 8y^2 - w^2$ (b) $\dfrac{\partial f}{\partial w} = -2xw$

   (c) $\dfrac{\partial f}{\partial y} = 16xy + 4z$ and $\dfrac{\partial f}{\partial y}(1, 2, 1, 3) = 16(2)(1) + 4(3) = 44$

   (d) $\dfrac{\partial f}{\partial z} = 4y$ and $\dfrac{\partial f}{\partial z}(0, 2, 1, 3) = 4(1) = 4$

4. $z_x = 12x^2y^4 + 4y$ and $z_y = 16x^3y^3 + 4x$

   (a) $z_{xx} = 24xy^4$ (b) $z_{yy} = 48x^3y^2$

   (c) $z_{xy} = 48x^2y^3 + 4$ (d) $z_{yx} = 48x^2y^3 + 4$

5. $z_x = 2x + 4e^{xy}(y) = 2x + 4ye^{xy}$

   Calculation of z_{xy} requires the Product Rule.

   $$z_{xy} = 0 + (4y)(e^{xy}x) + (e^{xy})(4) = 4xye^{xy} + 4e^{xy}$$

14.2 Exercises

1. If $z = x^4 - 5x^2 + 6x + 3y^3 - 5y + 7$,

   find $\dfrac{\partial z}{\partial x}$ and $\dfrac{\partial z}{\partial y}$.

2. If $z = x^5 - 6x + 4y^4 - y^2$, find $\dfrac{\partial z}{\partial x}$ and $\dfrac{\partial z}{\partial y}$.

3. If $z = x^3 + 4x^2y + 6y^2$, find z_x and z_y.

4. If $z = 3xy + y^2$, find z_x and z_y.

5. If $f(x, y) = (x^3 + 2y^2)^3$, find $\dfrac{\partial f}{\partial x}$ and $\dfrac{\partial f}{\partial y}$.

6. If $f(x, y) = (xy^3 + y)^2$, find $\dfrac{\partial f}{\partial x}$ and $\dfrac{\partial f}{\partial y}$.

7. If $f(x, y) = \sqrt{2x^2 - 5y^2}$, find f_x and f_y.

8. If $g(x, y) = \sqrt{xy - x}$, find g_x and g_y.

9. If $C(x, y) = 600 - 4xy + 10x^2y$, find $\dfrac{\partial C}{\partial x}$ and $\dfrac{\partial C}{\partial y}$.

10. If $C(x, y) = 1000 - 4x + xy^2$, find $\dfrac{\partial C}{\partial x}$ and $\dfrac{\partial C}{\partial y}$.

11. If $Q(s, t) = \dfrac{2s - 3t}{s^2 + t^2}$, find $\dfrac{\partial Q}{\partial s}$ and $\dfrac{\partial Q}{\partial t}$.

12. If $q = \dfrac{5p_1 + 4p_2}{p_1 + p_2}$, find $\dfrac{\partial q}{\partial p_1}$ and $\dfrac{\partial q}{\partial p_2}$.

13. If $z = e^{2x} + y \ln x$, find z_x and z_y.

14. If $z = \ln (1 + x^2y) - ye^{-x}$, find z_x and z_y.

15. If $f(x, y) = \ln (xy + 1)$, find $\dfrac{\partial f}{\partial x}$ and $\dfrac{\partial f}{\partial y}$.

16. If $f(x, y) = 100e^{xy}$, find $\dfrac{\partial f}{\partial x}$ and $\dfrac{\partial f}{\partial y}$.

17. Find the partial derivative of
$$f(x, y) = 4x^3 - 5xy + y^2$$
with respect to x at the point $(1, 2, -2)$.

18. Find the partial derivative of
$$f(x, y) = 3x^2 + 4x + 6xy$$
with respect to y at $x = 2$, $y = -1$.

19. Find the slope of the tangent in the positive x-direction to the surface $z = 5x^3 - 4xy$ at the point $(1, 2, -3)$.

20. Find the slope of the tangent in the positive y-direction to the surface $z = x^3 - 5xy$ at $(2, 1, -2)$.

21. Find the slope of the tangent in the positive y-direction to the surface $z = 3x + 2y - 7e^{xy}$ at $(3, 0, 2)$.

22. Find the slope of the tangent in the positive x-direction to the surface $z = 2xy + \ln (4x + 3y)$ at $(1, -1, -2)$.

23. If $u = f(w, x, y, z) = y^2 - x^2z + 4x$, find the following.
    (a) $\dfrac{\partial u}{\partial w}$ (b) $\dfrac{\partial u}{\partial x}$ (c) $\dfrac{\partial u}{\partial y}$ (d) $\dfrac{\partial u}{\partial z}$

24. If $u = x^2 + 3xy + xz$, find the following.
    (a) u_x (b) u_y (c) u_z

25. If $C(x_1, x_2, x_3) = 4x_1^2 + 5x_1x_2 + 6x_2^2 + x_3$, find the following.
    (a) $\dfrac{\partial C}{\partial x_1}$ (b) $\dfrac{\partial C}{\partial x_2}$ (c) $\dfrac{\partial C}{\partial x_3}$

26. If $f(x, y, z) = 2x\sqrt{yz - 1} + x^2z^3$, find the following.
    (a) $\dfrac{\partial f}{\partial x}$ (b) $\dfrac{\partial f}{\partial y}$ (c) $\dfrac{\partial f}{\partial z}$

27. If $z = x^2 + 4x - 5y^3$, find the following.
    (a) z_{xx} (b) z_{xy} (c) z_{yx} (d) z_{yy}

28. If $z = x^3 - 5y^2 + 4y + 1$, find the following.
    (a) z_{xx} (b) z_{xy} (c) z_{yx} (d) z_{yy}

29. If $z = x^2y - 4xy^2$, find the following.
    (a) z_{xx} (b) z_{xy} (c) z_{yx} (d) z_{yy}

30. If $z = xy^2 + 4xy - 5$, find the following.
    (a) z_{xx} (b) z_{xy} (c) z_{yx} (d) z_{yy}

31. If $f(x, y) = x^2 + e^{xy}$, find the following.
    (a) $\dfrac{\partial^2 f}{\partial x^2}$ (b) $\dfrac{\partial^2 f}{\partial y\, \partial x}$ (c) $\dfrac{\partial^2 f}{\partial x\, \partial y}$ (d) $\dfrac{\partial^2 f}{\partial y^2}$

32. If $z = xe^{xy}$, find the following.
    (a) z_{xx} (b) z_{yy} (c) z_{xy} (d) z_{yx}

33. If $f(x, y) = y^2 - \ln xy$, find the following.
    (a) $\dfrac{\partial^2 f}{\partial x^2}$ (b) $\dfrac{\partial^2 f}{\partial y\, \partial x}$ (c) $\dfrac{\partial^2 f}{\partial x\, \partial y}$ (d) $\dfrac{\partial^2 f}{\partial y^2}$

34. If $f(x, y) = x^3 + \ln (xy - 1)$, find the following.
    (a) $\dfrac{\partial^2 f}{\partial x^2}$ (b) $\dfrac{\partial^2 f}{\partial y^2}$ (c) $\dfrac{\partial^2 f}{\partial x\, \partial y}$ (d) $\dfrac{\partial^2 f}{\partial y\, \partial x}$

35. If $f(x, y) = x^3y + 4xy^4$, find $\left. \dfrac{\partial^2}{\partial x^2} f(x, y) \right|_{(1, -1)}$.

36. If $f(x, y) = x^4y^2 + 4xy$, find $\left. \dfrac{\partial^2}{\partial y^2} f(x, y) \right|_{(1, 2)}$.

37. If $f(x, y) = \dfrac{2x}{x^2 + y^2}$, find the following.
    (a) $\left. \dfrac{\partial^2 f}{\partial x^2} \right|_{(-1, 4)}$ (b) $\left. \dfrac{\partial^2 f}{\partial y^2} \right|_{(-1, 4)}$

38. If $f(x, y) = \dfrac{2y^2}{3xy + 4}$, find the following.
    (a) $\left. \dfrac{\partial^2 f}{\partial x^2} \right|_{(1, -2)}$ (b) $\left. \dfrac{\partial^2 f}{\partial y^2} \right|_{(1, -2)}$

39. If $z = x^2y + ye^{x^2}$, find $z_{yx}|_{(1, 2)}$.

40. If $z = xy^3 + x \ln y^2$, find $z_{xy}|_{(1, 2)}$.

41. If $z = x^2 - xy + 4y^3$, find z_{xyx}.

42. If $z = x^3 - 4x^2y + 5y^3$, find z_{yyx}.

43. If $w = 4x^3y + y^2z + z^3$, find the following.
    (a) w_{xxy} (b) w_{xyx} (c) w_{xyz}

44. If $w = 4xyz + x^3y^2z + x^3$, find the following.
    (a) w_{xyz} (b) w_{xzz} (c) w_{yyz}

APPLICATIONS

45. *Mortgage* When a homeowner has a 25-year variable-rate mortgage loan, the monthly payment R is a function of the amount of the loan A and the current interest rate i (as a percent); that is, $R = f(A, i)$. Interpret each of the following.
    (a) $f(100,000, 8) = 1289$
    (b) $\dfrac{\partial f}{\partial i}(100,000, 8) = 62.51$

46. *Mass transportation ridership* Suppose that in a certain city, the number of people N using the mass transportation system is a function of the fare f and the daily cost of downtown parking p, so that $N = N(f, p)$. Interpret each of the following.
    (a) $N(5, 10) = 6500$ (b) $\dfrac{\partial N}{\partial f}(5, 10) = -400$
    (c) $\dfrac{\partial N}{\partial p}(5, 10) = 250$

47. *Wilson's lot size formula* In economics, the most economical quantity Q of goods (TVs, dresses, gallons of paint, etc.) for a store to order is given by Wilson's lot size formula

$$Q = \sqrt{2KM/h}$$

where K is the cost of placing the order, M is the number of items sold per week, and h is the weekly holding costs for each item (the cost of storage space, utilities, taxes, security, etc.).

(a) Explain why $\dfrac{\partial Q}{\partial M}$ will be positive.

(b) Explain why $\dfrac{\partial Q}{\partial h}$ will be negative.

48. *Cost* Suppose that the total cost (in dollars) of producing a product is $C(x, y) = 25 + 2x^2 + 3y^2$, where x is the cost per pound for material and y is the cost per hour for labor.

(a) If material costs are held constant, at what rate will the total cost increase for each $1-per-hour increase in labor?

(b) If the labor costs are held constant, at what rate will the total cost increase for each increase of $1 in material cost?

49. *Pesticide* Suppose that the number of thousands of insects killed by two brands of pesticide is given by

$$f(x, y) = 10{,}000 - 6500e^{-0.01x} - 3500e^{-0.02y}$$

where x is the number of liters of brand 1 and y is the number of liters of brand 2. What is the rate of change of insect deaths with respect to the number of liters of brand 1 if 100 liters of each brand are currently being used? What does this mean?

50. *Profit* Suppose that the profit (in dollars) from the sale of Kisses and Kreams is given by

$$P(x, y) = 10x + 6.4y - 0.001x^2 - 0.025y^2$$

where x is the number of pounds of Kisses and y is the number of pounds of Kreams. Find $\partial P/\partial y$, and give the approximate rate of change of profit with respect to the number of pounds of Kreams that are sold if 100 pounds of Kisses and 16 pounds of Kreams are currently being sold. What does this mean?

51. *Utility* If $U = f(x, y)$ is the utility function for goods X and Y, the *marginal utility* of X is $\partial U/\partial x$ and the *marginal utility* of Y is $\partial U/\partial y$. If $U = x^2y^2$, find the marginal utility of

(a) X.   (b) Y.

52. *Utility* If the utility function for goods X and Y is $U = xy + y^2$, find the marginal utility of

(a) X.   (b) Y.

53. *Production* Suppose that the output Q (in units) of a certain company is $Q = 75K^{1/3}L^{2/3}$, where K is the capital expenditures in thousands of dollars and L is the number of labor hours. Find $\partial Q/\partial K$ and $\partial Q/\partial L$ when capital expenditures are $729,000 and the labor hours total 5832. Interpret each answer.

54. *Production* Suppose that the production Q (in gallons of paint) of a paint manufacturer can be modeled by $Q = 140K^{1/2}L^{1/2}$, where K is the company's capital expenditures in thousands of dollars and L is the size of the labor force (in hours worked). Find $\partial Q/\partial K$ and $\partial Q/\partial L$ when capital expenditures are $250,000 and the labor hours are 1225. Interpret each answer.

Wind chill **Dr. Paul Siple conducted studies testing the effect of wind on the formation of ice at various temperatures and developed the concept of wind chill, which we hear reported during winter weather reports. The wind chill temperatures for selected air temperatures and wind speeds are shown in the following table. For example, the table shows that an air temperature of 15°F together with a wind speed of 35 mph feels the same as an air temperature of −27°F when there is no wind.**

Wind Speed (mph)	Air Temperature (°F)						
	35	25	15	5	−5	−15	−25
5	33	21	12	0	−10	−21	−31
15	16	2	−11	−25	−38	−51	−65
25	8	−7	−22	−36	−51	−66	−81
35	4	−12	−27	−43	−58	−74	−89
45	2	−14	−30	−46	−62	−78	−93

Source: National Weather Service, NOAA, U.S. Department of Commerce

One form of the formula that meteorologists use to calculate wind chill temperature is

$$WC = 48.064 + 0.474t - 0.020ts - 1.85s + 0.304t\sqrt{s} - 27.74\sqrt{s}$$

where s is wind speed and t is the actual air temperature. Use this equation to answer Problems 55 and 56.

55. (a) To see how the wind chill temperature changes with wind speed, find $\partial WC/\partial s$.

(b) Find $\partial WC/\partial s$ when the temperature is 10°F and the wind speed is 25 mph. What does this mean?

56. (a) To see how wind chill temperature changes with temperature, find $\partial WC/\partial t$.

(b) Find $\partial WC/\partial t$ when the temperature is 10°F and the wind speed is 25 mph. What does this mean?

14.3

OBJECTIVES
- To evaluate cost functions at given levels of production
- To find marginal costs from total cost and joint cost functions
- To find marginal productivity for given production functions
- To find marginal demand functions from demand functions for a pair of related products

Applications of Functions of Two Variables in Business and Economics

◗ Application Preview

Suppose that the **joint cost function** for two commodities is

$$C = 50 + x^2 + 8xy + y^3$$

where x and y represent the quantities of each commodity and C is the total cost for the two commodities. We can take the partial derivative of C with respect to x to find the **marginal cost** of the first product and we can take the partial derivative of C with respect to y to find the marginal cost of the second product. (See Example 1.) This is one of three types of applications that we will consider in this section. In the second case, we consider production functions and revisit Cobb-Douglas production functions. Marginal productivity is introduced and its meaning is explained. Finally, we consider demand functions for two products in a competitive market. Partial derivatives are used to define marginal demands, and these marginals are used to classify the products as competitive or complementary.

Joint Cost and Marginal Cost

Suppose that a firm produces two products using the same inputs in different proportions. In such a case the **joint cost function** is of the form $C = Q(x, y)$, where x represents the quantity of product X and y represents the quantity of product Y. Then $\partial C/\partial x$ is the **marginal cost** of the first product and $\partial C/\partial y$ is the marginal cost of the second product.

◗ EXAMPLE 1 *Joint Cost* (Application Preview)

If the joint cost function for two products is

$$C = Q(x, y) = 50 + x^2 + 8xy + y^3$$

where x represents the quantity of product X and y represents the quantity of product Y, find the marginal cost with respect to the number of units of:

(a) Product X (b) Product Y (c) Product X at $(5, 3)$ (d) Product Y at $(5, 3)$

Solution

(a) The marginal cost with respect to the number of units of product X is $\partial C/\partial x = 2x + 8y$.

(b) The marginal cost with respect to the number of units of product Y is $\dfrac{\partial C}{\partial y} = 8x + 3y^2$.

(c) $\dfrac{\partial C}{\partial x}\bigg|_{(5, 3)} = 2(5) + 8(3) = 34$

Thus if 5 units of product X and 3 units of product Y are produced, the total cost will increase approximately \$34 for a unit increase in product X if y is held constant.

(d) $\dfrac{\partial C}{\partial y}\bigg|_{(5, 3)} = 8(5) + 3(3)^2 = 67$

Thus if 5 units of product X and 3 units of product Y are produced, the total cost will increase approximately \$67 for a unit increase in product Y if x is held constant.

● ***Checkpoint***

1. If the joint cost in dollars for two products is given by

$$C = 100 + 3x + 10xy + y^2$$

find the marginal cost with respect to (a) the number of units of Product X and (b) the number of units of Product Y at $(7, 3)$.

Production Functions

An important problem in economics concerns how the factors necessary for production determine the output of a product. For example, the output of a product depends on available labor, land, capital, material, and machines. If the amount of output z of a product depends on the amounts of two inputs x and y, then the quantity z is given by the **production function** $z = f(x, y)$.

● **EXAMPLE 2** *Crop Harvesting*

Suppose that it is known that z bushels of a crop can be harvested according to the function

$$z = (21)\frac{6xy + 4x^2 - 3y}{2x + 0.01y}$$

when $100x$ work-hours of labor are employed on y acres of land. What would be the output (in bushels) if 200 work-hours were used on 300 acres?

Solution
Because $z = f(x, y)$,

$$f(2, 300) = (21)\frac{6(2)(300) + 4(2)^2 - 3(300)}{2(2) + 3}$$

$$= (21)\frac{3600 + 16 - 900}{7} = 8148 \quad \text{(bushels)}$$

If $z = f(x, y)$ is a production function, $\partial z/\partial x$ represents the rate of change in the output z with respect to input x while input y remains constant. This partial derivative is called the **marginal productivity of x**. The partial derivative $\partial z/\partial y$ is the **marginal productivity of y** and measures the rate of change of z with respect to input y.

Marginal productivity (for either input) will be positive over a wide range of inputs, but it increases at a decreasing rate, and it may eventually reach a point where it no longer increases and begins to decrease.

● **EXAMPLE 3** *Production*

If a production function is given by $z = 5x^{1/2}y^{1/4}$, find the marginal productivity of

(a) x.   (b) y.

Solution

(a) $\dfrac{\partial z}{\partial x} = \dfrac{5}{2}x^{-1/2}y^{1/4}$ (b) $\dfrac{\partial z}{\partial y} = \dfrac{5}{4}x^{1/2}y^{-3/4}$

Note that the marginal productivity of x is positive for all values of x but that it decreases as x gets larger (because of the negative exponent). The same is true for the marginal productivity of y.

● *Checkpoint*

2. If the production function for a product is

$$P = 10x^{1/4}y^{1/2}$$

find the marginal productivity of x.

Calculator
Note If we have a production function and fix a value for one variable, then we can use a graphing calculator to analyze the marginal productivity with respect to the other variable. ■

● EXAMPLE 4 *Production*

Suppose the Cobb-Douglas production function for a company is given by

$$z = 100x^{1/4}y^{3/4}$$

where x is the company's capital investment (in dollars) and y is the size of the labor force (in work-hours).

(a) Find the marginal productivity of x.
(b) If the current labor force is 625 work-hours, substitute $y = 625$ in your answer to (a) and graph the result.
(c) From the graph in (b), what can be said about the effect on production of additional capital investment when the work-hours remain at 625?
(d) Find the marginal productivity of y.
(e) If current capital investment is \$10,000, substitute $x = 10,000$ in your answer to (d) and graph the result.
(f) From the graph in (e), what can be said about the effect on production of additional work-hours when capital investment remains at \$10,000?

Solution
(a) $z_x = 25x^{-3/4}y^{3/4}$
(b) If $y = 625$, then z_x becomes

$$z_x = 25x^{-3/4}(625)^{3/4} = 25\left(\frac{1}{x^{3/4}}\right)(125) = \frac{3125}{x^{3/4}}$$

The graph of z_x can be limited to Quadrant I because the capital investment is $x > 0$, and hence $z_x > 0$. Knowledge of asymptotes can help us determine range values for x and z_x that give an accurate graph. See Figure 14.6.

(c) Figure 14.6 shows that $z_x > 0$ for $x > 0$. This means that any increases in capital investment will result in increases in productivity. However, Figure 14.6 also shows that z_x is decreasing for $x > 0$, which means that as capital investment increases, productivity increases, but at a slower rate.

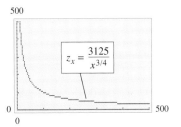

Figure 14.6

(d) $z_y = 75x^{1/4}y^{-1/4}$
(e) If $x = 10,000$, then z_y becomes

$$z_y = 75(10,000)^{1/4}\left(\frac{1}{y^{1/4}}\right) = \frac{750}{y^{1/4}}$$

The graph is shown in Figure 14.7 on the next page.

(f) Figure 14.7 also shows that $z_y > 0$ when $y > 0$, so increasing work-hours increases productivity. Note that z_y is decreasing for $y > 0$ but that it does so more slowly than z_x (try large values of x and y to confirm this). This indicates that increases in work-hours have a diminishing impact on productivity, but still a more significant one than increases in capital expenditures (as seen in Figure 14.6).

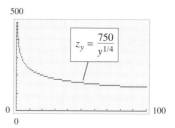

Figure 14.7

Demand Functions

Suppose that two products are sold at prices p_1 and p_2 (both in dollars) in a competitive market consisting of a fixed number of consumers with given tastes and incomes. Then the amount of each *one* of the products demanded by the consumers is dependent on the prices of *both* products on the market. If q_1 represents the demand for the number of units of the first product, then $q_1 = f(p_1, p_2)$ is the **demand function** for that product. The graph of such a function is called a **demand surface.** An example of a demand function in two variables is $q_1 = 400 - 2p_1 - 4p_2$. Here q_1 is a function of two variables p_1 and p_2. If $p_1 = \$10$ and $p_2 = \$20$, the demand would equal $400 - 2(10) - 4(20) = 300$ units.

● **EXAMPLE 5** *Demand*

The demand functions for two products are

$$q_1 = 50 - 5p_1 - 2p_2$$
$$q_2 = 100 - 3p_1 - 8p_2$$

where q_1 and q_2 are the numbers of units and p_1 and p_2 are in dollars.

(a) What is the demand for each of the products if the price of the first is $p_1 = \$5$ and the price of the second is $p_2 = \$8$?
(b) Find a pair of prices p_1 and p_2 such that the demands for product 1 and product 2 are equal.

Solution
(a)
$$q_1 = 50 - 5(5) - 2(8) = 9$$
$$q_2 = 100 - 3(5) - 8(8) = 21$$

Thus if these are the prices, the demand for product 2 is higher than the demand for product 1.

(b) We want q_1 to equal q_2. Setting $q_1 = q_2$, we see that

$$50 - 5p_1 - 2p_2 = 100 - 3p_1 - 8p_2$$
$$6p_2 - 50 = 2p_1$$
$$p_1 = 3p_2 - 25$$

Now, any pair of positive values that satisfies this equation will make the demands equal. Letting $p_2 = 10$, we see that $p_1 = 5$ will satisfy the equation. Thus the prices $p_1 = 5$ and $p_2 = 10$ will make the demands equal. The prices $p_1 = 2$ and $p_2 = 9$ will also make the demands equal. Many pairs of values (that is, all those satisfying $p_1 = 3p_2 - 25$) will equalize the demands.

If the demand functions for a pair of related products, product 1 and product 2, are $q_1 = f(p_1, p_2)$ and $q_2 = g(p_1, p_2)$, respectively, then the partial derivatives of q_1 and q_2 are called **marginal demand functions.**

$\dfrac{\partial q_1}{\partial p_1}$ is the marginal demand of q_1 with respect to p_1.

$\dfrac{\partial q_1}{\partial p_2}$ is the marginal demand of q_1 with respect to p_2.

$\dfrac{\partial q_2}{\partial p_1}$ is the marginal demand of q_2 with respect to p_1.

$\dfrac{\partial q_2}{\partial p_2}$ is the marginal demand of q_2 with respect to p_2.

For typical demand functions, if the price of product 2 is fixed, the demand for product 1 will decrease as its price p_1 increases. In this case the marginal demand of q_1 with respect to p_1 will be negative; that is, $\partial q_1/\partial p_1 < 0$. Similarly, $\partial q_2/\partial p_2 < 0$.

But what about $\partial q_2/\partial p_1$ and $\partial q_1/\partial p_2$? If $\partial q_2/\partial p_1$ and $\partial q_1/\partial p_2$ are both positive, the two products are **competitive** because an increase in price p_1 will result in an increase in demand for product 2 (q_2) if the price p_2 is held constant, and an increase in price p_2 will increase the demand for product 1 (q_1) if p_1 is held constant. Stated more simply, an increase in the price of one of the two products will result in an increased demand for the other, so the products are in competition. For example, an increase in the price of a Japanese automobile will result in an increase in demand for an American automobile if the price of the American automobile is held constant.

If $\partial q_2/\partial p_1$ and $\partial q_1/\partial p_2$ are both negative, the products are **complementary** because an increase in the price of one product will cause a decrease in demand for the other product if the price of the second product doesn't change. Under these conditions, a *decrease* in the price of product 1 will result in an *increase* in the demand for product 2, and a decrease in the price of product 2 will result in an increase in the demand for product 1. For example, a decrease in the price of gasoline will result in an increase in the demand for large automobiles.

If the signs of $\partial q_2/\partial p_1$ and $\partial q_1/\partial p_2$ are different, the products are neither competitive nor complementary. This situation rarely occurs but is possible.

● **EXAMPLE 6** *Demand*

The demand functions for two related products, product 1 and product 2, are given by

$$q_1 = 400 - 5p_1 + 6p_2 \qquad q_2 = 250 + 4p_1 - 5p_2$$

(a) Determine the four marginal demands.
(b) Are product 1 and product 2 complementary or competitive?

Solution

(a) $\dfrac{\partial q_1}{\partial p_1} = -5 \qquad \dfrac{\partial q_2}{\partial p_2} = -5 \qquad \dfrac{\partial q_1}{\partial p_2} = 6 \qquad \dfrac{\partial q_2}{\partial p_1} = 4$

(b) Because $\partial q_1/\partial p_2$ and $\partial q_2/\partial p_1$ are positive, products 1 and 2 are competitive.

● *Checkpoint*

3. If the demand functions for two products are

$$q_1 = 200 - 3p_1 - 4p_2 \quad \text{and} \quad q_2 = 50 - 6p_1 - 5p_2$$

find the marginal demand of
(a) q_1 with respect to p_1.   (b) q_2 with respect to p_2.

● *Checkpoint Solutions*

1. (a) $\dfrac{\partial C}{\partial x} = 3 + 10y$

   (b) $\dfrac{\partial C}{\partial y} = 10x + 2y$ $\dfrac{\partial C}{\partial y}(7, 3) = 10(7) + 2(3) = 76$ (dollars per unit)

2. $\dfrac{\partial P}{\partial x} = \dfrac{2.5y^{1/2}}{x^{3/4}}$

3. (a) $\dfrac{\partial q_1}{\partial p_1} = -3$ (b) $\dfrac{\partial q_2}{\partial p_2} = -5$

14.3 Exercises

JOINT COST AND MARGINAL COST

1. The cost (in dollars) of manufacturing one item is given by

$$C(x, y) = 30 + 3x + 5y$$

   where x is the cost of 1 hour of labor and y is the cost of 1 pound of material. If the hourly cost of labor is $10, and the material costs $3 per pound, what is the cost of manufacturing one of these items?

2. The manufacture of 1 unit of a product has a cost (in dollars) given by

$$C(x, y, z) = 10 + 8x + 3y + z$$

   where x is the cost of 1 pound of one raw material, y is the cost of 1 pound of a second raw material, and z is the cost of 1 work-hour of labor. If the cost of the first raw material is $16 per pound, the cost of the second raw material is $8 per pound, and labor costs $18 per work-hour, what will it cost to produce 1 unit of the product?

3. The total cost of producing 1 unit of a product is

$$C(x, y) = 30 + 2x + 4y + \frac{xy}{50} \quad \text{dollars}$$

   where x is the cost per pound of raw materials and y is the cost per hour of labor.
   (a) If labor costs are held constant, at what rate will the total cost increase for each increase of $1 per pound in material cost?
   (b) If material costs are held constant, at what rate will the total cost increase for each $1 per hour increase in labor costs?

4. The total cost of producing an item is

$$C(x, y) = 40 + 4x + 6y + \frac{x^2 y}{100} \quad \text{dollars}$$

   where x is the cost per pound of raw materials and y is the cost per hour for labor. How will an increase of

   (a) $1 per pound of raw materials affect the total cost?
   (b) $1 per hour in labor costs affect the total cost?

5. The total cost of producing 1 unit of a product is given by

$$C(x, y) = 20x + 70y + \frac{x^2}{1000} + \frac{xy^2}{100} \quad \text{dollars}$$

   where x represents the cost per pound of raw materials and y represents the hourly rate for labor. The present cost for raw materials is $10 per pound and the present hourly rate for labor is $12. How will an increase of
   (a) $1 per pound for raw materials affect the total cost?
   (b) $1 per hour in labor costs affect the total cost?

6. The total cost of producing 1 unit of a product is given by

$$C(x, y) = 30 + 10x^2 + 20y - xy \quad \text{dollars}$$

   where x is the hourly labor rate and y is the cost per pound of raw materials. The current hourly rate is $15, and the raw materials cost $6 per pound. How will an increase of
   (a) $1 per pound for the raw materials affect the total cost?
   (b) $1 in the hourly labor rate affect the total cost?

7. The joint cost (in dollars) for two products is given by

$$C(x, y) = 30 + x^2 + 3y + 2xy$$

   where x represents the quantity of product X produced and y represents the quantity of product Y produced.
   (a) Find and interpret the marginal cost with respect to x if 8 units of product X and 10 units of product Y are produced.
   (b) Find and interpret the marginal cost with respect to y if 8 units of product X and 10 units of product Y are produced.

8. The joint cost (in dollars) for products X and Y is given by

$$C(x, y) = 40 + 3x^2 + y^2 + xy$$

where x represents the quantity of X and y represents the quantity of Y.

(a) Find and interpret the marginal cost with respect to x if 20 units of product X and 15 units of product Y are produced.

(b) Find and interpret the marginal cost with respect to y if 20 units of X and 15 units of Y are produced.

9. If the joint cost function for two products is

$$C(x, y) = x\sqrt{y^2 + 1} \qquad \text{dollars}$$

(a) find the marginal cost (function) with respect to x.
(b) find the marginal cost with respect to y.

10. Suppose the joint cost function for x units of product X and y units of product Y is given by

$$C(x, y) = 2500\sqrt{xy + 1} \qquad \text{dollars}$$

Find the marginal cost with respect to
(a) x.   (b) y.

11. Suppose that the joint cost function for two products is

$$C(x, y) = 1200 \ln(xy + 1) + 10{,}000 \qquad \text{dollars}$$

Find the marginal cost with respect to
(a) x.   (b) y.

12. Suppose that the joint cost function for two products is

$$C(x, y) = y \ln(x + 1) \qquad \text{dollars}$$

Find the marginal cost with respect to
(a) x.   (b) y.

PRODUCTION FUNCTIONS

13. Suppose that the production function for a product is $z = \sqrt{4xy}$, where x represents the number of work-hours per month and y is the number of available machines. Determine the marginal productivity of
(a) x.   (b) y.

14. Suppose the production function for a product is

$$z = 60x^{2/5}y^{3/5}$$

where x is the capital expenditures and y is the number of work-hours. Find the marginal productivity of
(a) x.   (b) y.

15. Suppose that the production function for a product is $z = \sqrt{x} \ln(y + 1)$, where x represents the number of work-hours and y represents the available capital (per week). Find the marginal productivity of
(a) x.   (b) y.

16. Suppose that a company's production function for a certain product is

$$z = (x + 1)^{1/2} \ln(y^2 + 1)$$

where x is the number of work-hours of unskilled labor and y is the number of work-hours of skilled labor. Find the marginal productivity of
(a) x.   (b) y.

For Problems 17–19, suppose that the number of crates of an agricultural product is given by

$$z = \frac{11xy - 0.0002x^2 - 5y}{0.03x + 3y}$$

where x is the number of hours of labor and y is the number of acres of the crop.

17. Find the output when $x = 300$ and $y = 500$.
18. Find and interpret the marginal productivity of the number of acres of the crop (y) when $x = 300$ and $y = 500$.
19. Find and interpret the marginal productivity of the number of hours of labor (x) when $x = 300$ and $y = 500$.
20. If a production function is given by $z = 12x^{3/4}y^{1/3}$, find the marginal productivity of
(a) x.   (b) y.
21. Suppose the Cobb-Douglas production function for a company is given by

$$z = 400x^{3/5}y^{2/5}$$

where x is the company's capital investment and y is the size of the labor force (in work-hours).

(a) Find the marginal productivity of x.
(b) If the current labor force is 1024 work-hours, substitute $y = 1024$ in your answer to part (a) and graph the result.
(c) Find the marginal productivity of y.
(d) If the current capital investment is \$59,049, substitute $x = 59{,}049$ in your answer to part (c) and graph the result.
(e) Interpret the graphs in parts (b) and (d) with regard to what they say about the effects on productivity of an increased capital investment (part b) and of an increased labor force (part d).

22. Suppose the Cobb-Douglas production function for a company is given by

$$z = 300x^{2/3}y^{1/3}$$

where x is the company's capital investment and y is the size of the labor force (in work-hours).

(a) Find the marginal productivity of x.
(b) If the current labor force is 729 work-hours, substitute $y = 729$ in your answer to part (a) and graph the result.
(c) Find the marginal productivity of y.
(d) If the current capital investment is \$27,000, substitute $x = 27{,}000$ in your answer to part (c) and graph the result.
(e) Interpret the graphs in parts (b) and (d) with regard to what they say about the effects on productivity of an increased capital investment (part b) and of an increased labor force (part d).

DEMAND FUNCTIONS

In Problems 23–26, prices p_1 and p_2 are in dollars and q_1 and q_2 are numbers of units.

23. The demand functions for two products are given by

$$q_1 = 300 - 8p_1 - 4p_2$$
$$q_2 = 400 - 5p_1 - 10p_2$$

Find the demand for each of the products if the price of the first is $p_1 = \$10$ and the price of the second is $p_2 = \$8$.

24. The demand functions for two products are given by

$$q_1 = 900 - 9p_1 + 2p_2$$
$$q_2 = 1200 + 6p_1 - 10p_2$$

Find the demands q_1 and q_2 if $p_1 = \$10$ and $p_2 = \$12$.

25. Find a pair of prices p_1 and p_2 such that the demands for the two products in Problem 23 will be equal.

26. Find a pair of prices p_1 and p_2 such that the demands for the two products in Problem 24 will be equal.

In Problems 27–30, the demand functions for q_A and q_B units of two related products, A and B, are given.

Complete parts (a)–(e) for each problem. Assume p_A and p_B are in dollars.

(a) Find the marginal demand of q_A with respect to p_A.
(b) Find the marginal demand of q_A with respect to p_B.
(c) Find the marginal demand of q_B with respect to p_B.
(d) Find the marginal demand of q_B with respect to p_A.
(e) Are the two goods competitive or complementary?

27. $\begin{cases} q_A = 400 - 3p_A - 2p_B \\ q_B = 250 - 5p_A - 6p_B \end{cases}$

28. $\begin{cases} q_A = 600 - 4p_A + 6p_B \\ q_B = 1200 + 8p_A - 4p_B \end{cases}$

29. $\begin{cases} q_A = 5000 - 50p_A - \dfrac{600}{p_B + 1} \\ q_B = 10{,}000 - \dfrac{400}{p_A + 4} + \dfrac{400}{p_B + 4} \end{cases}$

30. $\begin{cases} q_A = 2500 + \dfrac{600}{p_A + 2} - 40p_B \\ q_B = 3000 - 100p_A + \dfrac{400}{p_B + 5} \end{cases}$

14.4   Maxima and Minima

OBJECTIVES

● To find relative maxima, minima, and saddle points of functions of two variables

● To apply linear regression formulas

◑ Application Preview

Adele Lighting manufactures 20-inch lamps and 31-inch lamps. Suppose that x is the number of thousands of 20-inch lamps and that the demand for these is given by $p_1 = 50 - x$, where p_1 is in dollars. Similarly, suppose that y is the number of thousands of 31-inch lamps and that the demand for these is given by $p_2 = 60 - 2y$, where p_2 is also in dollars. Adele Lighting's joint cost function for these lamps is $C = 2xy$ (in thousands of dollars). Therefore, Adele Lighting's profit (in thousands of dollars) is a function of the two variables x and y. In order to determine Adele's maximum profit, we need to develop methods for finding maximum values for a function of two variables. (See Example 4.)

In this section we will find relative maxima and minima of functions of two variables, and use them to solve applied optimization problems. We will also use minimization of functions of two variables to develop the linear regression formulas. Recall that we first used linear regression in Chapter 2.

In our study of differentiable functions of one variable, we saw that for a relative maximum or minimum to occur at a point, the tangent line to the curve had to be horizontal at that point. The function $z = f(x, y)$ describes a surface in three dimensions. If all partial derivatives of $f(x, y)$ exist, then there must be a horizontal plane tangent to the surface at a point in order for the surface to have a relative maximum at that point (see Figure 14.8a) or a minimum at that point (see Figure 14.8b). But if the plane tangent to the surface at the point is horizontal, then all the tangent lines to the surface at that point must also be horizontal, for they lie in the tangent plane. In particular, the tangent line in the direction of the *x*-axis will be horizontal, so $\partial z/\partial x = 0$ at the point; and the tangent line in the direction of the *y*-axis will be horizontal, so $\partial z/\partial y = 0$ at the point. Thus those points where *both* $\partial z/\partial x = 0$ and $\partial z/\partial y = 0$ are called **critical points** for the surface.

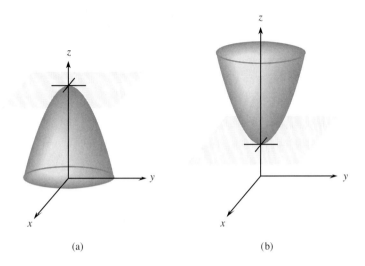

Figure 14.8   (a)   (b)

How can we determine whether a critical point is a relative maximum, a relative minimum, or neither of these? Finding that $\partial^2 z/\partial x^2 < 0$ and $\partial^2 z/\partial y^2 < 0$ is not enough to tell us that we have a relative maximum. The "second-derivative" test we must use involves the values of the second partial derivatives and the value of *D* at the critical point (a, b), where *D* is defined as follows:

$$D = \frac{\partial^2 z}{\partial x^2} \cdot \frac{\partial^2 z}{\partial y^2} - \left(\frac{\partial^2 z}{\partial x \partial y}\right)^2$$

We shall state, without proof, the result that determines whether there is a relative maximum, a relative minimum, or neither at the critical point (a, b).

Test for Maxima and Minima

Let $z = f(x, y)$ be a function for which both

$$\frac{\partial z}{\partial x} = 0 \quad \text{and} \quad \frac{\partial z}{\partial y} = 0 \quad \text{at a point } (a, b)$$

and suppose that all second partial derivatives are continuous there. Evaluate

$$D = \frac{\partial^2 z}{\partial x^2} \cdot \frac{\partial^2 z}{\partial y^2} - \left(\frac{\partial^2 z}{\partial x \partial y}\right)^2$$

at the critical point (a, b), and conclude the following:

a. If $D > 0$ and $\partial^2 z/\partial x^2 > 0$ at (a, b), then a relative minimum occurs at (a, b). In this case, $\partial^2 z/\partial y^2 > 0$ at (a, b) also.

b. If $D > 0$ and $\partial^2 z/\partial x^2 < 0$ at (a, b), then a relative maximum occurs at (a, b). In this case, $\partial^2 z/\partial y^2 < 0$ at (a, b) also.

c. If $D < 0$ at (a, b), there is neither a relative maximum nor a relative minimum at (a, b).

d. If $D = 0$ at (a, b), the test fails; investigate the function near the point.

We can test for relative maxima and minima by using the following procedure.

Maxima and Minima of $z = f(x, y)$

Procedure	Example
To find relative maxima and minima of $z = f(x, y)$:	Test $z = 4 - 4x^2 - y^2$ for relative maxima and minima.
1. Find $\partial z/\partial x$ and $\partial z/\partial y$.	1. $\dfrac{\partial z}{\partial x} = -8x;\ \dfrac{\partial z}{\partial y} = -2y$
2. Find the point(s) that satisfy *both* $\partial z/\partial x = 0$ and $\partial z/\partial y = 0$. These are the critical points.	2. $\dfrac{\partial z}{\partial x} = 0$ if $x = 0$. $\dfrac{\partial z}{\partial y} = 0$ if $y = 0$. The critical point is $(0, 0, 4)$.
3. Find all second partial derivatives.	3. $\dfrac{\partial^2 z}{\partial x^2} = -8;\ \dfrac{\partial^2 z}{\partial y^2} = -2;\ \dfrac{\partial^2 z}{\partial x\, \partial y} = \dfrac{\partial^2 z}{\partial y\, \partial x} = 0$
4. Evaluate D at each critical point.	4. At $(0, 0)$, $D = (-8)(-2) - 0^2 = 16$.
5. Use the test for maxima and minima to determine whether relative maxima or minima occur.	5. $D > 0$, $\partial^2 z/\partial x^2 < 0$, and $\partial^2 z/\partial y^2 < 0$. A relative maximum occurs at $(0, 0)$. See Figure 14.9.

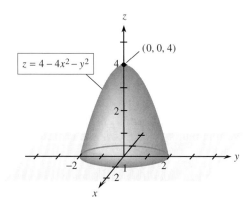

Figure 14.9

● **EXAMPLE 1** *Relative Minima*

Test $z = x^2 + y^2 - 2x + 1$ for relative maxima and minima.

Solution

1. $\dfrac{\partial z}{\partial x} = 2x - 2;\quad \dfrac{\partial z}{\partial y} = 2y$

2. $\dfrac{\partial z}{\partial x} = 0$ if $x = 1$. $\dfrac{\partial z}{\partial y} = 0$ if $y = 0$.

   Both are 0 if $x = 1$ *and* $y = 0$, so the critical point is $(1, 0, 0)$.

3. $\dfrac{\partial^2 z}{\partial x^2} = 2;\quad \dfrac{\partial^2 z}{\partial y^2} = 2;\quad \dfrac{\partial^2 z}{\partial x \partial y} = \dfrac{\partial^2 z}{\partial y\, \partial x} = 0$

4. At $(1, 0)$, $D = 2 \cdot 2 - 0^2 = 4$.

5. $D > 0$, $\partial^2 z/\partial x^2 > 0$, and $\partial^2 z/\partial y^2 > 0$.
   A relative minimum occurs at $(1, 0)$.
   (See Figure 14.10.)

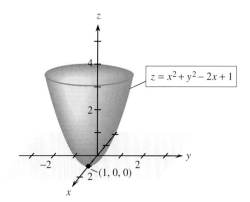

$z = x^2 + y^2 - 2x + 1$

Figure 14.10

● **EXAMPLE 2** *Saddle Points*

Test $z = y^2 - x^2$ for relative maxima and minima.

Solution

1. $\dfrac{\partial z}{\partial x} = -2x$; $\dfrac{\partial z}{\partial y} = 2y$

2. $\dfrac{\partial z}{\partial x} = 0$ if $x = 0$; $\dfrac{\partial z}{\partial y} = 0$ if $y = 0$.
   Thus both equal 0 if $x = 0$, $y = 0$. The critical point is $(0, 0, 0)$.

3. $\dfrac{\partial^2 z}{\partial x^2} = -2$; $\dfrac{\partial^2 z}{\partial y^2} = 2$; $\dfrac{\partial^2 z}{\partial x\, \partial y} = \dfrac{\partial^2 z}{\partial y\, \partial x} = 0$

4. $D = (-2)(2) - 0 = -4$

5. $D < 0$, so the critical point is neither a relative maximum nor a relative minimum. As Figure 14.11 shows, the surface formed has the shape of a saddle. For this reason, critical points that are neither relative maxima nor relative minima are called **saddle points.**

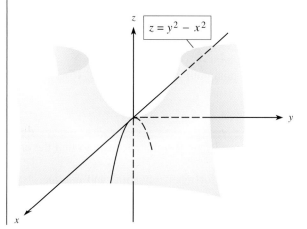

$z = y^2 - x^2$

Figure 14.11

The following example involves a surface with two critical points.

● **EXAMPLE 3** *Maxima and Minima*

Test $z = x^3 + y^2 + 6xy + 24x$ for maxima and minima.

Solution

1. $\dfrac{\partial z}{\partial x} = 3x^2 + 6y + 24$ and $\dfrac{\partial z}{\partial y} = 2y + 6x$

2. $\dfrac{\partial z}{\partial x} = 0$ if $0 = 3x^2 + 6y + 24$; that is, if $y = -\dfrac{1}{2}x^2 - 4$.

    $\dfrac{\partial z}{\partial y} = 0$ if $0 = 2y + 6x$; that is, if $y = -3x$.

    Because *both* conditions must be satisfied, we can equate these two expressions for y and obtain

    $$-3x = -\frac{1}{2}x^2 - 4$$

    $$\frac{1}{2}x^2 - 3x + 4 = 0 \quad \text{or} \quad x^2 - 6x + 8 = 0$$

    $$(x - 2)(x - 4) = 0 \quad \text{so} \quad x = 2 \text{ or } x = 4$$

    When $x = 2$, $y = -\dfrac{1}{2}(2)^2 - 4 = -6$ and $z = 2^3 + (-6)^2 + 6(2)(-6) + 24(2) = 20$, so one critical point is $(2, -6, 20)$.

    When $x = 4$, $y = -\dfrac{1}{2}(4)^2 - 4 = -12$ and $z = 4^3 + (-12)^2 + 6(4)(-12) + 24(4) = 16$, so the second critical point is $(4, -12, 16)$.

3. $\dfrac{\partial^2 z}{\partial x^2} = 6x$; $\dfrac{\partial^2 z}{\partial y^2} = 2$; $\dfrac{\partial^2 z}{\partial x\,\partial y} = \dfrac{\partial^2 z}{\partial y\,\partial x} = 6$

4. $D = (6x)(2) - 6^2$

5. At $(2, -6)$, $\dfrac{\partial^2 z}{\partial x^2} = 6 \cdot 2 = 12 > 0$, $\dfrac{\partial^2 z}{\partial y^2} = 2 > 0$, and $D = (6 \cdot 2)(2) - 6^2 = -12 < 0$, so a saddle point occurs at $(2, -6, 20)$.

    At $(4, -12)$, $\dfrac{\partial^2 z}{\partial x^2} = 6 \cdot 4 = 24 > 0$, $\dfrac{\partial^2 z}{\partial y^2} = 2 > 0$, and $D = (6 \cdot 4)(2) - 6^2 = 12 > 0$, so a relative minimum occurs at $(4, -12, 16)$.

● *Checkpoint*

Suppose that $z = 4 - x^2 - y^2 + 2x - 4y$.
1. Find z_x and z_y.
2. Solve $z_x = 0$ and $z_y = 0$ simultaneously to find the critical point(s) for the graph of this function.
3. Test the point(s) for relative maxima and minima.

◐ **EXAMPLE 4** *Maximum Profit* **(Application Preview)**

Adele Lighting manufactures 20-inch lamps and 31-inch lamps. Suppose that x is the number of thousands of 20-inch lamps and that the demand for these is given by $p_1 = 50 - x$, where p_1 is in dollars. Similarly, suppose that y is the number of thousands of 31-inch lamps and that the demand for these is given by $p_2 = 60 - 2y$, where p_2 is also in dollars. Adele Lighting's joint cost function for these lamps is $C = 2xy$ (in thousands of dollars). Therefore, Adele Lighting's profit (in thousands of dollars) is a function of the two variables x and y. Determine Adele's maximum profit.

Solution

The profit function is $P(x, y) = p_1x + p_2y - C(x, y)$. Thus,

$$P(x, y) = (50 - x)x + (60 - 2y)y - 2xy$$
$$= 50x - x^2 + 60y - 2y^2 - 2xy$$

gives the profit in thousands of dollars. To maximize the profit, we proceed as follows.

$$P_x = 50 - 2x - 2y \quad \text{and} \quad P_y = 60 - 4y - 2x$$

Solving $P_x = 0$ and $P_y = 0$ simultaneously, we have

$$\begin{cases} 0 = 50 - 2x - 2y \\ 0 = 60 - 2x - 4y \end{cases}$$

Subtraction gives $-10 + 2y = 0$, so $y = 5$. Thus $0 = 40 - 2x$, so $x = 20$. Now

$$P_{xx} = -2, \quad P_{yy} = -4, \quad \text{and} \quad P_{xy} = -2, \quad \text{and}$$
$$D = (P_{xx})(P_{yy}) - (P_{xy})^2 = (-2)(-4) - (-2)^2 = 4$$

Because $P_{xx} < 0$, $P_{yy} < 0$, and $D > 0$, the values $x = 20$ and $y = 5$ yield maximum profit. Therefore, when $x = 20$ and $y = 5$, $p_1 = 30$, $p_2 = 50$, and the maximum profit is

$$P(20, 5) = 600 + 250 - 200 = 650$$

That is, Adele Lighting's maximum profit is $650,000 when the company sells 20,000 of the 20-inch lamps at $30 each and 5000 of the 31-inch lamps at $50 each.

Linear Regression

We have used different types of functions to model cost, revenue, profit, demand, supply, and other real-world relationships. Sometimes we have used calculus to study the behavior of these functions, finding, for example, marginal cost, marginal revenue, producer's surplus, and so on. We now have the mathematical tools to understand and develop the formulas that graphing calculators and other technology use to find the equations for linear models.

The formulas used to find the equation of the straight line that is the best fit for a set of data are developed using max-min techniques for functions of two variables. This line is called the **regression line.** In Figure 14.12, we define line ℓ to be the best fit for the data points (that is, the regression line) if the sum of the squares of the differences between the actual y-values of the data points and the y-values of the points on the line is a minimum.

In general, to find the equation of the regression line, we assume that the relationship between x and y is approximately linear and that we can find a straight line with equation

$$\hat{y} = a + bx$$

where the values of $\hat{y}$ will approximate the y-values of the points we know. That is, for each given value of x, the point $(x, \hat{y})$ will be on the line. For any given x-value, x_i, we are interested in the deviation between the y-value of the data point (x_i, y_i) and the $\hat{y}$-value from the equation, $\hat{y}_i$, that results when x_i is substituted for x. These deviations are of the form

$$d_i = \hat{y}_i - y_i, \quad \text{for } i = 1, 2, \ldots, n$$

(See Figure 14.12 for a general case with the deviations exaggerated.)

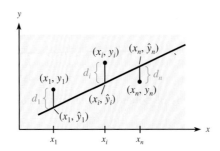

Figure 14.12

To measure the deviations in a way that accounts for the fact that some of the y-values will be above the line and some below, we will say that the line that is the best fit for the data is the one for which the sum of the squares of the deviations is a minimum. That is, we seek the a and b in the equation

$$\hat{y} = a + bx$$

such that the sum of the squares of the deviations,

$$S = \sum_{i=1}^{n} (\hat{y}_i - y_i)^2 = \sum_{i=1}^{n} [(bx_i + a) - y_i]^2$$

$$= (bx_1 + a - y_1)^2 + (bx_2 + a - y_2)^2 + \cdots + (bx_n + a - y_n)^2$$

is a minimum. The procedure for determining a and b is called the **method of least squares.**
We seek the values of b and a that make S a minimum, so we find the values that make

$$\frac{\partial S}{\partial b} = 0 \quad \text{and} \quad \frac{\partial S}{\partial a} = 0$$

where

$$\frac{\partial S}{\partial b} = 2(bx_1 + a - y_1)x_1 + 2(bx_2 + a - y_2)x_2 + \cdots + 2(bx_n + a - y_n)x_n$$

$$\frac{\partial S}{\partial a} = 2(bx_1 + a - y_1) + 2(bx_2 + a - y_2) + \cdots + 2(bx_n + a - y_n)$$

Setting each equation equal to 0, dividing by 2, and using sigma notation give

$$0 = b\sum_{i=1}^{n} x_i^2 + a\sum_{i=1}^{n} x_i - \sum_{i=1}^{n} x_i y_i \tag{1}$$

$$0 = b\sum_{i=1}^{n} x_i + a\sum_{i=1}^{n} 1 - \sum_{i=1}^{n} y_i \tag{2}$$

We can write equations (1) and (2) as follows:

$$\sum_{i=1}^{n} x_i y_i = a\sum_{i=1}^{n} x_i + b\sum_{i=1}^{n} x_i^2 \tag{3}$$

$$\sum_{i=1}^{n} y_i = an + b\sum_{i=1}^{n} x_i \tag{4}$$

Multiplying equation (3) by n and equation (4) by $\sum_{i=1}^{n} x_i$ permits us to begin to solve for b.

$$n\sum_{i=1}^{n} x_i y_i = na\sum_{i=1}^{n} x_i + nb\sum_{i=1}^{n} x_i^2 \tag{5}$$

$$\sum_{i=1}^{n} x_i \sum_{i=1}^{n} y_i = na\sum_{i=1}^{n} x_i + b\left(\sum_{i=1}^{n} x_i\right)^2 \tag{6}$$

Subtracting equation (5) from equation (6) gives

$$\sum_{i=1}^{n} x_i \sum_{i=1}^{n} y_i - n \sum_{i=1}^{n} x_i y_i = b\left(\sum_{i=1}^{n} x_i\right)^2 - nb \sum_{i=1}^{n} x_i^2$$

$$= b\left[\left(\sum_{i=1}^{n} x_i\right)^2 - n \sum_{i=1}^{n} x_i^2\right]$$

Thus

$$b = \frac{\sum_{i=1}^{n} x_i \sum_{i=1}^{n} y_i - n \sum_{i=1}^{n} x_i y_i}{\left(\sum_{i=1}^{n} x_i\right)^2 - n \sum_{i=1}^{n} x_i^2}$$

and, from equation (4),

$$a = \frac{\sum_{i=1}^{n} y_i - b \sum_{i=1}^{n} x_i}{n}$$

It can be shown that these values for b and a give a minimum value for S, so we have the following.

Linear Regression Equation

Given a set of data points $(x_1, y_1), (x_2, y_2), \ldots, (x_n, y_n)$, the equation of the line that is the best fit for these data is

$$\hat{y} = a + bx$$

where

$$b = \frac{\Sigma x \cdot \Sigma y - n\Sigma xy}{(\Sigma x)^2 - n\Sigma x^2}, \qquad a = \frac{\Sigma y - b\Sigma x}{n}$$

and each summation is taken over the entire data set (that is, from 1 to n).

● **EXAMPLE 5** *Inventory*

The following data show the relation between the diameter of a partial roll of blue denim material at MacGregor Mills and the actual number of yards remaining on the roll. Use linear regression to find the linear equation that gives the number of yards as a function of the diameter.

Diameter (inches)	Yards/Roll	Diameter (inches)	Yards/Roll
14.0	120	22.5	325
15.0	145	24.0	360
16.5	170	24.5	380
17.75	200	25.25	405
18.5	220	26.0	435
19.8	255	26.75	460
20.5	270	27.0	470
22.0	305	28.0	500

Solution

Let x be the diameter of the partial roll and y be the number of yards on the roll. Before finding the values for a and b, we evaluate some parts of the formulas:

$$n = 16$$
$$\Sigma x = 348.05$$
$$\Sigma x^2 = 7871.48$$
$$\Sigma y = 5020$$
$$\Sigma xy = 117{,}367.75$$

$$b = \frac{\Sigma x \Sigma y - n \Sigma xy}{(\Sigma x)^2 - n \Sigma x^2}$$

$$= \frac{(348.05)(5020) - 16(117{,}367.75)}{(348.05)^2 - 16(7871.48)} \approx 27.1959$$

$$a = \frac{\Sigma y - b \Sigma x}{n}$$

$$= \frac{(5020) - (27.1959)(348.05)}{16} \approx -277.8458$$

Thus the linear equation that can be used to estimate the number of yards of denim remaining on a roll is

$$\hat{y} = -277.85 + 27.20x$$

Note that if we use the linear regression capability of a graphing utility, we obtain exactly the same equation.

● **Checkpoint**

4. Use linear regression to write the equation of the line that is the best fit for the following points.

x	50	25	10	5
y	2	4	10	20

Finally, we note that formulas for models other than linear ones, such as power models ($y = ax^b$), exponential models ($y = ab^x$), and logarithmic models $[y = a + b \ln(x)]$, can also be developed with the least-squares method. That is, we apply max-min techniques for functions of two variables to minimize the sum of the squares of the deviations.

● **Checkpoint Solutions**

1. $z_x = -2x + 2$, $z_y = -2y - 4$
2. $-2x + 2 = 0$ gives $x = 1$.
   $-2y - 4 = 0$ gives $y = -2$.
   Thus the critical point is $(1, -2, 9)$.
3. $z_{xx} = -2$, $z_{yy} = -2$, and $z_{xy} = 0$, so

$$D(x, y) = (-2)(-2) - (0)^2 = 4$$

Hence, at $(1, -2, 9)$ we have $D > 0$ and $z_{xx} < 0$, so $(1, -2, 9)$ is a relative maximum.

4.
$$\Sigma x = 50 + 25 + 10 + 5 = 90$$
$$\Sigma x^2 = 2500 + 625 + 100 + 25 = 3250$$
$$\Sigma y = 2 + 4 + 10 + 20 = 36$$
$$\Sigma xy = 100 + 100 + 100 + 100 = 400$$

Then

$$b = \frac{\Sigma x \cdot \Sigma y - n\Sigma xy}{(\Sigma x)^2 - n\Sigma x^2} = \frac{90 \cdot 36 - 4 \cdot 400}{90^2 - 4 \cdot 3250} = \frac{1640}{-4900} \approx -0.33$$

and

$$a = \frac{\Sigma y - b\Sigma x}{n} = \frac{36 - (-0.33)(90)}{4} = \frac{65.7}{4} \approx 16.43$$

Thus the line that gives the best fit to these points is

$$\hat{y} = -0.33x + 16.43$$

14.4 Exercises

In Problems 1–16, test for relative maxima and minima.

1. $z = 9 - x^2 - y^2$
2. $z = 16 - 4x^2 - 9y^2$
3. $z = x^2 + y^2 + 4$
4. $z = x^2 + y^2 - 4$
5. $z = x^2 + y^2 - 2x + 4y + 5$
6. $z = 4x^2 + y^2 + 4x + 1$
7. $z = x^2 + 6xy + y^2 + 16x$
8. $z = x^2 - 4xy + y^2 - 6y$
9. $z = \dfrac{x^2 - y^2}{9}$
10. $z = \dfrac{y^2}{4} - \dfrac{x^2}{9}$
11. $z = 24 - x^2 + xy - y^2 + 36y$
12. $z = 46 - x^2 + 2xy - 4y^2$
13. $z = x^2 + xy + y^2 - 4y + 10x$
14. $z = x^2 + 5xy + 10y^2 + 8x - 40y$
15. $z = x^3 + y^3 - 6xy$
16. $z = x^3 + y^3 + 3xy$

In Problems 17 and 18, use the points given in the tables to write the equation of the line that is the best fit for the points.

17.
x	3	4	5	6
y	15	22	28	32

18.
x	10	20	30	40
y	2	6	5	6

APPLICATIONS

19. *Profit* Suppose that the profit from the sale of Kisses and Kreams is given by

$$P(x, y) = 10x + 6.4y - 0.001x^2 - 0.025y^2 \quad \text{dollars}$$

where x is the number of pounds of Kisses and y is the number of pounds of Kreams. Selling how many pounds of Kisses and Kreams will maximize profit? What is the maximum profit?

20. *Profit* The profit from the sales of two products is given by

$$P(x, y) = 20x + 70y - x^2 - y^2 \quad \text{dollars}$$

where x is the number of units of product 1 sold and y is the number of units of product 2. Selling how much of each product will maximize profit? What is the maximum profit?

21. *Nutrition* A new food is designed to add weight to mature beef cattle. The weight in pounds is given by $W = 13xy(20 - x - 2y)$, where x is the number of units of the first ingredient and y is the number of units of the second ingredient. How many units of each ingredient will maximize the weight? What is the maximum weight?

22. *Profit* The profit for a grain crop is related to fertilizer and labor. The profit per acre is

$$P = 100x + 40y - 5x^2 - 2y^2 \quad \text{dollars}$$

where x is the number of units of fertilizer and y is the number of work-hours. What values of x and y will maximize the profit? What is the maximum profit?

23. *Production* Suppose that

$$P = 3.78x^2 + 1.5y^2 - 0.09x^3 - 0.01y^3 \quad \text{tons}$$

is the production function for a product with x units of one input and y units of a second input. Find the values of x and y that will maximize production. What is the maximum production?

24. *Production* Suppose that x units of one input and y units of a second input result in

$$P = 40x + 50y - x^2 - y^2 - xy$$

units of a product. Determine the inputs x and y that will maximize P. What is the maximum production?

25. *Profit*   Suppose that a manufacturer produces two brands of a product, brand 1 and brand 2. Suppose the demand for brand 1 is $x = 70 - p_1$ thousand units and the demand for brand 2 is $y = 80 - p_2$ thousand units, where p_1 and p_2 are prices in dollars. If the joint cost function is $C = xy$, in thousands of dollars, how many of each brand should be produced to maximize profit? What is the maximum profit?

26. *Profit*   Suppose that a firm produces two products, A and B, that sell for $\$a$ and $\$b$, respectively, with the total cost of producing x units of A and y units of B equal to $C(x, y)$. Show that when the profit from these products is maximized,

$$\frac{\partial C}{\partial x}(x, y) = a \quad \text{and} \quad \frac{\partial C}{\partial y}(x, y) = b$$

27. *Manufacturing*   Find the values for each of the dimensions of an open-top box of length x, width y, and height $500,000/(xy)$ (in inches) such that the box requires the least amount of material to make.

28. *Manufacturing*   Find the values for each of the dimensions of a closed-top box of length x, width y, and height z (in inches) if the volume equals 27,000 cubic inches and the box requires the least amount of material to make. (*Hint:* First write the height in terms of x and y, as in Problem 27.)

29. *Profit*   A company manufactures two products, A and B. If x is the number of thousands of units of A and y is the number of thousands of units of B, then the cost and revenue in thousands of dollars are

$$C(x, y) = 2x^2 - 2xy + y^2 - 7x - 10y + 11$$
$$R(x, y) = 5x + 8y$$

Find the number of each type of product that should be manufactured to maximize profit. What is the maximum profit?

30. *Production*   Let x be the number of work-hours required and let y be the amount of capital required to produce z units of a product. Show that the average production per work-hour, z/x, is maximized when

$$\frac{\partial z}{\partial x} = \frac{z}{x}$$

Use $z = f(x, y)$ and assume that a maximum exists.

31. *Auto age*   The following figure shows the median age of automobiles and trucks on the road in the United States.

(a) Use linear regression to find the linear equation that is the best fit for the data, with x equal to the number of years past 1970.
(b) Use the equation to predict the median age of cars and trucks on the road in 2005.
(c) Discuss reasons why the median age is increasing.
(d) Write a sentence that interprets the slope of the linear regression line.

USA TODAY Snapshots®

Hanging on to the old buggy
The median age of automobiles and trucks on the road in the USA:

1970	4.9 years
1975	5.4 years
1980	6 years
1985	6.9 years
1990	6.5 years
1995	7.7 years
1998	8.3 years

by Keith Simmons, USA TODAY

Source: USA Today, September 27, 2000. Copyright © 2000. Reprinted by permission of *USA Today.*

32. *Retirement benefits*   The following table gives the approximate benefits for PepsiCo executives who earned an average of $250,000 per year during the last 5 years of service, based on the number of years of service, from 15 years to 45 years.

(a) Use linear regression to find the linear equation that is the best fit for the data.
(b) Use the equation to find the expected annual retirement benefits after 38 years of service.
(c) Write a sentence that interprets the slope of the linear regression line.

Years of Service	Annual Retirement
25	$109,280
30	121,130
35	132,990
40	145,490
45	160,790

Source: TRICON Salaried Employees Retirement Plan

33. *World population*   The following table gives the actual or projected world population in billions for selected years from 2000 to 2050.

(a) Use linear regression to find the linear equation that is the best fit for the data, with x equal to the number of years past 2000.

(b) Use the equation to predict the population in 2012.
(c) Write a sentence that interprets the slope of the linear regression line.

Years	Population (billions)
2000	6.08
2010	6.82
2020	7.54
2030	8.18
2040	8.72
2050	9.19

Source: U.S. Bureau of the Census,
International Data Base

 34. *Hourly earnings* The following table shows the average hourly earnings for full-time production workers in various industries for selected years.

(a) Find the linear regression equation for hourly earnings as a function of time (with $x = 0$ representing 1970).
(b) What does this model predict for the average hourly earnings in 2010?
(c) Write a sentence that interprets the slope of the linear regression equation.

Year	Hourly Earnings	Year	Hourly Earnings
1970	$3.40	1994	$11.32
1975	4.73	1996	12.03
1980	6.84	1998	13.00
1985	8.73	2000	14.00
1990	10.19	2001	14.53
1992	10.76	2002	14.95

Source: Bureau of Labor Statistics, U.S. Department of Labor

14.5 Maxima and Minima of Functions Subject to Constraints: Lagrange Multipliers

OBJECTIVE

● To find the maximum or minimum value of a function of two or more variables subject to a condition that constrains the variables

◗ Application Preview

Many practical problems require that a function of two or more variables be maximized or minimized subject to certain conditions, or constraints, that limit the variables involved. For example, a firm will want to maximize its profits within the limits (constraints) imposed by its production capacity. Similarly, a city planner may want to locate a new building to maximize access to public transportation yet may be constrained by the availability and cost of building sites.

Specifically, suppose that the utility function for commodities X and Y is given by $U = x^2y^2$, where x and y are the amounts of X and Y, respectively. If p_1 and p_2 represent the prices in dollars of X and Y, respectively, and I represents the consumer's income available to purchase these two commodities, the equation $p_1x + p_2y = I$ is called the *budget constraint*. If the price of X is $2, the price of Y is $4, and the income available is $40, then the budget constraint is $2x + 4y = 40$. Thus we seek to maximize the consumer's utility $U = x^2y^2$ subject to the budget constraint $2x + 4y = 40$. (See Example 4.) In this section we develop methods to solve this type of constrained maximum or minimum.

We can obtain maxima and minima for a function $z = f(x, y)$ subject to the constraint $g(x, y) = 0$ by using the method of **Lagrange multipliers,** named for the famous eighteenth-century mathematician Joseph Louis Lagrange. Lagrange multipliers can be used with functions of two or more variables when the constraints are given by an equation.

In order to find the critical values of a function $f(x, y)$ subject to the constraint $g(x, y) = 0$, we will use the new variable λ to form the **objective function**

$$F(x, y, \lambda) = f(x, y) + \lambda g(x, y)$$

It can be shown that the critical values of $F(x, y, \lambda)$ will satisfy the constraint $g(x, y)$ and will also be critical points of $f(x, y)$. Thus we need only find the critical points of $F(x, y, \lambda)$ to find the required critical points.

To find the critical points of $F(x, y, \lambda)$, we must find the points that make all the partial derivatives equal to 0. That is, the points must satisfy

$$\partial F/\partial x = 0, \quad \partial F/\partial y = 0, \quad \text{and} \quad \partial F/\partial \lambda = 0$$

Because $F(x, y, \lambda) = f(x, y) + \lambda g(x, y)$, these equations may be written as

$$\frac{\partial f}{\partial x} + \lambda \frac{\partial g}{\partial x} = 0$$

$$\frac{\partial f}{\partial y} + \lambda \frac{\partial g}{\partial y} = 0$$

$$g(x, y) = 0$$

Finding the values of x and y that satisfy these three equations simultaneously gives the critical values.

This method will not tell us whether the critical points correspond to maxima or minima, but this can be determined either from the physical setting for the problem or by testing according to a procedure similar to that used for unconstrained maxima and minima. The following examples illustrate the use of Lagrange multipliers.

● **EXAMPLE 1** *Maxima Subject to Constraints*

Find the maximum value of $z = x^2y$ subject to $x + y = 9, x \geq 0, y \geq 0$.

Solution
The function to be maximized is $f(x, y) = x^2y$. The constraint is $g(x, y) = 0$, where $g(x, y) = x + y - 9$. The objective function is

$$F(x, y, \lambda) = f(x, y) + \lambda g(x, y)$$

or

$$F(x, y, \lambda) = x^2y + \lambda(x + y - 9)$$

Thus

$$\frac{\partial F}{\partial x} = 2xy + \lambda(1) = 0, \quad \text{or} \quad 2xy + \lambda = 0$$

$$\frac{\partial F}{\partial y} = x^2 + \lambda(1) = 0, \quad \text{or} \quad x^2 + \lambda = 0$$

$$\frac{\partial F}{\partial \lambda} = 0 + 1(x + y - 9) = 0, \quad \text{or} \quad x + y - 9 = 0$$

Solving the first two equations for λ and substituting gives

$$\lambda = -2xy$$
$$\lambda = -x^2$$
$$2xy = x^2$$
$$2xy - x^2 = 0$$
$$x(2y - x) = 0$$

so

$$x = 0 \quad \text{or} \quad x = 2y$$

Because $x = 0$ could not make $z = x^2y$ a maximum, we substitute $x = 2y$ into $x + y - 9 = 0$.

$$2y + y = 9$$
$$y = 3$$
$$x = 6$$

Thus the function $z = x^2y$ is maximized at 108 when $x = 6$, $y = 3$, if the constraint is $x + y = 9$. Testing values near $x = 6$, $y = 3$, and satisfying the constraint shows that the function is maximized there. (Try $x = 5.5$, $y = 3.5$; $x = 7$, $y = 2$; and so on.)

● **EXAMPLE 2** *Minima Subject to Constraints*

Find the minimum value of the function $z = x^3 + y^3 + xy$ subject to the constraint $x + y - 4 = 0$.

Solution
The function to be minimized is $f(x, y) = x^3 + y^3 + xy$. The constraint function is $g(x, y) = x + y - 4$. The objective function is

$$F(x, y, \lambda) = f(x, y) + \lambda g(x, y)$$

or

$$F(x, y, \lambda) = x^3 + y^3 + xy + \lambda(x + y - 4)$$

Then

$$\frac{\partial F}{\partial x} = 3x^2 + y + \lambda = 0$$

$$\frac{\partial F}{\partial y} = 3y^2 + x + \lambda = 0$$

$$\frac{\partial F}{\partial \lambda} = x + y - 4 = 0$$

Solving the first two equations for λ and substituting, we get

$$\lambda = -(3x^2 + y)$$
$$\lambda = -(3y^2 + x)$$
$$3x^2 + y = 3y^2 + x$$

Solving $x + y - 4 = 0$ for y gives $y = 4 - x$. Substituting for y in the equation above, we get

$$3x^2 + (4 - x) = 3(4 - x)^2 + x$$
$$3x^2 + 4 - x = 48 - 24x + 3x^2 + x$$
$$22x = 44 \quad \text{or} \quad x = 2$$

Thus when $x + y - 4 = 0$, $x = 2$ and $y = 2$ give the minimum value $z = 20$ because other values that satisfy the constraint give larger z-values.

● **Checkpoint**

Find the minimum value of $f(x, y) = x^2 + y^2 - 4xy$, subject to the constraint $x + y = 10$, by:
1. forming the objective function $F(x, y, \lambda)$;
2. finding $\dfrac{\partial F}{\partial x}, \dfrac{\partial F}{\partial y}$, and $\dfrac{\partial F}{\partial \lambda}$;
3. setting the three partial derivatives (from Question 2) equal to 0, and solving the equations simultaneously for x and y;
4. finding the value of $f(x, y)$ at the critical values of x and y.

We can also use Lagrange multipliers to find the maxima and minima of functions of three (or more) variables, subject to two (or more) constraints. The method involves using two multipliers, one for each constraint, to form an objective function $F = f + \lambda g_1 + \mu g_2$. We leave further discussion for more advanced courses.

We can easily extend the method to functions of three or more variables, as the following example shows.

● **EXAMPLE 3** *Minima of Function of Three Variables*

Find the minimum value of the function $w = x + y^2 + z^2$, subject to the constraint $x + y + z = 1$.

Solution
The function to be minimized is $f(x, y, z) = x + y^2 + z^2$. The constraint is $g(x, y, z) = 0$, where $g(x, y, z) = x + y + z - 1$. The objective function is

$$F(x, y, z, \lambda) = f(x, y, z) + \lambda g(x, y, z)$$

or

$$F(x, y, z, \lambda) = x + y^2 + z^2 + \lambda(x + y + z - 1)$$

Then

$$\frac{\partial F}{\partial x} = 1 + \lambda = 0$$

$$\frac{\partial F}{\partial y} = 2y + \lambda = 0$$

$$\frac{\partial F}{\partial z} = 2z + \lambda = 0$$

$$\frac{\partial F}{\partial \lambda} = x + y + z - 1 = 0$$

Solving the first three equations simultaneously gives

$$\lambda = -1, \qquad y = \frac{1}{2}, \qquad z = \frac{1}{2}$$

Substituting these values in the fourth equation (which is the constraint), we get $x + \frac{1}{2} + \frac{1}{2} - 1 = 0$, so $x = 0, y = \frac{1}{2}, z = \frac{1}{2}$. Thus $w = \frac{1}{2}$ is the minimum value because other values of x, y, and z that satisfy $x + y + z = 1$ give larger values of w.

◑ **EXAMPLE 4** *Utility* **(Application Preview)**

Find x and y that maximize the utility function $U = x^2y^2$ subject to the budget constraint $2x + 4y = 40$.

Solution
First we rewrite the constraint as $2x + 4y - 40 = 0$. Then the objective function is

$$F(x, y, \lambda) = x^2y^2 + \lambda(2x + 4y - 40)$$

$$\frac{\partial F}{\partial x} = 2xy^2 + 2\lambda, \qquad \frac{\partial F}{\partial y} = 2x^2y + 4\lambda, \qquad \frac{\partial F}{\partial \lambda} = 2x + 4y - 40$$

Setting these partial derivatives equal to 0 and solving gives

$$-\lambda = xy^2 = x^2y/2, \quad \text{or} \quad xy^2 - x^2y/2 = 0$$

so

$$xy(y - x/2) = 0$$

yields $x = 0$, $y = 0$, or $x = 2y$. Neither $x = 0$ nor $y = 0$ maximizes utility. If $x = 2y$, we have

$$0 = 4y + 4y - 40$$

Thus $y = 5$ and $x = 10$.

Testing values near $x = 10$, $y = 5$ shows that these values maximize utility at $U = 2500$.

Figure 14.13 shows the budget constraint $2x + 4y = 40$ from Example 4 graphed with the indifference curves for $U = x^2y^2$ that correspond to $U = 500$, $U = 2500$, and $U = 5000$.

Whenever an indifference curve intersects the budget constraint, that utility level is attainable within the budget. Note that the highest attainable utility (such as $U = 2500$, found in Example 4) corresponds to the indifference curve that touches the budget constraint at exactly one point—that is, the curve that has the budget constraint as a tangent line. Note also that utility levels greater than $U = 2500$ are not attainable within the budget because the indifference curve "misses" the budget constraint line (as for $U = 5000$).

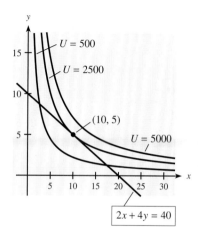

Figure 14.13

● **EXAMPLE 5** *Production*

Suppose that the Cobb-Douglas production function for a certain manufacturer gives the number of units of production z according to

$$z = f(x, y) = 100x^{4/5}y^{1/5}$$

where x is the number of units of labor and y is the number of units of capital. Suppose further that labor costs \$160 per unit, capital costs \$200 per unit, and the total cost for capital and labor is limited to \$100,000, so that production is constrained by

$$160x + 200y = 100,000$$

Find the number of units of labor and the number of units of capital that maximize production.

Solution
The objective function is

$$F(x, y, \lambda) = 100x^{4/5}y^{1/5} + \lambda(160x + 200y - 100,000)$$

$$\frac{\partial F}{\partial x} = 80x^{-1/5}y^{1/5} + 160\lambda, \qquad \frac{\partial F}{\partial y} = 20x^{4/5}y^{-4/5} + 200\lambda$$

$$\frac{\partial F}{\partial \lambda} = 160x + 200y - 100,000$$

Setting these partial derivatives equal to 0 and solving gives

$$\lambda = \frac{-80x^{-1/5}y^{1/5}}{160} = \frac{-20x^{4/5}y^{-4/5}}{200} \quad \text{or} \quad \frac{y^{1/5}}{2x^{1/5}} = \frac{x^{4/5}}{10y^{4/5}}$$

This means $5y = x$. Using this in $\frac{\partial F}{\partial \lambda} = 0$ gives

$$160(5y) + 200y - 100{,}000 = 0$$
$$1000y = 100{,}000$$
$$y = 100$$
$$x = 5y = 500$$

Thus production is maximized at $z = 100(500)^{4/5}(100)^{1/5} \approx 36{,}239$ when $x = 500$ (units of labor) and $y = 100$ (units of capital). See Figure 14.14.

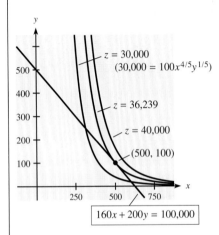

Figure 14.14

In problems of this type, economists call the value of $-\lambda$ the **marginal productivity of money.** In this case,

$$-\lambda = \frac{y^{1/5}}{2x^{1/5}} = \frac{(100)^{0.2}}{2(500)^{0.2}} \approx 0.362$$

This means that each additional dollar spent on production results in approximately 0.362 additional unit produced.

Finally, Figure 14.14 shows the graph of the constraint, together with some production function curves that correspond to different production levels.

Spreadsheet    The Excel tool called Solver can be used to find maxima and minima of functions subject
Note   to constraints. In Example 5, the numbers of units of labor and capital that maximize production if the production is given by the Cobb-Douglas function

$$P(x, y) = 100x^{4/5}y^{1/5}$$

and if capital and labor are constrained by

$$160x + 200y = 100{,}000$$

were found to be 500 and 100, respectively. The following Excel screen shows the entries for beginning the solution of this problem.

	A	B	C
1	Variables		
2			
3	units labor (x)	0	
4	units capital (y)	0	
5			
6	Objective		
7			
8	Maximize production	=100*B3^(4/5)*B4^(1/5)	
9			
10	Constraint		
11		Amount used	Available
12	Cost	=160*B3+200*B4	100000
13			

The solution, found with Solver, is shown below. For details of the solution process, see the *Excel Guide for Finite Mathematics and Applied Calculus,* which accompanies this text.

	A	B	C
1	Variables		
2			
3	units labor (x)	500	
4	units capital (y)	100	
5			
6	Objective		
7			
8	Maximize production	36238.98	
9			
10	Constraint		
11		Amount used	Available
12	Cost	100000	100000
13			

● **Checkpoint Solutions**

1. $F(x, y, \lambda) = x^2 + y^2 - 4xy + \lambda(10 - x - y)$

2. $\dfrac{\partial F}{\partial x} = 2x - 4y - \lambda, \qquad \dfrac{\partial F}{\partial y} = 2y - 4x - \lambda, \qquad \dfrac{\partial F}{\partial \lambda} = 10 - x - y$

3. $0 = 2x - 4y - \lambda \qquad (1)$
   $0 = 2y - 4x - \lambda \qquad (2)$
   $0 = 10 - x - y \qquad (3)$

   From equations (1) and (2) we have the following:

   $$\lambda = 2x - 4y \quad \text{and} \quad \lambda = 2y - 4x, \quad \text{so}$$
   $$2x - 4y = 2y - 4x$$
   $$6x = 6y, \quad \text{or} \quad x = y$$

   Using $x = y$ in equation (3) gives $0 = 10 - x - x$, or $2x = 10$. Thus $x = 5$ and $y = 5$.

4. $f(5, 5) = 25 + 25 - 100 = -50$ is the minimum because other values that satisfy the constraint give larger z-values.

14.5 Exercises

1. Find the minimum value of $z = x^2 + y^2$ subject to the condition $x + y = 6$.
2. Find the minimum value of $z = 4x^2 + y^2$ subject to the constraint $x + y = 5$.
3. Find the minimum value of $z = 3x^2 + 5y^2 - 2xy$ subject to the constraint $x + y = 5$.
4. Find the maximum value of $z = 2xy - 3x^2 - 5y^2$ subject to the constraint $x + y = 5$.
5. Find the maximum value of $z = x^2y$ subject to $x + y = 6, x \geq 0, y \geq 0$.
6. Find the maximum value of the function $z = x^3y^2$ subject to $x + y = 10, x \geq 0, y \geq 0$.
7. Find the maximum value of the function $z = 2xy - 2x^2 - 4y^2$ subject to the condition $x + 2y = 8$.
8. Find the minimum value of $z = 2x^2 + y^2 - xy$ subject to the constraint $2x + y = 8$.
9. Find the minimum value of $z = x^2 + y^2$ subject to the condition $2x + y + 1 = 0$.
10. Find the minimum value of $z = x^2 + y^2$ subject to the condition $xy = 1$.
11. Find the minimum value of $w = x^2 + y^2 + z^2$ subject to the constraint $x + y + z = 3$.
12. Find the minimum value of $w = x^2 + y^2 + z^2$ subject to the condition $2x - 4y + z = 21$.
13. Find the maximum value of $w = xz + y$ subject to the constraint $x^2 + y^2 + z^2 = 1$.
14. Find the maximum value of $w = x^2yz$ subject to the constraint $4x + y + z = 4, x \geq 0, y \geq 0$, and $z \geq 0$.

APPLICATIONS

15. *Utility*   Suppose that the utility function for two commodities is given by $U = xy^2$ and that the budget constraint is $3x + 6y = 18$. What values of x and y will maximize utility?

16. *Utility*   Suppose that the budget constraint in Problem 15 is $5x + 20y = 90$. What values of x and y will maximize $U = xy^2$?

17. *Utility*   Suppose that the utility function for two products is given by $U = x^2y$, and the budget constraint is $2x + 3y = 120$. Find the values of x and y that maximize utility. Check by graphing the budget constraint with the indifference curve for maximum utility and with two other indifference curves.

18. *Utility*   Suppose that the utility function for two commodities is given by $U = x^2y^3$, and the budget constraint is $10x + 15y = 250$. Find the values of x and y that maximize utility. Check by graphing the budget constraint with the indifference curve for maximum utility and with two other indifference curves.

19. *Production*   A company has the Cobb-Douglas production function

$$z = 400x^{0.6}y^{0.4}$$

where x is the number of units of labor, y is the number of units of capital, and z is the units of production. Suppose labor costs $150 per unit, capital costs $100 per unit, and the total cost of labor and capital is limited to $100,000.
(a) Find the number of units of labor and the number of units of capital that maximize production.
(b) Find the marginal productivity of money and interpret it.
(c) Graph the constraint with the optimal value for production and with two other z-values (one smaller than the optimal value and one larger).

20. *Production*   Suppose a company has the Cobb-Douglas production function

$$z = 100^{0.75}y^{0.25}$$

where x is the number of units of labor, y is the number of units of capital, and z is the units of production. Suppose further that labor costs $90 per unit, capital costs $150 per unit, and the total costs of labor and capital are limited to $90,000.
(a) Find the number of units of labor and the number of units of capital that maximize production.
(b) Find the marginal productivity of money and interpret it.
(c) Graph the constraint with the optimal value for production and with two other z-values (one smaller than the optimal value and one larger).

21. *Cost*   A firm has two plants, X and Y. Suppose that the cost of producing x units at plant X is $x^2 + 1200$ dollars and the cost of producing y units of the same product at plant Y is given by $3y^2 + 800$ dollars. If the firm has an order for 1200 units, how many should it produce at each plant to fill this order and minimize the cost of production?

22. *Cost*   Suppose that the cost of producing x units at plant X is $(3x + 4)x$ dollars and that the cost of producing y units of the same product at plant Y is $(2y + 8)y$ dollars. If the firm that owns the plants has an order for 149 units, how many should it produce at each plant to fill this order and minimize its cost of production?

23. *Revenue*   On the basis of past experience a company has determined that its sales revenue (in dollars) is related to its advertising according to the formula $s = 20x + y^2 + 4xy$, where x is the amount spent on

radio advertising and y is the amount spent on television advertising. If the company plans to spend \$30,000 on these two means of advertising, how much should it spend on each method to maximize its sales revenue?

24. *Manufacturing*   Find the dimensions x, y, and z (in inches) of the rectangular box with the largest volume that satisfies

$$3x + 4y + 12z = 12$$

25. *Manufacturing*   Find the dimensions (in centimeters) of the box with square base, open top, and volume 500,000 cubic centimeters that requires the least materials.

26. *Manufacturing*   Show that a box with a square base, an open top, and a fixed volume requires the least material to build if it has a height equal to one-half the length of one side of the base.

Key Terms and Formulas

Section	Key Terms	Formulas
14.1	Function of two variables	$z = f(x, y)$
	Variables: independent, dependent	independent: x and y
	Domain	dependent: z
	Coordinate planes	
	Utility	
	Indifference curve	
	Indifference map	
	Cobb-Douglas production function	$Q = AK^{\alpha}L^{1-\alpha}$
14.2	First-order partial derivative	
	With respect to x	$z_x = \dfrac{\partial z}{\partial x}$
	With respect to y	$z_y = \dfrac{\partial z}{\partial y}$
	Higher-order partial derivatives	
	Second partial derivatives	$z_{xx}, z_{yy}, z_{xy},$ and z_{yx}
14.3	Joint cost function	$C = Q(x, y)$
	Marginal cost	
	Marginal productivity	
	Demand function	
	Marginal demand function	
	Competitive products	
	Complementary products	
14.4	Critical values for maxima and minima	Solve simultaneously $\begin{cases} z_x = 0 \\ z_y = 0 \end{cases}$.
	Test for critical values	Use $D(x, y) = (z_{xx})(z_{yy}) - (z_{xy})^2$.
	Linear regression	$\hat{y} = a + bx$
		$b = \dfrac{\Sigma x \cdot \Sigma y - n\Sigma xy}{(\Sigma x)^2 - n\Sigma x^2}$
		$a = \dfrac{\Sigma y - b\Sigma x}{n}$
14.5	Maxima and minima subject to constraints	
	Lagrange multipliers	
	Objective function	$F(x, y, \lambda) = f(x, y) + \lambda g(x, y)$

Review Exercises

Section 14.1

1. What is the domain of $z = \dfrac{3}{2x - y}$?

2. What is the domain of $z = \dfrac{3x + 2\sqrt{y}}{x^2 + y^2}$?

3. If $w(x, y, z) = x^2 - 3yz$, find $w(2, 3, 1)$.

4. If $Q(K, L) = 70K^{2/3}L^{1/3}$, find $Q(64{,}000, 512)$.

Section 14.2

5. Find $\dfrac{\partial z}{\partial x}$ if $z = 5x^3 + 6xy + y^2$.

6. Find $\dfrac{\partial z}{\partial y}$ if $z = 12x^5 - 14x^3y^3 + 6y^4 - 1$.

In Problems 7–12, find z_x and z_y.

7. $z = 4x^2y^3 + \dfrac{x}{y}$ 8. $z = \sqrt{x^2 + 2y^2}$

9. $z = (xy + 1)^{-2}$ 10. $z = e^{x^2y^3}$

11. $z = e^{xy} + y \ln x$ 12. $z = e^{\ln xy}$

13. Find the partial derivative of $f(x, y) = 4x^3 - 5xy^2 + y^3$ with respect to x at the point $(1, 2, -8)$.

14. Find the slope of the tangent in the x-direction to the surface $z = 5x^4 - 3xy^2 + y^2$ at $(1, 2, -3)$.

In Problems 15–18, find the second partials.
(a) z_{xx} **(b)** z_{yy} **(c)** z_{xy} **(d)** z_{yx}

15. $z = x^2y - 3xy$

16. $z = 3x^3y^4 - \dfrac{x^2}{y^2}$

17. $z = x^2e^{y^2}$

18. $z = \ln(xy + 1)$

Section 14.4

19. Test $z = 16 - x^2 - xy - y^2 + 24y$ for maxima and minima.

20. Test $z = x^3 + y^3 - 12x - 27y$ for maxima and minima.

Section 14.5

21. Find the minimum value of $z = 4x^2 + y^2$ subject to the constraint $x + y = 10$.

22. Find the maximum value of $z = x^4y^2$ subject to the constraint $x + y = 9$, $x \geq 0$, $y \geq 0$.

APPLICATIONS

Section 14.1

23. *Utility*   Suppose that the utility function for two goods X and Y is given by $U = x^2y$, and a consumer purchases 6 units of X and 15 units of Y. If the consumer purchases 60 units of Y, how many units of X must be purchased to retain the same level of utility?

24. *Utility*   Suppose that an indifference curve for two products, X and Y, has the equation $xy = 1600$. If 80 units of X are purchased, how many units of Y must be purchased?

Section 14.2

25. *Concorde sonic booms*   The width of the region on the ground on either side of the path of France's Concorde jet in which people hear the sonic boom is given by

$$w = f(T, h, d) = 2\sqrt{Th/d}$$

where T is the air temperature at ground level in kelvins (K), h is the Concorde's altitude in kilometers, and d is the vertical temperature gradient (the temperature drop in kelvins per kilometer).*

(a) Suppose the Concorde approaches Washington, D.C., from Europe on a course that takes it south of Nantucket Island at an altitude of 16.8 km. If the surface temperature is 293 K and the vertical temperature gradient is 5 K/km, how far south of Nantucket must the plane pass to keep the sonic boom off the island?

(b) Interpret $f(287, 17.1, 4.9) \approx 63.3$.

(c) Find $\dfrac{\partial f}{\partial h}(293, 16.8, 5)$ and interpret the result.

(d) Find $\dfrac{\partial f}{\partial d}(293, 16.8, 5)$ and interpret the result.

Section 14.3

26. *Cost*   The joint cost, in dollars, for two products is given by $C(x, y) = x^2\sqrt{y^2 + 13}$. Find the marginal cost with respect to
(a) x if 20 units of x and 6 units of y are produced.
(b) y if 20 units of x and 6 units of y are produced.

27. *Production*   Suppose that the production function for a company is given by

$$Q = 80K^{1/4}L^{3/4}$$

where Q is the output (in hundreds of units), K is the capital expenditures (in thousands of dollars), and L is the work-hours. Find $\partial Q/\partial K$ and $\partial Q/\partial L$ when expenditures are \$625,000 and total work-hours are 4096. Interpret the results.

*N. K. Balachandra, W. L. Donn, and D. H. Rind, "Concorde Sonic Booms as an Atmospheric Probe," *Science,* 1 July 1977, Vol. 197, p. 47

28. *Marginal demand* The demand functions for two related products, product *A* and product *B*, are given by

$$q_A = 400 - 2p_A - 3p_B$$
$$q_B = 300 - 5p_A - 6p_B$$

where p_A and p_B are the respective prices in dollars.
(a) Find the marginal demand of q_A with respect to p_A.
(b) Find the marginal demand of q_B with respect to p_B.
(c) Are the products complementary or competitive?

29. *Marginal demand* Suppose that the demand functions for two related products, *A* and *B*, are given by

$$q_A = 800 - 40p_A - \frac{2}{p_B + 1}$$

$$q_B = 1000 - \frac{10}{p_A + 4} - 30p_B$$

where p_A and p_B are the respective prices in dollars. Determine whether the products are competitive or complementary.

Section 14.4

30. *Profit* The weekly profit (in dollars) from the sale of two products is given by $P(x, y) = 40x + 80y - x^2 - y^2$, where *x* is the number of units of product 1 and *y* is the number of units of product 2. Selling how much of each product will maximize profit? Find the maximum weekly profit.

Section 14.5

31. *Utility* If the utility function for two commodities is $U = x^2y$, and the budget constraint is $4x + 5y = 60$, find the values of *x* and *y* that maximize utility.

 32. *Production* Suppose a company has the Cobb-Douglas production function

$$z = 300x^{2/3}y^{1/3}$$

where *x* is the number of units of labor, *y* is the number of units of capital, and *z* is the units of production. Suppose labor costs are $50 per unit, capital costs are $50 per unit, and total costs are limited to $75,000.
(a) Find the number of units of labor and the number of units of capital that maximize production.
(b) Find the marginal productivity of money and interpret your result.
(c) Graph the constraint with the production function when $z = 180,000$, $z = 300,000$, and when the *z*-value is optimal.

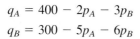 33. *Taxes* The following data show U.S. national personal income and personal taxes for selected years.
(a) Write the linear regression equation that best fits these data.
(b) Use the equation found in part (a) to predict the taxes when national personal income reaches $7000 billion.

Income (x) (billions)	Taxes (y) (billions)
$2285.7	$312.4
2560.4	360.2
2718.7	371.4
2891.7	369.3
3205.5	395.5
3439.6	437.7
3647.5	459.9
3877.3	514.2
4172.8	532.0
4489.3	594.9
4791.6	624.8
4968.5	624.8
5264.2	650.5
5480.1	689.9
5753.1	731.4
6150.8	795.1
6495.2	886.9

Source: Bureau of Economic Analysis, U.S. Commerce Department.

34. *Supply and demand* The following table gives the number of color television sets (in thousands) sold in 15 different years, along with the corresponding average price per set for the year. Use linear regression to find the best linear equation defining the demand function $q = f(p)$.

Price (p) (dollars)	Quantity Sold (q) (thousands)	Price (p) (dollars)	Quantity Sold (q) (thousands)
471.56	9793	509.86	7908
487.79	10236	524.96	6349
487.32	9107	514.10	4822
510.78	7700	515.43	5962
504.39	6485	520.82	5981
466.29	8411	525.01	5777
449.81	10071	462.32	5892
		560.09	2646

Source: U.S. Bureau of the Census, *Statistical Abstract of the United States,* Washington, D.C.

14 Chapter Test

1. Consider the function $f(x, y) = \dfrac{2x + 3y}{\sqrt{x^2 - y}}$.
   (a) Find the domain of $f(x, y)$.
   (b) Evaluate $f(-4, 12)$.

2. Find all first and second partial derivatives of
$$z = f(x, y) = 5x - 9y^2 + 2(xy + 1)^5$$

3. Let $z = 6x^2 + x^2y + y^2 - 4y + 9$. Find the pairs (x, y) that are critical points for z, and then classify each as a relative maximum, a relative minimum, or a saddle point.

4. Suppose a company's monthly production value Q, in thousands of dollars, is given by the Cobb-Douglas production function
$$Q = 10K^{0.45}L^{0.55}$$
where K is thousands of dollars of capital investment per month and L is the total hours of labor per month. Capital investment is currently \$10,000 per month and monthly work-hours of labor total 1590.
   (a) Find the monthly production value (to the nearest thousand dollars).
   (b) Find the marginal productivity with respect to capital investment, and interpret your result.
   (c) Find the marginal productivity with respect to total hours of labor, and interpret your result.

5. The monthly payment R on a loan is a function of the amount borrowed, A, in thousands of dollars; the length of the loan, n, in years; and the annual interest rate, r, as a percent. Thus $R = f(A, n, r)$. In parts (a) and (b), write a sentence that explains the practical meaning of each mathematical statement.
   (a) $f(94.5, 25, 7) = \$667.91$
   (b) $\dfrac{\partial f}{\partial r}(94.5, 25, 7) = \$49.76$
   (c) Would $\dfrac{\partial f}{\partial n}(94.5, 25, 7)$ be positive, negative, or zero? Explain.

6. Let $f(x, y) = 2e^{x^2y^2}$. Find $\dfrac{\partial^2 f}{\partial x \, \partial y}$.

7. Suppose the demand functions for two products are
$$q_1 = 300 - 2p_1 - 5p_2 \quad \text{and} \quad q_2 = 150 - 4p_1 - 7p_2$$
where q_1 and q_2 represent quantities demanded and p_1 and p_2 represent prices. What calculations enable us to decide whether the products are competitive or complementary? Are these products competitive or complementary?

8. Suppose a store sells two brands of disposable cameras and the profit for these is a function of their two selling prices. The type 1 camera sells for \$x, the type 2 sells for \$y, and profit is given by
$$P = 915x - 30x^2 - 45xy + 975y - 30y^2 - 3500$$
Find the selling prices that maximize profit. Find the maximum profit.

9. Find x and y that maximize the utility function $U = x^3y$ subject to the budget constraint $30x + 20y = 8000$.

10. For a middle-income family, the estimated annual expenditures for each year of raising a child through age 11 are given in the following table.
   (a) Find the linear regression line for these data.
   (b) Use the regression equation to project the annual expenditure associated with raising a child during his or her 14th year.
   (c) If this linear model had the form $E = f(x)$, where E is the expenditures and x is the age, would $f(35)$ make sense? Explain.

Age	Expenditures	Age	Expenditures
0	\$7,880	6	\$11,020
1	8,270	7	11,590
2	8,700	8	12,200
3	9,380	9	12,780
4	9,870	10	13,450
5	10,390	11	14,150

Source: Family Economics Research Group, U.S. Dept. of Agriculture

Extended Applications & Group Projects

I. Advertising

To model sales of its tires, the manufacturer of GRIPPER tires used the quadratic equation $S = a_0 + a_1x + a_2x^2 + b_1y$, where S is regional sales in millions of dollars, x is TV advertising expenditures in millions of dollars, and y is other promotional expenditures in millions of dollars. (See the Extended Application/Group Project "Marginal Return to Sales," in Chapter 9.)

Although this model represents the relationship between advertising and sales dollars for small changes in advertising expenditures, it is clear to the vice president of advertising that it does not apply to large expenditures for TV advertising on a national level. He knows from experience that increased expenditures for TV advertising result in more sales, but at a decreasing rate of return for the product.

The vice president is aware that some advertising agencies model the relationship between advertising and sales by the function

$$S = b_0 + b_1(1 - e^{-ax}) + c_1y$$

where $a > 0$, S is sales in millions of dollars, x is TV advertising expenditures in millions of dollars, and y is other promotional expenditures in millions of dollars.* The equation

$$S_n = 24.58 + 325.18(1 - e^{-x/14}) + b_1y$$

has the form mentioned previously as being used by some advertising agencies. For TV advertising expenditures up to $20 million, this equation closely approximates

$$S_1 = 30 + 20x - 0.4x^2 + b_1y$$

which, in the Extended Application/Group Project in Chapter 9, was used with fixed promotional expenses to describe advertising and sales in Region 1.

To help the vice president decide whether this is a better model for large expenditures, answer the following questions.

1. What is $\partial S_1/\partial x$? Does this indicate that sales might actually decline after some amount is spent on TV advertising? If so, what is this amount?
2. Does the quadratic model $S_1(x, y)$ indicate that sales will become negative after some amount is spent on TV advertising? Does this model cease to be useful in predicting sales after a certain point?
3. What is $\partial S_n/\partial x$? Does this indicate that sales will continue to rise if additional money is devoted to TV advertising? Is S_n growing at a rate that is increasing or decreasing when promotional sales are held constant? Is S_n a better model for large expenditures?
4. If this model does describe the relationship between advertising and sales, and if promotional expenditures are held constant at y_0, is there an upper limit to the sales, even if an unlimited amount of money is spent on TV advertising? If so, what is it?

*Edwin Mansfield, *Managerial Economics* (New York: Norton, 1990).

II. Competitive Pricing

Retailers often sell different brands of competing products. Depending on the joint demand for the products, the retailer may be able to set prices that regulate demand and, therefore, influence profits.

Suppose HOME-ALL, Inc., a national chain of home improvement retailers, sells two competing brands of interior flat paint, En-Dure 100 and Croyle & James, which the chain purchases for $8 per gallon and $10 per gallon, respectively. HOME-ALL's research department has determined the following two monthly demand equations for these paints:

$$D = 120 - 40d + 30c \quad \text{and} \quad C = 680 + 30d - 40c$$

where D is hundreds of gallons of En-Dure 100 demanded at d per gallon and C is hundreds of gallons of Croyle & James demanded at c per gallon. For what prices should HOME-ALL sell these paints in order to maximize its monthly profit on these items?

To answer this question, complete the following.

1. Recall that revenue is a product's selling price per item times the number of items sold. With this in mind, formulate HOME-ALL's total revenue function for the two paints as a function of their prices.
2. Form HOME-ALL's profit function for the two paints (in terms of their selling prices).
3. Determine the price of each type of paint that will maximize HOME-ALL's profit.
4. Write a brief report to management that details your pricing recommendations and justifies them.

Appendix

TABLE I Future Value of an Ordinary Annuity of $1 ($s_{\overline{n}|i}$)

Periods	1%	2%	3%	4%	5%	6%
1	1.000 000	1.000 000	1.000 000	1.000 000	1.000 000	1.000 000
2	2.010 000	2.020 000	2.030 000	2.040 000	2.050 000	2.060 000
3	3.030 100	3.060 400	3.090 900	3.121 600	3.152 500	3.183 600
4	4.060 401	4.121 608	4.183 627	4.246 464	4.310 125	4.374 616
5	5.101 005	5.204 040	5.309 136	5.416 323	5.525 631	5.637 093
6	6.152 015	6.308 121	6.468 410	6.632 975	6.801 913	6.975 319
7	7.213 535	7.434 284	7.662 462	7.898 294	8.142 008	8.393 838
8	8.285 670	8.582 969	8.892 336	9.214 226	9.549 109	9.897 468
9	9.368 527	9.754 629	10.159 106	10.582 795	11.026 564	11.491 316
10	10.462 212	10.949 721	11.463 879	12.006 107	12.577 893	13.180 795
11	11.566 834	12.168 716	12.807 796	13.486 351	14.206 787	14.971 643
12	12.682 503	13.412 090	14.192 030	15.025 805	15.917 126	16.869 941
13	13.809 328	14.680 332	15.617 790	16.626 838	17.712 983	18.882 138
14	14.947 421	15.973 939	17.086 324	18.291 911	19.598 632	21.015 066
15	16.096 895	17.293 418	18.598 914	20.023 588	21.578 564	23.275 970
16	17.257 864	18.639 286	20.156 881	21.824 531	23.657 492	25.672 529
17	18.430 443	20.012 072	21.761 588	23.697 512	25.840 366	28.212 880
18	19.614 747	21.412 313	23.414 435	26.645 413	28.132 385	30.905 653
19	20.810 895	22.840 559	25.116 869	27.671 229	30.539 004	33.759 992
20	22.019 004	24.297 371	26.870 375	29.778 079	33.065 954	36.787 592
21	23.239 194	25.783 318	28.676 486	31.969 202	35.719 252	39.992 727
22	24.471 586	27.298 985	30.536 780	34.247 970	38.505 214	43.392 291
23	25.716 301	28.844 964	32.452 884	36.617 889	41.430 475	46.995 829
24	26.973 464	30.421 864	34.426 470	39.082 604	44.501 999	50.815 578
25	28.243 199	32.030 301	36.459 264	41.645 908	47.727 099	54.864 513
26	29.525 631	33.670 907	38.553 042	44.311 745	51.113 454	59.156 384
27	30.820 887	35.344 325	40.709 634	47.084 214	54.669 126	63.705 767
28	32.129 096	37.051 212	42.930 923	49.967 583	58.402 583	68.528 113
29	33.450 387	38.792 236	45.218 850	52.966 286	62.322 712	73.639 800
30	34.784 891	40.568 081	47.575 416	56.084 938	66.438 847	79.058 188
31	36.132 740	42.379 443	50.002 678	59.328 335	70.760 790	84.801 679
32	37.494 067	44.227 031	52.502 759	62.701 469	75.298 829	90.889 780
33	38.869 008	46.111 572	55.077 842	66.209 528	80.063 771	97.343 167
34	40.257 698	48.033 804	57.730 177	69.857 909	85.066 959	104.183 757
35	41.660 275	49.994 480	60.462 082	73.652 225	90.320 307	111.434 783
36	43.076 878	51.994 369	63.275 945	77.598 314	95.836 323	119.120 870
37	44.507 646	54.034 257	66.174 223	81.702 247	101.628 139	127.268 122
38	45.952 723	56.114 942	69.159 450	85.970 336	107.709 546	135.904 209
39	47.412 250	58.237 241	72.234 233	90.409 150	114.095 023	145.058 462
40	48.886 373	60.401 986	75.401 260	95.025 516	120.799 774	154.761 970

(continued)

TABLE I Future Value of an Ordinary Annuity of $1 ($s_{\overline{n}|i}$) (*Continued*)

Periods	7%	8%	9%	10%	11%	12%
1	1.000 000	1.000 000	1.000 000	1.000 000	1.000 000	1.000 000
2	2.070 000	2.080 000	2.090 000	2.100 000	2.110 000	2.120 000
3	3.214 900	3.246 400	3.278 100	3.310 000	3.342 100	3.374 400
4	4.439 943	4.506 112	4.573 129	4.641 000	4.709 731	4.779 328
5	5.750 739	5.866 601	5.984 711	6.105 100	6.227 801	6.352 847
6	7.153 291	7.335 929	7.523 335	7.715 610	7.912 860	8.115 189
7	8.654 021	8.922 803	9.200 435	9.487 171	9.783 274	10.089 012
8	10.259 803	10.636 628	11.028 474	11.435 888	11.859 434	12.299 693
9	11.977 989	12.487 558	13.021 036	13.579 477	14.136 972	14.775 656
10	13.816 448	14.486 563	15.192 930	15.937 425	16.722 009	17.548 735
11	15.783 599	16.645 488	17.560 293	18.531 167	19.561 430	20.654 583
12	17.888 451	18.977 127	20.140 720	21.384 284	22.713 187	24.133 133
13	20.140 643	21.495 297	22.953 385	24.522 712	26.211 638	28.029 109
14	22.550 488	24.214 920	26.019 189	27.974 984	30.094 918	32.392 602
15	25.129 022	27.152 114	29.360 916	31.772 482	34.405 359	37.279 715
16	27.888 054	30.324 283	33.003 399	35.949 730	39.189 949	42.753 281
17	30.840 218	33.750 226	36.973 705	40.544 703	44.500 843	48.883 674
18	33.999 033	37.450 244	41.301 338	45.599 174	50.395 936	55.749 715
19	37.378 965	41.446 263	46.018 458	51.159 091	56.939 489	63.439 681
20	40.995 493	45.761 965	51.160 120	57.275 000	64.202 833	72.052 443
21	44.865 177	50.422 922	56.764 530	64.002 501	72.265 145	81.698 736
22	49.005 740	55.456 756	62.873 338	71.402 750	81.214 310	92.502 584
23	53.436 142	60.893 296	69.531 939	79.543 025	91.147 885	104.602 894
24	58.176 672	66.764 760	76.789 813	88.497 328	102.174 152	118.155 242
25	63.249 039	73.105 940	84.700 896	98.347 061	114.413 309	133.333 871
26	68.676 471	79.954 416	93.323 977	109.181 767	127.998 773	150.333 935
27	74.483 824	87.350 769	102.723 135	121.099 944	143.078 638	169.374 007
28	80.697 692	95.338 831	112.968 217	134.209 938	159.817 288	190.698 888
29	87.346 531	103.965 937	124.135 357	148.630 932	178.397 190	214.582 755
30	94.460 788	113.283 212	136.307 539	164.494 026	199.020 881	241.322 686
31	102.073 043	123.345 869	149.575 217	181.943 428	221.913 178	271.292 608
32	110.218 156	134.213 539	164.036 987	201.137 771	247.323 628	304.847 721
33	118.933 427	145.950 622	179.800 316	222.251 548	275.529 227	342.429 447
34	128.258 767	158.626 671	196.982 344	245.476 703	306.837 442	384.520 981
35	138.236 881	172.316 805	215.710 755	271.024 374	341.589 561	431.663 499
36	148.913 462	187.102 150	236.124 723	299.126 811	380.164 413	484.463 119
37	160.337 405	203.070 322	258.375 948	330.039 493	422.982 498	543.598 693
38	172.561 023	220.315 948	282.629 783	364.043 442	470.510 573	609.830 536
39	185.640 295	238.941 223	309.066 464	401.447 787	523.266 737	686.010 201
40	199.635 116	259.056 521	337.882 446	442.592 566	581.826 078	767.091 425

TABLE II Present Value of an Ordinary Annuity of $1 ($a_{\overline{n}|i}$)

Periods	1%	2%	3%	4%	5%	6%
1	0.990 099	0.980 392	0.970 874	0.961 538	0.952 381	0.943 396
2	1.970 395	1.941 561	1.913 470	1.886 095	1.859 410	1.833 393
3	2.940 985	2.883 883	2.828 611	2.775 091	2.723 248	2.673 012
4	3.901 965	3.807 729	3.717 098	3.629 895	3.545 951	3.465 106
5	4.853 431	4.713 460	4.579 707	4.451 822	4.329 477	4.212 364
6	5.795 476	5.601 431	5.417 191	5.242 137	5.075 692	4.917 324
7	6.728 194	6.471 991	6.230 283	6.002 055	5.786 373	5.582 381
8	7.651 678	7.325 482	7.019 692	6.732 745	6.463 213	6.209 794
9	8.566 017	8.162 237	7.786 109	7.435 332	7.107 822	6.801 692
10	9.471 304	8.982 585	8.530 203	8.110 896	7.721 735	7.360 087
11	10.367 628	9.786 848	9.252 624	8.760 477	8.306 414	7.886 875
12	11.255 077	10.575 342	9.954 004	9.385 074	8.863 252	8.383 844
13	12.133 740	11.348 374	10.634 955	9.985 648	9.393 573	8.852 683
14	13.003 703	12.106 249	11.296 073	10.563 123	9.898 641	9.294 984
15	13.865 052	12.849 264	11.937 935	11.118 387	10.379 658	9.712 249
16	14.717 874	13.577 710	12.561 102	11.652 296	10.837 770	10.105 895
17	15.562 251	14.291 872	13.166 118	12.165 669	11.274 066	10.477 260
18	16.398 268	14.992 032	13.753 513	12.659 297	11.689 587	10.827 604
19	17.226 008	15.678 462	14.323 799	13.133 939	12.085 321	11.158 117
20	18.045 553	16.351 434	14.877 475	13.590 326	12.462 210	11.469 921
21	18.856 983	17.011 210	15.415 024	14.029 160	12.821 153	11.764 077
22	18.660 379	17.658 049	15.936 917	14.451 115	13.163 003	12.041 582
23	20.455 821	18.292 205	16.443 608	14.856 842	13.488 574	12.303 379
24	21.243 387	18.913 926	16.935 542	15.246 963	13.798 642	12.550 358
25	22.023 155	19.523 457	17.413 148	15.622 080	14.093 945	12.783 356
26	22.795 203	20.121 036	17.876 842	15.982 769	14.375 185	13.003 166
27	23.559 607	20.706 898	18.327 032	16.329 586	14.643 034	13.210 534
28	24.316 443	21.281 273	18.764 108	16.663 063	14.898 127	13.406 164
29	25.065 785	21.844 385	19.188 455	16.983 715	15.141 074	13.590 721
30	25.807 708	22.396 456	19.600 441	17.292 033	15.372 451	13.764 831
31	26.542 285	22.937 702	20.000 429	17.588 494	15.592 810	13.929 086
32	27.269 589	23.468 335	20.388 766	17.873 552	15.802 677	14.084 043
33	27.989 692	23.988 564	20.765 792	18.147 646	16.002 549	14.230 230
34	28.702 666	24.498 592	21.131 837	18.411 198	16.192 904	14.368 141
35	28.408 580	24.998 620	21.487 220	18.664 613	16.374 194	14.498 246
36	30.107 505	25.488 843	21.832 253	18.908 282	16.546 852	14.620 987
37	30.799 510	25.969 454	22.167 235	19.142 579	16.711 287	14.736 780
38	31.484 663	26.440 641	22.492 462	19.367 864	16.867 893	14.846 019
39	32.163 033	26.902 589	22.808 215	19.584 485	17.017 041	14.949 075
40	32.834 686	27.355 480	23.114 772	19.792 774	17.159 086	15.046 297

(continued)

TABLE II Present Value of an Ordinary Annuity of $1 ($a_{\overline{n}|i}$) (*Continued*)

Periods	7%	8%	9%	10%	11%	12%
1	0.934 579	0.925 926	0.917 431	0.909 091	0.900 901	0.892 857
2	1.808 018	1.783 265	1.759 111	1.735 537	1.712 523	1.690 051
3	2.624 316	2.577 097	2.531 295	2.486 852	2.443 715	2.401 831
4	3.387 211	3.312 127	3.239 720	3.169 865	3.102 446	3.037 349
5	4.100 197	3.992 710	3.889 651	3.790 787	3.695 897	3.604 776
6	4.766 540	4.622 880	4.485 919	4.355 261	4.230 538	4.111 407
7	5.389 289	5.206 370	5.032 953	4.868 419	4.712 196	4.563 757
8	5.971 299	5.746 639	5.534 819	5.334 926	5.146 123	4.967 640
9	6.515 232	6.246 888	5.995 247	5.759 024	5.537 048	5.328 250
10	7.023 582	6.710 081	6.417 658	6.144 567	5.889 232	5.650 223
11	7.498 674	7.138 964	6.805 191	6.495 061	6.206 515	5.937 699
12	7.942 686	7.536 078	7.160 725	6.813 692	6.492 356	6.194 374
13	8.357 651	7.903 776	7.486 904	7.103 356	6.749 870	6.423 548
14	8.745 468	8.244 237	7.786 150	7.366 687	6.981 865	6.628 168
15	9.107 914	8.559 479	8.060 688	7.606 080	7.190 870	6.810 864
16	9.446 649	8.851 369	8.312 558	7.823 709	7.379 162	6.973 986
17	9.763 223	9.121 638	8.543 631	8.021 553	7.548 794	7.119 630
18	10.059 087	9.371 887	8.755 625	8.201 412	7.701 617	7.249 670
19	10.335 595	9.603 599	8.950 115	8.364 920	7.839 294	7.365 777
20	10.594 014	9.818 147	9.128 546	8.513 564	7.963 328	7.469 444
21	10.835 527	10.016 803	9.292 244	8.648 694	8.075 070	7.562 003
22	11.061 241	10.200 744	9.442 425	8.771 540	8.175 739	7.644 646
23	11.272 187	10.371 059	9.580 207	8.883 218	8.266 432	7.718 434
24	11.469 334	10.528 758	9.706 612	8.984 744	8.348 137	7.784 316
25	11.653 583	10.674 776	9.822 580	9.077 040	8.421 745	7.843 139
26	11.825 779	10.809 978	9.928 972	9.160 945	8.488 058	7.895 660
27	11.986 709	10.935 165	10.026 580	9.237 223	8.547 800	7.942 554
28	12.137 111	11.051 079	10.116 128	9.306 567	8.601 622	7.984 423
29	12.277 674	11.158 406	10.198 283	9.369 606	8.650 110	8.021 806
30	12.409 041	11.257 783	10.273 654	9.429 914	8.693 793	8.055 184
31	12.531 814	11.349 799	10.342 802	9.479 013	8.733 146	8.084 986
32	12.646 555	11.434 999	10.406 240	9.526 376	8.768 600	8.111 594
33	12.753 790	11.513 888	10.464 441	9.569 432	8.800 541	8.135 352
34	12.854 009	11.586 934	10.517 835	9.608 575	8.829 316	8.156 564
35	12.947 672	11.654 568	10.566 821	9.644 159	8.855 240	8.175 504
36	13.035 208	11.717 193	10.611 763	9.676 508	8.878 594	8.192 414
37	13.117 017	11.775 179	10.652 993	9.705 917	8.899 635	8.207 513
38	13.193 473	11.828 869	10.690 820	9.732 651	8.918 590	8.220 993
39	13.264 928	11.878 582	10.725 523	9.756 956	8.935 666	8.233 030
40	13.331 709	11.924 613	10.757 360	9.779 051	8.951 051	8.243 777

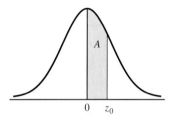

TABLE III Areas Under the Standard Normal Curve

The value of A is the area under the standard normal curve between $z = 0$ and $z = z_0$, for $z_0 \geq 0$. Areas for negative values of z_0 are obtained by symmetry.

z_0	A	z_0	A	z_0	A	z_0	A
0.00	0.0000	0.43	0.1664	0.86	0.3051	1.29	0.4015
0.01	0.0040	0.44	0.1700	0.87	0.3079	1.30	0.4032
0.02	0.0080	0.45	0.1736	0.88	0.3106	1.31	0.4049
0.03	0.0120	0.46	0.1772	0.89	0.3133	1.32	0.4066
0.04	0.0160	0.47	0.1808	0.90	0.3159	1.33	0.4082
0.05	0.0199	0.48	0.1844	0.91	0.3186	1.34	0.4099
0.06	0.0239	0.49	0.1879	0.92	0.3212	1.35	0.4115
0.07	0.0279	0.50	0.1915	0.93	0.3238	1.36	0.4131
0.08	0.0319	0.51	0.1950	0.94	0.3264	1.37	0.4147
0.09	0.0359	0.52	0.1985	0.95	0.3289	1.38	0.4162
0.10	0.0398	0.53	0.2019	0.96	0.3315	1.39	0.4177
0.11	0.0438	0.54	0.2054	0.97	0.3340	1.40	0.4192
0.12	0.0478	0.55	0.2088	0.98	0.3365	1.41	0.4207
0.13	0.0517	0.56	0.2123	0.99	0.3389	1.42	0.4222
0.14	0.0557	0.57	0.2157	1.00	0.3413	1.43	0.4236
0.15	0.0596	0.58	0.2190	1.01	0.3438	1.44	0.4251
0.16	0.0636	0.59	0.2224	1.02	0.3461	1.45	0.4265
0.17	0.0675	0.60	0.2258	1.03	0.3485	1.46	0.4279
0.18	0.0714	0.61	0.2291	1.04	0.3508	1.47	0.4292
0.19	0.0754	0.62	0.2324	1.05	0.3531	1.48	0.4306
0.20	0.0793	0.63	0.2357	1.06	0.3554	1.49	0.4319
0.21	0.0832	0.64	0.2389	1.07	0.3577	1.50	0.4332
0.22	0.0871	0.65	0.2422	1.08	0.3599	1.51	0.4345
0.23	0.0910	0.66	0.2454	1.09	0.3621	1.52	0.4357
0.24	0.0948	0.67	0.2486	1.10	0.3643	1.53	0.4370
0.25	0.0987	0.68	0.2518	1.11	0.3665	1.54	0.4382
0.26	0.1026	0.69	0.2549	1.12	0.3686	1.55	0.4394
0.27	0.1064	0.70	0.2580	1.13	0.3708	1.56	0.4406
0.28	0.1103	0.71	0.2612	1.14	0.3729	1.57	0.4418
0.29	0.1141	0.72	0.2642	1.15	0.3749	1.58	0.4430
0.30	0.1179	0.73	0.2673	1.16	0.3770	1.59	0.4441
0.31	0.1217	0.74	0.2704	1.17	0.3790	1.60	0.4452
0.32	0.1255	0.75	0.2734	1.18	0.3810	1.61	0.4463
0.33	0.1293	0.76	0.2764	1.19	0.3830	1.62	0.4474
0.34	0.1331	0.77	0.2794	1.20	0.3849	1.63	0.4485
0.35	0.1368	0.78	0.2823	1.21	0.3869	1.64	0.4495
0.36	0.1406	0.79	0.2852	1.22	0.3888	1.65	0.4505
0.37	0.1443	0.80	0.2881	1.23	0.3907	1.66	0.4515
0.38	0.1480	0.81	0.2910	1.24	0.3925	1.67	0.4525
0.39	0.1517	0.82	0.2939	1.25	0.3944	1.68	0.4535
0.40	0.1554	0.83	0.2967	1.26	0.3962	1.69	0.4545
0.41	0.1591	0.84	0.2996	1.27	0.3980	1.70	0.4554
0.42	0.1628	0.85	0.3023	1.28	0.3997	1.71	0.4564

(continued)

TABLE III   Areas Under the Standard Normal Curve (*Continued*)

z_0	A	z_0	A	z_0	A	z_0	A
1.72	0.4573	2.26	0.4881	2.80	0.4974	3.34	0.4996
1.73	0.4582	2.27	0.4884	2.81	0.4975	3.35	0.4996
1.74	0.4591	2.28	0.4887	2.82	0.4976	3.36	0.4996
1.75	0.4599	2.29	0.4890	2.83	0.4977	3.37	0.4996
1.76	0.4608	2.30	0.4893	2.84	0.4977	3.38	0.4996
1.77	0.4616	2.31	0.4896	2.85	0.4978	3.39	0.4997
1.78	0.4625	2.32	0.4898	2.86	0.4979	3.40	0.4997
1.79	0.4633	2.33	0.4901	2.87	0.4980	3.41	0.4997
1.80	0.4641	2.34	0.4904	2.88	0.4980	3.42	0.4997
1.81	0.4649	2.35	0.4906	2.89	0.4981	3.43	0.4997
1.82	0.4656	2.36	0.4909	2.90	0.4981	3.44	0.4997
1.83	0.4664	2.37	0.4911	2.91	0.4982	3.45	0.4997
1.84	0.4671	2.38	0.4913	2.92	0.4983	3.46	0.4997
1.85	0.4678	2.39	0.4916	2.93	0.4983	3.47	0.4997
1.86	0.4686	2.40	0.4918	2.94	0.4984	3.48	0.4998
1.87	0.4693	2.41	0.4920	2.95	0.4984	3.49	0.4998
1.88	0.4700	2.42	0.4922	2.96	0.4985	3.50	0.4998
1.89	0.4706	2.43	0.4925	2.97	0.4985	3.51	0.4998
1.90	0.4713	2.44	0.4927	2.98	0.4986	3.52	0.4998
1.91	0.4719	2.45	0.4929	2.99	0.4986	3.53	0.4998
1.92	0.4726	2.46	0.4931	3.00	0.4987	3.54	0.4998
1.93	0.4732	2.47	0.4932	3.01	0.4987	3.55	0.4998
1.94	0.4738	2.48	0.4934	3.02	0.4987	3.56	0.4998
1.95	0.4744	2.49	0.4936	3.03	0.4988	3.57	0.4998
1.96	0.4750	2.50	0.4938	3.04	0.4988	3.58	0.4998
1.97	0.4756	2.51	0.4940	3.05	0.4989	3.59	0.4998
1.98	0.4762	2.52	0.4941	3.06	0.4989	3.60	0.4998
1.99	0.4767	2.53	0.4943	3.07	0.4989	3.61	0.4999
2.00	0.4773	2.54	0.4945	3.08	0.4990	3.62	0.4999
2.01	0.4778	2.55	0.4946	3.09	0.4990	3.63	0.4999
2.02	0.4783	2.56	0.4948	3.10	0.4990	3.64	0.4999
2.03	0.4788	2.57	0.4949	3.11	0.4991	3.65	0.4999
2.04	0.4793	2.58	0.4951	3.12	0.4991	3.66	0.4999
2.05	0.4798	2.59	0.4952	3.13	0.4991	3.67	0.4999
2.06	0.4803	2.60	0.4953	3.14	0.4992	3.68	0.4999
2.07	0.4808	2.61	0.4955	3.15	0.4992	3.69	0.4999
2.08	0.4812	2.62	0.4956	3.16	0.4992	3.70	0.4999
2.09	0.4817	2.63	0.4957	3.17	0.4992	3.71	0.4999
2.10	0.4821	2.64	0.4959	3.18	0.4993	3.72	0.4999
2.11	0.4826	2.65	0.4960	3.19	0.4993	3.73	0.4999
2.12	0.4830	2.66	0.4961	3.20	0.4993	3.74	0.4999
2.13	0.4834	2.67	0.4962	3.21	0.4993	3.75	0.4999
2.14	0.4838	2.68	0.4963	3.22	0.4994	3.76	0.4999
2.15	0.4842	2.69	0.4964	3.23	0.4994	3.77	0.4999
2.16	0.4846	2.70	0.4965	3.24	0.4994	3.78	0.4999
2.17	0.4850	2.71	0.4966	3.25	0.4994	3.79	0.4999
2.18	0.4854	2.72	0.4967	3.26	0.4994	3.80	0.4999
2.19	0.4857	2.73	0.4968	3.27	0.4995	3.81	0.4999
2.20	0.4861	2.74	0.4969	3.28	0.4995	3.82	0.4999
2.21	0.4865	2.75	0.4970	3.29	0.4995	3.83	0.4999
2.22	0.4868	2.76	0.4971	3.30	0.4995	3.84	0.4999
2.23	0.4871	2.77	0.4972	3.31	0.4995	3.85	0.4999
2.24	0.4875	2.78	0.4973	3.32	0.4996	3.86	0.4999
2.25	0.4878	2.79	0.4974	3.33	0.4996		

Answers

Below are the answers to odd-numbered Section Exercises and all the Chapter Review and Chapter Test problems.

Exercise Set 0.1

1. $\in$ **3.** $\in$ **5.** $\notin$ **7.** $\{1, 2, 3, 4, 5, 6, 7\}$
9. $\{x: x$ is a natural number greater than 2 and less than 8$\}$
11. yes   **13.** no   **15.** $D \subseteq C$ **17.** $D \subseteq A$
19. $A \subseteq B$ or $B \subseteq A$ **21.** yes   **23.** no
25. A and B, B and D, C and D
27. $A \cap B = \{4, 6\}$ **29.** $A \cup B = \varnothing$
31. $A \cup B = \{1, 2, 3, 4, 5\}$
33. $A \cup B = \{1, 2, 3, 4\} = B$
35. $A' = \{4, 6, 9, 10\}$
37. $A \cap B' = \{1, 2, 5, 7\}$ **39.** $(A \cup B)' = \{6, 9\}$
41. $A' \cup B' = \{1, 2, 4, 5, 6, 7, 9, 10\}$
43. $\{1, 2, 3, 5, 7, 9\}$ **45.** $\{4, 6, 8, 10\}$
47. $A - B = \{1, 7\}$ **49.** $A - B = \varnothing$ or $\{\ \}$
51. **(a)** $L = \{99, 2000, 2001\}$
     $H = \{97, 98, 99, 2000, 2001, 2002\}$
     $C = \{95, 2001, 2002\}$
   (b) $L \subseteq H$
   (c) C' is the set of years when the percent change from low to high was 35% or less.
   (d) $B' = \{95, 96, 97, 98, 99, 2000\} =$ the set of years when the high was 7500 or less or the percent change was 35% or less.
   (e) $\{95, 2002\} =$ the set of years when the low was 7500 or less and the percent change exceeded 35%.
53. **(a)** 130   **(b)** 840   **(c)** 520
55.

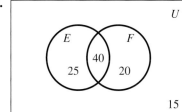

   (a) 40
   (b) 85
   (c) 25

57.

   (a) 25
   (b) 43
   (c) 53

59. **(a)** and **(b)**

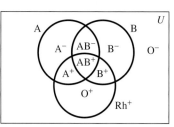

   (c) A$^+$: 34%; B$^+$: 9%; O$^+$: 38%; AB$^+$: 3%; O$^-$: 7%;
     A$^-$: 6%; B$^-$: 2%; AB$^-$: 1%

Exercise Set 0.2

1. **(a)** irrational   **(b)** rational, integer
   (c) rational, integer, natural   **(d)** meaningless
3. **(a)** Commutative   **(b)** Distributive
   (c) Multiplicative identity
5. $<$ **7.** $<$ **9.** $>$ **11.** 11   **13.** 4
15. 2   **17.** $\frac{-4}{3}$ **19.** 3   **21.** $\frac{17}{11}$ **23.** entire line
25. $(1, 3]$; half open   **27.** $(2, 10)$; open
29. $-3 \le x < 5$ **31.** $x > 4$
33. $(-3, 4)$
35. $(4, +\infty)$
37. $[-1, +\infty)$
39. $(-\infty, 0) \cup (7, +\infty)$
41. -0.000038585
43. 9122.387471   **45.** 3240.184509
47. **(a)** \$1088.91   **(b)** \$258.62   **(c)** \$627.20
49. **(a)** Formula (2) is more accurate; 1996: \$24.54; 1997: \$26.10
   (b) \$46.37
51. **(a)** $70{,}351 \le I \le 146{,}750$; $146{,}751 \le I \le 319{,}100$; $I > 319{,}100$
   (b) $T = \$4000$ for $I = \$29{,}050$; $T = \$14{,}325$ for $I = \$70{,}350$
   (c) $[4000, 14{,}325]$

Exercise Set 0.3

1. 256   **3.** -64 **5.** $\frac{1}{9}$ **7.** $-\frac{9}{4}$ **9.** 6^8
11. $\frac{1}{10}$ **13.** 3^9 **15.** $\left(\frac{3}{2}\right)^2 = \frac{9}{4}$ **17.** $1/x^6$
19. x/y^2 **21.** x^7 **23.** $x^{-2} = 1/x^2$ **25.** x^4
27. y^{12} **29.** x^{12} **31.** $x^2 y^2$ **33.** $16/x^{20}$

35. $x^8/(16y^4)$ **37.** $-16a^2/b^2$ **39.** $2/(xy^2)$
41. $1/(x^9y^6)$ **43.** $(a^{18}c^{12})/b^6$
45. (a) $1/(2x^4)$ (b) $1/(16x^4)$ (c) $1/x^4$ (d) 8
47. x^{-1} **49.** $8x^3$ **51.** $\frac{1}{4}x^{-2}$ **53.** $-\frac{1}{8}x^3$
55. 2.0736   **57.** 0.1316872428
59. $S = \$2114.81$; $I = \$914.81$
61. $S = \$9607.70$; $I = \$4607.70$
63. $7806.24
65. (a) 20, 45, 50, 52
   (b) $984.63, $5257.31, $7349.56, $8403.50
   (c) The boom of the 1990s ended; War on Terror and in Iraq
   (d) $14,363.39; too high based on the trend from 2002
67. (a) 162, 516, 1029
   (b) 198
   (c) Two possibilities might be more environmental protections and the fact that there are only a limited number of species.
   (d) There are only a limited number of species. Also, below some threshold level the ecological balance might be lost, perhaps resulting in an environmental catastrophe (which the equation could not predict). Upper limit = 1209
69. (a) 10   (b) $78.103 billion
   (c) $729.195 billion   (d) $3,801.290 billion

Exercise Set 0.4

1. (a) $\frac{16}{3}$ (b) 1.2
3. (a) 8   (b) not real   **5.** $\frac{9}{4}$
7. (a) 4   (b) $\frac{1}{4}$ **9.** $(6.12)^{4/9} \approx 2.237$
11. $m^{3/2}$ **13.** $(m^2n^5)^{1/4}$ **15.** $\sqrt[6]{x^7}$
17. $-1/(4\sqrt[4]{x^5})$ **19.** $y^{3/4}$ **21.** $z^{19/4}$ **23.** $1/y^{5/2}$
25. x **27.** $1/y^{21/10}$ **29.** $x^{1/2}$ **31.** $1/x$ **33.** $8x^2$
35. $8x^2y^2\sqrt{2y}$ **37.** $2x^2y\sqrt[3]{5x^2y^2}$ **39.** $6x^2y\sqrt{x}$
41. $42x^3y^2\sqrt{x}$ **43.** $2xy^5/3$ **45.** $2b\sqrt[4]{b}/(3a^2)$
47. $1/9$ **49.** 7   **51.** $\sqrt{6}/3$ **53.** $\sqrt{mx}/x$
55. $\sqrt[3]{mx^2}/x^2$ **57.** $-\frac{2}{3}x^{-2/3}$ **59.** $3x^{3/2}$
61. $(3\sqrt{x})/2$ **63.** $1/(2\sqrt{x})$
65. (a) $10^{8.5} = 10^{17/2} = \sqrt{10^{17}}$ (b) 316,227,766
   (c) 14.125
67. (a) $S = 1000\sqrt{\left(1 + \dfrac{r}{100}\right)^5}$
   (b) $1173.26 (nearest cent)
69. (a) $P = 0.924\sqrt[100]{t^{13}}$
   (b) 2000 to 2010; 1.1074 billion vs. 0.0209 billion. By 2040 and 2050 the population will be much larger than earlier in the 21st century, and there is a limited number of people that any land can support—in terms of both space and food.
71. 74 kg   **73.** 39,491   **75.** (a) 10   (b) 259

Exercise Set 0.5

1. (a) 2   (b) -1 (c) 10   (d) one
3. (a) 5   (b) -14 (c) 0   (d) several
5. (a) 5   (b) 0   (c) 2   (d) -5
7. -12 **9.** $\frac{-7}{31}$ **11.** 87.4654   **13.** $21pq - 2p^2$
15. $m^2 - 7n^2 - 3$ **17.** $3q + 12$
19. $x^2 - 1$ **21.** $35x^5$ **23.** $3rs$
25. $2ax^4 + a^2x^3 + a^2bx^2$ **27.** $6y^2 - y - 12$
29. $12 - 30x^2 + 12x^4$ **31.** $16x^2 + 24x + 9$
33. $x^4 - x^2 + \frac{1}{4}$ **35.** $36x^2 - 9$ **37.** $0.01 - 16x^2$
39. $0.1x^2 - 1.995x - 0.1$
41. $x^3 - 8$ **43.** $x^8 + 3x^6 - 10x^4 + 5x^3 + 25x$
45. (a) $9x^2 - 21x + 13$ (b) 5
47. $3 + m + 2m^2n$ **49.** $8x^3y^2/3 + 5/(3y) - 2x^2/(3y)$
51. $x^3 + 3x^2 + 3x + 1$ **53.** $8x^3 - 36x^2 + 54x - 27$
55. $x^2 - 2x + 5 - 11/(x + 2)$
57. $x^2 + 3x - 1 + (-4x + 2)/(x^2 + 1)$
59. $x + 2x^2$ **61.** $x - x^{1/2} - 2$ **63.** $x - 9$
65. $4x^2 + 4x$ **67.** $55x$
69. (a) $99.95 + 0.21x$ (b) $127.67
71. (a) $4000 - x$ (b) $0.10x$ (c) $0.08(4000 - x)$
   (d) $0.10x + 0.08(4000 - x)$
73. $(15 - 2x)(10 - 2x)x$
75. (a) $= A2 - 1$ (b) width $= 50 -$ length
   (c) $= 50 - A2$ (d) $= A2*B2$
   (e) length $= 33$, width $= 17$
77. (a)

	A	B
1	Year	Internet Households
2	1995	14.61
3	1996	20.35
4	1997	26.09
5	1998	31.83
6	1999	37.57
7	2000	43.31
8	2001	49.05
9	2002	54.79
10	2003	60.53
11	2004	66.27
12	2005	72.01
13	2006	77.75
14	2007	83.49
15	2008	89.23
16	2009	94.97
17	2010	100.71

   (b) 2006
   (c) when the percent p exceeds 100%

Exercise Set 0.6

1. $3b(3a - 4a^2 + 6b)$ **3.** $2x(2x + 4y^2 + y^3)$
5. $(7x^2 + 2)(x - 2)$ **7.** $(6 + y)(x - m)$
9. $(x + 2)(x + 6)$ **11.** $(x - 3)(x + 2)$
13. $(7x + 4)(x - 2)$ **15.** $(x - 5)^2$
17. $(7a + 12b)(7a - 12b)$
19. (a) $(3x - 1)(3x + 8)$ **(b)** $(9x + 4)(x + 2)$
21. $x(4x - 1)$ **23.** $(x^2 - 5)(x + 4)$
25. $2(x - 7)(x + 3)$ **27.** $2x(x - 2)^2$
29. $(2x - 3)(x + 2)$ **31.** $3(x + 4)(x - 3)$
33. $2x(x + 2)(x - 2)$ **35.** $(5x + 2)(2x + 3)$
37. $(5x - 1)(2x - 9)$
39. $(y^2 + 4x^2)(y + 2x)(y - 2x)$
41. $(x + 2)^2(x - 2)^2$
43. $(2x + 1)(2x - 1)(x + 1)(x - 1)$
45. $(x + 1)^3$ **47.** $(x - 4)^3$
49. $(x - 4)(x^2 + 4x + 16)$
51. $(3 + 2x)(9 - 6x + 4x^2)$ **53.** $x + 1$
55. $1 + x$ **57.** $7x - 3x^3$ **59.** $P(1 + rt)$
61. $m(c - m)$
63. (a) $p(10,000 - 100p)$; $x = 10,000 - 100p$
    (b) 6200
65. (a) $R = x(300 - x)$
    (b) $300 - x$

Exercise Set 0.7

1. $2y^3/z$ **3.** $\frac{1}{3}$ **5.** $(x - 1)/(x - 3)$
7. $(2xy^2 - 5)/(y + 3)$ **9.** $20x/y$ **11.** $\frac{32}{3}$
13. $3x + 9$ **15.** $\dfrac{-(x + 1)(x + 3)}{(x - 1)(x - 3)}$
17. $15bc^2/2$ **19.** $5y/(y - 3)$
21. $\dfrac{-x(x - 3)(x + 2)}{x + 3}$ **23.** $\dfrac{1}{x + 1}$ **25.** $\dfrac{4a - 4}{a(a - 2)}$
27. $\dfrac{-x^2 + x + 1}{x + 1}$ **29.** $\dfrac{16a + 15a^2}{12(x + 2)}$ **31.** $\dfrac{79x + 9}{30(x - 2)}$
33. $\dfrac{9x + 4}{(x - 2)(x + 2)(x + 1)}$
35. $(7x - 3x^3)/\sqrt{3 - x^2}$ **37.** $\frac{1}{6}$ **39.** xy
41. $\dfrac{x + 1}{x^2}$ **43.** $\dfrac{1}{\sqrt{a}} = \dfrac{\sqrt{a}}{a}$ **45.** $\dfrac{x - 2}{(x - 3)\sqrt{x^2 + 9}}$
47. (a) -12 **(b)** $\frac{25}{36}$ **49.** $2b^2 - a$
51. $(1 - 2\sqrt{x} + x)/(1 - x)$ **53.** $1/(\sqrt{x + h} + \sqrt{x})$
55. $(bc + ac + ab)/abc$
57. (a) $\dfrac{0.1x^2 + 55x + 4000}{x}$
    (b) $0.1x^2 + 55x + 4000$
59. $\dfrac{t^2 + 9t}{(t + 3)^2}$

Chapter 0 Review Exercises

1. yes   **2.** no   **3.** no   **4.** $\{1, 2, 3, 4, 9\}$
5. $\{5, 6, 7, 8, 10\}$ **6.** $\{1, 2, 3, 4, 9\}$
7. yes, $(A' \cup B') = \{1, 3\} = A \cap B$
8. (a) Commutative Property of Addition
    (b) Associative Property of Multiplication
    (c) Distributive Law
9. (a) irrational   **(b)** rational, integer
    (c) meaningless
10. (a) $>$ **(b)** $<$ **(c)** $>$ **11.** 6
12. 142 **13.** 10 **14.** $5/4$ **15.** 9 **16.** -29
17. $13/4$ **18.** -10.62857888
19. (a) $[0, 5]$, closed

(b) $[-3, 7)$, half-open

(c) $(-4, 0)$, open

20. (a) $-1 < x < 16$ **(b)** $-12 \leq x \leq 8$
    (c) $x < -1$
21. (a) 1 **(b)** $2^{-2} = 1/4$ **(c)** 4^6 **(d)** 7
22. (a) $1/x^2$ **(b)** x^{10} **(c)** x^9 **(d)** $1/y^8$ **(e)** y^6
23. $-x^2y^2/36$ **24.** $9y^8/(4x^4)$ **25.** $y^2/(4x^4)$
26. $-x^8z^4/y^4$ **27.** $3x/(y^7z)$ **28.** $x^5/(2y^3)$
29. (a) 4 **(b)** $2/7$ **(c)** 1.1
30. (a) $x^{1/2}$ **(b)** $x^{2/3}$ **(c)** $x^{-1/4}$
31. (a) $\sqrt[3]{x^2}$ **(b)** $1/\sqrt{x} = \sqrt{x}/x$ **(c)** $-x\sqrt{x}$
32. (a) $5y\sqrt{2x}/2$ **(b)** $\sqrt[3]{x^2y}/x^2$ **33.** $x^{5/6}$ **34.** y
35. $x^{17/4}$ **36.** $x^{11/3}$ **37.** $x^{2/5}$ **38.** x^2y^8
39. $2xy^2\sqrt{3xy}$ **40.** $25x^3y^4\sqrt{2y}$ **41.** $6x^2y^4\sqrt[3]{5x^2y^2}$
42. $8a^2b^4\sqrt{2a}$ **43.** $2xy$ **44.** $4x\sqrt{3xy}/(3y^4)$
45. $-x - 2$ **46.** $-x^2 - x$
47. $4x^3 + xy + 4y - 4$ **48.** $24x^5y^5$
49. $3x^2 - 7x + 4$ **50.** $3x^2 + 5x - 2$
51. $4x^2 - 7x - 2$ **52.** $6x^2 - 11x - 7$
53. $4x^2 - 12x + 9$ **54.** $16x^2 - 9$
55. $2x^4 + 2x^3 - 5x^2 + x - 3$
56. $8x^3 - 12x^2 + 6x - 1$
57. $x^3 - y^3$ **58.** $(2/y) - (3xy/2) - 3x^2$
59. $3x^2 + 2x - 3 + (-3x + 7)/(x^2 + 1)$
60. $x^3 - x^2 + 2x + 7 + 21/(x - 3)$
61. $x^2 - x$ **62.** $2x - a$ **63.** $x^3(2x - 1)$
64. $2(x^2 + 1)^2(1 + x)(1 - x)$ **65.** $(2x - 1)^2$
66. $(4 + 3x)(4 - 3x)$ **67.** $2x^2(x + 2)(x - 2)$
68. $(x - 7)(x + 3)$ **69.** $(3x + 2)(x - 1)$
70. $(x - 3)(x - 2)$ **71.** $(x - 12)(x + 2)$

72. $(4x + 3)(3x - 8)$ **73.** $(2x + 3)^2(2x - 3)^2$

74. $x^{2/3} + 1$ **75. (a)** $\dfrac{x}{(x + 2)}$ **(b)** $\dfrac{2xy(2 - 3xy)}{2x - 3y}$

76. $\dfrac{x^2 - 4}{x(x + 4)}$ **77.** $\dfrac{(x + 3)}{(x - 3)}$ **78.** $\dfrac{x^2(3x - 2)}{(x - 1)(x + 2)}$

79. $(6x^2 + 9x - 1)/(6x^2)$ **80.** $\dfrac{4x - x^2}{4(x - 2)}$

81. $-\dfrac{x^2 + 2x + 2}{x(x - 1)^2}$ **82.** $\dfrac{x(x - 4)}{(x - 2)(x + 1)(x - 3)}$

83. $\dfrac{(x - 1)^3}{x^2}$ **84.** $\dfrac{1 - x}{1 + x}$ **85.** $3(\sqrt{x} + 1)$

86. $2/(\sqrt{x} + \sqrt{x - 4})$

87. (a)

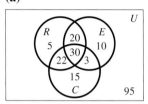

R: recognized
E: exercise
C: community involvement

(b) 10   **(c)** 100

88. \$72.96 billion   **89.** 16

90. (a) \$4115.27   **(b)** \$66,788.69

91. \$400 million; \$5078.676 million

92. (a) $10{,}000\left[\dfrac{(0.0065)(1.0065)^n}{(1.0056)^n - 1}\right]$

(b) \$243.19 (for both)

93. (a) $S = k\sqrt[3]{A}$ **(b)** $\sqrt[3]{2.25} \approx 1.31$

94. $26x - 300 - 0.001x^2$

95. $450{,}000 - 1125x$

96. $(50 + x)(12 - 0.5x)$

97. (a) $\dfrac{5400p}{100 - p}$

(b) \$0. It costs nothing if no effort is made to remove pollution.

(c) \$264,600

(d) Undefined. Removing 100% would be impossible, and the cost of getting close would be enormous.

98. $\dfrac{56x^2 + 1200x + 8000}{x}$

Chapter 0 Test

1. (a) $\{3, 4, 6, 8\}$ **(b)** $\{3, 4\}$; $\{3, 6\}$; or $\{4, 6\}$
(c) $\{6\}$ or $\{8\}$

2. 21

3. (a) 8   **(b)** 1   **(c)** $\frac{1}{2}$ **(d)** -10 **(e)** 30
(f) $\frac{5}{6}$ **(g)** $\frac{2}{3}$ **(h)** -3

4. (a) $\sqrt[5]{x}$ **(b)** $\dfrac{1}{\sqrt[4]{x^3}}$ **5. (a)** $\dfrac{1}{x^5}$ **(b)** $\dfrac{x^{21}}{y^6}$

6. (a) $\dfrac{\sqrt{5x}}{5}$ **(b)** $2a^2b^2\sqrt{6ab}$ **(c)** $\dfrac{1 - 2\sqrt{x} + x}{1 - x}$

7. (a) 5   **(b)** -8 **(c)** -5

8. $(-2, 3]$ (number line from -4 to 6, open circle at -2, closed dot at 3)

9. (a) $2x^2(4x - 1)$ **(b)** $(x - 4)(x - 6)$
(c) $(3x - 2)(2x - 3)$ **(d)** $2x^3(1 + 4x)(1 - 4x)$

10. (c); -2 **11.** $2x + 1 + \dfrac{2x - 6}{x^2 - 1}$

12. (a) $19y - 45$ **(b)** $-6t^6 + 9t^9$
(c) $4x^3 - 21x^2 + 13x - 2$ **(d)** $-18x^2 + 15x - 2$
(e) $4m^2 - 28m + 49$ **(f)** $\dfrac{x^4}{3x + 9}$ **(g)** $\dfrac{x^7}{81}$
(h) $\dfrac{6 - x}{x - 8}$ **(i)** $\dfrac{x^2 - 4x - 3}{x(x - 3)(x + 1)}$

13. $\dfrac{y - x}{y + xy^2}$ **14. (a)** 0   **(b)** 175

15. \$4875.44 (nearest cent)

Exercise Set 1.1

1. $x = -9/4$ **3.** $x = 0$ **5.** $x = -32$
7. $x = -29/2$ **9.** $x = 17/13$
11. $x = -1/3$ **13.** $x = 3$ **15.** $x = 5/4$
17. no solution   **19.** $x \approx -0.279$
21. $x \approx -1147.362$ **23.** $y = \frac{3}{4}x - \frac{15}{4}$
25. $y = -6x + \frac{22}{3}$ **27.** $t = \dfrac{S - P}{Pr}$
29. $x < 2$ **31.** $x < -4$ **33.** $x \leq -1$
35. $x < -3$ (number line, open circle at -3)
37. $x < -6$ (number line, open circle at -6)
39. $x < 2$ (number line, open circle at 2)
41. 145 months
43. \$3356.50   **45.** 440 packs, or 220,000 CDs
47. \$29,600   **49.** 78.68%   **51.** 96
53. \$90,000 at 9%; \$30,000 at 13%
55. \$2160/month; 8% increase
57. $x > 80$ (number line, open circle at 80)
59. $695 + 5.75x \leq 900$; 35 or fewer
61. (a) $t = 10$ **(b)** $t > 12.80$ **(c)** in 2008
63. (a) $0.479 \leq h \leq 1$; $h = 1$ means 100% humidity
(b) $0 \leq h \leq 0.237$

Exercise Set 1.2

1. (a) To each x-value there corresponds exactly one y-value.
(b) domain: $\{-7, -1, 0, 3, 4.2, 9, 11, 14, 18, 22\}$
range: $\{0, 1, 5, 9, 11, 22, 35, 60\}$
3. $f(0) = 1, f(11) = 35$
5. yes; to each x-value there corresponds exactly one y-value; domain $= \{1, 2, 3, 8, 9\}$, range $= \{-4, 5, 16\}$

7. The vertical-line test shows that graph (a) represents a function of *x*, but graph (b) does not.

9. yes   **11.** no

13. (a) -10 (b) 6   (c) -34 (d) 2.8

15. (a) -3 (b) 1   (c) 13   (d) 6

17. (a) 63/8   (b) 6   (c) -6

19. (a) no, $f(2 + 1) = f(3) = 13$ but $f(2) + f(1) = 10$
(b) $1 + x + h + x^2 + 2xh + h^2$
(c) no, $f(x) + f(h) = 2 + x + h + x^2 + h^2$
(d) no, $f(x) + h = 1 + x + x^2 + h$
(e) $1 + 2x + h$

21. (a) $-2x^2 - 4xh - 2h^2 + x + h$
(b) $-4x - 2h + 1$

23. (a) 10   (b) 6

25. (a) $b = a^2 - 4a$ (b) $(1, -3)$, yes
(c) $(3, -3)$, yes   (d) $x = 0, x = 4$, yes

27. domain: all reals; range: reals $y \geq 4$

29. domain: reals $x \geq -4$; range: reals $y \geq 0$

31. $x \geq 1, x \neq 2$ **33.** $-7 \leq x \leq 7$

35. (a) $3x + x^3$ (b) $3x - x^3$ (c) $3x^4$ (d) $\dfrac{3}{x^2}$

37. (a) $\sqrt{2x} + x^2$ (b) $\sqrt{2x} - x^2$
(c) $x^2\sqrt{2x}$ (d) $\dfrac{\sqrt{2x}}{x^2}$

39. (a) $-8x^3$ (b) $1 - 2(x - 1)^3$
(c) $[(x - 1)^3 - 1]^3$ (d) $(x - 1)^6$

41. (a) $2\sqrt{x^4 + 5}$ (b) $16x^2 + 5$
(c) $2\sqrt{2\sqrt{x}}$ (d) $4x$

43. (a) $f(20) = 103{,}000$ means that if \$103,000 is borrowed, it can be repaid in 20 years (of \$800-per-month payments).
(b) no; $f(5 + 5) = f(10) = 69{,}000$, but $f(5) + f(5) = 40{,}000 + 40{,}000 = 80{,}000$
(c) 15 years; $f(15) = 89{,}000$

45. (a) $f(1950) = 16.5$ means that in 1950 there were 16.5 workers supporting each Social Security beneficiary.
(b) 3.4
(c) Through 1995, the graph must be the same. After 1995, the graph might be the same.
(d) domain: $1950 \leq t \leq 2050$
range: $1.9 \leq n \leq 16.5$

47. (a) $f(95) = 1.0$; $g(95) = 0.6$
(b) $f(100) = 1.2$ means that at the end of 2000 there were 1.2 million persons in state prisons.
(c) $g(90) = 0.5$ means that at the end of 1990 there were 0.5 million persons on parole.
(d) $(f - g)(100) = f(100) - g(100) = 1.2 - 0.7 = 0.5$ means that at the end of 2000 there were 0.5 million more persons in state prisons than there were on parole.

(e) $(f - g)(98)$ is greater because for 1998 there is a greater distance between the graphs.

49. (a) $s \geq 0$
(b) $f(10) \approx -29.33$ means that if the air temperature is $-5°$F and there is a 10 mph wind, then the temperature feels like $-29.33°$F.
(c) $f(0) = 45.694$ from the formula, but $f(0)$ should equal the air temperature, $-5°$F.

51. (a) $C(10) = \$4210$
(b) $C(100)$ is the total cost of producing 100 items.
(c) $C(100) = \$32{,}200$

53. (a) $0 \leq p < 100$
(b) \$5972.73; to remove 45% of the particulate pollution would cost \$5972.73.
(c) \$65,700; to remove 90% of the particulate pollution would cost \$65,700.
(d) \$722,700; to remove 99% of the particulate pollution would cost \$722,700.
(e) \$1,817,700; to remove 99.6% of the particulate pollution would cost \$1,817,700.

55. (a) yes   (b) $A(2) = 96$; $A(30) = 600$
(c) $0 < x < 50$

57. (a) $(P \circ q)(t) = 180(1000 + 10t)$
$- \dfrac{(1000 + 10t)^2}{100} - 200$
(b) $x = 1150, P = \$193{,}575$

59. (a) yes; the output of g (customers) is the input for f.
(b) no; the output of f is revenue, and this is not the input for g.
(c) $f \circ g$: input (independent variable) is advertising dollars.
output (dependent variable) is revenue dollars.
$f \circ g$ shows how advertising dollars result in revenue dollars.

61. $L = 2x + 3200/x$

63. $R = (30 + x)(10 - 0.20x)$

Exercise Set 1.3

1. *x*-intercept 4
y-intercept 3

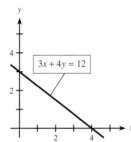

3. *x*-intercept 12
y-intercept -7.5

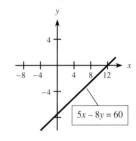

5. *x*-intercept 0
   y-intercept 0

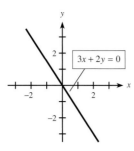

7. $m = 4$ **9.** $m = 0$ **11.** 0   **13.** 0
15. (a) negative   **(b)** undefined
17. $m = 7/3, b = -1/4$ **19.** $m = 0, b = 3$

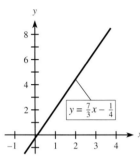

21. undefined slope,   **23.** $m = -2/3, b = 2$
   no *y*-intercept

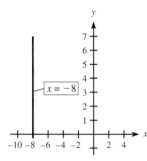 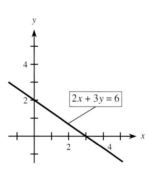

25.

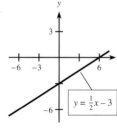

27.

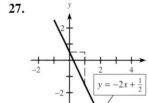

29.

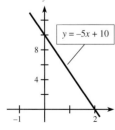

31.

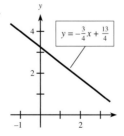

33.

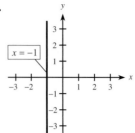

35. $y = 2x - 4$ **37.** $-x + 13y = 32$
39. perpendicular   **41.** neither; same line
43. $y = -\frac{3}{5}x - \frac{41}{5}$ **45.** $y = -\frac{6}{5}x + \frac{23}{5}$
47. (a)

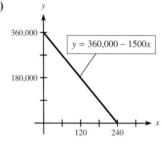

(b) 240 months
(c) After 60 months, the value of the building is $270,000.
49. (a) $m = 5.74; b = 14.61$
   (b) In 1995 (when $x = 0$), 14.61% of the U.S. population had Internet service.
   (c) The percentage of the U.S. population with Internet service is changing at the rate of 5.74% per year.
51. (a) $m = -0.762; b = 85.284$
   (b) The *M*-intercept represents the percent of unmarried women in 1950 who were married that year.
   (c) -0.762% per year. This means that after 1950, the percent of unmarried women who marry decreases by 0.762% per year.
53. (a) $m = 0.551$
   (b) This means that the average annual earnings of females increases $0.551 for each $1 increase in the average annual earnings of males.
   (c) $28,802
55. $y = 4.95 + 0.0838x$
57. (a) $B = 0.619W + 108.556$ **(b)** $634.71
59. (a) $y = 4.43x - 8689$
   (b) The CPI-U is changing at the rate of $4.43/year.
61. $p = 85,000 - 1700x$ **63.** $R = 3.2t - 0.2$
65. $y = 0.48x - 71$

Exercise Set 1.4

1.

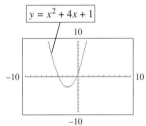

$y = x^2 + 4x + 1$

3.

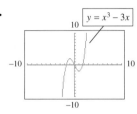

$y = x^3 - 3x$

5.

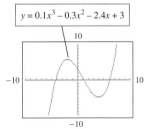

$y = 0.1x^3 - 0.3x^2 - 2.4x + 3$

7.

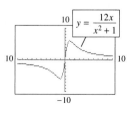

$y = \dfrac{12x}{x^2 + 1}$

9.

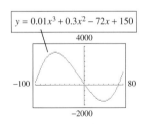

$y = x^3 - 12x - 1$

11. $y = (3x + 7)/(x^2 + 4)$

13. (a) $y = 0.01x^3 + 0.3x^2 - 72x + 150$

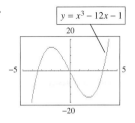

$y = 0.01x^3 + 0.3x^2 - 72x + 150$

(b) standard window

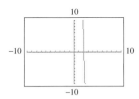

15. (a) $y = \dfrac{x + 15}{x^2 + 400}$

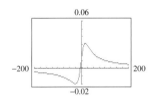

(b) standard window

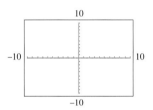

17. (a) x-intercept 30, y-intercept -0.03

(b)

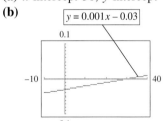

$y = 0.001x - 0.03$

19.

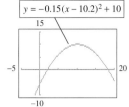

$y = -0.15(x - 10.2)^2 + 10$

21. $y = \dfrac{x^3 + 19x^2 - 62x - 840}{20}$

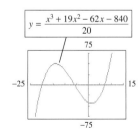

$y = \dfrac{x^3 + 19x^2 - 62x - 840}{20}$

23.

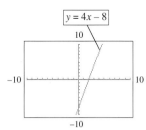

$y = 4x - 8$

Linear

25.

$y = \dfrac{5}{2} - 2x^2$

Not linear

27.

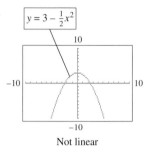

Not linear

29. $f(-2) = -18; f\left(\frac{3}{4}\right) = 0.734375$

31.

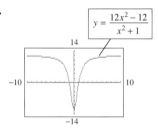

33.

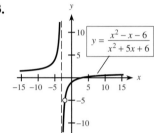

35. $x = 3.5$ **37.** Either $x = 5$ or $x = -2$

39. (a) $-1.1098, 8.1098$ **(b)** $-1.1098, 8.1098$

41. (a)

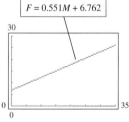

(b) When average annual earnings for males is $50,000, average annual earnings for females is $34,312.   **(c)** $41,199.50

43. (a)

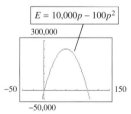

(b) $E \geq 0$ when $0 \leq p \leq 100$

45. (a)

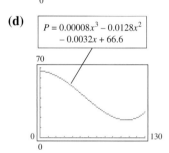

(b) Decreasing; as more people become aware of the product, there are fewer to learn about it, so the rate will decrease.

47. (a) x-min $= 0$, x-max $= 120$

(b) y-min $= 0$, y-max $= 70$

(c)

(d)

(e) The percent decreases before 2000 and increases after 2000.

49. (a)

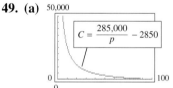

(b) Near $p = 0$, cost grows without bound.

(c) The coordinates of the point mean that obtaining stream water with 1% of the current pollution levels would cost $282,150.

(d) The p-intercept means that stream water with 100% of the current pollution levels would cost $0.

51. (a)

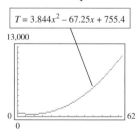

(b) increasing; the per capita federal tax burden is increasing.

Exercise Set 1.5

1. one solution; $(-1, -2)$
3. infinitely many solutions (each point on the line)
5. $x = 2, y = 2$ **7.** no solution
9. $x = 10/3, y = 2$ **11.** $x = 14/11, y = 6/11$
13. $x = 3, y = -2$ **15.** $x = 2, y = 1$
17. $x = -52/7, y = -128/7$ **19.** $x = 1, y = 1$
21. dependent
23. $x = 4, y = 2$ **25.** $x = -1, y = 1$
27. $x = -17, y = 7, z = 5$
29. $x = 4, y = 12, z = -1$
31. $x = 44, y = -9, z = -1/2$
33. $x \approx 2.455$; during 1998; amount $\approx \$842.67$ billion
35. (a) $x + y = 1800$ **(b)** $20x$ **(c)** $30y$
(d) $20x + 30y = 42,000$
(e) 1200 tickets at \$20; 600 tickets at \$30
37. \$68,000 at 18%; \$77,600 at 10%
39. \$135,000 at 10%; \$100,000 at 12%
41. 4 oz of A, $6\frac{2}{3}$ oz of B
43. 4550 of species A, 1500 of species B
45. 7 cc of 20%; 3 cc of 5%
47. 10,000 at \$20; 6000 at \$30
49. 80 cc   **51.** 5 oz of A, 1 oz of B, 5 oz of C
53. 200 type A, 100 type B, 200 type C

Exercise Set 1.6

1. (a) $P(x) = 17x - 3400$ **(b)** \$1700
3. (a) $P(x) = 37x - 1850$
(b) $-\$740$, loss of \$740   **(c)** 50
5. (a) $m = 5, b = 250$
(b) $\overline{MC} = 5$ means each additional unit produced costs \$5.
(c) slope = marginal cost; C-intercept = fixed costs
(d) 5, 5
7. (a) 27
(b) $\overline{MR} = 27$ means each additional unit sold brings in \$27.
(c) 27, 27
9. (a) $P(x) = 22x - 250$ **(b)** 22   **(c)** $\overline{MP} = 22$
(d) Each unit sold adds \$22 to profits at all levels of production, so produce and sell as much as possible.
11. $P = 58x - 8500$, $\overline{MP} = 58$
13. (a) $C(x) = 35x + 6600$ **(b)** $R(x) = 60x$
(c) $P(x) = 25x - 6600$
(d) $C(200) = 13,600$ dollars is the cost of producing 200 helmets.
$R(200) = 12,000$ dollars is the revenue from sale of 200 helmets.

$P(200) = -1600$ dollars; will lose \$1600 from production and sale of 200 helmets.
(e) $C(300) = 17,100$ dollars is the cost of producing 300 helmets.
$R(300) = 18,000$ dollars is the revenue from sale of 300 helmets.
$P(300) = 900$ dollars; will profit \$900 from production and sale of 300 helmets.
(f) $\overline{MP} = 25$ dollars per unit; each additional unit produced and sold increases profit by \$25.
15. (a) Revenue passes through the origin.
(b) \$2000   **(c)** 400 units
(d) $\overline{MC} = 2.5$; $\overline{MC} = 7.5$
17. 33
19. (a) $R(x) = 12x$; $C(x) = 8x + 1600$
(b) 400 units
21. (a) $P(x) = 4x - 1600$
(b) $x = 400$ units to break even
23. (a) $C(x) = 4.5x + 1045$
(b) $R(x) = 10x$
(c) $P(x) = 5.5x - 1045$
(d) 190 surge protectors
25. (a) $R(x) = 54.90x$ **(b)** $C(x) = 14.90x + 20,200$
(c) 505
27. (a) R starts at origin and is the steeper line.
FC is a horizontal line.
VC starts at origin and is not as steep as R.
(See figure.)

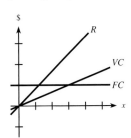

(b) C starts where FC meets the \$-axis and is parallel to VC. Where C meets R is the break-even point (BE).
P starts on the \$-axis at the negative of FC and crosses the x-axis at BE. (See figure.)

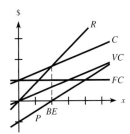

29. demand decreases

31. (a) 600 **(b)** 300 **(c)** shortage
33. 16 demanded, 25 supplied; surplus
35. $p = -2q/3 + 1060$
37. $p = 0.0001q + 0.5$
39. (a) demand falls; supply rises **(b)** (30, \$25)
41. (a) $q = 20$ **(b)** $q = 40$
  (c) shortage, 20 units short
43. shortage **45.** $q = 20$, $p = \$18$
47. $q = 10$, $p = \$180$ **49.** $q = 100$, $p = \$325$
51. (a) \$15 **(b)** $q = 100$, $p = \$100$
  (c) $q = 50$, $p = \$110$ **(d)** yes
53. $q = 8$; $p = \$188$ **55.** $q = 500$; $p = \$40$
57. $q = 1200$; $p = \$15$

Chapter 1 Review Exercises

1. $x = \frac{31}{3}$ **2.** $x = -13$ **3.** $x = -\frac{29}{8}$
4. $x = -\frac{1}{9}$ **5.** $x = 8$
6. no solution **7.** $y = -\frac{2}{3}x - \frac{4}{3}$
8. $x \le 3$

9. $x \ge -20/3$

10. $x \ge -15/13$

11. yes **12.** no **13.** yes
14. domain: reals $x \le 9$; range: reals $y \ge 0$
15. (a) 2 **(b)** 37 **(c)** 29/4
16. (a) 0 **(b)** 9/4 **(c)** 10.01
17. $9 - 2x - h$ **18.** yes **19.** no **20.** 4
21. $x = 0$, $x = 4$
22. (a) domain: $\{-2, -1, 0, 1, 3, 4\}$
  range: $\{-3, 2, 4, 7, 8\}$
  (b) 7 **(c)** $x = -1$, $x = 3$
  (d)

(e) No; for $y = 2$, there are two different x-values,
  -1 and 3.
23. (a) $x^2 + 3x + 5$ **(b)** $(3x + 5)/x^2$
  (c) $3x^2 + 5$ **(d)** $9x + 20$

24. x: 2, y: 5 **25.** x: 3/2, y: 9/5

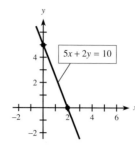
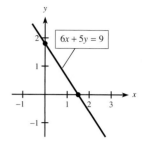

26. x: -2, y: none

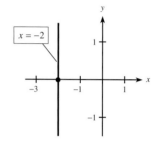

27. $m = 1$ **28.** undefined **29.** $m = -\frac{2}{5}$, $b = 2$
30. $m = -\frac{4}{3}$, $b = 2$ **31.** $y = 4x + 2$
32. $y = -\frac{1}{2}x + 3$ **33.** $y = \frac{2}{5}x + \frac{9}{5}$
34. $y = -\frac{11}{8}x + \frac{17}{4}$ **35.** $x = -1$ **36.** $y = 4x + 2$
37. $y = \frac{4}{3}x + \frac{10}{3}$
38. **39.**

40. (a) **(b)**

(c) The graph in (a) shows a complete graph. The
graph in (b) shows a piece that rises toward the
high point and a piece between the high and low
points.

41. (a)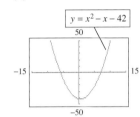

(b)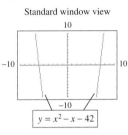

Standard window view

(c) The graph in (a) shows a complete graph. The one in (b) shows pieces that fall toward the minimum point and rise from it.

42. reals $x \geq -3$ with $x \neq 0$

43. $0.2857 \approx 2/7$ **44.** $x = 2$, $y = 1$

45. $x = 10$, $y = -1$ **46.** $x = 3$, $y = -2$

47. no solution **48.** $x = 10$, $y = -71$

49. $x = 1$, $y = -1$, $z = 2$

50. $x = 11$, $y = 10$, $z = 9$

51. (a) 1997 **(b)** 27 **(c)** $x = 30.94$; in 2011

52. 95%

53. 40,000 mi. He would normally drive more than 40,000 miles in 5 years, so he should buy diesel.

54. (a) yes **(b)** no **(c)** 4

55. (a) \$565.44

(b) The monthly payment on a \$70,000 loan is \$494.75.

56. (a) $(P \circ q)(t) = 180(1000 + 10t)$
$$-\frac{(1000 + 10t)^2}{100} - 200$$

(b) $x = 1150$, $P = \$193{,}575$

57. $(W \circ L)(t) = 0.02\left(50 - \dfrac{(t - 20)^2}{10}\right)^3$

58. (a)

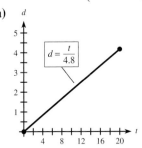

(b) When the time between seeing lightning and hearing thunder is 9.6 seconds, the storm is 2 miles away.

59.

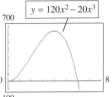

60. (a) $P = 58x - 8500$

(b) The profit increases by \$58 for each unit sold.

61. (a) yes

(b) $m = 0.219$, $b = 2.75$

(c) In 1950, national health care costs were 2.75% of the GDP.

(d) National health care costs (as a percent of the GDP) are changing at the rate of 0.219% per year.

62. $F = \frac{9}{5}C + 32$ or $C = \frac{5}{9}(F - 32)$

63. (a) 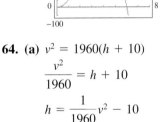 **(b)** $0 \leq x \leq 6$

64. (a) $v^2 = 1960(h + 10)$
$$\frac{v^2}{1960} = h + 10$$
$$h = \frac{1}{1960}v^2 - 10$$

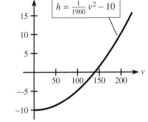

(b) 12.5 cm

65. \$100,000 at 9.5%; \$50,000 at 11%

66. 2.8 liters of 20%; 1.2 liters of 70%

67. (a) 12 supplied; 14 demanded **(b)** shortfall

(c) increase

68.

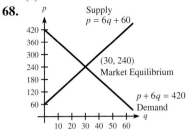

69. (a) 38.80 **(b)** 61.30 **(c)** 22.50 **(d)** 200

70. (a) $C(x) = 22x + 1500$ **(b)** $R(x) = 52x$

(c) $P(x) = 30x - 1500$ **(d)** $\overline{MC} = 22$

(e) $\overline{MR} = 52$ **(f)** $\overline{MP} = 30$ **(g)** $x = 50$

71. $q = 300$, $p = \$150$ **72.** $q = 700$, $p = \$80$

Chapter 1 Test

1. $x = 18/7$ **2.** $x = -3/7$ **3.** $x = -38$
4. $5 - 4x - 2h$
5. $t \geq -9$

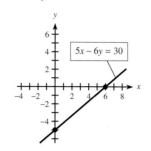

6. x: 6; y: -5 **7.** x: 3; y: 21/5

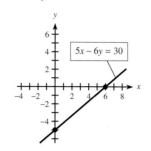
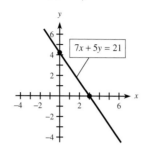

8. (a) domain: $x \geq -4$
      range: $f(x) \geq 0$
   (b) $2\sqrt{7}$ **(c)** 6
9. $y = -\frac{3}{2}x + \frac{1}{2}$ **10.** $m = -\frac{5}{4}$; $b = \frac{15}{4}$
11. (a) $x = -3$ **(b)** $y = -4x - 13$
12. (a) no; a vertical line intersects the curve twice.
   (b) yes; there is exactly one y-value for each x-value.
   (c) no; one value of x gives two y-values.
13. $x = -2$, $y = 2$
14. (a) $5x^3 + 2x^2 - 3x$ **(b)** $x + 2$
   (c) $5x^2 + 7x + 2$
15. (a) 30   **(b)** $P = 8x - 1200$ **(c)** 150
   (d) $\overline{MP} = 8$; the sale of each additional unit gives \$8 more profit.
16. (a) $R = 50x$
   (b) 19,000; it costs \$19,000 to produce 100 units.
   (c) 450
17. $q = 200$, $p = \$2500$
18. (a) 360,000; original value of the building
   (b) -1500; building depreciates \$1500 per month.
19. 400   **20.** 12,000 at 9%, 8000 at 6%

Exercise Set 2.1

1. $x^2 + 2x - 1 = 0$ **3.** $y^2 + 3y - 2 = 0$
5. $\frac{3}{2}, -\frac{3}{2}$ **7.** 0, 1   **9.** $-7, 3$ **11.** $\frac{1}{2}$
13. 8, -4 **15.** $-6, 2$ **17.** 8, 1   **19.** 1/2
21. (a) $2 + 2\sqrt{2}, 2 - 2\sqrt{2}$ **(b)** 4.83, -0.83
23. no real solutions
25. (a) $-\frac{7}{4}, \frac{3}{4}$ **(b)** $-1.75, 0.75$
27. (a) $(1 \pm \sqrt{31})/5$ **(b)** 1.31, -0.91
29. $\sqrt{7}, -\sqrt{7}$ **31.** no real solutions   **33.** 1, -9
35. $-9, -10$ **37.** $-2, 5$ **39.** $-300, 100$
41. 0.69, -0.06 **43.** $x = 20$ or $x = 70$

45. (a) $x = 10$ or $x = 345\frac{5}{9}$
   (b) yes; for any $x > 10$ and $x < 345\frac{5}{9}$
47. 6 sec
49. (a) ± 50
   (b) $s = 50$; there is no particulate pollution in the air above the plant.
51. 92.8 mph
53. (a) $0.034(7)^2 - 0.044(7) + 12.642 = 14$
   (b) $x = 7, x \approx -5.7$; 1997
55. when $t \approx 19.97$; 2000   **57.** \$80
59. almost 2004 (when $x \approx 13.98$)
61. (a) 18   **(b)** ≈ 31
   (c) speed triples, but K changes only by a factor of 1.72.

Exercise Set 2.2

1. (a) $(-1, -\frac{1}{2})$ **(b)** minimum   **(c)** -1 **(d)** $-\frac{1}{2}$
3. (a) $(1, 9)$ **(b)** maximum   **(c)** 1   **(d)** 9
5. (a) $(3, 9)$ **(b)** maximum   **(c)** 3   **(d)** 9
7. maximum, $(2, 1)$; zeros $(0, 0)$, $(4, 0)$

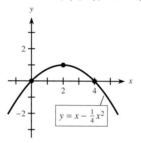

9. minimum, $(-2, 0)$; zero $(-2, 0)$

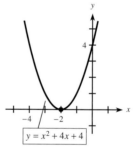

11. minimum, $(-1, -3\frac{1}{2})$; zeros $(-1 + \sqrt{7}, 0)$, $(-1 - \sqrt{7}, 0)$

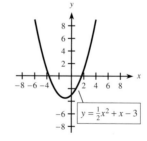

13. **(a)** 3 units to the right and 1 unit up

(b)

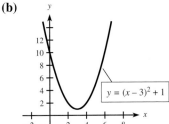

$y = (x - 3)^2 + 1$

15. $y = (x + 2)^2 - 2$

(a) 2 units to the left and 2 units down

(b)

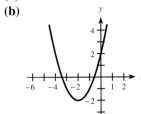

17. vertex $(1, -8)$; zeros $(-3, 0)$, $(5, 0)$

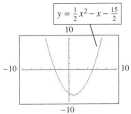

$y = \frac{1}{2}x^2 - x - \frac{15}{2}$

19. vertex $(-6, 3)$; no real zeros

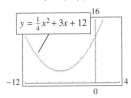

$y = \frac{1}{4}x^2 + 3x + 12$

21. -5

23. vertex $(10, 64)$; zeros $(90, 0)$, $(-70, 0)$

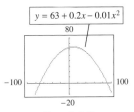

$y = 63 + 0.2x - 0.01x^2$

25. vertex $(0, -0.01)$; zeros $(10, 0)$, $(-10, 0)$

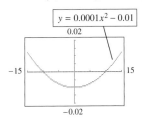

$y = 0.0001x^2 - 0.01$

27. **(a)** $(1, -24)$ **(b)** $x \approx -0.732, 2.732$

29. **(a)** $x = 2$ **(b)** $x - 2$

(c) $(x - 2)(3x - 2)$ **(d)** $x = 2, \frac{2}{3}$

31. **(a)** 80 units   **(b)** \$540   **33.** 400 trees

35. Dosage $= 500$ mg   **37.** Intensity $= 1.5$ lumens

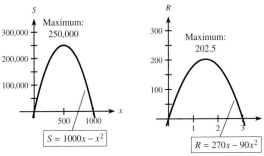

$S = 1000x - x^2$

$R = 270x - 90x^2$

39. Equation **(a)** $(384.62, 202.31)$ **(b)** $(54, 46)$

Projectile **(a)** goes higher.

41. **(a)** From b to c. The average rate is the same as the slope of the segment. Segment b to c is steeper.

(b) $d > b$ to make segment a to d be steeper (have greater slope).

43. **(a)**

Rent	Number Rented	Revenue
600	50	\$30,000
620	49	\$30,380
640	48	\$30,720

(b) increase

(c) $R = (50 - x)(600 + 20x)$

(d) \$800

45. **(a)** quadratic

(b) $a < 0$ because the graph opens downward.

(c) The vertex occurs after 2004 (when $x > 0$). Hence $-b/2a > 0$ and $a < 0$ means $b > 0$. The value $c = f(0)$, or the y-value in 2004. Hence $c > 0$.

47. **(a)**

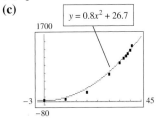

(b) quadratic

(c)

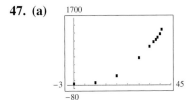

$y = 0.8x^2 + 26.7$

(Other answers could give a reasonable fit to the data.)

(d) \$2026.7 million (Answers will vary for other models.)

49.

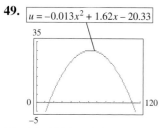

$u = -0.013x^2 + 1.62x - 20.33$

51. (a) 1914 and 2010   **(b)** When $u(x) < 0$

Exercise Set 2.3

1. (a) and **(b)**

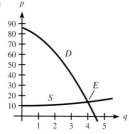

(c) See E on graph.   **(d)** $q = 4, p = \$14$

3. (a) and **(b)**

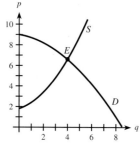

(c) See E on graph.   **(d)** $q = 4, p = \$6.60$
5. $q = 10, p = \$196$ **7.** $q = 216\frac{2}{3}, p = \$27.08$
9. $p = \$40, q = 30$ **11.** $q = 90, p = \$50$
13. $q = 70, p = \$62$ **15.** $x = 40$ units, $x = 50$ units
17. $x = 50$ units, $x = 300$ units
19. $x = 15$ units; reject $x = 100$ **21.** \$41,173.61
23. \$87.50   **25.** $x = 55, P(55) = \$2025$
27. (a)

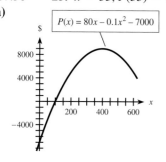

$P(x) = 80x - 0.1x^2 - 7000$

(b) (400, 9000); maximum point   **(c)** positive
(d) negative   **(e)** closer to 0 as a gets closer to 400
29. (a) $P(x) = -x^2 + 350x - 15,000$; maximum profit
is \$15,625
(b) no   **(c)** x-values agree
31. (a) $x = 28$ units, $x = 1000$ units
(b) \$651,041.67
(c) $P(x) = -x^2 + 1028x - 28,000$; maximum profit
is \$236,196
(d) \$941.60
33. (a) $t \approx 8$; 2000 $R = \$60.792$ billion
(b) The data show a smaller revenue,
$R = \$60.53$ billion, in 1999
(c)

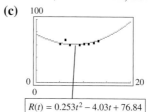

$R(t) = 0.253t^2 - 4.03t + 76.84$

(d) Except for 1998 ($t = 6$), the model fits the data
quite well.
35. (a) $P(t) = -0.019t^2 + 0.284t - 0.546$
(b) 1999
(c)

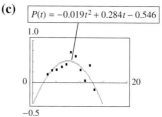

$P(t) = -0.019t^2 + 0.284t - 0.546$

(d) The model projects decreasing profits, and except
for 2004, the data support this.
(e) Management would be interested in increasing
revenues or reducing costs (or both) to improve
profits.

Exercise Set 2.4

1. j   **3.** b   **5.** f   **7.** g   **9.** i
11. k   **13. (a)** cubic   **(b)** quartic   **15.** b
17. g   **19.** a   **21.** d

A15

23.

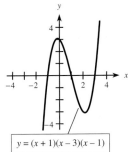

$y = (x + 1)(x - 3)(x - 1)$

25.

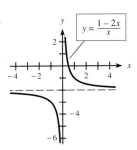

$y = \dfrac{1 - 2x}{x}$

27.

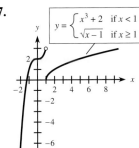

$y = \begin{cases} x^3 + 2 & \text{if } x < 1 \\ \sqrt{x - 1} & \text{if } x \geq 1 \end{cases}$

29. (a) 8/3   **(b)** 9.9   **(c)** -999.999 **(d)** no
31. (a) 64   **(b)** 1   **(c)** 1000   **(d)** 0.027
33. (a) 2   **(b)** 4   **(c)** 0   **(d)** 2
35. (a)

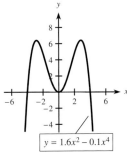

$y = 1.6x^2 - 0.1x^4$

(b) polynomial   **(c)** no asymptotes
(d) turning points at $x = 0$ and approximately
   $x = -2.8$ and $x = 2.8$

37. (a)

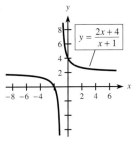

$y = \dfrac{2x + 4}{x + 1}$

(b) rational
(c) vertical: $x = -1$,
   horizontal: $y = 2$
(d) no turning points

39. (a)

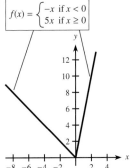

$f(x) = \begin{cases} -x & \text{if } x < 0 \\ 5x & \text{if } x \geq 0 \end{cases}$

(b) piecewise defined
(c) no asymptotes
(d) turning point at
   $x = 0$

41. (a) 6800; 11,200   **(b)** $0 < x < 27$
43. (a) upward   **(b)**

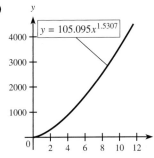

$y = 105.095x^{1.5307}$

(c) during 2000   **(d)** no; the value of investments
   declined after 9/11/01.
45. (a) $0 \leq p < 100$
(b) $5972.73; to remove 45% of the particulate
   pollution would cost $5972.73.
(c) $65,700; to remove 90% of the particulate
   pollution would cost $65,700.
(d) $722,700; to remove 99% of the particulate
   pollution would cost $722,700.
(e) $1,817,700; to remove 99.6% of the particulate
   pollution would cost $1,817,700.
(f) It increases without bound.
47. (a) $A(2) = 96; A(30) = 600$
(b) $0 < x < 50$
49. (a) $P(t) = \begin{cases} 1.965t - 5.65 & 5 \leq t \leq 20 \\ 0.095t^2 - 2.925t + 54.15 & 20 < t \leq 40 \end{cases}$

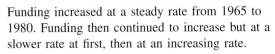

Funding increased at a steady rate from 1965 to
1980. Funding then continued to increase but at a
slower rate at first, then at an increasing rate.

(b) $33.65 billion   **(c)** $80.18 billion

51. (a) $P(x) = \begin{cases} 0.37 & 0 < x \le 1 \\ 0.60 & 1 < x \le 2 \\ 0.83 & 2 < x \le 3 \\ 1.06 & 3 < x \le 4 \end{cases}$

(b) $0.60; the postage for a 1.2-ounce letter

(c) domain = $\{x: 0 < x \le 4\}$
range = $\{0.37, 0.60, 0.83, 1.06\}$

(d) 2-ounce: $0.60; 2.01-ounce: $0.83

53. (a) $p = \dfrac{200}{2 + 0.1x}$

(b) Yes, when
$p = \$100$.

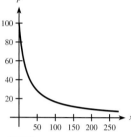

55. $C(x) = 30(x - 1) + \dfrac{3000}{x + 10}$

(a)

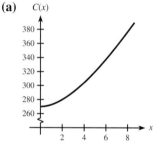

(b) Any turning point would indicate the minimum or the maximum cost. In this case, $x = 0$ gives a minimum.

(c) The y-intercept is the fixed cost of production.

Exercise Set 2.5

1. linear   **3.** quadratic   **5.** quartic

7. quadratic or power

9. $y = 2x - 3$

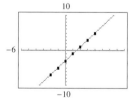

11. $y = 2x^2 - 1.5x - 4$

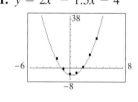

13. $y = x^3 - x^2 - 3x - 4$

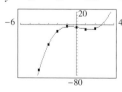

15. $y = 2x^{0.5}$

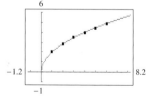

17. (a)

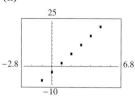

(b) linear

(c) $y = 5x - 3$

19. (a)

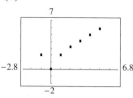

(b) quadratic

(c) $y = 0.09595x^2 + 0.4656x + 1.4758$

21. (a)

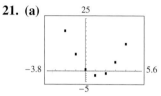

(b) quadratic

(c) $y = 2x^2 - 5x + 1$

23. (a)

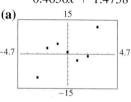

(b) cubic

(c) $y = x^3 - 5x + 1$

25. (a)

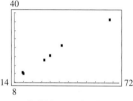

(b) $y = 0.518x + 2.775$

(c) $m = 0.518$; for each $1000 of annual earnings for males, females earned $518 per year.

27. (a) $y = -0.0620x^3 + 2.5848x^2 - 28.2315x + 103.9594$

(b) $x = 20.3$ (2001). According to the data, the year with the maximum number was 1999.

29. (a)

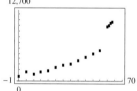

(b) $y = 149.931x - 683.033$

(c) $y = 0.1156x^3 - 7.6370x^2 + 186.4609x + 597.9660$

(d)

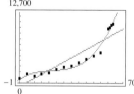

Cubic model is the better fit.

31. (a) $y = -7.888x^2 + 117.803x - 421.039$
$y = -1.974x^3 + 33.575x^2 - 106.491x - 205.432$
(b) The cubic model is a much better fit.
Cubic: $y \approx -912.0$; Quadratic: $y \approx -428.7$
33. (a) quadratic or power
(b) quadratic: $y = 1.0275x^2 - 29.598x + 299.31$
power: $y = 0.05193x^{2.5679}$
(c) quadratic (it predicts \$1392.2 billion)
(d) \$2152.4 billion
35. (a) $y = 0.3488x^{4.208}$ **(b)** 266,139 thousand
(c) It is nearly the entire population of the U.S.

Chapter 2 Review Exercises

1. $x = 0, x = -\frac{5}{3}$ **2.** $x = 0, x = \frac{4}{3}$
3. $x = -2, x = -3$
4. $x = (-5 + \sqrt{47})/2, x = (-5 - \sqrt{47})/2$
5. no real solutions   **6.** $x = \frac{\sqrt{3}}{2}, x = -\frac{\sqrt{3}}{2}$
7. $\frac{5}{7}, -\frac{4}{5}$ **8.** $(-1 + \sqrt{2})/4, (-1 - \sqrt{2})/4$
9. $7/2, 100$ **10.** $13/5, 90$
11. no real solutions   **12.** $z = -9, z = 3$
13. $x = 8, x = -2$ **14.** $x = 3, x = -1$
15. $x = (-a \pm \sqrt{a^2 - 4b})/2$
16. $r = (2a \pm \sqrt{4a^2 + x^3 c})/x$
17. $1.64, -7051.64$ **18.** $0.41, -2.38$
19. vertex $(-2, -2)$;   **20.** vertex $(0, 4)$;
zeros $(0, 0), (-4, 0)$ zeros $(4, 0), (-4, 0)$

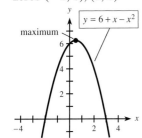   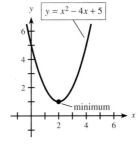

21. vertex $(\frac{1}{2}, \frac{25}{4})$;   **22.** vertex $(2, 1)$;
zeros $(-2, 0), (3, 0)$ no real zeros

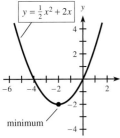

23. vertex $(-3, 0)$;   **24.** vertex $(\frac{3}{2}, 0)$;
zero $(-3, 0)$ zero $(\frac{3}{2}, 0)$

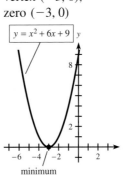

25. vertex $(0, -3)$;   **26.** vertex $(0, 2)$;
zeros $(-3, 0), (3, 0)$ no real zeros

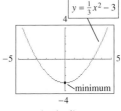

27. vertex $(-1, 4)$;   **28.** vertex $(\frac{7}{2}, \frac{9}{4})$;
no real zeros   zeros $(2, 0), (5, 0)$

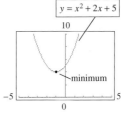

29. vertex $(100, 1000)$;   **30.** vertex $(75, -6.25)$;
zeros $(0, 0), (200, 0)$ zeros $(50, 0), (100, 0)$

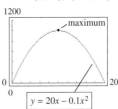   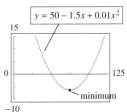

31. 20   **32.** 30
33. (a) $(1, -4\frac{1}{2})$ **(b)** $x = -2, x = 4$ **(c)** B
34. (a) $(0, 49)$ **(b)** $x = -7, x = 7$ **(c)** D
35. (a) $(7, 25)$ approximately, actual is $(7, 24\frac{1}{2})$
(b) $x = 0, x = 14$ **(c)** A
36. (a) $(-1, 9)$ **(b)** $x = -4, x = 2$ **(c)** C

37. (a)

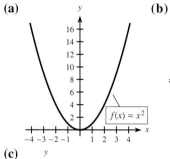

(b)

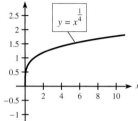

(c)

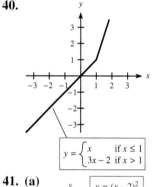

38. (a) 0   **(b)** 10,000   **(c)** −25   **(d)** 0.1

39. (a) −2   **(b)** 0   **(c)** 1   **(d)** 4

40.

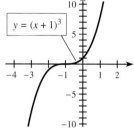

41. (a)

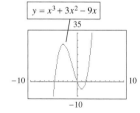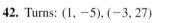

(b)

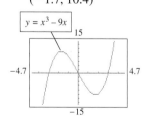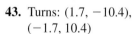

42. Turns: $(1, -5)$, $(-3, 27)$

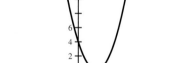

43. Turns: $(1.7, -10.4)$, $(-1.7, 10.4)$

44. VA: $x = 2$;
HA: $y = 0$

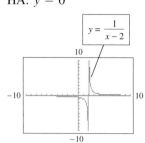

45. VA: $x = -3$;
HA: $y = 2$

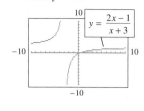

46. (a)

(b) $y = -2.1786x + 159.8571$
(c) $y = -0.0818x^2 - 0.2143x + 153.3095$

47. (a)

(b) $y = 2.1413x + 34.3913$
(c) $y = 22.2766x^{0.4259}$

48. (a) $t = -1.65$, $t = 3.65$ **(b)** Just $t = 3.65$
(c) At 3.65 seconds

49. $x = 20$, $x = 800$

50. (a) $x = 76.9$, in 1977; $x = 108$, in 2008
(b) $x = 92.5$, in 1993; 9.15%

51. (a) $x = 200$ **(b)** $A = 30,000$ square feet

52.

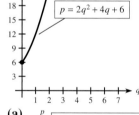

53.

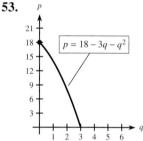

(b) $p = 41$,
$q = 20$

54. (a)

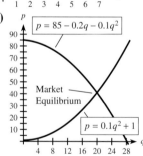

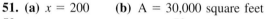

55. $p = 400, q = 10$ **56.** $p = 10, q = 20$
57. $x = 46 + 2\sqrt{89}, x = 46 - 2\sqrt{89}$
58. (15, 1275), (60, 2400)
59. max revenue = \$2500; max profit = \$506.25
60. max profit = 12.25; break-even points $x = 100, x = 30$

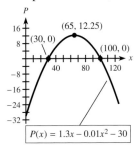

61. $x = 50, P(50) = 640$
62. **(a)** $C = 15{,}000 + 140x + 0.04x^2$;
   $R = 300x - 0.06x^2$
   (b) 100, 1500   **(c)** 2500
   (d) $P = 160x - 15{,}000 - 0.1x^2$; max at 800
   (e) at 2500: $P = -240{,}000$; at 800: $P = 49{,}000$
63. **(a)** power   **(b)** 5,599,885
   (c) 20.481364; this model predicts 2,048,136 cases
   in 1995.
64. **(a)**   **(b)** $0 \le x \le 6$

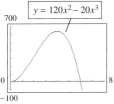

65. **(a)** rational   **(b)** $0 \le p < 100$
   (c) 0; it costs \$0 to remove no pollution.
   (d) \$475,200
66. **(a)** $x = 12; C(12) = \$18.68$
   (b) $x = 825; C(825) = \$909.70$
67. **(a)** $y = 17.3969x^{0.5094}$
   (b)   **(c)** 39.5 mph
   (d) 18.3 seconds

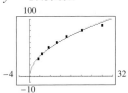

68. **(a)** $y = 1.8155x^2 - 11.1607x + 29.1845$
   (b) Use $x = 11$; 126 million cubic yards
69. **(a)** $y = 6.299x^{1.668}$
   (b) $y = 3.342x^2 - 86.763x + 981.894$
   (c) The quadratic model is better.

Chapter 2 Test

1. (a)   **(b)**

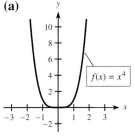

(c)   **(d)**

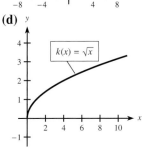

2. b; a   **3.**

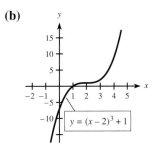

4. (a)

(b)

5. b; the function is cubic, $f(1) < 0$
6. **(a)** -10 **(b)** $-16\frac{1}{2}$ **(c)** -7

7.

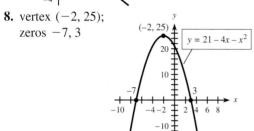

$g(x) = \begin{cases} x^2 \text{ if } x \le 1 \\ 4 - x \text{ if } x > 1 \end{cases}$

8. vertex $(-2, 25)$; zeros $-7, 3$

$y = 21 - 4x - x^2$

$(-2, 25)$

9. $x = 2, x = 1/3$

10. $x = \dfrac{-3 + 3\sqrt{3}}{2}, x = \dfrac{-3 - 3\sqrt{3}}{2}$

11. $x = 2/3$

12. c; $g(x)$ has a vertical asymptote at $x = -2$, as does graph c.

13. HA: $y = 0$; VA: $x = 5$

14. 42

15. (a) quartic (fourth-degree) **(b)** cubic

16. (a)

Model: $y = -0.3577x + 19.9227$

(b) $f(x) = 5.6$

(c) At $x = 55.7$

17. $q = 300, p = \$80$

18. (a) $P(x) = -x^2 + 250x - 15,000$

(b) 125 units, \$625 **(c)** 100 units, 150 units

19. (a) $f(15) = -31$ means that when the air temperature is 0°F and the wind speed is 15 mph, the air temperature feels like -31°F.

(b) -55°F

20. (a)

(b) quadratic

(c) $y = -0.313x^2 + 4.034x + 22.797$ **(d)** 26.1%

Exercise Set 3.1

1. 3 **3.** $\begin{bmatrix} -1 & -2 & -3 \\ 1 & 0 & -1 \\ -2 & 3 & 4 \end{bmatrix}$ **5.** A, C, D, F, Z

7. 1 **9.** $\begin{bmatrix} 1 & 3 & 4 \\ 0 & 2 & 0 \\ 2 & 1 & 3 \end{bmatrix}$ **11.** $\begin{bmatrix} 0 & 0 & 0 \\ 0 & 0 & 0 \\ 0 & 0 & 0 \end{bmatrix}$

13. $\begin{bmatrix} 9 & 5 \\ 4 & 7 \end{bmatrix}$ **15.** $\begin{bmatrix} 0 & -2 & -1 \\ 4 & 2 & 0 \\ 2 & 3 & 7 \end{bmatrix}$ **17.** $\begin{bmatrix} 2 & 3 & 6 \\ 3 & 4 & 1 \\ 6 & 1 & 6 \end{bmatrix}$

19. impossible **21.** $\begin{bmatrix} 3 & 3 & 9 & 0 \\ 12 & 6 & 3 & 3 \\ 9 & 6 & 0 & 3 \end{bmatrix}$

23. $\begin{bmatrix} 28 & 16 \\ 10 & 18 \end{bmatrix}$ **25.** impossible

27. $x = 3, y = 2, z = 3, w = 4$

29. $x = 4, y = 1, z = 3, w = 3$

31. $x = 2, y = 2, z = -3$

33. (a) $A = \begin{bmatrix} 65 & 78 & 14 & 12 & 71 \\ 9 & 14 & 22 & 9 & 44 \end{bmatrix}$

$B = \begin{bmatrix} 251 & 175 & 64 & 8 & 11 \\ 17 & 6 & 15 & 1 & 0 \end{bmatrix}$

(b) $A + B = \begin{bmatrix} 316 & 253 & 78 & 20 & 82 \\ 26 & 20 & 37 & 10 & 44 \end{bmatrix}$

(c) $\begin{bmatrix} 186 & 97 & 50 & -4 & -60 \\ 8 & -8 & -7 & -8 & -44 \end{bmatrix}$

more species in U.S.

35. (a) $\begin{bmatrix} 11,041.7 & 8978.4 & 6461 \\ 8739.8 & 9159.6 & 6877.3 \\ 9798.1 & 9086.7 & 6448.4 \\ 9696.6 & 8926.7 & 6109.5 \end{bmatrix}$

(b) air pollution

37. (a) $\begin{bmatrix} 825 & 580 & 1560 \\ 810 & 650 & 350 \end{bmatrix}$ **(b)** $\begin{bmatrix} -75 & 20 & -140 \\ 10 & -50 & 50 \end{bmatrix}$

39. $A = \begin{bmatrix} & M & F \\ & 54.4 & 55.6 \\ & 62.1 & 66.6 \\ & 67.4 & 74.1 \\ & 70.7 & 78.1 \\ & 72.9 & 79.4 \\ & 75.0 & 80.2 \end{bmatrix}$ $B = \begin{bmatrix} & M & F \\ & 45.5 & 45.2 \\ & 51.5 & 54.9 \\ & 61.1 & 67.4 \\ & 63.8 & 72.5 \\ & 64.5 & 73.6 \\ & 68.6 & 75.5 \end{bmatrix}$

$A - B = C = \begin{bmatrix} 8.9 & 10.4 \\ 10.6 & 11.7 \\ 6.3 & 6.7 \\ 6.9 & 5.6 \\ 8.4 & 5.8 \\ 6.4 & 4.7 \end{bmatrix}$

41. (a) $\begin{bmatrix} 376 & 342 \\ 603 & 493 \\ 731 & 520 \\ 777 & 565 \\ 738 & 505 \\ 537 & 378 \end{bmatrix}$ **(b)** $\begin{bmatrix} 451.20 & 410.40 \\ 723.60 & 591.60 \\ 877.20 & 624 \\ 932.40 & 678 \\ 885.60 & 606 \\ 644.40 & 453.60 \end{bmatrix}$

43. (a) $\begin{bmatrix} 46.20 & 84.00 & 210.00 & 10.50 \\ 42.00 & 84.00 & 42.00 & 0.00 \\ 58.80 & 147.00 & 94.50 & 0.00 \\ 31.50 & 147.00 & 42.00 & 21.00 \\ 42.00 & 0.00 & 210.00 & 10.50 \end{bmatrix}$

(b) $\begin{bmatrix} 48.40 & 88.00 & 220.00 & 11.00 \\ 44.00 & 88.00 & 44.00 & 0.00 \\ 61.60 & 154.00 & 99.00 & 0.00 \\ 33.00 & 154.00 & 44.00 & 22.00 \\ 44.00 & 0.00 & 220.00 & 11.00 \end{bmatrix}$

45. (a) $A = \begin{bmatrix} 0 & 1 & 1 & 0 \\ 1 & 0 & 1 & 1 \\ 0 & 0 & 0 & 1 \\ 0 & 1 & 1 & 0 \end{bmatrix}$

(b) $B = \begin{bmatrix} 0 & 1 & 1 & 1 \\ 1 & 0 & 1 & 1 \\ 0 & 1 & 0 & 1 \\ 1 & 1 & 1 & 0 \end{bmatrix}$ **(c)** person 2

47. (a) $\begin{bmatrix} 80 & 75 \\ 58 & 106 \end{bmatrix}$ **(b)** $\begin{bmatrix} 176 & 127 \\ 139 & 143 \end{bmatrix}$

(c) $\begin{bmatrix} 10 & 4 \\ 7 & 2 \end{bmatrix}$ **(d)** $\begin{bmatrix} -10 & 19 \\ -7 & 20 \end{bmatrix}$ Shortage, taken from inventory.

49. (a) 3, 4, 5, 6 **(b)** 1

51. Worker 1: 0.9625 Worker 2: 0.9375
Worker 3: 0.9125 Worker 4: 0.8875
Worker 5: 0.85 Worker 6: 0.875
Worker 7: 0.90 Worker 8: 0.925
Worker 9: 0.95
Worker 5 is least efficient; performs best at center 5

Exercise Set 3.2

1. $[32]$ **3.** $[11 \quad 17]$ **5.** $\begin{bmatrix} 29 & 25 \\ 10 & 12 \end{bmatrix}$

7. $\begin{bmatrix} 14 & 2 & 16 \\ 28 & 5 & 12 \end{bmatrix}$

9. $\begin{bmatrix} 7 & 5 & 3 & 2 \\ 14 & 9 & 11 & 3 \\ 13 & 10 & 12 & 3 \end{bmatrix}$ **11.** impossible

13. $\begin{bmatrix} 13 & 9 & 3 & 4 \\ 9 & 7 & 16 & 1 \end{bmatrix}$ **15.** $\begin{bmatrix} 9 & 7 & 16 \\ 5 & 17 & 20 \end{bmatrix}$

17. $\begin{bmatrix} 9 & 0 & 8 \\ 13 & 4 & 11 \\ 16 & 0 & 17 \end{bmatrix}$ **19.** $\begin{bmatrix} 161 & 126 \\ 42 & 35 \end{bmatrix}$ **21.** no

23. no **25.** $\begin{bmatrix} -55 & 88 & 0 \\ -42 & 67 & 0 \\ 28 & -44 & 1 \end{bmatrix}$ **27.** $\begin{bmatrix} 0 & 0 & 0 \\ 0 & 0 & 0 \\ 0 & 0 & 0 \end{bmatrix}$

29. $\begin{bmatrix} 0 & 0 & 0 \\ 0 & 0 & 0 \\ 0 & 0 & 0 \end{bmatrix}$ **31.** A **33.** Z

35. no (see Problem 27)

37. (a) $AB = \begin{bmatrix} 1 & 0 \\ 0 & 1 \end{bmatrix}$, $BA = \begin{bmatrix} 1 & 0 \\ 0 & 1 \end{bmatrix}$

(b) $ad - bc \neq 0$

39. $\begin{bmatrix} 2 - 2 + 2 \\ 6 - 4 - 4 \\ 4 + 0 - 2 \end{bmatrix} = \begin{bmatrix} 2 \\ -2 \\ 2 \end{bmatrix}$; solution

41. $\begin{bmatrix} 1 + 2 + 2 \\ 4 + 0 + 1 \\ 2 + 2 + 1 \end{bmatrix} = \begin{bmatrix} 5 \\ 5 \\ 5 \end{bmatrix}$; solution

43. $\begin{bmatrix} 1 & 0 & 0 & 0 & 0 \\ 0 & 1 & 0 & 0 & 0 \\ 0 & 0 & 1 & 0 & 0 \\ 0 & 0 & 0 & 1 & 0 \\ 0 & 0 & 0 & 0 & 1 \end{bmatrix}$ (Some entries may appear as decimal approximations of 0.)

45. $\begin{bmatrix} 31,600 & 36,960 \\ 42,500 & 47,600 \end{bmatrix}$

The entries represent the dealer's cost for each car.

47. (a) $A = \begin{bmatrix} 2703 & 2378 \\ 2383 & 2723 \\ 4376 & 5114 \\ 1443 & 1581 \\ 3152 & 2490 \\ 1289 & 1245 \\ 916 & 824 \end{bmatrix}$ $B = \begin{bmatrix} 96.1 & 131.0 \\ 48.1 & 47.8 \\ 79.4 & 86.8 \\ 99.0 & 118.0 \\ 77.3 & 107.0 \\ 65.8 & 64.2 \\ 89.5 & 96.3 \end{bmatrix}$

(b) 7×7 **(c)** the diagonal entries

49. $\begin{bmatrix} 1300 \\ 1520 \\ 1620 \end{bmatrix}$ Cost for female teens
Cost for single women 20–35
Cost for married women 35–60

51. 172, 208, 268, 327, 101, 123, 268, 327, 216, 263, 162, 195, 176, 215, 343, 417

53. $\begin{bmatrix} 2137 & 84 & 41.5 & 128.5 & 158.5 & 317 & 738 \\ 1285.5 & 64 & 30.5 & 115 & 136 & 229 & 590 \\ 969.5 & 46.5 & 24 & 98 & 139 & 224 & 476.5 \\ 809 & 44.5 & 22.5 & 99 & 141.5 & 240.5 & 455.5 \\ 687.5 & 33 & 17.5 & 80 & 104.5 & 203.5 & 430.5 \end{bmatrix}$

55. (a) $B = \begin{bmatrix} 0.7 & 8.5 & 10.2 & 1.1 & 5.6 & 3.6 \\ 0.5 & 0.2 & 6.1 & 1.3 & 0.2 & 1.0 \\ 2.2 & 0.4 & 8.8 & 1.2 & 1.2 & 4.8 \\ 251.8 & 63.4 & 81.6 & 35.2 & 54.3 & 144.2 \\ 30.0 & 1.0 & 1.0 & 1.0 & 1.0 & 1.0 \\ 788.9 & 0 & 0 & 0 & 0 & 0 \end{bmatrix}$ **(b)** $A = \begin{bmatrix} 1.11 & 0 & 0 & 0 & 0 & 0 \\ 0 & 0.95 & 0 & 0 & 0 & 0 \\ 0 & 0 & 1.11 & 0 & 0 & 0 \\ 0 & 0 & 0 & 1.11 & 0 & 0 \\ 0 & 0 & 0 & 0 & 0.95 & 0 \\ 0 & 0 & 0 & 0 & 0 & 0.95 \end{bmatrix}$

57. $\begin{bmatrix} 0.900 & 0.855 & 0.855 & 0.855 & 0.855 & 0.765 & 0.765 & 0.765 & 0.765 \\ 0.765 & 0.900 & 0.855 & 0.855 & 0.855 & 0.855 & 0.765 & 0.765 & 0.765 \\ 0.765 & 0.765 & 0.900 & 0.855 & 0.855 & 0.855 & 0.855 & 0.765 & 0.765 \\ 0.765 & 0.765 & 0.765 & 0.900 & 0.855 & 0.855 & 0.855 & 0.855 & 0.765 \\ 0.765 & 0.765 & 0.765 & 0.765 & 0.900 & 0.855 & 0.855 & 0.855 & 0.855 \\ 0.855 & 0.765 & 0.765 & 0.765 & 0.765 & 0.900 & 0.855 & 0.855 & 0.855 \\ 0.855 & 0.855 & 0.765 & 0.765 & 0.765 & 0.765 & 0.900 & 0.855 & 0.855 \\ 0.855 & 0.855 & 0.855 & 0.765 & 0.765 & 0.765 & 0.765 & 0.900 & 0.855 \\ 0.855 & 0.855 & 0.855 & 0.855 & 0.765 & 0.765 & 0.765 & 0.765 & 0.900 \end{bmatrix}$

Exercise Set 3.3

1. $\begin{bmatrix} 1 & -2 & -1 & | & -7 \\ 0 & 7 & 5 & | & 21 \\ 4 & 2 & 2 & | & 1 \end{bmatrix}$

3. $\begin{bmatrix} 1 & -3 & 4 & | & 2 \\ 2 & 0 & 2 & | & 1 \\ 1 & 2 & 1 & | & 1 \end{bmatrix}$ **5.** $x = 2, y = 1/2, z = -5$

7. $x = 18, y = 10, z = 0$ **9.** $x = -5, y = 2, z = 1$
11. $x = 4, y = 1, z = -2$ **13.** $x = 15, y = -13, z = 2$
15. $x = 15, y = 0, z = 2$ **17.** $x = 1, y = -1, z = 1$
19. $x_1 = 1, x_2 = 0, x_3 = 1, x_4 = 0$
21. $x = 1, y = 3, z = 1, w = 0$
23. no solution
25. (a) $x = (11 + 2z)/3, y = (-1 - z)/3,$
  $z = $ any real number
  (b) many possibilities, including $x = 11/3, y = -\frac{1}{3},$
  $z = 0$ and $x = 13/3, y = -2/3, z = 1$
27. If a row of the matrix has all 0's in the coefficient matrix and a nonzero number in the augment, there is no solution.
29. $x = 0, y = -z, z = $ any real number
31. $x = 3z - 2, y = 3 - 5z, z = $ any real number
33. $x = 1 - z, y = \frac{1}{2}z, z = $ any real number
35. $x = 2z - 2, y = 1 + z, z = $ any real number
37. $x = \frac{7}{2} - z, y = -\frac{1}{2}, z = $ any real number
39. $x = \frac{26}{5} - \frac{7}{5}z, y = \frac{4}{5} + \frac{2}{5}z, z = $ any real number
41. $x_1 = \frac{23}{38}x_3, x_2 = \frac{12}{19}x_3, x_3 = $ any real number
43. $x = 7/5, y = -3/5, z = w, w = $ any real number
45. no solution
47. $x_1 = 1 - 2x_4 - 3x_5, x_2 = 4 + 5x_4 + 7x_5,$
  $x_3 = -3 - 3x_4 - 5x_5, x_4 = $ any real number,
  $x_5 = $ any real number

49. $x = (b_2c_1 - b_1c_2)/(a_1b_2 - a_2b_1)$
51. (a) \$50,000 at 12%, \$85,000 at 10%, \$100,000 at 8%
  (b) \$6000 at 12%, \$8500 at 10%, \$8000 at 8%
53. beef: 2 cups; sirloin: 8 cups
55. AF: 2 oz, FP: 2 oz, NMG: 1 oz
57. 2 of portfolio I, 2 of portfolio II
59. $\frac{3}{8}$ pound of red meat, 6 slices of bread, 4 glasses of milk
61. Type I = 3 Type IV, Type II = 1000 − 2(Type IV), Type III = 500 − Type IV, Type IV = any integer satisfying 0 ≤ Type IV ≤ 500
63. Bacteria III = any amount satisfying
  1800 ≤ Bacteria III ≤ 2300
  Bacteria I = 6900 − 3 (Bacteria III)
  Bacteria II = $\frac{1}{2}$ (Bacteria III) − 900
65. (a) $C = 2800 + 0.6R$
  $U = 7000 - R$
  $R = $ any integer satisfying $0 \le R \le 7000$
  (b) $R = 1000$: $C = 3400$
    $U = 6000$
  $R = 2000$: $C = 4000$
    $U = 5000$
  (c) Min $C = 2800$
    when $R = 0$ and
    $U = 7000$
  (d) Max $C = 7000$ when $R = 7000$ and $U = 0$
67. There are three possibilities:
  (1) 4 of I and 2 of II
  (2) 5 of I, 1 of II, and 1 of III
  (3) 6 of I and 2 of III

Exercise Set 3.4

1. $\begin{bmatrix} 1 & 0 & 0 \\ 0 & 1 & 0 \\ 0 & 0 & 1 \end{bmatrix}$ **3.** yes **5.** $\begin{bmatrix} 2 & -7 \\ -1 & 4 \end{bmatrix}$

7. no inverse **9.** $\begin{bmatrix} \frac{5}{2} & -1 \\ -2 & 1 \end{bmatrix}$ **11.** $\begin{bmatrix} -\frac{1}{10} & \frac{7}{10} \\ \frac{1}{5} & -\frac{2}{5} \end{bmatrix}$

13. $\begin{bmatrix} \frac{1}{3} & 0 & 0 \\ 0 & \frac{1}{3} & 0 \\ 0 & 0 & \frac{1}{3} \end{bmatrix}$ **15.** $\begin{bmatrix} -1 & 1 & 0 \\ 1 & 0 & 0 \\ -1 & 0 & 1 \end{bmatrix}$

17. $\begin{bmatrix} \frac{1}{3} & -\frac{1}{3} & \frac{1}{3} \\ -\frac{2}{3} & -\frac{1}{3} & \frac{7}{3} \\ \frac{1}{3} & \frac{2}{3} & -\frac{5}{3} \end{bmatrix}$

19. no inverse **21.** no inverse

23. $\begin{bmatrix} 2 & 2 & 0 & 2 & 2 \\ 1 & 0 & 2 & 2 & 1 \\ 0 & 1 & 0 & 2 & 1 \\ 2 & 0 & 2 & 2 & 1 \\ 1 & 0 & 0 & 0 & 2 \end{bmatrix}$ **25.** $\begin{bmatrix} 13 \\ 5 \end{bmatrix}$ **27.** $\begin{bmatrix} 9 \\ 6 \\ 3 \end{bmatrix}$

29. $\begin{bmatrix} x \\ y \\ z \end{bmatrix} = \begin{bmatrix} 1 \\ 1 \\ 2 \end{bmatrix}$ **31.** $x = 2, y = 1$

33. $x = 1, y = 2$ **35.** $x = 1, y = 1, z = 1$
37. $x = 1, y = 3, z = 2$
39. $x_1 = 5.6, x_2 = 5.4, x_3 = 3.25, x_4 = 6.1, x_5 = 0.4$
41. -2 **43.** 14 **45.** -5 **47.** -19 **49.** yes
51. no **53.** Hang on **55.** Answers in back
57. $x_0 = 2400, y_0 = 1200$
59. (a) A = 5.5 mg and B = 8.8 mg for patient I;
  (b) A = 10 mg and B = 16 mg for patient II
61. $68,000 at 18%, $77,600 at 10%
63. 2 Deluxe, 8 Premium, 32 Ultimate
65. $200,000 at 6%, $300,000 at 8%, $500,000 at 10%
67. (a) $\begin{bmatrix} 0 & 1 & 0 \\ 0 & 0 & 1 \\ 1 & 1 & 1 \end{bmatrix}$ (b) 108

69. (a) $M = \begin{bmatrix} 0 & 1 & 0 \\ 0 & 0 & 1 \\ 1 & 1 & 1 \end{bmatrix}$ (b) 30

Exercise Set 3.5

1. (a) 15 (b) 4 **3.** 8 **5.** 40
7. most: raw materials; least: fuels
9. raw materials, manufacturing, service
11. 5 units of manufacturing; 10 units of agriculture;
  105 units of utilities

13. farm products = 200; machinery = 40
15. agricultural products = 100; oil = 700
17. utilities = 200; manufacturing = 400
19. mining = 310; manufacturing = 530
21. (a)

$$A = \begin{bmatrix} 0.3 & 0.6 \\ 0.2 & 0.2 \end{bmatrix} \begin{matrix} \text{Electronic components} \\ \text{Computers} \end{matrix}$$

with columns EC, C

  (b) electronic components = 1200; computers = 320
23. (a)

$$A = \begin{bmatrix} 0.30 & 0.04 \\ 0.35 & 0.10 \end{bmatrix} \begin{matrix} \text{Fishing} \\ \text{Oil} \end{matrix}$$

with columns F, O

  (b) fishing = 100; oil = 1250
25. development = $21,000; promotional = $12,000
27. engineering = $15,000; computer = $13,000
29. fishing = 400; agriculture = 500; mining = 400
31. electronics = 1240; steel = 1260; autos = 720
33. service = 90, manufacturing = 200, agriculture = 100
35. products = $\frac{7}{17}$ households;
  machinery = $\frac{1}{17}$ households
37. government = $\frac{10}{19}$ households;
  industry = $\frac{11}{19}$ households
39. (a)

$$A = \begin{bmatrix} 0.5 & 0.4 & 0.3 \\ 0.4 & 0.5 & 0.3 \\ 0.1 & 0.1 & 0.4 \end{bmatrix} \begin{matrix} \text{Manufacturing} \\ \text{Utilities} \\ \text{Households} \end{matrix}$$

with columns M, U, H

  (b) manufacturing = 3 households;
  utilities = 3 households

41. $\begin{bmatrix} 24 \\ 96 \\ 24 \\ 120 \\ 492 \\ 3456 \end{bmatrix}$ 3456 bolts, 492 braces, 120 panels

43. $\begin{bmatrix} 10 \\ 10 \\ 20 \\ 56 \\ 20 \\ 26 \\ 300 \end{bmatrix}$ 56 2 × 4's, 20 braces, 26 clamps, 300 nails

Chapter 3 Review Exercises

1. 4 **2.** 0 **3.** A, B **4.** none **5.** D, F, G, I
6. $\begin{bmatrix} -2 & 5 & 11 & -8 \\ -4 & 0 & 0 & -4 \\ 2 & 2 & -1 & -9 \end{bmatrix}$
7. Zero matrix **8.** Order

9. $\begin{bmatrix} 6 & -1 & -9 & 3 \\ 10 & 3 & -1 & 4 \\ -2 & -2 & -2 & 14 \end{bmatrix}$ **10.** $\begin{bmatrix} 3 & -3 \\ 4 & -1 \\ 2 & -6 \\ 1 & -2 \end{bmatrix}$

11. $\begin{bmatrix} 2 & 1 \\ 5 & 1 \end{bmatrix}$ **12.** $\begin{bmatrix} 12 & -6 \\ 15 & 0 \\ 18 & 0 \\ 3 & 9 \end{bmatrix}$ **13.** $\begin{bmatrix} 4 & 0 \\ 0 & 4 \end{bmatrix}$

14. $\begin{bmatrix} 2 & -12 \\ -8 & -22 \end{bmatrix}$ **15.** $\begin{bmatrix} 9 & 20 \\ 4 & 5 \end{bmatrix}$ **16.** $\begin{bmatrix} 5 & 16 \\ 6 & 15 \end{bmatrix}$

17. $\begin{bmatrix} 2 & 37 & 61 & -55 \\ -2 & 9 & -3 & -20 \\ 10 & 10 & -14 & -30 \end{bmatrix}$ **18.** $\begin{bmatrix} 43 & -23 \\ 33 & -12 \\ -13 & 15 \end{bmatrix}$

19. $\begin{bmatrix} 10 & 16 \\ 15 & 25 \\ 18 & 30 \\ 6 & 11 \end{bmatrix}$ **20.** $\begin{bmatrix} 17 & 73 \\ 7 & 28 \end{bmatrix}$ **21.** $\begin{bmatrix} 3 & 7 \\ 23 & 42 \end{bmatrix}$

22. F   **23.** F   **24.** $\begin{bmatrix} -19 & 12 \\ -8 & 5 \end{bmatrix}$ **25.** F

26. (a) Infinitely many solutions (bottom row of 0's)
   (b) $x = 6 + 2z$
     $y = 7 - 3z$
     $z =$ any real number
     Two specific solutions:
     If $z = 0$, then $x = 6$, $y = 7$.
     If $z = 1$, then $x = 8$, $y = 4$.

27. (a) No solution (last row says $0 = 1$)
   (b) No solution

28. (a) Unique coefficient matrix is I_3
   (b) $x = 0$, $y = -10$, $z = 14$

29. $(1, 2, 1)$ **30.** $x = 22$, $y = 9$

31. $x = -3$, $y = 3$, $z = 4$

32. $x = -\frac{3}{2}$, $y = 7$, $z = -\frac{11}{2}$ **33.** no solution

34. $x = 2 - 2z$; $y = -1 - 2z$, $z =$ any real number

35. $x = -2 + 8z$
   $y = -2 + 3z$
   $z =$ any real number

36. $x_1 = 1$, $x_2 = 11$, $x_3 = -4$, $x_4 = -5$ **37.** yes

38. $\begin{bmatrix} \frac{1}{2} & \frac{1}{4} \\ \frac{5}{2} & \frac{7}{4} \end{bmatrix}$ **39.** $\begin{bmatrix} -1 & -2 & 8 \\ 1 & 2 & -7 \\ 1 & 1 & -4 \end{bmatrix}$

40. $\begin{bmatrix} 2 & 1 & -2 \\ 7 & 5 & -8 \\ -13 & -9 & 15 \end{bmatrix}$

41. $x = -33$, $y = 30$, $z = 19$

42. $x = 4$, $y = 5$, $z = -13$

43. $A^{-1} = \begin{bmatrix} -41 & 32 & 5 \\ 17 & -13 & -2 \\ -9 & 7 & 1 \end{bmatrix}$; $x = 4$, $y = -2$, $z = 2$

44. no   **45. (a)** 16   **(b)** yes, det $\neq 0$

46. (a) 0   **(b)** no, det $= 0$

47. $\begin{bmatrix} 250 & 140 \\ 480 & 700 \end{bmatrix}$ **48.** $\begin{bmatrix} 1030 & 800 \\ 700 & 1200 \end{bmatrix}$

49. (a) higher in June   **(b)** higher in July

50. $\begin{matrix} \text{Men} & \text{Women} \\ \begin{bmatrix} 865 & 885 \\ 210 & 270 \end{bmatrix} & \begin{matrix} \text{Robes} \\ \text{Hoods} \end{matrix} \end{matrix}$ **51.** $\begin{bmatrix} 1750 \\ 480 \end{bmatrix} \begin{matrix} \text{Robes} \\ \text{Hoods} \end{matrix}$

52. (a) $\begin{bmatrix} 13{,}500 & 12{,}400 \\ 10{,}500 & 10{,}600 \end{bmatrix}$
   (b) Department A should buy from Kink; Department B should buy from Ace.

53. (a) $\begin{bmatrix} 0.20 & 0.30 & 0.50 \end{bmatrix}$
   (b) $\begin{bmatrix} 0.013469 \\ 0.013543 \\ 0.006504 \end{bmatrix}$
   (c) $\begin{bmatrix} 0.20 & 0.30 & 0.50 \end{bmatrix} \begin{bmatrix} 0.013469 \\ 0.013543 \\ 0.006504 \end{bmatrix} = 0.20(0.013469) +$
     $0.30(0.013543) + 0.50(0.006504) = 0.0100087$
   (d) The historical return of the portfolio, 0.0100087, is the estimated expected monthly return of the portfolio. This is roughly 1% per month.

54. 400 fast food, 700 software, 200 pharmaceutical

55. (a) A = 2C, B = 2000 − 4C, C = any integer satisfying $0 \leq C \leq 500$
   (b) yes; A = 500, B = 1000, C = 250
   (c) max A = 1000 when B = 0, C = 500

56. 3 passenger, 4 transport, 4 jumbo

57. shipping = 6000; agriculture = 800

58. (a) $A = \begin{matrix} & \begin{matrix} \text{S} & \text{C} \end{matrix} \\ \begin{bmatrix} 0.1 & 0.1 \\ 0.2 & 0.05 \end{bmatrix} & \begin{matrix} \text{Shoes} \\ \text{Cattle} \end{matrix} \end{matrix}$
   (b) shoes = 1000; cattle = 500

59. mining = 360, manufacturing = 320, fuels = 400

60. government = $\frac{64}{93}$ households; agriculture = $\frac{59}{93}$ households; manufacturing = $\frac{40}{93}$ households

Chapter 3 Test

1. $\begin{bmatrix} 3 & 1 & 5 \\ 1 & 3 & 6 \end{bmatrix}$ **2.** $\begin{bmatrix} -1 & 2 & 2 \\ 1 & -1 & 6 \end{bmatrix}$

3. $\begin{bmatrix} -12 & -16 & -155 \\ 5 & 12 & 87 \end{bmatrix}$ **4.** $\begin{bmatrix} 23 & 6 \\ 182 & 45 \\ 21 & 1 \end{bmatrix}$

5. $\begin{bmatrix} 0 & -7 \\ 26 & 1 \end{bmatrix}$ **6.** $\begin{bmatrix} -43 & -46 & -207 \\ 39 & 30 & -77 \\ 17 & 5 & -216 \end{bmatrix}$

7. $\begin{bmatrix} -2 & 3/2 \\ 1 & -1/2 \end{bmatrix}$ **8.** $\begin{bmatrix} -3 & 2 & 2 \\ 1 & 0 & -1 \\ 1/2 & -1/2 & 0 \end{bmatrix}$

9. $\begin{bmatrix} 5 \\ 14 \\ 15 \end{bmatrix}$ **10.** $x = -0.5, y = 0.5, z = 2.5$

11. $x = 4 - 1.8z, y = 0.2z, z =$ any real number

12. no solution **13.** $x = 2, y = 2, z = 0, w = -2$

14. $x = 6w - 0.5, y = 0.5 - w, z = 2.5 - 3w, w =$ any real number

15. (a) $B = \$45{,}000, E = \$40{,}000$
  (b) $\$0 \leq H \leq \$25{,}000$ (so $B \geq 0$)
  (c) min $E = \$20{,}000$ when $H = \$0$ and $B = \$75{,}000$

16. (a) $\begin{bmatrix} 0.08 & 0.22 & 0.12 \\ 0.10 & 0.08 & 0.19 \\ 0.05 & 0.07 & 0.09 \\ 0.10 & 0.26 & 0.15 \\ 0.12 & 0.04 & 0.24 \end{bmatrix}$

  (b) 0.08, 0.22, 0.12 consumed by carnivores 1, 2, 3
  (c) plant 5 by 1, plant 4 by 2, plant 5 by 3

17. (a) $\begin{bmatrix} 1000 & 4000 & 2000 & 1000 \end{bmatrix}$
  (b) $\begin{bmatrix} 45{,}000 & 55{,}000 & 90{,}000 & 70{,}000 \end{bmatrix}$
  (c) $\begin{bmatrix} 5 \\ 3 \\ 4 \\ 4 \end{bmatrix}$ **(d)** [\$1,030,000] **(e)** $\begin{array}{c} A \\ B \\ C \\ D \end{array} \begin{bmatrix} 65 \\ 145 \\ 125 \\ 135 \end{bmatrix}$ $\$$

18. (a) 121, 46, 247, 95, 261, 99, 287, 111, 179, 69, 169, 64
  (b) Frodo lives

19. growth, 2000; blue-chip, 400; utility, 400

20. agriculture: 315, minerals: 245

21. profit = households
  nonprofit = $\frac{2}{3}$ households

22. $\begin{array}{cccc} \text{Ag} & \text{M} & \text{F} & \text{S} \end{array}$
$\begin{bmatrix} 0.2 & 0.1 & 0.1 & 0.1 \\ 0.3 & 0.2 & 0.2 & 0.2 \\ 0.2 & 0.2 & 0.3 & 0.3 \\ 0.1 & 0.4 & 0.2 & 0.2 \end{bmatrix} \begin{array}{l} \text{Agriculture} \\ \text{Machinery} \\ \text{Fuel} \\ \text{Steel} \end{array}$

23. agriculture: 5000; machinery: 8000; fuel: 8000; steel: 7000

24. agriculture: $\frac{520}{699}$ households; steel: $\frac{236}{233}$ households; fuel: $\frac{159}{233}$ households

Exercise Set 4.1

1.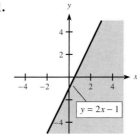

3.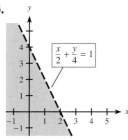

5.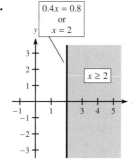

7. $(0, 0), (20, 10), (0, 15), (25, 0)$

9. $(5, 0), (15, 0), (6, 9), (2, 6)$

11. $(0, 5), (1, 2), (3, 1), (6, 0)$

13.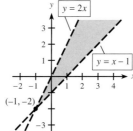

15.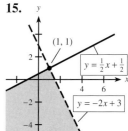

17.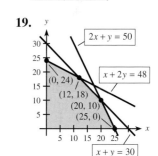

19.

21.

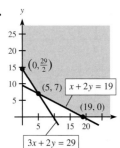

23.

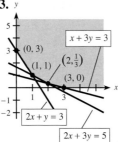

25.

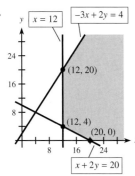

27. (a) Let x = the number of deluxe models and y = the number of economy models.

$$3x + 2y \le 24$$
$$\tfrac{1}{2}x + y \le 8$$
$$x \ge 0, y \ge 0$$

(b)

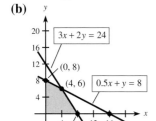

Corners: (0, 0), (8, 0), (4, 6), (0, 8)

29. (a) Let x = the number of cord-type trimmers and y = the number of cordless trimmers.

Constraints are
$$x + y \le 300$$
$$2x + 4y \le 800$$
$$x \ge 0, y \ge 0$$

(b)

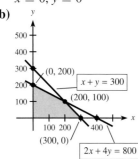

31. (a) Let x = the number of minutes on finance programs and y = the number of minutes on sports programs.

$$7x + 2y \ge 30$$
$$4x + 12y \ge 28$$
$$x \ge 0, y \ge 0$$

(b)

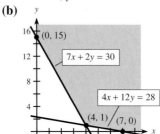

33. (a) Let x = the number of minutes of radio and y = the number of minutes of television.

Constraints are
$$x + y \ge 80$$
$$0.006x + 0.09y \ge 2.16$$
$$x \ge 0, y \ge 0$$

(b)

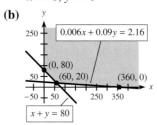

35. (a) Let x = the number of pounds of regular hot dogs and y = the number of pounds of all-beef hot dogs.

$$0.18x + 0.75y \le 1020$$
$$0.2x + 0.2y \ge 500$$
$$0.3x + \le 600$$

(b)

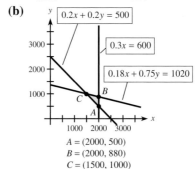

$A = (2000, 500)$
$B = (2000, 880)$
$C = (1500, 1000)$

Exercise Set 4.2

1. max = 20 at (4, 4); min = 0 at (0, 0)
3. max = 68 at (12, 4); min = 14 at (2, 2)
5. no max; min = 10 at (2, 1)
7. (0, 0), (0, 20), (10, 18), (15, 10), (20, 0); max = 66 at (10, 18); min = 0 at (0, 0)
9. (0, 60), (10, 30), (20, 20), (70, 0); min = 90 at (10, 30); no max

11. max = 80 at $x = 20$, $y = 10$

13. min = 36 at $x = 3$, $y = 0$ **15.** max = 22 at (2, 4)

17. max = 30 on line between (0, 5) and (3, 4)

19. min = 32 at (2, 3)   **21.** min = 9 at (2, 3)

23. max = 10 at (2, 4)   **25.** min = 3100 at (40, 60)

27. If x = the number of Deluxe models and
y = the number of Economy models, then
max = \$132 at (4, 6).

29. If x = the number of cord-type trimmers and
y = the number of cordless trimmers, then
max = \$9000 at any point with integer coordinates on
the segment joining (0, 200) and (200, 100), such as
(20, 190).

31. radio = 60, TV = 20, min C = \$16,000

33. inkjet = 45, laser = 25, max P = \$3300

35. 250 fish: 150 bass and 100 trout

37. (a) Max P = \$315,030 when corn = 3749.5 acres and
soybeans = 2251.5 acres
(b) Max P = \$315,240 when corn = 3746 acres and
soybeans = 2262 acres
(c) \$30/acre

39. 30 days for factory 1 and 20 days for factory 2;
minimum cost = \$700,000

41. 60 days for location I and 70 days for location II;
minimum cost = \$86,000

43. reg = 2000 lb; all-beef = 880 lb; maximum profit
= \$1328

45. From Pittsburgh: 20 to Blairsville, 40 to Youngstown;
From Erie: 15 to Blairsville, 0 to Youngstown;
minimum cost = \$1540

47. (a) R = \$366,000 with 6 satellite and 17 full-service
branches
(b) Branches: used 23 of 25 possible; 2 not used
(slack)
New employees: hired 120 of 120 possible; 0 not
hired (slack)
Budget: used all \$2.98 million; \$0 not used
(slack)
(c) Additional new employees and additional budget.
These items are completely used in the current
optimal solution; more could change and improve
the optimal solution.
(d) Additional branches. The current optimal solution
does not use all those allotted; more would just
add to the extras.

Exercise Set 4.3

1. $3x + 5y + s_1 = 15$, $3x + 6y + s_2 = 20$

3. $\begin{bmatrix} 1 & 5 & 1 & 0 & 0 & | & 200 \\ 2 & 3 & 0 & 1 & 0 & | & 134 \\ -4 & -9 & 0 & 0 & 1 & | & 0 \end{bmatrix}$

5. $\begin{bmatrix} 2 & 7 & 9 & 1 & 0 & 0 & 0 & | & 100 \\ 6 & 5 & 1 & 0 & 1 & 0 & 0 & | & 145 \\ 1 & 2 & 7 & 0 & 0 & 1 & 0 & | & 90 \\ -2 & -5 & -2 & 0 & 0 & 0 & 1 & | & 0 \end{bmatrix}$

7. one slack variable for each constraint row (above the
last row)

9. (a) $x_1 = 0$, $x_2 = 0$, $s_1 = 200$, $s_2 = 400$, $s_3 = 350$, $f = 0$
(b) not complete

(c) $\begin{bmatrix} 10 & ㉗ & 1 & 0 & 0 & 0 & | & 200 \\ 4 & 51 & 0 & 1 & 0 & 0 & | & 400 \\ 15 & 27 & 0 & 0 & 1 & 0 & | & 350 \\ -6 & -7 & 0 & 0 & 0 & 1 & | & 0 \end{bmatrix}$

$\frac{1}{27}R_1 \to R_1$, then $-51R_1 + R_2 \to R_2$,
$-27R_1 + R_3 \to R_3$, $7R_1 + R_4 \to R_4$

11. (a) $x_1 = 0$, $x_2 = 15$, $s_1 = 12$, $s_2 = 0$, $f = 15$
(b) not complete

(c) $\begin{bmatrix} 2 & 0 & 1 & -\frac{3}{4} & 0 & | & 12 \\ ③ & 1 & 0 & \frac{1}{3} & 0 & | & 15 \\ -4 & 0 & 0 & 3 & 1 & | & 15 \end{bmatrix}$

$\frac{1}{3}R_2 \to R_2$, then $-2R_2 + R_1 \to R_1$, $4R_2 + R_3 \to R_3$

13. (a) $x_1 = 0$, $x_2 = 14$, $x_3 = 11$, $s_1 = 6$, $s_2 = 0$, $s_3 = 0$,
$f = 525$
(b) complete (no part (c))

15. (a) $x_1 = 0$, $x_2 = 0$, $x_3 = 12$,
$s_1 = 4$, $s_2 = 6$, $s_3 = 0$, $f = 150$
(b) not complete

(c) $\begin{bmatrix} 4 & 4 & 1 & 0 & 0 & 2 & 0 & | & 12 \\ ② & ④ & 0 & 1 & 0 & -1 & 0 & | & 4 \\ -3 & -11 & 0 & 0 & 1 & -1 & 0 & | & 6 \\ -3 & -3 & 0 & 0 & 0 & 4 & 1 & | & 150 \end{bmatrix}$

Either circled number may act as the next pivot
entry, but only one of them. If 4 is used,
$\frac{1}{4}R_2 \to R_2$, then $-4R_2 + R_1 \to R_1$,
$11R_2 + R_3 \to R_3$, $3R_2 + R_4 \to R_4$. If 2 is used,
$\frac{1}{2}R_2 \to R_2$, then $-4R_2 + R_1 \to R_1$,
$3R_2 + R_3 \to R_3$, $3R_2 + R_4 \to R_4$.

17. (a) $x_1 = 0$, $x_2 = 0$, $x_3 = 12$, $s_1 = 5$, $s_2 = 0$, $s_3 = 6$,
$f = 120$
(b) no solution (no part (c))

19. $x = 11$, $y = 9$; $f = 20$

21. $x = 0$, $y = 14$, $z = 11$; $f = 525$

23. $x = 50, y = 10; f = 100$. Multiple solutions are possible.

Next pivot is circled.

$$\begin{bmatrix} 1 & 0 & 3 & 0 & 6 & 0 & | & 50 \\ 0 & 0 & 4 & 1 & -4 & 0 & | & 6 \\ 0 & 1 & -2 & 0 & ② & 0 & | & 10 \\ \hline 0 & 0 & 9 & 0 & 0 & 1 & | & 100 \end{bmatrix}$$

25. $x = 0, y = 5; f = 50$ **27.** $x = 4, y = 3; f = 17$

29. $x = 4, y = 3; f = 11$

31. $x = 0, y = 2, z = 5; f = 40$

33. $x = 15, y = 15, z = 25; f = 780$

35. $x = 6, y = 2, z = 26; f = 206$

37. $x_1 = 36, x_2 = 24, x_3 = 0, x_4 = 8; f = 1728$

39. $x = 8, y = 16; f = 32$

41. no solution

43. $x = 0, y = 50$ or $x = 40, y = 40; f = 600$

45. (a)
$$\begin{bmatrix} 1 & 1 & 1 & 0 & 0 & | & 60 \\ 1 & 3 & 0 & 1 & 0 & | & 120 \\ \hline -40 & -60 & 0 & 0 & 1 & | & 0 \end{bmatrix}$$

    (b) Maximum profit is \$3000 with 30 inkjet and 30 laser printers.

47. 300 style-891, 450 style-917, max $P = \$5175$

49. Maximum profit is \$3900 with 11 axles and 2 wheels

51. 500 tomatoes, 1800 peaches; maximum $P = \$4100$

53. 21 newspapers, 13 radio; 230,000 exposures

55. medium 1 = 10, medium 2 = 10, medium 3 = 12

57. \$1650 profit with 46 A, 20 B, 6 C

59. 8000 Regular, 0 Special, and 1000 Kitchen Magic; maximum profit = \$32,000

61. (a) 26 one-bedroom; 40 two-bedroom; 48 three-bedroom

    (b) \$100,200 per month

63. 19-in. TVs = 40, 27-in. TVs = 115, 35-in. table = 0, 35-in. floor = 38; Max $P = \$12,540$

Exercise Set 4.4

1. (a)
$$\begin{bmatrix} 5 & 2 & | & 10 \\ 1 & 2 & | & 6 \\ 3 & 1 & | & g \end{bmatrix} \text{ transpose} = \begin{bmatrix} 5 & 1 & | & 3 \\ 2 & 2 & | & 1 \\ 10 & 6 & | & g \end{bmatrix}$$

    (b) maximize $f = 10x_1 + 6x_2$ subject to $5x_1 + x_2 \leq 3, 2x_1 + 2x_2 \leq 1, x_1 \geq 0, x_2 \geq 0$.

3. (a)
$$\begin{bmatrix} 1 & 1 & | & 9 \\ 1 & 3 & | & 15 \\ 5 & 2 & | & g \end{bmatrix} \text{ transpose} = \begin{bmatrix} 1 & 1 & | & 5 \\ 1 & 3 & | & 2 \\ 9 & 15 & | & g \end{bmatrix}$$

    (b) maximize $f = 9x_1 + 15x_2$ subject to
$$x_1 + x_2 \leq 5$$
$$x_1 + 3x_2 \leq 2$$

5. (a) $y_1 = 8, y_2 = 2, y_3 = 0$; min: $g = 252$

    (b) $x_1 = 5, x_2 = 0, x_3 = 9$: max $f = 252$

7. maximize $f = 11x_1 + 11x_2 + 16x_3$ subject to
$$2x_1 + x_2 + x_3 \leq 2$$
$$x_1 + 3x_2 + 4x_3 \leq 10$$
primal: $y_1 = 16, y_2 = 0$; $g = 32$ (min)
dual: $x_1 = 0, x_2 = 0, x_3 = 2; f = 32$ (max)

9. maximize $f = 11x_1 + 12x_2 + 6x_3$ subject to
$$4x_1 + 3x_2 + 3x_3 \leq 3$$
$$x_1 + 2x_2 + x_3 \leq 1$$
primal: $y_1 = 2, y_2 = 3$; $g = 9$ (min)
dual: $x_1 = 3/5, x_2 = 1/5, x_3 = 0; f = 9$ (max)

11. min $= 28$ at $x = 2, y = 0, z = 1$

13. $y_1 = 2/5, y_2 = 1/5, y_3 = 1/5$; $g = 16$ (min)

15. (a) minimize $g = 120y_1 + 50y_2$ subject to
$$3y_1 + y_2 \geq 40$$
$$2y_1 + y_2 \geq 20$$
    (b) primal: $x_1 = 40, x_2 = 0; f = 1600$ (max)
dual: $y_1 = 40/3, y_2 = 0; g = 1600$ (min)

17. min $= 480$ at $y_1 = 0, y_2 = 0, y_3 = 16$

19. min $= 90$ at $y_1 = 0, y_2 = 3, y_3 = 1, y_4 = 0$

21. Atlanta $= 150$ hr, Ft. Worth $= 50$ hr; min $C = \$210,000$

23. line 1 for 4 hours, line 2 for 1 hour; \$1200

25. A = 12 weeks, B = 0 weeks, C = 0 weeks; cost = \$12,000

27. factory 1: 50 days, factory 2: 0 days; min cost \$500,000

29. 105 minutes on radio, nothing on TV; min cost \$10,500

31. (a) Georgia package = 10
Union package = 20
Pacific package = 5

    (b) \$4630

33. (a) min cost = \$16

    (b) Many solutions are possible; two are: 16 oz of food I, 0 oz of food II, 0 oz of food III and 11 oz of food I, 1 oz of food II, 0 oz of food III.

35. Mon. = 8, Tues. = 0, Wed. = 5, Thurs. = 4, Fri. = 5, Sat. = 0, Sun. = 3; min = 25

Exercise Set 4.5

1. $-3x + y \leq -5$ **3.** $-6x - y \leq -40$

5. (a) maximize $f = 2x + 3y$ subject to
$$7x + 4y \leq 28$$
$$3x - y \leq -2$$
$$x \geq 0, y \geq 0$$

    (b)
$$\begin{bmatrix} 7 & 4 & 1 & 0 & 0 & | & 28 \\ 3 & ⊖1 & 0 & 1 & 0 & | & -2 \\ \hline -2 & -3 & 0 & 0 & 1 & | & 0 \end{bmatrix}$$

7. (a) Maximize $-g = -3x - 8y$ subject to
$$4x - 5y \leq 50$$
$$x + y \leq 80$$
$$x - 2y \leq -4$$
$$x \geq 0, y \geq 0$$

$$(b) \begin{bmatrix} 4 & -5 & 1 & 0 & 0 & 0 & | & 50 \\ 1 & 1 & 0 & 1 & 0 & 0 & | & 80 \\ 1 & -2 & 0 & 0 & 1 & 0 & | & -4 \\ 3 & 8 & 0 & 0 & 0 & 1 & | & 0 \end{bmatrix}$$

9. $x = 6, y = 8, z = 12; f = 120$

11. $x = 10, y = 17; f = 57$

13. $x = 5, y = 7; f = 31$

15. $x = 5, y = 15; f = 45$

17. $x = 10, y = 20; f = 120$

19. $x = 20, y = 10, z = 0; f = 40$

21. $x = 5, y = 0, z = 3; f = 22$

23. $x = 70, y = 0, z = 40; f = 2100$

25. $x_1 = 20, x_2 = 10, x_3 = 20, x_4 = 80, x_5 = 10, x_6 = 10;$ $f = 3250$

27. 500 filters, 0 housing units; min cost = \$3300

29. regular = 2000 lb; beef = 880 lb; profit = \$1328

31. Produce 200 of each at Monaca; produce 300 commercial components and 550 domestic furnaces at Hamburg; profit = \$355,250

33. Produce 200 of each at Monaca; produce 300 commercial components and 550 domestic furnaces at Hamburg; cost = \$337,750

35. I = 3 million, II = 0, III = 3 million; cost = \$180,000

37. 2000 footballs, 0 soccer balls, 0 volleyballs; \$60,000

Chapter 4 Review Exercises

1. 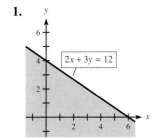 **2.**

3. 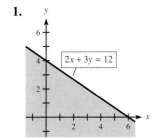 **4.**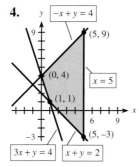

5. max = 25 at (5, 10); min = −12 at (12, 0)

6. max = 194 at (17, 23); min = 104 at (8, 14)

7. max = 140 at (20, 0); min = −52 at (20, 32)

8. min = 115 at (5, 7); no max exists, f can be made arbitrarily large

9. $f = 66$ at (6, 6) **10.** $f = 43$ at (7, 9)

11. $g = 24$ at (3, 3) **12.** $g = 84$ at (64, 4)

13. $f = 76$ at (12, 8) **14.** $f = 75$ at (15, 15)

15. $f = 168$ at (12, 7) **16.** $f = 260$ at (60, 20)

17. $f = 360$ at (40, 30) **18.** $f = 80$ at (20, 10)

19. $f = 270$ at (5, 3, 2) **20.** $f = 4380$ at (40, 10, 0, 0)

21. $f = 640$ on the line between (160, 0) and (90, 70)

22. no solution **23.** $g = 32$ at $y_1 = 2, y_2 = 3$

24. $g = 20$ at $y_1 = 4, y_2 = 2$

25. $g = 7$ at $y_1 = 1, y_2 = 5$

26. $g = 1180$ at $y_1 = 80, y_2 = 20$

27. $g = 140$ at $y_1 = 0, y_2 = 20, y_3 = 20$

28. $g = 1400$ at $y_1 = 0, y_2 = 100, y_3 = 100$

29. $f = 165$ at $x = 20, y = 21$

30. $f = 54$ at $x = 6, y = 5$

31. $f = 156$ at $x = 15, y = 2$

32. $f = 31$ at $x = 4, y = 5$

33. $f = 1000$ at $x_1 = 25, x_2 = 62.5, x_3 = 0, x_4 = 12.5$

34. $g = 2020$ at $x_1 = 0, x_2 = 100, x_3 = 80, x_4 = 20$

35. $P = \$14,750$ when 110 large and 75 small swing sets are made

36. $C = \$300,000$ when factory 1 operates 30 days, factory 2 operates 25 days

37. $P = \$320;$ I = 40, II = 20

38. $P = \$420;$ Jacob's ladders = 90, locomotive engines = 30

39. (a) Let $x_1 =$ the number of 27″ sets, $x_2 =$ the number of 32″ sets, $x_3 =$ the number of 42″ HD sets, $x_4 =$ the number of 42″ plasma sets.

(b) Maximize $P = 80x_1 + 120x_2 + 160x_3 + 200x_4$ subject to

$$8x_1 + 10x_2 + 12x_3 + 15x_4 \le 1870$$
$$2x_1 + 4x_2 + 4x_3 + 4x_4 \le 530$$
$$x_1 + x_2 + x_3 + x_4 \le 200$$
$$x_3 + x_4 \le 100$$
$$x_2 \le 120$$

(c) $x_1 = 15, x_2 = 25, x_3 = 0, x_4 = 100;$ max profit = \$24,200

40. food I = 0 oz, food II = 3 oz; $C = \$0.60$ (min)

41. cost = \$5.60; A = 40 lb, B = 0 lb

42. cost = \$8500; A = 20 days, B = 15 days, C = 0 days

43. pancake mix = 8000 lb; cake mix = 3000 lb; profit = \$3550

44. Texas: 55 desks, 65 computer tables; Louisiana: 75 desks, 65 computer tables; cost = \$4245

45. Midland: grade 1 = 486.5 tons, grade 2 = 0 tons; Donora: grade 1 = 13.5 tons, grade 2 = 450 tons; Cost = \$90,635

Chapter 4 Test

1. max $= 120$ at $(0, 24)$

2. (a) C;
$$\begin{bmatrix} 1 & 2 & 0 & 1 & 0 & -3/2 & 0 & | & 40 \\ 0 & ① & 0 & -2 & 1 & 1/2 & 0 & | & 15 \\ 0 & 3 & 1 & -1 & 0 & 1/4 & 0 & | & 60 \\ 0 & 0 & 0 & 4 & 0 & 6 & 1 & | & 220 \end{bmatrix}$$

$-2R_2 + R_1 \rightarrow R_1 \qquad -3R_2 + R_3 \rightarrow R_3$

(b) A; pivot column is column 3, but new pivot is undefined.

(c) B; $x_1 = 40, x_2 = 12, x_3 = 0, s_1 = 0,$
$s_2 = 20, s_3 = 0; f = 170$

3. (a) 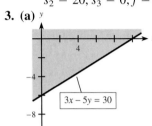 **(b)**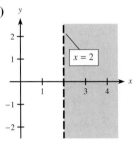

(c)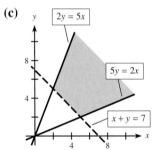

4. maximize $f = 100x_1 + 120x_2$ subject to
$3x_1 + 4x_2 \leq 2$
$5x_1 + 6x_2 \leq 3$
$x_1 + 3x_2 \leq 5$
$x_1 \geq 0, x_2 \geq 0$

5. min $= 21$ at $(1, 8)$; no max exists, f can be made arbitrarily large

6. min $= 136$ at $(28, 52)$

7. maximize $-g = -7x - 3y$ subject to
$x - 4y \leq -4$
$x - y \leq 5$
$2x + 3y \leq 30$

8. max: $x_1 = 17, x_2 = 15, x_3 = 0; f = 658$ (max)
min: $y_1 = 4, y_2 = 18, y_3 = 0; g = 658$ (min)

9. max $= 6300$ at $x = 90, y = 0$

10. max $= 1200$ at $x = 0, y = 16, z = 12$

11. If $x =$ the number of barrels of beer and $y =$ the number of barrels of ale, then maximize $P = 35x + 30y$
subject to
$3x + 2y \leq 1200$
$2x + 2y \leq 1000$
$P = \$16,000$ (max) at $x = 200, y = 300$

12. If $x =$ the number of day calls and $y =$ the number of evening calls, then minimize $C = 3x + 4y$ subject to
$0.3x + 0.3y \geq 150$
$0.1x + 0.3y \geq 120$
$x \leq 0.5 (x + y)$
$C = \$1850$ (min) at $x = 150, y = 350$

13. max profit $= \$5000$ when product 1 $= 25$ tons,
product 2 $= 62.5$ tons, product 3 $= 0$ tons,
product 4 $= 12.5$ tons

Exercise Set 5.1

1. 3.162278 **3.** 0.01296525 **5.** 1.44225

7. 7.3891 **9.**

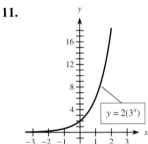

11. 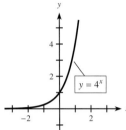 **13.**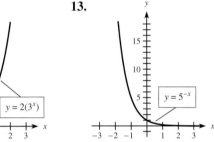

15. 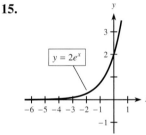 **17.**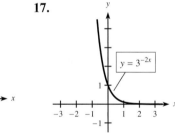

19. (a) $y = 3(2.5)^{-x}$

(b) Decay. They have the form $y = C \cdot a^x$ for
$0 < a < 1$ or $y = C \cdot b^{-x}$ for $b > 1$.

(c) The graphs are identical.

21. (a) and **(b)**

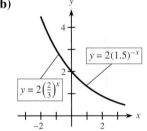

(c) $(1.5)^{-x} = \left(\dfrac{3}{2}\right)^{-x} = \left(\dfrac{2}{3}\right)^{x}$

23. All graphs have the same basic shape. For larger positive values of k, the graphs fall more sharply. For positive values of k nearer 0, the graphs fall more slowly.

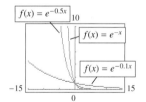
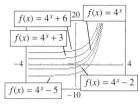

25. $y = f(x) + C$ is the same graph as $y = f(x)$ but shifted C units on the y-axis.

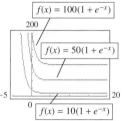

27. (a)

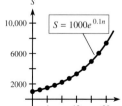

(b) As c changes, the y-intercept and the asymptote change.

29. \$1884.54

31.

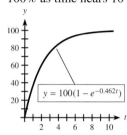

33. At time 0, the concentration is 0. The concentration rises rapidly for the first 4 minutes, then tends toward 100% as time nears 10 minutes.

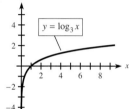

35.

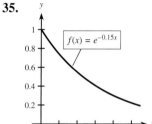

As the TV sets age (x increases), the fraction of sets still in service declines.

37.

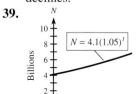

39.

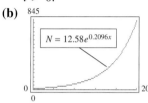

41. (a) Growth; $e > 1$ and the exponent is positive for $t > 0$.

(b)

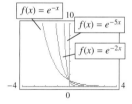

43. The linear model fails in 2007, giving a negative number of processors.

45. (a) $y = 434.24(1.0779)^x$
(b) \$18,478 billion

47. (a) $y = 2.58(1.043)^x$
(b) 264.79
(c) 2013

49. (a) $y = 116.83(0.85651)^x$
(b) decay; base satisfies $0 < b < 1$
(c) 1.12

Exercise Set 5.2

1. $2^4 = 16$ **3.** $4^{1/2} = 2$ **5.** $x = 8$ **7.** $x = \frac{1}{2}$
9. $x = 12$ **11.** $\log_2 32 = 5$ **13.** $\log_4\left(\frac{1}{4}\right) = -1$
15.

17.

19.

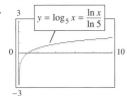

21. (a) 3   **(b)** −1   **23.** x **25.** 3
27. (a) 4.9   **(b)** 0.4   **(c)** 12.4   **(d)** 0.9
29. $\log x - \log (x + 1)$ **31.** $\log_7 x + \frac{1}{3}\log_7 (x + 4)$
33. $\ln (x/y)$ **35.** $\log_5 \left[x^{1/2}(x + 1) \right]$
37. equivalent; Properties V and III
39. not equivalent; $\log (\sqrt[3]{8}/5)$
41. (a)

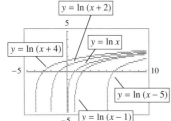

(b) For each c, the domain is $x > c$ and the vertical asymptote is at $x = c$.
(c) Each x-intercept is at $x = c + 1$.
(d) The graph of $y = f(x - c)$ is the graph of $y = f(x)$ shifted c units on the x-axis.
43. (a) 4.0875   **(b)** −0.1544
45.

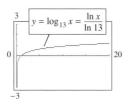

47.

49. If $\log_a M = u$ and $\log_a N = v$, then $a^u = M$ and $a^v = N$. Therefore, $\log_a (M/N) = \log_a (a^u/a^v) = \log_a (a^{u-v}) = u - v = \log_a M - \log_a N$.
51. $y = 3.818e^{-0.0506x}$ **53.** 7.9 times as severe
55. 14 times as severe   **57.** 40
59. $L = 10 \log (I/I_0)$

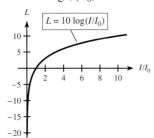

61. 0.1 and 1×10^{-14}
63. $\text{pH} = \log \dfrac{1}{[\text{H}^+]} = \log 1 - \log [\text{H}^+] = -\log [\text{H}^+]$

65. $\log_{1.02} 2 = 4t$; $t \approx 8.75$ years
67. $L(99) = 76.69$; $\ell(99) = 76.61$
$L(112) = 78.45$; $\ell(112) = 78.36$
The two models are quite similar.
69. (a)

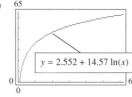

(b) 62.2%
(c) between $x = 51$ and $x = 52$; 2002
71. (a) $y = 24.926 + 14.504 \ln x$ **(b)** 74.3%
(c) The percent must be less than or equal to 100%.

Exercise Set 5.3

1. $\frac{5}{3}$ **3.** 2.943   **5.** 18.971
7. 151.413   **9.** 6.679
11. (a)

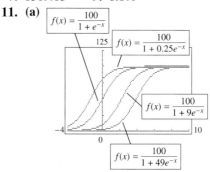

(b) Different c-values change the y-intercept and how the graph approaches the asymptote.
13. (a) 2038   **(b)** 4.9 months
15. (a) 13.86 years   **(b)** \$105,850.00
17. 24.5 years   **19.** 128,402
21. (a) $t \approx 50.6$; in 2011
(b) The intent of such plan would be to reduce future increases in health care expenditures. A new model might not be exponential, or, if it were, it would be one that rose more gently.
23. (a) \$4.98   **(b)** 8   **25.** \$502   **27.** \$420.09
29. \$2706.71   **31. (a)** \$10,100.31   **(b)** 6.03 years
33. (a) \$5469.03
(b) 7 years, 9 months (approximately)
35. (a) \$74.04 billion   **(b)** 5.3 years
(c) The September 11, 2001 terrorist attacks, the war in Iraq, larger federal deficits, and corporate scandals (such as Enron)
37. (a) $P(10) = 2.30$ means that a 1983 dollar was worth \$2.30 in 1970.
$P(70) = 0.11$ means that a 1983 dollar should be worth \$0.11 in 2030.
(b) 53.9 years (during 2014)
39. $x \approx 993.3$; about 993 units

41. **(a)** 600   **(b)** 2119   **(c)** 3000
   (d)

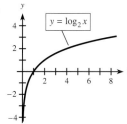

$N = 3000(0.2)^{0.6t}$

43. **(a)** 10   **(b)** 2.5 years
45. **(a)** 37   **(b)** 1.5 hours
47. **(a)** 52   **(b)** the 10th day
49. **(a)** 2%   **(b)** 20 months ($x = 19.6$)
51. **(a)** 0.23 km^3 **(b)** 5.9 years
53. **(a)** 6910 (approximately)
   (b) $t \approx 34.6$; in 2010
55. **(a)** $y = \dfrac{59.57}{1 + 5.22e^{-0.585x}}$
   (b) 56.8%
   (c) $x \approx 14.4$; in 2010

Chapter 5 Review Exercises

1. **(a)** $\log_2 y = x$ **(b)** $\log_3 2x = y$
2. **(a)** $7^{-2} = \frac{1}{49}$ **(b)** $4^{-1} = x$
3.

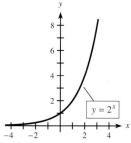

$y = e^x$

4.

$y = e^{-x}$

5.

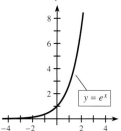

$y = \log_2 x$

6.

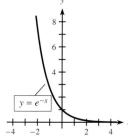

$y = 2^x$

7.

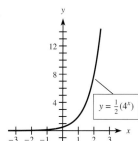

$y = \frac{1}{2}(4^x)$

8.

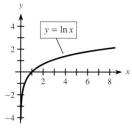

$y = \ln x$

9.

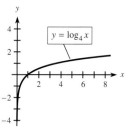

$y = \log_4 x$

10.

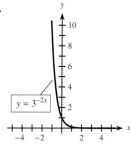

$y = 3^{-2x}$

11.

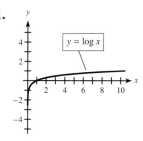

$y = \log x$

12.

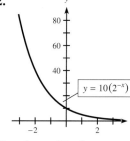

$y = 10(2^{-x})$

13. 0   **14.** 3   **15.** $\frac{1}{2}$ **16.** -1 **17.** 8
18. 1   **19.** 5   **20.** 3.15   **21.** -2.7
22. 0.6   **23.** 5.1   **24.** 15.6   **25.** $\log y + \log z$
26. $\frac{1}{2}\ln(x+1) - \frac{1}{2}\ln x$ **27.** no   **28.** -2
29. 5   **30.** 1   **31.** 0   **32.** 3.4939   **33.** -1.5845
34.

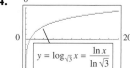

$y = \log_{\sqrt{3}} x = \dfrac{\ln x}{\ln \sqrt{3}}$

35.

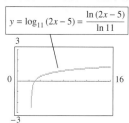

$y = \log_{11}(2x - 5) = \dfrac{\ln(2x-5)}{\ln 11}$

36. $x = 1.5$ **37.** $x \approx 51.224$
38. $x \approx 28.175$ **39.** $x \approx 40.236$
40. Growth exponential, because the general outline has the same shape as a growth exponential.
41. Decay exponential, because the general shape is similar to the graph of a decay exponential, and the number will diminish toward 0.
42. **(a)** \$32,627.66
   (b)

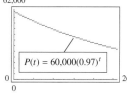

$P(t) = 60{,}000(0.97)^t$

43. **(a)** $y = 267.78(1.0729)^x$
   (b) \$1790.1 billion
   (c) \$3143.1 billion
   (d) too high, because the data seem to level off

44. (a) $10,383 (approximately)

(b)

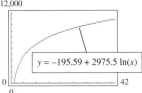

45. (a) $y = -31.5026 + 22.9434 \ln (x)$

(b) 65.97%

(c) Between $x = 103$ and $x = 104$; in 2044

46. (a) -3.9 **(b)** $0.14B_0$ **(c)** $0.004B_0$ **(d)** yes

47. (a) 27,441 **(b)** 12 weeks

48. 1366 **49.** 5.8 years

50. (a) $5532.77 **(b)** 5.13 years

51. logistic, because the graph begins like an exponential function but then grows at a slower rate

52. (a) 3000 **(b)** 8603 **(c)** 10,000

53. (a)

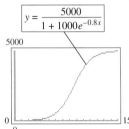

(b) 5 **(c)** 4970 **(d)** 10 days

Chapter 5 Test

1.

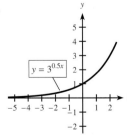

2.

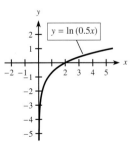

3.

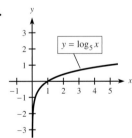

4.

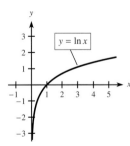

5.

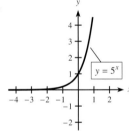

6.

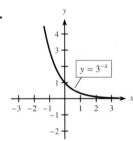

7.

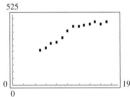

8.

9. 54.59815 **10.** 0.122456 **11.** 1.38629

12. 3.04452 **13.** $x = 17^{3.1} \approx 6522.163$

14. $\log_3 (27) = 2x$ **15.** 3

16. x^4 **17.** 3 **18.** x^2

19. $\ln M + \ln N$ **20.** $\ln (x^3 - 1) - \ln (x + 2)$

21. $\dfrac{\ln (x^3 + 1)}{\ln 4} \approx 0.721 \ln (x^3 + 1)$

22. $x \approx 38.679$

23. a decay exponential

24. With years on the horizontal axis, a growth exponential would probably be the best model.

25. (a) $2653.45 billion

(b) about 8.3 years

26. about 6.2 months

27. (a) about $12,002 billion

(b) between $x = 54$ and $x = 55$; in 2015

28. (a)

An exponential model is not appropriate; the shape of the scatter plot does not resemble an exponential function (neither growth nor decay).

(b) $y_1 = -54.846 + 192.7 \ln x$ and

$$y_2 = \dfrac{495}{1 + 3.9e^{-0.275x}}$$

(c) $y_1 \approx$ $565.43 billion is more optimistic than $y_2 \approx$ $493.01 billion.

Exercise Set 6.1

1. (a) $9600 **(b)** $19,600

3. (a) $30 **(b)** $1030

5. $864 **7.** $3850 **9.** 13%

11. (a) 5.13% **(b)** 4.29% **13.** $1631.07

15. $12,000 **17.** 10 years **19.** pay on time

21. (a) $2120 **(b)** $2068.29 (nearest cent)

23. 3, 6, 9, 12, 15, 18, 21, 24, 27, 30

25. $-\frac{1}{3}, \frac{1}{5}, -\frac{1}{7}, \frac{1}{9}, -\frac{1}{11}, \frac{1}{13}$ **27.** $-1, -\frac{1}{4}, -\frac{1}{15}, 0; a_{10} = \frac{1}{20}$

29. (a) $d = 3, a_1 = 2$ **(b)** 11, 14, 17

31. (a) $d = \frac{3}{2}, a_1 = 3$ **(b)** $\frac{15}{2}, 9, \frac{21}{2}$

33. -35 **35.** 203 **37.** 2185 **39.** 1907.5

41. $-15,862.5$ **43.** 21, 34, 55

45. $2400 **47.** the job starting at $20,000

49. (a) $3000 **(b)** $4500 **(c)** plan II, by $1500

(d) $10,000 **(e)** $13,500 **(f)** plan II, by $3500
(g) plan II

Exercise Set 6.2

(Minor differences may occur because of rounding.)
1. (a) 8% **(b)** 7 **(c)** 2% = 0.02 **(d)** 28
3. (a) 9% **(b)** 5 **(c)** $(\frac{9}{12})$% = 0.0075 **(d)** 60
5. $24,846.79 **7.** $4755.03 **9.** $3086.45
11. $6661.46 **13.** $5583.95
15. $7309.98 **17.** $502.47
19. (a) $12,245.64 **(b)** $11,080.32
(c) A $\frac{1}{2}$% increase in the interest rate reduces the amount required by $1165.32.
21. $50.26 more at 8% **23. (a)** 7.55% **(b)** 6.18%
25. 8% compounded monthly, 8% compounded quarterly, 8% compounded annually
27. 8.2% continuously yields 8.55%. 8.4% compounded quarterly yields 8.67% and so is better.
29. The higher graph is for continuous compounding because its yield (its effective annual rate) is higher.
31. 13.25% **33.** 3 years **35.** 4% **37.** $3996.02
39. (a) $554,021.04 **(b)** $125,173.66 more
41. 5.12 years (approximately) **43.** $13,916.24
45.

	A	B	C
1		Future Value	(Yearly)
2	End of Year	Quarterly	Monthly
3	0	$5000.00	$5000.00
4	1	$5322.52	$5324.26
5	2	$5665.84	$5669.54
6	3	$6031.31	$6037.22
7	4	$6420.36	$6428.74
8	5	$6834.50	$6845.65
9	6	$7275.35	$7289.60
10	7	$7744.64	$7762.34
11	8	$8244.20	$8265.74
12	9	$8775.99	$8801.79
13	10	$9342.07	$9372.59

(a) from quarterly and monthly spreadsheets: after $6\frac{1}{2}$ years (26 quarters or 78 months)
(b) See the spreadsheet.
47. (a) 24, 48, 96 **(b)** 24, 16, $\frac{32}{3}$
49. $10(2^{12})$ **51.** $4 \cdot (\frac{3}{2})^{15}$ **53.** $\frac{6(1 - 3^{17})}{-2}$
55. $\frac{3^{35} - 1}{2}$ **57.** $18[1 - (\frac{2}{3})^{18}]$
59. $350,850 (approx.) **61.** 24.4 million (approx.)
63. 35 years **65.** 40.5 ft **67.** $4096
69. 320,000 **71.** $7,231,366 **73.** $1801.14
75. 6; $5^6 = 15,625$ **77.** 305,175,780

Exercise Set 6.3

1. (a) The higher graph is $1120 per year.
(b) $R = \$1000$: $S = \$73,105.94$;
$R = \$1120$: $S = \$81,878.65$;
Difference = $8772.71

3. $7328.22 **5.** $1072.97 **7.** $4774.55
9. $4372.20 **11.** $1083.40 **13.** $741.47
15. $n \approx 27.1$; 28 quarters
17. A sinking fund is a savings plan, so the 10% rate
(a) is better.
19. $4651.61 **21.** $1180.78 **23.** $4152.32
25. $3787.92
27. (a) ordinary annuity **(b)** $266.10
29. (a) ordinary annuity **(b)** $n \approx 108.9$; 109 months
31. (a) annuity due **(b)** $235.16
33. (a) annuity due **(b)** $26,517.13
35. (a) ordinary annuity **(b)** $265.25
37. $53,677.40
39. (a) $n \approx 35$ quarters **(b)** $1,062,412 (nearest dollar)
41. The spreadsheet shows the amounts at the end of each of the first 12 months and at the end of the last 12 months. The amount after 10 years is shown.

	A	B	C
1		Future Value	
2	End of Month	Ordinary Ann.	Annuity Due
3	0	0	100
4	1	$100.00	$100.60
5	2	$200.60	$201.80
6	3	$301.80	$303.61
7	4	$403.61	$406.04
8	5	$506.04	$509.07
9	6	$609.07	$612.73
10	7	$712.73	$717.00
11	8	$817.00	$821.91
12	9	$921.91	$927.44
13	10	$1027.44	$1033.60
14	11	$1133.60	$1140.40
15	12	$1240.40	$1247.85
⋮	⋮	⋮	⋮
112	109	$15324.39	$15416.34
113	110	$15516.34	$15609.44
114	111	$15709.44	$15803.70
115	112	$15903.70	$15999.12
116	113	$16099.12	$16195.71
117	114	$16295.71	$16393.49
118	115	$16493.49	$16592.45
119	116	$16692.45	$16792.60
120	117	$16892.60	$16993.96
121	118	$17093.96	$17196.52
122	119	$17296.52	$17400.30
123	120	$17500.30	$17605.30

(a) $12,000
(b) Annuity due. Each payment for an annuity due earns 1 month's interest more than that for an ordinary annuity.

Exercise Set 6.4

1. $69,913.77
3. $2,128,391
5. $4595.46

7. taking $250,000 and $70,000 payments for the next 10 years has a slightly higher present value: $753,218.12

9. $11,810.24   11. $n \approx 73.8$; 74 quarters

13. (a) The higher graph corresponds to 8%.

(b) $1500 (approximately)

(c) With an interest rate of 10%, a present value of about $9000 is needed to purchase an annuity of $1000 for 25 years. If the interest rate is 8%, about $10,500 is needed.

15. Ordinary annuity—payments at the end of each period
Annuity due—payments at the beginning of each period

17. $69,632.02   19. $445,962.23   21. $2145.59

23. $146,235.06   25. $22,663.74

27. (a) ordinary annuity   (b) $5441.23

29. (a) annuity due   (b) $316,803.61

31. (a) ordinary annuity   (b) $27,590.62

33. (a) $30,078.99   (b) $16,900   (c) $607.02
(d) $36,421.20

35. (a) $4504.83   (b) $n \approx 21.9$; 22 withdrawals

37. $7957.86   39. $74,993.20   41. $59,768.91

43. $1317.98   45. $85,804.29

47. (a) The spreadsheet shows the payments for the first 12 months and the last 12 months. Full payments for $13\frac{1}{2}$ years.

	A	B	C	D
1	End of Month	Acct. Value	Payment	New Balance
2	0	$100000.00	$0.00	$100000.00
3	1	$100650.00	$1000.00	$99650.00
4	2	$100297.73	$1000.00	$99297.73
5	3	$99943.16	$1000.00	$98943.16
6	4	$99586.29	$1000.00	$98586.29
7	5	$99227.10	$1000.00	$98227.10
8	6	$98865.58	$1000.00	$97865.58
9	7	$98501.70	$1000.00	$97501.70
10	8	$98135.47	$1000.00	$97135.47
11	9	$97766.85	$1000.00	$96766.85
12	10	$97395.83	$1000.00	$96395.83
13	11	$97022.40	$1000.00	$96022.40
14	12	$96646.55	$1000.00	$95646.55
⋮	⋮	⋮	⋮	⋮
154	152	$10684.71	$1000.00	$9684.71
155	153	$9747.66	$1000.00	$8747.66
156	154	$8804.52	$1000.00	$7804.52
157	155	$7855.25	$1000.00	$6855.25
158	156	$6899.81	$1000.00	$5899.81
159	157	$5938.16	$1000.00	$4938.16
160	158	$4970.25	$1000.00	$3970.25
161	159	$3996.06	$1000.00	$2996.06
162	160	$3015.53	$1000.00	$2015.53
163	161	$2028.64	$1000.00	$1028.64
164	162	$1035.32	$1000.00	$35.32
165	163	$35.55	$35.55	$0.00

(b) The spreadsheet shows the payments for the first 12 months and the last 12 months. Full payments for almost 4 years.

	A	B	C	D
1	End of Month	Acct. Value	Payment	New Balance
2	0	$100000.00	$0.00	$100000.00
3	1	$100650.00	$2500.00	$98150.00
4	2	$98787.98	$2500.00	$96287.98
5	3	$96913.85	$2500.00	$94413.85
6	4	$95027.54	$2500.00	$92527.54
7	5	$93128.97	$2500.00	$90628.97
8	6	$91218.05	$2500.00	$88718.05
9	7	$89294.72	$2500.00	$86794.72
10	8	$87358.89	$2500.00	$84858.89
11	9	$85410.47	$2500.00	$82910.47
12	10	$83449.39	$2500.00	$80949.39
13	11	$81475.56	$2500.00	$78975.56
14	12	$79488.90	$2500.00	$76988.90
⋮	⋮	⋮	⋮	⋮
38	36	$27734.95	$2500.00	$25234.95
39	37	$25398.98	$2500.00	$22898.98
40	38	$23047.83	$2500.00	$20547.83
41	39	$20681.39	$2500.00	$18181.39
42	40	$18299.57	$2500.00	$15799.57
43	41	$15902.26	$2500.00	$13402.26
44	42	$13489.38	$2500.00	$10989.38
45	43	$11060.81	$2500.00	$8560.81
46	44	$8616.45	$2500.00	$6116.45
47	45	$6156.21	$2500.00	$3656.21
48	46	$3679.98	$2500.00	$1179.98
49	47	$1187.65	$1187.65	$0.00

Exercise Set 6.5

1. (a) the 10-year loan, because the loan must be paid more quickly

(b) the 25-year loan, because the loan is paid more slowly

3. $1288.29

5. $553.42   7. $10,345.11

9.

Period	Payment	Interest
1	$39,505.50	$9000.00
2	39,505.50	6254.51
3	39,505.43	3261.92
	118,516.43	18,516.43

Period	Balance Reduction	Unpaid Balance
		$100,000.00
1	$30,505.50	69,494.50
2	33,250.99	36,243.51
3	36,243.51	0.00
	100,000.00	

11.

Period	Payment	Interest
1	$5380.54	$600.00
2	5380.54	456.58
3	5380.54	308.87
4	5380.54	156.71
	21,522.16	1,522.16

Period	Balance Reduction	Unpaid Balance
		$20,000.00
1	$4780.54	15,219.46
2	4923.96	10,295.50
3	5071.67	5,223.83
4	5223.83	0.00
	20,000.00	

13. $8852.05
15. $5785.83
17. (a) $8718.46
(b) $174,369.20 + $75,000 = $249,369.20
(c) $74,369.20
19. (a) $1237.78
(b) $9902.24 + $2000 = $11,902.24
(c) $1902.24
21. $8903.25
23. (a) $148,495.66   (b) $141,787.65
25. (a) $89,120.53   (b) $6451.45
27. (a) $368.43; $383.43   (b) $n \approx 57.1$ (c) $211.95
29. (a) The line is the total amount paid ($644.30 per month × the number of months). The curve is the total amount paid toward the principal.
(b)

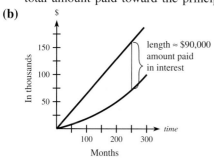

31.

	Rate	Payment	Total Interest
(a)	8%	$366.19	$2577.12
	8.5%	$369.72	$2746.56
(b)	6.75%	$518.88	$106,796.80
	7.25%	$545.74	$116,466.40

(c) The duration of the loan seems to have the greatest effect. It greatly influences payment size (for a $15,000 loan versus one for $80,000), and it also affects total interest paid.

33.

		Payment	Points	Total Paid
(a)	(i)	$738.99	–	$221,697
	(ii)	$722.81	$1000	$217,843
	(iii)	$706.78	$2000	$214,034

(b) The 7% loan with 2 points.
35. (a) $17,525.20   (b) $508.76
(c) $n \approx 33.2$; 34 quarters   (d) $471.57
37. The spreadsheet shows the amortization schedule for the first 12 and the last 12 payments.

	A	B	C	D	E
1	Period	Payment	Interest	Bal. Reduction	Unpaid Bal.
2	0				$16700.00
3	1	$409.27	$114.12	$295.15	$16404.85
4	2	$409.27	$112.10	$297.17	$16107.68
5	3	$409.27	$110.07	$299.20	$15808.48
6	4	$409.27	$108.02	$301.25	$15507.23
7	5	$409.27	$105.97	$303.30	$15203.93
8	6	$409.27	$103.89	$305.38	$14898.55
9	7	$409.27	$101.81	$307.46	$14591.09
10	8	$409.27	$99.71	$309.56	$14281.52
11	9	$409.27	$97.59	$311.68	$13969.84
12	10	$409.27	$95.46	$313.81	$13656.03
13	11	$409.27	$93.32	$315.95	$13340.08
14	12	$409.27	$91.16	$318.11	$13021.97
⋮	⋮	⋮	⋮	⋮	⋮
39	37	$409.27	$32.11	$377.16	$4322.49
40	38	$409.27	$29.54	$379.73	$3942.75
41	39	$409.27	$26.94	$382.33	$3560.43
42	40	$409.27	$24.33	$384.94	$3175.49
43	41	$409.27	$21.70	$387.57	$2787.91
44	42	$409.27	$19.05	$390.22	$2397.70
45	43	$409.27	$16.38	$392.89	$2004.81
46	44	$409.27	$13.70	$395.57	$1609.24
47	45	$409.27	$11.00	$398.27	$1210.97
48	46	$409.27	$8.27	$401.00	$809.97
49	47	$409.27	$5.53	$403.74	$406.24
50	48	$409.02	$2.78	$406.24	$0.00

Chapter 6 Review Exercises

1. $1, \frac{1}{4}, \frac{1}{9}, \frac{1}{16}$
2. Arithmetic: (a) and (c) (a) $d = -5$ (c) $d = \frac{1}{6}$
3. 235   **4.** 109   **5.** 315
6. Geometric: (a) and (b) (a) $r = 8$ (b) $r = -\frac{3}{4}$
7. 8   **8.** $2,391,484\frac{4}{9}$ **9.** $10,880   **10.** $6\frac{2}{3}\%$
11. $2941.18   **12.** $4650
13. the $20,000 job ($245,000 versus $234,000)
14. (a) 40   (b) 2% = 0.02
15. (a) $S = P(1 + i)^n$ (b) $S = Pe^{rt}$
16. (b) monthly   **17.** $372.79   **18.** $14,510.26
19. $1616.07   **20.** $21,299.21   **21.** 34.3 quarters
22. (a) 13.29%   (b) 14.21%
23. (a) 7.40%   (b) 7.47%
24. 2^{63} **25.** $2^{32} - 1$ **26.** $29,428.47
27. $6069.44   **28.** $31,194.18   **29.** $10,841.24
30. $n \approx 36$ quarters

31. $130,079.36   **32.** $12,007.09
33. (a) $11.828 million   (b) $161.5 million
34. $1726.85   **35.** $5390.77
36. $n \approx 16$ half-years (8 years)
37. $88.85   **38.** $3443.61   **39.** $114,597.40
40.

Payment Number	Payment Amount	Interest	Balance Reduction	Unpaid Balance
57	$699.22	$594.01	$105.21	$94,936.99
58	$699.22	$593.36	$105.86	$94,831.13

41. 16.32%   **42.** $213.81
43. (a) $1480   (b) $1601.03
44. $9319.64   **45.** 14.5 years   **46.** 79.4%
47. $21,474.08   **48.** $12,162.06
49. Quarterly APY = 6.68%. This rate is better for the bank; it pays less interest. Continuous APY = 6.69%. This rate is better for the consumer, who earns more interest.
50. $32,834.69   **51.** $3,466.64
52. (a) $592.76   (b) $177,828   (c) $85,828
   (d) $78,183.79
53. (a) $95,164.21   (b) $1300.14   **54.** $497.04
55. Future value of IRA = $86,485.66
   Present value needed = $1,160,760.05
   Future value needed from deposits = $1,074,274.39
   Deposits = $355.80
56. Regular payment = $64,337.43
   Unpaid balance = $2,365,237.24
   Number of $70,000 payments = $n \approx 46.1$
   Savings = $118,546.36

Chapter 6 Test

1. 25.3 years (approximately)   **2.** $840.75
3. 6.82%   **4.** $158,524.90   **5.** 33.53%
6. (a) $698.00   (b) $112,400
7. $2625   **8.** $7999.41   **9.** 8.73%
10. $119,912.92   **11.** $40,552.00
12. $32,488 (to the nearest dollar)   **13.** $6781.17
14. $n \approx 66.8$; 67 half-years
15. $1688.02
16. (a) $279,841.35   (b) $13,124.75
17. $23,381.82   **18.** $29,716.47
19. (a) The difference between successive terms is always -5.5.
   (b) 23.8   (c) 8226.3
20. 1000 mg (approximately)
21. (a) $145,585.54 (with the $2000);
      $147,585.54 (without the $2000)
   (b) $n \approx 318.8$ months
      Total interest = $243,738.13

(c) $n \approx 317.2$ months
   Total interest = $245,378.24
(d) Paying the $2000 is slightly better; it saves about $1640 in interest.

Exercise Set 7.1

1. (a) $\frac{2}{5}$ (b) 0   (c) 1   **3.** $\frac{1}{4}$ **5.** 1
7. (a) $\frac{3}{10}$ (b) $\frac{1}{2}$ (c) $\frac{1}{5}$ (d) $\frac{3}{5}$ (e) $\frac{7}{10}$
9. (a) $\frac{1}{13}$ (b) $\frac{1}{2}$ (c) $\frac{1}{4}$
11. {HH, HT, TH, TT};   (a) $\frac{1}{4}$ (b) $\frac{1}{2}$ (c) $\frac{1}{4}$
13. (a) $\frac{1}{12}$ (b) $\frac{1}{12}$ (c) $\frac{1}{36}$ **15.** (a) $\frac{1}{2}$ (b) $\frac{5}{12}$
17. (a) 431/1200
   (b) If fair, $\Pr(6) = \frac{1}{6}$; 431/1200 not close to $\frac{1}{6}$, so not a fair die
19. (a) 2:3   (b) 3:2   **21.** (a) $\frac{1}{21}$ (b) $\frac{20}{21}$
23. 0.46   **25.** (a) 63/425   (b) 32/425
27. (a) R: 0.63   D: 0.41   I: 0.51
   (b) Republican
29. (a) 1/3601   (b) 100/3601   (c) 3500/3601
   (d) 30%
31. (a) 0.193   (b) 0.316   (c) 0.089
33. $S = \{A+, A-, B+, B-, AB+, AB-, O+, O-\}$; No. Type O+ is the most frequently occurring blood type.
35. (a) 0.04   (b) 0.96   **37.** (a) 0.10   (b) 0.90
39. 0.03   **41.** 0.75   **43.** $\frac{1}{3}$ **45.** $\frac{1}{3}$
47. 0.22; yes, 0.39 is much higher than 0.22
49. $\frac{5}{13}$ **51.** $\frac{3}{8}$
53. (a) no   (b) {BB, BG, GB, GG}   (c) $\frac{1}{2}$
55. $\frac{1}{8}$ **57.** $\frac{3}{125}$
59. $\Pr(A) = 0.000019554$, or about 1.9 accidents per 100,000
   $\Pr(B) = 0.000035919$, or about 3.6 accidents per 100,000
   $\Pr(C) = 0.000037679$, or about 3.8 accidents per 100,000
   Intersection C is the most dangerous.
61. (a) 557/1200   (b) 11/120
63. (a) boy: 1/5; girl: 4/5
   (b) boy: 0.4946; girl: 0.5054   (c) part (b)

Exercise Set 7.2

1. $\frac{1}{6}$ **3.** $\frac{2}{3}$ **5.** $\frac{2}{5}$ **7.** (a) $\frac{1}{7}$ (b) $\frac{5}{7}$
9. $\frac{3}{4}$ **11.** $\frac{2}{3}$ **13.** $\frac{10}{17}$ **15.** $\frac{2}{3}$
17. (a) $\frac{1}{2}$ (b) $\frac{1}{3}$ (c) $\frac{8}{9}$ (d) $\frac{1}{9}$
19. 0.54   **21.** (a) 362/425   (b) $\frac{66}{85}$
23. $\frac{17}{50}$ **25.** (a) $\frac{5}{6}$ (b) $\frac{1}{6}$
27. (a) 0.35   (b) 0.08   (c) 0.83
29. (a) 0.336   (b) 0.261   (c) 0.295
31. (a) 0.169   (b) 0.248   (c) 0.380   (d) 0.912
33. (a) $\frac{11}{12}$ (b) $\frac{5}{6}$ **35.** (a) $\frac{1}{2}$ (b) $\frac{7}{8}$ (c) $\frac{3}{4}$
37. 0.56   **39.** 0.965

41. (a) 0.72   (b) 0.84   (c) 0.61
43. $\frac{31}{40}$ **45.** 0.13

Exercise Set 7.3

1. (a) $\frac{1}{2}$ (b) $\frac{1}{13}$ **3.** (a) $\frac{1}{3}$ (b) $\frac{1}{3}$ **5.** $\frac{4}{7}$
7. (a) $\frac{2}{3}$ (b) $\frac{4}{9}$ (c) $\frac{3}{5}$ **9.** (a) $\frac{1}{4}$ (b) $\frac{1}{2}$
11. $\frac{1}{36}$ **13.** (a) $\frac{1}{8}$ (b) $\frac{7}{8}$
15. (a) $\frac{3}{50}$ (b) $\frac{1}{15}$
   (c) The events in part (a) are independent because the result of the first draw does not affect the probability for the second draw.
17. (a) $\frac{4}{25}$ (b) $\frac{9}{25}$ (c) $\frac{6}{25}$ (d) 0   **19.** $\frac{5}{68}$
21. (a) $\frac{1}{5}$ (b) $\frac{3}{5}$ (c) 0   **23.** (a) $\frac{1}{17}$ (b) 13/204
25. (a) $\frac{13}{17}$ (b) $\frac{4}{17}$ (c) $\frac{8}{51}$ **27.** $\frac{31}{52}$ **29.** $\frac{25}{96}$
31. $\frac{43}{50}$ **33.** $\frac{65}{87}$ **35.** $35/435 = \frac{7}{87}$ **37.** $\frac{1}{10}$
39. 1/144,000,000   **41.** 0.004292   **43.** 0.06
45. 0.045   **47.** $(0.95)^5 = 0.774$ **49.** 0.06
51. (a) 0.366   (b) 0.634
53. (a) 0.4565   (b) 0.5435
55. (a) $\left(\frac{1}{3}\right)^3\left(\frac{1}{5}\right)^4 = 1/16{,}875$
   (b) $\left(\frac{2}{3}\right)^3\left(\frac{4}{5}\right)^4 = 2048/16{,}875$ (c) 14,827/16,875
57. 4/11; 4:7   **59.** (a) 364/365   (b) $\frac{1}{365}$
61. (a) 0.59   (b) 0.41

Exercise Set 7.4

1. $\frac{2}{5}$ **3.** $\frac{1}{325}$ **5.** (a) $\frac{2}{21}$ (b) $\frac{4}{21}$ (c) $\frac{23}{35}$
7. (a) $\frac{1}{30}$ (b) $\frac{1}{2}$ (c) $\frac{5}{6}$ **9.** $\frac{3}{5}$
11. (a) $\frac{6}{25}$ (b) $\frac{9}{25}$ (c) $\frac{12}{25}$ (d) $\frac{19}{25}$
13. $\frac{2}{3}$ **15.** $\frac{2}{3}$ **17.** 0.3095
19. (a) 81/10,000   (b) 1323/5000
21. (a) $\frac{6}{35}$ (b) $\frac{6}{35}$ (c) $\frac{12}{35}$
23. (a) $\frac{4}{7}$ (b) $\frac{5}{14}$ (c) $\frac{7}{10}$ (d) $\frac{16}{25}$ **25.** $\frac{17}{45}$
27. 0.079   **29.** (a) 49/100   (b) $\frac{12}{49}$

Exercise Set 7.5

1. 360   **3.** 151,200   **5.** 1
7. (a) $6 \cdot 5 \cdot 4 \cdot 3 = 360$ (b) $6^4 = 1296$ **9.** $n!$
11. $n + 1$ **13.** 16   **15.** 4950   **17.** 1   **19.** 1
21. 10   **23.** (a) 12   (b) 72   **25.** 604,800
27. 120   **29.** 24   **31.** 64   **33.** 720
35. $2^{10} = 1024$ **37.** $10! = 3{,}628{,}800$ **39.** 252
41. 30,045,015   **43.** 792   **45.** 210
47. 2,891,999,880   **49.** 3,700,000

Exercise Set 7.6

1. $\frac{1}{120}$ **3.** (a) 120   (b) $\frac{1}{120}$
5. 0.639   **7.** (a) 1/10,000   (b) 1/5040
9. $1/10^6$ **11.** 0.000048   **13.** $1/10! = 1/3{,}628{,}800$
15. (a) $\frac{1}{22}$ (b) $\frac{6}{11}$ (c) $\frac{9}{22}$
17. 0.098   **19.** $\dfrac{_{90}C_{28} \cdot {}_{10}C_2}{_{100}C_{30}}$

21. (a) 0.119   (b) 0.0476   (c) 0.476
23. 0.0238   **25.** (a) 0.721   (b) 0.262   (c) 0.279
27. (a) $\frac{1}{3}$ (b) $\frac{1}{6}$ **29.** $\dfrac{_{20}C_{10}}{_{80}C_{10}} = 0.00000011$
31. (a) 0.033   (b) 0.633
33. (a) 0.0005   (b) 0.002   **35.** 0.00748

Exercise Set 7.7

1. can   **3.** cannot, sum $\neq 1$
5. cannot, not square   **7.** can
9. [0.248   0.752]   **11.** [0.228   0.236   0.536]
13. $\left[\frac{1}{4} \quad \frac{3}{4}\right]$ **15.** $\left[\frac{1}{4} \quad \frac{1}{4} \quad \frac{1}{2}\right]$
17. [0.5   0.4   0.1]; [0.44   0.43   0.13];
   [0.431   0.43   0.139]; [0.4292   0.4291   0.1417]
19.
$$\begin{array}{c} R \\ N \end{array}\begin{bmatrix} 0.8 & 0.2 \\ 0.3 & 0.7 \end{bmatrix}$$
     $\begin{array}{cc} R & N \end{array}$ **21.** 0.45
23.
$$\begin{array}{c} A \\ F \\ V \end{array}\begin{bmatrix} 0 & 0.7 & 0.3 \\ 0.6 & 0 & 0.4 \\ 0.8 & 0.2 & 0 \end{bmatrix}$$
     $\begin{array}{ccc} A & F & V \end{array}$
25. [0.3928   0.37   0.2372]
27. [46/113   38/113   29/113]
29.
$$\begin{array}{c} r \\ u \end{array}\begin{bmatrix} 0.7 & 0.3 \\ 0.1 & 0.9 \end{bmatrix};\ \begin{bmatrix} 1/4 & 3/4 \end{bmatrix}$$
     $\begin{array}{cc} r & u \end{array}$
31. $\left[\frac{1}{14} \quad \frac{3}{14} \quad \frac{5}{7}\right]$ **33.** $\left[\frac{4}{7} \quad \frac{2}{7} \quad \frac{1}{7}\right]$
35. [49/100   42/100   9/100]

Chapter 7 Review Exercises

1. (a) $\frac{5}{9}$ (b) $\frac{1}{3}$ (c) $\frac{2}{9}$
2. (a) $\frac{3}{4}$ (b) $\frac{1}{2}$ (c) $\frac{1}{6}$ (d) $\frac{2}{3}$
3. (a) 3:4   (b) 4:3   **4.** (a) $\frac{1}{4}$ (b) $\frac{1}{2}$ (c) $\frac{1}{4}$
5. (a) $\frac{3}{8}$ (b) $\frac{1}{8}$ (c) $\frac{3}{8}$ **6.** $\frac{2}{13}$ **7.** 16/169
8. $\frac{3}{4}$ **9.** $\frac{2}{13}$ **10.** $\frac{7}{13}$ **11.** (a) $\frac{2}{9}$ (b) $\frac{2}{3}$ (c) $\frac{7}{9}$
12. $\frac{2}{7}$ **13.** $\frac{1}{2}$ **14.** 7/320
15. 7/342   **16.** 3/28   **17.** $\frac{8}{15}$
18. (a) $\frac{3}{14}$ (b) $\frac{4}{7}$ (c) $\frac{3}{8}$ **19.** 49/89
20. 30   **21.** 35   **22.** 26^3 **23.** 56
24. (a) Not square
   (b) The row sums are not 1.
25. [0.76   0.24], [0.496   0.504]
26. [0.2   0.8]
27. $\frac{5}{8}$ **28.** $\frac{1}{7}$ **29.** $\frac{29}{50}$
30. $\frac{5}{56}$ **31.** $\frac{33}{56}$ **32.** $\frac{15}{22}$ **33.** 0.72
34. (a) 63/2000   (b) $\frac{60}{63}$ **35.** 39/116   **36.** $4! = 24$
37. $_8P_4 = 1680$ **38.** $_{12}C_4 = 495$ **39.** $_8C_4 = 70$
40. (a) $_{12}C_2 = 66$ (b) $_{12}C_3 = 220$ **41.** 62,193,780
42. If her assumption about blood groups is accurate, there would be $4 \cdot 2 \cdot 4 \cdot 8 = 256$, not 288, unique groups.

43. $\frac{1}{24}$ **44.** $\frac{3}{500}$ **45.** $\frac{3}{1250}$
46. (a) 0.3398   (b) 0.1975   **47.** $\frac{1}{10}$
48. (a) $(_{10}C_5)(_2C_1)/_{12}C_6$

(b) $\dfrac{(_{10}C_5)(_2C_1) + (_{10}C_4)(_2C_2)}{_{12}C_6}$

49. [0.135   0.51   0.355], [0.09675   0.3305   0.57275], [0.0640875   0.288275   0.6476375]
50. [12/265   68/265   37/53]

Chapter 7 Test

1. (a) $\frac{4}{7}$ (b) $\frac{3}{7}$ **2.** (a) $\frac{2}{7}$ (b) $\frac{5}{7}$
3. (a) 0   (b) 1   **4.** $\frac{1}{7}$ **5.** $\frac{1}{7}$
6. (a) $\frac{2}{7}$ (b) $\frac{4}{7}$ **7.** $\frac{2}{7}$ **8.** $\frac{3}{7}$ **9.** $\frac{2}{3}$
10. 1/17,576   **11.** 0.2389   **12.** (a) $\frac{1}{5}$ (b) $\frac{1}{20}$
13. (a) $\frac{3}{95}$ (b) $\frac{6}{19}$ (c) $\frac{21}{38}$ (d) 0
14. (a) 5,245,786   (b) 1/5,245,786
15. (a) 2,118,760   (b) 1/2,118,760
16. 0.064   **17.** (a) 0.633   (b) 0.962   **18.** 0.229
19. (a) $\frac{1}{5}$ (b) $\frac{1}{14}$ (c) $\frac{13}{14}$ **20.** $\frac{3}{14}$

21. (a) 2^{10} (b) $\dfrac{1}{2^{10}}$ (c) $\frac{1}{3}$ (d) Change the code.

22. (a) $A = \begin{bmatrix} 0.80 & 0.20 \\ 0.07 & 0.93 \end{bmatrix}$ (b) $[0.25566 \quad 0.74434]$;

about 25.6%   (c) $\frac{7}{27}$; 25.9% of market

Exercise Set 8.1

1. 0.0595
3. (a) 1/6   (b) 5/6   (c) 18   (d) 0.045
5. (a) $\frac{1}{64}$ (b) $\frac{5}{16}$ (c) $\frac{15}{64}$ **7.** 0.0284
9. (a) 0.2304   (b) 0.0102   (c) 0.3174
11. 0.0585   **13.** 0.2759
15. (a) 0.375   (b) 0.0625
17. (a) 0.1157   (b) 0.4823
19. (a) $\frac{27}{64}$ (b) $\frac{27}{128}$ (c) $\frac{81}{256}$
21. (a) 0.0729   (b) 0.5905   (c) 0.9914
23. 0.2457   **25.** 0.0007
27. (a) 0.1323   (b) 0.0308
29. (a) 0.9044   (b) 0.0914   (c) 0.0043
31. (a) 0.8683   (b) 0.2099   **33.** 0.740

Exercise Set 8.2

1. 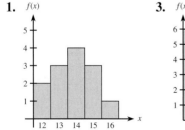   **3.**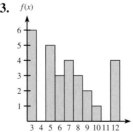

5. 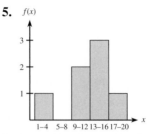   **7.**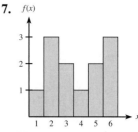

9. 3   **11.** 13   **13.** 2   **15.** 1
17. mode = 2, median = 4.5, mean = 6
19. mode = 17, median = 18.5, mean = 23.5
21. mode = 5.3, median = 5.3, mean = 5.32
23. 12.21, 14.5, 14.5   **25.** 9   **27.** 14
29. 4, 8.5714, 2.9277   **31.** 14, 4.6667, 2.1602
33. 2.73, 1.35   **35.** 6.75, 2.96
37. (a)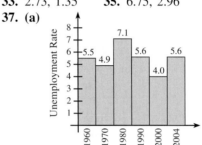

(b) $\bar{x} = 5.45$, $s = 1.02$
39. (a)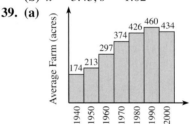

(b) $\bar{x} = 339.71$
41. The mean will give the highest measure.
43. The median will give the most representative average.
45. (a) 70.09%   (b) 8.79%
47.

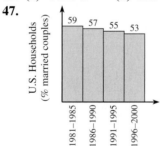

49. $\bar{x} = 3.32$ kg; $s = 0.677$ kg
51. (a) $60,000   (b) $36,000   (c) $32,000
53. (a) $23.325   (b) $5.139
55. (a) 7.88%   (b) 6.13%   (c) 7.00%   (d) 1.25%

57. (a)

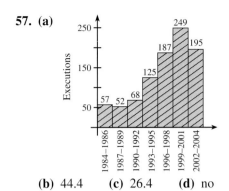

(b) 44.4   **(c)** 26.4   **(d)** no

Exercise Set 8.3

1. no; $\Pr(x) \not\geq 0$ **3.** yes   **5.** yes
7. no; $\Sigma\Pr(x) > 1$ **9.** $\frac{15}{8}$ **11.** 5
13. $\mu = \frac{13}{8}$, $\sigma^2 = 1.48$, $\sigma = 1.22$
15. $\mu = \frac{13}{3}$, $\sigma^2 = 2.22$, $\sigma = 1.49$
17. 3   **19.** 2
21. (a)

x	$\Pr(x)$
0	125/216
1	25/72
2	5/72
3	1/216

(b) $3\left(\frac{1}{6}\right) = \frac{1}{2}$ **(c)** $\sqrt{3\left(\frac{1}{6}\right)\left(\frac{5}{6}\right)} = \left(\frac{1}{6}\right)\sqrt{15}$
23. (a) 42   **(b)** 3.55
25. (a) 30   **(b)** 3.464
27. 125
29. $a^6 + 6a^5b + 15a^4b^2 + 20a^3b^3 + 15a^2b^4 + 6ab^5 + b^6$
31. $x^4 + 4x^3h + 6x^2h^2 + 4xh^3 + h^4$
33. 1.85
35. TV, 37,500; personal appearances, 35,300
37. $-$67.33
39. Expect to lose \$2 each time.
41. If he buys 0, his profit is 0. If he buys 100, his expected profit is $3(100)(0.25) + $3(50)(0.20) + $3(10)(0.55) + $(-1)50(0.20) + $(-1)90(0.55) = $62. If he buys 200, his expected profit is $3(180)(0.25) + $3(50)(0.20) + $3(10)(0.55) + (-$1)(20)(0.25) + (-$1)(150)(0.20) + (-$1)(190)(0.55) = $42. He should buy 100.
43. E(cost with policy) = $108, E(cost without policy) = $80; save \$28 by "taking the chance"
45. no; some pipes may be more than 0.01 in. from 2 in. even if average is 2 in.
47. (a) $100(0.10) = 10$ **(b)** $\sqrt{100(0.10)(0.90)} = 3$
49. (a) 60,000   **(b)** $\sqrt{24,000} = 154.919$
51. 59,690   **53. (a)** 4   **(b)** 1.79
55. 2, 1.41   **57.** 300

Exercise Set 8.4

1. 0.4641   **3.** 0.4641   **5.** 0.9153   **7.** 0.1070
9. 0.0166   **11.** 0.0227   **13.** 0.8849   **15.** 0.1915
17. 0.3944   **19.** 0.3830   **21.** 0.7745   **23.** 0.9773
25. 0.0668   **27. (a)** 0.3413   **(b)** 0.3944
29. 0.9876
31. (a) 0.4192   **(b)** 0.0227   **(c)** 0.0581   **(d)** 0.8965
33. (a) 0.0668   **(b)** 0.3085   **(c)** 0.3830
35. (a) 0.0475   **(b)** 0.2033   **(c)** 0.5934
37. (a) 0.0227   **(b)** 0.1587   **(c)** 0.8186

Chapter 8 Review Exercises

1. 0.0774   **2. (a)** 0.3545   **(b)** 0.5534
3. 0.407
4.

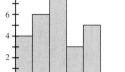

5. 3
6. $\frac{77}{26} \approx 2.96$
7. 3
8.

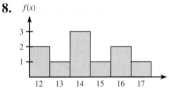

9. 14
10. 14
11. 14.3
12. $\bar{x} = 3.86$; $s^2 = 6.81$; $s = 2.61$
13. $\bar{x} = 2$; $s^2 = 2.44$; $s = 1.56$
14. 2.4   **15.** yes   **16.** no; $\Sigma\Pr(x) \neq 1$
17. yes   **18.** no; $\Pr(x) \not\geq 0$ **19.** 2
20. (a) 4.125   **(b)** 2.7344   **(c)** 1.654
21. (a) $\frac{37}{12}$ **(b)** 0.9097   **(c)** 0.9538
22. $\mu = 4$, $\sigma = (2\sqrt{3})/3$ **23.** 3
24. $x^5 + 5x^4y + 10x^3y^2 + 10x^2y^3 + 5xy^4 + y^5$
25. 0.9165   **26.** 0.1498   **27.** 0.1039   **28.** 0.3413
29. 0.6826   **30.** 0.1360   **31.** 0.297   **32.** 0.16308
33. 0.2048
34. (a) $(99,999/100,000)^{99,999} \approx 0.37$
   (b) $1 - (99,999/100,000)^{100,000} \approx 0.63$
35.

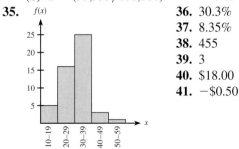

Class Marks

36. 30.3%
37. 8.35%
38. 455
39. 3
40. \$18.00
41. $-$0.50

42. (a) 1   **(b)** $\left(\frac{4}{5}\right)^4$
43. (a) 0.4773   **(b)** 0.1360   **(c)** 0.0227
44. 15%

Chapter 8 Test

1. (a) $\frac{40}{243}$ **(b)** $\frac{51}{243} = \frac{17}{81}$
2. (a) 4   **(b)** $\mu = 4, \sigma^2 = \frac{8}{3}, \sigma = \frac{2}{3}\sqrt{6}$
3. (i) For each $x, 0 \le \Pr(x) \le 1$ **(ii)** $\Sigma \Pr(x) = 1$
4. 5.1
5. $\mu = 16.7, \sigma^2 = 26.61, \sigma = 5.16$
6. $\bar{x} = 21.57$, median $= 21$, mode $= 21$
7. (a) 0.4706   **(b)** 0.8413   **(c)** 0.0669
8. (a) 0.3891   **(b)** 0.5418   **(c)** 0.1210
9.

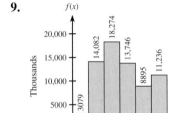

10. $\bar{x} = 46.48, s = 15.40$
11. (a) 35.98   **(b)** under 30; it would be lower
12. (a) 71.2   **(b)** It might be higher; cell use is spreading.
13. (a) 0.00003   **(b)** 30
14. 2 (1.8)   **15.** 5 (5.4)
16. 0 (0.054) with correct use
17. (a) 0.0158   **(b)** 0.0901   **(c)** 0.5383

Exercise Set 9.1

1. (a) -8 **(b)** -8
3. (a) 10   **(b)** does not exist
5. (a) 0   **(b)** -6
7. (a) does not exist $(+\infty)$ **(b)** does not exist $(+\infty)$
   (c) does not exist $(+\infty)$ **(d)** does not exist
9. (a) 3   **(b)** -6 **(c)** does not exist   **(d)** -6
11.

x	$f(x)$
0.9	-2.9
0.99	-2.99
0.999	-2.999
$\downarrow$	$\downarrow$
1	-3
$\uparrow$	$\uparrow$
1.001	-3.001
1.01	-3.01
1.1	-3.1

$\lim\limits_{x \to 1} f(x) = -3$

13.

x	$f(x)$
0.9	3.5
0.99	3.95
0.999	3.995
$\downarrow$	
1	4
$\uparrow$	5 } Different
1.001	4.995999
1.01	4.9599
1.1	4.59

$\lim\limits_{x \to 1} f(x)$ does not exist
15. -1 **17.** -4 **19.** -2 **21.** 6
23. 3/4   **25.** 0   **27.** does not exist
29. -3 **31.** does not exist   **33.** does not exist
35. $3x^2$ **37.** $\frac{1}{30}$ **39.** does not exist
41. -4 **43.** 9
45.

a	$(1 + a)^{1/a}$
0.1	2.5937
0.01	2.7048
0.001	2.7169
0.0001	2.7181
0.00001	2.71827
$\downarrow$	$\downarrow$
0	≈ 2.718

47. (a) 2   **(b)** 6   **(c)** -8 **(d)** $-\frac{1}{2}$
49. $150,000
51. (a) $32 (thousands)   **(b)** $55.04 (thousands)
53. (a) $2800   **(b)** $700   **(c)** $560
55. (a) 1.52 units/hr   **(b)** 0.85 units/hr   **(c)** lunch
57. (a) $-0; p \to 100^-$ means the water approaches not being treated (containing 100% or all of its impurities); the associated costs of nontreatment approach zero.   **(b)** ∞ **(c)** no, because $C(0)$ is undefined
59. (a) $4000   **(b)** $4000   **(c)** $4000
61. $C(x) = \begin{cases} 3.59 + 7.98x & \text{if } 0 \le x \le 10 \\ 83.39 + 6.78(x - 10) & \text{if } 10 < x \le 120 \\ 829.19 + 5.43(x - 120) & \text{if } x > 120 \end{cases}$
$\lim\limits_{x \to 10} C(x) = 83.39$
63. 10,216.54. This corresponds to the Dow Jones opening average on 10/5/2004.
65. (a) -0.21 (approximately)
   (b) This predicts the percent of U.S. workers in farm occupations as the year approaches 2010.
   (c) The model is inaccurate because the percent must be 0 or positive; it cannot be negative.

Exercise Set 9.2

1. (a) continuous
   (b) discontinuous; $f(1)$ does not exist
   (c) discontinuous; $\lim\limits_{x \to 3} f(x)$ does not exist
   (d) discontinuous; $f(0)$ does not exist and $\lim\limits_{x \to 0} f(x)$ does not exist

3. continuous

5. discontinuous; $f(-3)$ does not exist

7. discontinuous; $\lim_{x \to 2} f(x)$ does not exist

9. continuous

11. discontinuity at $x = -2$; $g(-2)$ and $\lim_{x \to -2} g(x)$ do not exist

13. continuous   **15.** continuous

17. discontinuity at $x = -1$; $f(-1)$ does not exist

19. discontinuity at $x = 3$; $\lim_{x \to 3} f(x)$ does not exist

21. vertical asymptote:
$x = -2$; $\lim_{x \to +\infty} f(x) = 0$; $\lim_{x \to -\infty} f(x) = 0$; $y = 0$

23. vertical asymptotes:
$x = -2, x = 3$; $\lim_{x \to +\infty} f(x) = 2$; $\lim_{x \to -\infty} f(x) = 2$; $y = 2$

25. (a) 0   **(b)** $y = 0$ is a horizontal asymptote

27. (a) 1   **(b)** $y = 1$ is a horizontal asymptote

29. (a) 5/3   **(c)** $y = 5/3$ is a horizontal asymptote

31. (a) does not exist $(+\infty)$
(b) No horizontal asymptotes

33. (a)

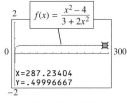

$$\lim_{x \to +\infty} f(x) = 0.5$$

(b) The table indicates $\lim_{x \to +\infty} f(x) = 0.5$.

35. (a) $x = -1000$ **(b)** 1000
(c) These values are so large that experimenting with windows may never locate them.

37. $\lim_{x \to \infty} \dfrac{a_n + \dfrac{a_{n-1}}{x} + \cdots + \dfrac{a_1}{x^{n-1}} + \dfrac{a_0}{x^n}}{b_n + \dfrac{b_{n-1}}{x} + \cdots + \dfrac{b_1}{x^{n-1}} + \dfrac{b_0}{x^n}}$
$= \dfrac{a_n + 0 + \cdots + 0 + 0}{b_n + 0 + \cdots + 0 + 0} = \dfrac{a_n}{b_n}$

39. (a) no, not at $p = -8$ **(b)** yes   **(c)** yes
(d) $p > 0$

41. (a) yes, $q = -1$ **(b)** yes

43. (a) R/i **(b)** \$10,000   **45.** yes, $0 \le p \le 100$

47. 100%; No, for p to approach 100% (as a limit) requires spending to increase without bound, which is impossible.

49. $R(x)$ is discontinuous at $x = 14{,}300$; $x = 58{,}100$; $x = 117{,}250$; $x = 178{,}650$; and $x = 319{,}100$.

51. (a) \$79.40   **(b)** $\lim_{x \to 100} C(x) = 19.40$;
$\lim_{x \to 500} C(x) = 49.40$ **(c)** yes

53. (a) $d(t) = -0.000019704t^4 + 0.0053192t^3 - 0.51299t^2 + 20.958t - 295.25$
(b) $d(109) \approx 1.48$; 1.48% of federal expenditures devoted to payment of interest in 2009

(c) $d(t) \to -\infty$ as $t \to +\infty$

(d) No, negative values for $d(t)$ do not make sense.

(e) The model is invalid when $d(t) < 0$—that is, for years before 1930 and for 2010 and beyond.

Exercise Set 9.3

1. (a) 6   **(b)** 8   **3. (a)** $\frac{10}{3}$ **(b)** -5

5. (a) $-3.9, -3.99$ **(b)** $-4.1, -4.01$ **(c)** -4

7. (a) 32   **(b)** 32   **(c)** $(4, 64)$

9. (a) verification   **(b)** -5 **(c)** -5 **(d)** $(-1, 3)$

11. (a) $P(1, 1), A(3, 0)$ **(b)** $-\frac{1}{2}$ **(c)** $-\frac{1}{2}$ **(d)** $-\frac{1}{2}$

13. (a) $P(1, 3), A(0, 3)$ **(b)** 0   **(c)** 0   **(d)** 0

15. (a) $f'(x) = 8x - 2$ **(b)** $8x - 2; -26$ **(c)** -26

17. (a) $p'(q) = 2q + 4$ **(b)** $2q + 4; 14$ **(c)** 14

19. (a) 89.000024   **(b)** 89.0072   **(c)** ≈ 89

21. (a) 294.000008   **(b)** 294.0084   **(c)** ≈ 294

23. -31

25. (a) At A the slope is positive; at B it is negative.
(b) $-1/3$

27. $f'(-3) = \frac{5}{2}; f(-3) = -9$ **29.** $y = 3x - 4$

31. (a) a, b, d **(b)** c **(c)** A, C, E

33. (a) A, B, C, D **(b)** A, D

35. (a) $f'(x) = 2x + 1$ **(b)** $f'(2) = 5$
(c) $y = 5x - 4$ **(d)**

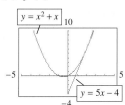

37. (a) $f'(x) = 3x^2$ **(b)** $f'(1) = 3$ **(c)** $y = 3x + 1$
(d)

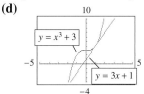

39. (a) 43 dollars per unit   **(b)** 95.50 dollars per unit
(c) The average cost per printer when 100 to 300 are produced (a) is \$43 per printer, and the average cost when 300 to 600 are produced (b) is \$95.50 per printer.

41. (a) $-100/3$ **(b)** $-4/3$

43. AB, AC, BC. Average rate is found from the slope of a segment; AB rises most slowly; BC is steepest.

45. (a) $R'(x) = \overline{MR} = 300 - 2x$
(b) 200; the predicted change in revenue from selling the 51st unit is about \$200.
(c) -100; the predicted change in revenue from the 201st unit is about -100 dollars.   **(d)** 0
(e) It changes from increasing to decreasing.

47. 200

49. (a) 100; the expected profit from the sale of the 201st car is $100.
  (b) −100; the expected profit from the sale of the 301st car is a loss of $100.
51. (a) 1.039
  (b) If humidity changes by 1%, the heat index will change by about 1.039°F.
53. (a) Marginal revenue is given by the slope of the tangent line, which is steeper at 300 cell phones. Hence marginal revenue is greater for 300 cell phones.
  (b) Marginal revenue predicts the revenue from the next unit sold. Hence, the 301st item brings in more revenue because the marginal revenue for 300 cell phones is greater than for 700.

Exercise Set 9.4

1. $y' = 0$ **3.** $y' = 1$ **5.** $f'(x) = 6x^2 - 5x^4$
7. $y' = 24x^3 - 10x + 1$
9. $g'(x) = 90x^8 - 25x^4 + 21x^2 + 5$
11. (a) 19   **(b)** 19   **13. (a)** 0   **(b)** 0
15. $y' = -5x^{-6} - 8x^{-9} = \frac{-5}{x^6} - \frac{8}{x^9}$
17. $y' = 11x^{8/3} - \frac{7}{2}x^{3/4} - \frac{1}{2}x^{-1/2}$

$$= 11\sqrt[3]{x^8} - \frac{7}{2}\sqrt[4]{x^3} - \frac{1}{2\sqrt{x}}$$

19. $f'(x) = -4x^{-9/5} - \frac{8}{3}x^{-7/3}$

$$= \frac{-4}{\sqrt[5]{x^9}} - \frac{8}{3\sqrt[3]{x^7}}$$

21. $g'(x) = -\frac{12}{x^5} - \frac{10}{x^6} + \frac{5}{3\sqrt[3]{x^2}}$

23. $y = -3x + 6$ **25.** $y = 3$
27. $(1, -1), (5, 31)$ **29.** $(0, 9), (3, -18)$
31. (a) $-1/2$ **(b)** -0.5000 (to four decimal places)
33. (a) $f'(x) = 6x^2 + 5$
  (b)

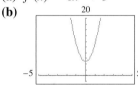

Graph of $f'(x)$ and numerical derivative of $f(x)$
35. (a) $h'(x) = -30x^{-4} + 4x^{-7/5} + 2x$

$$= \frac{-30}{x^4} + \frac{4}{\sqrt[5]{x^7}} + 2x$$

  (b)

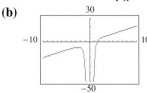

Graph of $h'(x)$ and numerical derivative of $h(x)$

37. (a) $y = 8x - 3$
  (b)

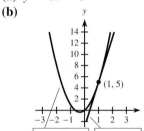

  (c) $x: 0.7 \rightarrow 1.6$; $y: 3.0 \rightarrow 7.9$

39. (a) $f'(x) = -2 - 2x$
  (b)

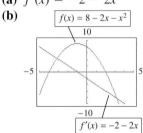

  (c) $f'(x) = 0$ at $x = -1$; $f'(x) > 0$ for $x < -1$; $f'(x) < 0$ for $x > -1$
  (d) $f(x)$ has a maximum when $x = -1$
    $f(x)$ rises for $x < -1$
    $f(x)$ falls for $x > -1$
41. (a) $f'(x) = 3x^2 - 12$
  (b)

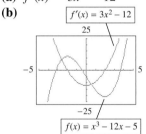

  (c) $f'(x) = 0$ at $x = -2$ and $x = 2$
    $f'(x) > 0$ for $x < -2$ and $x > 2$
    $f'(x) < 0$ for $-2 < x < 2$
  (d) $f(x)$ has a maximum when $x = -2$, a minimum when $x = 2$
    $f(x)$ rises when $x < -2$ and when $x > 2$
    $f(x)$ falls when $-2 < x < 2$
43. (a) 40; the expected change in revenue from the 301st unit is about $40
  (b) −20; the expected change in revenue from the 601st unit is about −20 dollars
45. (a) 920   **(b)** 926
47. (a) −4; if the price changes to $26, the quantity demanded will change by approximately −4 units
  (b) $-\frac{1}{2}$; if the price changes to $101, the quantity demanded will change by approximately $-\frac{1}{2}$ unit
49. (a) $\overline{C(x)}' = (-4000/x^2) + 0.1$ **(b)** 200
  (c) $C'(200) = \overline{C}(200) = 95$

51. (a) $-120,000$
 (b) If the impurities change from 1% to 2%, then the expected change in cost is $-120,000$ (dollars).
53. (a) $WC = 55.174 - 2.15s - 23.18\sqrt{s}$
 (b) -4.468
 (c) If the wind speed changes by $+1$ mph (to 26 mph), then the wind chill will change by approximately $-4.468°F$.
55. (a) $S(x) = 0.0003488x^{4.2077}$
 (b) 12.9734 million subscriberships per year
 (c) $S'(x) = 0.001468x^{3.2077}$
   $S'(22) \approx 29.704$ means that for the next year (2003), the number of subscriberships will change by about 29.704 million.
57. (a) $P(t) = -0.00025829t^3 + 0.035245t^2 + 1.42004t + 184.834$
 (b) $P'(t) = -0.00077487t^2 + 0.070490t + 1.42004$
 (c) 2000: $P'(40) \approx 3.00$ means that for 2001, the U.S. population will rise by about 3.00 million people.
   2025: $P'(65) \approx 2.73$ means that for 2026, the U.S. population is expected to rise by about 2.73 million people.

Exercise Set 9.5

1. $y' = 3x^2 + 2x - 6$
3. $f'(x) = (x^{12} + 3x^4 + 4)(12x^2) + (4x^3 - 1)(12x^{11} + 12x^3) = 60x^{14} - 12x^{11} + 84x^6 - 12x^3 + 48x^2$
5. $y' = (7x^6 - 5x^4 + 2x^2 - 1)(36x^8 + 21x^6 - 10x + 3) + (4x^9 + 3x^7 - 5x^2 + 3x)(42x^5 - 20x^3 + 4x)$
7. $y' = (x^2 + x + 1)(\frac{1}{3}x^{-2/3} - x^{-1/2}) + (x^{1/3} - 2x^{1/2} + 5)(2x + 1)$
9. (a) 40 **(b)** 40
11. $\dfrac{dp}{dq} = (q^2 - 4q - 1)/(q - 2)^2$
13. $\dfrac{dy}{dx} = \dfrac{4x^5 - 4x^3 - 16x}{(x^4 - 2x^2 + 5)^2}$
15. $\dfrac{dz}{dx} = 2x + \dfrac{2x - x^2}{(1 - x - 2x^2)^2}$
17. $\dfrac{dp}{dq} = \dfrac{2q + 1}{\sqrt[3]{q^2}(1 - q)^2}$ **19.** $y' = \dfrac{2x^3 - 6x^2 - 8}{(x - 2)^2}$
21. (a) $\frac{3}{5}$ **(b)** $\frac{3}{5}$ **23.** $y = 44x - 32$
25. $y = \frac{10}{3}x - \frac{10}{3}$ **27.** 104
29. 1.3333 (to four decimal places)
31. (a) $f'(x) = 3x^2 - 6x - 24$

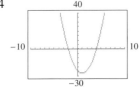

 Graph of both $f'(x)$ and numerical derivative of $f(x)$

(b) Horizontal tangents where $f'(x) = 0$; at $x = -2$ and $x = 4$
(c)

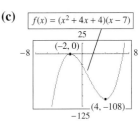

33. (a) $y' = \dfrac{x^2 - 4x}{(x - 2)^2}$

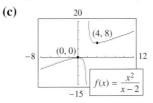

 Graph of both y' and the numerical derivative of y
 (b) Horizontal tangents where $y' = 0$; at $x = 0$ and $x = 4$
 (c)

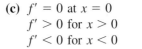

35. (a) $f'(x) = \dfrac{20x}{(x^2 + 1)^2}$ **(b)**

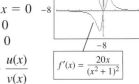

 (c) $f' = 0$ at $x = 0$
   $f' > 0$ for $x > 0$
   $f' < 0$ for $x < 0$
 (d) f has a minimum at $x = 0$
   f increasing for $x > 0$
   f decreasing for $x < 0$

37. $f'(x) = \lim\limits_{h \to 0} \dfrac{\dfrac{u(x + h)}{v(x + h)} - \dfrac{u(x)}{v(x)}}{h}$

$= \lim\limits_{h \to 0} \dfrac{u(x + h)v(x) - u(x)v(x + h)}{h \cdot v(x)v(x + h)}$

$= \lim\limits_{h \to 0} \dfrac{u(x + h)v(x) - u(x)v(x) + u(x)v(x) - u(x)v(x + h)}{h \cdot v(x)v(x + h)}$

$= \lim\limits_{h \to 0} \dfrac{v(x)\left[\dfrac{u(x + h) - u(x)}{h}\right] - u(x)\left[\dfrac{v(x + h) - v(x)}{h}\right]}{v(x)v(x + h)}$

$= \dfrac{v(x)u'(x) - u(x)v'(x)}{[v(x)]^2}$

39. $C'(p) = 810,000/(100 - p)^2$
41. $R'(49) \approx 30.00$ The expected revenue from the sale of the next unit (the 50th) is about \$30.00.
43. $R'(5) = -50$ As the group changes by 1 person (to 31), the revenue will drop by about \$50.
45. $S = 1000x - x^2$
47. $\dfrac{dR}{dn} = \dfrac{r(1 - r)}{[1 + (n - 1)r]^2}$

49. (a) $P'(6) \approx 0.045$ During the next (7th) month of the campaign, the proportion of voters who recognize the candidate will change by about 0.045, or 4.5%.

  (b) $P'(12) \approx -0.010$ During the next (13th) month of the campaign, the proportion of voters who recognize the candidate will drop by about 0.010, or 1%.

  (c) It is better for $P'(t)$ to be positive—that is, to have increasing recognition.

51. (a) $f'(20) \approx -0.79$

  (b) At 0°F, if the wind speed changes by 1 mph (to 21 mph), the wind chill will change by about -0.79°F.

53. (a) $B'(t) = (0.01t + 3)(0.04766t - 9.79)$ $+ (0.01)(0.02383t^2 - 9.79t + 3097.19)$

  (b) $B'(60) = 1.00634$ means that in 2010 the number of beneficiaries will be changing at the rate of 1.00634 million per year.

  (c) [2000, 2010] : 0.85,
  [2010, 2020] : 1.55,
  [2000, 2020] : 1.2
  The average rate over [2000, 2010] is best but is still off by almost 0.15 million per year.

55. (a) $f'(t) \approx 1000\left[\dfrac{9.6373t^2 - 4019.7t + 29,906.6}{(1.2882t^2 - 122.18t + 21,483)^2}\right]$

  (b) 1870: $f'(70) \approx -0.55$; 2000: $f'(200) \approx -0.16$

  (c) $f'(70)$ means that from 1870 to 1871, the model predicts a change of about -0.55% in U.S. workers in farm occupations. $f'(200)$ means that from 2000 to 2001, the model predicts a change of about -0.16% in U.S. workers in farm occupations.

Exercise Set 9.6

1. $\dfrac{dy}{du} = 3u^2, \dfrac{du}{dx} = 2x, \dfrac{dy}{dx} = 3u^2 \cdot 2x = 6x(x^2 + 1)^2$

3. $\dfrac{dy}{du} = 4u^3, \dfrac{du}{dx} = 8x - 1, \dfrac{dy}{dx} = 4u^3(8x - 1)$
  $= 4(8x - 1)(4x^2 - x + 8)^3$

5. $f'(x) = 25(2x^4 - 5)^{24}(8x^3) = 200x^3(2x^4 - 5)^{24}$

7. $h'(x) = \frac{16}{3}(x^6 + 3x^2 - 11)^7(6x^5 + 6x)$
  $= 32(x^5 + x)(x^6 + 3x^2 - 11)^7$

9. $g'(x) = -2(x^2 + 4x)^{-3}(2x + 4) = \dfrac{-4(x + 2)}{(x^2 + 4x)^3}$

11. $f'(x) = -3(x^2 + 2)^{-4}(2x) = \dfrac{-6x}{(x^2 + 2)^4}$

13. $g'(x) = -\dfrac{3}{4}(2x^3 + 3x + 5)^{-7/4}(6x^2 + 3)$
  $= \dfrac{-3(6x^2 + 3)}{4(2x^3 + 3x + 5)^{7/4}}$

15. $y' = \dfrac{1}{2}(x^2 + 4x + 5)^{-1/2}(2x + 4)$
  $= \dfrac{x + 2}{\sqrt{x^2 + 4x + 5}}$

17. $y' = 8(x^2 - 3)^4(2x) = 16x(x^2 - 3)^4$

19. $y' = \dfrac{15(3x + 1)^4 - 3}{7}$

21. (a) and **(b)** 96,768

23. (a) and **(b)** 2   **25.** $y = -3x + 4$

27. $9x - 5y = 2$

29. (a) $f'(x) = 6x(x^2 - 4)^2$

  (b)
  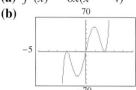

  (c) $x = 0, x = 2, x = -2$

  (d) $(0, -52), (2, 12), (-2, 12)$

  Graph of both $f'(x)$ and numerical derivative of $f(x)$

  (e)
  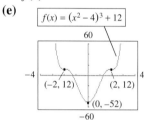

31. (a) $f'(x) = 8x(1 - x^2)^{1/3}$

  (b)
  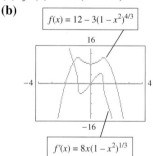

  (c) $f'(x) = 0$ at $x = -1, x = 0, x = 1$
  $f'(x) > 0$ for $x < -1$ and $0 < x < 1$
  $f'(x) < 0$ for $-1 < x < 0$ and $x > 1$

  (d) $f(x)$ has a maximum at $x = -1$ and $x = 1$, a minimum at $x = 0$
  $f(x)$ increasing for $x < -1$ and $0 < x < 1$
  $f(x)$ decreasing for $-1 < x < 0$ and $x > 1$

33. (a) $y' = 2x^2$ **(b)** $y' = -2/x^4$

  (c) $y' = 2(2x)^2$ **(d)** $y' = \dfrac{-18}{(3x)^4}$

35. 120 in./sec

37. $1499.85 (approximately); if a 101st unit is sold, revenue will change by about $1499.85

39. (a) -0.114 (approximately)

(b) If the price changes by \$1, to \$22, the weekly sales volume will change by approximately -0.114 thousand unit.

41. (a) $-\$3.20$ per unit

(b) If the quantity demanded changes from 49 to 50 units, the change in price will be about $-\$3.20$.

43. $\dfrac{dy}{dx} = \left(\dfrac{8k}{5}\right)(x - x_0)^{3/5}$

45. $\dfrac{dp}{dq} = -100(2q + 1)^{-3/2} = \dfrac{-100}{(2q + 1)^{3/2}}$

47. $\dfrac{dK_c}{dv} = 8(4v + 1)^{-1/2} = \dfrac{8}{\sqrt{4v + 1}}$

49. (a) \$658.75. If the interest changed from 6% to 7%, the amount of the investment would change by about \$658.75.

(b) \$2156.94. If the interest rate changed from 12% to 13%, the amount of the investment would change by about \$2156.94.

51. (a) 1980: $d'(50) \approx 0.68$; 2000: $d'(70) \approx -0.84$. These rates mean that the model predicts that the percent of federal expenditures devoted to payment of interest on the public debt will change by about 0.68% from 1980 to 1981 and by about -0.84% from 2000 to 2001.

(b) average rate 1975 to 1980: 0.58% is the best approximation

53. (a) 1970: $p'(10) \approx -0.41$; 1998: $p'(38) \approx -0.51$. These rates mean that the model predicts that the number of persons living below the poverty level is expected to change by about -0.41 million from 1970 to 1971 and by about -0.51 million from 1998 to 1999.

(b) 0.5133   **(c)** 1980; $p'(20) \approx 0.582$

Exercise Set 9.7

1. 0   **3.** $4(-4x^{-5})$; $-16/x^5$

5. $15x^2 + 4(-x^{-2})$; $15x^2 - 4/x^2$

7. $(x^2 - 2)1 + (x + 4)(2x)$; $3x^2 + 8x - 2$

9. $\dfrac{x^2(3x^2) - (x^3 + 1)(2x)}{(x^2)^2}$; $(x^3 - 2)/x^3$

11. $(3x^2 - 4)(x^3 - 4x)^9$

13. $\frac{5}{3}x^3[3(4x^5 - 5)^2(20x^4)] + (4x^5 - 5)^3(5x^2)$; $5x^2(4x^5 - 5)^2(24x^5 - 5)$

15. $(x - 1)^2(2x) + (x^2 + 1)2(x - 1)$; $2(x - 1)(2x^2 - x + 1)$

17. $\dfrac{(x^2 + 1)3(x^2 - 4)^2(2x) - (x^2 - 4)^3(2x)}{(x^2 + 1)^2}$;

$\dfrac{2x(x^2 - 4)^2(2x^2 + 7)}{(x^2 + 1)^2}$

19. $3[(q + 1)(q^3 - 3)]^2[(q + 1)3q^2 + (q^3 - 3)1]$; $3(4q^3 + 3q^2 - 3)[(q + 1)(q^3 - 3)]^2$

21. $4[x^2(x^2 + 3x)]^3[x^2(2x + 3) + (x^2 + 3x)(2x)]$; $4x^2(4x + 9)[x^2(x^2 + 3x)]^3$

23. $4\left(\dfrac{2x - 1}{x^2 + x}\right)^3\left[\dfrac{(x^2 + x)2 - (2x - 1)(2x + 1)}{(x^2 + x)^2}\right]$;

$\dfrac{4(-2x^2 + 2x + 1)(2x - 1)^3}{(x^2 + x)^5}$

25. $(8x^4 + 3)^2 3(x^3 - 4x)^2(3x^2 - 4) + (x^3 - 4x)^3 2(8x^4 + 3)(32x^3)$; $(8x^4 + 3)(x^3 - 4x)^2(136x^6 - 352x^4 + 27x^2 - 36)$

27. $\dfrac{(4 - x^2)\frac{1}{3}(x^2 + 5)^{-2/3}(2x) - (x^2 + 5)^{1/3}(-2x)}{(4 - x^2)^2}$;

$\dfrac{2x(2x^2 + 19)}{3\sqrt[3]{(x^2 + 5)^2}(4 - x^2)^2}$

29. $(x^2)\frac{1}{4}(4x - 3)^{-3/4}(4) + (4x - 3)^{1/4}(2x)$; $(9x^2 - 6x)/\sqrt[4]{(4x - 3)^3}$

31. $(2x)\frac{1}{2}(x^3 + 1)^{-1/2}(3x^2) + (x^3 + 1)^{1/2}(2)$; $(5x^3 + 2)/\sqrt{x^3 + 1}$

33. (a) $F_1'(x) = 12x^3(x^4 + 1)^4$

(b) $F_2'(x) = \dfrac{-12x^3}{(x^4 + 1)^6}$

(c) $F_3'(x) = 12x^3(3x^4 + 1)^4$

(d) $F_4'(x) = \dfrac{-300x^3}{(5x^4 + 1)^6}$

35. $dP/dx = 90(3x + 1)^2$

37. (a) \$59,900

(b) An 11th camper sold would change revenue by about \$59,900.

39. $dC/dy = 1/\sqrt{y + 1} + 0.4$

41. $dV/dx = 144 - 96x + 12x^2$

43. -1.6 This means that from the 9th to the 10th week, sales are expected to change by -1600 dollars (decrease).

45. (a)

$f'(t) \approx 1000\left[\dfrac{9.6373(t + 20)^2 - 4019.7(t + 20) + 29,906.6}{(1.2882(t + 20)^2 - 122.18(t + 20) + 21,483)^2}\right]$

(b) 1850: $f'(30) \approx -0.425$; 1950: $f'(130) \approx -0.345$

(c) These rates mean that the percent of U.S. workers in farm occupations is predicted to change by about -0.425% from 1850 to 1851 and by about -0.345% from 1950 to 1951.

Exercise Set 9.8

1. $180x^8 - 360x^3 - 72x$ **3.** $6x - 2x^{-3}$

5. $6x + \frac{1}{4}x^{-3/2}$ **7.** $60x^2 - 96$ **9.** $1008x^6 - 720x^3$

11. $-6/x^4$ **13.** $20x^3 + \frac{1}{4}x^{-3/2}$ **15.** $\frac{3}{8}(x + 1)^{-5/2}$

17. 0   **19.** $-15/(16x^{7/2})$ **21.** $24(4x - 1)^{-5/2}$

23. $-2(x + 1)^{-3}$ **25.** 26

27. 16.0000 (to four decimal places)

29. 0.0004261

31. (a) $f'(x) = 3x^2 - 6x$ $f''(x) = 6x - 6$

(b)
$\boxed{f'(x) = 3x^2 - 6x}$

$\boxed{f''(x) = 6x - 6}$
15

−3 ⎪ 5

−10
$\boxed{f(x) = x^3 - 3x^2 + 5}$

(c) $f''(x) = 0$ at $x = 1$
$f''(x) > 0$ for $x > 1$
$f''(x) < 0$ for $x < 1$

(d) $f'(x)$ has a minimum at $x = 1$
$f'(x)$ is increasing for $x > 1$
$f'(x)$ is decreasing for $x < 1$

(e) $f''(x) < 0$ **(f)** $f''(x) > 0$

33. (a) $f'(x) = -x^2 - 2x + 3$ $f''(x) = -2x - 2$

(b)

$\boxed{f(x) = -\frac{1}{3}x^3 - x^2 + 3x + 7}$
15

−5 ⎪ 5

−15
$\boxed{f''(x) = -2x - 2}$
$\boxed{f'(x) = -x^2 - 2x + 3}$

(c) $f''(x) = 0$ at $x = -1$
$f''(x) > 0$ for $x < -1$
$f''(x) < 0$ for $x > -1$

(d) $f'(x)$ has a maximum at $x = -1$
$f'(x)$ is increasing for $x < -1$
$f'(x)$ is decreasing for $x > -1$

(e) $f''(x) < 0$ **(f)** $f''(x) > 0$

35. $a = 0.12$ m/sec^2 **37.** -0.02 \$/unit per unit

39. (a) $\dfrac{dR}{dm} = mc - m^2$ **(b)** $\dfrac{d^2R}{dm^2} = c - 2m$

(c) second

41. (a) 0.0009 (approximately)

(b) When 1 more unit is sold (beyond 25), the marginal revenue will change by about 0.0009 thousand dollars per unit, or \$0.90 per unit.

43. (a) $S' = \dfrac{-3}{(t + 3)^2} + \dfrac{36}{(t + 3)^3}$ **(b)** $S''(15) = 0$

(c) After 15 weeks, the rate of change of the rate of sales is zero because the rate of sales reaches a minimum value.

45. (a) $p'(t) = -0.0057t^2 + 0.2710t - 2.542$

(b) $p''(t) = -0.0114t + 0.271$
1980: $p''(20) \approx 0.043$; 1998: $p''(38) \approx -0.1622$

(c) $p'(20) \approx 0.598$ means that in 1980, the number of people living below the poverty level was changing at the rate of 0.598 million people per year (the number is increasing). $p''(20) \approx 0.043$ means that in 1980, the rate of change of the number of people living below the poverty level was chang-

ing at the rate of 0.043 million people per year, per year (the rate was increasing). Thus, in 1980, $p(t)$ was increasing at an increasing rate.

47. (a) $p(t) = 2586t^{0.4077}$ **(b)** $p'(t) = 1054.3t^{-0.5923}$

(c) $p''(t) = -624.46t^{-1.5923}$
1995: $p''(15) \approx -8.372$; 2002: $p''(22) \approx -4.550$

(d) $p'(22) \approx 168.98$ means that in 2002 the poverty threshold income for a single person was changing at the rate of \$168.98 per year (the level was increasing). $p''(22) \approx -4.550$ means that in 2002, the rate of change of the poverty threshold income was changing at the rate of -4.550 dollars per year per year (the rate was decreasing). Thus, in 2002, $p(t)$ was increasing at a decreasing rate.

Exercise Set 9.9

1. (a) $\overline{MR} = 4$

(b) The sale of each additional item brings in \$4 revenue at all levels of production.

3. (a) \$3500; this is revenue from the sale of 100 units.

(b) $\overline{MR} = 36 - 0.02x$

(c) \$34; Revenue will increase by about \$34 if a 101st item is sold and by about \$102 if 3 additional units past 100 units are sold.

(d) Actual revenue from the sale of the 101st item is \$33.99.

5. (a) $R(x) = 80x - 0.4x^2$ (in hundreds of dollars)

(b) 7500 subscribers $(x = 75)$; $R = \$375,000$

(c) Lower the price per month.

(d) $\overline{MR} = R'(x) = 80 - 0.8x$; when $p = 50, x = 75$
$\overline{MR}(75) = 20$ means that if the number of customers increased from 75 to 76 (hundred), the revenue would increase by about 20 (hundred dollars), or \$2000. This means the company should try to increase subscribers by lowering its monthly charge.

7. (a)

$\overline{MR}$
36
24
12

−12 1200
−24
$\boxed{MR = 36 - 0.02x}$

(b) $x = 1800$
(c) \$32,400

9. $\overline{MC} = 8$ **11.** $\overline{MC} = 13 + 2x$

13. $\overline{MC} = 3x^2 - 12x + 24$ **15.** $\overline{MC} = 27 + 3x^2$

17. (a) \$10; the cost will increase by about \$10.

(b) \$11

19. \$46; the cost will increase by about \$46. For 3 additional units, the cost will increase by about \$138.

21.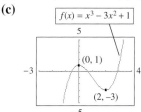

23. (a) The 101st item costs more. The tangent line slope is greater at $x = 100$ than at $x = 500$, and the slope of the tangent line gives the marginal cost and predicts the cost of the next item.
(b) More efficient. As x increases, the slopes of the tangents decrease. This means that the costs of additional items decrease as x increases.

25. $\overline{MP} = 5$ This means that for each additional unit sold, profit changes by \$5.

27. (a) \$0 **(b)** $\overline{MP} = 30 - 2x$
(c) -10; profit will decrease by about \$10 if a 21st unit is sold.
(d) -11; the sale of the 21st item results in a loss of \$11.

29. (a) $P(x) = R(x) - C(x)$, so profit is the distance between $R(x)$ and $C(x)$ (when $R(x)$ is above $C(x)$). $P(100) < P(700) < P(400)$; $P(100) < 0$, so there is a loss when 100 units are sold.
(b) This asks us to rank $\overline{MP}(100)$, $\overline{MP}(400)$, and $\overline{MP}(700)$. Because $\overline{MP} = \overline{MR} - \overline{MC}$, compare the slopes of the tangents to $R(x)$ and $C(x)$ at the three x-values. Thus $\overline{MP}(700) < \overline{MP}(400) < \overline{MP}(100)$. $\overline{MP}(700) < 0$ because $C(x)$ is steeper than $R(x)$ at $x = 700$. At $x = 100$, $R(x)$ is much steeper than $C(x)$.

31. (a) $A < B < C$. Amount of profit is the height of the graph. There is a loss at A.
(b) $C < B < A$. Marginal profit is the slope of the tangent to the graph. Marginals (slopes) are positive at all three points.

33. (a)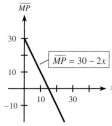
(b) 15
(c) 15
(d) \$25 thousand

35. 70 **37.** \$200

Chapter 9 Review Exercises

1. (a) 2 **(b)** 2 **2. (a)** 0 **(b)** 0
3. (a) 2 **(b)** 1 **4. (a)** 2 **(b)** does not exist
5. (a) does not exist **(b)** 2
6. (a) does not exist **(b)** does not exist

7. 55 **8.** 0 **9.** -2 **10.** 6 **11.** $\frac{1}{2}$ **12.** $\frac{1}{5}$
13. no limit **14.** 0 **15.** 4 **16.** no limit
17. 3 **18.** no limit **19.** $6x$ **20.** $1 - 4x$
21. -14 **22.** 5 **23. (a)** yes **(b)** no
24. (a) yes **(b)** no **25.** 2 **26.** no limit
27. 1 **28.** no **29.** yes **30.** yes
31. discontinuity at $x = 5$ **32.** discontinuity at $x = 2$
33. continuous **34.** discontinuity at $x = 1$
35. (a) $x = 0, x = 1$ **(b)** 0 **(c)** 0
36. (a) $x = -1, x = 0$ **(b)** $\frac{1}{2}$ **(c)** $\frac{1}{2}$
37. -2 **38.** 0 **39.** 7 **40.** true **41.** false
42. $f'(x) = 6x + 2$ **43.** $f'(x) = 1 - 2x$
44. $[-1, 0]$; the segment over this interval is steeper
45. (a) no **(b)** no **46. (a)** yes **(b)** no
47. (a) -5.9171 (to four decimal places) **(b)** -5.9
48. (a) $4.9/3$ **(b)** 7 **49.** about $-1/4$
50. $B, C, A: B < 0$ and $C < 0$; the tangent line at $x = 6$ falls more steeply than the segment over $[2, 10]$.
51. $20x^4 - 18x^2$ **52.** $8x$ **53.** 3 **54.** $1/(2\sqrt{x})$
55. 0 **56.** $-4/(3\sqrt[3]{x^4})$
57. $\dfrac{-1}{x^2} + \dfrac{1}{2\sqrt{x^3}}$ **58.** $\dfrac{-3}{x^3} - \dfrac{1}{3\sqrt[3]{x^2}}$
59. $y = 15x - 18$ **60.** $y = 34x - 48$
61. (a) $x = 0, x = 2$ **(b)** $(0, 1)\ (2, -3)$
(c)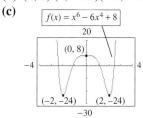
62. (a) $x = 0, x = 2, x = -2$
(b) $(0, 8)\ (2, -24)(-2, -24)$
(c)

63. $9x^2 - 26x + 4$ **64.** $15x^4 + 9x^2 + 2x$
65. $\dfrac{2(1 - q)}{q^3}$ **66.** $\dfrac{1 - 3t}{[2\sqrt{t}(3t + 1)^2]}$ **67.** $\dfrac{9x + 2}{2\sqrt{x}}$
68. $\dfrac{5x^6 + 2x^4 + 20x^3 - 3x^2 - 4x}{(x^3 + 1)^2}$
69. $(9x^2 - 24x)(x^3 - 4x^2)^2$
70. $6(30x^5 + 24x^3)(5x^6 + 6x^4 + 5)^5$
71. $72x^3(2x^4 - 9)^8$ **72.** $\dfrac{-(3x^2 - 4)}{2\sqrt{(x^3 - 4x)^3}}$
73. $2x(2x^4 + 5)^7(34x^4 + 5)$ **74.** $\dfrac{-2(3x + 1)(x + 12)}{(x^2 - 4)^2}$

75. $36[(3x + 1)(2x^3 - 1)]^{11}(8x^3 + 2x^2 - 1)$

76. $\dfrac{3}{(1 - x)^4}$ **77.** $\dfrac{(2x^2 - 4)}{\sqrt{x^2 - 4}}$ **78.** $\dfrac{2x - 1}{(3x - 1)^{4/3}}$

79. $y'' = \frac{-1}{4}x^{-3/2} - 2$ **80.** $y'' = 12x^2 - 2/x^3$

81. $\dfrac{d^5y}{dx^5} = 0$ **82.** $\dfrac{d^5y}{dx^5} = -30(1 - x)$

83. $\dfrac{d^3y}{dx^3} = -4/[(x^2 - 4)^{3/2}]$ **84.** $\dfrac{d^4y}{dx^4} = \dfrac{2x(x^2 - 3)}{(x^2 + 1)^3}$

85. **(a)** \$400,000   **(b)** \$310,000   **(c)** \$90,000

86. **(a)** \$70,000; this is fixed costs
   (b) \$0; $x = 1000$ is break-even

87. **(a)** \$140 per unit
   (b) $\overline{C}(x) \to +\infty$; the limit does not exist

88. **(a)** $C(x) \to +\infty$; the limit does not exist. As the number of units produced increases without bound, so does the total cost.
   (b) \$60 per unit. As more units are produced, the average cost of each unit approaches \$60.

89. The average annual percent change of **(a)** elderly men in the work force is -0.445% per year and of **(b)** elderly women in the work force is 0.017% per year.

90. **(a)** 1990–2000: Average rate $= 0.10\%$ per year for elderly men in the work force.
   (b) 1950–1960: Average rate $= 0.25\%$ per year for elderly women in the work force.

91. **(a)** $x'(10) = -1$ means if price changes from \$10 to \$11, the number of units demanded will change by about -1.
   (b) $x'(20) = -\frac{1}{4}$ means if price changes from \$20 to \$21, the number of units demanded will change by about $-\frac{1}{4}$.

92. $P'(t) \approx 3.695t^{-0.441}$. $P(65) \approx 68.2$ means that in 2005, 68.2% of U.S. roads and streets were paved.
$P'(65) \approx 0.59$ means that in 2005, the rate at which U.S. roads and streets were being paved was 0.59% per year.

93. The slope of the tangent at A gives $\overline{MR}(A)$. The tangent line at A is steeper (so has greater slope) than the tangent line at B. Hence, $\overline{MR}(A) > \overline{MR}(B)$, so the $(A + 1)$st unit will bring more revenue.

94. $R'(10) = 1570$. Raised. An 11th rent increase of \$30 (and hence an 11th vacancy) would change revenue by about \$1570.

95. **(a)** $P(20) = 23$ means productivity is 23 units per hour after 20 hours of training and experience.
   (b) $P'(20) \approx 1.4$ means that the 21st hour of training or experience will change productivity by about 1.4 units per hour.

96. $\dfrac{dq}{dp} = \dfrac{-p}{\sqrt{0.02p^2 + 500}}$

97. $x'(10) = \frac{1}{6}$ means if price changes from \$10 to \$11, the number of units supplied will change by about $\frac{1}{6}$.

98. $s''(t) = a = -2t^{-3/2}$; $s''(4) = -0.25$ ft/sec/sec

99. $P'(x) = 70 - 0.2x$; $P''(x) = -0.2$
$P'(300) = 10$ means that the 301st unit brings in about \$10 in profit.
$P''(300) = -0.2$ means that marginal profit ($P'(x)$) is changing at the rate of -0.2 dollars per unit, per unit.

100. **(a)** $\overline{MC} = 6x + 6$ **(b)** 186
   (c) If a 31st unit is produced, costs will change by about \$186.

101. $C'(4) = 53$ means that a 5th unit produced would change total costs by about \$53.

102. **(a)** $\overline{MR} = 40 - 0.04x$ **(b)** $x = 1000$ units

103. $\overline{MP}(10) = 48$ means that if an 11th unit is sold, profit will change by about \$48.

104. **(a)** $\overline{MR} = 80 - 0.08x$ **(b)** 72
   (c) If a 101st unit is sold, revenue will change by about \$72.

105. $\dfrac{120x(x + 1)}{(2x + 1)^2}$ **106.** $\overline{MP} = 4500 - 3x^2$

107. $\overline{MP} = 16 - 0.2x$

108. **(a)** A: Tangent line to $C(x)$ has smallest slope at A, so $\overline{MC}(A)$ is smallest and the next item at A will cost least.
   (b) B: $R(x) > C(x)$ at both B and C. Distance between $R(x)$ and $C(x)$ gives the amount of profit and is greatest at B.
   (c) A: $\overline{MR}$ greatest at A and $\overline{MC}$ least at A, as seen from the slopes of the tangents. Hence $\overline{MP}(A)$ is greatest, so the next item at A will give the greatest profit.
   (d) C: $\overline{MC}(C) > \overline{MR}(C)$, as seen from the slopes of the tangents. Hence $\overline{MP}(C) < 0$, so the next unit sold reduces profit.

Chapter 9 Test

1. **(a)** $\frac{3}{4}$ **(b)** $-8/5$ **(c)** $9/8$ **(d)** does not exist

2. **(a)** $f'(x) = \lim\limits_{h \to 0} \dfrac{f(x + h) - f(x)}{h}$
   (b) $f'(x) = 6x - 1$

3. $x = 0, x = 8$

4. **(a)** $\dfrac{dB}{dW} = 0.523$ **(b)** $p'(t) = 90t^9 - 42t^6 - 17$
   (c) $\dfrac{dy}{dx} = \dfrac{99x^2 - 24x^9}{(2x^7 + 11)^2}$
   (d) $f'(x) = (3x^5 - 2x + 3)(40x^9 + 40x^3) + (4x^{10} + 10x^4 - 17)(15x^4 - 2)$
   (e) $g'(x) = 9(10x^4 + 21x^2)(2x^5 + 7x^3 - 5)^{11}$
   (f) $y' = 2(8x^2 + 5x + 18)(2x + 5)^5$
   (g) $f'(x) = \dfrac{6}{\sqrt{x}} + \dfrac{20}{x^3}$

5. $\dfrac{d^3y}{dx^3} = 6 + 60x^{-6}$

6. (a) $y = -15x - 5$ **(b)** $(4, -90), (-2, 18)$

7. -15 **8. (a)** 2 **(b)** does not exist **(c)** -4

9. $g(-2) = 8$; $\lim\limits_{x \to -2^-} g(x) = 8$, $\lim\limits_{x \to -2^+} g(x) = -8$

∴ $\lim\limits_{x \to -2} g(x)$ does not exist and $g(x)$ is not continuous at $x = -2$.

10. (a) $\overline{MR} = R'(x) = 250 - 0.02x$

(b) $R(72) = 17{,}948.16$ means that when 72 units are sold, revenue is $17,948.16.
$R'(72) = 248.56$ means that the expected revenue from the 73rd unit is about $248.56.

11. (a) $P(x) = 50x - 0.01x^2 - 10{,}000$

(b) $\overline{MP} = 50 - 0.02x$

(c) $\overline{MP}(1000) = 30$ means that the predicted profit from the sale of the 1001st unit is approximately $30.

12. 104

13. (a) -5 **(b)** -1 **(c)** 4 **(d)** does not exist
(e) 2 **(f)** $3/2$ **(g)** $-4, 1, 3, 6$ **(h)** $-4, 3, 6$
(i) $f'(-2) <$ average rate over $[-2, 2] < f'(2)$

14. (a) $2/3$ **(b)** -4 **(c)** $2/3$

15. (a) B: $R(x) > C(x)$ at B, so there is profit. Distance between $R(x)$ and $C(x)$ gives the amount of profit.

(b) A: $C(x) > R(x)$

(c) A and B: slope of $R(x)$ is greater than the slope of $C(x)$. Hence $\overline{MR} > \overline{MC}$ and $\overline{MP} > 0$.

(d) C: Slope of $C(x)$ is greater than the slope of $R(x)$. Hence $\overline{MC} > \overline{MR}$ and $\overline{MP} < 0$.

Exercise Set 10.1

1. (a) $(1, 5)$ **(b)** $(4, 1)$ **(c)** $(-1, 2)$

3. (a) $(1, 5)$ **(b)** $(4, 1)$ **(c)** $(-1, 2)$

5. (a) $3, 7$ **(b)** $3 < x < 7$ **(c)** $x < 3, x > 7$
(d) 7 **(e)** 3

7. $x = 0, x = 4$

9.

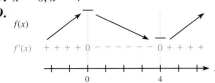

min: $(4, -58)$; max: $(0, 6)$

11. (a) max: $(-1, 6)$; min: $(1, 2)$

(b) $dy/dx = 3x^2 - 3$; $x = 1, x = -1$

(c) $(1, 2), (-1, 6)$

(d) yes

13. (a) HPI: $(-1, -3)$

(b) $dy/dx = 3x^2 + 6x + 3$; $x = -1$

(c) $(-1, -3)$

(d) yes

15. (a) $\dfrac{dy}{dx} = x - 1$ **(b)** $x = 1$ **(c)** $\left(1, -\tfrac{1}{2}\right)$

(d) decreasing: $x < 1$ **(e)**
increasing: $x > 1$

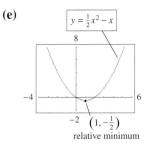

17. (a) $dy/dx = x^2 + x - 2$

(b) $x = -2, x = 1$ **(c)** $\left(-2, \tfrac{13}{3}\right), \left(1, -\tfrac{1}{6}\right)$

(d) increasing: $x < -2$ and $x > 1$
decreasing: $-2 < x < 1$

(e)

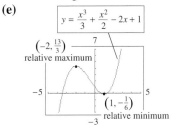

19. (a) $\dfrac{dy}{dx} = \dfrac{2}{3x^{1/3}}$ **(b)** $x = 0$ **(c)** $(0, 0)$

(d) decreasing: $x < 0$
increasing: $x > 0$

(e)

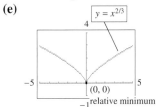

21. (a) $f'(x) = 0$ at $x = -\tfrac{1}{2}$
$f'(x) > 0$ for $x < -\tfrac{1}{2}$
$f'(x) < 0$ for $x > -\tfrac{1}{2}$

(b) $f'(x) = -1 - 2x$ verifies these conclusions.

23. (a) $f'(x) = 0$ at $x = 0, x = -3, x = 3$
$f'(x) > 0$ for $-3 < x < 3, x \neq 0$
$f'(x) < 0$ for $x < -3$ and $x > 3$

(b) $f'(x) = \tfrac{1}{3}x^2(9 - x^2)$ verifies these conclusions.

25. HPI $\left(1, \tfrac{4}{3}\right)$
no max or min

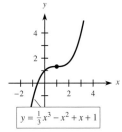

27. $(-6, 128)$ rel max;
$\left(4, -38\frac{2}{3}\right)$ rel min

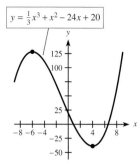

29. $(-1, 3)$ rel max;
$(1, -1)$ rel min;
HPI $(0, 1)$

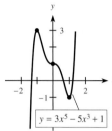

31. $(1, 1)$ rel max;
$(0, 0), (2, 0)$ rel min

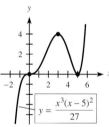

33. $(3, 4)$ rel max;
$(5, 0)$ rel min; HPI $(0, 0)$

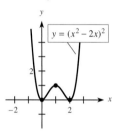

35. $(0, 0)$ rel max;
$(2, -4.8)$ rel min

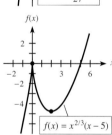

37. $(50, 300{,}500), (100, 238{,}000)$
$0 \le x \le 150, 0 \le y \le 301{,}000$

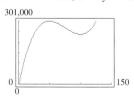

39. $(0, -40{,}000), (60, 4{,}280{,}000)$
$-20 \le x \le 90 \quad -50{,}000 \le y \le 5{,}000{,}000$

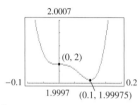

41. $(0, 2)$
$(0.1, 1.99975)$
$-0.1 \le x \le 0.2$
$1.9997 \le y \le 2.0007$

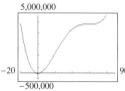

43. critical values: $x = -1, x = 2$
$f(x)$ increasing for $x < -1$ and $x > 2$
$f(x)$ decreasing for $-1 < x < 2$
rel max at $x = -1$; rel min at $x = 2$

possible graph of $f(x)$

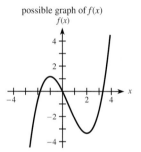

45. critical values: $x = 0, x = 3$
$f(x)$ increasing for $x > 3$
$f(x)$ decreasing for $x < 3, x \ne 0$
rel min at $x = 3$; HPI at $x = 0$

possible graph of $f(x)$

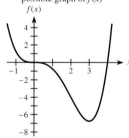

47. Graph on left is $f(x)$; on right is $f'(x)$ because $f(x)$ is increasing when $f'(x) > 0$ (i.e., above the x-axis) and $f(x)$ is decreasing when $f'(x) < 0$ (i.e., below the x-axis).

49. decreasing for $t \ge 0$

51. **(a)** $2 \pm \sqrt{13}$
(b) $2 + \sqrt{13} \approx 5.6$
(c) $0 \le t < 2 + \sqrt{13}$
(d)

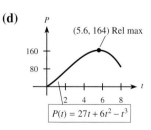

53. (a) $x = 5$ **(b)** $0 < x < 5$
 (c) increasing for $x > 5$
55. (a) at $x = 150$, increasing; at $x = 250$, changing from
 increasing to decreasing; at $x = 350$, decreasing
 (b) increasing for $x < 250$ **(c)** 250 units
57. (a) $t = 6$ **(b)** 6 weeks
59. (a) 10 **(b)** January 1
61. (a) $x = 10$, in 2000; yes
 (b) September 11, 2001 attack and wars in Afganistan
 and Iraq
63. (a) cubic
 (b) $y = -0.000298x^3 + 0.0163x^2 - 0.0511x + 4.273$
 (c) $x = 34.8$, in 1985

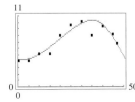

65. (a) third-degree (cubic)
 (b) $y = 6.164x^3 - 90.530x^2 + 373.598x + 202.867$
 (c) $(2.96, 675.39)$
 (d) The largest amount of money earmarked for
 colleges in the early 1990s was approximately
 $675 million, in 1993.
 (e) $1050 million, in 2000

Exercise Set 10.2

1. (a) concave down **(b)** concave up
3. (a) concave down **(b)** concave up
5. (a, c) and (d, e) **7.** (c, d) and (e, f) **9.** c, d, e
11. concave up when $x > 2$; concave down when $x < 2$;
 POI at $x = 2$
13. concave up when $x < -2$ and $x > 1$
 concave down when $-2 < x < 1$
 points of inflection at $x = -2$ and $x = 1$
15. no points of inflection;
 $(2, -2)$ min

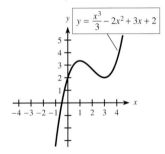

17. $\left(1, \frac{10}{3}\right)$ max; $(3, 2)$ min;
 $\left(2, \frac{8}{3}\right)$ point of inflection

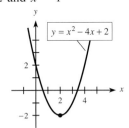

19. $(0, 0)$ rel max; $(2\sqrt{2}, -64)$, $(-2\sqrt{2}, -64)$ min;
 points of inflection: $(2\sqrt{6}/3, -320/9)$ and
 $(-2\sqrt{6}/3, -320/9)$

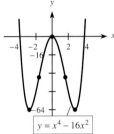

21. $(-2, 64)$ rel max; $(2, -64)$
 rel min; points of inflection:
 $(-\sqrt{2}, 39.6)$, $(0, 0)$, and
 $(\sqrt{2}, -39.6)$

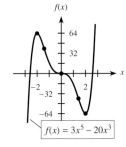

23. $(1, -3)$ min; points of
 inflection: $(-2, 7.6)$
 and $(0, 0)$

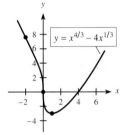

25. (a) $f''(x) = 0$ when $x = 1$
 $f''(x) > 0$ when $x < 1$
 $f''(x) < 0$ when $x > 1$
 (b) rel max for $f'(x)$ at $x = 1$; no rel min
 (c)

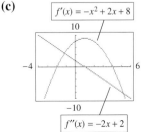

27. (a) concave up when $x < 2$; concave down when
 $x > 2$
 (b) point of inflection at $x = 2$
 (c)

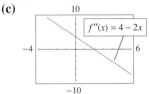

(d) possible graph of $f(x)$

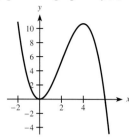

29. (a) G **(b)** C **(c)** F **(d)** H **(e)** I
31. (a) concave up when $x < 0$
   concave down when $x > 0$
   point of inflection at $x = 0$
   (b) concave up when $-1 < x < 1$
   concave down when $x < -1$ and $x > 1$
   POI at $x = -1$ and $x = 1$
   (c) concave up when $x > 0$
   concave down when $x < 0$
   point of inflection at $x = 0$
33. (a) $P'(t)$ **(b)** B **(c)** C
35. (a) C **(b)** right   **(c)** yes
37. (a) in an 8-hour shift, max when $t = 8$ **(b)** 4 hr
39. (a) 9 days   **(b)** 15 days
41. when $x \approx 14.3$; during 2000
43. (a) $y = 0.0000591x^3 - 0.00675x^2 + 0.0523x + 14.147$
   (b) $x = 72.048$, $y = 4.980$
   (c) The minimum percent is 4.98% in 1973.
45. (a) $y = -0.4052x^3 + 6.0161x^2 - 17.9605x + 52$
   (b) $x = 1.83$, during 1914
   (c) $x = 8.07$, during 1921

Exercise Set 10.3

1. min -6 at $x = 2$, max $3.\overline{481}$ at $x = -2/3$
3. min -1 at $x = -2$, max 2 at $x = -1$
5. (a) $x = 1800$ units, $R = \$32,400$
   (b) $x = 1500$ units, $R = \$31,500$
7. $x = 20$ units, $R = \$24,000$ **9.** 40 people
11. $p = \$15$, $R = \$22,500$
13. (a) max $= \$2100$ at $x = 10$
   (b) $\overline{R}(x) = \overline{MR}$ at $x = 10$
15. $x = 5$ units, $\overline{C} = \$23$
17. $x = 10$ units, $\overline{C} = \$20$
19. 200 units $(x = 2)$, $\overline{C} = \$108$ per 100 units
21. $\overline{C}(x)$ has its minimum and $\overline{C}(x) = \overline{MC}$ at $x = 5$
23. (a) A line from $(0, 0)$ to $(x, C(x))$ has slope $C(x)/x = \overline{C}(x)$; this is minimized when the line has the least rise—that is, when the line is tangent to $C(x)$.
   (b) $x = 600$ units

25. $x = 80$ units, $P = \$280,000$
27. $x = 10\sqrt{15} \approx 39$ units, $P \approx \$71,181$ (using $x = 39$)
29. $x = 1000$ units, $P = \$39,700$
31. (a) B **(b)** B **(c)** B **(d)** $\overline{MR} = \overline{MC}$
33. $\$860$ **35.** $x = 600$ units, $P = \$495,000$
37. (a) 60   **(b)** $\$570$ **(c)** $\$9000$
39. (a) 1000 units   **(b)** $\$8066.67$ (approximately)
41. 2000 units priced at $\$90$/unit; max profit is $\$90,000$/wk
43. (a) $y = 0.0002524x^3 - 0.02785x^2 + 1.6300x + 2.1550$
   (b) $(36.78, 36.99)$
   (c)

95
0   88
−10

The *rate* of change of the number of beneficiaries was decreasing until 1987, after which the rate has been increasing. Hence, since 1987 the number of beneficiaries has been increasing at an increasing rate.
45. (a) About mid-May
   (b) Just after September 11, when the terrorists' planes crashed into the World Trade Center and the Pentagon
47. (a) 16.5
   (b) 1.9
   (c) Rise. As the number of workers per beneficiary drops, either the amount contributed by each worker must rise or support must diminish.

Exercise Set 10.4

1. (a) $x_1 = \$25$ million, $x_2 = \$13.846$ million
   (b) $\$38.846$ million
3. 100 trees   **5. (a)** 5   **(b)** 237.5   **7.** $\$50$
9. $m = c$ **11.** 1 week   **13.** $t = 8$, $p = 45\%$
15. 240 ft   **17.** 300 ft $\times$ 150 ft
19. 20 ft long, $6\frac{2}{3}$ ft across (dividers run across)
21. 4 in. $\times$ 8 in. $\times$ 8 in. high   **23.** 30,000
25. 12,000   **27.** $x = 2$ **29.** 3 weeks from now
31. 25 plates   **33.** $t = 5.3$, 2005

Exercise Set 10.5

1. (a) $x = 2$ **(b)** 1   **(c)** $y = 1$ **(d)** 1
3. (a) $x = 2$, $x = -2$ **(b)** 3   **(c)** $y = 3$ **(d)** 3
5. HA: $y = 2$; VA: $x = 3$
7. HA: $y = 0$; VA: $x = -2$, $x = 2$
9. HA: none; VA: none

11. HA: $y = 2$; VA: $x = 3$
no max, min, or points of
inflection

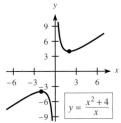

13. VA: $x = 0$;
$(-2, -4)$ rel max;
$(2, 4)$ rel min

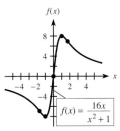

15. VA: $x = -1$; HA: $y = 0$;
$(0, 0)$ rel min; $(2, 4)$ rel max;

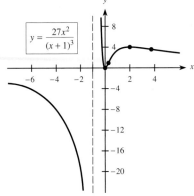

17. HA: $y = 0$; $(1, 8)$ rel max;
$(-1, -8)$ rel min;
points of inflection:
$(0, 0)$, $(-\sqrt{3}, -4\sqrt{3})$,
and $(\sqrt{3}, 4\sqrt{3})$

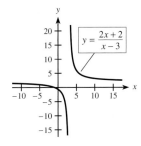

19. HA: $y = 0$; VA: $x = 1$; $\left(-1, -\frac{1}{4}\right)$ rel min; point of
inflection: $\left(-2, -\frac{2}{9}\right)$

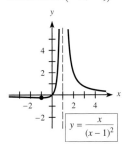

21. VA: $x = 3$;
$(2, -1)$ rel max;
$(4, 7)$ rel min

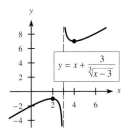

23. HA: $y = 0$;
VA: $x = 0$;
$(2, 0)$ rel min;
$(3, 1)$ rel max;
points of inflection:
$(1.87, 0.66)$,
$(4.13, 0.87)$

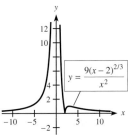

25. **(a)** HA: approx. $y = -2$; VA: approx. $x = 4$
(b) HA: $y = -\frac{9}{4}$; VA: $x = \frac{17}{4}$

27. **(a)** HA: approx. $y = 2$;
VA: approx. $x = 2.5$, $x = -2.5$
(b) HA: $y = \frac{20}{9}$; VA: $x = \frac{7}{3}$, $x = -\frac{7}{3}$

29. $f(x) = \dfrac{x + 25}{x^2 + 1400}$

(a)

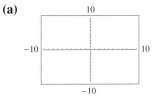

(b) HA: $y = 0$; rel min $(-70, -0.0071)$;
rel max $(20, 0.025)$
(c) x: -500 to 400
y: -0.01 to 0.03

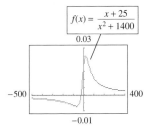

31. $f(x) = \dfrac{100(9 - x^2)}{x^2 + 100}$

(a)

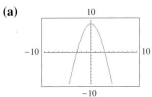

(b) HA: $y = -100$;
rel max $(0, 9)$

(c) x: -75 to 75
y: -120 to 20

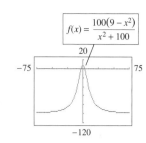

33. $f(x) = \dfrac{1000x - 4000}{x^2 - 10x - 2000}$

(a)

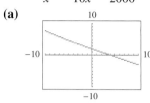

(b) HA: $y = 0$;
VA: $x = -40$,
$x = 50$;
no max or min

(c) x: -200 to 200
y: -200 to 200

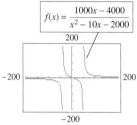

35. (a) none **(b)** $C \geq 0$ **(c)** $p = 100$ **(d)** no

37. (a)

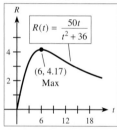

(b) 6 weeks
(c) 22 weeks after its release

39. (a) yes, $x = -1$
(b) no; domain is $x \geq 5$
(c) yes, $y = -58.5731$
(d) At $0°F$, as the wind speed increases, there is a limiting wind chill of about $-58.6°F$. This is meaningful because at high wind speeds, additional wind probably has little noticeable effect.

41. (a) $P = C$ **(b)** C **(c)** $P' = 0$ **(d)** 0

43. (a) 0
(b) As the years past 1800 increase, the percent of workers in farm occupations approaches 0.
(c) no

45. (a) No. Barometric pressure can drop off the scale (as shown), but it cannot decrease without bound. In fact, it must always be positive.
(b) See your library with regard to the "Storm of the Century" in March 1993.

Chapter 10 Review Exercises

1. $(0, 0)$ max **2.** $(2, -9)$ min **3.** HPI $(1, 0)$
4. $\left(1, \frac{3}{2}\right)$ max, $\left(-1, -\frac{3}{2}\right)$ min
5. (a) $\frac{1}{3}, -1$ **(b)** $(-1, 0)$ rel max, $\left(\frac{1}{3}, -\frac{32}{27}\right)$ rel min
(c) none **(d)**

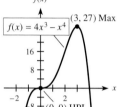

6. (a) $3, 0$ **(b)** $(3, 27)$ max **(c)** $(0, 0)$
(d)

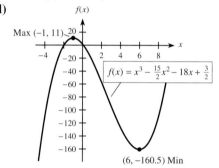

7. (a) $-1, 6$
(b) $(-1, 11)$ rel max, $(6, -160.5)$ rel min
(c) none **(d)**

8. (a) $0, \pm 1$ **(b)** $(-1, 1)$ rel max, $(1, -3)$ rel min
(c) $(0, -1)$ **(d)**

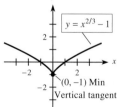

9. (a) 0 **(b)** $(0, -1)$ min **(c)** none
(d)

10. (a) 0, 1, 4
   (b) (0, 0) rel min, (1, 9) rel max, (4, 0) rel min
   (c) none   **(d)**

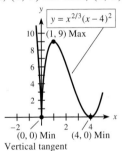

11. concave up

12. concave up when $x < -1$ and $x > 2$; concave down when $-1 < x < 2$; points of inflection at $(-1, -3)$ and $(2, -42)$

13. $(-1, 15)$ rel max; $(3, -17)$ rel min; point of inflection $(1, -1)$

14. $(-2, 16)$ rel max; $(2, -16)$ rel min; point of inflection $(0, 0)$

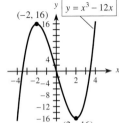

15. $(1, 4)$ rel max; $(-1, 0)$ rel min; points of inflection: $\left(\dfrac{1}{\sqrt{2}}, 2 + \dfrac{7}{4\sqrt{2}}\right)$, $(0, 2)$, and $\left(-\dfrac{1}{\sqrt{2}}, 2 - \dfrac{7}{4\sqrt{2}}\right)$

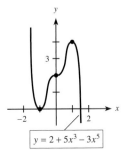

16. (a) (0, 0) absolute min; (140, 19,600) absolute max
   (b) (0, 0) absolute min; (100, 18,000) absolute max

17. (a) (50, 233,333) absolute max; (0, 0) absolute min
   (b) (64, 248,491) absolute max; (0, 0) absolute min

18. (a) $x = 1$ **(b)** $y = 0$ **(c)** 0   **(d)** 0

19. (a) $x = -1$ **(b)** $y = \frac{1}{2}$ **(c)** $\frac{1}{2}$ **(d)** $\frac{1}{2}$

20. HA: $y = \frac{3}{2}$; VA: $x = 2$

21. HA: $y = -1$; VA: $x = 1, x = -1$

22. (a) HA: $y = 3$; VA: $x = -2$

(b) no max or min   **(c)**

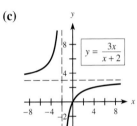

23. (a) HA: $y = 0$; VA: $x = 0$
   (b) (4, 1) max   **(c)**

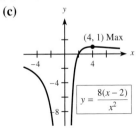

24. (a) HA: none; VA: $x = 1$
   (b) (0, 0) rel max; (2, 4) rel min
   (c)

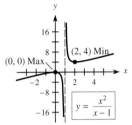

25. (a) $f'(x) > 0$ for $x < \frac{2}{3}$ (approximately) and $x > 2$
   $f'(x) < 0$ for about $\frac{2}{3} < x < 2$
   $f'(x) = 0$ at about $x = \frac{2}{3}$ and $x = 2$
   (b) $f''(x) > 0$ for $x > \frac{4}{3}$
   $f''(x) < 0$ for $x < \frac{4}{3}$
   $f''(x) = 0$ at $x = \frac{4}{3}$
   (c)

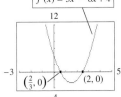

   (d)

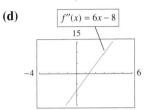

26. (a) $f'(x) > 0$ for about $-13 < x < 0$ and $x > 7$
   $f'(x) < 0$ for about $x < -13$ and $0 < x < 7$
   $f'(x) = 0$ at about $x = 0, x = -13, x = 7$

(b) $f''(x) > 0$ for about $x < -8$ and $x > 4$
$f''(x) < 0$ for about $-8 < x < 4$
$f''(x) = 0$ at about $x = -8$ and $x = 4$

(c)

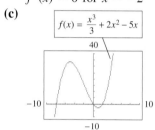

$$f'(x) = 0.01x^3 + 0.06x^2 - 0.96x + 0.08$$

(d)

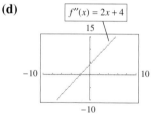

$$f''(x) = 0.03x^2 + 0.12x - 0.96$$

27. (a) $f(x)$ increasing for $x < -5$ and $x > 1$
$f(x)$ decreasing for $-5 < x < 1$
$f(x)$ has rel max at $x = -5$, rel min at $x = 1$

(b) $f''(x) > 0$ for $x > -2$ (where $f'(x)$ increases)
$f''(x) < 0$ for $x < -2$ (where $f'(x)$ decreases)
$f''(x) = 0$ for $x = -2$

(c)

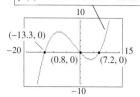

$$f(x) = \frac{x^3}{3} + 2x^2 - 5x$$

(d)

$$f''(x) = 2x + 4$$

28. (a) $f(x)$ increasing for $x < 6$, $x \neq 0$
$f(x)$ decreasing for $x > 6$
$f(x)$ has rel max at $x = 6$, point of inflection at $x = 0$

(b) $f''(x) > 0$ for $0 < x < 4$
$f''(x) < 0$ for $x < 0$ and $x > 4$
$f''(x) = 0$ at $x = 0$ and $x = 4$

(c)

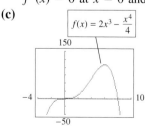

$$f(x) = 2x^3 - \frac{x^4}{4}$$

(d)

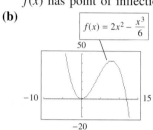

$$f''(x) = 12x - 3x^2$$

29. (a) $f(x)$ is concave up for $x < 4$
$f(x)$ is concave down for $x > 4$
$f(x)$ has point of inflection at $x = 4$

(b)

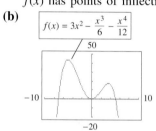

$$f(x) = 2x^2 - \frac{x^3}{6}$$

30. (a) $f(x)$ is concave up for $-3 < x < 2$
$f(x)$ is concave down for $x < -3$ and $x > 2$
$f(x)$ has points of inflection at $x = -3$ and $x = 2$

(b)

$$f(x) = 3x^2 - \frac{x^3}{6} - \frac{x^4}{12}$$

31. $x = 5$ units, $\overline{C} = \$45$ per unit

32. (a) $x = 1600$ units, $R = \$25,600$
(b) $x = 1500$ units, $R = \$25,500$

33. $P = \$54,000$ at $x = 100$ units   **34.** $x = 300$ units

35. $x = 150$ units   **36.** $x = 7$ units

37. $x = 500$ units, when $\overline{MP} = 0$ and changes from positive to negative.

38. 30 hours

39. (a) $I = 60$. The point of diminishing returns is located at the point of inflection (where bending changes).
(b) $m = f(I)/I =$ the average output
(c) The segment from $(0, 0)$ to $y = f(I)$ has maximum slope when it is tangent to $y = f(I)$, close to $I = 70$.

40. \$260 per bike   **41.** \$360 per bike

42. \$93,625 at 325 units

43. (a) 150   **(b)** \$650

44. \$208,490.67 at 64 units

45. $x = 1000$ mg   **46.** 10:00 A.M.

47. 325 in 2010   **48.** 20 mi from A, 10 mi from B

49. 4 ft $\times$ 4 ft   **50.** $8\frac{3}{4}$ in. $\times$ 10 in.

51. 500 mg

52. **(a)** $x \approx 13.04$; during 1994
   (b) point of inflection
53. 24,000
54. **(a)** vertical asymptote at $x = 0$
   (b)

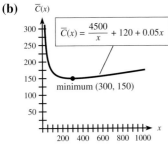

minimum (300, 150)

$$\bar{C}(x) = \frac{4500}{x} + 120 + 0.05x$$

55. **(a)** 3%
   (b) $y = 38$. The long-term market share approaches 38%.

Chapter 10 Test

1. max $(-3, 3)$; min $(-1, -1)$;
   POI $(-2, 1)$

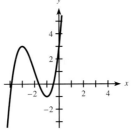

2. max $(3, 17)$; HPI $(0, -10)$; POI $(2, 6)$

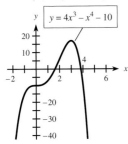

$y = 4x^3 - x^4 - 10$

3. max $(0, -3)$;
   min $(4, 5)$;
   vertical asymptote $x = 2$

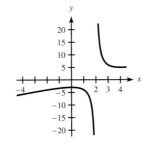

4. $\left(-\frac{1}{\sqrt{2}}, 0\right)$ and $\left(\frac{1}{\sqrt{2}}, \infty\right)$

5. $(0, 2)$, HPI; $\left(-\frac{1}{\sqrt{2}}, 3.237\right), \left(\frac{1}{\sqrt{2}}, 0.763\right)$

6. max $(-1, 4)$; min $(1, 0)$

7. max 67 at $x = 8$; min -122 at $x = 5$
8. horizontal asymptote $y = 200$; vertical asymptote $x = -300$
9.

Point	f	f'	f''
A	$-$	$+$	$-$
B	$+$	$-$	0
C	$+$	0	$+$

10. **(a)** -2 **(b)** $x = -1$
   (c) $y = 2$ (on right); $y = -2$ (on left)
11. local max at $(6, 10)$
12. **(a)** $x = 43.7, 1944$
   (b) The ratio of priests to Catholics was highest in 1944.
13. **(a)** $x = 7200$ **(b)** \$518,100   **14.** 100 units
15. \$250   **16.** $\frac{10}{3}$ centimeter   **17.** 28,000 units
18. **(a)** $y = 0.00253x^3 - 0.0570x^2 + 0.5842x + 3.110$
   (b) $x \approx 7.5$; during 1998
   (c) x-coordinate of the point of inflection

Exercise Set 11.1

1. $f'(x) = 4/x$ **3.** $y' = 1/x$

5. $y' = 4/x$ **7.** $f'(x) = \dfrac{4}{4x + 9}$

9. $y' = \dfrac{4x - 1}{2x^2 - x} + 3$ **11.** $dp/dq = 2q/(q^2 + 1)$

13. **(a)** $y' = \dfrac{1}{x} - \dfrac{1}{x - 1} = \dfrac{-1}{x(x - 1)}$

   (b) $y' = \dfrac{-1}{x(x - 1)}$; $\ln\left(\dfrac{x}{x - 1}\right) = \ln(x) - \ln(x - 1)$

15. **(a)** $y' = \dfrac{2x}{3(x^2 - 1)}$

   (b) $y' = \dfrac{2x}{3(x^2 - 1)}$; $\ln(x^2 - 1)^{1/3} = \frac{1}{3}\ln(x^2 - 1)$

17. **(a)** $y' = \dfrac{4}{4x - 1} - \dfrac{3}{x} = \dfrac{-8x + 3}{x(4x - 1)}$

   (b) $y' = \dfrac{-8x + 3}{x(4x - 1)}$;

   $\ln\left(\dfrac{4x - 1}{x^3}\right) = \ln(4x - 1) - 3\ln(x)$

19. $\dfrac{dp}{dq} = \dfrac{2q}{q^2 - 1} - \dfrac{1}{q} = \dfrac{q^2 + 1}{q(q^2 - 1)}$

21. $\dfrac{dy}{dt} = \dfrac{2t}{t^2 + 3} - \dfrac{1}{2}\left(\dfrac{-1}{1 - t}\right) = \dfrac{3 + 4t - 3t^2}{2(1 - t)(t^2 + 3)}$

23. $\dfrac{dy}{dx} = \dfrac{3}{x} + \dfrac{1}{2(x + 1)} = \dfrac{7x + 6}{2x(x + 1)}$

25. $y' = 1 - \dfrac{1}{x}$ **27.** $y' = (1 - \ln x)/x^2$

29. $y' = 8x^3/(x^4 + 3)$ **31.** $y' = \dfrac{4(\ln x)^3}{x}$

33. $y' = \dfrac{8x^3 \ln (x^4 + 3)}{x^4 + 3}$ **35.** $y' = \dfrac{1}{x \ln 4}$

37. $y' = \dfrac{4x^3 - 12x^2}{(x^4 - 4x^3 + 1) \ln 6}$

39. rel min $(e^{-1}, -e^{-1})$

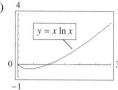

41. rel min $(2, 4 - 8 \ln 2)$

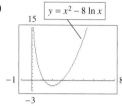

43. (a) $\overline{MC} = \dfrac{400}{2x + 1}$

(b) $\overline{MC} = \dfrac{400}{401} \approx 1.0$; the approximate cost of the 201st unit is $1.00

(c) $\overline{MC} > 0$. Yes

45. (a) $\overline{MR} = \dfrac{2500[(x + 1) \ln (10x + 10) - x]}{(x + 1) \ln^2 (10x + 10)}$

(b) 309.67; at 100 units, selling 1 additional unit yields $309.67.

47. (a) -5.23 **(b)** -1.89 **(c)** increasing

49. A/B **51.** $dR/dI = 1/(I \ln 10)$

53. (a) $y = -195.59 + 2975.50 \ln (x)$

(b) $99.18 per year

Exercise Set 11.2

1. $y' = 5e^x - 1$ **3.** $f'(x) = e^x - exe^{-1}$

5. $y' = 3x^2 e^{x^3}$ **7.** $y' = 36xe^{3x^2}$

9. $y' = 12x(x^2 + 1)^2 e^{(x^2+1)^3}$ **11.** $y' = 3x^2$

13. $y' = e^{-1/x}/x^2$ **15.** $y' = \dfrac{2}{x^3} e^{-1/x^2} - 2xe^{-x^2}$

17. $ds/dt = te^t(t + 2)$ **19.** $y' = 4x^3 e^{x^4} - 4e^{4x}$

21. $y' = \dfrac{4e^{4x}}{e^{4x} + 2}$ **23.** $y' = e^{-3x}/x - 3e^{-3x} \ln (2x)$

25. $y' = (2e^{5x} - 3)/e^{3x} = 2e^{2x} - 3e^{-3x}$

27. $y' = 30e^{3x}(e^{3x} + 4)^9$ **29.** $y' = 6^x \ln 6$

31. $y' = 4^{x^2}(2x \ln 4)$

33. (a) $y'(1) = 0$ **(b)** $y = e^{-1}$

35. (a) $z = 0$ **(b)**

37. rel min at $x = 1$, $y = e$
39. rel max at $x = 0$, $y = -1$
41. (a) $(0.1) Pe^{0.1n}$ **(b)** $(0.1) Pe^{0.1}$
(c) Yes, because $e^{0.1n} > 1$ for any $n \geq 1$.

43. (a) $\dfrac{dS}{dt} = -50,000e^{-0.5t}$

(b) The function is a decay exponential. The derivative is always negative.

45. $40e \approx 108.73$ dollars per unit

47. (a) $\dfrac{dy}{dt} = 46.2e^{-0.462t}$

(b) 29.107 percent per hour

49. $y' = \dfrac{98,990,100e^{-0.99t}}{(1 + 9999e^{-0.99t})^2}$

51. $\dfrac{dx}{dt} = -0.0684e^{-0.38t}$

53. (a) $P'(t) = \dfrac{1.2595e^{-0.029t}}{(1 + 3.97e^{-0.029t})^2}$

(b) $P'(100) \approx 0.0467$ means that in 2045 the population is expected to change at the rate of 0.0467 billion people per year.

(c) $P''(95) < 0$ means the rate is decreasing.

55. $\dfrac{dI}{dR} = 10^R \ln 10$ **57.** 301.8 ($billion/year)

59. (a) $d'(t) = 0.139104e^{0.0828t}$

(b) $d'(40) \approx 3.82 billion per year
$d'(100) \approx 548.65 billion per year

(c) 1999–2000: average rate = $17.9 billion per year
2000–2001: average rate = $133.3 billion per year
Average rates are much smaller than the model's prediction and suggest that the model may need to be changed.

(d) The terrorist attacks of September 11, 2001 and the subsequent wars on terrorism and in Iraq

61. (a) $P'(42) \approx -0.0229$

(b) This means that in 2002, the purchasing power of $1 was changing at the rate of -0.0229 dollars per year.

(c) For 2001–2002, average rate = -0.009 dollars per year.

63. (a) $y = 257.8(1.0693)^x$ (or $y = 257.8e^{0.067x}$)

(b) $y'(60) \approx 962.41 per year is the predicted rate at which U.S. federal tax per capita will change in 2010.

65. (a) $P(x) = \dfrac{1,817,670}{1 + 5.0135e^{-0.1375x}}$

(b) $P'(x) = \dfrac{1,253,022.175e^{-0.1375x}}{(1 + 5.0135e^{-0.1375x})^2}$

$P(20) \approx 1,376,500$ means the model predicts this number of prisoners in 2000.
$P'(20) \approx 45,938$ means the model predicts that in 2000, the prison population will be changing at the rate of 45,938 prisoners per year.

Exercise Set 11.3

1. $\frac{1}{2}$ **3.** $-\frac{1}{2}$ **5.** $-\frac{5}{3}$ **7.** $-x/(2y)$

9. $-(2x + 4)/(2y - 3)$ **11.** $y' = -x/y$

13. $y' = \dfrac{-y}{2(x - 1)}$ **15.** $\dfrac{dp}{dq} = \dfrac{p^2}{4 - 2pq}$

17. $\dfrac{dy}{dx} = \dfrac{x(3x^3 - 2)}{3y^2(1 + y^2)}$ **19.** $\dfrac{dy}{dx} = \dfrac{4x^3 + 6x^2y^2 - 1}{-4x^3y - 3y^2}$

21. $\dfrac{dy}{dx} = \dfrac{(4x^3 + 9x^2y^2 - 8x - 12y)}{(18y + 12x - 6x^3y + 10y^4)}$ **23.** undefined

25. 1   **27.** $y = \frac{1}{2}x + 1$ **29.** $y = 4x + 5$

31. $\dfrac{dy}{dx} = \dfrac{1}{2xy}$ **33.** $\dfrac{dy}{dx} = \dfrac{-y}{2x \ln x}$ **35.** -4

37. $-1/x$ **39.** $-y/x$

41. $ye^x/(1 - e^x)$ **43.** $\frac{1}{3}$ **45.** $y = 3 - x$

47. (a) $(2, \sqrt{2}), (2, -\sqrt{2})$

 (b) $(2 + 2\sqrt{2}, 0), (2 - 2\sqrt{2}, 0)$

49. (a) and **(b)** are verifications

 (c) yes, because $x^2 + y^2 = 4$

51. $1/(2x\sqrt{x})$

53. max at $(0, 3)$; min at $(0, -3)$

55. $\frac{1}{2}$, so an additional 1 (thousand dollars) of advertising yields about $\frac{1}{2}$ (thousand) additional units

57. $-\frac{243}{128}$ hours of skilled labor per hour of unskilled labor

59. At $p = \$80$, $q = 49$ and $dq/dp = -\frac{5}{16}$, which means that if the price is increased to $81, quantity demanded will decrease by approximately $\frac{5}{16}$ unit.

61. $-0.000436y$ **63.** $\dfrac{dh}{dt} = -\dfrac{3}{44} - \dfrac{h}{12}$

Exercise Set 11.4

1. 36   **3.** $\frac{1}{8}$ **5.** $-\frac{24}{5}$ **7.** $\frac{7}{6}$

9. -5 if $z = 5$, -10 if $z = -5$

11. -80 units/sec   **13.** 12π ft²/min

15. $\frac{16}{27}$ in./sec   **17.** $1798/day   **19.** $0.42/day

21. 430 units/month   **23.** 36π mm³/month

25. $\dfrac{\frac{dW}{dt}}{W} = 3\left(\dfrac{\frac{dL}{dt}}{L}\right)$ **27.** $\dfrac{\frac{dC}{dt}}{C} = 1.54\left(\dfrac{\frac{dW}{dt}}{W}\right)$

29. $\dfrac{1}{4\pi}$ micrometer/day   **31.** $1/(20\pi)$ in./min

33. 0.75 ft/sec   **35.** $-120\sqrt{6}$ mph ≈ -294 mph

37. approaching at 61.18 mph   **39.** $\frac{1}{25}$ ft/hr

Exercise Set 11.5

1. (a) 1   **(b)** no change

3. (a) 84   **(b)** Revenue will decrease.

5. (a) $\frac{100}{99}$ **(b)** elastic   **(c)** decrease

7. (a) 0.81   **(b)** inelastic   **(c)** increase

9. (a) $\eta = 11.1$ (approximately)   **(b)** elastic

11. (a) $\eta = \dfrac{375 - 3q}{q}$

 (b) unitary: $q = 93.75$; inelastic: $q > 93.75$; elastic: $q < 93.75$

 (c) As q increases over $0 < q < 93.75$, p decreases, so elastic demand means R increases. Similarly, R decreases for $q > 93.75$.

 (d) Maximum for R when $q = 93.75$; yes.

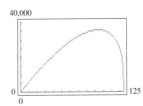

13. $12/item   **15.** $t = \$350$ **17.** $115/item

19. $483 per item; $40,100   **21.** $1100/item

Chapter 11 Review Exercises

1. $dy/dx = (6x - 1)e^{3x^2 - x}$ **2.** $y' = 2x$

3. $\dfrac{dp}{dq} = \dfrac{1}{q} - \dfrac{2q}{q^2 - 1}$

4. $dy/dx = e^{x^2}(2x^2 + 1)$

5. $f'(x) = 10e^{2x} + 4e^{-0.1x}$

6. $g'(x) = 18e^{3x + 1}(2e^{3x + 1} - 5)^2$

7. $\dfrac{dy}{dx} = \dfrac{12x^3 + 14x}{3x^4 + 7x^2 - 12}$

8. $\dfrac{ds}{dx} = \dfrac{9x^{11} - 6x^3}{x^{12} - 2x^4 + 5}$ **9.** $dy/dx = 3^{3x - 3} \ln 3$

10. $dy/dx = \dfrac{1}{\ln 8}\left(\dfrac{10}{x}\right)$ **11.** $\dfrac{dy}{dx} = \dfrac{1 - \ln x}{x^2}$

12. $dy/dx = -2e^{-x}/(1 - e^{-x})^2$

13. $y = 12ex - 8e$, or $y \approx 32.62x - 21.75$

14. $y = x - 1$ **15.** $\dfrac{dy}{dx} = \dfrac{y}{x(10y - \ln x)}$

16. $dy/dx = ye^{xy}/(1 - xe^{xy})$ **17.** $dy/dx = 2/y$

18. $\dfrac{dy}{dx} = \dfrac{2(x + 1)}{3(1 - 2y)}$ **19.** $\dfrac{dy}{dx} = \dfrac{6x(1 + xy^2)}{y(5y^3 - 4x^3)}$

20. $d^2y/dx^2 = -(x^2 + y^2)/y^3 = -1/y^3$ **21.** 5/9

22. $\left(-2, \pm\sqrt{\frac{2}{3}}\right)$ **23.** 3/4   **24.** 11 square units/min

25. (a) $y'(t) = \dfrac{2.62196}{t}$

 (b) $y(50) \approx 6.343$ is the predicted number of hectares of deforestation in 2000.
 $y'(50) \approx 0.05244$ hectares per year is the predicted rate of deforestation in 2000.

26. 135.3 dollars/year

27. (a) 152.5 dollars/year   **(b)** 1.13 times as fast

28. (a) $-0.00001438A_0$ units/year
 (b) $-0.00002876A_0$ units/year   **(c)** less

29. $1200e \approx \$3261.94$ per unit

30. −$603.48 per year   **31.** $-1/(25\pi)$ mm/min

32. $\frac{48}{25}$ ft/min   **33.** $\dfrac{dS/dt}{S} = \dfrac{1}{3}\left(\dfrac{dA/dt}{A}\right)$ **34.** yes

35. $t = \$1466.67, T \approx \$58,667$

36. $t = \$880, T = \$3520$

37. (a) 1   (b) no change   **38.** $\frac{25}{12}$, elastic   **39.** 1

40. (a)

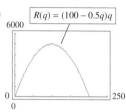

(b) $q = 100$

(c) max revenue at $q = 100$

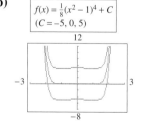

(d) Revenue is maximized where elasticity is unitary.

Chapter 11 Test

1. $y' = 15x^2 e^{x^3} + 2x$ **2.** $y' = \dfrac{12x^2}{x^3 + 1}$

3. $y' = \dfrac{12x^3}{x^4 + 1}$ **4.** $f'(x) = 20(3^{2x}) \ln 3$

5. $\dfrac{dS}{dt} = e^{t^4}(4t^4 - 1)$ **6.** $y' = \dfrac{e^{x^3+1}(3x^3 - 1)}{x^2}$

7. $y' = \dfrac{1 - \ln x}{x^2}$ **8.** $g'(x) = \dfrac{8}{(4x + 7) \ln 5}$

9. $y' = \dfrac{-3x^3}{y}$ **10.** $-\frac{3}{2}$ **11.** $y' = \dfrac{-e^y}{xe^y - 10}$

12. $1349.50 per week   **13.** $\eta = 3.71$; decreases

14. −0.05 unit per dollar   **15.** 586 units per day

16. (a) 1990: $y'(90) \approx 31.14$ million shares daily/year

(b) 2010: $y'(110) \approx 641.92$ million shares daily/year

17. $540

18. (a) $y = -31.5026 + 22.9434 \ln x$

(b) $y' = \dfrac{22.9434}{x}$

(c) $y(70) \approx 65.97\%$ means that this percent of roads and streets will be paved in 2010. $y'(70) \approx 0.328\%$ per year is the predicted rate of change of the percent of paved roads and streets in 2010.

19. $P'(50) \approx -0.0155$ means that in 2010 the purchasing power of a dollar is changing at the rate of $−0.0155 per year.

Exercise Set 12.1

1. $x^4 + C$ **3.** $\frac{1}{7}x^7 + C$ **5.** $\frac{1}{8}x^8 + C$

7. $\frac{4}{3}x^6 + C$ **9.** $27x + \frac{1}{14}x^{14} + C$

11. $3x - \frac{2}{5}x^{5/2} + C$ **13.** $\frac{1}{5}x^5 - 3x^3 + 3x + C$

15. $2x + \frac{4}{3}x\sqrt{x} + C$ **17.** $\frac{24}{5}x\sqrt[4]{x} + C$

19. $-5/(3x^3) + C$ **21.** $\frac{3}{2}\sqrt[3]{x} + C$

23. $\dfrac{1}{4}x^4 - 4x - \dfrac{1}{x^5} + C$

25. $\dfrac{1}{10}x^{10} + \dfrac{1}{2x^2} + 3x^{2/3} + C$

27. $2x^8 - \frac{4}{3}x^6 + \frac{1}{4}x^4 + C$ **29.** $-1/x - 1/(2x^2) + C$

31.

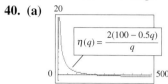

33. $f(x) = 18x^8 - 35x^4$ **35.** $\int (5 - \frac{1}{2}x)\,dx$

37. $\int (3x^2 - 6x)\,dx$ **39.** $R(x) = 30x - 0.2x^2$

41. $R(50) = \$22,125$ **43.** $P(t) = \frac{1}{4}t^4 + \frac{4}{3}t^3 + 6t$

45. (a) $x = t^{7/4}/1050$ (b) 0.96 ton

47. (a) $\overline{C}(x) = x/4 + 100/x + 30$ (b) $56 per unit

49. (a) $t \approx 3.9$ years past 2000

(b) $B = -1.0568t^2 + 8.2593t + 74.07$

(c) $t \approx 13$ years past 2000

51. (a) Yes; $\dfrac{dA}{dt} > 0$ for $t > 0$ so A is increasing and $\dfrac{dA}{dt}$ increases for increasing t.

(b) $A = 105.095t^{1.5307} - 0.012$

53. (a) $H = 0.0042t^3 + 0.766t^2 - 5.08t + 26.7$

(b) $H(50) \approx \$2212.7$ billion

Exercise Set 12.2

1. $du = 10x^4\,dx$ **3.** $\frac{1}{4}(x^2 + 3)^4 + C$

5. $(15x^2 + 10)^5/5 + C$ **7.** $\frac{1}{3}(3x - x^3)^3 + C$

9. $\frac{1}{8}(x^2 + 5)^4 + C$ **11.** $\frac{1}{4}(4x - 1)^7 + C$

13. $-\frac{1}{4}(x^2 + 1)^{-2} + C$ **15.** $\frac{1}{10}(x^2 - 2x + 5)^5 + C$

17. $-\frac{1}{8}(x^4 - 4x + 3)^{-4} + C$ **19.** $\frac{7}{6}(x^4 + 6)^{3/2} + C$

21. $\frac{3}{8}x^8 + \frac{6}{5}x^5 + \frac{3}{2}x^2 + C$ **23.** $\frac{72}{7}x^7 - \frac{48}{5}x^5 + \frac{8}{3}x^3 + C$

25. $\frac{2}{9}(x^3 - 3x)^{3/2} + C$ **27.** $\dfrac{-1}{[10(2x^5 - 5)^3]} + C$

29. $\dfrac{-1}{[8(x^4 - 4x)^2]} + C$ **31.** $\frac{2}{3}\sqrt{x^3 - 6x^2 + 2} + C$

33. $f(x) = 70(7x - 13)^9$

35. (a) $f(x) = \frac{1}{8}(x^2 - 1)^4 + C$

(b)

37. (a) $F(x) = \frac{15}{4}(2x - 1)^{2/5} + C$
(b)

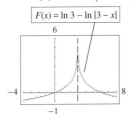

$F(x) = \frac{15}{4}(2x-1)^{2/5} - \frac{7}{4}$

(c) $x = \frac{1}{2}$
(d) vertical

39. $\displaystyle\int \frac{8x(x^2 - 1)^{1/3}}{3} \, dx$

41. (b) $\dfrac{-7}{3(x^3 + 4)} + C$
(d) $\int (x^2 + 5)^{-4} \, dx$ (Many answers are possible.)

43. $R(x) = \dfrac{15}{2x + 1} + 30x - 15$

45. 3720 bricks **47. (a)** $s = 10\sqrt{x + 1}$ **(b)** 50
49. (a) $A(t) = 100/(t + 10) - 1000/(t + 10)^2$
(b) 2.5 million
51. 7400
53. (a) $p(t) = -0.001905(t + 60)^3 + 0.4783(t + 60)^2$
$\qquad\qquad - 39.37t - 1270.785$

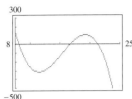

(b)

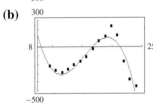

(c) The equation fits quite well overall, but is somewhat inaccurate for the mid-1990s and for 2002.
55. (a) $D(t) = -1975(0.1t - 1)^3 + 3357.4(0.1t - 1)^2$
$\qquad\qquad - 106.47t + 843.5$

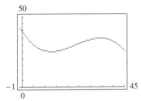

(b)

(c) The data fit quite well overall, with most inaccuracies occurring during the years of surplus.

Exercise Set 12.3

1. $e^{3x} + C$ **3.** $-e^{-x} + C$ **5.** $10{,}000e^{0.1x} + C$
7. $-1200e^{-0.7x} + C$ **9.** $\frac{1}{12}e^{3x^4} + C$
11. $-\frac{3}{2}e^{-2x} + C$ **13.** $\frac{1}{18}e^{3x^6 - 2} + C$
15. $\frac{1}{4}e^{4x} + 6/e^{x/2} + C$
17. $\ln|x^3 + 4| + C$ **19.** $\frac{1}{4}\ln|4z + 1| + C$
21. $\frac{1}{4}\ln|x^4 + 1| + C$ **23.** $2\ln|x^2 - 4| + C$
25. $\ln|x^3 - 2x| + C$ **27.** $\frac{1}{3}\ln|z^3 + 3z + 17| + C$
29. $\frac{1}{3}x^3 + \ln|x - 1| + C$ **31.** $x + \frac{1}{2}\ln|x^2 + 3| + C$
33. $f(x) = h(x), \int f(x)\,dx = g(x)$
35. $F(x) = -\ln|3 - x| + C$

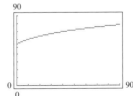

$F(x) = \ln 3 - \ln|3 - x|$

37. $f(x) = 1 + \dfrac{1}{x}; \displaystyle\int\left(1 + \frac{1}{x}\right)dx$

39. $f(x) = 5e^{-x} - 5xe^{-x}; \int (5e^{-x} - 5xe^{-x})\,dx$
41. (c) $\frac{1}{3}\ln|x^3 + 3x^2 + 7| + C;$ **(d)** $\frac{5}{8}e^{2x^4} + C$
43. \$1030.97 **45.** $n = n_0e^{-Kt}$ **47.** 55
49. (a) $S = Pe^{0.1n}$ **(b)** ≈ 7 years
51. (a) $p = 95e^{-0.491t}$ **(b)** ≈ 90.45
53. (a) $l(t) = 14.202\ln|t + 20| + 11.283$

(b)

(c) The model is an excellent fit.
55. (a) Yes, the exponential factor is always positive; hence -0.0548 times that factor is always negative. $P'(t) < 0$ means P is decreasing.
(b) $P(t) = 1.598e^{-0.0343t} + 0.0228$
(c) $P(45) \approx 0.36418$ and $P'(45) \approx -0.0117$ mean that for 2015 the model predicts the CPI will be \$0.36418 and changing at the rate of -0.0117 dollar per year.

Exercise Set 12.4

1. $C(x) = x^2 + 100x + 200$
3. $C(x) = 2x^2 + 2x + 80$ 5. $3750
7. (a) $x = 3$ units is optimal level
   (b) $P(x) = -4x^2 + 24x - 200$ (c) loss of $164
9. (a) profit of $3120   (b) 896 units
11. (a) $\overline{C}(x) = \frac{6}{x} + \frac{x}{6} + 8$ (b) $10.50
13. (a) and (b)

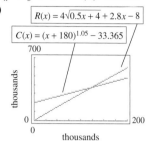

$R(x) = 4\sqrt{0.5x + 4} + 2.8x - 8$

$C(x) = (x + 180)^{1.05} - 33.365$

(c) Maximum profit is $114.743 thousand at $x = 200$ thousand units.
15. $C(y) = 0.80y + 7$
17. $C(y) = 0.3y + 0.4\sqrt{y} + 8$
19. $C(y) = 2\sqrt{y + 1} + 0.4y + 4$
21. $C(y) = 0.7y + 0.5e^{-2y} + 5.15$
23. $C(y) = 0.85y + 5.15$
25. $C(y) = 0.8y + \dfrac{2\sqrt{3y + 7}}{3} + 4.24$

Exercise Set 12.5

1. $4y - 2xy' = 4x^2 - 2x(2x) = 0$ ✓
3. $2y\,dx - x\,dy = 2(3x^2 + 1)\,dx - x(6x\,dx) = 2\,dx$ ✓
5. $y = \frac{1}{2}e^{x^2 + 1} + C$ 7. $y^2 = 2x^2 + C$
9. $y^3 = x^2 - x + C$ 11. $y = e^{x-3} - e^{-3} + 2$
13. $y = \ln|x| - \dfrac{x^2}{2} + \dfrac{1}{2}$ 15. $\dfrac{y^2}{2} = \dfrac{x^3}{3} + C$
17. $\dfrac{1}{2x^2} + \dfrac{y^2}{2} = C$ 19. $\dfrac{1}{x} + y + \dfrac{y^3}{3} = C$
21. $\dfrac{1}{y} + \ln|x| = C$ 23. $x^2 - y^2 = C$
25. $y = C(x + 1)$ 27. $y - \ln|y + 1| = -\frac{1}{2}e^{-2x} + C$
29. $3y^4 = 4x^3 - 1$
31. $2y = 3x + 4xy$ or $y = \dfrac{3x}{2 - 4x}$
33. $e^{2y} = x^2 - \dfrac{2}{x} + 2$ 35. $y^2 + 1 = 5x$
37. $y = Cx^k$
39. (a) $x = 10,000e^{0.06t}$ (b) $10,618.37; $13,498.59
    (c) 11.55 years
41. $P = 100,000e^{0.05t}$; 5%   43. ≈ 8.4 hours
45. $y = \dfrac{32}{(p + 8)^{2/5}}$ 47. ≈ 23,100 years
49. $x = 6(1 - e^{-0.05t})$ 51. $x = 20 - 10e^{-0.025t}$
53. $V = 1.86e^{2 - 2e^{-0.1t}}$ 55. $V = \dfrac{k^3t^3}{27}$

57. $t ≈ 4.5$ hours
59. (a) $E(t) = 0.13e^{0.0315t}$
    (b) The graph is a similar, but smooth, representation of the given data.

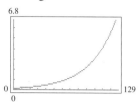

61. (a) $P = 50,000e^{-0.05t}$ (b) $11,156.50

Chapter 12 Review Exercises

1. $\frac{1}{7}x^7 + C$ 2. $\frac{2}{3}x^{3/2} + C$
3. $3x^4 - x^3 + 2x^2 + 5x + C$
4. $\frac{7}{5}x^5 - \frac{14}{3}x^3 + 7x + C$
5. $\frac{7}{6}(x^2 - 1)^3 + C$ 6. $\frac{1}{18}(x^3 - 3x^2)^6 + C$
7. $\frac{3}{8}x^8 + \frac{24}{5}x^5 + 24x^2 + C$ 8. $\frac{5}{63}(3x^3 + 7)^7 + C$
9. $\frac{1}{3}\ln|x^3 + 1| + C$ 10. $\dfrac{-1}{3(x^3 + 1)} + C$
11. $\frac{1}{2}(x^3 - 4)^{2/3} + C$ 12. $\frac{1}{3}\ln|x^3 - 4| + C$
13. $\dfrac{1}{2}x^2 - \dfrac{1}{x} + C$
14. $\frac{1}{3}x^3 + \frac{1}{2}x^2 - 2x - \ln|x - 1| + C$
15. $\frac{1}{3}e^{y^3} + C$
16. $\frac{1}{3}(x - 1)^3 + C$ or $x^3/3 - x^2 + x + C$
17. $\frac{1}{2}\ln|2x^3 - 7| + C$ 18. $\dfrac{-5}{4e^{4x}} + C$
19. $x^4/4 - e^{3x}/3 + C$ 20. $\frac{1}{2}e^{x^2 + 1} + C$
21. $\dfrac{-3}{40(5x^8 + 7)^2} + C$ 22. $-\frac{7}{2}\sqrt{1 - x^4} + C$
23. $\frac{1}{4}e^{2x} - e^{-2x} + C$ 24. $x^2/2 + 1/(x + 1) + C$
25. (a) $\frac{1}{10}(x^2 - 1)^5 + C$ (b) $\frac{1}{22}(x^2 - 1)^{11} + C$
    (c) $\frac{3}{16}(x^2 - 1)^8 + C$ (d) $\frac{3}{2}(x^2 - 1)^{1/3} + C$
26. (a) $\ln|x^2 - 1| + C$ (b) $\dfrac{-1}{x^2 - 1} + C$
    (c) $3\sqrt{x^2 - 1} + C$ (d) $\frac{3}{2}\ln|x^2 - 1| + C$
27. $y = C - 92e^{-0.05t}$
28. $y = 64x + 38x^2 - 12x^3 + C$
29. $(y - 3)^2 = 4x^2 + C$ 30. $(y + 1)^2 = 2\ln|t| + C$
31. $e^y = \dfrac{x^2}{2} + C$ 32. $y = Ct^4$
33. $3(y + 1)^2 = 2x^3 + 75$
34. $x^2 = y + y^2 + 4$ 35. $96   36. 472
37. $P(t) = 400[1 - 5/(t + 5) + 25/(t + 5)^2]$
38. $p = 1990.099 - 100,000/(t + 100)$
39. (a) $y = -60e^{-0.04t} + 60$ (b) 23%
40. $R(x) = 800\ln(x + 1)$
41. (a) $1000   (b) $C(x) = 3x^2 + 4x + 1000$
42. 80 units, $440

43. $C(y) = \sqrt{2y + 16} + 0.6y + 4.5$

44. $C(y) = 0.8y - 0.05e^{-2y} + 7.85$ **45.** $W = CL^3$

46. (a) $\ln |P| = kt + C_1$ (b) $P = Ce^{kt}$
   (c) $P = 50,000\, e^{0.1t}$ (d) The interest rate is
   $k = 0.10 = 10\%$

47. ≈ 10.7 million years   **48.** $x = 360(1 - e^{-t/30})$

49. $x = 600 - 500e^{-0.01t}$; ≈ 161 min

Chapter 12 Test

1. $2x^3 + 4x^2 - 7x + C$ **2.** $4x + \frac{2}{3}x\sqrt{x} + \frac{1}{x} + C$

3. $\dfrac{(4x^3 - 7)^{10}}{24} + C$ **4.** $-\frac{1}{6}(3x^2 - 6x + 1)^{-2} + C$

5. $\dfrac{\ln |2s^4 - 5|}{8} + C$ **6.** $-10,000e^{-0.01x} + C$

7. $\frac{5}{8}e^{2y^4 - 1} + C$ **8.** $e^x + 5\ln |x| - x + C$

9. $\dfrac{x^2}{2} - x + \ln |x + 1| + C$ **10.** $6x^2 - 1 + 5e^x$

11. $y = x^4 + x^3 + 4$ **12.** $y = \frac{1}{4}e^{4x} + \frac{7}{4}$

13. $y = \dfrac{4}{C - x^4}$ **14.** 157,498

15. $P(x) = 450x - 2x^2 - 300$

16. $C(y) = 0.78y + \sqrt{0.5y + 1} + 5.6$

17. 332.3 days

Exercise Set 13.1

1. 7 square units   **3.** 7.25 square units

5. 3 square units   **7.** 11.25 square units

9. $S_L(10) = 4.08$; $S_R(10) = 5.28$

11. Both equal 14/3.

13. It would lie between $S_L(10)$ and $S_R(10)$. It would equal 14/3.

15. 3   **17.** 42   **19.** −5   **21.** 180   **23.** 11,315

25. $3 - \dfrac{3(n + 1)}{n} + \dfrac{(n + 1)(2n + 1)}{2n^2} = \dfrac{2n^2 - 3n + 1}{2n^2}$

27. (a) $S = (n - 1)/n$ (b) 9/10   (c) 99/100
   (d) 999/1000   (e) 1

29. (a) $S = \dfrac{(n + 1)(2n + 1)}{6n^2}$

   (b) $77/200 = 0.385$ (c) $6767/20,000 \approx 0.3384$
   (d) $667,667/2,000,000 \approx 0.3338$ (e) $\frac{1}{3}$

31. $\frac{20}{3}$ **33.** $234 billion

35. There are approximately 90 squares under the curve, each representing 1 second by 10 mph, or

   $1\ \text{sec} \times \dfrac{10\ \text{mi}}{1\ \text{hr}} \times \dfrac{1\ \text{hr}}{3600\ \text{sec}} = \dfrac{1}{360}\ \text{mile}.$

   The area under the curve is approximately
   $90\left(\frac{1}{360}\ \text{mile}\right) = \frac{1}{4}\ \text{mile}.$

37. 1550 square feet

39. 466.1875 square units. This represents the total spent for wireless services between 2000 and 2005: $466.1875 billion.

Exercise Set 13.2

1. 18   **3.** 2   **5.** 60   **7.** $12\sqrt[3]{25}$ **9.** 98

11. $-\frac{1}{10}$ **13.** 12,960   **15.** 0   **17.** $\frac{49}{3}$

19. 2   **21.** $e^3/3 - 1/3$ **23.** 4   **25.** $\frac{8}{3}(1 - e^{-8})$

27. (a) $\frac{1}{6}\ln (112/31) \approx 0.2140853$ (b) 0.2140853

29. (a) $\frac{3}{2} + 3\ln 2 \approx 3.5794415$ (b) 3.5794415

31. (a) A, C (b) B

33. $\int_0^4 (2x - \frac{1}{2}x^2)\, dx$ (b) 16/3

35. (a) $\int_{-1}^0 (x^3 + 1)\, dx$ (b) 3/4

37. $\frac{1}{6}$ **39.** $\frac{1}{2}(e^9 - e)$

41. $\int_0^a g(x)\, dx > \int_0^a f(x)\, dx$; more area under $g(x)$

43. same absolute values, opposite signs

45. 6   **47.** 0   **49.** (a) $450,000   (b) $450,000

51. (a) $7007   (b) $19,649

53. $20,405.39

55. 15,233.6 billion dollars of revenue from 2000 to 2010

57. 0.04 cm³   **59.** 1222 (approximately)

61. 0.1808

63. (a) 0.5934   (b) 0.1733

65. (a) $E(t) = 0.001t^4 - 0.058t^3 + 0.979t^2$
   $- 6.77t + 204.1$

   (b) Total millions of short tons of CO emissions during the 1990s were about 1340.

Exercise Set 13.3

1. (a) $\int_0^2 (4 - x^2)\, dx$ (b) $\frac{16}{3}$

3. (a) $\int_1^8 \left[\sqrt[3]{x} - (2 - x)\right] dx$ (b) 28.75

5. (a) $\int_1^2 \left[(4 - x^2) - (\frac{1}{4}x^3 - 2)\right] dx$ (b) 131/48

7. (a) $(-1, 1), (2, 4)$ (b) $\int_{-1}^2 \left[(x + 2) - x^2\right] dx$
   (c) 9/2

9. (a) $(0, 0), \left(\frac{5}{2}, -\frac{15}{4}\right)$
   (b) $\int_0^{5/2} \left[(x - x^2) - (x^2 - 4x)\right] dx$ (c) $\frac{125}{24}$

11. (a) $(-2, -4), (0, 0), (2, 4)$
   (b) $\int_{-2}^0 \left[(x^3 - 2x) - 2x\right] dx + \int_0^2 \left[2x - (x^3 - 2x)\right] dx$
   (c) 8

13. $\frac{28}{3}$ **15.** $\frac{1}{4}$ **17.** $\frac{16}{3}$ **19.** $\frac{1}{3}$ **21.** $\frac{37}{12}$

23. $4 - 3\ln 3$ **25.** $\frac{8}{3}$ **27.** 6   **29.** 0   **31.** $-\frac{4}{9}$

33. $11.8\overline{3}$

35. average profit $= \dfrac{1}{x_1 - x_0} \displaystyle\int_{x_0}^{x_1} (R(x) - C(x))\, dx$

37. (a) $1402 per unit   (b) $535,333.33

39. (a) 102.5 units   (b) 100 units

41. $10,465 (nearest dollar)   **43.** 147 milligrams

45. 1988: 0.4034; 2000: 0.4264
   More equally distributed after Reagan. This is contrary to conventional wisdom.

47. Blacks: 0.4435; Asians: 0.4297
   Income was more nearly equally distributed among Asians, although both distributions were similar and not particularly equal.

Exercise Set 13.4

1. $120,000   **3.** $346,664 (nearest dollar)
5. $506,000 (nearest thousand)
7. $18,660 (nearest dollar)
9. $82,155 (nearest dollar)
11. $PV = \$265,781$ (nearest dollar), $FV = \$377,161$ (nearest dollar)
13. $PV = \$190,519$ (nearest dollar), $FV = \$347,148$ (nearest dollar)
15. Gift Shoppe, $151,024; Video Palace, $141,093. The gift shop is a better buy.
17. $83.33   **19.** $161.89   **21.** (5, 56); $83.33
23. $11.50   **25.** $204.17   **27.** $2766.67
29. $17,839.58   **31.** $133.33   **33.** $2.50
35. $103.35

Exercise Set 13.5

1. formula 5: $\frac{1}{8} \ln |(4 + x)/(4 - x)| + C$
3. formula 11: $\frac{1}{3} \ln [(3 + \sqrt{10})/2]$
5. formula 14: $w(\ln w - 1) + C$
7. formula 12: $\frac{1}{3} + \frac{1}{4} \ln \left(\frac{3}{7}\right)$
9. formula 13: $\frac{1}{8} \ln \left| \dfrac{v}{3v + 8} \right| + C$
11. formula 7: $\frac{1}{2}[7\sqrt{24} - 25 \ln (7 + \sqrt{24}) + 25 \ln 5]$
13. formula 16: $\dfrac{(6w - 5)(4w + 5)^{3/2}}{60} + C$
15. formula 3: $\frac{1}{2}(5^{x^2})\log_5 e + C$
17. formula 1: $\frac{1}{3}(13^{3/2} - 8)$
19. formula 9: $-\frac{5}{2} \ln \left| \dfrac{2 + \sqrt{4 - 9x^2}}{3x} \right| + C$
21. formula 10: $\frac{1}{3} \ln |3x + \sqrt{9x^2 - 4}| + C$
23. formula 15: $\frac{3}{4}\left[\ln |2x - 5| - \dfrac{5}{2x - 5} \right] + C$
25. formula 8: $\frac{1}{3} \ln |3x + 1 + \sqrt{(3x + 1)^2 + 1}| + C$
27. formula 6: $\frac{1}{4}[10\sqrt{109} - \sqrt{10} + 9 \ln (10 + \sqrt{109}) - 9 \ln (1 + \sqrt{10})]$
29. formula 2: $-\frac{1}{6} \ln |7 - 3x^2| + C$
31. formula 8: $\frac{1}{2} \ln |2x + \sqrt{4x^2 + 7}| + C$
33. $2(e^{\sqrt{2}} - e) \approx 2.7899$
35. $\frac{1}{32}[\ln (9/5) - 4/9] \approx 0.004479$ **37.** $3391.10
39. (a) $C = \frac{1}{2}x\sqrt{x^2 + 9} + \frac{9}{2} \ln |x + \sqrt{x^2 + 9}| + 300 - \frac{9}{2} \ln 3$
   (b) $314.94
41. $3882.9 thousand

Exercise Set 13.6

1. $\frac{1}{2}xe^{2x} - \frac{1}{4}e^{2x} + C$ **3.** $\frac{1}{3}x^3 \ln x - \frac{1}{9}x^3 + C$
5. $\dfrac{104\sqrt{2}}{15}$ **7.** $-(1 + \ln x)/x + C$ **9.** 1
11. $\dfrac{x^2}{2} \ln (2x - 3) - \frac{1}{4}x^2 - \frac{3}{4}x - \frac{9}{8} \ln (2x - 3) + C$
13. $\frac{1}{5}(q^2 - 3)^{3/2}(q^2 + 2) + C$ **15.** 282.4
17. $-e^{-x}(x^2 + 2x + 2) + C$ **19.** $(3e^4 + 1)/2$
21. $\frac{1}{4}x^4 \ln^2 x - \frac{1}{8}x^4 \ln x + \frac{1}{32}x^4 + C$
23. $\frac{2}{15}(e^x + 1)^{3/2}(3e^x - 2) + C$ **25.** II; $\frac{1}{2}e^{x^2} + C$
27. IV; $\frac{2}{3}(e^x + 1)^{3/2} + C$ **29.** I; $-5e^{-4} + 1$
31. $2794.46   **33.** $34,836.73   **35.** 0.264
37. $5641.3 billion

Exercise Set 13.7

1. $1/5$ **3.** 2   **5.** $1/e$ **7.** diverges   **9.** diverges
11. 10   **13.** diverges   **15.** diverges   **17.** diverges
19. 0   **21.** 0.5   **23.** $1/(2e)$ **25.** $\frac{3}{2}$
27. $\int_{-\infty}^{\infty} f(x) \, dx = 1$ **29.** $c = 1$ **31.** $c = \frac{1}{4}$
33. 20   **35.** area $= \frac{8}{3}$ **37.** $\int_0^{\infty} Ae^{-rt} \, dt = A/r$
39. $2,400,000   **41.** $700,000
43. (a) 0.368   (b) 0.018
45. 0.147
47. (a) $500 \left[\dfrac{e^{-0.03b} + 0.03b - 1}{0.0009} \right]$
   (b) the amount approaches ∞

Exercise Set 13.8

1. $h = \frac{1}{2}; x_0 = 0, x_1 = \frac{1}{2}, x_2 = 1, x_3 = \frac{3}{2}, x_4 = 2$
3. $h = \frac{1}{2}; x_0 = 1, x_1 = \frac{3}{2}, x_2 = 2, x_3 = \frac{5}{2}, x_4 = 3, x_5 = \frac{7}{2}, x_6 = 4$
5. $h = 1; x_0 = -1, x_1 = 0, x_2 = 1, x_3 = 2, x_4 = 3, x_5 = 4$
7. (a) 9.13   (b) 9.00   (c) 9   (d) Simpson's
9. (a) 0.51   (b) 0.50   (c) $\frac{1}{2}$ (d) Simpson's
11. (a) 5.27   (b) 5.30   (c) 5.33   (d) Simpson's
13. (a) 3.283   (b) 3.240
15. (a) 0.743   (b) 0.747
17. (a) 7.132   (b) 7.197   **19.** 7.8   **21.** 10.3
23. 119.58 ($119,580)   **25.** $32,389.76
27. $14,133.33   **29.** 1222.35 (1222 units)
31. 1990: 0.4348; 2003: 0.4416
   More equally distributed in 1990, but only slightly
33. 1586.67 ft^2

Chapter 13 Review Exercises

1. 212   **2.** $\dfrac{3(n + 1)}{2n^2}$ **3.** $\frac{91}{72}$ **4.** 1   **5.** 1
6. 14   **7.** $\frac{248}{5}$ **8.** $-\frac{205}{4}$ **9.** $\frac{825}{4}$ **10.** $\frac{4}{13}$
11. -2 **12.** $\frac{1}{6} \ln 47 - \frac{1}{6} \ln 9$ **13.** $\frac{9}{2}$

14. $\ln 4 + \frac{14}{3}$ **15.** $\frac{26}{3}$ **16.** $\frac{1}{2}\ln 2$
17. $(1 - e^{-2})/2$ **18.** $(e - 1)/2$ **19.** $95/2$
20. 36 **21.** $\frac{1}{4}$ **22.** $\frac{1}{2}$
23. $\frac{1}{2}x\sqrt{x^2 - 4} - 2\ln|x + \sqrt{x^2 - 4}| + C$
24. $2\log_3 e$ **25.** $\frac{1}{2}x^2(\ln x^2 - 1) + C$
26. $\frac{1}{2}\ln|x| - \frac{1}{2}\ln|3x + 2| + C$
27. $\frac{1}{6}x^6 \ln x - \frac{1}{36}x^6 + C$
28. $(-xe^{-2x}/2) - (e^{-2x}/4) + C$
29. $2x\sqrt{x + 5} - \frac{4}{3}(x + 5)^{3/2} + C$
30. 1 **31.** diverges **32.** -100
33. $\frac{5}{3}$ **34.** $-\frac{1}{2}$
35. (a) $\frac{8}{9} \approx 0.889$ **(b)** 1.004 **(c)** 0.909
36. 3.135 **37.** 3.9
38. (a) $n = 5$ **(b)** $n = 6$ **39.** \$28,000
40. $e^{-2.8} \approx 0.061$ **41.** \$1297.44 **42.** \$76.60
43. 1969: 0.3737; 2000: 0.4264; more equally distributed in 1969
44. (a) $(7, 6)$ **(b)** \$7.33 **45.** \$24.50
46. \$1,621,803 **47. (a)** \$403,609 **(b)** \$602,114
48. \$217.42 **49.** \$10,066 (nearest dollar)
50. \$86,557.41
51. $C(x) = 3x + 30(x + 1)^2 \ln(x + 1)$
$\qquad - 15(x + 1)^2 + 2015$
52. $e^{-1.4} \approx 0.247$
53. \$4000 thousand, or \$4 million
54. \$197,365 **55.** \$480,000

Chapter 13 Test

1. 3.496 (approximately)
2. (a) $5 - \dfrac{n + 1}{n}$ **(b)** 4
3. $\int_0^6 (12 + 4x - x^2)\, dx$; 72
4. (a) 4 **(b)** 3/4 **(c)** $\frac{5}{4}\ln 5$ **(d)** 7
(e) 0; limits of integration are the same
(f) $\frac{5}{6}(e^2 - 1)$
5. (a) $3xe^x - 3e^x + C$ **(b)** $\dfrac{x^2}{2}\ln(2x) - \dfrac{x^2}{4} + C$
6. -8
7. (a) $x[\ln(2x) - 1] + C$
(b) $\dfrac{2(9x + 14)(3x - 7)^{3/2}}{135} + C$
8. 16.089 **9. (a)** \$4000 **(b)** \$16,000/3
10. (a) \$961.18 thousand **(b)** \$655.68 thousand
(c) \$1062.5 thousand
11. 125/6 **12.**

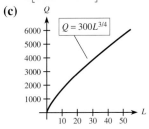

13. Before, 0.446; After, 0.19. The change decreases the difference in income.
14. (a) 1683.863 thousand barrels per day
(b) 1870.967 thousand barrels per day
15. 2.96655 **16.** 6800 ft^2

Exercise Set 14.1

1. $\{(x, y): x \text{ and } y \text{ are real numbers}\}$
3. $\{(x, y): x \text{ and } y \text{ are real numbers and } y \neq 0\}$
5. $\{(x, y): x \text{ and } y \text{ are real numbers and } 2x - y \neq 0\}$
7. $\{(p_1, p_2): p_1 \text{ and } p_2 \text{ are real numbers and } p_1 \geq 0\}$
9. -2 **11.** $\frac{5}{3}$ **13.** 2500 **15.** 36 **17.** 3
19. $\frac{1}{25}\ln(12)$ **21.** $\frac{13}{3}$
23. \$6640.23; the amount that results when \$2000 is invested for 20 years
25. 500; if the cost of placing an order is \$200, the number of items sold per week is 625, and the weekly holding cost per item is \$1, then the most economical order size is 500.
27. Max: $S \approx 112.5°F; A \approx 106.3°F$
Min: $S \approx 87.4°F; A \approx 77.6°F$
29. (a) \$752.80; when \$90,000 is borrowed for 20 years at 8%, the monthly payment is \$752.80.
(b) \$1622.82; when \$160,000 is borrowed for 15 years at 9%, the monthly payment is \$1622.82.
31. (a) $x = 4$ **(b)** $y = 2$
(c)

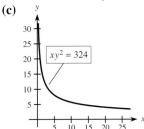

33. (a) 37,500 units
(b) $30(2K)^{1/4}(2L)^{3/4} = 30(2^{1/4})(2^{3/4})K^{1/4}L^{3/4} = 2[30K^{1/4}L^{3/4}]$
(c)

35. (a) 7200 units **(b)** 5000 units **37.** \$284,000

Exercise Set 14.2

1. $\dfrac{\partial z}{\partial x} = 4x^3 - 10x + 6$ $\qquad \dfrac{\partial z}{\partial y} = 9y^2 - 5$
3. $z_x = 3x^2 + 8xy$ $\qquad z_y = 4x^2 + 12y$

5. $\dfrac{\partial f}{\partial x} = 9x^2(x^3 + 2y^2)^2$ $\qquad$ $\dfrac{\partial f}{\partial y} = 12y(x^3 + 2y^2)^2$

7. $f_x = 2x(2x^2 - 5y^2)^{-1/2}$ $\qquad$ $f_y = -5y(2x^2 - 5y^2)^{-1/2}$

9. $\dfrac{\partial C}{\partial x} = -4y + 20xy$ $\qquad$ $\dfrac{\partial C}{\partial y} = -4x + 10x^2$

11. $\dfrac{\partial Q}{\partial s} = \dfrac{2(t^2 + 3st - s^2)}{(s^2 + t^2)^2}$ $\qquad$ $\dfrac{\partial Q}{\partial t} = \dfrac{3t^2 - 4st - 3s^2}{(s^2 + t^2)^2}$

13. $z_x = 2e^{2x} + \dfrac{y}{x}$ $\qquad$ $z_y = \ln x$

15. $\dfrac{\partial f}{\partial x} = \dfrac{y}{xy + 1}$ $\qquad$ $\dfrac{\partial f}{\partial y} = \dfrac{x}{xy + 1}$ $\qquad$ 17. 2

19. 7 $\qquad$ 21. -19

23. (a) 0 $\quad$ (b) $-2xz + 4$ $\quad$ (c) $2y$ $\quad$ (d) $-x^2$

25. (a) $8x_1 + 5x_2$ $\quad$ (b) $5x_1 + 12x_2$ $\quad$ (c) 1

27. (a) 2 $\quad$ (b) 0 $\quad$ (c) 0 $\quad$ (d) $-30y$

29. (a) $2y$ $\quad$ (b) $2x - 8y$ $\quad$ (c) $2x - 8y$ $\quad$ (d) $-8x$

31. (a) $2 + y^2e^{xy}$ $\quad$ (b) $xye^{xy} + e^{xy}$
(c) $xye^{xy} + e^{xy}$ $\quad$ (d) x^2e^{xy}

33. (a) $1/x^2$ $\quad$ (b) 0 $\quad$ (c) 0 $\quad$ (d) $2 + 1/y^2$

35. -6 $\qquad$ 37. (a) $\dfrac{188}{4913}$ $\quad$ (b) $\dfrac{-188}{4913}$ $\qquad$ 39. $2 + 2e$

41. 0 $\qquad$ 43. (a) $24x$ $\quad$ (b) $24x$ $\quad$ (c) 0

45. (a) For a mortgage of \$100,000 and an 8% interest rate, the monthly payment is \$1289.
(b) The rate of change of the payment with respect to the interest rate is \$62.51. That is, if the rate goes from 8% to 9% on a \$100,000 mortgage, the approximate increase in the monthly payment is \$62.51.

47. (a) If the number of items sold per week changes by 1, the most economical order quantity should also increase. $\dfrac{\partial Q}{\partial M} = \sqrt{\dfrac{K}{2Mh}} > 0$
(b) If the weekly storage costs change by 1, the most economical order quantity should decrease.
$\dfrac{\partial Q}{\partial h} = -\sqrt{\dfrac{KM}{2h^3}} < 0$

49. (a) 23.912; If brand 2 is held constant and brand 1 is increased from 100 to 101 liters, approximately 24,000 additional insects will be killed.

51. (a) $2xy^2$ $\quad$ (b) $2x^2y$

53. $\dfrac{\partial Q}{\partial K} = 100$; If labor hours are held constant at 5832 and K changes by \$1 (thousand) to \$730,000, Q will change by about 100 units. $\dfrac{\partial Q}{\partial L} = 25$; If capital expenditures are held constant at \$729,000 and L changes by 1 hour (to 5833), Q will change by about 25 units.

55. (a) $\dfrac{\partial WC}{\partial s} = -0.020t - 1.85 + \dfrac{0.152t}{\sqrt{s}} - \dfrac{13.87}{\sqrt{s}}$
(b) At $t = 10$, $s = 25$, $\dfrac{\partial WC}{\partial s} = -0.20 - 1.85 + 0.304 - 2.774 = -4.52$. This means that if wind speed changes by 1 mph (from 25 mph) while the temperature remains at 10°F, the wind chill temperature will change by about -4.52°F.

Exercise Set 14.3

1. \$75 $\qquad$ 3. (a) $2 + y/50$ $\qquad$ (b) $4 + x/50$

5. (a) \$21.46 $\qquad$ (b) \$72.40

7. (a) If y remains at 10, the expected change in cost for a 9th unit of X is \$36.
(b) If x remains at 8, the expected change in cost for an 11th unit of Y is \$19.

9. (a) $\sqrt{y^2 + 1}$ dollars per unit
(b) $xy/\sqrt{y^2 + 1}$ dollars per unit

11. (a) $1200y/(xy + 1)$ dollars per unit
(b) $1200x/(xy + 1)$ dollars per unit

13. (a) $\sqrt{y/x}$ $\qquad$ (b) $\sqrt{x/y}$

15. (a) $\ln(y + 1)/(2\sqrt{x})$ $\qquad$ (b) $\sqrt{x}/(y + 1)$

17. $z = 1092$ crates (approximately)

19. $z_x = 3.6$; If 500 acres are planted, the expected change in productivity from a 301st hour of labor is 3.6 crates.

21. (a) $z_x = \dfrac{240y^{2/5}}{x^{2/5}}$ $\qquad$ (b)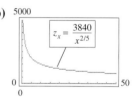

(c) $z_y = \dfrac{160x^{3/5}}{y^{3/5}}$ $\qquad$ (d)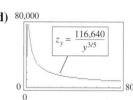

(e) Both z_x and z_y are positive, so increases in both capital investment and work-hours result in increases in productivity. However, both are decreasing, so such increases have a diminishing effect on productivity. Also, z_y decreases more slowly than z_x, so that increases in work-hours have a more significant impact on productivity than increases in capital investment.

23. $q_1 = 188$ units; $q_2 = 270$ units

25. any values for p_1 and p_2 that satisfy $6p_2 - 3p_1 = 100$ and that make q_1 and q_2 nonnegative, such as $p_1 = \$10, p_2 = \$21\frac{2}{3}$

27. **(a)** -3 units per dollar **(b)** -2 units per dollar **(c)** -6 units per dollar **(d)** -5 units per dollar **(e)** complementary

29. **(a)** -50 units per dollar **(b)** $600/(p_B + 1)^2$ units per dollar **(c)** $-400/(p_B + 4)^2$ units per dollar **(d)** $400/(p_A + 4)^2$ units per dollar **(e)** competitive

Exercise Set 14.4

1. max$(0, 0, 9)$ **3.** min$(0, 0, 4)$ **5.** min$(1, -2, 0)$
7. saddle$(1, -3, 8)$ **9.** saddle$(0, 0, 0)$
11. max$(12, 24, 456)$ **13.** min$(-8, 6, -52)$
15. saddle$(0, 0, 0)$; min$(2, 2, -8)$
17. $\hat{y} = 5.7x - 1.4$
19. $x = 5000, y = 128; P = \$25,409.60$
21. $x = \frac{20}{3}, y = \frac{10}{3}; W \approx 1926$ lb
23. $x = 28, y = 100; P = 5987.84$ tons
25. $x = 20$ thousand, $y = 30$ thousand; $P = \$1900$ thousand
27. length $= 100$ in., width $= 100$ in., height $= 50$ in.
29. $x = 15$ thousand, $y = 24$ thousand; $P = \$295$ thousand
31. **(a)** $y = 0.113x + 4.860$, x in years past 1970, $y =$ median age of autos and trucks
 (b) 8.815
 (c) The reasons include higher prices and better quality.
 (d) The median age of autos and trucks is changing at the rate of 0.113 year per year past 1970.
33. **(a)** $y = 0.06254x + 6.1914$, x in years past 2000, y in billions
 (b) 6.94188 billion
 (c) World population is changing at the rate of 0.06254 billion persons per year past 2000.

Exercise Set 14.5

1. 18 at $(3, 3)$ **3.** 35 at $(3, 2)$ **5.** 32 at $(4, 2)$
7. -28 at $\left(3, \frac{5}{2}\right)$ **9.** $\frac{1}{5}$ at $\left(-\frac{2}{5}, -\frac{1}{5}\right)$
11. 3 at $(1, 1, 1)$ **13.** 1 at $(0, 1, 0)$
15. $x = 2, y = 2$
17. $x = 40, y = \frac{40}{3}$

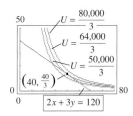

19. **(a)** $x = 400, y = 400$
 (b) $-\lambda = 1.6$ means that each additional dollar spent on production results in approximately 1.6 additional units produced.
 (c)

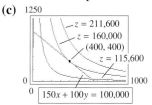

21. $x = 900, y = 300$; 900 units at plant X, 300 units at plant Y
23. $x = \$10,003.33, y = \$19,996.67$
25. length $= 100$ cm, width $= 100$ cm, height $= 50$ cm

Chapter 14 Review Exercises

1. $\{(x, y): x$ and y are real numbers and $y \neq 2x\}$
2. $\{(x, y): x$ and y are real numbers with $y \geq 0$ and $(x, y) \neq (0, 0)\}$
3. -5 **4.** 896,000
5. $15x^2 + 6y$ **6.** $24y^3 - 42x^3y^2$
7. $z_x = 8xy^3 + 1/y; z_y = 12x^2y^2 - x/y^2$
8. $z_x = x/\sqrt{x^2 + 2y^2}; z_y = 2y/\sqrt{x^2 + 2y^2}$
9. $z_x = -2y/(xy + 1)^3; z_y = -2x/(xy + 1)^3$
10. $z_x = 2xy^3e^{x^2y^3}; z_y = 3x^2y^2e^{x^2y^3}$
11. $z_x = ye^{xy} + y/x; z_y = xe^{xy} + \ln x$
12. $z_x = y; z_y = x$ **13.** -8 **14.** 8
15. **(a)** $2y$ **(b)** 0 **(c)** $2x - 3$ **(d)** $2x - 3$
16. **(a)** $18xy^4 - 2/y^2$ **(b)** $36x^3y^2 - 6x^2/y^4$
 (c) $36x^2y^3 + 4x/y^3$ **(d)** $36x^2y^3 + 4x/y^3$
17. **(a)** $2e^{y^2}$ **(b)** $4x^2y^2e^{y^2} + 2x^2e^{y^2}$
 (c) $4xye^{y^2}$ **(d)** $4xye^{y^2}$
18. **(a)** $-y^2/(xy + 1)^2$ **(b)** $-x^2/(xy + 1)^2$
 (c) $1/(xy + 1)^2$ **(d)** $1/(xy + 1)^2$
19. max$(-8, 16, 208)$
20. saddles at $(2, -3, 38)$ and $(-2, 3, -38)$; min at $(2, 3, -70)$; max at $(-2, -3, 70)$
21. 80 at $(2, 8)$ **22.** 11,664 at $(6, 3)$
23. 3 units **24.** 20 units
25. **(a)** 62.8 km (approximately)
 (b) If the surface temperature is 287 K, the Concorde's altitude is 17.1 km, and the vertical temperature gradient is 4.9 K/km, then the width of the sonic boom is about 63.3 km.
 (c) $\frac{\partial f}{\partial h} \approx 1.87$; If the surface temperature is 293 K and the temperature gradient is 5 K/km, then a change of 1 km in the Concorde's altitude would result in a change in the width of the sonic boom of about 1.87 km.

(d) $\dfrac{\partial f}{\partial d} \approx -6.28$; If the surface temperature is 293 K and the Concorde's altitude is 16.8 km, then a change of 1 K/km in the temperature gradient would result in a change in the width of the sonic boom of about −6.28 km.

26. (a) 280 dollars per unit of x
(b) 2400/7 dollars per unit of y

27. $\partial Q/\partial K = 81.92$ means that when capital expenditures increase by \$1000 (to \$626,000) and work-hours remain at 4096, output will change by about 8192 units; $\partial Q/\partial L = 37.5$ means that when labor hours change by 1 (to 4097) and capital expenditures remain at \$625,000, output will change by about 3750 units.

28. (a) −2   **(b)** −6   **(c)** complementary
29. competitive   **30.** $x = 20$, $y = 40$; $P = \$2000$
31. $x = 10$, $y = 4$
32. (a) $x = 1000$, $y = 500$
(b) $-\lambda \approx 3.17$ means that each additional dollar spent on production results in approximately 3 additional units.
(c)

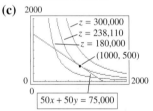

33. (a) $\hat{y} = 0.127x + 8.926$ **(b)** \$897.926 billion
34. $\hat{q} = 34{,}726 - 55.09p$

Chapter 14 Test

1. (a) all pairs (x, y) with $y < x^2$ **(b)** 14
2. $z_x = 5 + 10y(xy + 1)^4$ $z_y = -18y + 10x(xy + 1)^4$
$z_{xx} = 40y^2(xy + 1)^3$ $z_{yy} = -18 + 40x^2(xy + 1)^3$
$z_{xy} = z_{yx} = 10(5xy + 1)(xy + 1)^3$

3. (0, 2), a relative minimum; (4, −6) and (−4, −6), saddle points
4. (a) \$1625 thousand
(b) 73.11 means that if capital investment increases from \$10,000 to \$11,000, the expected change in monthly production value will be \$73.11 thousand, if labor hours remain at 1590.
(c) 0.56 means that if labor hours increase by 1 to 1591, the expected change in monthly production value will be \$0.56 thousand, if capital investment remains at \$10,000.
5. (a) When \$94,500 is borrowed for 25 years at 7%, the monthly payment is \$667.91
(b) If the percent goes from 7% to 8%, the expected change in the monthly payment is \$49.76, if the loan amount remains at \$94,500 for 25 years.
(c) Negative. If the loan amount remains at \$94,500 and the percent remains at 7%, increasing the time to pay off the loan will decrease the monthly payment, and vice versa.
6. $8xy\, e^{x^2y^2}(x^2y^2 + 1)$
7. Find $\dfrac{\partial q_1}{\partial p_2}$ and $\dfrac{\partial q_2}{\partial p_1}$ and compare their signs. Both positive means competitive. Both negative means complementary. These products are complementary.
8. $x = \$7$, $y = \$11$; $P = \$5065$
9. $x = 200$, $y = 100$
10. (a) $\hat{y} = 573.57x + 7652.05$
(b) \$15,682 (nearest dollar)
(c) No; a child cannot be 35 years old.

Index